General Pathology

Pathology of Simian Primates

Part I: General Pathology

Editor: R. N. T-W-Fiennes (London)

250 figures and 125 tables

S. Karger · Basel · München · Paris · London · New York · Sydney · 1972

S. Karger · Basel · München · Paris · London · New York · Sydney
Arnold-Böcklin-Strasse 25, CH–4000 Basel 11 (Switzerland)

Printed in Switzerland by Buchdruckerei National-Zeitung AG, Basel

Contents

Part I: General Pathology

Preface XXXIII
Introduction XXXV
Taxonomy and Nomenclature. Systematic List of Living Primates XXXVII

I. Natural History of Primates

LAPIN, B. A. (Sukhumi): The Rational Use of Primates in Biomedical Experiments 2

General 2
Infectious Diseases 4
Neuroses 4
Atherosclerosis 5
Pharmacology 6
Immunology 6
Reproduction 7
Oncology 7
References 9

GROVES, C. P. (Cambridge): Phylogeny and Classification of Primates 11

Tree Shrews 11
Prosimii 12
    Lemuriformes 14
    Lorisiformes 15
    Tarsiiformes 16

Anthropoidea 16
Platyrrhini 17
Catarrhini 18
Taxonomy of the Macaques 20
Short-tailed Macaques 21
Macaca seu Cynopithecus 22
Long-tailed Macaques 23
Taxonomy of the Patas Monkeys 27
Taxonomy of the Vervet Monkeys 28
Taxonomy of the Baboons 31
Taxonomy of the Geladas 33
Taxonomy of the Talapoins 35
Taxonomy of the Colobinae 38
Taxonomy of the Chimpanzees 39
Taxonomy of Marmosets 42
Family Callitrichidae 43
Taxonomy of other South American Monkeys 48
Classification of the Primates 53
References 56

JONES, C. (Washington, D.C.): Natural Diets of Wild Primates 58

Introduction 58
Methods 59
General Feeding Habits 60
Seasonal Variations 66
Arboreal and Terrestrial Feeding Habits 69
Geographic Variations 70
Feeding Habits and Body Size 71
Activity Rhythms and Feeding Habits 71
Natural Diets and Morphology 72
Summary 75
Acknowledgments 76
References 76

LAPIN, B. A. and CHERKOVICH, G. M. (Sukhumi): Biological Normals 78

Introduction 79
Blood 80
Blood Coagulation 80
Morphological Composition of Monkey Blood 83
Bone Marrow 88
Monkey Serum Proteins 89
Antigenic Properties of Serum Proteins 93
Physico-Chemical Values and Aminoacid Composition of Certain Serum Proteins 96

Genetic Polymorphism of Monkey Serum Proteins 96
Nonprotein Components of Monkey Serum 97
Circulation 104
Pulse Rate 104
ECG 108
Blood Pressure 117
Systemic Arterial 117
The Velocity of the Pulse Wave Distribution in Monkeys 120
Respiratory System 121
Oxygen Dissociation Curve of Blood 122
Digestive System 123
Saliva 123
Gastric Juice 127
Gastric Movements 132
Secretion and Movements of the Small Intestine 132
Metabolism and Body Temperature 133
Gaseous Exchange 133
Body Temperature 134
Renal Function 137
Electroencephalography 138
References 146

II. Development, Growth, Nutrition and Metabolism

SCHULTZ, ADOLPH H. (Zürich): Developmental Abnormalities 158

I. Introduction 158
II. Abnormal Development of Head and Skull 161
III. Abnormal Development of Vertebral Column and Thorax 167
IV. Abnormal Development of Limbs 171
V. Abnormal Development of Dentition 177
VI. Abnormal Development of other Features 184
VII. Summary 185
References 186

O'GARA, R. W. †[1] and ADAMSON, R. H. (Bethesda, Md.): Spontaneous and Induced Neoplasms in Nonhuman Primates 190

General 190
Skeletal System 191
Sex Organs 193
Skin and Subcutaneous Tissue 198

1 To the regret of his friends and to the loss of medicine science, Dr. O'GARA's death was reported early in 1971.

Liver and Biliary System 201
Pancreas and Lungs 204
Leukemia 204
Alimentary Tract 209
Kidney and Adrenal Gland 214
Recent Developments in Primate Carcinogenesis 216
References 231

WOLF, R. H. (Covington, La.): The Pathology of Nutritional Deficiencies in Nonhuman Primates 239

I. Introduction 240
II. Protein 241
III. Fat Excess and Resultant Pathology 243
IV. Vitamins 243
A. Fat Soluble Vitamins 243
1. Vitamin A Deficiency 243
2. Vitamin D Deficiency 244
3. Vitamin E Deficiency 248
4. Vitamin K Deficiency 249
B. Water Soluble Vitamins 249
1. Thiamine Deficiency 249
2. Riboflavin Deficiency 250
3. Pyridoxine Deficiency 252
4. Vitamin B_{12} Deficiency 253
5. Niacin Deficiency 254
6. Folic Acid Deficiency 255
7. Biotin Deficiency 255
8. Pantothenic Acid Deficiency 256
9. Choline Deficiency 256
10. Ascorbic Acid Deficiency 257
V. Water 259
VI. Minerals 259
A. Zinc Deficiency 259
B. Calcium Deficiency 259
C. Magnesium Deficiency 260
VII. Conclusion 260
References 261

MOOR-JANKOWSKI, J. and WIENER, A. S. (New York, N.Y.): Red Cell Antigens of Primates 270

Introduction 271
Human-Type Blood Groups 273
A-B-O Blood Groups 273
Man 273

Apes ... 275
Old World Monkeys ... 278
New World Monkeys ... 279
Lewis Types ... 282
M-N Types ... 284
Man vs. Non-Human Primates ... 284
Chimpanzees ... 286
Gorillas ... 286
Gibbons ... 286
Orang-Utans ... 287
Old World and New World Monkeys ... 287
Rh-System ... 288
Chimpanzees ... 289
Gorillas ... 291
Gibbons, Orang-Utans and Monkeys ... 292
The I-i Blood Factors ... 292
Xg^a Blood Factor ... 293
Other Human-Type Blood Factors ... 293
Simian-Type Blood Groups ... 296
Chimpanzees ... 296
Gibbons ... 297
Celebes Black Apes ... 299
Baboons ... 299
Rhesus Monkeys ... 301
Sero-Primatology ... 302
Chimpanzees ... 305
Gibbons ... 310
References ... 311

HAWKEY, CHRISTINE (London): Coagulation, Fibrinolysis and Platelet Function ... 318

General ... 318
Blood Coagulation ... 321
i) The Thrombin-Fibrinogen Reaction ... 321
ii) The Conversion of Prothrombin to Thrombin ... 322
Blood Platelets ... 324
Blood Fibrinolytic Activity ... 325
Pathology ... 327
References ... 329

HILL, C. W. and THOMSEN, C. E. (Covington, La.): Stress ... 331

The Concept of Stress ... 331
Stress – Metabolism Relations ... 332
Clinical Evidence ... 334
Experimental Evidence ... 335

Immunological Responses 335
Physiological Disorders 336
Endocrine Responses 337
Metabolic Processes 339
Summary 340
References 340

DAMIAN, R. T. and GREENE, N. D. (San Antonio, Tex.): The Immunology of Non-Human Primates 342

Introduction 343
Immunoglobulins 343
Major Classes (IgG, IgA, IgM; κ and λ Light-Chains) 344
Additional Studies on IgG 347
'Minor' Immunoglobulin Classes 349
Subclasses and other Forms of Heterogeneity 351
IgG Subclasses 351
IgM Heterogeneity 351
IgA Heterogeneity 352
Serum Levels of IgG, IgA, and IgM 352
Antibody Production 352
Characteristics of the Immune Response 352
Range of Immune Competency with Respect to Antigens 358
Haptens and Synthetic Polypeptides 358
Proteins 359
Infectious Agents 359
Immunological Ontogeny and Competency 359
Transplacental Transfer of Immunoglobulins 360
Transfer of Immunoglobulins by Lactating Mothers 361
Ontogeny of Immune Competency 361
Fetal Studies 361
Newborn Studies 363
Tolerogenesis 364
Natural 364
Induced 365
Immediate Hypersensitivity 365
Transfer of Human Reaginic Antibodies to Primates 366
Induction of Immediate Hypersensitivity in Primates 368
Nature of Antibodies Involved in Immediate Hypersensitivity in Primates 371
Summary 371
References 372

FELSENFELD, O. (Covington, La.): Autoimmune Diseases 380

General 380
Etiology 381

Autoimmunity Induced by Microbial Agents ... 385
Rheumatoid Disease ... 386
Sjögren's Syndrome ... 388
Lupus Erythematosus ... 388
Scleroderma and Related Lesions ... 389
Thyroiditis ... 390
Allergic Gastritis and Pernicious Anaemia ... 391
Autoallergic Phenomena of the Gut ... 391
Diseases of the Liver ... 392
Renal Diseases ... 393
Heart ... 393
Myasthenia gravis ... 393
Endocrine Glands ... 394
Ophthalmic Conditions ... 394
Central Nervous System ... 395
References ... 395

SPIES, H. G. and CLEGG, M. T. (Covington, La.): Diseases of the Endocrine System 399

Introduction ... 399
Hypothalamo-Hypophyseal Axis ... 402
Pancreas ... 403
Parathyroid ... 404
Thyroid ... 405
Adrenal Gland ... 407
References ... 409

FIENNES, R. N. T-W- (London): Poisons and Toxins ... 414

III. The Bodily Systems

FIENNES, R. N. T-W- (London): The Skin and Appendages, and Special Senses ... 418

Diseases of the Skin ... 418
I. General ... 418
II. Constitutional and Allergic Causes of Skin Disease ... 420
III. Traumatic Conditions ... 420
IV. Nutritional Causes of Skin Disease ... 421
V. Infectious Causes of Dermatitis ... 423
1. Arthropod Ectoparasites ... 423
2. Fungal Diseases ... 423
3. Bacterial Infections ... 424
4. Viral Diseases ... 424
5. Protozoal Diseases ... 425

6. Helminthic Diseases 425
VI. Conclusion 425
References 426
Special Senses 428
References 429

BARKER, M. J. M. and HERBERT, R. T. (London): Diseases of the Skeleton 433

I. Introduction 435
A. The Importance of Bone Disease of Primates 435
B. The History of Metabolic Bone Disease and Definition of Terms 436
1. Goundou 437
2. Paget's Disease of Bone or Osteitis Deformans 437
3. von Recklinghausen's Disease of Bone 438
4. Leontiasis Ossea 438
5. Craniotabes 438
6. Osteodystrophia Fibrosa 438
7. Renal Osteodystrophy 439
8. Vitamin D Resistant Rickets 439
9. Woolly Monkey Disease 441
10. Osteogenesis Imperfecta 441
11. Cage Paralysis 442
12. Conclusions 443
II. Pathology of Bone Diseases 445
A. Diseases of Nutritional Origin 445
1. Multifactorial Conditions 445
2. Rickets 447
3. Osteomalacia 456
4. Osteoporosis 462
5. Mineral and Protein Deficiency 464
6. Scurvy 466
7. Hypervitaminosis D 466
B. Diseases of Endocrine Origin 467
1. Parathyroid Syndromes 467
2. Thyroid Syndromes 469
C. Diseases Due to Mechanical Defects and Trauma 470
1. Congenital Deformities 470
2. Fractures 471
3. Cephalhaematoma 472
4. Atrophy of Disuse 472
D. Bone Tumours 474
E. Infections of Bone 475
1. Tuberculosis 475
2. Miscellaneous Infections 475
F. Osteoarthritis 476
III. Bone Structure and Physiology 479

A. Formation and Growth of Bone ... 479
1. Ossification in Membrane ... 479
2. Endochondral Ossification ... 479
B. The Elements of Normal Bone ... 481
1. Bone Matrix ... 481
2. Periosteum ... 481
3. The Cellular Components ... 482
C. Varieties of Bone and Nourishment of Bone Cells ... 484
D. Bone Activity ... 485
E. Mechanism of Calcification ... 485
F. Resorption of Bone ... 488
G. Calcium Homeostasis ... 489
1. Parathyroid Hormone ... 490
2. Calcitonin ... 490
3. Vitamin D ... 490
IV. Methods of Bone Study ... 491
A. General ... 491
1. Case History ... 491
2. Cage Design ... 491
3. Diet ... 494
4. Balance Studies ... 494
B. Radiology ... 496
1. Anaesthesia for X-Ray Examination ... 496
2. Main Features of the X-Ray Film ... 496
C. Histology ... 498
D. Histochemistry ... 500
E. Biochemistry ... 500
1. Interpretation of Data ... 500
2. Preparation of Samples ... 504
3. Analysis for and Significance of Parameters of Bone Disease ... 505
4. Calcium Kinetics ... 509
V. Conclusions ... 509
Acknowledgements ... 510
References ... 510

Golarz de Bourne, M. Nelly and Bourne, Geoffrey H. (Atlanta, Ga.): Skeletal Muscle Pathology in Non-Human Primates ... 520

I. General ... 520
II. *Macaca mulatta* ... 523
III. Gorilla ... 526
IV. Chimpanzee ... 528
V. Java Macaques ... 534
VI. Squirrel Monkeys ... 537
VII. Celebes Black Ape ... 537
VIII. Drill ... 538

IX. Other Groups 538
X. Discussion 538
Acknowledgement 540
References 540

BRIZZEE, K. R. (Covington, La.): Diseases of the Nervous System 542

Foreword by the Editor 543
Introduction 545
Congenital Abnormalities 545
Naturally Occurring 545
Experimentally Induced 548
Vincristine 548
Thalidomide 548
Hypoxia 548
Vascular Diseases 551
Demyelinating Diseases 552
Experimental 552
Cyanide Poisoning 552
Sodium Azide Poisoning 554
Barbiturate Poisoning 554
Hypoxia 555
Carbon Monoxide 555
Allergic Encephalomyelitis 556
Lead Poisoning 558
Naturally Occurring 558
Confluent Leukoencephalosis and Perivascular Myelosis 558
Isolated Papillomacular Degeneration 559
Infectious Diseases 560
Viral 560
Encephalitides 560
B Virus Infection 567
Acute Polioencephalitis 569
Diffuse Non-Suppurative Meningitides and Meningoencephalitides 575
Neurotropic Newcastle Infection 576
Neurotropic Dengue 576
Lymphocytic Choriomeningitis 577
Coxsackie and Echo Viruses 577
Equine Encephalitis 580
Venezuelan Equine Encephalitis 581
Encephalitis Affecting Basal Ganglia 582
Austrian Tick-Borne Encephalitis 583
Non-Suppurative Meningitis 584
Infectious Diseases-Bacterial 587
Suppurative Meningitis 587
Tuberculosis 591

Anthrax 599
Encephalopathies Related to Dysenteric Syndromes 600
Fungal Infections 602
Torulosis (Cryptococcus) 602
Other Fungal Infections 602
Degenerative Diseases 602
Diseases of Cerebellum 602
Diseases of Basal Ganglia 604
Manganese Poisoning 604
Carbon Disulfide Poisoning 606
Spontaneous Degeneration 606
Widespread Spontaneous Degeneration of CNS 607
Kuru 609
Pigmentation and Siderosis 610
Organic Arsenical Compounds 613
Acute Amaurotic Epilepsy 613
Acute Amaurosis without Epilepsy 614
Nutritional Deficiencies 614
Experimental 614
Spontaneous Occurrence 616
Peripheral Neuropathy 617
Parasitic Infections 617
Epilepsies 618
Tumors 619
Pathological Effects of Intrathecal Injection of Spinal Anesthetic and Vasoconstrictor Drugs 619
Pathological Effects of Ionizing Radiation 620
Pathological Effects of Hyperthermia 631
References 631

COHEN, B. and BOWEN, W. H. (London): Diseases of the Oral Cavity 639

General 639
Abnormalities in Post-Natal Development 639
Abnormalities in Morphology 641
Injuries to the Teeth 649
Malnutrition 652
Infectious Diseases 658
Tumours 669
References 670

FIENNES, R. N. T-W- (London); LAPIN, B. A.; DZHIKIDZE, E. K., and YAKOVLEVA, L. A. (Sukhumi): The Respiratory and Alimentary Systems 671

I. General Survey 671
No Pathogen Involved 672

Opportunist Pathogens 675
Primary Pathogens 677
II. Gastro-Intestinal Diseases of Simian Primates 681
1. Bacillary Dysentery 681
2. Salmonellosis 685
3. Enterocolitis Caused by the Organisms of the Proteus Group 688
4. Amoebiasis 689
5. Rare Conditions of the Alimentary Tract 689
6. Bezoars 690
7. Appendicitis 691
8. Hernia 693
9. Gallstone Disease 693
III. Diseases of the Respiratory Tract 694
1. Pneumonia 694
2. Lung Mites 701
3. Pulmonary Nocardiosis 704
References 704

KRUCKENBERG, S. M.; CORNELIUS, C. E., and COOK, J. E. (Manhattan, Kan.): Liver Function and Disease in Primates 711

Introduction 711
Classification of Liver Function Tests 712
Tests Measuring Hepatic Uptake, Conjugation and Excretion of Certain Organic Anions in Health and Disease 713
A. Bilirubin 713
B. BSP Clearance 713
C. Indocyanine Green (ICG) 718
Tests Dependent Upon Elevations of Serum Enzyme Activities Due to Hepatic Necrosis or Cholestasis 718
A. Serum Glutamic Pyruvic Transaminase (SGPT) and Serum Glutamic Oxaloacetic Transaminase (SGOT) 718
B. Serum Alkaline Phosphatase (SAP) 719
C. Miscellaneous Enzymes 719
Miscellaneous Tests 725
A. Serum Protein 725
B. Prothrombin Time 730
C. Thymol Turbidity 730
D. Cephalin-Cholesterol Flocculation 730
Compendium of Hepatic Diseases: Spontaneous and Experimental 732
A. Viral Diseases 732
B. Bacterial Diseases 736
C. Nutritional Deficiency Diseases 738
D. Parasites 740
E. Miscellaneous 743
References 748

McNulty, W. P. and Malinow, M. R. (Beaverton, Ore.): The Cardiovascular System 757

General 756
Heart 758
Endocardium and Valves 758
Myocardium 761
Diseases of Abnormal Depositions 761
Nutritional Deficiency Diseases 762
Myocarditis 764
Circulatory Insufficiency 767
Cardiac Hypertrophy 774
Congenital Heart Disease 775
Neoplasms 777
Pericardium 777
Blood Vessels 779
Arteries 779
Atherosclerosis 779
Hypertensive Arteriopathy 792
Medial Calcification 795
Necrotizing Arteritis (Periarteritis nodosa) 796
Aneurysms 796
Veins 799
References 801

Lapin, B. A. and Yakovleva, L. A. (Sukhumi): Blood Formation Disturbances in Monkeys 809

General 809
Disturbances of White Cell Production 809
The Effects of Radiation 815
Disturbances of Red Cell Production 817
References 819

Roberts, J. A. (Covington, La.): The Urinary System 821

Anatomy 821
Kidney 821
Ureter, Bladder, Urethra 822
Renal Physiology 823
Ureteral Physiology 825
Natural Pathology 826
Congenital Anomalies 826
Obstruction 827
Neoplasm 828

Urolithiasis 828
Vascular Disease 829
Chronic Renal Disease 830
Urinary Tract Infection 831
Bacterial 831
Specific Infections 836
References 838

KRAEMER, D. C. and VERA CRUZ, N. C. (San Antonio, Tex.): The Female Reproductive System 841

I. Anatomy 842
A. External Genitalia 842
B. Internal Genitalia 843
C. Mammary Glands 845
II. Physiology 846
A. Puberty 846
B. Breeding Season 846
C. Birth Season 849
D. Sexual Receptivity (Estrus) 849
E. Sexual (Menstrual) Cycle 850
1. Hormone Patterns 850
2. Ovarian Changes 851
3. Endometrial Changes 852
4. Changes in Cervical Mucus 852
5. Vaginal Changes 853
6. Perineal (Sex Skin) Changes 853
7. Changes in the External Genitalia 854
F. Gestation 854
G. Parturition 856
H. Lactation 856
III. Pathology 857
A. Prenatal and Neonatal Mortality 857
1. Embryo Mortality 857
2. Congenital Malformations 858
3. Placental Abnormalities 858
4. Abortions and Stillbirths 858
5. Dystocia 859
6. Neonatal Mortality 859
B. Noninfectious Gynecologic Pathology 859
1. Ovaries 859
2. Uterus and Fallopian Tubes 860
3. Vagina and External Genitalia 861
4. Mammary Glands 861
C. Infectious Gynecologic Pathology 862
1. Viral 862

2. Mycoplasma 864
3. Bacterial Infections 864
4. Mycotic Infections 866
5. Protozoan Infections 867
6. Parasitic Helminth Infections 867
References 868

ROBERTS, J. A. (Covington, La.): The Male Reproductive System 878

I. Anatomy and Physiology 878
1. A Male Sexual Cycle 878
2. The Testis 879
3. The Semen 880
4. The External Genitalia 880
5. The Accessory Glands 880
II. Pathology 882
1. Infertility 882
2. Cryptorchidism 884
3. Inflammatory Disease 884
4. Neoplasm 884
5. Miscellaneous 886
General References 886
References 887

Subject Index 889
Author Index 900

Part II: Infectious and Parasitic Diseases

Preface XXXIII
Introduction XXXV
Taxonomy and Nomenclature. Systematic List of Living Primates XXXVII

I. Protozoa

BURROWS, R.B. (Research Triangle Park, N.C.): Protozoa of the Intestinal Tract 2

General 2
Collection and Examination for Protozoa 3
Class Mastigophora 4
Family Bodonidae 4
Family Tetramitidae 6
Family Chilomastigidae 6
Family Trichomonadidae 7
Family Hexamitidae 9

Class Sarcodina 10
Family Endamoebidae 10
Class Sporozoa 17
Family Eimeriidae 18
Class Ciliata 20
Family Pycnothricidae 20
Family Bursariidae 21
Family Cycloposthiidae 24
Conclusions 24
References 25

BARKER, J.R. (Cambridge): Protozoa of Tissues and Blood (Other than the Haemosporina) 29

Classification 29
Diagnosis 30
Leishmania 31
Trypanosoma 32
Asia 32
Africa 33
South and Central America 37
Toxoplasma 42
Sarcocystis 45
Babesia 47
Acknowledgements 50
References 51

VOLLER, A. (London): Plasmodium and Hepatocystis 57

Plasmodiidae 57
Life Cycle of Primate Malaria Parasites 58
Disease Patterns 59
Symptoms 60
Haematology 60
Major Pathology 63
Preparation and Staining of Blood Films 63
Identification of the Malaria Parasites of Non-Human Primates 64
References 65
Appendix 65

II. Helminths

ORIHEL, T.C. and SEIBOLD, H.R. (Covington, La.): Nematodes of the Bowel and Tissues 76

I. General ... 76
II. Trichuroidea ... 77
III. Oxyuroidea ... 82
IV. Rhabdiasoidea ... 82
V. Ascaroidea ... 84
VI. Strongyloidea ... 86
VII. Spiruroidea ... 89
VIII. Filarioidea ... 94
References ... 99

KUNTZ, R.E. (San Antonio, Tex.): Trematodes of the Intestinal Tract and Biliary Passages ... 104

I. The Intestinal Tract and Biliary Passages ... 104
II. Taxonomic Representation of Trematodes Reported from Non-Human Primates ... 106
III. Zoonotic Significance of Primates ... 106
IV. Host-Parasite Relationships ... 111
  1. Somatic Trematodes ... 113
  2. Intestinal Trematodes ... 120
V. Recognition and Identification of Trematodes of Non-Human Primates ... 121
References ... 121

MYERS, BETTY J. (San Antonio, Tex.): Echinococcosis, Coenurosis, Cysticercosis, Sparganosis, etc. ... 124

I. General ... 124
II. Taxonomic Representation of Cestodes Reported from Non-Human Primates 125
III. Zoonotic Significance of Non-Human Primates ... 127
IV. Host-Parasite Relationships ... 130
  1. Somatic Cestodes ... 130
  2. Intestinal Forms ... 136
V. Recognition and Identification of Cestodes of Non-Human Primates ... 138
References ... 140

SCHMIDT, G.D. (Greeley, Col.): Acanthocephala of Captive Primates ... 144

General ... 144
Key to the Acanthocephala of Primates ... 145
Pathogenic Species ... 145
  Prosthenorchis elegans ... 145
  Prosthenorchis spirula ... 148
  Moniliformis moniliformis ... 149

Clinical Signs 149
Pathology 150
Diagnosis 154
Treatment and Control 155
References 155

III. Arthropods

FIENNES, R.N. T-W- (London): Ectoparasites and Vectors 158

Ectoparasites 158
Acari (Mites) 159
Insecta (Fleas, Flies, Lice) 160
Arthropod Vectors 163
Ixodoidea (Ticks) 163
Culicidae (Mosquitoes) 166
Transmission of Simian Malarias 168
Anthomyidae 169
Diptera (General) 172
Acknowledgements 173
References 173

INNES, J.R.M. (Kensington, Md.) and HULL, W. B. (Bethesda, Md.): Endoparasites 177

I. Introduction 177
II. Clinical Findings 178
III. Pathology 179
1. Macroscopic Findings 179
2. Microscopic Findings 179
IV. Comments 182
V. Acarology. The Causal Mites (Acarina: Halarachnidae) 184
References 191

SELF, J.T. and COSGROVE, G.E. (Oak Ridge, Tenn.): Pentastomida 194

General 194
Life History 194
Genera Infecting Primates 196
Pathology 198
Spread of Infection and Control 198
References 202

IV. Fungal and Bacterial Diseases

AL-DOORY, Y. (Washington, D.C.): Superficial Mycoses 206

General 206
Dermatomycoses 207
Piedra 210
References 211

AL-DOORY, Y. (Washington, D.C.): Intermediate Mycoses 214

Candidiasis 214
Phycomycosis 217
Aspergillosis 220
Penicilliosis 221
References 222

AL-DOORY, Y. (Washington, D.C.): Systemic Mycoses 224

Actinomycosis 224
Nocardiosis 226
Coccidioidomycosis 227
Histoplasmosis 231
African Histoplasmosis 234
Blastomycosis 235
Cryptococcosis 237
Sporotrichosis 239
References 240

PINKERTON, MARY (San Antonio, Tex.): Spirocheta, Spirillum, Leptospira 243

Rat-Bite Fever 243
Relapsing Fever 244
Leptospirosis 246
References 251

FRIBOURG-BLANC, A. (Paris): Treponema 255

Historical and General 255
Symptoms and Pathology 256
Geographical and Racial Distribution 256
Bacteriology 257
Experimental Pathogenesis 258
Serology 259
Problems and Conclusions 260
References 261

FIENNES, R.N. T-W- (London); PINKERTON, MARY (San Antonio, Tex.), and DZHIKIDZE, E.K. (Sukhumi): Enteropathogenic Organisms 263

General Introduction 263

Escherichia, Paracolon, Proteus, Vibrio 265
I. Escherichia 265
II. Paracolons 267
III. Proteus 267
Proteus morgani 268
IV. Vibrio 268
References 269
Shigelloses, Salmonelloses, and other Enterobacteriacea 270
Shigella 271
Salmonella 272
Escherichia Coli 273
Arizona 274
Providence Group 274
Proteus 274
References 274

FIENNES, R. N. T-W- (London) and DZHIKIDZE, E. K. (Sukhumi): Respiratory Pathogens and other Organisms 277

General Survey 277
Infections Caused by Cocci and other Miscellaneous Organisms 277
References 280
Cocci – Streptococcus, Staphylococcus, Diplococcus, Meningcoccus, Micrococcus 280
I. Staphylococci 280
II. Streptococci 280
III. Pneumococci 281
IV. Meningococci 281
References 281

PINKERTON, MARY (San Antonio, Tex.): Miscellaneous Organisms 283

I. General 283
II. Pasteurelloses 284
III. Brucellosis 289
IV. Tularemia 293
V. Melioidosis 296
VI. Bordetella 298
VII. Klebsiella 299
VIII. Pseudomonas 300
IX. Corynebacterium 302
X. Listeriosis 303
XI. Mycoplasma 304
References 304

FIENNES, R. N. T-W- (London): Tuberculosis 314

I. General 314
II. Incidence in Different Species 317
III. Strains of Tuberculosis 318
IV. Clinical Course 319
V. Pathology 320
VI. Diagnosis 321
Tuberculin Testing 322
Chest Radiology 324
Examination of Sputum, Gastric Washings, and Faeces 324
VII. Treatment 325
VIII. Prophylaxis and Prevention 325
Appendix 1 327
Suggested Standard Procedures for Detection of Tuberculosis 327
Appendix 2 331
Detection of Tuberculosis in Monkeys and Apes 331
Acknowledgement 333
References 333

THORPE, B.D. (Greeley, Col.): Anthrax, Fusiformis, Anaerobic Organisms 335

Bacillus SPP (Bacillosis) 335
References 337
Bacillus Anthracis (Anthrax) 337
General 337
Bacteriology 338
Distribution and Animals Affected 341
Transmission and Diagnosis 342
Treatment 343
Pathogenesis 343
Immunity 344
Anthrax in Primates 345
(a) Pathogenesis 345
(b) The Clinical Disease 347
(c) Pathology 348
References 355
Clostridium SPP (Anaerobic Toxemias, Tetanus, Botulism and Gas Gangrene) 358
Clostridium Tetani (Tetanus) 360
Clostridium Botulinum (Botulism, Allantiasis or Ichthyosismus) 362
Clostridial Induced Gaseous Gangrene 365
References 372
The Bacteroidoceae (Necrobacillosis and Bacteroidosis) 374
References 379

V. Viruses and Small Organisms of Indeterminate Position

KALTER, S.S. (San Antonio, Tex.): Identification and Study of Viruses 382

Introduction 383
Nature of Viruses 384
Virus Structure and Composition 385
Virus Classification 386
RNA Viruses 387
DNA Viruses 390
Other Viruses 392
Simian Viruses 392
Simian Virus Pathology 395
Viral Inclusion Bodies 396
Cytopathogenic Effect (CPE) 397
Importance of Viruses in Simian Pathology 398
Zoonotic and Ecological Considerations 398
Natural and Experimental Infections 403
Viral Contamination of Tissues 406
Carrier State 409
Detection and Recognition of Viruses 410
Collection and Shipment of Specimens 411
Types of Specimens for Routine Diagnosis 412
Handling and Shipment of Specimens 416
Storage of Specimens 420
Information Accompanying Specimens 420
Preparation of Specimens for Inoculation 420
Precautions 424
Laboratory Procedures 424
Cell Culture 425
Chick Embryo 430
Mouse Inoculation 432
Inoculation of Other Animals 433
Serology 434
Identification of an Isolate 452
Routine Laboratory Procedures for Handling Clinical Material 454
Interpretation of Results 456
Abbreviations 463
Acknowledgements 464
General References 464
References 464

KALTER, S.S. (San Antonio, Tex.): Serologic Surveys 469

Introduction 469
Adenoviruses 472

Arboviruses ... 474
Herpesviruses ... 476
Myxoviruses ... 477
Papovaviruses ... 480
Picornaviruses ... 480
Poxviruses ... 482
Reoviruses ... 485
Miscellaneous Viruses ... 486
Rubella ... 486
Viral Hepatitis ... 486
Marburg Virus ... 487
Simian Hemorrhagic Fever ... 488
Vesicular Stomatitis ... 488
Rabies ... 488
Rous Sarcoma ... 489
Rauscher Murine Leukemia Virus ... 489
SA11 ... 489
References ... 490

HEBERLING, R.L. (San Antonio, Tex.): The Simian Picornaviruses ... 497

The Simian Enteroviruses ... 497
Isolation and Pathogenicity ... 500
Serology ... 506
Diagnostic Virology ... 507
Conclusion ... 509
The Unclassified Simian Picornaviruses ... 510
The Rhinoviruses ... 511
Acknowledgements ... 512
References ... 512

HEBERLING, R.L. (San Antonio, Tex.): The Simian Reoviruses ... 516

Isolation and Pathogenicity ... 517
Serology ... 519
Diagnostic Virology ... 520
Conclusion ... 520
Acknowledgements ... 521
References ... 521

FELSENFELD, A.D. (Covington, La.): The Arboviruses ... 523

General and Technical ... 523
Arboviruses Causing Disease in Monkeys in Their Natural Habitats ... 525

Yellow Fever 525
Kyasanur Forest Disease 529
Non-Human Primates as Models for the Study of Arboviruses 533
Non-Human Primates as Sentinels 534
References 534

HSIUNG, G.D. (New Haven, Conn.) and SWACK, N.S. (West Haven, Conn.): Myxovirus and Pseudomyxovirus Groups 537

Introduction 538
General Properties 538
Morphology and Structure of the Virus 538
Effects of Physical and Chemical Agents 542
Growth Characteristics 542
Hemagglutination and Hemadsorption 542
Myxovirus Group with Special Reference to SV_5 543
Natural Infections in Monkeys 543
History of the Isolation of SV_5 and Related Virus Strains 545
Propagation in Tissue Culture 546
Cytopathic Effect and Hemadsorption 546
Plaque Formation and Autoinhibition 549
Interferon Production and Interference 552
Syncytial Cell Formation with or without Production of Infectious Virus 552
Pathogenesis and Persistence of SV_5 in Experimental Animals 553
Monkeys 553
Hamsters and Mice 554
Chick Embryos and Day-old Chicks 556
Antigenic Relationships Among the Parainfluenza-SV_5-mumps Newcastle Disease Viruses 556
Epidemiology and Control of SV_5 in Monkeys 558
Prevalence of SV_5 in Monkey Kidney Cell Cultures 558
Acquisition of SV_5 Infection in Monkeys 560
Distribution of SV_5 Antibody in Different Animal Species 561
Comparison of SV_5 Antibody Titers in Human Sera Collected from Different Geographical Areas 561
Pseudomyxovirus Group 562
Measles and Respiratory Syncytial Viruses 562
Foamy Viruses 563
Serological Types 564
Epidemiological Studies 565
Summary 566
Acknowledgements 566
References 566

HEBERLING, R.L. (San Antonio, Tex.): The Simian Adenoviruses 572

General 572

Isolation and Pathogenicity 574
Serology 577
Experimental Infections 581
Oncogenicity 582
Adenoviruses Associated Viruses 583
Diagnostic Virology 584
Conclusion 586
References 587

DANIEL, M. D.; MELÉNDEZ, L. V.; HUNT, R. D., and TRUM, B. F. (Southboro, Mass.): The Herpesvirus Group 592

Introduction 592
Herpesvirus in Old World Monkeys 593
- *Herpesvirus simiae* or B-virus 593
- SA8 Virus 596
- Liverpool Vervet Monkey Virus 597
- Patas Monkey *(Erythrocebus patas)* Virus 597

Herpesviruses in New World Monkeys 598
- *Herpesvirus T* 598
- *Herpesvirus saimiri* 601
- Squirrel Monkey Heart Isolate (SMHI) 602
- Marmoset Herpesvirus 602
- Owl Monkey Herpesvirus 603
- Owl Monkeys and *Herpesvirus simplex* 603
- Spider Monkey Herpesvirus 604
- Tree Shrew *(Tupaia glis)* Herpes-type Virus 604

Discussion 605
Note Added in Proof 607
Acknowledgements 607
References 607

MCNULTY, W. P. (Beaverton, Oreg.): Pox Diseases in Primates 612

I. Introduction 612
II. Vaccinia Group 615
- A. Variola and Alastrim 615
- B. Vaccinia 616
- C. Monkey-Pox 617

III. Paravaccinia Group 627
- A. Molluscum Contagiosum 628
- B. Orf 630

IV. Unclassified Group 630
- A. Yaba 630
- B. OrTeCa 635

V. Summary 642
References 643

FIENNES, R.N. T-W- (London): Rabies 646

General 646
Incidence of Rabies in Wild Monkeys 646
Experimental Infection 648
Rabies in Imported Monkeys 648
Incubation Period 649
Clinical Signs 650
The Autopsy 651
Pathological Changes in Simian Rabies 652
Modification of Lesions Following Intrasciatic Inoculation 657
Negri Bodies 659
Comparative Study of Lesions due to the Virus-fixe 659
Acknowledgements 661
References 661

SMETANA, H.F. (Covington, La.): Infectious Hepatitis 663

Clinical Course and Pathology 664
Experiences with Primate and Human Viral Hepatitis at the Delta Regional Primate Research Center of Tulane University 668
Fatal Spontaneous Viral Hepatitis in a Chimpanzee with Transmission of Disease to Several Human Beings 668
Viral Hepatitis in a Caretaker of 5 Chimpanzees Traceable to a Subsiding Viral Hepatitis in 2 Animals in his Care 669
Spontaneous Viral Hepatitis in Subhuman Primates 669
Apparent Transmission of the Disease from one Infected Animal to other Chimpanzees of the Colony 669
Spontaneous Viral Hepatitis in Patas Monkeys 670
Induced Viral Hepatitis in Subhuman Primates 670
Experimental Transfer of Viral Hepatitis to Patas Monkeys 675
Use of Serum 675
Use of 'P.M. Stool Suspension' 675
Induction of Viral Hepatitis with Materials from Human Cases of Homologous Serum Hepatitis 678
Thrombin (Cutter) 678
Hyland Laboratory Plasma 678
Pilot Study for Testing the Effect of Re-infection of Subhuman Primates Following an Attack of Viral Hepatitis 679
Experiments with Chimpanzees Having Recovered from Spontaneous Viral Hepatitis 679

Re-challenge of Chimpanzees Following Recovery from Induced Viral Hepatitis 679
Re-challenge of Patas Monkeys after Recovery from Induced Infectious Viral Hepatitis 679
Re-challenge of Patas Monkeys after Recovery from Induced Homologous Serum Hepatitis 679
Summary 680
References 680

HOLMES, A.W.; WOLFE, L., and DEINHARDT, F. (Chicago, Ill.): Infectious Hepatitis in Marmosets 684

I. Introduction 684
II. Structure and Function of the Normal Marmoset Liver 685
III. Transmission of Human Infectious Hepatitis to Marmosets 687
A. Experimental Method 687
B. Structural Alterations 688
C. Biochemical Alterations 691
D. Evidence for the Human Origin of the Hepatitis Seen in Marmosets 692
IV. Serial Passages of Hepatitis in Marmosets 695
V. Immunity of Marmosets Following Hepatitis 696
VI. Physico-Chemical Characteristics of the GB Agent 699
VII. Summary 700
References 700

SIMPSON, D.I.H. (Salisbury): Other Virus Diseases 702

Simian Haemorrhagic Fever 702
Marburg Virus Disease 706
Kuru and other Degenerative Conditions of the Central Nervous System 711
References 712

BARKER, L.F. (Bethesda, Md.): Chlamydiae and Rickettsiae 715

Introduction 715
Chlamydiae 717
Trachoma and Inclusion Conjunctivitis 717
Psittacosis 717
Rickettsiae 720
Typhus Group 720
Scrub Typhus 721
Spotted Fever Group 721
Q Fever 722
Trench Fever 723

Addendum 723
Acknowledgements 723
References 723

LAPIN, B.A. and YAKOVLEVA, L.A. (Sukhumi): Oncogenic Viruses 725

Introduction 725
Autologous Tumors of the Simian Host 726
Heterologous Host Tumors Induced in Simian Primates 729
Latent Infections with Oncogenic Viruses 732
Summary 734
References 735

Subject Index 739
Author Index 747

Preface[1]

Research with primates has increased so much during the past decade that it sometimes appears to have grown faster than the human population. Primate Centers and other major primate laboratories have been established in several countries during the past decade, and more are contemplated. Scientists have gained experience with many primate species not used appreciably before. New monkey diseases have been identified, some of which produce illness, and, possibly, even death in man, something which concerns us considerably. In the same period, new uses for monkeys have been identified, many of which relate to experimental and others to spontaneous illness. An important result of this expanded effort has been a proliferation of significant information published in numerous journals of many languages.

One of the functions of the American Regional Primate Centers is the dissemination of information about primates living in the laboratory and in their natural environment. The time appeared overdue, in contemplating the profusion of data a few years ago, to consider undertaking the challenging task of assembling the current information about primate pathology, and to analyze it, synthesize it, and present it in a way that would aid both the colony manager who is faced with an immediate problem and the researcher who is either experimentally investigating a disease or adaptive process or looking for an appropriate host animal.

1 The organization and compilation of the material in these volumes was supported by the Zoological Society of London and a grant, RR00164, from the National Institutes of Health. Research summarized in the chapters was supported from many sources, acknowledgement for which appeared in the original publications. We are grateful to these grantors for enabling the many scientists to produce the valuable data cited, to the original authors for making their data available, and to the chapter authors for their contributions. The involvement of all made these volumes possible.

The National Institutes of Health wisely incorporated a 'Visiting Scientists' program into the enabling grants for the operation of the Delta Center, and we used this mechanism to engage a visiting scientist who, freed from some of the routine duties in his home institution, could tackle the task of compiling the information. RICHARD FIENNES was our first and only candidate. He has both field and laboratory experience, and he has been interested in primate diseases for several years.

I won't tell you the circumstances or the spirit of our conversation that led to his acceptance of our invitation; I will tell you, however, that the Delta Center was as fortunate to get him for this responsibility as it was charmed by his wife, ALICE, who accompanied him. He devoted a year to the enormous job of planning the contents, scanning the literature, identifying expert authors for each chapter, and, handling the administrative and mechanical arrangements for the production of a complex pair of volumes. And he did some research at the Center as well!

It is fitting that the national origin of the editor is not the same as mine. Even though national origin was not a commanding factor in the selection of an editor, multiple nationality has characterized primate research from the beginning. In recognition of this, the International Primatological Society, created at the Delta Center in 1964, has grown to become a distinguished scientific association. The diverse nationalities of the chapter authors, a conspicuous characteristic of the two volumes, is a reflection of the interest in the world-wide use, and contributions to the use of primates in research.

ARTHUR J. RIOPELLE
Covington, La.

Introduction

The preparation of a work on the Pathology of Simian Primates was first envisaged by Dr. ARTHUR RIOPELLE, who arranged for me to spend a year at the Delta Primate Center to organise the work and to recruit contributors. I found in the States a keen awareness of the necessity for a work of this nature and – unexpectedly – considerable eagerness to participate. As a result, the task of recruiting a team of distinguished authors, with first hand practical knowledge of their subject, was rendered the easier.

To this circumstance, I must also attribute the unexpected length of the work. Originally, we thought of a much shorter work which would cover the various aspects of the subject in less depth. However, one after another authors preferred to give their subjects comprehensive coverage and presented the series of masterly and well documented theses which form this work. This was unforeseen but had to be accepted. The result is a work of much greater value which should stand the test of time as an authoritative reference for the future.

There are a few gaps, for which suitable authors were not found. These have been filled by short notes prepared by myself; the contributions on 'Toxins and Poisons', and on 'The Skin and Appendages, and Special Senses' are no more than stop-gaps; it is to be hoped that contributors with special knowledge of these subjects will be found when the work has to be revised for a new edition. The chapter on 'Ectoparasites and Insect Vectors' was to have been prepared by Dr. M. M. J. LAVOIPIERRE, a leading expert on the subject; owing to family difficulties, Dr. LAVOIPIERRE was unable to complete the task and the material in this chapter is largely taken from my own book 'Zoonoses of Primates' Weidenfeld & Nicolson, London; Cornell University Press, Ithaca, N.Y.). The chapter on Rabies, I reserved for myself, because it had become clear to me that in this same work I had underestimated the importance of simian rabies and I wished to correct the record.

I have attempted, in editing the work, to achieve as much uniformity as possible, and this has meant suggesting some re-arrangement of their material to some authors. In no case, has this been taken amiss, and I am grateful to the authors concerned for their co-operation. I have not attempted to standardise the spelling of scientific words. and the authors' own spelling of, say, estrus or oestrus, hematology or haematology, has been left to the dictates of geography. Some discrepancies will also be found in the handling of references and bibliographies; in some contributions, the numbered references are listed alphabetically; in others in order of appearance. This is, I think, relatively unimportant, but does offend a tidy mind. It was decided, however, not to alter this because of the danger of errors occurring when the numbers of the text references were transposed.

The high standard of presentation lies with the authors, who have so generously contributed and have given so much of their time to prepare this material in such depth. The organisers of the work owe them a special debt of gratitude. In preparation of the section on Parasitology, I have been assisted by Dr. TOM ORIHEL of the Delta Primate Center, as sub-editor. Dr. JOHN AYRES, also of the Delta Primate Center, has similarly assisted as sub-editor of the section of Virology. I am grateful to both my ex-colleagues for their help.

This work is not intended to supplant any existing work, such as 'Diseases of Laboratory Primates' by Dr. THEODORE RUCH, a work of proven value, which will retain its place as an important work of reference. Many other works on special aspects of simian medicine and husbandry are of especial value in their relative spheres. The present work is intended to supplement these as a comprehensive and authoritative source of information.

The clerical work involved in arranging and editing this work has fallen on the shoulders of Mrs. JACKIE FORBES at the Delta Primate Center, and of Mrs. ETHELYN HAZELL at the Nuffield Institute of Comparative Medicine. I am most grateful to both of them for the willing assistance they have afforded.

In conclusion, I would express my appreciation of the facilities afforded by the Delta Primate Center of Tulane University, and by the Nuffield Institute of Comparative Medicine of the Zoological Society of London. The work is the result of international collaboration and experts from USA, USSR, France, Switzerland and Britain have all contributed.

August 1970
RICHARD N. T-W-FIENNES, London

Taxonomy and Nomenclature

Systematic List of Living Primates[1]

J. R. Napier and P. H. Napier

Order PRIMATES Linnaeus, 1758
Suborder PROSIMII Illiger, 1811
Infraorder LEMURIFORMES Gregory, 1915
Superfamily Tupaioidea Dobson, 1882
Family Tupaiidae Mivart, 1868
Subfamily Tupaiinae Lyon, 1913
Genus *Tupaia* Raffles, 1821
Subgenus *Tupaia* Raffles, 1821
Tupaia glis Diard, 1820 (type species)
- *T. g. glis* Diard, 1820
- *T. g. belangeri* Wagner, 1841
- *T. g. dissimilis* Ellis, 1860
- *T. g. chinensis* Anderson, 1879
- *T. g. modesta* Allen, 1906
- *T. g. concolor* Bonhote, 1907
- *T. g. siccata* Thomas, 1914
- *T. g. laotum* Thomas, 1914
- *T. g. sinus* Kloss, 1916
- *T. g. clarissa* Thomas, 1917
- *T. g. cambodiana* Kloss, 1919
- *T. g. olivacea* Kloss, 1919
- *T. g. assamensis* Wroughton, 1921
- *T. g. cochinchinensis* Robinson and Kloss, 1922
- *T. g. annamensis* Robinson and Kloss, 1922

Tupaia glis Diard (continued)
- *T. g. versurae* Thomas, 1922
- *T. g. lepcha* Thomas, 1922
- *T. g. brunetta* Thomas, 1923
- *T. g. kohtauensis* Shamel, 1930
- *T. g. wilkinsoni* Robinson and Kloss, 1911
- *T. g. ferruginea* Raffles, 1821
- *T. g. operosa* Robinson and Kloss, 1914
- *T. g. ultima* Robinson and Kloss, 1914
- *T. g. longicauda* Kloss, 1911
- *T. g. obscura* Kloss, 1911
- *T. g. sordida* Miller, 1900
- *T. g. pemangilis* Lyon, 1911
- *T. g. pulonis* Miller, 1903
- *T. g. cognata* Chasen, 1940
- *T. g. umbratilis* Chasen, 1940
- *T. g. raviana* Lyon, 1911
- *T. g. lacernata* Thomas and Wroughton, 1909
- *T. g. demissa* Thomas, 1904
- *T. g. jacki* Robinson and Kloss, 1918
- *T. g. siaca* Lyon, 1908
- *T. g. castanea* Miller, 1903
- *T. g. batamana* Lyon, 1907

1 Reprinted with amendments from Napier, J. R. and Napier, P. H.: A Handbook of Living Primates (Academic Press, London 1967).

Tupaia glis Diard (continued)
T. g. redacta Robinson, 1916
T. g. phaeura Miller, 1902
T. g. discolor Lyon, 1906
T. g. tephrura Miller, 1903
T. g. chrysogaster Miller, 1903
T. g. siberu Chasen and Kloss, 1927
T. g. chrysomalla Miller, 1900
T. g. riabus Lyon, 1913
T. g. anambae Lyon, 1913
T. g. hypochrysa Thomas, 1895
T. g. longipes Thomas, 1893
T. g. salatana Lyon, 1913
Tupaia gracilis Thomas, 1893
T. g. gracilis Thomas, 1893
T. g. inflata Lyon, 1906
T. g. edarata Lyon, 1913
Tupaia javanica Horsfield, 1822
Tupaia minor Günther, 1876
T. m. minor Günther, 1876
T. m. malaccana Anderson, 1879
T. m. sincipis Lyon, 1911
T. m. caedis Chasen and Kloss, 1931
T. m. humeralis Robinson and Kloss, 1919
Tupaia montana Thomas, 1892
T. m. montana Thomas, 1892
T. m. baluensis Lyon, 1913
Tupaia muelleri Kohlbrugge, 1896
Tupaia nicobarica Zelebor, 1869
T. n. nicobarica Zelebor, 1869
T. n. surda Miller, 1902
Tupaia palawanensis Thomas, 1894
T. p. palawanensis Thomas, 1894
T. p. cuyonis Miller, 1910
T. p. moellendorffi Matschie, 1898
T. p. busuangae Sanborn, 1952
Tupaia picta Thomas, 1892
T. p. picta Thomas, 1892
T. p. fuscior Medway, 1965
Tupaia splendidula Gray, 1865
T. s. splendidula Gray, 1865
T. s. lucida Thomas and Hartert, 1895
T. s. natunae Lyon, 1911
T. s. carimatae Miller, 1906
Subgenus *Lyonogale* Conisbee, 1953
Tupaia tana Raffles, 1821 (type species)

Tupaia tana Raffles (continued)
T. t. tana Raffles, 1821
T. t. speciosa Wagner, 1840
T. t. besara Lyon, 1913
T. t. nitida Chasen, 1933
T. t. utara Lyon, 1913
T. t. chrysura Günther, 1876
T. t. paitana Lyon, 1913
T. t. banguei Chasen and Kloss, 1931
T. t. cervicalis Miller, 1903
T. t. masae Lyon, 1913
T. t. tuancus Lyon, 1913
T. t. lingae Lyon, 1913
T. t. sirhassenensis Miller, 1901
T. t. bunoae Miller, 1900
T. t. kelabit Davis, 1958
T. t. kretami Davis, 1962
Tupaia dorsalis Schlegel, 1857

Genus *Anathana* Lyon, 1913
Anathana ellioti Waterhouse, 1849
A. e. ellioti Waterhouse, 1849
A. e. wroughtoni Lyon, 1913
A. e. pallida Lyon, 1913

Genus *Urogale* Mearns, 1905
Urogale everetti Thomas, 1892

Genus *Dendrogale* Gray, 1848
Dendrogale murina Schlegel and Müller, 1844 (type species)
Dendrogale melanura Thomas, 1892
D. m. melanura Thomas, 1892
D. m. baluensis Lyon, 1913

Subfamily Ptilocercinae Lyon, 1913
Genus *Ptilocercus* Gray, 1848
Ptilocercus lowii Gray, 1848
Pt. l. lowii Gray, 1848
Pt. l. continentis Thomas, 1910

Superfamily Lemuroidea Mivart, 1864
Family Lemuridae Gray, 1821
Subfamily Lemurinae Mivart, 1864
Genus *Lemur* Linnaeus, 1758
Lemur catta Linnaeus, 1758 (type species)
Lemur variegatus Kerr, 1792

Lemur variegatus Kerr (continued)
L. v. variegatus Kerr, 1792
L. v. subcinctus A. Smith, 1833
L. r. ruber E. Geoffroy, 1812
Lemur macaco Linnaeus, 1766
L. m. macaco Linnaeus, 1766
L. m. albifrons E. Geoffroy, 1796
L. m. collaris E. Geoffroy, 1812
L. m. flavifrons Gray, 1867
L. m. fulvus E. Geoffroy, 1812
L. m. mayottensis Schlegel, 1886
L. m. rufus Audebert, 1800
L. m. sanfordi Archbold, 1932
Lemur mongoz Linnaeus, 1766
L. m. mongoz Linnaeus, 1766
L. m. coronatus Gray, 1842
Lemur rubriventer I. Geoffroy, 1850

Genus *Hapalemur* I. Geoffroy, 1851
Hapalemur griseus Link, 1795 (type species)
H. g. griseus Link, 1795
H. g. olivaceus I. Geoffroy, 1851
Hapalemur simus Gray, 1870

Genus *Lepilemur* I. Geoffroy, 1851
Lepilemur mustelinus I. Geoffroy, 1851
L. m. mustelinus I. Geoffroy, 1851
L. m. microdon Forsyth Major, 1894
L. m. ruficaudatus Grandidier, 1867
L. m. leucopus Forsyth Major, 1894
L. m. dorsalis Gray, 1870

Subfamily Cheirogaleinae Gregory, 1915
Genus *Cheirogaleus* E. Geoffroy, 1812
Cheirogaleus major E. Geoffroy, 1812 (type species)
Ch. m. major E. Geoffroy, 1812
Ch. m. crossleyi Grandidier, 1870
Cheirogaleus medius E. Geoffroy, 1812
Cheirogaleus trichotis Günther, 1875

Genus *Microcebus* E. Geoffroy, 1828
Microcebus murinus Miller, 1777 (type species)
M. m. murinus Miller, 1777
M. m. smithii Gray, 1842
Microcebus coquereli Grandidier, 1867

Genus *Phaner* Gray, 1870
Phaner furcifer Blainville, 1841

Family Indriidae Burnett, 1828
Genus *Indri* E. Geoffroy and G. Cuvier, 1795
Indri indri Gmelin, 1788

Genus *Avahi* Jourdan, 1834
Avahi laniger Gmelin, 1788
A. l. laniger Gmelin, 1788
A. l. occidentalis Lorenz, 1898

Genus *Propithecus* Bennett, 1832
Propithecus diadema Bennett, 1832 (type species)
P. d. diadema Bennett, 1832
P. d. candidus Grandidier, 1871
P. d. edwardsi Grandidier, 1871
P. d. holomelas Günther, 1875
P. d. perrieri Lavauden, 1931
Propithecus verreauxi Grandidier, 1867
P. v. verreauxi Grandidier, 1867
P. v. coquereli Milne-Edwards, 1867
P. v. coronatus Milne-Edwards, 1871
P. v. deckenii Peters, 1870
P. v. majori Rothschild, 1894

Superfamily Daubentonioidea Gill, 1872
Family Daubentoniidae Gray, 1870
Genus *Daubentonia* E. Geoffroy, 1795
Daubentonia madagascariensis Gmelin, 1788

Infraorder LORISFORMES Gregory, 1915
Family Lorisidae Gregory, 1915
Subfamily Lorisinae Flower and Lydekker, 1891
Genus *Loris* E. Geoffroy, 1796
Loris tardigradus Linnaeus, 1758
L. t. tardigradus Linnaeus, 1758
L. t. grandis Hill and Phillips, 1932
L. t. lydekkerianus Cabrera, 1908
L. t. malabaricus Wroughton, 1917
L. t. nordicus Hill, 1933
L. t. nycticeboides Hill, 1942

Genus *Nycticebus* E. Geoffroy, 1812
Nycticebus coucang Boddaert, 1785 (type species)
N. c. coucang, Boddaert 1785
N. c. bengalensis Fischer, 1804
N. c. tenasserimensis Elliot, 1913
N. c. insularis Robinson, 1917
N. c. hilleri Stone and Rehn, 1902
N. c. bancanus Lyon, 1906
N. c. borneanus Lyon, 1906
N. c. natunae Stone and Rehn, 1902
N. c. javanicus E. Geoffroy, 1812
Nycticebus pygmaeus Bonhote, 1907

Genus *Arctocebus* Gray, 1863
Arctocebus calabarensis J. A. Smith, 1860
A. c. calabarensis J. A. Smith, 1860
A. c. aureus de Winton, 1902

Genus *Perodicticus* Bennett, 1831
Perodicticus potto Müller, 1766
P. p. potto Müller, 1766
P. p. edwardsi Bouvier, 1879
P. p. faustus Thomas, 1910
P. p. ibeanus Thomas, 1910
P. p. juju Thomas, 1910

Subfamily Galaginae Mivart, 1864
Genus *Galago* E. Geoffroy, 1796
Subgenus *Galago* E. Geoffroy, 1796
Galago senegalensis E. Geoffroy, 1796 (type species)
G. s. senegalensis E. Geoffroy, 1796
G. s. albipes Dollman, 1909
G. s. braccatus Elliot, 1907
G. s. dunni Dollman, 1910
G. s. gallarum Thomas, 1901
G. s. granti Thomas and Wroughton, 1907
G. s. moholi A. Smith, 1836
G. s. sotikae Hollister, 1920
G. s. zanzibaricus Matschie, 1893
Galago crassicaudatus E. Geoffroy, 1812
G. c. crassicaudatus E. Geoffroy, 1812
G. c. agisymbanus Coquerel, 1859
G. c. argentatus Lönnberg, 1913
G. c. garnettii Ogilby, 1838

Galago crassicaudatus (continued)
G. c. kikuyuensis Lönnberg, 1912
G. c. lasiotis Peters, 1876
G. c. loennbergi Schwarz, 1930
G. c. monteiri Gray, 1863
G. c. panganiensis Matschie, 1906
G. c. umbrosus Thomas, 1917
Galago alleni Waterhouse, 1837

Subgenus *Euoticus* Gray, 1863
Galago elegantulus Le Conte, 1857 (type species)
G. e. elegantulus Le Conte, 1857
G. e. pallidus Gray, 1863
Galago inustus Schwarz, 1930

Subgenus *Galagoides* A. Smith, 1833
Galago demidovii Fischer, 1808
G. d. demidovii Fischer, 1808
G. d. anomurus Pousargues, 1893
G. d. orinus Lawrence and Washburn, 1936
G. d. phasma Cabrera and Ruxton, 1926
G. d. poensis Thomas, 1904
G. d. murinus Murray, 1859
G. d. thomasi Elliot, 1907

Infraorder TARSIIFORMES Gregory, 1915
Family Tarsiidae Gill, 1872
Genus *Tarsius* Storr, 1780
Tarsius spectrum Pallas, 1779 (type species)
T. s. spectrum Pallas, 1779
T. s. sangirensis Meyer, 1897
T. s. dentatus Miller and Hollister, 1921
T. s. pumilis Miller and Hollister, 1921
T. s. pelengensis Sody, 1949
Tarsius bancanus Horsfield, 1821
T. b. bancanus Horsfield, 1821
T. b. saltator Elliot, 1910
T. b. borneanus Elliot, 1910
T. b. natunensis Chasen, 1940
Tarsius syrichta Linnaeus, 1758
T. s. syrichta Linnaeus, 1758
T. s. fraterculus Miller, 1910
T. s. carbonarius Heude, 1898

Suborder ANTHROPOIDEA Mivart, 1864
Superfamily Ceboidea Simpson, 1931
Family Callitrichidae Thomas, 1903
Subfamily Callitrichinae Thomas, 1903
Genus *Callithrix* Erxleben, 1777
Callithrix jacchus Linnaeus, 1758 (type species)
C. j. jacchus Linnaeus, 1758
C. j. aurita E. Geoffroy, 1812
C. j. flaviceps Thomas, 1903
C. j. geoffroyi Humboldt, 1812
C. j. penicillata E. Geoffroy, 1812
Callithrix argentata Linnaeus, 1766
C. a. argentata Linnaeus, 1766
C. a. leucippe Thomas, 1922
C. a. melanura E. Geoffroy, 1812
Callithrix penicillata E. Geoffroy, 1812
C. h. humeralifer E. Geoffroy, 1812
C. h. chrysoleuca Wagner, 1842

Genus *Cebuella* Gray, 1866
Cebuella pygmaea Spix, 1823
C. p. pygmaea Spix, 1823
C. p. niveiventris Lönnberg, 1940

Genus *Saguinus* Hoffmannsegg, 1807
Saguinus midas group:
Saguinus midas Linnaeus, 1758
S. m. midas Linnaeus, 1758
S. m. tamarin Link, 1795
Saguinus nigricollis group:
Saguinus nigricollis Spix, 1823
Saguinus graellsi Jimenez de la Espada, 1870
Saguinus fuscicollis Spix, 1823
S. f. fuscicollis Spix, 1823
S. f. acrensis Carvalho, 1957
S. f. avilapiresi Hershkovitz, 1966
S. f. crandalli Hershkovitz, 1966
S. f. cruzlimai Hershkovitz, 1966
S. f. fuscus Lesson, 1840
S. f. illigeri Pucheran, 1845
S. f. lagonotus Jimenez de la Espada, 1870
S. f. leucogenys Gray, 1865
S. f. melanoleucus Ribiero, 1912
S. f. nigrifrons I. Geoffroy, 1851

Saguinus fuscicollis (continued)
S. f. tripartitus Milne-Edwards, 1878
S. f. weddelli Deville, 1849
Saguinus mystax group:
Saguinus mystax Spix, 1823
S. m. mystax Spix, 1823
S. m. pileatus I. Geoffroy and Deville, 1848
S. m. pluto Lönnberg, 1926
Saguinus labiatus E. Geoffroy, 1812
S. l. labiatus E. Geoffroy, 1812
S. l. thomasi Goeldi, 1907
Saguinus imperator Goeldi, 1907
Saguinus oedipus group:
Saguinus oedipus Linnaeus, 1758
S. oe. oedipus Linnaeus, 1758
S. oe. geoffroyi Pucheran, 1845
Saguinus leucopus Günther, 1876
Saguinus inustus Schwarz, 1951
Saguinus bicolor group:
Saguinus bicolor Spix, 1823
S. b. bicolor Spix, 1823
S. b. martinsi Thomas, 1912
S. b. ochraceus Hershkovitz, 1966

Genus *Leontopithecus* Lesson, 1840
Leontopithecus rosalia Linnaeus, 1766
L. r. rosalia Linnaeus, 1766
L. r. chrysomelas Kuhl, 1820
L. r. chrysopygus Mikan, 1823

Subfamily Callimiconinae Thomas, 1913
Genus *Callimico* Ribiero, 1911
Callimico goeldii Thomas, 1904

Family Cebidae Swainson, 1835
Subfamily Aotinae Elliot, 1913
Genus *Aotus* Illiger, 1811
Aotus trivirgatus Humboldt, 1812
A. t. trivirgatus Humboldt, 1812
A. t. boliviensis Elliot, 1907
A. t. griseimembra Elliot, 1912
A. t. lemurinus I. Geoffroy, 1843
A. t. microdon Dollman, 1909
A. t. nigriceps Dollman, 1909
A. t. roberti Dollman, 1909

Aotus trivirgatus Humboldt (continued)
- *A. t. bipunctatus* Bole, 1937
- *A. t. azarae* Humboldt, 1812

Genus *Callicebus* Thomas, 1903
Callicebus personatus E. Geoffroy, 1812 (type species)
- *C. p. personatus* E. Geoffroy, 1812
- *C. p. brunello* Thomas, 1913
- *C. p. nigrifons* Spix, 1823
- *C. p. melanochir* Kuhl, 1820

Callicebus moloch Hoffmannsegg, 1807
- *C. m. moloch* Hoffmannsegg, 1807
- *C. m. donacophilus* d'Orbigny, 1836
- *C. m. hoffmannsi* Thomas, 1908
- *C. m. brunneus* Wagner, 1842
- *C. m. cupreus* Spix, 1823
- *C. m. discolor* I. Geoffroy and Deville, 1848
- *C. m. ornatus* Gray, 1866

Callicebus torquatus Hoffmannsegg, 1807
- *C. t. torquatus* Hoffmannsegg, 1807
- *C. t. lugens* Humboldt, 1812
- *C. t. medemi* Hershkovitz, 1963

Subfamily Pitheciinae Mivart, 1865
Genus *Pithecia* Desmarest, 1804
Pithecia pithecia Linnaeus, 1766 (type species)
Pithecia monachus E. Geoffroy, 1812
- *P. m. monachus* E. Geoffroy, 1812
- *P. m. capillamentosa* Spix, 1823

Genus *Chiropotes* Lesson, 1840
Chiropotes satanas Hoffmannsegg, 1807 (type species)
- *Ch. s. satanas* Hoffmannsegg, 1807
- *Ch. s. chiropotes* Humboldt, 1812

Chiropotes albinasus I. Geoffroy and Deville, 1848

Genus *Cacajao* Lesson, 1840
Cacajao melanocephalus Humboldt, 1812 (type species)
Cacajao calvus I. Geoffroy, 1847
Cacajao rubicundus I. Geoffroy and Deville, 1848

Cacajao rubicundus (continued)
- *C. r. rubicundus* I. Geoffroy and Deville, 1848
- *C. r. ucayalii* Thomas, 1928

Subfamily Alouattinae Elliot, 1904
Genus *Alouatta* Lacépède, 1799
Alouatta belzebul Linnaeus, 1766 (type species)
- *A. b. belzebul* Linnaeus, 1766
- *A. b. discolor* Spix, 1823
- *A. b. mexianae* Hagmann, 1908
- *A. b. nigerrima* Lönnberg, 1941
- *A. b. ululata* Elliot, 1912

Alouatta villosa Gray, 1845
Alouatta palliata Gray, 1849
- *A. p. palliata* Gray, 1849
- *A. p. aequatorialis* Festa, 1903
- *A. p. coibensis* Thomas, 1902
- *A. p. mexicana* Merriam, 1902
- *A. p. trabeata* Lawrence, 1933

Alouatta seniculus Linnaeus, 1766
- *A. s. seniculus* Linnaeus, 1766
- *A. s. sara* Elliot, 1910
- *A. s. straminea* Humboldt, 1812
- *A. s. arctoidea* Cabrera, 1940
- *A. s. insulanus* Elliot, 1910

Alouatta caraya Humboldt, 1812
Alouatta fusca E. Geoffroy, 1812
- *A. f. fusca* E. Geoffroy, 1812
- *A. f. beniensis* Lönnberg, 1941
- *A. f. clamitans* Cabrera, 1940

Subfamily Cebinae Mivart, 1865
Genus *Cebus* Erxleben, 1777
Cebus capucinus Linnaeus, 1758 (type species)
- *C. c. capucinus* Linnaeus, 1758
- *C. c. curtus* Bangs, 1905
- *C. c. imitator* Thomas, 1903
- *C. c. limitaneus* Hollister, 1914
- *C. c. nigripectus* Elliot, 1909

Cebus albifrons Humboldt, 1812
- *C. a. albifrons* Humboldt, 1812
- *C. a. hypoleucus* Humboldt, 1812
- *C. a. cesarae* Hershkovitz, 1949

Cobus albifrons (continued)
C. a. pleei Hershkovitz, 1949
C. a. versicolor Pucheran, 1845
C. a. leucocephalus Gray, 1865
C. a. adustus Hershkovitz, 1949
C. a. unicolor Spix, 1823
C. a. yuracus Hershkovitz, 1949
C. a. cuscinus Thomas, 1901
C. a. aequatorialis Allen, 1914
C. a. trinitatis von Pusch, 1941
Cebus nigrivittatus Wagner, 1848
C. n. nigrivittatus Wagner, 1848
C. n. apiculatus Elliot, 1907
C. n. brunneus Allen, 1914
C. n. castaneus I. Geoffroy, 1851
C. n. olivaceus Schomburgk, 1848
Cebus apella Linnaeus, 1758
C. a. apella Linnaeus, 1758
C. a. margaritae Hollister, 1914
C. a. macrocephalus Spix, 1823
C. a. pallidus Gray, 1865
C. a. libidinosus Spix, 1823
C. a. xanthosternos Wied, 1820
C. a. robustus Kuhl, 1820
C. a. nigritus Goldfuss, 1809
C. a. cay Illiger, 1815
C. a. vellerosus I. Geoffroy, 1851
C. a. versutus Elliot, 1910

Genus *Saimiri* Voigt, 1831
Saimiri sciureus Linnaeus, 1758
S. s. sciureus Linnaeus, 1758
S. s. boliviensis d'Orbigny, 1834
S. s. collinsi Osgood, 1916
S. s. macrodon Elliot, 1907
S. s. nigriceps Thomas, 1902
S. s. ustus I. Geoffroy, 1844
Saimiri oerstedii Reinhardt, 1872
S. oe. oerstedii Reinhardt, 1872
S. oe. citrinellus Thomas, 1904

Subfamily Atelinae Miller, 1924
Genus *Ateles* E. Geoffroy, 1806
Ateles paniscus Linnaeus, 1758 (type species)
A. p. paniscus Linnaeus, 1758
A. p. chamek Humboldt, 1812
Ateles belzebuth E. Geoffroy, 1806
A. b. belzebuth E. Geoffroy, 1806
A. b. hybridus I. Geoffroy, 1829
A. b. marginatus E. Geoffroy, 1809
Ateles fusciceps Gray, 1866
A. f. fusciceps Gray, 1866
A. f. robustus J. A. Allen, 1914
Ateles geoffroyi Kuhl, 1820
A. g. geoffroyi Kuhl, 1820
A. g. vellerosus Gray, 1866
A. g. yucatanensis Kellogg and Goldman, 1944
A. g. pan Schlegel, 1876
A. g. frontatus Gray, 1842
A. g. ornatus Gray, 1870
A. g. panamensis Kellogg and Goldman, 1944
A. g. azuerensis Bole, 1937
A. g. grisescens Gray, 1866

Genus *Brachyteles* Spix, 1823
Brachyteles arachnoides E. Geoffroy, 1806

Genus *Lagothrix* E. Geoffroy, 1812
Lagothrix lagothricha Humboldt, 1812 (type species)
L. l. lagothricha Humboldt, 1812
L. l. lugens Elliot, 1907
L. l. poeppigii Schinz, 1844
L. l. cana E. Geoffroy, 1812
Lagothrix flavicauda Humboldt, 1812

Superfamily Cercopithecoidea Simpson, 1931
Family Cercopithecidae Gray, 1821
Subfamily Cercopithecinae Blanford, 1888
Genus *Macaca* Lacépède, 1799
Macaca sylvanus Linnaeus, 1758 (type species)
Macaca sinica Linnaeus, 1771
M. s. sinica Linnaeus, 1771
M. s. aurifrons Pocock, 1931
M. s. opisthomelas Hill, 1942
Macaca radiata E. Geoffroy, 1812
M. r. radiata E. Geoffroy, 1812
M. r. diluta Pocock, 1931
Macaca silenus Linnaeus, 1758

Macaca nemestrina Linnaeus, 1766
M. n. nemestrina Linnaeus, 1766
M. n. leonina Blyth, 1863
M. n. blythii Pocock, 1931
M. n. pagensis Miller, 1903
Macaca fascicularis Raffles, 1821
M. f. fascicularis Raffles, 1821
M. f. aurea E. Geoffroy, 1831
M. f. umbrosa Miller, 1902
M. f. valida Elliot, 1909
M. f. atriceps Kloss, 1919
M. f. argentimembris Kloss, 1911
M. f. capitalis Elliot, 1910
M. f. laeti Elliot, 1909
M. f. pumila Miller, 1900
M. f. fusca Miller, 1903
M. f. lasiae Lyon, 1916
M. f. phaeura Miller, 1903
M. f. mordax Thomas and Wroughton, 1909
M. f. cupida Elliot, 1910
M. f. baweana Elliot, 1910
M. f. philippinensis I. Geoffroy, 1843
M. f. mindanensis Mearns, 1905
M. f. mindora Hollister, 1913
M. f. tua Kellogg, 1944
M. f. limitis Schwarz, 1913
M. f. sublimitis Sody, 1933
Macaca mulatta Zimmerman, 1780
M. m. mulatta Zimmerman, 1780
M. m. vestita Milne-Edwards, 1892
M. m. villosa True, 1894
M. m. mcmahoni Pocock, 1932
Macaca assamensis M'Clelland, 1839
M. a. assamensis M'Clelland, 1839
M. a. pelops Hodgson, 1840
Macaca cyclopis Swinhoe, 1862
Macaca arctoides I. Geoffroy, 1831
M. a. arctoides I. Geoffroy, 1831
M. a. melanota Ogilby, 1839
Macaca thibetana Milne-Edwards, 1870
Macaca fuscata Blyth, 1875
M. f. fuscata Blyth, 1875
M. f. yakui Kuroda, 1941
Macaca nigra Desmarest, 1822
M. n. nigra Desmarest, 1822
M. n. maura Schinz, 1825
Macaca nigra (continued)
M. n. brunnescens, Matschie, 1901
M. n. ochreata Ogilby, 1840
M. n. nigrescens Temminck, 1849
M. n. hecki Matschie, 1901
M. n. tonkeana Heyer, 1899
Genus *Cercocebus* E. Geoffroy, 1812
Cercocebus albigena group:
Cercocebus albigena Gray, 1850
C. a. albigena Gray, 1850
C. a. johnstoni Lydekker, 1900
C. a. zenkeri Schwarz, 1910
Cercocebus aterrimus Oudemans, 1890
Cercocebus torquatus group:
Cercocebus torquatus Kerr, 1792
Cercocebus atys Audebert, 1797 (type species)
C. a. atys Audebert, 1797
C. a. lunulatus Temminck, 1853
Cerocebus galeritus Peters, 1879
C. g. galeritus Peters, 1879
C. g. agilis Rivière, 1886
C. g. chrysogaster Lydekker, 1900

Genus *Papio* Erxleben, 1777
Papio cynocephalus group:
Papio anubis Lesson, 1827
P. a. anubis Lesson, 1827
P. a. tibestianus Dekeyser and Derivot, 1960
P. a. heuglini Matschie, 1898
P. a. neumanni Matschie, 1897
Papio cynocephalus Linnaeus, 1766
P. c. cynocephalus Linnaeus, 1766
P. c. ibeanus Thomas, 1893
P. c. kindae Lönnberg, 1919
Papio papio Desmarest, 1820 (type species)
Papio ursinus Kerr, 1792
P. u. ursinus Kerr, 1792
P. u. orientalis Goldblatt, 1926
P. u. ruacana Shortridge, 1942
P. u. griseipes Pocock, 1911
Papio hamadryas group:
Papio hamadryas Linnaeus, 1758

Genus *Mandrillus* Ritgen, 1824
Mandrillus sphinx Linnaeus, 1758 (type species)

Mandrillus leucophaeus F. Cuvier, 1807
M. l. leucophaeus F. Cuvier, 1807
M. l. poensis Zukowsky, 1922

Genus *Theropithecus* I. Geoffroy, 1841
Theropithecus gelada Rüppell, 1835
Th. g. gelada Rüppell, 1835
Th. g. obscurus Heuglin, 1863

Genus *Cercopithecus* Linnaeus, 1758
Subgenus *Cercopithecus* Linnaeus, 1758
Cercopithecus aethiops group:
Cercopithecus aethiops Linnaeus, 1758
C. ae. aethiops Linnaeus, 1758
C. ae. hilgerti Neumann, 1902
C. ae. ellenbecki Neumann, 1902
C. ae. zavattarii de Beaux, 1943
C. ae. tantalus Ogilby, 1841
C. ae. budgetti Pocock, 1907
C. ae. marrensis Thomas and Wroughton, 1923
Cercopithecus pygerythrus F. Cuvier, 1821
C. p. pygerythrus F. Cuvier, 1821
C. p. arenarius Heller, 1913
C. p. centralis Neumann, 1900
C. p. callidus Hollister, 1912
C. p. johnstoni Pocock, 1907
C. p. excubitor Schwarz, 1926
C. p. nesiotes Schwarz, 1926
C. p. rufoviridis I. Geoffroy, 1842
C. p. whytei Pocock, 1907
C. p. cloeti Roberts, 1931
C. p. ngamiensis Roberts, 1932
C. p. marjoriae Bradfield, 1936
C. p. cynosuros Scopoli, 1786
Cercopithecus sabaeus Linnaeus, 1766
Cercopithecus cephus group:
Cercopithecus cephus Linnaeus, 1758
C. c. cephus Linnaeus, 1758
C. c. cephodes Pocock, 1907
Cercopithecus diana group:
Cercopithecus diana Linnaeus, 1758 (type species)
C. d. diana Linnaeus, 1758
C. d. roloway Schreber, 1774
C. d. dryas Schwarz, 1932
Cercopithecus lhoesti group:
Cercopithecus lhoesti Sclater, 1898
Cercopithecus preussi Matschie, 1898
Cercopithecus hamlyni group:
Cercopithecus hamlyni Pocock, 1907
Cercopithecus mitis group:
Cercopithecus mitis Wolf, 1822
C. m. mitis Wolf, 1822
C. m. boutourlinii Giglioli, 1887
C. m. stuhlmanni Matschie, 1893
C. m. doggetti Pocock, 1907
C. m. maesi Lönnberg, 1919
C. m. kandti Matschie, 1905
C. m. schoutedeni Schwarz, 1928
C. m. opisthostictus Sclater, 1893
Cercopithecus albogularis Sykes, 1831
C. a. albogularis Sykes, 1831
C. a. albotorquatus Pousargues, 1896
C. a. monoides I. Geoffroy, 1841
C. a. phylax Schwarz, 1927
C. a. kibonotensis Lönnberg, 1910
C. a. kolbi Neumann, 1902
C. a. moloneyi Sclater, 1893
C. a. francescae Thomas, 1902
C. a. nyasae Schwarz, 1928
C. a. erythrarchus Peters, 1852
C. a. schwarzi Roberts, 1931
C. a. labiatus I. Geoffroy, 1841
Cercopithecus mona group:
Cercopithecus mona Schreber, 1774
Cercopithecus campbelli Waterhouse, 1838
C. c. campbelli Waterhouse, 1838
C. c. lowei Thomas, 1923
Cercopithecus wolfi Meyer, 1891
C. w. wolfi Meyer, 1891
C. w. pyrogaster Lönnberg, 1919
C. w. elegans Dubois and Matschie, 1912
Cercopithecus denti Thomas, 1907
Cercopithecus pogonias Bennett, 1833
C. p. pogonias Bennett, 1833
C. p. grayi Fraser, 1850
C. p. nigripes du Chaillu, 1860
C. p. schwarzianus Schouteden, 1946
Cercopithecus neglectus group:
Cercopithecus neglectus Schlegel, 1876
Cercopithecus nictitans group:
Cercopithecus nictitans Linnaeus, 1766

C. n. nictitans Linnaeus, 1766
C. n. martini Waterhouse, 1841
C. n. stampflii Jentink, 1888
Cercopithecus petaurista Schreber, 1774
C. p. petaurista Schreber, 1774
C. p. buettikoferi Jentink, 1886
Cercopithecus ascanius Audebert, 1799
C. a. ascanius Audebert, 1799
C. a. katangae Lönnberg, 1919
C. a. whitesidei Thomas, 1909
C. a. montanus Lorenz, 1914
C. a. schmidti Matschie, 1892
Cercopithecus erythrotis Waterhouse, 1838
C. e. erythrotis Waterhouse, 1838
C. e. camerunensis Haymann, 1940
C. e. sclateri Pocock, 1904
Cercopithecus erythrogaster, Gray, 1866

Subgenus *Miopithecus* I. Geoffroy, 1842
Cercopithecus talapoin Schreber, 1774
C. t. talapoin Schreber, 1774
C. t. ansorgei Pocock, 1907
C. t. vleeschouwersi Poll, 1940
C. t. pilettei Lönnberg, 1919

Subgenus *Allenopithecus* Lang, 1923
Cercopithecus nigroviridis Pocock, 1907

Genus *Erythrocebus* Trouessart, 1897
Erythrocebus patas Schreber, 1774
E. p. patas Schreber, 1774
E. p. villiersi Dekeyser, 1950
E. p. pyrronotus Hemprich and Ehrenberg, 1829
E. p. baumstarki Matschie, 1905

Subfamily Colobinae Elliot, 1913
Genus *Presbytis* Eschscholtz, 1821
Presbytis aygula group:
Presbytis aygula Linnaeus, 1758 (type species)
P. a. aygula Linnaeus, 1758
P. a. fredericae Sody, 1930
P. a. hosei Thomas, 1889
P. a. canicrus Miller, 1934
P. a. nubila Miller, 1942
P. a. sabana Thomas, 1893
P. a. thomasi Collett, 1892
P. a. margae Hooijer, 1948
Presbytis melalophos Raffles, 1821
P. m. melalophos Raffles, 1821
P. m. femoralis Martin, 1838
P. m. australis Miller, 1913
P. m. siamensis Müller and Schlegel, 1841
P. m. nubigena Elliot, 1909
P. m. robinsoni Thomas, 1910
P. m. rhionis Miller, 1903
P. m. cana Miller, 1906
P. m. natunae Thomas and Hartert, 1894
P. m. catemana Lyon, 1908
P. m. percura Lyon, 1908
P. m. paenulata Chasen, 1940
P. m. sumatrana Müller and Schlegel, 1841
P. m. batuana Miller, 1903
P. m. ferruginea Schlegel, 1876
P. m. fluviatilis Chasen, 1940
P. m. fuscomurina Elliot, 1906
P. m. chrysomelas Müller, 1838
P. m. cruciger Thomas, 1892
Presbytis frontata Müller, 1838
P. f. frontata Müller, 1838
P. f. nudifrons Elliot, 1909
Presbytis rubicunda Müller, 1838
P. r. rubicunda Müller, 1838
P. r. rubida Lyon, 1911
P. r. ignita Dollman, 1909
P. r. carimatae Miller, 1906
P. r. chrysea Davis, 1962
Presbytis entellus group:
Presbytis entellus Dufresne, 1797
P. e. entellus Dufresne, 1797
P. e. schistacea Hodgson, 1840
P. e. hypoleucos Blyth, 1841
P. e. dussumieri I. Geoffroy, 1843
P. e. anchises Blyth, 1844
P. e. priam Blyth, 1844
P. e. thersites Blyth, 1847
P. e. lania Elliot, 1909

Presbytis entellus Dufresne (continued)
P. e. achilles Pocock, 1928
P. e. ajax Pocock, 1928
P. e. achates Pocock, 1928
P. e. iulus Pocock, 1928
P. e. aeneas Pocock, 1928
P. e. elissa Pocock, 1928
P. e. priamellus Pocock, 1928
Presbytis senex group:
Presbytis senex Erxleben, 1777
P. s. senex Erxleben, 1777
P. s. vetulus Erxleben, 1777
P. s. nestor Bennett, 1833
P. s. monticola Kelaart, 1850
P. s. harti Deraniyagala, 1954
Presbytis johnii Fischer, 1829
Presbytis cristata group:
Presbytis cristata Raffles, 1821
P. c. cristata Raffles, 1821
P. c. pyrrhus Horsfield, 1823
P. c. sondaica Robinson and Kloss, 1919
P. c. kohlbruggei Sody, 1931
P. c. ultima Elliot, 1910
P. c. vigilans Miller, 1913
P. c. germaini Milne-Edwards, 1876
P. c. atrior Pocock, 1928
Presbytis pileata Blyth, 1843
P. p. pileata Blyth, 1843
P. p. shortridgei Wroughton, 1915
P. p. brahma Wroughton, 1916
P. p. durga Wroughton, 1916
P. p. tenebrica Hinton, 1923
Presbytis geei Gee, 1956
Presbytis obscura Reid, 1837
P. o. obscura Reid, 1837
P. o. flavicauda Elliot, 1910
P. o. halonifer Cantor, 1845
P. o. carbo Thomas and Wroughton, 1909
P. o. styx Kloss, 1911
P. o. seimundi Chasen, 1940
P. o. sanctorum Elliot, 1910
Presbytis phayrei Blyth, 1847
P. p. phayrei Blyth, 1847
P. p. crepuscula Elliot, 1909
P. p. shanica Wroughton, 1917

Presbytis phayrei Blyth (continued)
P. p. ruhei Knottnerus-Meyer, 1933
Presbytis francoisi Pousargues, 1898
P. f. francoisi Pousargues, 1898
P. f. poliocephala Trouessart, 1911
P. f. laotum Thomas, 1921
P. f. delacouri Osgood, 1932
Presbytis potenziani Bonaparte, 1856
P. p. potenziani Bonaparte, 1856
P. p. siberu Chasen and Kloss, 1927

Genus *Rhinopithecus* Milne-Edwards, 1872
Rhinopithecus roxellanae Milne-Edwards, 1870 (type species)
Rh. r. roxellanae Milne-Edwards, 1870
Rh. r. bieti Milne-Edwards, 1897
Rh. r. brelichi Thomas, 1903
Rhinopithecus avunculus Dollman, 1912

Genus *Pygathrix* E. Geoffroy, 1812
Pygathrix nemaeus Linnaeus, 1771
P. n. nemaeus Linnaeus, 1771
P. n. nigripes Milne-Edwards, 1871

Genus *Nasalis* E. Geoffroy, 1812
Nasalis larvatus van Wurmb, 1781

Genus *Simias* Miller, 1903
Simias concolor Miller, 1903
S. c. concolor Miller, 1903
S. c. siberu Chasen and Kloss, 1927

Genus *Colobus* Illiger, 1811
Subgenus *Colobus* Illiger, 1811
Colobus polykomos Zimmerman, 1780 (type species)
C. p. polykomos Zimmerman, 1780
C. p. adolfifriederici Matschie, 1914
C. p. angolensis Sclater, 1860
C. p. cordieri Rahm, 1959
C. p. cottoni Lydekker, 1905
C. p. dollmani Schwarz, 1927
C. p. palliatus Peters, 1868
C. p. prigoginei Verheyen, 1959
C. p. ruwenzorii Thomas, 1901
C. p. satanas Waterhouse, 1838

Colobus polykomos (continued)
C. p. sharpei Thomas, 1902
C. p. vellerosus I. Geoffroy, 1834
Colobus guereza Rüppell, 1835
C. g. guereza Rüppell, 1835
C. g. caudatus Thomas, 1885
C. g. dodingae Matschie, 1913
C. g. gallarum Neumann, 1902
C. g. kikuyuensis Lönnberg, 1912
C. g. matschiei Neumann, 1899
C. g. occidentalis Rochebrune, 1886–7
C. g. percivali Heller, 1913
C. g. poliurus Thomas, 1901
C. g. uellensis Matschie, 1913

Subgenus *Procolobus* Rochebrune, 1886–7
Colobus verus van Beneden, 1838

Subgenus *Piliocolobus* Rochebrune, 1886–7
Colobus badius Kerr, 1792 (type species)
C. b. badius, Kerr 1792
C. b. bouvieri Rochebrune, 1887
C. b. ellioti Dollman, 1909
C. b. foai Poursargues, 1899
C. b. gordonorum Matschie, 1900
C. b. graueri Dollman, 1909
C. b. gudoviusi Matschie, 1914
C. b. langi J. A. Allen, 1925
C. b. metternichi Krumbiegel, 1942
C. b. nigrimanus Trouessart, 1906
C. b. oustaleti Trouessart, 1906
C. b. pennantii Waterhouse, 1838
C. b. powelli Matschie, 1913
C. b. preussi Matschie, 1900
C. b. rufomitratus Peters, 1879
C. b. temminckii Kuhl, 1820
C. b. tephrosceles Elliot, 1907
C. b. tholloni Rivière, 1886
C. b. waldroni Hayman, 1936
Colobus kirkii Gray, 1868

Superfamily Hominoidea Simpson, 1931
Family Hylobatidae Blyth, 1875
Genus *Hylobates* Illiger, 1811
Hylobates lar Linnaeus, 1771 (type species)
H. l. lar Linnaeus, 1771
H. l. longimanus Schreber, 1774
Hylobates pileatus Gray, 1861
Hylobates moloch Audebert, 1797–8
H. m. moloch Audebert, 1797–8
H. m. pongoalsoni Sody, 1949
H. m. muelleri Martin, 1841
H. m. albibarbis Lyon, 1911
H. m. abbotti Kloss, 1929
H. m. funereus I. Geoffroy, 1850
Hylobates agilis F. Cuvier, 1821
Hylobates hoolock Harlan, 1834
Hylobates concolor Harlan, 1826
H. c. concolor Harlan, 1826
H. c. hainanus Thomas, 1892
H. c. lu Delacour, 1951
H. c. leucogenys Ogilby, 1840
H. c. siki Delacour, 1951
H. c. gabriellae Thomas, 1909
Hylobates klossii Miller, 1903

Genus *Symphalangus* Gloger, 1841
Symphalangus syndactylus Raffles, 1821
S. s. syndactylus Raffles, 1821
S. s. continentis Thomas, 1908

Family Pongidae Elliot, 1913
Genus *Pongo* Lacépède, 1799
Pongo pygmaeus Linnaeus, 1760
P. p. pygmaeus Linnaeus, 1760
P. p. abelii Lesson, 1827

Genus *Pan* Oken, 1816
Pan troglodytes Blumenbach, 1779 (type species)
P. t. troglodytes Blumenbach, 1779
P. t. verus Schwarz, 1934
P. t. schweinfurthii Giglioli, 1872
Pan paniscus Schwarz, 1929

Genus *Gorilla* I. Geoffroy, 1852
Gorilla gorilla Savage and Wyman, 1847
G. g. gorilla Savage and Wyman, 1847
G. g. beringei Matschie, 1903
G. g. graueri Matschie, 1914
Family Hominidae Gray, 1825
Genus *Homo* Linnaeus, 1758
Homo sapiens Linnaeus, 1758

Some Common Synonyms

Alouatta guariba	=	*Alouatta fusca*
Anthropopithecus	=	*Pan*
Aotes	=	*Aotus*
Ateles ater	=	*Ateles paniscus*
Ateles melanochir	=	*Ateles geoffroyi*
Ateles variegatus	=	*Ateles belzebuth*
Bonobo	=	*Pan paniscus*
Brachitanytes	=	*Hylobates klossii*
Brachyurus	=	*Cacajao*
Callithrix santaremensis	=	*Callithrix humeralifer*
Cebus azarae	=	*Cebus apella*
Cebus cirrifer	=	*Cebus apella*
Cebus fatuellus	=	*Cebus apella*
Cebus hypoleucus	=	*Cebus capucinus*
Cercopithecus leucampyx	=	*Cercopithecus mitis*
Chaeropithecus	=	*Papio*
Cheirogaleus commersonii	=	*Cheirogaleus major*
Cheirogaleus milii	=	*Cheirogaleus major*
Cheirogaleus samati	=	*Cheirogaleus medius*
Chiromys	=	*Daubentonia*
Chlorocebus	=	*Cercopithecus*
Chrysothrix	=	*Saimiri*
Colobus abyssinicus	=	*Colobus guereza*
Comopithecus	=	*Papio hamadryas*
Corypithecus	=	*Presbytis*
Cynocephalus	=	*Papio*
Cynomolgus	=	*Macaca fascicularis*
Cynopithecus	=	*Macaca nigra*
Eriodes	=	*Brachyteles*
Galago pusillus	=	*Galago (Galagoides)*
Galeocebus	=	*Lepilemur*
Gastrimargus	=	*Lagothrix*
Gymnopyga	=	*Macaca nigra*
Hapale	=	*Callithrix*
Hemigalago	=	*Galago (Galagoides)*
Indri brevicaudatus	=	*Indri indri*
Inuus	=	*Macaca sylvanus*
Jacchus vulgaris	=	*Callithrix jacchus*
Kasi	=	*Presbytis*

Some Common Synonyms (continued)

Lagothrix hendeei	=	*Lagothrix flavicauda*
Lasiopyga	=	*Cercopithecus*
Leontideus	=	*Leontopithecus*
Leontocebus	=	*Leontopithecus* and/or *Saguinus*
Lepidolemur	=	*Lepilemur*
Lichanotus	=	*Avahi*
Lophocebus	=	*Cercocebus*
Lophopithecus	=	*Presbytis*
Lyssodes	=	*Macaca*
Macaca irus	=	*Macaca fascicularis*
Macaca rhesus	=	*Macaca mulatta*
Macaca speciosa	=	*Macaca arctoides*
Magus	=	*Macaca sylvanus*
Maimon	=	*Papio*
Marikina	=	*Saguinus*
Mico	=	*Callithrix argentata*
Microrhynchus	=	*Avahi*
Midas	=	*Saguinus*
Mixocebus	=	*Lepilemur*
Mycetes	=	*Alouatta*
Mystax	=	*Saguinus*
Mystax devillei	=	*Saguinus fuscicollis*
Neocothurus	=	*Cacajao*
Nomascus	=	*Hylobates concolor*
Nyctipithecus	=	*Aotus*
Oedipomidas oedipus	=	*Saguinus oedipus oedipus*
Oedipomidas spixii	=	*Saguinus oedipus geoffroyi*
Opolemur	=	*Cheirogaleus*
Otogale	=	*Galago (Galago)*
Otolemur	=	*Galago (Galago)*
Ouakaria	=	*Cacajao*
Papio comatus	=	*Papio ursinus*
Papio doguera	=	*Papio anubis*
Papio porcarius	=	*Papio ursinus*
Pithecia chrysocephala	=	*Pithecia pithecia*
Presbypithecus	=	*Presbytis*
Presbytis femoralis	=	*Presbytis melalophos*
Presbytis maura	=	*Presbytis cristata*
Presbytis mitrata	=	*Presbytis aygula*

Some Common Synonyms (continued)

Presbytis pyrrhus	=	*Presbytis cristata*
Presbytiscus	=	*Rhinopithecus avunculus*
Prolemur	=	*Hapalemur simus*
Rhesus	=	*Macaca*
Rhinostigma	=	*Cercopithecus hamlyni*
Semnocebus	=	*Cercocebus*
Semnopithecus	=	*Presbytis*
Seniocebus	=	*Saguinus bicolor*
Silenus	=	*Macaca*
Simia satyrus	=	*Pan troglodytes* or *Pongo pygmaeus*

Some Common Synonyms (continued)

Stenops	=	*Loris* and/or *Nycticebus*
Stentor	=	*Alouatta*
Tamarin	=	*Saguinus*
Tamarinus	=	*Saguinus*
Tana	=	*Tupaia (Lyonogale)*
Tardigradus	=	*Loris* and/or *Nycticebus*
Tarsius philippinensis	=	*Tarsius syrichta*
Trachypithecus	=	*Presbytis*
Troglodytes	=	*Pan*
Varecia	=	*Lemur variegatus*
Yarkea	=	*Pithecia*
Zati	=	*Macaca*

For more detailed synonymies, reference can be made to the following books and checklists:

ALLEN, G. M.: A checklist of Africa mammals. Bull. Mus. comp. Zool. Harv. *83:* 1–763 (1939).

CABRERA, A.: Catalogo de los mamiferos de America del Sur. Buenos Aires y Peru: Instituto Nacional de Investigacion de la Ciencias Naturales, Ciencio Zoologica. *IV:* No. 1 (1957).

CHASEN, F. N.: A handlist of Malaysian mammals. Bull. Raffles Mus. *15:* 1–209 (1940).

ELLERMAN, J. R. and MORRISON-SCOTT, T. C. S.: Checklist of Palearctic and Indian mammals 1758–1946. (Trustees of the British Museum, London 1951).

ELLERMAN, J. R.; MORRISON-SCOTT, T. C. S., and HAYMAN, R. W.: Southern African mammals 1758–1951: a reclassification (Trustees of the British Museum, London 1953).

HALL, E. R. and KELSON, K. R.: The mammals of North America, vol. I (The Ronald Press Company, New York 1959).

HILL, W. C. OSMAN: Primates. Comparative Anatomy and Taxonomy (University Press, Edinburgh 1953–1970). – Strepsirhini, vol. I (1953). – Haplorhini. Tarsioidae, vol. II (1955). Pithecoidea. Platyrrhini. Hapalidae, vol. III (1957). – Cebidae, part A, vol. IV (1960). – Cebidae, part B, vol. V (1962). – Catarrhini. Cercopithecoidea. Cercopithecinae, vol. VI (1966). – Cynopithecinae. Papio, Mandrillus, Theropithecus, vol. VIII (1970).

LAURIE, E. M. O. and HILL, J. E.: List of land mammals of New Guinea, Celebes and adjacent islands, 1758–1952. (Trustees of the British Museum, London 1954).

I. Natural History of Primates

Pathology of Simian Primates, Part I, pp. 2–10 (Karger, Basel 1972)

The Rational Use of Primates in Biomedical Experiments

B. A. LAPIN

Institute of Experimental Pathology and Therapy, USSR Academy of Medical Sciences, Sukhumi

Contents

General ... 2
Infectious Diseases ... 4
Neuroses ... 4
Atherosclerosis ... 5
Pharmacology ... 6
Immunology ... 6
Reproduction ... 7
Oncology ... 7
References ... 9

General

The use of primates in biomedical research has increased enormously during recent years. This has resulted in an increase of establishments in which monkeys are the main subjects of study and explains the growing use of monkeys as experimental subjects in nonspecialized scientific research laboratories.

The numbers of monkeys imported yearly from the natural habitat today reaches a few hundred thousands.

Everything mentioned above is explained by the great anatomo-physiologic similarity of monkeys to man in contrast with other animals.

The most characteristic models of many human diseases can be produced in monkeys and certain human diseases can be reproduced only in monkeys. Indeed, during the recent years some scientists have expressed an opinion that experimental biomedical research should be based entirely on monkeys.

However, this trend arouses apprehension in ecologists, primatologists and other scientists because the uncontrolled capture and destruction of an enormous amount of primates could finally result in the reduction of numbers, or even the extinction of these species in the wild. It should also be taken into consideration that experiments on monkeys are expensive, much more so than those on small laboratory rodents.

It seems that the use of monkeys in biomedical research is justified when the use of other animals for this purpose is impossible.

It is impossible to discuss monkeys in general, as even within one and the same genus different species and sub-species may differ sharply in those physiological systems which are important for the study of certain phenomena or pathologic syndromes. In choosing an experimental animal for the study of this or that pathologic process, we must first discover in each case to what extent the findings reveal the general regularities of a certain pathologic process, and to what extent they reflect morpho-physiologie peculiarities of the experimental monkey species used. The above is very important not only in the study of characteristic syndromes such as allergy, inflammation, dystrophy etc., but also in concrete nosologic forms of disease.

It is impossible to discuss all the aspects of the problem touched in this article; undoubtedly investigations oriented on monkeys such as studies of their normal biology, physiology, psychology etc. are of importance. These investigations have also a utilitarian practical side besides their general biologic importance. They provide data enabling us to indicate in detail the suitability of monkey species as experimental subjects. Investigations oriented on the study of 'spontaneous' pathology occurring outside the experimental situation in the wild, in the natural habitat, and in captivity, have a special significance from this point of view. For example, in the study of a certain pathologic process it is advisable to choose such experimental animals as are subject to this process spontaneously.

The incidence of 'spontaneous' diseases in monkeys maintained in captivity (colonies and zoos) is closely connected with the problem of adaptation to the new environment and regimen and acclimatization. This incidence may prove useful in revealing factors associated with the environment and the situations which cause this or that form of pathology.

Infectious Diseases

Infectious disease is one of the most widely spread subjects for which monkeys are used in medical experiments. As already mentioned, the greatest number of human infectious diseases can be produced in monkeys in a more characteristic form than in other animals. The production of certain human infectious diseases is possible only in monkeys. Bacterial infections such as dysentery and virus infections such as pox, measles, mumps, etc. are examples.

Interesting data on the transmissibility of hepatitis from chimpanzee to man and vice versa have appeared during the last years [14] [*vide* SMETANA, part II]. On the basis of experiments in Old World monkeys BEARCROFT [1] reached the conclusion that *Erythrocebus patas patas*, in which passages of the virus were possible without a decrease of its pathogenicity, showed the greatest sensitivity to the human virus of hepatitis. DEINHARDT'S [3] studies have shown that inoculation of marmosets with blood from human patients suffering from hepatitis has caused hepatitis in them; such studies are of extreme interest. In assessing the results of these experiments, if we suppose that the hepatitis in them is caused by their own activated virus, we cannot avoid the question why is the blood of human patients suffering from hepatitis complementary to the simian virus of hepatitis? If in DEINHARDT'S test hepatitis in marmosets is caused by their own virus, we must acknowledge a considerable antigenic similarity between the human and simian virus as of hepatitis, just as the sera of human convalescent patients in experiments *in vivo* neutralizes the virus in monkeys. [Personal communication.]

The studies revealing the etiology of diseases of obscure but presumably infectious nature are also highly promising. The use of monkeys made it possible to determine the viral nature of kuru [4, Studies in the chimpanzee]. These studies have led also to the suggestion of a viral etiology of amyotrophic lateral sclerosis as in the studies [18] in macaques.

Neuroses

The use of monkeys in studies on neuroses and their somatic manifestations is also very important. Such somatic disturbances as hypertension, coronary insufficiency, myocardial infarction, stable gastric achylia, long-term amenorrhea in young females have appeared against the background of neurosis in our tests with *Papio hamadryas* and *Macaca mulatta*.

The pathologic states mentioned above are of course not all the varieties of pathology caused by neurosis. However, they are the most frequent and obvious. Coronary insufficiency and myocardial infarction caused by neurosis in monkeys are not connected with organic changes in the coronary arteries and are thus fully caused by functional vasomotor disturbances of the coronary circulation.

The appearance of neurogenic states in monkeys is most frequently caused by 'social conflicts' (the disturbance of normal social relations); the similarity of somatic neurotic manifestations in man and monkeys makes us consider this aspect of experimental studies in monkeys to be one of the most promising. Our experimental tests on the production of neurosis in monkeys, and the resulting pathology, were begun only when it was already found that stable hypertension and coronary insufficiency was appearing in monkeys living in captivity for a long period of time and being used in different experiments other than those on neurosis. The assumption that these pathologic states were caused by neurosis connected with inadequate conditions of maintenance was confirmed by specially designed experiments. Thus this aspect of experimental use resulted from the study of 'spontaneous disease'.

The pecularities of the primate higher nervous activity, the functions of their brain, make tests of psychotropic substances in them rewarding.

Atherosclerosis

The studies on atherosclerosis in monkeys are of great interest. Only a few investigators managed to produce atherosclerosis in monkeys experimentally [10]; [16]; [9] though it is sometimes encountered 'spontaneously' in old monkeys (*M. mulatta*, *P. hamadryas*, etc.). Other published articles entitled 'Atherosclerosis in monkeys' have dealt with changes not related to clearly defined atherosclerosis; they describe changes in cholesterol and lipoprotein levels in blood [5]; [13]; [2]. These changes did not necessarily occur simultaneously with the development of atherosclerosis and did not reflect the extent to which the arteries were affected; the direct connection of lipoidosis with atherosclerosis is arguable. Some reports on atherosclerosis refer to the initial changes of its development which could be observed only microscopically or electron microscopically [6]; [15]. Such changes do not always develop into atherosclerosis in clearly defined form. When plaques are developed the number and extent of those changes does not usually exceed those occurring spontaneously in captivity. Sometimes such changes were detected both in experimental and control animals [12]; [19]. Some investigators also obtained negative results in general [7]; [8].

Under the same conditions and a similar cholesterol diet, we did not manage to produce strongly pronounced atherosclerotic manifestations with the exception of hypercholesterolemia and increased β-lipoprotein levels at the beginning of the experiment, and lipid spots occupying a considerable area of the intima, and single plaques in aorta and iliac arteries. This makes us widen the search for the factors involved in the development of clinical atherosclerosis against the background of TAYLOR's cholesterol diet.

Pharmacology

Studies on the pharmacological effect of different drugs have become extensive during recent years. For example, the pharmacological effect of lipid-soluble preparations is similar in monkeys and man in similar concentrations in the blood. But we must be very careful when directly transferring the preparation dosages from monkey to man as has already been stated in the literature many times.

The study of the possible teratogenic effect of new drugs on monkeys is also promising. Experiments on contergan, for instance, have shown that it is teratogenic only in monkeys and man.

Immunology

Studies in the area of noninfectious immunology and organ and tissue transplantation have been much developed recently. This tendency is based on the great immunologic similarity of human and monkey proteins.

Investigations of iso- and auto-immunization processes, reactions, 'graft against host', 'host against graft', immuno-depressant dosages, discovering the role of different protein allotypes and cell clones in reactions of incompatibility, have taken place under conditions closely copying those of the human organism. These investigations suggest an insignificant antigenic activity of monkey sera in relation to the human organism. This creates an additional aspect of the use of monkeys in the study of possible pratical applications of monkey plasma, albumins, and gamma-globulins (including immune) in medicine.

Reproduction

As only monkeys have an ovarian-menstrual cycle identical with human females, these animals become indispensible in the study of gynaecological and endocrine diseases of the female sexual sphere for the investigation of the problem of contraception.

Oncology

The advisability of the utilization of monkeys in oncologic experiments involving the use of radioactive and chemical carcinogens is disputable. The research on carcinogenic effects of chemical and radioactive substances in monkeys did not contribute any important new data to the understanding of human tumor pathogenesis up to the present time. These experiments in fact determined the possibility of experimental tumor production and also certain pecularities of their development. One of these pecularities is first of all a very long latent period from the moment of carcinogen administration to the time of tumor development [11]. This latent period is sometimes found to be more than 10 years. It was determined that there existed a definite dependance between the carcinogen dose and the length of the latent period. The latter was considerably reduced when the dose of carcinogen was increased.

This long latent period is probably due to the long life span of monkeys and seems to be a characteristic of the life cycle of the organism. However this long latent period hampers experimental research in the area of oncogenesis. The similarity of the endocrine systems of simians and humans opens wide prospects for studies in the interrelationship between hormonal disturbances and tendencies to tumor development. These tendencies are seen not only in the anatomo-physiologic similarity of human and simian endocrine systems but also in successful attempts to produce pretumor changes in mammary glands and uterus (endometriosis and fibrocystic mastopathy) in monkeys in the longterm administration of sex hormones [17].

The discovery that viral tumors can be produced in monkeys has also opened wide perspectives. In contrast with many animals usually used in oncologic experiments in which the percent of 'spontaneous' tumors is high, spontaneous tumors in monkeys are rare. This enables us to treat monkeys as animals with a low incidence of cancer. From this point of view if the production of the tumor in animals when injected with filtrates or other cell-free material is successful we can consider with certainty that the tumor is induced by aninoculated virus. At the same time the anatomo-physiological similarity of monkeys and man enhances the probability of adaptation of the presumed 'human tumor virus' and the induction of tumor processes in monkeys with its help. These prospects justify the widest use of monkeys in the study of the possible virus nature of human tumors.

Research oncologists mostly believe that the bulk of spontaneous tumors in animals is of viral nature, and that man seems to be an exception. The use of monkeys to investigate this is also highly promising, since we have failed in this field using other animals in our experiments.

It should not be understood from this that the inoculation of a few monkeys of this or that species with human tumor material (or its filtrates) can answer the question of whether a certain human tumor is of viral nature.

It is known that mature virus cannot for the most part be detected in animal tumors induced by DNA-containing viruses. Thus, if a certain human tumor is virus-associated and this is a DNA-containing virus, the induction of the tumor may not occur due to the absence of infectious virus in the material. But in case the tumor contains a mature virus, a long period of time during which the virus adapts to monkey tissues and causes a pathologic process may be expected. The possibility that this presumably human tumor virus may not cause a tumor but some other process in monkeys can also not be excluded. And finally, the different sensitivity of different monkey species to certain pathogenic factors, including tumor viruses, must always be taken into consideration.

Thus, in view of the extraordinary importance of determining a viral role in human oncogenesis, the wide use of different monkey species with a view to establishing the role of viruses in tumor induction in the primate order and in isolating such viruses from different human tumors is fully justified and well-reasoned.

All the apsects of the possible use of monkeys in experimental medicine touched on in this article determine only general and to our mind the most important directions of experimental primatology. However, we think it necessary to remind experimenters again of the great importance attached to the choice of monkeys as experimental subjects, especially when dealing with such expensive animals as apes.

And finally we should like to say that the use of monkeys born in a colony as experimental subjects is strongly recommended as the family tree, the character of disease and contacts and so on, are known. The latter also helps to regulate the capture of monkeys in their natural habitat and will lead to the preservation of these valuable species.

References

1 Bearcroft, W. G. C.: Hepatitis in African monkeys. Nature *197:* 806–807 (1963).

2 Chazov, E. I.; Ushkalov, A. F., and Klembovsky, A. I.: Early changes in arteries in experimental atherosclerosis. Arkhiv patol. *25:* 29–36 (1963).

3 Deinhardt, F.; Holmes, A. W.; Capps, R. B., and Popper, H.: Studies on the transmission of human viral hepatitis to marmoset monkeys. I. Transmission of disease serial passages and description of liver lesions. J. exp. Med. *125:* 673–688 (1967).

4 Gajdusek, D. C.: Kuru and experimental kuru in chimpanzees. Meeting of investigators on the genetics of primitive groups. Geneva 1967, 3–7 July, Geneva WHO (1967).

5 Gopalan, C.; Srikantia, S., and Ramanathan, K.: Effect of different fats on serum cholesterol levels in monkeys. Ind. J. med. Res. *48:* 495–502 (1960).

6 Gresham, G. A.; Howard, A. N.; McQueen, J., and Bowyer, D. E.: Atherosclerosis in primates. Brit. J. Pathol. *46:* 94–103 (1965).

7 Hueper, W. C.: Experimental studies in cardiovascular pathology. XIV. Experimental atherosclerosis in Macacus rhesus monkeys. Amer. J. Path. *22:* 1287 (1946).

8 Kawamura, R.: Neue Beiträge zur Morphologie und Physiologie der Cholesterinsteatose, p. 267 (Gustav Fisher, Jena 1927).

9 Malinow, M. R.; Maruffo, C. A., and Perley, A. M.: Experimental atherosclerosis in squirrel monkeys *(Saimiri sciurea)*. J. Path. Bact. *92:* 491–510 (1966).

10 Mann, G. V. and Andrus, S. B.: Xanthomatosis and atherosclerosis produced by diet in an adult rhesus monkey. J. Lab. clin. Med. *48:* 533–550 (1956).

11 Petrov, N. N.; Krotkina, N. A.; Barabadze, E. M.; Vadova, A. V.; Gelshtein, V. I.; Melnikov, R. A.; Postnikova, Z. A., and Smoilovskaya, E. Y.: Results of 18 years of work in Sukhumi on the induction of malignant tumours in monkeys. Vopr. onkologii *4:* 643–655 (1958).

12 Pickering, D.; Fisher, D.; Perley, A.; Basinger, G., and Moon, H.: Influence of dietary fatty acids on serum lipids. Amer. J. Dis. Child. *102:* 42–51 (1961).

13 Portman, O. W.; Hegsted, D. M.; Stare, F.; Bruno, D.; Murphy, R., and Sinisterra, L.: Effect of the level and type of dietary fat on the metabolism of cholesterol and beta lipoproteins in the cebus monkey. J. exp. Med. *104:* 817–822 (1956).

14 Riopelle, A. J.: Chimpanzee associated hepatitis. Proceedings of a conference on research with primates. Beaverton 19–26 (1963).

15 Scott, R. F.; Coulston, F., and Kroms, M.: Serum and aortic lipids in monkeys developing experimental atherosclerosis. Fed. Proc. *24:* 311 (1965).

16 Taylor, C. B.; Patton, D. E., and Cox, G. E.: Atherosclerosis in rhesus monkeys. VI. Fatal myocardial infarction in a monkey fed fat and cholesterol. Arch. Path. *76:* 404–412 (1963).

17 Vadova, A. V. and Smoilovskaya, E. Y.: Dyshormonal proliferation developing in monkey uterus and mammary grand after sinestrol injection. Vopr. onkologii *5:* 515–520 (1959).

18 Zilber, L. A.; Baidakova, Z. L.; Gardashian, A. H.; Konovalov, N. V.; Bunina, T. L., and Barabadze, E. M.: Possible viral etiology of a myotrophic lateral sclerosis. Vopr. virusol. *5:* 520–528 (1962).

19 Wissler, R. W.; Frazier, L. E.; Hughes, R. H., and Rasmussen, R. A.: Atherogenesis in the cebus monkey. I. Comparison of three food fats under controlled dietary conditions. Arch. Path. *74:* 312–322 (1962).

Author's address: Dr. B. A. Lapin, Institute of Experimental Pathology and Therapy, USSR Academy of Medical Sciences, *Sukhumi* (USSR)

Pathology of Simian Primates, Part I, pp. 11–57 (Karger, Basel 1972)

Phylogeny and Classification of Primates

C. P. GROVES

Duckworth Laboratory of Physical Anthropology, Faculty of Archaeology and Anthropology, Cambridge

Contents

Tree Shrews ... 11
Prosimii ... 12
Lemuriformes ... 14
Lorisiformes ... 15
Tarsiiformes ... 16
Anthropoidea ... 16
Platyrrhini ... 17
Catarrhini ... 18
Taxonomy of the Macaques ... 20
Short-tailed Macaques ... 21
Macaca seu Cynopithecus ... 22
Long-tailed Macaques ... 23
Taxonomy of the Patas Monkeys ... 27
Taxonomy of the Vervet Monkeys ... 28
Taxonomy of the Baboons ... 31
Taxonomy of the Geladas ... 33
Taxonomy of the Talapoins ... 35
Taxonomy of the Colobinae ... 38
Taxonomy of the Chimpanzees ... 39
Taxonomy of Marmosets ... 42
Family Callitrichidae ... 43
Taxonomy of other South American Monkeys ... 48
Classification of the Primates ... 53
References ... 56

Tree Shrews

Tree shrews are often considered to be either analogous to, or closely akin to, the most primitive Primates, because of several undeniably Primate characteristics: a large brain with a calcarine sulcus and extensive visual cortex; reduced olfactory bulbs; a certain degree of manipulatory ability in the hands; and, like some of the living lower Primates, a tympanic anulus enclosed within the auditory bulla.

VAN VALEN [1965] has reviewed the most important characteristics showed by tree shrews and Primates, and believes that they are due to convergent evolution, and the retention of common primitive features. Both tree shrews and Primates are arboreally adapted and thus depend more on the visual sense than on the olfactory. The specialized structure of the auditory region of the skull is due to convergence, because the bulla itself has a different origin in tree shrews and lemurs, being in the former entotympanic and in the latter an outgrowth of the petrosal. VAN VALEN's work has shown that the absence of an entotympanic element can be regarded as one of the basic characteristics of Primates.

MARTIN [1965] has shown that the tree shrews have reproductive habits different from Primates, and of the most primitive among mammals. The female suckles her young until they are bloated with milk, which forms about two-fifths of their weight. She then leaves them in their own nest and sleeps with the male in a separate nest, and does not return except at 48 h intervals to give suck. The young move around in the nest and maintain a constant body temperature. They give no distress calls, and the mother shows no response when they are disturbed. They reach sexual maturity at about four months, whereas the smallest of the Primates, the Mouse lemur, matures at about eight months. This process is very reminiscent of the Lagomorpha, which also give a minimum of parental care. The genital apparatus of tree shrews also resembles that of Lagomorpha and Marsupials; the testes are placed in front of the penis; there is no inguinal ring, there is no bone in either penis or clitoris; and the urethra opens far back in the vagina. Consequently, MARTIN agrees with VAN VALEN that the tree shrews should not be considered as Primates, but placed in an order, Tupaioidea, by themselves, as among the most primitive of living mammals.

Prosimii

The true Primates have been divided into two suborders, Prosimii and Anthropoidea. This grouping has the merit of convenience, but is hardly a natural one. None the less, because it is widely used and serves to separate the lower Primates into a single unit, even though their similarities are those of primitive retention, this arrangement will be adopted here.

SIMPSON'S [1945] classification divides the Prosimii into three infraorders – Lemuriformes, Lorisiformes and Tarsiiformes. The Lemuriformes and Lorisiformes have far more in common with each other than with the Tarsiiformes. Both have the highly characteristic 'dental comb', an arrangement where the lower incisors and canines are flattened, somewhat pliable and procumbent, so that they look and act like the teeth of a comb and are used in grooming and perhaps also in feeding.

The Lorisiformes are more like the higher Primates than the Lemuriformes. It seems likely that their similarities are primitive, not advanced, and that the Lemuriformes are more specialized.

In the Lorisiformes, as in Anthropoidea, the ethmoid bone is exposed in the orbit, separating the palatine from the lacrimal and the frontal from the maxilla. In the Lemuriformes, the ethmoid is not exposed and the palatine articulates with the lacrimal. On the margin of the orbit the lacrimal bone is separated from the malar by the maxilla, instead of articulating with it as in the Lemuriformes. In the auditory region, the tympanic ring is exposed and attached to the mouth of the bulla, instead of being enclosed in it as in the Lemuriformes. In the brain, a sulcus centralis separates the motor and sensory areas, unlike the Lemuriformes. The cartilaginous rings of the trachea are C-shaped, and incomplete behind, as in the Anthropoidea and indeed in most other mammals; in the Lemuriformes the rings are completely circular. However, recent work has indicated that the differences may not be quite as clear cut as this.

In other ways too, the Lorisiformes show special features, less specialized than the Lemuriformes but unlike the Anthropoidea. In the anatomy of the vascular system of the ear, the main artery, the entocarotid, divides into the promontory and stapedial arteries before reaching the skull; moreover the stapedial is very small, and the promontory very big: the latter does not pass through the tympanic cavity but enters the skull directly through the foramen lacerum. In the Lemuriformes the division takes place inside the tympanic cavity and the stapedial is the larger branch; in other Primates also the stapedial is the larger, but the promontory is less reduced, and the undivided artery enters the bulla near its posterior margin – not centrally as in the Lemuriformes.

Both groups have an M-shaped loop in the colon, the ansa coli. However, in the Lorisiformes this is twisted into an elaborate spiral.

Both Lemuriformes and Lorisiformes are popularly known as 'lemurs', although some prefer to restrict this term to those found on Madagascar, i.e. to the Lemuriformes. Madagascar seems to have become isolated from the mainland of Africa at some time during the Oligocene, although faunal interchange by 'rafting' since then cannot be ruled out.

Lemuriformes

1. Daubentoniidae. The Lemuriformes can be divided into four families. One of these, Daubentoniidae, is very aberrant and is superficially so unlike a lemur, that it was first described as a rodent. The family contains a single genus, *Daubentonia*, with only one living species, the Aye-aye, *D. madagascariensis*. In this form, as in rodents, the incisors are reduced to one pair in each jaw and are chisel-like; they grow from persistent pulps, and have enamel only on the anterior and posterior surfaces. The canines and lower premolars are absent, and the upper premolars are reduced to one, so that there is a wide gap, the diastema, between the front and back teeth. The Aye-aye is up to 40 in long, of which the bushy tail forms more than half, and is black; the short, woolly hairs of the coat are interspersed with long guard hairs. The eyes and ears are very large. It is a nocturnal animal and lives solitarily; listens for grubs under the bark of trees, then gnaws through the bark and inserts a long, wiry middle finger, to crush and remove its prey. Aye-ayes have become very rare as have many of the lemurs due to the extensive deforestation of Madagascar. A small group has been transferred to the island of Nosy-Bé, off the northwest coast, for conservation.

2. Cheirogaleidae. The second family, Cheirogaleidae, is recorded as a subfamily of Lemuridae, but has little in common with it. These small creatures are known as mouse-lemurs and dwarf-lemurs, and are often under a foot long. They have many primitive features; the vascularisation of the ear region is similar to the Lorisiformes. Some of the dwarf-lemurs store food in the tail and aestivate – becoming torpid for a week or more during exceptionally dry spells.

3. Lemuridae. The Lemuridae are larger animals; one species, the ruffed-lemur *(Lemur variegatus)*, reaches four feet in length. They are quadrupedal except for the little sportive-Lemur *(Lepilemur mustelinus)*, which clings to vertical stems and moves by jumping from one to another. The forms of the genus *Lemur*, of which there may be five to eight species, are not uncommon in captivity, and breed under captive conditions. They all tend to be crepuscular, except for the ring-tailed lemurs *(L. catta)*, which are diurnal and are found in troops of a dozen or so.

4. Indriidae. The Indriidae, like *Lepilemur*, are adapted to the locomotor pattern of vertical clinging and leaping; unlike Lemuridae, they do not normally move quadrupedally. They are rather large animals, lack the lower canine and a lower premolar; they are strictly vegetarian, feeding mainly on leaves, and are consequently difficult to keep in captivity.

Lorisiformes

1. Galaginae. The Lorisiformes are usually placed in a single family Lorisidae, with two subfamilies which only differ markedly in locomotor pattern. The subfamily Galaginae consists of the African galagos or bush-babies, small largely nocturnal creatures of jumping habit like the Indriidae. The commonest species, *G. senegalensis*, thrives in captivity and is often kept as a pet.

2. Lorisinae. The subfamily Lorisinae are slow-moving, quadrupedal animals, from Africa and southern Asia. The veins of the lower part of the limbs form a network of vessels, the rete mirabile, which allow the feet to remain clamped tightly round a branch for long periods of time: the slow movement seemingly aids in the capture of prey. In all forms the index finger is reduced to give a wider span to the hand, and in the African species, the potto and angwantibo *(Perodicticus, = Arctocebus)*, is almost absent.

Most Prosimii are seasonally polyestrous, the estrus cycles being restricted to certain seasons. Outside the breeding season, the vagina may be closed, as in the Cheirogaleidae; during the season, in the latter family, estrus lasts for 2–5 days, and the cycles occur at approximately monthly intervals. The season itself lasts for approximately three months. Cycling occurs in males, too; in the active season, there is enlargement of testes, epididymis, seminal vesicles, prostate and Cowper's glands. In any given area in Madagascar, the different lemur species breed and give birth at different times of the year – perhaps a mechanism for ensuring maximum survival in each. The stimulus may be temperature: in Madagascar the Mouse-lemur breeds from October to December; in the northern hemisphere, from March to June.

All lemurs bear a single young, except the Cheirogaleidae which may have two or three. They are born well-developed and cling to the mother's fur, except in Cheirogaleidae and in the ruffed-lemur, in which the young are born naked and blind, and are placed in nests built by the mother with vegetation and lined with fur from her own coat.

Tarsiiformes

Very different from the lemurs are the tarsiers, belonging to the infraorder Tarsiiformes. These are small, rat-like, leggy forms with huge eyes and ears and a short snout; the digits are very long with expanded pads at the end, toilet claws on pedal toes 2 and 3 (lemurs have them on the second toe only), and a highly opposable big toe. The tail is very long and thin, and scaly beneath. Like Galagos and Indriidae, the tarsiers are vertical clingers, leaping from one vertical stem to another; the scaly underside of the tail forms a friction surface, and presses against the support There is no dental comb. The teeth are primitive. The features of the skull are similar to Anthropoidea and Lorisiformes, except that the orbits are very large. Like Anthropoidea the nose is hairy, its lateral margins complete, and the upper lip is free from the gum; in contrast to lemurs in which there is a moist, naked muzzle, the lateral nasal margins are notched, and the lip is bound down to the gum. The arrangement in tarsiers results from reduction of the olfactory mechanism.

There are three species of tarsiers, placed in a single genus *Tarsius*, occurring in Indonesia and the Philippines. Because they are clearly the most primitive living true Primates, their classification has been continuously altered. They have been placed with Prosimii, and in Anthropoidea, or in a group peculiar to themselves. In reproduction they resemble Anthropoidea, having a monthly cycle with no anestrus, unlike most lemurs; the existence of menstruation has been claimed but not proven. The placenta is hemochorial like Anthropoidea and single like Hominoidea, instead of epitheliochorial like lemurs, and double like other higher Primates. Nevertheless, in certain features they are specialized, as for nocturnal existence: the orbits are very large, as stated, as is also the visual cortex and the retina consists entirely of rods. They are also solitary; in captivity, they do not form groups other than mated pairs and show little social interaction.

Anthropoidea

The Anthropoidea, or 'higher Primates', consist of two groups. All share certain features which are more advanced than those of the Prosimii – although they are independently acquired in the New and Old World groups. All have a large brain and short face; large orbits are separated from the temporal fossa by a nearly complete bony wall; the lacrimal foramen is inside the orbital margin; comparatively olfactory senses, and although there are sternal glands in smaller Platyrrhines which are used for marking along with urine, tactile senses arising from vibrissae are also reduced; palmar and plantar pads are little developed and fused. In the brain, the frontal and occipital lobes are very large; the corpus callosum is also enlarged, and the association areas are expanded. In the smallest-brained Anthropoidea (the marmosets) the brain is, compared to the body weight, three times the size of the brain in tarsiers or galagos.

Platyrrhini

The infraorder Platyrrhini consists of the South American monkeys. These monkeys show striking convergences with the Old World monkeys, but have a very distant common origin. The differences between them seem not very great for such long-separated forms. In the Platyrrhini there are three premolars, upper and lower; the Catarrhini (Old World monkeys and apes) have only two. The Platyrrhini have a bulla with the tympanic ring attached at its orifice; in the Catarrhini there is no bulla, the tympanic ring being elongated into a tube. In the Platyrrhini the thumb is not opposable; objects are often gripped between digits 2 and 3. Finally, the nostrils are far apart in the Platyrrhini, and look sideways; in the Catarrhini the septum is narrow and the nostrils look forward and down.

1. Callitrichidae. Traditionally the marmosets, the smallest of the South American monkeys, are separated from the rest into a family of their own, the Callitrichidae. They possess claws, and many arguments have raged over whether these are primitive or secondary. Alone among the Anthropoidea the marmosets possess carpal vibrissae, which all the lemurs possess. Most marmosets have only two molar teeth, upper and lower; but M^2 is missing as also in some other New World monkeys.

Marmosets have certain reproductive characteristics which set them apart from other Anthropoidea, with the exception of *Aotus* (the night monkey) and *Callicebus* (the titi), two genera of Cebidae. About 85% of all births are of twins, which the father carries around on his back, generally returning them to the mother only for nursing. In these three groups also the normal social group is a mated pair; the large troops of some other South American monkeys are not formed.

2. Cebidae. The Cebidae are a diverse group, with several very distinct and ancient lines. Some are difficult to keep in captivity, but the douroucouli or night monkey *(Aotus)* and the squirrel monkey *(Saimiri)* survive well in captivity. The former is nocturnal. Both are quadrupedal forms with some locomotor patterns rather reminiscent of the vertical clining lemurs. The larger forms – woolly monkeys *(Lagothrix)*, spider monkeys *(Ateles)* and capuchins *(Cebus)* – also thrive in captivity, but need a great deal of room as they are very active; these monkeys also have prehensile tails, which except in the capuchins are modified at the tip with a naked area of dermatoglyphics. Even in *Cebus* the body weight can be supported by the tail alone when hanging from a branch.

Catarrhini

Cercopithecoidea. The Catarrhini, or Old World higher primates, live in Africa, Asia and Europe. Two superfamilies are recognized: Cercopithecoidea for the Old World monkeys, and Hominoidea for the apes and man. In the former, the basic locomotion is quadrupedal; there is a tail, often long; both sexes have large, hard buttock pads, called ischial callosities, which develop well before birth, and in the male may be fused across the midline below the anus. The molar teeth are bilophodont (the cusps are united in pairs, transversely) making them very hard-wearing; and the males have long canine teeth with razor-sharp edges along the back, and a groove which extends to the root. The Cercopithecoidea tend to live in large social groups usually with a well-marked rank order among the males, and a less marked one among the females. There is an elaborate pattern of facial gestures of which some recur throughout the superfamily and some are shared even with the apes, while others are peculiar to one species or another.

Colobinae. To this superfamily belongs only a single family, Cercopithecidae, with two subfamilies. One of these Colobinae, contains forms specialized as leafeaters, with complex sacculated stomachs. Included are the colobus monkeys of Africa *(Colobus)*, and the langurs of Asia *(Presbytis)*. Except for the sacred langur *(Presbytis entellus)* of India, which has a diet less narrow than the rest, all the Colobinae are delicate and difficult to keep in captivity.

Cercopithecinae. The Cercopithecinae, the other subfamily, are omnivorous monkeys with simple stomachs but enormous cheek-pouches in which food is collected. Of the several recognized genera, all are suitable as laboratory animals.

The Baboons. Baboons have long, dog-like muzzles especially in the adult male. The savannah baboon *(Papio cynocephalus)* is common in south, east and west Africa in large troops sometimes numbering 50 or more, with a central coalition of dominant males. The highest (alpha) male mates with most of the females that are in estrus, and so fathers about 80% of the next generation. The hamadryas or mantled baboon *(P. hamadryas)* lives in the Somali desert and in S.W. Arabia; there is only one male in each troop, the surplus ones forming unisexual troops by themselves. Recent evidence has demonstrated intergrading between these two supposed species (vide infra). The drill and mandrill *(P. leucophaeus* and *P. sphinx)* live in a restricted area in the west-central African forests; they have short tails, and the mandrill has a brightly coloured red and blue face. Superficially like the baboons is the Gelada *(Theropithecus gelada)*, of the Simien mountains of Ethiopia. Like the hamadryas baboon the male gelada has a copious mane of hair; like it, too, the troops consist of one male with several females, which can come together to form sleeping and foraging groups several hundred strong. The gelada has strong, deep jaws and a snub nose.

Closely related to the baboons are the mangabeys *(Cercocebus)*, of which about seven species inhabit the African forests. They are essentially longtailed, lightly built arboreal baboons.

The Macaques. The macaques of Asia *(Macaca)* somewhat resemble baboons; one species is still found on the Rock of Gibraltar and so is the only Primate in Europe apart from man. The macaques are currently the most important laboratory Primates, and will be considered in detail in the next section.

The Old World monkeys considered hitherto are large-jawed forms in which the females tend to develop estrus swellings. In some of the macaques the onset of estrus is indicated only by a reddening of the perineal skin; but in most of them, as well as in all the baboons and mangabeys, the sexual skin of the female begins to swell after menstruation, reaching its maximum during the ovulation; it then recedes. Investigation has shown that the swelling contains mostly a watery fluid. In the gelada there is a variant on this pattern: the female develops a ring of bright red pustules around the bare patch on her chest.

The Guenons. The guenons *(Cercopithecus)*, consisting of twelve to twenty species, have short muzzles, small teeth, and mostly no sexual swellings. They are mainly arboreal, but one species, the green, grivet or vervet monkey *(C. aethiops)* is a terrestrial savanna-dweller. Two species develop sexual swellings; one of them, the talapoin *(C. talapoin)* is the smallest living catarrhine, with a total length of two to three feet, of which half is tail. Because of its relatively small size and docility it makes a good laboratory animal. Other members of this genus, e.g. the vervets, are also suitable as laboratory animals being readily available and docile. In general, cercopitheques do not breed so readily in captivity as macaques.

The last of the Old World genera is *Erythrocebus*, the patas monkey. This is a large, savanna-living monkey with a reddish coat. The male is twice as large as the female; troops contain a single male. Patas are becoming increasingly popular in laboratories, though they are rather large and may be aggressive.

Hominoidea. The Hominoidea are the second superfamily of Anthropoidea. They differ from the Cercopithecoidea in a number of ways: all are tailless, with the few remaining caudal vertebrae fused into an inturned bone, the coccyx; they have a broad thorax and skeletal proportions which seem to indicate that in the past, if not today, their mode of life has been that of a brachiator. The gibbons (*Hylobates* – the genus *Symphalangus*, often applied to the large siamang gibbon, is invalid, vide GROVES [1967]) are the only Hominoidea that habitually brachiate today. They are rather removed in morphology and behavior from the other apes (they live in pairs, not in large groups); and have sometimes been placed in a separate family, Hylobatidae; for the moment, however, we may accept them as Pongidae like the quadrumanous orang-utan *(Pongo pygmaeus)* and the terrestrial, knuckle-walking chimpanzee and gorilla *(Pan troglodytes* and *P. gorilla)*.

Taxonomy of the Macaques

Amongst the macaques, it is of fundamental importance to identify the species or subspecies, which is being used. Errors of identification and nomenclature have been made so often in the past that experiments have been duplicated and results have proved conflicting. A list of the various species, putative or real, is given below with their commonest vernacular names. It is certain that all of them are distinct entities in one way or another, but some may be only geographical races (subspecies) rather than full species.

---

a. *Tailless or short-tailed forms*

1. *Macaca sylvanus* (Gibraltar, N. Africa). Barbary ape; Gibraltar macaque
2. *Macaca fuscata* (Japan). Japanese monkey/ape/macaque
3. *Macaca thibetana* (S. China). Tibetan macaque
4. *Macaca arctoides* (Burma, Indo-China, S. China). Stumptailed monkey, Indo-Chinese bear macaque, '*Macaca speciosa*'

5. *Macaca ('Cynopithecus') nigra/maura* (Celebes). Black ape, Celebes ape, crested ape, moor macaque, ashy-black monkey

b. *Medium to long-tailed forms*

6. *Macaca silenus* (S.W. India). Lion-tailed macaque, wanderoo
7. *Macaca nemestrina* (Burma to Sumatra, Borneo). Pigtailed monkey/macaque
8. *Macaca assamensis* (Nepal, Assam, N. Burma, N. Indo-China). Assam macaque.
9. *Macaca mulatta* (India, N. of Godavari R.; Burma, N. Indo-China). Rhesus monkey/macaque
10. *Macaca cyclopis* (Taiwan). Formosan rock macaque
11. *Macaca fascicularis* (S. Indo-China to Indonesia, Philippines). Crab-eating monkey/macaque; Java monkey; Philippine macaque; Irus macaque; Cynomolgus monkey, '*Macaca irus*', '*Macaca cynomolgus*'
12. *Macaca sinica* (Ceylon). Toque macaque
13. *Macaca radiata* (India, S. of Godavari R.). Bonnet macaque

Short-tailed Macaques

The first group have either no tails, or little stumpy ones that hardly project from between the ischial callosities. In the first four the face is often some shade of red; the first three have a fringe of hair on either side, sometimes extended into a beard.

M. sylvanus has buff-yellow fur, tipped with black; the flanks and limbs are greyer. The facial fringe is grey-white, and there is a characteristic irregular black line from eye to ear. The face is rounded, the build stocky and rather pudgy. There is rarely any difficulty recognizing this species.

M. fuscata has darker fur, the face fringe being dark brown; the face and buttocks are red or pink; the face is long and narrow. In this species the tail is always visible, although very short. The head and body length is about the same as in *M. sylvanus* – 600–750 mm, the female being smaller – but the tail is longer, reaching 80 mm.

M. thibetana is a little-known species, generally confused with the next one until FOODEN [1967] showed they are clearly different. It is the largest species, reaching 830 mm head and body length in big males, with the tail up to 100 mm. The fur is dark brown; the male has a long grey-white face fringe and beard. The face is brownish to pale flesh-color. The fur is very long, reaching 80–90 mm. It is found from the mountains of Szechwan to the rocky hills of Fukien.

M. arctoides, the stumptail, is smaller than the foregoing, rarely above 600 mm in head and body length, and very stockily built. There is a whorl on the crown, in front of which the hair is very short, and old specimens go quite bald in front. The face is bright red, going dark in patches with exposure. It is distinguished from *M. fuscata*, with which it is most likely to be confused, by its baldness, its rounder build, and the sparseness of the hair on its belly, where the Japanese has a thick grey covering. The stumptail also has a very curious genital specialization: the glans penis is very long and tapering, with the urethral opening on its dorsal surface. It has recently been shown by FOODEN that the vagina is similarly modified, a 'lock and key' effect which would give a rather efficient mechanical isolation from other species in most cases. The name *Macaca speciosa*, often applied to this species at one time, cannot be used for this species as it was originally applied to the Japanese macaque!

The island of Celebes is inhabited by a variety of different forms of macaques, but only one in each locality. The two end-points, the one on the northern tip and the one on the southwestern tip, are so different in appearance that they have generally been placed in two different genera, and known as *Cynopithecus niger* and *Macaca maura* respectively, though they do in fact, shade into each other by intermediate forms. All are blackish in color, with a bare area around the ischial callosities, strong brow ridges and a tendency to develop bony ridges on the snout beside the nose, the jaws being rather elongate. The seven most clearly recognizable forms are classified by FOODEN [1969] as species as follows:

Macaca seu Cynopithecus

1. Niger. Strong maxillary ridges; nostrils at end of snout; a long, backwardly directed crest in middle of crown; tail at most 2 cm long; perianal area long-haired; ischial callosities orange colored, transversely folded (kidney-shaped); color entirely black. Northern tip of Celebes.

2. Nigrescens. Less strong ridges; nostrils at end of snout; long crest on crown; tail at most 2 cm long; perianal area long-haired; ischial callosities grey, not folded; color very dark brown, but limbs and underside black. A small area on the northern peninsula of Celebes.

3. Hecki. Ridges still less strong; nostrils partly on upper surface of snout; crest shorter, occupying whole breadth of crown; perianal area long-haired; ischial callosities grey, somewhat kidney-shaped; color black, but arms and underside brown, legs clear brown. Most of the northern peninsula of Celebes.

4. Tonkeana. No marked ridges; nostrils on upper surface of snout; a slightly elongated tuft on crown; tail 4–6 cm long; perianal area sparsely haired or bare; ischial callosities grey, unfolded; color black-brown or black, but underside brown and hind surface of thighs greyish. Central Celebes and nearby Togean Islands.

5. Ochreata. In general similar, but no elongation of hair on crown; underside and limbs greyish. South-eastern Celebes.

6. Brunnescens. Similar to the last, but brown instead of grey. Muna and Buton Islands.

7. Maura. Again similar, but again crown is short-haired; underside and limbs are black-brown or black like upper surface, only hind surface of thighs being greyish. South-western Celebes.

For the present, these forms may be accepted as species, since there is no evidence that they intergrade; the probability is that they will come to be classed as a single species, as intergradation seems likely.

All the Celebes macaques are of similar temperament; they tend to be docile, almost gentle, and easy to handle, and consequently are becoming more and more popular as laboratory animals. They have been bred in captivity.

Long-tailed Macaques

The other macaques are all rather long-tailed animals – longer tailed, at any rate, than the proceeding forms. *M. silenus* is a rare, odd-looking macaque, with a grey-white facial fringe and a black color. The hairs on the crown radiate from a whorl, and are also black. There is often a slight tuft on the tail. The species is restricted to the hill forests of southwestern India. Although superficially so unique as to be recognizable at a glance from all others, there is little doubt that it is very closely related to the well-known pig-tailed macaque, *M. nemestrina.* Although the latter is larger and more robust, as well as being brown in color, there are many points of resemblance. Both species have a whorl on the crown and a facial fringe of hair; the tail is not too long, and curves up and back in a loop, often curving forward again at the tip which rests lightly on top of the rump. Both species stand similarly, with the hands turned outwards; both species show a characteristic facial gesture, the pout, Flehmen face or simply pigtail face whose function is one of appeasement. Apart from rather small-sized races from Borneo and Bangka, *M. nemestrina* has two widely-recognized forms: the Malayan and Sumatran subspecies, *M. n. nemestrina,* which is deep brown with a blackish cap of hairs on top of the head and a short facial fringe; and the Burmese race, *M. n. leonina,* which is lighter brown with a cap only slightly darker, and a longer, greyer facial fringe. Looking at the Burmese subspecies, one could almost be looking at a pale-colored *M. silenus*. The Burmese pigtail has been introduced to the Andaman Islands; but there are no macaques of this group between the ranges of *M. n. leonina* and *M. silenus*.

The Assam macaque *(M. assemensis)* which superficially resembles a rhesus monkey, is robustly built like a pigtail and stands in the same way; it has a short facial fringe and generally the crown hairs radiate in a whorl. It is brown all over, except the cheeks which are grey. The canines are very large. It has long hair, which may reach 85 mm on the shoulders, and is a highland species.

The Assam macaque is often confused with the rhesus monkey, *M. mulatta*, currently the most commonly used laboratory animal. The rhesus is smaller then the pigtail or Assam macaques (head and body length about 400–550 mm) with a medium-length curly tail, hairs that sweep straight back from the brow, and bright reddish or orange hindparts which contrast strongly with the drab grey-brown foreparts. Related to the rhesus is the crab-eating macaque *(M. fascicularis)*, also common in laboratories. This latter is wholly brownish, or slightly olive in color, with a longer tail than the rhesus; the hair often rises to a little peak on the crown.

It was suggested by FOODEN that the rhesus and crabeater are conspecific. They share characters – like the pale face and the direction of crown hairs – which bring them close together, but do not overlap in range. Moreover, three skins from Ban Umphang in central Thailand seemed to him to show characters intermediate between the two, in both tail length and in color. It is certain that the two are closely related, and they do indeed share common patterns of geographic variation. For instance in tail length, there is a gradual increase from the north (China) to the south and east (Indonesia); To say that they form a single species is too far-reaching on the basis of three specimens[1].

M. mulatta, China: tail length as % of head and body	30
M. mulatta, N. India – N. Burma	35–50
Intermediates, Ban Umphang	55–96
M. fascicularis, S. Thailand – S. Burma	100–135
M. fascicularis, Indonesia – Philippines	130–150

1 A recent paper by FOODEN [1971] clarifies the *mulatta/fascicularis* situation in Thailand, suggesting on the basis of field study that the hybrids result from 'occasional fertile breeding between two species'.

For a number of years laboratory workers have called the crab-eater, *Macaca irus* or *Macaca cynomolgus*, names which are invalid taxonomically.

A rare species of macaque, closely related to *M. mulatta* and *fascicularis*, is *M. cyclopis* from Taiwan. This macaque has a woolly coat and a thickly-furred tail, carried behind it in a loop or question-mark, which is about 50–70% of the head and body length.

The last two species are both long-tailed like the crab-eater: the toque *(M. sinica)*, of Ceylon and the bonnet *(M. radiata)*, of southern India. They are closely related and might be better classified as one species, *M. sinica*. They are small, 450–600 mm long, with the tail 500–740 mm, and short-faced. Both have a whorl on the crown with copious straight hair radiating from it like a straw hat; in the bonnet this hair is short over the brows, leaving a nearly bare forehead reminiscent of *M. arctoides*. The two have a large penis with a swollen glans and well-developed corona; like *M. arctoides*, it would probably be difficult for these two to interbreed with any other species.

Sexual swellings occur in all macaque species except *sinica*, *radiata*, *fascicularis* and some *mulatta*. In *M. silenus* the swelling involves all the skin round the anus, genitalia and ischial callosities, with a transverse expansion lifting the root of the tail. In *M. nemestrina* and *M. arctoides* the part under the tail-root falls in thick transverse folds. In *M. sylvana* the swelling is purplish or grey instead of reddish, and is similar but with a high dome-shaped expansion above the anus. In the Celebes species the swelling covers the whole lateral rump area but does not affect the vulva. In *M. cyclopis* the swelling extends from the tail root down the back of the thigh. In the Chinese *M. mulatta sanctijohannis* there is a smaller swelling similar to *M. silenus*, but in other races of the same species the swelling is small or absent, and – like those species entirely without swellings – there is merely a reddening of the sexual skin towards estrus.

The most-used species of macaque is certainly the rhesus monkey *(M. mulatta)*, followed by the crab-eater *(M. fascicularis)*. These two species are still the most easily available, although as stated elsewhere in this chapter this may not continue to be the case for very long as wild stocks are fast being depleted. From the point of view of the animals' temperament, however, and ease of handling, these two species are far from being the most satisfactory. The bigger, more imposing looking pigtail *(M. nemestrina)* is relatively more docile, especially when young; on the other hand being so much larger it requires a larger cage and more food. Still more docile and cooperative is the stumptail *(M. arctoides)*; the males of this species, which are not quite as large as pigtails but more stockily built, will often display with tremendous gusto, shaking the bars and grimacing, on being approached by the handler; but after a minute the animal quietens down and is then easily handled.

Similarly, the Celebes macaques (including the 'black ape', *M. nigra)* are relatively unaggressive and even friendly. It is possible to walk into a colony of these macaques without unduly disturbing the animals, or even being threatened; of course this depends in part on the degree to which the animals are familiar with man, and on the behaviour of the individual human being, but it is true to say that the threshold of aggression of Celebes macaques is much higher than in any other species, and that they are preferable to any other macaque for these reasons. There is no difficulty in breeding macaques in captivity, and several laboratories are taking pains to set up breeding colonies of both black apes and stumptails: a wise precaution, since both these highly satisfying types occupy rather restricted ranges, and extensive exploitation of wild resources would quickly exterminate them.

Other species of macaque are less used in the lab. The Japanese monkey *(M. fuscata)* is not widely used: it is larger than the rhesus and shares some of the latter's disagreeable qualities. The bonnet/toque group *(M. sinica* and *M. radiata)* are the smallest macaques and somewhat more cooperative than rhesus, but are not readily available. The Assam macaque *(M. assamensis)* occasionally finds its way into a laboratory, where it is invariably mistaken for a large, rather docile rhesus, and only rarely recognised as something rather different. The lion-tail *(M. silenus)* and Tibetan *(M. thibetana)* are rare and should not be used for experiment even should they unexpectedly prove obtainable. The same is probably true for the stocky Barbary ape *(M. sylvanus)* of North Africa.

The demand for macaques, especially rhesus and crabeating macaques, for laboratories today threatens the wild population. In parts of India where rhesus monkeys used to be plentiful there are now few or none; many troops consist only of old animals, as the youngsters have been swept off into captivity. In India, as well as in Malaya, Thailand and the Philippines, the depletion in wild populations of macaques has been severe. A recent calculation estimated that a quarter of a million monkeys every year are used in terminal experiments; considering also losses during transportation, this represents a severe annual loss to wild populations. If importation of them into the USA and Europe continues at the present rate, the macaques – at least in southeast Asia and in India – will be threatened with extermination.

JOLLY [1966] has suggested some remedies. In the first place, monkeys must be used more economically in the laboratory: a monkey should not be used when a rat will suffice; unnecessary sacrifice of the animal should be avoided; and full use should be made of cadavers after sacrifice. A non-infectious cadaver should *on no account* be destroyed. Dr. R. H. G. VAN GELDER of the Mammalogy Department, American Museum of Natural History, New York, has offered facilities for permanent preservation of cadavers used for experimentation. In this way, the cadavers would be available not only to anatomists, but also to research workers interested in doing follow-ups or similar research. It would provide a means of checking identification – an important point if research is to be meaningful, and duplication and conflicting results avoided.

In the second place, every effort should be made to breed monkeys in the laboratory. If this is too costly, then breeding of monkeys in their country of origin should be undertaken.

In the third place, untapped sources should be utilized. In particular, monkeys which would be needlessly slaughtered for other purposes (as crop-raiders, or for tsetse control) should instead be captured and used as laboratory animals. Combined efforts of dealers and pest control officers could give rise to a saving of both money and animal resources. Nobody seems yet to have considered making use of the vast monkey drives which take place every year in Sierra Leone; only recently has anyone considered exploiting the talapoin, a good laboratory subject, which is actually regarded as a pest in Equatorial Guinea.

Monkeys are necessary for medical research, but without rational exploitation, both in the laboratory and in the field, the sources will cease to flow.

Taxonomy of the Patas Monkeys

There are few problems connected with the taxonomy of the patas monkey, another commonly used laboratory form. All belong to a single species, *Erythrocebus patas:* but there is some disagreement over whether they should be separated into a distinct genus or not, or whether they would not better be placed within the large genus *Cercopithecus.* The problem is that in the skull the patas monkey is remarkably similar to *Cercopithecus;* but in its limb morphology it is very different, being specialised for running in a savannah environment. Indeed, it is if possible a more thoroughly terrestrial animal than the baboons. Patas monkeys have long, slender limbs and a greyhound-like build; the tail is shorter than in *Cercopithecus* and just reaches the ground; the short fur is reddish (a colouration not found as a ground tint in *Cercopithecus*) except that the lower half of the limbs are white in males, pale fawn in female and young. The male's scrotum is turquoise blue.

OSMAN HILL [1966] lists four subspecies: *E. p. patas:* western savannah from Senegal to Nigeria. The face is pale pink and the nose is black; *E. p. villiersi:* Air (Asben) massif in the Sahara. Differs only in its smaller size from the nominate race; *E. p. pyrrhonotus:* Cameroun through southern Sudan to Kenya and Tanzania, to the west of the Rift valley. The whole face is blackish and adults develop a white nasal spot; there is a black fronto-temporal line; *E. p. baumstarki:* Ikoma district, Tanzania. Paler than *pyrrhonotus* without the fronto-temporal line.

In addition there are patas monkeys on the Athi plains, south-east of Nairobi and east of the Rift valley; these have a pink face with a blue nasal spot. No subspecific name has been applied to them as yet. East African patas monkeys are referred to as Nisnas monkeys.

Taxonomy of the Vervet Monkeys

The savannah and woodland-living guenons are all referable to a single species *Cercopithecus aethiops*, and go under such vernacular names as green, grivet, vervet, malbrouk etc. They have been referred to a varying number of species: DANDELOT [1959] recognises three and in this he is followed by OSMAN HILL [1966]; in his 1968 revision DANDELOT increases the number to five. The classification given below follows DANDELOT in the matter of subspecies, but recognises his five 'species' only as 'subspecies-groups' within a single species. Reasons for not recognising more than one species will be given after the descriptive section.

I. Subspecies-group sabaeus. Green monkey. Yellowish cheek-whiskers; a temporal whorl; a golden-yellow tip to the tail; no white brow-band; scrotum pale blue. Colour a bright gold-green.

1. C. a. sabaeus. West Africa, from Senegal to Volta river, Ghana.

II. Subspecies-group tantalus. Tantalus monkey. White cheek-whiskers pointing upwards, strongly contrasting with the olive-hued crown, and separated from white frontal band by a thin black line; white tail-tip; scrotum sky-blue, surrounded by orange hairs.

2. C. a. tantalus. Volta river, Ghana, to northwestern Congo (Kinshasa). Olive-green colour.

3. C. a. budgetti. Northeastern Congo (Kinshasa) into north and west Uganda. Fairly dark, olive-brown colour.

4. C. a. marrensis. Djebel Marra, Sudan. Large; light olive-fawn colour.

III. Subspecies-group aethiops. Grivet monkey. Long white falciform cheek-whiskers; white tail-tip; white brow-band inconspicuous; scrotum sky-blue but not surrounded by orange hairs.

5. C. a. aethiops. White Nile valley, Sudan, to Eritrea. Pale grey colour, more olive on back.

6. C. a. matschiei. Southwestern Ethiopia (west of Rift valley). Fawn colour, with russet tints.

7. C. a. hilgerti. Eastern Ethiopia (east of Rift valley). Fairly dark, olive-green coat; long woolly fur.

IV. Subspecies-group pygerythrus. Vervet monkey. White cheek-whiskers gradually blending with orange-hued crown; white frontal band; black tail-tuft; scrotum turquoise blue; a red patch under root of tail; black face.

8. C. a. arenarius. Low country of eastern Ethiopia, Somalia and northern Kenya, south to Tana river (a desert form). Olive-brown colour.

9. C. a. excubitor. Patta island, Kenya-Somalia border. As last; very small.

10. C. a. centralis. Kivu province, Congo (Kinshasa) to western shore of Lake Victoria in Uganda and Tanzania. Brownish fawn colour.

11. C. a. callidus. Kenya and northern Tanzania, west of Rift valley. Pale olive-yellow; black hands and feet.

12. C. a. johnstoni. Northeast Tanzania and southeast Kenya (south of Tana river), east of Rift valley. Pale olive-yellow; hands and feet less dark.

13. C. a. nesiotes. Zanzibar, Pemba and Mafia islands. Brownish fawn colour with light reddish flanks.

14. C. a. rufoviridis. Southern Tanzania, Malawi and Mozambique as far south as Zambesi river. Olive fawn colour with similar pale reddish flanks.

15. C. a. whytei. Lower course of Zambesi in Mozambique. Rather dark brownish green colour; intermediate between No. 14 and 16.

16. C. a. pygerythrus. Southeast Africa, from Limpopo to Cape province. Deep olive-green colour.

17. C. a. ngamiensis. Middle Zambesi, from Victoria Falls west through Botswana to Caprivi. Greenish colour, with pale hands and feet.

18. C. a. marjoriae. Lower course of Orange river. Pale ash-grey colour.

19. C. a. helvescens. Southwest Africa, south of Cunene river (Angola). Pale yellowish colour with pale hands and feet.

V. Subspecies-group cynosuros. Malbrouk monkey. White cheek-whiskers blending with orange crown; white frontal band; black tail-tip; scrotum 'lapis-lazuli blue'; dirty pink face. Colour olive-grey to olive fawn.

20. C. a. cynsuros. Southern Congo (Kinshasa); Zambia, east to Luangwa river.

There are thus twenty subspecies (according to DANDELOT), covering the whole savannah region of subsaharan Africa, and much of the sub-desert zones as well. The following points argue against the many-species concept of DANDELOT and HILL:

a) The geographic variation is of a clinal nature. Within each subspecies-group, the colouration nearest to the neighbouring group is found in the subspecies which abuts on the latter: thus within the *tantalus* group it is *C. a. tantalus*, geographically nearest to *C. a. sabaeus*, which shows the greenest colour, like *sabaeus*, and it is *C. a. marrensis*, geographically nearest to *C. a. aethiops*, which shows the lightest and greyest colour, like *C. a. aethiops*. The same similarity in colour combined with geographic proximity is shown between *C. a. hilgerti* (*aethiops*-group) and *arenarius* (*pygerythrus*-group); *C. a. budgetti* (*tantalus*-group) and *centralis* (*pygerythrus*-group); and *C. a. rufoviridis* and *ngamiensis* (*pygerythrus*-group) and *cynosuros*. This is a rather suggestive situation, if only indirect evidence.

b) One case of a hybrid swarm (i.e. continuous intergradation) between two of DANDELOT'S 'species' is known – recorded, as a matter of fact, by DANDELOT himself. The border between *C. a. budgetti* and *C. a. centralis* runs slantwise across the equator in Uganda, from Beni (Congo) to Kagera. In the contact zone, all types of intermediates are found: towards the *budgetti* side are both 'pure tantalus, and hybrids of tantalus predominance', and towards the *centralis* side are 'pure vervets, and hybrids of vervet predominance', i.e. both tantalus with russet fur under the tail like vervets, and vervets with orange fur round the scrotum like tantalus. This is not just F_1 hydridisation; it is true intergradation. Because of the indirect evidence cited under a) above, it seems a fairly safe bet that further hybrid swarms will be detected by more detailed study.

Vervets are increasingly kept in the laboratories; even though there is little likelihood that more than one species is involved, there are fairly conspicuous differences between the races, perhaps analogous to the case of the baboons. It becomes of importance to differentiate between them in publications.

Taxonomy of the Baboons

The overall picture of baboon classification is strikingly similar to that of vervets: namely, a single species inhabiting open country – savannah and desert – throughout subsaharan Africa, showing well-marked geographic variation but forming a continuously interbreeding whole. Even the highly differentiated Hamadryas, sacred or mantled baboon which has in its day been referred to a separate genus, *Comopithecus*, proves, on the strength of data shortly to be published by HANS KUMMER, to intergrade with the other more familiar savannah baboon.

Like the vervet monkeys, the baboons can be divided into five 'subspecies-groups'. The groups are, however, somewhat different geographically; they are more clearly differentiated morphologically than the vervet groups, but contain fewer recognisable subspecies within them. The same criteria as are used to unite all the vervets into one species can be employed in the baboons too: the morphological similarity of geographically neighbouring subspecies, and the existence of hybrid zones at any rate between the *hamadryas* and *anubis* groups (see above) and between the *anubis* and *cynocephalus* groups [1].

By the law of priority, which states that the species should be known by the name of the earliest described subspecies, the Savannah baboon has to be called *Papio hamadryas* LINNAEUS, 1758.

I. Subspecies-group hamadryas. Hamadryas, sacred or mantled baboon. Small- to medium-sized; skin of face and ischial callosities pink; tail rising and falling behind in a smooth curve. The skull has a deep-set nasal root, making the supraorbital torus appear more projecting. Females and young males are olive-fawn with no mane; males are grey with a huge mane on head and foreparts, accentuated by the very short hair on the hindparts.

1. P. h. hamadryas. Eastern Ethiopia and northern Somalia. Larger.

2. P. h. arabicus. Southwestern Arabia. Slightly smaller.

II. Subspecies-group papio. Guinea baboon. Size similar to last; skin of face blackish, of ischial callosities red; tail smooth-curved. A less-developed mane on shoulders and back of male, but not contrasting with short rump hairs; both sexes reddish-coloured, the individual hairs being closely ringed with alternate bands of black and reddish-yellow.

3. P. h. papio. Gambia, Guinea, coastal Senegal.

III. Subspecies-group anubis. Olive, doguera or anubis baboon. Usually considerably larger; skin black; tail kinked near base; nostrils protrude beyond plane of upper lip. Male maned, but no marked differentiation in hair-lengths (as in *papio*, unlike *hamadryas*); colour dark olive-green or olive-brown, often with blackish blotches, each hair being greyish with only one or two buffy rings, and the tip black.

4. P. h. anubis. Eastern Senegal to lowlands of western Ethiopia. Large, pale, with ocherous or yellowish tints.

5. P. h. tibestianus. Southern Tibesti; possibly Air. Small and pale.

6. P. h. doguera. Ethiopian highlands, at least the southern and western parts. Dark, olive, with black-blotched appearance.

7. P. h. tessellatus. Forest and forest borders of northern Congo (Kinshasa), western Uganda, Southern Sudan and eastern Central African Republic. Darker, browner, again rather blotched; often virtually indistinguishable from No. 6.

8. P. h. neumanni. Eastern Uganda, and Kenya west of Rift valley, south to Serengeti region of Tanzania. Fairly small, dark greenish-olive.

IV. Subspecies-group cynocephalus. Yellow baboon. Somewhat smaller; skin black; tail kinked near base; slenderly built and long-legged; nostrils not projecting. Mane much less developed, and less wavy than in previous group; colour yellow to grey-brown, each hair being either, plain yellow or with a narrow black ring or tip. Cheeks and underside whitish or plain buff.

9. P. h. ibeanus. Eastern Somalia, south to Tana river, Kenya. Dark brown-grey.

10. P. h. cynocephalus. Southern Kenya, west of Rift valley; most of Tanzania; Mozambique north of the Zambesi. Yellow-fawn, with long silvery tips to some of the hairs in adult males, giving it a sandy tinge.

11. P. h. pruinosus. Southern Malawi; Zambia east of R. Luangwa. Smaller; sandy-yellow colour. Size gets noticeably smaller from east to west.

12. P. h. kindae. Luangwa River, Zambia, west via Katanga to south-eastern Angola. Very small, but long-legged; short-faced; short smooth coat, rich ochre with long black tips on back.

V. Subspecies-group ursinus. Chacma baboon. The largest baboons; skin black; tail kinked near base; heavily built. Skull with muzzle considerably bent down on braincase. Mane little developed, as in previous group; colour olive-fawn to nearly black, with a single yellow or greyish ring on a dark grey hair (often black at tip); hair elongated on nape.

13. P. h. ruacana. Southwestern Angola through Southwest Africa to Orange R. Small for a chacma; blackish back, nape and crown with ocherous flanks; black hands, feet and tail. There is a distinct north-south cline in size in this race; southern specimens are as large as those from Cape Province.

14. P. h. griseipes. Rhodesia, southern Mozambique, northern Transvaal. Brown-yellow; hands, feet and tail not blackish.

15. P. h. orientalis. Coast of South Africa from Limpopo to Great Fish River, inland to Swaziland and Lesotho. Darker, more blackish than preceding.

16. P. h. ursinus. Cape Province north to Transvaal. Thickset, dark olive-grey (nearly black), with black hands, feet and tail.

C. J. JOLLY and I are at present collaborating in a study of geographic variation in baboons; consequently the above list should be taken as tentative only.

Baboons are not unlike long-nosed macaques to look at, and their large size and the general presence of a cape or mane make them look rather fearsome. Somewhat unexpectedly, however, they are distinctly more cooperative than rhesus monkeys, and except for their large size would be among the favourite research animals. Being considered vermin in many parts of their ranges, there are good possibilities for utilising this otherwise unappreciated resource of many African countries.

Taxonomy of the Geladas

The gelada *(Theropithecus gelada)* has often been referred to as 'gelada baboon', but studies by JOLLY have indicated that it must be clearly distinguished [contra BUETTNER-JANUSCH, 1966]. The face, though elongated like a baboon's, is elongated downwards rather than forwards, and is very deep with massive jaws and high-crowned teeth; the incisors and canines are relatively smaller than in true baboons. Since geladas sit down to feed – plucking seeds and roots and shuffling along on their bottoms as they move – a sexual signal on the perineal-vaginal area would be inconspicuous: consequently, the gelada's sexual skin is on the breast. Both sexes possess a naked, heart-shaped patch of skin on the breast, which in females becomes bright red during oestrus, with a chain of pustules which become engorged and stand out like ears of corn. Like baboons, the male gelada develops a mane; it is enormous, covering the foreparts and most of the forelegs. The ischial callosities are quite small and, unlike baboons, do not fuse across the midline in males; below them are a pair of fatty cushions. The nose is small and snubbed, with the nostrils not reaching the end of the snout. From this description it will be seen that the similarities between baboons and geladas are only superficial, probably in fact due to convergence; the two genera seem to have been separate for five million years.

Geladas vary geographically, although their range is limited to the Ethiopian highlands, and indeed does not extend further south than some 70 km north of Addis Ababa. According to Dr. JOHN CROOK [*in litt.*], who has studied wild geladas over much of their range, those living in the High Simien, on the northern rim of that part of the plateau, seemed to be more hairy and greyer than elsewhere; those at Debra Libanos in Shoa, much further to the south, appeared less hairy and browner in tone; these differences, as Crook remarks, make sense as the High Simien is colder and bleaker for parts of the year. Even more striking was the white fur surrounding the bare skin-patch of the adult male, which in geladas on Mt. Ambalimba was always 'persil-white', far more intense than the pale greyish-white colour in geladas of other areas.

Dr. CROOK writes: 'My general impression from visiting geladas in numerous sites in Ethiopia is that there is probably no clear division between a northern and a southern race. Far more likely is it that certain local and isolated populations on mountains and plateau areas may have diverged sufficiently in the genetic control of coat colour for contrasts in pelage to be apparent. These may also show some clinal variations in relation to climate'.

The importance of these data lies in the direct observation of numerous specimens. With the present large-scale depletion of many wild Primate populations, it becomes more and more essential to observe them in the wild, rather than collecting specimens and studying the remains. This too has to be done, in order to quantify and assess the characters one sees: but the relatively new possibilities of field taxonomy can extend traditional methods without reducing wild populations. In the present instance, the observations of at least three colour-types, to some extent at least separated geographically, fits in with the description of three 'species' of gelada described in the last century. These three supposed species, clearly conspecific according to modern taxonomic methods, fall into position as geographic races of *T. gelada* as follows:

1. T. g. gelada: this is the subspecies that CROOK records from the High Simien (or Semien); RÜPPELL, the describer, gave its range as 'Simien, Aksum and Haramat' (Haramat lies just to the east of Aksum). The general colour was given by RÜPPELL as being dark sooty-brown, with the shoulders, forearms, hands, feet and tail black, and the chest, upper arms and legs iron grey. The characterisation by CROOK of this as being 'grey' squares well with this; it remains to be seen whether it is more hairy.

2. T. g. senex: described by PUCHERAN in 1857, without a definite locality, as being pale yellow-brown on the sides of the head, neck and hindlimbs – lighter and browner than *gelada*, with no black on the tail. Although it is never very safe to link up two descriptions without geographic data, in this case PUCHERAN'S description is not unlike CROOK'S statement that the geladas around Debra Libanos are 'browner'. Until more exact information comes to light, therefore, I suggest the use of PUCHERAN'S name for the southern race which, as CROOK plausibly suggests, is probably the extreme of a north-south cline.

3. T. g. obscurus: HEUGLIN in 1863 described this form from the mountains at the sources of the Takazze river, in the districts of Lasta, Wadela, Talanta, Daund (= Dahont), Amhara and Worro, in the land of Jedju and Wollo-Galla. The colour of this race was described as darker, nearly black, with the legs pale yellow-brown and the tail pale brown; fleshcoloured rings round the eyes; and white hair around the chest patches. As the map shows (fig. 1), these localities are precisely in the Mount Ambalimba region where CROOK saw his 'persil-white'-chested males. BLANFORD recorded this race at Magdala; Dr. CLIFFORD JOLLY informs me that, as far as can be told from a rather small number of skins, this race seems valid on the characters assigned to it by its describer, in particular the white chest-hair.

Taxonomy of the Talapoins

This smallest of the Old World monkeys is becoming increasingly popular as a laboratory animal. It is small, not especially aggressive and so easy to handle, and can be kept in large groups – which can become self-perpetuating – in a comparatively small space. Moreover, in many parts of its range (such as Rio Muni) it is regarded as a pest and so the specimens which find their way into the possession of dealers are often those which would otherwise be slaughtered.

With the spread of cultivation, the talapoin (unlike all too many other African monkeys) has adapted itself to changed conditions and has in fact become more abundant. It lives near water and close to human settlement; such populations live mainly on the cassava which is left to soak in pools to leach out the toxic substances [GAUTIER-HION, 1966]. An obvious question poses itself: by what means do talapoins avoid poisoning themselves by eating unleached cassava? The answer to this question is apparently unknown, and would make a useful study for a laboratory possessing a talapoin colony.

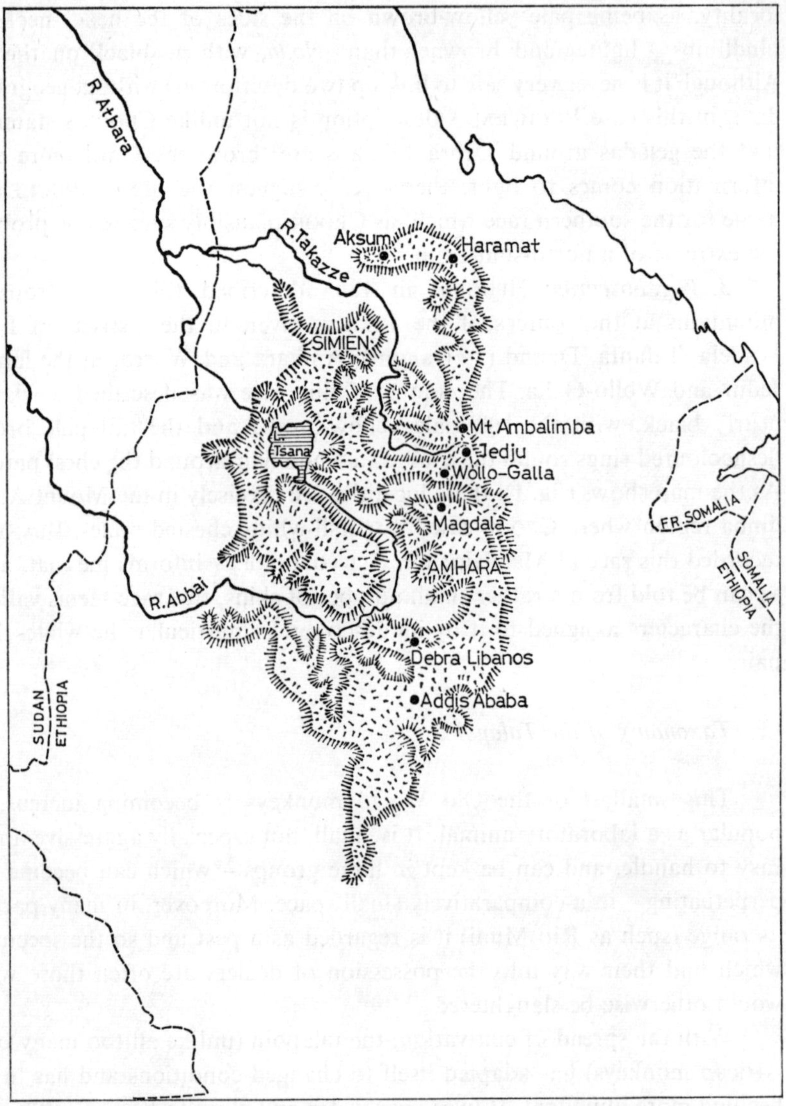

Fig. 1. Distribution of races of gelada.

The talapoin is generally placed in the genus *Cercopithecus,* although in recent years there has been an increasing tendency to allot it a separate genus, *Miopithecus.* The talapoin does differ from other guenons *(Cercopithecus* s.s.*)* in a number of features, notably the possession of cyclic sexual swellings in the female, and its greater range of facial gestures [VAN HOOFF, 1967]. There are also differences in the teeth. These differences make it possible that *Miopithecus* will eventually come to be recognised as a distinct genus somewhat intermediate between the guenons and the baboon/mangabey/macaque group; the swamp monkey *(Cercopithecus* [*Allenopithecus*] *nigroviridis)*, a much rarer species, shows some similar features and may prove congeneric. For the moment, however, the talapoin will be retained in the same genus as the other guenons.

There is only one species, *C. talapoin*. This is 340–400 mm in head and body length, with a rounded skull and a short, pale-yellowish face. The fur is greenish, varying in tone according to the width of the subterminal yellow band on each hair; the shoulders, hips and limbs are always much yellower than the back. On the cheeks are light whiskers, fanning out from just in front of the ear; a black band runs from eye to ear, and there are black infraorbital patches which are however absent in the young. HILL [1966] recognises four subspecies:

1. C. t. talapoin. Forests of Cameroun, south to mouth of R. Congo. Colour more olive due to yellow band's narrowness; cheek whiskers black-tipped, in general golden yellow; ears only slightly pigmented. Nose and upper lip covered with short black hairs. Head and body length 340–350 mm, tail 375–385 to 107–113% of head and body.

2. C. t. ansorgei. Forest of Angola, from south bank of lower Congo to Benguella district. Larger; more yellowish on back with broader yellow bands; whiskers whitish; less black on cheeks; nose and upper lip black-haired; ears black. Head and body 400 mm, tail 525–131% of head and body.

3. C. t. vleeschowersi. Lower Congo, from Kinshasa east to Kasai-Kwango confluence. Colour similar to last, but with short white hairs on the centre of the upper lip, golden yellow dark-tipped whiskers, and darker tail. Head and body 370, tail 360 – index between them 95–98%.

4. C. t. pilettei. Said to be from swamps at 2,500 m, Mt. Ruwenzori. Not unlike (3), but yellower, as in (1); tail grey below, not yellow as in all other races – this colour being sharply demarcated from that of the upper side. A shiny black patch on either side of the forehead, continuous laterally with the black eye-ear stripe. Head and body 365, tail 460, index between them 126%.

The validity of these races, all of which are based on just one or two specimens, is uncertain; since laboratory examples are mostly from Rio Muni, research workers will probably be able to estimate the likelihood of the validity of the races for themselves, by noting the variability within a single consignment. The subspecies *pilettei* would seem to be the most easily distinguishable from the description, but the correctness of its type locality, 850 miles to the northwest of the nearest confirmed talapoin range in the Congo-Kinshasa, has been doubted [RAHM, personal communication].

According to MACHADO [1969] the talapoin from north of the R. Congo (Cameroun, Rio Muni, Gabon, Congo-Brazzaville) is a distinct species from that found to the south of the Congo (northern Angola, Congo-Kinshasa). In the Angola species, as he describes it, the back is longer haired, a coarse mixture of greenish-yellow and black; the underside is white; the ears are big, black and very mobile; the nose is black, with dark lateral hairs bordering it forming a stripe down to the mouth; the tail is greyish underneath at the base; the scrotum is blue at the sides, ocherous-salmon on the most visible part. By contrast the Gabon species has a finer mixture of yellow and black, and shorter hair; the underside is light yellowish or greyish; the ears are small, not black, and less mobile; the nose is not so darkened and lacks the lateral darkening; the hollows in the cheek region are very prominent, outlined with a black line; the tail is yellowish underneath at the base; and the scrotum is entirely blue.

Moreover, the two species are said to have different behavioural characters. Apart from the difference in mobility of the ears, the Angola talapoin makes its threat grimace with the lips hardly open, while the Gabon species opens its mouth wide in threat.

A survey of the literature, and the photographs given by MACHADO from BUFFON and DAUBENTON's work on which SCHREBER's name *talapoin* was based, reveal that all names thus far given to talapoins belong to the Angolan species; so that *ansorgei* becomes an objective synonym of *talapoin*, and *pilettei* and *vleeschouwersi* are either synonyms or (perhaps) geographic races of the same species. The Gabon species lacks a name; MACHADO states that a formal description and naming of the new species is in preparation by himself and P. DANDELOT.

On the evidence given, MACHADO's claims seem well-founded. Whether or not the Angola species *C. talapoin* has any subspecies must await further specimens, but the two species should be carefully distinguished.

Taxonomy of the Colobinae

Members of the Old World Monkey subfamily Colobinae are sometimes kept in laboratories, but are highly unsatisfactory because of their specialised dietary requirements. In addition to this, some forms appear to be in general rather delicate and may refuse food of any kind. With two notable exceptions, the hanuman or entellus langur *(Presbytis entellus)* and the mantled guereza *(Colobus guereza)* and probably the other black-and-white colobs as well, all of the Colobinae are strictly leaf-eaters and cannot be persuaded to take other food. The hanuman langur does not in the wild restrict itself to leaves [RIPLEY, 1970] and is thus relatively easy to feed in captivity; accurate dietary data for black-and-white colobus are at present lacking, but the ease with which they can be adequately fed with other foodstuffs than leaves in captivity suggests that their diet in the wild may not be so very narrow.

It is not recommended, however, that even these two species should be used extensively for medical research. The black-and-white colobus is by no means abundant in many parts of Africa, and its numbers could not stand up to depletion such as occurred in the 1920s when its beautiful long silky fur became 'fashionable'. The hanuman langur is sacred to Hindus and thus protected in its native land, but nowhere does it build up such large populations as its compatriot the rhesus monkey (in the past at any rate).

Of the other Colobines, the spectacled langur *(Presbytis obscura)* has of late been added to the list of those which, though difficult, can be kept in captivity, and even bred, with success. The red and olive colobus have by and large proved impossible to confine, for psychological reasons as much as anything else. The proboscis monkey *(Nasalis larvatus)* has been bred in zoos at San Diego and Jogjakarta, and the extremely beautiful golden snub-nosed monkey *(Pygathrix [Rhinopithecus] roxellanae)* has been bred in the zoos of Pekin and Shanghai; neither of these species would be at all suitable for laboratories, nor are they likely to become widely available. The same probably applies to the other Asiatic Colobines.

Taxonomy of the Chimpanzees

Many authorities, for instance HILL [1967], recognise two species of chimpanzees. It will be shown below that the pygmy chimpanzee is perhaps better regarded as a rather distinct subspecies of the common chimpanzee. The four well-defined and easily distinguished subspecies are as follows:

1. Pan troglodytes verus. Masked chimpanzee. When young, a pale face with darker 'mask' round eyes; large tan spots may develop, and the face eventually turns black, but retaining the darker ocular region in many cases.

Ears large, always pale-coloured; palms and soles remain fairly pale. A bald frontal spot may develop; male has a grey-white beard, rather thick. Strong brow-ridges, narrow skull abruptly descending behind. Gambia east to Niger river.

2. Pan troglodytes troglodytes. Black-faced chimpanzee or Tschego. The palefaced young have no dark mask; the large tan spots grow and finally fuse to give a brown skin which later turns black – sometimes a very intense black. Ears smaller, black, as are palms and soles. Individuals tend to go quite bald, no white beard. Brow ridges strong, face more concave, braincase more rounded, less size difference between males and females. Niger river, south and east to Ubangui-Congo system.

3. Pan troglodytes schweinfurthii. Long-haired or Schweinfurth's chimpanzee. Face brown from an early age, becoming bronze then black without formation of tan spots; nose and upper lip turn black last. Body fur long, thick. Ears fairly large, bronzed. Bald spots develop; white beard present, longer but less copious than in No. 1, spreading to cheeks. Brow-ridges strong, but not raised as in previous races. Narrow skull, flat braincase, straight profile. North of Congo river, from Ubangui river east to Uganda and southern Sudan, Kivu province (Congo), Rwanda, Burundi, northwest Tanzania.

4. Pan troglodytes paniscus. Pygmy chimpanzee. Becomes black soon after birth, except for lips which remain flesh-coloured. Ears blackish, nearly hidden behind long cheek-whiskers. No balding, no white beard. No parting in hair, even when young, in marked contrast to other races especially No. 1. Brow ridges weak; skull narrow; face concave; braincase very high, rounded. Small – only 30 in high, compared to 36 for other races, and slenderly built; voice more shrill. Congo river south to R. Lukenie.

HILL [1967] recognises a further race which he calls *Pan troglodytes koolookamba.* The supposed range of this form falls entirely within that of the nominate race (No. 2 above): thus 'the koolookamba', if distinct at all, must be a full species. However HILL's description makes it clear that the claimed subspecies has all the essential characters of the nominate race, and differs only in such features as the gorilla-like nose and sagittal crest. Individuals answering to such a description are to be seen in any chimpanzee troop in the interior of the Gabon-Congo (Brazzaville) region [VAN BREE, personal communication]; any extra large animal will develop a sagittal crest, and gorilla-like noses turn up sporadically in black-faced chimpanzees. HILL's [1967] revision was based on living specimens of unknown locality; it demonstrates that the types of chimpanzees which SCHWARZ [24] showed to be geographic and named as subspecies can be picked out with relative ease from an unlocalised collection, and that there are additional characters by which they can be distinguished. SCHWARZ's geographic types were four in number (those above described); only a population study – i.e. based on *localised* specimens – can show whether there are others over and above this number. Such a study has not yet been conducted.

The pygmy chimpanzee, as can be seen from the descriptions, differs least from the nominate race (its black ears, lack of white beard, concave face and high rounded braincase are similar to, or accentuations of, the characters of *P. t. troglodytes*), and also has features recalling the long-haired chimpanzee (early darkening of skin, long cheek-whiskers, less marked brow ridges and narrow skull are in the same category). Its 'special features', that are not shared with other races, are the flesh-coloured lips, absolute lack of a parting, very high and rounded braincase and 'infantile' skull in general, small size and high-pitched voice. The first two characters are foreshadowed in individuals of other races; the rest are purely allometric. Consequently, all the differences between pygmy chimpanzees and the rest boil down to a matter of its smaller size. Since it follows the morphological pattern of its neighbours across the Congo river, in the absence of any intergradation (between *any* of the four subspecies, not just between the pygmy one and the rest!) all four forms may be conveniently classed as subspecies of a single species.

Studies on the serology and karyology of chimpanzees purporting to show the distinctiveness of the pygmy chimpanzee are inconclusive unless at the same time the three larger subspecies are differentiated. They are easily recognisable by a multiplicity of characteristics and there should be little difficulty about this. HILL and FINEG [in HILL, 1967] were able to agree on the classifications of almost all the chimpanzees at the Holloman Air-force Base independently. As a research animal, the chimpanzee has marked advantages and equally marked disadvantages. The clearest of its advantages, its phylogenetic nearness to man, makes it the ideal 'blueprint': experimental work carried out on chimpanzees are much more likely to be applicable to man than similar work carried out on monkeys. This statement would apply to gorillas and orang-utans as well, but these latter are endangered species and their import controlled; hence the burden falls on chimpanzees.

The disadvantages of using chimpanzees are moral as well as practical. It must be remembered that, with the possible exception of the gorilla, the chimpanzee is the most intelligent of the non-human primates (one is often tempted to omit the phrase 'non-human' in this context). Its intellectual requirements – companions, space, things to sustain its interest – are rather taxing to the resources of many laboratories; but boredom must not be allowed to creep in, for an unhappy chimpanzee is so often an unhealthy one, just as in human beings. This of course takes no account of the distress which one feels on seeing an unhappy chimpanzee: the laboratory personnel may become inured to the daily spectacle of a chimpanzee in cramped solitary confinement, but to visitors this is always a shock and among other considerations leaves a nasty taste where that particular laboratory is concerned.

Being a valuable and an intelligent animal, it is also incumbent on researchers not to sacrifice a chimpanzee needlessly. Just as a monkey should not be used where a rat will do, so one can add that a chimpanzee should not be used where a monkey will do.

The practical drawbacks to using chimpanzees are their need for space and variety, and their uncertain temper. When young, a chimpanzee will often be all too willing to cooperate in an experiment, even to insisting on the right to work the apparatus. Its volatile temperament may lead to sudden fright, or a rather destructive rampage. When adult, chimpanzees like all apes become crotchety and morose and often extremely uncooperative, and this places obvious limitations on their value as experimental animals.

Chimpanzees will breed in captivity if given the chance and inclination: yet another reason for adequate and humane housing, for the availability of wild-caught animals decreases daily. Being the only ape that is available for research, demand for specimens must not be allowed to increase so far that the species becomes rare in the wild. Every effort must therefore be made to breed them if they are to continue to be available.

Taxonomy of Marmosets

The family Callitrichidae, containing the marmosets and their relatives the tamarins, are the smallest of the New World monkeys, and for this reason as well as the comparative ease with which they may be kept in captivity they assume a great importance for laboratory studies. It should not, however, be forgotten they they are very primitive Primates: whether their claws which replace nails on their digits are primitive or secondary is one of the great disputes of primatologists, but in many other respects they are strikingly primitive: their brains are smooth and hardly convoluted at all, they possess well-defined pads on the hands and feet and well-developed vibrissae, and so on.

No modern revision of marmosets exists. However, HERSHKOVITZ is preparing a monograph of New World Primates and has published several preliminary papers on marmosets [HERSHKOVITZ, 1949, 1958, 1966a and b, 1968, 1969]. From these papers it is possible to see the way in which this authoritative worker is revising the family; his conclusions would result in a classification rather different from that in such works as HILL [1962]. The following classification may be taken as provisional until HERSHKOVITZ's monograph appears:

Family Callitrichidae

I. Subfamily Callimiconinae. [HERSHKOVITZ, 1966, makes this a full family]. M_3^3 present, claws less extreme in shape, upper incisors vertical, chin vertical. One genus, *Callimico,* with one species *C. goeldii* (Goeldi's marmoset): completely black with long silky hair forming a cape over the foreparts. Upper Amazon region.

II. Subfamiliy Callitrichinae. M_3^3 absent; claws well curved, sharp; upper incisors more sloping; chin receding. Four genera:

a) Genus *Cebuella.* The smallest marmoset; young *Callithrix* closely resemble adults of this genus. Lower incisors elongated, as long as canine, thus called 'short-tusked'; penis very short, projecting from anterior wall of scrotum; no pustules on scrotum. Skull short, rounded; hair of head slightly elongated into a cape; colour agouti (banded light and dark, usually brown, hairs). Head and body only 150 mm long; tail 190 mm. Upper Amazon; one species, *C. pygmaea* (pygmy marmoset).

b) Genus *Callithrix.* Larger, head and body length 200–230 mm, tail 300–320 mm. Similar but with less rounded skull, no cape except in young, and scrotum bears pustules which are glandular. Three species.

1. C. jacchus (common marmoset). Brazil coast. Marbled black and grey or black and brown; tail ringed with black and grey; ears with tufts growing in front (usually above and behind) of them or from inner surface of pinna. Face yellowish, hairy; a white blaze on forehead; rest of facial surround usually black. The subspecies – five recognised – are as follows:

C. j. jacchus. River Paraiba south to Bahia. Aural tufts white, fan-shaped. Forehead spot small.

C. j. penicillata. Bahia south to about Belmonte, and inland to Parana/ Tocantins watershed. Tufts black, less elongate but also fan-shaped.

C. j. geoffroyi. Belmonte south to Victoria. Similar, but whole facial surround is white. Tufts black.

C. j. flaviceps. Victoria to Rio de Janeiro. Head and facial surround light coloured; small yellowish tufts spring from inside pinna.

C. j. aurita. Rio de Janeiro south to R. Ribeirao. A large white forehead spot, but otherwise head is black; small white tufts from inside ear.

2. C. humeralifer (Santarem marmoset). Interior of Mato Grosso. Grizzled grey to white; tail weakly ringed, black at base; ears with thick tufts growing from pinna; face dark, with pink muzzle. Two subspecies:

C. h. humeralifer. West bank of lower R. Tapajóz. Grizzled brown-black, darker on hind parts; tail rings prominent.

C. h. chrysoleuca. Between rivers Canuma and lower Madeira. White; rump yellowish; tail only slightly ringed.

3. C. argentata (silvery marmoset). Interior of Mato Grosso. Brownish to silvery white; tail black; ears bare; face pink or reddish. Three races:

C. a. argentata. Between Tapajóz and Iriri rivers. White with black tail.

C. a. melanura. North and west of R. Tacuari. Brownish with black tail.

C. a. leucippe. Small area on east bank of R. Tapajóz. Completely white.

c) Genus *Saguinus.* Slightly larger than last; lower incisors not as long as canines ('long-tusked'); penis and clitoris long. Skull less rounded than in previous genera. Scrotum bears pustules. Ten species.

1. S. midas (black tamarin). Purely black marbled with buff, except for a brow-band in young, and sometimes orange hands. Two subspecies:

S. m. midas. North of lower Amazon, north to Guyana and west to Rio Negro. Hands and feet red-gold.

S. m. niger (synonym *tamarin*). South of Amazon estuary, between rivers Gurupi and Xingú. Hands and feet black like rest of body.

2. S. nigricollis (black-and-red tamarin). Anterior part of back coloured differently from posterior (including hindlimbs). Hairs around mouth white; genitalia pigmented; a pale brown patch on temples. Colombia-Ecuador-Brazil border. Two subspecies:

S. n. nigricollis. North of Amazon; south of rivers Putumayo/Iça. Anterior part of dorsum black, tapered behind; lower part reddish; below reddish mixed with black.

S. n. graellsi. west bank of Rio Napo. Forepart of dorsum agouti, ending more abruptly; hindpart and underparts brownish.

3. S. fuscicollis (saddle-back tamarin). Similar, but with 'saddle' of colour distinct from both anterior and posterior parts of back. No temporal patches. Thirteen subspecies are clearly recognisable:

S. f. fuscicollis. Between rivers Juruá and Javari. Upper side agouti hued, tail black.

S. f. avilapiresi. Between Juruá and Tapauá. Darker agouti; tail black.

S. f. fuscus. Bounded by rivers Caguán/Caquetá/Japurá in the north, and rivers Putumayo/Iça in south. Agouti; forehead and crown blackish.

S. f. nigrifrons. Between Juruá and Javari. Agouti; forehead and tail black.

S. f. illigeri. South of R. Maranon; between rivers Ucayali and Huallaga. Differs from last in its black crown; from *fuscus* in its black tail.

S. f. leucogenys. Between R. Huallaga and the Andes. Body more blackish.

S. f. lagonotus. South of R. Napo, north of R. Maranon. Similar, but nape black, forepart of back red.

S. f. tripartitus. South of Putumayo/Iça, north of R. Napo. A very beautiful race with black nape and orange-yellow foreparts, and a small white V-mark above the eyes.

S. f. weddelli. South of R. Purús, northwest of Madeira/Mamoré. White V-mark larger; whole of foreparts blackish.

S. f. cruzlimai. Range unknown: probably between Tapauá and Parús rivers. Similar to last, but whole body lighter with a reddish wash.

S. f. crandalli. Also unknown; probably between lower Solimoes, Tapauá and Madeira. Red replaced by a drab orange wash; tail drab-orange.

S. f. acrensis. Between upper Juruá and upper Tarauaca. Yellow wash.

S. f. melanoleucus. Between lower courses of same rivers. White.

4. S. labiatus (red-bellied tamarin). Skin beneath white circumbuccal hairs also depigmented; white extends to nasal septum. Genitalia unpigmented. Black-brown, marbled on loins; underparts red, sharply marked off; a reddish patch on crown. Two subspecies:

S. l. labiatus. Between rivers Purús and Maranon. Red of underparts extends to throat; on crown, very extensive red patch.

S. l. thomasi. A small area on north bank of R. Maranon. Red of chest does not extend to throat; on crown, restricted to vertex.

5. S. mystax (moustached tamarin). Similar to the last, but a much wider depigmented, white-haired area, round mouth and including nostrils. Back marbled, yellow-brown; underparts not sharply differentiated; tail black; usually no red patch on crown. Three subspecies:

S. m. mystax. South of R. Maranon, northwest of R. Juruá, east of R. Huallaga. Depigmented area of genitalia not extending under tail.

S. m. pluto. South of R. Solimoes, between Purús and Madeira. White in genital region extends along the underside of the tail.

S. m. pileatus. Between Juruá and Purús. Similar, but with a red crown mark.

6. S. imperator (emperor tamarin). Between upper Juruá and R. Madre de Dios. White hairs similar to last in extent, but elongated into a long sideswept moustache, and short beard. Body grey speckled with black; tail red with a black tip. A highly characteristic and unmistakeable species.

7. S. bicolor (pied tamarin). North of Amazon, between lower R. Negro and R. Erepecurú. Face and sides of head naked; foreparts whitish, hindparts dark; tail colour sharply distinct above and below. Genitalia only slightly pigmented. Three races:

S. b. bicolor. Between Rio Negro and Rio Uatuma. Neck, shoulders, arms and chest white; rest of body medium to dark grey-brown; underside orange; tail brown-black above, like underside below.

S. b. ochraceus. Probably, between R. Uatuma and lower R. Nhamundá. Anterior part of body cinnamon to buff, browner on hindparts; limbs ocherous; underside orange; upper side of tail brown.

S. b. martinsi. Between R. Nhamundá and R. Erepecurú. Darker, mainly brownish with strongly annulated hairs; less distinction between fore- and hindparts of back. Upper surface of tail dark brown.

8. S. inustus (Schwarz's tamarin). Brazil-Colombia border area, bounded by the upper Rio Negro, upper Orinoco, Guaviare and Japurá. Genitalia black; face bare, like the last, but with a short crest of hear on forehead; lips, cheeks and brows unpigmented; body black, becoming chocolate on mid-back, cinnamon on flanks; tail black.

9. S. leucopus (white-footed tamarin). Lowlands of northern Colombia between the Cauca and Magdalena rivers, near their confluence. Similar to last, with somewhat longer forehead crest and fully pigmented face; brown-grey, reddish below; tail black with a white tip. Hands and feet white.

10. S. oedipus (cottontop or pinché). Colombia west of the R. Cauca, and Panama. Crest more elongated; dark brown with buff speckling, belly and limbs white; tail brown at base, becoming lighter with a black tip. The lower margin of the ear pinna is deeply excavated. Two well-marked races:

S. o. geoffroyi. Colombia west of the Rio Atrato, the whole of Panama east of the canal, and the south coast to the west of the canal. Forehead crest comparatively short; brown colour of back strongly marbled with buff.

S. o. oedipus. Northwestern Colombia, near the coast between rivers Atrato and Cauca. Crest long, flowing like an Iroquois Indian; buffy marbling on back inconspicuous or absent.

d) Genus *Leontopithecus* (synonym *Leontideus*). Like the last but rather larger; scrotum lacking pustules; fingers very long and thin. There are also skull differences. These very beautiful marmosets (or tamarins) are rare; they are found in the sclerophyll forests of southeastern Brazil. All three species – which may ultimately come to be classed as races of a single species – are maned with dark faces, comparatively short tails (hardly longer than the head and body), and much gold in the body-colour.

1. L. rosalia (golden lion marmoset). Sierra do Mar. Completely red-gold.

2. L. chrysomelas (golden-maned tamarin). Sierra do Periperi. Black with golden mane, arms and upper surface of tail.

3. L. chrysopygus (golden-rumped tamarin). Mountains between Tiete and Panapanema. Black with a golden forehead, rump and inner sides of hindlimbs.

The above seventeen species of marmosets are all fairly suitable, as far as ease of maintenance goes, for laboratory work, although some tend to be aggressive both to handlers and to other marmosets of the same sex. However, some species are rare: members of the genera *Callimico*, *Cebuella* and *Leontopithecus* all rank as endangered species and, if they end up in a laboratory, attempts should be made to breed them – they should certainly not be experimented upon. *Saguinus* and *Callithrix*, on the other hand, especially certain species *(C. jacchus, S. fuscicollis, S. oedipus)* are quite often offered for sale; their identification down to species level and, if possible, down to subspecies level, is quite important. When identifying a marmoset, it is commonly the case that hair-tracts and tufts, and colour-pattern are more important as specific recognition characters than colour itself; this is true for mammals generally, but in marmosets it assumes greater significance than usual because three species have independently evolved white subspecies which can, however, be distinguished by the remnants of colour-pattern visible on them, and by teeth and ear-tufts (these three are *Callithrix humeralifer*, *C. argentata* and *Saguinus fuscicollis*). The geographic variation of the various species has been well-studied, and tends to be quite sharply defined, since borders of subspecies ranges tend to be large rivers, thus reducing gene-flow between them.

Taxonomy of other South American Monkeys

Both the night monkey *(Aotus)* and the capuchin monkey *(Cebus)* are not uncommon in laboratories, and both are fairly good, convenient animals to keep. However the former is nocturnal and if its psychological welfare is considered important it should not be disturbed by day; alternatively a daylight reversal scheme can be adopted, as for many prosimians.

The night monkey or douroucouli *(Aotus trivirgatus)* is the only species of its genus; numerous subspecies have, however, been described, some being of uncertain validity. The wide distribution extends from Colombia in the north to Paraguay in the south; the coat is thick, soft, brown or grey, and there are three stripes above the eyes, nearly meeting on the crown; between them are two white spots. Colouration and size are the characters that vary subspecifically.

The taxonomy of squirrel monkeys became of importance with the publication of a paper by MACLEAN [1964], which showed that two types of squirrel monkeys called the 'Gothic' and 'Roman' types differed in their contexts of penile display: the 'Gothic' type had a distant and a mirror display, whereas the 'Roman' type showed neither of these types. The Gothic type was said to come from Colombia or Peru; the Roman type, from Brazil. The question presents itself: are there really two distinct races of squirrel monkey, with the circumocular patch resembling a Gothic and a Roman arch respectively? If this is the case, then it is clear that there are distinct ethological differences between them; if not, then these differences fall within the range of individual variability.

CABRERA [1957] recognised a single species of squirrel monkey in South America, *Saimiri sciureus*, with six subspecies, commenting that the Central American form, *S. oerstedti*, was undoubtedly conspecific as well. A revision of the genus by THORINGTON is in press at this time, so CABRERA'S classification should be regarded as provisional:

1. S. s. oerstedti. Panama and Costa Rica. Colour reddish; crown of head black or nearly so. Shoulders and limbs grey-yellow; hands and feet ochraceous. Upper side of tail black mixed with yellow for basal two-thirds, with black tip.

2. S s macrodon. E. Colombia, Ecuador, northern Peru (east of the Andes). Colour light, rich tawny; crown grey; hands and feet darker.

3. S. s. sciureus. the Guianas, Venezuela, central and eastern Brazil as far south as the north of Goyaz state. Slightly greenish grey; crown grey; more golden colour in middle of back. Hands, feet and forearms gold or orange. Light parts of face are not completely white.

4. S. s. collinsi. Marajó island, Amazon estuary. Much paler and greyer, except on hindparts which are tawny; hands and feet darker (tawny). Smaller.

5. S. s. ustus. Banks of the Solimoes R. and its tributaries, N.W. Brazil. Redder or more orange colour, more or less overlaid with black. Basal two-thirds of tail blackish; circumocular region, as in the two previous races, not white.

6. S. s. nigriceps. Upper Jurua river in Brazil; banks of the Ucayali and Urubamba in E. Peru. Dark, grizzled colour, with blackish crown; hands and feet orange-rufous, not extending much onto forearms. White facial marks very conspicuous.

7. S. s. boliviensis. Eastern Bolivia, extending in the south to Paraguay. Grizzled olive as No. 3, but redder on hindparts; crown black and extensive. Hands, feet and forearms dark, burnt ocherous.

It seems that No. 6 may be MACLEAN'S typical 'Gothic' type, No. 3 his 'Roman'; but No. 1, 2, and 7 are all more or less 'Gothic', and No. 3, 4 and 5 are 'Roman'.

The capuchins are currently classified into four species, arranged in a tufted and an untufted group. The three untufted species replace one another geographically; the tufted capuchin overlaps each of the three untufted ones widely in range. They may be characterised as follows:

1. Tufted capuchin (*Cebus apella* LINNAEUS, 1758). The largest species;

dark brown with a broad blackish 'cap' on head, where the hairs are upright and tend to be elongated at the sides like horns; the blackish of the cap extends into long sideburns which reach under the chin. Many subspecies are recognised, from Colombia to Paraguay.

2. White-throated capuchin (*C. capucinus* LINNAEUS, 1758). Smaller, lacking the erect crown hairs and 'horns'. Coat black, except for face, chest, shoulders and upper part of arms, which are whitish. A mainly Central American distribution, from as far north as Honduras into Colombia, where it occurs only in the north and west, and western Ecuador, on the west side of the Cordillera Occidental.

3. White-fronted capuchin (*C. albifrons* HUMBOLDT, 1812). Similar to the last, but brown (often cinnamon-coloured) where the last is black, and buff where the last is white. Female may have a brush of long hairs on the brow, but there is never a true tuft as in the first species. Found in the forest of Colombia, western Venezuela, Ecuador, Peru, northern Bolivia and western Brazil, bounded to the west by the High Andes and to the east approximately by the rivers Orinoco, Negro and Tapajóz.

4. Weeper capuchin (*C. nigrivittatus* WAGNER, 1848). Brown – usually darker than the last – with white face and chest as in (2), but head much more white as cap is very narrow, V-shaped on the forehead. Found mainly to the east of the last, but apparently overlapping it in northern Colombia, and including Venezuela, Guyana, Surinam, Cayenne and Brazil north of the Amazon and east of the Rio Negro.

It should be noted that capuchins are highly intelligent. They are alert and inquisitive monkeys, and this has to be taken into account in keeping them in captivity. They are known in the pet trade as 'ringtails' because of the coiled tail which is prehensile, but not modified as in spider, woolly and howler monkeys.

A common but rather little-known South American monkey, the titi, has recently come into the laboratory world with a large colony now maintained in the Delta Primate Centre, where they have been bred. They are in many respects somewhat primitive Cebids, in dental and other anatomical features, although HERSHKOVITZ [1963] disputes that they are as primitive as formerly assumed. They are instantly recognised by the their short limbs, thick fur, long bushy tails and flat faces; the nails on hands and feet are strongly curved, rather clawlike. The two species are well-known through HERSHKOVITZ's work:

1. Widow monkey (*Callicebus torquatus* HOFFMANSEGG, 1807). Reddish-black in colour with a white to orange throat-patch, black tail and black

forearms, legs and feet. The hands are black in one race, whitish in the other two. Found on the tributaries of the upper Amazon, as far east as the Rio Negro, and on the Amazon-Orinoco watershed.

2. Dusky titi (*C. moloch* HOFFMANSEGG, 1807). A more variable species with eleven subspecies: the colour is dark brown or grey, often reddish; the underparts may be more reddish; no throat-patch but face often black, or with a black forehead band; the hands and feet are black or else coloured like the limbs (grey, black, red); the tail is red or grey. The wide range extends from the upper Amazon tributaries in Colombia and Ecuador, east along the southern bank of the Amazon, and southeast via Bolivia to the dry, hardwood forests of the southern Brazil littoral, from Sao Paulo to Bahía. The pale, black-footed masked titi, often classed as a separate species *(C. personatus)* is a race of this species.

Titis live in mated pairs which are territorial, and show aggression when they meet. Consequently it is unwise to cage them in groups of more than two, one of each sex.

The three genera – *Lagothrix*, *Ateles* and *Alouatta*, the woolly, spider and howler monkeys – are unsatisfactory as lab. animals because of the animals' agility and need for space in the case of the first two (especially spider monkeys) and because of the difficulty in keeping howlers under any conditions. For one reason or another (disease, wrong diet, mental factors) howler monkeys invariably die within a couple of years of leaving their native land; and it is really rather pointless as well as malicious to try to keep them, let alone perform experiments on them. Of the other two, there are two species in each case, one of which is rare and rather unlikely to be captured while the other is common enough and does fairly well in zoos. The common species of woolly monkey is HUMBOLDT'S which varies from pale grey or brown to nearly black, with often a black cap on the head; since this species is not uncommon in captivity, for better or worse, diagnoses may be given here of the four subspecies [after FOODEN, 1963]:

1. Lagothrix lagotricha lagotricha. North of the Amazon, from the Rio Negro west to the Andean foothills north of the Rio Napo, and north to southeastern Colombia. Uniform brown colour.

2. L. l. lugens. Ranges along the Cordillera Oriental to the Sierra de Merida, in Colombia. May be grey, grey with a black head, or completely matt black.

3. L. l. poeppigii. South of the Amazon and Rio Napo, southeast to the Rio Juruá, and west to the Andes. Brown, with a black head.

4. L. l. cana. From the R. Juruá in the west to the R. Tapajóz in the east, north to the Amazon and south into northern Bolivia. Olivaceous in colour, with a black head.

The other species is HENDEE'S woolly monkey *(L. flavicauda)* which is known only from a small area, 77° W and 6° S, on the Andes; it is mahogany brown with a sharp yellow stripe along the underside of the distal half of the tail. There is no likelihood of this species turning up in the lab., but if it does so then it should be treated with the care and circumspection its rarity deserves.

Spider monkeys (genus *Ateles*) are more difficult to classify, since there has been no modern revision of them since that by KELLOGG and GOLDMAN [1944], which was unsatisfactory in that it recognised too many species, somewhat arbitrarily allotted. All forms of true spider monkey replace one another; with a few exceptions, the differences concern colour only, and some supposed 'species-differences' are overlapped by individual variation. A review of the subspecies of the common spider monkey, *Ateles paniscus*, that seem to be valid is given below:

1. A. p. paniscus. Guyana, Cayenne, Surinam; Brazil north of the Amazon and east of the Rio Branco/R. Negro system. Long, silky jet-black hair; red face.

2. A. p. marginatus. Brazil, south of the Amazon between the rivers Tocantins and Tapajóz. Long silky black hair; face black, except around eyes and on muzzle where flesh-coloured. A white forehead band and side-whiskers.

3. A. p. chamek. Brazil, south of the Amazon and west of the Tapajóz; Bolivia, south as far as Santa Cruz; Peru, east of the Ucayali-Urubamba. Hair somewhat less long and silky, but still black; face black.

4. A. p. belzebuth. Venezuela and Colombia south of the Orinoco/Mxeta system, and east of the Cordillera Oriental; Brazil, north of the Amazon and west of the rivers Branco/Negro; Ecuador and Peru, east of the Andes and north of the Ucayali. Blackish, like the last, with a black face; often a pale triangular frontal diadem of upstanding hair; underparts much lighter, abruptly demarcated from back.

5. A. p. hybridus. Colombia and Venezuela, north of the Orinoco/Meta, east of the Magdalena. Lighter, wood-brown, but otherwise similar to last; frontal diadem distinct.

6. A. p. fusciceps (synonyms *robustus, dariensis*). Ecuador and Colombia, west of the Andes and east of the Rio Atrato. Blackish, with white hairs on the muzzle; tips of hairs on anterior half of crown brownish for a greater or lesser distance, sometimes over whole crown.

7. A. p. grisescens. Colombia, west of Rio Atrato, to eastern Panama. Hairs yellow at base, black at tips, giving a grizzled grey effect. Face often partly depigmented. Both 'hooded' and bare-forehead forms occur.

8. A. p. ornatus (synonyms *panamensis, azuerensis).* Western Panama, southern Costa Rica. Crown, neck and upper arms black-brown; cheeks whitish; back red-tawny to gold; underside paler, cinnamon. Face usually black, but often partly depigmented (round eyes and on muzzle).

9. A. p. geoffroyi (synonyms *frontatus, pan, vellerosus*). Northern Costa Rica, Nicaragua, Honduras, Salvador, Guatemala, and Mexico west to Oaxaca, north to Veracruz. Medium to dark brown; crown, neck and withers black; below buff or tawny. Face black or partly depigmented.

10. A. p. yucatanensis. Yucatan penisula only. Pale buff above (except crown, neck and upper arms which are black as usual); silvery-white below.

The woolly spider monkey was formerly referred to a distinct genus, *Brachyteles*, but is today more generally placed as a second species of the spider monkey genus *Ateles – A. arachnoides.* It is robustly built and with a rather thick, woolly coat, but like the common spider monkey it lacks a thumb and is rather pot-bellied. Its relationship to the common spider monkey is rather like that of HENDEE's woolly monkey to HUMBOLDT's woolly monkey: a highly distinctive allopatric species, but hardly deserving of generic rank. It is found in the littoral of southeastern Brazil, in the drier, more scrubby hardwood forests which have unfortunately been so cut down that its existence as a wild animal is endangered. Like howler monkeys, but for a different reason, it should not be kept in the lab.

Finally, the howler monkey *(Alouatta)* must be mentioned. Easily recognisable by its enormous jowls, covered by a shaggy beard and being the outward sign of the inflated hyoid bone which acts as a resonator for the voice, the howler is a totally unsatisfactory animal in every way for a laboratory. Six species exist, widespread from southern Mexico to Paraguay.

A recent review by PERKINS [1969] is extremely useful in its discussion of South American monkeys as laboratory animals, with observations on their behaviour. It should be noted, however, that the taxonomy differs from that used in the present chapter, and is based largely on HILL.

Classification of the Primates (order Primates)

Suborder Prosimii

  Infraorder Lemuriformes

Family Lemuridae
- Genus Lemur (lemurs) 5–8 species, Madagascar
  - Hapalemur (gentle lemur) 2 species, Madagascar
  - Lepilemur (sportive lemur) 1 species, Madagascar

Family Daubentoniidae
- Genus Daubentonia (aye-aye) 1 species, N. W. Madagascar

Family Cheirogaleidae
- Genus Cheirogaleus (dwarf lemurs) 2 species, Madagascar
  - Allocebus (hairy-eared dwarf lemur) 1 species, E. Madagascar
  - Phaner (fork-crown dwarf lemur) 1 species, W. Madagascar
  - Microcebus (mouse lemur) 2–4 species, Madagascar

Family Indriidae
- Genus Indri (indri) 1 species, N.E. Madagascar
  - Propithecus (sifaka), 2 species, Madagascar
  - Avahi (woolly indri), 1 species, Madagascar

Infraorder Lorisiformes

Family Lorisidae

Subfamily Lorisinae
- Genus Loris (slender loris) 1 species, Ceylon and S. India
  - Nycticebus (slow loris) 2 species, S.E. Asia
  - Perodicticus (potto and angwantibo) 2 species, Africa

Subfamily Galaginae
- Genus Galago (bushbabies) 5–6 species, Africa

Infraorder Tarsiiformes

Family Tarsiidae
- Genus Tarsius (tarsier) 3 species, Indonesia and Philippines

Suborder Anthropoidea

Infraorder Platyrrhini

Family Callitrichidae
- Genus Callithrix (common marmoset) 3 species, S. America
  - Cebuella (pygmy marmoset) 1 species, Upper Amazon
  - Saguinus (tamarins) 10–15 species, C. and S. America
  - Leontopithecus (lion marmoset) 3 species, S.E. Brazil
  - Callimico (Goeldi's marmoset) 1 species, Upper Amazon

Family Cebidae
Genus Aotus (night monkey) 1 species, S. America
Callicebus (titi) 2 species, S. America
Saimiri (squirrel monkey) 1 species, C. and S. America
Pithecia (saki) 4 species, northern S. America
Cacajao (uakari) 2 species, northern S. America
Cebus (capuchin monkeys) 4 species, northern S. America
Alouatta (howler monkeys) 5 species, C. and S. America
Lagothrix (woolly monkeys) 2 species, northern S. America
Ateles (spider monkeys) 2 species, C. and S. America
Infraorder Catarrhini
Superfamily Cercopithecoidea[1]
Family Cercopithecidae
Subfamily Colobinae
Genus Colobus (colobus monkeys) 4–8 species, Africa
Presbytis (langurs) 13 species, S.E. Asia
Pygathrix (including Rhinopithecus) (doud and snub-nosed monkeys) 4 species, Vietnam and S.W. China
Nasalis (proboscis monkeys) 2 species, Borneo and Mentawei Is.
Subfamily Cercopithecinae
Genus Cercopithecus (guenons) 12–20 species, Africa
Erythrocebus (patas monkeys) 1 species, Africa
Cercocebus (mangabeys) 4–7 species, Africa
Papio (baboons) 3–4 species, Africa
Theropithecus (gelada) 1 species, Ethiopia
Macaca (macaques) 10–16 species
Superfamily Hominoidea
Family Pongidae
Genus Hylobates (gibbons) 6 species, S.E. Asia
Pongo (orang-utan) 1 species, Borneo and Sumatra
Pan (gorilla and chimpanzee) 2 species, Africa
Family Hominidae
Genus Homo (man) 1 species, world-wide

1 For a recent collection of papers on this superfamily, with emphasis throughout on taxonomy, see NAPIER and NAPIER [1970].

References

BRAMBLETT, C. A.: (Personal communication. 1967).

BREE, P. J. H. VAN: (Personal communication. 1969).

BUETTNER-JANUSCH, J.: A problem in evolutionary systematics; nomenclature and classification of baboons, genus *Papio*. Folia primat. *4:* 288–308 (1966).

CABRERA, A.: Catalogo de los mamiferos de America del Sur (Buenos Aires 1957).

CROOK, J. H.: (*In litt*. 1970).

DANDELOT, P.: Note sur la classification des cercopithèques du groupe Aethiops. Mammalia *23:* 357–368 (1959).

FOODEN, J.: A revision of the woolly monkeys (Genus *Lagothrix*). J. mammal. *44:* 213–247 (1963).

FOODEN, J.: Identification of the stump-tailed monkey, *Macaca speciosa* I. GEOFFROY, 1826. Folia primat. *5:* 153–164 (1967).

FOODEN, J.: Taxonomy and evolution of the monkeys of Celebes. Bibl. primat. *10:* 1–148 (1969).

FOODEN, J.: Report on primates collected in Western Thailand, January–April, 1967. Fieldiana, Zoology, 59 *1:* 1–62 (1971).

GAUTIER-HION, A.: L'écologie et l'éthologie du Talapoin, *Miopithecus talapoin talapoin*. Biol. Gabon. *2:* 311–329 (1966).

HEUGLIN, M. T. VON: Beitrage zur Zoologie Afrikas; über einige Säugetiere des Bäschlo-Gebiets. Nova Acta Ac. Caes. Leop.-Car. *30:* 10–14 (1963).

HERSHKOVITZ, P.: Mammals of Northern Colombia. Preliminary Report No. 4: monkeys (Primates). Proc. US Nat. Mus. *98:* 323–427 (1949).

HERSHKOVITZ, P.: Type localities and nomenclature of some South American Primates. Proc. biol. Soc., Wash. *71:* 53–56 (1958).

HERSHKOVITZ, P.: A systematic and zoogeographic account of the monkeys of the genus *Callicebus* (Cebidae) of the Amazonas and Orinoco River basins. Mammalia *27:* 1–80 (1963).

HERSHKOVITZ, P.: Taxonomic notes on tamarins, genus *Saguinus*. (Callithricidae, Primates), with descriptions of four new forms. Folia primat. *4:* 381–395 (1966a).

HERSHKOVITZ, P.: On the identification of some marmosets, family Callithricidae. Mammalia *30:* 327–332 (1966b).

HERSHKOVITZ, P.: Metachromism or the principle of evolutionary change in mammalian tegumentary colours. Evolution *22:* 556–575 (1968).

HERSHKOVITZ, P.: The evolution of mammals on southern continents. VI. The recent mammals of the neotropical region; a zoogeographic and ecological review. Quart. Rev. Biol. *44:* 1–70 (1969).

HILL, W. C. O.: Primates; comparative anatomy and taxonomy. 3. Hapalidae. (Univ. Press, Edinburgh 1962).

HILL, W. C. O.: Primates; comparative anatomy and taxonomy. 6. Cercopithecoidea. (Univ. Press, Edinburgh 1966).

HILL, W. C. O.: Geographic variation in the chimpanzee, in STARCK, KUHN and SCHNEIDER Neue Ergebnisse der Primatologie (Frankfurt 1967).

HOOFF, J. A. R. A. M. VAN: The facial displays of the Catarrhine monkeys and apes; in MORRIS Primate ethology (London 1967).

JOLLY, C. J.: Introduction to the Cercopithecoidea, with notes on their use as laboratory animals. Symp. Zool. Soc. Lond. *17:* 427–457 (1966).

KELLOGG, R. and GOLDMAN, E. A.: Review of the Spider monkeys. Proc. US nat. Mus. *96:* 1–45 (1944).

KUMMER, H.: Cross-species modifications of social behaviour in baboons; in NAPIER The old world monkeys (New York 1970).

MACHADO, A. DE BARROS: Mamíferos de Angola ainda não citados ou pouco conhecidos. Publ. cult. Co. diam. Ang. Lisbon No. *46:* 93–232 (1969).

MACLEAN, P. D.: Responses to mirror image in two types of Squirrel monkey. Science *146:* 950 (1964).

MARTIN, R. D.: Reproduction and ontogeny in Tree-shrews (*Tupaia belangeri*) with reference to their general behaviour and taxonomic relationships. Z. Tierpsychol. *25:* 409–532 (1968).

NAPIER, J. R. and NAPIER, P. H.: The Old World Monkeys: Evolution, systematics and behavior (Academic Press, London/New-York).

PERKINS, E. M.: Monkeys of the New World. Primate News *7:* 4–15 (1969).

PUCHERAN, M.: Rev. Mag. Zool. (2) *9:* 243 (1857).

RAHM, U.: Some remarks on *Cercopithecus hamlyni, lhoesti* and *mitis* and on *Colobus polykomos;* in Napier Old world monkeys (1970).

RIPLEY, S.: Leaves and leaf-monkeys; in Napier Old world monkeys (1970).

RÜPPELL, E.: Neue Wirbeltiere zu der Fauna von Abyssinien gehörig. (Frankfurt 1835).

SCHWARZ, E.: On the local races of the chimpanzee. Ann. Mag. nat. Hist. *13:* 576–583 (1934).

SIMPSON, G. G.: The principles of classification and a classification of mammals. Bull. amer. Mus. nat. Hist. *85:* 1–350 (1945).

VALEN, L. VAN: Tree-shrews, Primates and fossils. Evolution *19:* 137–151 (1965).

Author's address: Prof. COLIN P. GROVES, Duckworth Laboratory of Physical Anthropology, Faculty of Archaeology and Anthropology, Downing Street, *Cambridge* (England)

Pathology of Simian Primates, Part I, pp. 58–77 (Karger, Basel 1972)

Natural Diets of Wild Primates

C. JONES

Tulane University, Delta Regional Primate Research Center, Covington, La.

Contents

Introduction 58
Methods 59
General Feeding Habits 60
Seasonal Variations 66
Arboreal and Terrestrial Feeding Habits 69
Geographic Variations 70
Feeding Habits and Body Size 71
Activity Rhythms and Feeding Habits 71
Natural Diets and Morphology 72
Summary 75
Acknowledgments 76
References 76

Introduction

There has been a great surge of interest and activity in various aspects of studies of nonhuman primates during the past decade; unfortunately, much remains unlearned with regard to natural diets and nutrition of primates. Data on various aspects of the natural diets of wild primates have been contributed by numerous authors. Information and summaries of published materials on natural diets of wild primates were presented by HILL [1953, 1955, 1957, 1960, 1962, 1966], WALKER [1964], ANDERSON [1967] and NAPIER and NAPIER [1967]. Many observations on natural diets have been incidental to studies of behavior or other aspects of the ecology of groups of primates and relatively few data published on natural diets were based on careful observations or thorough examinations of stomach contents of wild primates.

The purposes of this report are to summarize generally some of the information available on natural diets of wild primates, to discuss briefly some of the general relationships between primates with regard to natural diets, and to discuss and exemplify some of the variations of diets and some corresponding modifications of the gastro-intestinal tracts of wild primates.

Methods

It is not within the scope of this brief report to present either a complete list of food items consumed by various kinds of wild primates or a summary of the mass of literature containing information and notes on foods and feeding of wild primates. References are given that contain either additional information or examples pertinent to the aforementioned purposes of this paper.

Because teeth are recognized diagnostic features of the major groups of primates and are described and discussed in detail in taxonomic treatments of primates, discussions of teeth and correlations with natural diets of wild primates are not included in this paper. For photographs and discussions of the teeth of primates, see JAMES [1960].

Information on the natural diets of wild primates is obtained usually from careful observation and collection of specimens of food materials immediately after or while primates are feeding on them, from analyses of fecal materials, and from examinations of stomach contents. Examinations of stomach contents apparently provide the most accurate identifications of materials included in the natural diets of primates, but it is often either impractical or impossible to obtain specimens for these purposes. For brief discussions of some limitations in determining natural diets of monkeys from examinations of stomach contents, see THORINGTON [1968] and JONES [1970].

For the purposes of this report, wild primates may be classified generally with regard to natural foods and feeding behavior as follows:

Stenophagous, Monophagous – feed on a single kind of material.
Sarcophagous, Carnivorous – feed on the flesh of animals.
Insectivorous – feed on insects.
Carnivorous – feed on animals that they (feeders) kill.
Phytophagous, Vegetarian – feed on living plants.

Herbivorous – feed on leaves, stems, and soft parts of green plants.

Frugivorous – feed on fruits.

Graminivorous – feed on seeds and grains.

Euryphagous, Polyphagous, Omnivorous – feed on both plant and animal materials.

These categories are often not sharply defined because many species are not highly restricted in their natural diets; only a few species of primates feed only on certain foods, such as specific kinds or parts of plants. The many possibilities of food and various conditions for feeding present difficulties in tabulating accurately the natural diet of any primate. When foods are plentiful, or at least adequate, most primates apparently are selective. When foods are scarce, many edible substances are eaten. In addition, there are daily and seasonal shifts in the feeding habits of some primates. However, the above classification is useful in that it applies to both arboreal and terrestrial forms and to any taxonomic group of primates.

The names of primates used herein are mostly in accordance with the nomenclature given by NAPIER and NAPIER [1967]. In a few cases, some generally recognized names of subgenera and subspecies of primates are given for ease and clarity of presentation.

General Feeding Habits

In general, most primates are euryphagous in nature in that they eat a rather wide variety of foods. Summaries of some information available on natural diets of the families of primates considered herein are given in tables I–III. Two families (Indriidae, Tarsiidae) are stenophagous in the sense that the natural diets are rather restricted; plant materials are consumed by the former and animal matters are ingested by the latter. Two families (Daubentoniidae, Callithricidae) are completely euryphagous in the sense that both plant and animal materials are consumed. The seven remaining families (Lemuridae, Cebidae, Cercopithecidae, Lorisidae, Tupaiidae, Hylobatidae, Pongidae) are mostly euryphagous, but include some forms that are stenophagous in that they exhibit either phytophagy or sarcophagy with regard to natural diets.

Families of primates that have broad geographic distributions usually exhibit wide ecologic ranges with regard to the major general kinds of foods utilized (tables I–III). The families Lorisidae and Cercopithecidae, which are distributed widely across Africa and in southeast Asia, utilize a rather broad

Table I. Summary of general kinds of natural diets of families and genera of primates. Kinds of materials eaten most frequently are indicated by ● and kinds of other materials eaten occasionally are marked by ○. Only sources of information based on examinations of contents of stomachs are listed; other information presented is from HILL [1953, 1955, 1957, 1960, 1962, 1966]; NAPIER and NAPIER [1967] and WALKER [1964]

Genus	Kind of diet				Source
	Herbi-vorous	Frugi-vorous	Insecti-vorous	Carni-vorous	
Tupaiidae					
Anathana		○	●	○	
Dendrogale			●		
Ptilocercus		○	●	○	LIAT, 1967
Tupaia	○	○	●	○	DAVIS, 1962
Urogale		○	●	○	
Lemuridae					
Cheirogaleus	●	●	○	○	
Microcebus		○	●	○	
Phaner		○	●		
Hapalemur	●	●			
Lemur	○	●			
Lepilemur	●	○			
Indriidae					
Avahi	●	○			
Propithecus	●	○			
Indri	●	○			
Daubentoniidae					
Daubentonia	○	○	●	○	
Lorisidae					
Arctocebus			●		CHARLES-DOMINIQUE, 1966
Loris			●	○	
Nycticebus	○	○	●	○	
Perodicticus	○	○	●	○	CHARLES-DOMINIQUE, 1966
Galago		○	●	○	BOOTH, 1956; CHARLES-DOMINIQUE, 1966
Tarsiidae					
Tarsius			●	○	

Table I (continued)

Genus	Kind of diet Herbi-vorous	 Frugi-vorous	 Insecti-vorous	 Carni-vorous	Source
		Callithricidae			
Callimico	–	–	–	–	
Callithrix	○	○	●	○	KREIG, 1930
Cebuella		○	●	○	
Leontideus		○	●	○	
Saguinus	○	●	○	○	FOODEN, 1964
		Cebidae			
Aotus	○	○	●	○	
Callicebus	○	●	○	○	
Cacajao	○	●	○	○	
Chiropotes	○	●			FOODEN, 1964
Pithecia	○	●			FOODEN, 1964
Alouatta	●	○			FOODEN, 1964
Cebus		●	○		FOODEN, 1964
Saimiri		●	●		FOODEN, 1964; THORINGTON, 1968
Ateles	○	●			
Brachyteles	○	●			
Lagothrix	○	●			
		Cercopithecidae			
Cercocebus	○	●	○		JONES, 1970
Cercopithecus	○	●	○		JONES, 1970
Cynopithecus		●			
Erythrocebus	○	●	○	○	
Macaca	○	●	○	○	
Papio	○	●	○	○	
Theropithecus	○	●	○	○	
Colobus	●				BOOTH, 1956; JONES, 1970
Nasalis	●	○			
Presbytis	●	○	○		
Pygathrix	–	–	–	–	
Rhinopithecus	●	○			
Simias	–	–	–	–	
		Hylobatidae			
Hylobates	○	●	○	○	
Symphalangus	○	●	○	○	
		Pongidae			
Pongo	○	●	○	○	
Pan	○	●	○	○	
Gorilla	●	○			

Table II. Major categories of natural diets of families of primates. Except for numbers of genera in each family, numbers indicate percentages of genera in each family classified with regard to category of natural diet

Family	Genera	Stenophagous		Eury-phagous
		Phyto-phagous	Sarco-phagous	
Tupaiidae	5	0	20.0	80.0
Lemuridae	6	50.0	0	50.0
Indridae	3	100	0	0
Daubentoniidae	1	0	0	100
Lorisidae	5	0	40.0	60.0
Tarsiidae	1	0	100	0
Callithricidae	4	0	0	100
Cebidae	11	45.5	0	54.5
Cercopithecidae	11	36.4	0	63.6
Hylobatidae	2	0	0	100
Pongidae	3	33.3	0	66.6
Total	52	30.8	7.7	61.5

Table III. Numbers of major kinds of food materials utilized by families of primates. Except for numbers of genera in each family, numbers indicate percentage of genera in each family with regard to the numbers of major kinds of food materials included in the natural diets

Family	Genera	Numbers of kinds of foods utilized			
		1	2	3	4
Tupaiidae	5	20.0	0	60.0	20.0
Lemuridae	6	0	66.6	16.7	16.7
Indridae	3	0	100	0	0
Daubentoniidae	1	0	0	0	100
Lorisidae	5	20.0	20.0	20.0	40.0
Tarsiidae	1	0	100	0	0
Callithricidae	4	0	0	50.0	50.0
Cebidae	11	0	63.6	9.1	27.3
Cercopithecidae	11	18.2	18.2	27.3	36.3
Hylobatidae	2	0	0	0	100
Pongidae	3	0	33.3	0	66.6
Total	52	7.7	38.5	19.2	34.6

spectrum of food materials in their natural diets. Some families of primates with restricted geographic distributions are limited with regard to ecologic range in that natural diets are relatively restricted. The families Indriidae (found only on Madagascar) and Tarsiidae (distributed in a limited area of the Malay Archipelago) seemingly have the least variable natural diets of the families of primates considered herein.

On the basis of the data presented in table I, it is apparent that considerable variations in natural diets occur among the genera of some families of primates. Of 55 genera of primates listed, 32 genera have natural diets that include plant and animal materials, 16 genera are vegetarians in that natural diets consist of herbaceous materials and fruits, and 4 genera are carnivorous. For 3 genera, data on natural diets are unavailable. Only 4 genera, representatives of 3 families, apparently are almost completely monophagous in that a single kind of food material is ingested. *Dendrogale* (Tupaiidae) and *Arctocebus* (Lorisidae) are insect eaters, *Cynopithecus* (Cercopithecidae) consumes fruits, and *Colobus* (Cercopithecidae) ingests leaves. Although there are no reports confirming the presence of insects among stomach contents of *Colobus* [BOOTH, 1956; JONES, 1970], it seems likely that this species, as well as *Cynopithecus*, may eat some insects along with the vast quantities of plant materials that are consumed.

Table IV. Stomach contents of four species of lorisids from Gabon. Data are taken from CHARLES-DOMINIQUE [1966]

Species	Number examined	Animal materials in stomach contents (%)	
Arctocebus calabarensis	4	Mean	88.8
		Minimum	75.0
		Maximum	95.0
Perodicticus potto	6	Mean	39.2
		Minimum	5.0
		Maximum	90.0
Galago alleni	8	Mean	70.6
		Minimum	5.0
		Maximum	100.0
Galago elegantulus	14	Mean	93.9
		Minimum	60.0
		Maximum	100.0

Table V. Stomach contents of six species of cebids from South America. Data are taken from FOODEN [1964]

Specimens examined		Stomach contents						
Species	Number	Leaves only	Leaves and fruits	Fruits only	Insects only	Fruits and insects	Leaves, fruits, and insects	Leaves and insects
Chiropotes chiropotes	2	–	–	2	–	–	–	–
Pithecia pithecia	3	–	–	3	–	–	–	–
Alouatta seniculus	3	1	2	–	–	–	–	–
Cebus apella	2	–	–	2	–	–	–	–
Cebus nigrivittatus	1	–	–	–	–	1	–	–
Saimiri sciureus	12	–	–	10	1	1	–	–
Total	23	1	2	17	1	2	0	0

Results of examinations of stomach contents of some prosimians are presented in table IV. Although the four species of lorisids listed are primarily carnivorous, some plant materials are consumed by certain species [CHARLES-DOMINIQUE, 1966]. *Perodicticus* apparently is more omnivorous than the other lorisids that have been studied [CHARLES-DOMINIQUE, 1966; JONES, 1969].

Summaries of examinations of stomach contents of some South American monkeys are given in table V. Of the 6 genera of cebids studied, 1 genus *(Alouatta)* feeds mostly on leaves, 2 genera *(Chiropotes, Pithecia)* are predominantly frugivorous, and 2 genera *(Cebus, Saimiri)* eat insects, as well as fruits [FOODEN, 1964].

Some data available on stomach contents of several African primates are presented in table VI. Of the 3 genera of cercopithecids examined, 1 genus *(Cercocebus)* feeds on leaves and fruits, natural diets of 1 genus *(Cercopithecus)* are variable in that food materials include leaves, fruits, and insects, and 1 genus *(Colobus)* feeds exclusively on leaves [JONES, 1970].

In addition to the aforementioned intergeneric variations in natural diets of primates, some intrageneric variations in diets are apparent; for example, see the information on 5 species of *Cercopithecus* given in table VI. Leaf and fruit materials are eaten by 2 species *(C. neglectus, C. pogonias)* and leaves, fruits, and insects are consumed by 3 species *(C. cephus, C. nictitans,*

Table VI. Stomach contents of eight species of cercopithecids from West Africa. Data are taken from JONES [1970]

Specimens examined		Stomach contents						
Species	Number	Leaves only	Leaves and fruits	Fruits only	Insects only	Fruits and insects	Leaves, fruits and insects	Leaves and insects
Cercocebus albigena	2	1	1	–	–	–	–	–
Cercocebus torquatus	3	–	1	2	–	–	–	–
Cercopithecus cephus	3	–	1	1	–	–	–	1
Cercopithecus neglectus	2	1	1	–	–	–	–	–
Cercopithecus nictitans	3	–	1	–	–	1	1	–
Cercopithecus pogonias	6	–	6	–	–	–	–	–
Cercopithecus talapoin	9	–	2	2	–	5	–	–
Colobus polykomos	3	3	–	–	–	–	–	–
Total	31	5	13	5	0	6	1	1

C. talapoin). Of the species of *Cercopithecus* listed, *C. talapoin* has the most variable diet and seemingly is more insectivorous than the other species [JONES, 1970]. For additional information on natural diets, as well as comments on digestion of certain food materials, of *C. talapoin*, see GAUTIER-HION [1966].

Wide ranges of dietary habits are known to occur within the genus *Papio*. Foods consumed by baboons in southern Africa include marine foods and insects [HALL, 1961], birds and eggs [DEVORE and WASHBURN, 1963], as well as other wild mammals and even livestock [DART, 1963]. In addition, baboons are also frugivorous, herbivorous, and occasionally graminivorous in their natural diets [NAPIER and NAPIER, 1967].

Seasonal Variations

Reviews of the results of a few intensive studies of the natural diets of primates indicate that kinds and amounts of certain food materials included in the diets of some primates vary seasonally. This is true especially for some

Table VII. Food materials utilized by *Macaca fuscata* in winter at seven localities in Japan. Data are taken from SUZUKI [1965]

Food materials	Utilization (%)
Bark	24.80
Fiber of bark	7.50
Twigs	5.75
Winter buds	7.60
Leaves	12.75
Flowers	1.75
Miscanthus and *Carex*	15.75
Sasa leaves	2.95
Roots	3.75
Seeds	6.50
Fruits	3.35
Others	7.55

primates that occur in temperate regions, but is probably true generally for most primates that exhibit euryphagous habits of diets.

SUZUKI [1965] listed foods utilized by *Macaca fuscata* in winter at several localities in Japan (table VII). IZAWA and NISHIDA [1963] discussed the main items included in the summer and winter diets of *M. fuscata* on the Shimokita Peninsula, Japan, the northern limit of the distribution of nonhuman primates, and listed some additional foods consumed by this species during the season of snowfall (table VIII). General conclusions of the aforementioned authors were that natural diets of *M. fuscata* in winter reflected availability of food items during the times that severe winter climatic conditions prevailed.

GOODALL [1968] provided information on seasonal utilization of plant foods by chimpanzees in the Gombe Stream Reserve (table IX). YOSHIBA [1967] listed by month food plants that were utilized by *Presbytis entellus* in south India. Many plants in the areas studied were eaten by these animals; parts of some plants were eaten throughout much of the year, but parts of some plants were consumed only during a specific time of the year. The temporal utilizations of parts of plants, especially flowers and fruits, as foods seemingly reflect the seasonal availability of these specific materials.

On the basis of observations made in Tanzania [GOODALL, 1965, 1968] and in Equatorial Guinea [JONES and SABATER PI, 1969], it was reported that *Pan troglodytes* did not generally consume termites during the major seasons

Table VIII. Materials eaten by *Macaca fuscata* in the snowfall season on the Shimokita Peninsula, Japan. Data are taken from IZAWA and NISHIDA [1963]

Mollusca
Collisella grata
Collisella heroldi
Cellana dorsuosa
Cellana corenna
Volsalla dificilis
Seaweed
Sargassum fulvellum
Porphyra tenera
Zostera marina
Jetsam
Raphanus sativus
Brassica compestric
Brassica oleracea
Citrus unshiu

Table IX. Seasonal feeding on fruits by chimpanzees. Data are taken from GOODALL [1968]

	Month											
Plant	J	A	S	O	N	D	J	F	M	A	M	J
Elaeis	X	X			X	X	X	X	X	X	X	X
Ficus	X	X	X			X	X			X	X	
Tinospora								X	X			
Ampelidaceae									X	X	X	
Grewia	X	X									X	
Parinari											X	X
Uapaca	X	X										
Pseudospondias			X	X								
Garcinia					X	X						
Anisophyllea					X							
Annona						X	X	X				
Strychnos						X	X					
Saba						X	X					

of drought. GOODALL [1965, 1968] explained that termite passages were extended to the surface of the mound and swarming of fertile termites to form new colonies occurred in the early portions of the rainy season in Tanzania.

Arboreal and Terrestrial Feeding Habits

Some intraspecific differences in compositions of natural diets have been shown for some of the primates that feed on the ground and in trees. For indications of some differences between arboreal and terrestrial feedings of *P. entellus*, see figure 1. The differences between foods utilized by these langurs during arboreal and terrestrial feedings reflect the availability of suitable food materials in the trees and on the ground [YOSHIBA, 1967].

Differences have been observed between arboreal and terrestrial feedings of *P. troglodytes* where this species occurs in forest habitats. GOODALL [1965] reported that about 90% of the vegetable foods of chimpanzees was found growing on trees and that the animals seldom fed intensively on the ground. In some areas, chimpanzees consumed termites that were in nests on the ground [GOODALL, 1965, 1968; JONES and SABATER PI, 1969]. In other words, in some areas, chimpanzees were mostly frugivorous when feeding in

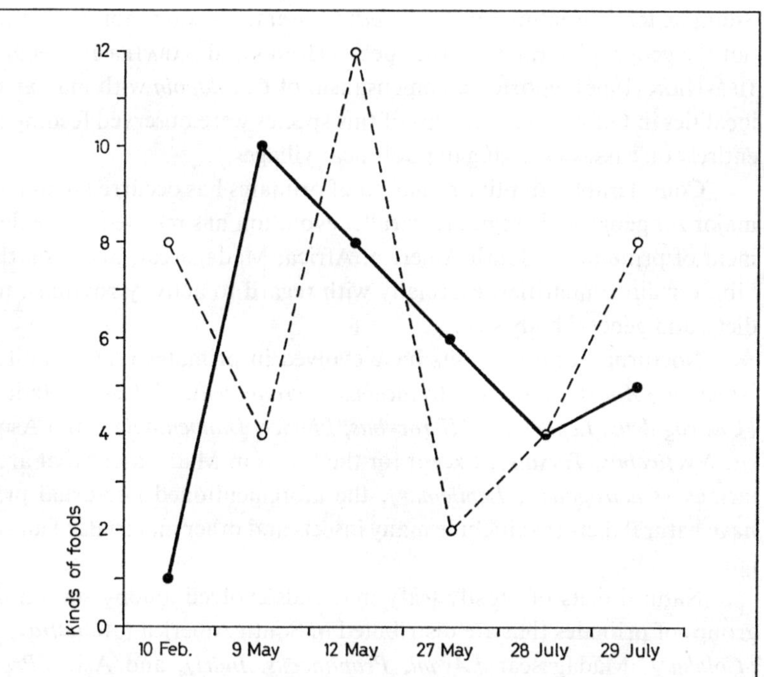

Fig. 1. Numbers of kinds of foods utilized by *Presbytis entellus* during arboreal (circles) and terrestrial (dots) feedings. Data are taken from YOSHIBA [1967].

the upper strata, but these animals were mainly insectivorous when feeding on the ground.

Great differences between arboreal and terrestrial feedings of primates occur within the species of primates that have become accustomed to raiding crops. Some crops, more or less ubiquitous in tropical and subtropical areas, probably raided most frequently by primates include *Arachis hypogea* (ground nut), *Musa* sp. (banana, plantain), *Saccharum officinarum* (sugar cane), and *Zea mays* (corn).

Geographic Variations

Some geographic variations in natural diets are known for some primates. Variations in diets, such as insectivorous and carnivorous feeding habits of chimpanzees, as well as some other primates, may be the results of learned behavior in local regions of the geographic range of the species. For example, termite eating by *P. troglodytes* seemingly does not occur throughout the geographic range of the species [JONES and SABATER PI, 1969]. GAUTIER-HION [1966] reported commensalism of *C. talapoin* with man at several localities in Gabon in the troops of this species were observed feeding almost entirely on cassava soaking in pools near villages.

Considerable adaptive radiation of primates has occurred in most of the major zoogeographic regions. Parallel evolution has resulted in the development of primates in South America, Africa, Madagascar, and Asia that exhibit certain similarities, especially with regard to activity rhythms, natural diets, and general body size.

Nocturnal feeding habits have evolved in primates that inhabit South America *(Aotus)*, Africa *(Arctocebus, Perodicticus, Galago)*, Madagascar *(Cheirogaleus, Lepilemur, Microcebus, Phaner, Daubentonia)*, and Asia *(Loris, Nycticebus, Tarsius)*. Except for the forms in Madagascar that are vegetarians *(Cheirogaleus, Lepilemur)*, the aforementioned nocturnal primates have natural diets that include many insects and other materials of animal origin.

Natural diets of mostly leafy materials evolved among several diverse groups of primates that are distributed in South America *(Alouatta)*, Africa *(Colobus)*, Madagascar *(Avahi, Propithecus, Indri)*, and Asia *(Presbytis, Rhinopithecus, Nasalis)*. One form from Madagascar *(Avahi)* is mostly nocturnal; the other leaf-eating forms mentioned exhibit diurnal activity rhythms.

Feeding Habits and Body Size

Examinations of the data available reveal some correlations between natural diet and body size. In general, small primates are insectivorous or carnivorous and large primates are mostly frugivorous or herbivorous (table I). The extremes of these correlations are exemplified by the smallest primates (length of body less than 200 mm) that are primarily carnivorous and the largest primates (gorillas) that, according to all available data, are completely vegetarian.

Within some groups of closely related primates, the smaller forms seemingly exhibit greater tendencies toward insectivorous diets than the larger species. Some examples are available from the data on five species of cebids (table X) presented by FOODEN [1964] and the information on five species of *Cercopithecus* (table XI) summarized by JONES [1970]. Of the species studied, the smallest cebid *(Saimiri sciureus)* and the smallest cercopithecid *(C. [Miopithecus] talapoin)* are more insectivorous in their natural diets than the larger species.

Activity Rhythms and Feeding Habits

Relationships between activity rhythms, body sizes, and natural diets of primates occurring on Madagascar are similar to the relationships of these features among the aforementioned groups studied in South America and Africa. In Madagascar, however, adaptive radiation has resulted in the evolution of several lemurs with nocturnal activity rhythms that consume considerable amounts of fruits and leafy materials. Of these nocturnal vegeta-

Table X. Body sizes of five species of cebids from South America. Data are taken from FOODEN [1964]

Specimens examined		Length of head and body		
Species	Number	Mean	Minimum	Maximum
Saimiri sciureus	12	266.6	225	286
Pithecia pithecia	3	346.3	342	349
Chiropotes chiropotes	2	406.5	406	407
Cebus nigrivittatus	1	428.0	–	–
Alouatta seniculus	3	537.3	512	564

Table XI. Body sizes of five species of *Cercopithecus* from West Africa. Data are taken from JONES [1970]

Specimens examined		Length of head and body		
Species	Number	Mean	Minimum	Maximum
Cercopithecus talapoin	9	295.4	212	357
Cercopithecus cephus	6	343.3	215	450
Cercopithecus pogonias	8	406.3	335	457
Cercopithecus neglectus	2	459.0	448	470
Cercopithecus nictitans	2	557.5	515	600

Table XII. Activity rhythms and major food items included in the natural diets of three families of primates from Madagascar

Genus	Activity rhythm	Major food item
	Lemuridae	
Microcebus	Nocturnal	Insects
Cheirogaleus	Nocturnal	Fruits
Phaner	Nocturnal	Insects
Hapalemur	Diurnal	Fruits
Lepilemur	Nocturnal	Leaves
Lemur	Diurnal	Fruits
	Daubentoniidae	
Daubentonia	Nocturnal	Insects
	Indriidae	
Avahi	Nocturnal	Leaves
Propithecus	Diurnal	Leaves
Indri	Diurnal	Leaves

rians *(Cheirogaleus, Lepilemur, Avahi)*, *Cheirogaleus* is a lemur with relatively small body size (table XII). On the basis of the available information on the natural diets of the primates of Madagascar, it is apparent that the natural diets of *Cheirogaleus* and *Daubentonia* are more variable than the diets of the other genera (table I).

Natural Diets and Morphology

Correlations are apparent between natural diets and some aspects of the morphology of gastro-intestinal tracts of primates. Species of primates with diets of mostly leaves have digestive tracts modified mostly by enlarged stomachs and by elongated intestines. Species of primates that are somewhat insectivorous have short gastro-intestinal tracts. For some correlations of data on natural diets with measurements of digestive tracts of seven species of New World monkeys, see FOODEN [1964]. For data on natural diets and corresponding modifications of the gastro-intestinal tracts of eight species of Old World monkeys, as well as some comparisons between natural diets and gastro-intestinal tracts of South American and African primates, see JONES [1970]. For notes on natural diets and brief comments on gastro-intestinal proportions of four species of West African lorisids, see JONES [1969]. A review of relative surface areas of digestive tracts and general morphology of intestinal villi of primates was presented by HLADIK [1967].

Relative proportions (measurements of organs/length of head and body) of some component parts of the gastro-intestinal tracts of two Old World monkeys, *C.* [*Miopithecus*] *talapoin* and *Colobus polykomos satanas*, and two New World primates, *Saimiri sciureus* and *Alouatta seniculus*, are depicted in figures 2–5. Comparisons are made between these species because *Cercopithecus* [*Miopithecus*] and *Saimiri* consume insects, as well as some fruits and other plant materials, but *Colobus* and *Alouatta* ingest mostly leaves.

Greatest differences between the groups of genera are seen in relative sizes of stomachs (fig. 2). Stomachs of *Colobus* and *Alouatta* on average are of much greater relative length than stomachs of *Miopithecus* and *Saimiri*.

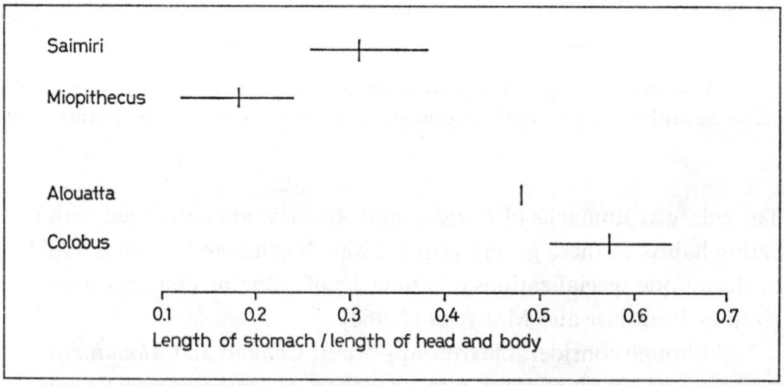

Fig. 2. Relative length of stomach of four species of primates. Vertical lines represent the arithmetic means and horizontal lines represent the ranges of measurements.

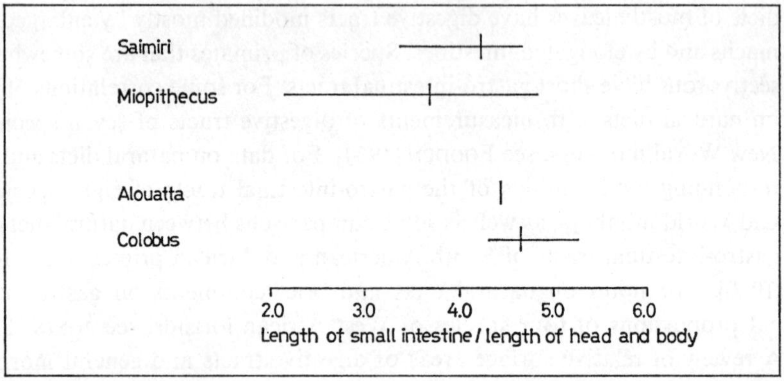

Fig. 3. Relative length of small intestine of four species of primates. Vertical lines represent the arithmetic means and horizontal lines indicate the ranges of measurements.

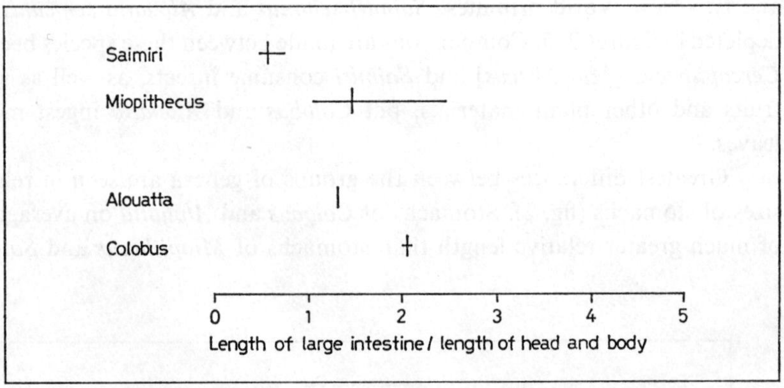

Fig. 4. Relative length of large intestine of four species of primates. Vertical lines indicate the arithmetic means and horizontal lines represent the ranges of measurements.

The enlarged stomachs of *Colobus* and *Alouatta* are correlated with the leaf-eating habits of these genera [HILL, 1966; NAPIER and NAPIER, 1967]. Some of the unique specializations of stomachs of colobine monkeys were summarized by BAUCHOP and MARTUCCI [1968].

Although considerable overlaps occur, *Colobus* and *Alouatta* on average have greater length of small intestines than *Miopithecus* and *Saimiri* (fig. 3). Relative lengths of large intestines are fairly constant among the genera for which data are given in figure 4; large intestines of *Saimiri* are smaller on av-

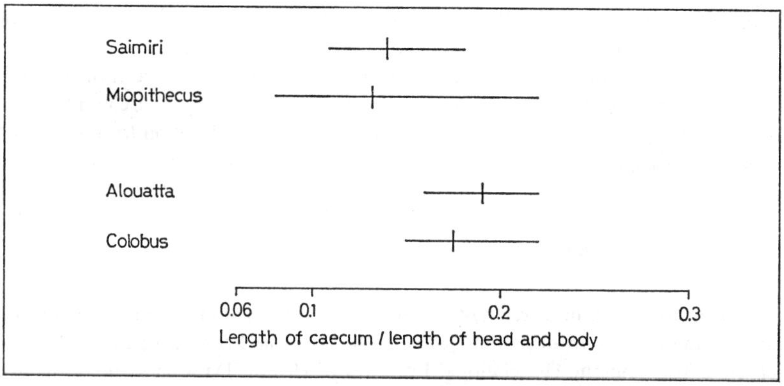

Fig. 5. Relative length of caecum of four species of primates. Vertical lines represent the arithmetic means and horizontal lines depict the ranges of measurements.

erage than those of other genera. In lengths of caeca (fig. 5), ranges of measurements overlap, but caeca of *Alouatta* and *Colobus* are on average greater in length than the caeca of *Saimiri* and *Miopithecus*.

An understanding of basic natural diets, variations of foods and feeding habits, as well as corresponding modifications of the digestive tracts of primates, would provide helpful information for most efficient and practical maintenance and utilization of primates kept in captivity for various research purposes. In addition, this information should be considered in the selection of particular species of primates for use in specific research projects.

Summary

Although some groups of primates have restricted diets, many primates generally eat a wide variety of foods. Major groups of primates with broad geographic distributions tend to have variable natural diets; groups of primates with restricted geographic distributions tend to have restricted natural diets.

Groups of primates that are similar with regard to kinds of food materials included in the natural diets occur in the major zoogeographic regions of the world. Within each major group of primates, variations in natural diets are correlated with geography and season, as well as activity rhythms and feeding levels of the animals in their natural habitats. Natural diets are correlated with body sizes of primates in that the smaller animals are mostly insectivorous and the larger primates generally exhibit greater vegetarian tendencies with regard to natural diets. Natural diets are correlated with some aspects of the morphology of the gastro-intestinal tracts of primates. Species of primates with diets of mostly leaves have digestive tracts modified by enlarged stomachs and elongated intestines, but species of primates that are somewhat insectivorous have short gastro-intestinal tracts.

Data on natural diets and morphologies of gastro-intestinal tracts of primates are considerations for selecting certain species of primates for use in research and are helpful for determining practices of maintenance for most efficient utilization for research of primates in captivity.

Acknowledgments

Some assistance in searching the literature for information presented herein was provided by the Primate Information Center of the Regional Primate Research Center, University of Washington, The National Institutes of Health. Data on some primates from West Africa were obtained during field studies supported by the National Geographic Society and the National Institutes of Health (FR00164).

References

ANDERSON, S.: Primates; in ANDERSON and JONES Recent mammals of the world (Ronald Press, New York 1967).

BAUCHOP, T. and MARTUCCI, R.: Ruminant-like digestion of the langur monkey. Science *161:* 698–700 (1968).

BOOTH, A. H.: The distribution of primates in the Gold Coast. J. West. afr. Sci. Assoc. *2:* 122–133 (1956).

CHARLES-DOMINIQUE, P.: Analyse de contenus stomacaux d'*Arctocebus calabarensis, Perodicticus potto, Galago alleni, Galago elegantulus, Galago demidoffi.* Biol. Gabon. *2:* 347–353 (1966).

DART, R.: The carnivorous propensity of baboons. Symp. zool. Soc. Lond. *10:* 49–56 (1963).

DAVIS, D.: Mammals of the lowland rain-forest of north Borneo. Bull. Nat. Mus. St. Singapore *31:* 1–129 (1962).

DEVORE, I. and WASHBURN, S.: Baboon ecology and human evolution; in HOWELL and BOURLIÈRE African ecology and human evolution (Aldine Publishing Company, Chicago 1963).

FINDLEY, J.: Insectivores and dermopterans; in ANDERSON and JONES Recent mammals of the world (Ronald Press, New York 1967).

FOODEN, J.: Stomach contents and gastro-intestinal proportions in wild-shot Guianian monkeys. Amer. J. Phys. Anthropol. *22:* 227–232 (1964).

GAUTIER-HION, A.: L'écologie et l'éthologie du talapoin, *Miopithecus talapoin talapotn.* Biol. Gabon. *2:* 311–329 (1966).

GOODALL, J.: Chimpanzees of the Gombe Stream Reserve; in DEVORE Primate behavior (Holt, Rinehart and Winston, New York 1965).

GOODALL, J.: The behaviour of free-living chimpanzees in the Gombe Stream Reserve. Anim. Behav. *1:* 161–311 (1968).

HALL, K.: Feeding habits of the chacma baboon. Adv. Science *17:* 559–567 (1961).

HILL, W. C. O.: Primates; Comparative anatomy and taxonomy, Vol. 1, Strepsirrhini (University Press, Edinburgh 1953).

HILL, W. C. O.: Primates; Comparative anatomy and taxonomy, Vol. 2, Haplorhini (University Press, Edinburgh 1955).

HILL, W. C. O.: Primates; Comparative anatomy and taxonomy, Vol. 3, Hapalidae (University Press, Edinburgh 1957).

HILL, W. C. O.: Primates; Comparative anatomy and taxonomy, Vol. 4, Cebidae, Part A (University Press, Edinburgh 1960).

HILL, W. C. O.: Primates; Comparative anatomy and taxonomy, Vol. 5, Cebidae, Part B (University Press, Edinburgh 1962).

HILL, W. C. O.: Primates; Comparative anatomy and taxonomy, Vol. 6, Cercopithecoidea (University Press, Edinburgh 1966).

HLADIK, C.: Surface relative du tractus digestif de quelques primates, morphologie des villosites intestinales et corrélations avec le régime alimentaire. Mammalia *31:* 120–147 (1967).

IZAWA, K. and NISHIDA, T.: Monkeys living in the northern limits of their distribution. Primates *4:* 67–88 (1963).

JAMES, W.: The jaws and teeth of primates (Pitman Medical Publishing, London 1960).

JONES, C.: Stomach contents and gastro-intestinal relationships of monkeys collected in Rio Muni, West Africa. Mammalia *34:* 107–117 (1970).

JONES, C.: Notes on ecological relationships of four species of lorisids in Rio Muni, West Africa. Folia primat. *11:* 255–267 (1969).

JONES, C. and SABATER PI, J.: Sticks used by chimpanzees in Rio Muni, West Africa. Nature *223:* 100–101 (1969).

KREIG, H.: Biologische Reisenstudien in Südamerika. Die Affen des Gran Chaco und seiner Grenzgebiete. Z. Morph. Okol. Tiere *18:* 760–785 (1930).

LIAT, L.: Notes on the food habits of *Ptilocercus lowii* Gray (pentail tree-shrew) and *Echinosorex gymnurus* (Raffles, moonrat) in Malaya with remarks on 'ecological labelling' by parasite patterns. J. Zool., Lond. *152:* 375–379 (1967).

NAPIER, J. and NAPIER, P.: A handbook of living primates (Academic Press, New York 1967).

SUZUKI, A.: An ecological study of wild Japanese monkeys in snowy areas. Primates *6:* 31–72 (1965).

THORINGTON, R., jr.: Observations of squirrel monkeys in a Colombian forest; in ROSENBLUM and COOPER The squirrel monkey (Academic Press, New York 1968).

WALKER, E.: Mammals of the world, Vol. 1 (Johns Hopkins Press, Baltimore 1964).

YOSHIBA, K.: An ecological study of hanuman langurs, *Presbytis entellus*. Primates *8:* 127–154 (1967).

Author's address: Dr. CLYDE JONES, Mammal Section, Bird and Mammal Laboratories, United States National Museum, *Washington, DC 20560* (USA)

Pathology of Simian Primates, Part I, pp. 78–156 (Karger, Basel 1972)

Biological Normals

B. A. LAPIN and G. M. CHERKOVICH

Institute of Experimental Pathology and Therapy, USSR Academy of Medical Sciences, Sukhumi

Contents

Introduction ... 79
Blood ... 80
  Blood Coagulation ... 80
  Morphological Composition of Monkey Blood[1] ... 83
  Bone Marrow[1] ... 88
  Monkey Serum Protein[2] ... 89
    Antigenic Properties of Serum Proteins ... 93
    Physico-Chemical Values and Aminoacid Composition of Certain Serum Proteins ... 96
    Genetic Polymorphism of Monkey Serum Proteins ... 96
  Nonprotein Components of Monkey Serum[2] ... 97
Circulation ... 104
  Pulse Rate ... 104
  ECG ... 108
  Blood Pressure ... 117
    Systemic Arterial ... 117
  The Velocity of the Pulse Waves Distribution in Monkeys ... 120
Respiratory System ... 121
  Oxygen Dissociation Curve of Blood ... 122
Digestive System ... 123
  Saliva ... 123
  Gastric Juice ... 127
  Gastric Movements ... 132
  Secretion and Movements of the Small Intestine ... 132
Metabolism and Body Temperature ... 133
  Gaseous Exchange ... 133

1 These two sections were written by M. I. KUKSOVA.
2 These two sections were written by G. A. ANNENKOV.

Body Temperature 134
Renal Function 137
Electroencephalography 138
References 146

Introduction

The important prospects for utilizing monkeys in biomedical experiments have brought on an extremely rapid growth in the number of research workers making extensive use of these animals.

However, insufficient knowledge of (1) primate ecology, of (2) a great number of normal physiological values and of (3) 'spontaneous' pathology, encountered outside the experimental situation, often stand in the way of the correct use of these rather expensive animals by the experimenter. This reference book was designed as a handy reference for clinicians and pathologists to help them identify disease and its causes, parasites and other agents of disease.

However, it was suggested that the section on pathology be prefaced with a relatively small chapter on normal physiological values. This was done to enable the experimenter to decide whether certain values should be considered normal or pathological.

In describing biological normals the authors have paid special attention to the conditions under which the recording of certain values was carried out. Physiological values obtained during the process of capture and restraint of the animals without preliminary tranquillization or anaesthesia may prove inadequate and in some cases they contrast sharply with those recorded in free ranging animals, especially when the experimenter is out of the animal's sight, and orienting and stress reactions are not induced. Radiotelemetric recording of functions, or long-term conditioning of animals to the surroundings under which recordings are made, give the best results.

The above is especially significant in the registration of such particularly labile values as those of the cardio-vascular and respiratory systems, cell composition of peripheral blood, etc.

Thus, data on biological normals in monkeys published by different authors without taking into consideration these conditions, must be accepted with reserve, since they may not reflect normal values in conditions of emotional tranquility, or at best only the upper limit of the norm or a physiological value determined by defensive reactions. Such values, owing to the great lability of physiological values in monkeys, can considerably surpass those obtained from animals in a completely tranquil state. Such data can serve as values for physiological hyper-activity and must presumably be treated as an initial background for a given situation. These findings have a limited value also in determining the possible limits of normal physiological values.

The material presented in this chapter is to a considerable extent based on long-term observations carried out by the associates of the Institute of Experimental Pathology and Therapy of the USSR Acad. Med. Sci. The material also brings together the data obtained from the most important literary sources such as the studies of MALINOW *et al.* [105, 115, 116, 117, 146, 164] and VAN CITTERS *et al.* [183] and others, who have made valuable contributions to the understanding of problems connected with the activity of the cardio-vascular system. This material is also based on current Primate References issued by the Regional Primate Research center of University of Washington. Due to limitations of space in this volume we have not presented data describing the functions of the endocrine system and certain other data. We have tried to describe the most widely used values and now it is the job of the reader to judge whether we have succeeded.

Blood

Blood volume is known from studies in *P. hamadryas* and in two macaca species. As in man it is 6–7% of the body weight. The data of investigations are summarized in table I.

Red-cell volume (hematocrit) for different species of macaca *(M. mulatta, M. nemestrina, M. fuscata, M. speciosa, M. radiata)* ranges as in man from 41.2±1.1% to 46.9±1.3% [140]. Somewhat lower values (38–45%) were presented by HAYDEN [78] for *M. mulatta*. Extracellular fluid volume measured by modification of the NaSCN dilution method showed statistically significant seasonal variations 184±25.2 ml/kg in winter and 250±40.7 ml/kg in summer for *M. mulatta* [128].

Blood Coagulation

Values of blood coagulation for macaques and baboons are presented in table II.

Besides these data, there is also information on blood coagulation for other species of macaques *(M. fuscata, M. speciosa, M. radiata, M. maurus,*

Table I. Red-cell, plasma and total blood volume

Red-cell volume	Plasma volume	Total blood volume	Monkey species	Authors	Comments
31.8 ml/kg calculated from total bl.v.-pl.v.	44.9±7.0 ml/kg	75.1±12.0 ml/kg calculated from pl.v. + 2 Hts	*M. mulatta*	OVERMAN, FELDMAN, 1947 [128]	23 animals of 2.4 — 4.40 kg. A modification of the dye dilution method T_{1824}
17.7±1.66 ml/kg, P^{32} red-cell tag	36.4±3.98 ml/kg, dye dilution method T_{1824}	54.1±4.72 ml/kg calculated from pl.v. + red-cell	*M. mulatta*	GREGERSEN *et al.*, 1959 [67]	18 animals of 3.4–7.5 kg
		60.9±0.96 ml/kg (49–71 ml/kg)	*M. mulatta*	BENDER, 1955 [16]	20 animals of 2.2–5.3 kg (10♂ and 10♀) J^{131} albumin tag
		89.5±6.0 ml/kg	*M. mulatta*	FORSYTH *et al.*, 1968 [54]	6 unanesthetized animals
		83±13 ml/kg ♂, 85±12 ml/kg ♀	*M. fuscata*	HIDEO NIGI, 1967 [80]	Dye dilution method T_{1824}, 35♂ of 5–8 kg, 17♀ of 5–8 kg
		77± 5 ml/kg ♂, 78±4 ml/kg ♀	*M. fuscata*	HIDEO NIGI, 1967 [80]	13 juveniles (6♂ and 7♀ of 1.8–5.2 kg)
		62.4 ml/kg ♂, 65 ml/kg ♀	*P. hamadryas*	WILLMAN, HANLON *et al.*, 1964 [192]	8 adult animals (4♂ of 16–21 kg and 4♀ of 9–12 kg)

M. silenus) [77, 146] and also for *Cebus apella, C. capucina, Cercopithecus ascanius Schmidti, C. patas, Cercocebus torquatus* and *Lagothrix lagotricha* [77]. However, these data were obtained from a very small number of animals (1–2) of each species. Due to this reason the authors consider these data as preliminary.

SEAMAN and MALINOW [146] consider that blood coagulation is more efficient in monkeys than in man as their prothrombin and thrombocyte counts are greater in monkeys in comparison with man (in *Saimiri* especially).

Table II. Some values of blood coagulation in monkeys

Prothrombin %	Fibrinogen mg %	Prothrombin time (sec) (Quick test)	Recalcification time (sec)	XIII factor %	Platelets $\times 10^3$/mm^3	Index of adhesiveness	Tolerance to heparin (min)	Fibrinolytic activity (min)	Monkey species	Authors	Comments
172±8	310±18	12.0±0.3	–	–	495±35	–	–	–	*M. mulatta*	SEAMAN and MALINOW [146]	10 unanaesthetized animals
94 (60–110)	249 (180–440)	–	124	106	225	1.3 (1.0–1.62)	4′5″	464	*M. mulatta*	CHERNIGOVSKAYA [1969, unpublished]	10 unanaesthetized animals
–	268±15.0	–	123 ±6.15	–	339.8 ±11.9	–	3′05″	–	*M. mulatta*	UVAROVA [1969, unpublished]	30 unanaesthetized animals
153±8	338±26	12.6±0.3	–	–	478±93	–	–	–	*M. nemestrina*	SEAMAN and MALINOW [146]	11 unanaesthetized animals
–	344	15.5	–	–	330	48%	–	585	*P. anubis*	HAWKEY and SYMONS [77]	4 animals anaesthetized with sernyl
95 70–105)	363	–	145	100	217	1.24	4′50″	385	*P. hamadryas*	CHERNIGOVSKAYA [1969, unpublished]	6 unanaesthetized animals
–	336.53 ±69.39	15.86 ±1.13	–	–	319±62	38 ±10.9%	–	–	Baboon sp. not stated	HAMPTON and MATHEWS [75]	16 animals anaesthetized with sernyl
157±6	268±15	8.6±0.2	–	–	700±41	–	–	–	*Saimiri sciurea*	SEAMAN and MALINOW [146]	11 animals anaesthetized with sernyl

According to the data of the same authors *Alouatta caraya* (a species of New World monkeys) is not very rich in thrombocytes but blood coagulation in this species is quicker than in man. HAMPTON and MATTHEWS [75] note a great similarity of human and baboon systems of blood coagulation but they still note that blood coagulation is quicker in baboons.

Morphological Composition of Monkey Blood

For the analysis of morphological composition, blood is best obtained by ear puncture. If necessary it is collected from the humeral vein, located superficially on the outer surface of the humerus. The values for peripheral blood in monkeys are labile, namely hemoglobin values and erythrocyte counts and to the greatest extent leucocyte counts. According to the data of various authors the total leucocyte count per 1 mm^3 in healthy monkeys ranges from 1,200 to 43,000 (tables III and IV) [3, 13, 17, 29, 86, 120].

The variations mentioned above are due to the fact that techniques of blood puncture were not uniform. Age, seasonal and diurnal peculiarities were sometimes not taken into consideration [29, 45, 68, 135, 140, 181]. However, these variations are very significant: the minimum hemoglobin and leucocyte count is observed at 4 pm, the maximum – from midnight to 4 am [55, 95, 98]. The erythrocyte count in the blood of *M. mulatta*, *P. hamadryas* and *C. aethiops* is increased during the summer and autumn from 5,000,000 to 6,500,000 per 1 mm^3. Hemoglobin is increased from 67–74 % to 87–90 %. The leucocyte count is decreased relatively [95, 98]. When swellings of the sexual skin are observed in ♀ *P. hamadryas* and *M. nemestrina* the erythrocyte count is increased from 4,000,000 to 5,900,000. Hemoglobin is increased on an average by 16% in adult females and 11% in young females. The changes of red blood in monkeys depend on the sex and age of the animals. The hemoglobin level is higher in males in comparison with females. During the first 24 h after birth the infants show high hemoglobin (100 ± 1.78 %) and erythrocyte values ($6{,}167{,}000 \pm 30{,}000$ per mm^3). Subsequently the hemoglobin content is reduced to 60 to 68% and in 4–5 month is increased again remaining on this level up to 1 year of age. Another increase of hemoglobin content is observed when the animals reach the age of sexual maturity (3–4 years).

Specific (as between macaques and baboons) and sexual differences are apparent in monkeys after 5 months of age. Males in comparison with females and baboons in contrast to macaques show higher hemoglobin values. Age

Table III. Normal values of red blood and sedimentation rate

Normal values	Monkey species	Authors	Comments
	Hemoglobin		
79.3±1.8% (60–92)	*M. mulatta*	VAN DEN BERGHE and BLITSTEIN [17]	
11.72±3.025 g (7.0–16.5)	*M. mulatta*	KRISE [94]	655 observations on 538 animals
♂ 69.0±1.74% ♀ 60.2±1.60 P>0.01 (mean 64.6±8.14)	*M. mulatta*	KUKSOVA [103]	♂ 41 unanaesthetized animals ♀ 48
♂ 81.0±1.97% ♀ 75.3±0.76 P>0.005 (mean 78.2±5.04)	*P. hamadryas*	KUKSOVA [103]	♂ 49 unanaesthetized animals ♀ 138
	Erythrocytes in mm^3		
4.69±0.105 mln (3.3–5.5)	*M. mulatta*	VAN DEN BERGHE *et al.* [17]	
5.57±73 mln (3.1–8.8)	*M. mulatta*	KRISE [94]	876 observations in 538 animals
♂5.83±0.26 mln ♀ 5.63±0.2 mln (5.73±0.18)	*M. mulatta*	KUKSOVA [103]	Sex difference is statistically insignificant. See above
♂ 5.78±0.18 mln ♀ 5.82±0.13 mln (5.8±0.04)	*P. hamadryas*	KUKSOVA [103]	Sex difference is statistically insignificant. See above
	Reticulocytes %		
8.4±0.84 (0.5–20)	*M. mulatta*	VAN DEN BERGHE [17]	
0.35±0.19 (0.1–3.4)	*M. mulatta*	KRISE [94]	76 observations
	Sedimentation rate mm/h		
0.90±2.2 (0.0–28)	*M. mulatta*	KRISE [94]	448 observations
2–5	*M. mulatta* *P. hamadryas*	KUKSOVA [103]	89 *M. mulatta* 187 *P. hamadryas*

Table IV. Normal values of leucocytes in different monkey

Normal values	Monkey species	Authors	Comments
	Leucocytes in mm³		
Thousands 16.67 ± 1.57 (5.0–40.0)	*M. mulatta*	VAN DEN BERGHE *et al.* [17]	
15,155 ± 5,981 (1,200–43,100)	*M. mulatta*	KRISE [94]	891 observations in 538 animals
♂ 10,266 ± 911 ♀ 11,540 ± 436 (mean 10,833 ± 1,163)	*M. mulatta*	KUKSOVA [103]	Sex difference is statistically not significant ♂ 111, ♀ 48
♂ 13,953 ± 646 ♀ 12,841 ± 241 (mean 13,397 ± 985)	*P. hamadryas*	KUKSOVA [103]	Sex difference is statistically not significant ♂ 49, ♀ 138
	Basophils %		
0.41 ± 0.10 (0–2)	*M. mulatta*	VAN DEN BERGHE [17]	
0.21 ± 0.18 (0–4)	*M. mulatta*	KRISE [94]	738 observations in 538 animals
0–1	*M. mulatta*, *P. hamadryas*	KUKSOVA [103]	89 *M. mulatta*, 187 *P. hamadryas*
	Eosinophils %		
1–6	*M. mulatta*, *P. hamadryas*	KUKSOVA [103]	See above
	Polymorphonuclears % Neutrophile metamyelocyte		
0–1	*M. mulatta*, *P. hamadryas*	KUKSOVA [103]	See above
	Neutrophile staff-like		
2–7	*M. mulatta*, *P. hamadryas*	KUKSOVA [103]	See above
	Neutrophile segmented		
45.46 ± 5.93 (15.2–76.0)	*M. mulatta*	VAN DEN BERGHE [17]	
35.79–16.70 (4–91)	*M. mulatta*	KRISE [94]	738 observations in 538 animals

Table IV (continued)

Normal values	Monkey species	Authors	Comments
♂ 48.0±5.01 ♀ 49.7±1.78 (mean 46.7±1.50)	*M. mulatta*	KUKSOVA [103]	See above
♂ 56.3±5.32 ♀ 54.7±1.08 (mean 55.5±1.42)	*P. hamadryas*	KUKSOVA [103]	See above
	Lymphocytes %		
46.36±3.63 (15.0–76.0)	*M. mulatta*	VAN DEN BERGHE [17]	
60.52±17.26 (7–95)	*M. mulatta*	KRISE [94]	738 observations in animals
♂ 36.5±5.32 ♀ 38.6±1.42 (mean 37.6±2.74)	*M. mulatta*	KUKSOVA [103]	Sex difference is statistically not significant. No. of animals. See above
♂ 32.3±5.90 ♀ 33.3±0.97 (mean 32.8±0,89)	*P. hamadryas*	KUKSOVA [103]	Sex difference is statistically not significant. No. of animals. See above
	Monocytes %		
2.53±0.36 (0–6.4)	*M. mulatta*	VAN DEN BERGHE [17]	
0.72±0.38 (0–8)	*M. mulatta*	KRISE [94]	738 observations in 538 animals
4–8	*M. mulatta*, *P. hamadryas*	KUKSOVA [103]	See above

variations of erythrocytes are not distinct except after the newborn period. Sexual differences in the concentration of hemoglobin and the erythrocyte count in adult animals were noted by MYERS and BLACKWELL [123].

Both unconditioned and conditioned digestive leucocytosis related to certain periods of feeding is manifested in monkeys [127].

It should be noted that the leucocyte count in the blood depends considerably on stress reactions. Such reactions can be induced by new surroundings (when the procedure of the experiment is changed, for instance). High leucocyte values in monkeys undergoing primary investigations testify to

Table V. Dynamics of the leucocytes count in repeated hematological analyses (UTKIN and KUKSOVA [182]

Number of animals	Number of analyses	Duration of the test (days)	Leucocytes count per 1 mm³ 1st day	Average count (the first 5 days)	Average count (the last 5 days)	The last day	The possibility of occasional divergence of average values obtained at the beginning and end of experiment (P)
1,367	16	23	15,800	13,400±870	10,800±830	9,400	0.011
1,270	25	62	15,200	17,560±1,600	9,480±930	10,000	0.003
1,183	11	29	17,400	15,640±2,100	9,480±840	7,400	0.043
1,291	42	93	17,400	15,040±800	13,160±590	13,400	0.081
1,275	43	71	14,800	13,500±860	8,850±1,020	8,400	0.009

physiological leucocytosis caused by the 'alarm reaction'. Regularly repeated blood punctures show that the leucocyte count can be halved and maintained on this level only if the procedure of the test is not changed [182] (table V).

Due to this the hematological analysis of leucocytes carried out only once is of no clinical significance.

Sedimentation rate is an accurate indicator of the state of health of the animal. This value is very sensitive and is to some extent specific, though some workers disagree on the point [125]. The normal value for sedimentation rate is 1–8 mm per h [97, 104]. The sedimentation rate is accelerated in summer.

In healthy animals normoblasts, polychromatophiles and anisocytosis are absent from blood smears. The appearance of pathological red cells in the blood should be treated as a certain manifestation of an anemic state.

The lymphocyte count is higher than that of neutrophils in infants to 1 year of age (as in man). In adult animals, as a rule neutrophiles predominate. The period of sexual maturation is an exception as different variants of the lymphocyte/neutrophile ratio can be observed within this period.

There exists a definite interrelationship between the total leucocyte count and lymphocyte/neutrophile segmented ratio. For instance, an increase of neutrophile-segmented cells is observed when the leucocyte cell count is relatively high – 15,000.

Monkey neutrophiles are characterized by a great number of lobes (7–8, sometimes 9–15). The basophiles and eosinophiles of *Papio* and *Macaca* have many granules. In *C. aethiops* fine granules in great number are observed.

It is inadvisable to be guided by average data in respect to one animal or a whole group due to the great lability of the total leucocyte count as well as various leucocyte forms. The results are best estimated from average individual values when the figures are arranged according to the response to a certain stimulus.

Bone Marrow

Bone marrow in monkeys is obtained by aspiration from the femoral bone [HALL and KNEZEVIC, 72], sternum [SUAREZ, DIAZ-RIVERA and HERNANDEZ-MORALES, 170], anterior and posterior crest of the pelvic bone [MOOR-JANKOWSKI *et al.*, 120; HUSER and BEARD, 85], ischial tuber [SWITZER, 169], tibial tuber [KUKSOVA, 99, 100], humeral condyle [COHEN, 73].

The ratio of bone marrow cells does not depend on the site of aspiration; the differences are, however, observed between values obtained from live animals and those obtained from dead animals [VAN DEN BERGHE and BLITSTEIN, 18], since the bone marrow obtained from live animals is sometimes diluted by peripheral blood [SUAREZ and DIAZ-RIVERA, 170; VAN DEN BERGHE and BLITSTEIN, 18; USACHEVA and RAEVA, 180]. This affects significantly the appearances of bone marrow haemopoiesis and the possibility to evaluate the results. To determine whether bone marrow samples are free from peripheral blood it is necessary to estimate the myelokaryocyte count in

Table VI. The influence of orientation response upon cell composition of the bone marrow (%) in *M. mulatta* – mean data obtained from 17 animals

Bone marrow cells	Days of aspiration		
	1	2	3
Promyelocytes, myelocytes; juvenile	12.0	13.7 $p = 0.002$	16.0
Staff-like neutrophiles	16.3	14.2	16.7
Segmented neutrophiles	23.8	18.5 $p = 0.108$	21.0
Red cell series	31.5	32.5 $p = 0.06$	25.9

the samples [KUKSOVA, 99]. Myelograms obtained from 100,000–350,000 nucleated cells per 1 mm^3 show normal rates of haemopoiesis. When the bone marrow sample contains 30,000–85,000 cells per 1 mm^3, the count of nucleated cells typical of peripheral blood (neutrophiles and lymphocytes) is increased considerably and the percentage of red cells and young cells of white blood series is reduced.

Also, in repeated (more than 3–4 times) punctures of one and the same haemopoietic area [VAN DEN BERGHE and BLITSTEIN, 18; KUKSOVA, 99] the bone marrow becomes diluted by peripheral blood. Its cell composition becomes low and the bone tissue is destroyed. The reduction of the latter takes place only in 9–12 months.

Stress (emotional) reactions also influence bone marrow values (table VI).

The extraction of the animal from the cage during 3 days for bone marrow puncture is accompanied by a statistically significant increase in the percentage of promyelocytes, myelocytes and juvenile cells and a decrease of red cell series. KUKSOVA [100] obtained data on normal bone marrow composition on the basis of the Sukhumi monkey colony for *P. hamadryas, M. mulatta* and *C. aethiops*. Since the data for haemopoietic values in males are more stable than in females, data for males are presented in table VII.

The red bone marrow stem is reduced in females in comparison with males; their leuco-erythroblastic ratio of 3.6 testifies to this.

The increase of erythrocytes at a considerable decrease of normoblasts is typical in females.

Bone marrow undergoes considerable changes in the process of growth and development of monkeys [KUKSOVA, VOLKOVA, VERBITSKY and ANDREEVA, 101].

Bone marrow values for healthy *P. hamadryas* infants during the first year of life, with regard for individual variations, are presented in table VIII. The increase of lymphocytes and their early stages (lymphoblasts, prolymphocytes) in the bone marrow to 60% by 4 months of age is typical of the majority of monkey infants and of human babies as well.

The picture of the bone marrow in infants becomes closer to that of adult animals by one year of age.

Monkey Serum Proteins

The number of works dedicated to monkey serum compounds is not large. A considerable part of those investigations was carried out on isolated

Table VII. Cell composition of the bone marrow in males. According to KUKSOVA [100]

Values	*P. hamadryas*		*M. mulatta*		*C. aethiops*	
	Mean	Variation range	Mean	Variation range	Mean	Variation range
Reticular cells	0.6	0.4–2.0	0.4	0.2–1.0	0.5	0.2–0.4
Hemohistioblasts	0.3	0.0–0.3	0.3	0.2–0.4	0.5	0.2–0.8
Hemocytoblasts	1.4	0.8–2.2	1.2	0.2–3.6	1.1	0.2–1.8
Myeloblasts	0.5	0.2–1.2	0.4	0.2–1.0	0.5	0.2–1.2
Promyelocytes	1.8	0.2–1.8	3.0	0.8–6.2	2.5	0.8–3.4
Myelocytes	7.1	3.2–8.4	7.1	4–13.0	7.0	3.2–12.2
Metamyelocytes	13.4	6.8–22.2	9.5	7.4–25.4	15.7	6.8–24.0
Staff-like neutrophiles	21.5	12.4–32.8	24.3	14–30.8	22.4	11.6–30.4
Segmented neutrophiles	10.7	6.2–22.8	13.5	8–23.8	11.7	4.6–18.0
Eosinophils	0.4	0.2–1.2	0.8	0.2–1.4	1.5	0.6–3.4
Basophils	0.2	0.0–0.2	0.4	0.2–0.8	0.3	0.2–0.4
Lymphoblasts	0.3	0.2–0.6	0.4	0.2–0.8	0.2	0–0.2
Prolymphocytes	0.3	0.2–0.6	0.6	0.4–1.0	0.3	0.2–0.4
Lymphocytes	10.8	7.2–14.2	8.0	4.4–13.2	8.7	3.4–12.8
Promonocytes	0.1	0.0–0.1	0.2	0.0–0.2	0.2	0.0–0.2
Monocytes	1.4	0.2–3.4	1.2	0.4–2.8	0.6	0.2–1.8
Proerythroblasts	1.2	0.2–2.8	1.1	0.6–2.2	0.7	0.4–1.4
Basophilic erythroblasts	5.6	4.6–9.8	4.7	1.4–10.0	4.0	1.6–7.0
Polychromatophilic erythroblasts	10.3	4.2–16.4	10.8	3.8–23.4	9.7	4–17.0
Orthochromatic erythroblasts	0.8	0.2–3.4	0.5	0.2–1.2	0.4	0.2–0.8
Polychromatophilic normoblasts	8.0	0.4–13.8	8.6	2.8–13.4	8.2	2.6–12.6
Orthochromatic normoblasts	2.0	1.4–3.6	1.6	0.2–2.6	1.6	0.4–2.8
Plasma cells	0.8	0.2–1.6	1.4	0.8–1.8	1.8	0.6–3.4
Red cell series mitosis	5:500	–	4:500	–	3:500	–
White cell series mitosis	2:500	–	3:500	–	3:500	–
Leukoerythrocytic relation	2.4	–	2.5	–	2.7	–
Index of neutrophilic maturity	0.6	–	0.5	–	0.7	–
Index of erythrocyte maturity	0.7	–	0.8	–	0.8	–

animals and therefore the findings often diverge significantly. The tables given below present data which are the most convincing as far as the number of the animals investigated, research technique and correlation of the data with that of other workers are concerned.

As seen from table IX, total serum protein content in all the primates is approximately the same. The protein spectrum of monkeys differs from that

Table VIII. Cell composition of the bone marrow in *P. hamadryas* during the first year of life (in %). According to KUKSOVA *et al.* [101]

Bone marrow cells	Age (in days)					
	1	10	20	30	120	360
Reticular cells	0.4–1.8	0.6–1	0.8–6.4	0.2–17.5	0.4–7.2	0–1.8
Hemohistioblasts	–	–	–	–	–	–
Hemocytoblasts	0–0.6	0–0.2	0–0.4	0–0.1	0–0.4	–
Myeloblasts	0–0.4	0–0.4	0–0.8	0–0.6	0–0.4	–
Promyelocytes	0.2–1.4	0.2–1.2	0.2–1.6	0.4–2.0	0.2–1.4	0–1.0
Myelocytes basophilic	–	–	0.0–0.2	–	–	–
Myelocytes eosinophilic	–	0.2–0.4	0.4–6.6	0–0.2	0–0.8	–
Myelocytes neutrophilic	2.4–9.4	2.4–4.8	–	1.2–7.2	0.8–8.0	2.0–7.0
Metamyelocytes basophilic	–	–	0.2–0.4	–	0–0.1	–
Metamyelocytes eosinophilic	0–0.2	0.2–0.4	1.0–9.0	0.2–2.2	0.2–1.6	–
Metamyelocytes neutrophilic	4.4–9.2	4.4–9.2	–	1.6–11.6	1.4–10.6	9.6
Basophile staff-like	–	–	0.2–0.8	–	–	–
Eosinophile staff-like	0–0.2	0–0.2	1.4–16.6	0.1–0.2	0.1–0.2	–
Neutrophile staff-like	4.4–13.6	6.0–18.6	14.6–29.0	1.6–19.2	2.4–18.0	8.0
Segmented neutrophiles	12.0–46.6	12.6–37.8	0.2–0.4	7.0–30.8	9.8–19.6	9.4
Basophils	0–0.2	0–0.2	0.4–3.4	0–0.5	–	–
Eosinophils	0.2–1.2	0.2–1.4	–	0.2–1.6	0.8–1.2	1.4
Lymphoblasts	–	–	0.2–5.6	–	0–0.5	–
Prolymphocytes	0–1.2	0.8–3.0	17.6–26.4	1.1–2.2	2.6–4.4	0.8
Lymphocytes	1.0–20.8	3.0–29.6	–	4.0–35.2	4.8–59.4	30.6
Promonocytes	–	–	1.8–6.4	–	–	–
Monocytes	0.2–6.8	2.6–9.4	0.2–1.4	1.0–6.9	2.0–4.8	4.6
Proerythroblasts	0.2–1.2	0.2–0.6	2.0–4.4	0.2–1.8	0.2–0.6	0.2
Erythroblasts basophilic	1.8–9.2	0.6–3.0	9.6–19.8	0.5–2.8	0.4–4.0	2.4
Erythroblasts polychromatophilic	11.4–32.6	9.2–21.4	–	6.2–22.6	3.2–16.4	12.2
Erythroblasts orthochromatic	–	–	–	–	0–0.1	–
Normoblasts polychromatophilic	3.6–16.2	5.6–15.8	6.4–14.2	6.3–16.0	1.8–14.2	10.8
Normoblasts orthochromatic	–	–	–	–	0–0.2	–
Megakaryocytes	+	+	+	+	+	+
Plasma cells	–	0–0.2	–	0–0.2	0.1–1.0	0.4
Bone marrow index of neutrophils	0.1–0.5	0.1–0.5	0.1–0.6	0.1–0.8	0.2–0.7	0.4
Leukoerythrocytic relation	0.9–3.1	2.0–5.8	2.4–3.4	1.1–5.0	2.0–9.5	2.8
Index of erythrocyte maturity	0.8–0.9	0.8–0.9	0.8–0.9	0.8–0.9	0.5–0.9	0.8

Table IX. Total protein (g%) and protein fraction (% to total protein) content in the M± m). According to ANNENKOV [5]

M. species	Sex	Number of analyses	Total protein	Albumins	Globulins	
					a_1	a_2
Man	M	15	7.5±0.09	59.8±0.79	4.5±0.23	7.4±0.88
	F	25	7.9±0.19	58.8±0.65	5.3±0.21	8.2±0.38
M. mulatta	♂	197	7.76±0.06	57.0±0.75	5.5±0.15	5.1±0.12
	♀	97	7.85±0.09	50.9±0.92	5.1±0.13	5.4±0.14
P. hamadryas	♂	67	7.40±0.10	58.9±1.04	5.3±0.16	6.5±0.12
	♀	122	7.54±0.06	53.6±0.62	4.6±0.08	7.4±0.17
C. aethiops	♂	46	7.26±0.18	56.3±2.28	4.6±0.33	5.6±0.33
	♀	43	7.45±0.13	50.0±1.73	5.6±0.44	6.7±0.43
Saimiri	♂	15	7.1±0.25	54.9±1.7	5.4±0.32	7.4±0.99
	♀	12	7.0±0.40	47.0±2.47	6.1±0.33	8.1±1.25
					a_1	a'_2
Saimiri [44]**	♂+♀	67	7.3±0.08	62.1±0.54	1.7±0.15	3.8±0.16

* Occurs in 10% of analyses and only in monkeys.
** Acetate cellulose electrophoresis.

of man in showing more examples of polymorphism, since 7 globulin fractions may be revealed, and lability, since significant individual variations occurs in the quantity of fractions and their protein content even in one and the same animal, as revealed by repeated analyses. The protein spectrum of *P. hamadryas* (males) show greater similarity to that of man than other species.

It is noteworthy that the albumin content in certain big animals *(Papio* and *M. mulatta)* exceeds 70% of the total protein [5]. No significant variations in total protein content, dependent on the time of the day, were detected [32]. Significant seasonal variations in the level of protein fractions were not detected either [161].

It is shown that the defence-protective reaction to the whole complex of various stimuli (capture, hand and cage fixation, venipuncture), which influence a number of biochemical values (sugar, cholesterol, residual nitrogen levels) did not affect considerably total protein and serum protein fractions.

The electrophoretic mobility of protein fractions in monkeys is higher than that of man (table X). The monkey serum glucoprotein spectrum also sera of man and of the most widely used in experiments species of monkeys (Shown as

Globulins				
α_3*	β_1	β_2	γ_1	γ_2
–	9.9±0.47		18.6±0.60	
–	9.5±0.40		18.2±0.35	
2.8	5.3±0.19	11.8±0.26	6.7±0.25	15.2±0.45
–	5.2±0.20	11.3±0.24	6.5±0.37	18.7±0.44
5.6	5.7±0.40	9.4±0.37	6.2±0.29	11.7±0.45
–	7.2±0.50	12.3±0.35	7.8±0.14	13.8±0.20
3.7	4.1±0.33	10.5±0.42	7.5±0.6	15.5±0.50
3.4	6.3±1.07	11.3±0.51	6.3±0.47	19.1±0.75
–	6.6±0.65		25.7±1.46	
–	7.7±1.71		26.8±1.97	
α_2	α''_2	β	γ	
6.5±0.21	4.1±0.15	6.4±0.17	15.3±0.45	

Total protein was determined refractometrically, fractions – by paper electrophoresis in horizontal cuvette in barbiturate buffer (pH 8.6; 1.5 mA/cm of the width of paper) during 6 h. Stained by bromphenolblue eluates were photometried.

shows marked divergence from the human spectrum. *Cercopithecus aethiops* shows less divergence in this respect (table XI). Monkey lipoprotein spectra are characterized by the quantitative prevalance of α-fractions whereas in man β-fractions are prevalent (table XI).

The ratio of β/α-lipoproteins in *M. nemestrina* shows most similarity to that of man.

Antigenic Properties of Serum Proteins

Many authors [81, 131, 133, 189, 191] determined a high degree of monkey and human antigenic similarity. The rate of immunoprecipitates produced with the help of antihuman serum [131] is as follows: man 100%, *Pan troglodytes* 94%, *P. anubis* 83%, *P. hamadryas* 80%, *Theropithecus gelada* 78%, *Macaca mulatta* 71%, *M. irus* 61%, *Cebus* 49% and *Galago* 36%. The investigation of structural albumin similarity by complement fixation test yielded a somewhat different sequence of dissimilarity indexes; man 1.0, go-

Table X. Electrophoretic analysis of monkey serum. According to ANNENKOV [5]

Fractions	Ranges of mobility	Baboons		Macaques		
		Hamad-ryas	Mandrill	*Mulatta*	*Speciosa*	*Nemes-trina*
		Protein quantity (% to total protein content)				
Prealbumin						
ς_1	8.0–8.2	–	0	1.1	0	0.6
ς_2	7.5–7.8	–	0	1.5	3.9	0.8
Albumin	7.0–7.1	60.4	58.4	60.0	51.0	58.6
Globulin						
α_1	5.8–6.0	2.3	0	5.3	2.7	2.2
α_2	5.4–5.6	4.4	5.7	0	0	4.4
α'_3	5.1–5.2	0	0	4.1	6.0	2.8
α''_3	4.9	6.9	5,8	0	0	0
Total α		13.6	11.5	9.4	8.7	9.4
β'_1	4.5–4.7	0	5.9	4.5	4.7	6.1
β''_1	4.2–4.3	7.7	0	0	0	0
β'_2	3.9–4.0	0	0	6.8	0	3.8
β''_2	3.6–3.8	0	6.5	0	8.2	0
Total β		7.7	12.4	11.3	12.9	9.9
γ_1	3.1–3.3	5.2	–	2.4	0	4.7
γ_2	1.6–1.8	13.1	17.7	14.3	23.5	16.0
Total γ		18.3	17.7	16.7	23.5	20.7

Note: Diaethylbarbiturate buffer pH = 8.6; 0.07,0° ascending boundary; peak mobility in $cm^2/sec^{-1}.10^5$; protein quantity in fractions – % to total protein.

rilla 1.0, chimpanzee 1.14, orang-utan 1.22, gibbon 1.28, macaques 2.23, baboon 2.44, vervet monkey 2.59, squirrel monkey 4.5, capuchin 5.0, galago 8.6 and lemur 18.0 [144].

Most of the monkey and human proteins do not show complete immunologic identity: when rhesus monkeys are immunized with human sera it is

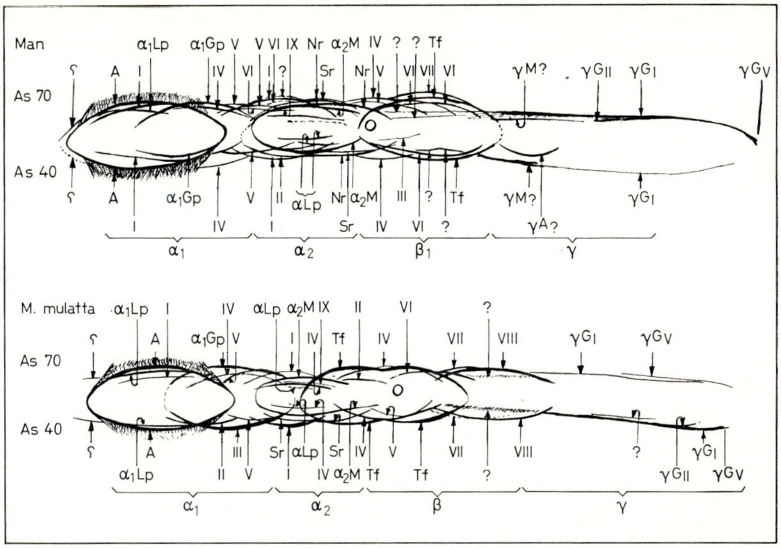

Fig. 1. Immunophoregrams of serum proteins in man (above) and *M. mulatta* (below), developed by antisera against blood proteins of man (AS_{40}) and of *M. mulatta* (AS_{70}). The arcs were identified according to table XII. Some arcs were not identified.

possible to get antibodies against all the proteins except albumin, ceruloplasmin, γG-globulin, a_0-cholinesterase [107]. Figure 1 shows a great immunologic similarity of human and *M. mulatta* proteins. The nomenclature for the proteins as used by different authors in immunophoregrams is not completely coincidental (table XII).

The differences in the immunophoretic human and monkey protein spectra consist predominantly in high prealbumin and α-lipoprotein mobility, accuracy with which they are revealed and the smaller mobility of transferrin in man. The characteristic dense arc β-8 and also arc β-7 located at the cathode end of β-zone is constantly and accurately revealed. Similar human arcs are not observed [5].

The divergence among the immunophoretic spectra of Old World monkeys is not pronounced [6]. One of the most marked differences between the spectra of genera *Macaca* and *Papio* is the absence of prealbumin-2 in the former [12, 62]. Specific differences are even less pronounced. They are comparable with the individual variations in the spectra depending on the allotypical protein polymorphism.

Table XI. Lipo- and glucoproteid spectra in the sera of certain monkeys and man. The black, glucoproteins were stained by Schiff-periodic acid. The extinction of eluate was

Species	Sex	Number of animals	Total protein	Albumins	Lipoproteins		
					A	B	C
Man	M	15	7.49±0.10	58.9±1.95	27.7±1.95	43.6±6.74	28.9±1.75
	F	16	7.84±0.22	54.9±1.09	28.3±2.90	48.8±2.75	22.8±2.63
M. mulatta	♂	25	8.25±0.18	53.2±1.3	45.0±2.5	30.3±1.5	25.0±1.64
	♀	25	8.17±0.22	48.0±2.04	44.5±2.30	30.8±1.571	25.3±2.25
P. hamadryas	♂	17	7.43±0.88	56.0±1.67	58.6±2.02	24.7±2.75	16.7±2.52
	♀	12	7.38±0.93	52.5±1.97	48.0±2.85	27.3±1.06	24.7±2.80
C. aethiops	♂	14	7.57±0.18	54.0±1.11	57.0±2.33	21.2±2.08	21.2±3.74
	♀	18	7.26±0.16	53.7±2.03	54.0±1.66	24.7±1.62	19.3±1.03
Saimiri	♂	15	7.13±0.25	54.9±1.70	–	–	–
	♀	11	7.03±0.40	47.0±2.47	–	–	–

Physico-Chemical Values and Aminoacid Composition of Certain Serum Proteins

The great similarity of human and monkey serum proteins is supported by the data on physico-chemical parameters and the amino acid composition of some of them (tables XIII and XIV). Unfortunately similar comparative data have not been widely published.

Genetic Polymorphism of Monkey Serum Proteins

Haptoglobin polymorphism is not prominent in monkeys while it is characteristic for man. Monkeys seem to lack the ^{2}Hp allele. The differences in the electrophoretic mobility of ^{1}Hp allotype in different monkey species is not noted by all authors. They are not pronounced and need further investigation. Transferrin polymorphism, on the other hand, is very extensive in monkeys (table XV). Thirteen allotypes of this protein have been revealed in macaques [63, 64, 136]. African monkeys differ greatly in this respect. The frequency of transferrin homozygotes is much higher in them.

HIRSCHFELD published data on the availability of Gc-proteins in the blood of *M. irus* [82]. He detected 4 phenotypes in 36 monkeys: 1–1, 2–1, 2–2 and 1F–2.

conditions of paper electrophoresis see in table IX. Lipoproteins were stained by Sudan estimated by photometry (M ± m are indicated). According to ANNENKOV [5]

	Glucoproteins				
Alb.	α_1	α_2	β	γ_1	γ_2
18.5±0.88	16.2±0.80	22.9±0.65	21.4±0.75	21.0±0.67	
14.3±1.59	16.7±0.98	26.2±0.66	22.8±0.81	20.9±1.18	
9.7±0.83	21.2±1.7	27.1±4.05	30.4±1.72	17.4±0.82	14.6±1.41
8.6±1.61	22.6±3.17	26.0±2.23	31.0±1.98	18.9±1.76	16.4±1.60
9.1±0.68	17.8±0.70	30.7±2.01	20.0±0.88	14.7±0.75	9.6±0.55
7.5±0.92	18.7±0.71	34.1±2.32	20.5±2.19	10.4±0.33	8.58±11.42
10.4±0.97	21.66±0.58	29.0±1.50	20.8±1.08	20.8±1.08	–
11.3±1.24	23.6±0.99	28.8±0.79	18.6±0.78	–	–
13.6±1.05	14.6±1.13	27.1±2.09	15.9±1.22	28.8±2.22	
12.6±1.54	14.9±1.10	29.9±3.18	19.5±2.72	13.0±2.26	

Genetically determined systems of prealbumins [12] and Lp (a)-factor [141] are known.

The variability of Gm-factors b, f and Z is noted in Cercopithecoidea; Gm-a, y, g and x have not been detected [108], nv(1)-factor was discovered in certain baboon populations [1]. Gm-factors have not been detected in South American capuchin monkeys, loris and *E. patas*. It seems that Gm-factor polymorphism of γG-globulin H-chain increases in non-human primates in proportion to their phylogenetic affinity to man.

Nonprotein Components of Monkey Serum

Monkeys differ from man considerably in a number of values.

These data seem to be indicative of the differences in rate, metabolic pathways and different clearances in human and monkey organs of excretion. The small numbers of animals investigated do not enable the data presented here to be completely representative but still they can prove very useful and helpful for a beginner who has started to investigate these values in monkeys for the first time (tables XVI, XVII and XVIII).

Table XII. The nomenclature of arcs in immunophoregramas of monkeys

ANNENKOV [5]		PICARD [133]		ANNENKOV [5]		PICARD [133]	
Zones	Arcs	Zones	Arcs	Zones	Arcs	Zones	Arcs
Albu-min	ς1 (prealbum.)	Preal-bumin	ς		β1(Tf)		β1-I (Tf)
					βII	β1	β1-II
	ς2 (prealbum.)		A		βIII		β1-III
	A (albumin)		α-LP	β	βIV		β1-IV
	α-LP (lipoprot.)		α1-I		βV		β1-V (Hp)
			α1-II (α_1GP)		βVI	β2	β2-I
α1	α1-I	α1	α1-III		βVII		β2-II
	α1-II		α1-IV		βVIII		β2-III
	α1-III (α1A, α1GP)		α1-V	γ	γ1-I(γG_1)		β2-IV
					γ1-II(γGII)		β2-V
	α1-IV				γ1-III(γGIII)		β-2VI(β2M)
	α1-V				γ1-IV(γGIV)		β2-VII
	α1-VI				γ1-V(γGV)		
α2	α2-I	α2	α2-I		γ2 (γA)		β2-VIII
	α2-II		α2-II		γ3 (γM)		β2-IX
	α2-III		α2-III(α2M)		γ4	γ	γ
	α2-IV		α2-IV				
	α2-V (Cp)		α2-V				
	α2-VI		α2-VI (αLP)				
	α2-VII (Hp)						
	α2-VIII (α2LP)						
	α2-IX						
	α2-X						
	α2-XI						
	α2-XII						

Table XIII. Certain physico-chemical ceruloplasmin values in *M. mulatta* and Man. According to SHAPOSHNICOV *et al.* [149]

Value	Man	*M. mulatta*
Serum content mg%	30–35	60–90
E 1%/1 cm at 610 μm	0.64–0.68	0.66
E 1%/1 cm at 280 μm	14.6	14.0
Cu atoms per 1 protein molecule	7.5–8.0	8.0
S 20, w $\times 10^{-13}$	7.1–7.6	7.4
D 20, w $\times 10^{-7}$ cm^2 sec^{-1}	4.4–4.5	4.5
Oxidase activity per 1 mg protein	11.6 ± 0.4	11.5
Molecular weight	151,000–160,000	160,000

Table XIV. The amino acid composition of human (C type) and cynomolgus transferrins (in μmol). According to PARKER *et al.* [132]

Amino acids 1	Man 2	Cynomolgus 3
Lysine	1.10	0.984
Histidine	0.353	0.351
NH_3	1.14	1.17
Arginine	0.508	0.469
Cysteic acid	0.732	–
Asparagic acid	1.50	1.52
Threonine	0.553	0.593
Serine	0.715	0.888
Glutamic acid	1.13	1.19
Proline	0.655	0.658
Glycine	0.932	0.896
Alanine	1.08	1.19
½-cystine	0.608	0.590
Valine	0.758	0.860
Methionine	0.166	0.189
Isoleucine	0.270	0.263
Tyrosine	0.477	0.496
Phenyl-alanine	0.522	0.549
Tryptophane	0.143	0.125
Glucosamine	0.095	0.059
Leucine	1.10	1.10

It is also necessary to note that sugar level in *P. hamadryas* and probably other catarrhine monkeys varies during 24 h, reaching the minimum at 8–9 pm and the maximum at 4–6 am. The variation may reach 20% of the maximum sugar level; the differences are statistically significant. During the winter period the sugar level is regularly reduced 20 to 30%, cholesterol content 50% or more and ATP increased from 1 to 3 mg%. The amount of residual nitrogen, inorganic and acid-soluble phosphorus was not observed to alter significantly [32, 38].

Of great interest is the fact [30, 38] that monkeys *(P. hamadryas, Cebus, C. aethiops* and *Theropithecus gelada)* show a stress reaction on the conditions of blood puncture which can induce an increase of sugar and cholesterol content in the blood. When blood puncture is repeated daily the reaction is suppressed. The differences between the analyses performed on the first day and the 10th to 15th days after the beginning of the experiment reached 20 mg% for sugar and 100 mg% for cholesterol.

Table XV. The frequency of 'Tf' allele in Cercopithecoidea (analyses by gel electrophoresis (after SMITHIES). Phenotypes were determined according to GOODMAN's standards. Most of the groups examined miscellaneous)

Monkey species	Number of examined animals	Transferrin alleles A	B	C	D′	D	E	F′	F	G	H′	H	Total	Authors
P. hamadryas	404	0.001	0.004	0.072	0.011	0.871	0.009		0.004			0.027	0.999	5
P. anubis	7	–	–	0.786	0.141	0.081	–						1.008	5
P. sphinx	7									0.072	0.928	–	1.000	5
C. aethiops	33				0.030	0.030	0.760	0.182	–	–			1.002	5
E. patas (Nigeria)	124				0.024	–	–	0.975	–	–			0.999	5
M. mulatta	244	0.027	0.049	0.475	0.027	0.172	0.061	0.023	0.051	0.133	–	0.004	1.000	5
M. assamensis	5				–	0.900	–	–		0.100			1.000	5
M. irus	9	0.055	–	0.167	–	0.278	0.278	–		0.222			1.000	5
M. nemestrina	15	–	–	0.206	0.069	0.380	–	0.069	0.035	0.241			1.000	5
M. speciosa	6	–	–	0.500	–	–	–	0.250	–	0.250			1.000	5
M. mulatta (Nepal border)	97	0.010	0.036	0.454	0.010	0.217	0.072	0.005	0.046	0.119		0.031	1.000	63
M. fuscata fuscata (Japan)	5								1.000				1.000	63
M. fuscata yakui (Japan)	5								1.000				1.000	63
M. radiata	2								1.000				1.000	63
M. irus (Thailand)	20		0.075	0.275		0.4000	0.025	0.075	0.025	0.050	0.075		1.000	63
M. speciosa (India)	21										1.000		1.000	63
M. speciosa (Thailand)	16							0.469			0.531		1.000	63

Table XVI. The content of certain nonprotein components in monkey serum (mg %)

Components	Monkey species	Number of tests	Normative M ± m	References	Method of determination
Sugar	*M. mulatta*	18	78 ± 13.7	31	Hagedorn-Iensen
Sugar	*P. hamadryas*	20	86.3 ± 14.5	31	Hagedorn-Iensen
Sugar	*C. aethiops*	9	98 ± 17.7	31	Hagedorn-Iensen
Sugar	Marmoset	20	94–466	83	Reinhold
Polysaccharides	*M. mulatta*	18	179 ± 46.4	31	Steppun
Polysaccharides	*P. hamadryas*	20	213 ± 58.9	31	Steppun
Polysaccharides	*C. aethiops*	9	137 ± 37.7	31	Steppun
Ergothionein	*M. mulatta*	10	0	9	Not indicated
Neuraminic acid	*M. mulatta*	10	73–101	9	Not indicated
Cholesterol	*M. mulatta*	153	141 ± 2.8 (100–200)	38	Pearson
Cholesterol	*M. radiata*	339	126 ± 8.73	163	Abell *et al.*
Cholesterol	*P. hamadryas*	158	110 ± 2.65 (90–150)	38	Pearson
Cholesterol	*C. aethiops*	9	119 ± 25.3	31	Levtchenko
Cholesterol	*M. mulatta*	5	139 ± 22	22	Schoenheimer
Cholesterol ether	*M. mulatta*	5	106 ± 39	22	Schoenheimer, Sperry
Phospholipids	*M. mulatta*	5	281 ± 39	22	Fiske-Subbarow
Phospholipids	*M. radiata*	238	140.8 ± 6.32	163	Fiske-Subbarow
Neutral lipid	*M. mulatta*	5	360 ± 54	22	Total lipids (cholesterol + phospholipids)
Neutral lipid	*M. mulatta*	18	365 ± 81.5	31	Bang-Pinkussen
Neutral lipid	*P. hamadryas*	20	317 ± 65.5	31	Bang-Pinkussen
Neutral lipid	*C. aethiops*	9	389 ± 44.4	31	Bang-Pinkussen
Total lipids	*M. mulatta*	5	780 ± 73	22	Bloor
Alkali reserve in volume %	*M. mulatta*	18	44.1 ± 5.7	31	Van Slyke
Alkali reserve in volume %	*P. hamadryas*	20	52.3 ± 6.2	31	Van Slyke
Alkali reserve in volume %	*C. aethiops*	2	33.9	31	Van Slyke

Table XVII. The content of nonprotein nitrogen components in monkey serum (mg %)

Components	Monkey species	Number of tests	Normative m ± m	Authors	Method of determination
Total nitrogen	*M. mulatta* and	27	1000.1 ± 20	4	Parnes
	P. hamadryas	10			
Residual nitrogen	Not indicated	2	26.5	110	Folin
Residual nitrogen	*M. mulatta* and	27	25.5 ± 0.95	4	Assel
	P. hamadryas	10			
Residual nitrogen	*Saimiri*	32	12.0–49.6	44	Micro-Kjeldahl
Urea	Not indicated	2	14.5	110	Folin
Urea	*M. mulatta*	10	8.9–14.6	9	Not indicated
Urea	*P. hamadryas*		8.57	29	
Urea	*M. mulatta*	27	13.8 ± 2.2	4	Gasanov
Urea	*P. hamadryas*	10			
Urea	Marmoset	24	5–21	83	Natelson
	Saimiri	67	11.4–42.2	44	Diacetyl monoxime
Uric acid	Not indicated	2	0.35	110	Folin
Uric acid	*M. mulatta*	10	0.28–0.65	9	Not indicated
Uric acid	*P. hamadryas*		3.35	29	
Uric acid	*M. mulatta* and	27	0.88 ± 0.34	4	Dubbs
	P. hamadryas	10			
Uric acid	Marmoset	24	0.63–7.68	83	Henry, Sobel, Kim
Uric acid	*Saimiri*	67	0.2–2.1	44	Conway
Creatinine	Not indicated	2	1.33	110	Folin
Creatinine	*M. mulatta*	10	0.9–2.3	9	Not indicated
Creatinine	*P. hamadryas*		0.9–1.6	29	
Creatinine	Marmoset		0.25–1.9	29	
Creatinine	Marmoset	24	0.41–1.95	83	Bonsnes, Taussky
Creatinine	*M. mulatta* and	27	1.5 ± 0.09	4	Kostir, Sonka
	P. hamadryas	10			
Creatin	*M. mulatta* and	27	1.57 ± 0.15	4	Kostir, Sonka
	P. hamadryas	10			
Aminoacid nitrogen	Not indicated	2	6.1	110	Folin
Aminoacid nitrogen	*M. mulatta*	10	5.9–8.3	9	Not indicated
Allantoin	*M. mulatta* and	27	3.24 ± 0.41	4	Joung
	P. hamadryas	10			
Glutathione reduced	*M. mulatta*	10	8.0–19.4	9	Not indicated

Table XVIII. The content of certain inorganic ions in monkey serum (in meq/l; *mg %; ** g %)

Components	Monkey species	Number of tests	Normative m ± m	References	Method
Sodium	*M. mulatta*	10	153.0 ± 1.7	188	
Sodium	*P. hamadryas*		137.5–145.5	29	
Sodium	Marmoset		122.2–157	29	
Sodium	Marmoset	90	150–187	83	Flame photometry
Potassium	*M. mulatta*	10	4.2 ± 0.3	188	
Potassium	*P. hamadryas*		3.05–4.85	29	
Potassium	Marmoset		3.44–6.67	29	
Potassium	Marmoset	95	3.3–8.6	83	Flame photometry
Chlorine	*M. mulatta*	10	110.2–1.0	188	Not indicated
Chlorine	Capuchin		110.2–122.8	8	
Chlorine	*P. hamadryas*		102.3–115.4	29	
Chlorine	Marmoset		95.4–123.6	29	
Chlorine	Marmoset	90	99–125	83	Cotlove, Nishi
Calcium	*M. mulatta*	4	9.9*	15	Copp
Calcium	*M. radiata*	18	12.5 ± 0.73	163	Clark-Collip
Calcium	*P. hamadryas*		3.3–5.9	29	
Calcium	Marmoset		3.6–5.0	29	
Calcium	Marmoset	79	7.83–13.05*	83	EDTA titration using Cal-Red
Calcium	*M. mulatta*	16	10.9 ± 0.16*	90	Clark
Magnesium	*M. mulatta*		2.1–2.7	8	Not indicated
Magnesium	*M. mulatta*	4	1.4*	15	Breen, Marshall
Magnesium	Marmoset	31	1.40–3.67	83	Sky-Peck
Zinc	*M. mulatta*		0.36*	9	Not indicated
Silicic	*M. mulatta*		20–26**	9	Not indicated
Nonprotein sulphur	*M. mulatta*		5.6–8.1*	9	Not indicated
Phosphorus	*M. mulatta*	4	3.75*	15	Fiske, Subbarow
Phosphorus	*P. hamadryas*		4.3–8.8*	29	
Phosphorus	Marmoset		2.26–8.94*	29	
Phosphorus	Marmoset	49	1.4–9.5*	83	Gomori
Phosphorus	*M. radiata*	18	4.6 ± 0.23*	163	Fiske, Subbarow
Phosphorus	*M. mulatta*	16	4.7 ± 0.33*	90	Fiske

The monkeys, predominantly Cercopithecoidea show few differences in biochemical values. This refers not only to protein but also to nonprotein components. Thus, the investigation of calcium content in different species of catarrhine monkeys by de Vaard technique did not reveal significant differences between them: *M. mulatta*, 10 mg%; *M. nemestrina*, 9.5; *M. irus*, 8.9, *M. speciosa*, 9.3; *M. assamensis*, 10.3; *P. hamadryas*, 9.3; *P. anubis*, 10.3; *C. aethiops*. 9.1. Sex differences have not been detected [93].

Circulation

Pulse Rate

A considerable number of papers (about 70) on the studies of ECG and pulse rates in different monkey species has been accumulated recently. The main bulk of the data refers to the last decade (1958–1969). However, the problem of normal pulse rate and ECG values has not been solved completely. We have included in the tables only the data extracted from the latest publications on certain monkey species and those dealing with the investigations carried out on an adequate number of animals.

As is seen from table XIX the pulse rate in monkeys in comparison with man is rather high, irrespective of the use of anaesthetics. It is noteworthy, that the pulse rate is very labile not only in one and the same monkey species but also in one and the same animal. Different data on pulse rates presented by different authors for unanaesthetized animals seem to depend on how much the animal is conditioned to the procedure of the test at the moment of observation; and in anaesthetized animals, on the depth of the anaesthesia and the properties of the anaesthetic used.

Thiopental-Na, pentobarbital-Na and sernylan are more frequently used as anaesthetics but there exists no common opinion of how the anaesthetic used influences the pulse rate. Thus, Kaminer [88] and Groover [69] consider that the pulse rate in anaesthetized animals is reduced parallel to the depth of anaesthesia. Van Citters *et al.* [183], contrarily, believed that the above mentioned substances increase the pulse rate.

Van Citters [183] is probably right in maintaining that anaesthetics should be avoided in order to establish normal values. However, we think it necessary to add that restraint of the animals and the presence of people in their field of vision should also be avoided since the pulse rate in monkeys is extraordinarily labile and depends significantly on the emotional state of the

Table XIX. Heart rate

Normal values	Monkey species	Authors	Comments
195 (150–240)/min	*M. mulatta*	KOKAYA [91, 92]	96 unanaesthetized animals
162 ± 3.56/min (120–218)	*M. mulatta*	CHERKOVICH [35]	Mean ± SD in 79 animals
160–333/min (260)	*M. mulatta*	ATTA and VANACE [10]	44 unanaesthetized animals
257 ± 31.2/min	*M. mulatta*	MALINOW [117]	351 unanaesthetized animals mean weight of 4.9 ± 1.7 kg
240/min	*M. cynomolgus*	BORUTAU [21]	
197 (140–260)/min	*M. radiata*	GOPALAN and PADMAVATHI [65]	28 observations in 22 animals of 3–16 lb
208/min	*M. speciosa*	SMITH and CROUT [157]	
170 (140–200)/min	*P. hamadryas*	KOKAYA [91, 92]	86 unanaesthetized animals
100–140/min	*P. ursinus*	KAMINER [88]	15 animals of 15–29 kg. Nembutal anaesthesia
74–138/min	*P. anubis*	GROOVER, SELJESKOG, HAGLIN and HITCHCOCK [70]	49 animals. Sernylan anaesthesia
160–180/min	*P. hamadryas*	FUFACHEVA [57]	230 unanaesthetized animals
125/min	Baboons (species not stated)	HERRMAN and WILLIAMS [79]	170 animals. Sernyl anaesthesia
201 ± 30/min	*Cynopithecus niger*	MALINOW and DE LANNOY [116]	25 unanaesthetized animals of 4–5 kg
230 (200–260)/min	Green monkey *Cercopithecus aethiops*	KOKAYA [92]	16 unanaesthetized animals
165/min	*Cebus apella*	WILBER [190]	
210/min	*Ateles ater* Spider	WILBER [190]	
260–306/min	*Saimiri sciurea*	BEISCHER and KNEPTON [14]	4 animals
300 (225–350)/min	*Saimiri sciurea*	REIS and OLIPHANT [138]	6 animals

animal and especially on the intensity of the orientation and defensive reactions. But even against this background one can observe the greatest heart rate in *Saimiri;* it is somewhat lower in macaques and lowest in baboons.

Even a whole year of conditioning the animal to the procedure of the test is not sufficient for the complete extinction of the defensive reaction heart component [FUFACHEVA, 57]; though the pulse rate becomes stable, it always exceeds 120/min.

Only MALINOW [166] in macaques and VAN CITTERS [183] in *Papio* succeded in recording the pulse rate less than 100 per min in unanaesthetized monkeys at rest. On these grounds they expressed an opinion that we always dealt with monkey tachycardia and the normal values vary between 80–90 per min or lower.

This opinion was supported by the new data of the ECG radiotelemetric technique by which it was possible to obtain ECG recordings from an unrestrained, free-moving monkey with no people in sight. [TATOYAN, 171]. In such conditions brought very close to natural, the pulse rate in *Papio* and macaques is 80–100 beats per min (table XX). However, as soon as a keeper to whom the animal is accustomed comes into view, the pulse rate immediately increases to 150.

The pulse rate in macaques is somewhat higher than in *Papio* (table XX). Thus, the normal value of the pulse rate in *Papio* and macaques is less than 100 beats per min, and a more frequent pulse usually recorded in restrained and even unrestrained animals in the presence of people should be treated as a consequence of the actual absence of extinction of the orientation and defensive reaction heart components (stress) (fig. 2).

The above mentioned normal values are observed when the ECG recording is performed during the day. Radiotelemetric techniques have enabled records to be made of the real diurnal dynamics of the pulse rate without waking the animal up at night. The night pulse rate is significantly reduced to 70–85 beats per min in adult *Papio* and 78–100 per min in *M. mulatta.* It is possible to observe sinus arrhythmia at such a low heart rate. When the pulse is increased to 100–120 beats per min sinus arrhythmia is usually not observed.

Monkeys show distinct changes of the pulse rate with age. Thus the pulse rate in newborn *M. mulatta* and *P. hamadryas* according to ECG data is 250–300 beats per min [CHERKOVICH, 35]. These values are probably not increased as the pulse rate recorded in the foetal *M. mulatta* in the intra- or extra-uterine space when the foetus is still in the amnion is 250–300 beats per min [DE GALLARDO *et al.*, 58].

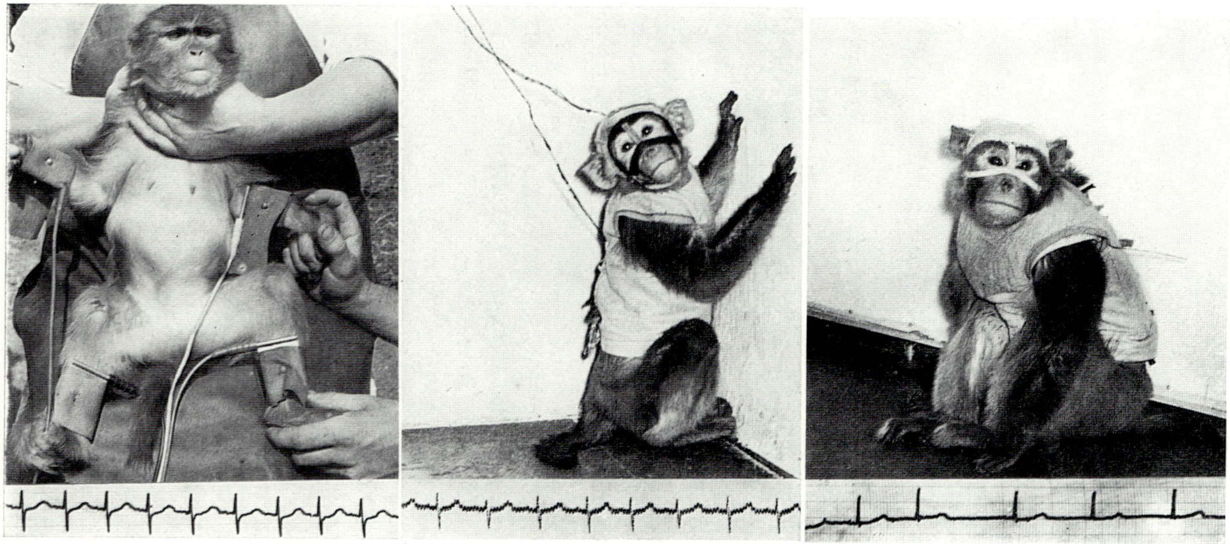

a *b* *c*

Fig. 2. A-ECG in monkeys restrained by hand, the heart rate is 250/min. B-ECG by hard-wine recording from free-moving animal in the presence of the experimenter: heart rate = 190/min. C-ECG by radio telemetry in the absence of experimenter: heart rate = 100/min with synus arrhythmia. A-lead V_3; B and C-Nehb leads. According to TATOYAN, CHERKOVICH, FUFACHEVA.

Table XX. Radiotelemetric ECG

Heart rate	Monkey species	Authors	Comments
Free-moving animals			
98–108/min	*M. mulatta*	TATOYAN [171]	10 unanaesthetized unrestrained animals
85–90/min	*P. hamadryas*	TATOYAN [171]	27 unanaesthetized unrestrained animals
Restrained animals			
170±20/min	*M. mulatta*	GRAYBIEL *et al.* [66]	9 observations in 1 restrained animal. No anaesthesia
253–280/min	*Saimiri sciurea*	GRAYBIEL *et al.* [66]	15 observations in 2 restrained animals. No anaesthesia

ECG

As it is seen from table XXI the duration of ECG segments is inversely dependent on pulse rate: the segments of least duration are observed in *Ateles ater*. Segments of greater duration are observed in macaques and *Papio*.

Cercopithecus aethiops has the greatest wave voltage which exceeds significantly that of macaques and *Papio* [KOKAYA, 92; CHERKOVICH, 36].

The R wave in standard leads in macaques and *Papio* has seldom exceeded 10–16 mm; in *C. aethiops* this wave is often 23–27 mm.

P wave in *M. mulatta* is high and sharp-pointed in 30 to 40% of cases [MALINOW, 117]. Such a P wave occurs in healthy monkeys, though it looks like P-pulmonale (fig. 3 [2]). The P wave in standard leads is always positive. In juvenile rhesus monkeys *(M. mulatta)* this wave is sometimes negative in the right precordial leads (fig. 3 [1]).

Q and S waves are often absent, but they sometimes reach 2–3 mm.

The changes of T wave and S-T segment observed in *Macaca mulatta* and especially in *Papio hamadryas* are most significant. HERRMAN and WILLIAMS [79] reported that 10% of 170 healthy *Papio* investigated showed a negative T wave in leads II, III and AVF. VAN CITTERS *et al.* [188] also note the presence of a negative T wave in a healthy *P. doguera* in leads II, III, AVF and even in the left precordial leads. When investigating 351 healthy *M. mu-*

Table XXI. Normal values of cardiac cycle intervals

Duration in sec					
P–R	QRS	Q–T	Monkey species	Authors	Comments
0.06–0.1	0.020–0.050	0.14–0.20	*M. mulatta*	RUSKIN and RIGDON [142]	27 animals of 2–7 kg. No anaesthesia
0.10 (0.07–0.13)	0.035 (0.03–0.04)	0.21 (0.15–0.26)	*M. mulatta*	KOKAYA [92]	96 unanaesthetized animals
0.064 (0.047–0.1)	0.027 (0.020–0.036)	0.148 (0.107–0.200)	*M. mulatta*	ATTA and VANACE [10]	44 unanaesthetized animals of 1.7–3.7 kg
0.089 (0.075–0.1)	0.045 (0.035–0.055)	–	*M. mulatta*	HAMLIN, ROBINSON and SMITH [73]	11 animals. Na-pentobarbital anaesthesia
0.07 ± 0.01 (0.06–0.08)	0.03 ± 0.007	0.14 ± 0.01	*M. mulatta*	MALINOW [116]	351 unanaesthetized animals. Mean weight 4.9 ± 1.7 kg
0.073 (0.054–0.088)	0.028 (0.020–0.044)	0.172 (0.140–0.220)	*M. irus*	DE WAART and STORM [185]	12 animals. Na-Evipan anaesthesia
0.06–0.14	0.04 (0.03–0.06)	0.2 (0.120–0.240)	*M. radiata*	GOPALAN and PADMAVATHI [65]	22 animals. Na-pentothal anaesthesia
0.11 (0.09–0.14)	0.04 (0.03–0.05)	0.22 (0.18–0.25)	*P. hamadryas*	KOKAYA [92]	86 animals. No anaesthesia
0.05–0.15	0.04 (0.01–0.07)	0.27 (0.2–0.3)	*P. ursinus*	KAMINER [88]	15 animals of 12–29 kg. Nembutal anaesthesia
0.118 ± 0.068	0.061 ± 0.008	0.26 ± 0.17	Baboons (species not stated)	HERRMAN and WILLIAMS [79]	170 animals. Sernylan or pentobarbital anaesthesia
0.12–0.16	0.06–0.08	0.24	*P. doguera*	VAN CITTERS and LASRY [183]	4 animals of 20–27 kg. Nembutal anaesthesia

Table XXI (continued)

Duration in sec					
P–R	QRS	Q–T	Monkey species	Authors	Comments
0.08 ± 0.01	0.03	0.17 ± 0.01	*Cynopithecus niger*	MALINOW and DE LANNOY [116]	25 animals
0.09 (0.08–0.1)	0.035 (0.03–0.04)	0.17 (0.15–0.20)	*Cercopithecus aethiops*	KOKAYA [92]	16 unanaesthetized animals
0.08	0.02	0.10	*Cebus apella*	WILBER [190]	
0.08	0.02	0.08	*Ateles ater*	WILBER [190]	

latta MALINOW noted the presence of the negative T wave in leads II, III and AVF in 6% of cases. We have also often observed the presence of the negative T wave in healthy *Papio hamadryas* (fig. 3 [3, 4, 5]).

S-T segment in *Papio hamadryas* is often shifted upwards or downwards in standard leads.

In macaques S-T segment is often shifted downwards in precordial leads (fig. 3 [6]). No histological data supporting this phenomenon were detected by thorough morphological analysis of such monkeys at death. Therefore it is suggested that the presence of the negative T wave in above mentioned leads in a certain percent of cases is a normal variation and is not indicative of pathology.

Radiotelemetric investigations have enabled us to observe conversion of the positive T wave into the negative and vice versa during short intervals of time [TATOYAN, 1969, 171].

Telemetric ECG recording (front Nehb lead) showed constantly the negative T wave in young *P. hamadryas* free ranging or restrained in a chair, against the background of low pulse rate and sinus arrhythmia provided man was out of sight. As soon as a person comes into the animal's view, the pulse rate immediately increases and the ECG recording shows the positive T wave (fig. 4 [1, 10]). The negative T wave and low heart rate are recorded again if the animals are not disturbed and there are no people in sight.

The results of these tests do not depend on the changes of body posture, as this phenomenon was also observed in monkeys restrained in a chair, where changes of body posture are not possible. A similar phenomenon was

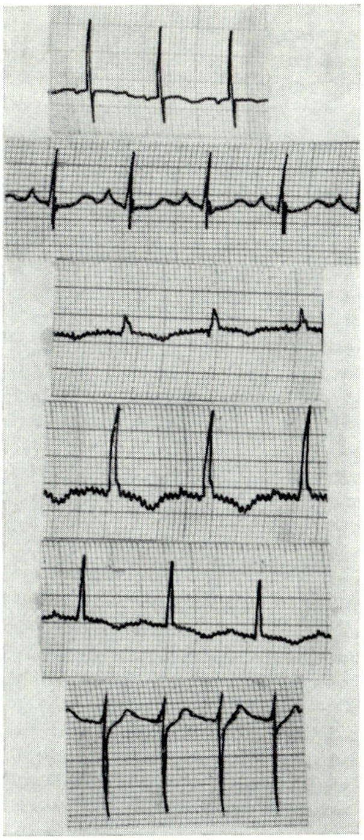

Fig. 3. 1-Negative P wave and 2-sharp P wave in a healthy *M. mulatta* 3.4.5-negative T waves in standard leads in a healthy *P. hamadryas;* 6-depressed S-T segment in lead V_2 in a healthy *M. mulatta.* According to CHERKOVICH.

Table XXII. Changes of electrical axis in various body postures. According to ATTA and VANACE [10]

Waves	Supine	Right semi-lateral 45°	Left semi-lateral 45°
P	+100°	+80°	+63°
QRS	+75°	+56°	+123°
T	+87°	+112°	−1°

not observed by us in *M. mulatta.* These data support the fact that repolarisation processes in monkeys are not always similar to those of man.

The wave voltage is decreased at night. This is demonstrated when ECG recordings are performed at night with the use of radiotelemetric technique.

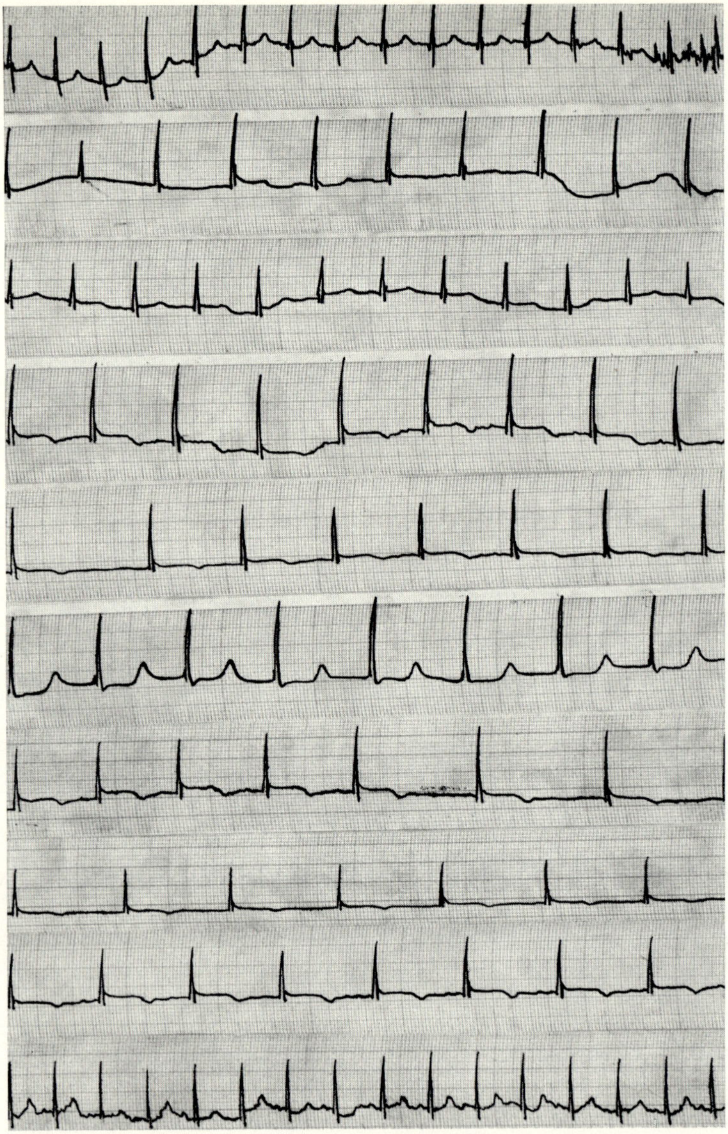

Fig. 4. Diurnal ECG changes in radiotelemetric recording. Recordings were made mostly every two hours. 1 ECG initial in restraint by hands at 10am; 2 at 12 (noon); 3 at 2pm (during eating); 4 at 4pm; 5 at 8 pm; 6 at12 (mid night) during storm; 7 at 2am; 8 at 4am; 9 at 8am; 10 at 10am in restraint by hands. According to Tatoyan.

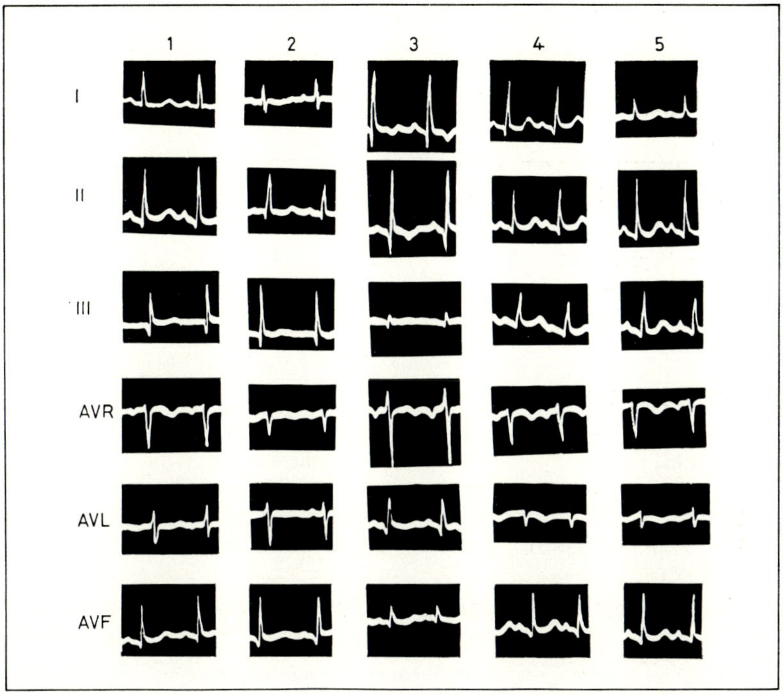

Fig. 5. ECG and the deviation of the electrical axis, recorded in *C. aethiops* in different postures; 1 – half lying on back (+56°), 2 – lying on the left side (+76°) 3 – lying on the right side (+36°), 4 – sitting (+54°), 5 – lying in prone position (+71°). According to KOKAYA.

According to HERRMAN and WILLIAMS [79] ECG zone of transition in *P. hamadryas* is often in the area $V_{1\text{-}2}$, $V_{2\text{-}3}$ or $V_{3\text{-}4}$. According to our data for *M. mulatta* and *P. hamadryas* it is found in the area $V_{3\text{-}4}$ or $V_{4\text{-}5}$.

In monkeys, especially young animals, the heart is more mobile in the thorax than in man. When the body posture is changed from upright to horizontal or the body is inclined on one side, characteristic ECG changes connected with the dislocation of the anatomical and electrical cardiac axis take place.

ATTA and VANACE [10] present data on P, Q, R, S and T axes deviation in *M. mulatta* induced by right and left lateral postures – 45° (table XXII).

Corresponding changes in ECG wave voltage in *M. mulatta* are shown by HAMLIN, ROBINSON and SMITH [73]. ECG changes at various postures in *Cercopithecus aethiops* are presented by KOKAYA [92] (fig. 5).

Table XXIII. Age alterations of the electrical axis. According to KOKAYA [92]

Age	Number of monkeys	Electrical axis deviation				
		0 to –90	0 to +40	+40 to +70	+70 to +90	+90
From 1 to 3–4 years	20	–	1	8	7	4
Over 18 years	12	6	2	3	1	–

Table XXIV. ECG Axes°

Normal values	Monkey species	Authors	Comments
+40° to +90° (QRS) (+55°)	*M. mulatta*	KOKAYA [92]	See above tables XIX and XXI
–41° to +127° (+55°)	*M. mulatta*	ATTA and VANACE [10]	See above tables XIX and XXI
+50° to +115°	*M. mulatta*	RUSKIN and RIGDON [142]	See above tables XIX and XXI
+50° to +100°	*M. mulatta*	MALINOW [116]	See above tables XIX and XXI
+63° (+50° to +75°)	*M. irus*	DE WAART and STORM [185]	See above tables XIX and XXI
+60° to +100° (19) –70° to –114° (9)	*M. radiata*	GOPALAN and PADMAVATHI [65]	See above tables XIX and XXI
+40° to +70° (14) 0° to +30° (2)	*Cercopithecus aethiops*	KOKAYA [92]	See above tables XIX and XXI
+36° (–30° to +90°)	*P. ursinus*	KAMINER [88]	See above tables XIX and XXI
+60,4° +29.6° (R)	Baboons	HERRMANN and WILLIAMS [79]	See above tables XIX and XXI
+35.5° (T)	Baboons	HERRMANN and WILLIAMS [79]	See above tables XIX and XXI
+40° to +70° (25) +70° to +90°, 0–30° (1)	*P. hamadryas*	KOKAYA [92]	See above tables XIX and XXI
+49° + 100°	*Cynopithecus niger*	MALINOW and DE LANNOY [116]	See above tables XIX and XXI

Table XXV. Phonocardiogram

Monkey species	Authors	Heart sounds (duration in sec)					Duration of systole in sec	Duration of diastole in sec
		Distance from QRS	I	II	III	IV		
M. mulatta (13 animals)	HAMLIN *et al.* [74]	0.036	0.074	0.038	–	–	0.19	0.157
P. doguera (4 animals)	VAN CITTERS [183]	0.03	0.04	0.03	Absent	Preceded the 1st sound by 0.04 sec	0.26	0.24

Apexcardiogram

The rapid-filling wave lasted 0.11 sec. The systolic wave was notched.

Cineangiography

(12 Observations in 3 animals with 70% Hypaque).

The maximum transverse thoracic diameter of the hearts was: 10.2; 10.0; 9.2 cm during diastole; 9.8; 9.4; 8.9 cm during systole. Diameter of the ascending aorta was about 2.0 cm.

Vectorcardiogram data for monkeys are presented in the papers by HAMLIN *et al.* [73]; BRISTOW and MALINOW [24]; CHERKOVICH and RUTTKAY-NEDECKY [37, 39]; RUTTKAY-NEDECKY and TSCHERKOVITSCH [143].

There exist definite age alterations in the electrical axis position. The electrical axis in adult monkeys is either normal or upright. Right-lateral deviation more often occurs in young monkeys while left-lateral deviation occurs in adult animals as in man [KOKAYA, 92] (table XXIII).

Data on electrical axis position in adult monkeys of different species are presented in table XXIV.

Data on the phonocardiogram (table XXV), apexcardiogram and cineangiogram carried out on one and the same animal *(P. doguera)* are presented in a paper by VAN CITTERS *et al.* [183]. Data presented above are extracted from this paper.

Data on cardiac output, stroke volume and peripheral resistance are important for the characteristics of cardiac activity. Those data are given in table XXVI and cited according to STAHL and MALINOW [164].

The study of the cardiac cycle phase structure yields valuable information on cardiac function. This information for *M. mulatta* was presented by

Table XXVI. Some data on the characteristics of cardiac activity

Normal values	Monkey species	Authors	Comments
	Cardiac output		
90–200 ml/kg/min	*M. mulatta*	LEES, MALINOW and PARER [105]	
1,086 ± 228 ml, 273 ± 62 ml/kg/min	*M. mulatta*	FORSYTH *et al.* [53]	13 unanaesthetized, restrained monkeys (sitting position) 3.1–6.8 kg (mean 4.1)
190 ml/kg/min, 2.84–5.22 l	*P. cynocephalus*	GUENTER *et al.* [71]	7 adult males (mean 4.1) anaesthetized with pentobarbital
	Stroke volume		
5.8 ± 2.4 ml	*M. mulatta*	LEES, MALINOW and PARER [105]	8 animals of 5–6 kg (Fick principle)
6.4 ± 1.4 ml	*M. mulatta*	FORSYTH *et al.* [53]	Dye dilution method T_{1824}
20–50 ml	*Papio cynocephalus*	GUENTER *et al.* [71]	7 adult males pentobarbital anaesthesia
	Peripheral resistance		
11,200 ± 3,700 dyne/s/cm^5	*M. mulatta*	HAYDEN [78]	8 animals mean weight 7.6 kg, pentobarbital anaesthesia
2,506 dyne/sec/cm^5	*P. cynocephalus*	GUENTER *et al.* [71]	See above
	Pulmonary vascular resistance		
849 ± 376 dyne/s/cm^5	*M. mulatta*	HAYDEN [78]	See above
mean 436 dyne/sec/cm^5	*P. cynocephalus*	GUENTER *et al.* [71]	See above
	Brain vascular resistance		
2.2 ± 0.4 (1.7–2.6) mm Hg/cm^3/100 g brain/min	*M. mulatta*	MEYER and HANDA [118]	7 animals of 3–4 kg, ether anaesthesia

BELKANIA 1969. No similar data are known from the existing literature on primates. Anaesthesia was not used. The heart rate in the experimental animals restrained supine ranged as usual. Data on duration of the main cardiac cycle phases are in table XXVII.

Table XXVII. Normal chronocardiogram for *M. mulatta* at a varying heart rate (average values in sec). According to BELKANIA 1969 (unpublished)

Phases	M ± m C 0.4–0.5 sec	C 0.3–0.39 sec	C 0.2–0.29 sec
Cardiac cycle (c)	0.423 ± 0.002	0.370 ± 0.002	0.263 ± 0.004
Asynchronous contraction phase (AC)	0.051 ± 0.001	0.044 ± 0.001	0.026 ± 0.002
Isometric contraction phase (ic)	0.038 ± 0.002	0.039 ± 0.001	0.028 ± 0.002
Tension period (T)	0.089 ± 0.001	0.083 ± 0.003	0.054 ± 0.002
Ejection phase (E)	0.159 ± 0.001	0.143 ± 0.002	0.095 ± 0.003
Mechanical systole (Sm)	0.197 ± 0.002	0.180 ± 0.001	0.123 ± 0.003
Joint systole (So)	0.248 ± 0.002	0.225 ± 0.002	0.148 ± 0.002
Protodiastolic phase (P)	0.014 ± 0.002	0.011 ± 0.001	0.014 ± 0.001
Ventricular diastole (D)	0.174 ± 0.003	0.147 ± 0.002	0.117 ± 0.002
Distance from 1st to 2nd sound	0.197 ± 0.002	0.180 ± 0.001	0.125 ± 0.002
Distance c–f	0.172 ± 0.001	0.155 ± 0.003	0.109 ± 0.004
Phase values			
Intrasystolic index (ISI)	80.7 ± 0.6%	77.9 ± 1.5%	76.7 ± 1.6%
Myocardial tension index (MTI)	35.7 ± 0.4%	37.0 ± 1.2%	36.4 ± 0.8%

Note: 16 unanaesthetized animals of 7–10 kg were used. The recordings of ECG (V_3), Phonocardiogram and Sphygmogram (a. axillaris) were carried out simultaneously. 6 or 10 successive cardiac cycles were analysed.

On the basis of the investigation of 6–10 subsequent cardiac cycles certain phase values were calculated. The Intrasystolic Index (ISI = E/Sm × 100%) ranged from 76.7 to 80.7%. Thus it took $^8/_{10}$ of the mechanical systole's duration to eject the blood from the left ventricle into aorta. The value of the Myocardial Tension Index (MTI = T/So × 100%) ranged from 35.7 to 37.0%. There exists a high degree of correlation between the different phases of the cardiac cycle ($p<0.001$) (table XXVIII).

Blood Pressure

Systemic Arterial

Data on diastolic and systolic blood pressure in monkeys, measured either by the indirect (cloth cuff) or direct method (vascular catheter) are presented in table XXIX.

Table XXVIII. Coefficients of phase correlation and some phase indexes of cardiac activity in *M. mulatta*

Designations	C					
T	0.993 $p<0.001$	T				
E	0.989 $p<0.001$	0.885 $p<0.001$	E			
So	0.982 $p<0.001$	0.926 $p<0.001$	0.981 $p<0.001$	So		
D	0.995 $p<0.001$	0.829 $p<0.001$	0.838 $p<0.001$	0.871 $p<0.001$	D	
ISI	0.004 $p>0.1$	0.144 $p>0.1$	0.219 $p>0.1$	0.047 $p>0.1$	0.011 $p>0.1$	ISI
MTI	0.145 $p>0.1$	0.199 $p>0.1$	–0.023 $p>0.1$	0.14 $p>0.1$	0.179 $p>0.1$	–0.619 $p>0.1$

Table XXIX. Blood pressure (in mm Hg)

Normal values	Monkey species	Authors	Comments
125/75	*M. mulatta*	MAGAKYAN, 1953 [111]	Indirect method (cloth cuff), unanaesthetized animals
158±18 systolic, 101±10 diastolic	*M. mulatta*	LEES, MALINOW and PARER, 1965 [105]	14 animals of 5.3 kg (mean weight), direct femoral catheter, phencyclidine analgesia
118.6±10.6–130.5±9.8 systol, 73.0±7.5–80.7±9.5 diastolic	*M. mulatta*	FORSYTH, 1968 [53]	8 animals of 3.5–5 kg
135/80	*P. hamadryas*	MAGAKYAN, 1953 [111]	Indirect method (cloth cuff), unanaesthetized animals
120/70	*Cercopithecus aethiops*	MAGAKYAN, 1953 [111]	Indirect method (cloth cuff), unanaesthetized animals

Table XXX. Blood pressure in cardiac cavities, (cardiac catheterization), aorta and pulmonary artery (in mm Hg)

	Right ventricle			Left ventricle			Pulmonary artery			
Right atrium	Systolic	Diastolic	Left atrium	Systolic	Diastolic	Aorta	Systolic	Diastolic	Monkey species	Authors
+1 mean (+2–2)	20	4	–	120–140	0–5	140/90	18–24	6–7	*P. doguera*	VAN CITTERS and LASRY [183]. 4 animals of 20–27 kg. Light nembutal anaesthesia
2±2 mean	23 ±11	2±2	–	121±21	4±3	Syst. 120±26 Diast. 84±12	21 ±8	9±4	*M. mulatta*	HAYDEN [78]. 8 animals mean weight 7.6 kg. Pentobarbital anaesthesia
3.5±2 mean	33±6	–	–	–	–	–	31 ±6	15 ±5	*M. mulatta*	LEES, MALINOW and PARER [105]. 14 animals of 7.6 kg mean weight. Phencyclidine analgesia

Blood pressure values in monkeys at rest are close to those of man, but diastolic pressure is somewhat higher than in man and in the majority of the cases is equal to 80–90 mm Hg. Blood pressure in monkeys refers to very labile functions and is increased sharply to 180–200/100–120 in healthy monkeys in periods of increased muscular activity (for example during restraint procedure).

Blood pressure can be also increased as a result of emotional overstrain when the animals are unconditioned to the procedure of a test. As soon as the animals get used to the procedure of the experiment and the defensive reflex becomes extinct, the values normalize LAPIN *et al.* [103]; FORSYTH and BAIREUTHER [52] also note that blood pressure, measured automatically every day (catheterization of abdominal aorta) in 13 restrained monkeys under a chronic experiment, was higher during the first week (138/86 mm Hg) and reduced in the course of the following days (125/78 mm Hg).

Table XXXI. The velocity of PWD along different vessels in monkeys and man in cm/sec. According to CHERKOVICH [40]

Arteries	*M. mulatta*		*P. hamadryas*		Man
	♂	♀	♂	♀	
Humeral artery	299.8±7.7 (57)	335.6±12.2 (56)	277.1±6.1 (52)	293.3±9.97 (30)	
Aorta	543.8±8.1 (63)	621.8±16.7 (60)	522.3±8.4 (59)	539.0±8.1 (37)	400–530
Segment aorta-shin artery	638.8±9.3 (34)	699.0±20.8 (32)	622.9±14.2 (23)	637.0±11.5 (24)	550–790
Shin artery	981.8±42.7 (34)	1061.8±21.8 (32)	99.1±43.3 (20)	883.5±37.3 (21)	820–1,100

Note: The number of investigations is given in brackets.

Blood pressure has a definite diurnal rhythm and is reduced during night hours. Material on circadian rhythms of blood pressure, heart rate and other physiological functions are published in 'Circadian Rhythms in Nonhuman Primates' [42].

Arterial pressure becomes greater with the increasing age of the animal.

VAN CITTERS [183] data on blood pressure in heart cavities are tabulated in table XXX.

The Velocity of the Pulse Wave Distribution in Monkeys

The velocity of the Pulse Wave Distribution (PWD) along different vessels in monkeys and man is given in table XXXI. The velocity of the PWD was recorded on a Visocard-Multivector 3-channel electrocardiograph recorder using special transducers of the same firm. The transducer was placed over a certain artery and the velocity of the PWD was recorded simultaneously with the ECG 2nd lead (fig. 6).

The velocity of the PWD in *M. mulatta* is somewhat higher than in *Papio hamadryas*. Sex differences are statistically insignificant.

The values of the velocity of the PWD in monkeys are similar to those of man but the ratio of the PWD velocity in the arteries of muscular type to that

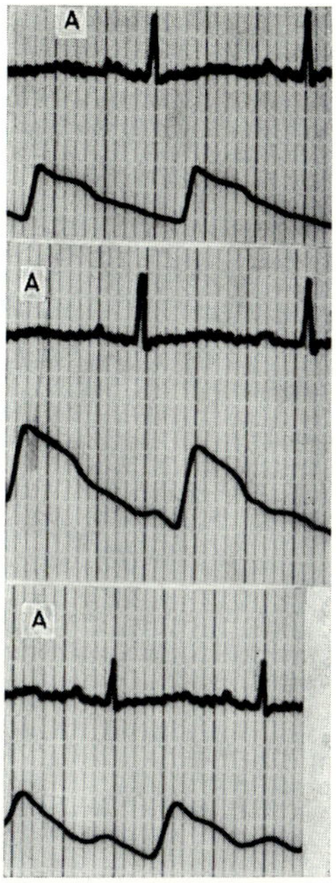

Fig. 6. Pulse wave recorded simultaneously with IInd ECG standard lead with transducers on a. brachialis (I), a. femoralis (II) and a. tibialis (III). According to CHERKOVICH.

in the arteries of elastic type is higher in monkeys (1.81–1.84) than in man (1.1–1.2), which may testify to a higher tone of vascular smooth muscle in comparison with man.

Respiratory System

The values of frequency are widely known for *Macaca* and *Papio*. Some data are presented in table XXXII. The values obtained from anaesthetized monkeys are higher than those obtained from unanaesthetized animals. The frequency is higher in young monkeys.

Data on the volume of residual air, dead space and lung compliance, etc. for *M. mulatta* are presented in a paper by STAHL and MALINOW [164]. The same paper contains data on arterial and venous blood gases for *M. mulatta*. Some of those and other data are given in table XXXIII.

Table XXXII. Frequency of respiration

Normal values	Monkey species	Authors	Comments
36/min	*M. mulatta*	MIYAMOTO, REYNOLDS, PATTERSON, CUGELL, KETTEL [119]	9 adult animals. Pentobarbital anaesthesia
42 ± 7/min	*M. mulatta*	LEES, MALINOW and PARER [105]	14 animals of 5.3 kg. Phencyclidine (sernylan) analgesia
30/min	*M. mulatta*	SHCHERBAKOVA [150]	6 adult unanaesthetized, unrestrained animals
31/min	*M. nemestrina*	SHCHERBAKOVA [150]	4 adult unanaesthetized, unrestrained animals
29/min	*P. hamadryas*	SHCHERBAKOVA [150]	12 adult unanaesthetized, unrestrained animals
88 ± 27/min	*P. papio* and *P. cynocephalus*	VICE, BRITTON, RATNER and KALTER [184]	21 young animals brought up in the nursery
18–22/min	*P. cynocephalus*	GUENTER *et al.* [71]	7 adult animals. Pentobarbital anaesthesia
Minute volume			
315 ml/kg	*P. hamadryas*	SLONIM and SHCHERBAKOVA [153]	3 animals of 5.9 kg (mean weight). No anaesthesia
1,791 ± 389 ml	*M. mulatta*	LEES, MALINOW and PARER [105]	14 animals of 5.3 kg mean weight
1.014 ± 40 cm^3, 383 ± 129 cm^3/kg	*M. mulatta*	KAREL and WESTON [89]	39 unanaesthetized animals of 2.63 kg mean weight
Tidal volume			
47 ml	*M. mulatta*	MIYAMOTO, REYNOLDS *et al.* [119]	4 adult animals
26 ± 8.7 cm^3	*M. mulatta*	KAREL, WESTON [89]	See above

Oxygen Dissociation Curve of Blood

According to the data given by BEHRMAN [13] for *M. mulatta* (females) the dissociation curves can be expressed by the following empirical equation: log p O_2 = 3.709–0.296 pH + 0.378 log (S/(100-S) where S stands for oxygen saturation of blood.

The haemoglobin of the foetus shows more affinity to oxygen. It has the following formula: log p O_2 = 4.823–0.484 pH + 0.416 log (S/100-S)).

Table XXXIII. Blood pH and blood gas constants

Normal values	Monkey species	Authors	Comments
pH of arterial blood			
7.4	*M. mulatta*	MIYAMOTO, REYNOLDS, PATTERSON, CUGELL, and KETTEL [119]	9 adult animals anaesthetized with pentobarbital
7.4	*M. mulatta*	BERHMAN [13]	19 animals
7.432±0.072	*M. mulatta*	LEES, MALINOW and PARER [105]	14 animals
7.4±0.169	*M. irus*	BANERJEE [11]	58 animals with sernylan analgesia
O_2 tension in arterial blood			
83 mm Hg	*M. mulatta*	MIYAMOTO *et al.* [119]	9 adult animals
101±7 mm Hg	*M. irus*	BANERJEE *et al.* [11]	See above
69 mm Hg	*P. cynocephalus*	GUENTER *et al.* [71]	7 baboons anaesthetized with pentobarbital
CO_2 tension arterial blood			
43 mm Hg	*M. mulatta*	MIYAMOTO *et al.* [119]	9 adult animals
42.9±8.3 mm Hg	*M. mulatta*	LEES, MALINOW and PARER [105]	See above
42±4 mm Hg	*M. irus*	BANERJEE [11]	See above
39 mm Hg	*P. cynocephalus*	GUENTER *et al.* [71]	See above

The values 'n' (from HILL's equation) and P_{50} with pH = 7.4 for females are equal to 2.65 and 32.9 mmHg respectively. The same values for the blood of the foetus are equal to 2.40 and 17.4 mm Hg. The paper also contains the nomogram for calculation of the dissociation curves HbO_2 of the mother and foetus *(M. mulatta)* according to pO_2 pH and S obtained experimentally.

Digestive System

Saliva

Saliva secretion in monkeys was studied with the help of fistulated parotid and submaxillary (sub-mandibular) glands. Parotid ducts in *M. mulatta* open opposite the first upper molar; in *Papio hamadryas*-opposite the 2nd to 3rd molar.

Submaxillary gland ducts open on the lateral side of the tongue frenulum. Sublingual ducts also open into the mouth cavity.

Buccal pouches do not possess salivary glands but saliva can flow into them from the mouth cavity. The presence of food in buccal pouches does not stimulate salivary secretion. The latter starts as soon as food penetrates the mouth cavity.

The question on the availability of spontaneous salivary secretion remains open until now.

Some investigators have failed to observe spontaneous salivary secretion in macaques [MALINOVSKY, 114]. SHESTOPALOVA *et al.* [151] observed it in small amounts: 1.0 to 20 mg per 5 min. UGOLEV, VOLKOVA and KORNEVITS [173] observed spontaneous salivary secretion only in restrained macaques and baboons.

Spontaneous salivary secretion in fasting monkeys shows a larger amount of secretion derived from the submaxillary gland than from the parotid gland. After feeding vice versa, the amount of secretion derived from the submaxillary glands is smaller than that produced by the parotid glands.

Saliva consists of a liquid part and solids (table XXXIV). The pH of saliva for different species is shown in table XXXV. Saliva of the submaxillary gland is more viscid (10–12 sec) than that of the parotid gland (5–6 sec) in comparison with water (5 sec), according to LAGUTINA [102]. The saliva of macaques and baboons contains a considerable amount of amylase.

Table XXXIV. The effect of different stimuli on the percentage of solids in saliva. According to LAGUTINA [102]

Stimuli	Parotid gland			Submaxillary gland		
	Solids	Organic	In-organic	Solids	Organic	In-organic
Cranberry	2.00	1.35	0.65	1.24	0.62	0.62
Apple juice (acid)	2.01	1.56	0.47	1.12	0.74	0.38
Oranges	2.13	1.71	0.42	1.17	0.92	0.25
Plum juice (sweet)	2.48	1.89	0.59	1.12	0.80	0.42
Sunflower seeds	2.48	1.90	0.53	1.31	1.05	0.36
Cream toffee	3.05	2.55	0.50	1.27	0.94	0.36
Green onion	2.41	1.90	0.51	1.75	1.28	0.42
Clay	3.29	2.79	0.50	1.67	1.24	0.37

Table XXXV. pH of saliva for different monkey species

Normal values		Monkey species	Authors	Comments
P	S			
8.2–8.5	–	*M. mulatta*	MALINOVSKY [114]	78 observations in 1 adult unanaesthetized male
8.04–8.8	8.2–8.07	*M. mulatta*	LAGUTINA [102]	1 adult unanaesthetized male
7.97–8.24	–	*M. irus*	BOWEN, EASTOE and MORRIS [23]	19 animals of 1.9–4.5 kg. Anaesthetized (N_2O, O_2 and fluothane). Salivation induced by pilocarpine (0.2–1.0 mg/kg)
8.0	8.0	*P. hamadryas*	SHESTOPALOVA [151]	4 unanaesthetized animals of 3 kg
Whole Saliva 8.16±0.055 (7.90–8.90)		*Callithrix jacchus*	DREIZEN, GOODRICH and LEVY [46]	30 adult animals of 200–500 g. Sernylan analgesia. Salivation induced by pilocarpine (0.5 mg/100 g)

P – *Gl. Parotis.*
S – *Gl. Submandibularis.*

The solids consist of organic and inorganic constituents. Mucin produced by the salivary glands in *Saimiri sciureus* and *M. mulatta* according to histo-chemical investigations [LEPPI and SPICER, 106] contains sulphurmucin which is predominantly secreted by the sublingual glands. Saliva of macaques also contains sialomucin.

The submaxillary gland also seems to secrete a certain growth hormone as there exist data [NARASIMHAN and GANLA, 124], showing that the ectomy of this gland inhibits growth as in the case of hypophysectomy.

The content of the solids in the saliva of baboons and macaques is distinguished from that of man and dog by a particularly great amount of organic constituents in the parotid saliva. The fermentative activity of the parotid saliva is 3–4 times higher than that of the submaxillary saliva [LAGUTINA, 102].

Spontaneous and conditioned reflex salivary secretion in baboons and macaques is not considerable in volume (0.1–0.2 ml per 5 min). Due to this, the technique of drop count is not convenient. UGOLEV *et al.* [173] proposed a

Table XXXVI. The changes of quantitative and qualitative composition of saliva from parotid and submaxillary glands in response to different food constituents in *Papio hamadryas*. According to SHESTOPALOVA [151]

Nature of the stimulus	Quantity of the stimulus (g)	Quantity of saliva per 30 min (cm^3)		pH		Amylolytic activity [acc. to VOLGEMUT]		Number of tests
		P	S	P	S	P	S	
In fasting animals	–	0.12	0.23	8	8	20,460	583	10
Candies	7	1.36	0.8	7.8	8.8	20,114	102	10
Apples	7	2.5	0.9	8	9	76,460	78	10
Cranberry	7	4.3	1.1	8	8.1	44,180	53	10
Milk	20	0.56	0.52	8	8.1	507	45	8
(Sweet) cherry	7	2.08	0.76	8.2	8.5	60,945	91.6	10
HCL 0.36%	15	0.88	0.54	9	8.6	40,010	60.7	5

Table XXXVII. Average data on the amount of secretion produced by salivary glands in response to different foods. According to LAGUTINA [102]

Stimuli	Quantity (g)	Secretion of saliva – cm^3 – per 3 min		Coefficient P/S	Number of tests
		Parotid gland (P)	Submaxillary gland (S)		
Bread (white)	2–4	0.65	0.14	4.5	21
Sunflower seeds (roasted)	2	0.65	0.21	3.0	13
Milk (sweet)	25	0.02	0.08	0.25	15
Cream toffee	2–5	1.25	0.44	2.8	6
Chocolates	2	0.86	0.31	2.6	10
Syrup filled bonbons (vitaminized)	2.5	0.95	0.34	2.8	14
Green onion	1–2	0.70	0.23	3.2	17
Tangerines	4–8	0.45	0.27	1.7	7
Oranges	10	0.46	0.20	2.3	5
Apples	2–4	0.83	0.26	3.2	3
Tinned apple juice	15–20	0.73	0.41	1.8	13
Lemon	2–4	1.22	0.62	2.0	13
Cranberry	2	1.34	0.49	2.7	18
Tinned plum juice	4	1.55	0.52	3.0	8
Chalk	1–2	0.60	0.20	3.0	8
Clay	2	0.82	0.33	2.5	7

weight technique. According to the latter, a sheet of filter paper weighed beforehand is applied to the fistulous opening and is immediately reweighed.

Saliva composition in response to different food stimuli in *M. mulatta* and *Papio* is shown in tables XXXVI and XXXVII.

The quantity of saliva and its composition are constant with one and the same food stimulus, however, they show much variation with different food stimuli.

The importance of mechanical and chemical food constituents is very well seen when solid food is eaten (hard candies). In this case a considerable amount of saliva (1.4 ml per 5 min) is secreted with a low content of solids (2.1%). The same candies softened in water induce a smaller amount of saliva (1 ml) with an increased content of solids and increased amylolytic activity [MALINOVSKY, 114].

Gastric Juice

In free-ranging monkeys as in man gastric juice is secreted continuously. The juice contains mucus, HCL and pepsin. pH optimum for monkey gastric juice pepsin secreted when protein is digested is equal to 1–1.5 [STARTSEV, 166].

It is inadvisable to restrain monkeys in order to collect juice of normal composition and quantity via the gastric fistula. It is necessary to condition the monkey to the following procedure. Before each collection of the juice the animal is taken in the arms, the juice poured out via the fistula and the animal is put back into the cage. In restrained monkeys the secretion of the HCL is inhibited and the gastric juice secreted is of a low acid, neutral or even alkaline reaction, and a low digestive power.

The quantity of juice secreted per unit of time and its acidity in free-ranging animals is greater than in restrained animals (tables XXXVII–XLII). When HCL is added to gastric juice the digestive power is restored. This inhibition of acid secretion in restrained monkeys has a neuro-reflex genesis and can be observed while the juice is extracted via a fistula or the 'Pavlov pouch' the vascular and nervous connections of which are left intact. When the 'Heidenhein denervated pouch' is employed, this phenomenon is not observed. Acid gastric juice of a great digestive power derived from this pouch can also be collected from restrained animals.

Similar data were published by MAHL [112]. He reported that in several rhesus monkeys unconditioned to the procedure of gastric juice collection,

Table XXXVIII. Volume of gastric juice in free-moving animals

Normal values (ml/h)	Monkey species	Authors	Comments
Old World monkeys			
11.8–19.2	*P. hamadryas*	STARTSEV [165, 166]	Range based on 30 observations in 5 animals
New World monkeys			
4.6 ± 1.2–14.8 ± 2.2	Spider	SMITH, BROOKS, DAVIS and ROTHMAN [159]	17 observations in 5 animals
13 ± 8.1	Spider	BROOKS, RIDLEY, ATTINGER and NEFF [27]	26 observations in 6 animals of unknown weight
Volume of gastric juice in restraint conditions			
Old World monkeys			
4.3–6.2	*P. hamadryas*	STARTSEV [165, 166]	See above
5.78	*M. mulatta*	STARTSEV [165, 166]	Mean based on 10 observations in 2 animals
3.5	*M. mulatta*	BRODIE and MARSHALL [25]	16 observations in 5 animals
9.6	*M. speciosa*	BRODIE and MARSHALL [25]	3 observations in 1 animal
5.49	*Cercopithecus aethiops*	STARTSEV [165, 166]	Mean based on 10 observations in 2 animals
New World monkeys			
0.1	*Tamarinus nigricollis*	BRODIE and MARSHALL [25]	1 observation in 1 animal
3.0	*Cebus capucinus*	BRODIE and MARSHALL [25]	17 observations in 5 animals
1.5	*Cebus apella*	BRODIE and MARSHALL [25]	10 observations in 3 animals
3.9 ± 0.9–5.8 ± 0.7	Spider	SMITH, BROOKS, DAVIS and ROTHMAN [159]	8 observations in 5 animals
1.7 (0.8–2.8)	Spider	BROOKS, RIDLEY, ATTINGER and NEFF [27]	21 observations in 6 monkeys

Table XXXIX. Free and total acid, pH of gastric juice in restraint fasted monkeys

pH	Free acid meq/l	Total acid meq/l	Monkey species	Authors	Comments
Old World monkeys					
–	0	–	*M. mulatta*	SCHENDORF and IVY [145]	45 animals 1–6 years of age. Peroral intubation
5–7	–	–	*M. mulatta*	FRENCH, LONGMIRE, PORTER and MOVIS [56]	31 animals
7.0	–	–	*M. mulatta*	STARTSEV [165, 166]	10 observations in 2 animals. Chronic gastric fistula
3.5	1.0	18	*M. mulatta*	BRODIE and MARSHALL [25]	16 observations in 5 animals of 1.9 kg. Thomas gastric canula
1.9 ± 0.5	49.2 ± 31.6	–	*M. mulatta*	SMITH, MASON and JACOBSON [160]	6 animals conditioned to restraint in a chair. Chronic gastric fistula
–	0	20	*M. speciosa*	BRODIE and MARSHALL [25]	3 observations in 1 animal of 4.3 kg
6.7	0	28	*P. hamadryas*	STARTSEV [165, 166]	30 observations in 5 animals
5.1	–	–	*Cercopithecus aethiops*	STARTSEV [165, 166]	10 observations in 2 animals
–	6	35	*Cercopithecus aethiops*	BRODIE and MARSHALL [25]	14 observations in 5 animals of 2.2 kg
New World monkeys					
–	47.0 ± 10.6 –88.0 ± 8.0	66.0 ± 14.4 –117.0 ± 7.0	Spider	SMITH, BROOKS, DAVIS and ROTHMAN [160]	18 observations in 5 animals
–	68 ± 36	92 ± 34	Spider	BROOKS, RIDLEY, ATTINGER and NEFF [27]	6 animals
–	56.9 8.0–102.0	81.7 38.6–118.3	*Saimiri sciureus*	BRODIE and MARSHALL [26]	56 observations in 23 animals. Thomas gastric fistula
–	44	88	*Tamarinus nigrocollis*	BRODIE and MARSHALL [25]	1 observation in 1 animal of 0.280 kg
–	35	72	*Cebus capucinus*	BRODIE and MARSHALL [25]	17 observations in 5 animals of 0.876 kg
–	7	39	*Cebus apella*	BRODIE and MARSHALL [25]	100 observations in 3 animals of 0.853 kg

Table XL. Free and total acid, pH of gastric juice in free-moving fasted monkeys

pH	Free acid	Total acid	Monkey species	Authors	Comments
New World monkeys					
–	7.8±2.5–102.0±5.5 meq/l	99.0±11.9 –127.0± 12.5 meq/l	Spider	SMITH, BROOKS, DAVIS and ROTHMAN [159]	17 observations in 5 animals
–	94±22 meq/l	113± 21 meq/l	Spider	BROOKS, RIDLEY, ATTINGER and NEFF [27]	30 observations in 6 animals
Old World monkeys					
1–2.2	36 titration units	65 titration units	*P. hamadryas*	STARTSEV [165, 166]	30 observations in 5 animals
1.8	41.8 meq/l	–	*M. mulatta*	SMITH, MASON and JACOBSON [160]	6 animals with chronic gastric fistula

Table XLI. Gastric juice pepsin

Pepsin digestive power	Monkey species	Authors	Comments
Free monkeys			
7.2 mm/h by Mett	*P. hamadryas*	STARTSEV [166]	See above
167.0±22.6–448.0±180.0 $PU^{Hb} \times 10^{-4}$/ml	Spider	SMITH, BROOKS, DAVIS and ROTHMAN [159]	See above
407±275	Spider	BROOKS, RIDLEY, ATTINGER and NEFF [27]	26 observations in 6 animals
Restrained monkeys			
2 mm/h by Mett	*M. mulatta*	STARTSEV [166]	See above
0	*P. hamadryas*	STARTSEV [166]	See above
2.3 mm/h	*Cercopithecus aethiops*	STARTSEV [166]	See above
148.0±64.9–400.0±45 $PU^{Hb} \times \pm 10^{-4}$/ml	Spider	SMITH, BROOKS, DAVIS and ROTHMAN [159]	See above
563±398 $PU^{Hb} \times 10^{-4}$/ml	Spider	BROOKS, RIDLEY, ATTINGER and NEFF [27]	40 observations in 6 animals

Table XLII. Fasting gastric juice electrolyte content

K^+ meq/l	Na^+ meq/l	Cl^- meq/l	Monkey species	Authors	Comments
Restraint conditions					
14.4±0.8 – 20.5±1.7	44.8±7.0 – 70.8±2.4	128±3.1 – 150.2±1.7	Spider	SMITH, BROOKS, DAVIS and ROTHMAN [159]	See above
11.5 (7.6–21.0)	60.9 (26.5–103.0)	155.9 (105.0–181.6)	Squirrel	BRODIE and MARSHALL [26]	See above
Free monkey					
14.2±1.0 – 19.2±0.3	36.8±14.0 – 68.7±14.0	138.0±9.3 – 149.0±6.9	Spider	SMITH, BROOKS, DAVIS and ROTHMAN [159]	See above

the juice was of a low acidity while in one rhesus monkey conditioned to the procedure the juice was acid. In experiments with restrained spider monkeys continuing for 3 h the acidity and secretion of the gastric juice was considerably reduced [BROOKS, RIDLEY *et al.*, 27].

Diurnal variations were observed only in free-ranging monkeys and manifested themselves in the reduction of juice volume and acidity at night. BRODIE and MARSHALL [26] noted inverse dependence in respect to New World monkeys: the volume of the juice secreted and the amount of free HCL in restrained monkeys were greater than in free-ranging animals. It should be noted that the above mentioned animals were conditioned to the chair and therefore it seems that the inhibition of HCL secretion was not manifested in them. SMITH *et al.* [160] also collected acid gastric juice from macaques, conditioned to the chair. According to the data of the same authors BRODIE and MARSHALL [26] the juice volume and its acidity was reduced in squirrel monkeys restrained for 24 h.

The fact that fixation appears to be a significant stress for monkeys is supported by the data reported by STARTSEV [166]: a healthy monkey *(P. hamadryas)* died soon after being restrained in the chair. A similar case with two healthy animals (Spider monkey and Woolly monkey) is described by BROOKS *et al.* [27].

But there exist data [SMITH, MASON and JACOBSON, 160] which disagree with the findings of the above mentioned authors. These investigators did not detect significant differences in the gastric juice acidity in restrained and free-ranging monkeys. The juice was always acid. Seasonal alterations of the gastric juice were not detected.

Gastric Movements

Gastric movements in *M. mulatta, M. irus* and *P. hamadryas* in free-ranging behaviour bear a continuous character and are accompanied by acid gastric juice secretion. The disturbance of the normal contractile gastric movements and their complete inhibition are observed in restrained *M. mulatta, P. hamadryas* and *C. aethiops*.

Sometimes the interchange of periodic and continuous gastric movements of the fasting stomach (as in the dog) can be observed. Continuous gastric movements last from 4 to 6 h; active periods last from 30 to 60 min; passive from 0.5 to 3 h.

Secretion and Movements of the Small Intestine

The monkey's small intestine is covered with villi of various forms. Predominant in the duodenum are broad ramifying villi. The villi of the jejunum and ileum are mainly foliaceous and sometimes lingual [SEHGAL, CHAKRAVARTI *et al.*, 147].

The juice of the intestine in *M. mulatta* investigated with the help of the Tiri fistula looks like a yellowish opalescent fluid with lumps (clots) of mucus, degenerative epithelial cells and lymphocytes; the pH = 7–9. The amount of the juice as well as the percentage of the solids vary widely in different individuals and also in one and the same individual. The amount of juice which can be secreted during 4 h varies from 0.22 to 1.62 g; the percentage of solids from 1–20%. Juice secretion induced by mechanical stimulation reaches 8 g during the same period of time.

The average activity of enterokinase is 2,450 units (the range of activity varies from 900 to 10,000 units). The average activity of alkaline phosphatase (352 units) varies from 150 to 3,015 units. Saccharase ranges from 20 to 161 units, amylase from 0 to 5 units, lipase from traces to 42 units.

Alkaline phosphatase is distributed evenly along the whole length of the small intestine. Enterokinase is predominantly concentrated in the duodenum. Its contents reduce sharply at the beginning of the jejunum and is furthermore reduced at the end of the small intestine. The movements of the intestine in green monkeys consist of continuous contractions of the ileum and caecum during 5–6 h of observation. Certain contractions last for about 1–2 min. Such contractions are not connected with the character of the gastric movements and are not dependent on it. Restraint of monkeys does not inhibit intestinal movements while gastric movements are inhibited in restrained animals [STARTSEV, 165].

Metabolism and Body Temperature

Gaseous Exchange

The chamber technique is most advisable for the investigation of gaseous exchange in the lungs and basal metabolism, since the animals get used to the mask with great difficulty [SLONIM *et al.*, 154]. Data on O_2 consumption for different monkey species are presented in table XLIII.

Table XLIII. Oxygen consumption

Normal values	Monkey species	Authors	Comments
8.5 ml/kg/min	*M. mulatta*	SLONIM and KALMIKOVA [154]	20 unanesthetized monkeys of 6 kg (mean weight)
4.8–10.5 ml/kg/min	*M. mulatta*	LEES, MALINOW and PARER [105]	
12.5 ml/kg/min 15.6 l/kg/per day	*M. nemestrina*	SLONIM and KALMIKOVA [154]	11 monkeys of 4.0 kg (mean weight)
14 l/kg/per day	*M. nemestrina*	PACE, HANSEN *et al.* [129]	Restrained animals of 8–12 kg
8.7 ml/kg/min	*P. hamadryas*	SLONIM and SHCHERBAKOVA [153]	37 unanaesthetized animals of 7 kg
827.4±44.0–829.4±41.7 ml/kg/h	*Saimiri sciurea*	MALINOW and WAGNER [115]	11 unanaesthetized animals of 800–900 g
0.80 ml/g/h	*Callithrix jacchus*	MORRISON and SIMOES [121]	Diurnal monkey
0.45 ml/g/h	*Aotus trivirgatus*	MORRISON and SIMOES [121]	Nocturnal monkey
Cerebral metabolic rate of oxygen			
3.4±0.8 cm^3/100 g Brain/min	*M. mulatta*	MEYER and HANDA [118]	16 animals of 3–4 kg. Ether inhalation

Table XLIV. Metabolic rate

Normal values	Monkey species	Authors	Comments
748 kcal/m²/24 h 46.5 kcal/kg/24 h	Baboon	BRUHN [28]	7 sleeping animals of 5.3–7.6 kg (between 9 pm and 3 am). Chamber technique
847 kcal/m²/24 h 82.54 kcal/kg/24 h	*P. hamadryas*	SLONIM and SHCHERBAKOVA [153]	37 unanaesthetized animals of 7 kg (mean weight)
884 kcal/m²/24 h 65.4/kcal/kg/24 h	*M. mulatta*	SLONIM and SHCHERBAKOVA [153]	20 unanaesthetized animals of 6 kg (mean weight)
610–675 kcal/m²/24 h 48.4–49.3 kcal/kg/24 h	*M. mulatta*	SPECTOR [162]	Animals of 3.2–4.2 kg at rest
653 kcal/m²/24 h 49.3 kcal/kg/24 h	Mangabey	BRUHN [28]	6 animals of 3.4–4.1 kg
100 kcal/kg/24 h	*Saimiri sciurea*	MALINOV and WAGNER [115]	11 unanaesthetized animals of 800–900 g

The respiratory quotient in monkeys is usually rather low. SLONIM *et al.* [154] presented data for *P. hamadryas* and *M. mulatta* which had an average respiratory quotient of 0.7. According to LEES, MALINOV and PARER [105] the respiratory quotient in *M. mulatta* is 0.77 ± 0.03.

Body Temperature

The body temperature measured in the rectum of different *M.* species is not constant and has significant diurnal variations reaching 3° C (table XLV).

The diurnal amplitude decreases with increasing t° of the surroundings. Due to this the indication of the body t° must be accompanied by the indication of the time of the day (when t° was measured) and t° of the surroundings. Diurnal monkeys show maximum body t° in the day time, nocturnal animals at night. It should be also taken into consideration that increased motor activity, during restraint, for instance, causes an increase in body t° from 37° C to 40° C.

Table XLV. Body temperature (rectal)

Time and t° of the body	Monkey species	Authors	Comments
max 12 (noon) to 4 pm 39.0° C min 2 am–6 am 36.0° C	*M. mulatta*	SLONIM [156], SHCHERBAKOVA [150], CHERKOVICH [34]	20 unanaesthetized animals air t° 17–23° C
max 39.9° C min 37.5° C	*M. mulatta*	WAHLE [186]	8 animals
max 38.84 ± 0.47 at 4 pm min 37.42 ± 0.56 at 4 am Diurnal amplitude 1.42° C	*M. irus*	HONJO, FUJIWARA *et al.* [84]	10 unanaesthetized animals of 2–3 kg
max 12(n)–4 pm 39.0° C min 2 am–6 am 36° C Diurnal amplitude 2.5–3.0° C	*P. hamadryas*	SLONIM [156], SHCHERBAKOVA [150], CHERKOVICH [34]	See above
max at 11 pm 38.5° C min 7 am–15 pm 37.3° C Diurnal amplitude 1.4°	*Aotus trivirgatus*	MORRISON and SIMOES [121]	Nocturnal monkey
mean 39.1° C (day time) max 39.7° at 9.30 am mean 36.2° C (at night) min 35.4 C° at 1.30 am Diurnal amplitude 4.3°	*Callithrix jacchus*	MORRISON and SIMOES [121]	Diurnal monkey

Among homoiothermal animals the most labile body t° is observed in monkeys. As has already been mentioned, their diurnal amplitude can reach 3° C whereas in man it does not exceed 1 °C; in cat, 0.5°; in birds, 0.7° and in dog, 0.2–0.3 °C. The body t° in young *Papio* and *Macaca* is lower than in adult animals.

The diurnal amplitude was also high when measured vaginally: 1.3 to 3.0° F (2.16 + 1.47) [ERIKSON, 48].

Skin temperature in *Papio* and *Macaca* was investigated by ARHANGELSKAJA [7]. Data on it are presented in figure 7.

The highest t° is observed in the area of the chest, it is somewhat lower in the forehead, face and neck. The lowest skin t° is that of the palms and soles; in *M. mulatta* it is somewhat higher than in *P. hamadryas.*

During the first days after birth the diurnal variations of the rectal t° in *Macaca* and *Papio* are insignificant and irregular. They vary from 35.5° to

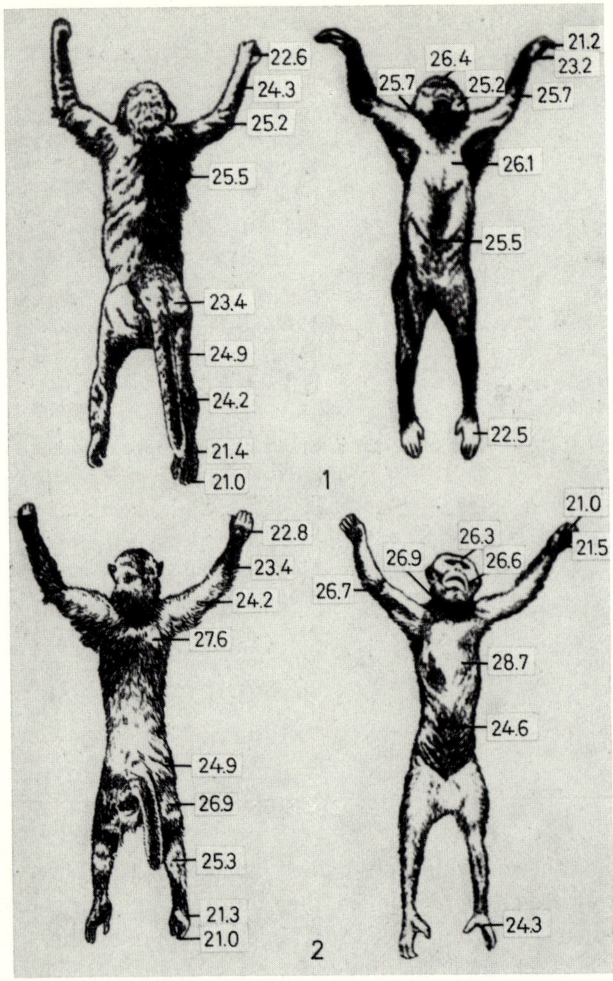

Fig. 7. Skin temperature in adult monkeys. 1 – *P. hamadryas*, 2 – *M. mulatta*. According to ARHANGELSKAYA.

36.5° C. At one and a half month the rectal t° of the infant acquires a definite diurnal rhythm with an amplitude of 2.5–3.0° C but the absolute values are lower than in adult animals, thus, when the surrounding t° is 18–19° C, rectal t° ranges from 34° C to 37° C; at surrounding t° 23–25° C from 35.5 to 38.5° C [CHERKOVICH, 34].

Table XLVI. Voiding in *M. mulatta* [FELDMAN *et al.*, 49]

Monkey weight kg	Urinary output ml/24 h	Urinary output ml/m²	Frequency of voiding	Nonvoiding period	Urine volume (1st morning voiding)	% of total output
5.2 (3.1–9.6)	262 (105.0–504.0)	741 (316.0–1,330.0)	4.5 (1.0–15.7)	17.8 10.6–23.7)	150 (44.0–270.0)	62 (13.0–100.0)

Table XLVII. Comparative thickness of kidney cortex and medulla. According to FISHER *et al.* [172]

Species	The thickness of:				Urinary concentrating ability
	Cortex	Medulla	Outer medulla	Inner medulla	
Monkey	8 mm	7 mm	7 mm	1 mm	1,800 mOsm
Man	15 mm	15 mm	6 mm	9 mm	1,400 mOsm

Newborn *P. papio* and *P. cynocephalus* show a similar dependence on the t° of the surroundings [VICE, BRITTON, RATNER and KALTER, 184] when the surrounding t° is 80–85 °F, the rectal t° is 95.1 °F, at 88–92 °F it is 97.8 °F.

Renal Function

The amount of urine excreted in *M. mulatta* during 24 h is presented in table XLVI [FELDMAN, SMITH and LEVENTHAL, 49]. The investigation was carried out on 20 animals (16 females) which consumed 125 g of food and 410 ml of water per day 1,179 ml per m² of surface). Average data on voiding in the conditions mentioned above are presented in table XLVI.

Voiding has a definite diurnal rhythm: the greatest amount of urine is excreted in the morning (45–85% of the urine excreted during 24 h). Later in the day voiding of small amounts of urine was observed 1 or 2–3 times. Voiding was not observed at night, though the animals wake up sometimes and drink water. A seasonal rhythm of voiding is also observed. In summer the first voiding occurs before 6 am, in winter after 6 am. There are monkeys which void 11–23 times after the principal voiding in the morning and do not void at night. The prementioned refers to monkeys conditioned to their surroundings. Fear and excitement can induce a voiding at abnormal times.

The detailed paper by STHAL and MALINOW [164] concerning the normals of *M. mulatta* contains data on urine formation rate (70–80 ml/kg/per day) presented by ALTMAN and DITTMER [2]. Urine volume of other non-human primates is given in special tables published by MORROW and TERRY [122] (Seattle Primate Information Center).

According to the data of TISHER, ROSEN and TESCHAN [172], urinary concentrating ability in *M. mulatta* is equivalent to that of man though the length of the monkey medulla is less than that of man due to a poorly developed inner medulla, the thickness of which was measured on sagittal kidney sections.

It is accounted for the fact that the ascending limb of Henle's loop in monkeys is composed of cuboidal epithelium but in man it is lined by squamous epithelium.

Inulin, endogenous creatinine, PAH and isotops: hippuran J^{131}; B_{12}; Co^{57} were used to determine the clearance in monkeys.

These data for different monkey species are presented in table XLVIII.

Most of the authors note that the values characterizing the renal function in monkeys resemble those of man more than those of rabbits, rats, and dogs. PICKERING *et al.* [134] did not detect any age alterations in glomerular filtration rates of *M. mulatta.* SKEITH *et al.* [152] found out that glomerular filtration rate decreased with advancing age in Cebus monkeys.

The amino acid composition of urine for 112 individuals belonging to 27 different species is given in a work by FOODEN [51].

The investigation of urine formation rate in 6 foetuses *(M. mulatta)* of an almost complete pregnancy term (2 weeks before delivery) carried out by CHEZ, SMITH and HUTCHINSON [41] showed that the average rate is equal to 0.03 ml/min and the osmotic pressure of urine is similar to that in the blood of the foetus.

Electroencephalography

Electric activity of the brain was studied in 20 *P. hamadryas*, 20 *Cercopithecus aethiops* and 31 *M. mulatta* [URMANCHEEVA and CHERKOVICH, 174, 175]. During the first days after birth irregular waves of different duration with the prevalence of slow waves were observed (fig. 8). The amplitude of

Table XLVIII. Renal function parameters

Normal values	Monkey species	Authors	Comments
Inulin clearance			
1.96 ± 445 ml/min/kg (1.18–3.03)	*M. mulatta*	PICKERING and SUSSMAN [134]	40 observations in 7 mature reserpine treated animals
2.18 ± 0.680 ml/min/kg (1.00–3.60)	*M. mulatta*	PICKERING and SUSSMAN [134]	42 observations in 6 unsedated immature animals
3.1 ± 0.5 ml/min/kg (2.2–3.8)	*M. cynomolgus*	SWEET, LEVITT and HODES [168]	12 unanaesthetized animals of 3.6–5.7 kg
35 ± 12.7 ml/min/m² (21.7–53.0)	*Cebus albifrons*	SKEITH and HEALEY [152]	5 adult animals of 1.5–2.5 kg. Sernylan + Pentobarbital sodium anaesthesia
0.59 ± 0.120 ml/min/g kidney weight	*Saimiri sciurea*	SELKURT and WATHEN [148]	7 animals of 450–850 g. Pentobarbital anaesthesia. Had not been given water since the previous day
Creatinine clearance			
3.08 ± 0.939 ml/min/kg (1.73–5.22)	*M. mulatta*	PICKERING and SUSSMAN [134]	See above
41.1 ± 13.5 ml/min	Baboon	JOHNSTON *et al.* [87]	174 animals
42.8 ± 13.3 ml/min/m² (30.8–43.9)	*Cebus albifrons*	SKEITH and HEALEY [152]	See above
PAH Clearance			
8.06 ± 1.220 ml/min/kg (6.71–10.90)	*M. mulatta*	PICKERING and SUSSMAN [134]	See above
1.03 ± 0.15 ml/min/g kidney weight	*Saimiri sciurea*	SELKURT and WATHEN [148]	See above
Isotope Clearance			
3.29 ml/min/kg (2.19–4.34) Hypaque J^{131}	*M. speciosa*	O'DELL, RADWIN, BERNSTEIN and SCHLEGEL [126]	6 females of 6.8–8.9 kg. Innovar-vet treated
2.07 ml/min/kg (0.93–3.30) B_{12} Co^{57}	*M. mulatta*	WEIGEL, HANSEN and SMITH [182]	8 unanaesthetized animals
40 ± 12.4 ml/min/m² (32.8–57.3) B_{12} Co^{57}	*Cebus albifrons*	SKEITH and HEALEY [152]	See above

Table XLVIII (continued)

Normal values	Monkey species	Authors	Comments
Creatinine-to-Inulin Clearance ratio			
I.14	*M. mulatta*	SMITH and CLARKE [158]	
1.06–1.56	*Cebus albifrons*	SKEITH and HEALEY [152]	3 animals Sernylan + Pentobarbital anaesthesia
Renal Plasma Flow			
11.9 ml/min/kg (9.6–14.4) Hippuran J^{131}	*M. mulatta*	WEIGEL, HANSEN and SMITH [187]	See above
19.8 ml/min/kg (13.1–25.2) Hippuran J^{131}	*M. mulatta*	WEIGEL, HANSEN and SMITH [187]	Pentobarbital sodium anaesthesia
15.1 ml/min/kg (13.8–17.7) Hippuran J^{131}	*M. speciosa*	O'DELL, RADWIN, BERNSTEIN and SCHLEGEL [126]	See above
140.1 + 69.2 ml/min	Baboon	JOHNSTON *et al.* [87]	See above
Filtration Fraction			
0.243	*M. mulatta*	PICKERING and SUSSMAN [134]	See above
0.173	*M. mulatta*	WEIGEL, HANSEN and SMITH [187]	See above
0.218	*M. speciosa*	O'DELL, RADWIN, BERNSTEIN and SCHLEGEL [126]	See above

those waves had increased by the age of one month and reached the maximum by 6–8 months. Later, the amplitude decreased and stabilized at the age of 2–2.5 years with a marked prevalence of the more frequent EEG components (8–12 and 16–32 per sec). Distinct regular rhythms appeared for the first time at the age of 4–6 months. In old age further reduction of the amplitude of all the EEG waves was noted. From the first days of life and during the whole first year after birth distinct specific differences appeared in the monkey EEG (fig. 8 and 9).

The greatest frequency of EEG components was observed in *Cercopithecus aethiops*. Distinct regular driving response to flashing light at a rate of

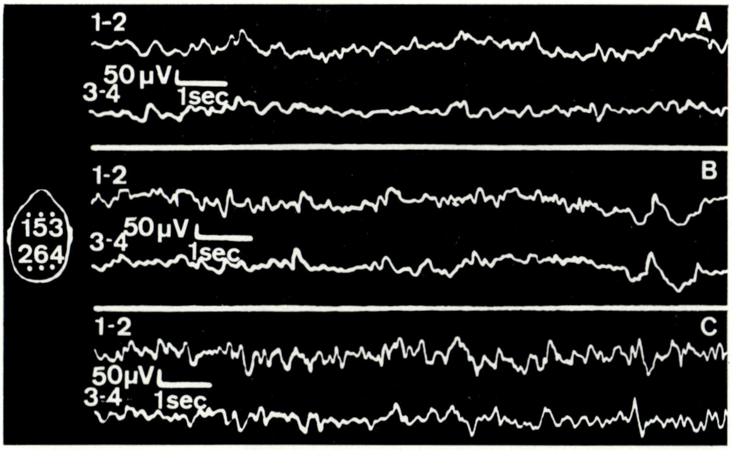

Fig. 8. EEG in an infant (A) *P. hamadryas*, (B) *M. mulatta* and (C) *C. aethiops*. On the left: 6 points from which ECG was recorded. According to URMANCHEEVA and CHERKOVICH.

5–10 per sec and a feeble driving response of 15 per sec were observed in the EEG of the occipital region in *M. mulatta* and *P. hamadryas* during the first days of life. The driving response in *Cercopithecus aethiops* corresponded to a more frequent rhythm (5–20 per sec) during the same period of life. Beginning with the second week of life all the monkey species examined showed a driving response to flashing light at a rate of 5 to 34 per sec. The greatest amplitude and regularity of waves in the induced rhythm were observed at a rate of 5–10 light flashes per sec. All components of the driving response increased at 6–7 months of age. The dependence of the kind of response to flashing light on background activity was observed from 3–4 to 8–10 months of age. The amplitude of the induced rhythm was decreased at 6–8 months of age. At 2 years of age the induced rhythm became stable. The components of the driving response to flashing light within all periods of age were higher in *C. aethiops* than in *P. hamadryas* and *M. mulatta*. At 15–20 years of age the driving response in the EEG became considerably weaker and disappeared completely at 27 years of age.

A detailed study of age alterations in the EEG in *M. mulatta* and the peculiarities of EEG during natural sleep, drowsiness and awakening were fully presented in a paper by ROBERT DE RAMIREZ DE ARELLANO [137] and in an excellent atlas on EEG by CAVENESS [33].

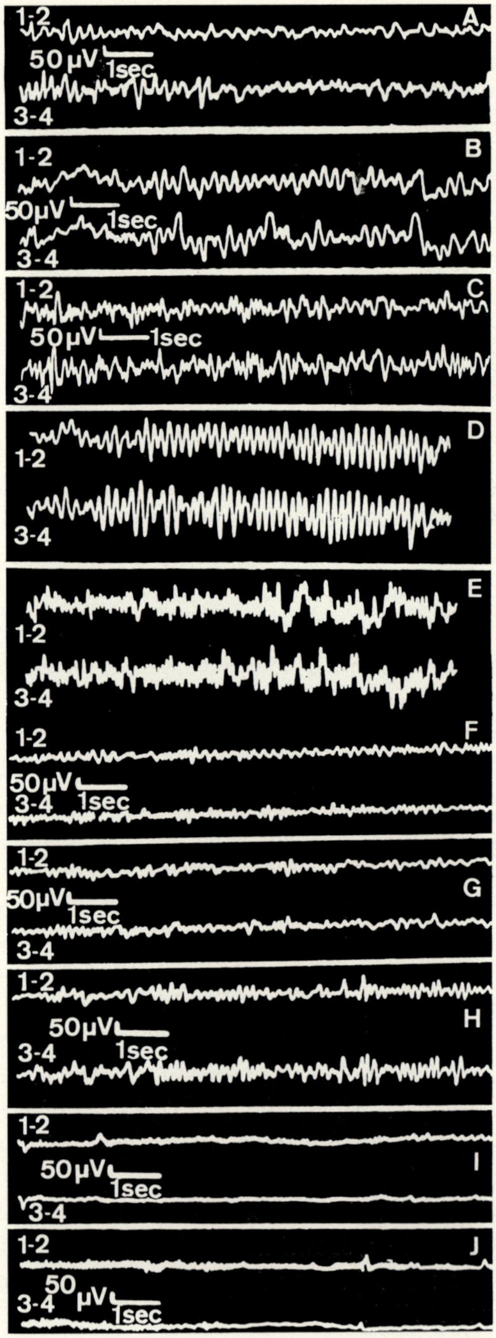
1-2
A
50 µV
1sec
3-4
1-2
B
50µV
1sec
3-4
1-2
C
50µV
1sec
3-4
D
1-2
3-4
E
1-2
3-4
1-2
F
50µV
3-4
1sec
1-2
50µV
G
1sec
3-4
1-2
H
50µV
3-4
1sec
1-2
50µV
I
1sec
3-4
1-2
J
50 µV
3-4
1sec

The examination of the foetal EEG in *M. mulatta* revealed spontaneous electric activity of the brain at 112 days of gestation (the full term being 168 days). This activity revealed irregular slow waves (0.5–1.0 per sec) of a low amplitude (8–16 μV). More frequent waves were superimposed. A burst of waves of 18 per sec frequency appeared as a result of fetal movements occurring at 125 days of gestation (fetal movement artifact). In other respects the basic activity of the foetus EEG in *M. mulatta* was the same during the whole gestation term. A marked change took place in the spontaneous EEG when the first gasp occurred. It resulted in an increase of both wave frequency and amplitude [DE GALLARDO *et al.*, 58].

A study of the electrical activity of the cortex and subcortical stem formations by means of implanted electrodes was carried out by URMANCHEEVA [177] in *M. mulatta* and *P. hamadryas* during the first year of life (fig. 10). This study showed that the fluctuations of potentials in deep structures of the brain during the first weeks of life were represented by irregular waves of varying duration. Waves of a frequency 2–4, 5–6, 8–12 and 18–24 per sec were recorded in the deep structures of the brain as in the EEG. Their amplitude usually ranged between 50–75 μV and was periodically increased to 100–150 μV in certain wave groups of short duration (fig. 11).

The picture of ESCoG in different subcortical stem structures is homogeneous and similar to ECoG of the front cortex areas of the brain.

In the ECoG of the occipital cortex the amplitude of waves was greater and the frequent components were more regular than in the deep structures of the brain. The prevalence of regular waves of a frequency 5–6 or 6–7 per sec was noted in infants of 3 months of age during wakefulness. The waves were modulated according to the amplitude and synchronized in all the leads. These peculiarities were connected with the reduction of the functional lability of cortical neurons (difficulties in revealing induced potentials and driving response to flashing light). A gradual reduction of the amplitude of waves was observed in infants from 5–6 months of age. The regularity of waves remained high. The ESCoG frequency-amplitude characteristics in 1-year-old *M. mulatta* did not differ significantly from that of adult animals.

Fig. 9. EEG in (A) *M. mulatta,* (B) *P. hamadryas* and (C) *C. aethiops* at 1 month of age EEG in (D) *P. hamadryas* and (E) *C. aethiops* at 7 months of age. EEG in (F) *P. hamadryas,* (G) *M. mulatta* and (H) *C. aethiops* at 2 years of age, and ECG in (I) *M. mulatta* and (J) *P. hamadryas* at 21 years of age. According to URMANCHEEVA and CHERKOVICH.

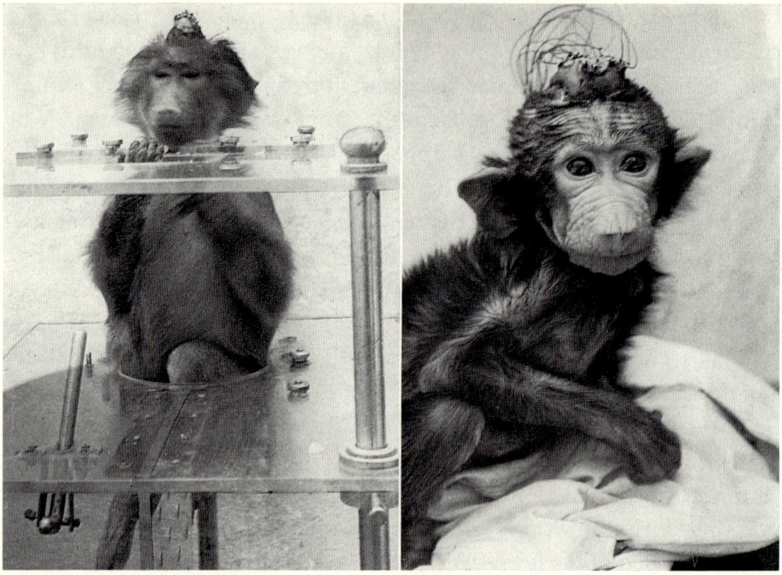

Fig. 10. A – an adult *P. hamadryas* and B – an infant with electrodes implanted in different structures. According to URMANCHEEVA.

Similar waves were observed in subcortical structures in adult *M. mulatta* (fig. 12). However, these waves were irregular and their amplitude characteristics recorded in *M. mulatta* and man by means of implanted electrodes [URMANCHEEVA, DIAKONOVA, 176] did not reveal any differences.

The electrophysiologic study in *M. mulatta* revealed significant peculiarities in electrographic responses of the hippocampus. In contrast to other mammals (rabbits, cats), which in response to sensory stimuli in the hippocampus showed a regular θ rhythm (5–7 per sec), the hippocampus of *M. mulatta* under the same conditions showed desynchronization response as it was usually observed in the neocortex [URMANCHEEVA, 178]. Similar data were given for saimiri [GERGEN, 60, 61], chimpanzee [RHODES *et al.*, 139] and man [PAGNIC and MARASSERO, 130]. In *M. mulatta* and *P. hamadryas* as in man, the propagation of hippocampus seizure activity (during hippocampal stimulation) to other brain areas is more restricted than in other animals [URMANCHEEVA, 178, 179; HASABOV and URMANCHEEVA, 76].

BERT and COLLOMB [19, 20] and FISCHER-WILLIAMS *et al.* [50] obtained a high regularity of waves of 5–7 and 8–10 cycles per sec in baboons ECoG.

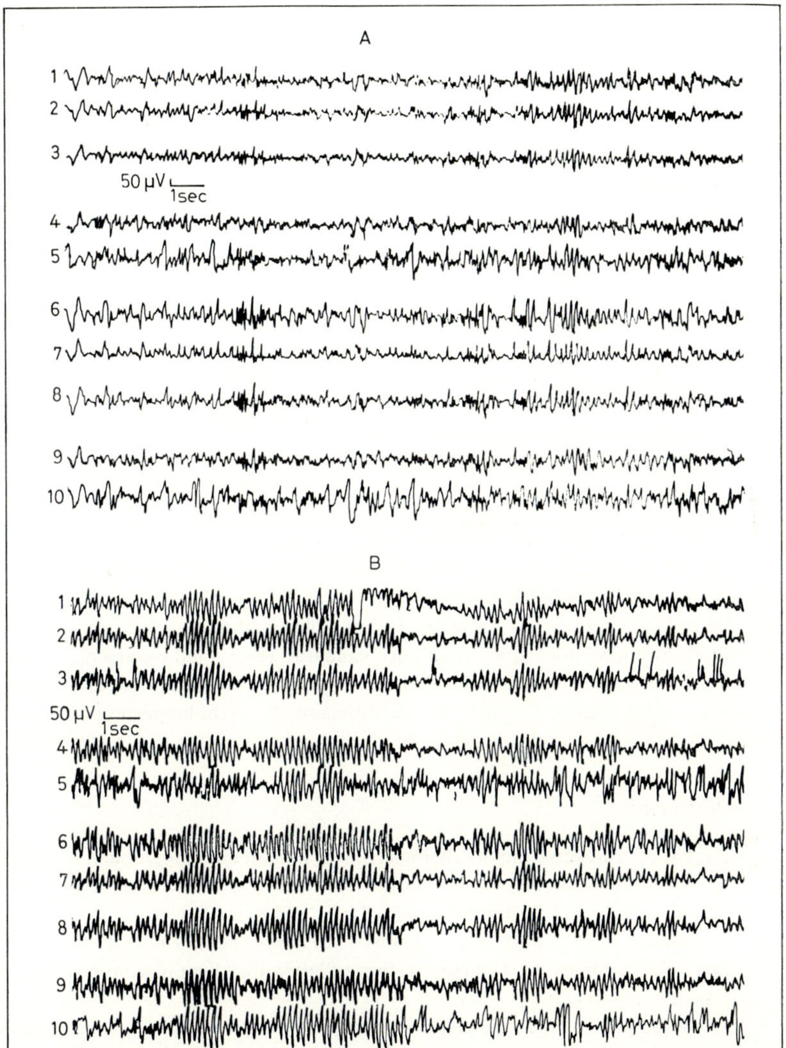

Fig. 11. ECoG and ESCoG in an infant *M. mulatta* (A) at 4 weeks of age and (B) at 13 weeks of age. During the recording, infant sits quietly with eyes open. Channel designation: 1 – putamen. 2 – thalamus, 3 – left hippocampus, 4 – cortex of the gyrus centralis post, 5 – occipital cortex, 6 – caput nuclei caudati, 7 – right hippocampus, 8 – globus pallidus, 9 – cortex of the gyrus centralis post, 10 – occipital cortex. According to URMANCHEEVA.

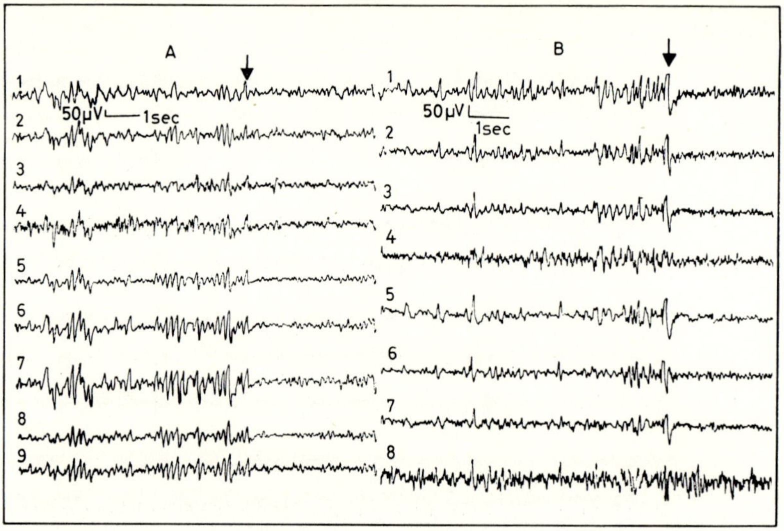

Fig. 12. Electrical activity of different brain structures in *M. mulatta* at rest and during orienting reaction. Channel designation: (A) 1 – left cortex parietalis, 2 and 3 – gyrus centralis post, 4 – right cortex parietalis, 5 – left globus pallidus, 6 – left nucl. caudatus, 7 – left hippocampus, 8 – fornix, 9 – right nucl. caudatus. (B) 1 – left putamen, 2 – corpus n. laudati (left), 3 – nucl. ventrolateralis thalami, 4 – occipital cortex, 5 – caput n. caudati, 6 – corpus n. caudati (right), 7 – nucl. reticularis thalami, 8 – right hippocampus. The switchcng on of the sound stimulus is designated by arrows. According to URMANCHEEVA.

References

1 ALEPA, P.: Antigenic factors characteristic of human immunoglobulin G, detected in the sera of non-human primates. Primates in Medicine *1:* 1 (Karger, Basel 1968).

2 ALTMAN, R. L. and DITTMER, D.: cites STAHL and MALINOW.

3 ANDERSON, E. T.; LEWIS, J. P.; POSSOVOY, M., and TROBAUGH, F. E.: Marmosets as laboratory animals. II. The Hematology of laboratory kept marmosets. Lab. Anim. Care *17:* 30–40 (1967).

4 ANNENKOV, G. A.: The content of certain terminal products of nitrogen metabolism in blood and urine of *M. mulatta* and *Papio hamadryas*. Vopr. physiol. exp. pathol. i radiobiol., Sukhumi, pp. 99–103 (1965).

5 ANNENKOV, G. A.: Serum protein spectra of certain monkeys and man. Diss., Sukhumi (1967).

6 ANNENKOV, G. A.: Immunophoretic spectra of blood serum of lower monkeys. The comparison of protein spectra of some monkeys and man. Vopr. Antropol. *32:* 153 (1969).

7 ARHANGELSKAJA, N. A.: cites SLONIM (1952).

8 ASATIANI, V. S.: Comparative characteristics of mineral, nitrogen and lipid contents in human and monkey blood. Soobshch. AN Grus., SSR *20:* 423–427 (1958).

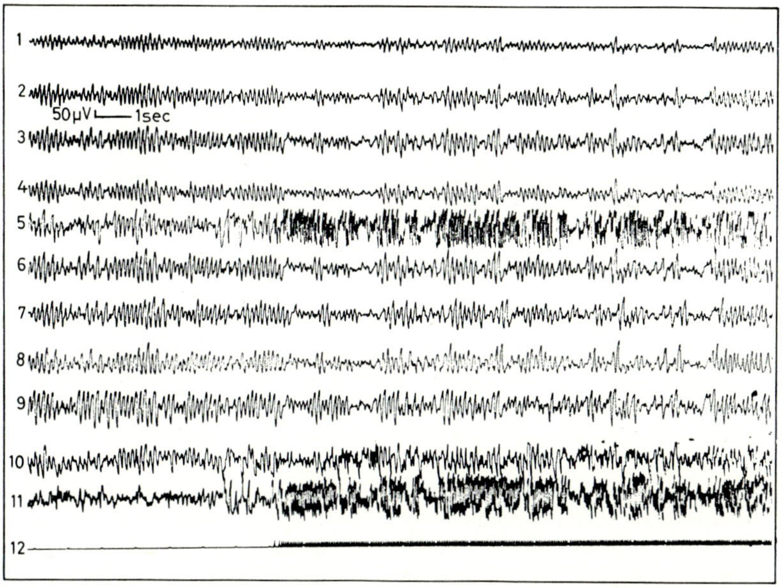

Fig. 13. Background electrical activity and driving response to a flashing light in the waking *P. hamadryas.* Channel designation: 1 – left putamen, 2 – left globus pallidus, 3 – left hippocampus, 4 – left area prefrontalis, 5 – occipital cortex, 6 – right putamen, 7 – right globus pallidus, 8 – right hippocampus, 9 – right area prefrontalis, 10 – gyrus centralis anterior, 11 – left occipital cortex. 12 – light flashes (20 per sec.).

9 ASATIANI, V. S.; PYCHKAYA, T. P., and AGIEVA, A. K.: Certain indices of blood chemistry in lower monkeys. Bull. exp. Biol. i. Med. *2:* 66–73 (1959).

10 ATTA, A. G. and VANACE, P. W.: Electrocardiographic studies in the *Macaca mulatta* monkey. Ann. N.Y. Acad. Sci. *85:* 811–818 (1960).

11 BANNERJEE, C. M.; WOOLARD, M., and ALARIE, Y.: Respiratory gas tensions and pH of arterial blood in cynomolgus monkeys *(M. irus).* Toxicol. appl. Pharmacol. *12:* 299–300 (1968).

12 BARNICOT, N.; JOLLY, C. Y.; HUEHUS, E. K., and DANCE, N.: Red cell and serum protein variants in baboons. First International Symposium on the Baboon and its use as an Experimental animal. San Antonio, pp. 323–338 (1963).

13 BEHRMAN, K. E.: *In vivo* oxygen dissociation curves for maternal and fetal rhesus monkeys. J. appl. Physiol. *25:* 224–229 (1968).

14 BEISCHER, D. E. and KNEPTON, J. C.: Influence of strong magnetic fields on the ECG of squirrel monkeys *(Saimiri sciureus).* Aerospace Med. *35:* 939–944 (1964).

15 BELL, N.; BARRETT, R., and PATTERSON, R.: Effects of porcine thyrocalcitonin on serum calcium, phosphorus and magnesium in monkey and man. Proc. Soc. exp. Biol. Med., N.Y. *123:* 114–118 (1966).

16 BENDER, M. A.: Blood volume of the rhesus monkey. Science *3158:* 156–156 (1955).

17 Berghe, L. van den et Blitstein, J.: Etudes hématologiques chez des singes. I. Compositions cytologique du sang. Ann. Soc. belge Méd. trop. *33:* 709–729 (1953).

18 Berghe, L. van den et Blitstein, J.: Critique des méthodes d'examen de la moelle osseuse. Rev. belge Path. *23:* 289–310 (1954).

19 Bert, J. et Collomb, H.: Diversité de l'électroencéphalogramme des primates cynomorphes. Electroenceph. clin. Neurophysiol. *17:* 485–616 (1964).

20 Bert, J. et Collomb, H.: Effet de l'immobilisation sur l'EEG du babouin. Comparaison avec les résultats obtenus par télémétrie chez l'animal en liberté. Comp. rend. Soc. biol. *159:* 1202–1204 (1965).

21 Borutau, H.: Das Affen-Electrocardiogramm. Zbl. Physiol. *32:* 1–3 (1918).

22 Bossak, E.; Wang, Ch.-I., and Adlersberg, D.: Comparative studies of Lipoproteins in various species by paper electrophoresis. Proc. Soc. exp. Biol. Med., N.Y. *87:* 637–643 (1964).

23 Bowen, N. H.; Eastoe, J. E., and Morris, G. C. R.: The parotid saliva of *M. irus.* Proc. of the Physiol. Soc. *171:* 16–17 (1964).

24 Bristow, J. D. and Malinow, M. R.: Spontaneous bundle branch block in rhesus. Circulat. Res. *16:* 210–220 (1965).

25 Brodie, D. A. and Marshall, R. W.: Gastric content of fasted primates: a survey. Science *141:* 174–175 (1963).

26 Brodie, D. A. and Marshall, R. W.: Fasting gastric content of the squirrel monkey. Amer. J. Physiol. *204:* 681–685 (1963).

27 Brooks, F. P.; Ridley, F.; Attinger, F., and Neff, K.: Effect of restraint on fasting gastric content of spider monkeys. Amer. J. Physiol. *205:* 1093–1095 (1963).

28 Bruhn, J. M.: The respiratory metabolism of infra-human primates. Amer. J. Physiol. *110:* 477–484 (1935).

29 Burns, K.; Fergusson, F., and Hampton, S.: Compendium of normal blood values for baboons, chimpanzees and marmosets. Amer. J. clin. Path. *48:* 484–494 (1967).

30 Butnev, Y. P.: Concerning the dynamics of the biological norms in monkeys. Biul. exp. Biol. i Med., Moscow *9:* 27–30 (1957).

31 Butnev, Y. P.: Chemical composition of blood and its dynamics in monkeys. Vop. Physiol. i Pathol. obez., Sukhumi 9–14 (1961).

32 Butnev, Y. P.: 24-h and seasonal variations in the biochemical blood indices in lower monkeys. Biul. exp. Biol. i Med., Moscow *1:* 51 (1962).

33 Caveness, W. F.: Atlas of electroencephalography in the developing monkey (Addison-Wosley 1962).

34 Cherkovich, G. M.: Experimental study of the diurnal rhythm of certain physiological functions in monkeys. Diss. Leningrad (1950).

35 Cherkovich, G. M.: Age-induced variations in frequency of the heart beat in normal monkeys and bradycardia with sinus arrhythmia as a result of experimental neurosis. Biul. exp. Biol. i Med., Moscow *8:* 46–49 (1960).

36 Cherkovich, G. M.: cites Lapin, Norkina, Cherkovich, Yakovleva, Kuksova, Alekseeva, Fufacheva and Startzev (1963).

37 Cherkovich, G. M. and Ruttkay-Nedecky, I.: Ontogenetic changes in the QRS of the baboon vectocardiogram. Physiol. Bohemoslovaca *16:* 264–271 (1967).

38 Cherkovich, G. M. and Usunian, L. A.: 'Serum cholesterol value in baboons and macaques, its seasonal variations and its relation to basal metabolic rate.' Zeitschr. für Versuchstierk. Vol. 9, pp. 236–242 (1967).

39 Cherkovich, G. M. und Ruttkay-Nedecky: Vectorcardiographic Study of the Unrestrained Monkey (Baboon and Macaque). Aktuelle Probleme der Vectorkardiographie, 143–145 (Thieme, Stuttgart 1968).

40 Cherkovich, G. M.: The velocity of the pulse wave distribution and arterial pjesography in adult Hamadryas baboons and rhesus monkeys. Zeitschr. für Versuchstierkunde Vol. *11:* 117–122 (1969).

41 Chez, R. A.; Smith, F. G., and Hutchinson, D. L.: Renal function in the intrauterine primate fetus. I. Experimental technique, rate of formation and chemical composition of urine. Amer. J. Obstet. Gynec. *90:* 128–131 (1964).

42 Circadian Rhythms in Non-Human Primates (Basel, New York 1969).

43 Cohen, B. S.: Bone marrow aspiration in the monkey *(Macacus rhesus)* Blood *8:* 661–663 (1953).

44 'Complication from literature of values for *S. sciurea.*' Univ. Washington, Regional Primate Center, Primate Inform. Center (1969).

45 Day, P.; Langston, W. C., and Shukers, C.: Leukopenia and anemia in the monkey resulting from vitamin deficiency. J. Nutr. *9:* 637–644 (1935).

46 Dreizen, S.; Goodrich, J. S., and Levy, B. M.: Comparison of salivary glands and salivary electrolytes in man and marmoset. Arch. oral Biol. *2:* 229–237 (1968).

47 Eberl-Rothe, G.: Die mikroskopische Anatomie der Primatenmilz. Primatologia. Handbuch der Primatenkunde, vol. 3, pp. 71–87 (1960).

48 Erikson, L. B.: Diurnal temperature variations in the rhesus monkey under normal and experimental conditions. Nature *186:* 83–84 (1960).

49 Feldman, A. Z.; Smith, W. K., and Leventhal, C. M.: Diurnal patterns of micturition and drinking in rhesus monkeys; in Care and disease of the research monkey. Ann. N.Y. Acad. Sci. *85:* 828–841 (1960).

50 Fischer-Williams, M.; Poncet, M.; Riche, D., and Naquet, R.: Light-induced epilepsy in the baboon, *Papio papio:* cortical and depth recordings. Electroenceph. clin. Neurophysiol. *25:* 557–569 (1968).

51 Fooden, J.: Urinary amino-acids of non-human primates. Zoologica (USA) *46:* 167–180 (1961).

52 Forsyth, R. P. and Baireuther, R.: Systemic arterial blood pressure and pulse rate in chronically restrained rhesus monkeys. Amer. J. Physiol. *212:* 1461–1463 (1967).

53 Forsyth, R. P.: Blood pressure and avoidance conditioning. Psychosom. Med. *30:* 125–135 (1968).

54 Forsyth, R. P.; Nies, A. S.; Wyler, F.; Neutze, J., and Melmon, K. L.: Normal distribution of cardiac output in the unanesthetized, restrained rhesus monkey. J. Appl. Physiol. *25:* 736–741 (1968).

55 Fox, H.: The blood of *M. rhesus.* Folia Haematol. *35:* 272–279 (1927).

56 Frensch, S. D.; Longmire, R. L.; Porter, R. W., and Movius, H. J.: Extra vagal influences on gastric hydrochloric acid secretion induced by stress stimuli. Surgery *34:* 621–632 (1953).

57 Fufacheva, A. A.: On the neural regulation of heart activity in monkeys. Dissert. Leningrad (1965).

58 Gallardo, F. O. E. de; Fleishman, R. W., and de Ramirez de Arellano, M. T. R.: EEG of the monkey fetus *in utero*, and changes in it at birth. Exp. Neurol. *9:* 73–84 (1964).

59 Gendon, U. Z.: Study of the relation of protein fraction content and antibody devel-

opment in monkeys immunized with poliomyelitis agent. Zh. Gigieny, epid., microbiol., immunol. (USSR) *5:* 499–507 (1961).

60 GERGEN, J. A. and MACLEAN, P. D.: Hippocampal seizures in squirrel monkey. Electroenceph. clin. Neurophysiol. *13:* 316–317 (1961).

61 GERGEN, J. A.: Functional properties of the hippocampus in the subhuman primate. Progress in Brain Res. *27:* 442–461 (1967).

62 GOODMAN, M.; POULIK, E., and POULIK, M.: Variations in the serum specificities of higher primates detected by two dimensional starch-gel electrophoresis. Nature *188:* 78–79 (1960).

63 GOODMAN, M.; KULKARN, A.; POULIK, E., and REKEYS, E.: Species and geographic differences in the transferrin polymorphism of macaques. Science *147:* 884–886 (1965).

64 GOODMAN, M.: Evolution of the catarrhine primates at the macromolecular level. Primates in Medicine *1:* 10 (Karger, Basel/New York 1968).

65 GOPALAN, C. and PADMAVATHI, S.: Electrocardiographic studies on normal and malnourished monkeys *(Macacus radiata)*. Indian J. med. Res. *45:* 423–430 (1957).

66 GRAYBIEL, A.; HOLMES, R. H. *et al.:* An account of experiments during which two monkeys returned to Earth safe and sound after a flight in a space ballistic rocket. Vopr. kosm. med., pp. 243–322 (a collection of articles by foreign authors) (Moscow 1962).

67 GREGERSEN, M. J.; SEAR, H.; RAWSON, R. A.; CHIEN SHU, and SAIGER, G. L.: Cell volume, plasma volume, total blood volume and F cells factor in the rhesus monkeys. Amer. J. Physiol. *196:* 184–187 (1959).

68 GRIGOROVA, O. P. and NESTURH, M. F.: Blood morphology of lower catarrhine monkeys from Cercopithecinae subfamily. Proc. of Moscow Zoo *3:* 95–122 (1946).

69 GROOVER, M. E.: Myocardial necrosis without atherosclerosis and produced by vagus stimulation in the wild Kenya Baboon *(P. anubis, P. doguera)*. Proc. of the 1st Symp. on the Baboon and its use as an Exp. anim.(Univ. Press, San Antonio 1965).

70 GROOVER, M. E.; SELJESKOG, E.L.; HAGLIN, J. J., and HITCHCOCK, C. R.: The relationship of coronary disease to myocardial infarction in the baboon. 1st Int. Symp. on the baboon and its use as an exp. anim. (Univ. Press, San Antonio 1965).

71 GUENTER, C. A.; MCCAFFREE, D. R.; DAVIS, L. J., and SMITH, V. S.: Hemodynamic characteristics and blood gas exchange in the normal baboon. J. Appl. Physiol. *25:* 507–510 (1968).

72 HALL, A. S. and KNEZEVIC, A.: Bone marrow sampling in monkeys. J. Amer. vet. med. Ass. *147:* 1075–1076 (1965).

73 HAMLIN, R. L.; ROBINSON, F. R., and SMITH, C. R.: ECG and Vectorcardiogram of *M. mulatta* in various postures. Amer. J. Physiol. *201:* 1083 (1961).

74 HAMLIN, R. L.; ROBINSON, F. R.; SMITH, C. R., and MARSLAND, W. P.: Heart sounds of healthy *M. mulatta*. J. appl. Physiol. *17:* 199–200 (1962).

75 HAMPTON, J. W. and MATTHEWS, C.: Similarities between baboon and human blood clotting. J. appl. Physiol. *21:* 1713–1716 (1966).

76 HASABOV, G. A. and URMANCHEEVA, T. G.: Propagation of seizure activity from neocortex and hippocampus in rabbits and monkeys. Evolut. neurophys. neurochem. Leningrad 76–88 (1967).

77 HAWKEY, CH. and SYMONS, C.: Preliminary report of studies on platelet aggregation, blood coagulation and fibrinolysis in non-human primates. Some recent development in compar. med. 213–222 (1966).

78 HAYDEN, R. O.: cites STAHL and MALINOW.

79 HERRMAN, G. R. and WILLIAMS, A. H.: The ECG patterns in 170 baboons in the domestic and African colonies at the Primate Center of the Southwest Foundation for Research and Education; in VAGRBORG The baboon in medical research. Proc. of the first Symp. on the baboon and its use as an exp. animal, pp. 251–264 (University of Texas Press, San Antonio 1965).

80 HIDEO, NIGI: Circulating blood volume of the Japanese monkey *(Macaca fuscata)* by means of the dye T1824. Primates *8:* 23–28 (1967).

81 HIRSCHFELD, J.: The use of immunoelectrophoresis in the analysis of normal sera and in studies of the inheritance of certain serum proteins. Sci. Tools *8:* 17 (1962).

82 HIRSCHFELD, J.: The Gc-system. Immunoelectrophoretic studies of normal human sera with special reference to a new genetically determined serum system (Gc). Progr. Allergy *6:* 155–186 (Karger, Basel/New York 1962).

83 HOLMES, A.; POSSOVOY, M., and CAPPS, R.: Marmosets as laboratory animals. III. Lab. Anim. Care *17:* 41–47 (1967).

84 HONJO, S.; FUJIWARA, T.; TAKASAKA, M.; SUZUKI, Y., and IMAIZUMI, K.: Observations on the diurnal temperature variation of cynomolgus monkeys *(Macaca irus)* and on the effect of changes in the routine lighting upon this variation. Japanese J. Med. Sci. Biol. *16:* 189–198 (1963).

85 HUSER, H.-J. and BEARD, M. E. J.: Studies on folate and vitamin B_{12} metabolism in primates. I. Blood and bone marrow morphology, folate and vitamin B_{12} levels. Folia Primat. *10:* 172–180 (1969).

86 INOYE, M.; ITAKURA, C.; TAKEMURA, N., and HAYAMA, S.: Peripheral blood of wild Japanese monkeys *(Macaca fuscata* and *M. Jyakui)*. Primates *5:* 75–112 (1964).

87 JOHNSTON, G. S.; HEERDEN, P. D.; VAN ZYL, J. A.; VAN ZYL, J. J. W.; RETIEF, C. P., and MURPHY, G. P.: Renal function studies in the intact baboon. Invest. Urol. *6:* 125–128 (1968).

88 KAMINER, B.: ECG of the baboon *Papio ursinus* with a note on the anatomy of the heart. S. Afr. J. Med. Sci. *23:* 231–240 (1958).

89 KAREL, L. and WESTON, R. E.: Respiration in *Macaca mulatta.* Proc. Soc. exp. Biol., N.Y. *61:* 291–296 (1946).

90 KING, T. and GARGUS, J.: Normal blood values of the adult female monkey *(M. mulatta)*. Lab. Animal Care *17:* 391 (1967).

91 KOKAYA, G. I.: The ECG of monkeys in normal and pathologic states. Biul. exp. Biol. i Med., Moscow *12:* 23–26 (1954).

92 KOKAYA, G. I.: The ECG of healthy monkeys of different species and age. Dissert. (Sukhumi 1958).

93 KOSTUCHENKO, L. V.: Calcium content in the blood of cattarrhine monkeys in comparison with other mammals. Vopr. Antropol. *24:* 137–142 (1966).

94 KRISE, G. M.: Hematology of the normal monkey. Ann. N.Y. Acad. Sci. *85:* 803–810 (1960).

95 KUKSOVA, M. I.: Seasonal and daily fluctuations of red blood in monkeys. Theoretical problems of medicine and biology in experiments on monkeys, pp. 98–106 (Moscow 1956).

96 KUKSOVA, M. I.: Concerning redistribution of the blood in the blood vessels of lower monkeys. Biul. exp. Biol. i Med., Moscow *45:* 31–35 (1958).

97 KUKSOVA, M. I.: Ontogenetic development of blood picture in rhesus monkeys and *Papio hamadryas*. Sov. Antropol. *2:* 35–53 (1958).

98 KUKSOVA, M. I.: Seasonal and daily fluctuations on white blood in lower monkeys. Zh. Obshch. Biol. *1;* 73–80 (1959).

99 KUKSOVA, M. I.: Investigation of the cellular structure of the bone marrow in monkeys in repeated punctures. Biul. exp. Biol. i Med., Moscow *8:* 112–115 (1960).

100 KUKSOVA, M. I. and DIKOVENKO, E. A.: The leucocyte reaction dynamics in lower monkeys. Vop. Physiol. i Patologii Obesjan, pp. 25–34 (Sukhumi 1961).

101 KUKSOVA, M. I.; VOLKOVA, L. S.; VERBITSKIY, M. SH., and ANDREEVA, A. V.: Peculiarities of blood formation in healthy infants *Papio hamadryas* in the first year of their life. Vop. Antrop., Iss. 30.M. (1968).

102 LAGUTINA, N. I.: Appraisal of unconditioned and of natural conditioned salivation responses in a *Macaca rhesus monkey*. Physiologich. Zh., SSSR *45:* 415–422 (1959).

103 LAPIN, B. A.; NORKINA, L. N.; CHERKOVICH, G. M.; YAKOVLEVA, L. A.; KUKSOVA, M. I.; ALEKSEEVA, L. V.; FUFACHEVA, A. A., and STARTSEV, V. G.: The monkey as an object of biomedical experiments (Sukhumi 1963).

104 LAPIN, B. A.; DZHIKIDZE, E. K., and YAKOVLEVA, L. A.: Monkeys diseases in natural conditions of Viet-Nam (Moscow 1965).

105 LEES, M. H.; MALINOW, M. R., and PARER, J. T.: cite STAHL and MALINOW (1967).

106 LEPPI, T. J. and SPICER, S. S.: The histochemistry of mucin in certain salivary glands. Amer. J. Anat. *118:* 833–850 (1966).

107 LICHTER, E. and DRAY, S.: Immunoelectrophoretic characterization of human serum proteins with primate antisera. J. Immunol. *92:* 91–99 (1964).

108 LITWIN, S.: Phylogenetic differences among the Gm factors of non-human primates. Nature *216:* 268–269 (1967).

109 LOGHEM, E.; SHUSTER, Y., and FUDENLERG, H.: Gm factors in non-human primates. Vox. Sang. *14:* 80–94 (1968).

110 MADDOCK, S. and SVEDBERG, A.: The effect of the total removal of the liver of the monkey. Amer. J. Physiol. *121:* 203–208 (1938).

111 MAGAKYAN, G. O.: Studies on arterial pressure in monkeys. Proceed. of the second scientific session of the Institute of Clinical and Experimental Cardiology., 107–111. Acad. Sci. Georg. SSR (Tbilisi 1953).

112 MAHL, C. T.: Relationship between acute and chronic fear and the gastric acidity and blood sugar levels in *M. mulatta* monkey. Psychosom. Med. *14:* 182–210 (1952).

113 MAKARCHENKO, A. F.; BOITRUB, B. A., and ZLATIN, R. S.: Changes of protein macrostructure in central nervous system stimulation. DAN SSSR *160:* 731–733 (1965).

114 MALINOVSKY, C. V.: Unconditioned secretion of parotid gland in the rhesus monkey. Physiologich. Zh. SSSR *39:* 47–51 (1953).

115 MALINOW, M. R. and WAGNER, R.: Oxygen uptake in squirrel monkey *(Saimiri sciurea)*. Lab. Anim. Care *16:* 105–108 (1966).

116 MALINOW, M. R. and LANNOY C. W. DE: The ECG of *Cynopithecus niger*. Folia Primat. *4:* 66–73 (1966).

117 MALINOW, M. R.: An Electrocardiographic study of *Macaca mulatta*. Folia Primat. *4:* 51–65 (1966).

118 MEYER, J. S. and HANDA, J.: Cerebral blood flow and metabolism during experimental Hyperthermia (fever). Minnesota Medicine *56:* 37–44 (1967).

119 MIYAMOTO, T.; REYNOLDS, L. B.; PUTTERSON, R.; CUGELL, D. W., and KETTEL, L. J.: Respiratory changes in passively sensitized dogs and monkeys as models of allergic asthma. Amer. Rev. resp. Dis. *97:* 76–88 (1968).

120 MOOR-JANKOWSKI, J.; HUSER, H. J.; WIENER, A.; KALTER, S. S.; PALLOTTA, A. J., and GUTHRIE, C. B.: Hematology, blood groups, serum isoantigens, and preservation of blood of the baboon. The baboon in medical research. Proc. of the 1st Intern. Symp. on the baboon and its use as an experimental animal, 363–405 (Austin 1965).

121 MORRISON, P. and SIMOES, J.: Body temperature in two brazilian primates. Bd. Fac. filos. ciênce letras. Univ. S. Paulo Zool. *24:* 167–177 (1962).

122 MORROW, A. C. and TERRY, M. W.: Urine volume in non-human primates. A tabulation from the literature. Seattle Primate Information centre (1968).

123 MYERS, J. H. and BLACKWELL, L. H.: Cardiovascular physiology and fluid volume studies in the baboon. 2nd Intern. Congr. of Primatology (Abstr. of scientific papers). Yerkes Regional Primate Research Center Emory Univ. Atlanta, Georgia (1968).

124 NARASIMHAN, M. J. and GANLA: The influence of the submandibular salivary gland on growth. Curr. Sci. *35:* 542–543 (1966).

125 NIGI, H.; TANAKA, T., and NOGUCHI, Y.: Hematological analyses of the Japanese monkey *(Macaca fuscata)*. Primates *8:* 107–120 (1967).

126 O'DELL, R. M.; RADWIN, H. M.; BERNSTEIN, L. H., and SCHEGEL, J. H.: Renal function in *Macaca speciosa*. J. appl. Physiol. *24:* 366–368 (1968).

127 ORLOVA, S. D.: The role of nervous-humoral factors in digestive leucocytosis. Arkhiv Biol. Nauk *44:* 123–139 (1937).

128 OVERMAN, R. P. and FELDMAN, H. A.: Circulatory and fluid compartment physiology in the normal monkey with especial reference to seasonal variations. Amer. J. Physiol. *148:* 455–459 (1947).

129 PACE, N.; HANSEN, L. T.; REHLMANN, D. F.; BARNSTEIN, N. J., and CANNON, M. D.: Preliminary observations of some physiological characteristics of the pigtailed monkey, *M. nemestrina.* Aerospace Med. *35:* 118–121 (1964).

130 PAGNIC, A. and MAROSSERO, F.: Some observations on the human rhinencephalon. A stereoelectroencephalographic study. Electroenceph. Clin. Neurophys. *18:* 260–271 (1965).

131 PALUSKA, E. and KORINEK, J.: Studium antigenni pribuznost krevnich proteini cloveka a nekterych primatu novymi metodami immunobiologickymi, i kvalitativni rozbor vysledku, ziskany testem dvijite difuze v agarovem gelu. Cs. Epidem. *9:* 185–195 (1960).

132 PARKER, W. and BEARN, A.: Studies on the transferrins of adult serum, cord serum, and cerebrospinal fluid. J. exp. Med. *115:* 83–105 (1962).

133 PICARD, P.: Les proteines sériques de *Macacus irus* (Cynomolgus). Rev. belge Path. *27:* 321 (1960).

134 PICKERING, D. E. and SUSSMAN, H. H.: Renal function in monkeys *(M. mulatta)*. Amer. J. vet. Res. *23:* 667–672 (1962).

135 PONDER, E. J.; GEAGER, F., and CHARIPPER, H. A.: Haematology of the primates. Zoologica *11:* 9–18 (1929).

136 PRICHOD'KO, W.; GOODMAN, M.; POULIK, E.; MIKI, T., and TANAKA, T.: Geographic variations of transferrin allelic frequencies in continental and insular macaque populations. II. Int. Congress of primatology Abstr. of Sci. Papers, Emory Univ. Atlanta 29 (1968).

137 DE RAMINEZ DE ARELLANO, R. M. Y.: Maturational changes in the electroencephalogram of normal monkeys. Exp. Neurol. *3:* 209–224 (1961).

138 REIS, D. J. and CLIPHANT, M. C.: Bradycardia and tachycardia following electrical stimulation of the amygdaloid region in monkey. J. Neurophysiol. *27:* 893–912 (1964).

139 RHODES, J. M.; ADEY, W. R.; KADO, R. T., and REITE, M. L.: Different EEG sleep stages and their response to tone stimulation in the chimp. Physiologist *6:* 264 (1963).

140 RINEHART, J. and GREENBERG, L. D.: Vitamin B_6 deficiency in the Rhesus monkey with particular reference to the occurrence of atherosclerosis, dental caries and hepatic cirrhosis. Amer. J. clin. Nutr. *4:* 318–328 (1956).

141 RITTNER, C.: Zum Nachweis der B-Lipoprotein-Polymorphismen Ag und Lp bei subhumanen Primaten. Humangenetic *2:* 219 (1966).

142 RUSKIN, A. and RIGDON, R. H.: The electrocardiogram of normal and malaria infected monkeys. J. Lab. clin. Med. *34:* 1105–1108 (1949).

143 RUTTKAY-NEDECKY, T. und TSCHERKOWITSCH, G. M.: Topographie des Proximitatseffektes im kardioelektrischen Felde beim Affen *(Macaca mulatta)*. Neue Ergebnisse der Elektrokardiologie (VEB Gustav Fischer Verlag, Jena 1965).

144 SARICH, V. and WILSON, A.: Quantitative immunochemistry and the evolution of primate albumins: microcomplement fixation. Science *3576:* 1563–1566 (1966).

145 SCHNEDORF, J. G. and IVY, A. C.: The incidence and permanence of unexplained gastric anacidity in the rhesus monkey after histamine and mecholyl, with hematologic studies. Amer. J. digest Dis. *4:* 423–432 (1937).

146 SEAMAN, A. J. and MALINOW, M. R.: Blood clotting in non-human primates. Lab. Anim. Care *18:* 80–84 (1968).

147 SEHGAL, A. K.; CHAKRAVARTI, R. N.; MYSOREKAR, N. R., and CHHUTTANI, P. N.: Intestinal villi in rhesus monkeys. Nature *210:* 972–973 (1966).

148 SELKURT, E. E. and WATHEN, R. L.: Renal concentrating mechanism of the squirref monkey. Amer. J. Physiol. *213:* 191–197 (1967).

149 SHAPOSHNIKOV, A. I.; MUKHA, C. V., and ZDRODOVSKAYA, E. P.: On the similarity of properties of ceruloplasmin in monkey and man. Biochemistry *32:* 908–912 (1967).

150 SHCHERBAKOVA, O. P.: Diurnal rhythm of physiological functions in certain orders ol mammals; The experience of the investigation of periodic changes in the physiological functions of the organism, 15–41 (Moscow 1949).

151 SHESTOPALOVA, S. K.: Unconditioned reflectory salivation in *Papio hamadryas*. Physiol. Zh. SSSR *50:* 812–819 (1964).

152 SKEITH, M. D. and HEALEY, L. A.: Urate clearance in Cebus monkeys. Amer. J. Physiol. *214:* 582–584 (1968).

153 SLONIM, A. D. and SHCHERBAKOVA, O. P.: Basal metabolism in monkeys. I. Respiratory gaseous exchange and basal metabolism in *P. hamadryas*. Proceed. of the Sukhumi biological station, Moscow, 5–17 (1949).

154 SLONIM, A. D. and KALMIKOVA, K. P.: Basal metabolism in monkeys. II. Basal metabolism in macaques. Proceed. of the Sukhumi biological station, Moscow 18–21 (1949).

155 SLONIM, A. D. and SHCHERBAKOVA, O. P.: Observations on the night sleep of monkeys. The experience of the investigation of periodic changes in the physiological functions of the organism, Moscow 156 (1949).

156 SLONIM, A. D.: Animal heat and its regulation in the mammalian organism (Moscow-Leningrad 1952).

157 SMITH, C. C. and CROUT, D.: Arterial blood pressure in *M. speciosa*. Lab. Primate Newsletter *3:* 1–2 (1964).

158 SMITH, H. W. and CLARKE, R. W.: The excretion of inulin and creatinine by the anthropoid apes and other infrahuman primates. Amer. J. Physiol. *122:* 132–139 (1938).

159 SMITH, C. P.; BROOKS, F. P.; DAVIS, R. A., and ROTHMAN, S. S.: Fasting gastric contents in the spider monkey. Amer. J. Physiol. *199:* 889–892 (1960).

160 SMITH, C. P.; MASON, J. W., and JACOBSON, E. D.: Fasting gastric content in conscious Macaca. Amer. J. Physiol. *24:* 629–633 (1966).

161 SOLOPAEV, B. P.; ANNENKOV, G. A.; SOLOPAEVA, I. M., and SOLOVIOVA, G. A.: Experimental study of the reversibility of liver pathological changes in monkey. Proc. of the Inst. of surgery named after A. V. Vishnevski. USSR Acad. Med. Sci., Moscow 1963.

162 SPECTOR, W. S.: cites STAHL and MALINOW (1956).

163 SRIKANTIA, S. and GOPALAN, C.: Some biochemical and physiological features of normal monkeys *(Macaca radiata)*. J. appl. Physiol. *18:* 1231–1233 (1963).

164 STAHL, W. R. and MALINOW, M. R.: A survey of physiological measurements in *M. mulatta*. Folia primat. *7:* 12–33 (1967).

165 STARTSEV, V. G.: On the natural food conditioned gastric reflexes in monkeys. Physiolog. i Patolog. VND obezjan, Sukhumi 71–82 (1960).

166 STARTSEV, V. G.: cites LAPIN, NORKINA, CHERKOVICH, YAKOVLEVA, KUKSOVA, ALEKSEEVA, FUFACHEVA and STARTSEV 1963.

167 STARTSEV, V. G.: Role of central nervous mechanism in chronic impairment of gastric secretory functional achylia in monkeys. Physiologich. Zh. SSSR *51:* 243–250 (1965).

168 SWEET, A. Y.; LEVITT, M. R., and HODES, H. L.: Kidney function, body fluid compartments and water and electrolyte metabolism in the monkey. Amer. J. Physiol. *201:* 975–979 (1961).

169 SWITZER, J. W.: Bone marrow composition in the adult rhesus monkey *(Macaca mulatta)*. J. amer. vet. med. Ass. *151:* 823–828 (1967).

170 SUAREZ, R. M.; DIAZ-RIVERA, R. S., and HERNANDEZ-MORALES, F.: Aspirated bone marrow studies in normal macacus rhesus monkeys. Amer. J. med. Sci. *205:* 581–586 (1943).

171 TATOYAN, S. H.: Normal ECG picture (in the front Nehb lead) in baboons and macaques according to the data of the telemetric investigation. Dissert., Moscow (1969).

172 TISCHER, C. C.; ROSEN, S., and TESCHAN, P.: Morphology of Henle's loop and its relation to urinary concentrating ability in monkey and man. Fed. Proc. *272:* 741–741 (1968).

173 UGOLEV, A. M.; VOLKOVA, E. M., and KORNEVITZ, G. V.: Natural food conditioned reflexes in monkeys. The experience of studies on the regulation of physiological functions under natural conditions *2:* 147–151 (Moscow-Leningrad 1953).

174 URMANCHEEVA, T. G. and CHERKOVICH, G. M.: Electrical activity of the brain cortex of monkeys in ontogenesis. Biul. exp. Biol. i Med., Moscow *12:* 3–7 (1962).

175 URMANCHEEVA, T. G. and CHERKOVICH, G. M.: Electrical activity of the hemisphere cortex in monkeys of different age at rhythmical light flashing. Physiologich. Zh. SSSR *57:* 431–440 (1963).

176 URMANCHEEVA, T. G. and DIAKONOVA, S. N.: Electrophysiological investigation of certain subcortical formations in humans with implanted electrodes. Physiolog. Zh. USSR *51:* 909–917 (1965).

177 URMANCHEEVA, T. G.: Electrographic patterns of development of cortical-subcortical interactions during ontogenesis in lower monkeys. Med. Primatology, Tbilisi 27–36 (1967).

178 URMANCHEEVA, T. G.: Electrophysiological character of hippocampus in mammals. 5th conference of evolut. physiol. in memory of acad. L. A. Orbeli. Leningrad, pp. 257–258 (1968).

179 URMANCHEEVA, T. G.: Interactions of hippocampus with neocortex and subcortical formations. Physiology and Pathology of the limbico-reticular complex, Moscow, pp. 31–32 (1968).

180 USACHEVA, I. N. and RAEVA, N. V.: Normal peripheral blood and bone marrow indices of monkeys *(Macacus rhesus)*. Biul. exp. Biol. i Med.. Moscow *11:* 106–108 (1962).

181 USPENSKY, S. S.: Hematology of *Papio hamadryas*. A collection of the students investigations (Moscow State University) *13:* 51–64 (1940).

182 UTKIN, I. A. and KUKSOVA, M. I.: Leucocyte count change during the repeated blood examination in monkeys in the same conditions. Dokladi Akad. Nauk SSSR *108:* 981–984 (1956).

183 CITTERS, R. L. VAN and LASRY, J. E.: Cardiovascular function in adult baboons as indicated by standard diagnostic tests. Folia primat. *3:* 13–21 (1965).

184 VICE, T. E.; BRITTON, H.A.; RATNER, I. A., and KALTER, S. S.: Care and raising of newborn baboons. Lab. Anim. Care *16:* 12 (1966).

185 WAART, A. DE and STORM, C. J.: Electrocardiographic observations on javanese monkeys. Arch. Neerland Physiol. *20:* 255–277 (1935).

186 WAHLE, H.: Beitrag zum Verhalten der Blutsenkung des Blutbildes und der Körpertemperatur beim gesunden nicht infizierten Rhesus-Affen. Folia haemat. *75:* 81–91 (1957).

187 WEIGEL, W. W.; HANSEN, H. A., and SMITH, C. C.: Application of isotopic clearance procedures in measuring renal function in rhesus monkeys. Fed. Proc. *25:* 198–198 (1966).

188 WHITE, R.; MCCARTY, C.; GRINDLAY, Y., and BOLLMAN, Y.: Hematocrit readings and values of plasma electrolytes in the rhesus monkey. Proc. Statt Mayo Clin. *35:* 114–117 (1960).

189 WIENER, A.; GORDON, E., and MOOR-JANKOWSKI, J.: Immunological relationship between serum globulins of man and of other primates, revealed by a serological inhibition test. Transfusion *4:* 347 (1964).

190 WILBER, CH. G.: Response of heart in two New World monkeys to hypothermia. Comp. Bioch. Physiol. *11:* 323–327 (1964).

191 WILLIAMS, C. and WEMYSS, C.: Experimental and evolutionary significance of similarities among serum protein antigens of man and the lower primates. Ann. N. Y. Acad. Sci. *94:* 77 (1961).

192 WILLMAN, V. L.; HANLON, C. R.; KAISER, G. C., and COOPER, T.: Chronic total extrinsic denervation of the heart in primates. The Physiologist *7:* 288 (1964).

Author's address: Prof. B. A. LAPIN, Institute of Experimental Pathology and Therapy, USSR Academy of Medical Sciences, *Sukhumi* (USSR)

II. Development, Growth, Nutrition and Metabolism

Pathology of Simian Primates, Part I, pp. 158–189 (Karger, Basel 1972)

Developmental Abnormalities

ADOLPH H. SCHULTZ

Anthropologisches Institut, Universität Zürich, Zürich

Contents

I.	Introduction	158
II.	Abnormal Development of Head and Skull	161
III.	Abnormal Development of Vertebral Column and Thorax	167
IV.	Abnormal Development of Limbs	171
V.	Abnormal Development of Dentition	177
VI.	Abnormal Development of other Features	184
VII.	Summary	185
References		186

I. Introduction

This chapter deals with the occurrence of manifestations of abnormal development in primates besides man. It must be stated at once that this remains a very incomplete and uneven survey on account of the scarcity of prenatal and infantile specimens in most collections in contrast to the far more extensive material of older age, which is rarely reached by individuals with severe defects. Furthermore, many authors have recorded single incidences of malformations without any indication of the number of normal specimens among which they were encountered. For these reasons it is often impossible to consider the relative frequency of abnormal development in different structures as well as species.

Congenital malformations are usually thought of as appearing already before birth, but the abnormal nature of some skeletal and dental conditions in particular, can be recognized only during their postnatal development. Inasmuch as normal development is properly defined as that which exists in the majority of cases, any abnormal development must really be exceptional for a given species. This seemingly platitudinous statement should be recalled whenever it is found that some obvious abnormalities of one species have become normal characters in certain other species, as is clearly the case in numerical reductions of nipples or of some digits and teeth. The often difficult or even arbitrary distinction between localized developmental abnormalities and extreme, *normal* variations can only be based on a knowledge of the typical variability of the structure under investigation in a particular species. For instance, cervical ribs in a primate are explained as resulting from abnormal development on account of their great rarity, whereas the lack of one or both ribs on the last thoracic vertebra is probably best interpreted as merely one extreme of a continuous series of variations in costal formation at the caudal border of the thorax, since such variations are very common in many primates. That there is at times no precise border between normal and abnormal, which can be clearly defined and consistently applied, is also shown by the example of the third molars which vary in man as well as some other primates through all degrees of degeneration to their complete absence. Whether the congenital lack of these teeth should be listed under 'developmental abnormalities' or merely as 'variations' depends on their prevailing tendency toward reduction, which differs widely among primates.

Severe malformations, leading to intrauterine or neonatal death in wild primates rarely become known, and even among captive ones such occurrences are not regularly investigated and recorded. Since developmental abnormalities are particularly frequent among miscarried fetuses of man, it seems most likely that the same is true in regard to nonhuman primates of which at least some species do show a similarly high prenatal mortality. For instance, according to HARTMAN [1] only about two out of every three conceptions survived beyond birth in a colony of captive macaques. In the far larger breeding colony of these monkeys at the Bionetics Research Laboratories at Bethesda it was more recently found that approximately 15% of females that conceive can be expected to abort or produce a still-birth [2]. YERKES [3] had listed 17 abortions and 5 still-births among the 66 pregnancies in his chimpanzee colony, or a mortality of one in three. According to the latest survey by GUILLOUD [4] 53 of a total of 170 conceptions among captive chimpanzees had died *in utero* (= 31%), 18 of which had reached less than 60 days. Relatively smaller losses have been reported from some other research institutes, such as the Russian primate center at Sukhumi where according to LAPIN and YAKOVLEVA [5] among 775 pregnancies of Old World monkeys only 9% ended in abortions and another 13% died at birth. Extremely low figures are given by PICKERING [6], who found merely 1.2% of abortions and 2.4% of still-births among 250 pregnancies in a carefully supervised colony of rhesus monkeys. KOFORD *et al.* [7] also report that among 1,003 births in free-ranging rhesus monkeys in Puerto Rico only 4% were represented by aborted and still-born fetuses. In general, however, prenatal deaths are probably not such rare exceptions in nonhuman primates and not often caused by traumatic accidents, but mostly due to some developmental disturbance.

That monkeys and apes in their natural habitat can at times manage to survive with even surprisingly profound malformations is evident from some examples to be mentioned here. The majority of the abnormal conditions, however, consists merely of localized defects without any immediate selective significance. Such cases of a single kind may be quite common in some species or local population of that species, while totally absent in others. The causes of malformations in wild primates are not known. Undoubtedly a great many represent mutations, but their precise hereditary nature is still obscure. At times there may be a connection with chromosomal abnormalities, of which some have actually been found in various monkeys [e.g. EGOZCUE *et al.*, 8]. A preliminary report has recently been published [9] on a captive-born chimpanzee with an extra chromosome and showing very marked physical and mental retardation as well as other symptoms of mongolism. It is also quite likely that some virus infections of mothers can lead to malformations in the offspring of wild as well as captive primates. Inasmuch as the experimental administration of thalidomide to monkeys during early stages of their pregnancy has been repeatedly demonstrated to produce various deformations in the fetus [e.g., BARROW *et al.*, 10; WILSON and FRADKIN, 11], it is not impossible that one or another chemical substance in plants eaten by wild primate mothers may interfere with the normal development of their descendants and be responsible for some kinds of birth defects. Vitamin deficiency especially of vitamin *A* during pregnancy is regarded by LAPIN and YAKOVLEVA [5] as a possible and perhaps even main cause of fetal abnormalities among captive monkeys. Finally, there is no reason to doubt that not only in man, but among all primates, delayed or faulty implantation and inadequate placental function can at times interfere with the delicate processes of embryonic differentiation in their normal sequence.

A great variety of conditions, equivalent to limited, minor malformations, has commonly been regarded as individual recurrences of ancestral characters, known as atavisms. This interpretation seems to be quite reasonable in such cases as supernumerary incisors or premolars, more than two mammary glands, metopism, or even hypertrichosis in higher primates, since they are conditions which resemble those normally existing in more primitive forms. Actually, however, all such supposedly atavistic occurrences, as well as innumerable other teratological conditions can be readily and more consistently explained as results of abnormal development, including retarded or arrested developmental change. Duplication in development seems to be especially prevalent and accounts not only for the occurrence of single-ovum twins, but also as limited twinning for 'double monsters', supernumerary teeth of all kinds, polydactylism, extra nipples, etc.

The accumulated literature on malformations in nonhuman primates has become so extensive that only selected examples can be considered here. Many years ago the writer [12] had attempted to survey the available information on teratological conditions in primates besides man and it is this former paper, together with many new data and references, which has formed the basis for the present revision.

II. Abnormal Development of Head and Skull

Comparatively few cases of major malformations of the head of nonhuman primates have become known, though more or less marked congenital defects of the skull are quite common in many species. Two instances of so-called double monsters, involving the head, have been found in dead-born macaques, the specimen of Bolk [13] representing a much more complete duplication than the one of Hartman [14], which is shown in figure 1. A case of anencephaly in one twin of *Macaca arctoides*, occurring among 22 births of these monkeys, has been reported by Christie [15]. Another macaque, which had lived five months at the time of its description by Koford *et al.* [7], had hemiatrophy of the right side of the face, palate and tongue accompanied by anophthalmos and gross malformation of the nose on the same side as well as a deformed right wrist and interdigital webbing. In the early infantile capuchin monkey, shown in figure 2, the embryonic naso-medial process had incompletely developed, leaving an incapacitating gap in the intermaxillary region and at the nostrils. A case of harelip in a mandrill and one of cleft palate in a newborn marmoset twin have been recorded by Hill [16, 17] and the occurrence of hydrocephaly in an infantile guenon by Fox [18].

Besides the great variety of mostly quite limited cranial malformations among primates, which had been collected in two descriptive monographs by LeDouble [19] more than 60 years ago, a great many further ones have since been published. Exceptionally rare is the case of an abnormally short skull

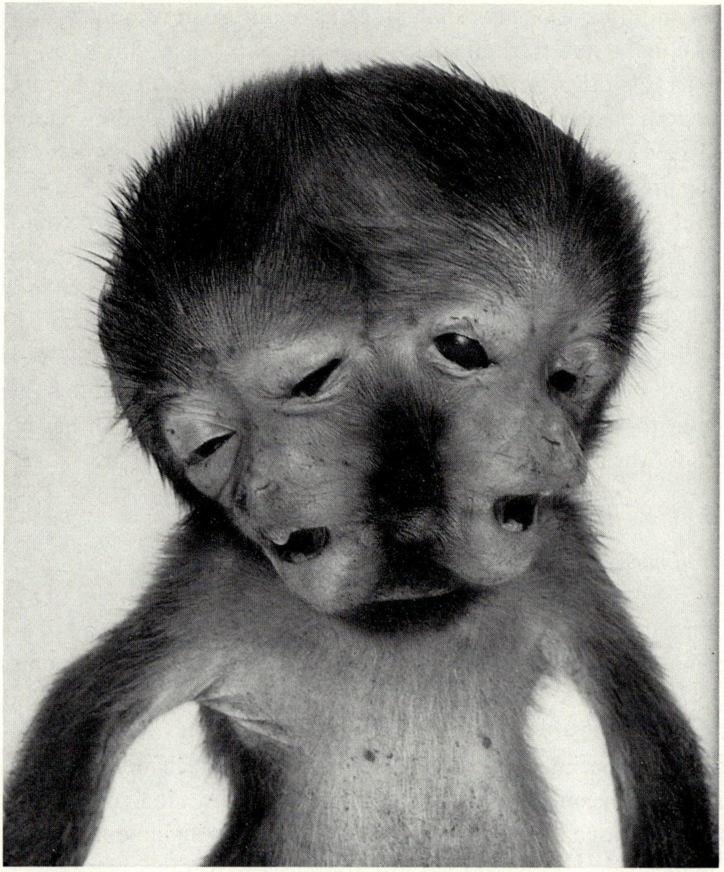

Fig. 1. Partial duplication of the head in a deadborn *Macaca mulatta* from collection of Carnegie Laboratory of Embryology.

base with lacking basal suture in a wild adult mangabey and the occurrence of a very strongly impressed cranial base in a wild adult guereza reported by VOGEL [20]. Previously known only in man was the syndrome of acrocephaly combined with digital defects which the writer [21] found in a wild infantile chimpanzee. This male ape was exceptionally small for its stage of development (complete first dentition); it had merely a short central part of the coronal suture, due to the bilaterally abnormal juxtaposition of the frontal and parietal ossification centers; this has resulted in retarded growth in length and a compensatory increase in the relative height of the brain-case, as shown in

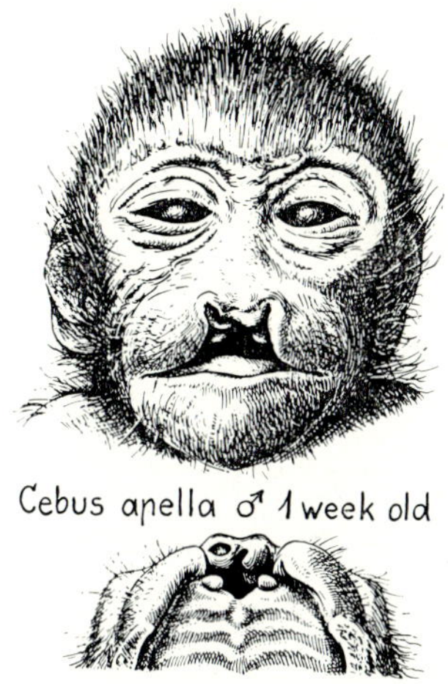

Fig. 2. Frontal and palatine views of incompletely developed upper lip and intermaxillary region of an infantile capuchin monkey [SCHULTZ, 12].

figure 3. This specimen also lacked both thumbs, while both first toes were merely represented by small metatarsal rudiments.

The interfrontal suture, which normally remains open to adult life in nearly all prosimians but disappears during early infantile development in all normal Simiae, can abnormally persist as an open suture among the latter and then usually produces a slight increase in frontal width. Such so-called metopism is extremely rare in nearly all monkeys and apes, but an outstanding exception has been found in the isolated subspecies of *Colobus badius kirkii* in which 77% of the cases examined by REMANE [22] show this condition. In man persisting metopic sutures occur in all races with widely varying frequency, reaching in some brachycephalics as much as 25%. Among platyrrhines the *Ateles* shown in figure 4 is the only incidence of real metopism of adult New World monkeys known to the author, but he has found this condition in every one of the man-like apes as very rare exceptions. A transverse suture dividing the occipital squama and producing a so-called Inca bone in man with varying frequency has been encountered in only a few apes; among monkeys exclusively in *Saimiri* in which it was found in 10 of 61 wild specimens [12], one being shown in figure 4. It is of interest that this is the only nonhuman prima ə with a very prominent occiput, projecting far behind the

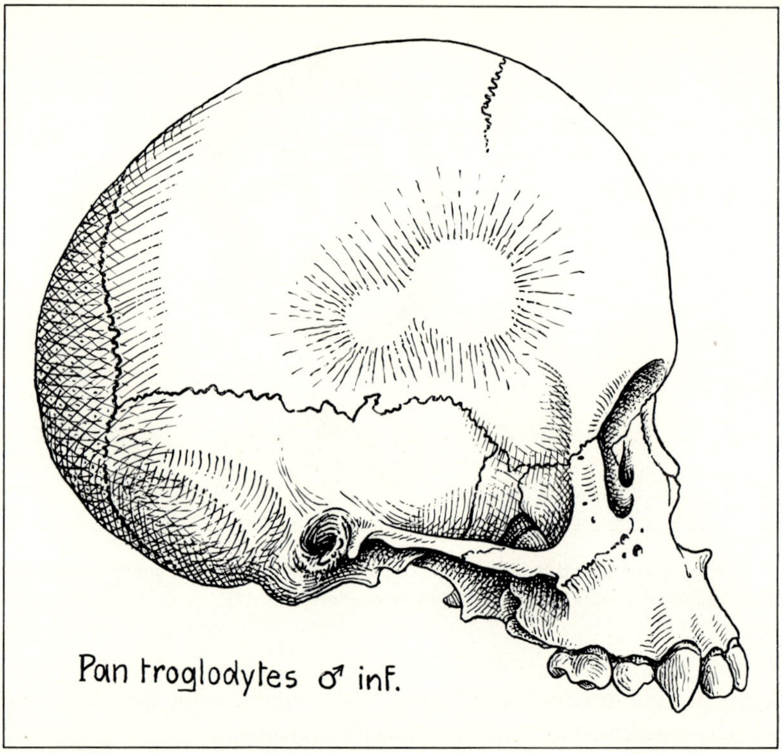

Fig. 3. Skull of an acrocephalic chimpanzee. The abnormal location of the ossification centers for the frontal and parietal is indicated by the circular zones surrounded by radiating lines [see also SCHULTZ, 21].

occipital condyles, since in man with Inca bones the occiput is usually also abnormally large and bulging.

Deviations from the normal growth and final shape of the neurocranium are mostly due to abnormal, early ossification and the consequent changes from the typical number or position of some sutures or the age of their obliteration. This is evident not only from the above quoted example of acrocephaly, but also from the less rare occurrence of plagiocephaly, characterized by the *unilateral* lack or premature closure of the coronal suture, connected with marked asymmetry of the entire cranial vault. One such case is shown in figure 4 and precisely the same condition has been found in two other wild gibbons [23], three spider monkeys and nine howler monkeys [24], besides a few single cases in various Old World monkeys. The frequency of plagiocephaly can differ greatly according to species. For instance, in very large series of wild specimens it existed in 2.4% of *Alouatta palliata* and in 1.5% of *Ateles geoffroyi*, but was never found in *Cebus capucinus*. The lack of the sagittal suture, producing an abnormally narrow skull, known as scaphocephaly, is fairly common in man, but has so far been reported for only one wild juvenile capuchin monkey and one wild juvenile howler monkey [24]. The parietal bones develop normally from single ossification centers, but more than one center on one or both sides do occur and produce anomalous, divided parietals. A clear case of this kind in a captive macaque is shown in figure 4. HRDLIČKA [25] found completely or incompletely divided parietals among several hundred monkeys in nearly 13% of the cases. The writer [12, 23, 26] has reported additional instances in various monkeys and concluded that among the apes this abnormality is most frequent in orang-utans. The latter ape is also distinguished by the highest incidence of divided zygomatic bones, an anomaly which was found in 11 of 166 orang-utan skulls [26], but only as very rare exceptions in a few other species of nonhuman primates.

The various fontanelles can become closed through ossification from abnormal single or multiple centers which then form fontanellar bones of occasionally very large size, as shown by two examples in figure 4. That there must be a hereditary basis for the occurrence of such anomalous bones is highly probable from their widely differing frequency according to species, as well as even the particular fontanelles involved. For instance, so-called bregmatic fontanelle bones were found among 251 skulls of *Ateles geoffroyi* in 42% of the cases, among 167 skulls of *Cebus capucinus* in 16%, but among 493 skulls of *Alouatta palliata* in only one single case. In other large cranial series such bones existed in 12% of *Colobus badius*, whereas not in any *Colobus polykomos*, and in 15% of *Hylobates lar*, but never in *Hylobates moloch* [23, 24]. Lambdoid, asteric and pteric fontanelle bones are much rarer than bregmatic ones in all the above named and many other species, but in the great apes and man this relation is often reversed. For instance, among gorillas the writer [12] had found very large abnormal bones in the region of the pteric fontanelles in 6 skulls and has since seen several more cases of the same kind. In a series of 170 skulls of chimpanzees of ages prior to general sutural obliteration, abnormal fontanellar bones existed at the pterion in 2%, at the bregma in 3%, at the asterion in 4%, but at the lambda in 24% of the cases [27]. These surprisingly common lambdoid bones of chimpanzees ranged from a single, small, so-called os apicis to multiple bones, at times even extending along the lambdoid and sagittal sutures, as in the case shown in figure 4.

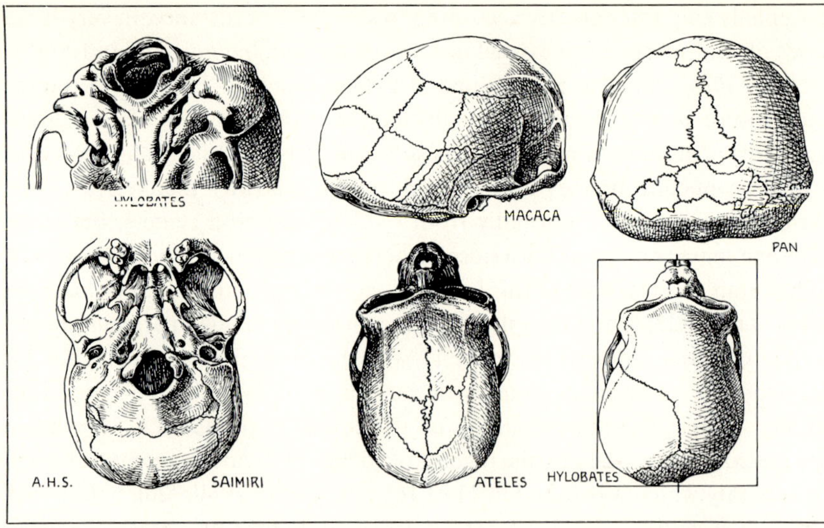

Fig. 4. Incomplete separation of atlas and occiput in an adult wild gibbon. Divided parietals (left 2, right 5 parts) in an adult captive macaque. Multiple abnormal ossicles along sagittal and lambdoid sutures of a wild juvenile chimpanzee. Abnormal transverse occipital suture in a wild adult squirrel monkey. Persisting metopic suture, large bregmatic fontanelle bones and obliteration of left coronal suture in a wild adult spider monkey (from a photograph kindly sent by Dr. M. A. SCHÖN). Premature closure or lack of right coronal suture, producing marked asymmetry, in a wild juvenile gibbon.

A great variety of still other kinds of abnormal cranial development in nonhuman primates has been recorded in the literature. The failure of some early embryonic conditions to disappear with advancing age is exemplified by the very rare persistence of a so-called canalis basilaris chordae in the basioccipital bone of juvenile and even adult specimens and by the partial or complete presence of a cranio-pharyngeal canal in the basisphenoid bone long after its usual obliteration. The latter canal persists with remarkably high frequency in chimpanzees, is also quite common in gibbons and spider monkeys, but totally unknown in many other primates [23, 24, 27]. Partial occlusion of the external auditory meatus, incomplete formation of zygomatic arches and doubling of mandibular condyles, evidently not caused by any accident or disease, have been encountered by the writer in occasional monkeys and apes as rare congenital malformations. The lack of nasal bones is another abnormality which is even fairly frequent in orang-utans and gibbons, though very exceptional among other primates, including man.

III. Abnormal Development of Vertebral Column and Thorax

The incomplete separation between the occiput and the atlas, clearly distinguished from arthritic fusion between the two structures, has been found in a gibbon (shown in fig. 4) as well as in a gorilla and a baboon [12]. That the atlas is more often defectively formed than the other cervical vertebrae is known from human skeletons and has also been found in a variety of monkeys and, particularly, in a series of 233 wild gibbons which contained 20 cases with incomplete development of the dorsal and/or ventral arches; one case is shown in figure 5a. Complete 'cervical' ribs, transforming the seventh

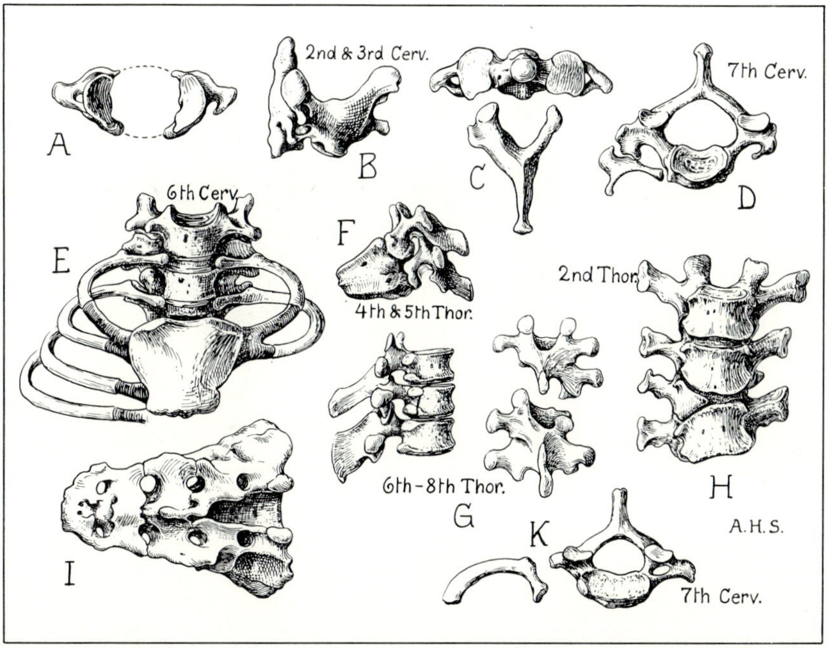

Fig. 5. Developmental abnormalities in vertebrae of apes: *a* = congenital lack of ven tral and dorsal arches of the atlas of an adult gibbon; *b* = fused second and third vertebrae of an adult gibbon; *c* = separate dorsal arch of epistropheus of an adult orang-utan; *d* = cervical rib of an adult gibbon; *e* = complete ribs from 7th vertebra in a gibbon; *f* = fused bodies of 4th and 5th thoracic vertebrae of a juvenile orang-utan; *g* = lateral and dorsal views of asymmetrical fusion of thoracic vertebrae of a juvenile chimpanzee; *h* = wedge-shaped vertebra with incomplete rib of an adult chimpanzee; *i* = incomplete dorsal arches of first two sacral vertebrae of an adult orang-utan; *k* = cervical rib of a juvenile chimpanzee. All, except g and k, are wild specimens [SCHULTZ, 12].

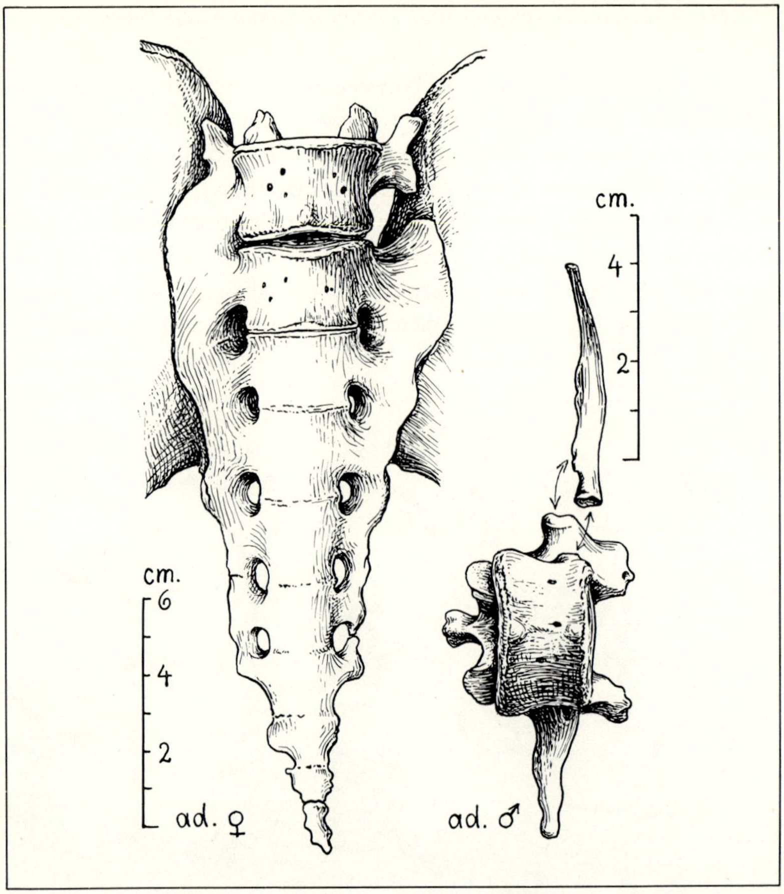

Fig. 6. Asymmetrical lumbo-sacral and thoraco-lumbar transition vertebrae of adult chimpanzees [SCHULTZ, 27].

cervical segment unilaterally or bilaterally into a thoracic one, are occasionally met with in all major groups of primates as rare deviations from normal development. One such case is illustrated in figure 5e, showing also a ventral contact between the abnormal ribs and the normal ones, belonging to the eighth segment. The latter condition existed also in an orang-utan and in two *Nycticebus* with complete 'cervical' ribs, which the writer [26, 28] reported elsewhere together with corresponding occurrences in still further primates. The cases of unilateral incomplete and complete ribs of the seventh vertebra in figure 5d and k serve as additional examples of such abnormal development in especially the seventh segment. The congenital lack of ribs coming from the eighth vertebra has also been found, but not nearly as often as the abnormal presence of ribs on the seventh vertebra [28]. As has already been mentioned, some primate species show such a marked instability of the costal development at the thoraco-lumbar border, that the unilateral presence of a last rib can be included among variations rather than malformations. Equally common in many species are asymmetrical formations at the lumbo-sacral border, represented by a transitional vertebra with a sacral wing on one side only, as shown by the example in figure 6. Such a lumbo-sacral vertebra has been found by OTTOW [29] to be connected with a strong lumbar scoliosis in an adult wild gorilla, but usually any asymmetrical development of the last lumbar segment has no noteworthy effect on the other parts of the spinal column. Clear vertebral malformations are much rarer in the middle of the different spinal regions than at their borders, but a considerable variety of them has nevertheless been encountered among prosimians, monkeys and especially apes. Decidedly asymmetrical development of some malformed thoracic vertebrae exists in the examples shown in figure 5g and h as well as in figure 7. Other types of vertebral anomalies are represented by incomplete separation of adjoining vertebrae, as in figure 5b and f, and by incomplete development of dorsal arches in c and i. Further cases of a very similar nature have been found in various apes and some monkeys [12, 28]. In a still-born baboon, described by HILL [30], the dorsal arches of four vertebrae above the root of the tail were incompletely formed and the overlying soft parts were also wide open.

In the caudal region of hominoids occur at times congenital abnormalities on the surface which seem to be due to the exceptional persistence of embryonic conditions. One such case is shown in figure 8 and others have been described in detail in former papers by the author [12, 26, 31]. The so-called fovea coccygea is a shallow pit or groove of varying size, entirely hairless and resembling umbilical scars, being located in the skin over the coccygeal vertebrae. Such scars existed in 2 fetal and 3 infantile orang-utans and in 14 older fetuses, infants and even juvenile chimpanzees examined by the writer, being evidently less rare in these apes than in man. By their location and appearance these coccygeal foveas may be remaining traces of imperfectly closed posterior neuropores. Before the embryonic outer tail of hominoids becomes completely overgrown by surrounding tissues there normally remains for a short time a tubercle-like prominence over the tip of the coccyx. The latter tubercle can abnormally persist on the surface after the coccyx itself has withdrawn from it and then usually contains merely soft tissue, but no bone or

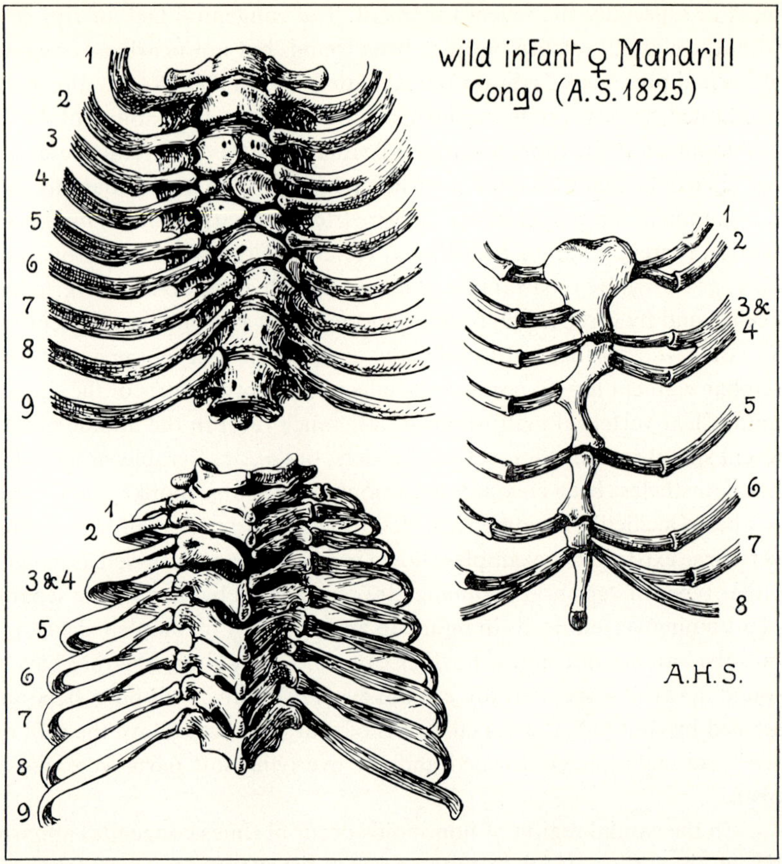

Fig. 7. Multiple congenital malformations of the thorax of a young mandrill [SCHULTZ, 12].

cartilage. Such coccygeal tubercles are very rarely seen in human newborns, but among 57 mostly very young chimpanzees, examined by the writer, good-sized tubercles existed in 7 specimens as abnormal embryonic remnants. So far this structure has not been found in any other apes, but the mere vestige of the bone-less outer tail of *Macaca sylvana* seems to be comparable, except that in this species it has become a constant and hence normal feature.

Ribs can develop abnormally through incomplete formation as well as by being bifurcated at their dorsal or ventral ends, as in the deformed mandrill of figure 7 and in numerous gibbons, several orang-utans and a chim-

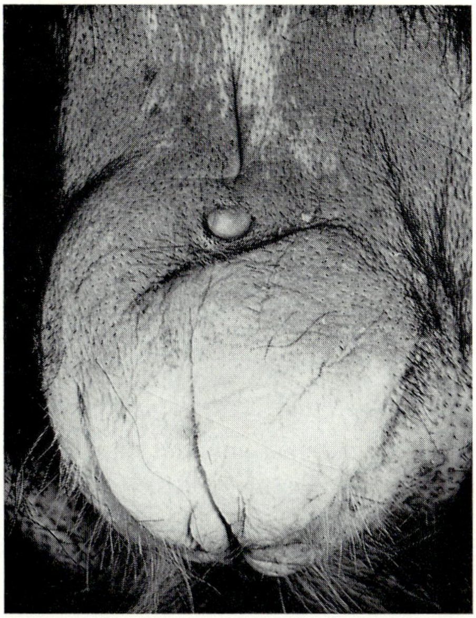

Fig. 8. Lower part of back of an infantile chimpanzee with hair shaved to show the abnormal, long coccygeal pit and large coccygeal tubercle [SCHULTZ, 12].

panzee, described by the writer in previous papers [23, 26, 28]. The sternum has been found to develop with remarkably many irregularities among the hominoids, probably due to its mostly double row of ossification centers in these primates. For instance, a series of 200 wild gibbons contained no less than 15 specimens with abnormal perforations in the corpus sterni [23] and marked asymmetries are very common especially in the sterna of orang-utans [26].

IV. Abnormal Development of Limbs

Developmental malformations are comparatively rare in the proximal and middle parts of the limbs, but surprisingly common in the distal parts, especially the digits, of not only man, but many other primates as well. Of the few extensive congenital defects of the extremities of apes the most noteworthy are the lack of one upper limb distal to the middle of the upper arm in an adult wild gibbon and the total absence of the distal half of one forearm in a freshly captured chimpanzee. In both these cases the proximal parts of the affected limbs were greatly underdeveloped, as described in other papers by the author [12, 23]. In a report on the capture of 56 chimpanzees in the Congo, RAHM [32] has briefly mentioned that 'As probably congenital deformations we noticed one missing forearm'. These conditions in wild apes represent most likely what in the teratology of man is recorded as unilateral peromelia and they are comparable to the many minor cases of agenesia at the distal ends of the extremities, such as the lack of first digits in the acrocephalic chimpanzee mentioned in a previous part of this paper. At least as extensive congenital defects of extremities have been found in a large troop of Japanese macaques by ITANI *et al.* [33], who enumerate not only various malformations of the hands and feet, including their total absence, but also 'some monkeys which lacked a part of their limbs from the forearm or from the upper arm down', an infant 'without half of the upper arm and the thigh' and a newborn whose 'both thighs were completely lacking and there were only traces of both arms'. VALERIO *et al.* [2] picture the fetus of an imported macaque with amelia of both arms and hemimelia of both legs and WILSON and FRADKIN [11] mention that thalidomide given to pregnant monkeys has produced phocomelia, hemimelia and syndactyly in some of the offspring.

Abnormal embryonic development of the hands and feet can result in a surprisingly great variety of defects in especially the digits, of which some may be partly or totally lacking, others more or less completely fused or bifurcated and even accessory digits can sprout almost anywhere from normal ones. The latter condition has been found, e.g., by BOLAU [34] in an orang-utan with polydactyly in the form of a small supernumerary digit arising from the fifth basal phalanx on both hands. In his comprehensive treatise on variation in relation to evolution BATESON [35] had described a most unusual incidence of nine toes on one foot of an adult macaque which also showed some incomplete doubling of the long bones and hipbone of the corresponding extremity. According to the same author a gibbon possessed a well-formed supernumerary digit sprouting from the lateral side of the fifth metacarpal on both hands. The writer has encountered many further cases of similar kind, namely among apes in an adult chimpanzee with an extra digit attached to the base of the right fourth finger, in a *Hylobates agilis* with two supernumerary toes on the left foot (and toe III abnormally short) and in no less than 10 of 113 wild *Hylobates lar* with accessory digits (6 times on feet, 4 times on hands) of which some are shown in figure 9. Among Old World monkeys ad-

---

Fig. 9. Developmental defects in the fingers and toes of wild adult gibbons *(Hylobates lar)* [SCHULTZ, 23].

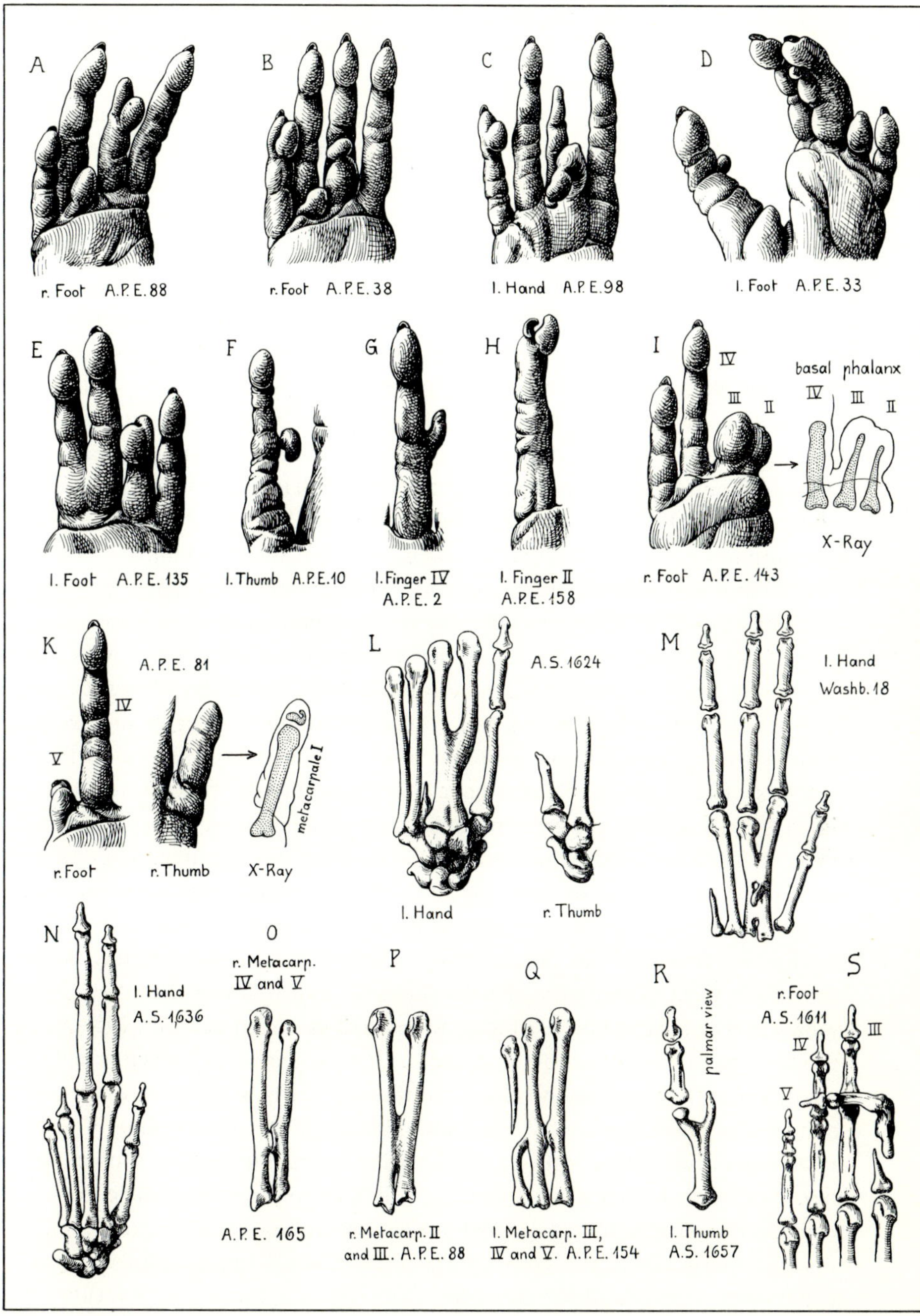
A
r. Foot A.P.E.88
B
r. Foot A.P.E.38
C
l. Hand A.P.E.98
D
l. Foot A.P.E.33
E
l. Foot A.P.E.135
F
l. Thumb A.P.E.10
G
l. Finger IV
A.P.E.2
H
l. Finger II
A.P.E.158
I
IV
III II
basal phalanx
IV III II
X-Ray
r. Foot A.P.E.143
K
A.P.E. 81
IV
V
metacarpale I
r. Foot
r. Thumb
X-Ray
L
A.S.1624
l. Hand
r. Thumb
M
l. Hand
Washb.18
N
l. Hand
A.S.1636
O
r. Metacarp.
IV and V
A.P.E. 165
P
r. Metacarp. II
and III. A.P.E.88
Q
l. Metacarp. III,
IV and V. A.P.E.154
R
palmar view
l. Thumb
A.S.1657
S
r. Foot
A.S.1611
IV
III
V

ditional digits were found in one of 51 proboscis monkeys, one of 83 langurs and three of 152 macaques examined by the author. While the larger of these sprouts contain phalanges and at times also nails, most of the smaller ones are mere grape-like appendages without any bone; all branch from the ventral or lateral sides of other digits, but never from the dorsal side.

In many instances such supernumerary digits have developed in association with reductions of other fingers or toes. Congenital deformities, in man known as 'cleft-foot' and 'lobster-claw', have been recorded in various nonhuman primates. In a chimpanzee, e.g., the toes II, III and IV of one foot were very incompletely developed [36] and in another chimpanzee the middle fingers of both hands had remained abnormally short from stunted growth of their basal phalanges [12]. In a further chimpanzee the fingers II to V of one hand lacked all phalanges [37], as shown in figure 10. The macaque in the same figure has extensive, though quite unequal, developmental defects in both hands and both feet, while the writer's mandrill in figure 10 has only one hand grossly malformed. A captive-born infant drill has also been found to have both feet extremely anomalous, even in the tarsal region [38]. The widely differing developmental defects of the 18 wild gibbons in figure 9 contain reductions as well as additions singly or in combination. In the author's [23] series of 113 gibbons from one limited area these anomalies occurred with the remarkable frequency of 16%. An additional series of 120 *Hylobates lar*, of which only the skeletons could be examined, contained 11 cases of digital anomalies and among the skeletons of other gibbon species several further instances of very similar defects have been encountered, though not in as high a frequency as in the gibbon population of the mountainous district of northern Thailand from which the first mentioned series came. The latter seems to suffer from an exceptional amount of developmental irregularities in many bodily parts and especially in the distal portions of the limbs [23]. The highest frequency of congenitally defective hands and feet has been found in a local population of wild Japanese macaques by FURUYA [39], namely in 41 of a total of 209 births which had occurred between 1957 and 1965 after having begun with a single case. Of these malformations it is merely stated that they correspond closely to the not nearly as frequent ones, encountered in another troop of Japanese monkeys (33), by consisting of various kinds of fusions, reductions and losses of digits, but no supernumerary ones. The hands are more often defective than the feet. Among 900 captive-born *Macaca mulatta* there occurred one case of clubfoot and the great many imported macaques of the same laboratory contained single cases of oligodactyly, polydactyly and lobster-claw of the hands according to VALERIO *et*

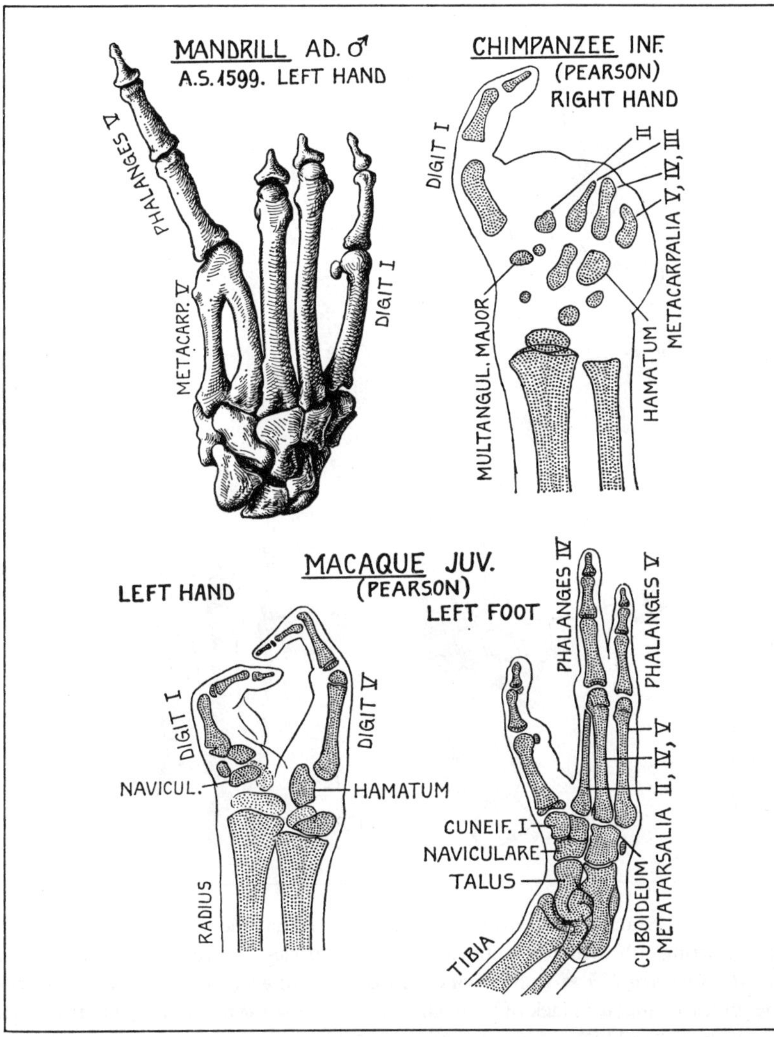

Fig. 10. Digital anomalies in the hands of a mandrill, a chimpanzee and a macaque and in the foot of the same macaque [SCHULTZ, 23; last two specimens redrawn from PEARSON, 35].

al. [2]. In one of 99 wild spider monkeys, examined by the writer in Panama, both first toes were incompletely developed, much like the thumbs of this species usually are.

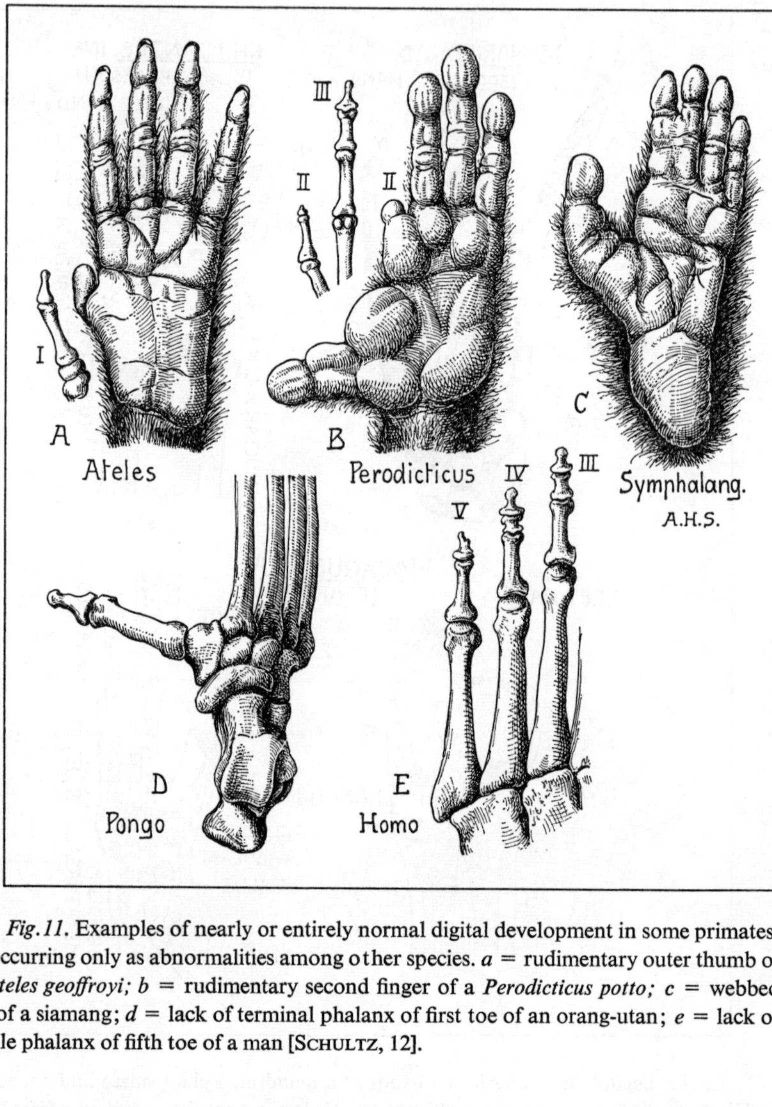

Fig. 11. Examples of nearly or entirely normal digital development in some primates, but occurring only as abnormalities among other species. *a* = rudimentary outer thumb of an *Ateles geoffroyi; b* = rudimentary second finger of a *Perodicticus potto; c* = webbed toes of a siamang; *d* = lack of terminal phalanx of first toe of an orang-utan; *e* = lack of middle phalanx of fifth toe of a man [SCHULTZ, 12].

It is of special interest to find that during evolution some primate species or even entire genera have acquired certain digital reductions which would unquestionably have to be regarded as developmental defects if they had remained limited to only few individuals. Among prosimians the phalangeal part of the second finger is greatly shortened in *Nycticebus*, even more reduced in *Perodicticus* (fig. 11b) and has totally disappeared from the outside of the hand in *Arctocebus*. Among platyrrhines the free part of the thumb has become lost in spider monkeys except for occasional useless vestiges in varying percentages of individuals (fig. 11a). The same elimination of the thumb distinguishes the guerezas among catarrhines and again some traces of this digit can persist even with a minute phalanx and nail. A reduction in the usual digital development exists among hominoids in orang-utans whose first toes have lost one phalanx and the nail in well over half of all cases (fig. 11d), this being much more common in females than in males [40]. An analogous digital degeneration has also appeared in the fifth toes of man which lack the middle phalanges with surprising frequency in some populations (fig. 11e).

The normal embryonic webbing between digits can abnormally persist especially between the second and third toes of man as a rare hereditary anomaly, known as zygodactyly. Exactly the same condition of arrested development is less rarely found among chimpanzees and gibbons, while in siamangs it has become a very nearly constant species character (fig. 11c). Again it is seen that certain congenital malformations of various species have changed by their greatly increased frequency into normal features of some other species.

V. Abnormal Development of Dentition

Developmental abnormalities in the permanent dentition of especially the man-like apes are extraordinarily common, as is evident from the great number and variety of such cases recorded in a widely scattered literature. The most extensive study of dental anomalies in nonhuman primates has been published by COLYER [41] and detailed descriptions of many further cases have been contributed especially by BATESON [35]; BENNEJEANT [42, 43] and the writer [12, 23, 24, 26, 44, 45]. The following brief review of the nature and incidence of developmental defects in primate dentitions will deal with (1) increase in the number of teeth, (2) decrease in the size and number of teeth and (3) crowded, misplaced and malformed teeth.

Supernumerary teeth result most likely from localized twinning in the early development of tooth buds, similar to the occasional duplication of digits or of nipples. That 'extra' teeth represent congenital anomalies, and are not simply atavisms, is evident from the fact that even the canines have been found in pairs and as many as six molars on one side of one jaw can develop in very rare instances. Furthermore, such twinning may be limited and produce merely a double crown on a single root, as the writer [12] has recorded

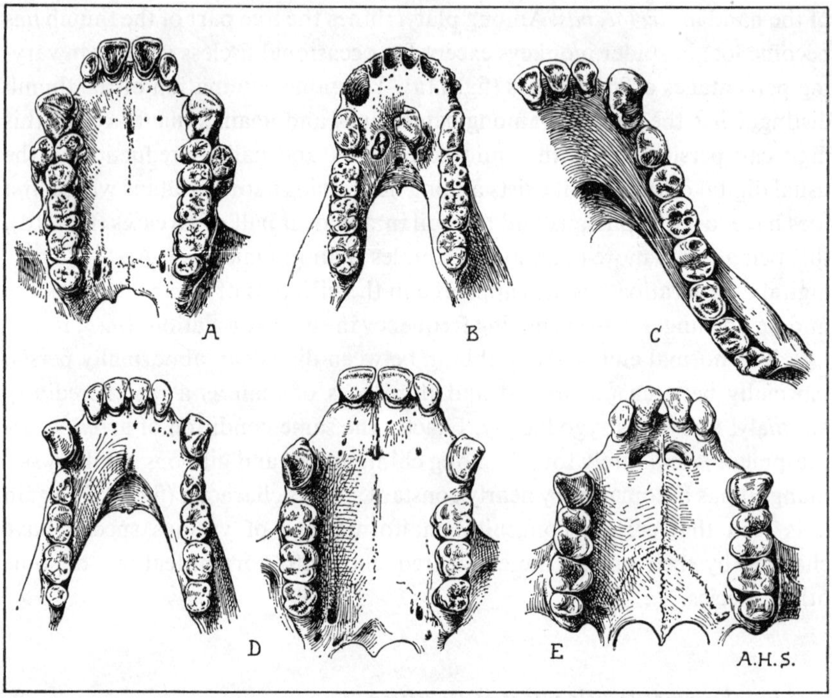

Fig. 12. Some examples of abnormal development in the permanent dentition of wild primates. *a* = duplication of right and left first premolars and reduced third molars of a male orang-utan; *b* = male gorilla with both lower second premolars displaced lingually (left impacted); *c* = male orang-utan with five molars on right (four on left); *d* = lower and upper jaws of a female chimpanzee with congenital lack of two lower and all upper premolars and of one upper incisor and with reduced lower left third molar; *e* = female marmoset with congenital lack of one premolar on both sides [SCHULTZ, 12].

for a lower deciduous canine of a wild chimpanzee and for both lower permanent canines of a wild gibbon. In an adult wild howler monkey one lower canine was found to have a crown divided into four unequal parts, but its root showed barely an indication of doubling [24]. Completely formed supernumerary canines can erupt behind or on the side of the normal ones. For instance, one adult gibbon has two upper left canines of nearly equal size in good alignment [46], but in an adult baboon the extra upper right and left canines stand lingual to the corresponding normal ones [47].

Supernumerary incisors are rare among most primates and so far only one such tooth per specimen has been encountered. COLYER [41] found extra incisors in 6 of 2,446 platyrrhines, in 14 of 5,900 catarrhine monkeys, in 2 of

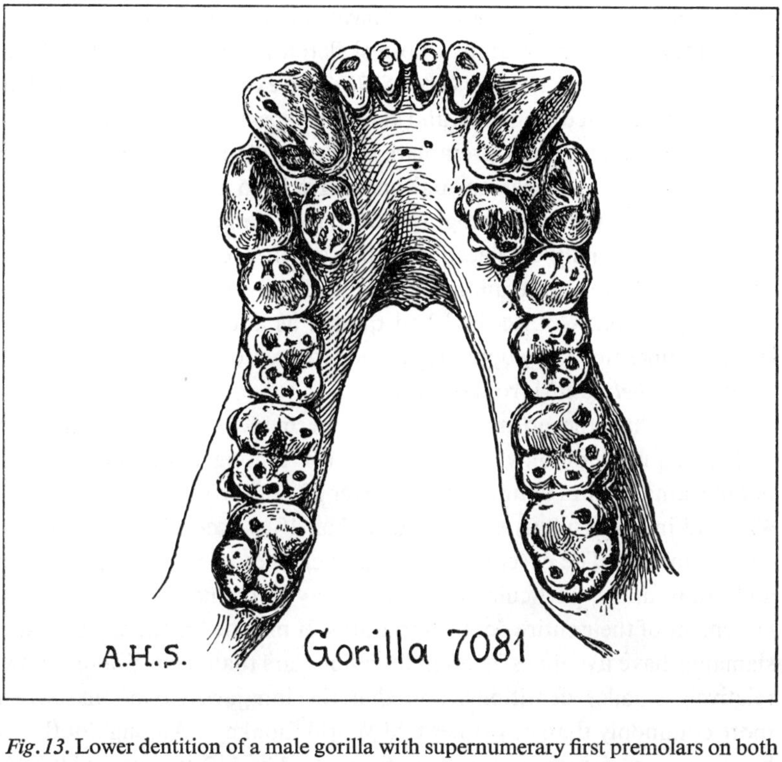

Fig. 13. Lower dentition of a male gorilla with supernumerary first premolars on both sides and exceptionally large teeth [SCHULTZ, 45].

391 hylobatids, in 3 of 467 chimpanzees, in 6 of 546 gorillas and in none of 229 orang-utans. According to the writer's [23, 24, 26, 44, 48] own records there existed no supernumerary incisors among over 900 Central American Cebidae, 400 Colobinae and 212 chimpanzees, but five incisors in one jaw were present in 3 of 550 gibbons, in 2 of 217 orang-utans and in 2 of 323 gorillas, the extra teeth occurring slightly more often in the upper than in the lower jaw.

Supernumerary premolars are even rarer than extra incisors. In the above mentioned large series of COLYER [41] only 4 platyrrhine and 4 catarrhine monkeys had more than the normal number of premolars and in his total of 1,633 skulls (with mandibles) of man-like apes such abnormal teeth existed merely in 1 siamang and 1 chimpanzee. In all of the writer's material extra premolars had been found in only 2 howler monkeys, 1 guereza, 1 gibbon, 1 orang-utan, 2 gorillas and no chimpanzee. As shown by figure 12a the abnormal premolars of the orang-utan have been crowded labially, whereas another adult male orang-utan, whose skeleton is on exhibit at the American Museum of Natural History (no. 15970), the extra upper right and left premolars had erupted lingually and the same specimen has also 5 lower incisors. The supernumerary premolars of the gorilla in figure 13 also stand far lingually while the larger normal ones have been pushed labially as well as forward. From the same geographical region came an adult female gorilla with also a lingually displaced extra lower left premolar [45], showing that such generally very rare anomalies can suddenly appear repeatedly in some local population. COLYER [41] had quoted as another incidence the occurrence of supernumerary upper right and left premolars in 3 of 30 *Ateles* '*marginatus*' *(= belzebuth)* from one locality.

More than the normal number of permanent molars generally occur with a frequency of only a fraction of 1% among monkeys and gibbons, but in siamangs and chimpanzees in over 2% of the series, in gorillas in over 4%, and in orang-utans even in more than 10% according to the combined data in table I. The exceptionally high incidence of extra molars in gorillas and orang-utans, particularly in males, may be connected with the great enlargement of their entire dental apparatus. It may be for the same reason that siamangs have five times as often fourth molars than have gibbons with their relatively smaller dentitions, and that the longjawed baboons have them more commonly than any other Old World monkeys. Among South African baboons, e.g., fourth molars were found in 4 of 112 specimens [49]. On the other hand, in all New World monkeys extra molars are least rare in *Ateles*, even though its dentition is of quite modest size and merely exceptionally variable at its aboral end. Supernumerary molars can often develop in such good alignment as in the orang-utan of figure 12c, or be displaced, impacted and reduced to mere pegs with a single root. They may appear on one side of one jaw only or may exist bilaterally in both jaws.

The congenital lack of teeth is among primates generally somewhat to very much rarer than are supernumerary teeth, but in man, gibbons and some platyrrhines this relation is usually reversed. In the mostly very large series examined by COLYER [41] these two categories of dental anomalies have been found in the proportions listed in table II. The much higher frequency of missing than of additional teeth among gibbons has been confirmed in the writer's [23] extensive material of these apes which contains 42 absent teeth in a total of 27 specimens, whereas only 16 extra teeth in 8 specimens. In the platyrrhine genera *Ateles*, *Alouatta* and *Cebus*, studied by the writer [24], there were 54 congenitally lacking teeth in 38 skulls, but only 10 extra teeth in

Table I. Number of specimens with supernumerary molars, total of adults examined and percentage of former in series according to different authors

Primates	COLYER	SCHULTZ	Other authors	Total
Platyrrhines	16:2,446 = 0.7 (41)	3:388 = 0.8 (24, 44)	6:409 = 1.5 (35)	19:2,834 = 0.7
Catarrh. monkeys	20:5,900 = 0.3 (41)	3:1,171 = 0.3 (44, 48, 50)	2:419 = 0.5 (35) 4:112 = 3.6 (49)	23:7,071 = 0.3
Hylobates	1:306 = 0.3 (41)	5:833 = 0.6 (23, 44)	0:51 = 0 (35)	6:1,139 = 0.5
Symphalangus	2:85 = 2.4 (41)	1:31 = 3.2 (new)	–	3:116 = 2.6
Pongo	16:229 = 7.0 (41)	9:142 = 6.3 (26)	38:194 = 19.6 (51)	63:565 = 11.1
Pan	14:560 = 2.5 (41)	6:239 = 2.5 (44 + new)	0:50 = 0 (52)	20:849 = 2.4
Gorilla	23:675 = 3.4 (41)	16:289 = 5.5 (54 + new)	8:98 = 8.2 (52) 14:336 = 4.2 (53)	61:1,398 = 4.4

Number of publication in parentheses. Authors besides the first two: BATESON [35][1], SHAW (for *Papio* only) [49], SELENKA [51, 52], RANDALL [53].
1 The data of BATESON are not included in the totals since COLYER had partly used the same material.

Table II. Total numbers of congenitally lacking and of supernumerary teeth in the primates of COLYER [41] and percentage relations between the former and the latter

Primates	Lacking teeth	Supernumerary teeth	Percentage
Platyrrhines	52	36	144
Catarrh. monkeys	49	55	89
Hylobates	21	4	525
Symphalangus	1	5	20
Pongo	10	32	31
Pan	15	28	54
Gorilla	6	58	10

5 other skulls. Such an exceptionally high preponderance of missing teeth is due most of all to the strong trend to suppress the last molars among gibbons as well as spider monkeys. Incisors are lacking surpsisingly often in *Alouatta palliatta*, namely in 18 of 248 specimens (= 7.3%) from Central America [24]. It is of special interest also that only the upper lateral or the lower middle incisors are usually the ones which are suppressed or at least much reduced, just as they are in man, in whom such cases are well-known to occur repeatedly in the same families. Among other simian primates the congenital lack of incisors is very rare, but the prosimian *Nycticebus* has quite often some or even all of its upper incisors undeveloped [12]. Abnormal reduction in the number of premolars has been found sporadically in a great variety of monkeys and apes, but only as comparatively rare exceptions. The existence of merely two premolars on both sides of the upper jaw in the marmoset of figure 12e and on one side of the lower jaw in the spider monkey of figure 14 suggests that the reduction of premolars, which is typical for all catarrhines, might have originated by way of the same kind of abnormal dental development. The suppression of a premolar in the *Ateles* of figure 14 is accompanied by abnormal occlusion of the lower canine, exactly as in two chimpanzees of the writer's [27] collection which also lack a lower premolar. The unique skull of a wild chimpanzee, shown in figure 12d, has a total of six premolars congenitally missing as well as one incisor and a last molar greatly reduced.

As already mentioned, the third molars of spider monkeys and of gibbons are specially prone to developmental failure. In *Ateles geoffroyi*, e.g., anywhere from one to four last molars were lacking in 3.2% of specimens from Panama and in even 15.4% of those from one locality in Nicaragua [24]. Among capuchin monkeys missing last molars were recorded in 3.2% of one series and in 2.5% of another series [24, 41]. The lack of third molars, combined with the marked reduction of even the second molars, shown in the *Cebus* in figure 14, is strikingly like the condition in many typical Callithricidae which, except for *Callimico*, have normally eliminated third molars and often have also their second molars reduced in varying degrees. Among all other monkeys the congenital absence of molars is much rarer, though it has been recorded for a considerable variety of species. In gibbons one or more third molars are lacking in 2.3% of the specimens of COLYER [41] and in 3.8% of those studied by the writer [23]. Among 105 siamangs (85 of COLYER and 20 seen by the author) only one possessed this abnormality. In a total of 371 orang-utans [26, 41] the absence of last molars occurred in 1.6% of the specimens and among altogether 706 chimpanzees [27, 41] in 1.7%. Of all manlike apes the gorilla lacks third molars most rarely, only a single case having

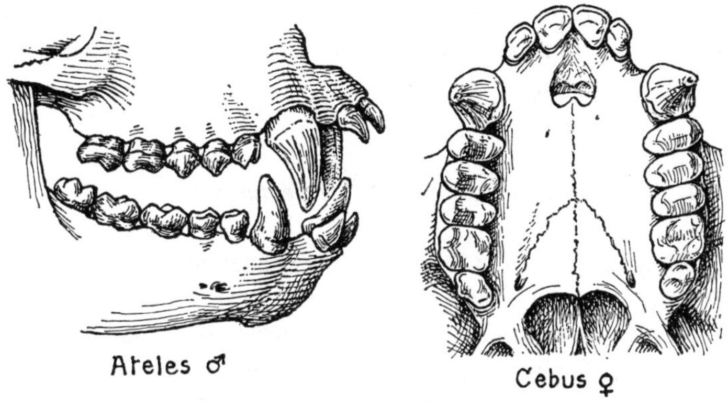

Fig. 14. Wild adult *Ateles geoffroyi* with congenital lack of both upper third molars and of lower right first premolar and with lower right canine occluding behind upper canine. Wild adult *Cebus albifrons* with congenital lack of all third molars and upper second molars much reduced.

been found in series of skulls totalling over one thousand [41, 53, 54]. In recent man, finally, congenitally missing third molars have been reported from all races with widely varying frequencies, being generally least common among prognathous races, but occurring in anywhere from about 4 to 10% of some European populations and even in 27% of a very large series of skulls from a pre-European Hawaiian population [55]. That the last molars are especially prone to undergo abnormal development not only in man, but also in some other primates, is evident in addition from the frequent cases in which these teeth are reduced in size, delayed in eruption or have remained impacted.

Some instances of marked displacement of teeth and of malformed crowns have already been mentioned. Dental abnormalities of this kind are surprisingly common in many primate species and a great variety of such cases has been described and illustrated particularly by COLYER [41] and the writer [23, 24, 26, 44, 48]. That supernumerary teeth are frequently displaced or kept from full eruption is readily explained by inadequate space for good alignment. Crowding and rotating of normal permanent teeth is often associated with the abnormal retention of the corresponding deciduous teeth or by a disproportion between the size of the jaws and that of the dentition. An extreme example of the latter kind is shown in figure 12b. In conclusion it is to be noted that all these varied defects in the complicated development of the dental apparatus are generally more frequent among the man-like apes than other nonhuman primates and more common in the permanent than the deciduous dentition.

VI. Abnormal Development of other Features

Congenital defects in the cutaneous pigmentation and especially its extreme form of total albinism have been recorded for many different primate species in widely scattered publications of which some had been enumerated by the writer [12]. Quite recently an albino gorilla [56] has been added to the list of such cases as well as partial albinism in a wild mandrill [17].

Supernumerary mammary glands or nipples in man represent rare teratological conditions, whereas as many as six nipples develop in some prosimians as perfectly normal species characters. In a great variety of monkeys and apes more than the normal two nipples have been encountered with at times considerable frequency, though usually they seem to be about as rare as in man [57 and numerous more recent case reports]. It is noteworthy that accessory nipples are occasionally associated with other anomalies in the same individuals not only in man, but also, e. g., in a macaque as well as a gibbon, both of which possessed supernumerary digits.

Only few cases of congenital hernias in nonhuman primates have been published so far, namely a large inguinal hernia in an adult captive orangutan [58], an umbilical one in an adult wild gibbon [23], a diaphragmatic one in a baboon fetus [59] and an 'internal hernia with death from strangulation' in a mona monkey [quoted in 60].

True hermaphroditism in a rhesus monkey has been thoroughly described [61] as an exceedingly rare occurrence among nonhuman primates. Cryptorchism is another developmental abnormality which seems to be very exceptional in most monkeys and apes, though the writer [62] found it unilaterally in an adult wild proboscis monkey and even in three of 62 wild gibbons. Congenital defects of the urinary organs have so far not been recorded for more than a few nonhuman primates. HILL [63] has contributed a detailed study of a congenital cystic kidney in a dwarf lemur and a unilateral renal agenesia in a marmoset and elsewhere [60] refers to a paper by HADDOW on renal agenesia in a wild guenon. Accessory spleens of varying number and size are fairly common in man and apparently also in some other primates. For instance, in a large series of wild howler monkeys these congenital abnormalities existed in 30 specimens [64]. A surprising variety of developmental anomalies of the cardio-vascular system is known to occur among monkeys and apes [5, 65, 66, a. o.] and may be even more frequent in the orang-utan than in man [67].

Many different prosimians and the marmosets regularly have multiple offspring as a normal condition, but in all other primates twinning is so exceptional that it can be regarded as belonging to developmental abnormalities. With a survey of the relevant literature and his own records the writer [57] had already demonstrated that twins do occur in a great variety of wild or captive monkeys and apes with frequencies which probably are similar to that found in man. From presently available data for births of mostly captive primates it can tentatively be concluded that twinning may be specially rare among Old World monkeys and fairly common in chimpanzees. For instance, among 900 births of rhesus monkeys there were only three cases of twins [2], whereas of 170 births of chimpanzees seven were multiple ones, showing the surprisingly high frequency of 4.1% [4]. New cases of twins of man-like apes have recently been reported for an orang-utan (♀ ♂) [68] a gorilla (♀ ♀) [69], a chimpanzee (♀ ♂) [70] and even triplets (♀ ♀ ♂) for another chimpanzee [71]. The type of twinning could be determined with certainty in only few instances, but these suffice to state definitely that not only double-ovum twins occur, as they do regularly in the Callithricidae, but also single-ovum twins.

VII. Summary

A great many different forms of developmental anomalies can and do occur among captive as well as wild nonhuman primates with widely varying frequency. The present enumeration of such abnormalities remains fragmentary and gives merely an indication of the real variety and prevalence of congenital defects on account of the very uneven availability of material for investigation. Far larger series of skeletons and dentitions have been examined for anomalies than have been entire cadavers and comparatively little has become known so far in regard to prenatal development in primates. Only few general conclusions are warranted as yet from the data collected in this report. The most frequent malformations have been found in the ossification of the cranial vault, in the vertebrae at the borders of spinal regions, in the last molar teeth and in the digits of the hands and feet, all of which structures seem to be particularly susceptible to irregular development. These anomalies are in general more common among the man-like apes than at least in the other catarrhines. There is much evidence supporting the assumption that some forms of cranial, vertebral, dental and digital malformations have a genetic basis. Various other causes, however, may be responsible for the sporadic occurrence of some or most of the abnormally developed different conditions, which have been described in this paper.

References

1 HARTMAN, C. G.: Studies in the reproduction of the monkey *Macacus (Pithecus) rhesus*, with special reference to menstruation and pregnancy. Carnegie Inst. Wash. Pub. 433, Contrib. to Embryol. *23:* 1–161 (1932).

2 VALERIO, D. A.; MILLER, R. L.; INNES, J. R. M.; COURTNEY, K. D.; PALLOTTA, A. J., and GUTTMACHER, R. M.: *Macaca mulatta*. Management of a Laboratory Breeding Colony (Academic Press, London/New York 1969).

3 YERKES, R. M.; Chimpanzees; A laboratory colony (Yale Univ. Press, New Haven 1943).

4 GUILLOUD, N. B.: The breeding of apes: Experience at the Yerkes Regional Primate Research Center. Ann. N.Y. Acad. Sci. *162:* 297–300 (1969).

5 LAPIN, B. A. und YAKOVLEVA, L. A.: Vergleichende Pathologie der Affen (Fischer, Jena 1964).

6 PICKERING, D. E.: Reproduction characteristics in a colony of laboratory confined mulatta macaque monkeys. Folia primat. *8:* 169–179 (1968).

7 KOFORD, C. B.; FARBER, P. A., and WINDLE, W. F.: Twins and teratisms in rhesus monkeys. Folia primat. *4:* 221–226 (1966).

8 EGOZCUE, J.; PERKINS, E. M.; HAGEMENAS, F., and FORD, D. M.: The chromosomes of some platyrrhini *(Callicebus, Ateles* and *Saimiri)*. Folia primat. *11:* 17–27 (1969).

9 SELLERS, T.: The 49th chromosome. Yerkes Newsletter, *6:* 1–4 (1969).

10 BARROW, M. V.; STEFFEK, A. J., and KING, C. T. G.: Thalidomide syndrome in rhesus monkeys *(Macaca mulatta)*. Folia primat. *10:* 195–203 (1969).

11 WILSON, J. G. and FRADKIN, R.: Teratogeny in nonhuman primates, with notes on breeding procedures in *Macaca mulatta*. Ann. N.Y. Acad. Sci. *162:* 267–277 (1969).

12 SCHULTZ, A. H.: The occurrence and frequency of pathological and teratological conditions and of twinning among nonhuman primates. Primatologia *1:* 965–1014 (Karger, Basel 1956).

13 BOLK, L.: Die Doppelbildung eines Affen. Beitr. path. Anat. *76:* 238–253 (1926).

14 HARTMAN, C. G.: Birth of a two-headed monster in the rhesus monkey. Science *98:* 449 (1943).

15 CHRISTIE, R. J.: An occurrence of monozygotic twinning and anencephaly in *Macaca arctoides*. Lab. Anim. Care *19:* 531–532 (1969).

16 HILL, W. C. O.: Report of the society's prosector for the year 1953. Proc. zool. Soc., Lond. *124:* 303–311 (1954).

17 HILL, W. C. O. and SABATER PI, J.: Notes on two anomalies in mandrills *(Mandrillus sphinx* Linn.*)*. Folia primat. *12:* 290–295 (1970).

18 FOX, H.: Report from the Penrose Research Laboratory, 14–25 (Philadelphia 1941).

19 LEDOUBLE, F.: Traité des variations des os du crâne de l'homme (1903) et Traité des variations des os de la face de l'homme (Vigot, Paris 1906).

20 VOGEL, C.: Über einige seltene pathologische Erscheinungsbilder am Schädel von Primaten aus freier Wildbahn; in STARCK, SCHNEIDER und KUHN Neue Ergebnisse der Primatologie, 128–136 (Fischer, Stuttgart 1967).

21 SCHULTZ, A. H.: Acrocephalo-oligodactylism in a wild chimpanzee. J. Anat. *92:* 568–579 (1958).

22 REMANE, A.: Das Stirnnaht-Problem. Z. Morph. Anthrop. *23:* 153–178 (1925).

23 Schultz, A. H.: Age changes and variability in gibbons. Amer. J. Phys. Anthrop. n.s. *2:* 1–129 (1944).
24 Schultz, A. H.: Age changes and variability in the skulls and teeth of the Central American monkeys *Alouatta*, *Cebus* and *Ateles*. Proc. zoolog. Soc., Lond. *133:* 337–390 (1960).
25 Hrdlička, A.: Divisions of the parietal bone in man and other mammals. Bull. Amer. Mus. Nat. Hist. *19:* 231–386 (1903).
26 Schultz, A. H.: Growth and development of the orang-utan. Carnegie Inst. Wash. Pub. 525, Contrib. to Embryol. *29:* 57–110 (1941).
27 Schultz, A. H.: The skeleton of the chimpanzee; in Bourne The Chimpanzee, 1, pp. 50–103 (Karger, Basel/New York 1969).
28 Schultz, A. H.: Vertebral column and thorax. Primatologia *4:* Liefer. *5:* 1–66 (Karger, Basel 1961).
29 Ottow, B.: Spondylosis deformans bei Anthropoiden, namentlich dem Gorilla. Z. Morphol. Antrop. *43:* 206–221 (1952).
30 Hill, W. C. O.: Spina bifida in a sacred baboon *(Papio hamadryas)*. Ceylon J. of Science, Sec. D, Med. Sc. V *1:* 9–15 (1939).
31 Schultz, A. H.: Chimpanzee fetuses. Amer. J. Phys. Anthrop. *18:* 61–79 (1933).
32 Rahm, U.: Observations during chimpanzee captures in the Congo; in Starck, Schneider und Kuhn Neue Ergebnisse d. Primatologie, pp. 195–207 (Fischer, Stuttgart 1967).
33 Itani, J.; Tokuda, K.; Furuya, Y.; Kano, K., and Shin, Y.: The social construction of natural troops of Japanese monkeys in Takasakiyama. Primates *4:* 1–42 (1963).
34 Bolau, H.: Über den Orang-Utan des Zoologischen Gartens in Hamburg. Verh. naturwiss. Ver. Hamburg-Altona, N. F. *1:* 119–121 (1877).
35 Bateson, W.: Materials for the study of variation treated with special regard to discontinuity in the origin of species (Macmillan, London 1894).
36 Goldschmidt, W.: Über einen Fall von Spaltfussbildung bei *Anthropopithecus troglodytes*. Anat. Anz. *37:* 246–249 (1910).
37 Pearson, K.: On the existence of the digital deformity – so-called lobster-claw in the apes. Ann. Eugen. *4:* 339–340 (1931).
38 Hill, W. C. O.: Lobster-claw deformity in a drill *(Mandrillus leucophaeus* F. Cuv.*)*. Bibl. primat. *1:* 239–251 (1962).
39 Furuya, Y.: On the malformation occurred in the Gagyusan troop of wild Japanese monkeys. Primates. *7:* 488–492 (1966).
40 Schultz, A. H.: The recent hominoid primates; in Washburn and Jay Perspectives on Human Evolution, pp. 122–195 (Holt, Rinehart & Winston, New York 1968).
41 Colyer, F.: Variations and diseases of the teeth of animals (Bale & Danielsson, London 1936).
42 Bennejeant, C.: Anomalies et variations dentaires chez les primates (Vallier, Clermont-Ferrand 1936).
43 Bennejeant, C.: Les anomalies dentaires numériques des primates. Bull. et Mém. Soc. d'Anthrop., Paris *5:* 154–197 (1954).
44 Schultz, A. H.: Eruption and decay of the permanent teeth in primates. Amer. J. Phys. Anthrop. *19:* 489–581 (1935).

45 SCHULTZ, A. H.: A gorilla with exceptionally large teeth and supernumerary premolars. Folia primat. *2:* 149–160 (1964).

46 RIESENFELD, A.: A rare case of canine variation. Amer. J. Phys. Anthrop. n. s. *13:* 541–544 (1955).

47 FREEDMAN, L.: A biometric study of *Papio cynocephalus* skulls from northern Rhodesia and Nyasaland. J. Mammalogy *44:* 24–43 (1963).

48 SCHULTZ, A. H.: Cranial and dental variability in Colobus monkeys. Proc. zool. Soc., Lond. *130:* 79–105 (1958).

49 SHAW, J. C. M.: Four cases of fourth molar teeth in South African baboons. J. Anat. *62:* 79–85 (1927).

50 SCHULTZ, A. H.: Growth and development of the proboscis monkey. Bull. Mus. Comp. Zool. Harvard *89:* 277–314 (1942).

51 SELENKA, E.: Rassen, Schädel und Bezahnung des Orang-Utan. Studien über Entwicklungsgesch. d. Tiere, H. 6. Menschenaffen, Lief. *1:* 1–91 (Wiesbaden 1898).

52 SELENKA, E.: Schädel des Gorilla und Schimpansen. Studien über Entwicklungsgesch. d. Tiere, H. 7. Menschenaffen, Lief. *2:* 93–160 (Wiesbaden 1899).

53 RANDALL, F.: The skeletal and dental development and variability of the gorilla. Human Biol. *15:* 236–254, 307–337; *16:* 23–76 (1943–1944).

54 SCHULTZ, A. H.: Morphological observations on gorillas; in GREGORY Raven Memorial Volume, pp. 227–254 (Columbia Univ. Press, New York 1950).

55 SNOW, C. E.: An old Hawaiian population on Oahu. Amer. J. phys. Anthrop. *20:* 69–70 (1962).

56 SABATER PI, J.: An albino lowland gorilla from Rio Muni, West Africa. Folia primat. *7:* 155–160 (1967).

57 SCHULTZ, A. H.: The number of young at a birth and the number of nipples in primates. Amer. J. Phys. Anthrop. n. s. *6:* 1–24 (1948).

58 STRAUS, W. L., jr.: The thoracic and abdominal viscera of primates, with special reference to the orang-utan. Proc. amer. Phil. Soc. *76:* 1–85 (1936).

59 HENDRICKX, A. G. and GASSER, R. F.: A description of a diaphragmatic hernia in a sixteen week baboon fetus *(Papio* sp.*)*. Folia primat. *7:* 66–74 (1967).

60 HILL, W. C. O.: Primates. 6. Cercopithecoidea, Cercopithecinae (Univ. Press, Edinburgh 1966).

61 SULLIVAN, D. J. and DROBECK, H. P.: True hermaphroditism in a rhesus monkey. Folia primat. *4:* 309–317 (1966).

62 SCHULTZ, A. H.: The relative weight of the testes in primates. Anat. Rec. *72:* 387–394 (1938).

63 HILL, W. C. O.: Congenital abnormalities of the urinary tract in primates. Folia primat. *2:* 111–118 (1964).

64 MARUFFO, C. A.: General pathology; in MALINOW Biology of the Howler Monkey. Bibl. primat. *7:* 151–174 (1968).

65 DEGARIS, C. F.: Pericardial patency and partial ectocardia in a newborn orang-utan. Anat. Rec. *59:* 69–82 (1934).

66 STRAUS, W. L., jr.: The visceral anatomy of an infant chimpanzee. J. Mammalogy *18:* 501–507 (1937).

67 CHASE, R. E. and DEGARIS, C. F.: Anomalies of venae cavae superiores in an orang. Amer. J. phys. Anthrop. *24:* 61–65 (1938).

68 HEINRICHS, L. and DILLINGHAM, L. A.: Bornean orang-utan twins born in captivity. Folia primat. *13:* 150–154 (1970).
69 FLEISCHHACKER, H.: Die Frankfurter Gorilla-Zwillinge. Anthropolog. Anz. *31:* 55–69 (1968).
HEDIGER, H. und ZWEIFEL, F.: Primaten-ethologische Schnappschüsse aus dem
70 Zürcher Zoo. Bibl. primat. *1:* 252–276 (1962).
BOURNE, G. H.: Premature chimpanzee triplets at Yerkes. Lab. Prim. Newsletter *6:*
71 14–15 (1967).

Author's address: Prof. A. H. SCHULTZ, Anthropologisches Institut, Universität Zürich, Künstlergasse 15, *Zürich 1* (Switzerland)

Pathology of Simian Primates, Part I, pp. 190–238 (Karger, Basel 1972)

Spontaneous and Induced Neoplasms in Nonhuman Primates[1]

R. W. O'GARA[2] and R. H. ADAMSON

Laboratory of Pathology and Laboratory of Chemical Pharmacology, National Cancer Institute, Bethesda, Md.

Contents

General ... 190
Skeletal System ... 191
Sex Organs ... 193
Skin and Subcutaneous Tissue ... 198
Liver and Biliary System ... 201
Pancreas and Lungs ... 204
Leukemia ... 204
Alimentary Tract ... 209
Kidney and Adrenal Gland ... 214
Recent Developments in Primate Carcinogenesis ... 216
References ... 231

General

Nearly 200 spontaneous tumors have been found in subhuman primates according to data reported in the literature (tables I–VII) although many of them are of doubtful neoplastic nature. RUCH lists one hundred and twenty tumors (including 5 previously unpublished) found prior to 1959, and LUTHER, covering most of the same material, found 132 spontaneous neoplasms reported in subhuman primates through 1961.

1 A part of the work reported in this chapter was supported by contract No. PH-43-66-84. Literature search for material contained herein was completed in 1969.

2 Dr. O'GARA's early death in January 1971 has deprived Primatology of one of the few leading experts in the sphere of Oncology. His loss is deeply regretted by those of us who counted him among our friends. *Ed.*

The word tumor is used here in a broad sense to include all neoplasms as well as all borderline lesions which might be confused with neoplasms, such as polyps, cysts, endometriosis and leukoplakia, and which need to be considered in the differential diagnosis. The word neoplasm is limited to those tumors which show an actual or potential growth pattern beyond what is usually expected for that tissue by competent observers. Whether a neoplasm is benign or malignant depends on the clinical behavior of the neoplasm, and clinical experience with neoplasms in animals has not yet reached the stage where the distinction is as meaningful as in humans. Hence the main distinction in this manuscript is made between neoplasms and non-neoplastic reactive lesions, although whenever possible the original investigator's diagnosis has been allowed to stand.

Skeletal System

Ten spontaneous tumors of the skeletal system have been reported in subhuman primates at ages ranging from 3 to 9 years (table I). Most of these have occurred in the jaws or skull, in contrast to humans where most osseous tumors occur in the long bones. Numerous successful efforts have been made to induce tumors of bone in primates. Notable is the work of Soviet investigators using radioactive substances. PETROV, KROTKINA and co-workers [1951] induced osteogenic sarcomas of the tibia in 3 out of 7 monkeys and baboons by the implantation of radioactive ore. In 1957, KROTKINA and BARABADZE reported the induction of 4 additional bone sarcomas from implants of radium bromide and radium ore. Subsequent manuscripts [PETROV *et al.*, 1958; KROTKINA and BARABADZE, 1964] have reported 4 more osteogenic sarcomas induced by radium compounds. MELNIKOV and BARABADZE also described the induction of fibrosarcomas of the maxilla in 3 rhesus monkeys bearing implants of ^{110}Ag or ^{60}Co in the antrum of HIGHMORE [1960 and 1968]. One of these monkeys also developed a squamous cell carcinoma of the tongue. The latent period for the induction of bone tumors by radioactive substances was 2–3 years for the higher doses of radium and about the same time or a little less for ^{60}Co and ^{110}Ag implants. Malignant tumors have been induced by external irradiation as well. KENT and PICKERING induced fibrosarcoma of periorbital tissue in 2 *Macaca mulatta* monkeys who had received 2,000–3,000 r of ^{60}Co gamma radiation to the region of the right eye 16 to 32 months previously. The maxilla and zygomatic bones were invaded but it is not clear whether the fibrosarcomas arose from the bone or from periorbital soft tissue.

Table I. Spontaneous tumors of bone in primates

Location	Type of tumor	Species	Sex	Age	Remarks	Reference
Skull	Osteoma	*Gorilla gorilla*	–	unknown		PETTIT, 1909
Jaw	Fibrosarcoma	Cochin-Chinese monkey	–	5 years		PEZET, 1920
Mandible	Sarcoma	*Macaca mulatta*	F	adult		BAGG, 1931
Ulna	Fibrosarcoma (with metastasis)	*Papio porcarius*	M	9 years 1 month	Giant cell tumor	RATCLIFFE, 1933
Maxilla	Fibro-osteoma	*Macaca mulatta*	M	6 years	Bilateral	RANKOW, 1947
Ulna	Osteosarcoma	*Lemur catta*	F	c 5 years		WARWICK, 1951
Humerus	Sarcoma	*Macaca mulatta*	F	c 3 years		VADOVA and GEL'SHTEIN, 1960
Maxilla	Osteosarcoma	Monkey	–	–		COHEN and GOLDMAN, 1960
Maxilla	Osteosarcoma	*Macaca mulatta*	M	c 3 years		KENT and PICKERING, 1958
Tibia	Osteoma	Hamadryad baboon	F	10 years		LAPIN and YAKOVLEVA, 1963

A fibrosarcoma of the mandible was observed by GOWGIEL in a *Macaca rhesus* monkey 25 months after he had received 7,500 r of external fractional irradiation to the head. This monkey was 28 months old at the time of irradiation, and was the only one of a group of 14 similarly treated monkeys to develop a malignant tumor. Another bone tumor, an osteogenic sarcoma, was reported by VALERIO and co-workers in a 10½-year-old *Macaca mulatta* monkey who had received 608 rep of whole body neutron. irradiation over a period of 39 months. The latent period was about 9 years In addition to the osteogenic sarcoma, the monkey also had a transitional cell carcinoma of the right renal pelvis, a papillary cystadenocarcinoma of the left kidney, and a basal cell carcinoma of the abdominal skin.

One pleomorphic sarcoma of the tibia has also been induced [KROTKINA, 1948] in a rhesus monkey by the intramedullary implantation of a paraffin pellet containing 30 mg of 3-methylcholanthrene (3-MC), after a latent peri, od of 3½ years. Three other monkeys similarly treated, as well as 5 who had received 15 mg of 9:10 dimethyl 1,2-benzanthracene (DMBA) and one who had received 50 mg of 3-MC did not develop tumors. Earlier experiments on 11 rhesus monkeys using implants of smaller amounts of DMBA and 3-MC (5–10 mg) also failed to induce sarcomas in the tibia, although 6 of these survived for more than 7 years after implantation.

It seems clear that ionizing radiation is as well established as a cancerogenic agent in subhuman primates as it is in man, and that contrary to usual opinion the time required to induce tumors in primates is not excessively long if a sufficiently strong stimulus is used. The latent period for both internal and external ionizing irradiation has been as short as 16–19 months (table IXa).

Sex Organs

Forty one tumors of female sex organs have been reported in 39 primates. Twenty four were tumors of the uterus including 12 of the uterine body (mostly fibroleiomyomas and polyps) and 12 of the uterine cervix (table II). In addition there have been observed 8 tumors of the ovary (fig. 1), 6 of the mammary gland, and 3 of the vagina. Two monkeys had double primary tumors: fibromyomata of uterus and granulosa cell tumor of ovary; adenoma of uterine cervix and ovarian cyst; and one had three separate tumors: adenocarcinoma of breast, renal adenoma, and endometrial cyst of ovary. At least 12 of these uterine tumors however are questionably neoplastic. Attempts to induce neoplasia in sex related organs in the female primate have been unsuccessful except for one tumor of the breast induced by implantation of ^{110}Ag (table IXa). OVERHOLSER and ALLEN [1935] reported atypical epithelial growth in the cervix of 4 ovariectomized monkeys induced by repeated cervical trauma and large doses of estrin or estrin combined with corpus luteum extract for 20 to 90 days. However, only two monkeys which received estrin and corpus luteum extract for over 80 days showed cervical lesions which resembled neoplasms. Subsequently numerous investigators treated monkeys intensively with estrogens but no tumors were induced [HISAW and LENDRUM, 1936; WAGENEN, 1937; ZUCKERMAN, 1937; ENGLE *et al.*, 1943; VARGAS, 1943; VADOVA and SMOYLOVSKAYA, 1959; GESCHICKTER and

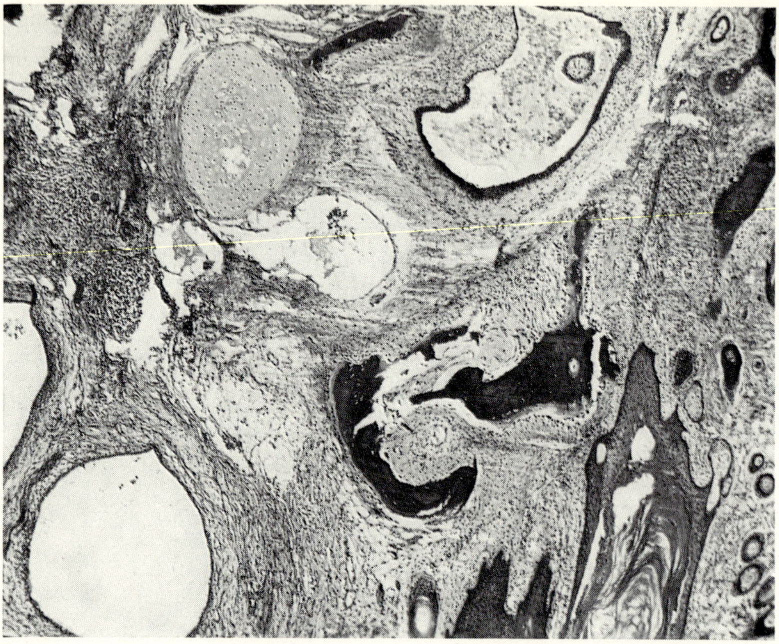

Fig. 1. Dermoid tumor of the ovary in a young rhesus monkey. Islands of cartilage and squamous epithelium can be seen as well as hair follicles and cysts lined by squamous or cuboidal epithelium. Hematoxylin and eosin. ×55. Courtesy of Hazleton Laboratories, Falls Church, Virginia [CREWS *et al.*, 1967].

HARTMAN, 1959]. PFEIFFER and ALLEN made a heroic attempt to induce neoplasia in rhesus monkeys with polycyclic hydrocarbons and estrogens. Estrogenic compounds were injected or implanted subcutaneously, intramuscularly or directly in the breast. Carcinogenic polycyclic hydrocarbons were injected or implanted subcutaneously, painted on the skin, or dusted as crystals into the peritoneal cavity; some was sewed into omental pockets or placed directly on the uterus; and pellets of carcinogens were implanted in the myometrium, in the cervix, in the submucosa of the vagina, in the mammary glands or at other sites. Methylcholanthrene was also given intravenously and orally in large doses. Out of 50 monkeys, some of which were treated over a period of 11 years, none developed cancer although 12 monkeys survived 3 years of treatment, and 5 survived for 5 to 11 years. Metaplasia of mammary ducts and cervical lymph nodes occurred with both estrogen and carcinogenic hydrocarbons given alone or in combination.

Table II. Spontaneous tumors of the female and male sex organs in primates

Location	Type of tumor	Species	Sex	Age	Remarks	References
Uterus	Myoma	Monkey	F	unknown		Sutton, 1885
Uterus	Leiomyoma	Black lemur	F			Schurmann, 1915
Uterus	Leiomyofibroma	*Lemur macaco*	F	unknown		Fox, 1923
Abdominal cavity	Endometriosis	*Macaca rhesus*	F	adult		Fraser, 1929
Uterus	Endometrial cystadenoma	*Papio cynocephalus*	F	>17½ years		Ratcliffe, 1932–1933
Uterus	Adenomyoma	*Papio cynocephalus*	F	very old		Hamerton, 1933
Uterus	Fibroleiomyoma	Long-tailed monkey	F	mature		Oshima, 1937
Uterus	Fibromyomata	*Saimiri sciurea*	F	aged	Also granulosa cell tumors of ovaries	Rewell, 1950
Uterus	Papillomatous excrescences	*Macaca mulatta*	F	>1 year	Pregnancy during year	Hill, 1954
Uterus and ovary	Carcinoma	*Macaca philippinensis*	F			Newberne and Robinson, 1960
Uterus	Leiomyoma	Long-tailed monkey (Meerkatze)	F			Kronberger, 1962
Uterus	Endometrial polyp	Hamadryad baboon	F	10 yr		Lapin and Yakovleva, 1963
Uterine cervix	Adenoma or carcinoma	Monkey	F			Hutchinson, 1901
Uterine cervix	Papilloma	*Macaca sinica*	F	>3½ years		Ratcliffe, 1933

Table II (continued)

Location	Type of tumor	Species	Sex	Age	Remarks	References
Uterine cervix	Carcinoma	*Macaca sinica*	F	unknown		FOX, 1936
Uterine cervix	Carcinoma with metastases	*Macaca mulatta*	F	old		HISAW and HISAW, 1958
Uterine cervix	Adenoma	*Macaca philippinensis*	F	young adult	Also simple ovarian cyst	RUCH, 1959
Uterine cervix	Carcinoma-*in-situ*	*Macaca mulatta*	F			STERNBERG, 1961
Uterine cervix	Polyps	Macaques (6)	F	8–22 years	6 cases	LAPIN and YAKOVLEVA, 1963
Vagina	Leukoplasia	*Cercopithecus mona*	F	unknown		PETTIT, 1904
Vagina	Papillomatosis	*Macaca mulatta*	F	3 years 5 months		VADOVA and GEL'SHTEIN, 1956
Vagina	Papillomatosis	*Macaca mulatta*	F	9 yr 8 months		VADOVA and GEL'SHTEIN, 1956
Ovary	Squam. carcinoma	*Macaca cynomolgus*	F	young	Metastasis to lung	JUNG, 1924
Ovary	Adenocarcinoma	*Macaca irus*	F	>13 years		FOX, 1937
Ovary	Granulosa cell tumors	*Saimiri sciurea*	F	aged	also fibromyomata of uterus	REWELL, 1950
Ovary	Simple cyst	*Macaca mulatta*	F	16 years		VADOVA and GEL'SHTEIN, 1956
Ovary	Simple cyst	*Macaca philippinensis*	F	young adult	also adenoma of cervix	RUCH, 1959

Ovary	Sarcoma	Anubis Baboon	F	20 years	Metastases	LAPIN and YAKOVLEVA, 1963
Ovary	Serous cystadenoma	Rhesus	F			FLINN, 1967
Ovary	Dermoid cyst and papillary serous cystadenoma	Rhesus	F	adult	weighed 168 g	CREWS *et al.*, 1967
Breast	Fibrosarcoma	Green monkey				O'CONNER, 1947
Breast	Carcinoma	*Macaca mulatta*	F	older adult		RUCH, 1959
Breast	Adenocarcinoma	*Macaca mulatta*	F	10 years	also had renal adenoma and endometrial cyst of ovary	VADOVA and GEL'SHTEIN, 1956
Breast	Adenocarcinoma	*Tupaia glis*	F		pregnant	ELLIOT *et al.*, 1966
Breast	Carcinoma	*Pongo pygmalus* (Orang-utan)	F	15 years		BRACK, 1966
Breast	Carcinoma	*M. mulatta*	F	9 years	X-irradiated	CHAPMAN, 1968
Testes	Fat bodies	Monkey	M	unknown	hermaphroditic	SUTTON, 1885
Testes	Adenocarcinoma	*Cynocephalus hamadryas*	M			VORONOFF, 1949
Testes	Seminoma	Orangutan	M			VORONOFF, 1949
Testes	Seminoma	Howler (*Alouatta caraya*)	M	adult		MARUFFO and MALINOW, 1966
Prostate	Papillomatous adenoma	*Lemur catta*	M	unknown		FOX, 1923
Prostate	Adenoma	*Lemur mongoz*	M	8 years 3 months		RATCLIFFE, 1933
Prostate	Carcinoma	*Macaca mulatta*	M	very old		ENGLE and STOUT, 1940

Tumors of the sex organs in the male primate are fewer than in the female. Only 4 spontaneous tumors of the testis and 3 of the prostate have been reported, and no claims of induced tumors have been made. In view of the fact that coal tar, mineral oil, and chimney soot have been incriminated in cancer of the scrotum in man, it is surprising that so little work has been done with these substances in subhuman primates.

Skin and Subcutaneous Tissue

Spontaneous tumors of the skin and subcutaneous tissue have been seen in a few monkeys (table III). One of these was a papilloma in a brown Cebus monkey which LUCKÉ and collaborators were able to transmit to 11 other monkeys of the same and different species. BEARCROFT and JAMIESON [1958] observed an epidemic of subcutaneous tumors of the face and limbs in 20 rhesus monkeys. The tumors were of a pleomorphic cell type and reached a diameter of 2–4 cm before spontaneously regressing. The tumor cells contained eosinophilic inclusions in the cytoplasm and the tumor was transmissible to other monkeys inoculated with lightly centrifuged emulsions of tumor cells. Subcutaneous tumors have also been induced in monkeys by injections of ROUS chicken sarcoma virus (especially the SCHMIDT-RUPIN and CARR-ZILBER strains) and have grown to considerable size (16–20 g). Tumors have also appeared after the injection of tissue culture cells derived from monkey heart [SALK and WARD, 1957] but these soon regressed. We have already mentioned the induction of carcinomas of the skin and mucous membranes by ^{110}Ag and by radiation, and in addition there has been one melanoblastoma induced by ^{110}Ag (table IXa). Many attempts to induce skin or subcutaneous tumors by chemical agents (table IXb) have resulted only in granulomas, epithelial hyperplasia and acanthosis (fig. 2). Tumors have appeared at the site of treatment but they do not have the characteristics of true neoplasms and regress after the stimulus is removed. BONNE [1930] reported carcinomas of the skin in 20 *Macacus cynomolgus* monkeys painted twice weekly with coal tar; however the 'carcinomas' promptly regressed when the painting was stopped. PFEIFFER and ALLEN produced hyperkeratinization, and papillomatous masses or wart-like excrescences by painting polycyclic hydrocarbons on the monkey's skin, and fibrotic masses developed about implanted pellets. Similar results were noted more recently by other investigators upon intradermal injection of carcinogenic hydrocarbons [O'GARA and KELLY, 1965] or by skin painting [ADACHI *et al.*, 1969]. In all cases the reactions

Table III. Spontaneous tumors of skin, subcutaneous tissue and miscellaneous sites

Location	Type of tumor	Species	Sex	Age	Remarks	Reference
Skin	Papilloma	*Cebus apella fatuellus*		unknown	Transmitted to 11 other monkeys	LUCKÉ *et al.*, 1950
Skin of rump	Carcinoma	*Macacus maurus*	M	9 years +		RATCLIFFE, 1955
Sexual skin	Squamous cell carcinoma	Hamadryad baboon	F	19 years		LAPIN and YAKOVLEVA, 1963
Scalp	Sarcoma	*Cercopitheque*		unknown		PLIMMER, 1914
Subcutaneous	Pleomorphic cell tumor	*Macaca mulatta*	M and F	varied	20 monkeys (Yaba virus)	BEARCROFT and JAMIESON, 1958
Subcutaneous	Hemangioendothelioma	*Macaca mulatta*	M	5 months		KENT and PICKERING, 1958
Subcutaneous	Lipoma and 2 fibromas	*M. mulatta* and *M. nemestrina*	M and F	5–15 years	3 monkeys	CHAPMAN, 1968
Brain		*Pan* sp.	M	unknown		SELLHEIM, 1934
Brain	Gliosarcoma	*Papio doguera*	M	6 years		SELLHEIM, 1936
Brain	Angioma	*Macaca rhesus*		unknown	out of 4,000 monkeys examined	UNTERHARNSCHEIDT, 1964
Peripheral nerve	Melanoma or spindle-cell carcinoma	*Macaca mulatta*	M	5½ years		BLOKHIN *et al.*, 1955
Eyeball	Glioma	*Macaca radiata*		young		SUTTON, 1885
Ocular conjunctiva	Fibroma	*Cercopithecus aethiops sabaeus*				WADSWORTH, 1953
Pituitary and optic tract	Carcinoma	Baboon		unknown		SUTTON, 1885
Pituitary	Adenoma	*Macaca mulatta*	M	3½ years		KENT and PICKERING, 1958
Thyroid	Adenoma	*Papio comatus*	M	30 years }	Double tumor	FOX, 1936
Omentum	Fibroma	*Papio comatus*	M	30 years }	Double tumor	FOX, 1936
Abdominal wall	Lipoma	*Cercopithecus diana*	F	adult	Postpartum	RATCLIFFE, 1942
Nose	Fibromyxoma	*Pan* sp.	M	11 years	Surgically excised	NICHOLS, 1939
Heart	Growth	*Lemur catta*		unknown		SCOTT, 1927
Pericardium	Adenocarcinoma	*Cebus apella*	F	2 years +		IGLESIAS and LIPSCHUTZ, 1947

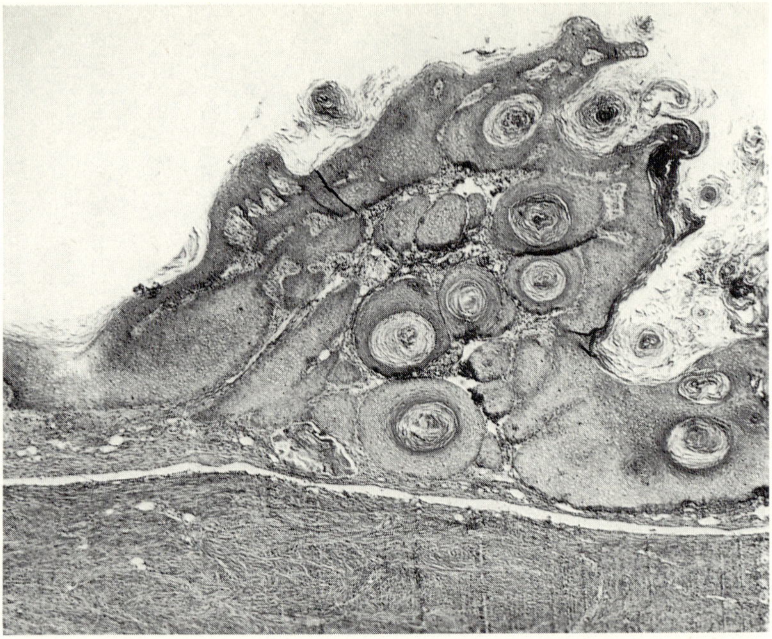

Fig. 2. Keratoacanthoma of skin in a 5-month-old monkey six weeks after three intradermal injections of 3-methylcholanthrene. Hematoxylin and eosin. ×60.

subsided upon removal of the stimulus. However, SUGIURA *et al.* [1956] were successful in inducing true squamous cell carcinoma of the skin in 3 of 6 rhesus monkeys painted with a high-boiling catalytically cracked oil. The induction time in this experiment was long (4–5 years) but it showed that tumors could be induced if the right carcinogen was used.

Recent investigations indicate that there may be a decided difference between the reaction of prosimian primates and higher subhuman primates to the application of carcinogenic hydrocarbons. Subcutaneous fibrosarcomas were obtained in 3 of 4 *Tupaia glis* (tree shrews) by NOYES [1968] in 6 months following the injection of 10 mg of benzpyrene (table IXb). Other investigators have induced fibrosarcomas in prosimian primates with polycyclic hydrocarbons but have not yet published their results. These include 3 sarcomas in tree shrews [ADAMSON *et al.*] (fig. 3) and a fibrosarcoma in galago [COOPER and KELLY]. A small fibrosarcoma has also been induced in a marmoset *(Tamarinus nigricollis)* by the injection of 2 mg of 3-methylcholanthrene [LEVY, 1963], and 2 sarcomas were induced in a cottontop marmo-

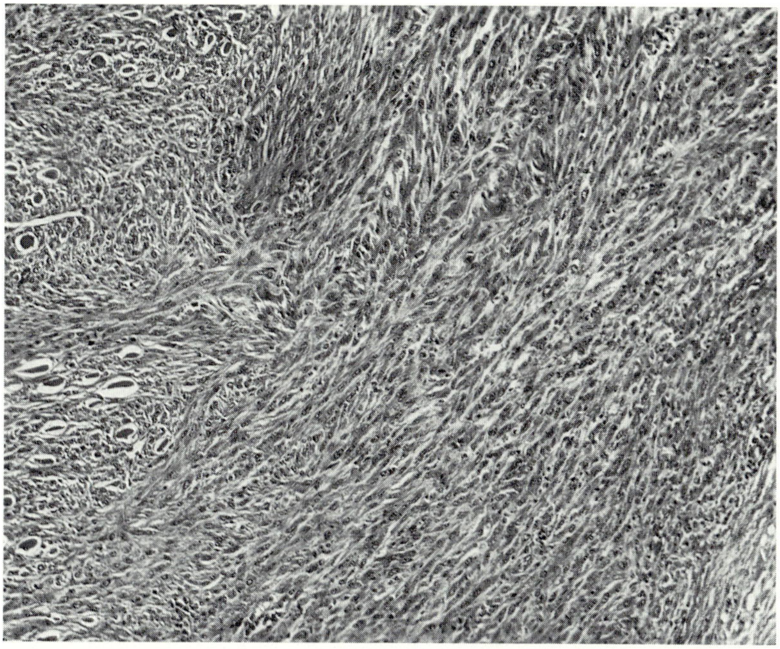

Fig. 3. Spindle cell fibrosarcoma of the subcutaneous tissue of a tree shrew *(Tupaia glis)* 20 months after an injection of 3-methylcholanthrene. Hematoxylin and eosin. ×135.

set *(Saguinus oedipus)* by NOYES [1969] by the subcutaneous injections of benzo(a)pyrene and 7,12-dimethylbenz(a)anthracene respectively.

Liver and Biliary System

Spontaneous tumors of the liver and biliary system are rarely seen in subhuman primates (table IV). Only 4 spontaneous tumors of the liver and 3 of the gall bladder are reported (fig. 4). However hepatomas and hepatic cell carcinomas of the liver are easily induced in several species of monkey (rhesus, cynomolgus, African greens, and South American capuchins) by the administration of diethylnitrosamine (DENA) either in the diet or by intraperitoneal injections. Approximately 50 hepatic cell neoplasms (fig. 5, 6) and 2 hemangioendotheliomas of liver or spleen (fig. 7) have been induced in monkeys during the past 8 years by DENA and one hepatoma by 1-Nitroso-piperidine [KELLY *et al.*, 1966; and ADAMSON *et al.* unpublished]. The tumors have

Table IV. Spontaneous tumors of the pancreas, liver and lung

Location	Type of tumor	Species	Sex	Age	Remarks	Reference
Pancreas	Adenocarcinoma	*Cercopithecus aethiops sabaeus*	F	14 years +	metastasized	RATCLIFFE, 1933
Pancreas	Adenocarcinoma	*Papio cynocephalus*	F	17 years +	also papillary cystadenoma of uterus	RATCLIFFE, 1933
Near pancreas	Fibroadenoma	Hoolock gibbon	F	15 years +		WADSWORTH, 1961
Liver	Adenoma of bile ducts	*Macaca fuscatus*	M			RATCLIFFE, 1932, 1933
Liver	Adenoma	*Presbytis entellus*		unknown		ZVETAEVA, 1941
Liver	Hepatoma	*Lemur macaco*	F	unknown		BRYGOO *et al.*, 1964
Liver	Hemangioendothelioma	*Macaca mulatta*	M	young	X-radiation (800 rads)	WOODRUFF and JOHNSON, 1968
Gall bladder	Adenocarcinoma	*Papio papio*	M	24 years +		FOX, 1938
Gall bladder	Adenocarcinoma	*Papio papio*	F			LOMBARD and WITTE, 1959
Gall bladder	Adenocarcinoma	*Macaca mulatta*	F	ca 10 years	metastasized	O'GARA *et al.*, 1967
Lung	Carcinoma	*Cercopithecus diana*	F			RATCLIFFE, 1942
Lung	Carcinoma	Squirrel monkey (*Saimiri sciurea*)	M			LOMBARD and WITTE, 1959
Lung	Alveolar carcinoma	Howler (*Alouatta caraya*)	F			MARUFFO, 1967

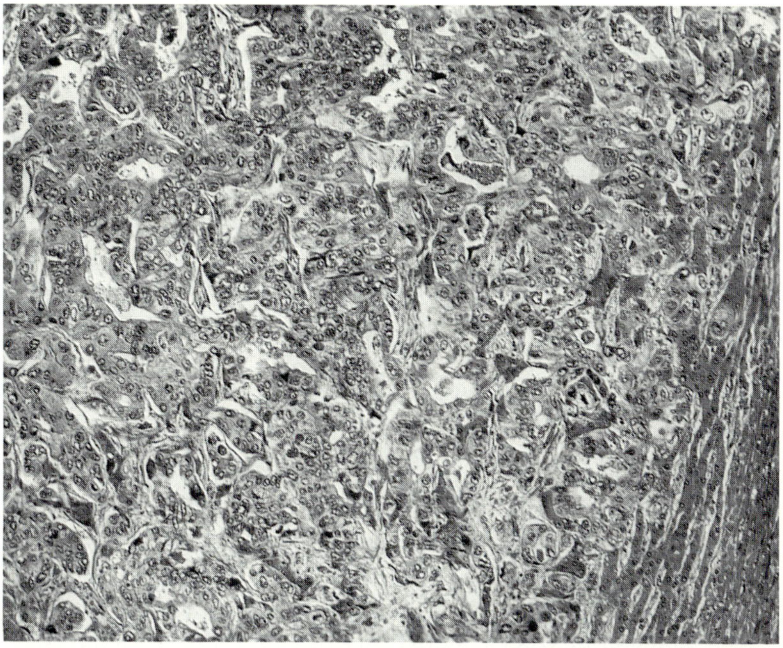

Fig. 4. Spontaneous carcinoma of the gall bladder in a 10-year-old *M. mulatta.* A portion of invaded liver is visible in the lower right corner. Hematoxylin and eosin. ×135.

appeared in an average of 22 months when DENA is given by the oral route and about 17 months when given parenterally. One of these tumors has been grown successfully in tissue culture for long periods of time [DAWE *et al.*, 1968] and repeated successful transplantations have been made intracerebrally into newborn monkeys both from tissue culture, and from monkey to monkey [KELLY *et al.*, 1967; WALKER, unpublished]. An interesting feature of the primary induced liver tumors has been the reappearance of a specific fetal protein in the serum of a high proportion of monkeys bearing such tumors [HULL *et al.*, 1969]. The specific protein, and α-fetoglobulin, is one which is produced by the immature liver cells of the fetus but disappears shortly after birth, and reappears only with the development of hepatic cell tumors. A similar protein has been described in the serum of humans with tumors of hepatic cell origin and with certain embryonal tumors of the testis.

The influence of other well known carcinogens on the monkey liver has also been studied over a relatively long period of time. The primate liver has proven refractory to the effects of carcinogenic azo dyes, acetylaminofluorene, and urethane, when administered for a period of up to 8 years. Morphologic changes in the liver of monkeys receiving these carcinogenic compounds has been minimal and no tumors have been induced. Feeding or intraperitoneal injections of aflatoxin B_1 results in striking histological changes in the liver of monkeys for a few months, but these toxic changes later disappear as the monkey apparently becomes more tolerant to the aflatoxin. Several groups of investigators have fed aflatoxin to monkeys for periods of up to 30 months without obtaining tumors [CUTHBERTSON *et al.*, 1967; ADAMSON *et al.*, unpublished]. The addition of cycasin (an ingredient of the cycad nut) to the diet of monkeys also produces toxic hepatic changes in the liver but not as severe as with aflatoxin [ADAMSON *et al.*, unpublished]. It is still too early to evaluate the eventual effect of long-term administration of aflatoxin and cycasin on the liver of primates.

Pancreas and Lungs

Only 3 spontaneous tumors of the pancreas have been seen, all in monkeys at least 14 years old, and no induced tumors of this organ have been reported. Spontaneous tumors of the lung are rarely seen in the subhuman primate. Three spontaneous tumors have been reported and a fourth tumor was found in a macaque which had been exposed to beryllium inhalation [SCHEPERS, 1964]. The tumor was a 0.3 cm alveolar carcinoma which had both a papillomatous and acinar component. In addition, VORWALD [1966] reported the induction of six cancers of the lung in macaque monkeys by bronchial intubation with beryllium.

Leukemia

Fifteen cases of spontaneous tumors of the leukemia-lymphoma type have been reported, distributed rather evenly between the green and macaque monkey, the gibbon and the baboon (table V). Reliable reports on in-

---

Fig. 5. a Hepatic cell carcinoma of the liver in a 12-month-old *Cercopithecus aethiops* (African Green monkey) treated with multiple intraperitoneal injections of nitrosodiethylamine. *b* The tumor cells are arranged in trabeculae which are several cells thick and frequently contain a central channel. Between the trabeculae, there are blood channels, lined by endothelial cells and Kupffer cells, which often contain foci of hematopoietic cells (upper center). Metastases were found in the lungs. Hematoxylin and eosin. ×135.

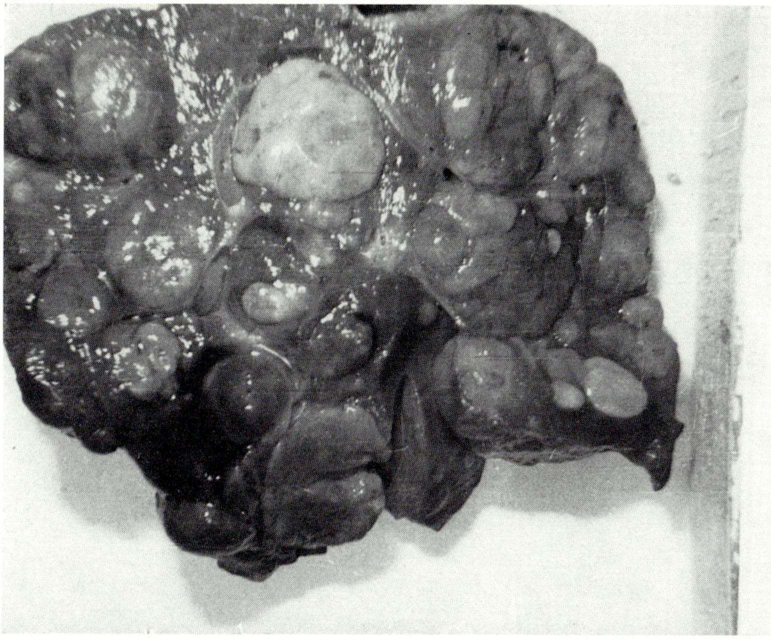

a

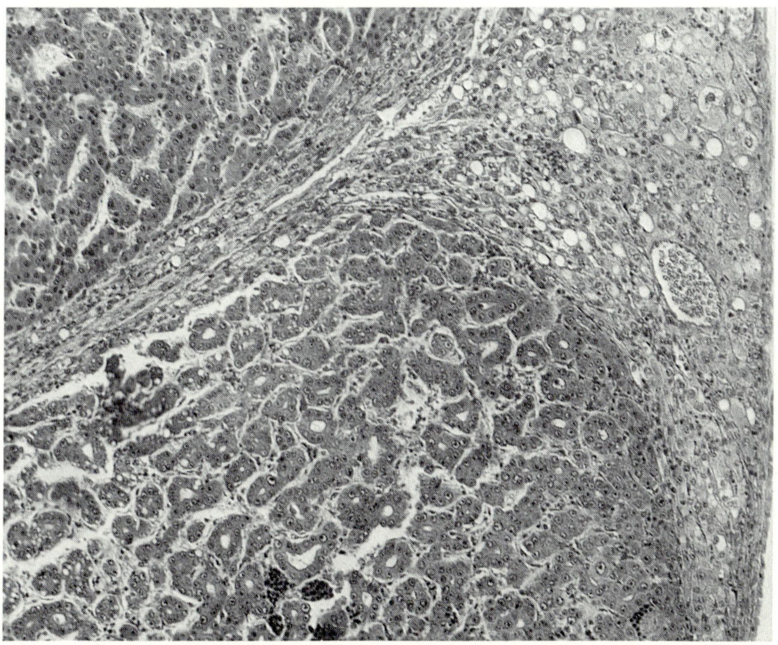

b

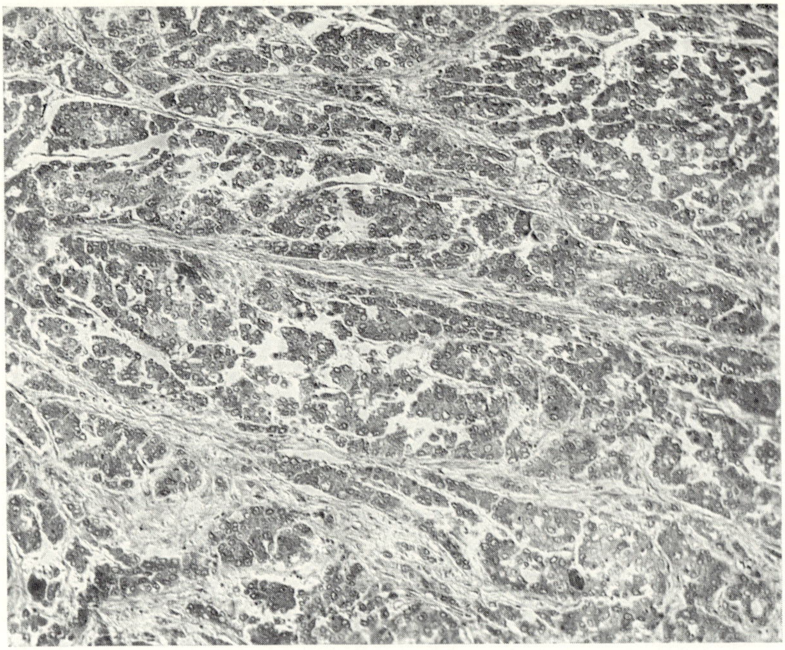

Fig. 6. Hepatic cell carcinoma of the liver in a 29-month-old Cynomolgus monkey *(Macaca mulatta fascicularis)* who received nitrosodiethylamine orally since birth. The tumor cells are poorly differentiated but some evidence of trabecular formation can be seen. Widespread metastases were found in lungs, pleura, diaphragm, spleen, omentum, peritoneum, and inguinal and pancreatic lymph nodes. Hematoxylin and eosin. ×125.

duced leukemia in primates have been relatively rare (table IX). BUNTING and YATES [1913–1914] described early HODGKIN's-like changes in lymph nodes of rhesus monkeys treated with bacteria isolated from human Hodgkin's dis-

---

Fig. 7. Hemangioendothelioma of the spleen in a 3-year-old *Macaca mulatta* monkey who had received nitrosodiethylamine by intraperitoneal injection for 22 months. There are numerous proliferating endothelial cells forming abortive small vascular channels. A portion of spleen can be seen in the lower left corner. This monkey also had hepatic cell carcinoma of the liver with metastases to lung and abdominal lymph nodes. Two courses of therapy with 6-mercaptopurine were ineffective in arresting the course of the hepatic tumor. Hematoxylin and eosin. ×135.

Fig. 8. Peripheral blood smear. Acute myelogenous leukemia in a 17-month-old female *Macaca mulatta* treated with procarbazine hydrochloride since birth. Nucleated cells in the peripheral blood rose to 352,000 on the day of death, mostly immature cells of the myeloid series. Wrights stain. ×690.

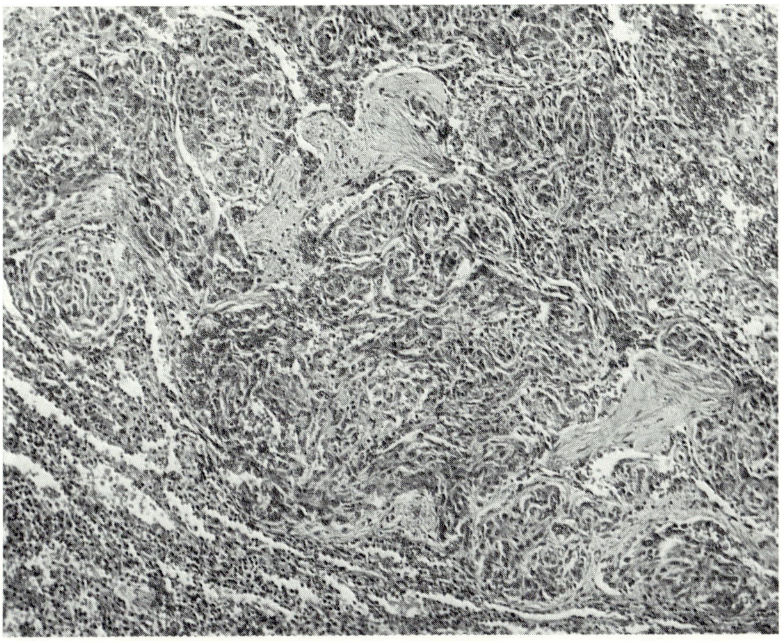

7

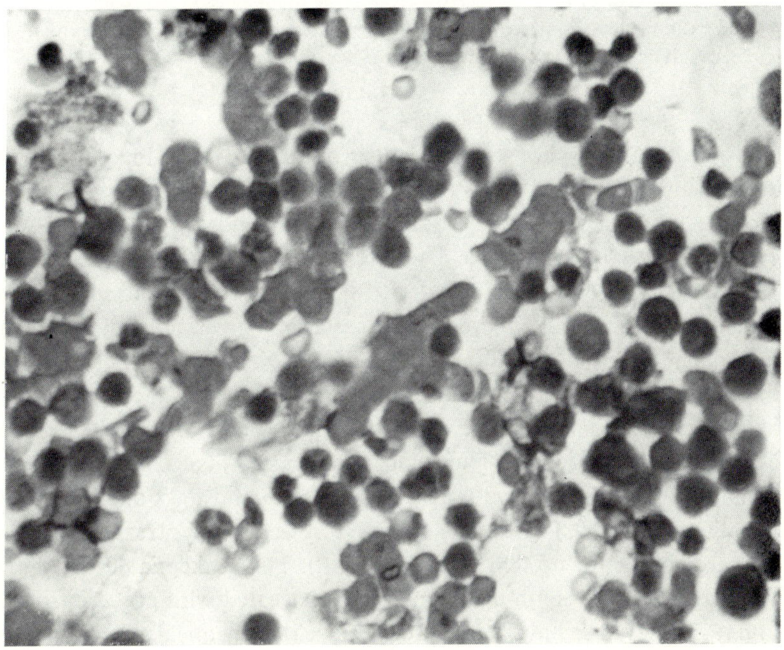

8

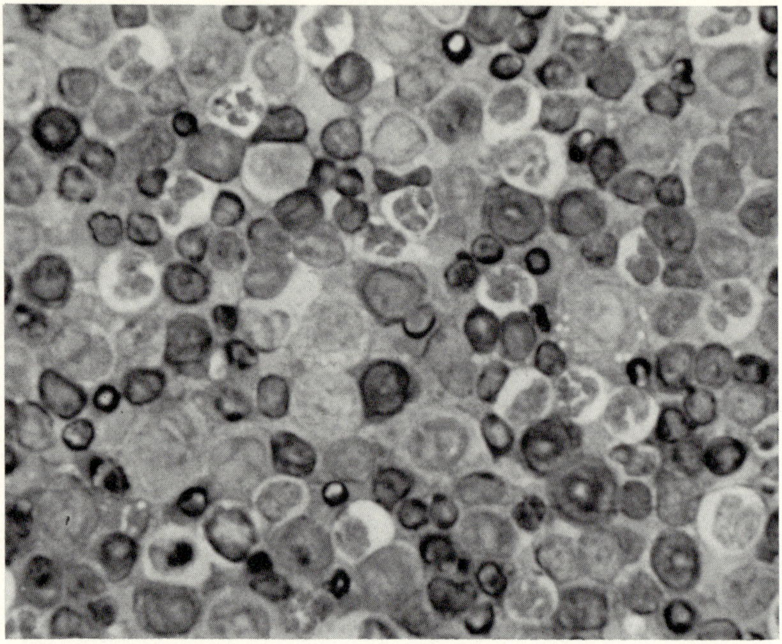

Fig. 9. Smear of femoral marrow from same monkey as in figure 8. The marrow is packed with myelocytes, promyelocytes, and myeloblasts. A few more differentiated myeloid cells can be seen as well as a few erythroblasts. Giemsa stain. ×690.

ease (table IXC). However, since then numerous other investigators have been unsuccessful in attempts to transmit Hodgkin's disease from humans to monkeys [LONGCOPE, 1907; TORREY, 1916; CUNNINGHAM and MCALPINE, 1923; M. J. STEWART, 1924; DELEON and REYES, 1929; H. L. STEWART, 1932]. KROTKINA and BARABADZE [1958] described a myeloid sarcoma and a reticulocytic sarcoma induced in a rhesus and a green monkey by the intramedullary injection of radioactive material. ZALUSKY and co-workers induced a generalized myelocytic leukemia in one of 5 *Macaca mulatta* monkeys given fractional doses of whole body neutron irradiation (608 rep) after a latent period of 7 years. Numerous recent attempts to induce leukemia in primates with viruses have not been successful and only one leukemia has been seen in the NCI Primate Colony No. 1 in monkeys treated with chemical carcinogens for up to 8 years (fig. 8–10). This leukemia occurred several years ago in a 14-month-old monkey being treated with procarbazine HCl, and it is believed to have been induced since this chemical is highly leukemo-

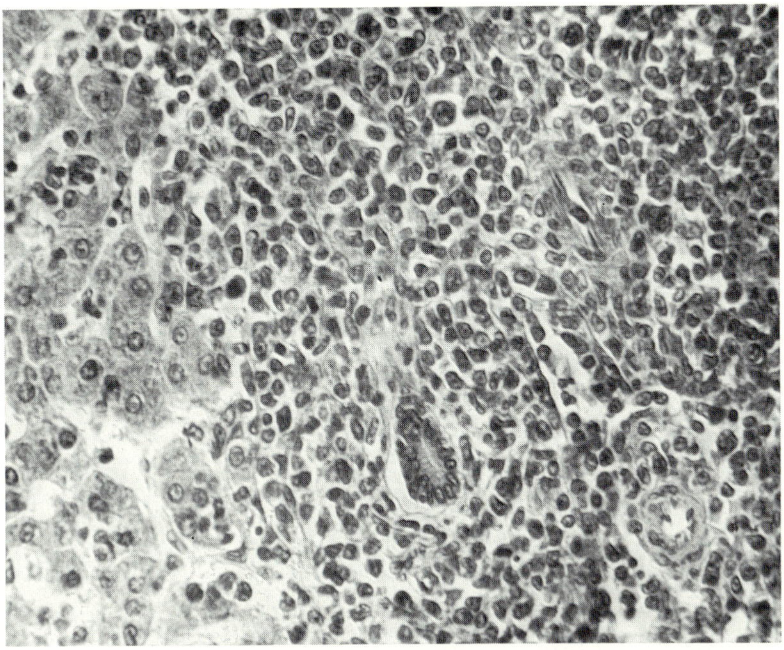

Fig. 10. Leukemic infiltration of the liver in the same monkey. A residual bile duct and hepatic artery can be seen in the lower right and residual hepatic cells on the left. There was also leukemic infiltration in the heart, spleen, lymph nodes, kidneys, skin, diaphragm, and meninges. Hematoxylin and eosin. ×430.

genic in mice. However, no additional leukemias have been seen since in this group, although one or two other monkeys have had hyperplastic bone marrows. Another group of monkeys in this colony has been treated with urethane orally plus fractional doses of whole-body X-radiation for a total dose of 500 r, with no leukemias induced in periods up to 6 years post-radiation.

Alimentary Tract

Thirteen spontaneous tumors of the oral cavity have been reported, exclusive of bone tumors. All except 2 have been carcinomas of squamous cell type (table VI and fig. 11). Five of these tumors were carcinomas of the tongue and 3 of these occurred in a single laboratory within a period of 6 years [STEINER *et al.*, 1942; KLÜVER and BRUNSCHWIG, 1947], but no causative agent or condition was ever established. Tumors of the stomach repres-

Table V. Spontaneous leukemias and lymphomas in primates

Location	Type of tumor	Species	Sex	Age	Remarks	Reference
Blood. Enlarged spleen and lymph nodes	Chronic lymphocytic leukemia	*Cynocephalus sphinx*			ave. WBC 36,000 with 82% lymphocytes and monocytes	MASSAGLIA, 1923
	Chronic lymphocytic leukemia	Green monkey				CORSON-WHITE, 1929
Lymph nodes	Lymphosarcoma	Chacma baboon	M			NOBACK, 1934
Ribs, pleura, mediastinum, lymph nodes and spleen	Multiple myeloma or myelogenous leukemia[1]	Mottled fox monkey		adult		OSHIMA, 1937
	Lymphocytic leukemia	Stairs monkey *(Cercopithecus albogularis Stairsi)*		5½ + years		HAMERTON, 1942
	HODGKIN's sarcoma	Baboon		unknown		KELLY, 1948
Abdominal viscera and lymph nodes	Reticulosarcoma	*Papio ursinus*	F	6 years +	subtotal thyroidectomized	GILLMAN and GILBERT, 1954
	Leukemia (HODGKIN)	Green monkeys			3 cases	LAPIN and YAKOVLEVA, 1956, 1963
	Malignant lymphoma	*M. mulatta*	F	7 years		KENT and PICKERING, 1958
Lower lip	Lymphosarcoma	*M. mulatta*				GORLIN. Quoted in COHEN and GOLDMAN, 1960
Liver, kidneys, spleen and lymph nodes	Malignant lymphoma	Gibbon *(Hylobates sp.)*	M			NEWBERNE and ROBINSON, 1960
Mesentery, liver, spleen and lymph nodes	Burkitt lymphoma	White handed gibbon *(Hylobates lar)*	F	4½ years		DIGIACOMO, 1967
Generalized	Acute lymphocytic leukemia	White-cheeked gibbon *(Hylobates concolor)*	M	1½ years	massive cerebral involvement	DE PAOLI and GARNER, 1968

1 Can be translated from Japanese as either myeloma or myeloid leukemia.

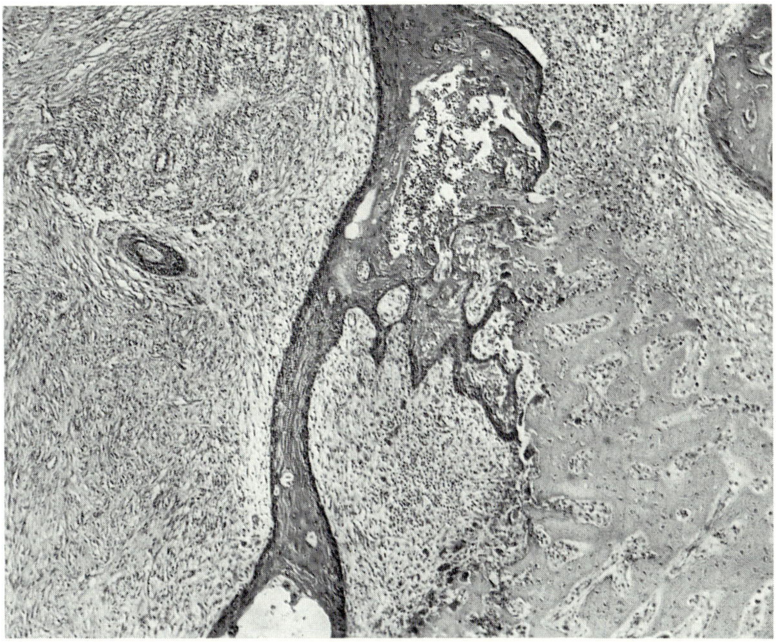

Fig. 11. Epithelial tumor of the mandible of a 22-year-old blackfaced spider monkey. There is a mass of loose fibrous tissue containing a strand of neoplastic squamous cells which is invading adjacent bone. Hematoxylin and eosin. ×65 (Courtesy of Dr. J. A. CRAN, University of Adelaide, South Australia).

ent the largest group of spontaneous tumors in nonhuman primates. There are 31 cases, all except five in macaques. However in 14 of these cases the tumors were associated with the presence of the nematode, *Nochtia nochti*, and probably were more of a hyperplastic reaction than a genuine neoplasm. Later BONNE and SANDGROUND [1939] were able to induce similar lesions in the stomach by administering adult worms orally to two normal monkeys. Most of the spontaneous tumors not associated with parasites were also benign and have been classified as adenomas, papillomas or polyps. Only 4 were considered malignant and only one of these metastasized. LUSHBAUGH [1947, 1949] reported the induction of infiltrating adenomatous lesions of the stomach and colon of rhesus monkeys by spraying their bodies or by feeding of diesel motor lubricating oil. These lesions resembled the lesions produced by BONNE and SANDGROUND in the stomachs of Javanese monkeys by parasitic nematodes and were not considered to be malignant neoplasms. Indeed pho-

Table VI. Spontaneous tumors in primates; alimentary tract

Location	Type of tumor	Species	Sex	Age	Remarks	Reference
Mouth	Epithelioma	*Macaca mulatta*	M	adult		ZUCKERMAN, 1930; HAMERTON, 1930
Tongue	Squam. Carcinoma	*Macaca irus*	M	16 years		STEINER *et al.*, 1942
Tongue	Squam. Carcinoma	*Macaca mulatta*	M	16 years		STEINER *et al.*, 1942
Tongue	Squam. Carcinoma	*Macaca mulatta*	M	14 years		KLÜVER and BRUNSCHWIG, 1947
Cheek	Squam. Carcinoma	*Saimiri sciurea*	F	8 years		KLÜVER and BRUNSCHWIG, 1947
Tongue	Squam. Carcinoma	*Macaca mulatta*	M	25 years		KROTKINA, 1956
Tongue	Carcinoma	Macaque	–			HEMMENS, unpublished (see RUCH)
Gingiva	Squam. carcinoma	Baboons (2)			2 cases	COHEN and GOLDMAN, 1960
Maxilla	Odontoma	Chimpanzee				COHEN and GOLDMAN, 1960
Mandible	Squam. carcinoma	*Macaca mulatta*	F	9 years	Local metastases	SASAKI *et al.*, 1961–1962
Mandible	Squam. carcinoma	Spider monkey		22 years		CRAN, 1969
Lip	Fibroma	Baboon				COHEN and GOLDMAN, 1960
Pharynx	Squam. papilloma	*Hylobates lar*				FOX, 1934
Esophagus	Epidermoid carcinoma	*Macaca fuscata*	M	15 years +		FOX, 1932; RATCLIFFE, 1932
Esophagus	Leiomyoma	*Macaca mulatta*	M	28 years +	also adenocarcinoma kidney and gastric polyp	RATCLIFFE, 1940
Stomach	Adenocarcinoma	*Cercopithecus rufoviridis*		8–10 years		SCHMEY, 1914

Stomach	Carcinoma	*Macaca sinica*	F	adult		McCarrison, 1919
Stomach	Papillary Adenoma	*Comopithecus hamadryaus*	M	2 years +		Fox, 1923; Ratcliffe, 1933
Stomach	Benign Adenoma	*Macaca mulatta*				Hamerton, 1930
Stomach	Fibroadenomatous polyps	*Macaca irus*			8 cases (worms)	Finkeldey, 1931
Stomach	Adenopapillomas	*Macaca mordax*	M and F		6 cases (worms)	Bonne and Sandground, 1939
Stomach	Polyadenoma	*Cynocephalus hamadryas*			observed in 1,100 autopsied monkeys	Voronoff, 1949
Stomach	Polyadenoma	*Cynocephalus hamadryas*			observed in 1,100 autopsied monkeys	Voronoff, 1949
Stomach	Adenoma	*Macaca mulatta*	F	4 years		Vadova and Gel'shtein, 1956
Stomach	Adenoma	*Macaca mulatta*	M	5 years		Vadova and Gel'shtein, 1956
Stomach	Adenoma	*Comopithecus hamadryas*	M	4 years		Vadova and Gel'shtein, 1956
Stomach	Adenoma	*Macaca nemestrina*	F	5 years		Vadova and Gel'shtein, 1956
Stomach	Multiple papillomata	*Macaca mulatta*	M	3½ years		Kent and Pickering, 1958
Stomach	Polyp	*Macaca rhesus*		2–19 years	4 cases	Lapin and Yakovleva, 1960, 1963
Stomach	Adenocarcinoma	*Macaca mulatta*	F	10 years +	fed DDT for 6 years	Kimbrough, 1966
Stomach	Carcinoma	*Macaca mulatta*	M		metastases	Klüver, unpublished (see Ruch)
Small intestine	Adenocarcinoma	Ringtail monkey				O'Conner, 1947
Sigmoid gyrus	Polyps				several animals	Kennard, 1941
Colon	Adenocarcinoma	*Macaca mulatta*			local metastases	Plentl *et al.*, 1968

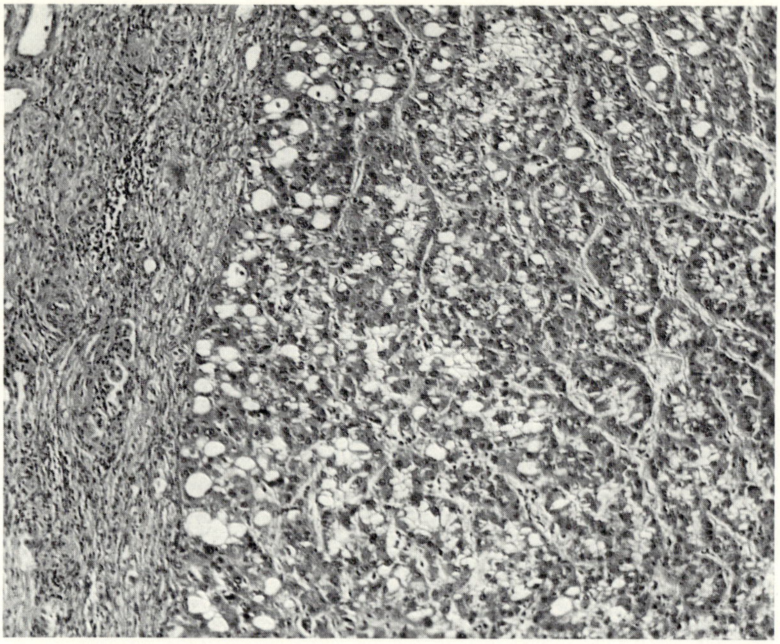

Fig. 12. Adenocarcinoma of the kidney in a 21-year-old *Macaca mulatta* monkey. The tumors were bilateral and had metastasized to the lung. Hematoxylin and eosin. ×120 (Courtesy of Drs. H. L. STEWART and K. C. SNELL, National Cancer Institute, Bethesda, Maryland).

tomicrographs of the lesions suggest a hyperplastic reaction of the mucosa to multiple ulcerations rather than neoplasia. LAPIN and YAKOVLEVA [1959] reported the induction of polyposis of the stomach and intestine and one carcinoma of the stomach in 4 of 10 rhesus monkeys administered cholesterin daily for 1 ½ to 2 years.

Kidney and Adrenal Gland

Renal tumors are among the most frequent spontaneous neoplasms of subhuman primates (table VII and fig. 12). Fourteen such tumors have been seen in the past 50 years, 8 of which were in rhesus monkeys (*Macaca mulatta*). All were epithelial tumors and nearly all were considered malignant, including 3 which had metastasized. Only one tumor occurred in a monkey younger than 10 years of age. Four of the renal tumors occurred in a single

Table VII. Spontaneous tumors in primates – kidneys and adrenal glands

Location	Type of tumor	Species	Sex	Age	Remarks	References
Kidney	Adenocarcinoma	*Lemur sp*		unknown		Plimmer, 1910
Kidney	Adenocarcinoma	*Cacajao sp*		unknown		Plimmer, 1915
Kidney	Carcinoma	*Cebus apella*			metastasized	Scott, 1928
Kidney	Adenocarcinoma	*Macaca mulatta*	M	28 years +	familial-father	Fox, 1934; Ratcliffe, 1940
Kidney	Adenocarcinoma	*Macaca mulatta*	F	16 years +	familial-offspring	Fox, 1935; Ratcliffe, 1940
Kidney	Adenocarcinoma	*Macaca mulatta*	M	10 years +	familial-offspring	Fox, 1935; Ratcliffe, 1940
Kidney	Adenocarcinoma	*Macaca mulatta*	M	10 years +	familial-offspring (metastasized)	Fox, 1935; Ratcliffe, 1940
Renal cortex	Simple adenoma	*Papio papio*	M	15 years +		Fox, 1937
Kidney	Papillary carcinoma	*Macaca rhesus*	F	16 years		Antonov, 1956
Kidney	Papillary carcinoma	*Macaca rhesus*	M	1 year		Antonov, 1956
Kidney	Malignant adenoma	*Macaca mulatta*	M	27 years		Vadova and Gel'shtein, 1956
Kidney	Adenoma of medulla	*Alouatta caraya*				Maruffo, 1967
Kidney	Tumor	*Papio papio*	M			Klüver, unpublished (see Ruch)
Kidney	Adenocarcinoma	*Macaca mulatta*	M	21 years	bilateral tumors with metastases to lungs	Stewart and Snell, unpublished
Adrenal gland	Hypernephroma	*Cebus apella fatuellus*	M	4 years +		Fox, 1923; Ratcliffe, 1933
Adrenal gland	Cortical adenoma	*Macaca irus*	M			Fox, 1933
Adrenal gland	Fibroma of medulla	*Macaca mulatta*	M			Kent and Pickering, 1958
Adrenal gland	Cortical adenoma	Macaque		18 years		Lapin and Yakovleva, 1963
Adrenal gland	Cortical adenomas	*Alouatta caraya*		unknown	11 wild Howler monkeys out of 292 autopsied	Maruffo, 1967

family of rhesus monkeys, a father and three of his offspring by two different mothers. The father also had a submucous leiomyoma which partly obstructed the esophagus, and a gastric polyp. Despite the relative frequency of spontaneous renal tumors, only one monkey exposed to carcinogenic influences is reported to have developed neoplasms of the kidney [VALERIO *et al.;* table IXa].

Fifteen spontaneous tumors of the adrenal gland have been reported, and all except one were apparently benign. Eleven of these tumors were discovered by one investigator in a series of autopsies on 292 howler monkeys.

Recent Developments in Primate Carcinogenesis

Until recently intensive studies of carcinogenesis in monkeys by a broad spectrum of chemicals has not been done either because of the difficulty and expense of maintaining primates for long periods, or from lack of interest in the subject, or because of the belief based on limited experience that monkeys were refractory to the known chemical carcinogens.

Study of table IXb shows that at least 31 genuine neoplasms were induced in primates by chemicals in the past 40 years. Seven of these were fibrosarcomas induced in prosimian primates by polycyclic hydrocarbons which have failed notably to induce tumors in higher primates, except in two isolated instances [KROTKINA, 1948; LEVY, 1963]. In rhesus monkeys, lung tumors induced by beryllium [SCHEPERS; VORWALD], liver tumors induced by diethylnitrosamine, and skin cancers induced by a high-boiling fraction of cracked oil have been the only unquestionably induced neoplasms. Experiments were initiated about 8 years ago by a group of investigators at the National Cancer Institute to study the effect of important classes of chemical compounds on subhuman primates. These compounds were all powerful carcinogens in mice and rats. Since then over 500 monkeys (rhesus, cynomolgus, cercopithicus and capuchin) have been treated by various routes with carcinogenic chemicals at maximum tolerated doses. Most of these monkeys were started on treatment while they were newborns and were continued on it for 5 years unless tumors developed earlier or the monkey became moribund. About 300 of these monkeys are still alive 1 to 8 years after the start of treatment, and the remaining 200 have been autopsied during the course of experimental studies or following death due to other causes (table VIII).

It is obvious from table VIII that only one of these rodent carcinogens, diethylnitrosamine (DENA), has thus far proved to be carcinogenic in pri-

Table VIII. Reaction of monkeys to different classes of chemical carcinogens

Compound	Route	Live monkeys – years since start of treatment[1]			Number of monkeys autopsied 1961–1968	Neoplasms and other significant lesions in monkeys surviving 1 month of treatment
		0–3 years	4–7 years	8 years		
N, N-Nitrosodiethylamine (DENA)	Oral	13	9		67	41 hepatomas and hepatic cell carcinomas in autopsied monkeys with metastases in 22 cases. Two fibrosarcomas of liver and 3 hemangiomas or hemangioendotheliomas of spleen and liver. In addition 7 living monkeys have confirmed hepatic tumors
	IP	22				
1-Nitrosopiperidine	Oral	2	8		2	1 hepatoma
	IP	10				
N, N′-dimethyl-p-(m-tolylazo)-anliine (3′methyl DAB)	Oral		9		19	Pulmonary thrombosis and periarteritis (5 cases). Gastritis, cecitis or ileocolitis (3 cases)
N,N′-dimethyl-p-phenylazo-analine (DAB)	Oral			2		
N-2-fluorenylacetamide (2-FAA)	Oral		8	5	7	Periportal lymphocytic infiltration of liver
N, N-2,7-fluorenylenebisacetamide (2,7-FAA)	Oral		12			
Copper chelate of N-hydroxy-2-fluorenylacetamide	S.C.		8		0	Foreign body granulomas on biopsy

Table VIII (continued)

Compound	Route	Live monkeys – years since start of treatment[1]			Number of monkeys autopsied 1961–1968	Neoplasms and other significant lesions in monkeys surviving 1 month of treatment
		0–3 years	4–7 years	8 years		
Ethyl carbamate (urethane)	Oral		22	2	21	Variety of lesions not clearly related to treatment
3-Methylcholanthrene (3-MC)	Oral		13		22	Papillomas and keratoacanthomas of skin. Periarteritis and periarteriolitis of pulmonary arteries
	S.C. and ID		5			
3,4,9,10-dibenzpyrene	S.C. and ID	4	11		3	
Aflatoxin B_1	Oral	25	1		24	Toxic hepatitis with bile plugs in small canaliculi and proliferation of pseudotubules or small bile ducts
	IP	15				
Cycasin (and its aglycone)	Oral	7	7		16	Centrolobular necrosis of liver and periportal fibrosis
	IP	10				
Procarbazine – HCl	Oral	17	12		15	1 acute myelogenous leukemia
Methylnitrosourea	Oral	25			0	
	IV	10				

1 Treatment is stopped after a maximum period of 5 years.

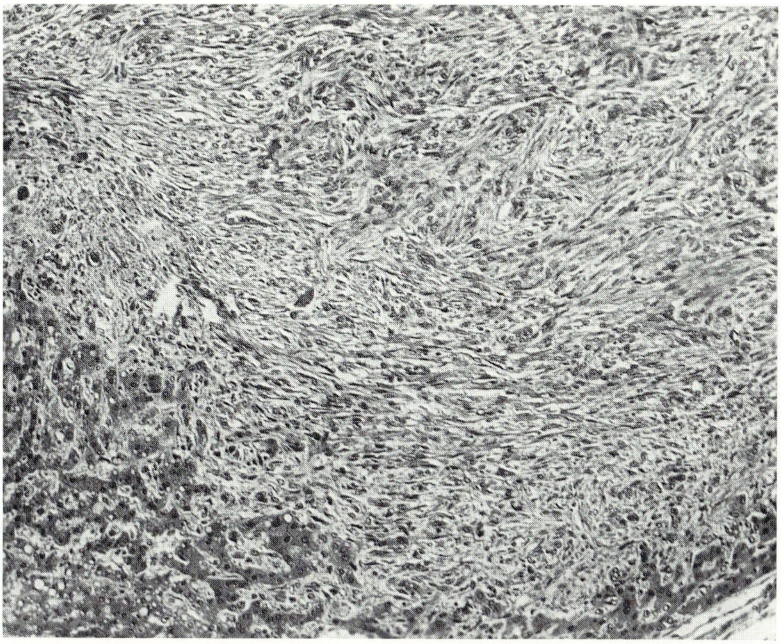

Fig. 13. Fibrosarcoma of the liver in an adult female Capuchin monkey *(Cebus capucinus)* who received nitrosodiethylamine orally for 31 months. This large tumor nodule was composed of a non-circumscribed mass of proliferating spindle cells. MASSON's trichrome stain showed considerable associated collagen, and WILDER's reticulum stain showed a typical sarcomatous pattern. Elsewhere the liver showed hepatic cell carcinoma, but no metastasis were found from either tumor. Hematoxylin and eosin. ×135.

mates. Hepatomas and hepatic cell carcinomas have been induced in about 50 monkeys with this chemical. Three of these monkeys also had hemangioendotheliomas of the liver or spleen, and two had fibrosarcomas in the liver (fig. 13). Over 80% of monkeys surviving for 2 years or longer on DENA have developed hepatic cell neoplasms when the carcinogen was incorporated in the diet and nearly 100% when the carcinogen was given by intraperitoneal injections. Liver tumors induced by intraperitoneal injections had a greater frequency of metastases to lung and lymph nodes than did those induced by the oral route, and the latent period was much shorter. The earliest tumors were induced in 10 months by intraperitoneal injections and in 14 months by oral administration. The average time for tumor induction was 22 months for oral administration and 17 months for intraperitoneal injections. One of the first induced liver tumors was placed in tissue culture and was suc-

Table IXa. Induced tumors in subhuman primates – radiation

Treatment	Type of tumor	Location	Species	Dose	Number	Latent period	Reference
Implants of radium ore	Osteogenic sarcoma	Tibia	*Macaca mulatta*	2.0–5.5 μg radium	2/4	8–10 years	PETROV *et al.*, 1951
			Hamadryad baboon		1/3		
Radium bromide or radium ore	Osteogenic, myeloid and reticulocytic sarcomas	Tibia	*Macaca rhesus* (3)	11.8–21 μg	4/16	23 months, 39 months, 34 months, 62 months (green)	KROTKINA and BARABADZE, 1957
			C. Aethiops (1) (green)	8.9 μg radium			
Radioactive substance	Osteogenic sarcomas				2		PETROV, KROTKINA *et al.*, 1958
Radium bromide ($RaBr_2$)	Sarcomas	Tibia	*Macaca rhesus* ape, Pavian baboon		2	5–6½ years	KROTKINA and BARABADZE, 1964
Local ionizing radiation	Fibrosarcoma	Maxilla	*Macaca mulatta*	2–3,000 r	2	474–892 days	KENT and PICKERING, 1958
	Glioblastoma	Brain	*Macaca mulatta*		1		
^{110}AG + progesterone and synoestrol	Carcinosarcoma	Breast	*Macaca mulatta*	606 μc 900,000 rad	1	3½ years after ^{110}Ag	SMOYLOVSKAIA *et al.*, 1960a

	Malignant melanoblastoma with metastasis	Skin	*Macaca mulatta*	300,000 r 795 μc	1	3 years 3 months	Smoylovskaia *et al.*, 1960b
^{110}Ag	Sarcoma	Maxilla	*Macaca rhesus*	4,380–5,350 r	2	892–961 days	Melnikov and Barabadze, 1960, 1968
	Squam. Carcinoma	Tongue	*Macaca rhesus*	5,350 rad	1		
^{60}Co	Fibrosarcoma	Maxilla	*Macaca rhesus*	397 rad	1	569–833 days	Melnikov and Barabadze, 1960, 1968
	Squam. Carcinoma	Maxilla	*Macaca rhesus*	397 rad	1		
External fractional X-radiation	Fibrosarcoma	Mandible	*Macaca rhesus*	7,500 r	1/14	25 months	Gowgiel, 1965
Whole body neutron radiation	Myeloid leukemia adenoma	Generalized bronchus	*Macaca mulatta*	608 rep	1	6 years	Zalusky *et al.*, 1965
X-radiation	Leukemia		*Papio hamadryas*	2,800 r	1	2 years	Dzhikidze and Yakovleva, 1965
Whole-body proton irradiation	Acute granulocytic leukemia	Generalized	*Macaca mulatta*	100 rad	1	2½ years	Siegal *et al.*, 1968
Whole body neutron radiation	Transitional cell carcinoma	Renal pelvis	*Macaca mulatta*	608 rep	1	9 years	Valerio *et al.*, 1968
	Cystadenocarcinoma	Kidney					
	Basal cell carcinoma	Skin					
	Osteogenic sarcoma	Femur					

Table IXb. Induced tumors in subhuman primates – chemical

Treatment	Type of tumor	Location	Species	Dose	Number	Latent period	Reference
Skin painting with coal tar	Carcinomas with regression	Skin	*Macacus cynomolgus*		20	2 years	BONNÉ *et al.*, 1930
Estrin and trauma	Atypical epithelial hyperplasia	Cervix	Rhesus	1660–5490 units	4		OVERHOLSER and ALLEN, 1935
Tar injections into tibia	Bone tumor and leukemoid state	Tibia	*Macacus sinicus*		1	6 months	BERNARD, 1935
Shale lubricating oil	Infiltrating adenomatous lesions	Stomach and colon	*Macaca mulatta*		8	3–7 months	LUSHBAUGH, 1947, 1949
Intramedullary 3-methylcholanthrene	Polymorphonuclear sarcoma	Tibia	Rhesus	30 mg	1	3½ years	KROTKINA, 1948
Estrogens and	Papillomatous masses and	Skin	Rhesus	large	12/50	survived 3 years	PFEIFFER and ALLEN, 1948
carcinogenic hydrocarbons	Wart-like excrescences			quantities	3/50	survived 9 years	
High-boiling cracked oil	Squamous carcinomas	Skin	Rhesus		3/6	4–5 years	SUGIURA *et al.*, 1956
Cholestrin	Polyposis	Stomach and intestine	*Macaca rhesus*	490–700 g	4/10	18–24 months	LAPIN and YAKOVLEVA, 1959
Polycyclic hydrocarbons	Papillomas and keratoacanthomas	Skin	Rhesus and cynomolgus	10–150 mg	31	1 week	KELLY and O'GARA, 1962; O'GARA and KELLY, 1965

3-methylcholanthrene	Fibrosarcoma	Subcutaneous	Marmoset (*Tamarinus nigricollis*)	2mg	1/2	10 months	Levy, 1963
Uracil	Colloid goiter	Thyroid	*Cebus*	0.5–0.75% in diet	4	86 weeks	Andrus *et al.*, 1964
Beryllium inhalation	Alveolar carcinoma	Lung	*Macaca mulatta*		1	92 days	Schepers, 1964
Beryllium intubation	Cancer	Lung	*Macaca mulatta*		6/20	4½–8 years	Vorwald, 1966
N-nitrosodiethylamine	Carcinomas	Liver	Macaques, *cebus* and *Cercopithicus*	1.6–26.0g	8/17	14–27 months	Kelly *et al.*, 1966
2-Naphthylamine	Carcinoma	Bladder	Rhesus	200mg/kg	3	33–46 months	Conzelman *et al.*, 1967
3,4 Benzpyrene	Fibrosarcoma	Subcutaneous	*Tupaia glis* (Tree shrew)	10mg	3/4	6 months	Noyes, 1968
7.12-dimethylbenz(a) anthracene	Papilloma and kerato-acanthomas	Skin	*Galago Crassicaudatus*	2½% sol	3/8	8 months	Adachi *et al.*, 1969
3.4 Benzpyrene	Fibrosarcoma	Shoulder	*Galago crassicaudatus* (Greater bush baby)	3mg	1	26 months	Cooper and Kelly (unpublished)
3-methylcholanthrene	Fibrosarcoma	Thigh	*Tupaia glis*	10mg	3	14–20 months	Adamson *et al.*, 1968 (unpublished) 1969
Benzo(a)pyrene	Fibrosarcoma	Subcutaneous	*Saguinus oedipus* (cottontop marmoset)	10mg	1/2	16 months	Noyes, 1969
Dimethylbenz(a)anthracene	Rhabdomyosarcoma			10mg			

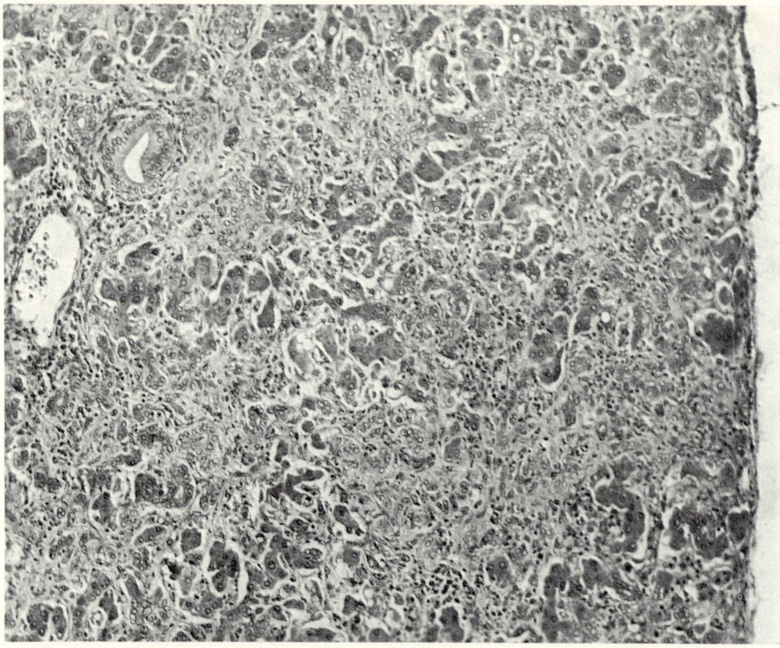

Fig. 14. Toxic hepatitis in a 2½-month-old rhesus monkey fed aflatoxin since birth. There is necrosis and atrophy of liver cords with regeneration of pale epithelial cells which sometimes form pseudotubules. Diffuse infiltration by lymphocytes and neutrophils. A portal vein and a bile duct are seen in the left upper corner. Hematoxylin and eosin. ×135.

cessfully grown *in vitro* for several years [DAWE *et al.*, 1968]. After 2 years in tissue culture, growth of the cells was sufficiently vigorous to allow transplantation back into monkeys. In the first attempt nine infant monkeys received intracerebral inoculations of tumor cells suspended in broth and 5 of the monkeys developed large intracerebral tumors which resembled the original hepatic tumor [KELLY *et al.*, 1967]. Subsequently additional successful intracerebral transplants were completed from tissue culture cells and also by monkey to monkey transplantation [WALKER, unpublished]. Use of immunosuppressive drugs has increased the transplantability of the tumor in adult monkeys as well as newborns and has made possible monkey to monkey transplantation for as long as 6 to 8 generations.

About 2 years ago it was discovered that most of these hepatic tumors were synthesizing an *α* fetoprotein similar to that produced by hepatic tumors in humans. This *α* fetoprotein is a serum globulin, which is normally

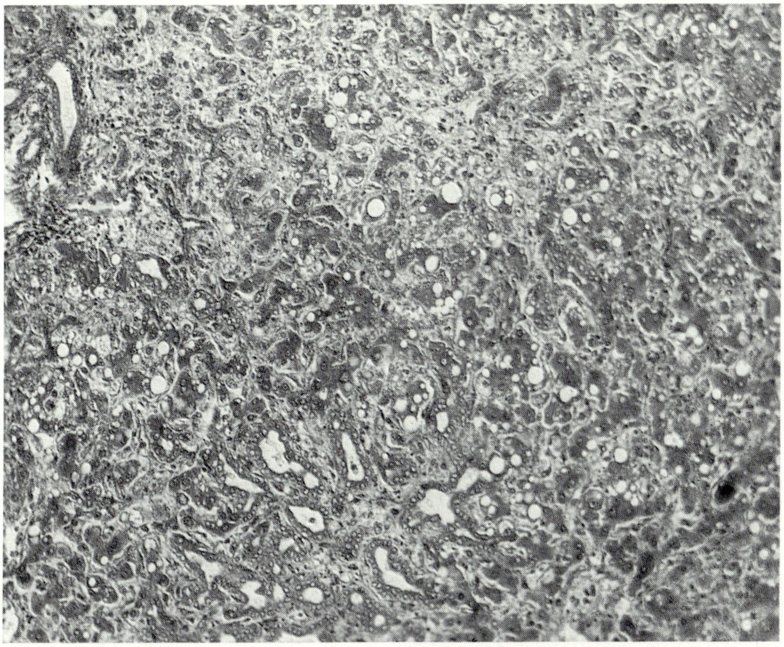

Fig. 15. Toxic hepatitis in a 3-month-old Cynomolgus monkey fed nitrosodiethylamine since birth. The histologic appearance is similar to that seen in figure 14. The liver shows atrophy of hepatic cords and regeneration of pale epithelial cells which frequently form pseudotubules. A portal triad with a pre-existing bile duct is seen in the left upper corner. Hematoxylin and eosin. ×135.

produced by the fetal liver but disappears from the serum shortly after birth. It is never found in the human adult except in the presence of a tumor of hepatic cell origin or in certain embryonal carcinomas of the testis. It is never found in humans associated with pure cholangio-carcinomas of the liver, nor with cirrhosis or other liver diseases. In monkeys the protein usually appears simultaneously with or slightly before the appearance of recognizable hepatic tumors, although in a few cases the appearance of the abnormal protein has been delayed for as long as a year. The protein has never been found in monkeys with livers damaged by other hepatotoxic chemicals, in which tumors have not appeared.

Administration of aflatoxin to infant monkeys was begun about 3 to 4 years ago, but the original material consisted of a crude mixture of fungal products. More recently pure aflatoxin B_1 has become available and is being

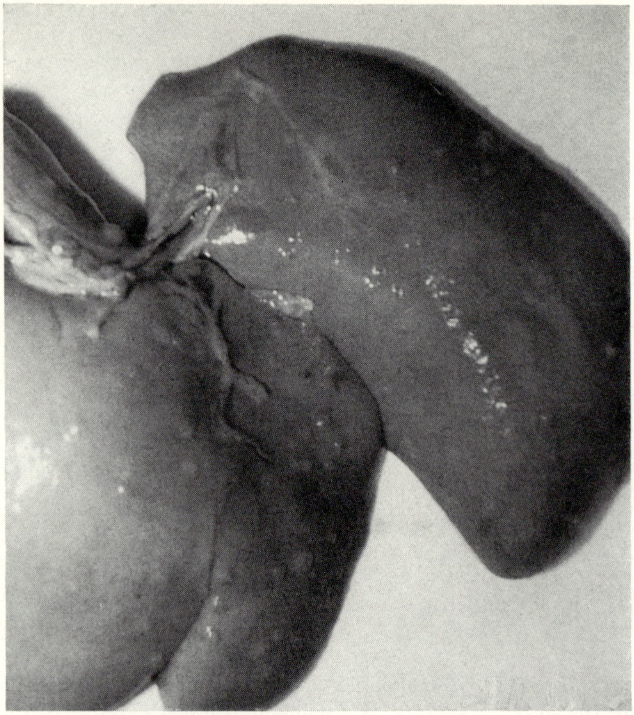

Fig. 16. Liver of a 17-month-old male rhesus monkey who had received intraperitoneal injections of aflatoxin B_1 for 7 months. There are numerous gray slightly raised nodules.

given both orally and by intraperitoneal injections at maximal tolerated doses. Aflatoxin is very toxic to the simian liver, especially in newborn animals. It produces a type of necrosis which closely mimics that produced by diethylnitrosamine (fig. 14–15). It is characterized by destruction and atrophy of most of the liver cells, especially those about the central part of the lobules and by marked alteration in the remaining hepatic cells so that they form a type of pseudotubules, bearing a superficial resemblance to small bile ducts. Preexisting bile ducts in the portal area, however, show no discernible reaction to the compound. No tumors have appeared in monkeys receiving this compound but the pure aflatoxin B_1 has been under test for only 18 months. Drs. CUTHBERTSON, LAURSEN and PRATT of Glaxo Research Ltd., England, have also fed aflatoxin to monkeys for long periods of time. One monkey sur-

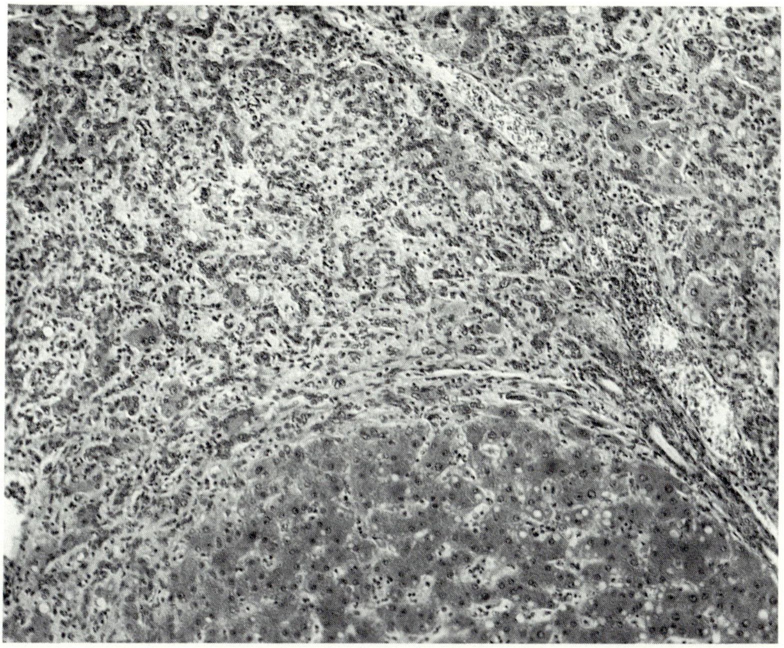

Fig. 17. Histologic appearance of the liver seen in figure 16. A nodule of regenerating liver is seen (lower margin) above which there are atrophic liver cords, proliferating pseudotubules, and diffuse infiltration by neutrophilic leukocytes. Hematoxylin and eosin. ×135.

vived for 3 years and showed a nodular cirrhosis but no tumors. However, the resemblance of the early toxic changes in the liver to those produced by diethylnitrosamine suggests that hepatic tumors will develop eventually if treatment is persistent. The liver of two monkeys dying after 7 months of intraperitoneal injections of aflatoxin showed extensive areas of necrosis and large hyperplastic nodules, which appeared grossly as small tumor nodules (fig. 16–17).

Cycasin, an ingredient of the cycad nut and a probable contaminant of human food in several parts of the world, has been fed to monkeys for several years. This chemical causes carcinomas of the liver, kidney, and intestinal tract in rats [LAQUEUR *et al.*], but so far has not been responsible for any tumors in primates. Recently the aglycone of cycasin has become available and has been administered to monkeys by intraperitoneal injection, since the aglycone does not require the action of intestinal bacteria for its activity. The

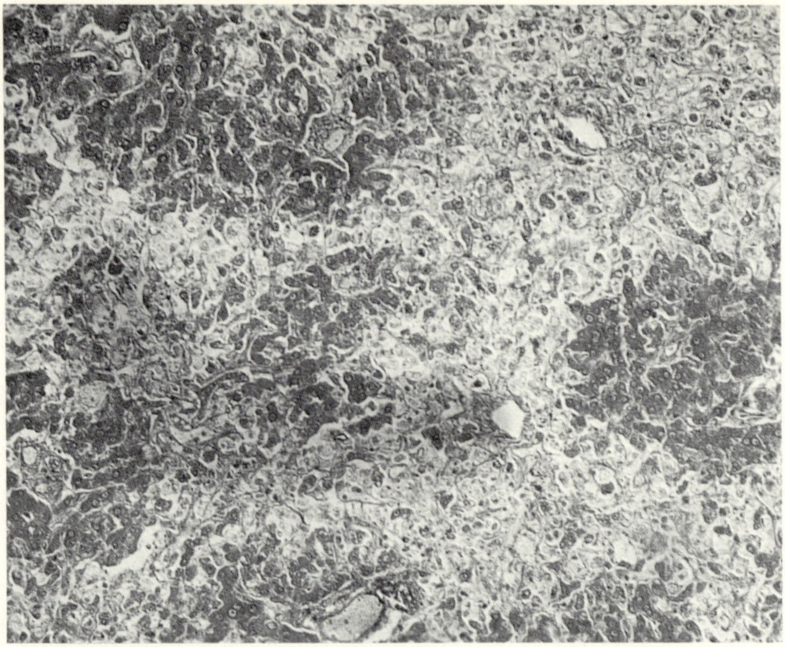

Fig. 18. Toxic hepatitis in a 2½-year-old male Cynomolgus monkey fed cycasin since birth. Compare with figures 14 and 15. The liver shows centrolobular necrosis but no proliferation of pseudotubules. Two portal triads are seen at the lower left and two central veins in the right center. Hematoxylin and eosin. ×135.

toxic effect of cycasin on the liver appears to differ from that of aflatoxin inasmuch as the necrosis is not as specific for the centrolobular area and there is less proliferation of pseudotubules (fig. 18).

Monkeys fed 2- and 2,7-fluorenylacetamide for 5 years and observed for up to 8 years, have not developed tumors nor have those autopsied shown any lesions suggestive of developing neoplasia. In rats, these compounds cause tumors of the liver, ear ducts and gastrointestinal tract probably due to the metabolic product, N-OH fluorenylacetamide. The fluorenylacetamides are rapidly excreted in the urine and feces following oral administration and in addition the monkey converts much less of the fluorenylacetamide to them N-OH derivative and more to other less carcinogenic products [DYER *et al.*, 1966]. However, 8 monkeys have been injected subcutaneously with the copper chelate of N-OH fluorenylacetamide and none have developed tumors during a subsequent 6 years observation period. The feeding of azo dyes

Table IXc. Induced tumors in subhuman primates – biological

Treatment	Type of tumor	Location	Species	Dose	Number	Latent period	Reference
Cultures from human Hodgkin's disease	Early Hodgkin's disease	Lymph nodes	Rhesus		1		Bunting and Yates, 1913
Cultures from human Hodgkin's disease	Giant-celled nuclei	Lymph nodes	Rhesus		5		Bunting and Yates, 1914
Adult *Nochtia nochti* nematodes	Adenopapillomas	Stomach	*Macacus mordax*	16–22 adult worms	2	2–3 months	Bonné and Sanground, 1939
Human papilloma suspension	Small tumors	Conjunctiva	*Macacus rhesus*		3		Green *et al.*, 1940
Papilloma cell suspension	Papillomas	Skin	*Cebus*		11/13	2 weeks	Lucké *et al.*, 1950
Monkey heart tissue	Tumors	Subcutaneous	Rhesus and *Cynomolgus*		300	1–2 weeks	Salk and Ward, 1957
Poxvirus (Yaba)	Pleomorphic fibrocytic tumors	Subcutaneous	Rhesus and *Cynomolgus*		16	1–3 weeks	Niven *et al.*, 1961
Rous sarcoma virus	Fibrosarcoma	Subcutaneous	Rhesus		7/8	2–6 weeks	Munroe and Windle, 1963

(N,N^1-dimethyl-p-phenylazoaniline (DAB), and N,N^1-dimethyl-p-(m-tolylazo) aniline (3′ Me DAB)) to monkeys in conjunction with a rice diet low in riboflavin has not induced any neoplastic changes during an observation period of 4 to 8 years.

Ethyl carbamate (urethan) has been administered orally 5 times weekly to a group of 24 monkeys for 4 to 5 years. Half of these animals also had whole body radiation with a total of 500 r of X-ray, divided into 10 weekly doses of 50 r each. Urethan induces lung tumors, leukemia and hepatomas in rodents but no monkeys have developed leukemia or other neoplastic lesions although they have been observed for 4 to 8 years since the start of treatment.

Procarbazine hydrochloride (N-isopropyl-*α*-(2-methylhydrazino)-p-toluamide, hydrochloride) has been administered by weekly subcutaneous injection and more recently by the oral route. This compound induces a high incidence of leukemia and pulmonary tumors in mice and mammary tumors in rats. It has also been used in the chemotherapy of HODGKIN's disease and related conditions. One case of acute myelogenous leukemia has appeared in a monkey after 15 months of treatment with this drug (fig. 8–10). However, judgment on the leukemogenic properties of this drug in primates must be reserved pending the outcome in other monkeys still under test.

Finally 3-methylcholanthrene and 3,4,9,10-dibenzpyrene have been administered either orally or by injections to rhesus and cynomologus monkeys. Single or multiple subcutaneous or intradermal injections have produced keratoacanthomas, papillomas, and giant cell granulomas. These animals have been under observation for 3 to 7 years now and no sarcomas or carcinomas have developed. Fourteen newborn monkeys given repeated injections of 3-methylcholanthrene died from arteriolitis and periarteritis of the pulmonary vessels, and four had cor pulmonale [O'GARA and KELLY, 1965]. In contrast single subcutaneous injections of 3-methylcholanthrene in the prosimian tree shrew *(Tupaia glis)* have induced fibrosarcoma in three of these primates within 14 to 20 months [ADAMSON *et al.*, unpublished]. Similar results in the tree shrew using 3,4-benzpyrene were reported by NOYES [1968]. In addition COOPER (at the San Diego Zoo) has been successful in inducing a subcutaneous fibrosarcoma in a *Galago crassicaudatus* with a single injection of 3,4-benzpyrene [COOPER and KELLY, unpublished]. It seems probable, therefore, that some of the prosimian primates are more closely related to the rodent than to the higher primates in their reaction to carcinogenic chemicals. Further studies along these lines are needed before a complete understanding of phylogenetic differences can be achieved.

References

ADACHI, K.; YAMASAWA, S., and MONTAGNA, W.: Epidermal hyperplasia induced by 7,12-dimethylbenz(a)anthracene in prosimians. J. nat. Cancer Inst. *42:* 61–68 (1969).

ADAMSON, R. H., *et al.:* unpublished.

ANDRUS, S. B.; ROACH, A. M., and FILLIOS, L. C.: Experimental production of colloid goiter in the Cebus monkey. Proc. Soc. exp. Biol., N.Y. *116:* 963–967 (1964).

ANTONOV, A. M.: Spontanye opukholi pochek u obez'ian *(Macacus rhesus).* (Spontaneous kidney tumors in monkeys). Vop. Onkol. *2:* 198–201 (1956).

BAGG, H. J.: Neoplasms in the lower primates, with a description of an osteogenic sarcoma of the jaw in a *Macacus rhesus.* Amer. J. Cancer *15:* 2143–2148 (1931).

BEARCROFT, W. G. C. and JAMIESON, M. F.: An outbreak of subcutaneous tumors in rhesus monkeys. Nature *182:* 195–196 (1958).

BERNARD, J.: Etat leucémoide et tuméfaction osseuse provoqués chez le singe par les injections intra-médullaires de goudron. Sang *9:* 790–808 (1935).

BLOKHIN, N. N.; VASILYEY, Y. M., and POGOSYANTS, Y. Y.: A new case of spontaneous malignant tumor in a monkey *(Macacus rhesus).* Vop. Onkol. *1:* 91–95 (1955).

BONNE, C.; LODDER, J. und STREEF, G. M.: Das Terrcarcinom beim Affen *(Macacus cynomolgus).* Z. Krebsforsch. *32:* 310–326 (1930).

BONNE, C. and SANDGROUND, J. H.: On the production of gastric tumors, bordering on malignancy, in Javanese monkeys through the agency of *Nochtia nochti,* a parasitic nematoda. Amer. J. Cancer *37:* 173–185 (1939).

BRACK, M.: Carcinoma solidum simplex Mammae bei einem Orang-Utan *(Pongo pygmalus).* Zbl. allg. Path. path. Anat. *109:* 474–480 (1966).

BRYGOO, E. R.; LEVADITI, J.; DESTOMBES, P., and GUILLON, J. C.: Adenocarcinoma with 'bronzed cirrhosis' observed in Madagascar in a Lemur *M. Macaco* L 1766. Bull. Soc. Path. exot. *57:* 228–233 (1964).

BUNTING, C. H. and YATES, J. L.: An etiologic study of Hodgkin's disease; preliminary note. J. amer. med. Ass. *61:* 1803–1804 (1913).

BUNTING, C. H. and YATES, J. L.: An etiologic study of Hodgkin's disease; second note. J. amer. med. Ass. *62:* 516–517 (1914).

CHAPMAN, W. L.: Neoplasia in nonhuman primates. J. amer. vet. med. Ass. *153:* 872–878 (1968).

COHEN, D. W. and GOLDMAN, H. M.: Oral disease in primates. Ann. N.Y. Acad. Sci. *5:* 889–909 (1960).

CONZELMAN, G. M., jr.; FLANDERS, L. E. III; SPRINGER, K., and CROUT, D. W.: Induction of transitional cell carcinoma in the urinary bladder of the rhesus monkey. Proc. amer. Ass. Cancer Res. *8:* 11 (1967).

CONZELMAN, G. M., jr.; MOULTON, J. E.; FLANDERS, L. E., III, SPRINGER, K., and CROUT, D. W.: Induction of transitional cell carcinomas of the urinary bladder in monkeys fed 2-naphthylamine. J. Nat. Cancer Inst. *42:* 825–836 (1969).

COOPER, R. W. and KELLY, M. G.: unpublished.

CORSON-WHITE, E. P.: Chronic lymphatic leukemia in a green monkey. Arch. Pathol. *8:* 1019 (1929).

CRAN, J. A.: An epithelial tumor occurring in the jaws of a black-faced spider monkey *(Ateles geoffroyi).* Oral Surg. *27:* 494–498 (1969).

CREWS, L. M.; KERBER, W. T., and FEINMAN, H.: An ovarian tumor of dual nature in a rhesus monkey. Path. vet. *4:* 157–161 (1967).

CUNNINGHAM, W. F. and MCALPIN, K.: Experiments with Hodgkin's disease; an attempt to produce it in anthropoids and other monkeys. Arch. intern. Med. *32:* 353–358 (1923).

CUTHBERTSON, W. F. J.; LAURSEN, A. C., and PRATT, D. A. H.: Effect of groundnut meal containing aflatoxin on Cynomolgus monkeys. Brit. J. Nutr. *21:* 893–908 (1967).

DAWE, C. J.; WANG-PENG, J.; MORGAN, W. D.; O'GARA, R. W., and KELLY, M. G.: Culture of a cell line (NCLP-6) derived from a hepatocarcinoma induced in *Macaca mulatta* by N-nitrosodiethylamine. J. nat. Cancer Inst. *40:* 1167–1193 (1968).

DELEON, W. and REYES, C.: Experimental transplantation of Hodgkin's disease in monkeys. J. Philipp. Isls. Med. Ass. *9:* 9–11 (1929).

DEPAOLI, A. and GARNER, F. M.: Acute lymphocytic leukemia in a white-cheeked Gibbon *(Hylobate concolor)*. Cancer Res. *28:* 2559–2561 (1968).

DIGIACOMO, R. F.: BURKITT's lymphoma in a white-handed Gibbon *(Hylobates lar)*. Cancer Res. *27:* 1178–1179 (1967).

DOBSON, R. L. and PINTO, J. S.: Cutaneous carcinogenesis in simian primates; in MONTAGNA and DOBSON Advances in biology of skin; Carcinogenesis, vol. 7, pp. 345–354 (Pergamon Press, New York 1965).

DVIZHKOV, P. P. and ZVETAEVA, N. P.: A contribution to the comparative pathology of tumors in animals. Trud. Moskov. Zoo. *3:* 166–198 (1946).

DYER, H. M.; KELLY, M. G., and O'GARA, R. W.: Lack of carcinogenic activity and metabolic fate of fluorenylacetamides in monkeys. J. nat. Cancer Inst. *36:* 305–322 (1966).

DZHIKIDZE, E. K. and YAKOVLEVA, L. A.: A case of leukemia in a monkey subjected to X-irradiation for a prolonged period of time. Probl. Gematol. Pereliv. Krovi. *10:* 58–63 (1965).

ELLIOT, O. S. and ELLIOT, M. W.: Breast cancer in a tree shrew *(Tupaia Glis)*. Nature *211:* 1105 (1966).

ENGLE, E. T.; KRAKOWER, C., and HAAGENSEN, C. D.: Estrogen administration to aged female monkeys with no resultant tumors. Cancer Res. *3:* 858–866 (1943).

ENGLE, E. T. and STOUT, A. P.: Spontaneous primary carcinoma of the prostate in a monkey *(Macaca mulatta)*. Amer. J. Cancer *39:* 334–337 (1940).

FINKELDEY, W.: Pathologisch-anatomische Befunde bei der Oesophagostomiasis der Javeneraffen. Z. Infektkr. Haustiere *40:* 146–164 (1931).

FLINN, R. M.: Serous cystadenoma of the ovary in a rhesus monkey. J. Path. Bact. *94:* 451–452 (1967).

FOX, H.: Disease in captive wild mammals and birds. Incidence, description, comparison (Lippincott, Philadelphia 1923).

FOX, H.: Notes on special animals. Rep. Lab. Comp. Path., Philadelphia, pp. 11–15 (1926).

FOX, H.: Special subjects of pathological interest. Rep. Lab. Comp. Path., Philadelphia, pp. 16–22 (1932).

FOX, H.: Subjects of pathological interest. Rep. Lab. Comp. Path., Philadelphia, pp. 16–22 (1933).

FOX, H.: Subjects of pathological interest. Rep. Lab. Comp. Path., Philadelphia, pp. 17–30 (1934).

FOX, H.: Mortality and pathological matters. Rep. Lab. Comp. Path., Philadelphia, pp. 12–18 (1935).

FOX, H.: Mortality and matters of pathological interest. Rep. Penrose Res. Lab., pp. 14–19 (1936).

FOX, H.: Mortality and matters of pathological interest. Rep. Penrose Res. Lab., pp. 13–17 (1937).

FOX, H.: Matters of pathological interest. Rep. Penrose Res. Lab., pp. 17–26 (1938).

FRASER, A. D.: Ectopic endometrium in a Macacus rhesus. J. Obstet. Gynaec. brit. Cwlth *36:* 590–591 (1929).

FUKUDA, N.; ASAKURA, S., and NAKAGAWA, S.: Two cases of esophageal cancer in monkeys. Jap. J. vet. Sci. *20:* 303 (1958).

GESCHICKTER, C. F. and HARTMAN, C. G.: Mammary response to prolonged estrogenic stimulation in the monkey. Cancer *12:* 767–781 (1959).

GILLMAN, J. and GILBERT, C.: Some connective tissue diseases (amyloidosis, arthritis, reticulosarcoma) in the baboon *(Papio ursinus)*. S. Afr. J. Med. Sci. *19:* 112 (1954).

GOWGIEL, J. M.: A sarcoma observed in the irradiated monkey. Radiat. Res. *24:* 446–451 (1965).

GREEN, R. G.; GOODLOW, R. J.; EVANS, C. A.; PAYTON, W. T., and TITRUD, L. A.: Transmission of a human papilloma to monkeys. Amer. J. Cancer *39:* 161–171 (1940).

HAMERTON, A. E.: Report on the deaths occurring in the Society's gardens during the year 1929. Proc. zool. Soc., Lond., pp. 357–380 (1930).

HAMERTON, A. E.: Report on the deaths occurring in the Society's gardens during the year 1932. Proc. zool. Soc., Lond., pp. 451–482 (1933).

HAMERTON, A. E.: Report on the deaths occurring in the Society's gardens during the year 1936. Proc. zool. Soc., Lond., pp. 443–474 (1937).

HAMERTON, A. E.: Primary degeneration of the spinal cord in monkeys, a study in comparative pathology. Brain *65:* 193–204 (1942).

HILL, W. C. O.: Report of the Society's prosecutor for the year 1953. Proc. zool. Soc., Lond. *124:* 303–311 (1954).

HISAW, F. L. and LENDRUM, F. C.: Squamous metaplasia in the cervical glands of the monkey following oestrin administration. Endocrinology *20:* 228–229 (1936).

HISAW, F. L. and HISAW, F. L.: Spontaneous carcinoma of the cervix uteri in a monkey *(Macaca mulatta)*. Cancer *11:* 810–816 (1958).

HULL, E. W.; CARBONE, P. P.; GITLIN, D.; O'GARA, R. W., and KELLY, M. G.: *a* Fetoprotein in monkeys with hepatoma. J. nat. Cancer Inst. *42:* 1035–1044 (1969).

HUTCHINSON, W.: 'Studies in human and comparative pathology', chapter 10; Pathological parodies; sarcoma; cancer, p. 258 (GLAISHER, London 1901).

INGLESIAS, R. and LIPSCHUTZ, A.: Effect of prolonged oestrogen administration in female New World monkeys with observations on a pericardial neoplasm. J. Endocrin. *5:* 88–98 (1947).

JUNG, G.: Carcinombildung an den Geschlechtsorganen bei einem jugendlichen weiblichen Affen. Z. Krebsforsch. *21:* 227-229 (1924).

KELLY, A. L.: Annual report of the hospital and research committee. Zoonooz Res. *21:* 7 (1948).

KELLY, M. G. and O'GARA, R. W.: Chemical carcinogenesis in primates: induction of squamous cell papillomas in young primates by 3-methylcholanthrene. Proc. amer. Ass. Cancer Res. *3:* 333 (1962).

KELLY, M. G.; O'GARA, R. W.; ADAMSON, R. H.; GADEKAR, K.; BOTKIN, C. C.; REESE, W.

H., and KERBER, W. T.: Induction of hepatic cell carcinomas in monkeys with N-Nitrosodiethylamine. J. nat. Cancer Inst. *36:* 323–351 (1966).

KELLY, M. G.; O'GARA, R. W.; WALKER, M. D.; DAWE, C. J.; MORGAN, W. D., and KERBER, W. T.: Intracerebral transplantation into rhesus monkeys of cell cultures (NCLP-6) derived from a chemically induced primate hepatocarcinoma. J. nat. Cancer Inst. *39:* 153–169 (1967).

KENNARD, M. A.: Abnormal findings in 246 consecutive autopsies on monkeys. Yale J. Biol. Med. *13:* 701–712 (1941).

KENT, S. P. and PICKERING, J. E.: Neoplasms in monkeys *(Macaca mulatta)*. Spontaneous and irradiation induced. Cancer *11:* 138–147 (1958).

KIMBROUGH, R.: Spontaneous malignant gastric tumor in a rhesus monkey *(Macaca mulatta)*. Arch. Path. *81:* 343–346 (1966).

KLÜVER, H. and BRUNSCHWIG, A.: Oral carcinoma in a monkey colony; a report of two additional cases. Cancer Res. *7:* 627–633 (1947).

KLÜVER, H. and WEIL, A.: Carcinoma of the tongue in monkeys and pathologic changes in the central nervous system. J. Neuropath. exp. Neurol. *7:* 144–153 (1948).

KRONENBERGER, H.: Tumors in zoo animals. Nord. Vet. Med. *14* (Suppl. I)*:* 297 (1962).

KROTKINA, N. A.: A case of malignant tumor of the bone in monkey induced by methylcholanthrene. Arkh. Pat. *10:* 38–42 (1948).

KROTKINA, N. A.: A case of spontaneous cancer of the tongue in a monkey. Vop. Onkol. *2:* 748–749 (1956).

KROTKINA, N. A. and BARABADZE, E. M.: On experimental carcinogenesis in the long bones of monkeys, pp. 70–71. Tez. Dok. Ras. Zas. Bur. med.-biol. Nauk AMN USSR Sukhumi (1957).

KROTKINA, N. A. and BARABADZE, E. M.: On the induction of bone tumors in monkeys. Pat. Fiziol. Eksp. Ter. *3:* 14–18 (1958).

KROTKINA, N. A. and BARABADZE, E. M.: Two sarcomas of the extremities induced in apes. Vopr. Onkol. *10:* 87–92 (1964).

LAPIN, B. A.: A case of lymphogranulomatosis in a green marmoset; in UTKIN Theoretical and practical questions of medicine and biology in experiments on monkeys, pp. 159–163 (Pergamon Press, New York 1960).

LAPIN, B. A. and YAKOVLEVA, L. A.: Gastrointestinal tumors in monkeys with prolonged cholesterin administration. Arkh. Pat. *21:* 25–30 (1959).

LAPIN, B. A. and YAKOVLEVA, L. A.: Comparative pathology in monkeys (Moscow, 1960; translation published by Thomas, Springfield 1963).

LAQUEUR, G. L.; MICKELSEN, O.; WHITING, M. G., and KURLAND, L. T.: Carcinogenic properties of nuts from Cycas Circinalis L. indigenous to Guam. J. nat. Cancer Inst. *31:* 919–951 (1963).

LEVY, B. M.: Induction of fibrosarcoma in the primate *Tamarinus nigricollis*. Nature *200:* 182–183 (1963).

LOMBARD, L. S. and WITTE, E. J.: Frequency and types of tumors in mammals and birds of the Philadelphia Zoological Garden. Cancer Res. *19:* 127–141 (1959).

LONGCOPE, W. T.: Notes on experimental inoculations of monkey glands from cases of Hodgkin's disease. Bull. Ayer Clin. Lab. *4:* 18–21 (1907).

LUCKÉ, B.; RATCLIFFE, H., and BREEDIS, C.: Transmissible papilloma in monkeys. Fed. Proc. *9:* 337 (1950).

LÜDKÉ, H.: Über die experimentelle Erzeugung leukämieähnlicher Blutbilder. Dtsch. Arch. klin. Med. *100:* 552–566 (1910).

LUSHBAUGH, C. C.: Experimental hyperplastic gastritis and gastric polyposis in monkeys. J. nat. Cancer Inst. *7:* 313–320 (1947).

LUSHBAUGH, C. C.: Infiltrating adenomatous lesions of the stomach, cecum, and rectum of monkeys similar to early human carcinoma and carcinoma *in situ*. Cancer Res. *9:* 385–394 (1949).

LUTHER, M.: Cancer in Subhuman Primates. US Public Health Service Publication No. 1138 (Washington 1962).

MCCARRISON, R.: The pathogenesis of deficiency disease; No. 9. On the occurrence of recently developed cancer of the stomach in a monkey fed on food deficient in vitamins. Indian J. med. Res. *7:* 342–345 (1919).

MARUFFO, C. A. and MALINOW, M. R.: Seminoma in a Howler monkey (Alouatta Caraya). J. Path. Bact. *91:* 280–282 (1966).

MARUFFO, C. A.: Spontaneous tumours in Howler monkeys. Nature *213:* 521 (1967).

MASSAGLIA, A. C.: Leukaemia in the monkey. Note on a syndrome in a monkey *(Cynocephalus sphinx)* similar to that of chronic lymphatic leukaemia in man. Lancet *204:* 1056–1057 (1923).

MELINKOV, R. A. and BARABADZE, E. M.: Cancer of the tongue in monkeys, provoked by introduction of a radioactive isotope (^{110}Ag) into the antrum of Highmore. Vop. Onkol. *6:* 69–72 (1960). Translated in Prob. Oncol. *6:* 390–393 (1960).

MELINKOV, R. A. and BARABADZE, E. M.: Malignant tumors in monkeys. English summary, pp. 197–202. Academy of Sciences, USSR (1968).

MUNROE, J. S. and WINDLE, W. F.: Tumors induced in primates by chicken sarcoma virus. Science *140:* 1415–1416 (1963).

NEWBERNE, I. W. and ROBINSON, V. B.: Spontaneous tumors in primates; a report of two cases with notes on the apparent low incidence of neoplasms in subhuman primates. Amer. J. vet. Res. *21:* 150–155 (1960).

NICHOLS, R. E.: Nasal polypus in a chimpanzee. J. Amer. vet. med. Ass. *47:* 56–57 (1939).

NIVEN, J. S. F.; ARMSTRONG, J. A.; ANDREWES, C. H.; PEREIRA, H. G., and VALENTINE, R. C.: Subcutaneous 'growths' in monkeys produced by a poxvirus. J. Path. Bact. *81:* 1–14 (1961).

NOBACK, C. V.: Report of the veterinary. N.Y. zool. Soc. Ann. Rept. *38:* 39–43 (1934).

NOYES, W. F.: Carcinogen-induced sarcoma in the primitive primate, *Tupaia glis*. Proc. Soc. exp. Biol. Med. *127:* 594–596 (1968).

NOYES, W. F.: Carcinogen-induced neoplasia with metastasis in a South American primate, *Saguinus oedipus*. Proc. Soc. exp. Biol. Med. *131:* 223–225 (1969).

O'CONNER, P.: Occurrence of tumors in zoo animals. Animaland *14:* 2–4 (1947).

O'GARA, R. W. and KELLY, M. G.: Pulmonary arteriolitis in infant monkeys. Produced by subcutaneous injections of polycyclic hydrocarbons. Arch. Path. *79:* 475–483 (1965).

O'GARA, R. W.; KELLY, M. G., and KERBER, W. T.: A spontaneous metastasizing carcinoma of the gallbladder in a rhesus monkey. Folia primat. *6:* 284–291 (1967).

OSHIMA, F.: Studies on tumors in wild animals. Gann *31:* 220–223 (1937).

OVERHOLSER, M. D. and ALLEN, E.: Atypical growth induced in cervical epithelium of the monkey by prolonged injections of ovarian hormone combined with chronic trauma. Surg. Gynec. Obstet. *60:* 129–136 (1935).

PETROV, N. N.; KROTKINA, N. A.; VADOVA, A. V., and POSTNIKOV, Z. A.: Dynamics of the origin and development of malignant growths in monkeys (Academy of Medical Sciences, Moscow 1951). Summarized by Shimkin and Shope; Some observations on cancer research in the Soviet Union. Cancer Res. *16:* 915–917 (1956).

PETROV, N. N.; KROTKINA, N. A.; BARABADZE, E. M.; VADOVA, A. V.; GEL'SHTEIN, V. I.; MEL'NIKOV, R. A.; POSTNIKOVA, Z. A., and SMOLOVSKAIA, E. I. A.: The results of 18 years of work in Sukhumi on the induction of malignant tumors in monkeys. Vop. Onkol. *4:* 643–655 (1958). (Translated in Prob. Oncol. *4:* 1–6 1958).

PETROV, N.; KROTKINA, N.; VADOVA, A.; BARABADZE, E.; GUELSTEINE, V.; MELINKOV, R. et SMOILOVSKAYA, E.: L'induction expérimentale des tumeurs malignes chez les singes. Acta Un. int. Cancer *15:* 119–123 (1959).

PETTIT, A.: Sur un cas de leucoplasie vaginale chez la guenon mone (*Cercocebus* mona Schreb). Bull. Mus. Hist. Nat., Paris *10:* 281 (1904).

PETTIT, A.: Lésions osseuses chez deux singes (*Cebus fatuellus*, L. et *Gorilla gorilla*, WYMANN). Bull. Soc. Path. exot. *2:* 220–222 (1909).

PEZET, M.: Tumeur de la machoire chez un singe. Rec. Méd. vét. *96:* 530–534 (1920).

PFEIFFER, C. A. and ALLEN, E.: Attempts to produce cancer in Rhesus monkeys with carcinogenic hydrocarbons and estrogens. Cancer Res. *8:* 97–127 (1948).

PLENTL, A. A.; DEDE, J. A., and GREY, R. M.: Adenocarcinoma of the large intestine in a pregnant rhesus monkey *(Macaca mulatta)*. Folia primat. *8:* 307–313 (1968).

PLIMMER, H. G.: Report on the deaths which occurred in the Zoological Gardens during 1909. Proc. zool. Soc., Lond. *1:* 131–136 (1910).

PLIMMER, H. G.: Report on the deaths which occurred in the Zoological Gardens during 1913, together with a list of blood-parasites found during the year. Proc. zool. Soc., Lond. *1:* 181–190 (1914).

PLIMMER, H. G.: Report on the deaths which occurred in the Zoological Gardens during 1914, together with a list of blood-parasites found during the year. Proc. zool. Soc., Lond. *1:* 123–130 (1915).

RANKOW, R. M: Bilateral fibro-osteoma of the maxilla in a monkey. J. dent. Res. *26:* 333–336 (1947).

RATCLIFFE, H. L.: Tumors in captive primates with a description of a giant cell tumor in a chacma baboon, *Papio porcarius*. J. Cancer Res. *14:* 453–460 (1930).

RATCLIFFE, H. L.: Cancer in animals. Trans. Coll. Physicians *54:* 152–157 (1932a).

RATCLIFFE, H. L.: Tumors in captive primates. Report of two cases. Amer. J. Path. *8:* 117– 121 (1932b).

RARCLIFFE, H. L.: Incidence and nature of tumors in captive wild mammals and birds Amer. J. Cancer *17:* 116–135, 172 (1933).

RATCLIFFE, H. L.: Familial occurrence of renal carcinoma in rhesus monkeys *(Macaca mulatta)*. Amer. J. Path. *16:* 619–624 (1940).

RATCLIFFE, H. L.: Deaths and important diseases. Rep. Penrose Res. Lab., pp. 11–25 1942).

RATCLIFFE, H. L.: Causes of death in the animal collection. Rep. Penrose Res. Lab., pp. 7–19 (1955).

REWELL, R. E.: Report of the society's pathologist for the year 1949. Proc. zool. Soc. Lond. *120:* 485–495 (1950).

RUCH, T. C.: Diseases of Laboratory Primates (Saunders, Philadelphia 1959).

SALK, J. E. and WARD, E. N.: Some characteristics of a continuously propagating cell derived from monkey heart tissue. Science *126:* 1338–1339 (1957).

SASAKI, T.; HIROKAWA, M., and USIZIMA, H.: A spontaneous squamous cell carcinoma of the lower jaw in *Macaca mulatta.* Primates (Jour. Primatology) *3:* 82–87 (1961–1962).

SCHEPERS, G. W. H.: Biological action of beryllium. Reaction of the monkey to inhaled aerosols. Indust. Med. Surg. *33:* 1–16 (1964).

SCHMEY, M.: Das Magenkarzinom bei Säugetieren. Dtsch. tierärztl. Wschr. *22:* 377–380 (1914).

SCHURMAN, E. A.: A note upon lesions of the female genitalia in wild animals. J. comp. Path. Therap. *28:* 330–334 (1915).

SCOTT, H. H.: Report on the deaths occurring in the Society's Gardens during the year 1926. Proc. zool. Soc. Lond. *1:* 173–198 (1926).

SCOTT, H. H.: Report on the deaths occurring in the Society's Gardens during the year 1927. Proc. zool. Soc. Lond. *1:* 81–119 (1928).

SELLHEIM, A. P.; Monkeys with singular attacks. English summary, pp. 98–99. Trud. Vaesoyuz. Inst. Eksp. Med. *1:* 93–98 (1934).

SELLHEIM, A. P.: Une tumeur cérébrale chez un singe. J. belge. Neurol. Psychiat. 240–241 (1936).

SIEGAL, A. M.; CASEY, H. W.; BOWMAN, R. W., and TRAYNOR, J. E.: Leukemia in a rhesus monkey *(Macaca mulatta)* following exposure to whole-body proton irradiation. Blood *32:* 989–996 (1968).

SMOYLOVSKAYA, E. J.; VADOVA, A. V.; PODVALNAYA, M. J., and CHACHIBAYA, I. A.: Carcinosarcoma of the mammary gland arising in a monkey after hyperoestrinization and administration of radioactive silver (^{110}Ag). Vop. Oncol. *6:* 35–42 (1960a). (Translated in: Prob. Oncol. *6:* 666–675 1960).

SMOYLOVAKAYA, E. J.; VADOVA, A. V.; PODVALNAYA, M. J., and CHACHIBAYA, I. A.: Induction of melanoblastoma in a monkey. Vop. Onkol. *6:* 69–74 (1960b). (Translated in: Prob. Oncol. *6:* 1477–1484 1960).

STEINER, P. E.; KLÜVER, H., and BRUNSCHWIG, A.: Three carcinomas of the tongue in two monkeys. Cancer Res. *2:* 704–709 (1942).

STEINMULLER, D.; DILLINGHAM, L. A., and PREHN, R. T.: Lack of carcinogenic activity of methylcholanthrene in a squirrel monkey. Proc. amer. Ass. Cancer Res. *10:* 87 (1969).

STERNBERG, S. S.: Carcinoma in situ of the cervix in a monkey *(Macaca mulatta).* Amer. J. Obstet. Gynec. *82:* 96–98 (1961).

STEWART, H. L.: Etiologic studies in Hodgkin's disease. J. Lab. clin. Med. *18:* 281–287 (1932).

STEWART, H. L. and SNELL, K. C.: Unpublished data.

STEWART, M. J. and DOBSON, J. F.: Inoculation and implantation experiments in monkeys with glands from cases of Hodgkin's disease. Brit. J. exp. Path. *5:* 65–68 (1924).

SUGIURA, K.; SMITH, W. E., and SUNDERLAND, D. A.: Experimental production of carcinoma in rhesus monkeys. Cancer Res. *16:* 951–955 (1956).

SUTTON, J. B.: Tumors in animals. J. Anat. *19:* 415–475 (1885).

TORREY, J. C.: Bacteria associated with certain types of abnormal lymph glands. J. Med. Res. *34:* 65–80 (1916).

UNTERHARNSCHEIDT, F.: Intracerebral cavernous angioma in a monkey *(Macacus rhesus).* Acta neuropath. *3:* 295–296 (1964).

VADOVA, A. V. and GEL'SHTEIN, V. I.: Prolonged use of synestrol in spontaneous mammary carcinoma of monkey. Vop. Onkol. *2:* 391–396 (1956a).

VADOVA, A. V. and GEL'SHTEIN, V. I.: Spontaneous tumors in catarrhine monkeys according to the data obtained in the monkey colony of the Sukhumi medico-biological station (Medgiz, Moskva 1956b). Translated in Utkin Theoretical and practical questions of medicine and biology in experiments on monkeys, pp. 137–158 (Pergamon Press, New York 1960).

VADOVA, A. V. and SMOYLOVSKAYA, E. J.: Dyshormonal proliferation developing in monkey uterus and mammary gland after sinestrol injection. Vop. Onkol. *5:* 515–520 (1959). Translated in Prob. Oncol. *5:* 1–6 (1959).

VALERIO, M.; LANDON, J. C., and INNES, J. R. M.: Neoplastic diseases in simians. J. nat. Cancer Inst. *40:* 751–756 (1968).

VARGAS, L.: Attempt to induce formation of fibroids with estrogen, in the castrated female rhesus monkey. Bull. Johns Hopk. Hosp. *73:* 22–30 (1943).

VORONOFF, S.: Tumeurs spontanées chez les singes; in Groupes sanguins chez les singes; La greffe du cancer humain aux singes, pp. 63–126 (Doin, Paris 1949).

VORWALD, A. J.: The induction of experimental pulmonary cancer in the primate, pp. 125. Ninth International Cancer Congress, Tokyo (1966).

WADSWORTH, J. R.: Tumors in zoo animals; in Animaland *20:* 2–3 (1953).

WADSWORTH, J. R.: Neoplasia of captive zoo species. Vet. Med. *56:* 25–26 (1961).

WAGENEN, G. VAN: The effects of oestrin on the urogenital tract of the male monkey. Anat. Rec. *63:* 387–403 (1937).

WALKER, M. D.: Unpublished data.

WARWICK, R. A.: Sarcoma of bone in a ring-tailed lemur *(Lemur catta)*. J. Path. Bact. *63:* 499–501 (1951).

WOODRUFF, J. M. and JOHNSON, D. K.: Hepatic hemangioendothelioma in a rhesus monkey. Path. vet. *5:* 327–332 (1968).

ZALUSKY, R.; GHIDONI, J. J.; MCKINLEY, J.; LEFFINGWELL, T. P., and MELVILLE, G. S.: Leukemia in a rhesus monkey *(Macaca mulatta)* exposed to wholebody neutron irradiation. Radiat. Res. *25:* 410–416 (1965).

ZUCKERMAN, S.: A rhesus macaque *(Macaca mulatta)* with carcinoma of the mouth. Proc. zool. Soc., Lond. *1:* 59–61 (1930).

ZUCKERMAN, S.: Effects of prolonged oestrin-stimulation on the cervix uteri. Lancet *232:* 435–437 (1937).

ZVETAEVA, N. P.: Tumors. Trud. Moskov. Zoo *2:* 94–101 (1941).

Author's address: Dr. RICHARD H. ADAMSON, Laboratory of Chemical Pharmacology, National Cancer Institute, *Bethesda, MD 20214* (USA)

Pathology of Simian Primates, Part I, pp. 239–269 (Karger, Basel 1972)

The Pathology of Nutritional Deficiencies in Nonhuman Primates

R. H. Wolf

Tulane University, Delta Regional Primate Research Center, Animal Care Division, Covington, La.

Contents

I. Introduction 240
II. Protein 241
III. Fat Excess and Resultant Pathology 243
IV. Vitamins 243
A. Fat Soluble Vitamins 243
1. Vitamin A Deficiency 243
2. Vitamin D Deficiency 244
3. Vitamin E Deficiency 248
4. Vitamin K Deficiency 249
B. Water Soluble Vitamins 249
1. Thiamine Deficiency 249
2. Riboflavin Deficiency 250
3. Pyridoxine Deficiency 252
4. Vitamin B_{12} Deficiency 253
5. Niacin Deficiency 254
6. Folic Acid Deficiency 255
7. Biotin Deficiency 255
8. Pantothenic Acid Deficiency 256
9. Choline Deficiency 256
10. Ascorbic Acid Deficiency 257
V. Water 259
VI. Minerals 259
A. Zinc Deficiency 259
B. Calcium Deficiency 259
C. Magnesium Deficiency 260
VII. Conclusion 260
References 261

I. Introduction

The excellent compilations of the literature by DAY [25] and later by RUCH [124], leave little for one to add to a new chapter on nutritional disease. Since, however, DAY'S summary is out of date, and RUCH'S book is out of print and so not generally available, portions of the material covered in those studies will be reconsidered here. This chapter will describe the effects of various nutritional deficiencies in the nonhuman primate and will report clinical, gross necropsy and histologic alterations produced by the deficiencies. Data reviewed in this study were gathered primarily from publications of investigations in the production of single or 'pure' nutritional deficiencies or excesses using the nonhuman primate as an experimental model. FOLLIS' *Deficiency disease* [40] or similar text should be consulted for explanations of the underlying biochemical defects induced by the deficiency or excess of a nutritional element. A contemporary and complete summary of the nutritional requirements of the nonhuman primate has been made recently by PORTMAN [106].

Before discussing specific nutritional disorders, a few general statements are in order. Reports often include a statement that as a clinical feature of the disease the animals become anorexic or that food consumption diminished. This must certainly be considered as an important influence on the appearance of lesions described. An understanding of those changes occurring in inanition such as loss of body weight, loss of body fat, disturbances in organ growth, and atrophy, hypoplasia or hyperplasia of certain organ systems is necessary in considering the effects of a specific disorder on the pathophysiology of an organism [41]. Of similar importance is an awareness that the expression of a disease can be seriously altered when several nutrients are deficient simultaneously [42].

It is probable that many primates imported from various regions of the world, when trapped, are in a marginal nutritional state or are deprived of adequate or even marginal diet by the feeding practices of the trappers and other personnel involved in the transporting of the primate to the end user. Animals purchased directly from exporters without intermediate conditioning may well arrive at the laboratory malnourished [114, 152].

RUCH'S [129] summary of *Studies of mixed deficiency diseases* includes the findings of RAO [113] and FOLLIS [39] each of whom fed *Macaca* species animals drastically inadequate diets such as milled rice with a few vegetables or ground yellow corn and water. Common to both studies were the facts that the animals deteriorated quickly, developed a great variety of deficiency signs and developed intercurrent infectious processes.

Brandom *et al.* [8] found that by withholding food from *Macaca speciosa (arctoides)* for 96 h with water available *ad libitum* would markedly reduce the secondary response to cholera toxin. Significantly fewer IgA and IgG producing immunocytes were found in the intestines of the starved animals than in the animals fed an adequate diet twice daily. Also, tritiated thymidine incorporation by the intestinal epithelium was reduced in the starved animals. This study indicates that it is necessary to provide captive primates with an adequately balanced diet on a regular and frequent basis.

II. Protein

Of the important nutritional elements, protein is most likely to be deficient in newly-imported primates [114, 152]. Several reasons for this might include the fact that (1) the areas of the world from which primates are exported are for the most part areas where human diets have marginal or totally inadequate protein levels, (2) protein-rich foods are expensive, and (3) the animals have more than likely been held in captivity before exportation by trappers totally ignorant of the animals dietary requirements.

Numerous studies have revealed many of the effects of severe protein deficiency on *Macaca mulatta*. Rhesus monkeys fed a diet low in protein but adequate in calories will develop a kwashiorkor-like syndrome [27–30, 97, 108, 110, 111, 112, 139]. This is a progressive disease, though to a point it is reversible with proper diet. Clinical findings in *Macaca mulatta* are an initial loss of weight [27–30, 97, 108, 110–112, 139], reduction of serum protein and albumin levels with a rise in γ-globulins [27–30, 97, 108, 110–112], hypocholesterolemia [27–30, 97, 110], loose stools [27–30, 108], decreases in serum iron-binding capacity [110, 112] variable fasting plasma iron levels [110, 112, 139], decreased ^{58}Co cyanocobalamin uptake [28], depressed ^{59}Fe absorption [139] and decreased serum alkaline phosphatase activity [97]. A moderate normocytic, normochromic anemia [30, 111, 139], and normoblastic, hypocellular bone marrow [30] develop in rhesus monkeys fed a diet (by gavage) with adequate calories and negligible protein (0.6%). These changes are reversible by the return of protein to the diet. Bulky stools were a finding on one longterm feeding study [108].

Necropsy findings were reproducible, fast to develop (3–8 weeks) and similar to those seen in kwashiorkor. Atrophy of the intestinal and gastric mucosa [28–30, 111] occurs, apparently due to a slowing of the migration of crypt epithelial cells to tips of the villi [29]. It was demonstrated by autoradiographic techniques that the protein deficiency caused a cellular proliferation disturbance in the jejunal epithelial cells [29]. Tritiated thymidine labeling was heaviest in the jejunal epithelial cells of the deficient animals and the mitotic index was lower, indicating that DNA synthesis was slowed. Cellular DNA content was reported to remain normal; therefore, the total number of epithelial cells was reduced resulting in hypoplasia of the mucosa [29]. The livers were soft and large, grossly mottled yellow and red. Periportal fatty infiltration developed early in deprivation and spread into the lobule though the centrilobular region was spared [27]. Fat content of the liver reached 40–45% [110] but was reversible on protein supplementation. Microscopically, severe portal fatty infiltration with nuclear displacement uncomplicated by necrosis, cirrhosis, or inflammatory response was seen [27, 30, 111]. High fat levels in the diet tended to accentuate the hepatic fatty changes [27]. Other significant lesions included pancreatic atrophy with loss of zymogen granules, atrophy of the acini with a relative proliferation of the pancreatic ducts [27, 29, 30], atrophic changes in the salivary glands resembling the pancreatic lesions, marked atrophy and lymphoid depletion of the spleen and lymph nodes, growth arrest in epiphyseal cartilage plates, and cessation of osteoblastic activity and osteoid formation; atrophic changes of the skin, and wasting and atrophy of skeletal muscle [30]. In addition to the skeletal muscle, the smooth muscle of the intestine and the cardiac muscle showed a marked reduction in myocyte cytoplasmic mass with approximation of sarcolemmal nuclei [30]. The bone marrow was hypocellular and normoblastic [30].

Bradycardia and reduction of ecg-complex amplitudes in *Macaca radiata* on both low protein-low calorie and low protein-high calorie diets was reported by GOPALAN and PADMAVATHI [51]. Histopathologic lesions included moderate to severe periportal fatty metamorphosis, severe pancreatic atrophy, with loss of zymogen granules and atrophy of the myocardial and skeletal muscle fibers.

WOLF *et al.* [170] have shown that *Erythrocebus patas* makes a markedly reduced immune response to *Escherichia coli* 055 infections when fed 3.8 g of protein/kilo/day. Caloric and vitamin levels were calculated to be adequate and no deficiency signs were seen. The animals fed at a higher level (7.5 g protein/kilo/day) gained a slight amount of weight over the experimental course whereas the low-protein group lost slightly. Bacteria were excreted for a longer time by patas given the low-protein diet and agglutinin, opsonin, and antibody formation were also reduced. Total serum protein, and serum albumin levels declined in the low-protein patas. Significantly less increase in immunoglobulins G, A and M levels in the spleen, mesenteric lymph nodes and intestines occurred after infection in the low-protein group than in the controls. Although the deprived animals in this study were not restricted severely enough to develop a kwashiorkor-like syndrome, it is important to note that the immune mechanism was markedly depressed or retarded in the face of an enteric infection; such infections are, of course, an important problem in laboratory primates.

Several investigators [21, 49, 66, 133, 138, 162, 163] have produced nutritional cirrhosis by feeding nonhuman primates low-protein, high-fat, and low-choline diets. The deficiency of choline appears to cause the fatty livers which developed in low-protein – and low-protein, high-fat – diets to progress to cirrhosis. This is described elsewhere in this chapter under choline deficiency.

III. Fat Excess and Resultant Pathology

Various levels of dietary fat have been fed to nonhuman primates in what are considered balanced or adequate diets. Dietary fat levels used in studies evaluated for other sections of this chapter ranged from little or no fat, to levels of 4–15%. Apparently nonhuman primates can tolerate wide variations in dietary fat levels at the lower ranges without obvious pathology.

However, when a variety of nonhuman primates are fed high-fat diets with added cholesterol [1–3, 10, 11, 14, 20, 68, 76, 82a, 82b, 84, 85, 87–89, 95, 104, 105, 135, 136, 141, 144–146, 166–168, 171], marked elevation of serum lipids (cholesterol and lipoproteins) occurs and atherosclerosis of the aorta, coronary arteries and systemic arteries develops. Xanthomata of the skin and tendon sheaths [2, 88, 144], cerebral aneurysms [1], and myocardial infarction [1, 145] may occur on this type of dietary regimen. Kim [74] found that essential fatty acid deficiency retarded growth rate in *Macaca mulatta.*

IV. Vitamins

A. Fat Soluble Vitamins

1. Vitamin A Deficiency

Studies of vitamin A deficiency have been made in *Macaca mulatta* [58, 59b, 64, 122, 147, 150, 153] and *Macaca sinica* [109]. Of these, the more detailed studies concerned ocular changes primarily [109, 122]. Few animals in the general studies showed minimal opthalmic pathology. Clinical signs included anorexia, diarrhea, asthenia, loss of weight, apathy, and roughened pelage. Xerophthalmia was reported in a few animals.

RAMALINGASWAMI *et al.* [109] reported xerophthalmia, and patchy loss of conjunctival pigment, which progressed to opaque foamy lesions in 2 of 3 *Macaca sinica* on a vitamin A deficient diet for 3 months. Corneal lesions varied from a cloudy dryness to keratomalacia. The third animal showed conjunctival lesions after 8 months on the diet, but no corneal alterations. Serum vitamin A levels were markedly lower (mean – 53 IU/100 ml) in the deficient animal compared to the control animal levels (159 to 168 IU/100 ml). Impairment of dark adaptation ranged from equivocal to marked in the deficient animals.

Histopathologic evidence of ocular damage was found in all three deficient animals. Corneal epithelium showed changes ranging from incipient keratinization to xerophthalmia. Alterations in the substantia propria included irregularities in the collagen matrix and leucocytic infiltration. Elastin was reduced in Descement's membrane, and the cytoplasm of Descement's endothelium was vacuolated. Retinal changes varied in severity according to the magnitude of the corneal lesions. The ganglion and bipolar cells were uniformly spared, but the rods in 2 of the 3 animals were missing in the macular region, to about 1 mm from the fovea and were damaged 2 mm out. The cones in the macular area had enlarged nuclei and were more irregular than usual. The neck sections were swollen, and the outer segments were damaged. Spiral-like structures were absent from the eyes of deficient animals. Nuclear pycnosis in the macular region extended out 2 mm from the fovea. Damage to pigment cells extended over a much wider area than the rod and cone damage.

RODGERS *et al.* [122] fed young rhesus monkeys a purified diet deficient in vitamin A after loading the animals with vitamin A for 2 months. No xerophthalmia was seen after 5 months of deprivation. Hemeralopia was reported as equivocal due to difficulty in assessment. Retinal changes included clumping of pigment granules within the cell bodies of the pigmented epithelium, so that the nuclei were concealed and the processes denuded. Other necropsy findings were cachexia, and keratinization of the epithelium of eyes, ears, maxillary sinuses, turbinates, salivary glands, kidneys, trachea, reproductive tract, bladder, and ureters.

2. Vitamin D Deficiency

The historical development of the study of primate bone diseases has been admirably summarized by RUCH [132]. In this work, the varied descriptions and exotic eponyms have been properly correlated and classed as manifestations of rickets and/or osteomalacia and discussed in relation to bone physiology. The awareness of the role of vitamin D in calcium metabolism resulted in its addition to primate diets. This supplement has, for the most part, eliminated rickets as a problem in most colonies of Old World monkeys. GERSTENBERGER has described the deficiency lesions in *Macaca mulatta* [50].

In spite of this knowledge, bone disease has persisted in captive primates, most notably in New World monkeys. A tabulation of the pathogenesis of rickets and osteomalacia has been made by FOLLIS [44]. Historically this disease in New World primates has been attributed to an imbalance in Ca:P metabolism [132], high phosphorus intake [77], and a viral infection [36].

Demineralizing bone disease and rickets have been reported in *Saimiri sciureus* [9, 15, 80], *Ateles sp.* [80], *Cebus albifrons* [69], *Saguinus oedipus* [69], *Saguinus nigricollis* [69], *Saguinus mystax* [69] and *Lagothrix sp.* [140] fed diets containing vitamin D_2 as the sole source of vitamin D. In addition to these species, *Callicebus moloch* housed indoors without access to sunlight or vitamin D_3 developed the disorder. The present author has seen animals that were reluctant to climb or jump, and that walked with an awkward gait[1]. The reluctance to move is probably due to the fact that these conditions are extremely painful, extrapolating from data on man [12]. Marked demineralization and spontaneous fractures of the long bones were found on radiographic examination, and vertebral collapse was a common sequela of the demineralization. The femurs and caudal vertebrae were easily bent manually. Rapid clinical improvement resulted after the feeding of 10 IU vitamin D_3 in corn oil daily. Mineralization of the metaphysis occurred within 3 weeks after institution of the vitamin D_3 supplement. Residua included thoracic deformities (pigeon breast), kyphosis and bending of the long bones of the limbs[2].

The pathology of this specific deficiency has been well described in young *Saimiri sciureus* by LEHNER *et al.* [80]. This group produced demineralizing bone disease with widening of the epiphyseal plates and fraying, widening and cupping of the metaphysis by feeding synthetic diets containing no vitamin D or vitamin D_2. The cartilaginous epiphyseal plates were hypertrophied and excess osteoid formation accompanied the demineralization. In addition, hypophosphatemia was observed. A 3rd group of *Saimiri sciureus* fed vitamin D_3 in the basic diet developed no bone disease.

1 WOLF, R. H., unpublished data.
2 WOLF, R. H., unpublished data.

The same deficiency produced in adult *Cebus albifrons*, and in *Saguinus oedipus*, *Saguinus nigricollis*, and *Saguinus mystax* a somewhat different type of bone disease designated osteodystrophia fibrosa [69]. Marked demineralization of cortical bones, demonstrated by radiographic techniques, resulted in kyphosis and fractures. Epiphyseal alterations were not reported, probably due to the age of the animals studied. Histopathologic lesions included marked demineralization of the bones along with active trabecular osteoid formation and fibrosis of the marrow. Treatment with vitamin D_3 caused regression of these lesions.

The differences in the lesions between the two studies probably reflect a difference in age rather than a specific response of the host; i.e. rickets in young or growing animals and osteomalacia in adults [12]. The *Callicebus* examined had features common to and intermediate to the animals reported in each of the studies, probably because they were of intermediate age between the 2 groups. The lesions included spontaneous fractures without callus formation, marked osseous demineralization, fibrous replacement of the marrow, epiphyseal plate thickening and rachitic rosary[3].

Bone disease of the types described in New World primates were induced by a deficiency of vitamin D_3 and regressed after supplementation with vitamin D_3. Additionally, certain New World monkeys [69, 80, 140] appear unable to use vitamin D_2 to any appreciable extent. Of interest, COWDRY and SCOTT [19] produced no evidence of toxicity including soft tissue calcification in 2 *Cebus apella* given over 27 million XI units of vitamin D over a 10-day period. Therefore, it is probable that deficiency of vitamin D_3 is responsible for much of the bone diseases plaguing New World primate colonies.

Toxicity due to hypervitaminosis D has been seen in *Macaca mulatta* accidentally fed a diet containing levels that would provide each animal 163,000 USP units vitamin D/day [73].

The earliest notable findings were an increase in respiratory and enteric infections. This was followed by slight loss of weight, decreased hemoglobin levels and red blood cell levels, slight elevation of BUN, equivocal elevation of serum calcium levels and electrocardiographic alterations including 'rounded' T-waves resembling those seen in hypercalcemia. Peaked hyperkalemic-type T-waves were also seen. Most of the clinical alterations reversed spontaneously after the diet was changed.

The most consistent lesions were mineral deposits in the renal paren-

3 WOLF, R. H., unpublished data.

chyma, especially the distal convoluted tubules, proximal tubules, collecting tubules and the thin loop tubules. The deposits were grossly visible in extreme cases. Calcification of the wall of the small ducts of the submaxillary and less often the parotid gland and the stomach were consistent findings. Accompanying iron deposits were seen. Deposits of iron and calcium were found in the mucosa of bronchi and alveolar ducts. In the more severe cases, incrustations of iron and calcium covered the alveolar walls. Myocardial lesions presented as a focal yellow-gray mottling of the myocardium. The left atrium and ventricle were affected most extensively. The major microscopic cardiac lesion was a focal deposition of calcium and iron along dead or degenerating muscle fibers. In addition, mineralization of the media of the aorta and intramyocardial arteries was seen. The severity of the lesions correlated positively with the duration of feeding and inversely with time after withdrawal of the toxic feed.

COWDRY and SCOTT [19] fed 19 rhesus monkeys apparently higher doses of vitamin D than that in the previous study. In this study soft tissue mineralization was found but was minimal. Two *Cebus apella fatuellus* were fed approximately 27 million XI units of Viosterol over 200 h developed no lesions.

Studies of hypervitaminosis D_2 and D_3 have been made on *Saimiri sciureus* and *Cebus albifrons* by HUNT, GARCIA and HEGSTED [70]. Four groups of 3 squirrel monkeys each maintained on a commercial monkey diet were supplemented with 50,000 IU vitamin D_3 *per os* daily. In addition, 4 *Cebus albifrons* were fed a purified diet containing no vitamin D. Two *Cebus albifrons* were fed 100,000 IU vitamin D_2 and another 2, 100,000 IU vitamin D_3 daily.

All animals receiving vitamin D_3 died. Death occurred earlier in the squirrel monkeys receiving 100,000 IU/day (20–26 days) than in the animals receiving 50,000 IU (28–55 days). The *Cebus* monkeys died at 52 and 89 days. None of the animals receiving D_2 at any dose level died.

The only microscopic lesions found in the squirrel monkeys, excluding 1 animal fed 100,000 IU D_3/day, were minimal nephrocalcinosis and occasional laminated tubular calculi. In the one animal of the high dose group the nephrocalcinosis was much more extensive and involved numerous medullary tubules. The lesions of the group, excluding the 100,000 IU D_3/day animal, which had marked nephrocalcinosis, were not different than those in the vitamin D_2 group.

Vitamin D_3-treated *Cebus albifrons* developed extensive granular or homogeneous mineral deposits staining purple in the presence of hematoxylin and eosin, positive with von Kossa and alizarin red S techniques and to the periodic acid Schiff reaction in renal tubular systems (especially cortical and medullary collecting tubules), renal parenchyma, bronchial and bronchiolar smooth muscle cells, capillary walls of the alveolar septa, the endocardium of all cardiac chambers, occasional ventricular myofibers, gastric mucosal epithelial cells, and arterial walls. Medial and intimal mineralization was found in the aorta, pulmonary artery, coronary arteries, gastric arteries, and arteries of the tongue, pancreas, adrenals and mesentery. Calcium oxalate calculi, and calculi resembling the tissue mineral deposits were found in the collecting tubules. The *Cebus* monkeys fed vitamin D_2 developed no lesions.

Serum calcium levels and BUN levels were elevated in all animals receiving vitamin D_3 but not in those receiving vitamin D_2. Serum phosphorus levels were elevated terminally in *Saimiri sciureus* receiving vitamin D_3. Serum alkaline phosphatase levels were not altered in any of the animals. Apparently renal failure was the cause of death in these animals, probably due to a direct nephrotoxic effect other than mineralization, at least in *Saimiri sciureus* [69].

3. Vitamin E Deficiency

Deficiency of the fat soluble vitamin E has been studied from a variety of aspects in *Macaca* species [26, 31a, 31b, 32, 37, 89, 90, 103, 161].

Production of deficiency states requires dietary deprivation ranging from 6 months to 3 years. This may be influenced by the quality of protein fed [32]. Progressive muscular weakness is one of the first signs of the deficiency [26, 31a, 37, 90], and is reversible with α-tocopherol.

Two studies [89, 90] report extensive cellular pigment accumulation. This pigment was described as a fuchsinophilic [89] and acid fast [90] material in the smooth muscle cells of the blood vessels, stomach, small intestine, gallbladder, and bronchi, in the epithelium of the urinary bladder, and oviduct, in reticuloendothelial cells of lymphnodes and the spleen, and in tissue macrophages.

A vitamin E responsive anemia was reported in a number of studies examining this organ system [32, 103]. The anemia was characterized by low hemoglobin values [103], insignificant increases in reticulocytes [103], red cell values below 1 million [32], red cells of normal size and hemoglobin content [32, 103], and normal gluthathione stability [103]. DINNING [32] reports that the anemia is not hemolytic, but probably due to impaired erythropoiesis. Bone marrow examination revealed moderate erythroid hyperplasia and a relative myeloid depression [103]. Many multinucleate erythroid precursors were present [32, 103].

FILER *et al.* [37] made electrocardiographic studies of vitamin E deficient monkeys. The heart rate, P-R interval, QRS interval, QT_c interval and axes were not affected. The voltages of R_I and R_{II} and R_{V4} were reduced in the deficient animals. R_{aVL} and R_{V3} were affected variably. The T waves of leads II and aVL were inverted in 75% of the deficient animals and 25% of the control animals. It was also found that the ejection time of the heart was reduced in the deficient group.

An evaluation of vitamin E neuropathy was attempted by WECHSLER and GLOBUS [161]. Diffuse degeneration of the anterior horn cells and lateral columns was found along with alopecia, wrist drop, and fasciculations in 1 *Macaca irus* [161]. Most of the animals on the diet died of intercurrent infection.

4. Vitamin K Deficiency

The effect of vitamin K deficiency has been studied in *Macaca radiata* [94] and *Macaca mulatta* [65]. No change was observed in prothrombin time in 10 *Macaca radiata* on a synthetic vitamin K deficient diet high in sugar and supplemented with aureomycin and a sulfa drug. The diet itself provided small quantities of vitamin K and the authors report only that coprophagy was not observed [94]. Prolongation of prothrombin time was observed in *Macaca mulatta* [65] fed a soyabean protein diet supplemented with methionine, irradiated beef, neomycin and oxytetracycline. These diets would not support rats in their vitamin K requirements. The prothrombin times could be shortened by vitamin K_4 administration.

B. Water Soluble Vitamins

1. Thiamine Deficiency

Thiamine deficiency has been induced in *Macaca mulatta* by the use of synthetic diets low in, or devoid of, thiamine [34, 115, 117–119, 159]. This condition was reported as a spontaneous deficiency in *Macaca sylvanus* fed a diet of bread, vegetables, noodles or white barley meal [79]. RUCH summarised the clinical disease in man [125].

The clinical course began as a loss of weight [34, 115, 117–119, 159] or decrease in weight gain [115, 117–119], followed by frank anorexia, general weakness and loss of various reflexes, ataxia, incoordination, convulsions, tremor, signs of cardiac insufficiency, occasional ptosis, retching, and vomiting [79, 115–118, 159]. Blood thiamine levels and tissue thiamine (skeletal muscle, brain, cardiac muscle, kidney and liver) levels dropped [115]. Blood pyruvate, on the other hand, increased [115]. Hematology values were normal [159] or reduced slightly [115]. Untreated, the deficiency resulted in death.

Autopsy and histopathologic findings of this condition must be carefully distinguished from those of inanition [43]. All the reports concerning *M. mulatta* [33, 114–117, 159] and the report on *M. sylvanus* [79] mention anorexia as a constant feature of the deficiency state. No inanition controls were used in these studies. Gross lesions were not strongly emphasized in any of the studies although RINEHART *et al.* [115, 119] reported right-sided dilatation of the heart with small foci of myocardial necrosis and hydropic degeneration of the myocardial fibers. Nonspecific electrocardiographic alterations including bradycardia, decreased R-wave amplitude, and T-wave inversion were found by WAISMANN and MCCALL [159]. The central nervous system lesions were studied by several investigators [34, 118]. Bilaterally symmetrical areas of degeneration were found in the corpus striatum, corpora quadragemina, putamen, caudate nucleus, inferior colliculus, cerebellar vermix, superior colliculus, mammillary bodies, and cerebral cortex nuclei corresponding to the 3rd, 6th, 8th, and 10th cranial nerves. The pathology ranged in severity from early focal edema and separation of glial nerve fibers with fragmentation and disintegration of myelin sheaths to disintegration of the axis cylinder, degeneration of glia, and the accumulation of microglia.

Proliferation and dilatation of cerebral capillaries accompanied by endothelial proliferation in *Macaca mulatta* reached a degree of severity resembling Wernicke's disease in man, though the lesions were more widespread in the monkey [34].

2. Riboflavin Deficiency

Riboflavin deficiency has been studied in *Macaca mulatta* [16, 56, 157], *Cebus apella* [86] and *Papio anubis* [45, 46, 42]. The deficiency state appears to manifest itself with certain species variation in its effects.

Riboflavin deficiency is reported to result in anemia, variable leucopenia, anorexia, impairment of reflexes, incoordination, ataxia, blindness, loss of weight, alopecia and a characteristic dermatitis in *Macaca mulatta* [16, 56, 157].

The dermatitis manifests itself in the rhesus monkey as a red freckling which first appears on the face and groin and later, and variably, in other areas, especially the arms and legs. The dermatitis is described as consisting of small, dry, red, foci which enlarge and coalesce, becoming dark scabs covering large areas [16, 56, 157]. GREENBERG [56] reports ulceration of the plantar surfaces and ischial callosities. A hypochromic-normocytic anemia occurred in the animals studied by COOPERMAN *et al.* [16] and WAISMAN [157] reported that hemoglobin and RBC levels were decreased. White blood cell levels were also variably or slightly reduced [16, 157]. In COOPERMAN'S report [16], riboflavin supplementation supported weight gain and partially reversed the anemia, but liver was needed to restore completely the blood cellular elements.

Alterations of locomotor activity reported include ataxia, incoordination and faulty grasping reflex [157]. Blindness reported by GREENBERG [56] was accompanied by atrophy of the neuronal axis cylinder with gliosis and reduction of ganglion cell population at the posterior pole of the optic nerve. The bipolar and receptor cells of the retina were not damaged. Fatty yellow livers were found at necropsy [157].

MANN *et al.* [86] reported that riboflavin deficiency causes diffuse alopecia with a flaky dryness of the skin in *Cebus apella*. A crusty seborrheic dermatitis was found on the hands of 2 of 4 animals. The physical condition of the animal deteriorated and plasma flavin decreased, but the animals were able to maintain the cellular elements of the blood at a normal level. Terminal ataxia was noted. Autopsy findings included peripheral demyelination, epithelial atrophy in the gastrointestinal tract and bladder, thymic involution, and medial fibrosis and intimal hyperplasia of the small arteries of the heart and kidney.

FOY *et al.* [45–47] used synthetic diets with adequate protein and calorie levels but devoid of riboflavin to produce the deficiency in *Papio anubis*. According to the authors, all became 'anorexic, asthenic, anemic, and lost weight' [45]. Certain animals developed bloody diarrhea, but bacteriological studies, if made, were not reported. Severe hyperkeratosis and parakeratosis or seborrheic dermatitis developed on the face, especially on the nose, eyebrows and eyelids, and on the axillae, scrotum and penis [45–47]. Swollen bleeding gums developed but no corneal or lingual changes were found [45, 46]. These lesions disappeared upon readministration of riboflavin. Adrenal hyperplasia occurred early, but was followed by adrenal depletion.

Examination of bone marrow in these animals revealed marked hypoplasia or aplasia of the erythroid elements [45–47]. Little change occurred in the myeloid series unless intercurrent infection developed. Giant stab cells, but no megaloblasts, were found in the deficient animals. The hypoplastic bone marrow contained about 10.3% red cell precursors which, after the administration of riboflavin increased to about 47.4%. Likewise, red cell ^{59}Fe uptake was markedly reduced in the deficient animals.

A decrease in the peripheral reticulocyte, platelet, and hemoglobin levels was found in the deficient animals following the bone marrow aplasia [47]. Increased urinary excretion of xanthurenic acid was found in the deficient baboons [47]. The authors suggest that the riboflavin induced erythroid aplasia is related to impaired adrenal cortical function [47].

3. Pyridoxine Deficiency

The summaries of the literature on pyridoxine deficiency by RUCH [126] and by GREENBERG [57] have described adequately the clinical course and pathology of this deficiency in *Macaca mulatta*.

Clinically, rhesus monkeys fed a B_6 deficient diet develop a hypochromic anemia [74, 93, 96, 102, 116, 121]; become anorexic [93, 102, 121] and lose weight [53, 74, 93, 116, 121]; and develop thinning, loss, and graying of the fur [73, 102, 116, 121]. They also develop low blood pyridoxine levels [53, 121]. The anemia is described generally as hypochromic [93, 102, 121]. However, cell size classification is somewhat confused as one author reports a normocytic anemia [93] while others report that mean cell diameter increased [102, 121]. POPPEN *et al.* [102] report also that microcytosis is common. Target cells [121], increased resistance to hemolysis [121], marked anisocytosis [102], poikilocytosis [102], and polychromasia are common [102]. The bone marrow undergoes normoblastic hyperplasia with a reversal of M/E ratio [102]. Leucopenia with a decrease in the granulocytic and mononuclear series sometimes occurs [55, 74, 93, 102, 121].

Dental lesions [57] including reduction in size of teeth, alterations in the shape of the teeth, defects in enamel and deformities were reported by BERDJIS *et al.* [5] and POPPEN *et al.* [102]. The teeth were soft and rapidly underwent attrition. Atrophic glossitis and acute and chronic gingivitis occurred along with the dental pathology.

Necropsy and histopathologic findings include arteriosclerosis [55, 57, 74, 116, 121], with intimal accumulations of mucoid metachromatic material [116] and proliferation of connective tissue in the intima [96] as well as intimal hyperplasia and edema [116]. Arteries most commonly involved were the abdominal aorta, iliac and femoral arteries [117] and the small and medium branches of the renal arteries [116, 120]. Other sites reported include the coronary arteries [96] and the small arteries of pancreas, testes, ovary, uterus, thymus, liver, adrenal, colon, and lung [96, 116]. The development of atherosclerosis appeared to be augmented by essential fatty acid deficiency [74].

Hepatic pathology [57] ranging from fatty metaplasia to marked cirrhosis was reported by RINEHART and GREENBERG [121], and KIM [74]. Corroborating data were found by WIZGIRD *et al.* [169]. Acute deprivation resulted in a smooth, pale liver. In chronic deprivation the liver showed changes ranging from diffuse cirrhosis to scarring with regenerating nodules [116, 118]. These changes (including fibrosis) are centrilobular initially but become generalized eventually [116, 120].

Other findings include acute and chronic gingivitis [5], dental defects [5], and atrophy of spleen, thymus and lymph nodes [96].

VICTOR and ADAMS [151] found peripheral nerve demyalination and degeneration of the cells of Betz and pyramidal cells of the motor cortex in *Macaca mulatta* dying from pyridoxine deficiency.

4. Vitamin B_{12} Deficiency

Central and peripheral nervous system lesions resembling those found in man affected with pernicious anemia have been found in various nonhuman primates [128, 131]. A high incidence of alterations in the spinal cord, motor activity, oral mucosa, and blood constituents have been reported in nonhuman primates that developed low serum vitamin B_{12} levels after having been fed vegetarian diets [38, 75, 98–101, 165]. This did not occur if animal-origin foodstuffs and vitamin B_{12}-supplemented diets were fed [101].

Paralysis seen in *Macaca mulatta* with low serum vitamin B_{12} levels was described as flaccid, involving the hind legs and tail. Gradations of motor impairment ranged from posterior ataxia or clumsiness to posterior paralysis [100]. Loss of knee, ankle and plantar reflexes were observed in these animals [100] as well as reduction of loss of those of the arm. Signs of hyperesthesia of plantar surfaces were reported to be elicited by manipulation of this area [100].

Microscopic lesions in the central nervous system of these animals consisted of varying degrees of sudanophilic, spongiform demyelinization of the dorsal and lateral columns of the spinal cord with axonal degeneration. Spongiform demyelinization of the cerebral white matter and severe sudanophilic demyelinization of the peripheral nerves were lesions frequently accompanying the spinal cord degeneration [100, 101]. Clinical improvement was reported in paralysed animals after parenteral administration of vitamin B_{12} [101].

Megaloblastic anemia and macrocytosis were not seen in any of the rhesus monkeys studied by OXNARD and SMITH [100]. The nuclei of polymorphonuclear leucocytes of man while in a state of hypovitaminosis B_{12} (and other states) are hyperlobulated and they resemble those seen in many nonhuman primate species. Hyperlobulation is not a deficiency sign in *Pan satyrus, Gorilla gorilla, Macaca mulatta, Macaca fascicularis, Macaca nemestrina, Cynopithecus niger, Papio anubis* and *Papio cynocephalus* but rather is a species-specific characteristic occurring in the presence of adequate vitamin B_{12} and folate intake [71]. Also, giant metamyelocytes are found in the bone marrow of these same animals, probably reflecting again a species-specific character [71].

Alterations of the epithelium resulting in smoothness and a shiny appearance of the surface of the tongue and an apparent increase in binucleate and multinucleate buccal epithelial cells (with accompanying chromatin alterations and mitotic reductions) were observed by OXNARD [99] in primates with low serum vitamin B_{12} levels. These changes, seen in otherwise normal animals, disappeared after the parenteral administration of vitamin B_{12}.

Reduced reproductive efficiency, or at least infertility, has been correlated with low serum vitamin B_{12} levels [100]. The parenteral administration of vitamin B_{12} to infant rhesus monkeys was reported to promote an apparent increase in growth rate over untreated infants on the same dietary regimen [75]. However, since food consumption was not controlled, such comparisons should be regarded cautiously.

The interaction of vitamin B_{12}, folic acid and ascorbic acid has been discussed elsewhere in this chapter, but worth comment is the fact that vitamin B_{12} administration will not correct the megaloblastosis of folic acid deficiency. Further, serum vitamin B_{12} levels may be normal in the presence of severe folic acid deficiency [46].

5. Niacin Deficiency

Little data is available on the pathology of niacin deficiency. The early studies [13, 18, 23, 24, 61, 62, 149] of niacin deficiency are probably multiple-B vitamin deficiencies considering the constituents of the diet and because elements in addition to niacin were required to correct the lesions which developed [18].

Niacin deficiency was studied in the rhesus monkey by feeding a purified diet low in niacin [143]. Gradual loss of weight and anemia were reported [143].

6. Folic Acid Deficiency

The detailed discussion of folic acid deficiency and the interrelationship of folic acid and vitamin C metabolism by RUCH [32] leaves little to be added. A review of this summary indicates that folic acid deficiency will cause in macaques a macrocytic anemia or cytopenia [17, 22, 24, 164], increased susceptibility to infection (especially dysentery) [22, 72, 156, 158], mucoid to bloody diarrhea [22, 24, 57, 72], leucopenia [22, 156], loss of weight [156] and anorexia [156]. Folic acid and vitamin C deficient *Macaca mulatta* develop a megaloblastic anemia [91, 92, 107, 142]. Folic acid supplementation would prevent megaloblastosis; however, scurvy and anemia would develop. Ascorbic acid would prevent the anemia and scurvy.

The feeding of a synthetic diet devoid of folic acid to *Callithrix jacchus* resulted in a syndrome resembling that described in *Macaca mulatta* [33]. The administration of a folic acid antagonist (4-amino-N_{10}-methylpteroylglutamic acid) at a rate of 0.1 mg 3 times a week speeded the development of the lesions and hurried somewhat the time of the animal's death. The syndrome was characterized by weight loss, anorexia, lassitude, diarrhea, alopecia, leucopenia, and granulocytopenia. Bilateral angular cheilosis was a frequent finding in the deficiency and deficient/antagonist groups. The cheilosis presented as a bilateral, progressive stomatitis and cheilitis originating at the mucocutaneous junctions near the labial commissures and progressing through all degrees of involvement from reddening to maceration and denuding of the adjacent epithelium. The histologic picture began as a leucocytic infiltration and progressed to degeneration and erosion of the epithelium at the mucocutaneous junction. *Callithrix* fed the same diet with folic acid supplemented did not develop cheilosis or any of the accompanying signs of the syndrome.

ROSENBLUM [123] reports that the supplementation of pregnant *Saimiri sciureus* with folic acid late in pregnancy improves the chances of viable births. A megaloblastic anemia resulting from folic acid deficiency is reported to develop in *Saimiri sciureus* in the stage of pregnancy where fetal growth is maximal.

7. Biotin Deficiency

The effects of acute and chronic biotin deficiencies in *Macaca mulatta* have been described by LEASE *et al.* [78], WAISMAN and ELVEHJEM [156] and WAISMAN *et al.* [159]. Although biotin is synthesized in the intestinal tract, a deficiency may result from the longterm administration of sulfonamides which suppresses synthesis of by the feeding or raw egg white which contains an antivitamin substance (avidin).

Clinical signs in the acute deficiency include a severe dermatitis on the face [156, 159], hands [156, 159] and feet [156, 159] and blepharitis with crusting of the eyelids and excessive lacrimation [159]. Thinning of the hair-coat, listlessness, anorexia, and loss of weight followed the appearance of the dermatitis [159]. The chronic deficiency [159] is characterised by patchy alopecia or thinning of the haircoat, hypochromotrichia, a rust-colored porphyrin deposit on the abdomen and groin, gingival lesions [78], and a dry dermatitis. Leucopenia and cachexia appeared, progressing to death. Lowered resistance to infection was noted in the study by LEASE *et al.* [78].

8. Pantothenic Acid Deficiency

Lack of growth, thinning and graying of the fur, ataxia, anemia, cachexia and diarrhea were reported as deficiency signs in *Macaca mulatta* [93]. Supplements of liver as well as pantothenate were required to correct fully the symptoms. Therefore, probably other elements were deficient or pantothenic acid deficiency increases the requirement for other nutrients.

9. Choline Deficiency

A number of authors [21, 49, 138, 162, 163] have shown that the feeding of high fat, low protein, choline deficient diets to *Macaca mulatta* will cause the development of florid hepatic cirrhosis. On this regimen the animals developed alopecia, lost weight, and became apathetic. Clinical chemistry examinations revealed elevated SGOT, SGTP, serum isocitrate dehydrogenase, serum cholesterol and bilirubin levels. Hepatic damage was found to progress from fatty metamorphosis to degeneration, with varying degrees of hyalinization of parenchymal cells, and the fibrosis followed by the development of regenerative nodules [138].

WILGRAM *et al.* [162] and WILGRAM [163] studied choline deficiency in 4 *Macaca mulatta* and 4 *Cebus* on a synthetic diet of soya protein 10%, alcohol extract of peanut meal 10%, sugar 33%, starch 13%, dextrose 13%, lard 18%, and salts 3%. The author stated that methionine was too low; however, both the deficient and test animal grew and maintained weight although the choline-supplemented animal did better. The livers of the deficient animals had a two- to threefold increase in extractable lipids, although serum lipids were lowered. At laparotomy, the livers of the deficient animals were slightly enlarged, firm, and yellow, while the livers of choline-supplemented animals were normal. Histologic examination of biopsy specimens revealed fine lipid droplets throughout the entire lobe, though in the periportal area the droplets coalesced into globules, pushing the cytoplasm to the rim of the cell and the nucleus to a corner of the cell.

WILGRAM fed the same diet to 2 *Cebus* monkeys for 4 months, the protein was then reduced to 5% for 6 months. Clinically, the animals maintained a moderate intake, but lost weight, became emaciated and unstable, lost hair, and developed small skin ulcers. Noticeable ascites developed, peripheral venous distention became prominent, and the livers became palpable. At this stage the animals were icteric, and scratched constantly. At laparotomy the liver was tawny and firm with nodules, and the spleen was firm, 'sugar-coated' and enlarged. Biopsy specimens examined microscopically showed changes ranging from moderate fatty metaplasia to florid cirrhosis. Inflammatory and fibroplastic changes were especially prominent in the periportal areas, and some of the parenchymal cells were degenerate or necrotic. Gross changes regressed after choline supplementation. Microscopically, however, the regression was not complete.

Other parameters evaluated showed a normochromic anemia, decreased A/G ratio, decreased serum albumin and an increase in the globulin fractions. The inclusion of ethanol in a similar feeding study did not enhance the pathology [21].

Baboons *(Papio anubis)* on a choline-deficient diet for 2 months developed a mild centrilobular fatty metamorphosis which was reversible by choline supplementation [66].

10. Ascorbic Acid Deficiency

RUCH'S [129] and DAY'S [25] summaries of the early literature on spontaneous and induced ascorbic acid deficiencies describe well the oral, soft tissue and osseous pathology. These studies were based primarily on the rhesus monkey, *Macaca mulatta* with its known requirement of exogenous ascorbic acid [25]. The interaction between vitamin C and folic acid is discussed under the folic acid section. Detailed description of the pathology of hypovitaminosis C has been made recently in *Saimiri sciureus* [81].

General clinical signs reported included loss of weight [47a, 47b, 59a, 60, 81, 137]; anorexia, or reduced forced intake [59a, 60, 81, 137]; muscular weakness, reluctance to move or arthralgia (especially of the knees and hind limbs) [48a, 48b, 59a, 60, 137]; anemia [54, 81] and alterations of the skin [48a, 48b, 81] and adnexal structures [48a, 48b, 58a, 60, 81, 137, 149]. In addition, chemical determinations of ascorbic acid revealed decreases in serum ascorbic acid levels [54, 67, 81] in vitamin C deficient or scorbutic primates.

Autopsy lesions included gingivitis of all degrees varying from mild reddening to hemorrhage and necrosis [59a, 60, 67, 137, 148], and dental abnormalities, including heavy tartar accumulations [48a, 48b, 137], loosened or missing teeth, and periodontal osseous defects or deficits [48a, 48b, 59a, 60, 67, 81, 137, 148].

Extensive hemorrhage into the subcutaneous and muscular tissue [59a, 60, 148] around the wrist, knees and ankles and into the subperiosteal and subendosteal spaces [58a, 60, 81, 148] of the long bones, palate, and vertex and parietal areas of the skull and ribs have been reported as prominent gross lesions. Hemorrhage into the wrist, temporomandibular hip, knee and elbow joints, were reported by TOMLINSON [148]. Osseous pathology reported included metaphyseal fractures [81], epiphyseal fractures [148] and resorption at the ends of the long bones [59a, 60]. LEHNER, BULLOCK and CLARKSON [81] reported that in *Saimiri sciureus*, 'the epiphysis and periosteum of the long bones were easily detachable from the shafts' of the long bones' and that 'the bones of the skull were easily separable by gentle traction.' Periosteal and epiphyseal separation, and impaction of the long bones were also reported for *Macaca mulatta* [59a, 60, 148]. Additional lesions reported as erosions or resorptions of the ends of the long bones were described by several authors [59a, 60, 137, 148].

Histopathologic examination of the epiphyseal plates and costochondral junctions revealed a rarefaction of cortical bone in *Saimiri sciureus* [81] and lack of bone deposition on the cartilage spicules [81] or osteoid formation [81]. Long cartilaginous spicules were reported to extend from the epiphyseal plate into loose medullary connective tissue just beneath the epiphyseal plate in both *Saimiri sciureus* [81] and *Macaca mulatta* [59a, 60]. LEHNER *et al.* [81] further reported that the subperiosteal hematomata were poorly organized, even in the parietal areas of the skull where hemorrhage was observed 1 month prior to necropsy examination.

Two authors [67, 113] investigated the role of ascorbic acid deficiency on infection. Thirteen of 21 *Macaca mulatta* on an ascorbic-deficient diet died of acute infection (5 of pneumonia) before gross signs of scurvy were evident [134]. HONJO *et al.* [67] found that intercurrent ascorbic acid deficiency reduced the quantity of *Shigella flexneri* 2a organisms needed to infect 3 of 7 *Macaca irus* from 10^9 to 10^{10} viable counts/animal to 7.1×10^6 viable counts/animal.

V. Water

Water should certainly be mentioned in this chapter though no formal data has been found on the pathology of water deprivation in the nonhuman primate. It would probably be safe to surmise that a nonhuman primate which has been dehydrated, whether from excessive excretion of fluids (commonly, dysentery) or inadequate intake (malfunctioning automatic watering system, etc.), would be a fair example of inadequate water intake. In such situations one might see, depending on severity, ataxia, incoordination or coma, sunken eyeballs, increased plasticity of the skin, dry mucous membranes, and clinical laboratory values indicating hemoconcentration and increased serum solids[4].

VI. Minerals

Little research has been done on the mineral requirements or deficiency responses of the nonhuman primate, very likely because of the difficulties in this area of nutritional research.

A. Zinc Deficiency

The feeding of a purified diet containing extremely low levels of zinc (0.5 ppm) to young *Saimiri sciureus* resulted in retardation of growth rate, and an unthrifty or unkempt appearance with varying degrees of alopecia and parakeratosis of the lingual epithelium [4, 83]. The parakeratosis was most pronounced on the anterior dorsum of the tongue though these changes were found on the ventral surface as well [4]. Coinciding with the areas of parakeratosis were regions of keratohyalin granule loss or reduction.

Erythrocytic values remained in a normal range although serum albumin levels were reduced [83]. Zinc concentrations in the hair, spleen, liver, cardiac muscle and pancreas were reduced in the deprived animals although adrenal levels were increased, and bone and testicular levels were normal [83].

B. Calcium Deficiency

It has been reported that the rhesus monkey metabolized calcium like man [63]. Therefore, deficiency of this element would probably result in the

4 Wolf, R. H., unpublished data.

neuromuscular and skeletal abnormalities. DAY [25] concludes that prolonged calcium deficiency results in 'irritability, anorexia, weakness, and muscular atrophy; analysis of bones revealed a marked reduction in calcium content'. Calcium deficiency, and calcium deficiency concomitant with ascorbic acid deficiency has been shown to cause a greater reduction of bone calcium in rhesus monkeys than with scurvy only [48a, 48b, 148, 149].

C. Magnesium Deficiency

The production of magnesium deficiency in *Cebus albifrons* by the use of a magnesium deficient diet resulted in loss of weight, hyperirritability, convulsions, electrocardiographic alterations including S-T segment depression and T-wave peaking, and an increased sensitivity to cardiac glycosides [155]. Approximately half the dose of K-strophanthidine required to produce extrasystoles in the control animals would do so in the deficient animals.

Sudanophilic vascular deposits and dense fibrous connective tissue plaques in the ascending and descending aorta found at necropsy were reported to be much more frequent in the test animal than in the controls.

VII. Conclusion

Of as much importance probably as the consistent though non-specific lesions produced in nonhuman primates by the exclusion of a specific nutrient is the fact that the great majority of the reports reviewed here indicate that increased susceptibility to infection, intercurrent infection and exacerbation of low grade or chronic infection become evident during the course of the experiment. Certainly a primate fed an inadequate diet is not a suitable subject for most types of biomedical research. A number of commercially produced, reasonably adequate diets are available to aid in this problem of properly feeding laboratory primates as are diets developed and used by various zoological gardens. Unfortunately, much of the rationale in the development of these diets is empirical or at least extrapolated from *Macaca mulatta* for which there are most data. Experimental production of deficiency syndromes has not been made, or at least reported, for a great number of nutrients. Investigations into the deficiency states and requirements of these elements would be of much benefit in the nutritional management of captive primates and should be encouraged.

References

1 ANDRUS, S. B.; PORTMAN, O. W., and RIOPELLE, A. J.: Comparative studies of spontaneous and experimental atherosclerosis in primates. II. Lesions in chimpanzees including myocardial infarction and cerebral aneurysms. Progr. biochem. Pharmacol. vol. 4, pp. 393–419 (Karger, Basel 1968).

2 ARMSTRONG, M. L.; CONNOR, W. E., and WARNER, E. D.: Xanthomatosis in rhesus monkeys fed a hypercholesterolemic diet. Arch. Path. *84:* 227–237 (1967).

3 ARMSTRONG, M. L.; CONNOR, W. E., and WARNER, E. D.: Tissue cholesterol concentration in the hypercholesterolemic rhesus monkey. Arch. Path. *87:* 87–92 (1969).

4 BARNEY, G. H.; MACAPINLAC, M. P.; PEARSON, W. N., and DARBY, W. J.: Parakeratosis of the tongue. A unique histopathologic lesion in the zinc deficient squirrel monkey. J. Nutr. *93:* 511–517 (1967).

5 BERDJIS, C. C.; GREENBERG, L. D.; RINEHART, J. F., and FITZGERALD, G.: Oral and dental lesions in vitamin B_6 deficient rhesus monkeys. Brit. J. exp. Path. *41:* 198–205 (1960).

6 BLOND, C. P. LE: *vide* LE BLOND, C. P.

7 BOULAY, G. H. DU and CRAWFORD, M. A.: Nutritional bone disease in captive primates. Symp. zool. Soc., Lond. *21:* 223–236 (1968).

8 BRANDOM, W. F.; FELSENFELD, O., and GREER, W. E.: Proliferation of immunologically active cells in the intestines of primates during secondary response to cholera toxin. Ann. trop. Med. Parasit. *63:* 129–138 (1968).

9 BULLOCK, B. C. and BOWEN, J. A.: Rickets and osteomalacia in squirrel monkeys *(Saimiri sciureus)*. Fed. Proc. *25:* 533 (1966).

10 BULLOCK, B. C.: Effect of age and diet on coronary artery atherosclerosis in *Cebus albifrons*. Fed. Proc. *26:* 371 (1967).

11 BULLOCK, B. C.; CLARKSON, T. B.; LEHNER, N. D. M.; LOFLAND, H. B., jr., and ST. CLAIR, R. W.: Atherosclerosis in *Cebus albifrons* monkeys. III. Clinical and pathologic studies. Exp. mol. Path. *10:* 39–62 (1969).

12 BULLOCK, B. C.: Personal communication (1970).

13 CHICK, H. and HUME, E. M.: The production in monkeys of symptoms closely resembling those of pellagra, by prolonged feeding of a diet of low protein content. Biochem. J. *1:* 135–146 (1920).

14 CLARKSON, T. B.; BULLOCK, B. C., and LEHNER, N. D. M.: Pathologic characteristics of atherosclerosis in New World monkeys. Progr. biochem. Pharmacol., vol. 4, pp. 420–428 (Karger, Basel 1968).

15 CLARKSON, T. B.; BULLOCK, B. C.; LEHNER, N. D. M., and FELDNER, M. A.: The squirrel monkey as a laboratory animal. Publ. 1594 (National Academy of Science, Washington 1968).

16 COOPERMAN, J. M.; WAISMAN, H. A.; MCCALL, K. B., and ELVEHJEM, C. A.: Studies on the requirements of the monkey for riboflavin and a new factor found in liver. J. Nutr. *30:* 45–57 (1945).

17 COOPERMAN, J. M.; ELVEHJEM, C. A.; MCCALL, K. B., and RUEGAMER, W. R.: 'Folic acid' active compounds in the nutrition of the monkey. Proc. Soc. exp. Biol. Med. *61:* 92–97 (1946).

18 COOPERMAN, J. M.; MCCALL, K. B.; RUEGAMER, W. R., and ELVEHJEM, C. A.:

Attempts to produce a niacin deficiency in the rhesus monkey. J. Nutr. *32:* 37–45 (1946).

19 COWDRY, E. V. and SCOTT, G. H.: Effect on monkeys of small doses of a concentrated preparation of viosterol. Arch. Path. *2:* 71–105 (1944).

20 COX, G. C.; TAYLOR, C. B.; COX, L. G., and COUNTS, M. A.: Atherosclerosis in rhesus monkeys. Hypercholesterolemia induced by dietary fat and cholesterol. Arch. Path. *66:* 32–52 (1958).

21 CUETO, J.; TAJEN, N.; GILBERT, E., and CURRIE, R. A.: Experimental liver injury in the rhesus monkey. I. Effects of cirrhogenic diet and ethanol. Ann. Surg. *166:* 19–28 (1967).

22 DAY, P. L.; LANGSTON, W. C., and SHUKERS, C. F.: Leukopenia and anemia in the monkey resulting from vitamin deficiency. J. Nutr. *9:* 637–644 (1935).

23 DAY, P. L.; LANGSTON, W. C., and DARBY, W. J.: Failure of nicotinic acid to prevent nutritional cytopenia in the monkey. Proc. Soc. exp. Biol. Med. *38:* 861–863 (1938).

24 DAY, P. L.; LANGSTON, W. C.; DARBY, W. J.; WAHLIN, J. H., and MIMS, V.: Nutritional cytopenia in monkeys receiving the Goldberger diet. J. exp. Med. *72:* 463–477 (1940).

25 DAY, P. L.: The nutrient requirements of primates other than man. Vitamins Hormones, N.Y. *2:* 71–105 (1944).

26 DAY, P. L. and DINNING, J. S.: Nutritional muscular dystrophy in monkeys receiving a diet deficient in both vitamins B_6 and E. Amer. J. clin. Nutr. *4:* 386–389 (1956).

27 DEO, M. G. and RAMALINGASWAMI, V.: Production of periportal fatty infiltration of the liver in the rhesus monkey by a protein deficient diet. Lab. Invest. *9:* 319–329 (1960).

28 DEO, M. G. and RAMALINGASWAMI, V.: Absorption of ^{58}CO labelled cyanocobalamin in protein deficiency. An experimental study in the rhesus monkey. Gastroenterology *46:* 167–174 (1964).

29 DEO, M. G. and RAMALINGASWAMI, V.: Reaction of the small intestine to induced protein malnutrition in rhesus monkeys. A study of cell population kinetics in the jejunum. Gastroenterology *49:* 150–157 (1965).

30 DEO, M. G.; SOOD, S. K., and RAMALINGASWAMI, V.: Experimental protein deficiency. Pathological features in the rhesus monkey. Arch. Path. *80:* 14–23 (1965).

31a DINNING, J. S. and DAY, P. L.: Vitamin E deficiency in the monkey. J. exp. Med. *105:* 395–402 (1957).

31b DINNING, J. S. and DAY, P. L.: Vitamin E deficiency in the monkey. II. Tissue concentration of nucleic acid and creatine. J. Nutr. *63:* 393–397 (1957).

32 DINNING, J. S.: Vitamin E responsive anemia in monkeys and man. Nutr. Rev. *21:* 289–291 (1963).

33 DREIZEN, S. and LEVY, B. M.: Histopathology of experimentally induced nutritional deficiency cheilosis in the marmoset *(Callithrix jacchus)*. Arch. oral Biol. *14:* 577–582 (1969).

34 DREYFUS, P. M. and VICTOR, M.: Effects of thiamine deficiency on the central nervous system. Amer. J. clin. Nutr. *9:* 414–425 (1961).

35 DU BOULAY, G. H.: *vide* BOULAY, G. H. DU

36 FIENNES, R. N. T- W-: Cystic bone disease in South American primates. II. Proc. zool. Soc., Lond. *143:* 521–523 (1964).

37 FILER, L. J., jr.; RUMERY, R. E.; YU, P. N. G., and MASON, K. E.: Studies on vitamin E deficiency in the monkey. Ann. N.Y. Acad. Sci. *52:* 284–291 (1949).

38 FLINN, R. M. and OXNARD, C. E.: The relationship between growth and the administration of cyanocobalamin in the rhesus monkey. Folia primat. *4:* 432–437 (1966).

39 FOLLIS, R. H., jr.: A kwashiorkor-like syndrome observed in monkeys fed maize. Proc. Soc. exp. Biol. Med. *96:* 523–528 (1957).

40 FOLLIS, R. H., jr.: Deficiency disease (Thomas, Springfield 1958).

41 FOLLIS, R. H., jr.: Deficiency disease, pp. 11–13 (Thomas, Springfield 1958).

42 FOLLIS, R. H., jr.: Deficiency disease, pp. 15–16 (Thomas, Springfield 1958).

43 FOLLIS, R. H., jr.: Deficiency disease, p. 197 (Thomas, Springfield 1958).

44 FOLLIS, R. H., jr.: Deficiency disease, pp. 362 (Thomas, Springfield 1958).

45 FOY, H.; KONDI, A., and MBAYA, V.: Effects of riboflavine deficiency on the bone marrow function and protein metabolism in baboons. Preliminary report. Brit. J. Nutr. *18:* 307–318 (1964).

46 FOY, H.; KONDI, A., and MBAYA, V.: Serum vitamin B_{12} and folate levels in normal and riboflavin-deficient baboons *(Papio anubis)*. Brit. J. Haemat. *12:* 239–245 (1966).

47 FOY, H. and KONDI, A.: Comparison between erythroid aplasia in marasmus and kwashiorkor and the experimentally induced erythroid aplasia in baboons by riboflavin deficiency. Vitamins Hormones, N.Y. *28:* 653–681 (1968).

48a FRASER, H. F.: A chronic deficiency of (1) calcium, (2) vitamin C, and (3) both calcium and vitamin C in monkeys. Publ. Hlth. Rep. Wash. *57:* 959–967 (1942).

48b FRASER, H. F.: Mouth lesions in monkeys associated with a chronic deficiency of (1) calcium, (2) vitamin C, and (3) both calcium and vitamin C. Publ. Hlth Rep., Wash. *57:* 968–973 (1942).

49 GAISFORD, W. D. and ZUIDEMA, G. D.: Nutritional Laennec's cirrhosis in the *Macaca mulatta* monkey. J. surg. Res. *5:* 220–235 (1965).

50 GERSTENBERGER, H. J.: Rickets in monkeys *(Macaca rhesus)*. Amer. J. Dis. Child. *56:* 694–695 (1938).

51 GOPALAN, C. and PADMAVATHI, S.: Electrocardiographic studies on normal and malnourished monkeys *(Macacus radiata)*. Indian J. med. Res. *45:* 423–430 (1957).

52 GRAHAM-JONES, O.: Cystic bone disease in South American primates. I. Proc. zool. Soc., Lond. *143:* 517–519 (1964).

53 GREENBERG, L. D. and RINEHART, J. F.: Serum iron levels in rhesus monkeys with chronic vitamin B_6 deficiency. Proc. exp. Soc. Biol. Med. *70:* 20–25 (1949).

54 GREENBERG, L. D. and RINEHART, J. F.: Serum iron levels in rhesus monkeys with chronic vitamin C deficiency. Proc. Soc. exp. Biol. Med. *88:* 325–328 (1955).

55 GREENBERG, L. D.; MCIVOR, B., and MOON, H. D.: Experimental arteriosclerosis in pyridoxine-deficient rhesus monkeys. Amer. J. clin. Nutr. *6:* 635–637 (1958).

56 GREENBERG, L. D. and MOON, H. D.: Riboflavin deficiency in the monkey. Fed. Proc. *22:* 592 (1963).

57 GREENBERG, L. D.: Arteriosclerotic, dental, and hepatic lesions in pyridoxine-deficient monkeys. Vitamins Hormones, N.Y. *22:* 677–694 (1964).

58 GRINKER, R. B. and KANDEL, E.: Experimental vitamin (A, B, B_2 and B complex) deficiency. Histologic changes in the central nervous system. Arch. Neurol. Psychiat. *30:* 1287–1297 (1933).

59a HARDEN, A. and ZILVA, S. S.: Experimental scurvy in monkeys. J. Path. Bact. *22:* 246–251 (1918–1919).

59b HARDEN, A. and ZILVA, S. S.: Edema observed in a monkey fed on a diet free from the fat-soluble 'A' accessory food factor and low in fat. Lancet *ii:* 780–781 (1919).

60 HARDEN, A. and ZILVA, S. S.: The antiscorbutic requirements of the monkey. Biochem. J. *1:* 131–134 (1920).

61 HARRIS, L. J.: The vitamin B_2 complex. Differentiation of the anti-blacktongue and P.-P. factors from lactoflavin and vitamin B_6. VII. Experiments with monkey and other species. Biochem. J. *37:* 1414–1421 (1937).

62 HARRIS, L. J.: CXCV. The vitamin B_2 complex. VIII. Further notes on 'Monkey pellagra' and its cure with nicotinic acid. Biochem. J. *32:* 1479–1481 (1938).

63 HARRIS, R. S.; MOOR, J. R., and WANNER, R. L.: Calcium metabolism of the normal rhesus monkey. J. clin. Invest. *40:* 1766–1774 (1961).

64 HETLER, R. A.: The development of xerophthalmia and the keratinization of epithelial tissue on withdrawal of vitamin A from the diet of the monkey *(Macaca rhesus)*, guinea pig, rabbit, and adult albino rat. J. Nutr. *8:* 75–96 (1934).

65 HILL, R. B.; SCHENDEL, H. W.; RAMA RAO, P. B., and JOHNSON, B. C.: Effect of dietary vitamin K on the nutritional status of the rhesus monkey. J. Nutr. *84:* 259–264 (1964).

66 HOFFBAUER, F. W. and ZAKI, F. G.: Choline deficiency in baboon and rat compared. Arch. Path. *79:* 364–369 (1965).

67 HONJO, S.; TAKASAKA, M.; FUJIWARA, T.; IMAIZUMI, K., and OGAWA, H.: Shigellosis in cynomolgus monkeys *(Macaca irus)*. VII. Experimental production of dysentery with a relatively small dose of *Shigella flexneri* 2a in ascorbic acid deficient monkeys. Jap. J. med. Sci. Biol. *22:* 149–162 (1969).

68 HOWARD, A. N.; GRESHAM, G. A.; BOWYER, D. E., and LINDGREN, F. T.: Aortic and coronary atherosclerosis in baboons, Progr. biochem. Pharmacol., vol. 4, pp. 438–444 (Karger, Basel 1968).

69 HUNT, R. D.; GARCIA, F. G., and HEGSTED, D. M.: A comparison of vitamin D_2 and D_3 in New World primates. I. Production and regression of osteodystrophia fibrosa. Lab. anim. Care *17:* 222–234 (1967).

70 HUNT, R. D.; GARCIA, F. G., and HEGSTED, D. M.: Hypervitaminosis D in New World monkeys. Lab. anim. Care *22:* 358–366 (1969).

71 HUSER, H. J. and BEARD, M. E. J.: Studies on folate and vitamin B_{12} metabolism in primates. I. Blood and bone marrow morphology, folate and vitamin B_{12} levels. Folia primat. *10:* 172–180 (1969).

72 JANOTA, M. and DACK, G. M.: Bacillary dysentery developing on a 'vitamin M' deficient diet. J. infect. Dis. *65:* 219–224 (1939).

73 KENT, S. P.; VAWTER, G. W.; DOWBEN, R. M., and BENSON, R. E.: Hypervitaminosis D in monkeys. Amer. J. Path. *34:* 37–59 (1958).

74 KIM, D. S.: Experimental production of arteriosclerosis in rhesus monkeys by deficiency of pyridoxine hydrochloride (vitamin B-6) and essential fatty acids. Yonsei med. J. *2:* 42–47 (1961).

75 KROHN, P. L.; OXNARD, C. E., and CHALMERS, J. N. M.: Vitamin B_{12} levels in the serum of rhesus monkeys. Nature, Lond. *197:* 186 (1963).

76 KRAMSCH, D. M. and HOLLANDER, W.: Occlusive atherosclerotic disease of the coronary arteries in monkey *(Macaca irus)* induced by diet. Exp. molec. Path. *9:* 1–22 (1968).

77 Krook, L. and Barrett, R. B.: Simian bone disease. A secondary hyperparathyroidism. Cornell Vet. *52:* 459–492 (1962).

78 Lease, J. G.; Parson, Helen T., and Kelly, Eunice: A comparison in five types of animals of the effects of dietary egg white and a specific factor given orally or parenterally. Biochem. J. *31:* 433–457 (1937).

79 Le Blond, C. P. and Chaulin-Serviniere, J.: Spontaneous beriberi of the monkey as compared with experimental avitaminosis. Amer. J. med. Sci. *203:* 100–109 (1942).

80 Lehner, N. D. M.; Bullock, B. C.; Clarkson, T. B., and Lofland, H. B., jr.: Biological activities of vitamins D_2 and D_3 for growing squirrel monkeys. Lab. anim. Care *17:* 483–493 (1967).

81 Lehner, N. D. M.; Bullock, B. C., and Clarkson, T. B.: Ascorbic acid deficiency in the squirrel monkey. Proc. Soc. exp. Biol. Med. *128:* 512–000 (1968).

82a Lofland, H. B., jr.; Clarkson, T. B.; St. Clair, R. W.; Lehner, N. D. M., and Bullock, B. C.: Atherosclerosis in *Cebus albifrons* monkeys. I. Sterol metabolism. Exp. molec. Path. *8:* 302–313 (1968).

82b Lofland, H. B., jr.; Clarkson, T. B.; St. Clair, R. W.; Lehner, N. D. M., and Bullock, B. C.: Atherosclerosis in *Cebus* monkeys. II. Arterial metabolism. Exp. molec. Path. *9:* 57–70 (1968).

83 Macapinlac, M. P.; Barney, G. H.; Pearson, W. N., and Darby, W. J.: Product of zinc deficiency in the squirrel monkey *(Saimiri sciureus)*. J. Nutr. *93:* 499–510 (1967).

84 MacNintch, J. E.; St. Clair, R. W.; Lehner, N. D. M; Clarkson, T. B., and Lofland, H. B., jr.: Cholesterol metabolism and atherosclerosis in *Cebus* monkeys in relation to age. Lab. Invest. *16:* 444–452 (1967).

85 Malinow, M. R.; Maruffo, C. A., and Perley, A. M.: Experimental atherosclerosis in *Cebus* monkeys. J. Path. Bact. *92:* 491–510 (1966).

86 Mann, G. V.; Watson, D. L.; McNally, A., and Goddard, J.: Primate nutrition. II. Riboflavin deficiency in the *Cebus* monkey and its diagnosis. J. Nutr. *47:* 225–241 (1952).

87 Mann, G. V.; Andrus, S. B.; McNally, A., and Stare, F. J.: Experimental atherosclerosis in Cebus monkeys. J. exp. Med. *98:* 195–218 (1953).

88 Mann, G. V. and Andrus, S. B.: Xanthomatosis and atherosclerosis produced by diet in an adult rhesus monkey. J. Lab. clin. Med. *48:* 533–550 (1956).

89 Mason, D. E. and Telford, I. R.: Vitamin E deficiency in the monkey. Anat. Rec. *94:* 522 (1946).

90 Mason, K. E. and Telford, I. R.: Some manifestations of vitamin E deficiency in the monkey. Arch. Path. *43:* 363–373 (1947).

91 May, C. D.; Sundberg, R. D.; Schaar, F., and Lowe, C. U.: Experimental nutritional megaloblastic anemia. Relationship of ascorbic acid and pteroylglutamic acid. Amer. J. Dis. Child. *82:* 282–309 (1951).

92 May, C. D.; Hamilton, A., and Stewart, C. T.: Experimental megaloblastic anemia and scurvy in the monkey. IV. Vitamin B_{12} and folic acid compounds in the diet, liver, urine, feces and effect of therapy. Blood *7:* 978–991 (1952).

93 McCall, K. B.; Waisman, H. A.; Elvehjem, C. A., and Jones, E. S.: A study of pyridoxine and pantothenic acid deficiency in the monkey *(Macaca mulatta)*. J. Nutr. *31:* 685–697 (1946).

94 METTA, V. C. and GOPALAN, C.: A study of the vitamin K requirements of the monkey *(Macacus radiata)*. Indian J. med. Res. *51:* 512–515 (1963).

95 MIDDLETON, C. C.; CLARKSON, T. B.; LOFLAND, H. B., jr., and PRICHARD, R. W.: Diet and atherosclerosis of squirrel monkeys. Arch. Path. *83:* 145–153 (1967).

96 MUSHETT, C. W.; STEBBINS, R. S., and BARTON, M. N.: Studies in the pathologic effects produced by two analogues of pyridoxine. Trans. N.Y. Acad. Sci. *9:* 291–296 (1947).

97 ORDY, J. M.; SAMORAJSKI, T.; ZIMMERMAN, R. R., and RADY, P. M.: Effects of postnatal protein deficiency on weight gain, serum protein, enzymes, cholesterol, and liver ultrastructure in a subhuman primate *(Macaca mulatta)*. Amer. J. Path. *48:* 769–782 (1966).

98 OXNARD, C. E.: Some variations in the amount of vitamin B_{12} nutrition in some primates in captivity. Nature, Lond. *201:* 1188–1191 (1964).

99 OXNARD, C. E.: Vitamin B_{12} nutrition in some primates in captivity. Folia primat. *4:* 424–431 (1966).

100 OXNARD, C. E. and SMITH, W. T.: Neurological degeneration and reduced serum vitamin B_{12} levels in captive monkeys. Nature, Lond. *210:* 507–509 (1966).

101 OXNARD, C. E.; SMITH, W. T., and TORRES, I.: Vitamin B_{12} deficiency in captive monkeys and its effect on the nervous system and the blood. Lab. Anim. *4:* 1–12 (1970).

102 POPPEN, K. J.; GREENBERG, L. D., and RINEHART, J. F.: The blood picture of pyridoxine deficiency in the monkey. Blood *7:* 436–444 (1952).

103 PORTER, F. S.; FITCH, C. D., and DINNING, J. S.: Vitamin E deficiency in the monkey. IV. Further studies on the anemia with emphasis on bone marrow morphology. Blood *20:* 471–477 (1962).

104 PORTMAN, O. W. and ANDRUS, S. B.: Comparative evaluation of three species of New World monkeys for studies of dietary factors, tissue lipids and atherogenesis. J. Nutr. *87:* 429–438 (1965).

105 PORTMAN, O. W.; ALEXANDER, M., and MARUFFO, C. A.: Nutritional control of arterial lipid composition in squirrel monkeys. Major ester classes and types of phospholipids. J. Nutr. *91:* 35–45 (1967).

106 PORTMAN, O. W.: Nutrient requirements of monkeys. NRC Bull. 990 (in press).

107 PROEHL, E. C. and MAY, C. D.: Experimental nutritional megaloblastic anemia and scurvy in the monkey. III. Protoporphyrin, coproporphyrin, urobilinogen, and iron in blood and excreta. Blood *7:* 671–682 (1952).

108 RACELA, A. S.; GRADY, H. J.; HIGGINSON, J., and SVOBODA, D. J.: Protein deficiency in the rhesus monkey. Amer. J. Path. *49:* 419–443 (1966).

109 RAMALINGASWAMI, V.; LEACH, E. H., and SRIRAMACHARI, S.: Ocular structure in vitamin A deficiency in the monkey. Quart. J. exp. Physiol. *40:* 337–347 (1955).

110 RAMALINGASWAMI, V.; DEO, M. G., and SOOD, S. K.: Protein deficiency in the rhesus monkey, in meeting protein needs of infants and children. NAC Pub. No. 843, pp. 365–375 (1961).

111 RAMALINGASWAMI, V.: Perspectives in protein malnutrition. Nature, Lond. *201:* 546–551 (1964).

112 RAMALINGASWAMI, V.: Interface of protein nutrition and medicine in tropics. Lancet *1969:* 733–736.

113 RAO, M. V. R.: Intestinal changes in monkeys fed on poor rice diets. Indian J. med. Res. *30:* 273–284 (1942).

114 RATCLIFFE, H. L.: Disease control and colony management of infrahuman primates at the Philadelphia Zoological Gardens. Proc. Anim. care Panel *5:* 8–15 (1954).

115 RINEHART, J. F.; GREENBERG, L. D., and FRIEDMAN, M.: Experimental thiamine deficiency in the rhesus monkey. Amer. J. Path. *23:* 879–881 (1947).

116 RINEHART, J. F. and GREENBERG, L. D.: Arteriosclerotic lesions in pyridoxine-deficient monkeys. Fed. Proc. *7:* 278 (1948).

117 RINEHART, J. F.; GREENBERG, L. D., and GINZTON, L. L.: Thiamine deficiency in the rhesus monkey. Clinical, metabolic and hematologic observations. Blood *3:* 1453–1459 (1948).

118 RINEHART, J. F.; FRIEDMAN, M., and GREENBERG, L. D.: Effect of experimental thiamine deficiency on the nervous system of the rhesus monkey. Arch. Path. *48:* 129–139 (1949).

119 RINEHART, J. F. and GREENBERG, L. D.: Effect of experimental thiamine deficiency on the heart of the rhesus monkey. Arch. Path. *48:* 89–95 (1949).

120 RINEHART, J. F. and GREENBERG, L. D.: Pathogenesis of experimental atherosclerosis in pyridoxine deficiency with notes on similarities to human atherosclerosis. Arch. Path. *51:* 12–18 (1951).

121 RINEHART, J. F. and GREENBERG, L. D.: Vitamin B_6 deficiency in the rhesus monkey. Amer. J. clin. Nutr. *4:* 318–325 (1956).

122 RODGERS, F. C.; GROVER, A. D., and FAZAL, H.: Experimental hemaralopia uncomplicated by xerophthalmia in *Macacus rhesus*. Brit. J. Ophthal. *45:* 96–108 (1961).

123 ROSENBLUM, L. A.: Some aspects of female reproductive physiology in the squirrel monkey; in ROSENBLUM and COOPER The squirrel monkey, pp. 163–166 (Academic Press, New York 1968).

124 RUCH, T. C.: Diseases of laboratory primates (Saunders, Philadelphia 1959).

125 RUCH, T. C.: Diseases of laboratory primates, pp. 7–9 (Saunders, Philadelphia 1959).

126 RUCH, T. C.: Diseases of laboratory primates, pp. 11–15 (Saunders, Philadelphia 1959).

127 RUCH, T. C.: Diseases of laboratory primates, p. 18 (Saunders, Philadelphia 1959).

128 RUCH, T. C.: Diseases of laboratory primates, p. 19 (Saunders, Philadelphia 1959).

129 RUCH, T. C.: Diseases of laboratory primates, pp. 21–24 (Saunders, Philadelphia 1959).

130 RUCH, T. C.: Diseases of laboratory primates, pp. 34–36 (Saunders, Philadelphia 1959).

131 RUCH, T. C.: Diseases of laboratory primates, pp. 388–393 (Saunders, Philadelphia 1959).

132 RUCH, T. C.: Diseases of laboratory primates, pp. 474–500 (Saunders, Philadelphia 1959).

133 RUTHERFORD, R. B.; BOITNOTT, J. D.; DONOHOO, J. S.; MARGOLIS, S.; SEBOR, J., and ZUIDEMA, G. D.: Production of nutritional cirrhosis in *Macaca mulatta* monkeys. Arch. Surg., Chicago *98:* 720–730 (1969).

134 SABIN, A. B.: Vitamin C in relation to experimental poliomyelitis with incidental observations on certain manifestations in *Macaca rhesus* monkeys on a scorbutic diet. J. exp. Med. *69:* 507–516 (1939).

135 SCOTT, R. F.; JONES, R.; DAOUD, A. S.; ZUMBO, O.; COULSTON, F., and THOMAS,

W. A.: Experimental atherosclerosis in rhesus monkeys. II. Cellular elements of proliferative lesions and possible role of cytoplasmic degeneration in pathogenesis as studies by electron microscopy. Exp. molec. Path. *7:* 34–57 (1957).

136 Scott, R. F.; Morrison, E. S.; Jarmolych, J.; Nam, S. C.; Kroms, M., and Coulston, F.: Experimental atherosclerosis in rhesus monkeys. I. Gross and light microscopy features and lipid values in serum and aorta. Exp. molec. Path. *7:* 11–33 (1967).

137 Shaw, J. H.; Phillips, P. H., and Elvehjem, C. A.: Acute and chronic ascorbic acid deficiencies in the rhesus monkey. J. Nutr. *29:* 365–372 (1945).

138 Smetana, H. F. and Greer, W. E.: Experimental study of the histogenesis of nutritional cirrhosis in primates. Proc. 3rd Congress of Gastroenterology *3:* 25–29 (1966).

139 Sood, S. K.; Deo, M. G., and Ramalingaswami, V.: Anemia in experimental protein deficiency in the rhesus monkey with special reference to iron metabolism. Blood *26:* 421–432 (1965).

140 Stare, F. J.; Andrus, S. B., and Portman, O. W.: Primates in medical research with special reference to New World monkey; in Pickering Proceedings of a conference on research with primates. Tecktronix Foundation. Beaverton, Ore. (1966).

141 Strong, J. P. and McGill, H. C.: Diet and experimental atherosclerosis in baboons. Amer. J. Path. *50:* 669–690 (1967).

142 Sundberg, R. D.; Schaar, F., and May, C. D.: Experimental nutritional megaloblastic anemia. II. Hematology. Blood *7:* 1143–1181 (1952).

143 Tappan, D. V.; Lewis, U. J.; Register, U. D., and Elvehjem, C. A.: Niacin deficiency in the rhesus monkey. J. Nutr. *46:* 75–83 (1952).

144 Taylor, C. B.; Manalo-Estrella, P., and Cox, G. E.: Atherosclerosis in rhesus monkeys. V. Marked diet-induced hypercholesterolemia with xanthomatosis and severe atherosclerosis. Arch. Path. *76:* 239–244 (1963).

145 Taylor, C. B.; Patton, D., and Cox, G. E.: Atherosclerosis in rhesus monkeys. VI. Fatal myocardial infarction in a monkey fed fat and cholesterol. Arch. Path. *76:* 404–412 (1963).

146 Taylor, B.: Experimentally induced arteriosclerosis in nonhuman primates; in Roberts and Straus Comparative atherosclerosis, pp. 215–243 (Harper, New York 1965).

147 Tilden, E. B. and Miller, E. G., jr.: The response of the monkey *(Macacus rhesus)* to withdrawal of vitamin A from the diet. J. Nutr. *3:* 121–139 (1930).

148 Tomlinson, T. H., jr.: Pathology of artificially induced scurvy in the monkey. With and without chronic calcium deficiency. Publ. Hlth Rep., Wash. *57:* 987–993 (1942).

149 Topping, N. H. and Fraser, H. F.: Mouth lesions associated with dietary deficiencies in monkeys. Publ. Hlth Rep., Wash. *54:* 416–431 (1939).

150 Turner, R. G. and Loew, E. R.: Effect of vitamin withdrawal on the monkey *(Macacus rhesus)*. J. Nutr. *5:* 29–34 (1932).

151 T- W- Fiennes, R. N.: *vide* Fiennes, R. N. T- W-

152 Ulmer, F. A., jr.: The monkey. A comparison of the natural environment with observations in captivity. Ann. N.Y. Acad. Sci. *85:* 737–746 (1960).

153 Verder, E. and Petran, E.: Vitamin A deficiency in rhesus monkeys. Studies on the gastro-intestinal tract, the blood, and nervous system. J. infect. Dis. *60:* 193–208 (1937).

154 VICTOR, M. and ADAMS, R. D.: The neuropathology of experimental vitamin B_6 deficiency in monkeys. Amer. J. clin. Nutr. *4:* 346–353 (1956).
155 VITALE, J. J.; VELEZ, H.; GUZMAN, C., and CORREA, P.: Magnesium deficiency in the *Cebus* monkey. Circulat. Res. *12:* 642–649 (1963).
156 WAISMAN, H. A. and ELVEHJEM, C. A.: The role of biotin and 'folic acid' in the nutrition of the rhesus monkey. J. Nutr. *26:* 361–375 (1943).
157 WAISMAN, H. A.: Production of riboflavin deficiency in the monkey. Proc. Soc. exp. Biol. Med. *55:* 69–71 (1944).
158 WAISMAN, H. A.; RASMUSSEN, A. F., jr.; ELVEHJEM, C. A., and CLARK, P. F.: Studies on the nutritional requirements of the rhesus monkey. J. Nutr. *26:* 205–217 (1943).
159 WAISMAN, H. A.; MCCALL, K. B., and ELVEHJEM, C. A.: Acute and chronic biotin deficiencies in the monkey, *Macaca mulatta.* J. Nutr. *29:* 1–11 (1945).
160 WAISMAN, H. A. and MCCALL, K. B.: A study of thiamin deficiency in the monkey *(Macaca mulatta).* Arch. Biochem. *4:* 265–279 (1949).
161 WECHSLER, I. S. and GLOBUS, J. H.: Histologic observations on the nervous system of monkeys experimentally deprived of vitamin E. J. Neuropath. exp. Neurol. *1:* 430–437 (1942).
162 WILGRAM, G. F.; LUCAS, C. C., and BEST, C. H.: Kwashiorkor type of fatty liver in primates. J. exp. Med. *108:* 361–370 (1958).
163 WILGRAM, G. F.: Experimental Laennec type of cirrhosis in monkeys. Ann. intern. Med. *51:* 1134–1158 (1959).
164 WILLS, L. and BILIMORIA, H. S.: Studies in pernicious anaemia of pregnancy. V. Production of a macrocytic anaemia in monkeys by deficient feedings. Indian J. med. Res. *20:* 391–403 (1932).
165 WILSON, H. E. and PITNEY, W. R.: Serum concentrations of vitamin B_{12} in normal and nutritionally deficient monkeys. J. Lab. clin. Med. *45:* 590–598 (1951).
166 WISSLER, R. W.; FRAZIER, L. E.; HUGHES, R. H., and RASMUSSEN, R. A.: Atherogenesis in the *Cebus* monkey. I. A comparison of three food fats under controlled dietary conditions. Arch. Path. *24:* 312–322 (1962).
167 WISSLER, R. W.; GETZ, G. S.; VESSELINOVITCH, D.; FRAZIER, L. E., and HUGHES, R. H.: Acute severe experimental atherosclerosis in rhesus monkeys. Fed. Proc. *25:* 597 (1966).
168 WISSLER, R. W.; VESSELINOVITCH, D.; GETZ, G. S., and HUGHES, R. H.: Aortic lesions and blood lipids in rhesus monkeys fed three food fats. Fed. Proc. *26:* 371 (1967).
169 WITZGIRD, J. P.; GREENBERG, L. D., and MOON, H. D.: Hepatic lesions in pyridoxine deficient monkeys. Arch. Path. *79:* 317–323 (1965).
170 WOLF, R. H.; FELSENFELD, O.; BRANNON, R. B., and GREER, W. E.: Low protein intake and response to *Escherichia coli* 055 infection in patas monkeys (in press).
171 YOUNGER, R. K.; SCOTT, H. W., jr.; BUTTS, W. H., and STEPHENSON, S. E., jr.: Rapid production of experimental hypercholesterolemia and atherosclerosis in the rhesus monkey. Comparison of five dietary regimens. J. surg. Res. *9:* 263–271 (1969).

Author's address: Dr. ROBERT H. WOLF, Tulane University Delta Regional Primate Research Center, Animal Care Division, *Covington, LA 70433* (USA)

Pathology of Simian Primates, Part I, pp. 270–317 (Karger, Basel 1972)

Red Cell Antigens of Primates[1]

J. MOOR-JANKOWSKI and A. S. WIENER

Department of Forensic Medicine, New York University School of Medicine; Laboratory for Experimental Medicine and Surgery in Primates (LEMSIP), New York University Medical Center, and Office of the Chief Medical Examiner of New York City, New York, N.Y.

Contents

Introduction ... 271
Human-Type Blood Groups ... 273
- A-B-O Blood Groups ... 273
  - Man ... 273
  - Apes ... 275
  - Old World Monkeys ... 278
  - New World Monkeys ... 279
- Lewis Types ... 282
- M-N Types ... 284
  - Man vs. Non-Human Primates ... 284
  - Chimpanzees ... 286
  - Gorillas ... 286
  - Gibbons ... 286
  - Orang-Utans ... 287
  - Old World and New World Monkeys ... 287
- Rh-System ... 288
  - Chimpanzees ... 289
  - Gorillas ... 291
  - Gibbons, Orang-Utans and Monkeys ... 292
- The I-i Blood Factors ... 292
- Xg^a Blood Factor ... 293
- Other Human-Type Blood Factors ... 293

Simian-Type Blood Groups ... 296
- Chimpanzees ... 296
- Gibbons ... 297
- Celebes Black Apes ... 299
- Baboons ... 299

1 Supported by USPHS, NIH grant GM 12074.

Rhesus Monkeys ... 301
Sero-Primatology ... 302
Chimpanzees ... 305
Gibbons ... 310
References ... 311

Introduction

Paralleling the manifest morphological differences that distinguish one species of primate from another, there are biochemical differences in the blood. Similarly, just as there are fine morphological features that distinguish individuals within a species, so are there fine biochemical differences among the individuals of a species. In this chapter will be discussed those biochemical differences in red cells, recognized by serological tests, that are designated blood groups [63, 77].

Because, morphologically, it is difficult or impossible to distinguish the red cells of one primate species from those of another, or even from the red cells of infraprimate species, it was believed for some time that animal blood could be used safely for blood transfusions to man. The untoward results of such experiments were explained at the end of the nineteenth century, when LANDOIS demonstrated that the serum of animals of one species could agglutinate and hemolyze the red cells of animals of another species due to the usual presence in sera of many species of naturally occurring antibodies for red cells of other species – the so-called 'heteroagglutinins'. These 'normal' heteroagglutinins vary considerably in strength, e.g. whereas normal human sera generally agglutinate red cells of all chimpanzees, the heteroagglutinating activity of normal chimpanzee sera against human red cells is weak or absent[2].

Following LANDSTEINER's rule of 'immunological perspective', heteroagglutinins produced by immunization in distantly-related species demonstrate lesser specificity than heteroagglutinins produced in closely-related species. For example, rabbits which are phylogenetically distant from primates, when immunized with human erythrocytes, produce antisera which react to almost the same titer with red cells of apes as with those of man, and to only a moderately lower titer with red cells of Old World monkeys (table I). Chimpanzees immunized with red cells of man, their close relative, pro-

2 However, as shown by BORDET, the heteroagglutinins can be greatly increased in avidity and titer by immunization so that, for example, injections of human blood can cause chimpanzees to develop potent immune heteroagglutinins for red cells of all human beings.

Table I. Cross-reactions of human, chimpanzee, macaque, and baboon erythrocytes with antisera prepared by immunization of rabbits[1]

Red cells of	Titers with rabbit anti-human erythrocyte serum (1/20 dilution)			Titers with rabbit anti-macaque erythrocyte serum (1/20 dilution)		
	Unabsorbed	Absorbed with macaque erythrocytes	Absorbed with human erythrocytes	Unabsorbed	Absorbed with macaque erythrocytes	Absorbed with human erythrocytes
Human	400	400	< 20	80	< 20	< 20
Chimpanzee	200	200	< 20	80	< 20	< 20
Baboon	80	< 20	< 20	600	< 20	800
Macaque	80	< 20	< 20	400	< 20	800

1 Abridged and modified from LANDSTEINER and MILLER [26].

duce antisera which react much stronger with human red cells than with those of gorilla, orang-utan, and gibbon, and only very weakly with red cells of monkeys. On the other hand, the heteroagglutinating rabbit antisera can be fractionated by absorption with red cells of monkeys and apes, yielding reagents reactive only with human cells; other antibody fractions are cross-reactive with red cells of chimpanzee as well as human red cells, but not reactive with monkey red cells [26].

The degree of cross-reactivity of an antiserum with red cells of species other than that whose red cells were used for immunizing the experimental animal generally parallels the degree of taxonomic relationship among the species. Similar cross-reactivity among sera paralleling taxonomic relationship was described as early as 1904 by NUTTALL [58]. He constructed quite a complete mammalian taxonomic system based on precipitating reactions of sera of rabbits immunized with sera of other species. This work was later confirmed and extended, especially for primates, by MOLLISON [36–39], and more recently by GOODMAN [15, 16].

A different kind of cross-reactivity that cuts across taxonomic lines was first described by FORSSMAN [12]. The classic Forssman, or heterophile, antibody characteristically is produced in rabbits by injections of guinea-pig kidney. It reacts with red cells of sheep and of some other, but not all, animal species. Using this reagent mammalian species have been classified as Forssman positive or Forssman negative.

At the turn of the present century KARL LANDSTEINER [22], seeking evidence of individual differences in human blood, tested blood specimens from members of his laboratory staff, including himself, by mixing the serum from each individual with the red cells of others. Instead of the negative or weak reactions that might have been expected, he found that in some combinations the red cells were strongly agglutinated, while in others the red cells remained unaffected. This phenomenon of isoagglutination he pointed out, explained why, after the use of animal blood had been abandoned for blood transfusions in man, severe hemolytic reactions continued to occur despite the use of human donors. LANDSTEINER also showed that the human isoagglutinins occurred naturally (i.e. without prior immunization) and regularly, and were of 2 sharply different specificities, so that in combination they distinguished 4 distinct blood groups: O, A, B and AB. In this way the field of blood grouping was born, and for this discovery of such fundamental importance in clinical medicine, KARL LANDSTEINER was awarded the Nobel Prize in 1930 [23].

The first part of the present paper deals with human-type blood groups of non-human primates, i.e. blood groups demonstrated with reagents originally produced for typing human red cells and then further purified for tests in apes and monkeys.

Human-Type Blood Groups

A-B-O Blood Groups

Man

Because the chemical nature of the 2 antigens on the red cell surface was unknown when LANDSTEINER discovered them, he arbitrarily named them A and B. There are 4 possible combinations of these 2 antigens (also called agglutinogens), and all 4 blood groups, indeed, exist. They have been named O, A, B and AB. According to LANDSTEINER's rule, those agglutinins – and only those agglutinins – are present (except in newborn babies) in the serum for which the corresponding agglutinogens are absent from the red cells. Modern blood typing techniques do not rely on the relatively low-titered natural anti-**A** and anti-**B** antisera[3]. Potent reagents are produced by immunizing human beings with material containing the group-specific substances A and B;

3 To avoid ambiguity, **boldface** type is used for symbols for specificities (blood factors) and their corresponding antibodies, regular type for symbols for agglutinogens and phenotypes, and *italics* for symbols for genes and genotypes.

Table II. The 4 blood groups of the human A-B-O systems and their heredity[1]

Phenotype	Red cell antigens (agglutinogens)	Serum isoagglutinins	Possible genotypes[2]
O	None	Anti-**A** and anti-**B**	*OO*
A	A	Anti-**B**	*AA* and *AO*
B	B	Anti-**A**	*BB* and *BO*
AB	A and B	None	*AB*

1 [116].
2 Gene *O* is an amorph, i.e. it determines nothing.

obviously, group B individuals produce anti-**A** reagents, whereas those whose blood is group A produce anti-**B**.

BERNSTEIN proved that the 4 blood groups are inherited by multiple alleles [5], *O*, *A* and *B*, so that there are 6 possible genotypes corresponding to the 4 blood groups as shown in table II. The heredity of the blood groups is so clear-cut that it is regularly applied in medicolegal cases of disputed paternity [77, 116].

The isoagglutinins are not homogeneous but each is made up of a spectrum of different antibody molecules which can be fractionated by suitable absorption. Thus, absorption of group B (anti-**A**) serum with some particular group A red cells (subgroup A_2) will leave behind a fraction of the antibodies capable of clumping most other group A red cells (subgroup A_1) though no longer reactive with the absorbing cells. The fraction remaining has been designated anti-$\mathbf{A_1}$, and group A and AB cells clumped by the reagent are designated subgroups A_1 and A_1B, respectively; those that fail to react are of subgroups A_2 and A_2B. The subgroup differences are genetically determined, thus increasing the number of alleles from 3 to 4, *O*, A^1, A^2, and *B*, and increasing the number of genotypes from 6 to 10, determining the 6 blood groups and subgroups [116].

Reagents giving reactions that parallel anti-**A** and anti-**B** have been extracted from seeds (so-called lectins) [108] and snails [83]. For example, lima bean *(Phaseolus vulgaris)* extracts have the specificity anti-**A**, whereas anti-$\mathbf{A_1}$ lectin derived from *Dolichos biflorus* is in regular use for subgrouping group A blood. A particularly useful reagent is the lectin from *Ulex europeus* which has the specificity designated anti-**H** [106]. It detects the agglutinogen H, inherited independently of A and B, for which almost all individuals are homozygous *HH*. Anti-**H** strongly and regularly clumps red cells of group O and subgroup A_2, but gives weaker or negative reactions with red cells of the other groups. It has been postulated that the negative association of H with A_1 and B is caused by competition between the respective genes for a common substrate from which each allele produces its corresponding gene product [106].

Most, but not all men are so-called secretors, i.e. their body fluids and secretions contain blood group substances in solution, which correspond to their A, B and H red cell antigens. Saliva is the most readily available secretion and is, therefore, usually used for testing. The tests are carried out by mixing the agglutinins with the saliva and determining the ability of the saliva to inhibit the agglutination of group A cells by anti-**A** reagents, of group B cells by anti-**B** and of group O cells by anti-**H.** The ability to secrete the group substances is determined by a pair of allelic genes *Se* and *se* [69]. Thus, a group A secretor, i.e. genotype *AA* or *AO*, genotype *SeSe* or *Sese*, will have saliva capable of inhibiting anti-**A** and anti-**H**, but not anti-**B**. Group O secretor saliva inhibits only anti-**H**. The group-specific substances in secretions are mucopolysaccharides, whereas the red cell antigens are lipopolysaccharides [21]. In both cases, however, the group-specific reactions have been shown to be due to the following identical terminal non-reducing simple sugars on the antigen molecules: N-acetyl-D-galactosamine for antigen A, D-galactose for antigen B, and L-fucose for antigen H [73].

Apes

Attempts were made early to demonstrate in animals blood groups comparable to those of man. However, with the exception of man and the anthropoid apes, isoagglutination generally does not occur when red cells of one animal are tested with the serum of another animal of the same species. Of the great apes, chimpanzees have been most extensively studied; only blood groups O and A occur in that species[4]. Of a total of 386 chimpanzees (table III) there were 56 group O and 330 group A, corresponding to the gene fre-

4 Human blood grouping sera must be treated to remove the heteroagglutinins they contain for red cells of non-human primates before they can be used for typing ape blood. For this purpose an ingenious but tedious method using reagents obtained by elution was devised by LANDSTEINER and MILLER [27]. In our own studies, we have used the simpler method of absorbing heteroagglutinins from our blood grouping reagents with chimpanzee group O red cells [91]. Failure to appreciate the need of removing heteroagglutinins from human sera used in testing primate animal red cells had led to the publication of some erroneous reports [4, 7].

Table III. Human-type A-B-O blood groups in anthropoid apes[1]

Species		Number of animals in groups					Estimated gene frequencies		
		O	A	B	AB	Totals	*O*	*A*	*B*
Chimpanzees									
Pan troglodytes	Wiener, Moor-Jankowski and Gordon [46, 78, 87]	33	195	0	0	228	38.0	62.0	0
	Other investigators[1]	23	135	0	0	158	38.1	61.9	0
	Totals Number	56	330	0	0	386	38.1	61.9	0
	%	14.5	85.5	0	0	100			
Pan paniscus	Schmitt *et al.* [70]	1	5	0	0	6			
	André *et al.* [3]	2	15	0	0	17			
Gibbons									
Hylobates lar[2]	Wiener, Moor-Jankowski and Gordon [46, 97]	0	10	24	22	56[2]	0	37.5	62.5
	Other investigators	0	2	10	2	14			
Hylobates hoolock	Wiener, Moor-Jankowski and Gordon [97]	0	1	0	0	1			
Siamangs									
Symphalangus brachytanites	Wiener, Moor-Jankowski and Gordon [97]	0	0	2	0	2			
Orang-utans									
Pongo pygmaeus	Wiener, Moor-Jankowski and Gordon [46]	0	22	1	3	26			
	Other investigators	0	12	11	11	34			
	Totals	0	34	12	14	60			
Gorillas									
Gorilla gorilla gorilla	Candela, Wiener and Goss [8]	0	0	9	0	9			
	Wiener and Moor-Jankowski [94]	0	0	14	0	14			
	Totals	0	0	23	0	23			

Table III (continued)

Species		Number of animals in groups					Estimated gene frequencies		
		O	A	B	AB	Totals	*O*	*A*	*B*
Gorilla gorilla berengei	WIENER, CANDELA and GOSS [84]	0	2	0	0	2			

1 After SCHMITT [70]. The series of 158 *Pan troglodytes* tested by ANDRÉ *et al.* [3] is not included, because many of the specimens were hemolyzed and were grouped by the absorption method, so that the findings may not be accurate.
2 Includes 52 *Hylobates lar lar* and 4 *Hylobates lar pileatus;* the latter 4 were all group B.

quencies O = 0.381 and A = 0.619. However, these results are somewhat misleading in that, according to their physical characteristics, a number of subspecies, or races of chimpanzees can be distinguished, and the results given here are for a mixture of these different varieties of chimpanzees. This subject will be discussed in detail in the chapter on sero-primatology.

Of the rare dwarf chimpanzees, *Pan paniscus*, only 23 have been blood grouped [70], and the results obtained are similar to those for *Pan troglodytes.*

Group A chimpanzees can be subgrouped (about one-tenth are subgroup A_1^{Ch}), but the distinction between A_1^{Ch} and A_2^{Ch} is not as sharp as between A_1 and A_2 in man [78]. Group O chimpanzees' red cells are clumped less strongly by anti-**H** lectin than are human group O red cells, and erythrocytes of some group O chimpanzees fail to react at all, thus behaving like human group O red cells of newborn babies [106]. Group A chimpanzee red cells fail to clump in anti-**H** lectin, or react weakly. Because of the difficulties in obtaining saliva for testing, it was not possible to determine the secretor type for all the blood-grouped chimpanzees, but more than 100 chimpanzees from whom saliva was obtained and tested by us proved to be secretors.

In gibbons and orang-utans, as in man and chimpanzees, isoagglutination is a normal and regular phenomenon, but in the gibbon and orang-utan only groups A, B and AB have been found (table III). It is noteworthy that group B red cells of gibbons are strongly and regularly agglutinated by anti-**H** lectin, and to the same titer as human group O red cells [106]. Of all the gorillas, chimpanzees, gibbons and orang-utans we tested, only one orang-utan was not a secretor [98].

The red cells of gorillas do not give clear-cut reactions in agglutination tests [98]. In this respect the reactions of red cells of gorillas resemble those of monkeys, as described below, rather than those of the other apes and man. However, it was possible to determine their A-B-O group by inhibition tests on saliva, urine, or on extracts of salivary glands from dead animals [8, 84, 94]. For easily understandable reasons, the studies on this species of anthropoid apes have been the most limited; of 23 lowland gorillas, all proved to be group B, while both of the 2 highland gorillas tested were group A.

Old World Monkeys (Catarrhina)

Unlike anthropoid apes, except the gorilla, tests for isoagglutination performed by us in numerous species of monkeys generally gave no, or at the most only very weak and irregular, reactions. Moreover, the red cells of Old World monkeys are not agglutinated by anti-**A**, anti-**B** or anti-**H** reagents suitable for grouping human or ape blood. Nevertheless, as first shown by WIENER *et al.* [8, 84], the existence of 4 A-B-O blood groups corresponding to the human blood groups can be demonstrated by testing monkey saliva for its ability to inhibit anti-**A** and anti-**B** reagents, and by testing monkey sera after absorption with human O red cells, for their ability to agglutinate human group A and group B red cells. Thus, LANDSTEINER's rule holds in monkeys also, but the reciprocal relationship is between agglutinins in the serum and group-specific substances in the saliva, rather than on the red cells. Of the hundreds of monkeys we have tested, all have proved to be secretors, and, irrespective of their A-B-O group, their saliva regularly inhibits anti-**H** lectin.

In *Catarrhina*, our most extensive studies have been carried out on baboons [47, 93]. The initial results indicated the presence of groups A, B and AB only. However, it is now well established that group O also occurs in baboons, but in low frequencies (table IV). By methods of population genetics, it could be shown that the distribution of the 4 A-B-O groups indicates inheritance by triple allelic genes, *O*, *A* and *B*, as in man, despite the failure of the red cells to agglutinate in anti-**A**, anti-**B** and anti-**H** reagents. It is noteworthy that the distribution of the A-B-O blood groups is not the same in all species of baboons, viz. *Papio ursinus* (South Africa) had the lowest frequency of group A and the highest of group B among the various species of baboons we have tested.

Of the other species of Old World monkeys, the largest numbers (more than 100) of animals tested have been rhesus monkeys, *Macaca mulatta*, all

Table IV. Human type A-B-O blood groups of baboons[1]

Species of baboons		Blood group distribution in the animals tested					Estimated gene frequencies		
		O	A	B	AB	Totals	*O*	*A*	*B*
Papio cynocephalus	Number	0	18	20	22	60			
	%	0	30	33.3	36.7	100	8.9	44	47.1
Papio anubis	Number	0	53	56	65	174			
	%	0	30.5	32.2	37.4	100	8.4	45	46.6
Papio ursinus,	Number	0	4	59	26	89			
South Africa	%	0	4.5	66.3	29.2	100	1.8	18.8	79.3
Papio papio, Senegal	Number	2	27	93	66	188			
	%	1.1	14.4	49.5	35.1		10.6	29.9	59.5
Papio (species	Number	1	42	65	65	173			
undetermined)	%	0.6	24.3	37.6	37.6		10	39	51

1 Modified after WIENER and MOOR-JANKOWSKI [93].

of which gave reactions corresponding to the human-type group B. Four *Macaca speciosa* also gave the reactions of group B. In contrast, among 48 *Macaca fascicularis*, a distribution of the A-B-O groups similar to that of baboons was found[5]. Tests on the *Macaca nemestrina* yielded unclear reactions [105]. In 26 patas monkeys *(Erythrocebus patas)*, 4 drills *(Mandrillus leucophaeus)* and 4 African green monkeys *(Cercopithecus sabaeus)* only group A was found: whereas among 8 vervet monkeys *(Cercopithecus pygerythrus)* 7 were group A, and 1 group B[6]. Remarkably, the saliva of all the 28 gelada monkeys *(Theropithecus gelada)* gave the reactions of group O, but the expected anti-**A** and anti-**B** agglutinins were often missing from the sera.

New World Monkeys (Platyrrhina)

Fewer New World monkeys than Old World monkeys have been tested by us. The red cells of all Platyrrhina tested failed to react with anti-**A** and

5 However, saliva from *M. fascicularis* gave only about one-eighth of the inhibition titers of baboon saliva with anti-**B** and anti-**H** reagents, though equal titer with anti-**A**.

6 The 2 groups of *Cercopithecus* were of different origin and were tested separately. The taxonomical classification was made by W. C. OSMAN HILL.

anti-**H** reagents but were weakly clumped by anti-**B** reagents, as first demonstrated in 1925 by LANDSTEINER and MILLER [27], thus indicating the presence of a B-like antigen on *Platyrrhina* red cells. This B-like antigen appears to be independent of the animal's A-B-O blood group as determined by tests on the saliva and serum. A similar B-like antigen can also be demonstrated on red cells of some non-primate species, as it is distributed among them in an apparently haphazard way, comparable to that of the Forssman antigen. The ubiquitous character of B-like antigens is confirmed by the presence of a B-like antigen in certain strains of *E. coli* and other bacteria [72]. A similar finding was that of indications of an A-like antigen on red cells from baboons (but not from geladas) which was unrelated to their blood group determined by tests on saliva [108].

In this connection it should be noted that injections of group A human red cells into some rabbits will elicit antibodies which are also capable of lysing sheep red cells, whereas injection of sheep cells can elicit potent anti-**A** [68]. This has been attributed to a Forssman-like antigen in human A cells, the so-called $\mathbf{F_A}$ blood factor. It is evident that contact with A-like and B-like antigens is inevitable since such antigens are ubiquitous in nature, and we presume that this accounts for the regular presence of anti-**A** and anti-**B** agglutinins in the serum of those primates and other animals in which the corresponding group-specific substances A and B are absent [76]. It is necessary to emphasize that the A-like and B-like antigens on monkey red cells are quite unrelated to the A-B-O blood groups of these monkeys, which are determined by reactions of their saliva and serum.

As for the A-B-O blood groups of New World monkeys, tests on saliva and sera confirm their presence, though the reciprocal relationship between group-substance in saliva and antibody in serum is not as regular as in man [102][7].

The largest series of *Platyrrhina* tested by us consisted of 31 marmosets of various species; all the animals were group A but the reactions of the saliva were atypical in that they failed to inhibit or only weakly inhibited anti-**H** lectin, and often lacked the expected anti-**B** agglutinin [107]. These data were in agreement with previously published results of an independent study by GENGOZIAN [14]. Among 15 spider monkeys of various species, groups O, A and B were found; 3 squirrel monkeys *(Saimiri sciureus)* were group A and 1 was group O [102].

7 Exceptions to Landsteiner's rule of reciprocity are always infrequent in primates; however, they do occur very rarely in man, rarely in chimpanzee, more often in Old World monkeys and still more often in New World monkeys.

Table V. Human-type A-B-O blood groups of Catarrhina and Platyrrhina[1]

Species	Blood group distribution in the animals tested				Totals
	O	A	B	AB	
Old World monkeys (Catarrhina)					
Baboons (see table IV)					
Drills					
Mandrillus leucophaeus	0	4	0	0	4
Geladas					
Theropithecus gelada[2]	18	0	0	0	18
Celebes black apes					
Cynopithecus niger[3]	1(?)	23	2	0	26
Patas monkeys					
Erythrocebus patas	0	26	0	0	26
Vervet monkeys					
Cercopithecus pygerythrus	0	7	1	0	8
Macaques					
Macaca mulatta[4]	0	0	46	0	46
Macaca fascicularis	1	10	5	4	20
Macaca nemestrina[5]	3	0	7(?)	0	10
Macaca speciosa	0	0	4	0	4
New World monkeys (Platyrrhina)					
Spider monkeys					
Ateles marginatus	0	3	0	0	3
Ateles paniscus	0	1	0	0	1
Ateles ater	0	6	0	0	6
Ateles cucullatus	1	0	4	0	5
Capuchin monkeys					
Cebus albifrons	1	0	3	0	4
Squirrel monkeys					
Saimiri sciurea[6]	1	3	0	0	4
Marmosets					
Various species[7]	0	31	0	0	31

1 Taxonomic determination of the animals tested kindly provided by W. C. OSMAN HILL [59].

2 These animals are unique in that their saliva inhibits neither anti-**A** nor anti-**B** though anti-**H** is inhibited; moreover, the serum lacks one or both of the expected agglutinins, anti-**A** and anti-**B**.

3 The expected anti-B is absent from the serum of the majority of these monkeys.

4 The salivary B antigen was poorly developed in about half of these animals.

5 Salivary B substance is only weakly reactive.

6 Red cells have a B-like antigen, and there is no anti-**A** or anti-**B** in the serum.

7 The salivary H substance is poorly developed. The expected anti-**B** is generally absent from serum.

Lewis Types

The Lewis types of human blood are inherited by a pair of allelic genes, *Le* and *le*, independent of the genes *A-B-O*, *H-h*, and the secretor genes *Se-se*. Despite this, the Lewis types are closely associated with the A-B-O blood groups and A-B-H secretor types [17]. The reason for this has been shown to be that the enzyme produced by the Le-le genes acts on the same substrate on which the A-B-O and H-h transferases act. The specificity Le appears to depend on a terminal non-reducing L-fucose monosaccharide. The fucose, however, is attached to a different point on the blood group mucopolysaccharide from that of the L-fucose determined by the *H*-gene [73].

The Lewis blood types are unique in that the transferases determined by the *Le* gene produce only soluble group substances in the plasma and secretions which are absorbed onto the red-cell surface secondarily. Thus, red cells of all newborn babies are non-reactive at birth, but become agglutinable by anti-**Le** sera later during the 1st year of life. As the Lewis genes and A-B-H secretor genes are independent [9], 4 phenotypes are demonstrable in tests on secretions such as saliva, designated as Sec Les, nS Les, Sec nL and nS nL, respectively, where Sec indicates the presence of H (and A and B) in the secretions and nS their absence, while Les indicates the presence of Lewis substance and nL its absence. The close association between the Les and Sec types,

Table VI. Lewis types in man[1]

Genotypes	Four saliva types	Three red cell types			
		Phenotypes	Reaction with		
			Anti-**Le**$_1$	Anti-**Le**^{H}	Anti-**Le**
sese LeLe *sese Lele*	nS Les	Le$_1$	+	–	+
SeSe LeLe *Sese LeLe* *SeSe Lele* *Sese Lele*	Sec Les	Le$_2$	–	+	+
SeSe lele *Sese lele*	Sec nL	le	–	–	–
sese lele	nS nL	le	–	–	–

1 After WIENER [79].

despite their genetic independence, is evident from the fact that when Le and H are both present together a hybrid specificity $\mathbf{Le^H}$ appears for which there is a specific antibody. Whereas saliva tests disclose 4 types, tests on red cells show only 3 types, the reactions of which with antisera of the 3 main Lewis specificities are shown in table VI.

The Lewis antibodies in freshly-separated human antisera can produce hemolysis of the appropriate red cells. For this reason they have been the cause of serious hemolytic transfusion reactions. Such reactions are most common among negroes who have the highest frequencies of genes *le* and *se*, and therefore also have the highest percentages of the blood types Le_1 and le [90].

Tests for the Lewis types of red cells of non-human primates have not been possible because of the lack of potent antisera free of non-specific hetero-

Table VII. Lewis types of saliva from apes and monkeys[1]

Species	Lewis types		Totals
	Les	nL	
Anthropoid apes			
Chimpanzees *(Pan troglodytes)*	8	22	30
Gibbons *(Hylobates lar)*	0	8	8
Orang-utans *(Pongo pygmaeus)*	26	0	26
Old World monkeys			
Baboons *(Papio cynocephalus)*	31	8	39
Drills *(Mandrillus leucophaeus)*	0	3	3
Celebes black apes *(Cynopithecus niger)*	26	0	26
Patas monkeys *(Erythrocebus patas)*	16	10	26
Vervet monkeys *(Cercopithecus pygerythrus)*	3	1	4
Rhesus monkeys *(Macaca mulatta)*	38	8	46
Crab-eating macaques *(Macaca fascicularis)*	8	0	8
Pig-tailed macaques *(Macaca nemestrina)*	10	0	10
New World monkeys			
Spider monkeys (Ateles, various species)	0	5	5
Cinnamon capuchins *(Cebus albifrons)*	0	4	4
Squirrel monkeys *(Saimiri sciureus)*	0	4	4
Marmosets (various species)	0	31	31

1 Modified after WIENER *et al.* [90] and MOOR-JANKOWSKI and WIENER [46].

agglutinins. However, as tests on the secretions are carried out by the inhibition technique, the presence of heteroagglutinins in the reagents does not interfere with those tests. As shown in table VII, all of 26 orang-utans and 8 of 30 chimpanzees proved to be Lewis-positive while all of 8 gibbons were Lewis-negative. Moreover, whereas most of the Old World monkeys tested were Lewis-positive, all of the 44 New World monkeys were Lewis-negative.

M-N Types

Man vs. Non-Human Primates

In 1926 LANDSTEINER and LEVINE [25] injected rabbits with group O human red cells and found that the antisera, after absorption with appropriate human red cells, detected hitherto unknown specificities of human red cells. Two of the specificities proved to be antithetical, and were designated **M** and **N**, from the word 'iMmuNe'. Tests with anti-**M** and anti-**N** reagents determine only 3 types of blood, M, N and MN: blood lacking both properties M and N was not found. This was explained when heredity was shown to be by a pair of allelic genes, so that only 3 genotypes are possible, namely, genotype *MM*, determining type M, genotype *NN* for type N and genotype *MN* for type MN [24]. The MN types determine a separate blood-group system, which is now known to be very complex and includes more than 20 distinct blood factors. Due to the unavailability of the other antisera, however, the routine blood-grouping work is still largely limited to the original 3 MN types of LANDSTEINER and LEVINE.

In studies on the MN specificities of red cells of non-human primates, it was soon found that reagents prepared from different rabbit antisera, though of identical specificity in tests on human red cells, differed in their reactions with red cells of apes and monkeys [28, 74]. Thus, the 6 selected anti-**M** sera shown in table VIII all reacted differently with red cells of non-human primates, although they could be used interchangeably for typing human blood[8]. These results show that there is not a one-to-one correspondence between antigen and antibody. Instead, each antigen can elicit a multiplicity of antibodies, each of which has a different specificity. Therefore, when tests are carried out on non-human primate red cells using a single reagent, a very limited view results which can be misleading (comparable to a photograph of an

8 It is of interest to note that similar differences in the selective reaction of standard anti-**M** blood typing sera were demonstrated by the mixed-agglutination technique in human tissue cells in culture [42].

Table VIII. The reactions of red cells of apes and monkeys with anti-M reagents[1]

Red cells	Anti-M reagents					
	No. 1	No. 2	No. 3	No. 4	No. 5	No. 6
Human M	+++	+++	++±	++±	++±	++±
Human N	–	–	–	–	–	–
Human MN	+++	+++	++±	++±	++±	++±
Chimpanzee *Pan troglodytes*	+++	+++	++±	++±	++±	–
Rhesus monkey *Macaca mulatta*	++±	++±	++	+	–	–
Green monkey *Cercopithecus pygerythrus*	+++	+++	–	–	–	–
Black spider monkey *Ateles ater*	+++	–	–	–	–	–
White spider monkey *Ateles*	+	–	–	–	–	–
Capuchin monkey *Cebus capucinus*	–	–	–	–	–	–

1 Abridged and modified from WIENER (Latin species identification by A. S. WIENER [94]) [74].

object from a single viewpoint only, which may omit important characteristics) [80]. Some of the earlier investigators carried out their tests on red cells of rhesus monkeys with a single anti-**M** reagent, and therefore conflicting results were obtained. A few workers concluded that red cells of rhesus monkeys were M-positive; others insisted that they were M-negative. The correct answer, of course, is that red cells of all rhesus monkeys do have M-like antigens, but these M-like antigens share only some – not all – of the serological specificities (blood factors) of human M-blood and, therefore, react only with some, but not all, anti-human M-sera. In fact table VIII demonstrates the presence of M-like agglutinogens in apes, Old World monkeys, and some species of New World monkeys. As can be seen, the number of specificities which red cells from non-human primates share with human M-cells varies with the species, and is correlated with the taxonomical position of the species in relation to man.

The results of tests on red cells of non-human primates with anti-**N** rabbit sera are analogous to those obtained with anti-**M** reagents [74]. However, the tests with anti-**N** reagents were less satisfactory because the available anti-**N** reagents in general are of lower titer and have a greater tendency to produce non-specific reactions. The discovery by Ottensooser and Silberschmidt [60] that extracts of seeds of *Vicia graminea* have anti-**N** specificity has proved most helpful in this connection, and this anti-$\mathbf{N}^V$ lectin has been used by us routinely in our studies on non-human primates.

Chimpanzees

Red cells of all chimpanzees have M-like agglutinogens; most but not all anti-**M** reagents agglutinate chimpanzee red cells. Similarly some anti-**N** rabbit sera agglutinate red cells of all chimpanzees, although many anti-**N** sera fail to react. With anti-$\mathbf{N}^V$ lectin *(Vicia graminea)* chimpanzees can be divided into two types. Almost half have red cells which react with this reagent and are designated $(MN_1)^{Ch}$. The remainder have red cells that do not react and are type $(MN_2)^{Ch}$ [48, 82]. More recently, we have found some rare anti-**N** rabbit sera which give reactions with chimpanzee red cells paralleling those of anti-$\mathbf{N}^V$ lectin.

Gorillas

Tests on the red cells of 12 lowland gorillas showed all to react strongly with anti-$\mathbf{N}^V$ lectin *(Vicia graminea)* so that all these gorillas evidently had an N-like agglutinogen on their red cells. In tests with suitably absorbed anti-**M** reagents on the other hand, most, but not all of the gorilla bloods gave positive reactions [94].

Gibbons

Some anti-**M** (and some anti-**N**) reagents clump red cells of all gibbons: whereas other similar reagents do not react with the red cells of *any* gibbons. Moreover, certain anti-**M** sera agglutinate red cells of some, but not other, gibbons [112], and tests with these reagents and with anti-$\mathbf{N}^V$ lectin define 3 blood types in gibbons, $(M)^{Gi}$, $(N)^{Gi}$ and $(MN)^{Gi}$ [97]. The distribution of these 3 types satisfied the Hardy-Weinberg law, indicating that heredity is by 2 contrasting alleles, as it is in man. Thus, despite the serological differences, the population genetic studies demonstrate that the M^{Gi}-N^{Gi} types of gibbons are the counterparts of the human MN types. Significantly, in gibbons as well as in chimpanzees, treatment of the red cells with ficin and other proteolytic enzymes destroys their reactivity with MN reagents, as is the case for human red cells [75].

Table IX. Human-type M and N blood factors in apes[1]

Species	M-like factors		Totals
	present	absent	
Chimpanzees			
Pan troglodytes	130	0	130
Gibbons			
Hylobates lar lar	21	31	52
Hylobates lar pileatus	4	0	4
Orang-utans			
Pongo pygmaeus	12	12	24
Gorilla			
Gorilla gorilla gorilla	9	3	12
	N^V factor *(Vicia graminea)*		
Chimpanzees			
Pan troglodytes	42	62	104
Gibbons			
Hylobates lar lar	43	9	52
Hylobates lar pileatus	0	4	4
Orang-utans			
Pongo pygmaeus	0	24	24
Gorillas			
Gorilla gorilla gorilla	12	0	12

1 [94].

Orang-Utans

Red cells of this species are difficult to test with rabbit antisera because rabbit serum normally contains potent heteroagglutinins for red cells of all orang-utans. After absorption with selected orang-utan red cells, however, anti-**M** reagents can be prepared which give type-specific reactions with orang-utan blood [100]. Because anti-$\mathbf{N}^V$ lectin does not agglutinate orang-utan erythrocytes, only 2 MN types have been demonstrated, designated $(M)^{Or}$ and $(m)^{Or}$ (table IX).

Old World and New World Monkeys

The studies on MN types in monkeys have not advanced far beyond the original observations made in 1937 [28]. Red cells of monkeys are not agglutinated by anti-$\mathbf{N}^V$ lectin, and their varying reactions with anti-**M** sera are shown in table VIII. The close relationship among the M-factors of man, apes and monkeys has been demonstrated by the production of anti-**M** reagents by immunization of rabbits with red cells of rhesus monkeys, baboons, and chimpanzees as well as with human type M blood [28, 99]. Despite the manifest serological differences among the reagents produced, they can be used interchangeably for typing human blood. As described below, it was the study of such anti-rhesus-monkey rabbit sera that led to the discovery of the Rh-factor.

Rh-System

The previously mentioned observation that anti-**A** reagents can be produced from the sera of rabbits immunized with sheep blood [68] and that anti-**M** reagents can be produced from sera of rabbits immunized with rhesus monkey blood [28] led LANDSTEINER and WIENER [29] to examine their anti-rhesus rabbit sera for antibodies of other specificities. Indeed, they found that suitably absorbed anti-rhesus sera agglutinated the red cells of 85% of Caucasians, independently of the other known blood groups. Persons whose red cells were agglutinated were said to be Rh-positive, those whose red cells were not agglutinated were said to be Rh-negative. The Rh-negative type, occurring with 15% frequency in Caucasians, is less common in Negroes and virtually absent in Mongoloids. The great importance of the Rh-factor in blood transfusion was soon recognized by WIENER and PETERS [113] and was followed by the detection of its role in problems of maternal-fetal incompatibility [32].

LANDSTEINER and WIENER found that better anti-**Rh** reagents could be produced by immunizing guinea pigs instead of rabbits [30]. More recently, we have found that similar reagents can be produced by injecting guinea pigs with baboon red cells [96]. Remarkably, guinea pigs do not produce Rh-antibodies when injected with human Rh-positive blood. As FISK and FOORD [11] showed, anti-rhesus guinea pig sera or anti-rhesus rabbit sera strongly agglutinate all human umbilical cord bloods, but after absorption with Rh-negative cord blood, the sera can be used also for Rh-typing of newborns. More recently LEVINE *et al.* [31] have found that absorption of the anti-rhesus guinea pig sera with blood from certain rare individuals yields a reagent that agglutinates the red cells of all but less than 1% of human beings. Since this reagent was produced in a way similar to LANDSTEINER and WIENER's original anti-**Rh** serum, LEVINE graciously assigned the symbol LW to his new blood factor. The relationship of the anti-**LW** serum of LEVINE to the anti-rhesus monkey sera of LANDSTEINER and WIENER was discussed in detail in a recent paper [96].

In summary, the following antibodies have been isolated from antisera produced by rabbits and guinea pigs in response to injections of blood from rhesus monkeys and baboons (cf. also table X): (1) antibodies reactive only for red cells of rhesus monkeys and baboons; (2) antibodies reactive not only for rhesus monkey blood, but also cross-reactive for red cells of all human beings; (3) antibodies reactive not only for rhesus monkey blood, but also cross-reactive for red cells from newborn humans (cord blood); (4) antibodies reactive not only for rhesus monkey blood, but also cross-reactive for human M blood; (5) antibodies reactive not only for rhesus monkey blood, but also cross-reactive for human Rh-positive blood (anti-**Rh** of LANDSTEINER and WIENER), and (6) antibodies reactive not only for rhesus monkey blood, but also cross-reactive for human blood having LEVINE's LW-factor.

The antisera produced by individual rabbits and guinea pigs are of course not all identical in that they may contain any combination of one or more of the above enumerated 6 antibodies.

From sera of isosensitized human patients, or from male Rh-negative volunteer donors immunized with Rh-positive blood, potent Rh-reagents can be prepared [86]. These anti-$\mathbf{Rh_0}$ sera do not clump red cells from baboons or rhesus monkeys, which proves that they have a different specificity from LANDSTEINER and WIENER's anti-**Rh**, even though the 2 kinds of reagent detect the same agglutinogen and can be used interchangeably for typing human blood. This is comparable to the situation for the MN types, in which reagents giving parallel reactions on human blood differ in their reactions on non-human primate blood, as has been described.

Studies on sera obtained from cases of hemolytic transfusion reaction and from mothers of erythroblastotic babies have yielded a series of antibodies (more than 30 by the latest count) defining specificities related to the original Rh-factor of LANDSTEINER and WIENER [117]. The human blood types defined by these reagents make up the so-called Rh-Hr system. Because of limited availability of suitable reagents, it has been necessary to limit our studies on non-human primates principally to the following few of the more important of these antisera: anti-$\mathbf{Rh_0}$, anti-**rh'**, anti-**rh''**, anti-**hr'**, and anti-**hr''** (table XI) [88].

Chimpanzees

Red cells of chimpanzees are not agglutinated by anti-rhesus rabbit or anti-rhesus guinea pig sera. However, all chimpanzee red cells are agglutinat-

Table X. Some absorption experiments with anti-rhesus guinea pig sera[1]

Guinea pig anti-rhesus serum (1:5)	Tested against red cells of									
	human cord blood		human adult blood		LW negative		Rh_{null}	Chimpanzee	Rhesus monkey	Guinea pig
	Rh+	Rh−	Rh+	Rh−	Rh+	Rh−				
Unabsorbed	+++	+++	+++	+++	+++	+++	++±	++	+++	−
Absorbed with group AB, Rh-negative cells	+++	++	++	−	−	−	n.d.	−	+++	−
Absorbed with group AB, Rh-positive cells	+++	+++	++	−	−	−	n.d.	−	+++	−
Absorbed with Rh_{null} red cells	+++	++±	++±	++	++±	+±	−	−	+++	−
Absorbed with chimpanzee red cells	+++	++±	++±	++	++	++	n.d.	−	+++	−

1 From [96].
n.d. = not done.

Table XI. Human-type Rh-Hr blood factors in non-human primates[1]

Species	Rh-Hr blood factors					Designation of phenotype
	Rh₀	**rh′**	**rh″**	**hr′**	**hr″**	
Chimpanzees						
Pan troglodytes	+	–	–	+	–	$\overline{\mathfrak{R}}h_0^{Ch}$
Gibbons						
Hylobates lar lar						
Hylobates lar pileatus	–	–	–	+	–	$\bar{r}h^{Gi}$
Orang-utans						
Pongo pygmaeus	w	–	–	w	–	$\overline{\mathfrak{R}}h_0^{Or}$
Gorilla	+	–	–	+	–	$\overline{R}h_0^{Go}$
Gorilla gorilla gorilla	w or –	–	–	+	–	$\overline{\mathfrak{R}}h_0^{Go}$
Rhesus monkeys						
Macaca mulatta	–	–	–	–	–	

1 [94].
w = weak.

ed by anti-**Rh_0** human sera, provided the proper technique is used. This proves that red cells of all chimpanzees have an agglutinogen closely related to, but not identical with, the Rh-agglutinogen of human blood. Interestingly, if human anti-**Rh_0** sera are absorbed with chimpanzee red cells this will generally remove the reactivity for chimpanzee blood, but a fraction of antibodies still specifically reactive for human Rh-positive blood will remain in the sera [101].

Chimpanzee red cells do not react with available anti-**rh′** and anti-**rh″** sera which makes them correspond to the human type Rh_0. They are agglutinated by anti-**hr′**, as must be expected in view of the fact that **rh′** is absent, but both factors of the contrasting pair, **rh″** and **hr″**, are absent. This makes it possible to define further the Rh-type of chimpanzee blood as indicated by the symbol $\overline{\mathfrak{R}}h_0^{Ch}$, the Germanic $\mathfrak{R}$ being used to emphasize that the **Rh_0** factor of chimpanzee and man are not identical, and the bar above the $\mathfrak{R}$ representing the absence of the contrasting factor pair **rh″-hr″**.

Gorillas

In the past, studies on gorillas have been unsatisfactory because only single animals could be tested precluding comparative studies. In a recent test on blood specimens from 11 gorillas *(Gorilla gorilla gorilla)* it was observed

[94] that the red cells of all the animals were agglutinated, both by saline agglutinating anti-$\mathbf{Rh_0}$ sera and by anti-$\mathbf{Rh_0}$ reagents clumping only ficin-treated red cells. It was noticed that 3 of 11 gorilla blood samples reacted in somewhat lower titers. Absorption of the anti-$\mathbf{Rh_0}$ sera with more weakly-reacting gorilla red cells produced a reagent still highly reactive with human Rh-positive red cells as well as with red cells of the more strongly reactive gorillas, but negative with red cells of the formerly weakly-reacting gorillas. On the other hand, absorptions of the antisera with human Rh-positive cells removed all activity for gorilla red cells as well as for human Rh-positive red cells.

The red cells of all gorillas reacted with anti-**hr**′ reagents to the same titer as human hr′-positive red cells, but no reactions occurred with anti-**rh**″ and anti-**hr**″ reagents. Thus the majority of gorillas could be assigned to the type Rh_0^{Go} corresponding to the human type Rh_0 except for the absence of **hr**″ factor. The remaining 3 gorillas belonged to type $\mathfrak{R}h_0^{Go}$ corresponding to the low-grade human Rh_0 variant type [117].

Gibbons, Orang-Utans and Monkeys

Tests on red cells of gibbons gave negative reactions with human anti-$\mathbf{Rh_0}$, anti-**rh**′ and anti-**rh**″ reagents, which would make them correspond to the human type rh. Tests with anti-**hr**′ give positive reactions, as was expected in the absence of **rh**′. However both factors of the contrasting pair **rh**″-**hr**″ are absent from gibbon blood as well as from chimpanzee blood. For these reasons the gibbon Rh type has been assigned the symbol $\overline{r}h^{Gi}$.

Red cells of orang-utans react as do those of chimpanzees, but in lower titer. Red cells of all the species of Old World and New World monkeys tested by us for human-type A-B-O and MN blood groups were also tested with our Rh-Hr antisera; however, no agglutination was observed.

The I-i Blood Factors

The I-i blood factors, first described in 1956 [89], are unique in that they are detected by cold-agglutinating high-titered sera generally obtained from patients with hemolytic anemia. The I agglutinogen is also unique in that its presence depends on the age of the individual; it is absent in the newborn in which the contrasting agglutinogen i is present in its place. During the first year of age the agglutinogen i gradually disappears and is replaced by the adult agglutinogen I [34]. Certain extremely rare individuals fail to form the agglutinogen I, and are said to be of type i. Our studies on the I-i blood factors in non-human primates have included a few newborn chimpanzees and baboons, but were otherwise limited to adult animals.

The results of tests for I-i factors on a series of non-human primates and non-primate animals [49, 109] are summarized in table XII. As shown, the distribution of the I-i blood factors in these species is comparable to that of the heterophile, or Forssman, antigens in that it cuts across taxonomic lines. For example, factor **I** is strongly developed in rabbit but absent in rat, factor **I** is present and **i** absent in capuchin monkeys, while **I** is absent and **i** present in squirrel monkeys, etc. There are some regularities in that all species of Old World monkeys tested to date had factor **i** but lacked **I**. However, among the anthropoid apes the chimpanzees, both adult and newborn, had only factor **i**. Gibbons lacked both factors **I** and **i**.

Xg$^{\mathbf{a}}$ Blood Factor

This human blood factor, which has been studied extensively in the laboratory of Drs. RACE and SANGER, is unique because of its sex-linked inheritance, i.e. the gene for Xg^a is carried on the X-chromosome [33]. Tests by RACE and SANGER on a large series of chimpanzees, orang-utans, a few gorillas, gibbons, numerous baboons, rhesus monkeys, other species of primates, and many species of non-primate animals, revealed the presence of Xg^a only in gibbons, where its distribution indicated sex linkage, as is the case in man [64].

Other Human-Type Blood Factors

The low titer and avidity of the available reagents for blood factors of human blood group systems other than described above, complicated by the presence of interfering heteroagglutinins, has discouraged us from attempting any large-scale investigations of those blood factors in non-human primates.

In the original study by LANDSTEINER and LEVINE on the agglutinogen-P [24], tests were carried out for this factor on 10 chimpanzees with negative results. Whenever potent enough anti-**P** sera became available to us, we used them to test chimpanzee red cells, always with negative results. Interestingly, on one occasion we detected in the serum of a non-immunized gibbon an agglutinin having the specificity anti-**P** [94].

Table XII. Human-type blood factors I and i in primates and lower an imals

Species	Titers[2] with anti-I	anti-i
Man		
Newborn	0	64
Adult, typical	32	0
Adult, type i	0	16
Apes		
Chimpanzees *Pan troglodytes*	0	16
Gibbons *Hylobates lar lar* and *Hylobates lar pileatus*	0	0
Orang-utans *Pongo pygmaeus*	0	4
Old World monkeys		
Baboons *Papio cynocephalus* and *Papio anubis*	0	64
Geladas *Theropithecus gelada*	0	64
Patas monkeys *Erythrocebus patas*	1	32
Vervet monkeys *Cercopithecus pygerythrus*	1	32
Sykes monkeys *Cercopithecus albogularis*	0	32
Mangabeys Various species	2	64
New World monkeys		
Capuchin monkeys Various species	128	1
Squirrel monkeys *Saimiri sciureus*	0	16
Spider monkeys Various species	2	32
Marmosets Various species	32	64
Prosimians		
Potto *Perodicticus potto*	0	0
Slow loris *Nycticebus coucang*	0	32
Lower mammals		
Sheep	0	0
Goat	0	0
Cat	0	0
Dog	0	0
Rabbit	512	0
Rat	0	8
Deer	0	0

1 From [44].

2 The titers reported here are the reciprocals of the highest dilution of the antiserum giving a positive reaction, e.g. a titer of 64 means that clumping occurs with the reagent in dilutions up to 1:64; while a titer of 1 means clumping only with the undiluted reagent.

With regard to other human-type blood factors, we are not prepared to draw any conclusions, even though some reports from other laboratories have appeared in the literature.

Table XIII. Simian-type blood factors of chimpanzees *(Pan troglodytes)* and their frequencies[1]

Blood factors	Source of reagent		Positive	Negative	Totals
$\mathbf{V_1}$	Cross-immune serum[2]	Number	56	116	172
		%	32.6	67.4	
$\mathbf{N^V}$ (V)	*Vicia graminea* lectin	Number	58	63	121
		%	47.9	52.1	
$\mathbf{A^c}$	Isoimmune serum	Number	72	86	158
		%	45.6	54.5	
$\mathbf{B^c}$	Isoimmune serum	Number	104	54	158
		%	65.8	34.2	
$\mathbf{C^c}$	Isoimmune serum	Number	86	72	158
		%	54.4	45.6	
$\mathbf{E^c}$	Isoimmune serum	Number	27	71	98
		%	27.6	72.4	
$\mathbf{F^c}$	Isoimmune serum	Number	52	46	98
		%	53.1	46.9	
$\mathbf{G^c}$	Isoimmune serum	Number	73	25	98
		%	74.5	25.5	
$\mathbf{H^c}$	Cross-immune serum[2]	Number	75	23	98
		%	76.5	23.5	
$\mathbf{I^c}$	Isoimmune serum	Number	32	9	41
		%	78.5	21.5	
$\mathbf{K^c}$	Cross-immune serum[2]	Number	132	5	137
		%	96.4	3.6	
$\mathbf{L^c}$	Isoimmune serum	Number	108	20	128
		%	84.4	15.6	
$\mathbf{c^c}$	Isoimmune serum	Number	90	15	105
		%	85.5	14.5	
$\mathbf{M^c}$	Cross-immune serum[3]	Number	5	20	25
		%	20.0	80.0	

1 Blood factor $\mathbf{I^c}$ has been omitted from the table because statistical data as to its distribution are not available.
2 Serum prepared in chimpanzees immunized with human red cells.
3 Serum prepared in chimpanzee immunized with gibbon red cells.

Simian-Type Blood Groups

Heretofore, only those blood groups of non-human primates have been described which could be demonstrated using reagents originally intended for typing human blood and then adapted for use with other species. The simian-type blood groups to be discussed hereafter are those demonstrated with reagents prepared by immunization with red cells of non-human primates. For this purpose, 3 methods of immunization have been used: (1) isoimmunization, i.e. injection of primates with red cells from animals of the same species; (2) heteroimmunization, i.e. injection of rabbits and other experimental animals with primate red cells, and (3) cross-immunization, defined by us as immunization between closely related species, e.g. immunization of chimpanzees with red cells of man or gibbon.

Chimpanzees

Currently, as many as 15 distinct simian-type blood factors have been identified by us in chimpanzees (table XIII). Our initial experiments, intended to demonstrate the existence of simian-type blood groups, proved successful in 1964 after a course of isoimmunization that lasted 18 months [57]. The 2 different reagents obtained were designated with the first letters of the alphabet as anti-$\mathbf{A}^c$ and anti-$\mathbf{B}^c$[9]. Statistical analysis of data on the distribution of the blood factors among chimpanzees showed that an association existed between $\mathbf{A}^c$, $\mathbf{B}^c$, and the factor detected by anti-$\mathbf{N}^V$ lectin extracted from *Vicia graminea*, already discussed. Thus, the 3 factors evidently belonged to a single chimpanzee blood group system. The factor detected by anti-$\mathbf{N}^V$ lectin was renamed V (table XIII), for *Vicia*, and the blood group system was called the V-A-B system [92]. Shortly thereafter, by cross-immunization of chimpanzees with human blood, a reagent was obtained which gave reactions with chimpanzee red cells, almost parallel with those of anti-$\mathbf{N}^V$ lectin. The new reagent, which originally had been called anti-$\mathbf{D}^c$ [103], was therefore designated anti-$\mathbf{V}_1$ [104]. The fact that anti-$\mathbf{N}^V$ detects the agglutinogen N of human blood indicated that the chimpanzee V-A-B system could be the counterpart of the human M-N-S system. This hypothesis is supported by the fact that ficin destroys the activity of the V-A-B antigens just as is the case for the human MN antigen.

9 The superscript 'c' stands for chimpanzee in order to avoid confusion with the human-type A-B-O blood group system.

A parallel series of immunizations on another group of chimpanzees yielded 3 antisera defining 3 other factors of chimpanzee blood designated $\mathbf{C}^c$, $\mathbf{E}^c$ and $\mathbf{F}^c$, respectively. The factors $\mathbf{C}^c$, $\mathbf{E}^c$ and $\mathbf{F}^c$ proved to be associated with one another, but independent of the V-A-B system. Therefore they define another blood group system which was designated as the C-E-F system [114]. The immunochemical behavior of the antigens indicates that the C-E-F system of chimpanzees is the counterpart of the human Rh-Hr system, because in both cases treatment of the red cells with ficin improves the reactions. A few years after our discovery of the C-E-F system, stimulated by the report of MASOUREDIS *et al.* [35], we obtained direct serological evidence to support this idea. In 1967, MASOUREDIS *et al.*, by testing blood from our chimpanzees with radioactively-labelled fractions of anti-$\mathbf{Rh}_0$ serum, were able to demonstrate individual differences in chimpanzee red cells. We were able to confirm MASOUREDIS' conclusions by testing chimpanzee red cells with antisera prepared by absorbing our human anti-$\mathbf{Rh}_0$ sera with blood of chimpanzees designated as negative by MASOUREDIS' tests. At about the same time, we had produced in a chimpanzee the isoantibody designated as anti-$\mathbf{c}^c$, because the factor $\mathbf{c}^c$ detected by it was reciprocally related to the factor $\mathbf{C}^c$, i.e. like the factors **M** and **N** in man [111]. We then found that the reactions of our isoimmune anti-$\mathbf{c}^c$ reagent with chimpanzee red cells paralleled those obtained with MASOUREDIS' radioactive labeled anti-$\mathbf{Rh}_0$ serum, as well as those obtained with our absorbed human anti-$\mathbf{Rh}_0$ sera. This provided further evidence of the close relationship between the chimpanzee C-E-F system and the human Rh-Hr system.

As shown in table XIV, additional simian-type factors of chimpanzee red cells have been identified, totalling 15 factors to date. Among them $\mathbf{H}^c$ and $\mathbf{L}^c$ appear to be part of the C-E-F system [118], judging from the population genetic studies as well as from immunochemical evidence. Others, e.g., $\mathbf{G}^c$ and $\mathbf{K}^c$ [85], appear to define new blood group systems. It is noteworthy that 3 of the reagents were produced by cross-immunizing chimpanzees with human blood and 1 by cross-immunization with gibbon blood [56]. In each case, the antisera produced react with red cells from all individuals of the donor animal's species, but show individual differences in the recipient species, i.e. chimpanzees.

Gibbons

Except for chimpanzees, our studies on simian-type blood factors of apes have been limited almost entirely to gibbons. To be sure, isoimmuniza-

Table XIV. Distribution of the A^p-B^p blood types in various species of baboons[1]

Source of animals	Species	Totals		Blood types			
				O^p	A^p	B^p	A^pB^p
LEMSIP[2]	*Papio cynocephalus* and others[6]	Number	136	38	30	44	24
		%	100	28	22.1	32.3	17.6
SFRE[3]	*Papio anubis*	Number	175	78	71	17	9
		%	100	44.5	40.6	9.7	5.2
Dr. G. P. MURPHY[4]	*Papio ursinus*	Number	97	97	0	0	0
		%	100	100	0	0	0
Dr. J. BERT[5], Dakar, Senegal, Africa	*Papio papio*	Number	184	20	56	12	96
		%	100	10.9	30.4	6.5	52.2

1 From WIENER *et al.* [115].
2 Laboratory for Experimental Medicine and Surgery in Primates, New York University Medical Center, New York, N.Y.
3 Southwest Foundation for Research and Education, San Antonio, Texas.
4 University of Stellenbosch, Bellville, Cape Province, South Africa.
5 Current address: 60, Avenue Cantini, 13 Marseille 80, France.
6 Some of the animals tested were of the species *Papio anubis*.

tion experiments were started in orang-utans at the Yerkes Regional Primate Research Center in 1963, but just at the point where one of the animals began to show antibody response, they became unavailable to us.

In gibbons, antibodies of 3 distinct specificities have so far been produced, and were designated anti-$\mathbf{A}^g$, and $\mathbf{B}^g$ and anti-$\mathbf{C}^g$, respectively. Two of the antisera, anti-$\mathbf{A}^g$ and anti-$\mathbf{B}^g$, appear to be on the same plane [50], as they react by all 3 serological methods used, i.e. the saline agglutination, antiglobulin and the ficinated-red-cell methods. Anti-$\mathbf{A}^g$ was produced by 2 gibbons of the subspecies *Hylobates lar pileatus* who had been injected with blood of the subspecies *Hylobates lar lar;* the factor A^g detected by the antisera appears to be characteristic for all animals of the subspecies *H. lar lar* (cf. chapter of sero-primatology). On the other hand, factor B^g appears to be characteristic of the subspecies *H. lar pileatus.*

The blood factor C^g belongs to a different category [110]. The anti-$\mathbf{C}^g$ sera which were produced in 2 gibbons, *H. lar lar*, define differences within their subspecies *H. lar lar*. Anti-$\mathbf{C}^g$ reacts in high titer by the antiglobulin method and hardly at all by the saline or ficin methods.

Celebes Black Apes *(Cynopithecus niger)*

By immunization of rabbits with red cells of Celebes black apes, an antiserum was obtained which, after absorption to remove species-specific heteroagglutinins, defined individual differences in the red cells of that species [53]. The blood factor detected by the reagent has been designated $\mathbf{A}^{ba}$, where ^{ba} stands for 'black ape', and the corresponding antibody is designated anti-$\mathbf{A}^{ba}$. Four of 11 Celebes black apes tested proved to have the blood factor $\mathbf{A}^{ba}$. The relative ease with which the anti-$\mathbf{A}^{ba}$ serum was produced in rabbits is comparable to the original experiments of LANDSTEINER and LEVINE in which they discovered the human MN types [24] and of OWEN and ANDERSON's experiments with rhesus monkey blood [61]. The main difficulty of the method is that the initial experiments must be done blind, in that not only the blood used for the immunizations but also that used for the absorptions has to be selected at random. Therefore, in our experiments on primate blood groups we have most often used the method of isoimmunization, as in the studies on baboons and rhesus monkeys described below.

Baboons

No naturally-occurring isoagglutinins have been found by us in screening tests on large numbers of baboons *(Papio cynocephalus* and *Papio anubis)* [47, 91]. Our attempts to produce typing sera by immunization of rabbits with red cells of baboons have been unsuccessful because we were unable to find at random the appropriate baboon red cells which would remove the non-type specific heteroagglutinins from the immune rabbit sera. We have, however, succeeded in producing typing reagents by isoimmunization.

Four baboons *(P. cynocephalus)* were isoimmunized by injecting intramuscularly whole blood mixed with an equal amount of complete Freund's adjuvant. Two animals responded by producing antisera giving parallel reactions for a blood factor designated $\mathbf{A}^{P}$, where the superscript ^{P} stands for 'Papio'. A third baboon produced an antibody of a different specificity $\mathbf{B}^{P}$. The factors $\mathbf{A}^{P}$ and $\mathbf{B}^{P}$ together determine 4 types of baboon blood [51] (table XIV). Very limited family studies have been done, but the distribution of the types in the baboon population show an association between $\mathbf{A}^{P}$ and $\mathbf{B}^{P}$, indicating that they probably belong to the same blood group system. This hypothesis is supported by immunochemical evidence, because the 2 antisera, anti-$\mathbf{A}^{P}$ and anti-$\mathbf{B}^{P}$, have similar serological properties in that they react

Table XV. Comparison of specificities of 7 rhesus isoimmune sera[1,2]

Test cells	Anti-A^{rh}	Anti-B^{rh}	Anti-C^{rh}	Anti-D^{rh}	Anti-H	Anti-E: Anti-globulin	Anti-E: Ficin method	Anti-K
Rhesus monkeys *(Macaca mulatta)*								
Standard LEMSIP test panel								
No. 1 Caesar	–	+	+	–	–	+	–	+
No. 2 Eve	+	–	+	+	–	+	–	+
No. 3 Eleanor	+	–	–	+	–	–	+	+
No. 4 Antigone	–	+	+	–	–	+	±	+
Average titers[3]								
Saline agglutination	0	6	1	4	0		0	0
Antiglobulin	70	8	90	150	640		48	24
Ficinated red cell method	0	3	80	70	48		20	0
Rhesus monkeys *(Macaca mulatta)*								
Dealer's stock animals								
Number positive	47	39	47	32	n.t.	n.t.	n.t.	n.t.
Number negative	0	0	0	15	n.t.	n.t.	n.t.	n.t.
Crab-eating macaques *(Macaca fascicularis)*								
Positive	8	2	12	10	n.t.	n.t.	n.t.	n.t.
Negative	4	1	0	2	n.t.	n.t.	n.t.	n.t.
Stump-tail macaques *(Macaca speciosa)*								
Positive	3	0	3	0	n.t.	n.t.	n.t.	n.t.
Negative	0	2	0	3	n.t.	n.t.	n.t.	n.t.
Baboons *(Papio cynocephalus)*								
Positive	20	3	20	0	n.t.	n.t.	n.t.	n.t.
Negative	0	0	0	20	n.t.	n.t.	n.t.	n.t.

1 Modified from WIENER *et al.* [95] and MOOR-JANKOWSKI and WIENER [52].

2 Results by antiglobulin method except for serum E, where method used is indicated.

3 The titers reported here are the reciprocals of the highest dilution of the antiserum giving a positive reaction, e.g. a titer of 64 means that clumping occurs with the reagent in dilutions up to 1:64; while a titer of 1 means clumping only with the undiluted reagent.

n.t. = Not tested for this study.

weakly by the saline agglutination method, do not agglutinate ficinated baboon red cells, and produce the clearest reactions by the antiglobulin method. The failure of the reagents to clump ficinated red cells suggests that the simian type A^p-B^p system of baboons may be the counterpart of the human MN system and of the chimpanzee V-A-B system. The A^p-B^p blood types show strikingly different distributions in the different species of baboons [115].

The 4th of the 4 original isoimmunized baboons has also produced isoantibodies, but the reactions are not sharply defined enough to yield a satisfactory diagnostic reagent. Moreover, our anti-$\mathbf{A}^p$ and anti-$\mathbf{B}^p$ sera contain antibodies other than those specific for blood factors $\mathbf{A}^p$ and $\mathbf{B}^p$ [94]. This indicates that further studies will shed more light on the complexities of the baboon blood types.

Rhesus Monkeys

An isoimmune serum was produced in rhesus monkeys as early as 1932 by FISCHER and KLINKHART [10]; it clumped the red cells of 20 of a total of 30 rhesus monkeys, but the experiments were not continued and the antiserum is no longer available. Subsequent attempts to isoimmunize rhesus monkeys were unsuccessful, but OWEN and ANDERSON [61] were able to demonstrate individual differences in rhesus monkey blood by immunizing rabbits with rhesus blood and then absorbing the antisera with selected rhesus red cells to remove non-specific heteroagglutinins.

Our own investigations on simian-type blood groups of rhesus monkeys were begun in 1966. Six rhesus monkeys were repeatedly injected intravenously with blood from one another, and when no antibodies were elicited, additional injections were given intramuscularly of blood mixed with complete Freund's adjuvant. Four of the 6 animals then responded, and as the 4 resulting antisera were different, they were designated as anti-$\mathbf{A}^{rh}$, anti-$\mathbf{B}^{rh}$, anti-$\mathbf{C}^{rh}$ and anti-$\mathbf{D}^{rh}$ (table XV) [52]. All 4 reagents gave best reactions by the antiglobulin technique. The weakest serum anti-$\mathbf{B}^{rh}$, reacted well by all 3 methods. Anti-$\mathbf{C}^{rh}$ and anti-$\mathbf{D}^{rh}$ were alike in that they both reacted well by the ficinated cell technique and the antiglobulin technique; anti-$\mathbf{A}^{rh}$, despite its high titer, failed to clump ficinated red cells. Presumably, therefore, factor $\mathbf{A}^{rh}$ belongs to a different blood group system from $\mathbf{C}^{rh}$ and $\mathbf{D}^{rh}$, while $\mathbf{C}^{rh}$ and $\mathbf{D}^{rh}$ may well belong to a single system. All 4 reagents showed polymorphism, of course, with the initial LEMSIP panel of rhesus monkeys, but anti-$\mathbf{A}^{rh}$, anti-$\mathbf{B}^{rh}$ and anti-$\mathbf{C}^{rh}$ agglutinated the red cells of all other rhesus monkeys obtained by us for testing purposes; only anti-$\mathbf{D}^{rh}$ showed good polymorphism (32 positive out of 47) and was therefore the most useful as a typing reagent. On the other hand, anti-$\mathbf{A}^{rh}$ and anti-$\mathbf{B}^{rh}$ showed polymorphism among the small number of crab-eating macaques *(Macaca fascicularis)* which were tested by us in the initial experiment (table XIV). The possibility that the 3 reagents anti-$\mathbf{A}^{rh}$, anti-$\mathbf{B}^{rh}$ and anti-$\mathbf{C}^{rh}$ might each contain a mixture of antibodies could not be excluded, although attempts to fractionate them by absorption proved unsuccessful.

In a subsequent study [95], 3 isoantisera, H, E and K, prepared by Dr. HANS BALNER by combined skin transplantation and blood injections were studied (table XV). These proved to be different from the original 4 studied by us, and all showed good polymorphism. Serum K, like anti-$\mathbf{A}^{rh}$, failed to agglutinate ficinated red cells, so that the factor detected by serum K and factor $\mathbf{A}^{rh}$ may well belong to one and the same blood group system, presumably the counterpart of the human M-N-S system. Serum E was of greatest interest in that it clearly contained at least 2 different antibodies, one demonstrable by the antiglobulin technique and the other reactive only for ficinated red cells.

Recently, one of the immunized animals from our original panel of 6 that had previously failed to respond, showed the presence of antibodies reactive for the red cells of the donor monkey. The antibody failed to agglutinate the red cells of the 25 rhesus monkeys that have been tested so far [94].

As of 1970, we have studied isoantibodies of at least 9 different specificities, determining blood factors belonging to at least 2 different blood group systems of rhesus monkeys [94].

Finally, attention is called to the recent report of HIROSE and BALNER [18] of an analysis of the reactions of a large number of antisera produced by them in rhesus monkeys. This study was carried out using the dextran agglutination technique. It confirms the high complexity of the blood types in rhesus monkeys, further study of which promises to be rewarding.

Sero-Primatology

A review of blood groups of non-human primates would be incomplete, were it not to include consideration of the variations in the distribution of serological characteristics among different species, as well as among population groups within the species.

In man, the differences in the distribution of blood group frequencies among various nationalities was first noted in 1919 by LUDWIK and HANNA HIRSZFELD [19]. Their discovery led to the establishment of the science of *sero-anthropology*. Subsequent investigations have provided a quite complete mapping of blood group distribution in practically all human populations [40, 41, 67, 81].

We now know that while no blood group is peculiar to any particular human ethnic group, the distributions of blood groups do vary according to population group. Among American whites, a typical distribution would be. O, 43.7%; A_1, 29%; A_2, 8.9%; B, 13.5%; A_1B, 3.5%; and A_2B, 1.4%. This distribution corresponds to the gene frequencies O = 0.650, A^1 = 0.183, A^2 = 0.065, B = 0.102 [81]. In Eurasia, the frequencies of group B increase from west to east, so that among Hindus group B is more common than group A. Moreover, among the mongoloids, American Indians and in some of the Walser isolates in Switzerland [43], subgroup A_2 does not occur; and some tribes of American Indians are exclusively, or almost exclusively, group O. Among negroes, approximately the same distribution of the A-B-O blood groups is found as among whites, though red cells of group A react somewhat differently with the subgrouping reagents anti-$\mathbf{A_1}$ and anti-**H**. The frequency of non-secretors in caucasoids and mongoloids is about 20 to 25%; the frequency is higher in negroids (about 35%). For further details on the situation in man, the reader is referred to the work of MOURANT [40, 41], RUFFIÉ [67], and WIENER [81].

In an analogy to sero-anthropology, the term *sero-primatology* has been coined by A. S. WIENER [54] to define the characterization of simian populations by differences in the distribution of their serological properties. For such purpose, sufficient data has become available only recently through our large scale surveys of various species of apes and monkeys [45]. Whereas, in general, methods of population genetics used in anthropological investigation of human population have been used also in our sero-primatological surveys, some important considerations which may influence the evaluation of the data obtained in tests of ape and monkey populations are as follows: (1) in contrast to the situation in most non-human primates, in man family, subfamily, genus and species are equivalent. In the simians, sero-primatological studies may be begun at the level of major taxonomical subdivisions. Thus, Catarrhina and Platyrrhina constitute the largest serologically distinct subdivisions, as shown by the previously described differences of their human-type A-B-O blood group characteristics. Among the Catarrhina, the Simoids form a separate serologically distinct group characterized by the overall similarity of their human type A-B-O, MN and Rh-Hr blood groups to those of man, although some species may differ in the characteristics of one of the blood group systems (e.g. gorilla, the A-B-O system of which is similar to that of catarrhine monkeys, but the Rh-Hr system of which is more like that of man than of all other primates). (2) On the level of species, enough serological information is available on simians that it may be used as one of the criteria of taxonomic classification. On the basis of the human type A-B-O blood groups alone, *M. mulatta* may be differentiated from *M. fascicularis; Theropithecus gelada* from all species of *Papio;* and *Hylobates lar lar* from *Hylobates lar pileatus*, to cite only a few striking examples. Moreover, antisera can be produced in one primate species which react selectively with red cells of another species in what is called by immunologists a 'species specific' reaction, as has been shown in gibbons. (3) Population genetic studies of human populations are based on the Hardy-Weinberg law, which assumes panmixia, i.e. random mating within each population. The results of such population genetic inferences in man have been confirmed by family studies which are, as yet, not possible in apes and monkeys because their breeding in captivity was started only recently. Therefore, in all our investigations of simians we have applied the human population genetic formulae, although we know that the breeding pattern varies from species to species and has not been sufficiently investigated in any of them. As a working hypothesis, we assume panmixia when studying large enough samples. Studies in man [43, 66] have demonstrated that many European populations consist of large numbers of small endogenous isolates, yet for the entire population the Hardy-Weinberg rule still holds. Our working hypothesis is that the same applies for simian populations in the wild. (4) Of all the varieties of sero-primatological differentiations, it is the determination of *population groups within a single species* which bears the closest resemblance to sero-anthropology of man.

In 1964 such a difference in the distribution of serological characteristics between 2 populations of the same species was observed for the first time by comparing the human-type A-B-O blood group frequencies of a colony of interbred[10] baboons with those of freshly-captured baboons at the Southwest Foundation for Research and Education, San Antonio, Texas [47].

More recently, we have been also using the simian-type blood groups for sero-anthropological investigations. These blood groups appear to be more

10 Recent observations seem to indicate that the interbred colony might have an admixture of baboons of a species different from that of the freshly captured *P. cynocephalus* and *P. anubis* used in the above-mentioned study.

specific for the species for which they were produced than are the human-type blood groups. Thus, antisera developed by immunization of one group of *M. mulatta* may show polymorphism in the original group and in some other random groups of their species, whereas they may react with all (cf. table XV 'dealer's stock') or with none of the animals of other randomly captured groups [94]. The same applies to the simian types of baboons [94].

To date, the differences in distribution of serological properties in population groups or races within primate species has been most thoroughly demonstrated for chimpanzees and gibbons by using both their human-type and simian-type blood groups and have been confirmed by studies in other areas of molecular biology.

Chimpanzees

In our investigations of the human-type and simian-type blood groups of chimpanzees, high χ^2 values were encountered in certain 2×2 contingency tests, suggesting associations between blood factors which, according to serological evidence, did not belong to the same blood group system [54]. We then postulated that these high χ^2 values might have been caused by non-homogeneity of the investigated chimpanzee population, even though all the animals tested by us belonged to the same species, *Pan troglodytes*. Upon further inquiry, Dr. W. C. OSMAN HILL called our attention to the existence of morphologically and geographically defined subgroups within the species *P.*

Table XVI. Distribution of the human-type A-B-O blood groups among 4 subspecies of chimpanzees[1]

Subspecies	O		A		Totals
	Number	%	Number	%	
P. troglodytes troglodytes	2	14.3	12	85.7	14
P. troglodytes schweinfurthi[2]	13	39.4	20	60.6	33
P. troglodytes verus[2]	4	9.5	38	90.5	42
P. troglodytes koola-kamba	1	–	1	–	2
Totals	20		71		91

1 From [45].
2 For the 2 species *P. troglodytes schweinfurthi* and *P. troglodytes verus*, $\chi^2 = 9.2$, $n = 1, 0.001 < P < 0.005$.

Table XVII. Distribution of the simian-type C-E-F system among 4 subspecies of chimpanzees[1]

Subspecies		c	E	F	EF	C	Totals
P. troglodytes troglodytes	Number	2	1	7	3	1	14
	%	14.3	7.1	50.0	21.4	7.1	
P. troglodytes schweinfurthi[2]	Number	2	5	16	10	0	33
	%	6.0	15.2	48.5	30.3	0	
P. troglodytes verus[2]	Number	27	3	8	4	0	42
	%	64.3	7.1	19.1	9.5	0	
P. troglodytes koola-kamba	Number	0	0	2	0	0	2
Totals		31	9	31	17	0	98

1 From [45].
2 For the 2 species *P. troglodytes schweinfurthi* and *P. troglodytes verus*, $\chi^2 = 26.6$, $n = 3$, $P<0.0001$.

Table XVIII. Distribution of the H^c factor associated with the simian-type C-E-F among 4 subspecies of chimpanzees[1]

Subspecies	h		H		Totals
	Number	%	Number	%	
P. troglodytes troglodytes	2	14.3	12	85.7	19
P. troglodytes schweinfurthi[2]	1	3.3	32	96.7	33
P. troglodytes verus[2]	17	40.5	25	59.5	42
P. troglodytes koola-kamba	0	–	2	–	2
Totals	20		71		91

1 From [45].
2 For the 2 species *P. troglodytes schweinfurthi* and *P. troglodytes verus*, $\chi^2 = 14.1$, $n = 1$, $P<0.0001$.

troglodytes, described by him [59]. When the chimpanzees tested were classified according to their morphological characteristics, as defined by OSMAN HILL, population groups homogeneous for their blood group distribution were obtained.

Table XIX. Distribution of the blood factor V of the simian-type V-A-B system among 4 subspecies of chimpanzees[1]

Subspecies	v		V		Totals
	Number	%	Number	%	
P. troglodytes troglodytes	9	64.3	5	35.7	14
P. troglodytes schweinfurthi[2]	24	72.8	9	27.2	33
P. troglodytes verus[2]	25	59.5	17	40.5	42
P. troglodytes koola-kamba	2	–	0	–	2
Totals	60		31		91

1 From [45].
2 For the 2 species *P. troglodytes schweinfurthi* and *P. troglodytes verus*, $\chi^2 = 1.38$, n = 1, 0.2<P<0.3.

Table XX. Distribution of the A^c-B^c types of the simian-type V-A-B system among 4 subspecies of chimpanzees[1]

Subspecies		O^c	A^c	B^c	A^cB^c	Totals
P. troglodytes troglodytes	Number	0	1	9	4	14
	%	0	7.1	64.3	28.6	
P. troglodytes schweinfurthi[2]	Number	3	4	2.1	5	33
	%	9.1	12.1	63.7	15.1	
P. troglodytes verus[2]	Number	5	14	14	9	42
	%	11.9	33.3	33.3	21.4	
P. troglodytes koola-kamba	Number	0	0	2	0	2
Totals		8	19	46	28	91

1 From [45].
2 For the 2 species *P. troglodytes schweinfurthi* and *P. troglodytes verus*, $\chi^2 = 7.77$, n = 3, P = 0.05 approximately.

The serological findings on chimpanzee population groups tested by us so far are summarized in tables XVI to XXI. Despite the relatively small size of the series, statistically significant differences in distribution both of human-type and of simian-type blood groups have been demonstrated for the

Table XXI. Distribution of the G^c factor of the simian-type G system among 4 subspecies of chimpanzees[1]

Subspecies	g		G		Totals
	Number	%	Number	%	
P. troglodytes troglodytes	5	35.7	9	64.3	14
P. troglodytes schweinfurthi[2]	6	18.2	27	81.8	33
P. troglodytes verus[2]	13	31	29	69	42
P. troglodytes koola-kamba	0	–	2	–	2
Totals	24		67		91

1 From [45].
2 For the 2 species *P. troglodytes schweinfurthi* and *P. troglodytes verus*, $\chi^2 = 1.63$, n = 1, P = 0.20, approximately.

populations of *P. troglodytes schweinfurthi* and *P. troglodytes verus*, which comprise the great majority of the animals tested, and at the same time show the most striking differences in blood group distribution. The frequency of group O was found to range from only 9.5% among *P. troglodytes verus* to as high as 39% in *P. troglodytes schweinfurthi* (table XVI) [45, 54]. Also, the differences in distribution of the simian-type C-E-F blood group system and of the blood factor $\mathbf{H}^c$ associated with this system are statistically significant (tables XVII and XVIII). The frequency of type c ranged from 6% among *P. troglodytes schweinfurthi* to 64.3% among *P. troglodytes verus*, and parallel variations were observed in the distribution of blood factor $\mathbf{H}^c$.

On the other hand, no significant differences were demonstrated in the distribution of the simian-type V-A-B blood group system (table XIX and XX) or of the simian-type blood factor $\mathbf{G}^c$ (table XXI).

Thus, the blood group data so far obtained confirm the subdivisions of the species *P. troglodytes* by Dr. OSMAN HILL, based on morphological characteristics, for at least 2 of the 4 races of chimpanzees. These findings were confirmed by studies on the distribution of serum groups. Working with Dr. PAUL ALEPA, we determined races of chimpanzees tested by him for genetic factors of human immunoglobulins of Gm and Inv types.

All the sera tested for Gm and Inv types formed a part of a larger collection of blood samples which served to demonstrate the differences in the distribution of blood groups [54] among the races of chimpanzees. The sera

Table XXII. Summary of results of Gm(6) and Inv(1) testing on sera of chimpanzee subspecies[1]

Chimpanzee (*P. troglodytes*) subspecies	Gm(6)			Inv(1)		
	Number positive	Number negative	Total number tested	Number positive	Number negative	Total number tested
P. troglodytes troglodytes	3	15	18	6	12	18
P. troglodytes schweinfurthi	4	21	25	1	23	24
P. troglodytes koola-kamba	0	3	3	0	3	3
P. troglodytes verus	19	18	37	6	30	36
Totals	26	57	83	13	68	81

1 [1, 2].

were coded and typed without prior knowledge of the classification of the donors. The antisera and the testing methods used have been previously reported [1, 2]. Of the 111 samples used in the initial study, 107 were tested for Gm(6) and 96 for Inv(1). However, several of the samples were obtained from animals that died before their morphological racial characteristics could be determined and whose geographical origins were unknown, or from hybrid animals bred in captivity. As a result, only 83 of the 107 animals tested for Gm(6), and 81 of the 96 tested for Inv(1) could be used in this study. The findings are summarized in table XXII. Despite the relatively small size of the series, statistically significant differences were demonstrated for the distribution of Gm(6). Among the chimpanzee races, *P. troglodytes schweinfurthi* and *P. troglodytes verus* show the most striking differences in the distribution of Gm(6) and, at the same time, comprise the great majority of animals tested ($\chi^2 = 6.63, p = 0.01$). These are the 2 races which have also shown significant differences in the distribution of their blood groups.

The differences in the distribution of Inv(1) were not statistically significant. At the present time, additional samples from morphologically classified chimpanzees are being collected in order to extend the statistical analysis.

Table XXIII. Comparison of the distribution of the A-B-O and MN types in 2 subspecies of gibbons[1]

Subspecies of gibbons[2,3]	Number of animals of blood group							Totals
	A_1	B	A_1B	A_2B	$(M)^{Gi}$	$(N)^{Gi}$	$(MN)^{Gi}$	
Hylobates lar lar	10	20	20	2	9	31	12	52
Hylobates lar pileatus	0	4	0	0	4	0	0	4

1 [94].
2 In *Hylobates lar lar*, about two-fifths are group B, so that the chance of observing 4 group B animals in succession, as in *Hylobates lar pileatus*, is only $2/5^4$ or only about 1 in 40.
3 In *Hylobates lar lar* only 9 among 52 animals are type M, so that the chance of observing 4 animals of type M in succession, as in *Hylobates lar pileatus*, is only $9/52^4$ or about one in 1,000.
The superscript ^{Gi} for Gibbon is used to emphasize the difference between the gibbon and human MN types.

The demonstration that the frequency of certain Gm-factors can vary in the races of chimpanzees as it does in the races of man, suggests a more general nature of this phenomenon in primates. It seems no longer justifiable to test samples from primates without a more precise determination of their origins and other polymorphic markers such as blood groups. 'Racial' differences could account for certain of the discrepancies in the reported frequencies of Gm and Inv types among chimpanzees [6, 62, 65, 71].

Gibbons

Human-type A-B-O and MN blood groups were determined in gibbons maintained at the Delta Regional Primate Research Center of Tulane University, Covington, Louisiana, the South East Asia Treaty Organization (SEATO) Research Laboratory, APO San Francisco, California, and at LEMSIP. Four of the animals belonged to the subspecies *H. lar pileatus*, and the remaining 52 to the subspecies *H. lar lar*. Statistically significant differences in the distribution of the human-type blood groups were observed for the 2 subspecies as shown in table XXIII [97].

It is of interest to note that similar differences between the above 2 groups of gibbons were observed in the same animals also for the distribution of the human-type, sex-linked blood group antigen Xg^a [13]. It may be appropriate to point out here again that of all species of primates tested, only man and gibbon share this antigen.

Our sero-primatological findings have been further confirmed by recent electrophoretic investigation of hemoglobin in the same group of gibbons. This work, carried out in collaboration with Dr. HAROLD HOFFMAN, revealed that both gibbon populations exhibit a single hemoglobin component which has a different mobility in each of the 2 populations [20].

References

1 ALEPA, E. P.: Genetic factors and polypeptide chain subclasses of human immunoglobulin G detected in the sera of non-human primates. Proc. 1st Congr. Int. Soc. Primat., Frankfort 1966, pp. 411–416 (Fischer, Stuttgart 1967).

2 ALEPA, F. P. and TERRY, W. D.: Genetic factors and polypeptide chain subclasses of human immunoglobulin detected in chimpanzee serums. Science *150:* 1293 (1965).

3 ANDRÉ, A.; COURTOIS, G.; LENNES, G.; NINANE, G. et OSTERRIETH, P. M.: Mise en évidence d'antigènes de groupes sanguins A, B, O et Rh chez les singes chimpanzés. Ann. Inst. Pasteur *101:* 82 (1961).

4 BEREZNAY, Y.: Composition du sang des singes anthropoides par rapport au sang humain. Bull. Soc. roy. Zool., Anvers, No. 10 (1959).

5 BERNSTEIN, F.: Zusammenfassende Betrachtungen über die erblichen Blutstrukturen des Menschen. Z. indukt. Abstamm. Vererbungsl. *37:* 223 (1925).

6 BOYER, S. H. and YOUNG, W. J.: Gammaglobulin (Gm group) heterogenicity in chimpanzees. Science *133:* 583–584 (1961).

7 BUETTNER-JANUSCH, J.: Antihuman sera and primate erythrocytes. Ann. N.Y. Acad. Sci. *97:* 9 (1962).

8 CANDELA, P. B.; WIENER, A. S., and GOSS, L. J.: New observations on the blood group factors in Simidii and Cercopithecidae. Zoologica *25:* 513 (1940).

9 CEPPELLINI, R.: On the genetics of secretor and Lewis characters; a family study. Proc. 5th Congr. Int. Soc. Blood Transfusion, Paris, 1955, pp. 207–211 (Karger, Basel 1956).

10 FISCHER, W. und KLINKHART, G.: Über Hämagglutination und Hämolyse bei *Macaca cynomologus* und *Simia rhesus*. Z. ImmunForsch. *75:* 513 (1932).

11 FISK, R. T. and FOORD, R. G.: Observations on the Rh-agglutinogen of human blood. Amer. J. clin. Path. *12:* 545 (1942).

12 FORSSMAN, J.: Die Herstellung hochwertiger spezifischer Schafhämolysine ohne Verwendung von Schafsblut. Biochem. Z. *37:* 78 (1911).

13 GAVIN, J.; NOADES, J.; TIPPETT, P.; SANGER, R., and RACE, R. R.: Blood group antigen Xg^a in gibbons. Nature, Lond. *204:* 1322 (1964).

14 GENOGOZIAN, N.: Human A- and B-like antigens on red cells of marmosets. Progr. Med. *117:* 858–861 (1964).

15 GOODMAN, M.: Evolution of the immunologic species specificity of human serum proteins. Human biol. *34:* 104–150 (1962).

16 Goodman, M.: Phylogenetic relationships among primates from immunodiffusion (Ouchterlony) data; in Goldsmith and Moor-Jankowski Medical primatology 1970 (in press).

17 Grubb, R.: Observations on the human blood group system Lewis. Acta path. microbiol. scand. *28:* 61 (1951).

18 Hirose, Y. and Balner, H.: Red cell iso-antigens of rhesus monkeys. Blood *34:* 661 (1969).

19 Hirszfeld, L. and Hirszfeld, H.: Serological differences between the blood of different races. Lancet *ii:* 675–679 (1919).

20 Hoffman, H. A. and Gottlieb, A. J.: Hemoglobins of chimpanzees and gibbons. Prim. Med., vol. 1, pp. 27–34 (Karger, Basel 1968).

21 Kabat, E. A.: Blood group substances (Academic Press, New York 1956).

22 Landsteiner, K.: Über Agglutinationserscheinungen normalen menschlichen Blutes. Klin. Wschr. *14:* 1132 (1901).

23 Landsteiner, K.: Individual differences in human blood. Science *74:* 403 (1931).

24 Landsteiner, K. and Levine, P.: On the inheritance of agglutinogens of human blood demonstrable by immune agglutinins. J. exp. Med. *48:* 731 (1928).

25 Landsteiner, K. and Levine, P.: On individual differences in human blood. J. exp. Med. *47:* 757 (1928).

26 Landsteiner, K. and Miller, C. P., jr.: Serological studies on the blood of primates. I. The differentiation of human and anthropoid blood. J. exp. Med. *42:* 841 (1925).

27 Landsteiner, K. and Miller, C. P., jr.: Serological studies on the blood of the primates. III. Distribution of serological factors related to human isoagglutinogens in the blood of lower monkeys. J. exp. Med. *42:* 863 (1925).

28 Landsteiner, K. and Wiener, A. S.: The presence of M-agglutinogens in the blood of monkeys. J. Immunol. *33:* 1 (1937).

29 Landsteiner, K. and Wiener, A. S.: An agglutinable factor in human blood recognized by immune sera for rhesus blood. Proc. Soc. exp. Biol. Med. *43:* 223 (1940).

30 Landsteiner, K. and Wiener, A. S.: Tests for Rh-factor with guinea pig immune sera. Proc. Soc. exp. Biol. Med. *51:* 313 (1942).

31 Levine, P. and Celano, M. J.: Agglutinating specificity for LW factor in guinea pig and rabbit anti-Rh serums. Science *156:* 1, 744 (1967).

32 Levine, P.; Katzin, E. M., and Burnham, L. J.: Isoimmunization in pregnancy; its possible bearing on the etiology of erythroblastosis fetalis. J. amer. med. Ass. *116:* 825 (1941).

33 Mann, J. D.; Cahan, A.; Gelb, A. G.; Fisher, N.; Hamper, J.; Tippett, P.; Sanger, R., and Race, R. R.: A sex-linked blood group. Lancet *i:* 8 (1962).

34 Marsh, W. L.: Anti-i, a cold autoantibody defining the I-i relationship in human red cells. Brit. J. Haemat. *7:* 200 (1961).

35 Masouredis, S. P.; Dupuy, M. E., and Elliot, M.: Distribution of the human Rh_0(D) antigen in the red cells of non-human primates. J. Immunol. *28:* 8 (1947).

36 Mollison, T.: Die biologische Eiweissdifferenzierung als Hilfsmittel zur Erkenntnis der verwandtschaftlichen Beziehungen der Wirbeltiere, besonders des Menschen. S.B. Isis, Dresden *1912:* 3–6.

37 Mollison, T.: Die Präzipitinreaktion als Zeugnis für die anthropomorphe Verwandtschaft des Menschen. Korresp. Bl. dtsch. Ges. Anthrop. *43:* 132–134 (1912).

38 MOLLISON, T.: Über den Begriff der Differenzierung im morphologischen und biochemischen Sinne. Verh. dtsch. Ges. Rassenforsch. *9:* 38–42 (1938).

39 MOLLISON, T.: Die epigenetische Entwicklung des Arteiweisses in der Ontogenie. Anthrop. Anz. *15:* 205–210 (1938).

40 MOURANT, A. E.: The distribution of the human blood groups (Blackwell, Oxford 1954).

41 MOURANT, A. E.; KOPEC, A. C., and DOMANIEWSKA-SOBCZAK, K.: The A-B-O blood groups (Blackwell, Oxford 1958).

42 MOOR-JANKOWSKI, J.: Immunological identification of cells in culture. Symp. on characterization and uses of human diploid cells strains. Int. Ass. Microbiol. Soc. Opatija 1963, pp. 161–182 (Inst. Immunology, Zagreb 1964).

43 MOOR-JANKOWSKI, J. and HUSER, H. J.: Sero-anthropological investigations in the Walser and Romansh isolates of the Swiss Alps. 6th Congr. Int. Soc. Blood Transfusion, Boston 1956, pp. 215–219 (Karger, Basel 1958).

44 MOOR-JANKOWSKI, J. and WIENER, A. S.: Blood groups of non-human primates, summary of the currently available information. Prim. Med. *1:* 49–67 (1968).

45 MOOR-JANKOWSKI, J. and WIENER, A. S.: Sero-primatology, a new discipline. 1st Congr. Int. Primat. Soc., Frankfort, 1966. Progress in primatology, pp. 378–381 (Fischer, Stuttgart 1967).

46 MOOR-JANKOWSKI, J. and WIENER, A. S.: Blood groups of apes and monkeys; human-type and simian-type. 1st Congr. Int. Primat. Soc., Frankfort, 1966. Progress in primatology, pp. 384–410 (Fischer, Stuttgart 1967).

47 MOOR-JANKOWSKI, J.; WIENER, A. S., and GORDON, E. B.: Blood groups of apes and monkeys. I. The A-B-O blood groups in baboons. Transfusion *4:* 92 (1964).

48 MOOR-JANKOWSKI, J.; WIENER, A. S., and GORDON, E. B.: Blood groups of apes and monkeys. III. The M-N blood factors of apes. Folia primat. *2:* 129 (1964).

49 MOOR-JANKOWSKI, J.; WIENER, A. S., and GORDON, E. B.: Blood antigens and cross-reacting antibodies in primates, including man. III. Heterophile-like behavior of the blood factor I. Exp. Med. Surg. *22:* 308 (1964).

50 MOOR-JANKOWSKI, J.; WIENER, A. S., and GORDON, E. B.: Simian blood groups. Two new blood factors of gibbon blood, $\mathbf{A^g}$ and $\mathbf{B^g}$. Transfusion *5:* 235 (1965).

51 MOOR-JANKOWSKI, J.; WIENER, A. S.; GORDON, E. B., and DAVIS, J. H.: Blood groups of baboons; demonstrated with isoimmune sera. Nature, Lond. *214:* 181 (1967).

52 MOOR-JANKOWSKI, J.; WIENER, A. S.; GORDON, E. B., and DAVIS, J. H.: Blood groups of monkeys; demonstrated with isoimmune rhesus monkey sera. Int. Arch. Allergy *30:* 33 (1967).

53 MOOR-JANKOWSKI, J.; WIENER, A. S.; GORDON, E. B., and GUTHRIE, C. B.: Simian blood groups. A new blood factor $\mathbf{A^{ba}}$ of Celebes black apes demonstrated with rabbit antisera. Folia primat. *3:* 245 (1965).

54 MOOR-JANKOWSKI, J.; WIENER, A. S.; KRATOCHVIL, C. H., and FINEG, J.: Seroprimatology of chimpanzee blood groups; blood group distribution as racial characteristic. Science *152:* 219–220 (1966).

55 MOOR-JANKOWSKI, J.; WIENER, A. S.; KRATOCHVIL, C. H., and FINEG, J.: Chimpanzee blood groups; demonstrated with isoimmune and cross-immune sera. Int. Arch. Allergy *29:* 397–403 (1966).

56 Moor-Jankowski, J.; Wiener, A. S., and McRitchie, R.: Isoantibodies produced in chimpanzees by cross-immunization with gibbon red cells. Int. Arch. Allergy *34:* 386 (1968).

57 Moor-Jankowski, J.; Wiener, A. S., and Rogers, C.: Blood groups of chimpanzees; demonstrated with iso-immune sera. Science *145:* 1441 (1964).

58 Nuttall, G. H. F.: Blood immunity and blood relationships (Cambridge University Press, Cambridge 1904).

59 Osman Hill, W. C.: Chimpanzees; Encyclopedia Britannica, vol. 5, p. 5 (Encyclopedia Britannica, Chicago 1966).

60 Ottensooser, F. and Silberschmidt, K.: Haemagglutinin anti-N in plant seeds. Nature, Lond. *172:* 914 (1953).

61 Owen, R. D. and Anderson, D. R.: Blood groups in rhesus monkeys. Ann. N.Y. Acad. Sci. *97:* 1 (1962).

62 Podliachouk, L.: Le facteur sérique Gm^{a} dans le sérum d'animaux de différentes espèces. Ann. Inst. Pasteur *96:* 362–364 (1959).

63 Prokop, O. and Uhlenbruck, G.: Human blood and serum groups (Wiley Interscience, New York 1968).

64 Race, R. R. and Sanger, R.: Blood groups in man (Davis, Philadelphia 1968).

65 Ritter, H.; Boshaar, E. und Putz, Z.: Zur Phylogenetik des Gammaglobulinpolymorphismus. Gm. Humangenetik *1:* 571–575 (1965).

66 Rosin, S.: Die Verteilung der ABO-Blutgruppen in der Schweiz. Arch. J. Klaus-Stiftung *31:* 17 (1956).

67 Ruffié, J. et Bernard, J.: Hémotypologie et évolution du groupe humain. Monographies du Centre d'Hémotypologie du CNRS (Herrmann, Paris 1966).

68 Schiff, F. und Adelsberger, L.: Über blutgruppenspezifische Antikörper und Antigene. Z. ImmunForsch. *40:* 335 (1924).

69 Schiff, F. und Sasaki, H.: Der Ausscheidungstypus, ein auf serologischem Wege nachweisbares mendelndes Merkmal. Klin. Wschr. *11:* 1426 (1932).

70 Schmitt, J.: Immunbiologische Untersuchungen bei Primaten. Ein Beitrag zur Evolution der Blut- und Serumgruppen. Bibl. primat. *8:* 1–146 (Karger, Basel 1968).

71 Schmitt, J.; Krupe, M. und Deicher, N.: Untersuchungen über Immunoglobulin-Erbmerkmale (Gm- und Inv-Gruppen) bei Subhuman-Primaten. Humangenetik *1:* 571–573 (1965).

72 Springer, G. F.: Occurrence of A-B-H-like substances in microorganisms and plants; in Wiener Advances in blood grouping, III (Grune & Stratton, New York 1970).

73 Watkins, W. M.: Blood-group substances; in Wiener Advances in blood grouping, III (Grune and Stratton, New York 1970).

74 Wiener, A. S.: The agglutinogens M and N in anthropoid apes. J. Immunol. *34:* 11 (1938).

75 Wiener, A. S.: Natural anti-M agglutinins in the sera of male negroid twin infants, with a comparison of different methods of treating anemia. Acta. genet. med. gemellol. *1:* 170 (1952).

76 Wiener, A. S.: Origin of naturally occurring hemagglutinins and hemolysins; a review. J. Immunol. *66:* 287 (1961).

77 Wiener, A. S.: Blood groups and transfusion, 3rd ed. (Thomas, Springfield 1943; reprinted by Hafner, New York 1962).

78 WIENER, A. S.: Blood groups of chimpanzees and other non-human primates; their implications for the human blood groups. Trans. N.Y. Acad. Sci. Ser. II, *27:* 488 (1965).

79 WIENER, A. S.: Lewis blood types. Theoretical implications and practical applications. Amer. J. clin. Path. *43:* 388 (1965).

80 WIENER, A. S.: The law of specificity, with special reference to the blood groups. Exp. Med. Surg. *24:* 134 (1966).

81 WIENER, A. S.: Problems and pitfalls in blood grouping tests for non-parentage. I. Distribution of the blood groups; in WIENER Advances in blood grouping, III (Grune & Stratton, New York 1970).

82 WIENER, A. S.; BALDWIN, M., and GORDON, E. B.: Blood groups in chimpanzees. Exp. Med. Surg. *31:* 159 (1963).

83 WIENER, A. S.; BRAIN, P., and GORDON, E. B.: Further observations on the hemagglutinin of the snail, *Achatina granulata*. Haematologia, Budapest *3:* 9 (1969).

84 WIENER, A. S.; CANDELA, P. B., and GOSS, L. J.: Blood group factors in the blood, organs and secretions of primates. J. Immunol. *45:* 225 (1942).

85 WIENER, A. S.; HANLY, C.; MOOR-JANKOWSKI, J., and GORDON, E. B.: Chimpanzee red cell isoantibody of a 'new' specificity, anti-K^c, produced by cross-immunization of chimpanzees with human blood. J. Immunol. *98:* 1011 (1967).

86 WIENER, A. S. and GORDON, E. B.: Simple method of preparing anti-Rh serum in normal male donors. Amer. J. clin. Path. *17:* 67 (1947).

87 WIENER, A. S. and GORDON, E. B.: The blood groups of chimpanzees; A-B-O and M-N types. Amer. J. phys. Anthrop. *19:* 35 (1961).

88 WIENER, A. S. and GORDON, E. B.: The blood groups of chimpanzees; the Rh-Hr (CDE/cde) blood types. Amer. J. phys. Anthrop. *19:* 35 (1961).

89 WIENER, A. S.; GORDON, L. J.; COHEN, L., and FELDMAN, J.: Type-specific cold autoantibodies as a cause of acquired hemolytic anemia and transfusion reactions; biologic test with bovine red cells. Ann. intern. Med. *44:* 221 (1956).

90 WIENER, A. S.; GORDON, E. B., and MOOR-JANKOWSKI, J.: The Lewis blood groups in man. A review with supporting data on non-human primates. J. forensic Med. *11:* 67 (1964).

91 WIENER, A. S. and MOOR-JANKOWSKI, J.: Blood groups in anthropoid apes and baboons. Science *142:* 67–69 (1963).

92 WIENER, A. S. and MOOR-JANKOWSKI, J.: The V-A-B blood group system of chimpanzees. A paradox in the application of the 2 × 2 contingency test. Transfusion *5:* 64 (1965).

93 WIENER, A. S. and MOOR-JANKOWSKI, J.: The A-B-O blood groups of baboons. Amer. J. phys. Anthrop. *30:* 117 (1969).

94 WIENER, A. S. and MOOR-JANKOWSKI, J.: Unpublished data (1970).

95 WIENER, A. S.; MOOR-JANKOWSKI, J.; BALNER, H., and GORDON, E. B.: Blood groups of monkeys; demonstrated with antisera produced by combined skin transplantation and isoimmunization with red cells. Int. Arch. Allergy *34:* 386 (1968).

96 WIENER, A. S.; MOOR-JANKOWSKI, J., and BRANCATO, G. J.: LW factor. Haematologia, Budapest *3:* 385 (1969).

97 WIENER, A. S.; MOOR-JANKOWSKI, J.; CADIGAN, F. O., Jr., and GORDON, E. B.: Comparison of the A-B-H blood group specificities and the M-N types in man, gibbons *(Hylobates)* and siamangs *(Symphalangus)*. Transfusion *8:* 235 (1968).

98 WIENER, A. S.; MOOR-JANKOWSKI, J., and GORDON, E. B.: Blood groups of apes and monkeys. II. The A-B-O blood groups, secretor and Lewis types in apes. Amer. J. phys. Anthrop. *21:* 271 (1963).

99 WIENER, A. S.; MOOR-JANKOWSKI, J., and GORDON, E. B.: Blood groups antigens and cross-reacting antibodies in primates including man. I. Production of antisera for agglutinogen M by immunization with blood other than human type M blood. J. Immunol. *92:* 391 (1964).

100 WIENER, A. S.; MOOR-JANKOWSKI, J., and GORDON, E. B.: Blood group antigens and cross-reacting antibodies in primates including man. II. Studies on the M-N types of orang-utans. J. Immunol. *93:* 101 (1964).

101 WIENER, A. S.; MOOR-JANKOWSKI, J., and GORDON, E. B.: Blood groups of apes and monkeys. IV. The Rh-Hr blood types of anthropoid apes. Amer. J. hum. Genet. *16:* 246 (1964).

102 WIENER, A. S.; MOOR-JANKOWSKI, J., and GORDON, E. B.: Blood groups of apes and monkeys. V. Studies on the human blood group factors **A**, **B**, **H** and **Le** in Old and New World monkeys. Amer. J. phys. Anthrop. *22:* 175 (1964).

103 WIENER, A. S.; MOOR-JANKOWSKI, J., and GORDON, E. B.: Blood group antigens and cross-reacting antibodies in primates including man. IV. The blood factor $\mathbf{D}^{c}$ and its relation to the V-A-B blood group system of chimpanzees. Exp. Med. Surg. *23:* 256 (1965).

104 WIENER, A. S.; MOOR-JANKOWSKI, J., and GORDON, E. B.: Blood group antigens and cross-reacting antibodies in primates including man. V. Production of antisera for a 'new' blood factor $\mathbf{V}_1$ by immunization of chimpanzees with human blood. Exp. Med. Surg. *23:* 264 (1965).

105 WIENER, A. S.; MOOR-JANKOWSKI, J., and GORDON, E. B.: Blood groups of apes and monkeys. VI. Further studies on the human-type blood group factors **A, B, H** and **Lewis** in monkeys. Folia primat. *4:* 81 (1966).

106 WIENER, A. S.; MOOR-JANKOWSKI, J., and GORDON, E. B.: The relationship of the H substance to the A-B-O blood groups. Int. Arch. Allergy *29:* 82 (1966).

107 WIENER, A. S.; MOOR-JANKOWSKI, J., and GORDON, E. B.: Marmosets as laboratory animals. V. Blood groups in marmosets. Lab. Anim. Care *17:* 71 (1967).

108 WIENER, A. S.; MOOR-JANKOWSKI, J., and GORDON, E. B.: The specificity of hemagglutinating bean and seed extracts (lectins); implications for the nature of A-B-H agglutinins. Int. Arch. Allergy *36:* 582 (1969).

109 WIENER, A. S.; MOOR-JANKOWSKI, J.; GORDON, E. B., and DAVIS, J.: The blood factors **I** and **i** in primates including man and in lower species. Amer. J. phys. Anthrop. *23:* 389 (1965).

110 WIENER, A. S.; MOOR-JANKOWSKI, J.; GORDON, E. B.; DAUMY, O. M., and DAVIS, J. H.: Blood groups of gibbons; further observations. Int. Arch. Allergy *30:* 466 (1966).

111 WIENER, A. S.; MOOR-JANKOWSKI, J.; GORDON, E. B., and KRATOCHVIL, C. H.: Individual differences in chimpanzee blood demonstrable with absorbed human anti-$\mathbf{Rh}_0$ sera. Proc. nat. Acad. Sci., Wash. *56:* 458 (1966).

112 WIENER, A. S.; MOOR-JANKOWSKI, J.; GORDON, E. B., and SHELL, N. F.: Human-type blood factors in gibbons, with special reference to the multiplicity of serological specificities of human type M blood. Transfusion *6:* 313 (1966).

113 WIENER, A. S. and PETERS, H. R.: Hemolytic reactions following transfusions of

homologous group with three cases in which the same agglutinogen was responsible. Ann. intern. Med. *13:* 2306 (1940).

114 WIENER, A. S.; MOOR-JANKOWSKI, J.; RIOPELLE, A. J., and SHELL, W. F.: Simian blood groups. Another blood group system, C-E-F, in chimpanzees. Transfusion *5:* 508 (1965).

115 WIENER, A. S.; SOCHA, W. W.; MOOR-JANKOWSKI, J., and GORDON, E. B.: The A^p-B^p blood types of baboons. Amer. J. Phys. Anthrop. *33:* 433 (1970).

116 WIENER, A. S. and WEXLER, I. B.: Heredity of the blood groups (Grune & Stratton, New York 1958).

117 WIENER, A. S. and WEXLER, I. B.: An Rh-Hr syllabus, 2nd ed. (Grune & Stratton, New York 1963).

118 WIENER, A. S.; WISECUP, W., and MOOR-JANKOWSKI, J.: A 'new' simian-type blood factor, L^c, associated with the C-E-F blood group system of chimpanzees. Transfusion *7:* 351 (1967).

Authors' address: Dr. J. MOOR-JANKOWSKI and Dr. A. S. WIENER, Department of Forensic Medicine, New York University School of Medicine, 520 First Avenue, *New York, NY 10016* (USA)

Pathology of Simian Primates, Part I, pp. 318–330 (Karger, Basel 1972)

Coagulation, Fibrinolysis and Platelet Function

CHRISTINE HAWKEY

Zoological Society of London, London

Contents

General 318
Blood Coagulation 321
i) The Thrombin-Fibrinogen Reaction 321
ii) The Conversion of Prothrombin to Thrombin 322
Blood Platelets 324
Blood Fibrinolytic Activity 325
Pathology 327
References 329

General

Published reports on the blood coagulation mechanism in non-human primates are confined to studies of relatively small numbers of *Macaca mulatta* [1, 2], *M. nemestrina* [1], *Erythrocebus patas* [2], *Saimiri sciureus* [1], *Cebus capucinus* [2], *Pan troglodytes* [2] and various species of Papio [2, 3, 4]. Blood fibrinolytic activity has been measured in a small series of Cebidae, Cercopithecidae and Pongidae [3] and ADP-induced platelet aggregation in a group of Cercopithecidae and Pongidae [4]. Thus there is very little information available on blood coagulation and related mechanisms in non-human primates and, for most species, normal values for the factors concerned with these mechanisms are not known. This is because it is difficult to obtain and test enough animals of each species to permit statistical analysis of the results and this situation is unlikely to be remedied for some time. Although levels of coagulation and fibrinolytic factors vary slightly in different species [2], there is an overall similarity in results of coagulation and fibrinolytic tests in mon-

	Synonyms for clotting factors
Factor I	Fibrinogen
Factor II	Prothrombin
Factor III	Thromboplastin
Factor IV	Calcium
Factor V	Labile factor, proaccelerin
Factor VII	Proconvertin, stable factor, serum prothrombin conversion accelerator (SPCA)
Factor VIII	Antihaemophilic factor (AHF) Antihaemophilic globulin (AHG)
Factor IX	Christmas factor, plasma thromboplastin component (PTC), antihaemophilic globulin B
Factor X	Stuart Prower factor
Factor XI	Plasma thromboplastin antecedent (PTA)
Factor XII	Hageman factor
Factor XIII	Fibrin stabilising factor (FSF) Laki-Lorand factor

keys classified in the same family. Monkeys are, therefore, grouped in families for the consideration of results, and the normal range for a given family is at present accepted as a guide for that family.

Unfortunately, even at family level, the total number of animals for which data are available is still very small and does not yet permit conclusions to be drawn regarding the influence of variables such as age and sex. We have found, however, that, by carefully controlling all technical details which may affect our tests, a fair amount of reliance can be placed on figures obtained from relatively small numbers of animals.

The results discussed in this section have been collected during a survey of the haematology of overtly healthy non-human primates in the collection of the Zoological Society of London (table I). Each monkey received a tranquillizing dose of phencyclidine hydrochloride (Sernylan 1–1.5 mg/kg) before venepuncture to minimise the effects of stress; this drug has no direct effect on coagulation, fibrinolytic or platelet factors. Blood samples were obtained by puncture of a femoral vein using disposable needles and plastic syringes. Criteria for collection, preparation and storage of the blood samples were rigidly controlled at all times [6]. Tests on normal human plasma were always made in parallel with the non-human subjects, although the human subjects received no tranquillising drug. The human studies provided controls for all test procedures used and also permitted a comparison of results between human and non-human primates.

Table I. Animals tested

Family	Total	Species	Total
Prosimii Lemuridae	6	*Lemur catta* (Ring-tailed lemur)	6
Anthropoidea Cebidae	62	*Cebus capucinus* (White-throated capuchin)	4
		C. nigrivittatus (Weeper capuchin)	15
		C. apella (Black-capped capuchin)	6
		Aotus trivirgatus (Owl monkey, Douroucouli)	25
		Saimiri sciureus (Squirrel monkey)	6
		Lagothrix lagothricha (Woolly monkey)	4
		Ateles paniscus (Spider monkey)	2
Cercopithecidae	137	*Macaca sylvanus* (Barbary ape)	4
		M. silenus (Lion-tailed monkey)	1
		M. nemestrina (Pig-tailed monkey)	6
		M. irus (Crab-eating monkey)	2
		M. mulatta (Rhesus monkey)	25
		M. assamensis (Assamese macaque)	2
		M. speciosa (Stump-tailed macaque)	5
		M. maurus (Moor monkey)	6
		Cynopithecus niger (Black ape)	2
		Cercocebus albigena (Grey-cheeked mangabey)	1
		C. torquatus (White-collared mangabey)	3
		Papio cynocephalus (Yellow baboon)	7
		P. anubis (Olive baboon)	7
		P. papio (Guinea baboon)	6
		P. hamadryas (Sacred baboon)	2
		Mandrillus sphinx (Mandrill)	4
		M. leucophaeus (Drill)	2
		Cercopithecus pygerythrus (Vervet monkey)	3
		C. sabaeus (Green monkey)	2
		C. diana (Diana monkey)	1
		C. l'hoesti (L'Hoests monkey)	1
		C. albogularis (Sykes monkey)	1
		C. mona (Mona monkey)	2
		C. ascanius (Spotnose monkey)	1
		C. sclateri (Sclater's spotnose monkey)	1
		C. talapoin (Talapoin monkey)	3
		Erythrocebus patas (Patas monkey)	37
Pongidae	24	*Pongo pygmaeus* (Orang-utan)	21
		Pan troglodytes (Chimpanzee)	2
		Gorilla gorilla (Gorilla)	1

Table II. Cross reactivity between brain extract and plasma from different primate species

Source of plasma	Prothrombin time in s with brain extract from					
	M. mulatta	*M. maurus*	*E. patas*	*Cercopithecus* sp.	*Cebus* sp.	*H. sapiens*
M. mulatta	14.0	13.5	13.5	14.0	14.0	15.0
M. maurus	14.0	14.0	14.5	14.0	14.0	14.5
E. patas	14.0	13.5	14.0	14.5	13.5	14.0
Cercopithecus sp.	15.0	14.0	15.0	15.0	13.0	14.5
Cebus sp.	13.5	14.5	14.0	14.0	14.0	14.0
H. sapiens	14.5	14.0	14.0	14.0	13.5	14.0

Laboratory methods routinely used in human coagulation studies can be applied directly to non-human primates as the problem of species specificity of reagents, which often hampers general comparative coagulation surveys, does not arise between different primate species. Thus human brain thromboplastin (table II) and cephalin react equally well with human and monkey plasma, and human plasma deficient in specific clotting factors can be used as substrate for assays of these factors in monkeys. It is, however, important to specify exactly the methods used as comparison of results obtained by different methods is not always valid.

Blood Coagulation

i) The Thrombin-Fibrinogen Reaction

Fibrinogen, measured as clottable protein [7], falls within the normal human range in each group of monkeys examined (table III). Occasionally, particularly in members of the Cebidae and Pongidae, unusually high fibrinogen levels are found and there is evidence that levels of greater than 500 mg% are associated with chronic pathological conditions such as bacterial infections, parasitic infestations and bone dyscrasias in monkeys as they are in man. The relatively high levels seen in some apparently healthy Cebidae may be a manifestation of chronic bone disease which has been recorded frequently in these animals in captivity [8, 9].

The rate of conversion of fibrinogen to fibrin by bovine thrombin [10] does not differ significantly in the Lemuridae, Pongidae and man. Cebidae

Table III. Tests on factors involved in the thrombin-fibrinogen reaction

Test	Prosimii	Anthropoidea			
	Lemuridae	Cebidae	Cercopithecidae	Pongidae	Hominidae
Fibrinogen [7] mg/100 ml	343 (3) 293–407	451 (34) 194–908	350 (102) 95–624	437 (21) 182–843	150–450
Thrombin time [10] sec (50 µm/ml)	6.7 (5) 6.0–7.5	9.5 (51) 6.0–16.0	8.2 (120) 5.0–16.5	6.8 (23) 4.0–9.0	6.5 (69) 4.0–8.5
Thrombin time [10] sec (5 µm/ml)	20.3 (5) 12.0–25.0	85.2 (51) 16.0–300	38.1 (120) 15.0–164	43.3 (23) 25.0–115	22.9 (69) 16.0–25
Factor XIII [11]	+ (5)	+ (39)	+ (102)	+ (21)	+
Antithrombin [20] units	350 (1)	1052 (5) 460–1610	452 (27) 300–680	516 (7) 390–725	529 (11) 350–720

Figures in brackets denote number of individuals tested.

and Cercopithecidae show a slightly retarded rate of clotting which becomes more evident when dilute thrombin is used (table II). This finding is difficult to interpret but may, in the Cebidae, be due to the high levels of plasma antithrombin often found in these animals, or from the presence in their plasma of anticoagulant fibrinogen breakdown products as these animals also often have high fibrinolytic activity (table VIII). Factor XIII, the factor responsible for stabilization of fibrin clots, has been found in all monkeys examined (table III). This protein has been assessed qualitatively by the method of Josso *et al.* [11]. No quantitative tests have yet been undertaken.

ii) The Conversion of Prothrombin to Thrombin

The prothrombin (factor II) level [12], expressed as a percentage of the amount found in normal human plasma, is low on average in some Cercopithecidae and Pongidae and occasionally in individuals of these families the level approaches the haemorrhagic range in man (table IV), although the monkeys do not suffer from a bleeding diathesis. The rate of conversion of prothrombin to thrombin by tissue factor, measured by the one-stage prothrombin test [13] is decreased in the Cercopithecidae and Pongidae and increased in Lemurs and Cebus monkeys compared with man (table V). Aver-

Table IV. Assays of plasma clotting factors

Clotting factor %	Prosimii	Anthropoidea			
	Lemuridae	Cebidae	Cercopithecidae	Pongidae	Hominidae
Factor 11 [12] (Prothrombin)	100 (1)	117 (18) 70–180	78 (43) 22–270	76 (20) 41–170	100 66–145
Factor V [21]	250 (1)	787 (26) 43–1,000	194 (75) 100–1,000	102 (16) 43–230	100 50–200
Factor VII [22]		802 (4) 210–1,000	325 (26) 115–1,000	244 (7) 45–1,000	100 60–150
Factor VIII [23]	690 (2) 380–1,000	550 (12) 180–1,000	433 (44) 110–1,000	625 (12) 115–1,000	100 50–200
Factor IX [6]		400 (7) 100–1,000	42 (27) 11–130	70 (7) 12–140	100 50–200
Factor X [24]	250 (1)	210 (25) 40–800	104 (55) 28–250	118 (17) 27–250	100 50–200
Contact activation index [16, 17] (Factors XI and XII)		41 (4) 36–44	50 (28) 23–72	54 (4) 43–60	46 (8) 34–61

Figures in brackets denote number of animals tested.

age levels of factors V, VII and X, which are the plasma clotting factors required for this reaction, are generally higher in monkeys than in man (table IV). Again, some individuals show surprisingly low levels of these factors but haemorrhagic symptoms have not been recorded.

It should be noted that the conversion of prothrombin to thrombin by Russell's viper venom (Stypven) is slow in some monkeys, particularly the Cercopithecidae (table V) and the Stypven test does not reflect the time levels of clotting factors in these monkeys as it does in man [14]. This test should not, therefore, be used in screening monkeys for coagulation disorders.

Generation of intrinsic (plasma) prothrombin activator is very rapid in the Lemuridae and all monkeys show a faster average rate of generation than man (table V). The cephalin time [15] (partial thromboplastin test) by which this pathway of prothrombin activation has been assessed, varies directly with the plasma factor VIII concentration and this factor is consistently high in monkeys compared with man (table IV). Factor IX which is also

Table V. Tests measuring the conversion of prothrombin to thrombin

Test	Prosimii	Anthropoidea			
	Lemuridae	Cebidae	Cercopithecidae	Pongidae	Hominidae
Cephalin time [15]	28.5 (6)	48.1 (54)	55.9 (132)	51.0 (24)	69.2 (85)
sec	23.0–37.0	31.5–95.0	38.0–88.0	30.0–73.0	40.0–105.0
Prothrombin time [13]	12.8 (4)	11.4 (51)	15.2 (129)	16.6 (24)	13.9 (100)
sec	11.5–15.0	7.0–16.5	11.0–22.0	10.0–28.0	12.5–15.5
Stypven time	8.0 (4)	11.4 (32)	15.4 (125)	10.5 (23)	8.2 (74)
sec	7.0–10.0	6.5–18.3	8.0–40.0	7.0–17.5	6.0–12.5

Figures in brackets denote number of individuals tested.

concerned with intrinsic prothrombin activation, is asymptomatically low in some Cercopithecidae and Pongidae. Contact factors (factors XI and XII) have not yet been assayed directly in monkeys but results obtained with a modification [16] of the contact activation test [17] suggest that in Cebidae these factors are not so active as in other members of the Anthropoidea (table IV). This aspect of intrinsic prothrombin activation requires a great deal of further study.

It should be mentioned that the low levels of prothrombin and factors VII, IX and X found in some individuals could be a result of sub-acute vitamin K deficiency or chronic liver disease. These possibilities have not yet been investigated but acute vitamin K deficiency has been encountered in *Aotus trivirgatus* characterised by prothrombin times of longer than 40 sec and levels of factors II, IV, IX and X of less than 10%. Monkeys with this condition suffered massive gastro-intestinal, sub-cutaneous and intramuscular haemorrhages. The deficiency responded rapidly to intramuscular injection of 5 mg vitamin K (Konakion), the prothrombin time returning to normal within 3–6 h.

Blood Platelets

Platelet counts do not vary significantly from the normal human range in any of the groups of monkeys examined (table VI). Clot retraction is also equal to that of man in the few monkeys in which it has been measured (table

Table VI. Platelet tests

Test	Prosimii Lemuridae	Anthropoidea				
		Cebidae	Cercopithecidae	Hylobatidae	Pongidae	Hominidae
Platelet count [18]	223 (12)	266 (64)	269 (126)	249 (4)	234 (20)	
$\times 10^3/mm^3$	100–350	108–558	115–498	139–278	129–318	150–500
Clot retraction			64 (17)	61 (1)	66 (5)	57–68
%			50–71		66–67	75
Platelet factor [3]		61 (4)	68 (15)		66 (2)	81 (17)
released [16]		54–65	56–81			71–92
Platelet factor [3]		96 (4)	103 (15)		124 (2)	107 (17)
Total [16]		93–98	80–120		116–133	100–127

Figures in brackets denote number of individuals tested.

VI). The total amount of platelet factor 3 and the amount released from a standard number of platelets by kaolin treatment [16] is greater in man than in monkeys. Platelets of all monkeys aggregate rapidly on addition to platelet-rich plasma of adenosine diphosphate (ADP). There is, however, a marked difference in the reversal of ADP-induced aggregation in different species (fig. 1) which is due to variable levels of ADP-inhibitor [5] in the plasma. The level of this inhibitor is high in baboons *(Papio cynocephalus* and *P. anubis),* patas monkeys and chimpanzees, and low in rhesus monkeys, weeper capuchins and man (fig. 2). Family averages are not helpful in respect of ADP inhibitor. The adhesiveness of platelets to glass has been measured, in a small number of monkeys, with results comparable to man in most instances [3, 4].

Blood Fibrinolytic Activity

Free plasmin is not found in the circulating blood of monkeys or man (table VII); as in man, the reason is probably that the little formed under normal circumstances is linked to plasma antiplasmins. The average level o-plasminogen is higher in monkeys than in man, particularly high levels ocf curring in the Cebidae and Cercopithecidae, suggesting a potentially more active fibrinolytic mechanism (table VI). Plasminogen of all primates is activated by human urokinase and of all except baboons by streptokinase. This

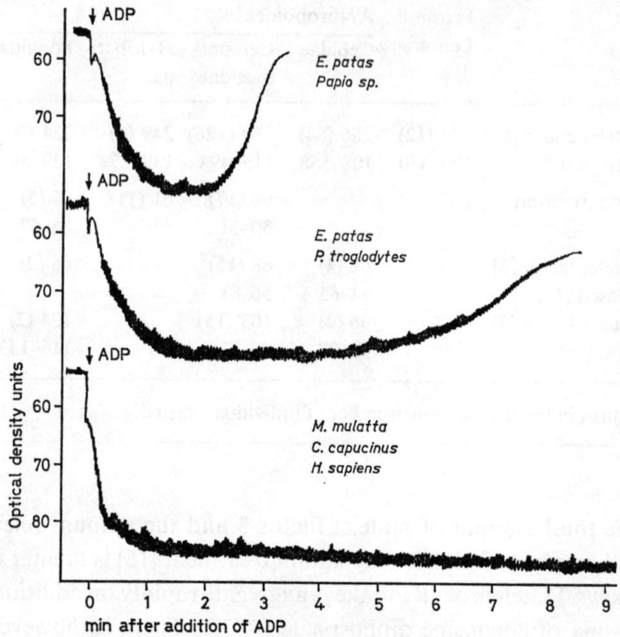

Fig. 1. Patterns of ADP-induced platelet aggregation found in different primate species.

Table VII. Fibrinolytic factors

	Anthropoidea			
	Cebidae	Cercopitheci-dae	Pongidae	Hominidae
Plasminogen [10] Units/ml	5.47 (5) 3.7–6.8	5.46 (28) 3.4–7.05	3.94 (5) 2.7–5.2	3.16 (20) 1.9–4.0
Plasmin [10] Units/ml	0 (5)	0 (10)	0 (10)	0 (20)
Fibrinolytic Inhibitors % [25]	100 (5) 48–116	104 (18) 70–140	122 (11) 80–150	100 (20) 80–120

Fig. 2. Variation in plasma ADP-inhibitor in different primate species. Shaded area denotes normal human range.

difference in reactivity of baboon plasminogen suggests a difference in the structure of the plasminogen molecule in baboons [19].

Blood activator, measured by the euglobulin lysis test [10] is within the normal human range in the Cercopithecidae and Pongidae but was found to be low in lemurs (table VIII). Some Cebidae have very short euglobulin lysis times indicating relatively high levels of blood activator. Fibrinolytic inhibitors have not varied significantly (table VII). Fibrinogen breakdown products have not been measured in monkey blood.

Pathology

Congenital deficiencies of coagulation factors, giving rise to severe haemorrhagic conditions, occur in man and in some inbred domestic mammals

Table VIII. Blood activator estimated by euglobulin lysis time

Euglobulin lysis time [10] (h)	Number of animals				
	Lemuridae	Cebidae	Cercopithecidae	Pongidae	Hominidae
less than 1	0	*16*	9	5	1
1–4	0	2	*64*	*16*	*56*
4–8	0	5	18	1	8
8–24	7	1	10	0	2
more than 24	0	2	0	0	0

such as dogs, horses and pigs. Factor VIII (deficiency haemophilia) is the most common of these; in man the incidence of haemophilia is 3–4 in 100,000 and it has been estimated that one in every ten cases arise as a spontaneous mutation. Such congenital deficiencies have not been found in monkeys or indeed in any wild mammal, almost certainly because they would be rapidly lethal in the natural habitat but it is conceivable that in the conditions of inbreeding and protected environment provided by large primate centres, cases may occur in monkeys in the future. There seems little justification at present for regarding the monkeys in which low levels of prothrombin and factors VII, IX and X have been recorded as being congenitally abnormal. The possible influence of chronic liver disease or vitamin K deficiency has already been discussed and receives support because factors V and VIII, which are not dependent on vitamin K or depressed in chronic liver disease, are always high in monkeys (table IV). In our experience, only in severe cases of proven vitamin K deficiency have haemorrhagic symptoms been encountered in monkeys. In these cases it was not possible to determine the cause of the vitamin deficit which could have been due to incorrect diet or imbalance of the intestinal flora.

Primary thrombocytopenia or functional platelet defects are also as yet unknown in monkeys but progressive thrombocytopenia, accompanied by reduction of some clotting factors has been described in rhesus [26] and aotus [27] monkeys experimentally infected with malaria. This condition, found in human malarial patients [28], is thought to result from red cell breakdown in the circulating blood; as a consequence there is released a phospholipid substance which activates the coagulation mechanism, and adenosine diphosphate which causes platelets to aggregate. In its intermediary stages this situation can give rise to the formation of platelet and fibrin microthrombi. These are usually cleared rapidly by the reticuloendothelial and fibrinolytic systems and may not be detectable clinically or histologically. Their production, however, results in depletion of circulating levels of platelets and plasma clotting factors, giving rise to a defective haemostatic mechanism. This condition can be reversed by the administration of heparin [27, 29].

References

1 SEAMAN, A. J. and MALINOW, M. R.: Blood clotting in non-human primates. Lab. anim. care *18:* 80 (1968).

2 HAWKEY, C. and SYMONS, C.: A comparative study of blood coagulation in human and non-human primates. Proc. 2nd Int. Cong. on Primatology, Atlanta, 1968 (Karger, Basel/München/New York 1969).

3 HAWKEY, C. and SYMONS, C.: Preliminary report of studies on platelet aggregation, blood coagulation and fibrinolysis in non-human primates. Proc. zool. Soc., Lond. *17:* 213 (1966).

4 HAMPTON, J. W. and MATTHEWS, C.: Similarities between human, baboon and blood clotting. J. appl. Physiol. *21:* 1713 (1966).

5 HAWKEY, C. and SYMONS, C.: Variation in ADP-induced platelet aggregation *in vitro* in primates as a result of differences in plasma ADP-inhibitor levels. Thromb. Diath. haemorrh. *19:* 29 (1968).

6 HARDISTY, R. M. and INGRAM, G. I. C.: Bleeding disorders. Investigation and management (Blackwell Scientific Publications, Oxford 1965).

7 RATNOFF, O. D. and MENZIE, C.: A new method for the determination of fibrin in small samples of plasma. J. Lab. clin. Med. *37:* 316 (1951).

8 RUCH, T. C.: Diseases of laboratory primates (Saunders, Philadelphia 1959).

9 BOULAY, G. H. DU and CRAWFORD, M. A.: Nutritional bone disease in captive primates. Proc. zool. Soc., Lond. *21:* 223 (1968).

10 HAWKEY, C. and HOWELL, M.: The laboratory control of thrombolytic therapy. J. clin. Path. *17:* 287 (1964).

11 JOSSO, F.; PROU-WARTELLE, O.; ALAGILLE, D. et SOULIER, J. P.: Le déficit congénital en facteur stabilisant de la fibrine (facteur XIII). Etude de deux cas. Nouv. Rev. franç. Hémat. *4:* 267 (1964).

12 PECHET, L.: One-stage specific determination of Factor II (Prothrombin); in TOCANTINS and KAZAL Blood Coagulation, Haemorrhage and Thrombosis (Grune and Stratton, London/New York 1964).

13 QUICK, A. J.: Haemorrhagic diseases (Kimpton, London 1967).

14 HAWKEY, C. and SYMONS, C.: Coagulation of primate blood by Russell's viper venom. Nature, Lond. *210:* 141 (1966).

15 RODMAN, N. F.; BARROW, E. M., and GRAHAM, J. B.: Diagnosis and control of the haemophilioid states with the partial thromboplastin time test. Amer. J. clin. Path. *29:* 525 (1958).

16 HAWKEY, C.: Unpublished (1970).

17 MARGOLIS, J.: Initiation of blood coagulation by glass and related surfaces. J. Physiol. *137:* 95 (1957).

18 BRECHER, G. and CRONKITE, E. P.: Morphology and enumeration of human blood platelets. J. appl. Physiol. *3:* 365 (1950).

19 HAMPTON, J. W.: The phylogeny of baboon fibrin-fibrinolysin. Zoologica et Pathologica *48:* 59 (1969).

20 BIGGS, R. and MACFARLANE, R. G.: Human blood coagulation and its disorders (Blackwell, Oxford 1962).

21 WOLF, P.: A modification of Stefanini's method of estimating factor V activity using human oxalated plasma. J. clin. Path. *6:* 34 (1953).

22 DISCHE, F. and BENFIELD, V.: Congenital factor VII deficiency. Haematological and genetic aspects. Acta haemat. *21:* 257 (1959).

23 HARDISTY, R. and MACPHERSON, J.: A one-stage factor VIII (antihaemophilic factor) assay and its use on venous and capillary plasma. Thromb. Diath. haemorrh. *7:* 215 (1962).

24 DENSON, K. W.: The specific assay of Prower-Stuart factor and factor VII. Acta haemat. *25:* 105 (1961).

25 BLIX, S.: The quantitative determination of fibrinolytic inhibitors in plasma or serum by means of the fibrin plate. Scand. J. clin. Lab. Invest. *16:* 403 (1964).

26 DENNIS, L. H.; EICHELBERGER, J. W.; DOENHOFF, A. VON, and CONRAD, M.: A coagulation defect and its treatment with heparin in *Plasmodium knowlesi* malaria in Rhesus monkeys. Milit. med. Supplement 1966.

27 VOLLER, A.; RICHARDS, W. H. G.; HAWKEY, C. M., and RIDLEY, D.: Human malaria *(Plasmodium falciparum)* in owl monkeys *(Aotus trivirgatus)*. J. trop. med. Hyg. *72:* 153 (1969).

28 DENNIS, L. H.; EICHELBERGER, J.; INMAN, M., and CONRAD, M. E.: Depletion of coagulation factors in drug-resistant *Plasmodium falciparum* malaria. Blood *29:* 713 (1967).

Author's address: Dr. CHRISTINE HAWKEY, Zoological Society of London, Regent's Park, *London, N. W. 1* (England)

Pathology of Simian Primates, Part I, pp. 331–341 (Karger, Basel 1972)

Stress

C. W. HILL and C. E. THOMSEN

Tulane University, Delta Regional Primate Research Center, Covington, La.

Contents

The Concept of Stress 331
Stress – Metabolism Relations 332
Clinical Evidence 334
Experimental Evidence 335
Immunological Responses 335
Physiological Disorders 336
Endocrine Responses 337
Metabolic Processes 339
Summary 340
References 340

The Concept of Stress

The term stress has been used in reference to environmental conditions and organismic reactions, both physiological and behavioral. Thus, stress as a noun is best applied to an entire situation, and terms such as stress stimuli or conditions and stress responses or reactions should be used for the various components of the stress situation. Furthermore, the operational definition of a stress situation requires a specification of events on both sides of the environmental-organismic relationship. It is not sufficient to know only that the stimulus conditions are stressful or only that the physiological and/or behavioral responses are stressed.

Stress conditions may be divided relatively discretely into two categories, psychological and physical. Psychological stress conditions produce their unique impact upon the organism by way of the sensory systems, while physical stress conditions rather directly result in shock and trauma. TEICHNER [1968] distinguished between physical and symbolic stimuli, in which the former produce direct physical effects by the force of their energies while the latter act through the central nervous system. Physical stress conditions include a large variety of traumatic events which may be arbitrarily classified in terms of injuries, shock, infections, confinement, and so forth [FIENNES, 1968; 1970]. Psychological stress conditions have been described in terms of novelty, high intensity, surprise, stimulus deprivation, averseness, and conflict [APPLEY and TRUMBULL, 1967; Ch.1]. Informational overload, i.e., more stimuli than the organism can handle at one time, has also been proposed as a cause of stress [TEICHNER, 1968; p. 273]. In this chapter, more emphasis will be given to psychological stress conditions, although physical stress will not be excluded.

Stress conditions in general tend to produce reactions of increased tension and emotion, increased sympathetic nervous system activity, physiological disorders and pathologies, various changes in endocrine functioning, and alterations in certain metabolic processes. More specifically stress reactions have been assessed by means of (a) emotional behaviors such as exaggerated and uncoordinated movements, freezing and immobility, defecation and urination, (b) physiological responses such as heart rate, respiration, and galvanic skin response, and (c) hormonal levels, particularly the steroids and catecholamines [LACEY, 1967; TEICHNER, 1968]. TEICHNER further emphasizes the importance of measuring the compensatory activity within the homeostatic regulating systems of the body, as well as the more peripheral physiological and behavioral responses. As he points out, organismic stress reactions may be completely or partially concealed by the controlling events of regulatory systems.

Stress – Metabolism Relations

The most complete theoretical model for stress situations in general has been developed by SELYE [1956] in which three phases (acute, chronic, and terminal) are proposed: (1) the initial reactions to stress conditions are compensatory adjustments with transitory, sub-clinical effects, (2) persistent stress conditions produce sustained reactions during which compensatory mechanisms gradually break down and irreversible abnormalities develop, and (3) finally there is a state of exhaustion, collapse, or death. FIENNES [1968, 1970] has utilized SELYE's model in descriptions of animal health problems in the wild and in the laboratory, and emphasized the paucity of information available on the physiological processes involved.

The sequence of events from stress conditions to basic metabolic processes is complex and incompletely known, but it may be diagrammed in the following over-simplified manner: stress conditions → physical impact → initial organismic response → hypothalamic and autonomic nervous system activity → hormonal secretions → metabolic alterations. When the stress conditions are psychological, the physical impact can be specified as stress stimuli, the organismic response as sensory system activity, and the involvement of the hypothalamus and autonomic nervous system is clear. These early stages cannot be described so confidently in the case of physical stress conditions. If MASON's suggestion [1968c; pp.798–800] is correct, the physical conditions may also be shown as leading in part to stimuli and consequent sensory system activity. The relatively direct effects of sensory system activity upon the functioning of the endocrine system by way of the hypothalamus and autonomic nervous system are well established as illustrated in figure 1. More recently, the effects of various hormones on metabolic functioning, especially in regard to the integrated action of the entire endocrine system, have been summarized by MASON [1968c; pp.793–795]: (1) each hormone has the capability for influencing a number of different metabolic processes, (2) each me-

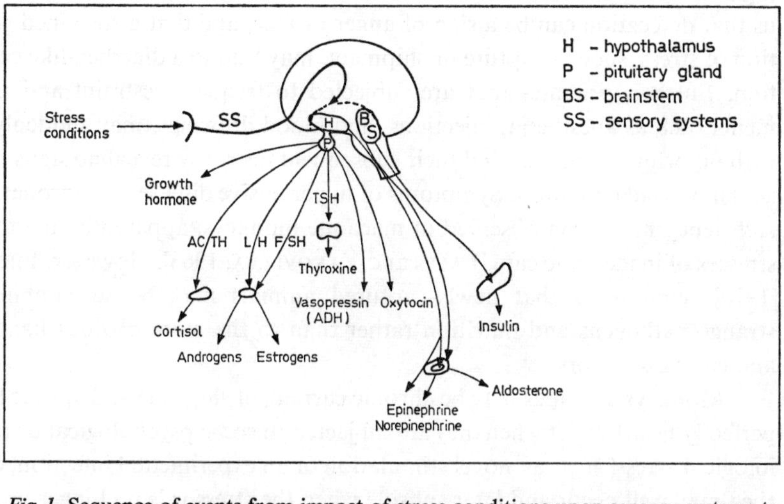

Fig. 1. Sequence of events from impact of stress conditions upon sensory systems to multiple responses within the endocrine system. (From fig. 1 in MASON 1968a)

tabolic process can be affected by several hormones in similar or divergent ways, and (3) the various hormones controlling a particular metabolic process may combine or oppose each others' effects, or their resultant may lie somewhere in between these two extremes.

Although the effects of stress conditions upon metabolic processes have not been studied extensively as yet, considerable evidence has been accumulated, both clinical and experimental, for certain sub-sets of relationships within the total sequence. Representative data will be reported below in support of the hypothesis that stress conditions influence the functioning of the neuroendocrine system and the operation of certain physiological compensating mechanisms. In addition, selected evidence will be presented linking marked changes in hormonal levels, physiological compensating activities, and certain peripheral behaviors with various metabolic processes. The conclusive demonstration of the direct effects of stress conditions upon metabolic processes must await future research.

Clinical Evidence

Many years of clinical experience with primates have shown that the stresses of handling, shipment, and strange housing can contribute to the appearance of symptoms of disease, and that mortality rates are usually at their highest during the first few weeks in quarantine. RUCH [1959; p.76] reminds us that defecation can be a sign of anger or fear, and that a sustained situation of stress, such as capture or shipment, may lead to a diarrhea-like condition. Further, primates that are subjected to frequent restraint and treatments such as anesthesia, injections, and blood drawings, may suddenly die with no prior symptoms, and their subsequent necropsy reveal no signs of an actual morbid condition. Symptoms of hypertensive disease and coronary insufficiency have been observed in macaque monkeys, apparently due to the stresses of inadequate care [LAPIN and YAKOVLEVA, 1963]. However, FIENNES [1970] reminds us that newly-captured animals may be succumbing to strange pathogens and nutrition rather than to the more obvious handling and caging stressors.

Monkeys and apes may be chronic carriers of shigellae and appear to be perfectly healthy; yet when they are subjected to some psychological or physiological stress, such as novel stimulation or an experimental infection, diarrhea may well occur and later subside when the stressor is no longer present [RUCH, 1959; p. 93]. Gastric ulcers, although not frequently found at necropsy, may be induced by the persistent stresses of captivity with its attendant handlings and treatments. Confirming evidence from research studies will be reported later. Latent malarial and other dormant infections may also be activated by the stresses of life in captivity [RUCH, 1959; p. 257].

Experimental Evidence

Research studies relevant to psychological stress situations may be divided into 5 categories based upon the kinds of responses (dependent variables) that were measured: (1) transient changes in physiological and behavioral responses, such as heart rate, respiration, bodily activity, and defecation; (2) immunological responses, e.g., changes in the rate and extent of antibody development; (3) relatively persistent physiological disorders and pathological conditions; (4) significant variations in specific hormonal levels; and (5) alterations in certain metabolic processes. Some studies, of course, have included measurements in more than one category, but none appears to have involved them all.

Although the majority of stress studies fall in the first category, it will not be reviewed because these response measures are not good indicators of persistent metabolic alterations. As BRADY [1964; p. 272] has pointed out, these effects are relatively transitory and apt to be conditioned to specific aspects of the experimental environment. In SELYE's model [1956], the physiological and behavioral responses of category 1 appear in all three stages of the general adaptation cycle, with little differentiation among them. The first stage of initial reactions and homeostatic adjustment would not be likely to result in metabolic derangement. In the second and third stages, however, the persistent high-level activity of the sympathetic nervous system together with its collaborative hormones, followed by the excessive compensatory reactions of the parasympathetic system and its endocrine allies, often produce serious damage or disruption within one or more of the physiological systems under autonomic control.

Immunological Responses

Monkeys as well as animals below the primate level have become either more or less susceptible to a variety of infectious agents when subjected to a number of different psychological stressors. For example, MARSH *et al.* [1963]

found that rhesus monkeys in a SIDMAN avoidance situation were less susceptible than their controls to intravenous injections of polio virus. On the other hand, in two related experiments [FELSENFELD *et al.*, 1966; HILL *et al.*, 1967], psychological stress produced by a variety of stimuli, including noise, light, movements, and disruptions of routine, delayed the onset of antibody development to both *Vibrio cholerae* and bovine serum albumin in African Green Monkeys, reduced the final level attained, and increased the output of cortisol. Although all the factors and their interrelations have still not been completely elucidated, it is quite probable that these divergent results are attributable to variations in the 'pattern and duration of stress exposure, or time of infection in relation to stress exposure' [MARSH, 1964; pp. 499–500].

The alternative ways by which the adrenal cortical steroids might influence antibody formation have been reviewed by MCMASTER and FRANZL [1961]. They concluded that there may be 'either a direct inhibition of antibody protein synthesis or an interference with the effective action of the specific antigenic stimulus required for the production of antibody'. They further speculate that the corticoids may (a) produce changes in lymphoid tissues, (b) increase the proportion of the antibody proteins in the cells at the expense of the amount in the blood, (c) prevent the formation of antigenic stimulus, and/or (d) block the stimulation of antibody synthesizing cells.

Physiological Disorders

BRADY and his colleagues [BRADY, 1963, 1964; BRADY *et al.*, 1958] have demonstrated the influence of certain kinds of psychological stress upon the development of peptic ulcers, together with related steroid and gastric acid levels. When rhesus monkeys were required to press a lever periodically in order to avoid shock, both their steroid and acid levels were elevated when the work and rest periods alternated every six hours, the steroid and acid levels remained high and peptic ulcers were likely to develop within several weeks. However, when the work period was extended to 72 h with four days off, both steroid and acid responses gradually decreased over extensive work periods and no ulcers developed. These findings were supplemented by the work of SMITH *et al.* [1964] in which an inhibition of gastric secretion was found at the beginning of a shock avoidance schedule with rhesus monkeys, followed by some variation around normal levels during the latter stages of a 72 h period. These authors emphasized the lack of relationship between the gastric acid response and the number of shocks received.

The induction of atherosclerosis by psychological stress was demonstrated in squirrel monkeys by LANG [1967]. Over a period of 25 months, 12 squirrel monkeys were transferred from their home cages to smaller restraint boxes for 2 h a day, 5 days a week, and 6 of them were also subjected to shock avoidance training. These experimental animals had higher levels of urinary 17 ketosteroids and serum cholesterol than those in a control group. In addition atherosclerosis was found in the coronary arteries of 9 of these animals whereas there were no such cases in the control group. There were no differences in any of the stress effects between the monkeys which worked on the avoidance task and those which were simply placed in the restraint box. On the other hand, LAPIN *et al.* [1966] concluded, on the basis of the results of others as well as their own studies, that atherosclerosis was extremely difficult to induce experimentally in monkeys, even when certain special psychological stressors such as shock and conflict were added to the routine stressors of captivity.

In an experimental stress situation with baboons, both hypertensive disease and coronary insufficiency developed in some animals, while in others only one of the conditions was produced [LAPIN *et al.*, 1966]. The stressors included (a) unusual social groupings, (b) conflicts between sexual and food incentives, and (c) the disruption of normal physiological rhythms. In a number of cases, the coronary insufficiency eventually ended in myocardial infarction and/or brain hemmorrhage.

Endocrine Responses

Changes in hormonal levels produced by psychological stressors in primates have been summarized and interpreted by MASON [1968b] in an excellent and comprehensive review. In addition to emphasizing the multiple influences of individual hormones and the various patterns of multiple hormone actions upon particular metabolic processes, the existence of a catabolic-anabolic cycle of endocrine responses was elaborated. In a typical sequence, energy-releasing hormones, such as the catecholamines and the corticosteroids, are secreted primarily during the period in which the stressors are active, while those hormones concerned with metabolic synthesis, such as insulin and the estrogens and androgens, increase to their highest levels after the stressors have ceased to function (fig. 2). It seems clear that these two groups of hormones do not respond in a fixed pattern to different stressors. Within each group there is some variation in the timing of the responses following the onset of a stressful stimulus.

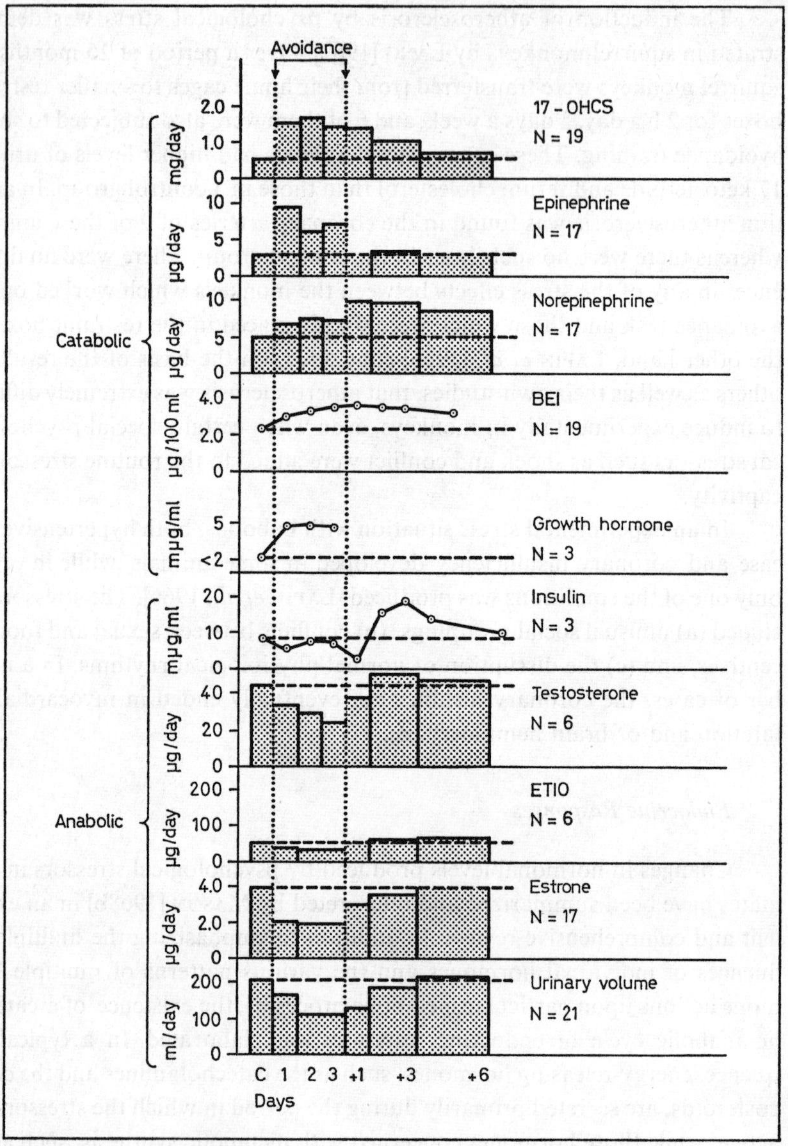

Fig. 2. **Pattern of multiple hormonal responses to stress conditions illustrating biphasic catabolic-anabolic cycle. (From fig. 1 in Mason 1968b)**

The timing and the extent of the various responses may also vary with the kind of stressor. The same general sequence of endocrine events has been found both in the avoidance situation and in the primate chair, yet different patterns of epinephrine-norepinephrine levels depend on whether the stress situation is relatively routine or novel and unpredictable [MASON, 1964; pp. 385–386].

Metabolic Processes

The possible disruption of metabolic processes due to the excessive secretion of adrenocorticosteroids was discussed by MAURICE GOLDSTEIN and others following a paper by BUESCHER [1964, pp. 481–487]. Although the specific processes and sites of action could not be identified, considerable emphasis was placed upon the presence or absence of some requirement that would utilize the product of these activities. Earlier studies of DWIGHT INGLE and NORMAN EPSTEIN were referred to in support of the hypothesis that when psychological stress was accompanied by physiological trauma, such as tissue damage or infection, the stress-produced steroids would be absorbed in the adaptive response to the trauma rather than contributing to metabolic derangement. In other words, there must be a complete integration, both in amount and in timing, of all the elements within SELYE's general adaptation response to stress, peripheral metabolic processes as well as central neural and hormonal activities, in order that the high steroid levels may be utilized in anabolic repair instead of excessive catabolic breakdown.

The metabolic processes of greatest relevance are those concerned with the mobilization and synthesis of proteins, fatty acids, and glucose [MASON, 1968b; pp. 777–779]. The catecholamines, corticosteroids, growth hormone, and thyroxin have in common the catabolic capability for promoting: (a) the release of free fatty acids, (b) hyperglycemia, or (c) both. On the other hand, the estrogens, androgens, and insulin share the anabolic tendency to stimulate protein synthesis and glycogenesis.

High levels of adrenal cortical hormones have been found to inhibit growth, suppress the rate of nitrogen release, and deplete the available protein in many tissues [RENOLD and ASHMORE, pp. 204–207]. The latter effect appears to be clearly the direct product of 'increased protein catabolism... resulting from excess glucocorticoid activity'. Although the specific mechanisms by which other metabolic processes are disrupted have not yet been determined, the available evidence seems to indicate that the corticosteroids enhance physiological responsiveness in general by increasing 'the amino acid pool at the expense of protein' [RENOLD and ASHMORE, 1960; p. 212].

Summary

An attempt has been made to describe the complex biological events, especially those at the metabolic level, produced within an organism at the primate level when placed in a stress situation. The concept of stress was analyzed in terms of stress conditions and reactions, and the functional components of its operational definition were specified. Psychological stress was emphasized throughout, but physical stress was not excluded.

A general outline of the sequence of events leading from stress conditions to metabolic processes has been described, with the endocrine system serving as the principal mediating mechanism. Clinical evidence was presented in support of the general conclusion that psychological stress conditions tend to produce observable symptoms of otherwise latent infections. Experimental evidence was sampled in 4 categories of response measurement: (1) immunological, (2) physiological, (3) hormonal, and (4) metabolic. In addition to emphasizing the multiple influences of individual hormones and the various patterns of multiple hormone actions upon particular metabolic processes, the existence of a catabolic – anabolic cycle of endocrine responses was indicated.

The possible disruption of metabolic processes by excessive secretion of adrenocorticosteroids was suggested. There must be a synchrony of neural, hormonal, and metabolic activity in order that high steroid levels may be absorbed in anabolic synthesis rather than remaining free to contribute to excessive catabolic breakdown. High levels of adrenal cortical hormones inhibit growth, suppress the rate of nitrogen release, and increase the amino acid pool at the expense of available protein.

References

APPLEY, M. H. and TRUMBULL, R.: On the concept of psychological stress; in APPLEY and TRUMBULL Psychological stress, pp. 1–13 (Appleton-Century-Crofts, New York 1967).

BRADY, J. V.: Further comments on the gastrointestinal system and avoidance behavior. Psychol. Rep. *12:* 742 (1963) – Experimental studies of psychophysiological responses to stressful situations; in Symposium on medical aspects of stress in the military climate, pp. 271–289 (Walter Reed Army Institute of Research, Washington 1964).

BRADY, J. V.; PORTER, R. W.; CONRAD, D. G., and MASON, J. W.: Avoidance behavior and the development of gastroduodenal ulcers. J. exp. Anal. Behav. *1:* 69–72 (1958).

BUESCHER, E. L.: Psychophysiological factors in susceptibility and response: A survey; in Symposium on medical aspects of stress in the military climate, pp. 475–480 (Walter Reed Army Institute of Research, Washington 1964).

FELSENFELD, O.; HILL, C. W., and GREER, W. E.: Response of *Cercopithecus aethiops* to cholera vibrio lipopolysaccharide and psychological stress. Proc. roy. Soc. trop. Med. Hyg. *60:* 514–518 (1966).

FIENNES, R. N. T-W-: Ecological concepts of stress in relation to medical conditions in captive wild animals. Proc. roy. Soc. Med. *61:* 161–162 (1968). – Stress in a crowded world. New Scientist *19:* 595–596 (1963). – Ecological concepts of stress in relation to medical conditions in captive wild animals. Proc. Soc. Med. *61:* 161–162 (1968). – Stress and the latent pathogen; in BALNER and BEVERIDGE 'Infections and immunosuppression in subhuman primates'. Proceedings of the International Symposium on Infections and Immunosuppression in Subhuman Primates, Rijswijk 1969 (Munksgaard, Copenhagen 1970).

HILL, C. W.; GREER, W.E., and FELSENFELD, O.: Psychological stress, early response to foreign protein, and blood cortisol in vervets. Psychosom. Med. *29:* 279–283 (1967).

LACEY, J. I.: Somatic response patterning and stress: Some revisions of activation theory; in APPLEY and TRUMBULL Psychological stress, pp. 14–37 (Appleton-Century-Crofts, New York 1967).

LANG, C. M.: Effects of psychic stress on atherosclerosis in the squirrel monkey *(Saimiri sciureus)*. Proc. Soc. exp. Biol., N. Y. *126:* 30–34 (1967).

LAPIN, B. A. and YAKOVLEVA, L. A.: Comparative pathology in monkeys (Thomas, Springfield 1963).

LAPIN, B. A.; YAKOVLEVA, L. A., and CHERKOVICH, G. M.: Use of non-human primates in medical research, especially in the study of cardiovascular pathology and oncology; in R. N. T. W. FIENNES Some recent developments in comparative medicine, pp. 195–212(Academic Press, London 1966).

MCMASTER, P. D. and FRANZL, R. E.: The effects of adrenocortical steroids upon antibody formation. Metabolism *10:* 990–1005 (1961).

MARSH, J. T.: Experimental studies on response to viral infections; in Symposium on medical aspects of stress in the military climate, pp. 489–501 (Walter Reed Army Institute of Research, Washington 1964).

MARSH, J. T.; LAVENDER, J. F.; CHANG, S., and RASMUSSEN, A. F.: Poliomyelitis in monkeys: decreased susceptibility after avoidance stress. Science *140:* 1414–1415 (1963).

MASON, J. W.: Psychoendocrine approaches in stress research; in Symposium on medical aspects of stress in the military climate, pp. 375–417 (Walter Reed Army Institute of Research, Washington 1964). The scope of psychoendocrine research. Psychosom. Med. *30:* 565–575 (1968a). – Organization of the multiple endocrine responses to avoidance in the monkey. Psychosom. Med. *30:* 774–790 (1968b). Overall hormonal balance as a key to endocrine organization. Psychosom. Med. *30:* 791–808 (1968c).

RENOLD, A. E. and ASHMORE, J.: Metabolic effects of adrenal corticosteroids; in WILLIAMS Diabetes, pp. 194–215 (Hoeber, New York 1960).

RUCH, T. C.: Diseases of laboratory primates (Saunders, Philadelphia 1959).

SELYE, H.: The stress of life (McGraw-Hill, New York 1956).

SMITH, G. P.; BOREN, J. J., and MCHUGH, P. R.: The gastric secretory response to acute environmental stress; in Symposium on medical aspects of stress in the military climate pp. 353–365 (Walter Reed Army Institute of Research, Washington 1964).

TEICHNER, W. H.: Interaction of behavioral and physiological stress reactions. Psychol. Rev. *75:* 271–291 (1968).

Authors' address: Dr. CHARLES W. HILL and Dr. CURTIS E. THOMSEN, Tulane University, Delta Regional Primate Research Center, *Covington, LA 70433* (USA)

Pathology of Simian Primates, Part I, pp. 342–379 (Karger, Basel 1972)

The Immunology of Non-Human Primates[1]

R. T. Damian and N. D. Greene

Division of Microbiology and Infectious Diseases, Southwest Foundation for Research and Education, San Antonio, Tex.

Contents

Introduction ... 343
Immunoglobulins ... 343
Major Classes (IgG, IgA, IgM; κ and λ Light-Chains) ... 344
Additional Studies on IgG ... 347
'Minor' Immunoglobulin Classes ... 349
Subclasses and other Forms of Heterogeneity ... 351
IgG Subclasses ... 351
IgM Heterogeneity ... 351
IgA Heterogeneity ... 352
Serum Levels of IgG, IgA, and IgM ... 352
Antibody Production ... 352
Characteristics of the Immune Response ... 352
Range of Immune Competency with Respect to Antigens ... 358
Haptens and Synthetic Polypeptides ... 358
Proteins ... 359
Infectious Agents ... 359
Immunological Ontogeny and Competency ... 359
Transplacental Transfer of Immunoglobulins ... 360
Transfer of Immunoglobulins by Lactating Mothers ... 361
Ontogeny of Immune Competency ... 361
Fetal Studies ... 361
Newborn Studies ... 363
Tolerogenesis ... 364
Natural ... 364

1 This review was aided by USPHS Grant 9-P06-FR00451 and by the United States – Japan Cooperative Medical Science Program administered by the National Institute of Allergy and Infectious Diseases of the National Institutes of Health, Department of Health, Education and Welfare (Grant No. PHS 5 R22 AI-08398).

Induced ... 365
Immediate Hypersensitivity ... 365
Transfer of Human Reaginic Antibodies to Primates ... 366
Induction of Immediate Hypersensitivity in Primates ... 368
Nature of Antibodies Involved in Immediate Hypersensitivity in Primates ... 371
Summary ... 371
References ... 372

Introduction

Despite the great popularity of non-human primates for biomedical research, there is surprisingly little known about these animals relative to basic immunology. However, this situation is rapidly correcting itself and more and more papers on this subject are appearing.

Our purpose in this review is to bring together the more modern knowledge of non-human primate immunology as a point of departure for future studies concerned with various aspects of infectious disease research and immunopathology. We have tried not to slight the very great intrinsic interest of immunology for its own sake, but have pointed out, where appropriate, the contributions and implications of studies on primate immunology to those areas of more general immunological concern.

In the ensuing discussion, we have, for the sake of brevity, equated the word 'primate' with the phrase 'non-human primate'. NAPIER and NAPIER [1] have been freely consulted as a source of common and scientific names.

Immunoglobulins

The intent of the authors in this section is to review studies on the occurrence, heterogeneity, and structure of immunoglobulins in various primates. We are not particularly concerned here with immuno-taxonomic research as it relates to phylogeny, beyond what information it provides toward our purpose. Studies on primate immuno-taxonomy are extensive and have been adequately reviewed elsewhere [2, 3, 4]. The reader is also referred to recent comprehensive compendia for details on immunoglobulin structure [5, 6].

Studies on the nature of primate immunoglobulins must of necessity be based upon knowledge of human immunoglobulins. This is so because human immunoglobulins are the most fully characterized of any species, a consequence not only of our anthropocentrism – but more cogently because of the existence and recognition in man of neoplastic diseases affecting the immunoglobulin synthetic systems. These diseases include multiple myeloma, Waldenstrom's macroglobulinemia, idiopathic monoclonal gammapathies, and the heavy-chain (Franklin's) diseases. The common characteristic among them is the frequent overproduction of highly homogeneous immunoglobulins or immunoglobulin subunits. The existence of these 'abnormal' proteins has allowed for their characterization by the methods of protein chemistry, an approach stubbornly resisted by natural, highly heterogeneous populations of antibodies or immunoglobulins. Also, the discovery of quantitatively minor classes or subclasses of immunoglobulins has been made possible, or at least tremendously facilitated by the availability of the corresponding homogeneous protein (myeloma protein, etc.). A case in point is the discovery of IgD [7].

In mammals other than primates, the establishment of strict homologies with the human varieties of immunoglobulins has often been difficult (for a general discussion of this problem, see [8]). A major molecular feature allowing immunoglobulin identification is antigenicity, i.e. the classes and subclasses are characterized by unique heavy-chain antigenic determinants. Phylogenetic affinity is correlated with antigenic similarity (cross-reactivity) and therefore, within the order Primates, such difficulties are minimized as compared with other orders of mammals. Consequently, the earliest identification of the various human immunoglobulin counterparts in the sera of primates was immunological, with the use of anti-human reagents. As the technique of immunoelectrophoresis [9] has played a major role in the differentiation of human immunoglobulin classes and subclasses, it is not surprising that this technique has also resulted in similar information on primates.

Major Classes (IgG, IgA, IgM; κ and λ Light-Chains)

The most thoroughly systematic immunoelectrophoretic studies have been those of PICARD, HEREMANS and VANDEBROEK [9–12], who utilized not only anti-whole human serum produced in several species of animals, but also anti-cynomolgus *(Macaca fascicularis)* and anti-lemur *(Lemur mongoz)*. These antisera were judiciously chosen since they covered the primate group in such a way that most cross-reactions of interest could be brought out. Homologies were rigorously established by means of analysis with homologous and heterologous antisera, use of the interrupted trough technique and specific, absorbed antisera to individual immunoglobulin classes. A good representation of species from the order was studied, and the results regard-

Table I. Immunoglobulin class occurrence in various primates (10–12)

Species	IgG	IgA	IgM
Chimpanzee *(Pan satyrus)*	+	+	+
Gorilla *(Corilla gorilla)*	+	+	+
Orang-utan *(Pongo pygmeus)*	+	+	+
Cynomolgus monkey *(Macaca fascicularis)*	+	+	+
Vervet monkey *(Cercopithecus aethiops)*	+	+	+
Capuchin monkey *(Cebus apella)*	+	+	+
Marmoset *(Hapale jacchus)*	+	–	+
Lemur *(Lemur mongoz)*	+	–	+
Bushbaby *(Galago* sp.*)*	+	–	+

ing the distribution of the major classes, IgG, IgA, and IgM, are summarized in table I.

More recent studies from FUDENBERG's laboratory [13, 14] have made use of the refinements of modern knowledge concerning immunoglobulin polypeptide chains. These workers produced antisera specific for various heavy-chains and the two major light-chain types (κ and λ) of human immunoglobulins by immunizing and absorbing with appropriate myeloma and other homogeneous immunoglobulins. Their assay system was a sensitive inhibition of precipitation technique using an isolated, radiolabelled reference standard human immunoglobulin and its corresponding specific rabbit antiserum. Dilutions of primate sera were added and the percent inhibition of the homologous system served as the measure of cross-reactivity. Although human and primate immunoglobulins were shown to be closely related, generally paralleling taxonomic relationships, some interesting differences in the degree of cross-reactivity were noted among the various types of chains examined. For example, α-chains (IgA heavy-chains) showed no lower than 82% cross-reactivity. In contrast to PICARD *et al.* [10, 11], they demonstrated IgA in both *Lemur* and *Galago*. γ-chains (IgG) were clustered at from 70–99% cross-reactivity, with two exceptions; galago at 20% and strangely, only 60% for the gibbon. Regarding γ-chains, another author, POULIK, found an antihuman γ-chain serum to cross-react very well with chimpanzee IgG, very weakly with rhesus IgG, and not at all with slow loris IgG by gel diffusion analysis [15]. SHUSTER *et al.* [14] suggested that μ-chains (IgM) have diverged the most in primate evolution. The chimpanzee showed 90% cross-reactivity but the gorilla and orang-utan were at about the 28% level. Gibbons again gave a surprising result, having only 6% cross-reactivity with man, a virtually negative result as it was within the experimental error of the technique. Cercopithecid monkeys ranged from 43–81%, while New World monkeys were not examined. Among the prosimians, the potto showed 9% cross-reactivity and the galago showed 12%. Lemur serum did not inhibit the μ-system, although PICARD *et al.* [10] were able to identify IgM in the lemur.

Light-chain inhibition of radioimmune-precipitation assays [13, 14] showed clustering of cross-reactivity correlated with phylogenetic position. For example, κ-chains from members of the Hominoidea were in the range of 87–100%, and λ-chains were 70–89% cross-reactive with human chains. Among Old World monkeys, χ cross-reactivity was 61–89% and λ was 36–45%. Prosimian κ-chains were 22–28% and λ-chains were 9 and 12% for two species, but 51% for the lemur. POULIK [15], using gel diffusion analysis and anti-human antisera, found good cross-reactivity between human and chimpanzee light-chains, weak cross-reactivity between human and rhesus monkey, and no cross-reactivity between human and slow loris light-chains.

Chemical evidence for the correspondence of human and simian (as well as other mammalian species) light-chains was provided by COHEN and PORTER [16], who analyzed baboon light chains electrophoretically in urea-glycine starch gels. This procedure resolved the light-chains into multiple electrophoretic subfractions, and the baboon proved to be similar to the other species tested (man, bovine, horse, guinea-pig, and rabbit) both in number and mobilities of the subfractions. HOOD *et al.* [17], in a comparative analysis of mammalian and avian light-chains based upon chemical κ and λ typing stated, 'all animals have light-chains analogous to L (λ) and/or $\underline{k}$ (κ) classes, though the relative amounts vary widely'. To illustrate the latter point, they found the horse to be completely devoid of κ-chains. Among the mammals tested were man, baboon, and 'monkey'. These three species have similar light-chains, as based upon peptide mapping techniques, and furthermore, their κ:λ ratios are similar. From these results [17], we would expect all primates to possess κ and λ chains as part of their immunoglobulin structure, although an occasional species might be found to be deficient in one or the other class.

In summary, it is clear that all the main groups of primates possess the major types of immunoglobulin polypetide chains, γ, a, μ, κ and λ. The Tupaioidea have not been studied with respect to their immunoglobulins. There are other, smaller gaps in knowledge. For example, gibbons have not been shown to possess IgM, at least IgM cross-reacting with human IgM. IgA has yet to be demonstrated in any member of the New World family Callitrichidae.

It is obvious from the foregoing that immunoglobulin cross-reactivity determinations and identification are strongly influenced by the nature of the antiserum used, and to a lesser degree by the sensitivity of the immunological tests employed. If an immunoglobulin class cannot be identified with an anti-human antiserum, success can probably be had by using an antiserum made to the corresponding immunoglobulin of a lower primate.

Additional Studies on IgG

Structurally, IgG is the best known of the major immunoglobulin classes. The pioneering studies of PORTER and of EDELMAN on controlled immunoglobulin degradation leading to the multi-chain structure concept were performed on rabbit and human IgG, and a human IgG was the first to have its primary structure (amino acid sequence) completely elucidated.

We consider here some studies that go beyond the identification of IgG in one or another primate, but rather provide structural information and additional comparative data.

Papain digestion of IgG [18] has yielded the classical fragments Fc and Fab in various species: chimpanzee [15], gibbon [19], rhesus monkey [15, 20], baboon [19. 21], and slow loris [15]. These fragments were identified immunologically or electrophoretically. POULIK [15] found rhesus Fc and Fab to cross-react but weakly with anti-human fragment specific antisera, and slow loris Fc and Fab gave no cross-reaction. DAMIAN *et al.* [21], using specific anti-baboon IgG antiserum, identified Fc and Fab in baboon IgG papain digests by means of immunoelectrophoresis. They isolated baboon Fc and Fab by chromatography on DEAE-cellulose following methods developed for human fragment isolation.

Besides Fc and Fab, another fragment termed F′c and known to be a product of further papain action on Fc [22] has been recognized in primate IgG. POULIK [15, 22] found F′c immunologically (anti-human F′c) and electrophoretically (starch gel) in chimpanzee IgG papain digest. Chimpanzee F′c is electrophoretically slower than human F′c at pH 8.6. Neither rhesus nor slow loris F′c reacted with the antihuman F′c in immunoelectrophoresis, although starch gel electrophoresis revealed F′c fragments in the papain digestion mixtures of the IgG of both species. F′c was recognized in the papain digestion mixture of baboon IgG by an anti-baboon IgG serum [21]. MALLEY and MACKLER [20] used the appearance of F′c as an indicator of the rate of papain digestion of specifically purified rhesus monkey antihapten (trinitrophenol) antibody which was of the IgG class. The F′c was identified by immunoelectrophoresis and sedimentation velocity (2S) in this study, and was found to increase in quantity with digestion times in excess of two hours.

POULIK [15] carried the analysis of papain IgG digestion products further with his technique of two-dimensional electrophoresis in starch gel containing urea and mercaptoethanol, a technique designed to facilitate comparison of related proteins by peptide mapping. Differences in human, chimpanzee, rhesus, and slow loris IgG papain digests, scarcely noticeable on conventional starch gel electrophoresis, were accentuated. It was evident that human and chimpanzee IgG are related, and that rhesus and slow loris IgG are very different from them, as well as being very different from each other by this technique.

It would appear that further insight into the structure of primate IgG can be gained through the use of naturally occurring antiglobulin factors. For example, a rheumatoid factor reacted with both rhesus and human IgG, forming similar amounts of 22S complex [23]. The reactivity with this particular rheumatoid factor was thought to be related to common sensitivity to papain of human and rhesus IgG, since papain-resistant IgG of other species showed low or no reactivity. Another report [24] was concerned with the reactivity for the IgG of various representative primates of a Waldenström macroglobulin antibody specific for the Fc portion of human IgG. This antibody was not able to distinguish between the IgG from the Hominoidea and the IgG from the Cercopithecoidea, and was unreactive with IgG from the Ceboidea and the Lemuriformes. It did react with IgG from the lorisiform species *Potto potto* and *Galago crassicaudatus*. This is interesting in relation to the report of SHUSTER *et al.* [14], mentioned above, in which low cross-reactivity (20%) between human and galago γ-chains was found, and serves to underscore the importance of the nature of the reagent used to investigate immunoglobulin relationships. Yet another factor which reacts with human IgG Fc is staphylococcal protein A, and a wide variety of animals' sera or isolated IgG was screened for reactivity with it [25]. All mammals tested, including a good series of primates, gave positive reactions as evidenced either by direct precipitation or by the inhibition of precipitation between protein A and a human IgG myeloma protein. Among the primates, *Tupaia glis* serum reacted by inhibition, whereas all others, including the lemur, galago, New World and Old World monkeys, and great apes reacted by precipitation with protein A. The authors suggested that the difference between inhibition and precipitation could be related to slightly differing specificities. This particular protein A reactive determinant was not related to determinants reactive with most of the rheumatoid factors which they tested, although one rheumatoid factor did have a reactivity which paralleled reactivity with protein A. The value of protein A in this regard seems limited in studies within the order Primates as compared to studies with a broader phylogenetic scope.

WANG *et al.* [26, 27] have performed some interesting studies on certain γ-chain peptides which are related to a serological specificity of the human IgGl molecules, the Gm (a) factor. It is known that IgGl molecules are either Gm(a +) or Gm(a–), and that IgG molecules of the other three subclasses are all Gm(a–). A unique peptide, the Gm 'a' peptide, could always be demonstrated in the tryptic peptide maps of degraded Gm(a+) immunoglobulins whereas another unique peptide, Gm 'non-a', appeared in similar preparations from Gm(a–) immunoglobulins. The above authors studied the phylogenetic distribution of the 'a' and 'non-a' peptides within the primates and found that chimpanzees and gorillas, which are Gm(a+) show the 'a' peptide. Gm(a–) primates, the rhesus monkey and baboon lacked the 'non-a' peptide and instead possessed yet another, designated the Old World monkey (OWM) peptide. WANG *et al.* [26, 27] postulated that both the 'a' and 'non-a' peptides arose from the OWM peptide by single-step point mutations, and that these peptides ('a' and 'non-a') appeared after the Hominoidea and Cercopithecoidea had separated as independent lines of evolution. These evolutionary considerations formed part of the basis upon which WANG and FUDENBERG [27] devised a test for the single locus [28] versus gene complex [29] hypotheses of the genetic control of γ-chain synthesis. Although the details are beyond the scope of this review, it should be mentioned that the latter hypothesis found support in this study.

'Minor' Immunoglobulin Classes

The first indications that primates possess an immunoglobulin homologous with human IgE emerged from studies on the passive transfer of human reaginic sera to various primates (for details, see section on transfer of human reaginic antibodies to primates).

Later, it was found that intradermal injection of antiserum specific for the Fc portion of human IgE into *Macaca irus (fascicularis)* resulted in the extravasation of previously injected Evans blue dye [30]. The amount of anti-IgE required for positive reaction was extremely low, in the range of 10^{-3} to 3×10^{-4} μg N/ml. This reversed type of skin reaction was interpreted by the authors to be due to the presence on target cells of an immunoglobulin which reacts with anti-human IgE antibody, releasing the mediators of local anaphylaxis. The reaction could be prevented by prior treatment of the anti-human IgE serum with either the serum of ragweed sensitive humans or normal monkey serum. When normal monkey serum or serum from monkeys hypersensitive to *Ascaris* antigen was fractionated by ion-exchange chromatography and gel filtration, the component responsible for removing the anti-IgE activity had the same elution characteristics as human IgE [30, 31].

Further proof of the existence of monkey IgE was provided by the isolation of a reagin-rich fraction from monkeys *(Macaca irus)* hypersensitive to *Ascaris* or infected with parasitic mites [31]. The protein reacted with anti-light chain antiserum showing that it was in fact, an immunoglobulin. It also gave a visible precipitin arc against anti-human IgE in gel diffusion analysis, and human IgE spurred over the monkey protein when both were set against the anti-human IgE, indicating partial identity between the two proteins. More evidence that the monkey and human protein are not identical was provided when rhesus monkeys responded to antigenic stimulation with human reagin-rich serum fractions or an IgE myeloma protein with production of specific anti-human IgE antibodies [32]. These antibodies exhibited functional activities associated with anti-IgE produced in other animals such as rabbits or guinea pigs, i.e. absorption of reaginic activity from hypersensitive human serum and production of wheal and flare reactions in human skin.

Physiochemically, the *Macaca* IgE is very similar to that of man [31]. Characterisitics examined included electrophoretic mobility, behavior upon gel-filtration and ion-exchange chromatography, and sedimentation velocity (which was 8.0S). For a full discussion of the characteristics of human IgE, the reader is referred to the review by ISHIZAKA and ISHIZAKA [33].

In addition to *Macaca* species, the baboon *(Papio cynocephalus)* also appears to possess IgE, as has been demonstrated by production of classical wheal and flare reactions upon intradermal injection of anti-human IgE serum [21].

From the suitability of a wide variety of primates including prosimians as recipients of passively transferred human reaginic sera in skin tests, reviewed in detail in a following section, it appears likely that the whole order possesses immunoglobulins homologous to the IgE of man.

IgD has not yet been reported from primates, and its existence in these animals is unproved, although we have seen a precipitin line in gel diffusion tests between normal baboon serum and a commercially prepared anti-human IgD serum (unpublished).

Low molecular weight antibodies, representing either a new class of immunoglobulins or intermediates in the synthesis or degradation of the established classes, have been identified in rhesus monkey antisera prepared against *Escherichia coli* endotoxin [34]. These antibodies were detected by a bentonite flocculation test, suggesting multivalency. Their molecular weight was estimated to be slightly larger than that of lysozyme (14,000 to 18,000) and the sedimentation constant was less than 2S. They migrated in the slow α to fast β regions in electrophoresis at pH 8.6. Human antibodies of this type were co-precipitated with anti-IgM antisera, suggesting a sharing of antigenic determinants with IgM. However, this was not examined for the monkey.

Subclasses and other Forms of Heterogeneity

IgG Subclasses

That the gamma-globulin of primates is antigenically heterogeneous was clearly shown by PICARD *et al.* [11], who found multiple gamma lines in immunoelectrophoretic analysis of primate sera developed by anti-cynomolgus sera.

It is now well established that the IgG of man is separable into four subclasses by heavy-chain differences [35, 36]. Several workers, using antisera specific for the various human subclasses have demonstrated their counterparts in the sera of chimpanzees, gorillas, and orang-utans [13, 37, 38]. However, gibbons and all lower primates were found to lack the particular antigenic determinants differentiating the four human subclasses. Baboons *(Papio cynocephalus)* were shown to also possess four subclasses of IgG, based upon γ-chain differences [21]. The antigenic determinants permitting their recognition were further localized on the Fc fragment of the heavy-chains. Absorption experiments with human globulin showed that they were not shared with human IgG. Further studies [39] demonstrated that other cercopithecid monkeys including the mandrill *(P. sphinx)*, the gelada *(Theropithecus gelada)*, and the talapoin *(Cercopithecus talapoin)* all share IgG subclass determinants with the baboon. Similar determinants in prosimians and ceboids could not be found, suggesting that these species either lack IgG subclasses altogether or more probably possess their own unique determinants; distinctive from those of man and the great apes on the one hand, and cercopithecid monkeys on the other. Gibbons displayed a cercopithecid affinity, sharing certain of these determinants with the baboon.

IgM Heterogeneity

LAKIN *et al.* [40] made the interesting discovery of two distinct types of IgM in the rhesus monkey immunized with bovine serum albumin, one characteristic of the primary anti-protein response and termed IgM-p and the other found in hyperimmune sera (secondary response) and termed IgM-h. Passive hemagglutinating activity in the former was sensitive to reduction by 2-mercaptoethanol (2-ME) whereas the hemagglutinating ability of IgM-h (as well as IgG) was not diminished after 2-ME reduction. This was in contrast to the findings of SIBAL *et al.* [41] that both primary and secondary response (to purified Rauscher murine leukemia virus) IgM in the rhesus monkey is sensitive to reduction by 2-ME. LAKIN *et al.* [40] were able to establish that this treatment did, in fact, result in the break-up of the IgM-h molecule into five 8.0S subunits and that these subunits retained hemagglutinating ability, which provided confirmation for the presence of ten or, at least, potentially ten combining sites on the native IgM molecule.

IgA Heterogeneity

Secretory IgA has not yet been positively identified in primates. However, an IgA sharing certain physiochemical characteristics with human secretory IgA and isolated following procedures devised for the isolation of human secretory IgA was obtained from baboon colostrum [21]. As the presence of secretory piece has not yet been shown for this baboon IgA, its relationship to secretory IgA remains speculative.

Serum Levels of IgG, IgA, and IgM

Scattered data are available in the literature on normal levels of the three major immunoglobulin classes in the plasma or serum of primates. In many cases quantitation has been by means of radial immunodiffusion [42] and the exact nature of the specific antiserum used has not been made clear. Presumably, in most cases the values were obtained with the use of anti-human immunoglobulin class-specific sera, which at best would appear to give but a rough estimate of the true values, in view of the above discussion on human and primate immunoglobulin cross-reactivity. The data are brought together in table II.

Antibody Production

Characteristics of the Immune Response

LAKIN *et al.* [41, 51–53] have broadly studied antibody characteristics and responses of rhesus monkeys serially immunized with heterologous serum proteins (bovine serum albumin, BSA and bovine gamma-globulin,

Table II. Normal serum immunoglobulin levels in adult primates, including man[1]

Species	Number of subjects	Method of quantitation	IgG	IgA	IgM	Reference
Chimpanzee	64	Zinc turbidity	6.4–13.8	–	–	37
Chimpanzee	55	Radial immunodiffusion[2]	8–13	0.8–2.9	0.2–0.8	38
Orang-utan	9	Radial immunodiffusion[2]	10.5–13.3	0.9–1.9	0.2–0.9	38
Baboon	25	Radial immunodiffusion[2]	11–14	1.1–4.1	0.3–1.8	38
Gelada baboon	6	Sharpless and LoGrippo immunodiffusion	7.9–8.6	1.0–1.3	2.2–3.0	43
Rhesus monkey	13	Radial immunodiffusion[2]	12–30	0.3–0.9	0.8–2.0	44
Stump-tailed macaque	4	Radial immunodiffusion[2]	8.5–11.1	3.1–4.3	1.0–1.2	45
Vervet monkey	9	Oudin tube with anti-monkey and anti-human globulins	10.0–10.2	4.0–4.2	1.4–1.6	46
Patas monkey	22	Method not stated	9.2–10.8	1.2–2.6	0.9–1.1	47
Patas monkey	20	Radial immunodiffusion[3]	8.8–9.3[4]	0.6–0.8[4]	1.0–1.5[4]	48
Human	30	Radial immunodiffusion	8.5–14.6[4]	1.4–2.6[4]	0.7–1.3[4]	49
Human	30	Radial immunodiffusion	7.7–13.5[4]	1.0–2.8[4]	0.6–1.5[4]	50

1 All values converted to mg/ml, to nearest 0.1 mg.
2 Nature of antisera unknown (anti-human or anti-monkey).
3 Anti-human globulins used.
4 Range includes ±1 SD from mean.

BGG) and the dinitrophenyl (DNP) antigenic determinant. Monkeys immunized with DNP_{17}/-BSA had made precipitating antibodies to both BSA determinants and the DNP determinant when examined nine weeks post-immunization. Using radio immunoelectrophoresis, they localized anti-BSA antibody in two types of IgG (suggested by them to represent κ and λ classes), in IgA, and in IgM. Activity in IgM was shown only after isolation and concentration of the macroglobulin fraction. Similarly, anti-DNP antibody was demonstrated to reside in IgG and IgM. IgA antibody was not detected by the technique. Anti-BSA IgM was also detected in hyperimmunized monkeys. Some special characteristics of rhesus monkey hyperimmune IgM are discussed in a previous section.

Quantitative precipitin analyses of the rhesus BSA-anti-BSA and DNP-anti-DNP systems (nine week post-immunization sera) revealed additional antibody characteristics. LAKIN *et al.* [51] obtained classical type precipitin curves for these two systems: the BSA system being characterized by a narrow equivalence zone, the DNP system having a broader zone of equivalence. The antibody:antigen combining ratios was 7:1 to 9:1 on a weight basis, or 3:1 to 4:1 on a molar basis for BSA-anti-BSA and the amount of precipitating antibody in the sera of two monkeys was 273 and 480 μg antibody N/ml. In the DNP system the combining ratio of antibody to antigen was 3:2 on both weight and molar bases, and the amount of precipitating antibody measured in two monkeys' antisera was 172 and 250 μg antibody N/ml. The absolute level of anti-DNP antibodies was found to diminish with time, despite the administration of more immunogen.

These authors [51] also utilized the direct radioactive antigen precipitin (P_{80}) technique of TALMAGE and MAURER [54] to study further rhesus antisera. Among their more interesting results was the finding of soluble complexes in the zone of antibody excess, formed in both nine week post-immunization and later hyperimmune anti-DNP-BSA sera. These complexes were demonstrated by co-precipitation with a rabbit anti-rhesus antiglobulin serum in the P_{80} test. Although the presence of non-precipitating antibodies was considered as a possible reason for the formation of soluble complexes, the authors favored physicochemical factors influencing the combination of DNP-BGG and potentially precipitating antibodies as being responsible for the phenomenon.

Later LAKIN *et al.* [52] purified IgG and IgM from rhesus monkeys immunized with BSA and compared the relative efficiencies of antibodies in these two classes in various *in vivo* and *in vitro* immunological reactions. They studied passive hemagglutinating, precipitating, and skin sensitizing properties. IgG was purified by DEAE-chromatography and IgM by a combination of DEAE-chromatography and gel-filtration. The basis for their comparisons was the measurement of BSA binding capacity of these preparations by the method of FARR [55]. Normalization of precipitin (P_{80}) and hemagglutination values was accomplished by dividing these values by the maximum BSA-binding activity of each immunoglobulin fraction. In this way, the authors concluded that rhesus IgM was 23 times more efficient than IgG in passive hemagglutination, whereas it had less than 2.3% of the precipitating capacity of rhesus IgG. These values are in line with relative agglutinating and precipitating efficiencies of IgM and IgG as reported for other species [56]. Rhesus IgG was capable of fixing to guinea-pig skin in passive cutaneous anaphylaxis (PCA) and IgM was not active in this respect.

Anti-protein responses of the rhesus monkey were studied in greater depth by means of an $^{131\text{-}125}$I double-label technique [53]. Monkeys were injected intravenously with a mixture of ^{125}I-BSA and ^{131}I-BGG. An initial injection of this material was followed after 57 days by a second injection containing a tenfold increase over the first in protein content for both BSA and BGG. Blood samples were taken periodically after antigen administration, and several aspects of the anti-protein responses were examined; including the kinetics of antigen elimination (blood clearance) in both non-immune and immune phases, the immunoglobulin class characteristics of the synthesized antibodies, and the temporal relationships of antibody concentrations. A wide range of responses to both foreign proteins was observed in the four immunized monkeys. Regarding the anti-BSA responses, one monkey failed to make a primary response which was recognizable by any of the following methods: measurement of immune elimination, passive hemagglutination, ammonium sulfate co-precipitation (technique of FARR [55] modified), or radio immunoelectrophoresis. Upon reinjection with BSA this monkey responded in a fashion strikingly similar to the primary responses of the other three monkeys in the series, suggesting unresponsiveness to the first injection of antigen followed by a primary response to the second. In those three monkeys demonstrating primary responses to the first injection of antigen the onset of the immune elimination phase averaged around 13 days, in contrast to a five-day latent period for two monkeys showing a secondary response. IgG antibody only was detectable by radioimmunoelectrophoresis in two of the three and both IgG and IgM antibodies were detected in the third. In this case, the IgM antibody was transient in appearance and paralleled the IgG antibody. The monkey giving a 'primary' response to the second antigen injection also formed antibodies of the IgG class only. The immunoglobulin class identification of secondary response antibodies was made in a similar fashion: one monkey had only IgG antibodies and a second had IgG antibodies along with other antibodies presumed but not proved to reside in the IgA fraction.

Another feature of anti-BSA responses in the rhesus monkey which emerged from this study [53] was a general lack of correlation between the amounts of hemagglutinating antibody measured and the total antigen binding capacity as determined by the FARR technique. Since IgG was the predominant species of antibody produced, this discrepancy led the authors to postulate an IgG intra-class heterogeneity although whether they were referring to a structural subclass-like heterogeneity or to what has been termed the 'maturation of the immune response' [57] is unclear from their discussion. Finally, these authors found both primary and secondary anti-BSA responses to be biphasic, usually showing cyclic passive hemagglutination titer levels with two waves of activity, the first lasting from six to ten days and the second from seven to eighteen days.

Anti-BGG responses in these monkeys were generally similar to their anti-BSA responses, both primary and secondary. The half-life of radiolabelled BGG in the circulation during the non-immune phase was lower than that for radiolabelled BSA: 2.7 ± 0.6 days as opposed to 7.5 ± 0.7 days. Circulating antibody in the primary response was first detected around 13 days after antigenic stimulation, and the latent period in the appearance of detectable secondary response antibodies averaged around eight days. As with the anti-BSA antibodies, IgG was the predominant class of primary and secondary response anti-BGG antibodies. The lone exception was the same monkey showing a transient IgM primary antibody response to BSA. A similar transient IgM anti-BGG antibody response was noted in this animal, which also went along with its primary IgG anti-BGG response. Again, hemagglutinating antibodies directed to BGG showed biphasic titration curves similar to those obtained in measuring anti-BSA antibodies by passive hemagglutination.

These extensive studies by Lakin *et al.* on rhesus monkey immune responses to heterologous serum proteins and their haptenized (DNP) derivatives may be briefly summarized as follows: Antibody activity appeared in IgG (which seemed heterogeneous), IgA and IgM. Primary responses were most often confined to IgG, although one monkey showed a mixed IgG-IgM response, *pari passu*, to two different protein antigens. IgM antibodies were also being synthesized in the hyperimmune state, and these differed physicochemically from early IgM antibodies (see IgM heterogeneity). Classical quantitative precipitin curves for anti-BSA and anti-DNP antibodies were obtained, with the DNP-anti-DNP system showing a broader equivalence zone than the other. Soluble antigen-antibody complexes were demonstrated in the zone of antibody excess for the DNP system. IgM antibodies were more efficient than IgG antibodies in hemagglutination and the reverse was true for their relative efficiency in precipitation. The secondary response was characterized by a shorter latent period and a higher antibody level than the primary response and both responses exhibited cyclic waves of antibody activity. Most of these results indicate that the rhesus monkey is not remarkably different from other, more intensively studied animals such as the rabbit and guinea-pig [58] in its antibody characteristics and responses. However, although LAKIN *et al.* [40] reported upon physiochemical comparisons of primary response IgM (details of its appearance with regard to their immunization schedule were not given) and hyperimmune IgM antibodies; in their later, more detailed studies of primary and secondary responses [53], there seemed to be a notable lack of an early IgM antibody response [cf. 56, 59].

Other authors have reported a more consistent appearance of IgM antibodies in the primary response to various antigens. For example, SIBAL *et al.* [41], in their study of the antibody responses of ten adult rhesus monkeys immunized with a purified virus (Rauscher murine leukemia virus) found an initial 19S antibody response followed by 7S antibody. After secondary stimulation, the predominant antibody was 7S but was accompanied by a transient, minimal 19S response. They used sucrose density gradient ultracentrifugation, passive hemagglutination, and micro-immunodiffusion methods for the isolation and assay of antibodies of the two groups. Both primary response and secondary response 19S antibodies were sensitive to 2-mercaptoethanol [cf. 40]. Similar results on the early appearance of 2-ME sensitive 19S antibodies in rhesus monkeys experimentally infected with coxsackievirus were obtained by SCHMIDT *et al.* [60]. Another study of anti-viral (mumps) responses in the rhesus monkey showed 2-ME sensitive 19S antibody to be present in the sera of all of four monkeys a short time after primary mumps infection [61]. SILVERSTEIN *et al.* [62] also studied the early primary immune response of the rhesus monkey, in this case to sheep erythrocytes. Their work is discussed in more detail in the section on fetal studies, but certain aspects are relevant here. Using the hemolytic plaque assay for antibody producing cells, they showed a peak reactivity of IgM antibody-forming cells in both fetal and adult animals six days after primary stimulus.

It would appear from these reports that the rhesus monkey is not generally deficient in its ability to form IgM antibodies during the primary immune response. The paucity of IgM primary responses found by Lakin *et al.* [53] therefore seems to be a peculiarity either to the particular antigens which they used or else to some other aspects of their system.

Felsenfeld *et al.* have published extensively on immunological studies after cholera and shigella immunizations in several species of primates, including patas monkeys [47, 48, 63], vervet monkeys [46, 64], stumptailed macaques [45], gelada baboons [43], and chimpanzees [63, 65, 66]. Their original reports should be studied for details on serum and tissue immunoglobulin and antibody content and immunoglobulin producing cell distribution following immunization by various routes.

Range of Immune Competency with Respect to Antigens

There are many reports in the literature of antibodies in primates to a great variety of antigens, introduced either naturally such as by infection or stimulated artificially by investigators for various purposes. An exhaustive review is not intended but it would be useful to give an indication of the diversity of antigenic stimuli to which primates have been shown to respond by making specific antibodies.

We have taken the liberty of introducing along with the discussion of certain antigens, additional miscellaneous information of immunological interest which would have been difficult to incorporate elsewhere.

Haptens and Synthetic Polypeptides

Rhesus monkey antibodies produced to the dinitrophenyl [51] and trinitrophenyl [20] antigenic determinants have been mentioned previously.

Gill *et al.* have intensively examined antibody responses of various animals to synthetic polypeptide antigens, and have included primates in several of these studies [67–69].

In *Cebus albifrons*, it was shown [67] that poly $Glu^{15}Lys^{33}Tyr^{16}$ was strongly immunogenic; in fact, eliciting a greater antibody level (600 μg antibody N/ml) than in rabbits (150 μg antibody N/ml). Furthermore, once established, the antibody response in the monkeys was maintained at a relatively high, stable level for the entire study period of 315 days. After the antibody response had reached stable levels, the effects of lysine and total protein deficiencies on antibody production were ascertained. The antibody response was found to be quite resistant, even while the animals were losing weight because of the nutritionally inadequate diets.

Gill and Gershoff [68] later successfully immunized *Cebus albifrons* monkeys with another synthetic polypeptide, poly $Glu^{52}Lys^{33}Tyr^{15}$. These immunizations actually formed the basis for studies on certain nutritional aspects of antibody production, in this case the effects of methionine deficiency and excess along with its antagonist, ethionine. This paper also yields interesting information on the secondary antibody response. The monkeys were found to fall into two well-defined groups regarding the time of maximal antibody response after secondary stimulation, the 'early responders' having their peak at 11 days and the 'late responders' peaking at 22 days. The maximum amounts of antibody produced by both groups were similar, being approximately 600 μg antibody N/ml.

A synthetic macromolecule composed entirely of D-amino acids, poly D-Glu^{55}D-Lys^{39}D-Tyr^{6} was also immunogenic in *Cebus albifrons* [69]. However, in this case immune paralysis was shown to occur after repeated antigen injections, both in the monkey and in rabbits.

The primate model developed by Gill *et al.* appears to have great potential usefulness for studying the important problems of the molecular bases of antigenicity [70] and nutritional effects on the development of immunity in a system closer to man.

Proteins

Antiprotein responses have been reviewed in the section on characteristics of the immune response. Hill *et al.* [71] have reported inhibitory effects of psychological stress on the development of anti-BSA titers in vervet monkeys.

Infectious Agents

Antibody responses against certain viruses were discussed previously (pp. 352–358). The development of antiviral antibodies is discussed by Kalter in part II. Other infectious organisms shown to be capable of stimulating antibodies in primates include bacteria [47, 72], mycoplasmas [73], fungi [74], chlamydiae [75], and parasites [76, 77].

Immunological Ontogeny and Competency

In this section we consider together several areas of primate immunology which are broadly concerned with developmental aspects. Three of these areas, passive transfer of immunoglobulins from mother to offspring, ontogenesis of immune capabilities, and the special case of marmoset chimerism are included on obvious grounds. The fourth, tolerogenesis in immunocompetent animals, for which but few data on primates are available, is included here on more tenuous grounds, although developmental overtones on the cellular level [78] justify its inclusion.

Transplacental Transfer of Immunoglobulins

As BRAMBELL pointed out in his review of the subject [79], the young mammal may receive antibodies from the mother either before or after birth, or both, depending upon the species. That human newborns have demonstrable immunity, that circulating antibody levels in newborns could be comparable to maternal levels, that humans have a hemochorial placenta, which by its structure was presumed to facilitate transplacental passage of molecules, and that the newborn infant does not absorb antibodies from colostrum [80] all led to the assumption that passive immunization in primates is the result of transplacental immunoglobulin passage. However, this assumption ignored the possibility of transfer via the gut of the fetus as a result of amniotic fluid swallowing, or even transfer via the naso-pharynx and respiratory tract.

The proof of transplacental passage of immunoglobulins in primates appeared in the work of BANGHAM *et al.* [81, 82] on the rhesus monkey. They labelled maternal serum proteins with ^{131}I and reinjected them into the circulation of pregnant monkeys. Proteins in samples of maternal and fetal serum and of amniotic fluid, taken 8 and 24 h after injection, were fractionated and checked for radioactivity. The fetal serum contained labelled γ-globulin and a small amount of labelled albumin, suggesting selectivity of transfer. Albumin alone reached the amniotic fluid. Most of the labelled globulin was in the γ-fraction, with little in the α- or β-globulin fractions. The role of amniotic fluid and fetal swallowing was examined by injecting labelled serum proteins directly into the amniotic cavity. Although labelled proteins appeared in the fetal gut, they did not reach the circulation. Thus, the route of γ-globulin and albumin transfer from mother to fetus was shown to be transplacental and selective. Later, BANGHAM *et al.* [83], using a technique of indwelling cannulation of the fetus, demonstrated that the exchange of γ-globulin and albumin also occurred transplacentally in the opposite direction, i.e. from fetus to mother. Again, the transfer was selective, and the amniotic fluid was not involved. QUINLIVAN [84] later confirmed these results, with the added refinement that he labelled isolated γ-globulin instead of whole serum, thereby eliminating the possibility that the results of BANGHAM were due to transplacental passage of free ^{131}I followed by re-iodination in the fetal circulation of serum proteins occurring in the same proportion as the original preparation of iodinated serum proteins.

In man, it is well established that transplacental passage of immunoglobulins occurs for the IgG, but not IgA, IgM, IgD, and IgE classes [85, 86].

Transfer of Immunoglobulins by Lactating Mothers

QUINLIVAN, in his study of immunoglobulin transfer between mother and offspring in the rhesus monkey [84] also injected ^{131}I-labelled γ-globulin into lactating mothers. Labelled γ-globulin was not recovered from the blood of the nursing infants. However, the three lactating females which were injected had been delivered of their infants 3–12 days previous to the injections which could conceivably have exceeded the period of time in which newborn monkeys might be capable of absorbing immunoglobulins through their guts [79]. Also, QUINLIVAN did not look for radiolabelled γ-globulin in the milk of these mothers.

Ontogeny of Immune Competency

Fetal Studies

BANGHAM *et al.* [87] presented evidence for the competency of the fetal rhesus monkey to reject allografts in as early as the ninth week of gestation. They injected fetuses *in utero* between the ninth and thirteenth week of gestation (of a normal gestation time of approximately 160 days) with nucleated allogeneic adult bone marrow cells in doses previously shown to be capable of inducing secondary disease (graft versus host reaction) in irradiated adult recipients. Fetuses which survived the procedure to term were not runted. The absence of secondary disease was confirmed by challenge at 5–30 days after birth with thoracic duct lymphocytes and/or spleen cells obtained from the original marrow cell donors. Again, no secondary disease was evident, supporting the conclusion that the fetal rhesus monkey is immunologically competent to reject allografts by the ninth week of gestation.

Fetal immune competency to a protein antigen was also established for the rhesus monkey by the studies of COTES *et al.* [88]. They gave intramuscular injections of bovine serum albumin (BSA) to 10–16 week fetuses *in utero*, and tested for antibody content in fetal serum by the antigen-binding capacity test of FARR [55] or by measuring the rate of elimination from the circulation of a tracer dose of ^{131}I-BSA in newborn animals. Fetuses injected with 2 mg BSA in Freund's complete adjuvant on the 66th through 87th day of gestation were tested and found positive for anti-BSA antibodies six weeks after injection, showing that the fetuses were able to form anti-BSA antibodies at some time in the period between 66 days and 129 days of gestation (the period between earliest injection and testing). Another fetal monkey, when injected at around day 70 with 2 mg of BSA in adjuvant, showed an immune elimination curve when challenged at 93 and 274 days after birth with tracer doses of labelled BSA. This experiment added no information on the onset of competency in fetal life but did argue against passive immunization from the mother.

The most thoroughly studied animal from the point of view of fetal immunology is the lamb. This knowledge is the result of efforts by SILVERSTEIN *et al.* (for reviews see [78 and 89]). The lamb was chosen primarly because the sheep placenta does not allow the passage of immunoglobulins between mother and fetus, and therefore antibodies found in the fetal circulation are synthesized by the fetus. These studies on the fetal lamb have established several important points concerning fetal immune responses: (1) attainment of competence is a stepwise process with respect to antigens, the ability to respond to some antigens appearing earlier than to others; (2) birth itself is irrelevant to immune competency, but rather competence to various antigens appears before or after birth, depending upon the species; (3) once competence to an antigen is attained, the initial response to the antigen is not quantitatively immature or weak.

More recently, these workers have turned their attention to the rhesus monkey, which has the disadvantage of an immunoglobulin-'permeable' placenta but offers the important advantage of being closer to man. Many of the results on the rhesus monkey have been offered, to date, in a preliminary form only [78, 89]. The intrauterine fetal surgical procedures needed in these studies, including hysterotomy and fetal exposure, thymectomy, and splenectomy have, however, been presented in detail [90]. From SILVERSTEIN's summaries, it appears that the rhesus monkey fetus, as the lamb fetus, shows a stepwise maturation of the ability to respond to various antigens. Fetal antibody responses were induced to the bacteriophage ΦX174, ovalbumin, and ferritin. Diphtheria toxoid was not immunogenic in fetal monkeys. Rejection of homografts occurred earlier in the fetal rhesus monkey than in the fetal lamb; as early as the 58th day of gestation for the monkey as compared to 80 days for the lamb.

More recently, SILVERSTEIN *et al.* [62] have published a report on antibody production by the fetal rhesus monkey in which antibody-producing cells and not circulating antibodies were assayed. At appropriate intervals after the injection of sheep erythrocytes intravenously into fetuses of 73–127 days gestation, the fetuses were either splenectomized *in utero* or delivered by caesarean section for their spleens. The number of anti-sheep erythrocyte antibody producing cells in the spleens was assayed by the direct hemolytic plaque technique for IgM antibody or by an antiglobulin facilitation modification of the plaque technique for IgG. The cellular kinetics of similarly immunized adults were followed for purposes of comparison. Disparities between total sizes and levels of lymphoid maturation between fetal and adult spleens were balanced by assessing the proportion of *available* lymphoid cells which were making antibodies. The results showed that fetal and adult primary IgM responses were quantitatively similar, confirming the maturation of the fetal system. A striking difference was noted in the declining phase of the primary response. The adult response decays slowly whereas the fetal response falls precipitously to an absolutely zero level on days 8 and 9. This lack of antibody forming cells, both of IgM and IgG type, was confirmed on additional animals. Several possible explanations for this interesting phenomenon are offered by the authors, but they favor either the idea of greater fetal sensitivity to the immunosuppressive effects of circulating antibody or a temporary sequestration of antigen coupled with a postulated rigid requirement for the presence of antigen for antibody synthesis in the fetus.

Newborn Studies

In addition to their work on the fetal immune response to BSA mentioned above, COTES *et al.* [88] also examined the responses of newborn rhesus monkeys. They treated one monkey one day after birth, and another seven days after with 100 mg BSA. The monkeys were challenged 20 and 55 days later, respectively, with a tracer dose of ^{131}I-BSA for determination of elimination rates. As might be expected from their demonstration of fetal immune responses to BSA, both of these monkeys showed immune elimination curves.

Inoculation of fetal and young rhesus monkeys with Rous sarcoma virus has also demonstrated the ability of young monkeys to form antibodies [91]. In one monkey inoculated *in utero* and born with a tumor, complement-fixing (COFAL) antibodies appeared three weeks after birth. This monkey also had virus neutralizing antibodies in its serum when tested soon after birth, although whether the antibody was also present during fetal life was not determined. Another monkey, inoculated with the oncogenic virus at the post-natal age of one month developed neutralizing antibodies in another month after inoculation, preceding tumor development by at least one week. The tumor and COFAL antibodies then appeared simultaneously.

Tolerogenesis

Natural

Marmosets (Callitrichidae) are unusual among primates in that fraternal twinning is the rule, occurring in about 80% of births [92, 93]. HILL [94] and WISLOCKI [92] described placental vascular anastomoses between twins, attributed to early fusion of the blastocysts. The female partner of heterosexual twins is never a freemartin [92]. BENIRSCHKE *et al.* [95] noted marrow chimerism on the basis of sex chromosome analysis in two species, *Callithrix jacchus* and *Leontocebus rosalia*, establishing as for cattle that the placental anastomoses are functional. GENGOZIAN *et al.* [96] showed that, in contrast to cattle, chimerism is the rule in marmosets. This phenomenon has more recently been studied in greater detail in animals of known sibling relationships [97].

PORTER and GENGOZIAN [98] grafted skin reciprocally between partners of six adult sets of marmoset twins. Four sets of twins were heterosexual, and lymphocyte sex chromosome chimerism could be demonstrated in them. A chimeric state could not be proved by sex chromosome analysis in the remaining two sets of isosexual twins. At the time of the writing of their paper, all twelve of the monkeys had accepted their twins' first-set skin grafts. Second- and third-set skin grafts were also accepted. Some of the grafts had been viable for over two years. That marmosets are immunologically competent to reject incompatible skin grafts was shown by reciprocal grafting between members of two different subspecies *(Saguinus fuscicollis illigeri* and *S. f. lagonotus)*. The grafts survived only two to four weeks and the rejection reactions were acute. It was of interest that two series of grafts within a subspecies *(S.f. illigeri to S.f. illigeri* and *S. oedipus oedipus* to *S.o. oedipus)* showed mean survival times approximately twice that of the intersubspecies grafts, suggesting a 'population histocompatibility tolerance' in marmosets. The extended survival of chimeragrafts and the inability of second- and third-set grafts to provoke a rejection response argues strongly for a state of immunological tolerance between chimeric twins resulting from prenatal parabiosis in marmosets. This condition would appear to offer an excellent model for the study of various immunological phenomena [99].

Induced

Fetal

Berman *et al.* [91], in experiments on *in utero* inoculation of rhesus monkeys with Rous sarcoma virus, found one monkey to be devoid of antibody activity to leukosis antigens, even though it bore a large tumor after birth. They postulated tolerance as an explanation for this observation, although eventual tumor regression suggested that the cellular homograft rejection mechanism was unimpaired.

Cotes *et al.* [88] found five of six rhesus monkeys which had been injected *in utero* with 100 mg BSA to be tolerant to BSA, as determined by tracer dose elimination of ^{131}I-BSA from the circulation after a challenge dose of 1 mg BSA in adjuvant. The duration of the tolerant state was as long as 345 days after birth.

Young and Adult Animals

Another monkey in the series of Berman *et al.* [91], injected with Rous sarcoma virus postnatally, failed to develop complement-fixing or virus neutralizing antibodies even though a small tumor had been induced. As with the prenatally inoculated non-responsive animal, tolerance was invoked as a possible explanation.

Cotes *et al.* [88], in their studies of BSA immunogenesis and tolerogenesis in prenatal and neonatal monkeys, found one monkey treated at three days of age with 250 mg BSA intravenously, and at 31 days of age with an additional 100 mgm BSA intravenously, to be tolerant at least to the age of 72 days.

Howard *et al.* [100] were able to induce partial tolerance in older rhesus monkeys (one-half and one-year old) to human gamma globulin (HGG) which had been ultracentrifuged clear of aggregates and administered intravenously. The tolerant state was recognized by passive hemagglutination and gel-diffusion techniques. Although titers to HGG were demonstrable in treated animals, they were substantially lower, both in the primary and secondary responses, than those of untreated control monkeys.

Immediate Hypersensitivity

The state of immediate hypersensitivity is characterized by the presence in serum of certain antibodies which have been variously referred to as reagins, skin-sensitizing, homocytotropic, and homologous anaphylactic antibodies, etc. The major properties distinguishing this type of antibody from others include heat lability, inability to precipitate, inability to fix complement, and affinity for the skin and other tissues of the homologous or closely related species with persistence in their attachment for long periods of time [101]. This tissue affinity of reaginic antibodies accounts for many of the manifestations of allergic disease, and secondarily, provides a means for their detection. The combination of tissue-fixed reagin with its specific antigen (allergen) results in the release from target cells of mediators of anaphylaxis, i.e. histamine and slow reacting substance of anaphylaxis (SRS-A).

In man, the reaginic activity is largely associated with the class of immunoglobulins designated IgE. In this section, we review some of the investigations elucidating the nature and mode of action of human IgE in which primates have played a significant role, as well as studies on the occurrence and properties of the homologous immunoglobulin in primates.

Transfer of Human Reaginic Antibodies to Primates

Since the initial successful attempts by CAULFIELD *et al.* [102] and STRAUS [103] to passively sensitize rhesus monkey skin with the serum of atopic humans, many investigators have searched for primate substitutes for human volunteers in routine allergy testing. The possibility of transmitting infectious agents via the passively transferred serum provided further incentive. Using the techniques of direct Prausnitz-Küstner (P-K) and passive cutaneous anaphylaxis (PCA) tests, LAYTON *et al.* [104–107] showed that a wide variety of primates could serve as efficient recipients of passively transferred human reaginic antibodies. Animals considered to be most satisfactory included the Asian macaques, baboons, capuchins, marmosets, lemurs, and galagos. The squirrel monkey failed to give a typical reaction and the dark pigmentation of the spider monkey interfered with visualization of the results. Tree shrews were negative.

LAYTON's results were independently confirmed for cynomolgus and rhesus monkeys, and extended to include a study of some of the properties of the antibody involved [108]. It was found that the heat lability of the responsible PCA antibody essentially paralleled that of the human P-K antibody, since heating at 56°C for 30 min diminished the reactivity while heating for two or four hours abolished it completely. The antibody was also shown to have a sedimentation velocity ranging between 7S and 11S. In a comparison of the reactivity of monkey and human skin, it was shown that human atopic sera giving only a weak reaction at low dilutions in human skin failed to sensitize rhesus monkey skin. However, higher titered human sera gave essentially similar results in both human and monkey skin.

Confirmation of reduced responsiveness of *Macaca mulatta* and *M. speciosa* skin upon passive transfer of human reaginic serum was provided by FEINBERG *et al.* [109]. It was suggested that vascular permeability induced by histamine and other mediators is less in monkeys than in man, accounting for the reduced reactivity of monkey skin.

The use of primate tissues other than skin was first reported by ARBESMAN *et al.* [110]. Strips of ileum from *M. irus*, *M. mulatta*, and *Saimiri sciureus* were exposed to dilutions of sera from patients sensitive to one or more of a variety of allergens. Upon subsequent exposure to the specific allergen, the anaphylactic contraction was measured (Schultz-Dale reaction). Quantitation was achieved with known concentrations of histamine. The reaction was shown to be specific, indicating the presence of specific antibody in the atopic sera. However, there was no correlation with hemagglutination titers. This antibody was labile after heating at 56°C for two hours. The same sera were examined using guinea-pig ileum, and with this tissue, tissue fixation was correlated with hemagglutination titer and heat stability. This suggested the presence of more than one type of tissue-sensitizing antibody in the atopic patients studied by ARBESMAN *et al.*, a possibility supported by evidence from other sources [111].

Sensitization of primate lung tissue with atopic human serum or homologous monkey skin-fixing antibodies has been described [112, 113]. The released histamine was measured by its effect upon guinea-pig ileum [112] or by fluorometry [113]. In a comparative evaluation of human and monkey (rhesus) tissues for detection of human reagins, good correlation was seen in the reactivity of human and monkey lung [114]. However, considerable variation may occur within individual tissue donors of both species. The chopped lung technique was found to be inferior in sensitivity to the Schultz-Dale technique on monkey ileum, but due to the reproducibility of results with the former, it was recommended as a valuable test in reaginic hypersensitivity research. Evidence was presented that the same antibody probably participates in the passive sensitization of human skin, human and monkey lung, and monkey ileum. Thus, the activity in all cases was heat labile and not correlated with the passive hemagglutination titers of the tested sera.

The use of monkey lung fragments in studying the mechanism of mediator release has been described [115]. Tissue passively sensitized with human serum rich in reagin, or with IgE myeloma protein, released histamine and SRS-A when exposed to specific allergen in direct anaphylaxis. A reversed anaphylactic release of these mediators also occurred after exposure to 7S anti-IgE or its $F(ab')_2$ fragments but not to its monomer Fab′ fragments. The authors therefore suggested that in the reversed release of histamine and SRS-A, bridging of two cell-bound IgE molecules could be the initial reaction step.

Induction of Immediate Hypersensitivity in Primates

The induction of homologous skin-sensitizing (reagin-like) antibodies with properties similar to human reagins in animals infected with various parasites has been documented [116–120]. In rhesus monkeys experimentally infected with *Schistosoma mansoni*, homologous skin-sensitizing antibodies first appeared between 6 and 10 weeks after exposure to the parasite [116, 117, 119]. Their initial appearance was generally transitory. Upon subsequent challenge with the parasite, the antibodies regularly appeared within one to two weeks and persisted for up to 20 weeks. Attempts to demonstrate a protective function for the reagin-like antibody, as could be shown with rats infected with the same parasite [116], were unsuccessful [117]. It was possible to stimulate reagin production only with a living infection, the injection of antigens extracted from the parasites failing to give a positive response. However, the same antigens were shown to react in passive cutaneous anaphylaxis with the reagins produced by infection, indicating that the production of reagins must depend upon some subtle combination of factors involving the chemical state of the allergen and its route of presentation, and perhaps other unknown factors.

Chimpanzees infected with *S. mansoni* likewise develop an antibody which has the property of prolonged skin fixation [118]. In this case rhesus monkeys served as recipients of the passively transferred serum, in a manner similar to the transfer of human sera to lower primates as discussed in the preceding section. Unlike the infection in rhesus monkeys, however, the reagin-like antibody was detectable only after the 12th week and persisted throughout the observation period of 7 months. A peak of reactivity was observed 5 to 6 months after infection. In *M. mulatta* infected with the related *S. japonicum*, skin-sensitizing antibodies appeared at irregular intervals, varying with the strain of the parasite used [120]. Thus, with the zoophilic Formosan strain, positive passive cutaneous anaphylaxis (PCA) was obtained with the sera of some monkeys within 4 to 5 weeks, while infections with the human Japanese strain resulted in positive PCA only after 17 to 23 weeks. However, in view of the transitory nature of the initial appearance of the antibody in schistosome infected rhesus monkeys shown by OGILVIE *et al.* [117] the irregular sampling schedule employed by HSÜ and HSÜ [120] makes the data difficult to interpret. The lack of correlation of the presence of reagin-like antibodies with the immune state of the animals, and the inability of passively transferred serum from PCA positive animals to provide protection from a challenge infection confirms the conclusion of others [117] that the antibody is non-functional in immunity to the parasite.

Rhesus monkeys experimentally infected with the nematode parasite *Toxocara canis* also became hypersensitive, as demonstrated by direct skin tests using homologous antigen [121]. Skin reactivity developed between the second and fifth week and persisted for at least 36 weeks.

Indications that primates can be *artificially* stimulated to produce reaginic antibodies began to emerge from the laboratory of PATTERSON [51, 122]. Rhesus monkeys immunized with BSA or DNP-BSA produced antibodies which reacted both in active cutaneous anaphylaxis (ACA) and in passive transfer to normal monkey skin (P-K test). The ACA test, in which BSA was injected intradermally into BSA-immunized animals, produced Arthus-type reactions; i.e. local induration, hemorrhage, and subsequent necrosis. However, passive transfer of anti-BSA hyperimmune serum and subsequent challenge with BSA (P-K test) resulted in skin reactions consistent with immediate hypersensitivity, as did direct skin challenge of DNP-BSA immunized animals with DNP-haptenized bovine gamma globulin (DNP-BGG). The latter reaction appeared to be specific for the DNP haptenic determinant, since no reaction was elicited with non-haptenized BGG. In one case, the anti-DNP hypersensitivity could be passively transferred [51]. WEISZER *et al.* [122] detected immediate skin reactivity to *Ascaris* antigen in *M. mulatta*, presumably due to cross-reacting antibodies directed to other nematodes naturally acquired by the animals. During the course of weekly intradermal injections of the antigen, a 10- to 100 fold increase in skin titer was observed in the animals with the lowest initial reactivity and was interpreted as being due to hypersensitization to the injected antigen. Repeated intravenous injections had a similar effect, leading to an increase in both skin and systemic reactivity. In the case of both skin and systemic sensitization, the increased reactivity returned to initial levels two to three weeks after the sensitizing injections were stopped.

The use of more highly purified antigens for the induction of *Ascaris* hypersensitivity in *M. mulatta* was reported by MALLEY *et al.* [123]. The active component of a crude *Ascaris* extract could be separated from inactive and non-specifically toxic components by powder block zone electrophoresis. In animals sensitized with the crude antigen, positive skin tests were seen three to four weeks after the initial injection, the response being dose-dependent. The purified antigen elicited a positive response within five to six weeks. The reactivity stimulated by crude antigen persisted for 40 to 90 days and could be readily re-stimulated to levels comparable to the initial response. Duration of reactivity stimulated by purified antigen was not followed beyond the 35th day.

These results in stimulation of hypersensitivity in primates using parasite antigens have been followed by several reports of similar immunization with allergens implicated in human atopic hypersensitivity [124–126]. 'Monkeys' (species unknown but presumably *M. mulatta)* immunized by inhalation of dry ragweed pollen developed antibodies capable of sensitizing monkey or guinea-pig ileum for histamine release upon subsequent antigen challenge, but sensitization of monkey skin could not be demonstrated [124]. Parenteral injection of the antigen in Freund's adjuvant gave rise to antibodies which, in addition to sensitizing homologous ileum, could also be detected in gel diffusion, hemagglutination, and Schultz-Dale reactions with guinea-pig ileum. Specific contraction of the ileum from the inhalant-immunized group upon direct exposure of the tissue to antigen could be demonstrated. Similar production of both reaginic and blocking antibody has been shown with a purified timothy pollen allergen in humans, although in the rhesus monkey only reaginic antibodies were produced [125]. These were detected by their ability to sensitize homologous lung tissue; attempts to passively sensitize skin of normal rhesus monkeys were unsuccessful. In *Cercopithecus aethiops* the dialysable component of timothy pollen extract, when injected initially in complete Freund's adjuvant with subsequent booster injections in incomplete adjuvant, elicits the production of a skin sensitizing antibody as well as antibodies with blocking characteristics [126]. In this case, the antibody responsible for skin sensitization could be passively transferred to normal monkey skin.

Utilizing rhesus monkeys both passively sensitized with human reaginic serum and actively sensitized with antigens cross-reactive with *Ascaris suum* antigens, PATTERSON *et al.* have developed a model system for the study of allergic respiratory disease [122, 127]. For a full consideration of the methods involved and the responses under investigation, the reader is referred to recent publications by PATTERSON [128–130].

Nature of Antibodies Involved in Immediate Hypersensitivity in Primates

In an earlier section, immunochemical evidence for the occurrence in primates of IgE, homologous to human IgE, was reviewed. In humans, it has been convincingly demonstrated that the bulk of reaginic or immediate type hypersensitivity is associated with this immunoglobulin class [33]. We now review evidence showing a similar association between IgE and reaginic activity in monkeys.

In the examples of hypersensitivity in primates induced either by infection with parasites or by antigen injection, discussed above, the characteristics of the antibody responsible for the allergic reactions were studied in several cases. Heat lability of the antibody produced in schistosome infections in rhesus monkeys [119, 120] and chimpanzees [118] was demonstrated. Further characterization in the case of rhesus monkeys infected with *Schistosoma mansoni* disclosed many properties shared with human IgE; i.e. charge properties as shown by ion-exchange chromatography and electrophoresis, and molecular size as indicated by gel-filtration [119]. Prolonged fixation to homologous but not to heterologous (guinea pig) skin was also shown. Gel-filtration of serum of *Ascaris*-sensitive *M. mulatta*, whose sensitivity was acquired either by prior infection with other parasites [122, 131] or by antigen injection [123], located the reaginic activity in the ascending portion of the 7S peak. This indicates a molecular size intermediate between the 7S and 19S serum components, and is consistent with the size characteristics of human IgE [33]. Further fractionation of the 7S peak by ion-exchange chromatography separated the reaginic activity from IgG [122]. Absorption of the reagin-rich fractions with specific anti-IgA, -IgG, and -IgM had no effect upon the reaginic properties; i.e., capacity to passively sensitize homologous skin [131] or lung [122]. Proof that the immunoglobulin responsible for *Ascaris* hypersensitivity in monkeys is homologous to human IgE was provided by ISHIZAKA *et al.* [31] who showed that specific anti-human IgE serum reacts in gel diffusion analysis with the monkey protein and absorption with anti-IgE removes the reaginic activity. The sedimentation velocity of the molecule was established at 8.0S, which is in agreement with the size of the human IgE molecule.

Summary

We are at a point in time where the new-found fundamental knowledge of immunology, gained mainly from studies on human material and conventional laboratory animals, is beginning to be channeled into research with primates. The lag in base-line immunological knowledge of these choicest of models for the study of human diseases is disappearing. Research in primate immunology can be expected to shed new light on basic immune mechanisms, as well. Comparative evolutionary studies on immunoglobulin structure is an obvious example, but other areas, some of which have been pointed out above, can be expected to benefit equally.

The approximately 200 species of living primates [1], in a continuum of phylogenetic distance from man that is truly remarkable, would appear to offer tremendous benefits to immunological research if exploited properly. In this regard, it is encouraging to see the accelerated replacement of the appellation 'monkey' with more precise identifications in experimental papers.

Besides offering unique situations such as the natural high-frequency chimerism of marmosets for biomedical research, this large assemblage of species could play a role in the understanding of those complex immune responses, both on the afferent and efferent limbs, in which a series of substances and events interact in definite and exact sequences. If, as we have tried to point out in this chapter, primates are not qualitatively vastly different from man in their immunological apparatus; it would seem reasonable to assume that various species have quantitative genetic adjustments that could give real insights, if understood, into subtle disturbances in human immune responses leading to great disturbances in human 'fitness'.

References

1 Napier, J. R. and Napier, P. H.: A handbook of living primates (Academic Press, London/New York 1967).

2 Goodman, M.: Immunochemistry of the primates and primate evolution. Ann. N.Y. Acad. Sci. *102:* 219–234 (1962).

3 Williams, C. A., jr.: Immunochemical analysis of serum proteins of the primates: a study in molecular evolution; in Buettner-Janusch Evolutionary and genetic biology of primates, vol. II, pp. 25–74 (Academic Press, New York/London 1964).

4 Lichter, E. A.: Immunological studies of serum proteins in nonhuman primates. Ann. N.Y. Acad. Sci. *162:* 43–49 (1969).

5 Cold Spr. Harb. Symp. quant. Biol. vol. XXXII. Antibodies (Cold Spr. Harbor Lab. Quant. Biol., New York 1967).

6 Killander, J. (ed.): Gamma globulins; structure and control of biosynthesis. 3rd Nobel Symp. Södergarn, 1967 (Almgvist & Wiksell, Stockholm 1967).

7 Rowe, D. A. and Fahey, J. L.: A new class of human immunoglobulins. I. A unique myeloma protein. J. exp. Med. *121:* 171–184 (1965).

8 Tomasi, T. B. and Bienenstock, J.: Secretory immunoglobulins. Adv. Immunol. *9:* 1–96 (1968).

9 Grabar, P. et Williams, C. A., jr.: Méthode permettant l'étude conjuguée des propriétés électrophorétiques et immunochimiques d'un mélange des protéines. Biochim. biophys. Acta *10:* 193–194 (1953).

10 Picard, J.; Heremans, J., and Vandebroek, G.: Serum proteins found in primates. I. Proteins of man, *Macacus irus* and *Lemur mongoz*. Vox Sang. *7:* 190–213 (1962).

11 Picard, J.; Heremans, J., and Vandebroek, G.: Serum proteins found in primates. II. Serum proteins of some other primate species. Vox Sang. *7:* 425–448 (1962).

12 Picard, J.; Heremans, J., and Vandebroek, G.: Serum proteins found in primates; comparative analyses of the antigenic structure of several proteins. Mammalia *27:* 285–299 (1963).

13 Wang, A.-C.; Shuster, J.; Epstein, A., and Fudenberg, H. H.: Evolution of antigenic determinants of transferrin and other serum proteins in primates. Biochem. Genet. *1:* 347–358 (1968).

14 Shuster, J.; Warner, N. L., and Fudenberg, H. H.: Cross-reactivity of primate immunoglobulins. Ann. N.Y. Acad. Sci. *162:* 195–201 (1969).

15 Poulik, M. D.: Immunological and structural studies of immunoglobulins of human, chimpanzee and other non-human primates; primates in med., vol. 1, pp. 68–94 (Karger, Basel/New York 1968).

16 Cohen, S. and Porter, R. R.: Heterogeneity of the peptide chains of γ-globulin. Biochem. J. *90:* 278–284 (1964).

17 Hood, L.; Gray, W. R.; Sanders, B. G., and Dreyer, W. J.: Light chain evolution. Cold Spr. Harb. Symp. quant. Biol. *32:* 133–146 (1967).

18 Porter, R. R.: Hydrolysis of rabbit gamma globulin and antibodies with crystalline papain. Biochem. J. *73:* 119–126 (1959).

19 Litwin, S. D.: The expression of human genetic factors (Gm) on primate immunoglobulins. Ann. N.Y. Acad. Sci. *162:* 177–188 (1969).

20 Malley, A. and Mackler, B.: Skin-fixing properties of papain digested rhesus antihapten antibody. Proc. Soc. exp. Biol., N.Y. *131:* 398–402 (1969).

21 Damian, R. T.; Greene, N. D., and Kalter, S. S.: IgG subclasses in the baboon *(Papio cynocephalus)*. J. Immunol. (in press).

22 Poulik, M. D.: F′c fragment of immunoglobulins. Nature, Lond. *210:* 133–134 (1966).

23 Normansell, D. E. and Stanworth, D. R.: Interactions between rheumatoid factor and native γG-globulins studied in the ultracentrifuge. Immunology, Lond. *15:* 549–560 (1968).

24 Stone, M. J. and Metzger, H.: The specificity of a monoclonal macroglobulin (γM) antibody; reactivity with primate γG-immunoglobulins. J. Immunol. *102:* 222–228 (1969).

25 Kronvall, G.; Seal, U. S.; Finstad, J., and Williams, R. C., jr.: Phylogenetic insight into evolution of mammalian Fc fragment of γG-globulin using staphylococcal protein A. J. Immunol. *104:* 140–147 (1970).

26 Wang, A.-C.; Shuster, J., and Fudenberg, H. H.: Evolutionary origin of the Gm 'a' peptide of immunoglobulins. J. molec. Biol. *41:* 83–86 (1969).

27 Wang, A.-C. and Fudenberg, H. H.: Genetic control of gamma chain synthesis. A chemical and evolutionary study of the Gm (a) factor of immunoglobulins. J. molec. Biol. *44:* 493–500 (1969).

28 Steinberg, A. G.: Genetic variations in human immunoglobulins; the Gm and Inv types; in Greenwalt Symposium on immunogenetics, pp. 75–98 (Lippincott, Philadelphia 1966).

29 Kunkel, H. G.; Allen, J. C.: Grey, H. M.; Mårtensson, L., and Grubb, R.: A relationship between the H chain groups of 7S γ-globulin and the Gm system. Nature, Lond. *203:* 413–414 (1964).

30 ISHIZAKA, K. and ISHIZAKA, T.: Reversed-type allergic skin reactions by anti-γE-globulin antibodies in humans and monkeys. J. Immunol. *100:* 554–562 (1968).

31 ISHIZAKA, K.; ISHIZAKA, T., and TADA, T.: Immungolobulin E in the monkey. J. Immunol. *103:* 445–452 (1969).

32 ITO, K.; WICHER, K., and ARBESMAN, C. E.: Antibodies to IgE produced in monkeys. J. Allergy *45:* 123 (1970).

33 ISHIZAKA, K. and ISHIZAKA, T.: Human reaginic antibodies and immunoglobulin E. J. Allergy *42:* 330–363 (1968).

34 ROSSEN, R. D.; WOLFF, S. M.; BUTLER, W. T., and VANNIER, W. E.: The identification of low molecular weight bentonite flocculating antibodies in the serum of the rabbit, monkey and man. J. Immunol. *98:* 764–777 (1967).

35 GREY, H. M. and KUNKEL, H. G.: H chain subgroups of myeloma proteins and normal 7S γ-globulin. J. exp. Med. *120:* 253–266 (1964).

36 TERRY, W. D. and FAHEY, J. L.: Subclasses of human γ_2-globulin based on differences in the heavy polypeptide chains. Science *146:* 400–401 (1964).

37 ALEPA, F. P. and TERRY, W. D.: Genetic factors and polypeptide chain subclasses of human immunoglobulin G detected in chimpanzee serums. Science *150:* 1293–1294 (1965).

38 ALEPA, F. P.: Antigenic factors characteristic of human immunoglobulin G, detected in the sera of non-human primates; primates in med., vol. 1, pp. 1–9 (Karger, Basel/New York 1968).

39 DAMIAN, R. T.; LUKER, M. F.; GREENE, N. D., and KALTER, S. S.: in preparation.

40 LAKIN, J. D.; PATTERSON, R., and PRUZANSKY, J. J.: Immunoglobulins of the rhesus monkey. III. Structure and activity of rhesus IgG and IgM antibodies synthesized in the primary response and in the hyperimmune state. J. Immunol. *102:* 975–985 (1969).

41 SIBAL, L. R.; FINK, M. A.; MCCUNE, C. L., and COWLES, C. A.: Antibody response of rhesus monkeys to a density gradient purified murine leukemia virus (Rauscher). J. Immunol. *98:* 368–373 (1967).

42 MANCINI, G.; CARBONARA, A. O., and HEREMANS, J. F.: Immunochemical quantitation of antigens by single radial immunodiffusion. Immunochemistry *2:* 235–254 (1965).

43 FELSENFELD, O. and GREER, W. E.: Vibriocidal activity, immune globulin producing cells and immune globulin levels in *Theropithecus gelada* after administration of a *Vibrio cholerae* antigen. Immunology, Lond. *14:* 319–324 (1968).

44 HOUBA, V.; BROWN, K. N., and ALLISON, A. C.: Heterophile antibodies, M-antiglobulins and immunoglobulins in experimental trypansomiasis. Clin. exp. Immunol. *4:* 113–123 (1969).

45 BRANDOM, W. F.; FELSENFELD, O., and GREER, W. E.: Proliferation of immunologically active cells in the intestines of primates during secondary response to cholera toxin. Ann. trop. Med. Parasit. *63:* 129–138 (1969).

46 FELSENFELD, O.; FELSENFELD, A. D.; GREER, W. E., and HILL, C. W.: Relationship of some vibrio antibodies to serum immune globulins in man and *Cercopithecus aethiops*. J. infect. Dis. *116:* 329–334 (1966).

47 GREER, W. E. and FELSENFELD, O.: Comparison of early responses to phenolized cholera vaccine and to vibrio lipopolysaccharide in patas monkeys *(Erythrocebus patas)*. Ann. trop. Med. Parasit. *60:* 417–422 (1966).

48 Felsenfeld, O.; Greer, W. E., and Jiřička, J.: Early immunoglobulin formation and hemagglutinating and bactericidal titers in *Erythrocebus patas* after oral administration of a weakly pathogenic shigella strain. Lab. Invest. *19:* 146–152 (1968).

49 Stiehm, E. R. and Fudenberg, H. H.: Serum levels of immune globulins in health and disease; a survey. Pediatrics *37:* 715–727 (1966).

50 Stoop, J. W.; Zegers, B. J. M.; Sander, P. C., and Ballieux, R. E.: Serum immunoglobulin levels in healthy children and adults. Clin. exp. Immunol. *4:* 101–112 (1969).

51 Lakin, J. D.; Patterson, R., and Pruzansky, J. J.: Immunoglobulins of the rhesus monkey *(Macaca mulatta)*. I. Antiserum to bovine serum albumin and the dinitrophenyl antigenic determinant. J. Immunol. *98:* 745–756 (1967).

52 Lakin, J. D.; Patterson, R., and Pruzansky, J. J.: Immunoglobulins of the rhesus monkey. II. Properties of rhesus IgG and IgM antibody in hemagglutination, precipitation and skin sensitization. J. Immunol. *102:* 253–262 (1969).

53 Lakin, J. D.; Patterson, R., and Pruzansky, J. J.: Immunoglobulins of the rhesus monkey. IV. Characteristics of the primary and secondary responses of the rhesus monkey to heterologus serum protein antigens. J. Immunol. *103:* 577–587 (1969).

54 Talmage, D. W. and Maurer, P. H.: ^{131}I-labelled antigen precipitation as a measure of quantity and quality of antibody. J. infect. Dis. *92:* 288–300 (1953).

55 Farr, R. S.: A quantitative immunochemical measure of the primary interaction between I*BSA and antibody. J. infect. Dis. *103:* 239–262 (1958).

56 Pike, R. M.: Antibody heterogeneity and serological reactions. Bact. Rev. *31:* 157–174 (1967).

57 Siskind, G. W. and Benacerraf, B.: Cell selection by antigen in the immune response. Adv. Immunol. *10:* 1–50 (1969).

58 Uhr, J. W. and Finkelstein, M. S.: The kinetics of antibody formation. Progr. Allergy, vol 10, pp. 37–83 (Karger, Basel/New York 1967).

59 Freeman, M. J. and Stavitsky, A. B.: Radioimmunoelectrophoretic study of rabbit anti-protein antibodies during the primary response. J. Immunol. *95:* 981–990 (1965).

60 Schmidt, N. J.; Lennette, E. H., and Dennis, J.: Characterization of antibodies produced in natural and experimental coxsackievirus infections. J. Immunol. *100:* 99–106 (1968).

61 Heffner, R. R., jr. and Schluederberg, A.: Specificity of the primary and secondary antibody responses to myxoviruses. J. Immunol. *98:* 668–672 (1967).

62 Silverstein, A. M.; Prendergast, R. A., and Parshall, C. J., jr.: Cellular kinetics of the antibody response by the fetal rhesus monkey. J. Immunol. *104:* 269–271 (1970).

63 Felsenfeld, O.; Burrows, W.; Kasai, G. J.; Greer, W. E., and Jiřička, Z.: The cellular immune response of non-human primates to crude type 2 cholera toxin. J. infect. Dis. *118:* 491–499 (1968).

64 Felsenfeld, O.; Hill, C. W., and Greer, W. E.: Response of *Cercopithecus aethiops* to cholera vibrio lipopolysaccharide and psychological stress. Trans. roy. Soc. trop. Med. Hyg. *60:* 514–518 (1966).

65 Felsenfeld, O.; Greer, W. E.; Kirtley, B., and Jiřička, Z.: Cholera studies in non-human primates. Part 8. Comparison of methods for the enumeration of immunologically active cells and early immune globulin, precipitin, vibriocidin and antitoxin formation. Trans. roy. Soc. trop. Med. Hyg. *62:* 278–284 (1968).

66 FELSENFELD, O.; FELSENFELD, A. D., and GREER, W. E.: Avidity of cholera antibody. Trans. roy. Soc. trop. Med. Hyg. *63:* 846–853 (1969).

67 GILL, T. J., III and GERSHOFF, S. N.: The antigenicity of synthetic polypeptides in primates. Biochem. biophys. Acta *97:* 373–376 (1965).

68 GILL, T. J., III and GERSHOFF, S. N.: The effects of methionine and ethionine on antibody production in primates. J. Immunol. *99:* 883–893 (1967).

69 GILL, T. J., III; KUNZ, H. W., and PAPERMASTER, D. S.: Studies on synthetic polypeptide antigens. XVIII. The role of composition, charge, and optical isomerism in the immunogenicity of synthetic polypeptides. J. biol. Chem. *242:* 3308–3318 (1967).

70 SELA, M.: Immunological studies with synthetic polypeptides. Adv. Immunol. *5:* 29–129 (1966).

71 HILL, C. W.; GREER, W. E., and FELSENFELD, O.: Psychological stress, early response to foreign protein, and blood cortisol in vervets. Psychosom. Med. *29:* 279–283 (1967).

72 RATNER, J. J.; PINKERTON, M. E.; KALTER, S. S., and GUILLOUD, N. B.: Antibodies in non-human primates to selected bacterial antigens. Z. Versuchstierk. *11:* 146–154 (1969).

73 KENDE, M.: Antibody response of animals to *Mycoplasma pneumoniae* measured by latex agglutination. Appl. Microbiol. *17:* 275–279 (1969).

74 TAYLOR, R. L.: Serological response of rhesus monkeys to *Histoplasma*, *Blastomyces*, and *Coccidioides* antigens. J. Bact. *94:* 1–5 (1967).

75 ISA, A. M.: Susceptibility of anti-chlamydial IgG antibodies to reduction and alkylation. J. Immunol. *102:* 1332–1335 (1969).

76 KALTER, S. S.; KAGAN, I. G., and KUNTZ, R. E.: Antibodies to parasites in Kenya baboons. Trans. roy. Soc. trop. Med. Hyg. *63:* 684–685 (1969).

77 HILLYER, G. V. and RITCHIE, L. S.: Immunoprecipitins in *Schistosoma mansoni* infections. II. *Cercopithecus sabaeus* and *Macaca mulatta* monkeys. Exp. Parasit. *20:* 326–333 (1967).

78 STERZL, J. and SILVERSTEIN, A. M.: Developmental aspects of immunity. Adv. Immunol. *6:* 337–459 (1967).

79 BRAMBELL, F. W. R.: The passive immunity of the young mammal. Biol. Rev. *33:* 488–531 (1958).

80 KUTTNER, A. and RATNER, B.: The importance of colostrum to the new-born infant. Amer. J. Dis. Child. *25:* 413–434 (1923).

81 BANGHAM, D. R.; HOBBS, K. R., and TERRY, R. J.: Selective placental transfer of serum-proteins in the rhesus. Lancet *ii:* 351–354 (1958).

82 BANGHAM, D. R.: The transmission of homologous serum proteins to the foetus and to the amniotic fluid in the rhesus monkey. J. Physiol., Lond. *153:* 265–289 (1960).

83 BANGHAM, D. R.; HOBBS, K. R., and TEE, D. E. H.: Transmission of serum proteins from foetus to mother in the rhesus monkey; indwelling cannulation of foetus without interruption of pregnancy. Lancet *ii:* 1173–1174 (1960).

84 QUINLIVAN, W. L. G.: Gamma globulin-^{131}I transfer between mother and offspring in the rhesus monkey. Amer. J. Physiol. *212:* 324–328 (1967).

85 COHEN, S. and MILSTEIN, C.: Structure and biological properties of immunoglobulins. Adv. Immunol. *7:* 1–89 (1967).

86 JOHANSSON, S. G. O.: Serum IgND levels in healthy children and adults. Int. Arch. Allergy *34:* 1–8 (1968).

87 Bangham, D. R.; Cotes, P. M.; Hobbs, K. R., and Tee, D. E. H.: An attempt to determine the age at which 'cellular immunological maturity' develops in the rhesus monkey. Proc. Int. Symp. on bone marrow therapy and chemical protection in irradiated primates, Leiden 1962, pp. 187–192 (Radiobiol. Inst. Org. Hlth Res. TNO, Rijswijk, Netherlands 1962).

88 Cotes, P. M.; Hobbs, K. R., and Bangham, D. R.: Development of the immune response in the foetal and newborn rhesus monkey. I. Response to bovine serum albumin. Immunology, Lond. *11:* 185–198 (1966).

89 Silverstein, A. M. and Kraner, K. L.: Studies on the ontogenesis of the immune response; in Sterzl Molecular and cellular basis of antibody formation, pp. 341–348 (Academic Press, New York/London 1965).

90 Parshall, C. J., jr. and Silverstein, A. M.: Surgical approaches to the study of fetal immunology in primate animals. Ann. N.Y. Acad. Sci. *162:* 254–266 (1969).

91 Berman, L. D.; Cotes, P. M., and Simons, P. J.: Tumors and immunological response in rhesus monkeys born after *in utero* inoculation with Rous sarcoma virus. J. nat. Cancer Inst. *39:* 119–131 (1967).

92 Wislocki, G. B.: Observations on twinning in marmosets. Amer. J. Anat. *64:* 445–483 (1939).

93 Hampton, J. K., jr. and Hampton, S. H.: Marmosets (Hapalidae); breeding seasons, twinning, and sex of offspring. Science *150:* 915–917 (1965).

94 Hill, J. P.: The developmental history of the primates. Philos. Trans. (B) *221:* 45–178 (1932).

95 Benirschke, K.; Anderson, J. M., and Brownhill, L. E.: Marrow chimerism in marmosets. Science *138:* 513–515 (1962).

96 Gengozian, N.; Batson, J. S., and Eide, P.: Hematologic and cytogenetic evidence for hematopoietic chimerism in the marmoset *Tamarinus nigricollis.* Cytogenetics *3:* 384–393 (1964).

97 Gengozian, N.; Batson, J. S.; Greene, C. T., and Goslee, D. G.: Hemopoietic chimerism in imported and laboratory-bred marmosets. Transplantation *8:* 633–652 (1969).

98 Porter, R. P. and Gengozian, N.: Immunological tolerance and rejection of skin allografts in the marmoset. Transplant. Bull. *8:* 653–665 (1969).

99 Gengozian, N.: Marmosets: Their potential in experimental medicine. Ann. N.Y. Acad. Sci. *162:* 336–362 (1969).

100 Howard, R. J.; Landon, J. C.; Dougherty, S. F.; Notkins, A. L., and Mergenhagen, S. E.: Induction of tolerance in immunologically competent primates. J. Immunol. *102:* 266–268 (1969).

101 Stanworth, D. R.: Reaginic antibodies. Adv. Immunol. *3:* 181–260 (1963).

102 Caulfield, A. H. W.; Brown, M. H., and Waters, E. T.: Suitability of the monkey *(Macacus rhesus)* as a recipient for the Prausnitz Kustner reaction. Proc. Soc. exp. Biol., N.Y. *35:* 109–112 (1936).

103 Straus, H. W.: Studies in experimental hypersensitiveness in rhesus monkeys; passive local cutaneous sensitization with human reaginic sera. J. Immunol. *32:* 251–269 (1937).

104 Layton, L. L.; Lee, S., and DeEds, F.: Diagnosis of human allergy utilizing passive skin-sensitization in the monkey *Macaca irus.* Proc. Soc. exp. Biol., N.Y. *108:* 623–626 (1961).

105 Layton, L. L.; Lee, S.; Yamanaka, E.; Greene, F. C., and Greene, T. W.: Allergy skin tests upon castor bean sensitive human and passively sensitized cynomolgus monkeys. Int. Arch. Allergy 20: 257–261 (1962).

106 Layton, L. L.; Greer, W. E.; Greene, F. C., and Yamanaka, E.: Passive transfer of human atopic allergies to catarrhine and platyrrhine primates of Suborder Anthropoidea. Int. Arch. Allergy *23:* 176–187 (1963).

107 Layton, L. L.: Passive transfer of human atopic allergies into lemurs, lorises, pottos and galagos; possible primate-ordinal specificity of acceptance of passive sensitization by human atopic reagin. J. Allergy *36:* 523–531 (1965).

108 Rose, N. R.; Kent, J. H.; Reisman, R. E., and Arbesman, C. E.: Demonstration of human reagin in the monkey. I. Passive sensitization of monkey skin with sera of untreated atopic patients. J. Allergy *35:* 520–534 (1964).

109 Feinberg, S. M.; Feinberg, A. R., and Lee, F.: Hypersensitivity responses in monkeys; histamine skin response, 48–80, passive transfer (man to monkey, monkey to monkey) and other features related to human reactivity. Ann. Allergy *26:* 99–106 (1968).

110 Arbesman, E. C.; Girard, P., and Rose, N. R.: Demonstration of human reagin in the monkey. II. *In vitro* passive sensitization of monkey ileum with sera of untreated atopic patients. J. Allergy *35:* 535–546 (1964).

111 Reid, R. T.; Minden, P., and Farr, R. S.: Reaginic activity associated with IgG immunoglobulin. J. exp. Med. *123:* 845–858 (1966).

112 Goodfriend, L.; Kovacs, B. A., and Rose, B.: *In vitro* sensitization of monkey lung fragments with human ragweed atopic serum. Int. Arch. Allergy *30:* 511–518 (1966).

113 Malley, A. and Harris, R. L., jr.: Passive sensitization of monkey lung fragments with sera of timothy-sensitive patients. I. Spectrofluorometric analysis of histamine release. J. Immunol. *100:* 915–920 (1968).

114 McGerity, J. L.; Arbesman, C. E., and Wicher, K.: Comparison of passive sensitization of skin and lung of human being and ileum and lung of monkey with sera of ragweed sensitive patients. J. Allergy *43:* 200–208 (1969).

115 Ishizaka, T.; Ishizaka, K.; Orange, R. P., and Austen, K. F.: The capacity of human immunoglobulin E to mediate the release of histamine and slow reacting substance of anaphylaxis (SRS-A) from monkey lung. J. Immunol. *104:* 335–343 (1970).

116 Ogilvie, B. M.: Reagin-like antibodies in animals immune to helminth parasites. Nature, Lond. *204:* 91–92 (1964).

117 Ogilvie, B. M.; Smithers, S. R., and Terry, R. J.: Reagin-like antibodies in experimental infections of *Schistosoma mansoni* and the passive transfer of resistance. Nature, Lond. *209:* 1221–1223 (1966).

118 Sadun, E. H.; Lichtenberg, E. von; Hickman, R. L.; Bruce, J. I.; Smith, J. H., and Schoenbechler, M. J.: Schistosomiasis mansoni in the chimpanzee: parasitologic, clinical, serologic, pathologic and radiologic observations. Amer. J. trop. Med. Hyg. *15:* 496–506 (1966).

119 Edwards, A. J.; Jones, V. E.; Smithers, S. R., and Terry, R. J.: The occurrence and properties of reagins in rhesus monkeys infected with *Schistosoma mansoni*. Ann. trop. Med. Parasit. *61:* 280–293 (1967).

120 Hsü, H. F. and Hsü, S. L. Y.: Reagin-like antibodies in rhesus monkeys immune to *Schistosoma japonicum*. Z. Tropenmed. Parasit. *17:* 166–176 (1966).

121 WISEMAN, R. A. and WOODRUFF, A. W.: Toxocara skin sensitivity tests and other observations in animals experimentally infected with *Toxocara canis*. Trans. roy. Soc. trop. Med. Hyg. *61:* 827–833 (1967).

122 WEISZER, I.; PATTERSON, R., and PRUZANSKY, J. J.: Ascaris hypersensitivity in the rhesus monkey. I. A model for the study of immediate type hypersensitivity in the primate. J. Allergy *41:* 14–22 (1968).

123 MALLEY, A.; AMKRAUT, A. A.; STREJAN, G., and CAMPBELL, D. H.: Hypersensitivity to Ascaris antigens. III. Atopic-type hypersensitivity induced in rhesus monkeys. J. Immunol. *101:* 292–300 (1968).

124 WICHER, R.; NEY, R.; REISMAN, R., and ARBESMAN, C. E.: Immunologic studies on active sensitization of monkeys with ragweed pollen. Fed. Proc. *27:* 367 (1968).

125 MALLEY, A. and PEARLMAN, F.: Induction of both reaginic and blocking antibodies with a low molecular weight fraction of timothy pollen. J. Allergy *43:* 59–64 (1969).

126 GIRARD, J. P. and HAMPAI, A.: Production of reaginic antibody in monkeys following immunization with dialyzable constitutents of timothy pollen extracts. Int. Arch. Allergy *37:* 561–574 (1970).

127 PATTERSON, R.; FINK, J. N.; WENNEMARK, J.; BAUM, J.; PRUZANSKY, J. J., and NISHIMURA, E. T.: The biologic consequences of the immediate type hypersensitivity transferred from man to monkey. J. Allergy *37:* 295–310 (1966).

128 PATTERSON, R.: Laboratory models of reaginic allergy. Progr. Allergy, vol. 13, pp. 332–407 (Karger, Basel/New York 1969).

129 PATTERSON, R.: Animal models demonstrating respiratory responses to inhalant antigens. Yale J. Biol. Med. *40:* 495–500 (1968).

130 PATTERSON, R. and TALBUT, C. H.: Respiratory responses in subhuman primates with immediate type hypersensitivity. J. Lab. clin. Med. *73:* 924–933 (1969).

131 PATTERSON, R.; ROBERTS, M., and PRUZANSKY, J. J.: Comparisons of reaginic antibodies from three species. J. Immunol. *102:* 466–475 (1969).

Authors' address: Dr. R. T. DAMIAN and Dr. N. D. GREENE, Division of Microbiology and Infectious Diseases, Southwest Foundation for Research and Education, *San Antonio, TX 78206* (USA)

Pathology of Simian Primates, Part I, pp. 380–398 (Karger, Basel 1972)

Autoimmune Diseases

O. Felsenfeld

Division of Communicable Diseases, Tulane University, Delta Regional Primate Research Center, Covington, La.

Contents

General ... 380
Etiology ... 381
Autoimmunity Induced by Microbial Agents ... 385
Rheumatoid Disease ... 386
Sjögren's Syndrome ... 388
Lupus Erythematosus ... 388
Scleroderma and Related Lesions ... 389
Thyroiditis ... 390
Allergic Gastritis and Pernicious Anaemia ... 391
Autoallergic Phenomena of the Gut ... 391
Diseases of the Liver ... 392
Renal Diseases ... 393
Heart ... 393
Myasthenia gravis ... 393
Endocrine Glands ... 394
Ophthalmic Conditions ... 394
Central Nervous System ... 395
References ... 395

General

The so-called autoimmune diseases have been discussed by numerous authors. Excellent reviews have been published by Condie and Good [1959]; Paterson [1959, 1966]; Kunkel and Tan [1964]; Burch and Burwell [1965]; Dumonde [1966]; Burnet [1966]; Nossal [1966]; Mackay [1966]; Pirofsky [1969]; Glynn and Holobrow [1969]; and others, dealing with one or more facets of this complex problem, which begins right with the designation of this group of sicknesses. The word 'immune' stood originally for 'exempt from military service' in Latin but modern usage gave it a different meaning. 'Autoimmune' could be translated as 'self-safe'; and 'autoallergic' as 'altered reactivity to self', principally when one takes into account the position of the term 'self' occupied in immunology as a result of the outstanding work of Sir MacFarlane Burnet on 'self-recognition' and related basic concepts. Autoallergy will not be discussed in this chapter. Neither will inroads be made into fundamental immunological considerations which are related in a different part of this volume. It is necessary, however, to summarize a few salient points that require consideration in the analysis of autoimmunity.

Etiology

The introduction of a suitable material (antigen) into the body sets up a sequence of events that are demonstrable by immunoserological tests and also often by the observation of histological changes. If very young animals are immunized (receive antigen), tolerance or unresponsiveness may develop. In immune responses, the antigen impresses a pattern of formation of antibodies on the cells called immunocytes. Several theories are available to explain this process. They can be divided into two groups. The template (direct or indirect) theories are based on the presence of an antigen that leads to a suitable formation of a template which will carry the regulatory message to the immunologically competent cells (immunocytes) that will manufacture immune globulins with antibodies. The clonal theories presume the presence of genetically endowed immunologically competent cell clones that form antibody if and when stimulated. Burwell [1963] strongly emphasized a recessive X-linked factor as part of this system, and saw in the immunocytes (which are lymphatic and plasma cells) a system committed to regulate tissues throughout the body. It is generally believed that the thymus is the organ where precursors of lymphoid-plasma cells (stem cells) formed in the bone marrow are primed for immunocytic activity. Good *et al.* [1966] demonstrated other regional centres, analogous to the *Bursa fabricii* of aves, where proimmunocyte and immunocyte regulation takes place. We were able to show the presence of such a regional centre in the caecum of *Erythrocebus patas* monkeys [Brandom *et al.*, 1969] by stimulation with cholera vibrio antigen. Undoubtedly, similar regional centres exist also in other sectors of the body, as. e.g. the respiratory and the urogenital tract. The cell-bound antibodies which are part of the structure or configuration of immune globulins, principally those designated IgG, IgM, IgA and IgE, are present in immunocytes but there are significant differences between the lymphocytes apparently governed by the thymus alone and playing a role in transplantation immunity, graft *versus* host reaction, and autoimmunity without plasmacytic involvement and often also without demonstrable circulating antibodies, and the plasma cells responding to infective agents. These factors are interwoven in autoimmune diseases.

BURNET [1966] defined autoimmune disease as a condition resulting from the interaction of antibodies or immunocytes and antigenic determinants carried by normal body components. The specificity of antibodies is regulated by the configuration of its combining site(s). Specificity is determined by the genetic formation carried and phenotypically expressed by the immunocyte concerned. After antibody-antigen contact, destructive processes may set in with liberation of pharmacologically active agents, or conversion of the lymphocytes, or proliferation of germinal centres and production of plasma cell clones, and antibody formation. Multiresponsive phenomena also may occur. BURNET considered immunological tolerance a natural phenomenon and believed that the thymus eliminates cells carrying immune patterns ('forbidden' clones) capable of reacting with potential antigenic determinants present in their natural environment. This concept leads to the possible consideration of neoplasms as conditioned autoimmune phenomena related to 'forbidden' clones, and to regarding hypergammaglobulinaemia the result of the formation of an enormous pool of immunocytes producing globulin, as well as to the understanding of autoimmune Coombs-positive haemolytic anaemia, in which red blood cells are coated by autoantibodies. Pathologic clones in autoimmune disease may appear as a result of random selection but the requirement for genotypic susceptibility remains a governing factor also in this group of diseases.

MACKAY [1966], a former co-worker of BURNET, also regarded autoimmune phenomena as a fault of recognition and tolerance for self-constituents. In autoimmune disease, the antibody-forming tissue may be aberrant, and homoeostatic elimination of anti-self immunocytes may not take place. Whereas in allergy or hypersensitivity a damaging immune reaction to a usually recognizable extrinsic antigen develops, in autoimmune disease there are hypergammaglobulinaemia, high titres of circulating antibodies, deposition of denaturated gamma-globulins, strong lymphoid-plasma cell hyperplasia, and responsiveness to immunosuppressive agents as well as corticosteroids. These phenomena coexist with other lesions attributable to the immunological response. Both conditions, allergy and autoimmune phenomena, often coexist in autoimmune disease, as e.g. in lupus erythematosus.

GRAS [1969] called attention to the immunological equilibrium that is indispensible for normal cell response. This may be inhibited by prolonged antigen administration (hyperimmunization) but hyperimmunization does not necessarily lead to autoimmune phenomena even though anti-organ antibodies may circulate after such treatment, as was shown also by our group in *Cercopithecus aethiops* hyperimmunized with conventional cholera vaccine [FELSENFELD and GREER, 1967]. Transient antibodies developed against extracts of the primate's own liver but disappeared after a few weeks.

BURCH and BURWELL [1965] believed that there is a specific mitotic control protein particular to the lymphoid system in all organs, which is the effector of symmetrical cell proliferation, as well as a protein tissue coding factor that is located on cell membranes and identifies the target tissue. This concept explains the presence of an afferent signal in the immunological feedback control circuit that may fail in autoimmune disease. They postulated also that symmetrical mitosis is normally inhibited by specific contact relationships between precursor cells of similar differential determinants bearing similar tissue coding factors. This concept would allow understanding of self-recognition on a molecular level, and explain the effect of genetic mutation in the lymphoid mitotic control cells as converting agents of control protein into autoantibody, with complementary relationship to the associated tissue coding factor.

NOSSAL [1966] emphasized sequestered antigens as incitants of autoantibody formation, and summarized the difficulties arising in animal experiments when Freund's adjuvant is employed. It stimulates but also disrupts the function of the lymphoid system. An adjuvant is never tolerogenic, even though there is a zone of antigenic dosis that may cause tolerance which is altered when Freund's adjuvant is administered. Focal histological changes and tissue breakdown as a result of such an injection was emphasized also by MACKAY and BURNET [1963] who pointed out that sequestered antigens may be liberated by trauma and infection. This will be discussed later.

NOSSAL [1966] did not believe that the immunological response consisting of germinal centre formation, increased immunocytopoiesis, and a varying degree of reticulum cell and histiocyte hyperplasia, differs in autoimmune diseases but noted that the site of these activities may vary.

WALFORD [1969] reviewed the question of the influence of age on autoimmune phenomena and their relationship to his own observations. He confirmed the effect of age on the occurrence of such diseases. WALFORD thought that perhaps also such markers of old age as loss of teeth, graying of the hair, as well as Dupuytren contracture may be expressions of autoimmunity. This is in agreement with BURNET's postulatum that immunological tolerance is breaking down with age, hence autoantibody formation increases. On the other hand, it is possible that cell alterations occur under the influence of physical agents (X-rays, heat, cold, etc.) or repeated trauma and numerous infections suffered during a long life-span unmask antigenic factors, release specific components, or that a central failure of responsiveness develops with aging.

The differences between the immunological response of normal and autoimmune *(sit venia verbo)* animals is not always remarkable. A transient rise in autoantibodies directed against various organs may be observed in numerous infectious diseases [DUMONDE, 1966]. KUNKEL and TAN [1964] suggested that autoantibodies be divided into the more infrequent 'pathogenical' and the more common 'physiogenical' types. Among the latter belong immunoconglutinins that appear to be autoantibodies directed against complement which is helpful in removing microbial invaders [COOMBS *et al.*, 1961] as well as in the phagocytosis of immune globulin-coated bacteria. Antiglobulins, especially against IgM (19S), are present in normal sera and appear to be harmless but their titre increases considerably in rheumatoid arthritis. Anticytoplasmic antibodies to liver, spleen and kidney are not a rarity, and their levels become higher during immunization [MACKAY and BURNET, 1963]. The complement-fixation test permits their detection. This test shows high titres also in macroglobulinaemia, perhaps due to aggregation of the immune globulins or their constituent components.

The essentials of autoimmune phenomena were presented by ASHERSON [1968] as an antigen-mediated depression of the immunological response that may also be called selective inhibition or immune deviation. Delayed hypersensitivity does not characterise all autoimmune diseases but if present, the immunological response in adults is more easily suppressed. Autoimmune disease or autoimmune response depends on autoreactive hypersensitivity that is synergetic with circulating antibodies. Delayed hypersensitivity also demonstrates greater antigenic specificity. Autoantibodies are directed against intracellular components or against cell or membrane surfaces. Many attach themselves to proteins activated after cell damage or by proteolytic enzymes in the intercellular spaces. It is possible that the surface antigens of pathogenic microorganisms undergo adaptive changes and acquire similarity to some of the host cells or membranes.

Summarizing, in autoimmune disease the antibody producing system may be abnormal, revealing a disturbance of immunological tolerance. There may be merely an alteration of partial tolerance [DUMONDE, 1966]. The same consequences may follow the presence of unusual or altered antigens including own proteins, foreign proteins, or the unmasking of antigenic determinants normally hidden from contact with extracellular elements. The reaction itself resembles the response to artificial hapten and protein conjugates, e.g. lipopolysaccharideprotein complexes, or brain lipoid hapten reactions. Organ specificity may be lacking or present, and is sometimes pathognomical. Genetic factors, alterations of self-recognition and environmental influences may stand in the foreground. In animal experiments, adjuvant or carrier protein enhance autoimmune responses but may modify them. The tissue damage may be minimal, organ-specific, or more generalized.

There is a large number of autoimmune diseases discussed in text-books and reference articles quoted previously. Here we will limit ourselves to those that are of interest in nonhuman primates, particularly in the use of the primate model in the study of human disease.

Autoimmunity Induced by Microbial Agents

BURNET [1962] and ACHERSON [1968] discussed these conditions in man and animals.

Microorganisms may liberate normal or damaged tissue components; modify tissues enzymatically; combine with tissue; may be chemically related to tissue factors; or influence the union of antibody with complement, giving rise to the so-called rheumatoid factor and/or conglutinin.

Many microbes harbour antigenic components similar to human blood antigens. Enterobacteriaceae and *Pasteurella* are best known in this group. *Escherichia coli* 086, for instance, is rich in blood-group *B*, pseudomonads in D, *Pneumococcus* type XIV in O and A, *Mycoplasma* and *Listeria* in I, *Ascaris* in P_1, echinococcal cyst fluid in P blood substances.

Tubercle and leprosy bacilli, *Bordetella*, *Corynebacteria*, moulds and yeasts, and (even endotoxin) have been enumerated among the microorganisms giving rise to autoimmunity.

Streptococci and the heart muscle, some *Escherichia coli* (principally serotype 014) and the colon, beta-haemolytic streptococci (principally type 12) and the kidney possess similar antigenic factors. Thus infection with such organisms may result in autoimmune responses. Plasmodia contain Wassermann antigen but the pathological significance of the phenomenon is not yet clear. The Wassermann antigen is a diphosphatide glycerol that is present in numerous normal primate tissues containing linoleic acid principally in the mitochondria.

The substance causing autoimmunity is usually very simple in its chemistry, as glycogen, other polysaccharides, lipids and nucleic acids but seldom (if ever) proteinic. It is important that such compounds have to be coupled to proteins or relatively high-molecular polypeptide building stones to achieve strong antigenicity. WILLIAMS *et al.* [1968], discussing autoimmunity and viruses, called attention to the morphological similarities of a chronic viral disease of the Aleutian mink and juvenile autoimmune plasmacytic hepatitis. These authors posed the question if the virus-carrying cell coated with immune globulin does not lead to the perception of the globulin as an antigen.

Kuru, successfully transmitted to chimpanzees by GAJDUSEK *et al.* [1966, 1967] has characteristics compatible with autoimmune disease with a strong genetic accent, and a hitherto not fully explored immunological mechanism involving the central nervous system.

In polyoma infection, transformation of the invaded cells with antigenic modification develops.

VAN LOGHEM [1965], reviewing the relationship of viruses and acquired haemolytic anemia, pointed out Newcastle disease as an example of an autoimmune response. Some Rickettsiae and *Anaplasma* also cause such reactions.

One may theorise as to whether the frequent occurrence of Rh-positive blood in Africa and Asia, accompanied by antiglobulin antibodies, may be ascribed to malaria. Moreover, as stated, Plasmodia contain also Wassermann antigen. The frequently elevated immune globulin M (19S) in Africa has been thought to be due to trypanosomiasis, malaria, genetic factors or helminthic infections. The need for further investigation and experimentation is evident in this field, especially as the origin and course of blackwater fever shows several features [cf. FELSENFELD, 1965] compatible with failure of immunological tolerance and 'normal' antibody response.

Rheumatoid Disease

We will consider here only the pathologic condition in which the so-called 'rheumatoid factor' (RF) is involved. Autoimmune globulins of the 19S type (IgM) play a role in the antibody formation and are specific for typical antigen-antibody complexes in this disease. Genetic factors are evident in the formation of the participating antibodies. Gm and InV loci carry groups of alleles in man and according to HESS and BÜTLER [1962] and ALEPA and STERNBERG [1964] also in nonhuman primates. The rheumatoid agglutinator (Ragg) shows lower specificity for Gm(a +) globulin. Aggregates of Gm(a–) globulin are considerably more inhibitory in the Ragg system but if the two factors are modified, the difference in their behaviour diminishes. In rheumatoid arthritis, Ragg is iso- or auto-specific, whereas the agglutinator present also in some normal sera is only iso-specific. The normal serum agglutinator becomes demonstrable after gamma-globulin injection; may be transmitted by the mother to her offspring; or appears after blood transfusion. It is non-precipitable by routinely employed procedures. Children form anti-Gm antibodies with ease, principally when the mother is Gm-positive, and the offspring Gm-negative. Since the antibody is of the 19S type that does not pass the placental barrier, no complications arise *in utero*. No anti-InV autoantibody has yet been recorded.

Altered antibodies may be present in rheumatoid arthritis. A non-specific factor (F) directed against a fragment of immune globulins also may be demonstrable. This does not seem to be in discord with the concept that altered autologous gamma globulins produce antibodies regulated by genetically governed agents. Immune globulin complexing is not rare. A labile 22S with weak affinity has been described [KUNKEL and TAN, 1964]. IgG also has the ability of complexing. The great variety of antibody coating of red blood cells coated with incomplete Rh antibodies under these circumstances raises the question of the greater readiness of heterozygous individuals to form such antibodies. The nature of antibody-antigen complexes has not yet been fully clarified nor is clear their relationship to the RF. When serum containing RF is injected into man or chimpanzee, no symptoms develop but polymorphonuclear leukocytes may carry the RF. These cells may show gamma-globulin inclusions, often related to IgM [VAUGHAN *et al.*, 1968a, 1968b]. Such inclusions are seen, however, also in other diseases. These leukocytes may be scavengers, or may cause cell and tissue destruction. It is of interest that the amount of complement is reduced when gamma-globulin inclusion body containing polymorphonuclear cells are numerous in the blood. The same holds true about joint fluid.

High RF levels have been found in lupus erythematosus, tuberculosis, leprosy, chronic bronchitis of undetermined nature, subacute bacterial endocarditis, infective hepatitis, trypanosomiasis, and a number of other conditions, including Felty's syndrome (splenomegaly with leukopenia and anaemia).

Perhaps the complex formed by RF-anti-RF antibody is causing vascular damage. SVARTZ *et al.* [1969] demonstrated an anti-RF factor, which is a new type of immune globulin G. This finding is opening new vistas for the study of rheumatoid arthritis. The abnormal streptolysin S present in the same disorder also requires further investigation before its relationship to rheumatoid arthritis can be fully established. Nonhuman primates offer an excellent model for such tests.

Sjögren's Syndrome

This condition, connected with dryness of the salivary, lacrimal and other glands, with hyperglobulinaemia and elevated rheumatoid factor levels, may be regarded as the connecting link between rheumatoid arthritis and lupus erythematosus. Multiple antibody formation takes place, and antibodies against salivary tissue have been demonstrated [BERTRAM and HALBERG, 1969]. ASHERSON [1968] believed that the immunological process is directed by nucleosides attached to proteins, the former having acquired immunogenicity because of this union. We have been able to produce transient antinuclear response in *Macaca speciosa* by repeated injections of autologous salivary gland tissue but evidently, as in other autoimmune diseases, the genetic constitution was unfavourable for a lasting response, and there was no underlying tissue damage prior to the immunization.

Lupus Erythematosus

MACKAY [1966] called attention to the fact that forbidden clones react with a wide variety of autoantigens. The role of genetic factors in this disease has been proved by the study of identical twins. This author believed that antibody-antigen complexes formed in lupus erythematosus cause vascular damage, including glomelural lesions of the so-called lupus nephritis type. Immunocytes invade the affected tissues. Cell-surface damage causes the liberation of intracellular substances as lysosomal enzymes, heparin and histamine, that are pharmacologically active. The lupus erythematosus cell (LE) factor, a 7S globulin, reacts with nucleoprotein. Complement is fixed in this process, and further tissue damage ensues. DUBOIS and ARTERBERRY [1966] reviewed another aspect of the problem, emphasising that hereditary dysproteinaemia coupled with some other defect of metabolic or infectious nature may be the underlying cause of lupus erythematosus if a triggering factor as extreme temperature, drugs or infection set off the chain of events leading to this condition.

Lupus erythematosus is characterised by multiple autoantibody formation. Several species of bacteria, as Lactobacilli, *E. coli* and *Brucella*, whole organisms as well as their nucleoprotein extracts, caused lupus erythematosus-like processes in laboratory rodents. The principal immunological sign of lupus erythematosus is the antinuclear factor reacting with DNA. It appears that the antibodies are directed against the purine and pyrimidine bases of DNA but a wide variety of them is aimed at different sites on the DNA molecule, as well as against histones of nucleoproteins. KUNKEL and TAN [1964] pointed out that the antibodies may be directed against binding sites on the nucleoproteins.

The LE cell factor is closely related to antibodies reacting with nucleoprotein but are independent on circulating immune globulin levels. Hyperimmunization with nucleoprotein did not produce lupus erythematosus in rabbits in the experiments of CHRISTIAN *et al.* [1963], nor in several species of nonhuman primates in our hands.

There is a cytoplasmic factor involved in lupus erythematosus that may be demonstrated in the kidneys, as well as an antigen attached to the basement membrane of the glomeruli. There is no evidence of antikidney antibodies. Tissue breakdown, however, may incite further antibody formation.

Globulin is present in the glomeruli, as well as complement, which is low in the blood serum. Steroid therapy often restores the balance but does not break down globulin aggregates in the kidneys.

The role of complement appears important to many authors in autoimmune disease, particularly in lupus erythematosus. Cryoglobulin that fixes complement is frequently found in the kidneys. The dermal-epidermal surface and the epidermal cells but not hyperkeratotic or necrotic skin contain much gamma globulin. Considerable amounts of antibody and components of the complement are utilized during this disease also in the tegument. The consequences are obvious.

The disease is often associated with rheumatoid arthritis.

Scleroderma and Related Lesions

Some forms of scleroderma are of intermediate character and cannot be distinguished from dermal lesions in lupus erythematosus. There are different patterns of nuclear staining, and a precipitin reaction identical with that observed in Sjögren's syndrome [BECK *et al.*, 1963]. The antigens appear to be cytoplasmic.

Dermatomyositis often accompanies malignancy, and improves when the growth is treated. It has also been seen together with benign tumours. One may believe that dermatomyositis is an expression of generalized sensitivity to malignancy, with the involvement of factors common to autoimmune disease and tumour growth.

Dermatitis herpetiformis appears to be evoked by antigens derived from the epidermal basement membrane. BEUTNER and JORDON [1964] succeeded in inducing it in *Macaca rhesus.*

Thyroiditis

One type of chronic thyroiditis, called also lymphadenoid goiter and Hashimoto's disease, is of autoimmune genesis. Thyroglobulin appears to be implicated. Antigen has been found in microsomal fractions, principally of the acinar cells. Fluorescent microscopical investigations revealed antigen in the follicular colloid. The antigens may form autoprecipitates, and are found in rheumatoid arthritis, lupus erythematosus, hypergammaglobulinaemia, myxoedema, thyrotoxicosis, pernicious anaemia, leprosy, and other conditions. Their chemical nature is anchored to the thyroxyl group that has haptenical properties. Perhaps abnormal coupling of thyroid hormones also take place. DUMONDE [1966] described successful experiments on *Macaca rhesus* with thyroglobulin-induced autoimmune state, accompanied by 7S antibody formation and an Arthus-type skin reaction but other serological tests gave results different from those observed in man. Autologous thyroid tissue, administered parenterally with Freund's adjuvant, produced in animals damage of the protective basement membrane. X-ray irradiation or the use of heterologous antigens facilitated the induction of the disease, as in most other autoimmune conditions.

ROSE *et al.* [1965, 1966]; KITE *et al.* [1965, 1966]; DOEBBLER and ROSE [1966], and ANDRADA *et al.* [1968] experimented on *Macaca mulatta* to a considerable extent, and found this species feasible for the study of autoimmune thyroiditis. The spontaneous occurrence of this disease in nonhuman primates has not yet been established beyond reasonable doubt.

Allergic Gastritis and Pernicious Anaemia

RAPP *et al.* [1964] distinguished autoantigenic factors in the stomach as anti-tissue, anti-intrinsic, and anti-gastric parietal cell factors. Undoubtedly, numerous components participate in autoimmune processes. BERNIER and HYNES [1967] analysed circulating (blocking) antibodies that inhibit the attachment of the intrinsic factor to B_{12} vitamin, and precipitating antibodies which form a complex with the intrinsic factor-B_{12} union. These authors defined autoantibodies in pernicious anaemia as predominantly of the 7S type, with heterogeneous side chains. SAINT MARTIN and EYQUEM [1967] found similarities between the cytoplasmic antigen in pernicious anaemia and autoallergic gastritis. Heterogeneous gastric antigen prepared from crude gastric extracts caused autoimmune disease in *Macaca rhesus.*

The serum inhibitor of B_{12} absorption in pernicious anaemia is not related to cytoplasmic extracts of gastric parietal cells. The former is localized in the gut and not in the stomach. The latter is located in microsomes and has a wide reactivity in man and nonhuman primates, appearing also in atrophic gastritis, thyroiditis and thyroidisms, diabetes, iron deficiency anaemia and other conditions. ANDRADA *et al.* [1969] administered pooled gastric mucosa with Freund's complete adjuvant to rhesus monkeys. Autoantibodies developed. Multifocal gastritis with monocytic infiltrates in the gastric mucosa with atrophy of the gastric glands in the neighbourhood of the lesions developed. Autoantibodies were present in the cytoplasm of the gastric parietal cells. Thyroidal, adrenal and testicular extracts did not provoke a similar reaction. These experiments confirm the close similarity of the immunological responses elicited in the human and nonhuman primate's gastroenterical system.

Autoallergic Phenomena of the Gut

EIDELMAN [1968] reviewed the problems of the small intestine, some of which have bearing on autoimmunity. The principal immune globulin of the intestinal tract, IgA, may be substituted by the large-molecular IgM. Hypogammaglobulinemia is often associated with steatorrhoea. The question of sprue is still unsettled, and local dysfunction may be instrumental in its origin. Moniliasis, decades ago considered the cause of sprue by Castellani, is able to produce autoimmune responses. The numerous factors involved in sprue disease have not yet been brought on a common denominator.

Chronic idiopathic ulcerative colitis has been a subject of extensive studies, more recently by KIRSNER, PERLMANN, BROBERGER and their co-workers [reviewed by DUMONDE, 1966]. It has been mentioned previously that some *Escherichia coli* strains, particularly serotype 014, contain a polysaccharide related to that of the colon. Such an agent may induce an autoimmune response, principally in the presence of genetic factors favourable for its development. It is not certain that only the 014 serotype may be involved, or that this organism has to be present at the time when the first clinical manifestations of the disease become apparent. Judging from our own experiments on chimpanzees, one may suspect the 'original sin' coming into play, similar to that in viral infections when antibodies against long past infections appear if invasion by another strain takes place. Responses to common antigen and to interaction between colonic bacteria and mucosa are more often manifest in the young, who are also more frequently victims of ulcerative colitis.

Autoantibody to colon may be demonstrated also in Crohn's disease and sometimes in amoebiasis, principally in the goblet cells, with the aid of fluorescent microscopy.

Lesions resembling chronic ulcerative colitis have been observed in gibbons [STOUT and SNYDER, 1969]. Psychological factors appear to enhance the development of this malady in gibbons [ENGEL, 1969] as in man.

Diseases of the Liver

MACKAY and BURNET [1963] emphasized the appearance of cytoplasmic antigens in the liver. There seem to be several antigens, related to various cytoplasmic components, as was evident also in our observation [FELSENFELD and GREER, 1967]. The rheumatoid factor is not infrequent in various liver diseases. Hepatic autoallergic phenomena have been recorded in schistosomiasis and in *Eimeria* infections of rodents.

DONIACH and WALKER [1969] demonstrated the overlapping of autoantibodies in juvenile cirrhosis, primary biliary cirrhosis and so-called cryptogenic (silent or overt) cirrhosis of the liver. According to these authors, one single process may damage hepatocytes or bile ductules and, according to the structure affected, cause active hepatitis and cirrhosis, or primary biliary disease and eventual cirrhosis, or the silent stage of cryptogenic cirrhosis. The latter syndrome represents a heterogeneous group of pathological states. Genetic predisposition appears to play a significant role, and tissue antibodies are markers but not the cause of these illnesses as in other autoallergic diseases

Renal Diseases

Those allied with lupus erythematosus have been discussed.

It appears that in 'nephrotoxic nephritis' the glomelural cortex is the site of the autoimmune process. Protein-polysaccharide complexes are formed on the basement membrane but possibly also in glomelural cells. In other instances the renal tubular cells are involved. Similar autoimmune processes take place in other organs but not in avascular tissues [DUMONDE, 1966]. Antibodies against glomeruli (in the broader sense) are usually accompanied by the formation of complexes between immune globulins, glomeruli and complement. The beta-haemolytic streptococci most frequently implicated in triggering renal autoimmune processes belong (in decreasing order) to serotypes 1, 4, 12 and 49.

STEBLAY [1963] injected *Macaca* with homologous glomerular basement membrane extracts. Autoantibodies could be demonstrated but local tissue lesions were rare. This observation fortifies the opinion that the administration of the antigen alone is not sufficient to induce autoimmune disease.

Heart

DUMONDE [1962] stated that autoantigens in man are rather of microsomal than of mitochondrial origin. Fluorescent antibody studies demonstratc ed sarcolemmal, subsarcolemmal, intermyofibrillar and diffuse sarcoplasmic antigens which were present also in skeletal muscles. Streptococci are often active agents. Organisms of the same *genus* are involved in subacute bacterial endocarditis but autoantibodies disappear in the latter after a favourable termination of the disease.

There has been a considerable output of reports on the relationship of immunity and myocardial diseases, especially infarction. The experiments hitherto reported show divergent results and are, therefore, not discussed at this time.

Myasthenia gravis

It has been proposed [SIMPSON, 1960] that in this disease antibodies against the neuro-muscular end-plates occur and act as competitive blocking agents. The complement level fluctuates in this disease, becoming lower during exacerbations. Therefore, a destructive autoimmune mechanism had been postulated. Fluorescent antibody techniques [STRAUSS *et al.*, 1960] revealed several cardiac autoantigens in skeletal muscles, with or without involvement of the sarcolemma and the thymus. The autoantibodies seem to be transmitted from mother to offspring but disappear from the child's blood. This confirms the 7S nature of such autoantibodies.

Injections of extracts of autologous thymus into rhesus monkeys did not cause autoantibody formation against components of skeletal muscles in our hands.

Endocrine Glands

Autoimmune thyroid disease was described under 'Thyroiditis'.

FEDERLIN [1969] reviewed the present status of investigations in this field. He believed that the triad Addison's disease, thyroiditis and hypoparathyreoidism is of genetic origin. Autoimmune polyendocrinopathia, particularly the type associated with moniliasis, is not a rare phenomenon.

Pancreatic lesions with isoantibody formation are difficult to study because of the considerable diversity of the antigens, including the so-called pancreatic juice. METZGAR [1964a, 1964b] succeeded in introducing autoimmune responses related to the pancreas in nonhuman primates.

Idiopathic adrenal atrophy may be due to autoimmune reactions but the adrenal is seldom the only organ affected in this condition.

Ophthalmic Conditions

Uveitis has been reviewed by LUNTZ [1968]. Autologous ocular tissue was shown to produce autoimmune phenomena in monkeys by VOJTIŠKOVÁ *et al.* [1962] and BROWN *et al.* [1963].

Sympathetic ophthalmia belongs in this group. Uveal and lens proteins may play a role in it. Endogenous uveitis and lens-induced uveitis show high autoantibody titres. Perhaps the synovium and/or the iris are sensitised by the same antigen. Low levels of anti-lens antibodies have been found in traumatic iritis. These antibodies may occur also in normal persons, in rheumatoid disease and other conditions. Anti-uveal antibodies, however, have not been found in healthy individuals. A broad field is open here for further studies.

Central Nervous System

Allergic encephalomyelitis is probably caused by a proteolipid antigen, or a collagen-like protein, or by a basic protein [PATERSON, 1966]. Myelin has encephalopathogenic activity resembling mitochondrial damage after parenteral administration. The 'encephalopathogenic protein' may be a constituent of myelin. The hypersensitivity can be transmitted to other animals by sensitised cells. The antibodies are not of the conventional type, and cross-react with uveal tissue. The commonest form of allergic encephalomyelitis is that developping after rabies vaccination in which rabbit nervous tissue is injected. The disease has been studied to a considerable extent [PATERSON, 1966] in nonhuman primates and findings resembling those in man have been recorded.

SIVADON [1968] analysed the relationship of mental pathology and autoimmune disease. Following the studies of HEATH, he injected sera of patients with schizophrenia into rhesus monkeys. This caused abnormal electroencephalograms in the recipients. No references could be found on similar studies in other mental diseases of non-infectious origin.

References

ALEPA, F. P. and STEINBERG, A. G.: Production of anti-Gm reagents by rhesus monkeys. Vox Sang. *7:* 231 (1964).

ANDRADA, J. A.; ROSE, N. R., and ANDRADA, E. C.: Experimental autoimmune gastritis in the rhesus monkey. Clin. exp. Immunol. *4:* 283 (1969).

ANDRADA, J. A.; ROSE, N.R., and KITE, J. H., jr.: Experimental thyroiditis in the rhesus monkey. Clin. exp. Immunol. *3:* 133 (1968).

ASHERSON, G. L.: The role of micro-organism in autoimmune responses. Progr. Allergy *12:* 192 (1968).

BECK, J. S.; ANDERSON, J. R.; GRAY, K. G., and ROWELL, N. R.: Antinuclear and precipitating antibodies in progressive systematic sclerosis. Lancet *ii:* 1188 (1963).

BERNIER, G. M. and HYNES, J. D.: Immunologic heterogeneity of autoantibodies in patients with pernicious anemia. New Engl. J. Med. *277:* 1386 (1967).

BERTRAM, U. and HALBERG, P.: Specific antibody in sera from patients with Sjögren's syndrome. Acta allerg. *19:* 458 (1964).

BEUTNER, E. H. and JORDON, R. E.: Demonstration of skin antibodies in sera of pemphigus vulgaris patients. Proc. Soc. exp. Biol., N.Y. *117:* 505 (1964).

BRANDOM, W. F.; FELSENFELD, O., and GREER, W. E.: Proliferation of immunologically active cells in the intestines. Ann. trop. Med. Parasit. *63:* 1129 (1969).

BROWN, P. C.; GLYNN, L. E., and HOLBOROW, E. J.: Pathogenesis of allergic orchitis and uveitis in guinea pigs. J. Path. Bact. *85:* 505 (1963).

BURCH, P. R. J. and BURWELL, R. G. Self and not self. A clonal approach to immunology. Quart. Rev. Biol. *40:* 252 (1965).

BURNET, F. M.: Autoimmune disease. Proc. roy. Soc. Med. *55:* 619 (1962).

BURWELL, R. G.: The role of lymphoid tissue in morphostasis. Lancet *ii:* 69 (1963).

CHRISTIAN, C. L.; DESIMONE, A. R., and ABRUZZO, L. J. L.: Anti-DNA antibodies in hyperimmunized rabbits. Arthr. Rheumat. *6:* 766 (1963).

CONDIE, R. M. and GOOD, R. A.: Immunologic aspects; in KOREY Biology of Myelin, 321 (Harper, New York 1959).

COOMBS, R. R.; COOMBS, A. M., and INGRAM, D. G.: The serology of conglutination and its relationship to disease (Thomas, Springfield 1961).

DOEBBLER, T. K. and ROSE, N. R.: Experimental thyroiditis in the rhesus monkey. Clin. exp. Immunol. *1:* 159 (1966).

DONIACH, D. and WALKER, J. G.: A unified concept of autoimmune hepatitis. Lancet *i:* 813 (1969).

DUBOIS, E. L. and ARTERBERRY, J. D.: Etiology; in DUBOIS Lupus Erythematosus, 66 (Blakiston, London/New York 1966).

DUMONDE, D. C.: Tissue specific antigens. Adv. Immunol. *5:* 245 (1966).

EIDELMAN, S.: Some immunological factors in small intestine disease. Amer. J. clin. Nutr. *21:* 1110 (1969).

ENGEL, G. L.: Psychological factors in ulcerative colitis in man and gibbon. Gastroenterol. *57:* 362 (1969).

FEDERLIN, K.: Autoimmunophänomene bei Erkrankungen der endokrinen Drüsen. Klin. Wschr. *47:* 337 (1969).

FELSENFELD, O.: Synopsis of Clinical Tropical Medicine (Mosby, St. Louis 1965).

FELSENFELD, O.; GREER, W. E., and FELSENFELD, A. D.: Transient antibodies against homologous hepatic microsomal and mitochondrial antigens. Ann. Allergy *25:* 6 (1967).

GAJDUSEK, D. C.; GIBBS, C. J., jr., and ALPERS, M.: Experimental transmission of a Kuru-like syndrome to chimpanzees. Nature, Lond. *209:* 794 (1966).

GAJDUSEK, D. C.; GIBBS, C. J., jr., and ALPERS, M.: Transmission and passage of experimental Kuru in chimpanzees. Science *155:* 794 (1967).

GLYNN, L. E. and HOLBROW, E. J.: Autoimmunity and Disease. 2nd ed. (Blackwell, Oxford 1969).

GOOD, R. A.: The Thymus. Ciba Symposium, London, 231 (1966).

GRAS, J.: L'équilibre immunologique. Rev. franç. Et. clin. biol. *14:* 520 (1969).

HESS, M. and BÜTLER, C.: Anti-Gm specificities of sera of rhesus monkeys immunized with human gamma-globulin. Vox Sang. *7:* 93 (1962).

KITE, J. H.; ARGUE, H., and ROSE, N. R.: Experimental thyroiditis in the rhesus monkey. Clin. exp. Immunol. *1:* 139 (1966).

KITE, J. H.; ROSE, N. R.; KANO, K., and WITEBSKY, E.: Cytotoxicity of human thyroid antibodies. Ann. N.Y. Acad. Sci. *124:* 201 (1969).

KUNKEL, H. G. and TAN, E. M.: Autoantibodies and disease. Adv. Immunol. *4:* 351 (1964).

LUNTZ, M. H.: Anti-uveal and anti-lens antibodies in uveitis and their significance. Exp. Eye Res. *7:* 561 (1968).

MACKAY, I. R.: Autoimmune aspects of systemic lupus erythematosus and the forbidden clone hypothesis; in DUBOIS Lupus Erythematosus, 85 (Blakiston, London/New York 1966).

MACKAY, I. R. and BURNET, M. F.: Autoimmune Disease (Thomas, Springfield 1963).

METZGAR, R. S.: Immunologic studies of pancreas-specific isoantigens. J. Immunol. *93:* 176 (1964a).

METZGAR, R. S.: Human pancreas-specific isoantigens. Nature, Lond. *203:* 660 (1964b).

NOSSAL, G. J. V.: Principles of antibody formation. Proc. 11th Congr. Internat. Soc. Blood Transfus. 436 (1966).

PATERSON, P. Y.: Allergic encephalomyelitis; in LAWRENCE Cellular and Humoral Aspects of the Hypersensitive States, 469 (Harper-Hoeber, New York 1959).

PIROFSKY, B.: Autoimmunization and the Autoimmune Haemolytic Anaemias (Karger, Basel/München/New York 1969).

RAPP, W.; ARONSON, S. B.; BURTIN, B., and GRABAR, P.: Constituents and antigens of normal human gastric mucosa. J. Immunol. *92:* 579 (1964).

ROSE, N. R.; KITE, J. H.; DOEBBLER, T. K.; SPIER, R.; SKELTON, F. R., and WITEBSKY, E.: Studies on experimental thyroiditis. Ann. N.Y. Acad. Sci. *124:* 201 (1965).

ROSE, N. R.; SKELTON, F. R.; KITE, J. H., and WITEBSKY, E.: Experimental thyroiditis in the rhesus monkey. Clin. exp. Immunol. *1:* 171 (1966).

SAINT MARTIN, J. et EYQUEM, A.: Apparition d'iso- et d'auto-anticorps contre le tractus digestif. Ann. Inst. Pasteur *113:* 591 (1967).

SIMPSON, J. A.: Myasthenia gravis: a new hypothesis. Scot. med. J. *5:* 419 (1960).

SIVADON, P.: Processus auto-immunitaires et pathologie mentale. Rev. Immunol. *32:* 193 (1968).

STEBLAY, R. W.: Attempts to introduce autoimmune renal disease. Proc. Int. Symp. Immunopathol. La Jolla, 252 (1963).

STEINBERG, A. G. and WILSON, J. A.: Hereditary globulin factors and immune tolerance in man. Science *140:* 303 (1963).

STOUT, C. and SNYDER, R. L.: Ulcerative colitis-like lesions in gibbons. Gasteroenterol. *57:* 256 (1969).

STRAUSS, A. J. L.; SEEGAR, B. C.; HSU, K. C.; BURKHOLDER, P. M.; NASTUK, W. L., and OSSERMAN, K. E.: Immunofluorescence demonstration of a muscle-binding, CF serum globulin fraction in myasthenia gravis. Proc. Soc. exp. Biol., N.Y. *105:* 184 (1960).

SVARTZ, N.; HEDMAN, S., and SÖDERBERG, O.: A 7S gammaglobulin completely neutralizing the rheumatoid factor macroglobulin. Acta med. scand. *185:* 21 (1969).

LOGHEM, J. J. VAN: Viral infection and autoimmune disease. Vox. Sang. *10:* 1 (1965).

VAUGHAN, J. H.; BARNETT, E. V.; SOBEL, M. V., and JACOX, R. F.: Intracytoplasmic inclusions of immunoglobulins. Arthr. Rheumat. *11:* 125 (1968a).

VAUGHAN, J. H.; JACOX, R. F., and NOEL, P.: Relation of intracytoplasmic inclusions in joint fluid leukocytes to anti-γ G globulins. Arth. Rheumat. *11:* 135 (1968b).

VOJTIŠKOVÁ, M.; CHUTNÁ, J.; RYCHLIKOVÁ, M., and POKORNÁ, Z.: Thymus transplantation and uveitis. Folia Biol., Praha *8:* 207 (1962).

WAKSMAN, B. H.: Experimental allergic encephalomyelitis and the autoallergic diseases. Int. Arch. Allergy *14:* 1 (1959).

WAKSMAN, B. H.: Delayed hypersensitive reactions. J. Allergy *31:* 468 (1960a).

WAKSMAN, B. H.: Distribution of experimental autoallergic lesions. Amer. J. Path. *37:* 673 (1960b).

WAKSMAN, B. H.: Auto-immunization and the lesions in autoimmunity. Medicine, Balt. *41:* 93 (1962).

WALFORD, R. L.: Immunologische Aspekte des Alterns. Klin. Wschr.: *47:* 599 (1969).
WILLIAMS, R. C., jr.; KENYON, A. J., and HUNTLEY, C. C.: Immunoglobulins, viruses and speculation on their interrelationship. Blood *31:* 522 (1968).

Author's address: Dr. OSCAR FELSENFELD, Chief, Division of Communicable Diseases, Tulane University, Delta Regional Primate Research Center, *Covington, LA 70433* (USA)

Pathology of Simian Primates, Part I, pp. 399–413 (Karger, Basel 1972)

Diseases of the Endocrine System

H. G. SPIES and M. T. CLEGG

Division of Reproductive Physiology, Delta Regional Primate Research Center, Covington, La.

Contents

Introduction ... 399
Hypothalamo-Hypophyseal Axis ... 402
Pancreas ... 403
Parathyroid ... 404
Thyroid ... 405
Adrenal Gland ... 407
References ... 409

Introduction

The endocrine glands play an essential role in the regulation of almost all the life processes ranging from reproduction to adjustment to environmental changes. The proper function of these systems requires delicate control mechanisms designed to maintain a reasonably static internal environment of the total organism. This is accomplished in several ways. One is by a system of feed-back controls, in which the functional activity of one endocrine gland is controlled by a hormone from another gland, which in turn is regulated by the circulating level of hormone produced by the first gland. For example: the secretion of thyroxin by the thyroid is stimulated by thyroid stimulating hormone (TSH) from the anterior pituitary. As the circulating level of thyroxin increases, this hormone inhibits further release of TSH; thyroid activity is diminished and the level of thyroxin drops. A second method of regulation is by way of changes in circulating levels of metabolites. For example, when blood glucose is high the secretion of insulin from the pancreatic Islets of LANGERHANS is increased. Similarly the circulating level of calcium controls the secretion of parathyroid hormone or thyrocalcitonin. A final method is by way of neural stimuli. Examples of this method are the stimulation of adrenocorticotropic hormone (ACTH) release by environmental stresses or the influence of photic stimuli on gonadotropin release.

Because of the close interrelationship existing between the endocrine glands and the necessity for integrating the proper balance of hormones, a deficiency of one hormone may manifest itself in a variety of metabolic disturbances. For purposes of convenience, it has been useful to categorize the hormones with respect to their principal metabolic effects, namely, those influencing water and electrolyte metabolism, such as mineralocorticoids (al-, dosterone) and antidiuretic hormone (vasotocin, pitressin or ADH); those influencing mineral metabolism, such as parathormone (PTH) and thyrocalcitonin; those influencing carbohydrate, fat and protein metabolism, such as growth hormone, insulin, epinephrine, and the adrenal glucocorticoids; and those having a general metabolic effect such as thyroxin and triiodothyronine. Metabolic disturbances relating to the malfunction of one endocrine gland, therefore may show a great variety of symptoms.

The classification and description of the more commonly known endocrine diseases have largely been supplied by human studies and comparative information on sub-human primates still requires considerable amplification. It is useful, therefore, to list some of the more important endocrinopathies described in human clinical literature.

- Diseases of the hypophysis
  - Anterior pituitary
  - Tumors of the anterior lobe causing hyperfunction
    - Acromegaly, gigantism
    - Cushing's syndrome
  - Hypopituitarism, dwarfism
  - Posterior pituitary and hypothalamic disease
    - Diabetes insipidus
    - Dystrophia adiposogetalis
    - Laurence-Moon-Biedl syndrome
- Diseases of the adrenal cortex
  - Adrenocortical insufficiency
  - Addison's disease
  - Adrenocortical hyperfunction
    - Adrenogenital syndrome

- Adrenal virilism
- Cushing's syndrome
- Primary aldosteronism

Diseases of the adrenal medulla
- Pheochromocytoma

Diseases of the pancreas
- Diabetes mellitus

Diseases of the thyroid
- Cretinism, juvenile
- Myxedema, adult
- Gull's disease
- Hyperthyroidism (Graves disease)
  - Diffuse and nodular toxic goiter
  - Exophthalmos
  - Colloid goiter endemic, spontaneous
  - Thyrotoxicosis

Diseases of the parathyroid
- Hypoparathyroidism
- Primary hyperparathyroidism
- von Recklinghausen's disease of bone
- Secondary hyperparathyroidism

Diseases of the gonads as described under by ROBERTS.

The apparent similarity between man and other primates makes the latter an obvious choice for studies of normal and pathological endocrine conditions. Valid comparisons are not easy, however, as data which depict the normal physiological endocrine patterns in monkeys and apes are sparse. Furthermore, reports of spontaneous anomalies of the endocrine glands in monkeys are rare [RUCH, 1959; KENT, 1960; NEWBERNE and ROBINSON, 1960]. In addition, baseline data of normal endocrine patterns are yet to be determined. With the development of assay procedures of sufficient sensitivity to measure hormones in peripheral plasma, however, it is anticipated that much valuable information will soon be forthcoming. Anatomical and histological criteria vary normally among species of monkeys so widely that abnormalities based on these criteria are difficult and sometimes misleading. Pathological conditions of animals with unknown previous histories observed following captivity have allowed only the most gross anomalies to be recognized. The causes of these anomalies have been mostly speculative; therefore, it is evident that a great deal of future effort is needed.

Hypothalamo-Hypophyseal Axis

No reported spontaneous anomalies of hypothalamic origin in monkeys were found in the literature. However, in 1913 Camus and Roussy noted that lesions in the hypothalamus, produced experimentally, resulted in diabetes insipidus. Subsequently, Fisher *et al.* [1938] have discussed the structural and functional relationship between the hypothalamus and posterior pituitary and the hormonal control of water balance in monkeys. Magoon *et al.* [1939] reported gross polyuria in 4 of 17 rhesus monkeys *(Macaca mulatta)* with lesions destroying the supraoptic-paraventricular area of the hypothalamus. Some years later, Holmes [1963] noted that section of the pituitary stalk in rhesus monkeys resulted in interruption of the menstrual cycles and diabetes insipidus. The latter was not permanent, however, since the source of ADH (supraoptic-paraventricular region) was not destroyed. Diabetes insipidus routinely occurs in squirrel monkeys *(Saimiri sciureus)* during the last trimester of pregnancy [Clewe, 1969].

In the squirrel monkey hypothalamic lesions in the anterior median eminence blocks the elevated response of growth hormone secretion which is prompted by insulin [Abrams *et al.*, 1966]. They concluded that this area was a part of the neural link by which hypoglycemia triggered growth hormone response. Brown *et al.* [1967], based on data in nine squirrel monkeys, suggested that the hypothalamus contained glycoreceptor neurons which controlled growth hormone secretion. However, Meyer and Knobil [1967] have suggested that hypoglycemia acts as a non-specific stress in the activation of growth hormone secretion in monkeys. Certainly minor changes in environment, such as stress, diet, glucose administration and fasting, influence plasma growth hormone levels. Also, amino acid metabolism seems to be linked to growth hormone secretion since infusion of several amino acids or ingestion of a protein rich food will prompt a discharge of growth hormone from the adenohypophysis. Squirrel monkeys have an elevation in growth hormone following handling or upon placement in primate chairs [Brown *et al.*, 1967]. Ether anesthesia also results in an elevation of growth hormone as do such physical stresses as insulin, induced hypoglycemia, and hemorrhage. In the rhesus monkey, pitressin and epinephrine are potent stimulators of growth hormone secretion [Meyer and Knobil, 1966 and 1967] but in the squirrel monkey epinephrine has no effect. There is little doubt that the hypothalamus influences secretion of pituitary growth hormone in primates, as has been conclusively shown in laboratory mammals, since electrical stimulation of the anterior ventral median eminence causes a rise in plasma growth hormone levels in chair adapted monkeys [BROWN *et al.*, 1967]. In addition, injections of crude extracts of ovine, porcine or bovine hypothalamus into monkeys produce elevation of plasma growth hormone levels [GARCIA and GESCHWIND, 1966; SMITH *et al.*, 1967; KNOBIL *et al.*, 1968]. Very little information concerning the influence of the hypothalamus on other pituitary hormones has been reported for monkeys. However, KUMAR and KNOWLES [1967] have described specialized ependymal cells in the tuber cinereum of rhesus monkeys which send long processes to the region of the pars tuberalis. These specialized ependymal cells have an affinity for estrogen since high concentrations of this labeled hormone were observed in them following injection into ovariectomized animals. Furthermore, there is a greater glandular concentration of these ependymal cells in adult female rhesus as compared to males. In adult females these cells are fully laden with granules during the pre-ovulatory stage of the menstrual cycle, while in females killed at menstruation the cells are degranulated [KUMAR, 1968].

Pituitary anomalies ranging from an adenoma in a capuchin monkey to cystic enlargement in a *Macaca sinica* [FOX, 1927] and gorilla [KOCH, 1937] have been noted. SUTTON [1885] noted one incidence of a carcinoma of the pituitary extending into the optic chiasm in a baboon *(Papio cynocephalus)* and KELLY [1948] called attention to HODGKINS sarcoma in a baboon. Also, an inflamed pituitary following meningitis [KENNARD and WILLNER, 1941] and intracellular inclusions of the pars distalis in a rhesus macaque [COWDRY and SCOTT, 1935] have been reported. The significance of these conditions relative to pituitary function were not disclosed. KENT and PICKERING [1958] noted a pituitary tumor in a male rhesus of three and one-half years of age following thyroid atrophy, but since the animal had been exposed to wholebody irradiation, the spontaneity of this occurrence can be questioned.

Pancreas

RATCLIFFE [1942] reported an adenocarcinoma of the pancreas in an 18 year old female baboon *(Papio cynocephalus)*. Diabetes mellitus has been described in monkeys [KELLY, 1947] and a number of cases of animals dying while in coma have been cited.

The relationship of diabetes mellitus to obesity in humans has lead to the suggestion that excessive food intake might precipitate the disease as a result of islet cell exhaustion. In support of this concept, HAMILTON and BRO-

BECK [1965] induced hyperphagia and obesity in male *Macaca mulatta* monkeys by bilateral lesions of the ventromedial nuclei. At the height of obesity as a result of increased food intake, both hyperglycemia and glycosuria developed. These conditions could be controlled by the administration of insulin. One male developed cataracts 27 months after the operation. When sacrificed, degeneration of the islets of the pancreas was observed. They also reported on the occasional occurrence of transient glycosuria in normal animals; this, they attributed to excessive food intake and the high carbohydrate constituent of the laboratory diet. GIBBS *et al.* [1966] reported the development of glomerulosclerosis in long-term alloxan diabetic monkeys. Cataracts and dwarfism also appeared and was more marked in animals on high carbohydrate-low fat diets than in animals on normal diets. LANE [1968] reported that in squirrel monkeys *(Saimiri sciureus)* following recent captivity there was an impairment in carbohydrate metabolism. Upon stress some animals developed overt diabetes, showed a rapid loss of weight and death occurred in a few days. Two animals had glucose levels as high as 510 mg percent and 750 mg percent which was accompanied by glycosuria and polyuria. The pancreas of both monkeys showed some degranulation of the islets upon histological examination. A high incidence of asclerotic changes in squirrel monkeys was noted by ST. CLAIRE and his associates [1967] following recent capture. The impairment of carbohydrate metabolism seems to be more prevalent in adults than in juveniles and more prevalent in males than in females. BECK [1969] recently noted that progesterone and mestronal, an oral contraceptive [BECK and WELLS, 1968], could be a diabetogenic factor during pregnancy in rhesus monkeys. Treatment with progesterone for three weeks produced enhanced plasma insulin response to iv glucose without altering glucose tolerance, perhaps because of a marked compensatory enhancement of the initial rate of insulin release following glucose stimulation. MINTZ *et al.* [1969] reported a bidirectional placental transfer of insulin from mother and fetus in monkeys. They noted that glucose injections into the fetus did not alter fetal plasma insulin concentrations whereas in the neonate glucose induced a delayed rise.

Parathyroid

Simian bone disease was first reported by MACLEOD in 1895 [RUCH, 1959]. Between 1908 and 1911 three different reports on the baboon, capuchin monkey and guenon monkey were reported. These workers noted hypertrophy of the skull and other skeletal bones [ROQUES and BOUFFARD, 1908; PETTIT, 1909; MARCHOUX and MESNIL, 1911]. CORSON-WHITE [1922] suggested three cases of bone disease similar to those described in man. CHRISTELLER [1923] classified lesions found in four young animals as hyperostotic sclerotic osteitis fibrosa. He tabulated all his cases of simian bone disease as osteitis fibrosis of the VON RECKLINGHAUSEN's type later shown to be primary hyperparathyroidism. HAMERTON [1939] reported a few cases of simian bone disease but did not discuss the etiology of the disease. In material from 79 simian necropsies compiled by KOCH and DIEMEL [1952] simian bone disease was listed as the third most common disease preceded only by enteritis and tuberculosis. In 1957 CORDY described simian bone disease in three young baboons in captivity. These animals had been fed a diet of 0.5 gm calcium and 1.8 gm phosphorus at the ratio of 1:3.6. ARNDT [1923] described the histology of the parathyroid gland in normal and bone diseased monkeys and noted only one case of simian bone disease in which the parathyroids were sclerotic. In this case the connective tissue made up $^1/_3$ to ½ of the sections. Both CHRISTELLER and CORDY had reported normal parathyroid glands earlier in their cases of simian bone disease. KROOK and BARRETT [1962] reported bone disease in a spider monkey *(Ateles)* and a capuchin monkey. The two diseased monkeys had been in a pet shop and appeared at their clinic for treatment of swollen gums and joints. Both animals showed roentgenographic evidence of skeletal lesions characteristic of simian bone disease. The functional portion of the parathyroid glands in both these animals exceeded that of a normal monkey by 10 to 30 times. They concluded the disease was a manifestation of secondary hyperparathyroidism caused by hypocalcemia which in turn was a secondary response to excessive phosphorus feeding. Whereas von RECKLINGHAUSEN'S disease was ascribed by definition as primary hyperparathyroidism caused by parathyroid neoplasm, they diagnosed the disease as one brought about by a secondary role of a hyperactive parathyroid. (*Vide* also BARKER and HERBERT, pp. 433 *et seq.*)

Thyroid

KATZBERG, in 1968, studied thyroids of 42 baboons from East Africa. He utilized as some additional observations animals from a colony at San Antonio, Texas. His studies included all ages and the full spectrum of weights from both sexes. He reported that at birth the average diameter of 50% of the follicles of the thyroid were 35 μm while in older, heavier animals 50% of the follicles were up to 300 μm. Relative volume of the stroma and epithelium declined when compared to the volume of the colloid. In newborns there were 36 sq mm of follicular surface per area of cubic mm of thyroid tissue. This declined during maturation till the largest animals had a surface area reduced to approximately 20 sq mm or 60% of that in the newborn animal. By actual count the number of follicles per unit volume was greater in females than males. The actual volume was larger in males. He suggested a decreased number of follicles with age and reduced surface area indicative of a depletion of the epithelial cell population. The effects of extrinsic factors influencing these observations were not known nor was the actual secretion rate of thyroid hormone from the changing cell populations. PICKERING [1968] has reported a number of studies on levels of thyroid function in fetal, newborn and young rhesus macacas. He indicated that follicle formation and colloid synthesis were apparent in the gland by conception age 50 days. These thyroid follicles undergo vigorous proliferation and vascularization and have large increases in total iodine content by 75 days of fetal development. ^{131}I uptake per mg thyroid tissue was 8–12 times greater in the fetus than in the mother; thus, he concluded that under normal conditions the amounts of iodine passing the placenta was in healthy excess of amounts required by fetal tissue. On the other hand, in the neonate, PICKERING and his co-workers report that congenital athyrosis can be readily simulated by experimental thyroid ablation. These thyroid deficiencies caused abnormalities of the nervous system ranging from abnormal vascularity, deficient myelinization and atrophic neuronal and interstitial components. These could be counteracted by giving thyroxin. The relationship of thyroid function to that of other endocrine glands is not clearly known, but CARLSON [1952] reported that in man the thyroid reached a maximum production rate at puberty and then decreased with a gradual reduction of the basal metabolic rate in man from birth to old age. According to KAR and SRIDASTAVA [1966], hyperthyroidism does not disturb sexual development in the prepuberal male or female rhesus macaca monkey. In prepuberal male or female rhesus monkeys, hypothyroidism for 180 days failed to produce any detectible changes on gonad development. There was a suggestion, most pronounced in the male, of increased sensitivity of the accessory sex glands to exogenous gonadotropins as evidenced by an increase in weight when compared to euthyroid, gonadotropin treated controls. However, hypothyroidism produced by thyroidectomy or thiouracil feeding results in amenorrhea which can be corrected by thyroid therapy. KITE *et al.*, reported that monkeys *(Macaca mulatta)* which had been treated with thyroid extract (thyroprotein) in FREUND's Adjuvant resulted in thyroid regression and subsequent regeneration in one of two monkeys over an 18-month period of time. No behavioral or physiological changes in either of the two rhesus monkeys were reported. Subsequent studies by the same group of the ultrastructure of thyroid tissue from monkeys immunized against thyroid tissue showed vacuolization of the endoplasmic reticulum and increased colloid droplets. A necrotic process in follicular epithelium occurred with basement membrane disruption under severe follicle damage [THEMANN *et al.*, 1968].

It is well established that elevated serum cholesterol levels are associated with thyroid deficiency in monkeys, a relationship which can be demonstrated experimentally. Thus the experimental induction of hypothyroidism in cebus monkeys by the ingestion of thiouracil is followed by elevated serum cholesterol levels [ANDRUS *et al.*, 1964]. Because of the established association between hypercholesteremia induced by a high cholesterol diet and atherosclerosis [MANN *et al.*, 1953; TAYLOR *et al.*, 1963], the functional state of the thyroid takes on additional significance when dealing with pathological findings. That a hypothyroid state may exacerbate the effects produced by a high cholesterol intake, has been experimentally verified. YOUNGER *et al.* [1969] induced hypothyroidism by ^{131}I administration in monkeys maintained on a 5 to 10 percent cholesterol diet. They reported that atherosclerotic symptoms occurred much sooner in the ^{131}I treated animals than in the non-treated high cholesterol controls. Heavy lesions in the abdominal aorta and iliac arteries were produced. Coronary lesions were also common, but no myocardial infarctions were observed. There was also an elevation of serum urea. Two females showed large colloid goiters while two males presented a picture of colloid goiter mixed with colloid hyperplasia. In one male this could not be distinguished from an early carcinoma. WARREN and MEISNER [1953] have described an adenomatous goiter in one monkey. The nodules were multiple, poorly encapsulated and variable in structure.

Adrenal Gland

Although the reported incidence of benign and malignant tumors is low in simian species [LOMBARD and WITTE, 1959; FOX, 1937; KENT, 1960], the most commonly observed tumors of endocrine origin have been in the adrenal gland [NEWBERNE and ROBINSON, 1960]. DEINHARDT and associates [1967] reported on a pathological study of 93 monkeys of which nine had adrenal necrosis. These monkeys were collected in Columbia and Peru and their normal habitats ranged from central Mexico to southern Brazil. MARUFFO

[1967] reported 14 tumors in 292 free ranging howler monkeys of which 11 were corticoadenomas of the adrenal varying in size from 1–20 mm in diameter. The cells of these adenomas were arranged in a columnar pattern similar to the zona fasciculata. None of these tumors were found in prepuberal animals. HOUSER *et al.* [1962] have reported nine cases of adrenal adenomas in howler *(Alouatta villosa)* monkeys of the Panama area. Other incidences of adrenal tumors have been noted and reviewed by RUCH [1959]. SCHMIDT [1956] has noted the occurrence of calcium deposits in the adrenal as well as one case of an accessory adrenal cortex embedded in the peripheral renal cortex of one rhesus monkey.

The wide variability of adrenal structure in simian species was disclosed in a report by HOUSER *et al.* [1962] on adrenal histology of the howler, night monkey, spider, capuchin and tamarin species. In the howler, the adrenal is oval with a broad zona granulosa composed of straight cords of large cells going into the zona fasciculata. Cortical adenomas appeared most common in this species. The zona reticularis was indistinct and the medulla had large cells arranged in cords. In the night monkey *(Aotus trivirgatus)* triangular adrenals were observed containing a glomerulosa of balls of cells devoid of lipid and separated from the zona fasciculata by a layer of small fat cells. The spider monkey *(Ateles)* had an oval shaped adrenal with a narrow zona glomerulosa composed of small cells, distinguished sharply from the zona fasciculata, and clearly separated from the reticularis. The medulla was the largest in the spider monkey of all the five species compared. The capuchin monkey *(Cebus capucinus)* had folded adrenals with an intermediate zone of cells next to the fasciculata. The adrenal medulla was completely absent in one monkey of this species. In the tamarin *(Saguinus geoffroyi)* the adrenals were folded with a large cortex, thin poorly developed glomerulosa of small cells which were rich in lipid. The zona fasciculata was not distinct and the zona reticularis was well developed. These data point to the variability in adrenal histology and the need for greater knowledge of corticoid levels in the blood which could be associated with functional changes of the species of interest. NAGEL *et al.* [1965] noted that both surgery and anesthesia increased plasma 17-hydroxy corticoid levels and resulted in a fall in total eosinophils in the majority of pigtail macacas so treated. Every monkey had a 50% decrease in eosinophil count 4 h following intramuscular ACTH. They showed lethargy, failure to eat, diarrhea and diuresis similar to the etiology of the stress syndrome which is frequently encountered following captivity. MASON *et al.* [1961, 1968a–g] published a series of papers relating the corticoid levels and adrenal medullary hormones to emotional stress. They have utilized rhesus monkeys trained on an avoidance response procedure over a 72-h test period. With this procedure the animals receive an electrical shock every 20 s unless they press a lever for its elimination. Such animals are preconditioned to the procedure, allowed to adjust and then subjected to the 72-h test period. Under these conditions they have observed a marked increase in both urinary and plasma 17-hydroxy corticosteroid levels during the 72-h test; an increase in urinary epinephrine (urinary norepinephrine was less consistently elevated); an increase of 50% over baseline in plasma extractable iodine which remained elevated well into a 2–3 week recovery period; decreased urinary estrogens during the 72-h followed by a rebound to normal; decreased urinary androgens followed by elevation above baseline levels after the 72-h test; a rapid decrease in plasma insulin levels during the first hours of the test followed by increases above baseline levels which remained high for long periods beyond the test; and variable changes in plasma growth hormone levels depending upon the levels they observed initially. In their earlier reports they noted gastric duodenal ulcerations in 4 (2 males and 2 females) monkeys on this test. They speculated that changes in growth hormone, corticoids and pepsinogen levels along with sleep deprivation may have contributed to the incidence of the ulcers.

References

1 Abrams, R. L.; Parker, M. L.; Blanco, S.; Reichlin, S., and Daughaday, W. H.: Hypothalamic regulation of growth hormone secretion. Endocrinology, Springfield *78:* 605–613 (1966).

2 Andrus, S. B.; Roach, A. M., and Fillios, L. C.: Exptl. production of colloid goiter in the cebus monkey. Proc. Soc. exp. Biol., N.Y. *116:* 963–967 (1964).

3 Arndt, H. J.: Über Epithelkörperchenbefunde bei einem Schimpansen; zugleich ein Beitrag zur Frage der pathogenetischen Zusammenhänge von Epithelkörperchen und Ostitis fibrosa. Virchows Arch. path. Anat. *247:* 454 (1923).

4 Beck, P. and Wells, S.: Plasma insulin responses to glucose in estradiol- and mestranol-treated rhesus monkeys. Clin. Res. *16:* 338 (1968).

5 Beck, P.: Progestin enhancement of the plasma insulin response to glucose in rhesus monkeys. Diabetes, N.Y. *18:* 146–152 (1969).

6 Brown, G. M.; Schalch, D. S., and Reichlin, S.: Growth hormone response to stress in the squirrel monkey. Fed. Proc. *26:* 585 (1967).

7 Camus, J. et Roussy, G.: Hypophysectomie et polyurie expérimentales. C. R. Soc. Biol. *75:* 483–486 (1913).

8 Carlson, A. J.: The thyroid, pancreatic islets, parathyroid, adrenals, thymus and pituitary. Crowdy's problems of aging. Biological and medical aspects, pp. 358–380 (Williams and Wilkins, Baltimore 1952).

9 Christeller, E.: Die Formen der Ostitis fibrosa und der verwandten Knochener-

krankungen der Säugetiere; zugleich ein Beitrag zur Frage der «Rachitis» der Affen. Ergebn. allg. Path. Abt 2, *20:* 1 (1923).

10 CLEWE, T. H.: Observations on reproduction of squirrel monkeys in captivity. J. Reprod. Fertil. *6:* 151–156 (1969).

11 CORDY, D. R.: Osteodystrophia fibrosa accompanied by visceral accumulation of lead. Cornell Vet. *47:* 480–490 (1957).

12 CORSON-WHITE, E. P.: Osteitis deformans in monkeys. Arch. intern. Med. *30:* 790 (1922).

13 COWDRY, E. V. and SCOTT, G. H.: Nuclear inclusions in the kidneys of macacus rhesus monkeys. Amer. J. Path. *11:* 659–668 (1935).

14 DEINHARDT, J. B.; DEVINE, J.; PASSOVOYR, M.; POHLAMAN, R., and DEINHARDT, F.: Marmosets as laboratory animals. Care of Marmosets in the laboratory. Pathology outline of statistical evaluation of data. Lab. Animal Care *17:* 11–29 (1967).

15 FISHER, C. W. R.; INGRAM, and RANSON, S. W.: Diabetes insipidus and a neural hormonal control of water balance; a contribution to the structure and function of the hypothalamic hypophysial system. Ann Arbor (Edwards 1938).

16 FOX, H.: Mortality and matters of pathological interests. Dept. Pemrose Research Lab. 13–17 (1937).

17 FOX, H.: Pathologic anatomy in the hypophysis cerebri in wild animals. Arch. Neurol., Chicago *17:* 471–480 (1927).

18 GARCIA, J. F. and GESCHWIND, I. I.: Increase in plasma growth hormone level in the monkey following the administration of sheep hypothalamic extract. Nature, Lond. *211:* 372–374 (1966).

19 GIBBS, G. E.; WILSON, R. B., and GIFFORD, H.: Glomerulosclerosis in the long-term alloxan diabetic monkey. Diabetes *15:* 258–261 (1966).

20 HAMERTON, A E.: Review of mortality rates and report on the deaths occurring in the society's garden during the year 1938. Proc. zool. Soc., Lond., p.281 (1939).

21 HAMILTON, C. L. and BROBECK, J. R.: Control of food intake in normal and obese monkeys. Ann. N.Y. Acad. Sci. *131:* 583–592 (1965).

22 HAMILTON, C. L. and BROBECK, J. R.: Diabetes mellitus in hyperphagic monkeys. Endocrinology, Springfield *73:* 512–515 (1963).

23 HOLMES, R. L.: The pituitary stalk in water metabolism in the monkey. J. Endocrin. *28:* 107–118 (1963).

24 HOUSER, R. G.; HARTMAN, F. A.; KNOUFF, R. A., and MCCOY, F. W.: Adrenals in some panama monkeys. Anat. Rec. *142:* 41–52 (1962).

25 KAR, A. B. and SRIDASTAVA, N. K.: Induced hypothyroidism and sexual development in prepuberal rhesus monkeys. Acta biol. med. J. *17:* 193–206 (1966).

26 KAR, A. B. and DATTA, J. K.: Induced hyperthyroidism and sexual development in prepuberal female rhesus monkeys. Arch. Anat. micr. Morph. exp. *54:* 119–128 (1965).

27 KATZBERG, A. A.: Changing dynamics in the thyroid follicles of the baboon, papio and anubis. Folia primat. *8:* 58–65 (1968).

28 KELLY, A. L.: Annual report of the hospital and research laboratory. Zoonoses Res. *21:* 5–6 (1948).

29 KENNARD, M. A. and WILLNER, M. D.: Findings at autopsies of seventy anthropoid apes. Endocrinology, Springfield *28:* 967–976 (1941).

30 KENT, S. P. and PICKERING, D. E.: Neoplasms in monkeys *(Macaca mulatta)*; spontaneous and irradiation-induced. Cancer *11:* 138–147 (1958).

31 KENT, S. P.: Spontaneous and induced malignant neoplasms in monkeys. Ann. N. Y. Acad. Sci. *85:* 819–827 (1960).

32 KERR, G. R.: Serum hormonal iodine values during and following pregnancy in monkey *(Macaca mulatta)*. J. Endocrin. *24:* 137–142 (1962).

33 KITE, J. H., jr.; ARGUR, HELEN, and ROSE, N. R.: Exptl. thyroiditis in the rhesus monkey; Course of disease. Clin. exp. Immunol. *1:* 171–188 (1966).

34 KNOBIL, E.; MEYER, V., and SCHALLY, A. V.: In PECILE and MÜLLER Growth Hormone. Excerpta Med. *1:* 1234 (1968).

35 KOCH, W.: Bericht über das Ergebnis der Obduktion des Gorilla «Bobby» des Zoologischen Gartens zu Berlin. Ein Beitrag zur vergleichenden Konstitution.

36 KOCH, W. und DIEMEL, B.: Über Krankheiten der Menschenaffen. Veröffentl. Morph. Path. *13:* 1 (1952).

37 KROOK, L. and BARRETT, R. B.: Simian bone disease; a secondary hyperparathyroidism. Cornell Vet. *52:* 457–492 (1962).

38 KUMAR, T. C. A. and KNOWLES, F. G. W.: A system linking the third ventricle with the pars tuberalis of the rhesus monkey. Nature, Lond. *215:* 54–55 (1967).

39 KUMAR, T. C. A.: Modified ependymal cells in the ventral hypothalamus of the rhesus monkey and their possible role in the hypothalamic regulation of anterior pituitary function. J. Endocrin. *41:* 17–18 (1968).

40 LANG, C. M.: Impaired carbohydrate metabolism. Penn med. J., pp. 73–74 (1968).

41 LOMBARD, L. S. and WITTE, E. J.: Frequency and types of tumors in mammals and birds of the Philadelphia Zoological Gardens. Cancer Res. *19:* 127–141 (1959).

42 MAGOON, H. W.; FISHER, C., and RANSON, S. W.: The neurohypophysis and water exchange in the monkey. Endocrinology, Springfield *25:* 161–174 (1939).

43 MANN, G. V.; ANDRUS, S. B.; MCNALLY, A., and STARE, F. J.: Experimental atherosclerosis in cebus monkeys. J. exp. Med. *98:* 195–218 (1953).

44 MANN, G. V. and ANDRUS, S. B.: Xanthomatosis and atherosclerosis produced by diet in an adult rhesus monkey. J. Lab. clin. Med. *48:* 533–550 (1956).

45 MARUFFO, C. A.: Spontaneous tumors in howler monkeys. Nature *213:* 521 (1967).

46 MARCHOUX, E. et MESNIL, F.: Ostéite hypertrophique généralisée des singes avec lésions rapellant le goundou. Bull. Soc. Path. exot. *4:* 150–155 (1911).

47 MASON, J. W.; BRADY, J. V.; POLISH, E.; BAUER, JOANN, A.; ROBINSON, J. A.; ROSE, R. M., and TAYLOR, ELISZABETH, D.: Patterns of corticosteroids and pepsinogen change related to emotional stressed captive monkeys. Science *133:* 1596–1598 (1961).

48 MASON, J. W.; BRADY, J. V., and TOLLIVER, G. A.: Plasma and urinary 17-hydroxy corticosteroid responses to 72-h avoidance sessions in the monkey. Psychosom. Med. *30:* 608–630 (1968a).

49 MASON, J. W.; TOLSON, W. W.; BRADY, J. V.; TOLLIVER, G. A., and GILMORE, L. I.: Urinary epinephrine and norepinephrine response to 72-h avoidance sessions in the monkey. Psychosom. Med. *30:* 654–665 (1968b).

50 MASON, J. W.; MOUGEY, E. H.; BRADY, J. V., and TOLLIVER, G. A.: Thyroid (Plasma Butanol-Extractable Iodine) responses to 72-h avoidance sessions in the monkey. Psychosom. Med. *30:* 682–695 (1968c).

51 MASON, J. W.; TAYLOR, E. D.; BRADY, J. V., and TOLLIVER, G. A.: Urinary estrone,

estradiol and estriol responses to 72-h avoidance sessions in the monkey. Psychosom. Med. *30:* 696–709 (1968d).

52 MASON, J. W.; TOLSON, W. W.; ROBINSON, J. A.; BRADY, J. V.; TOLLIVER, G. A., and JOHNSON, T. A.: Urinary androsterone, eticholanolone and dehydroepiandrosterone responses to 72-h avoidance sessions in the monkey. Psychosom. Med. *30:* 710–732 (1968e).

53 MASON, J. W.; WHERRY, F. E.; BRADY, J. V.; BEER, B.; PENNINGTON, L. L., and GOODAMAN, A. C.: Plasma insulin response to 72-h avoidance sessions in the monkey. Psychosom. Med. *30:* 746–759 (1968f).

54 MASON, J. W.; WOOL, M. S.; WHERRY, F. E.; PENNINGTON, L. L.; BRADY, J. V., and BEER, B.: Plasma growth hormone response to avoidance sessions in the monkey. Psychosom. Med. *30:* 760–773 (1968g).

55 MEYER, V. and KNOBIL, E.: Stimulation of growth hormone secretion by vasopressin in the rhesus monkey. Endocrinology, Springfield *79:* 1016–1017 (1966).

56 MEYER, V. and KNOBIL, E.: Growth hormone secretion in the unanesthetized rhesus monkey in response to noxious stimuli. Endocrinology, Springfield *80:* 163–171 (1967).

57 MINTZ, D. H.; CHEZ, R. A., and HORGER, E. O. III.: Fetal insulin and growth hormone metabolism in the subhuman primate. J. clin. Invest. *48:* 176–186 (1969).

58 NAGLE, J. P.; CAMMOCK, E. E.; NYHUS, L. M., and HARKENS, H. N.: Evidence for adrenal insufficiency in acutely stressed captive monkeys. J. appl. Physiol. *20:* 131–133 (1965).

59 NEWBERNE, J. W. and ROBINSON, V. B.: Spontaneous tumors in primates. A report of two cases with notes on the apparent low incidence of neoplasms in subhuman primates. Amer. J. vet. Res. *21:* 150–155 (1960).

60 PETTIT, A.: Lésions osseuses chez deux singes. Bull. Soc. Path. Exot. *2:* 220 (1909).

61 PICKERING, D. E. and KONTAXIS, N. E.: Thyroid function in the fetus of the macaque monkey *(Macaca mulatta)*. II. Chemical and morphological characteristics of the fetal thyroid gland. J. Endocrin. *23:* 267–275 (1961).

62 PICKERING, D. E.; KERR, G. R., and KONTAXIS, N. E.: Studies of fetal thyroid gland function in the mulatta macaque monkey. Proc. of conf. on Res. with Primates, pp. 81–88 (Tektronix Foundation, Beaverton 1963).

63 PICKERING, D. E.: Thyroid physiology in the developing monkey fetus *(Macaca mulatta)*. Gener. comp. Endocr. *10:* 182–190 (1968).

64 RATCLIFFE, H. L.: Deaths and important diseases. Rep. Penrose Res. Lab., pp. 11–25 (1942).

65 ROQUES, R. et BOUFFARD, B.: Un cas de goundou chez le cynocéphale. Bull. Soc. Path. exot. *1:* 295 (1908).

66 RUCH, T. C.: Diseases of Laboratory Primates (Saunders, Philadelphia 1959).

67 SCHMIDT, IRA G.: An accessory adrenal cortex in the kidney of a rhesus monkey and the effects of pyrimethamine on this tissue. Endocrinology, Springfield *59:* 454–457 (1956).

68 SCHMIDT, IRA G.: Bone formation in the adrenal gland of a rhesus monkey. Endocrinology, Springfield *61:* 780–782 (1957).

69 SMITH, G. S.; KATZ; ROOT, A.; DHARIWAL, A.; BONGIOVANNI, A.; EBERLEIN, W., and MCCANN, S. M.: Effect of thioglycollate or thypsin on GH-RF activity of ovine SME in rhesus monkeys. Fed. Proc. *26:* 316 (1967).

70 St. Clair, R. W.; MacNintch, J. E.; Middleton, C. C.; Clarkson, T. B., and Lofland, H. B.: Changes in serum cholesterol levels of squirrel monkeys during importation and acclimatization. Lab. Invest. *16:* 828–832 (1967).

71 Sutton, J. B.: Tumors in animals. J. Anat., Lond. *19:* 415–475 (1885).

72 Taylor, C. B.; Trueharrt, R. E., and Cox, G. E.: Atherosclerosis in rhesus monkeys. III. Role of increased thickness of arterial walls in atherogenesis. Arch. Path. *76:* 15–28 (1963).

73 Themann, H.; Andrada, J. A.; Rose, N. R.; Andrada, E. C., and Witebsky, E.: Experimental thyroiditis in the rhesus monkey. V. Electron microscopic investigations. Clin. exp. Immunol. *3:* 491–508 (1968).

74 Vadova, A. V. and Gelshtein, V. I.: Spontaneous tumors in lower catarrhine monkeys according to the data of the monkey breeding station of the Sukhumi Medico-Biological Station, pp. 137–163; in Utkin theoretical and practical questions of medicine and biology in experiments on monkeys (Pergamon, London 1960).

75 Warren, S. and Meissner, W. A.: Tumors of the thyroid gland. *4:* 14 (Armed Forces Institute of Pathology, Washington 1953).

76 Younger, R. K.; Scott, H. W., jr.; Butts, W. H., and Stephenson, S. E., jr.: Rapid production of experimental hypercholesterolemia and atherosclerosis in the rhesus monkey. Comparison of five dietary regimens. J. surg. Res. *9:* 263–271 (1969).

Authors' address: Dr. H. G. Spies and Dr. M. T. Clegg, Division of Reproductive Physiology, Delta Regional Primate Research Center, *Covington, LA 70433* (USA)

Pathology of Simian Primates, Part I, pp. 414–415 (Karger, Basel 1972)

Poisons and Toxins

R. N. T-W-FIENNES

Department of Pathology, Nuffield Institute of Comparative Medicine, Zoological Society of London, London

Simian primates play an important part in the evaluation of the toxicology of new drugs before they are used in human patients. No attempt is made in this work to review the vast literature on this subject. The latest reviews of the subject are MILLER's of 1966 [2] and that of GOLDSMITH and MOOR-JANKOWSKI 1969 [1].

The literature on clinical conditions arising from toxins and poisons is extremely sparse. RUCH [3], for instance, makes two passing references to the toxicity of lead in relation to neural and bone pathology. Lead poisoning occurred mainly in former times, when water systems were designed with lead pipes, and paints contained a lead base. These hazards are not important today.

Probably the most important source of toxic disease in primates arises from mouldy feed pellets. This may not always be diagnosed and the pathologist should be on the alert for signs in bowel and liver. In the smaller monkeys and prosimians, hepatitis arising from gammexane (BHC) has been encountered some years ago at the London Zoo [FIENNES, unpublished data].

Beyond these few remarks, the subject of toxins and poisons is not pursued, because of the dearth of information. Pathologists should, however, always bear in mind the possibility that toxic causes might be involved in simian deaths.

References

1 GOLDSMITH, E. I. and MOOR-JANKOWSKI, J.: Ann. N.Y. Acad. Sci. *162:* (1) 604 (1969).

2 MILLER, C. O.: Proceedings of conference on non-human primate toxicology. Department of Health, Education and Welfare, Virginia (1966).
3 RUCH, T. C.: Diseases of laboratory primates (Saunders, Philadelphia 1959).

Author's address: R. N. T-W-FIENNES, M. A., M.R.C.V.S., Head, Department of Pathology, Nuffield Institute of Comparative Medicine, Zoological Society of London, Regent's Park, *London, N.W.1* (England)

III. The Bodily Systems

Pathology of Simian Primates, Part I, pp. 418–432 (Karger, Basel 1972)

The Skin and Appendages, and Special Senses

R. N. T-W-Fiennes

Department of Pathology, Nuffield Institude of Comparative Medicine Zoological Society of London, London

Contents

Diseases of the Skin 418
I. General 418
II. Constitutional and Allergic Causes of Skin Disease 420
III. Traumatic Conditions 420
IV. Nutritional Causes of Skin Disease 421
V. Infectious Causes of Dermatitis 423
1. Arthropod Ectoparasites 423
2. Fungal Diseases 423
3. Bacterial Infections 424
4. Viral Diseases 424
5. Protozoal Diseases 425
6. Helminthic Diseases 425
VI. Conclusion 425
References 426
Special Senses 428
References 429

Diseases of the Skin

I. General

Very few systematic studies have been made on the pathology of primate skins and, unfortunately, it has not been possible to interest any of the few primate dermatologists in contributing a specialised chapter on the subject. Skin pathology is reviewed in an able chapter in Dr. Theodore Ruch's 'diseases of laboratory primates', and few significant advances in the pathological sense have been reported since the date of its publication in 1959. A selected bibliography is appended to this chapter, but most of the titles refer to specialised experimental approaches, which have little interest for the general primate clinician or pathologist. Most of the systematic work on primate skins has been done at the Oregon Research Primate Center's Department of Dermatology, and a series of papers on normal skin morphology has been published in the American Journal of Physical Anthropology by various authors. The present contribution, therefore, written by a general pathologist, not by a dermatologist, lacks the specialised and scholarly approach of other contributors, serving merely to survey the various influences which may lead to pathological skin conditions and to guide those who have the care of simian primates to a correct diagnosis.

In healthy primates having at least one other companion, also in primates in the wild, the skins are remarkably clean and free from parasites. This arises from the unique habit in this group of animals of grooming each other. This habit is so in-built into their natural instincts as to have social and almost ritual significance. A friendly chimpanzee, for example, will not only groom other chimpanzees, but will also work over the skin of its master or keeper, picking off any small excrescences it may find. One of the main signs that may guide the simian's keeper to know that its skin is in an unhealthy and irritable condition is that of scratching. However, scratching also has some sort of social significance, and a monkey will often scratch itself simply because it is apparently shy or embarrassed.

Skin lesions are only apparent on the hairless parts of the body, and their extent will not be evident unless there is accompanying hair loss. With some diseases, such as some forms of ring-worm or generalised scabies, hair loss will occur and this may help the diagnostician to discover the cause of the condition. In many infectious conditions, however, there is no loss of hair and the animal, whether alive or dead, must be clipped in order that the nature and extent of the lesions may be traced. Such occurs particularly in the pox diseases, in measles, and in herpes. Since some of these diseases are infectious to man, either by aerosol from skin debris or by contact, care must be taken both to wear adequate protective equipment and to dispose of the hair-clippings in a hygienic manner.

Even when hair loss occurs, this does not always indicate that a pathological skin condition exists. Hair loss may arise in conditions of stress, particularly in the first month or two following capture, in which case it will grow again once the animal has been conditioned to his new surroundings. It is thus by no means easy in many cases to determine the cause of hair loss, skin discomfort, or lesions, and thus to discover what remedial treatment should be given. In general, lesions affecting the integument arise from the following causes: constitutional; allergic; traumatic; nutritional and infectious. The infectious causes of skin disease may arise from: arthropod ectoparasites; fungal, bacterial, or viral causes; and finally from helminth parasites.

II. Constitutional and Allergic Causes of Skin Disease

Loss of hair from the scalp frequently occurs in ageing chimpanzees and sometimes also in macaques. Probably it is commoner in males than females. This form of alopecia is equivalent to the common baldness of man and appears to be associated with a seborrhoeic condition of the skin. The hair follicles become progressively atrophied and the skin of the scalp is converted virtually to a sebaceous secreting surface. As in man, the cause is unknown, and there is no cure for the condition.

Temporary patchy loss of hair from the tail or body may be due to stressful circumstances, as stated above, in which case it will return within 2 to 3 months. If it does not return, other causes of the condition should be sought as described under.

The general non-infectious skin diseases of man, such as psoriasis, have not been described in simian primates and do not, therefore, constitute a problem. A few papers describe the induction of allergic skin conditions, experimentally induced, but naturally occurring cases have not, so far, been encountered. It is, however, well for the investigator to be alive to the possibility of allergic causes of skin conditions, and the same may also be said of photosensitisation. The latter condition has also not been described, but theoretically could easily occur in monkeys exposed to sunlight, which had been fed on a diet suitable for its development.

III. Traumatic Conditions

Primate skins have good regenerative and healing powers and accidental and surgical wounds usually heal without complications. Suppurative conditions are rare even though lacerated wounds may occur beneath the hair cover, where they are not observed and thus escape treatment. This is in marked contrast to what occurs with some other groups of animals, such as bears, where wounds readily go septic. However, two forms of traumatic lesion require special mention, since they may be troublesome. These are frost-bite, and tail wounds.

With the possible exception of some South American primate groups, most simians are extremely cold-resistant, and in most temperate areas of the world can be permitted to exercise outside even in the winter when temperatures are below freezing. Under such conditions, the coat thickens, and the animals appear to enjoy the fresh air and exercise, even playing with the snow. The only requirement is mild heat, such as that of an infra-red bulb in the sleeping quarters. They are, however, liable under such conditions, to suffer from severe frost-bite which may affect the hands and feet. This is specially the case when metal perches or hand-holds are provided; these should be made of wood or be plastic-covered. The condition takes the usual form of a dry gangrene, which may lead to loss of fingers or toes.

Tail injuries take a somewhat similar form with a dry gangrene of the skin, leading to the loss of one or more coccygeal vertebrae with a raw bleeding area at the tip. Affected animals worry at the tail, sucking and biting at it with the result that the lesion becomes progressive. Surgical amputation is thought by some workers to be contra-indicated, because this merely intensifies the trouble so that in the end the whole tail is lost and the monkey must be sacrificed. Sometimes, if the tail is left alone, the end becomes dry and healing occurs. Where the condition is not due to trauma, it appears to result from draughty conditions in the house when the tail hanging down from the perch becomes chilled.

In addition to the above, the pathologist may sometimes be puzzled by extensive subcutaneous effusions in dead experimental monkeys at autopsy. These are usually the result of irritant subcutaneous injections or of giving large volumes of fluid for treatment of dehydration.

IV. Nutritional Causes of Skin Disease

Skin lesions commonly occur in deficiency states, and where a monkey colony is chronically affected their nutritional status should be examined. This can easily, though roughly, be done by determining the average intake of different food items, and estimating from food tables the probable intake of the items which may be concerned. Those which merit particular study are the essential fatty acids, vitamin A and members of the B-group vitamins. It should, however, be remembered that human requirements of vitamins are a poor guide to those which may be needed by other primates. For instance, vitamin D intake for the prevention of skeletal disease is totally different for man and other primates and differs even between different primate groups. The author had evidence of greatly enhanced requirements of some B-vitamins by Callicebus monkeys, lacking which they were subject to many different ailments as a result of which some died, and the survivors did not thrive well.

There appears to be no knowledge as to primate requirements of essential fatty acids. Deficiency in rats leads to scaliness of the feet and tail and even tail necrosis. In human infants, essential infant eczema is quickly cured by the addition of essential fatty acids to the diet. It is more than probable that some derangements, from which monkeys suffer in captivity, are associated with inadequate essential fatty acid intake. This is especially the case where the food, including pelleted foods, is stale or spoilt. Oxidation and deterioration of fats is progressive, and may continue even in foods stored at low temperature. At high temperatures, degradation can be very rapid indeed.

As is well known, an adequate intake of vitamin A is required for the maintenance of epithelial cells, and the vitamin provides the stimulus for cell division and new cell-growth. In excess, vitamin A can cause drying and peeling of the skin, general pruritus and loss of hair. In deficiency, the skin becomes dry and rough and papular eruptions are noted. There is also follicular hyperkeratosis.

Of the B-vitamins, deficiency of B_2 (riboflavin) leads to changes of the epidermis and its appendages. Baboons with chronic riboflavin deficiency exhibit massive hypertrophy of the skin of the face. Absence of niacin (nicotinic acid) is associated with the human disease of pellagra. Amongst the other serious symptoms associated with this disease, the skin suffers from atrophic changes, affecting the corium and associated with vaso-dilation, faulty keratinization of the epithelium, and atrophy of the sebaceous glands. The early stages are marked by an extensive erythema, aggravated by sunlight and heat. Pyridoxine (B_6) deficiency is also associated with skin changes, somewhat similar to those of unsaturated fatty acid deficiencies but accompanied by oedema. Biotin deficiency is also a cause of skin lesions of a seborrhoeic type, with generalised erythema, followed by scaling and alopecia; the shed scales are brown and greasy. When pantothenic acid is deficient, the hair becomes grey and tends to fall, and there are generalised scaling, erythema, and random foci of eczema.

It is plainly impossible to differentiate between all these lesions on clinical grounds. Where nutritional deficiencies are suspected, some estimate should be formed of the intake of factors which might be involved and then the diet should be supplemented with those which appear likely to be deficient. Where one factor is deficient, probably others are also, particularly in the case of B-vitamins, so that general supplementation with several factors may be desirable. Where deficiency states are present, supplementation of the diet should lead to improvement within a week or two, but an infectious cause should always be sought.

V. Infectious Causes of Dermatitis

1. Arthropod Ectoparasites

Primates rarely suffer from ticks, lice and fleas, due to their meticulous grooming habits. When animals groom themselves or each other, it is not a sign that they are picking off ectoparasites; usually, they pick off scurf, scales or blemishes. Where an animal is so affected, it is almost invariably in one that is caged by itself; if given a companion, the parasites will quickly be eliminated. Both apes and monkeys, however, are very susceptible to human scabies, which can be contracted from human sources and transmitted to human patients. Often the disease is progressive and may ultimately cause the death of the affected animal, in which the skin may also suffer from secondary infections and be covered with suppurating sores. The responsible parasite is the human *Sarcoptes scabiei*, though two other mange parasites have been described, *S. mutans* and *S. pitheci*. It is possible, even probable, that simian primates occasionally suffer from mange in the wild; if so, the disease is very rare and captured animals are unlikely to be suffering from it. It would appear that affected animals almost invariably acquire their infection from human sources. There have also been rare accounts of follicular mange in primates contracted from dogs.

2. Fungal Diseases

The dermatomycoses are well described by YOUSEF AL-DOORY in his chapter on mycoses. A typical case of ring-worm will be easily diagnosed and differs in no essential points from such diseases in man and other animals. Sometimes, however, the typical lesions of ring-worm are not seen and the disease takes the form of a generalised dermatitis, in which the hairs as well as the skin are infected. In such cases, the skin is affected by intense irritation with continual scratching and biting of the tail. There is a patchy alopecia which starts at the base of the tail and may come to affect the greater part of this organ. The body-hair falls out in patches and the lesions spread to the head and over the arms and legs. In the ultimate result, the animal may become almost naked with small tufts of hair left on the scalp, the limbs and other parts of the body. In these cases, both skin scrapings and hairs should be microscopically examined, when the fungus, usually of the *Trichophyton mentagrophytes* group, will be detected particularly in the shafts of infected hairs. The fungus grows well on Saubouraud's medium, at room or incubator temperatures. Both epidermal scales and an infected hair should be placed on the plate. This disease is unsightly and distressing to the affected animal, which improves greatly in condition as soon as treatment is instituted. The condition responds readily to oral therapy by griseofulvin; within a week or two, the diseased hairs are being shed and new hair-growth begins to cover naked parts of the skin.

3. Bacterial Infections

Bacterial infections of the skin are rare in primates and cause very little trouble, though cases of impetigo, associated with streptococci, have been described. In the chapter on alimentary diseases, Dzhikidze makes the point that primates affected with salmonellosis frequently have erythematous skin lesions, probably arising as an allergic phenomenon. Middle-ear disease is occasionally encountered at autopsy, due to cocci and other pyogenic organisms, such as Pseudomonas, the importance being that an extension of the infection to the inner ear and brain may sometimes occur causing death from meningitis.

4. Viral Diseases

Of the viral diseases, warts, Yaba and Yaba-like infections are described in sections on virus diseases and oncogenic viruses. There are reports also that monkeys have been experimentally infected with *Molluscum contagiosum*. The Yaba and Yaba-like diseases cause skin tumours which are usually regressive, but which may spread quickly through a colony.

A generalised 'rash' on primate skins will usually be associated with systemic diseases such as herpes, pox, or measles. In such cases, the animals are suffering from potentially serious epidemic disease, which may be dangerous to man, and the greatest care must be taken in handling the animals and their tissues. Skin biopsies or material taken from autopsy specimens, should be examined under the light microscope for determination of the type of lesion present and to locate inclusion-bodies. Specimens containing early and late lesions should be fixed in formalin and stained with haematoxylin and eosin. Herpes lesions consist of small sub-epidermal necrotic foci, in which typical Cowdry-type A intra-nuclear inclusion bodies can readily be found, especially in early lesions. These are diagnostic.

In pox cases, the lesions are typically vesicular and intracytoplasmic inclusion bodies can be seen. The lesions of measles are characterised by the presence of multi-nucleated cells and there are eosinophilic inclusion bodies in both nuclei and cytoplasm.

5. Protozoal Diseases

Protozoal diseases which affect the skin may be virtually ignored. The only disease concerned would be cutaneous leishmaniasis, to which experimentally some species of simian primates are certainly susceptible; under natural conditions or in captivity the occurrence of the disease has not been reported.

6. Helminthic Diseases

Helminthic diseases of the skin are of little importance. The larvae of hookworms or of strongyloides penetrate the skin and may set up temporary pruritus while they penetrate to the subcutaneous tissues. Occasionally larvae entering the simian as an unnatural host become localised in the skin especially overlying the joints and in these sites cause pustules or blisters. Guinea-worm infections have been described and one nematode, *Trichosoma*, is occasionally found in skin sections causing small blisters.

VI. Conclusion

Skin conditions of simian primates have not been extensively studied, and this is the reason for the difficulty in finding an author with specialised knowledge of the subject. Simian skin in most species does not closely resemble human skin, which has primitive features, supposedly due to 'neoteny'. Possibly for this reason, primates are not especially suitable as models of the more difficult human dermatological problems. However, as with man, the cause of diseases affecting the skin is often difficult to find and may involve some prolonged studies. As with man, too, skin involvement is often, perhaps usually, a sign of more deep-seated psychological or infectious disorder; it should thus never be neglected.

References

ADACHI, K. and YAMASAWA, S.: Aminopeptidase in primate skin. J. invest. Derm. *50:* 360 (1968).

AL-DOORY, Y.: The Mycoflora of sub-human primates. III. The flora of the skin of the baboon, vervet and gelada in captivity. Folia primat. *7:* 292–298 (1967).

ARAO, T. and PERKINS, E.: Skin of primates. 43. Further observations on Philippine tarsier *(Tarsius syrichta)*. Amer. J. phys. Anthrop. *31:* 93 (1969).

ARITA, I. and HENDERSON, D. A.: Smallpox and monkeypox in primates. Prim. medicine *3:* 122–123 (1969).

BELL, M.: Electromicroscopy and histochemistry of the Langerhans cells of sub-human primates. Adv. Biol. Skin *8:* 115–134 (1967).

BENDA, R.; MOSKOV, F. S., and DANES, L.: Visualization by immunofluorescence of elementary bodies and their aggregates in materials from skin lesions in poxvirus infections. Acta Virol., Praha *18:* 229–232 (1968).

CRAN, T. A.: An epithelial tumour occurring in jaws of a black-faced spider monkey *(Ateles geoffroyi)*. Oral Surg. *27:* 494 (1969).

DEMIS, D. J.: Allergy and drug sensitivity of skin. Annu. Rev. Pharmacol. *9:* 457–482 (1969).

DOBSON, R. M. and PINTO, J. S.: Cutaneous carcinogenesis in simian primates. Adv. Biol. Skin *7:* 345–354 (1966).

DOUGLAS, J. D.; TANNER, K. N.; PRIME, J. R.; RIPER, D. C. VAN, and DERWELIS, S. K.: *Molluscum contagiosum* in chimpanzees. J. amer. vet. med. Ass. *151:* 901–904 (1967).

DREIZEN, S. and LEVY, B. M.: Histopathology of experimentally induced nutritional deficiency cheilosis in marmosets *(Callithrix jacchus)*. Arch. oral. Biol. *14:* 577 (1969).

DUKE, B. O.: Experiments on the chemoprophylaxis of *Onchocerea volvulus* infection. Bull. Wld Hlth Org. *39:* 179–186 (1968).

FEINBERG, S. M.; FEINBERG, A. R., and LEE, F.: Hypersensitivity responses in monkeys. II. A comparison of several indicators in visualisation of immediate skin response to histamine 48–80 and antigen challenge. J. Allergy *43:* 65 (1969).

FEINBERG, A. R.; FEINBERG, S. M., and LEE, F.: Hypersensitivity responses in monkeys. III. Influence of various tranquillizers and other drugs on immediate skin reaction. J. Allergy *43:* 81 (1969).

FLATT, R. E. and PATTON, N. M.: A mite infestation in squirrel monkeys. J. amer. vet. med. Ass. *155:* 1233 (1969).

GIACOMETTI, L.: The healing of skin wounds in primates. I. The Kinetics of cell proliferation. J. invest. Derm. *48:* 133–137 (1967).

GIACOMETTI, L. and MONTAGNA, W.: Healing of skin wounds in primates. II. Proliferation of epidermal cell melanocytes. J. invest. Derm. *50:* 273 (1968).

GIACOMETTI, L.: The Healing of skin wounds in primates. III. Behaviour of the cells of Langerhans. J. invest. Derm. *53:* 151–154 (1969).

GIACOMETTI, L.: Behaviour of Langerhans cells in wound healing. Austr. Rec. *163:* 188 (Abstr. only 1969).

IGGO, A.: Electrophysiological and histological studies of cutaneous mechanoreceptors; in KENSHALO The skin diseases, ch. 5 (Thomas, Springfield 1968).

IGGO, A.: Cutaneous thermoreceptors in primates and sub-primates. J. Physiol., Lond. *200:* 403–430 (1969).

IGGO, A. and MUIR, A. R.: The structure and function of a slowly adapting touch corpuscle in hairy skin. J. Physiol., Lond. *200:* 763–796 (1969).

KADUR, B. M.; GULATI, S. M., and TALLAR, J. A.: Low molecular dextran in the management of frostbite in monkeys. Indian J. med. Res. *56:* 1675–1681 (1968).

KEBBER, W. T.; REESE, W. H., and HATTA, J. VAN: Balanitis, paronychia and onychia in a rhesus monkey. Lab. Anim. Care *18:* 506–507 (1968).

MACHIDA, H. and GIACOMETTI, L.: The skin. Bibl. primat. *7:* 126–140 (1968).

MACHIDA, H.; PERKINS, E., and HU, F.: The skin of primates. XXXV. The skin of the squirrel monkey *(Saimiri sciureus)*. Amer. J. phys. Anthrop. *26:* 45–54 (1967).

MACHIDA, H. and PERKINS, E. M., jr.: The distribution of melanotic melanocytes in the skin of sub-human primates. Adv. Skin Biol. *8:* 41–58 (1967).

MACKENZIE, J. C.: Ordered structure of the stratum corneum of mammalian skin. Nature *222:* 881–882 (1969).

MALLEY, A.: Skin manifestation of drug toxicity as revealed in non-human primates; in MILLER Conference on non-human primate toxicology, pp. 141–144 (Warrenton, Virginia 1968).

MONTAGNA, W.; MACHIDA, H., and PERKINS, E. M.: The skin of primates. XXXIII. The skin of the angwantibo *(Arctocebus calabarensis)*. Amer. J. phys. Anthrop. *25:* 272–290 (1966).

MONTAGNA, W. H.: Observation on the melanocytes of selected primates. Adv. Skin Biol. *8:* 59–88 (1967).

MONTAGNA, W. and UNO, H.: Baldness in non-human primates. J. Soc. cosmet. Chem. *19:* 173–185 (1968).

PERKINS, E. M., jr. and MACHIDA, H.: The skin of primates. XXXIV. The skin of the golden spider monkey *(Ateles geoffroyi)*. Amer. J. phys. Anthrop. *26:* 35–44 (1967).

PERKINS, E.; ARAO, T., and DOLNICK, C. H.: Skin of primates. XXXVII. Skin of pig-tail macaque *(M. nemestrina)*. Amer. J. phys. Anthrop. *28:* (5)75 (1968).

PERKINS, E. M.: Skin of primates. XXXVI. Skin of pygmy marmoset. *(Callithrix* [= *Cebuella*] *pygmaeas)*. Amer. J. phys. Anthrop. *29:* (5) 349 (1968).

PERKINS, E.; ARAO, T., and UNO, H.: Skin of primates. XXXVIII. Skin of red uakari *(Cacajao rubicundus)* (1968).

PERKINS, E. M.: Skin of primates. XLIV. The skin of Goeldi's marmoset *(Callinico goeldii)*. Amer. J. phys. Anthrop. *30:* 231–250 (1969).

PERKINS, E. M. and FORD, D. M.: Skin of primates. XXXIX. Skin of white-browed capuchin *(Cebus albifrons)*. Amer. J. phys. Anthrop. *30:* (5) 1 (1969).

PERKINS, E.M. and FORD, D.M.: Skin of cotton-top pinché *(Saguinus [=Oedopomidas] oedipus)*. Amer. J. phys. Anthrop. *30:* (5) (1969).

PERKINS, E.M.: The skin of primates. XLI. The skin of the silver marmoset *(Callithrix argentatus)*. Amer. J. phys. Anthrop. *30:* (5) 360–388 (1969).

PSEMICKA, P.: A contribution to the study of skin innervation of *Macaca mulatta.* Folia Morph., Praha *16:* 19–23 (1968).

SCHUMAKER, H.R. and MAHNO, G.: Thorns in the skin and joints of the monkey. Arch. Path. *84:* 536–538 (1967).

UNO, N.; ALLEGRA, F.; ADACHI, K., and MONTAGNA, W.: Studies of common baldness of the stumptailed macaque. I. Distribution of the follicles. J. invest. Derm. *49:* 288–296 (1967).

WALLACH, J.D.: Angioedema associated with strawberry ingestion by a gorilla. J. amer. vet. med. Ass. *153:* 879–880 (1968).

WENNER, W. L.; BOLANO, C.R.; CHO, C.T. *et al.:* Studies in pathogenesis of Monkey Pox. 3. Histopathological lesions and sites in immuno-fluorescence. Arch. Virusforsch. *27:* 13 (1969).

WICHER, K.; HAY, R.; REISMAN, R., and ARBESHAN, C.E.: Immunologic studies on active sensitization of monkey with ragweed pollen. Fed. Proc. *27:* (Abstr. only 1968).

Special Senses

As with diseases of the skin, no author has been found to make a specialist contribution on conditions affecting sight, hearing, and other senses. It does not, in fact, appear that specialist studies have been made in these areas. A bibliography of recent references is appended, but these mostly refer to experimental studies in relation to behavioural responses.

References to nutritional (e.g. vitamin and deficiency), nervous (e.g. amaurotic epilepsy), and endocrine (e.g. diabetes) conditions, which may affect sight, will be found in the appropriate sections.

To the clinician and pathologist, the most important condition affecting the sense organs is undoubtedly middle ear disease due to infectious causes, usually *Pseudomonas aeruginosa.* This is easily overlooked both in the living animal and at autopsy. This may accompany infection of the nasal sinuses and often spreads to the meninges with fatal results. In animals dying from meningitis, both nasal sinuses and the middle ears should be examined.

Blind monkeys have an uncanny way of concealing their disability and appear able to get around their cages and feed themselves so well that the fact of blindness often escapes detection. When there are no discernible lesions of the eyes, these cases are usually due to lesions of the brain and optic nerves.

Conjunctivitis and corneal ulcers, as with all animals, occasionally affect simians, and are treated as in any other animal.

In old monkeys, both Old World and New World, cataracts may affect one or both lenses. One case of naturally occurring detached retina has been described in an owl monkey.

There is little more that can usefully be said about the special senses, until further studies have been made on them or until more critical examinations have been made at autopsy. It is, however, appropriate to draw attention to the drastic rearrangement of the architecture of the nasal fossa in primates that has come about as a result of the enlargement of the brain and orbits in these animals. These are described by CAVE [1967] and reference to these papers will be rewarding. The characteristic primate changes are found in all Platyrrhine and Catarrhine genera, also in Tarsius. They are not found in Tupaia, whose primate status CAVE is reluctant to admit for this reason. It is also absent from both the lorisiform and lemuriform lemurs, and their standing in the primates is, therefore, also in doubt. CAVE writes, 'the distinctively primate type of nasal fossa results from the dethronement of olfaction and its replacement by vision as the dominant special sense'.

Almost nothing is known of olfactory and skin senses in primates, though specialised nerve endings and responses of primate skins are reviewed in KENSHALLO's 'The skin sense', pp. 45–46, 84 *et seq.*, 114 *et seq.*

References

Hearing and Scent

BRAGG, V. C. and DREHER, D. E.: A shock-avoidance technic for determining audiologic thresholds in the Cebus monkey. USAF Sch. Aerospace med. Tech. Rep. No. 5. SAM-TR-69-58 (Brooks, Texas 1969).

CARR, C. D.; SONNENWIRTH, A. C., and SENTURIA, B. N.: External otitis: production of *Pseudomonas aeruginosa* antibodies. Ann. Otol. Rhinol. Laryng., St. Louis *76:* 393–408 (1967).

CAVE, A. J. E.: Observations on the platyrrhine nasal fossa. Amer. J. phys. Anthrop. No. 5 *26:* 277–288 (1967).

CHORAZYMA, H. and DESMEDT, J. E.: Auditory discrimination in the baboon; in STARCK, SCHNEIDER, and KUHA Progress in primatology, pp. 320–321 (Gustav Fischer, Stuttgart 1967).

DEWSON, J. H. III and GOWEY, A.: Discrimination of auditory sequences by monkeys. Nature *222:* 695–697 (1969).

DREWER, D. E. and BRAGG, V. C.: Discovery of hearing loss in a *Cebus capucinus* monkey by a shock-avoidance technic. USAF Sch. Aerospace Med. Tech. Rep. No. SAM TR-6957 (1969).

ENGSTROM, H.; LINDEMAN, H. H., and ADES, H. W.: Anatomical features of the auricular sensory organ; in 2nd Sym. on the role of the vestibular organ in space exploration, pp. 43–45, US NASA Rep. No. NASA SP-115, Amer. Res. Center, Moffatt Field, Calif (1966).

HAWKINS, J. and JOHNSSON, L. G.: Light microscopic observations of inner ear in man and monkey. Ann. Otol. Rhinol. Laryng., St. Louis *77:* 608 (1968).

HEFFNER, H. E.; THURSTON, E., and MASTERTON, B.: Hearing in prosimians. J. acoust. Soc. Amer. *45:* 299 (1969).

HINCHCLIFFE, R. and PYE, A.: Variations in the middle ear of the mammalia. J. Zool. *157:* 277–278 (1969).

IGARASHI, M.; MAHON, R. C. J., and KONISHI, S.: Comparative movements of cochlea apparatus. J. Speech Res. *11:* 229–235 (1968).

JOHNSSON, L. G. and HAWKINS, J. E. J.: A direct approach to cochlea anatomy and pathology in man. Arch. oto-Laryng., Chicago *85:* 599–613 (1967).

KELLOG, W. H.: Communication and language in the houseraised chimpanzee. Science *162:* 423–427 (1968).

MASTERTON, B.; HEFFNER, H., and RANIZZA, R.: The evolution of human hearing. J. acoust. Soc. Amer. *45:* 966–985 (1969). – Abstr. Biol. Abstr. *1969:* 50, No. 95756.

MILLER, J. M.; MOODY, D. B., and STEBBINS, W. C.: Evoked potential and auditory reaction times in monkeys. Science *163:* 592–594 (1969).

PAPARELLA, M. M. and SUGIURA, S.: The pathology of suppurative labyrinthitis. Trans. amer. otol. Soc. *55:* 132–171 (1967).

PYE, A. and HINCHCLIFFE, R.: Structural varieties in the mammalian middle ear. Med. Biol. illus. *18:* 122–127 (1968).

REYNOLDS, P. C.: Evolution of primate vocal-auditory communication systems. Amer. Anthrop. *70:* 300 (1968).

STEBBINS, W. C.; JOHNSON, L. G., and HAWKINS, J., jr.: Relation of hearing loss to cochlea histopathology produced by antibiotics in the monkey. J. acoust. Soc. Amer. *44:* 357 (Abstr. only 1968).

STEBBINS, W. C.; MILLER, J. M.; JOHNSSON, L. G., and HAWKINS, J. E.: Ototoxic hearing loss and cochlea pathology in monkey. Ann. Otol. Rhinol. Laryng., St. Louis *78:* 1007 (1969).

STRUHSAKER, T. T.: Auditory communication among vervet monkeys *(C. aethiops)*; in ALTMANN Social communication among primates, ch. 16 (Chicago Univ. Press., Chicago 1967).

Vision

AABERG, T. M. and MACHENER, R.: Naturally occurring retinal detachment in owl mon key. Invest. opthal. *8:* 459 (Abstr. only 1969).

ADAMS, C. K. and JONES, A. E.: Spectral sensitivity of the sooty mangabey. Percep. Psychophys. *2:* 419–422 (1967).

ADAMS, C. K.; BRYAN, A. H., and JONES, A. E.: Electroretinographic determination of the spectral sensitivity of the baboon *(Papio anubis)*. Vision Res. *8:* 1399–1405 (1968).

BLAKEMORE, C.; HODGKINSON, R. G., and COWEY, A.: Retinal lesions in monkeys; recovery from misreaching. Vision Res. *8:* 883–888 (1968).

BROM, A. J.: Anterior corneal mosaic. Brit. J. Ophthal. *52:* 659–669 (1968).

BRUCHEA, J. M.: The frontal eye field of the monkey. Int. J. Neurol., Montevideo *5:* 262–281 (1966).

CAMPBELL, F. P.: Retina and optic nerve. Arch. Ophthal., Chicago *77:* 818–841 (1967).

CARPENTER, M. L.: A histological study of the chimpanzee eye. 6571st Aeromed. Res. Lab. Tec. Rep. No. 679 (Holloman, New Mexico 1967).

COOGAN, P. S. and MORRIS, F.: An improved histologic technic for studying primate retina. USAF Sch. Aerospace Med. Tech. Rep. No. SAM-TR-6953 (1969).

COOPER, G. F. and ROBSON, J. G.: The yellow colour of the lens of man and other primates. J. Physiol., Lond. *203:* 411–417 (1969).

COWEY, A. and ELLIS, C. M.: Visual acuity of rhesus and squirrel monkeys. J. comp. Physiol. Psychol. *64:* 80–84 (1967).

DE OLIVEIRA, L. F. and RIPPS, H.: The 'area centraneous' of the owl monkey *(Aotus trivirgatus)*. Vision Res. *8:* 223–228 (1968).

DE VALOIS, R. L. and JACOBS, G. H.: Primate colour vision. Science *162:* 533–540 (1968).

DOBELLE, W. H.; MARKS, W. B., and MACNICHOLL, E. F., jr.: Visual pigment density in single primate foveal cones. Science *166:* 1508–1510 (1969).

DOLLING, J. E. and BOYCOTT, B. B.: Organization of the primate retina; electron microscopy. Proc. roy. Soc. Edinb. B. *166:* 80–111 (1966).

GLICKSTEIN, M.: Organization of the visual pathways. Science *164:* 917–926 (1969).

GOLDSTEIN, E. B.: Visual pigments and the early receptor potential of frog and monkey retina. Exp. Eye Res. *8:* 247–248 (1969).

GOLDSTEIN, E. B.: Contribution of cones to the early receptor potential in the rhesus monkey. Nature *222:* 1273–1274 (1969).

GRAHAM, E. S.: MCVEAN, G. W., and FARRER, D. N.: Near and far visual acuity in rhesus monkeys *(M. mulatta)*. Percept. Mot. Skills *26:* 1067 (1968).

GREEN, K. F.; MOORE, J. W., and SARGENT, T. D.: Color preference in squirrel monkeys. Psychon. Sci. *4:* 367–368 (1966).

HELLING, C. S. and TURNER, B. C.: Auto-immune chorio-retinitis in rhesus monkeys. Science *162:* 561–563 (1968).

HOBACK, C. R. and SHRIVER, J. E.: Color vision in Tupaia and the phylogeny of color vision in the primates. Amer. J. phys. Anthrop. *27:* 241 (Abstr. only 1967).

HUBEL, D. H. and WIESEL, T. N.: Stereoscopic vision in macaque monkey. Nature *225:* 41–42 (1970).

JOHNSON, G. L.: Ophthalmoscopic studies on the eyes of animals. Philos. Trans. 1901. B. *194:* 1–82 (1968).

KILLAM, E. K.; STARK, L. G., and KILLAM, K. F.: Photo-stimulation in the species of baboons. Life Sci. *6:* 1569–1574 (1967).

LANDON, J. C. and BENNETT, D. G.: Viral induced simian conjunctivitis. Nature *222:* 683–684 (1969).

LATIES, A.: Histological features of retinal photoreceptors. Anat. Rec. *157:* 276 (Abstr. only 1967).

LATIES, A. M.; LIEBMAN, P. A., and CAMPBELL, C. E. M.: Photoreceptor orientation in the primate eye. Nature *218:* 172–173 (1968).

LEINFELDER, J. T.: Ocular histoplasmosis – a survey. Survey Ophthal. *12:* 103–104 (1967).

MARKOVA, A. Y.: 'Images of objects in outside world as perceptible to monkeys'. Vop. Psikhol. *1967:* (3) 98–109.

MARKOVA, A. Y.: 'Perception of outside world and its biological dependence'. Vop. Antrop. *25:* 45–54 (1967).

Massopust, L. C.; Wolin, L. R., and Kaboya, S.: Differential colour responses in the visual cortex of the squirrel monkey. Vision Res. *9:* 355–473 (1969).

Medin, D. L. and Davis, R. T.: Colour discrimination by rhesus monkeys. Psychon. Sci. *7:* 33–34 (1967).

Moreland, J. D.: Retinal topography and the blue-aves phenomenon. Vision Res. *9:* 965–976 (1969).

Ordy, J. M.: Visual acuity and the morphologic organization of the retina among primates. Anat. Rec. *157:* 294–295 (Abstr. only 1967).

Ripps, N. and Vaughan, H. G.: The spectral sensitivity of evoked potentials from the retina and vortex of nocturnal and diurnal monkeys. Vision Res. *9:* 895–907 (1969).

Samorajski, T.; Ordy, J. M., and Keefe, J. R.: Structural organization of the retina in the tree shrew *(Tupaia glis)*. J. Cell. biol. *28:* 489–504 (1966).

Schachter, J.; Arnstein, P., and Dawson, C. A.: Human follicular conjunctivitis caused by infection with a psittacosis agent. Proc. Soc. exp. biol., N.Y. *127:* 292–295 (1968).

Schrier, A. M. and Blough, D. S.: Photopic spectral sensitivity of macaque monkeys. J. comp. Physiol. Psychol. *62:* 457–458 (1966).

Serbanescu, T.; Bert, J.; Guillon, R., and Naquet, R.: Etude de la photosensibilité du *Papio papio* de Sénégal oriental. J. Physiol., Paris *60:* 399–403 (1968).

Shaffer, A. N.: Glaucoma research conference. Amer. J. Ophthal. *61:* 564–567 (1966).

Shriver, J. E. and Noback, C. R.: Color vision in the tree shrew *(Tupaia glis)* Folia primat. *6:* 161–169 (1967).

Sidley, M. A. and Sperling, H. G.: Photopic spectral sensitivity in the rhesus monkey. J. Opt. Soc. Amer. *57:* 816–818 (1967).

Stark, L.; Joy, R. M.; Hance, A. J., and Killam, K. F.: Further studies of photo stimulation in sub-human primates. Life Sci. *7:* 1037 (1968).

Stollnitz, F.: Can monkeys attend selectively to dimensions of colors? Psychon. Sci. *17:* 50 (1969).

Tigges, J. and Tigges, M.: The accessory optic system in Galago (Primates). Anat. Rec. *160:* 441 (Abstr. only 1968).

Törqvist, G.: Accommodation in monkeys. Acta ophthal., Kb *45:* 429 (1967).

Weale, R. A.: Why does the human retina possess 'fovea'. Nature *212:* 255–256 (1966).

Weale, R. A.: Why is there a fovea in the human retina. Proc. roy. Soc. Med. *62:* 573–574 (1969).

Wolin, L. R. and Massopust, L. C.: Characteristics of the ocular fundus in primates. J. Anat., Lond. *101:* 693–699 (1967).

Wolin, L. R. and Massopust, L. C., jr.: Ocular fundus in mammals. Anat. Rec. *160:* 519 (Abstr. only 1968).

Young, F. A. and Leary, G. A.: A comparison of the development of myopia in humans and chimpanzees. Invest. Ophthal. *7:* 114 (1968).

Young, F. A.: The development of simple myopia. Eye, Ear, Nose Throat Mon. *47:* 171–174 (1968).

Young, F. A. and Leary, G. A.: Comparative optical characteristics of major primates. Invest. Ophthal. *8:* 234 (1969).

Author's address: R. N. T-W-Fiennes, M. A., M.R.C.V.S., Head, Department of Pathology, Nuffield Institute of Comparative Medicine, Zoological Society of London, Regent's Park, *London, N.W.1* (England)

Pathology of Simian Primates, Part I, pp. 433–519 (Karger, Basel 1972)

Diseases of the Skeleton

M. J. M. BARKER and R. T. HERBERT

Department of Pathology, Nuffield Institute of Comparative Medicine, Zoological Society of London, London

Contents

I. Introduction 435
A. The Importance of Bone Disease of Primates 435
B. The History of Metabolic Bone Disease and Definition of Terms 436
1. Goundou 437
2. Paget's Disease of Bone or Osteitis Deformans 437
3. von Recklinghausen's Disease of Bone 438
4. Leontiasis Ossea 438
5. Craniotabes 438
6. Osteodystrophia Fibrosa 438
7. Renal Osteodystrophy 439
8. Vitamin D Resistant Rickets 439
9. Woolly Monkey Disease 441
10. Osteogenesis Imperfecta 441
11. Cage Paralysis 442
12. Conclusions 443
II. Pathology of Bone Diseases 445
A. Diseases of Nutritional Origin 445
1. Multifactorial Conditions 445
2. Rickets 447
3. Osteomalacia 456
4. Osteoporosis 462
5. Mineral and Protein Deficiency 464
6. Scurvy 466
7. Hypervitaminosis D 466
B. Diseases of Endocrine Origin 467

1. Parathyroid Syndromes 467
2. Thyroid Syndromes 469
C. Diseases Due to Mechanical Defects and Trauma 470
1. Congenital Deformities 470
2. Fractures 471
3. Cephalhaematoma 472
4. Atrophy of Disuse 472
D. Bone Tumours 474
E. Infections of Bone 475
1. Tuberculosis 475
2. Miscellaneous Infections 475
F. Osteoarthritis 476
III. Bone Structure and Physiology 479
A. Formation and Growth of Bone 479
1. Ossification in Membrane 479
2. Endochondral Ossification 479
B. The Elements of Normal Bone 481
1. Bone Matrix 481
2. Periosteum 481
3. The Cellular Components 482
C. Varieties of Bone and Nourishment of Bone Cells 484
D. Bone Activity 485
E. Mechanism of Calcification 485
F. Resorption of Bone 488
G. Calcium Homeostasis 489
1. Parathyroid Hormone 490
2. Calcitonin 490
3. Vitamin D 490
IV. Methods of Bone Study 491
A. General 491
1. Case History 491
2. Cage Design 491
3. Diet 494
4. Balance Studies 494
B. Radiology 496
1. Anaesthesia for X-Ray Examination 496
2. Main Features of the X-Ray Film 496
C. Histology 498
D. Histochemistry 500
E. Biochemistry 500
1. Interpretation of Data 500
2. Preparation of Samples 504
3. Analysis for and Significance of Parameters of Bone Disease 505
4. Calcium Kinetics 509
V. Conclusions 509
Acknowledgements 510
References 510

I. Introduction

Since Bland-Sutton's classical studies in 1884 [16], osteodystrophies of captive primates have continued to attract attention. Despite this, the field is still confused. The confusion is due, in part, to a lack of definition of terms and partly because primates are rarely chosen as experimental animals for the basic study of bone metabolism.

In this chapter we have attempted to set out clear definitions of the terms in use. We have reviewed the pathology of simian bone disease and have broadly outlined the physiology of bone so as to be able to highlight areas where the study of primate bone metabolism might be rewarding. Finally, we have outlined the methods by which we study bone diseases of simian primates and their experimental induction.

Whether primates are studied for their own intrinsic interest or for comparison with man, their bone metabolism, normal or pathological, presents an interesting challenge, that could lead to advances in the understanding and treatment of many human conditions.

A. The Importance of Bone Disease of Primates

The importance of bone disease of primates is twofold. Firstly these animals are becoming progressively more difficult and expensive to obtain from wild sources. The establishment of healthy colonies for breeding in order to provide a regular source of fit animals for research programmes therefore becomes increasingly more important. Until comparatively recently, bone disease was exceedingly frequent in captive primates. Unless attention is paid to the requirements of different species, bone disease can prove to be a major cause of ill-health and failure to breed in primate colonies. Secondly, bone diseases of primates are important in the possibilities for use of these animals as a model system to gain fundamental knowledge into the pathogenesis of several serious and incompletely understood bone diseases of man. For example, osteoporosis of humans is a common clinical condition which may arise from many causes, e.g. malabsorption or excessive loss of protein, calcium deficiency, disuse, senility, and hormonal disturbances such as lack of oestrogens and steroid therapy. There are still many uncertainties regarding the precise role of various factors in the aetiology and treatment of the young and old age groups, for example the effect of calcium treatment at all ages and the significance of exercise and weight-bearing in the elderly. The larger non-human primates, such as baboons and chimpanzees, may frequently suffer from a degree of osteoporosis in captivity and they would appear to be a logical choice of animal to investigate various aspects of this disease.

The current vast amount of work progressing on the effects of parathyroid hormone, calcitonin and vitamin D metabolism, suggests possibilities for the use of primates to investigate the direct effects of these compounds in an animal of the same order as man. Calcitonin is currently being used in the treatment of osteoporosis and Paget's disease [8, 89] as it inhibits bone resorption and alleviates the pain of Paget's disease. This latter disease has not so far been certainly diagnosed in non-human primates because, presumably, there is some unknown fundamental difference in the bone metabolism and environment of these animals compared with humans.

The precise mechanism of bone resorption still remains uncertain particularly regarding the relative roles of osteocytes, osteoclasts and parathyroid hormone. Detailed study of the demineralizing bone diseases which primates are prone to develop could add some more useful information to this problem.

Before reliable information can be obtained from these or any other research programmes using primates as a model system it is essential to understand the mechanism of production of the diseases from which they suffer and how they can be treated or prevented from occurring. Bone diseases of primates can be considered under the two general headings of metabolic bone disease and non-metabolic bone disease. The former, particularly rickets, osteomalacia and secondary hyperparathyroidism affects predominantly New World species, mainly at the time between weaning and becoming adult, and appears to be a disease solely of captivity, probably due to poor nutrition. Non-metabolic bone disease includes congenital lesions, tumours, fractures, infections and osteoarthritis. These conditions also show some species variability particularly regarding the relative incidence of fractures and osteoarthritis. The variation between species in this case possibly reflects differences in the social organisation and ecology of the animals rather than a fundamental difference in their bone metabolism.

B. The History of Metabolic Bone Disease and Definition of Terms

The most frequently occurring class of bone diseases of primates, and the most important regarding morbidity and mortality, is metabolic bone disease, particularly rickets, osteomalacia and secondary hyperparathyroidism.

The complex and confused history of this group of diseases was well reviewed by RUCH in 1959 [163] and by KROOK and BARRETT in 1962 [112]. Reference should be made to these works for a detailed account of the controversy.

Metabolic bone disease of primates has in the past been the subject of many conflicting diagnoses and there is still some uncertainty surrounding several terms. The most important of these are: Goundou; Paget's disease; von Recklinghausen's disease; leontiasis ossea; craniotabes; osteodystrophia fibrosa; renal osteodystrophy; vitamin D resistant rickets; woolly monkey disease; osteogenesis imperfecta; and cage paralysis.

1. Goundou

Goundou is not now thought to be a disease which causes bone disorders in simian primates. It is the name normally given to a condition of hypertrophic osteitis occurring in humans and connected in some uncertain way with yaws, i.e. an infection with the spirochaete *Treponema pertenue* distributed mainly in Central Africa. Confusion arose because of the appearance of maxillary swellings in both Goundou and severe osteomalacia. The basic difference, however, is that the bony lesions of the maxilla in the case of Goundou are essentially the primary change in the skull which may later spread to involve the hard palate and other skull bones. If maxillary swelling occurs in severe osteomalacia of primates it does so as a late result of the general thickening of the whole skull. The long bones of the leg are also bowed anteriorly in both conditions, the deformed osteomalacic or rachitic legs of the animals being likened to the 'boomerang' leg of yaws in humans [163].

2. Paget's Disease of Bone or Osteitis Deformans

This is a disease of unknown aetiology which has so far only been certainly diagnosed in humans. The disease produces bone pain, thickening of the skull and typically bowed tibias known in this case as 'sabre' tibias. The affected bones are highly vascular and there is a small but definite risk of sarcomatous change. The thick skull and bowed legs of the severely rachitic or osteomalacic animals were likened to this disease particularly by CORSON-WHITE [33] who reported 3 cases in 6,570 consecutive autopsies. Macroscopically the bones of Paget's disease and of osteomalacia may appear similar, both showing irregular thickening and softening.

Microscopically there are, however, differences. Paget's disease is characterised by a mosaic of cement lines. This arises because of the irregular formation of new bone occurring as a result of greatly increased osteoblastic activity. Osteoclasts are also present but their activity is not so marked. Between the bone trabeculae there are large vascular channels with much fibrous tissue.

In contrast the bones, in cases of osteomalacia, are less vascular with thin trabeculae, active osteoclasts, wide osteoid seams and little mineralised bone.

3. von Recklinghausen's Disease of Bone

This is also known as osteitis fibrosa cystica, and is caused in humans by a primary disorder of the parathyroid glands. This is usually an adenoma, adeno-carcinoma or occasionally primary hyperplasia, i.e. the condition is primary hyperparathyroidism.

No definite cases of this disease have so far been reported in non-human primates. The limb bones may be fibrous and cystic in other conditions, e.g. severe metabolic bone diseases. Unless a primary disorder of the parathyroids can be shown to be the cause, then the condition cannot correctly be called osteitis fibrosa cystica as described by VON RECKLINGHAUSEN.

KROOK and BARRETT in 1962 [112] referred to generalised osteitis fibrosa cystica as a radiological diagnosis of cases of supposed nutritional secondary hyperparathyroidism due to calcium-phosphorus imbalance. In this context osteitis fibrosa cystica is clearly not intended to mean von Recklinghausen's disease but is used in a purely descriptive sense.

Confusion can easily arise with the incorrect use of this term and careful definition is important.

4. Leontiasis Ossea

This term has been used in a descriptive sense in connection with skull deformities in primates and other animals [6, 163]. It is usually now used, at least in humans, to describe the appearance of the face and skull resulting from chronic inflammatory disease of the maxilla.

5. Craniotabes

In 1884, BLAND SUTTON [16] used this term to describe the thinning and softening of the bones of the base of the skull associated with 'tabetic' patches over the orbits. Col. A. S. HAMERTON who was pathologist at the London Zoo for 18 years considered the term to mean excessively thin cranial bones.

It is now generally used to describe the bossed appearance of the skull in rickets and is due to the failure of the suture lines and fontanelles to close.

6. Osteodystrophia Fibrosa

This term was first introduced by Sir ARNOLD THEILER to describe lesions of the skeleton in horses due to Ca/P imbalance. It has been applied to bone changes found in several genera of South American primates, (marmosets, tamarins and capuchins) of differing ages, fed a diet containing 2,000 IU vitamin D_2/kg diet [95].

The bone changes described were extensive osteoclast activity, resorption of trabecular and compact bone, active osteoblasts, new trabecular and subperiosteal osteoid and fibrous replacement of bone marrow.

The predominant biochemical changes consisted of high serum alkaline phosphatase activity and a tendency towards a low serum phosphorus.

When vitamin D_3 (2,000 IU/kg diet) was substituted for vitamin D_2 these changes were reversed. There was also a marginal increase in the serum calcium and an increase in the serum inorganic phosphate.

These features are very similar to those described by LEHNER *et al.* [114] who showed that up to 1,000 IU vitamin D_2/kg diet does not prevent rickets in young squirrel monkeys *(Saimiri sciureus)*.

As this condition was produced by feeding these animals a form of vitamin D that does not appear to be effective in New World monkeys, this is comparable to an inadequate supply of vitamin D. This consideration and the radiological, histological and biochemical features lead us to suggest that osteodystrophia fibrosa is a descriptive term for a form of rickets or osteomalacia depending on the age of the animal.

7. Renal Osteodystrophy

Bilateral renal disease affecting tubules and glomeruli and associated with chronic uraemia can result in severe skeletal lesions. Impaired renal function produces a diminished excretion of hydrogen ion and impaired reabsorption of bicarbonate, i.e. a state of metabolic acidosis. Plasma phosphate is raised and the plasma calcium may be low or normal. Plasma calcium may be lowered because retained organic ions combine with ionized calcium. Hence compensating secondary hyperparathyroidism may occur. The lesions in the bone vary with age and might be expected to range from generalized loss of density, to rickets, osteomalacia or secondary hyperparathyroidism. Some degree of resistance to the action of vitamin D is a constant finding in cases of renal osteodystrophy of humans. We know of no definite cases of this condition in non-human primates.

8. Vitamin D Resistant Rickets

In a review of the mode of action of vitamin D, NORMAN [142] has discussed the use of this term. DENT [43], FANCONI and GIRARDET [56] regard the condition as primarily a renal tubular disorder resulting in reduced renal reabsorption of phosphate and hypophosphataemia. WINTERS *et al.* [191]

have reviewed many cases of vitamin D resistant rickets and conclude that the disease is nearly always inherited and is usually congenital. NORMAN [142] suggested that this view supports the idea that vitamin D may act at the gene level. A genetic mutation could thus have an effect on an individual's ability to respond to vitamin D.

In recent years a considerable amount of work has been done on the metabolism of vitamin D and the results of some of this work have been outlined in the section on calcium homeostasis

AVIOLI *et al.* [5] have investigated the metabolism of vitamin D_3-^3H in vitamin D resistant rickets and tentatively concluded that abnormalities in vitamin D metabolism may account for observed osseous and biochemical changes in patients with vitamin D resistant rickets and familial hypophosphataemia.

FRASER and KODICEK [67] suggest, on the basis of their work on the biosynthesis by the kidney of a biologically active vitamin D metabolite, that in familial vitamin D resistant rickets a genetic defect may result in the absence of an enzyme responsible for the conversion of 25-hydroxycholecalciferol (25-HCC) to '1-oxygenated 25-HCC' known as 'compound P', which is 2.5 times as active as 25-HCC [111]. BORDIER *et al.* [21] have used 25-HCC to treat successfully a young woman with vitamin D resistant rickets.

As yet, no case of vitamin D resistant rickets that fits any of the above definitions has been recorded in simian primates. The apparent inability of New World monkeys to utilize vitamin D_2 is perhaps an analogous situation but we are not aware of any work using vitamin D_3-^3H or vitamin D_2-^3H in non-human primates that would give any indication as to whether the apparent molecular specificity of New World primates is due to an inability to convert vitamin D_2 to 25-HCC or 25-HCC to 'compound P'.

Neither is it clear whether these animals really do not metabolize vitamin D_2 or whether they require higher levels of the vitamin than the Old World primates. It should, perhaps, be noted in this context that vitamins D_2 and D_3 show different potency in developing rats and chicks. Vitamin D_3 has equal potency in rats and chicks; vitamin D_2 has equal potency to vitamin D_3 in rats; but vitamin D_2 has only one-tenth the activity of vitamin D_3 in chicks [29]. The relative effectiveness of vitamin D_2 and D_3 have not yet been fully determined for primates. Whilst the approximate amounts of vitamin D needed to keep the bones of various primates healthy are now reasonably well known [94, 95, 114], the minimum requirements for physiological activity have not been investigated to the same degree as they have for laboratory animals such as chicks and rats.

9. Woolly Monkey Disease

Metabolic bone disease seems to affect members of the genera *Lagothrix* and *Cebus* more severely than other New World primates. The great thickening of the skull which occurs as a late result of the disease has led some people to suggest that there is a separate woolly monkey disease [52] (fig. 1).

Thickening of the skull is not normally seen in human osteomalacia. However, rather than implicate a specific pathology for the woolly monkey, we believe that the thick skull is an end-result of severe osteomalacia complicated by secondary hyperparathyroidism, which is rarely seen to such a degree in other animals. The genus *Lagothrix* contains some of the largest South American primates. It is possible that their requirements for vitamin D_3 are seldom met completely in captivity by diets suitable for smaller animals, thus leaving them very prone to metabolic bone disease.

10. Osteogenesis Imperfecta

This is a congenital disease of humans resulting in abnormally thin bones with thin trabeculae although mineralization is normal. There are often, however, multiple fractures, hyperplastic callus and deformities. Hence

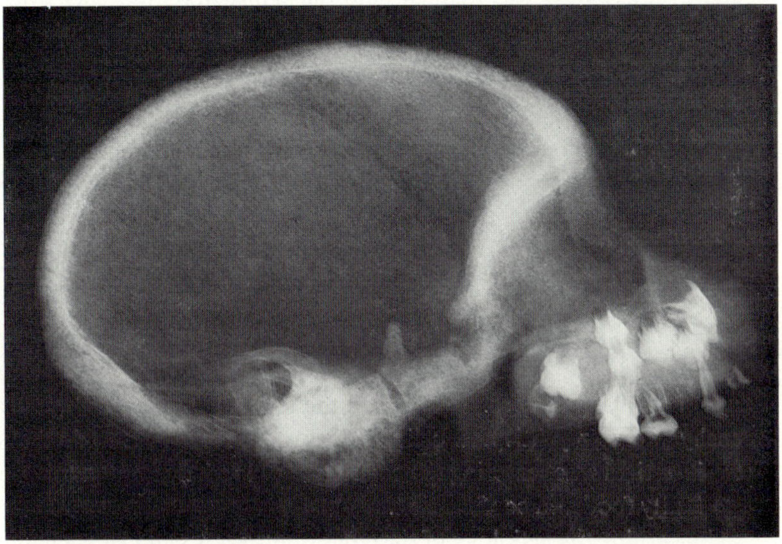

Fig. 1. X-ray appearance of woolly monkey skull *(Lagothrix lagotricha)* showing effects of osteomalacia and secondary hyperparathyroidism. Note thick skull and displacement of tooth buds.

severe rickets and osteomalacia of primates have been given this name in the past.

The basic abnormality in osteogenesis imperfecta lies in the osteoblasts and may be associated with thin sclerae which appear blue and otosclerosis leading to deafness.

Fragilitas ossium is the name sometimes given to humans with abnormally brittle bones due to osteogenesis imperfecta. This condition has not been known to occur so far in non-human primates.

11. Cage Paralysis

Describing this syndrome in 1905, BROOKS and BLAIR [24] report it as the commonest cause of death of primates in captivity, excluding tuberculosis. The symptoms of the disease as they described it consisted of decreased general activity, increasing stiffness and loss of the animal's normal interest in its surroundings. The limbs were frequently fractured and the arms were used as crutches for locomotion. Reflexes were depressed and there was often partial paralysis of the hind limbs with atrophy of the limb muscles. The muscles of the trunk were also affected resulting in the characteristic protuberant abdomen.

Dorsal kyphosis was frequent and the skulls were thickened and soft. Ultimately those symptoms progressed to complete paraplegia with loss of sphincter control of faeces and urine.

Distortion of the chest, typically a 'pigeon breast' deformity and of the thoracic spine commonly led to terminal bronchitis and hypostatic pneumonia due to poor respiratory movements and inadequate ventilation of the lungs.

The occurrence of lesions in the central nervous system (CNS) originally suggested to BROOKS [24] that the disease was primarily of the CNS. These lesions were later shown by them to be secondary to the deformed spinal column producing degeneration of direct and crossed pyramidal tracts.

CAMPBELL and CLELAND [27] describing this condition also showed changes in the spinal cord. They called the lesions a subacute degenerative parenchymatous myelitis and believed them to be due to a toxin from the gastro-intestinal tract entering the cord via the bloodstream; the toxin being due to faulty food and confinement.

Concerning the aetiology of the condition BROOKS and BLAIR [24] make three basic points. Firstly the possible deficiency of 'earthy salts'. Secondly, an infection and thirdly, a congenital predisposition to disease aggravated by ill-ventilated, small dark cages. They noticed that the condition occurred less frequently in larger lighted cages and they observed no case in a large cage exposed to much sunlight. Hence they recognised the importance of lack of minerals, complicating infections, and ultra-violet light.

BROOKS and BLAIR [24] regarded the prognosis of the established syndrome as serious and made several important observations regarding treatment. They recommended bone dust, lime water, sunlight, dry well-ventilated cages, a varied diet and a balance of organic and inorganic foods.

They therefore recognised the essential nature of 'cage paralysis' which they identified with osteomalacia, as being primarily a nutritional disturbance aggravated by insufficient ultra-violet light or as shown later insufficient vitamin D_2 or D_3 depending on the species. All these factors led them to conclude that 'cage paralysis as seen in primates in captivity is osteomalacia'.

This subsequently proved to be rather a sweeping statement as the term came to be used in a far wider sense and osteomalacia was not the only cause of 'cage paralysis'.

HAMERTON remarked in 1936 [78] that 'Cage paralysis has been used to mean any form of palsy occurring among animals in the menagerie'. The term was then a collective name for a range of diseases not all of them primarily skeletal, which was non-specific and implied that the cause was captivity in a cage that was too small.

RUCH [163] makes the point that 'cage paralysis' has occurred in large cages and with apparently normal skeletons. Possibly these cases were due to meningitis or other viral infection, or to a true paralysis due to vitamin B_1 deficiency in the diet producing a motor nerve degeneration [58]. The term 'cage paralysis' has been used, therefore, for a clinical picture which can be caused by several diseases.

Unwillingness or inability to move about the cage in a normal manner is a symptom of disease. The term 'cage paralysis' is, therefore, best avoided and a specific diagnosis made.

12. Conclusions

RUCH [163] concludes his section on the historical aspects of metabolic bone disease by raising the point that non-human primates may differ from humans in some fundamental way regarding endocrine function or bone metabolism. There are certainly some features of metabolic bone disease of these animals which are not commonly seen in their counterpart in human diseases, e.g. frequent skull lesions, severe fibrous replacement of bone with relatively few osteoclasts and high serum alkaline phosphatase (e.g. $>$1,000 IU/l).

Table I. Major details of semi-synthetic diets for New World primates that have (A)

	Reference	Animals used	Protein, %	Oil, %	Carbo-hydrate, %	Fibre, %	Mineral mix, %
A	125	*Cebus capucinus*	25.0 casein	8.0 Mazola corn oil	60.6 sucrose	–	4.0 salt mix IV (85)
A	114	*Saimiri sciureus*	15.0 vitamin-free casein	5.0 corn oil	20.0 white dextrin 20.0 sucrose	13.3 cellulose	2.0 USP XIV
A/B	BARKER and HERBERT	*Cebus albifrons* *Saimiri sciureus* *Lagothrix lagotricha*	27.0 casein	5.0 corn oil	10.0 glucose 40.0 corn starch	15.6 cellulose powder	2.0
B	95	*Cebus albifrons*	25.0 casein	8.0 corn oil	62.0 sucrose	–	4.0 salt mix IV [85]

This may occur as a result of the very advanced state that the disease rapidly reaches in these animals because of their high dietary requirement for protein, and for vitamins D_2 and D_3 (table I). Alternatively, the differences may be explained by some more fundamental metabolic differences.

Although it is possible that the disorder could be classified with the fibrous osteodystrophies of unknown origin [163], the evidence leads us to consider metabolic bone disease of primates as basically rickets or osteomalacia often complicated by secondary hyperparathyroidism. These diseases, however, may take an atypical form in these animals. Rickets is the disease produced by an insufficient quantity of the appropriate vitamin D available to the growing animal. Osteomalacia is the disease produced by an insufficient quantity of the appropriate vitamin D available to the adult animal.

maintained healthy bones, and (B) corrected abnormal bones

Vitamin mix, %	Vitamin D_3, IU/g	Ca, %	P, %	Ca/P	Calorific value, cal/g	Notes
0.4	2.0 approx. as 2% cod liver oil USP	0.56	0.30	1.8	2.50	+ supplement of additional minerals
2.2	1.25–10.0	0.24	0.24	1.0	2.68	+ 20% apple sauce + 2.5% agar + 1 oz reconstituted frozen orange juice daily
0.4	8.0	0.27	0.15	1.9	2.50	deionized-distilled water, *ad libitum*
1.0	2.0	0.56	0.30	1.8	2.50	

II. Pathology of Bone Diseases

A. Diseases of Nutritional Origin

1. Multifactorial Conditions

In the past, diets given to captive primates have been grossly imbalanced in many respects. Du Boulay and Crawford [52] have shown that the traditional vegetable and fruit diet fed to primates in captivity is not only low in protein and vitamin D but also has a very low calcium phosphorus ratio due to low calcium and high phosphorus content.

Such diets have produced severe bone disease particularly in the New World species (Cebidae). The apparently greater susceptibility of these animals to bone diseases when fed this type of diet may reflect their higher protein requirement and their apparent specificity for vitamin D_3 rather than vitamin D_2. Vitamin D requirements have been discussed by DAY [37] and WOLF (pp. 244–248).

The typical syndrome produced is a combination of demineralizing bone disease (rickets or osteomalacia) with a variable degree of secondary hyperparathyroidism and osteoporosis.

Table I gives details of the approximate composition of some of the diets that have been fed to a variety of Cebidae as treatment for metabolic bone disease or used to maintain healthy bone in these animals.

The combination of vitamin D deficiency and calcium-phosphorus imbalance is most common today among pet animals. Protein deficiency is likely to occur also. Although commercial monkey chows are manufactured, they are not always readily available to pet owners.

A typical case was that of a female woolly monkey *(Lagothrix lagotricha)* 18 months old, weighing 2 kg, quite unable to move or feed herself. She sat huddled in a crouched position typical for sick primates. A thoracic rosary was easily palpable and there was a protuberant abdomen associated with general muscular hypotonia. The coat was dull and sparse and the animal shrieked if its limbs were palpated.

Many limb deformities were obvious and radiological examination revealed the following: folding displaced fracture of right radius and ulna; folding fracture of left radius and ulna; fracture of midshaft of right and left humerus; fracture of left femur; fracture of right tibia and fibula, and dislocation of clavicular-scapular joints. Bowing of the limbs was also evident and total lack of callus formation with generalised decrease in bone density (fig. 2).

The changes of rickets at the epiphyses of the long bones were not as pronounced as might be expected from the evidence of the severity of the disease elsewhere, possibly because the predominant deficiency was of calcium and protein.

Treatment with 5,000 IU vitamin D_3 i.m./kg body wt/week and a high calcium and protein diet rapidly produced great improvement. Within 2 weeks she could walk and feed herself. Weight increased and she started to climb about her cage even though no fixation of the fractures had been attempted.

Repeat radiological examination after two months treatment, showed that all the fractures had united and there was much callus formation around the fracture sites (fig. 3).

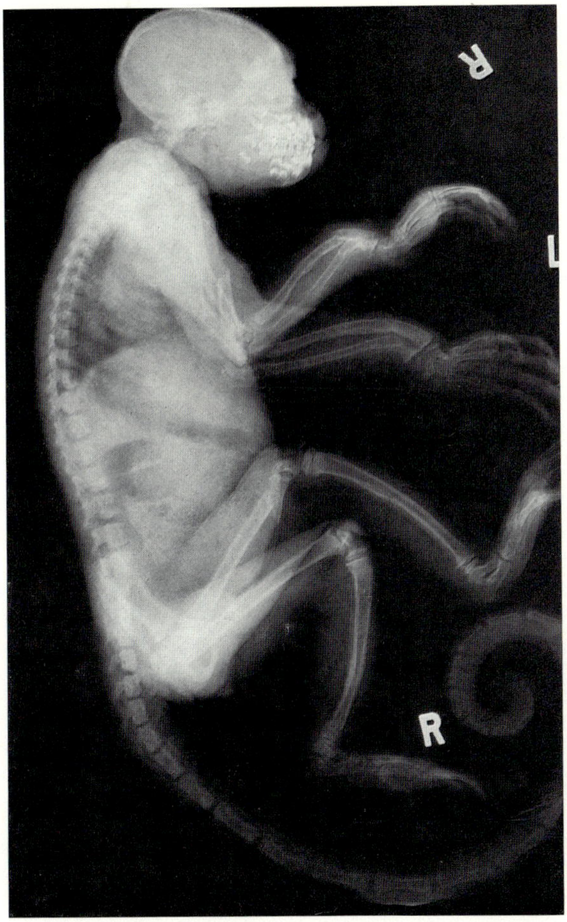

Fig. 2. X-ray appearance of woolly monkey skeleton *(Lagothrix lagotricha)* showing effects of severe demineralising metabolic bone disease. Female animal, weight 2 kg, age 2 years. Note features of rickets at epiphyses, frequent fractures, displaced teeth and low density bone, i.e. rickets and calcium deficiency.

2. Rickets

a) Introduction

Rickets is the disease produced by an insufficient quantity of the appropriate vitamin D available to the growing animal. It is virtually always an acquired metabolic bone disease due to the failure of calcification to keep pace

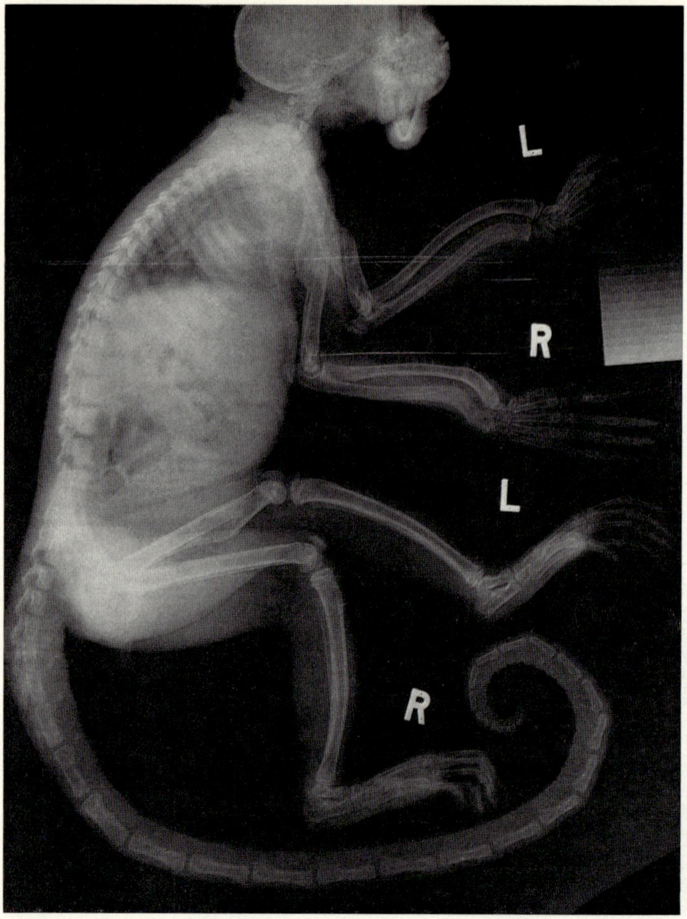

Fig. 3. X-ray of same woolly monkey as figure 2 after treatment with vitamin D_3 and calcium. Note callus formation around fractures and increased density of bone cortices.

with growth. Rickets and osteomalacia of animals represent the same basic abnormality, differences being due to age.

New World primates appear to require vitamin D_3 to maintain healthy bones [95, 114] or possibly a very high level of vitamin D_2 although extremely large doses have been given with no apparent effect [34, 96]. Old World species can apparently maintain healthy bones on vitamin D_2.

Lack of calcium results in a demineralised or poorly mineralised skeleton but not rickets unless accompanied by vitamin D deficiency.

b) Aetiology

Factors implicated in the past as causing rickets in primates in addition to malnutrition, include too early weaning, lack of exercise and poor climate [16].

In an assessment of nutritional bone disease in captive primates, DU BOULAY and CRAWFORD [52] drew attention to the practice of feeding these animals a predominantly vegetarian diet with a low calcium and high phytic acid content. Phytic acid (the hexaphosphate salt of *myo*-inositol) forms insoluble salts with calcium in the gut. If phytic acid is hydrolysed in the gut, phosphorus would be released, effectively altering the calcium-phosphorus ratios. The overall effect of either process is interference with the absorption of calcium. Also, they discussed the significance of low protein and calcium-phosphorus imbalance which, they believed, added to the severity of the effects produced by inadequate vitamin D.

However, the most important factor producing rickets in primates is undoubtedly inadequate dietary vitamin D or lack of exposure to ultra-violet light [93, 118, 119]. Ultra-violet light acts on 7-dehydrocholesterol to produce vitamin D_3, cholecalciferol, and acts on ergosterol to produce vitamin D_2, ergocalciferol (also known as calciferol). This reaction is thought to take place in the skin. The effect of these deficiencies may, as already stated, be aggravated by unbalanced calcium-phosphorus ratio in favour of phosphorus. The accepted normal for health is 1.0 [98, 112, 163].

c) Congenital Rickets

Marginal osteomalacia in a mother or a slight degree of rickets in an infant may pass undetected by standard whole skeleton X-rays without biochemistry or histology. It is, however, usually stated that congenital rickets does not occur, at least in humans, unless the mother is suffering from osteomalacia [63]. Certainly BLAND SUTTON did not observe any cases in his survey of over 200 primates.

In the years 1962–1965, three young were born to an orang-utan 'Tulie' *(Pongo pygmaeus)* in the London Zoo. Two were born alive, both with a disease resembling rickets. One died shortly after birth and one was treated and is still alive. The mother died in 1965, two days after giving birth to the third which died from a fractured cranium. Radiological examination of the third infant revealed imperfect ossification of the bones of the cranium resembling rickets. *Post mortem* examination of the mother revealed that she was thin (5.6 kg) and that death was primarily the result of *Pseudomonas aeruginosa* septicaemia. She also suffered from bilateral pyelonephritis and the bowel

4 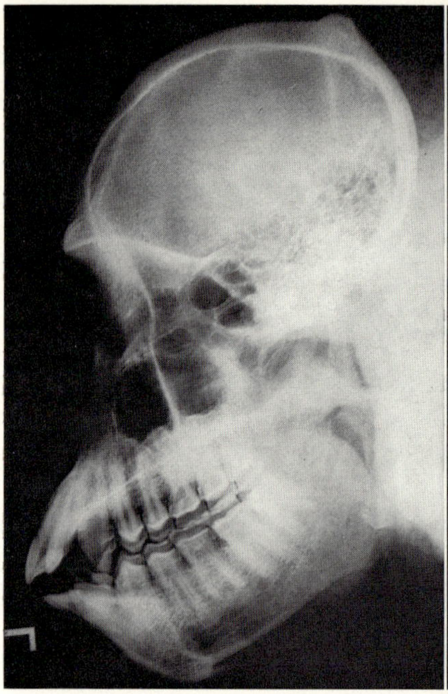

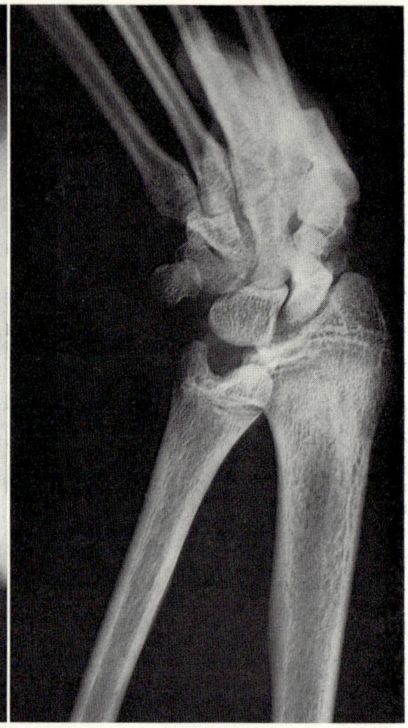 5

Fig. 4. X-ray of skull of female orang-utan 'Tulie' *(Pongo pygmaeus)*. No evidence of metabolic bone disease.

Fig. 5. X-ray of forearm and wrist of 'Tulie'. No evidence of old rickets or metabolic bone disease.

was haemorrhagic. No skeletal disease was apparent either radiologically or histologically [59] (fig. 4 and 5). Hence it is possible that at least in this case congenital rickets occurred in the absence of demonstrable maternal skeletal disease.

In 1967, HUNT *et al.* [95] observed 2 cases of neonatal rickets in stillborn squirrel monkeys *(Saimiri sciureus)* born to mothers fed on vitamin D_2, thereby underlining the apparent ineffectiveness of vitamin D_2 in these New World species.

d) Signs of Rickets

The main features of rickets in living primates as described by BLAND SUTTON in 1884 [16] were reduced activity, reluctance to move, pain, paralysis of the lower limbs, deformity, paraplegia, priapism, incontinence of urine and faeces, kyphosis and a deformed chest. The latter frequently resulted in bronchitis and death from pneumonia. This spectrum of features, now rare in zoos and scientific institutes, is not uncommon among pet animals.

In 1964, GRAHAM-JONES [74] described several other features stressing the changes in dentition, notably fibrous hyperplasia of the gums, disengaged primary dentition and the inability of the secondary dentition to erupt through the fibrous gums. Occlusal problems were frequent, particularly in adolescent animals at the time of the tooth change which is at approximately 1 year of age in South American primates. He also states that the hard palate may become vaulted and the animal suffers further from chronic diarrhoea, loss of weight, pneumonia (associated with scoliosis of the rib cage) and possibly also hyperaesthesia.

In 1967, REED and KRIEWALDT [155], describing the changes in a 6-month-old female baboon, noted inability to walk, curved diaphysis of the humerus but normal dentition.

e) Radiological Appearances of Rickets

Metaphyseal changes. The early changes of rickets are seen in the metaphyses of long bones, particularly distally in the ulna and in the costo-chondral junction. The cartilage columns are lengthened so that the gap on an X-ray film between the metaphysis and the epiphysis appears wide and irregular. The metaphysis itself is widened and irregular because of disordered endochondral ossification. Spikes of ossification may be noticeable on either side of the metaphysis because the central zones are more severely affected. Consequently the metaphysis is concave towards the epiphysis. In the epiphyses the centres of ossification are frequently late appearing and may have an uneven outline.

Diaphyseal changes. The shafts of long bones are frequently bowed and thinned, anterior bowing of the tibias being most often seen. Fractures and dislocation, particularly of the clavicles, are not uncommon when the disease is severe. Cortical bone becomes thin and less opaque and cancellous bone shows poorly defined trabeculae with an overall decrease in density (fig. 6).

Skull and jaw changes. In simian primates, rickets leads frequently to an irregular thickening of the skull bones with a decrease in density, probably due to secondary hyperparathyroidism. The mandible and maxilla are affected in most cases with gross thickening and retention of milk teeth. Displaced eruption of the secondary dentition may then follow.

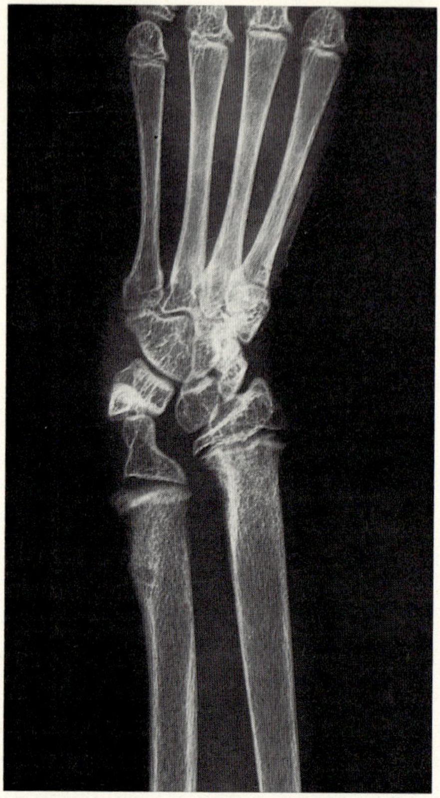

Fig. 6. X-ray of black spider monkey *(Ateles paniscus)* showing effects of rickets and osteomalacia. Note wide irregular epiphyseal line, poor definition of bone cortices and coarse trabecular pattern.

Histological abnormalities. The most characteristic feature of rickets is the great widening of the epiphyseal disc of cartilage, due to retarded removal of cartilage rather than to cellular hypertrophy. The cupped appearance of the metaphysis is a result of this retardation being more pronounced near the centre of the metaphysis than at the periphery [45] (fig. 7).

According to HAM and LEESON [77] the mature cartilage cells persist, because in the absence of adequate vitamin D the cells do not become surrounded by calcified cartilage matrix. The mature cells are still capable of secreting alkaline phosphatase and the hypertrophic cartilage cells persist instead of degenerating.

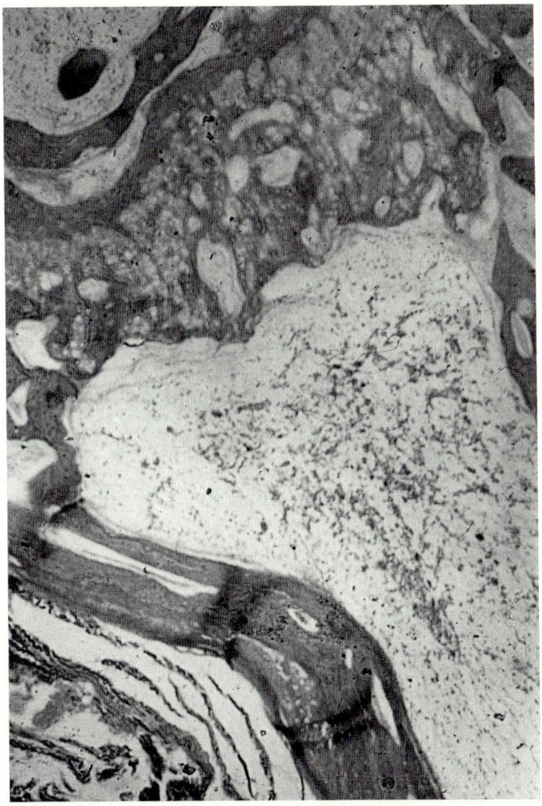

Fig. 7. Section of marmoset tibia *(Hapale jacchus)* showing appearance of true rickets. Note wide epiphysis with disordered cartilage columns and irregular capillary invasion. Also scalloping and distortion of metaphysis. ×28.

As the columns of these mature cells continue to proliferate and lengthen they become compressed and swollen [144], eventually disordered and arranged in irregular clumps; hence, normal orderly invasion of this zone by capillaries from the metaphysis does not occur.

The zone of invasion by capillaries is known as the zone of provisional calcification as it is usually here that the cartilage matrix becomes calcified. In rickets the invading blood vessels spread out laterally around and into the clumps of hypertrophic cells producing bushy groups of capillaries. Osteoblasts accompany these capillaries but in this abnormal environment the bone matrix cannot be mineralised. Osteoid material is, therefore, laid down around the disordered cartilage columns producing chondro-osteoid trabeculae.

Thus the main features of rickets at the epiphysis are: the failure of calcification of the cartilage matrix, and lack of resorption of the zone of mature cartilage cells. When rickets is severe, however, multiplication and growth of cartilage ceases resulting in a static state at the epiphysis [45].

Rickets affects all the growing bones, not only at the epiphysis but at all sites of bone mineralization, i.e. where osteoblasts are actively laying down bone, on the surfaces of trabeculae and osteons beneath the periosteum. The end-result is the production of excess osteoid tissue and the typical softening and deformities of the skeleton, e.g. bowed limbs and chest wall deformities. The 'rickety rosary' of the chest is due to swelling at the costo-chondral junction because of excess chondro-osteoid trabeculae and sub-periosteal osteoid.

Membrane bones are also affected and the skull is generally thickened. The suture lines may be late in uniting, producing a bossed appearance. The permanent teeth are usually late erupting, because of the thickening of the dental follicle.

Treatment and healing. In 1927 LUCAS and HUME treated and cured rickets and osteomalacia in a variety of New and Old World primates, by exposing them to ultra-violet light. Recurrence of the bone disease was prevented by prophylactic exposure to natural ultra-violet light using vitaglass [93, 118, 119]. GERSTENBERGER [71] succeeded in curing 81 primates suffering from rickets by exposing them to one-half erythema doses of ultraviolet light 6 times per week with no additional vitamin D.

The effective treatment of rickets in primates seems to depend not only upon adequate vitamin D or ultra-violet light but also upon sufficient dietary protein and a correct calcium-phosphorus ratio.

Very high doses of vitamin D are often necessary to produce a reversal of the rachitic process. We have treated New World primates suffering from rickets with vitamin D_3 5,000 IU i.m./kg body wt./week. This has been in addition to a high protein diet and a calcium-phosphorus ratio of at least unity. This regime is continued until there is clear evidence of healing radiologically, i.e. the appearance of a line of calcification in the rachitic cartilage zone and increased density of the metaphysis of the affected bones.

Although lack of calcium does not produce rickets a supplement of calcium is often necessary in treatment to provide for mineralisation of the skeleton and to avoid hypocalcaemia which could occur in the early stages of treatment.

The daily requirement of vitamin D_3 for New World species is not yet certain. However, a range from 2.0–8.0 IU D_3/g diet has been used successfully for a variety of these animals (table I). Old World primates seem less susceptible to rickets and diets containing approximately 2.0 IU D_3/g diet appear adequate.

The healing process of rickets has been extensively studied in rats by DODDS and CAMERON [45, 46, 47, 48]. The first sign of healing in their animals was calcification in the rachitic metaphysis near the edge of the epiphyseal cartilage. From here calcification spread through the metaphysis towards the shaft and later into the cartilage towards the end of the bone. The disordered cartilage was then reorganized by increasing calcification of the exposed cartilage (or chondro-osteoid) trabeculae. Correction of the shape of the bone was the last to take place. Although there are differences between the bones of rodents and the bones of primates, as we have discussed below (p. 508) the basic healing process occurs in the same manner.

In the widened epiphysis, healing is evidenced radiologically by the appearance of a line of calcification. In the metaphysis, there is an increase in density. This is greater than normal in the early stages, reverting to normal as healing progresses (fig. 8).

Simultaneously callus is formed around any fracture sites and the bone cortices increase in density and thickness. The trabecular pattern of the bones becomes more distinct, and the metaphysis eventually loses its concave shape as the whole bone is remodelled.

Histologically the earliest sign of healing is an increase in the proportion of osteoid tissue, which shows a calcification front [20, 128]. This is best studied in undecalcified sections, since the healing process is evident from the location of the mineral. Gradually the cartilage columns of the epiphyses return to normal size and shape as calcification occurs around the mature cells. The marrow, which tends to be fibrotic, is restored to a normal pattern. The bone is more metabolically active than normal at this stage while deposition of new mineral and reorganization of the diseased bone is occurring.

Prevention. The avoidance of rickets should not now be a serious problem. Monkey chows are available commercially designed to suit New and Old World species and maintain optimum health without supplementary feeding. If these diets are difficult to obtain, 15 min exposure per day to ultraviolet light either from sunlight or lamp is a useful prophylactic in some cases [79]. Protection of the eyes is not necessary as the animals do not expose them to the ultra-violet rays.

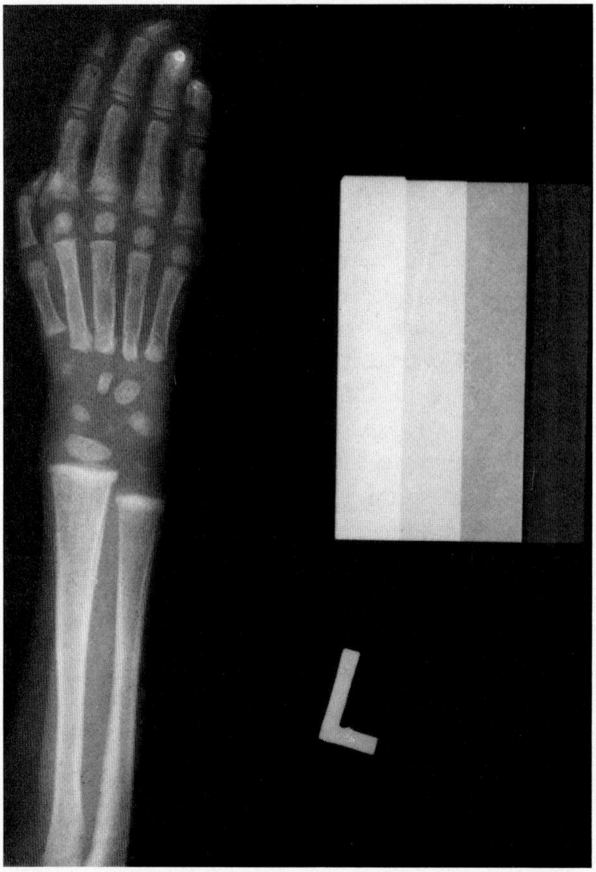

Fig. 8. X-ray of rickets when treated with vitamin D. Baboon *(Papio hamadryas)*. Note dense metaphysical lines and spikes at lateral borders of ulna.

3. Osteomalacia

a) Introduction

Osteomalacia is the disease produced by an insufficient quantity of the appropriate vitamin D available to the adult animal. It is essentially the same process as rickets. In the adult animal the epiphyses are fused so that the effects of vitamin D deficiency are most marked in the shafts of the long bones. Not uncommonly, however, osteomalacia is superimposed on a skeleton which has suffered a degree of rickets in the past. The normal plasma calcium concentration is maintained in spite of inadequate vitamin, by resorption of bone mineral probably stimulated by parathyroid hormone.

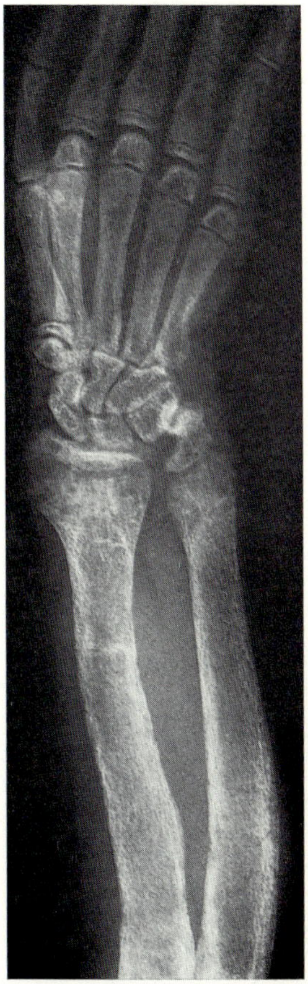

Fig. 9. X-ray of capuchin monkey *(Cebus nigrivittatus)* showing effects of severe metabolic bone disease. Features of old rickets, osteomalacia and secondary hyperparathyroidism are apparent. Note wide irregular metaphysis, bowing of radius and ulna, poor definition of cortices with low density bone and coarse trabeculation.

b) Aetiology

The possibility of osteomalacia being a secondary result of an infection of the nervous system was suggested by Todd [179] in 1913. The animals he observed (vervets, macaques and baboons), were undoubtedly suffering from osteomalacia. They used their forelimbs as crutches, became stiff and reluctant to move and frequently suffered from broncho-pneumonia. Their diet, however, consisted of milk pudding, carrots, onions, potatoes, greens, ap-

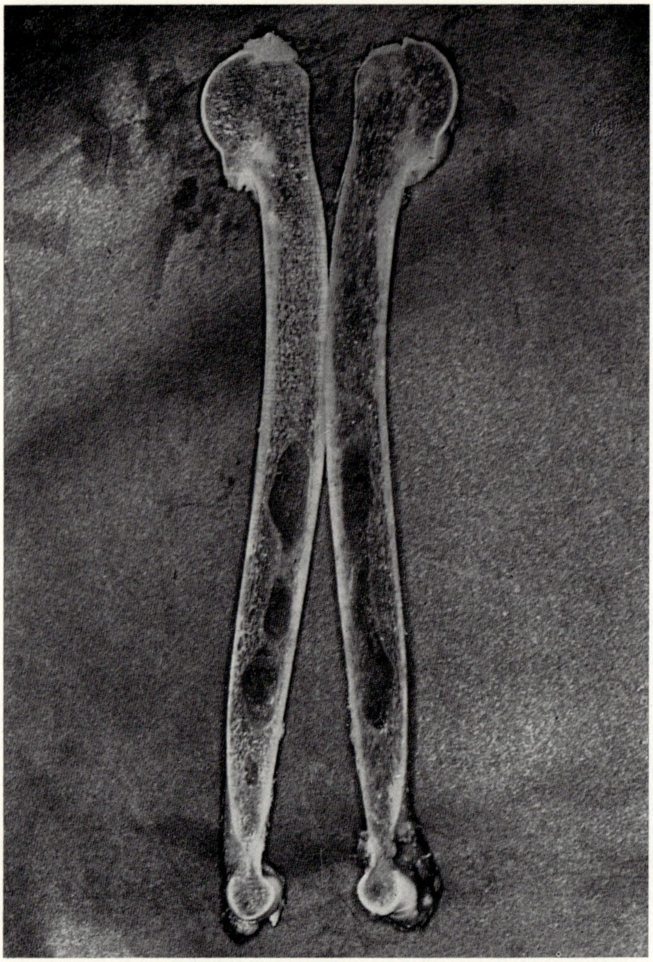

Fig. 10. Gross appearance of osteomalacia and secondary hyperparathyroidism affecting humerus of woolly monkey *(Lagothrix lagotricha)*. Longitudinal section. Note bowing and cystic features of bone marrow.

ples, bananas and occasionally meat. Combined with lack of sunlight they must have received negligible vitamin D.

In 1922 CORSON-WHITE [32] suggested that a low protein diet rich in carbohydrate resulted in a predominantly acid residue. She said that as a result, the alkalinity of the blood was lowered as the alkali was used to neutralize this excessive production of acid. The bones became soft, as the lime salts were more soluble as the alkalinity of the blood fell. She also suggested that imbalanced inorganic dietary substances and a loss of power to absorb calcium and phosphorus were of additonal importance.

Today it is recognized that the same dietary defects which cause rickets in young animals can cause osteomalacia in adults. There may also be varying degrees of secondary hyperparathyroidism and osteoporosis (fig. 9 and 10).

c) Signs of Osteomalacia

An adult primate suffering from osteomalacia shows many of the general signs already described for rickets, i.e. loss of weight, reluctance to move, progressive stiffness, hunched posture and bowing of the long bones. Pathological fractures of the long bones are frequent, although the limbs may not always be obviously painful when palpated. Collapse of vertebrae is also common. The effects on the spine and chest result in poor respiratory movements and fatal chest infections frequently ensue.

Osteomalacia also severely affects the shape, and hence function, of the female pelvis. In the past this has produced extreme distortion which led to stillbirth of young or death of the mother from dystocia.

In 1935, Ashley Montagu [3] described the effects of osteomalacia on the skull of an adult baboon associated with severe dental caries. This case was remarkable for the effect of the disease on the jaws and teeth. There was severe rarefaction of the bone of the skull with breakdown of the normal Haversian systems. Rarefaction of the mandibular bone was so extreme that the teeth (molars and canines) were everted so far as to be virtually horizontal at the front. The apices of the teeth were surrounded by broken dentine.

In this case osteomalacia was stated by Ashley Montagu [3] to be produced by a deficient diet and inadequate sunlight, and had progressed to such an extent that it had produced malocclusion.

d) Radiological Features

Bowing of limbs, fractures, pseudo-fractures, collapse of vertebrae and decreased density of bone are all indicative of osteomalacia. The pelvis may be compressed and distorted because of pressure from the femoral heads.

Additional changes are those of secondary parathyroid involvement which is present to varying degrees in many cases. These changes consist of subperiosteal erosions, pseudocysts, i.e. sharply defined areas of translucency, and irregular thickening, particularly of the skull and loss of trabecular pattern of the bone. The lamina dura around the roots of the teeth may be absent, a feature indicative of secondary hyperparathyroidism. Looser's

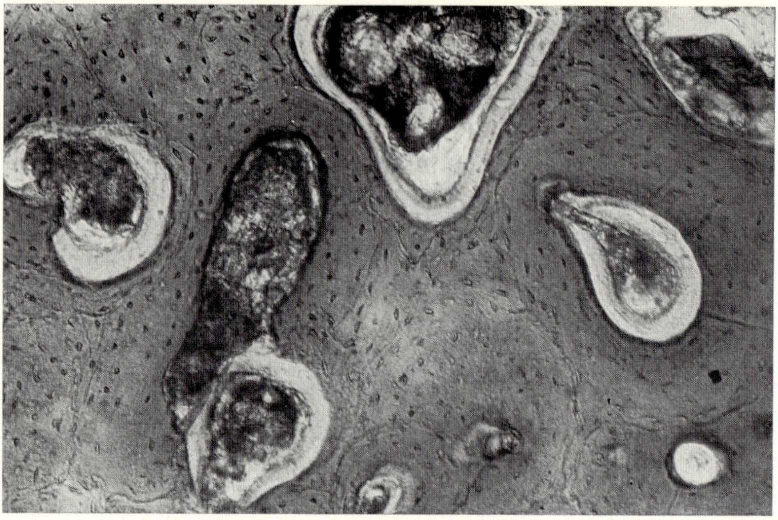

Fig. 11. Undecalcified ground section of femur of a woolly monkey *(Lagothrix lagotricha)* with signs of osteomalacia. Note wide osteoid seams on all surfaces. 20 μm thick. H and E. ×70.

zones, as described in humans, are usually symmetrical and can occur at the neck of the humerus and femur, pubic rami, ribs and axillary borders of the scapulae.

e) Histological Abnormalities

The most immediately striking microscopic feature of osteomalacia, is usually the abnormally increased amount of osteoid tissue present, as wide irregular seams on the surfaces of bone trabeculae (fig. 11). However, it is not limited to these sites and is also identifiable on the other bone surfaces, i.e. beneath the periosteum and lining resorption cavities and osteocyte lacunae. Undecalcified sections also reveal much low density bone surrounding osteocytes and occurring at sites of active bone resorption. In areas where bone surfaces are not covered by osteoid, osteoclasts may be prominent immediately adjacent to eroded bone. The absence of increased numbers of osteoclasts does not necessarily mean that active bone resorption is not taking place. The character of the bone surfaces and the structure of the bone is a more reliable guide.

Osteomalacic bone is also characterized by an increase in numbers of plump active osteoblasts indicative of the raised metabolic activity of the bone (fig. 12 and 13).

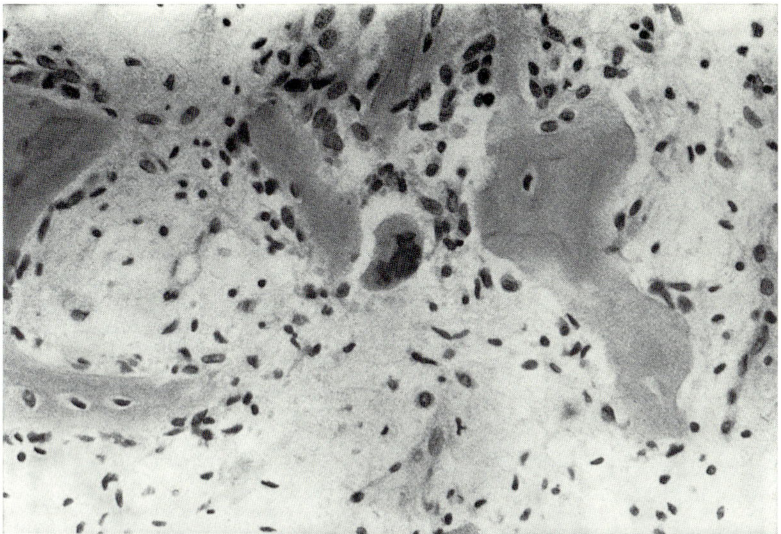

12

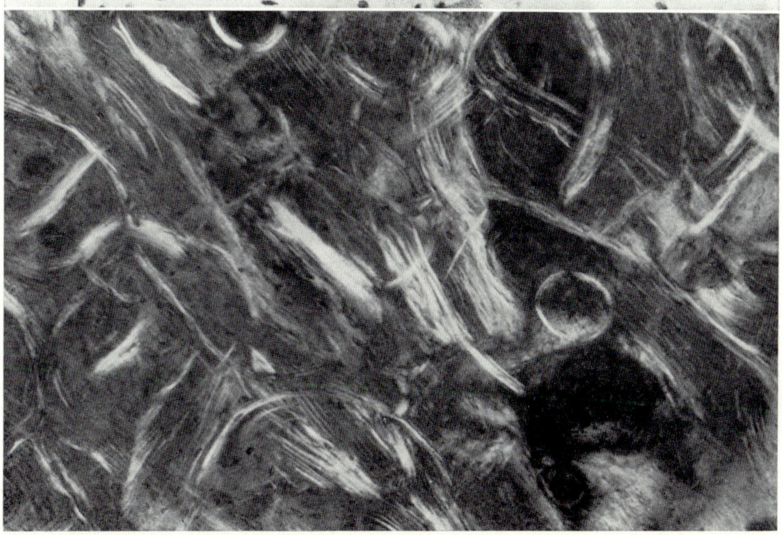

13

Fig. 12. Section of femur of woolly monkey *(Lagothrix lagotricha)* showing effect of severe osteomalacia. Note fibrous tissue, osteoclast adjacent to bone surface (centre) and many osteoblasts. Decalcified section 6 μm. H and E, ×175.

Fig. 13. Section of femur of woolly monkey *(Lagothrix lagotricha)* showing appearance of severe osteomalacia when viewed by polarised light. Note disruption of normal regular arrangement of lamellae, and compare with figure 20. Undecalcified section 20 μm. ×70.

The proportion of osteoid seams showing a calcification front is decreased in osteomalacia [20]. Looser's zones are areas of bone which are sites of much osteoid replacement. They were originally described in humans as occurring at stress sites where there is much resorption of bone, e.g. long bones, pelvis and scapula. In osteomalacia there is inability to replace the resorbed bone with mineralized tissue, osteoid being the only substance formed. Pseudo-fractures may occur at these sites as there is much decalcification and softening of the bones. The precise site of these zones may well be different from humans in non-human primates because of the differing postures adopted and consequently different sites of maximal stress.

The excessively fibrotic bone and fibro-cellular bone marrow commonly found in osteomalacia of adult and rickets of young primates is probably due to the action of parathyroid hormone (fig. 14 and 15).

Marmosets suffering from osteomalacia and experimental hyperparathyroidism have been studied by radiological and histological means by DREIZEN *et al.* [50]. This study showed firstly, that the periodontal bone was extremely sensitive to vitamin D deficiency and to the action of parathyroid hormone. A fall in bone density and loss of continuity with replacement of alveolar bone marrow by fibrous tissue were described. Secondly, there was great similarity between the changes in periodontal bone affected by osteomalacia and bone of animals injected with parathyroid hormone. These facts led to the suggestion that secondary hyperparathyroidism contributed to the periodontal changes in vitamin D deficient animals.

f) Treatment and Healing

As in rickets, vitamin D, correct calcium-phosphorus ratio and ultraviolet light all play an important part in treatment.

HAMPTON *et al.* in 1966 [79] cured osteomalacia of marmosets in 1 month by treatment with 1,000 IU vitamin D_3 orally/animal/day. They prevented recurrence by providing a diet supplying 500 IU D_3/day and 25% protein.

Healing is evidenced by increased activity of the animal, increase in bone density, reappearance of calcification at bone surfaces, i.e. the calcification front of osteoid seams and a normal amount of callus formation around fracture sites.

4. Osteoporosis

This term is used in humans to mean bones that are thin compared with the bones of normal young persons [134].

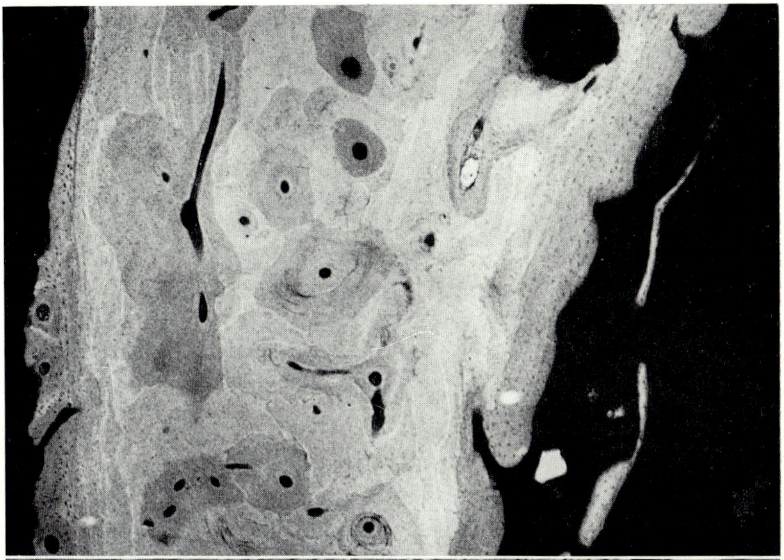

14

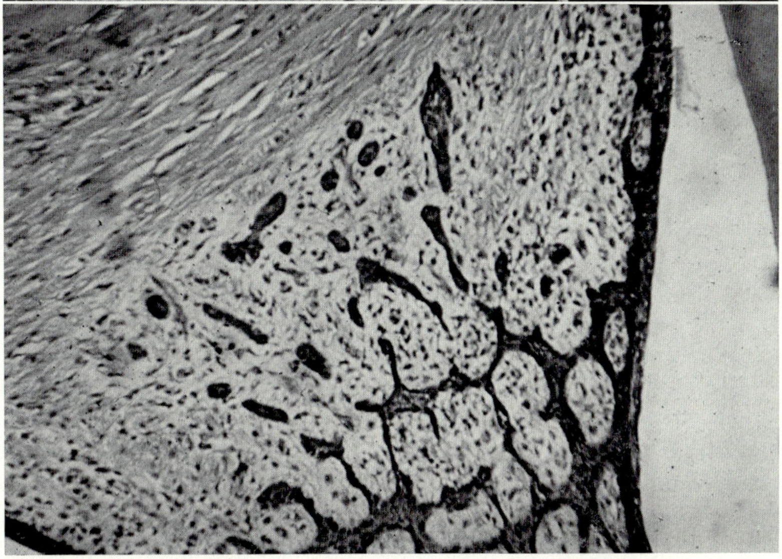

15

Fig. 14. Microradiograph of femur of woolly monkey *(Lagothrix lagotricha)* showing effects of osteomalacia. Note irregular distorted osteons and excessive low density bone. Undecalcified ground section 20 μm. ×28.

Fig. 15. Section of marmoset tibia *(Hapale jacchus)*. Animal injected with 800 U of parathyroid hormone. Note fibrosis and resorption of bone. ×28.

In non-human primates the majority of osteoporosis arises because of excessive loss of bone due to rickets or osteomalacia with secondary hyperparathyroidism and due to the effect of ageing and consequent senile atrophy. Low dietary calcium may itself cause osteoporosis [141].

DAMMRICH in 1967 [36] suggested that senile osteoporosis of primates starts relatively later than in other animals such as cattle and sheep. The onset of osteoporosis is, however, difficult to determine because of the wide variation in the amount of bone normally present in a particular species.

Radiologically, osteoporotic bone is more translucent than normal and the bone cortex may be obviously thin. Affected bones may show pathological fractures, and wedge-shaped vertebrae due to anterior collapse are frequent (fig. 16 and 17).

Resorption of bone of the mandible with ageing is common and so the teeth may be particularly prominent (fig. 17). Histologically there is typically a reduction in the number and width of trabeculae of cancellous bone and a thin cortex. In compact bone there may in addition be an abnormally large number of empty osteocyte lacunae.

5. *Mineral and Protein Deficiency*

a) Calcium Deficiency

A deficient intake of calcium can be expected to produce a combination of skeletal disorders. Loss of bone mass and density, i.e. osteoporosis, may be the dominant feature but if the deficiency is severe the parathyroid glands are likely to become hyperplastic and the changes of secondary hyperparathyroidism may be superimposed. The bones, therefore, show increased resorption and thinning.

b) Protein Deficiency

Deficiency purely of protein in young primates appears to have characteristic effects on bone growth. JHA *et al.* in 1968 [99] have studied rhesus monkeys *(Macaca mulatta)* fed a diet low in protein but with adequate vitamin D and calories. The result of this diet was loss of weight, reduction in the concentration of serum proteins and reduced endochondral bone formation. All the zones of cartilage at the epiphyses were affected, deficiency of protein resulting in poor proliferation of the cartilage cells. Osteoblast activity was notably reduced, these cells having a high protein turnover. Because of these effects the metaphyseal trabeculae were stunted and there was little remodelling activity and few osteoclasts.

The rate of appositional bone growth fell, although the shaft of the bone appeared to be normal. No changes in the concentration of serum calcium,

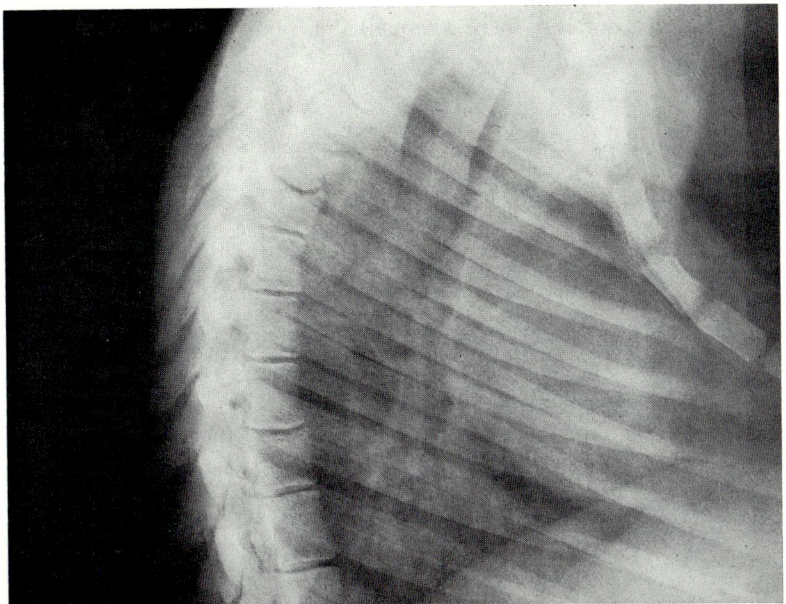

16

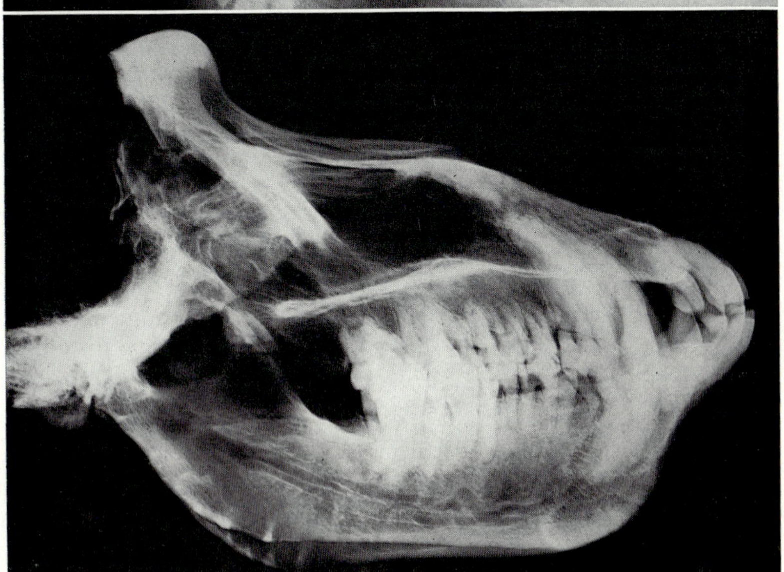

17

Fig. 16. X-ray of woolly monkey *(Lagothrix lagotricha)* showing effects of osteoporosis. Note collapse and irregular outline of upper thoracic vertebrae.

Fig. 17. X-ray of skull of baboon *(Papio hamadryas)* over 40 years old. Note resorption and reshaping of mandible as a result of ageing. Particularly obvious anteriorly.

phosphorus or the activity of serum alkaline phosphatase were found. Hence the fall in the rate of bone growth in protein deficiency appears to be due to the effect on collagen and to the reduced rate of osteoid formation. Uncomplicated protein deficiency is unlikely to occur except experimentally, since most commonly a low protein diet is also deficient in vitamin D and a mixed syndrome results.

6. Scurvy

Vitamin C deficiency is rarely found in animals fed a traditional fruit and vegetable diet. Semi-synthetic diets usually contain adequate vitamin C unless oxidation has occurred during storage. Lack of vitamin C results in reduced formation of collagen and osteoid tissue; blood vessels are abnormal due to weak intercellular cement substance resulting in frequent haemorrhages. The abnormal capillaries result in frequent haemorrhages in the metaphyses and beneath the periosteum (see WOLF, pp. 257–258).

As formation of osteoid tissue is depressed, bone matrix formation is abnormal and so the skeleton becomes somewhat osteoporotic. At the epiphyses the zone of provisional calcification persists and becomes abnormally wide due to a failure of bone resorption. Calcification is unaffected by vitamin C deficiency and so this wide zone is calcified in a normal manner, shown radiologically as a characteristically broad white line.

7. Hypervitaminosis D

This disorder is rare and has only been recorded due to accidental errors in feeding or as a result of deliberate experiment. COWDRY and SCOTT in 1936 [34] showed salivary gland changes in 2 mature *Cebus fatuellus* given 2.7×10^7 USP XI units vitamin D_2 in 200 h. (Recommended maximum dose for treatment of 1- to 2-year-old children at that time was 5.0×10^3 USP XI units/day [82].) The implications of this observation are obscured by the massive dose involved.

In 1958, KENT *et al.* [107] accidentally fed 558 rhesus monkeys a diet providing 1.63×10^5 USP XI units of vitamin D_2/animal/day. The diet contained 0.35% calcium and 0.29% phosphorus. This diet produced widespread signs. The general features were an increased incidence of upper respiratory tract infections, diarrhoea, weight loss and anaemia. There was also a rise in the concentration of blood urea suggesting a degree of renal failure. Calcium and phosphorus concentrations in the blood increased and calcium and iron deposits occurred in various organs notably the heart, liver, kidneys, lungs, salivary glands and stomach. The distribution of these lesions led them to conclude that these deposits occurred mostly in alkaline tissue (i.e. acid secreting).

HUNT *et al.* [96] induced hypervitaminosis D in squirrel and cebus monkeys by daily oral doses of between $0.5–1.0 \times 10^5$ IU vitamin D_3/animal. Vitamin D_2 at the same level produced no noticeable signs. The effects produced in the squirrel monkeys were increased blood calcium, phosphorus and urea concentrations without metastatic calcification. The cebus monkeys, however, did show widespread calcification. Neither species were shown to have any obvious bone lesions. It was suggested that the signs of hypervitaminosis D are probably related to the duration of overdosage and that vitamin D may have a direct effect on renal function, producing renal failure without calcification.

Bone lesions due to excess vitamin D have been produced experimentally in other animals. The changes consist of increased density of the bone shafts and microscopically, excess osteoid tissue.

B. Diseases of Endocrine Origin

1. Parathyroid Syndromes

Although occurring in human pathology, hyperparathyroidism due to neoplasm or primary chief cell hyperplasia has not so far been reported as occurring in non-human primates.

The only parathyroid disorders which have been reported are those associated with severe demineralizing bone disease of nutritional origin in captive primates. Conventional histology of the parathyroid glands in such cases has shown light chief cell hyperplasia with little connective tissue and occasional small water clear cells which may be the most active form [112] (fig. 18 and 19). The same survey included some quantitative measurements based on the principle of point counting and revealed an increased cytoplasm: nucleus ratio and an increased nuclear surface as measured by planimeter using a camera lucida. It was concluded that the increase in size of the nuclear surface in cases of parathyroid hyperplasia was due to the increased activity of the nucleus in synthesizing parathyroid hormone.

Radiological features which have been described as indicative of secondary hyperparathyroidism in primates include thickening of the skull bones, loss of trabecular pattern, cortical erosions and pseudocysts of bones [52]. Hyperostosis of the maxilla with retention of deciduous teeth and lack of eruption of the permanent teeth has also been reported [112].

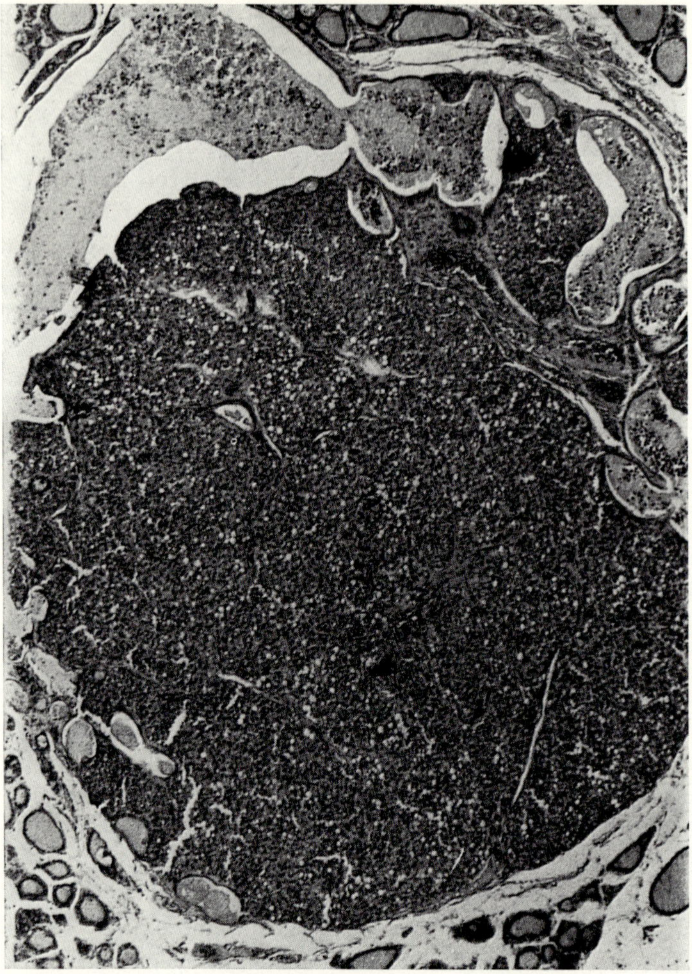

Fig. 18. Section of parathyroid gland of adult male woolly monkey *(Lagothrix lagotricha)* with evidence of bone disease, surrounded by thyroid tissue. Cystic peripheral areas are probably not significant. No distinct pattern, predominantly light chief and water clear cells with fatty changes probably due to age. H and E. $\times 34$.

It is possible for a gland which has become hyperplastic due to chronic low plasma ionized calcium to continue to secrete increased levels of parathyroid hormone even when the stimulus to increased secretion is removed, i.e. adenoma formation in a gland which is already overactive [63]. This has not so far been reported in non-human primates.

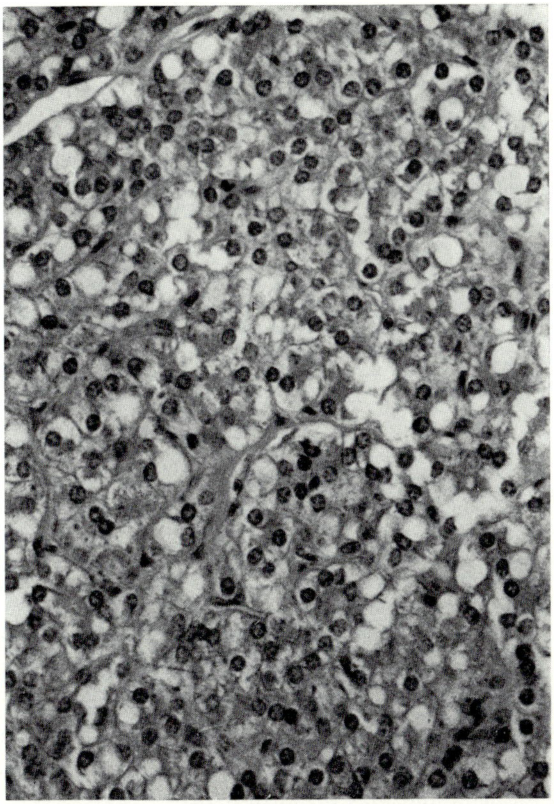

Fig. 19. Parathyroid gland of same animal as figure 15. Note predominantly water clear and light chief cells. Fatty changes and fibrous tissue also apparent. H and E. ×70.

2. Thyroid Syndromes

Lack of thyroxine in primates produces profound effects on the skeletal system. The most severe effects are produced in young animals when there is gross retardation of skeletal development. Bone growth, dental development and closure of sutures, are all affected.

A radiological study of young rhesus monkeys *(M. mulatta)* whose thyroids were ablated with ^{131}I has demonstrated these effects [121]. The animals in this study developed all the characters of human cretins (thick skin, dull hair, protuberant abdomen, etc.). The skeleton studied by radiology and histology showed decreased rate of linear bone growth, increased bone density and delayed appearance of secondary centres of ossification. Closure of the skull sutures was delayed and the eruption of secondary dentition retarded.

Metabolically there was reduced retention of calcium and evidence of secondary hyperpituitarism (enlarged sella turcica radiologically). In addition it has been demonstrated that thyroid gland destruction results in a generalized retardation of skeletal development affecting both linear and circumferential bone growth and reconstruction [35]. Cortical and medullary bone is affected equally, the retardation not being confined to one type or group of bones.

These effects of thyroxine lack are exactly as would be expected from studies in humans.

C. Diseases Due to Mechanical Defects and Trauma

1. Congenital Deformities

a) Spontaneous

Spontaneously arising congenital malformations appear to occur less frequently than in man. (13 out of 2,950 breeding animals in a recent study [190].) There is also some difference in incidence of the range of abnormalities which occur. For example developmental skull defects such as fontanellar bones, persistent sutures and premature closure of sutures are frequent. An interesting example of congenital cranial suture abnormality was decribed by FLOWER in 1882 [61]. In this case, a young chimpanzee, the coronal and sagittal sutures showed premature consolidation, a defect described by FLOWER as acrocephaly. This is apparently due to pathological changes at an early stage of development and may occur in man.

Webbing of the toes (zygodactyly) which is rare in man has been found in 8 cases of a study of 109 chimpanzees [168] (see SCHULTZ, pp. 159–189).

The survey conducted by SCHULTZ in 1956 [168] also revealed digital anomalies as being particularly common in gibbons (18 of 113 *Hylobates lar*). Many congenital defects of vertebrae and ribs, including cervical rib had resulted in scoliosis. One fully grown wild gibbon showed no limb distal to the middle of one upper arm and a juvenile chimpanzee had no development from the middle of one forearm [164], i.e. 2 cases of unilateral peromelia.

Spontaneously arising cleft palate has been described in marmoset twins [88].

A further case of spontaneous congenital skeletal deformity which has been described is that of the Arnold-Chiari malformation [26]. This malformation consists basically of herniation of part of the cerebellum, 4th ventricle and medulla into the cervical spinal canal. The case reported occurred in a newborn sacred baboon and was associated with a sacrococcygeal spina bifida (incomplete fusion of spinal neural arches) and a myelocoele, i.e. the spinal cord lying exposed at the site of the spina bifida. Spina bifida has also been reported by OSMAN HILL in 1939 [87] (see also K. R. BRIZZEE, pp. 545–546).

b) Induced

It has been shown that macaques and baboons respond in the same way as humans to Thalidomide, rubella and androgens [190]. Rhesus monkeys also respond as humans to Thalidomide [39]. A single dose of Thalidomide given between the 25th and 30th days of gestation has produced abnormal limb formation. Given earlier or later no effect has been demonstrated. WILSON *et al.*, in 1967 [190] have also shown a positive dose response and a cephalocaudal gradient within the 5-day period, i.e. an early dose producing predominantly forelimb abnormalities and later treatment affecting the hind limbs.

Deformities produced by maternal Thalidomide include phocomelia (hands and feet each attached to the trunk by one small bone), amelia (complete absence of all 4 extremities), hemimelia (distal parts of extremities absent) and digital abnormalities, hydrocephalus, facial capillary haemangioma and anotia (hypogenesis of ear auricles) [39].

Dosage of 5 mg/kg between the 18th and 20th days of gestation has also produced deformities, viz. amelia, phocomelia, spina bifida, tail shortening and bilateral cleft lip.

2. Fractures

A fracture is usually described as either complete or partial, i.e. one cortex of the affected bone remaining intact (greenstick). Also depending on the nature of the break the fracture may be simple, compound, comminuted or impacted.

Fractures may arise as a result of direct or indirect trauma or occur through an area of bone weakened by disease, e.g. a cyst, tumour or area of rarefaction due to metabolic bone disease. In this case normal stresses and strains on the bone may be enough to cause a fracture. These are pathological fractures.

a) Fractures Due to Trauma

The vast majority of fractures occurring in the wild state are probably traumatic in origin and appear to be particularly common among the larger primates, e.g. 12 out of 60 adult orang utans from the wild showed healed fractures [166]. Forty-two of 118 gibbons showed healed fractures [165] and all of 11 wild baboons over 13 years old had fractures [22]. Gorillas are also particularly liable to fractures in the wild state and are prone to their own particular fractures, e.g. great sagittal crest [53].

A further survey of gorilla skeletons [152] revealed 3.3% of skulls showing evidence of healed fractures and 5.6% of skeletons also with healed fractures. This series shows an apparent sex incidence, the male skeletons having fractures 2–3 times as often as the female. This is not surprising in view of the larger body bulk of the male.

The common sites of fractures appears to vary from species to species. This variation is presumably associated with the animals' different modes of life especially their methods of locomotion, e.g. baboons have frequent fractures of the axial skeleton, skull, neck of femur and pectoral girdle [22]. Gibbons show frequent fractures of humerus, femur and phalanges [163] and the humerus was a common site in orang utans [166].

b) Pathological Fractures

This form of fracture which may be of any degree of severity is most commonly associated with severe metabolic bone disease of New World primates (fig. 2). Other possible sources of weakness are tumours or infections, e.g. osteomyelitis and tuberculosis.

c) Healing of Fractures

In general, fractures of primates which have been observed in the wild appear to have healed well with few cases of complicating infections, excess callus or failure to unite [53, 165]. This is possibly due to the death of the severe cases rather than a reflection of the efficiency of the fracture-healing processes in wild monkeys.

3. Cephalhaematoma

A subperiosteal haematoma of the skull (cephalhaematoma) arising as a result of trauma, may be large and can calcify. The result is a hard swelling over the cranial bones which may be sharply defined by the suture lines (fig. 20 and 21). Any large cephalhaematoma, especially if calcified, should prompt investigation into the possibility of previously unsuspected disease, such as vitamin C deficiency. After correction of such a deficiency the swelling gradually diminishes as the animal grows.

4. Atrophy of Disuse

Due to the lack of the normal stresses and strains of weight bearing and activity an immobile limb will eventually show thinning and loss of density.

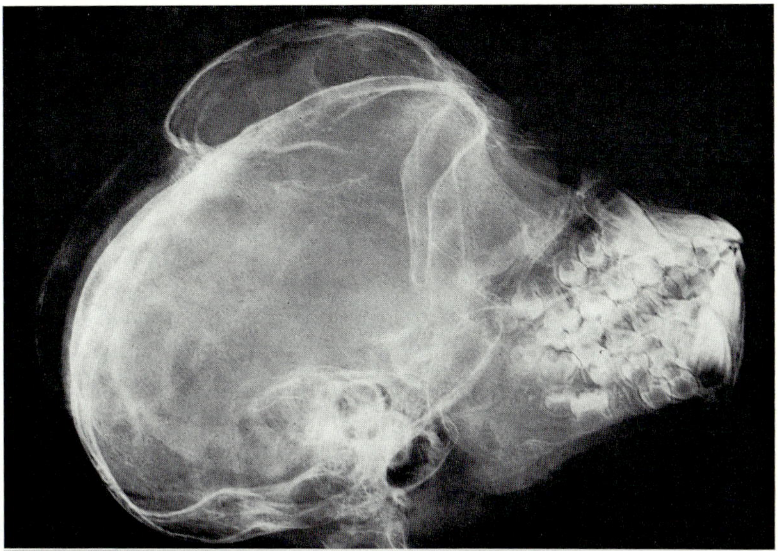

20

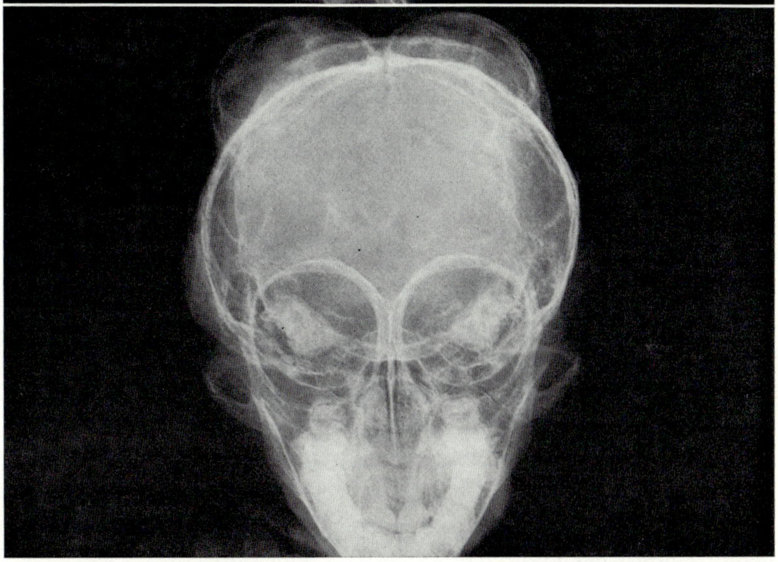

21

Fig. 20. X-ray of skull of capuchin monkey *(Cebus nigrivittatus)* showing a calcified cephalhaematoma. Lateral view. Note outer shell of calcification limited by suture lines.

Fig. 21. X-ray of calcified cephalhaematoma. Same animal as figure 17. Frontal view.

This disuse atrophy has been reported in primates that have developed poliomyelitis and a subsequent flaccid paralysis [75].

Disease due to chronic suppuration affecting the left leg of a chimpanzee resulted in the left femur weighing 60 g less than the normal right leg [65].

D. Bone Tumours

a) Spontaneous

Spontaneous bone tumours appear to be uncommon occurrences in non-human primates. Of those tumours which do occur there appears to be a preponderance of malignant growths – osteosarcomas. Reference should be made to pp. 191–193, O'GARA and ADAMSON, for a detailed list. One of their tables shows 10 reported cases of spontaneous tumours. This reveals that of these 10, 6 tumours were of the skull and 4 of the long bones. They have drawn attention to this apparent difference from humans in whom a predominance of tumours in the long bones would be expected.

A further report [28] concerns an osteogenic sarcoma, again of the skull, this time of the frontal bone, occurring in an old female rhesus monkeys in association with several neoplasms of soft tissues. These were probably not secondaries but further separate primaries. The osteogenic sarcoma consisted mostly of fibrous connective tissue and osteoid and recurred after surgery. The typical features of an osteosarcoma are, generally speaking, hyperchromatic neoplastic osteoblasts, fragments of osteoid tissue and occasionally bone. Benign osteomas and fibro-osteomas have been described (see O'GARA and ADAMSON, pp. 191–193).

In the case of a ten-year-old baboon, described by LAPIN and YAKOVLEVA in 1963 [113], the tumour occurred on the left thigh and consisted of compact bone merging with the bone of the femur and contained no malignant features. Bony exostoses, probably benign osteomas, have been reported in the zygomatic bones, scapula, humerus and radii of gorillas [152].

As far as species incidence is known, no clear picture emerges, spontaneous tumours having been described in a wide range of species.

b) Experimentally Induced Tumours

Bone tumours have been induced by various substances injected into or implanted in bones. Substances used have been radioactive ores, tar, radium bromide, ^{10}Ag, ^{60}Co and 3-methylcholanthrene (3-MC) all of which led to malignant tumours, osteosarcoma or fibrosarcoma usually after 2–3 years.

It has been shown using radium that higher doses result in a shorter time lag before the appearance of neoplasia [147].

E. Infections of Bone

1. Tuberculosis

The skeleton rarely appears to be affected by known tuberculosis. A survey of spontaneous disease in 600 animals revealed only 1 case of skeletal tuberculosis [55].

MILLER and KINARD in 1964 [131] reviewed the main cases reported up to that time [13, 64, 106 174,]. They added a report of widespread skeletal tuberculosis occurring in a juvenile female rhesus monkey *(Macaca mulatta)* which had shown a negative reaction to 8 tuberculin tests carried out at 2-week intervals. This case revealed numerous tubercles in the flat and long bones. The lesions were translucent on X-rays and occurred both in the marrow and cancellous bone. Microscopically the predominant finding was ofcaseous necrosis; epithelioid cells, lymphocytes and fibrosis being relatively inconspicuous.

Tuberculosis of the vertebral bodies (Potts' disease) has been described in rhesus monkeys [127, 174]. The case described by MARTIN *et al.*, in 1968 [127] presented with flaccid paralysis and reduced response in the hind limbs associated with a thoracic kyphosis. The lesions consisted of collapse of the tenth thoracic vertebra due to a tuberculous abscess and an encapsulated paravertebral mass extending from T9–T11. This mass was made up of caseous material and multiple granulomas composed of necrotic centres, epithelioid cells, lymphocytes, giant cells and fibrous tissues. The spinal canal was greatly narrowed between T9 and T11. Microscopically the collapsed tenth thoracic vertebra showed central fragmentation, necrosis, granuloma formation in the marrow and some periosteal new bone formation.

These findings are completely consistent with the disease as it occurs in humans. In man the spine is generally held to be affected in 30–50% of all skeletal tuberculosis and particularly the eleventh thoracic vertebra.

Tuberculous coxitis has been reported in a macaque [175] and tuberculous osteomyelitis of ribs of a Mexican spider monkey *(Ateles geoffroyi bellensus)* has occurred in an animal kept for over 13 years in captivity [57].

2. Miscellaneous Infections

Bone infections of primates are not common but of those which have been recorded, the majority have occurred in the skull bones, notably in the

Table II. Summary of reported cases of miscellaneous bone infections

Reference	Animal	Lesion and number of cases	Number of skeletons examined
66	Chimpanzee	chronic suppuration and periostitis, leg ×1	1
165	Chimpanzee	infected maxillary sinuses and drainage holes, ×3	35
166	Orang utan	perforation in orbit, ×1 abscess in clavicle, ×2 abscess in manubrium, ×1 abscess in cranial vault, ×1	92
152	Gorilla	orbital perforations, ×10 palatal perforations, ×1 frontal sinus perforations, ×1 scapula perforations, ×5 iliac crest perforations, ×4	180
169	Colobus	infected auditory meatus, ×1 infected orbital fossa, ×1 infected maxillary sinus, ×4	309
22	Baboon	acute osteomyelitis of foot, ×1 acute osteomyelitis of femur, ×1	60

sinuses and the orbits, often with perforations, presumably drainage holes (see table II).

F. Osteoarthritis

Describing chronic arthritis in mammals in 1939, Fox [66] showed the frequent occurrence of this condition in the wild state. Sixteen of his 20 cases of osteoarthritis of primates were found in wild animals. Of 389 primates examined, Fox noted a relatively high incidence in gorillas and incriminated body mass as a major cause of the condition through jolting and jarring. The arthritis he observed was mostly of the hypertrophic type. Vertebrae and the bones and joints of the hands and feet were the most affected showing, in varying degrees, lipping fibrosis, osteophyte formation and eburnation.

Severe spinal arthritis also seems to be particularly common in baboons [22] (and see fig. 22). In addition to arthritis of the spine, arthritic changes of

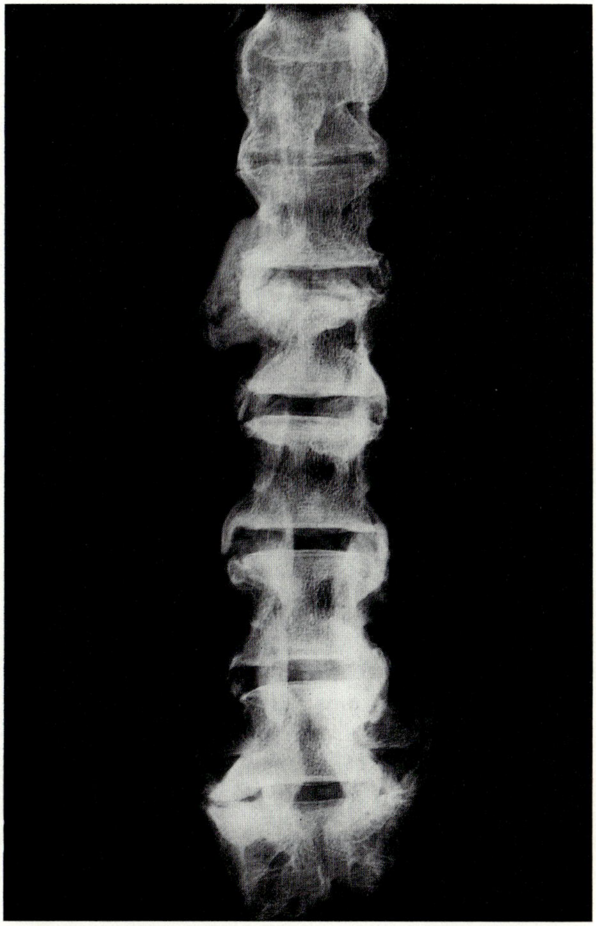

Fig. 22. X-ray of baboon *(Papio hamadryas)* showing effects of severe spinal arthritis. Note severe lipping, osteophyte formation and narrowing of joint space.

the wrists, hip, long bones, sternoclavicular joints and tempero-mandibular joints often with exostoses and ankylosis have all been described [165, 170] (fig. 23).

A similarity between rheumatoid arthritis in man and arthritis in other mammals has been suggested by Fox [66]. Two gorillas and one mandrill were described as showing swelling and demineralization of joints with periarticular ostosis suggestive, to him, of rheumatoid arthritis.

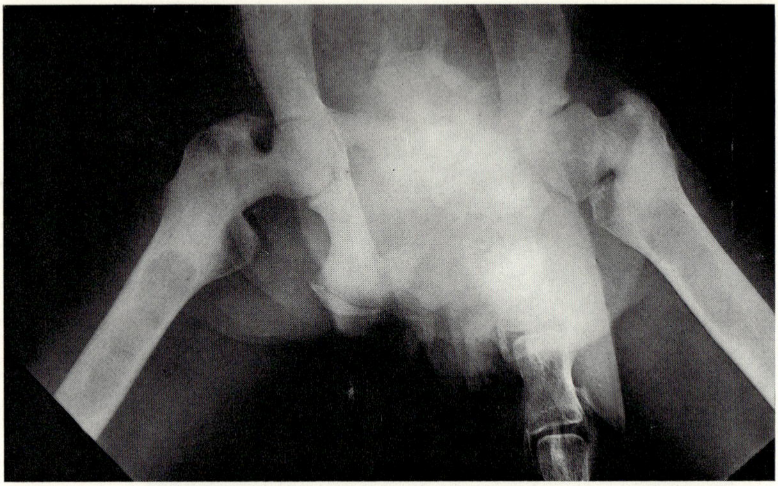

Fig. 23. X-ray of woolly monkey *(Lagothrix lagotricha)* showing osteoarthritis of the left hip. Note irregular destruction of femoral head and loss of joint space.

In man, rheumatoid arthritis presents as a chronic inflammatory disorder of connective tissues in which the joints are usually the most affected, although skin, muscle, heart, blood vessels, lungs and lymphoid tissue may all show specific lesions. The joints are usually painful, swollen and tender with much thickening and excessive growth of the synovial tissue. The result is usually a significant degree of fibrosis, deformity and limitation of movement.

In 1968, SOKOLOFF *et al.* [173] described 3 cases of spinal ankylosis in old rhesus monkeys. One animal was an 18½-year-old male which was killed because of inability to move. The vertebral column was flexed and rigid and the vertebral arches, joints and dorsal processes were found to be united by a continuous sheet of bone. Syndesmophytes (i.e. bridges of bone on vertebral bodies) were found on several lumbar vertebrae.

Microscopically the picture was of an active, non-specific, chronic inflammatory reaction with much vascular granulation tissue, polymorphs, plasma cells and osteoclasts. The knee joints also showed destructive changes and eburnation. This case they described as one of ankylosing spondylitis.

Ankylosing spondylitis of humans may be regarded as a form of rheumatoid arthritis affecting the spine and occasionally other joints. The disease progresses from synovial overgrowth of the spinal joints to fibrous and bony ankylosis with extreme rigidity of the spine. Human ankylosis spondylitis however, usually starts in young males. Apart from this age difference the disease described in rhesus monkeys is very similar.

One of the other cases described by SOKOLOFF *et al.* [173] was a male killed because of weight loss and poor general condition. This animal was still capable of movement. Examination of its spine showed bulbous osteophyte formation on the ventrolateral aspect of the tenth thoracic vertebra and a layer of compact bone in the position of the anterior longitudinal ligament. This layer of bone was up to 3 mm thick and continued over the next 4 vertebral bodies. There was further osteophyte formation over other thoracic and lumbar vertebrae. Also the right twelfth rib was partially fused to the spine and the sacro-iliac joints were ankylosed. The knees showed mild changes of osteoarthritis. These changes are consistent with osteoarthritis of a particularly hypertrophic form.

Sacro-iliac ankylosis and lumbar vertebral ankylosis has also been described in a radiological survey of Egyptian mummified baboons conducted by MOODIE in 1931 [133].

III. Bone Structure and Physiology

A. Formation and Growth of Bone

Bone formation and growth occurs in two ways. These are ossification in membrane and endochondral ossification.

1. Ossification in Membrane

The essential feature of this process is the formation of bone directly from connective tissue. In this case there is no intermediate cartilage model stage. The mesenchymal cells of connective tissue differentiate directly to osteoblasts.

Osteoid tissue is then laid down and subsequently mineralized to form trabeculae of bone in ossification centres. The majority of the skull bones are formed in this manner.

2. Endochondral Ossification

This is the process by which a cartilage model is converted to bone from centres of ossification and epiphyseal lines. Most of the skeleton is formed in this way and the long bones grow in length by this means.

The cartilage model of a bone is formed by differentiation of mesenchymal cells. The model grows by division and hypertrophy of mature cartilage cells, the chondrocytes. This is growth by the interstitial mechanism. Mature chondrocytes secrete alkaline phosphatase and are capable of initiating calcification of the intercellular substance of cartilage.

The cartilage model is surrounded by the perichondrium consisting of fibroblasts and undifferentiated cells. These cells are capable of differentiating into osteoblasts and osteocytes, so forming the periosteum of bone with its layer of osteogenic cells.

When osteogenic cells and capillaries reach the interior of the cartilage model they form a centre of ossification for the shaft (diaphysis). The osteogenic cells form osteoblasts, which also secrete alkaline phosphatase, and lay down the intercellular substance of bone around the calcified cartilage produced by the mature chondrocytes.

Similar centres of ossification arise in the ends of the bones, known as epiphyseal centres.

Between the bone arising from the ossification centre in the shaft and that from the epiphysis is a transverse disc of growing cartilage. This is the epiphyseal disc. This disc of cartilage is responsible for the growth in length of the bone and is replaced by bone when growth ceases. The cells of the disc are continually growing but the disc does not increase in thickness because of the calcification and replacement by bone which occurs on the diaphyseal side.

Several zones of cartilage can be recognised in the disc [77]. These are firstly the zone of resting cartilage which is adjacent to the bone of the epiphysis. Then the zone of proliferating cartilage where the cells can be seen to be mitotic. The cells in the third zone, of maturing cartilage, normally grow in regular columns, accumulate glycogen and produce alkaline phosphatase. Calcification occurs around these cells producing the fourth zone, that of calcified cartilage. The cells here are mostly dead or degenerate as they are surrounded by calcified intercellular substance.

The next stage in the formation of bone on the diaphyseal side of the disc is the invasion of the calcified cartilage by osteoblasts and capillaries from the shaft of the bone. The osteoblasts also contain alkaline phosphatase and initiate formation of the intercellular substance of bone around these cores of calcified cartilage.

This whole process is continuous as long as the cartilage columns are growing. The irregular trabecular framework of calcified cartilage and bone substance that is first formed in this way is known as the primary spongiosa. This is in direct continuity with the cartilage columns.

This irregular structure is then reshaped (modelled) by osteoclast activity to form the secondary spongiosa which is much thinner and is the zone where most of the calcified cartilage has disappeared and consists almost entirely of newly-formed bone.

Further growth in diameter of bones is achieved by new bone being deposited by osteoblast activity in direct apposition to the bone already formed. [For further details see 77.]

B. The Elements of Normal Bone

Normal bone in an adult animal consists basically of bone cells and an intercellular matrix impregnated with bone mineral which provides rigidity.

1. Bone Matrix

The bone matrix consists of collagen fibres and a ground substance of polysaccharides and a mucoprotein substance, impregnated with bone mineral. The physical and chemical properties of the collagen fibres of bone vary with species [124]. Chondroitin sulphate and keratin sulphate form the majority of the polysaccharide ground substance of bone.

The formation of bone mineral will be described in the section on calcification (III, E). It consists mainly of calcium and phosphate in the form of tricalcium phosphate and hydroxyapatite depending on the age of the animal [148]. There are also age and species differences in the amount of bone mineral [15, 103]. Other ions such as magnesium, bicarbonate, fluoride and citrate are also involved [10]. Osteoid tissue is bone matrix which is not mineralised. It is most noticeable when active formation of new lamellar bone is occurring, as for example in rickets and osteomalacia [68]. An osteoid seam denotes the layer of this unmineralized matrix found on the surfaces of active bone.

Osteoid seams of greater width than 10 μm are often accepted as pathological, although Frost has cast doubt upon their significance in man and has shown that the occasional wide seam occurs even in normal human subjects [70] (fig. 24).

2. Periosteum

The periosteum covers the outer surface of bones and has two layers. An outer tough fibrous layer of collagen fibres and fibroblasts and an inner layer which is capable of bone formation. The inner layer consists of osteoblasts joined together by thin cytoplasmic processes. The capillaries of the inner layer of the periosteum are connected with the blood vessels of the Haversian system by Volkmann's canals.

a) Endosteum

The endosteum lines the walls of the marrow cavity and trabeculae of cancellous bone. It also lines the interior of the Haversian canals of compact bone. It is usually a thin layer of connective tissue cells which can react as marrow cells or differentiate to form osteoblasts.

3. The Cellular Components

The cells of bone are all closely related and derived from connective tissue cells.

a) Osteogenic Cells

Healthy bone is lined by rather flat cells which are capable of differentiating into osteoblasts, osteoclasts or cartilage cells depending on circumstances. These are the osteogenic or stem cells of bone. Mature osteoblasts or osteocytes are incapable of division and so increased numbers in response to the need for further bone deposition as, for instance, at a fracture site, can only arise through the multiplication and differentiation of the osteogenic cells.

b) Osteoblasts

Osteoblasts are basophilic cuboidal, and mononuclear and are also situated on bone surfaces. The surfaces of bone consist of periosteum, endosteum, the vascular channels of cortical bone and the surfaces of the trabeculae of cancellous bone. Osteoblasts secrete alkaline phosphatase when active. The role of this enzyme in calcification will be discussed in section III E (Mechanisms of calcification). Osteoblasts may also be concerned with the supply of calcium ions to the bone matrix [145].

c) Osteocytes

Osteocytes are derived from osteoblasts and are found in lacunae embedded in the bone matrix. They are less strongly basophilic than osteoblasts and have long cytoplasmic processes which extend from their lacunae in canaliculi. Some of these canaliculi open onto the bone surface and are important in the nourishment of the osteocytes.

The osteocytes, once thought to be relatively inactive, are in fact intimately concerned with the metabolic activity of bone and the transport of substances between blood and bone. They are probably also involved in the control of resorption of bone surrounding them and the liberation of mineral [11, 12]. The precise function of osteocytes is not yet clear but it is possible that there is a difference in activity between the small newly formed osteocyte and the larger mature cell. Possibly the small osteocyte being recently incorporated into the organic matrix continues the process of the bone deposition started by the osteoblast.

The mature osteocyte may be the cell, responsive to parathyroid hormone, which modifies the surrounding matrix for release of bone mineral [184].

d) Osteoclasts

Osteoclasts are very large cells with many nuclei. Most commonly there are between 10 and 20 nuclei visible although there may be any number between 2 and 100 [80].

They are usually situated on eroded surfaces of bone, known as Howship's lacunae, which are sites of bone resorption. However, they are mobile cells and may be found apparently free from bone in the marrow spaces. Also, as they are large cells, the plane in which the section is cut may pass through a part of the cell away from the bone surface. Likewise only a few nuclei of a multinucleate cell may be cut.

When an osteoclast is in close contact with the bone surface a brush border of fine processes is usually visible under the light microscope. Electron microscopy shows this border to be made of folds of cytoplasm communicating with the interior of the cell and the site of bone resorption [7].

The main function of osteoclasts appears to be the resorption of bone, particularly the continuous process of remodelling and the liberation of bone mineral. Resorption by osteoclasts probably affects bone matrix and mineral together, although it is possible that calcium is removed first and matrix later [156].

Osteoclasts are responsive to many stimuli [80], parathyroid hormone and calcitonin being the two most fundamental hormonal influences. The osteoclasts increase in numbers and activity in response to parathyroid hormone raising the plasma calcium by liberating mineral from the bone. Calcitonin probably inhibits bone resorption and causes a decrease in the numbers of osteoclasts.

Other influences on the osteoclasts are cortisone, thyroxine, vitamin A and vitamin D [184].

C. Varieties of Bone and Nourishment of Bone Cells

Normal bone (i.e. cells and matrix) in an adult animal consists of two fundamental types, compact and cancellous bone. Shafts of long bones are examples of compact bone and the bone of the iliac crest or vertebrae is cancellous or trabecular bone.

Normally these two types are arranged in a regular lamellar pattern, i.e. regular sheets of matrix consisting of parallel fibres of collagen impregnated with bone mineral.

Bone which does not follow this regular pattern only occurs in young animals or in pathological states of the adult, e.g. fractures. It is then known as non-lamellar, coarsely bundled or woven bone.

In compact bone the fibres of the lamellae are arranged concentrically around longitudinally-running blood vessels in Haversian canals. This unit is an osteon and its outer limit is defined by a cement line which is stained deeply basophilic by haemotoxylin and eosin. The osteocytes are regularly situated in the osteon and communicate with each other via their canaliculi. Between the osteons of compact bone the osteocytes and bone matrix are still regularly arranged in parallel lamellae.

The osteocytes receive their nourishment from the blood vessels running in the Haversian canals via their canaliculi and diffusion of substances into the bone matrix.

Trabecular or cancellous bone contains fewer Haversian systems and the osteocytes do not follow quite such a regular pattern.

The bone is arranged as a slender network of trabeculae and the bone cells are maintained by substances from the bone surfaces.

All bone when first formed is an irregular pattern of trabeculae of woven bone which is reorganised by osteoblasts and osteoclasts to form compact or cancellous bone depending on the site of bone formation.

The Haversian systems of compact bone are produced as a result of osteoblasts laying down successive lamellae and reducing the irregular arrangement of newly-formed trabecular bone and vascular spaces to a regular pattern of blood vessels in narrow canals. The pattern of Haversian canals which is formed first in compact bone is known as the primary Haversian system.

In all large animals further secondary Haversian systems are formed from resorption cavities being roofed over by new bone lamellae. The resorption cavities are a result of osteoclast activity and the new bone lamellae which are then laid down are due to the osteoblasts. Hence both these cells are important at these sites. These secondary systems may cross earlier concentric Haversian systems so forming interstitial or circumferential lamellae. This occurs for example at the surfaces of bone.

D. Bone Activity

The metabolic activity of bone is a function of the activity of the bone cells lining the surfaces and lying in lacunae in its matrix.

When considering the activity of bone, the size of its surface area should be remembered for this is the area exposed and responsive to changes in its environment. For example 40 cm^3 of iliac crest bone has a surface area of about 1,600 cm^2 [172]. Trabecular bone as in the iliac crest is more metabolically active than compact bone possible because of its relatively greater surface area. See also POSNER [148]. The net turnover of bone tissue in the human clavicle has been calculated as 16% per year and as 0.95% per year in the tibia [69]. Hence there is considerable variation in activity of bone depending on its site. Similarly FROST has shown a net total skeletal turnover of approximately 2% per year in man [69].

Bone turnover in man and animals is achieved by the processes of growth in length and diameter and by the replacement of primary osteons by secondary osteons formed from resorption cavities.

The size and shape of the secondary osteon varies with species and with age, the smaller the size of the animal the smaller are the osteons. There is also a difference in mineral density with species. It has been further suggested that when the remodelling of bone which occurs with longitudinal growth ceases, turnover of bone continues because of the formation of secondary osteons [101].

Small laboratory animals such as rats do not completely close their epiphyses during life and form no secondary osteons.

Hence, when studying human bone disease, an animal with a similar bone structure to man such as a non-human primate is obviously more appropriate than a rat which, although convenient and cheap, differs markedly in bone structure and activity.

E. Mechanism of Calcification

It would be inappropriate to attempt here a review of the many experimental details and theoretical problems involved in calcification. For a historical review, WEIDMAN 1963 [186] should be consulted.

Cell differentiation and matrix formation are now considered to be the initial stages of bone formation, tissue mineralisation being regarded as the final step. Bone mineral consists mainly of salts of calcium and phosphate. Extracellular fluid is said to be supersaturated in calcium and phosphate ions with respect to the bone apatite but apparently unsaturated with respect to its formation [140, 148, 176].

Tissue calcification theories have been divided by POSNER into 3 groups [148].

a) Those theories that propose mechanisms by which the concentrations of calcium and phosphate are increased to the point where their solubility product is exceeded and mineral precipitated. This type of mechanism was first described by ROBISON in 1923 [159], and involved the extra-cellular hydrolysis of some phosphate ester by alkaline phosphatase, to produce locally high concentrations of phosphate ions thus causing the formation of bone mineral.

b) Those theories which propose mechanisms that would create nucleating sites or remove barriers to these sites, at locally calcifying areas. Apatite would then form spontaneously from the normal levels of calcium and phosphate. This type of theory has been discussed by POSNER [148], and an epitaxis system has been discussed by NEUMAN and NEUMAN [140].

c) Those theories that propose that apatite does not precipitate directly from extracellular fluid but that a less basic calcium phosphate salt is initially formed which then hydrolyses to apatite. POSNER [148] also discusses these theories.

The common objection to all 3 groups of theories is the ubiquity of collagen and alkaline phosphatase. This makes it difficult to understand how an extracellular process could result in the ossification of a specific tissue rather than in general calcification.

Recent information concerning the nature of calcium salts in bone has helped to make a synthesis of these theories. MOLNAR, 1959 [132], by means of electron-microscopy, HARPER and POSNER, 1966 [83], using X-ray diffraction techniques and TERMINE and POSNER, 1966 [177], using infra-red spectroscopy, have shown that bone mineral appears to be deposited first as amorphous tricalcium phosphate which then hydrolyses to minute hydroxyapatite crystals that POSNER [148] describes as calcium deficient hydroxyapatite on account of their low calcium-phosphate molar ratios. These crystals are thought to attach to collagen fibres and act as seeding centres for secondary deposits of hydroxyapatite. The proportions of the amorphous and crystalline forms vary with the animals' age [177] and the part of the bone examined. TERMINE and POSNER, 1967 [178], have shown that 'the epiphyseal region of a chick ulna is considerably lower in percent crystallinity than the metaphysis of the same limb'. The molar ratios of hydroxyapatite possibly also change with the age of the bone as the apatite tends towards true hydroxyapatite instead of a calcium deficient one. The extent to which other ions such as carbonate and magnesium are involved is as yet unknown.

The amorphous-crystalline ratio also varies with nutritional deficiencies. In rickets there is an increase in the amorphous content resulting in less mature bone for a given animal age, whilst osteoporosis does not affect the ratio [177].

POSNER [148] has pointed out that electron micrographs of amorphous bone mineral are difficult to obtain because hydrolysis to apatite may occur when the tissues are kept standing in aqueous solutions. The anhydrous fixation methods of MOLNAR [132] are, therefore, recommended.

It has been suggested that from the present data it is reasonable to believe that bone cells are involved in the calcification process. The primary factor responsible for limiting apatite formation to those areas that normally calcify is not then alkaline phosphatase nor pyrophosphatase but cellularly derived amorphous calcium phosphate. POSNER then goes further and says that the bone cell untimately governs the whole calcification process.

Whilst POSNER has come down heavily on the side of an intracellular initial step in mineralisation he does not provide a reasonable mechanism for this. LEHNINGER, in the 5th Jubilee Lecture to the Biochemical Society, has discussed the role of mitochondria in calcium ion transport [116]. He has proposed a series of steps in the mineralisation process that start with the concentration of calcium and phosphate within the mitochondria in an energy-linked process.

LEHNINGER describes the enormous amounts of calcium phosphate that can be sequestered by mitochondria, up to 3.0 μmol/mg of protein which represents an increase of over 25 % of their dry weight. This increase would exceed the solubility product of calcium and phosphate in the mitochondrial matrix water and precipitation of a salt would follow. This salt has been shown to be amorphous tricalcium phosphate.

The overall mechanism proposed by LEHNINGER is as follows:

Calcium and phosphate ions in the extracellular fluid and cytoplasm are pumped into mitochondria by a respiration-linked process. The solubility product of calcium phosphate in the mitochondria is exceeded and amorphous tricalcium phosphate precipitated. Micropackets of this amorphous mineral pass out of the mitochondria and attach to the collagen fibres. Nonreversible hydrolysis to hydroxyapatite follows.

There are still many gaps in the overall process but calcification now has a reasonable basis for the initial step and some of the areas requiring further investigation are becoming more clear. The change of emphasis from an extracellular process to an intracellular process for the initial step is especially important and adds weight to POSNER's conclusion that the bone cell ultimately governs the whole calcification process [148].

F. Resorption of Bone

Resorption is continuous throughout life at differing rates and is essential for the maintenance of healthy bone. Modelling and remodelling of growing bones, turnover of mature lamellar bone and excessive resorption associated with pathological disturbances of the control systems of the skeleton are all essentially cellular phenomena. The activities of the bone cells concerned are largely controlled by changes in their hormonal environment, however, the precise mechanisms of bone resorption are still uncertain. The most striking histological feature of active bone resorption is the increase in osteoclast activity. These large cells are situated in most instances close to eroded bone surfaces (Howship's lacunae). Studies of these areas with the electron microscope have shown that in the region of the ruffled border of the osteoclast apatite crystals have been removed from the collagen fibres of the bone [81]. The removal of apatite crystals from collagen fibres probably depends upon enzymes secreted by the osteoclasts on to the bone surface. This is suggested by the demonstration of many enzymes in these cells including acid phosphatase, carbonic anhydrase, oxidases, glucuronidase and leucine amino-peptidase [81] and by work demonstrating the secretion of ^{3}H-glucosamine by osteoclasts on to the bone surface [143]. Collagenase may be present in bone at sites of resorption for the subsequent breakdown of collagen fibres [193].

Osteoclast activity is broadly controlled by the level of circulating parathyroid hormone. The hormone probably stimulates the metabolic activity of osteoclasts already present, increases their production of enzymes and stimulates the stem cells of bone to form more osteoclasts [180, 184]. These effects of parathyroid hormone are reduced in vitamin D deficiency and the degree of bone resorption in the presence of a constant level of parathyroid hormone varies with the concentration of calcium and phosphate at least in experiments *in vitro* [151].

Calcitonin acts conversely to parathyroid hormone specifically to inhibit bone resorption by osteoclasts and to inhibit formation of new osteoclasts [62, 156, 157].

Areas of bone resorption also show large osteocytes lying in lacunae with irregular outlines and surrounded by bone which stains metachromatically with toluidine blue and is of low density. These large osteocytes appear to have a different function from the small osteocytes which are surrounded by smooth walled lacunae and dense bone. It is possible that the low density areas and large osteocytes represent areas of resorption taking place away from the exposed bone surfaces, i.e. osteolysis.

The osteocytes are also directly influenced by parathyroid hormone secretion which increases their activity. These cells secrete alkaline phosphatase and protease and so may modify the bone matrix to release calcium and break down collagen [11].

Bone resorption is also enhanced by excess cortisone, thyroxine and vitamin A and a fall in oestrogen concentration. The relationships between these additional factors and between them and the dominant hormones, parathyroid hormone and calcitonin, are still being investigated [184].

In summary, bone resorption is mediated directly through osteoclasts which separate bone mineral from collagen fibres probably by enzymic action under the general control of parathyroid hormone. The increase in numbers and activity in response to parathyroid hormone is also influenced by other factors such as vitamins A and D, cortisone, oestrogen, calcitonin and the concentration of circulating calcium and phosphate. Osteocytes also respond to parathyroid hormone to modify bone matrix and release mineral. It is possible that the modification of bone matrix by osteocytes and other factors that occur in pathological states results in abnormal skeletal material which also stimulates osteoclastic bone resorption.

G. Calcium Homeostasis

The physiology and pathology of bone are intimately linked with calcium homeostasis. The concentration of plasma calcium is controlled by feed-back mechanisms involving parathyroid hormone, calcitonin and vitamin D. The mechanisms of action of these 3 factors have been extensively investigated but a complete explanation is not yet available. A review of the literature related to these factors is inappropriate here, so a general summary of calcium homeostasis is given to provide a basis for understanding the pathology of bone disease. Parathyroid hormone has been reviewed by Ar-

NAUD *et al.* [2]; calcitonin by HIRSCH and MUNSON [90] and MACINTYRE [122, 123], and vitamin D by DE LUCA [42] and NORMAN [142]. Parathormone, calcitonin and calcium homeostasis have been reviewed together by COPP [31].

1. Parathyroid Hormone

Reduction in the concentration of plasma ionized calcium stimulates the secretion, by the parathyroid glands, of parathyroid hormone (PTH). This is a single-chain polypeptide of approximately 8,500 molecular weight which has 3 main sites of action; the gastrointestinal tract, kidney and bone.

The predominant effect of PTH on the gastrointestinal tract is to increase the absorption of calcium thus increasing the plasma calcium concentration. Vitamin D is considered to be necessary for PTH to have an effect at this site [153, 154].

The effects of PTH on the kidney are the increase of urinary excretion of phosphate and an increase of calcium resorption. Thus, the plasma phosphate tends to be lowered and the plasma calcium concentration increased. Vitamin D is not considered to be necessary for the action of PTH on the kidney [1].

VAUGHAN [184] has summarised the effects of PTH on bone as: (a) to promote resorption by osteoclasts; (b) to promote osteolysis by osteocytes, and (c) both to inhibit and stimulate osteoblastic activity. The overall effect is the resorption of bone mineral and matrix and the release of calcium and phosphate ions together with products from the matrix. Vitamin D is said to be necessary for PTH to have its effect on bone [1].

2. Calcitonin

Calcitonin is a linear peptide with a molecular weight of about 3,600. It is secreted by the 'C' cells of the thyroid. Calcitonin produces an increase in the urine phosphate excretion [158] but how important this action is compared to the direct effect of calcitonin on bone is uncertain [122]. The action on bone is to inhibit resorption with a consequent reduction in plasma calcium and phosphate.

3. Vitamin D

As a result of the work of KODICEK and of DE LUCA, it is now known that vitamin D is converted to a functional form before it acts in the control of calcium metabolism. The first of 2 steps involves the hydroxylation of the side chain of the molecule at carbon-25 to give 25-hydroxycholecalciferol (25-HCC). This reaction has been shown to take place in the liver and 25-

HCC to be the major form of the vitamin circulating in the blood plasma of rats, chickens and man.

A second step has recently been shown to take place in the kidney to yield a metabolite that has greater biological activity than either cholecalciferol or 25-HCC [67, 111]. It appears that this step involves an oxygen insertion at carbon-1 of 25 HCC. The exact nature of the molecule known as compound 'P' is, as yet, unknown[1]. FRASER and KODICEK [67] have outlined the following scheme for the intermediary metabolism of vitamin D. Cholecalciferol is metabolised to 25-HCC by the liver; 25-HCC passes to the blood and constitutes a circulating reserve; in the kidney oxygenation of 25 HCC occurs to yield compound P which then acts on the target tissues, small intestine, bone and skeletal muscle.

IV. Methods of Bone Study

A. General

The difficulties in knowing what is normal and what is abnormal have led us to develop a system of examination of animals with suspected bone disease that combines radiological, histological and biochemical techniques. By making sequential studies on the animal the initial diagnosis and treatment can be rejected or confirmed, and the treatment altered accordingly. Some of the techniques included are more suitable for research than routine use but we have dealt with them together for the sake of clarity.

1. Case History

As always, the case history of the animal is important and where bone disease is concerned details of the animals' diet, vitamin supplement and the amount of ultra violet light given (if any) are particularly important.

2. Cage Design

Metabolic studies are difficult and costly in time and money even in man where a certain degree of co-operation can be elicited. In simians they require considerable care if the results are to be reliable. The primary requirement is an effective metabolic cage system. We found it necessary to design our own unit which is illustrated in figures 24 and 25. Although the system shown here

1 Compound 'P' has now been identified by KODICEK as 1,25-dihydroxycholecalciferol [197].

Fig. 24. Unit of 8 stainless steel metabolic cages.

was designed for small South American monkeys we have successfully used a larger version for a woolly monkey *(Lagothrix lagotricha)* and for an Old World monkey *(Erythrocebus patas)*.

The cage system consists of 8 sub-units supported on a mobile tubular steel frame. The sub-units are made completely of stainless steel for ease of cleaning and minimum contamination. Each sub-unit consists of a cage and a collecting funnel. Each cage fits on to

Fig. 25. Close-up of individual cage/collecting funnel sub-unit.

the frame by means of L-shaped brackets on the cage, which slide on to inverted L-shaped brackets attached to the frame. The collecting funnel fits on to the frame in the same way. The cage and the funnel are secured to the frame by ring-clips. The funnel is designed so that the sides overlap the bottom of the cage thus avoiding a gap between the bottom of the cage and the collecting apparatus. The cages are 2 × 2 × 1 ft wide. The sides and back are made of sheet stainless steel. The top, front and floor are of stainless steel mesh. There is a sliding front door on to which the food hopper and water valve fit. We found it necessary to add a second sliding front of $^{3}/_{8}$ in Perspex sheet to prevent the animals urinating out of the front of the cages. Each cage has a 1″ square stainless steel perch set 1 ft from the roof and 3 in from the back of the cage. This was found to be necessary to prevent callus formation caused by the animal permanently sitting or standing on the mesh floor. The bottom of the cage is held in position by the sliding front door. The floor, door and Perspex sheet slide out for cleaning and this makes the cage itself very easy to clean. Each cage is light enough to be handled by 1 person if necessary.

The collecting funnel is steep-sided with a fine mesh stainless steel grid which collects faeces and uneaten food, set 1 ft from the top. This makes it difficult for the animal to regain faeces or uneaten food. The grid is held in position at the ends by 2 plates attached with wing nuts. By removing 1 plate the grid can be removed for cleaning. Urine is collected into screw-topped bottles supported on a wide cross-member shelf.

The animals are supplied with deionised, distilled water from a 30-litre reservoir via an automatic watering valve on each cage (Cramex Ltd.) locked into position on the front of the cage so as to be easily detachable from the outside. The water valve passes through a hole in the Perspex sheet, thus locking the sheet in position. The sheet can be lifted above the valve to allow the animal to feed but when in position it covers the food hopper hole avoiding losses of urine.

The whole unit was constructed by North Kent Cages Limited to our design, and further modifications have been made by us as the need has become apparent.

The unit described here has been in use for some time now and has proved to be very effective.

3. Diet

The second requirement of this type of investigation is an adequate diet of known composition, single or multiple constituents of which can be altered at will. We have developed a semi-synthetic diet which we mix and pellet ourselves. The approximate composition is given in table I.

Weighed quantities of the major constituents and the vitamin and mineral mixtures are added to 1 litre of deionised-distilled water, per kg of dry material. Chromic oxide (0.5 g/kg dry material) is then added. These constituents are then mixed for 10 min in a Hobart food mixer. The mixture is extruded through a mincer with a five-eight inch tube attached to the above machine. The resulting pellets are broken into appropriate lengths and dried in a Gallenkamp hot air oven at 40 °C for 12 h.

In this way we can regulate the composition of the diet and by regular assays of the major constituents of the finished pellets, control the quality.

We do not give any supplements, and have found this diet, combined with deionised-distilled water, to be quite adequate.

4. Balance Studies

WHITBY and LANG [187] and ROSE [161] have discussed the necessity for using a marker substance in balance studies on human patients. They gave chromium sesquioxide in gelatine capsules, 0,5 g 3 times a day. Analysis of the chromium content of the faeces over a 6-day period was used to assess the completeness of the faecal collections. We use chromium the opposite way round. As we know how much of the dietary constituents are added per gram of chromium, then from the amount of chromium excreted per day (assuming no absorption) we can calculate the quantity of calcium, magnesium, etc. that the animal has consumed. We can also calculate the weight of food the animal has eaten.

ROSE [161] considers that chromium sesquioxide or some other marker that is not absorbed is essential for all calcium balance work if a high degree of accuracy is required. We considered using phenol red but WIGGINS and DAWSON [188] have shown that it partitions in the gut. Polyethylene glycol was also considered, but WHITBY and LANG [187] have pointed out that it is a mixture of polymers which do not all behave identically in the method used for their estimation.

As far as we know, chromium sesquioxide fulfils the requirements for an inert marker substance, set out by WHITBY and LANG [187], as well in simian primates as it does in humans. So far we have no evidence of toxic effects in *Cebus*, *Lagothrix* or *Saimiri*. MERTZ [129] in a review of chromium in biological systems records that no reports of oral toxicity of trivalent chromium are known to him.

The balance study routine that we follow involves seven, 24-hour periods. The animals are first established on our normal diet for 3–4 days and are fed from 10.30–11.30 h. After the acclimatisation period the cages are cleaned and rinsed with deionised-distilled water. The balance starts with the feed at 10.30. At 11.30 the uneaten food, faeces and urine are discarded. The collecting funnels are washed and rinsed with deionised-distilled water. The next day the same procedure is followed except the faeces and urine are retained. This regime has proved adequate for the maintenance of body weight and for carrying out effective balances.

After the seven 24-hour balance samples have been collected but before the final feed the animals are anaesthetised, weighed, blood samples taken from the femoral vein without stasis, (using a Yale 23G. needle) X-rayed and a bone hiopsy taken if required. A tetracycline label is given before the animal recovers from the anaesthetic and on the 2 subsequent days, without anaesthesia.

The choice of anaesthesia for animals with possible bone disease is important, as an injectable anaesthetic which involves crushing the animal could cause damage. It also means that the cages must have crush bars which means that there would be gaps in the cage through which faeces and urine could be lost, and food stolen from the adjoining animals.

We train the animals to enter a small catching box and then give them fluothane, nitrous oxide and oxygen by essentially the same method as described by COHEN [30], except that we use only a face mask. On being returned to the cage the animal is fed.

The investigation of bone pathology of primates depends ideally upon the combination of observation of the animal, radiology, histology and biochemistry. Some methods are described below with notes on their interpretation.

The specific changes observed in particular diseases have been discussed when describing those diseases.

B. Radiology

This is probably the most frequently used screening examination if bone pathology is suspected during life or at autopsy.

However, the early changes of metabolic bone disease are best detected by histological or biochemical methods, as the disease process must be further established to be visible on a standard X-ray film. Earlier changes in bone mineralisation can be demonstrated radiologically by more specialised techniques such as autoradiographs, microradiographs (fig. 26) and densitometry. Densitometry requires films of a limb immersed in water with a calibrated standard, usually of aluminium [49, 100, 102, 105, 184].

1. Anaesthesia for X-Ray Examination

For routine whole skeleton X-ray examination the animals should be anaesthetised. In this way the animal is subject to less stress and physical trauma. Also complete relaxation is easily obtained and it is simple to position the animal for the examination. Anaesthesia is particularly advantageous in cases of severe metabolic bone disease or fractures as the animal is far less likely to injure itself than when forcibly restrained.

2. Main Features of the X-Ray Film

When examining the skeleton the main points to look for are:

a) Straight spine and long bones. There may, for example, be considerable deformity in rickets and osteomalacia.

b) Abnormal epiphyses, e.g. cupping and enlarged epiphyseal cartilage lines in cases of rickets.

c) Fractures, callus formation and displacement of bones in cases of metabolic bone disease and trauma.

d) Diminution of joint space as in osteoarthritis.

e) Density of bones. Changes of density of bones may be localised or generalised and increased or decreased. Osteomalacia complicated by secon-

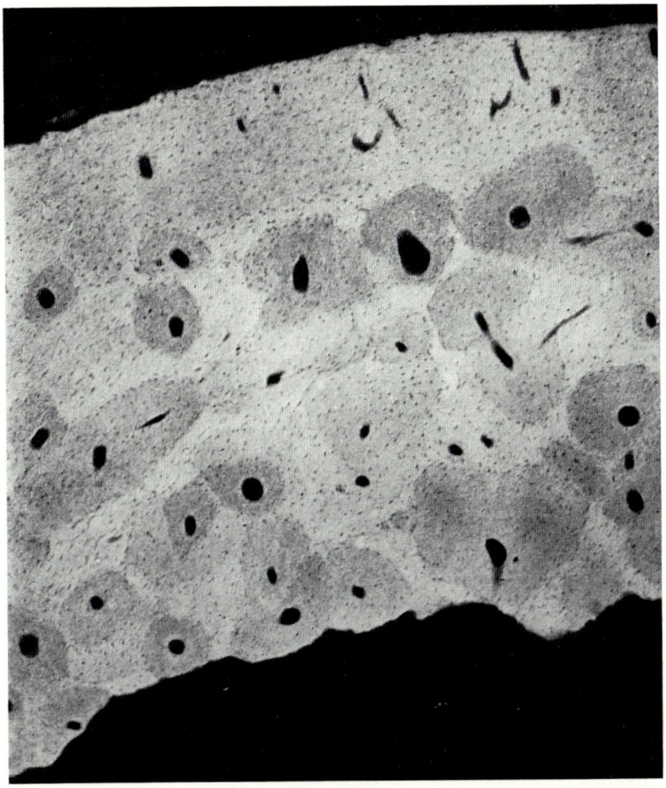

Fig. 26. Microradiograph of normal woolly monkey femur *(Lagothrix lagotricha)*. Note regular pattern of concentric and interstitial lamellae. Recently-formed bone is less dense and so appears grey.

dary hyperparathyroidism may produce a general fall in density and cystic spaces show as localised decrease in density in the long bones.

Osteoporosis alone usually produces a general fall in density, although localised osteoporosis may occur in an infected or paralysed limb.

f) Dentition. Metabolic bone disease can result in displacement or retention of the erupting secondary dentition. The lamina dura may disappear in cases of secondary hyperparathyroidism. The lamina dura is a thin layer of compact bone lining the tooth socket and its radiological appearance should be interpreted carefully in conjunction with any other evidence of skeletal disease. It is possible that its appearance of increased density is an artifact as its mineral content has been shown to be the same as that of neighbouring bone [126].

g) Trabecular pattern. Coarsening is evident in osteoporosis.

Bearing these points in mind any significant bone pathology should be detectable.

C. Histology

In life, the histology of bone hinges upon an investigation of a bone biopsy usually obtained from the iliac crest. This site is chosen because the method is straightforward and cancellous bone being more metabolically active than compact bone reveals abnormalities at an earlier stage.

We have used various sizes of modified versions of a simple stainless steel sleeve, trocar and cannula based on the full-thickness iliac crest biopsy instrument described by BYERS and SMITH in 1967 [25].

The small size of the samples obtained necessitates cutting undecalcified sections embedded in methacrylate on a Jung microtome to obtain the maximum number of sections and most information.

An alternative to iliac crest biopsy is removal of one or more of the terminal tail vertebrae of suitable animals such as the squirrel monkey *(Saimiri sciureus)* as suggested by LEHNER *et al.* [115].

We have found that removal of tail vertebrae under general anaesthesia is a practical procedure for squirrel and capuchin monkeys. To date we have amputated vertebrae at intervals of 2 months to a total of 5 vertebrae per tail and the technique has proved entirely straightforward. Infiltration of the amputation stump with 1% lignocaine reduces post-operative interference with the skin sutures by the animal.

The size of biopsy sample obtained in this way is large enough to provide a representative index of the state of health of the animal's skeleton, is repeatable and is sufficient for the preparation of both decalcified and undecalcified sections.

Routine examination *post mortem* for suspected bone pathology should include samples from the costo-chondral junction, femur, epiphysis of a suitable long bone, vertebra and whole skull.

Soft tissues are also important particularly the thyroid and parathyroids, kidney, liver, lymph glands, small and large intestine and lung.

The situation of the parathyroid glands in simians is variable. A survey of 50 *Macaca mulatta* (rhesus), and 50 *M. fascicularis* (cynomolgus) [104] showed 52% of the glands of rhesus and 56% of the cynomolgus to be within the thyroids. The remainder being lateral to the thyroids or partly inside and partly outside the thyroid capsule.

These glands are most important in cases of bone disease of primates and should be examined histologically whenever possible for the changes of secondary hyperparathyroidism, i.e. chief cell hyperplasia and for any tumour producing primary hyperparathyroidism.

The cervical lymphatic glands are often enlarged in metabolic bone disease of woolly monkeys perhaps due to a viral or other coincident infection [58]. Necrotic patches in the liver and pyelonephritis in the kidneys are also frequent. The villi of the small intestine may be atrophied and the lung often reveals a terminal pneumonia especially if there is any degree of chest or spine deformity.

Bone specimens should be cleaned preferably leaving the marrow intact and fixed in 10% formol saline. The specimens should then be divided into 2 and 1 set decalcified in formic acid or ethylenediaminetetra acetic acid (EDTA) which chelates with calcium. The other set should be dehydrated and embedded in methyl methacrylate.

The end-point for decalcification is best determined on a X-ray film and depending on the size of the specimen takes several weeks. The decalcified specimens should be processed and embedded in paraffin wax or preferably double-embedded in celloidin and wax [51].

Thin sections of approximately 6 μm can then be cut for routine examination the most useful microtome being the Leitz base sledge model No. 1300. The undecalcified specimens, after embedding in methacrylate, can be cut to a thickness of 6–8 μm for routine stains, using the Jung KM microtome. The sections can also be ground on a glass plate to 20 μm for microradiography and for examination by polarised light when indicated (fig. 27).

Bone formation rate and aspects of bone growth can be studied by administration of tetracycline which is incorporated into areas of bone activity and fluoresces under ultra violet light [69, 130] (fig. 28 and 29).

Decalcified sections stained routinely with haematoxylin and eosin provide information about osteoclast and osteoblast activity, cellularity of bone, trabecular size and pattern, fibrosis, cartilage cells and columns and the character of the bone surfaces. Osteoid seams although usually visible to some extent on a suitably stained decalcified section are best examined on an undecalcified section. Undecalcified sections are stained routinely with haematoxylin and eosin and von Kossa's silver nitrate method. In this way osteoid seam width can be measured and mineral composition determined. Further useful stains are Masson's trichrome stain and van Gieson stain, both of which show the pattern of the collagen fibres and osteoid seams well.

Staining sections with solochrome cyanine is most useful as an indication of early vitamin D deficiency or early healing of metabolic bone disease [128]. This stain reveals the granular appearance of osteoid which is about to calcify ('calcification front'), a decrease in the proportion of osteoid with a calcification front being evidence of early vitamin D deficiency.

Fig. 27. Section of normal compact bone seen by polarised light. Collagen fibres are birefringent showing up pattern of concentric and interstitial lamellae. Undecalcified ground section 20 μm. ×70.

Similarly as vitamin D is necessary for the proper formation of the calcification front an increase in the proportion of osteoid with a calcification front in a biopsy specimen is good evidence of improvement in the disease [20].

D. Histochemistry

The methods described above can be applied to bone for the demonstration of acid and alkaline phosphatase. The most satisfactory decalcification and enzyme preservation is EDTA at 4 °C [146].

E. Biochemistry

1. Interpretation of Data

Biochemical details of bone diseases of simian primates are relatively sparse. There is, however, a considerable amount of information in the form of tables of normal values (e.g. LAPIN). These tables usually cover a range

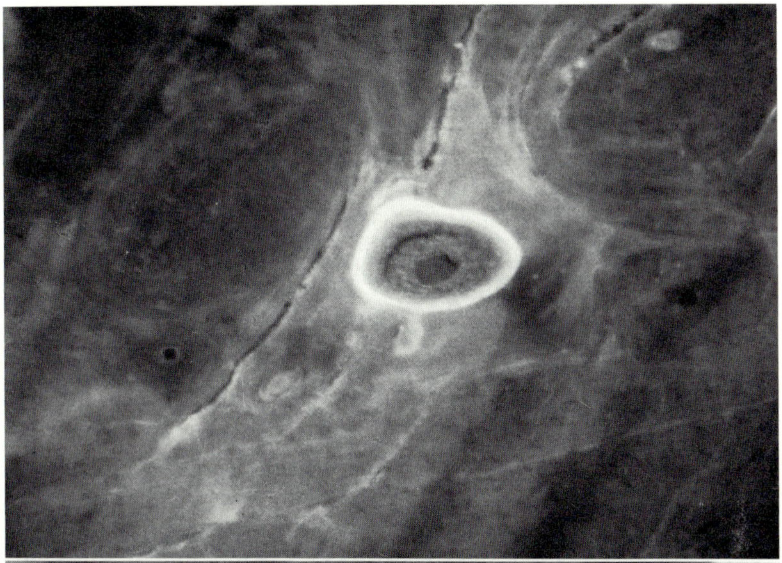

28

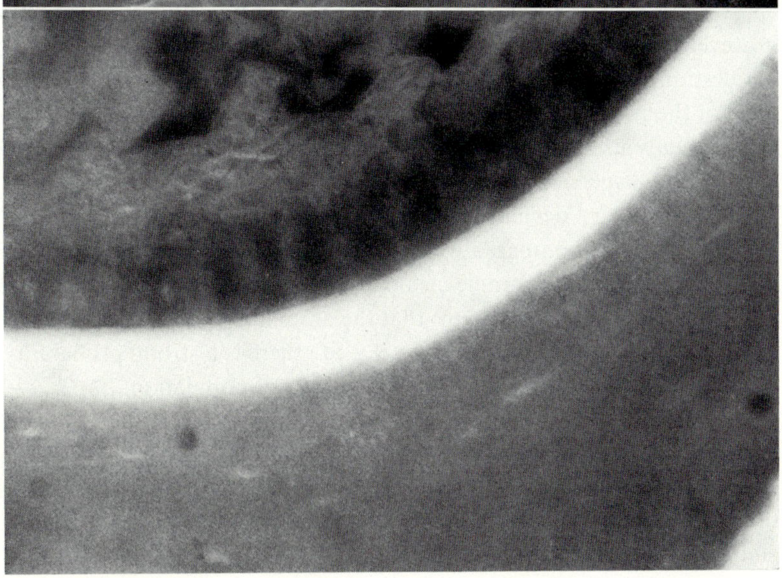

29

Fig. 28. Section of normal compact bone from a baboon *(Papio hamadryas)* injected 2 days before death with tetracycline (20 mg/kg). Note single labelled osteon with 1 fluorescent band. Undecalcified ground section 20 μm. Ultra violet light illumination. ×70.

Fig. 29. Same animal as figure 28. High-power view of labelled osteon. Sharp band of fluorescence denotes region of active bone mineralisation at time of labelling. Undecalcified ground section 20 μm. Ultra violet illumination. ×175.

of parameters, although not all of them are related to bone disease. They also deal with many primate species but it is often difficult to find tables relating to the particular animal that one is investigating, due in part to the wide range of journals in which normal values are published. There are also limitations to the use of this type of data for establishing a diagnosis. Some of these limitations are discussed below. It is hoped that this discussion will help in making the decisions as to whether an animal is normal or not with respect to bone disease somewhat earlier. With this aim in mind we also have outlined some of the diagnostic methods that we have found useful.

First, what is meant by 'normal' value? There are many possible meanings for the term depending on the use to which it is put. The biological and statistical meanings are not always the same, as highlighted by GRASBECK [73] who points out that to a statistician it is normal to be senile and even dead by the age of 80! To a biologist normal value often means the ideal or healthy value [150].

Many ways of obtaining normal values from data have been proposed, e.g. WOOTTON *et al.* [195], HOFFMAN [91], NEUMANN [139]. Normal values are, however, usually obtained by measuring the concentration of the constituent in question, in a homogeneous sample of animals considered to be representative of the population and drawn from it. The arithmetic mean of the sample is usually used as a measure of central tendency. There are a variety of measures of dispersion of the values about the mean, the usual one employed being the standard deviation. If the value for the particular constituent of a random animal is compared with the statistics from the representative sample and falls within the sample means ± 2 SD, one can be reasonably sure (assuming a Gaussian distribution) that the animal is similar, i.e. normal. If the value falls outside the sample mean ± 2 SD there is a strong probability that the animal is either abnormal or from a different population.

Before making a diagnosis (and possibly starting treatment) on the basis of the first conclusion that the animal is abnormal it is necessary to exclude the second possibility, i.e. that the animal is from a different population to that from which the normal data was drawn. There are many aspects to excluding the second possibility and some of them are discussed below.

a) Identification

Whilst published data normally carry the specific name of the animals used, it is essential that a reference to the work used to make the identification is included. The identity of the animal under study can then be checked against this reference before comparison with the data is made.

b) Sex of the Animal

The range and normal values for males and females are not always identical, e.g. COWDRY and SCOTT [34], DE LA PEÑA and GOLDZIEHER [40]. If the animal is female details of its reproductive state are also important. Factors such as oestrus, pregnancy, lactation, have effects on calcium metabolism which could invalidate comparison with published data. The reproductive cycle in male animals such as found in *Saimiri sciureus* should also be considered [54].

c) Age of the Animal

It is difficult to assess an animal's age unless it has been bred in captivity. An approximate age can be determined by examining whole body radiographs for such features as fusion of cranial sutures, the condition of the epiphyses of bones and the state of development of dentition [76], or by means of a formula such as that proposed by VAN WAGENEN and CATCHPOLE [182].

The age of the animal is especially important when comparison is to be made with data concerning parameters which possibly vary with age as, for example, serum alkaline phosphatase which is higher in growing animals than in normal adults [40].

d) The Nutritional State of the Animals

The dietary requirements of particular species of simian primates are gradually being determined and it is now possible to assess the adequacy of diets fed to animals used to obtain normal values. Although this is possible, the vague dietary details given in some papers make it rather difficult to obtain anything other than an approximate idea of whether the animals were adequately fed or not. The effects of a particular diet, if it is inadequate, especially if it has been fed to animals over a long period, can invalidate comparison with data obtained from these animals [9].

e) The Effects of Anaesthesia

The anaesthetic or restraint used obviously depends on the animal and the reasons for anaesthetizing it. The type of anaesthetic or restraint should, however, be noted when making comparison.

f) Method of Taking Sample

Venous-stasis should be avoided when taking blood for calcium determinations. The local increase in blood pressure has an ultrafiltration effect artificially raising the plasma calcium concentration (G. A. ROSE, personal communication).

If urine or faeces are to be collected considerable care has to be taken to obtain meaningful results. The type of caging system required has been described in detail above.

g) Biochemical Units

Comparison with tables of normal values is complicated by the use of different methodology and the consequent different units. This is particularly so with measurements of enzyme activity. An international enzyme unit (IU) is defined as the activity of an enzyme which transforms 1 μmol of substrate in 1 min under defined conditions. If the molecular weight of the substrate is unknown then it is defined in terms of micromoles of product. Conversion of enzyme values to international units (IU) does not facilitate comparison as the different methods employ different substrates under different conditions [183] and the normal ranges, converted to IU are not the same. The best solution is probably to determine the particular enzyme activity in the animal under study by the same method as that used in obtaining the normal values.

h) Statistics

The amount of confidence one has in the extent to which a given animal varies from normal depends on how good the estimates of population parameters are. Such factors as the sample size, the number of assays carried out on each sample, the number of samples taken from each animal and the extent of any statistical treatment used, are all pieces of evidence that are necessary in deciding whether the animal is abnormal.

2. Preparation of Samples

The choice of method of preparing samples for analysis depends on both the type of sample and the method of analysis. We have chosen analytical methods that enable a single preparation procedure to be used for each type of sample. This saves time, labour and reduces systematic errors.

a) Food and Faeces

For the analysis of the mineral content of food pellets, approximately 10 g are crushed in a pestle and mortar. A known weight of food powder is then transferred to a Kjeldahl flask. For the analysis of faecal minerals the total 24-hour collection of faeces is placed in the flask. The wet digestion procedure outlined by WOOTTON [194] using concentrated nitric acid and perchloric acid is then followed for both food and faecal samples. Particular care should be taken with this technique to avoid explosions. The final clear digestion fluid is washed into an appropriate sized volumetric flask and made up to volume with deionised-distilled water. Samples of this solution can then be assayed.

The concentration of a particular component per ml of this solution can then be related to the accurately known weight of food powder originally taken and hence the concentration of the component per gram of food can be calculated. The concentration of a component per ml can be used to calculate the total faecal excretion of that component simply by multiplying by the volume of the volumetric flask originally used.

b) Urine

Ten mls of concentrated hydrochloric acid are added to the 24-hour sample of urine, the solution shaken and the volume recorded. A 10-ml sample of this solution is then sealed in a stoppered tube and stored at 4 °C until analysed. Some authors recommend the addition of either concentrated hydrochloric acid to the collecting bottles under the animals' cages. This seems unwise if, as with our cage system, there are 2 tiers of cages. The accidental spilling of a container would pour concentrated acid over the lower animals.

c) Blood

Blood samples are taken and put into prepared tubes containing heparin, mixed by repeated gentle inversion and then centrifuged at 3,000 rpm for 10 min. The blood plasma is then separated from the cells and either stored in sealed tubes at –20 °C or assayed immediately.

3. Analysis for and Significance of Parameters of Bone Disease

a) Calcium and Magnesium

There are many methods of analysis for these two ions. Metabolic work involves a large number of samples, and a quick reliable assay system is essential. Atomic absorption spectroscopy or an automated titrimetric method are the most suitable.

Low plasma calcium values are found in hyperparathyroidism and occasionally in osteomalacia. In rickets the plasma calcium concentration is usually at the lower limit of the normal range, or else normal. Hyperparathyroidism can result in raised plasma calcium values but this is usually only the case when there is a parathyroid tumour. Hypervitaminosis-D may also result in raised plasma calcium [34, 96]. Diseases in which calcium absorption is impaired tend to produce only marginal alterations in the plasma calcium due to the release of calcium from bone by the action of parathyroid hormone [95, 97, 114].

Hypomagnesaemia has been produced experimentally in capuchin monkeys by feeding a purified diet deficient in magnesium, to young growing animals. The result was loss of weight, reduced plasma magnesium and potassium concentrations, and in some animals convulsions [185].

Urinary excretion of calcium is raised in primary hyperparathyroidism and hypervitaminosis-D and lowered in malabsorption diseases such as rickets. Faecal excretion of calcium is increased in malabsorption diseases contributing to the negative calcium balance found in these conditions.

b) Chromium

If chromic oxide has been used as a marker and is present in either the food or faecal samples, it will be oxidised to dichromate by the wet digestion procedure outlined in section E 2, a. The dichromate can then be estimated by the addition of excess standard ferrous ammonium sulphate solution and back titration with standard potassium dichromate solution using N-phenylanthranilic acid as an indicator [187, 194]. Gorsuch [72] using radioactive chromium, has estimated that 2% of the chromium in the original digestion mixture is lost, probably as volatile chromyl chloride.

c) Phosphorus

The method described by Delsal and Manhouri [41] is suitable for blood plasma, urine and food and faecal digest liquid. If dichromate is present in the digest liquid it will interfere with the assay.

The dichromate can be removed by acidifying part of the diluted digest mixture with a few drops of concentrated hydrochloric acid and shaking it in a separating funnel with an equal volume of 4-methylpentan-2-one. The dichromate passes into the organic phase which can be discarded. The aqueous phase can then be used for the assay [194].

Low plasma inorganic phosphorus levels are found in rickets, osteomalacia and to some extent in hyperparathyroidism. There is a moderate increase in hyperparathyroidism but it should be borne in mind that plasma inorganic phosphorus also increases in chronic nephritis.

The amount of phosphorus excreted in urine depends largely on the amount of phosphorus in the diet. There is an increased excretion of phosphorus in hyperparathyroidism and reduced excretion in hypoparathyroidism, and malabsorption diseases, e.g. rickets. Treatment of rickets with either ultra violet light or vitamin D leads to an increase in the urine excretion of phosphorus.

The faecal excretion of phosphorus depends on both the intake and the amount absorbed. In malabsorption conditions such as rickets and osteomalacia a negative balance is frequently found.

d) Hydroxyproline

The determination of hydroxyproline in urine and blood plasma though time consuming is relatively straightforward. The method we have found successful is that of WOESSNER [192].

Hydroxyproline accounts for approximately 14% of the total amino acid content of collagen and except for small amounts in elastin no significant amounts of the amino acid are found in any other body protein [4]. The urine excretion of hydroxyproline is known to increase in several diseases including hyperparathyroidism. Although such a rise is not a specific indication of increased parathyroid activity, it is a good guide to collagen degradation. In combination with other determinations, urinary hydroxyproline excretion is extremely useful for assessing parathyroid activity.

AVIOLI and PROCKOP [4] injected 300 μCi of uniformly-labelled proline-^{14}C, intravenously, into adult rhesus monkeys. Twenty-seven days after the initial injection of proline-^{14}C and on each of the subsequent 17 days, 2 of the animals were given intramuscular injections of 200 U of commercial parathyroid extract. Injection of parathyroid extract was followed by an increase in the urine excretion of hydroxyproline-^{14}C. From these and other results [149] it has been suggested that the effect of parathyroid hormone on collagen metabolism is limited to bone collagen [4].

HUECKEL and ROGERS [92] have shown that feeding gelatin to a variety of animals, including man, rhesus and capuchin monkeys, produced different effects. Human subjects showed essentially no change in the excretion of the amount of free hydroxyproline, 5½ h after ingestion of gelatin, while bound hydroxyproline was increased 40-fold. The rhesus and capuchin monkeys showed no change in the amount of either free or bound hydroxyproline excreted. HUECKEL and ROGERS concluded that as the simian primates used did not excrete bound hydroxyproline after the ingestion of gelatin they either have an enzyme that can readily hydrolyse the hydroxyproline-containing peptides or else they do not absorb such peptides. Whichever is the case, these results indicate a possible difference in the digestion and absorption of hydroxyproline-containing peptides between these simian primates and man.

e) Alkaline Phosphatase

There are many methods for determining the activity of this enzyme. The choice of method depends to a large extent on the method used to determine the normal data with which the results are to be compared, and the personal preference of the investigator. In the United Kingdom, the method of

KING and ARMSTRONG [108, 109, 110] is frequently used, whilst in the United States of America the method of BODANSKY [17, 18, 19] has been favoured.

The plasma alkaline phosphatase activity of normal young animals is higher than the adult range. In rickets, which by definition only occurs in growing animals, there is a marked increase in plasma alkaline phosphatase activity. In osteomalacia which occurs in older animals the increase is not so marked. Plasma alkaline phosphatase activity is also raised in hyperparathyroidism and in the multifactorial condition such as described as occurring in a woolly monkey (p. 447) extremely high values are often found. We are not aware of any record of a case of hypophosphatasia in simians comparable to the condition sometimes seen in human children.

Abnormalities of liver can result in a raised plasma alkaline phosphatase activity. This can complicate the diagnosis of bone disease unless measurements of liver function, such as transaminase determinations, are made.

The occurrence in blood of several iso-enzymes of alkaline phosphatase has been studied in humans [135, 136, 160], in the hope that separation of the isoenzymes associated with bone and those associated with liver would assist diagnosis. A similar study has been started in our laboratory using simian material.

f) Creatinine and Urea

The plasma inorganic phosphorus concentration is raised in chronic nephritis. The possibility of kidney malfunction should be considered as part of the differential diagnosis and creatinine clearance data and blood urea concentrations are useful monitors of kidney function.

Twenty-four hour urinary creatinine excretion data also has the advantage that it can be used to check the completeness of 24-hour urine collections.

g) Parathyroid Hormone

There are 2 basic assay systems for parathyroid hormone. The bioassay system [137] and the radioimmunoassay system [14]. Both are expensive and time-consuming to set up and use.

h) Calcitonin

As with parathyroid hormone there are 2 major assay systems. The bioassay system [171] and the radioimmunoassay system [38, 117]. VAUGHAN [184] has pointed out that the differences already found between the amino acid sequence of porcine calcitonin and human calcitonin means that antibodies to porcine material cannot provide a reliable basis for immunoassay of the human hormone. Similarly there would be problems using antibodies to porcine material for the assay of simian hormone. It seems doubtful whether reliable radioimmuno-assays for either parathyroid hormone or calcitonin in simian primates are practical at the moment.

i) Vitamin D

As with the two hormones vitamin D can be assayed by either a bioassay technique, e.g. SCHACHTER *et al.* [164] or by chemical techniques such as DE VRIES [44]. NORMAN [142] has pointed out that there is at the moment no really satisfactory assay system for vitamin D.

4. Calcium Kinetics

Although radionuclides of calcium have been used to investigate calcium metabolism in a variety of simian primates [84, 196], we are not aware of a comparative study of simian calcium kinetics such as BRONNER's study [23] of rat and man. HARRIS [84] has suggested as a result of investigation using ^{45}Ca in rhesus monkeys that 'the monkey is the best substitute for human beings in studies of calcium metabolism'.

The effects of vitamins D_2 and D_3 on the intestinal absorption of ^{47}Ca by capuchin monkeys have been investigated by HUNT *et al.* [97]. They demonstrate that vitamin D_2 does not increase the intestinal absorption of ^{47}Ca above that seen in monkeys deficient in vitamin D whilst vitamin D_3 promotes absorption of the nuclide.

V. Conclusions

The extent of our knowledge of the aetiology and pathology of simian bone diseases is a reflection of advances made in the understanding of the physiology and biochemistry of many of the factors involved, such as parathyroid hormone, calcitonin and vitamin D. It also reflects the use of non-primates in preference to primates for many basic studies and the paucity of studies on simian bone metabolism. This is necessarily the case considering the cost and difficulties involved in obtaining and maintaining colonies of experimental primates.

The discrepancies between results obtained from the different laboratory animals should countenance caution before extrapolation to man is made. We view the place of simian primates in studies on bone physiology as second in the sequence; non-primates (e.g. rat, dog, etc.) → simian primates → man. Thus simian primates could have a useful role as a necessary buffer between non-primates and man. The development of this role would provide continually more information about simian bone physiology which would both clarify bone diseases in these animals and enable a suitable simian model system to be selected for comparison with man.

Acknowledgements

We are grateful to Prof. H. BAUM, Dr. H. A. SISSONS for advice, and Mr. M. G. HART for technical assistance. We are also grateful to Dr. G. H. DU BOULAY for permission to use some of his radiographs, to Miss E. M. MONSON for help in taking radiographs, and to Mr. D. G. TAYLOR for preparation of the photographic plates. Our thanks are also due to Mrs. E. R. HAZELL for clerical assistance.

M. J. M. BARKER is supported by a grant from the Wellcome Trust, and R. T. HERBERT by a grant from the Universities Federation for Animal Welfare. Additional financial Assistance has been obtained through the generosity of the World Health Organization.

References

1 ARNAUD, C.; RASMUSSEN, H., and ANAST, C.: Further studies on the relationship between parathyroid hormone and vitamin D. J. clin. Invest. *45* (12): 1955–1964 (1966).

2 ARNAUD, C.; TENENHOUSE, A. M., and RASMUSSEN, H.: Parathyroid hormone. Annu. Rev. Physiol. *29:* 349–372 (1967).

3 ASHLEY MONTAGU, M. F.: A rare case of osteomalacia with severe caries of the maxillary canine teeth in a captive baboon. Int. J. orthod. Dent. Child. *21:* 672–681 (1935).

4 AVIOLI, L. V. and PROCKOP, D. J.: Collagen degradation and the response to parathyroid extract in the intact rhesus monkey. J. clin. Invest. *46* (2): 217–224 (1967).

5 AVIOLI, L. V.; WILLIAMS, T. F.; LUND, J., and LUCA, H. F. DE: Metabolism of vitamin D-^{3}H in vitamin D resistant rickets and familial hypophosphataemia. J. clin. Invest. *46* (12): 1907–1915 (1967).

6 BASSI, R.: Sopra un caso di leontiasi delle osoa della testa in una scimia *(Cynocephalus sphinx)*. Med. Vet. *3* (4): 289–297 (1874).

7 BAUD, C. A.: The fine structure of normal and parathormone-treated bone cells; in GALLIARD, VAN DEN HOOF and STEENDYK 4th European Symposium on Calcified Tissues, pp. 4–6, Excerpta med. int. Congr. Ser. No. 120 (1967).

8 BAIRD, C. A.; SITBENTHAL, J.; LANGER, B.; TUFLING, M. R., and MACK, R. S.: The effects of prolonged administration of thyrocalcitonin in human senile osteoporosis. In: Calcitonin 1969. Proc. 2nd Int. Symp., London 1969 (Heinemann, London 1970).

9 BEASLEY, J. W.: Effects of age and rearing conditions of *M. mulatta* on blood proteins and protein bound carbohydrates. Clin. Chem. *14:* 1107–1111 (1968).

10 BECKER, R. O.; SPADARO, J. A., and BERG, E. W.: Trace element of human bone. J. Bone Jt. Surg. *50-A* (2): 326–334 (1968).

11 BELANGER, L. F.: Osteolysis. An outlook on its mechanism and causation; in GALLIARD, TALMAGE and BUDY The parathyroid glands (University of Chicago Press, Chicago 1965).

12 BELANGER, L. F.: Osteocytic osteolysis. Calc. Tissue Res. *4:* 1–12 (1969).

13 BENSON, R. E.; FREMMING, B. D., and YOUNG, R. J.: A tuberculosis outbreak in a *Macaca mulatta* colony. Amer. Rev. Tuberc. Pulmon. Dis. *72:* 204–209 (1955).

14 BERSON, S. A.; YALOW, R. S.; AURBACH, G. D., and POTTS, J. T.: Immunoassay of bovine and human parathyroid hormone. Proc. nat. Acad. Sci., Wash. *49:* 613–617 (1963).

15 BILTZ, R. M. and PELLEGRINO, E. D.: The chemical anatomy of bone. I. A comparative study of bone composition in sixteen vertebrates. J. Bone Jt. Surg. *51-A* (3): 456–466 (1969).

16 BLAND SUTTON, J. B.: Observations on rickets etc. in wild animals. J. Anat., Lond. *18:* 363–397 (1884).

17 BODANSKY, A.: Phosphatase studies. I. Determination of inorganic phosphate. Beer's law and interfering substances in the Kuttner-Lichtenstein method. J. biol. Chem. *99:* 197–206 (1932).

18 BODANSKY, A.: Phosphatase studies. II. Determination of serum phosphatase. Factors influencing the accuracy of the determination. J. biol. Chem. *101:* 93–101 (1933).

19 BODANSKY, A.: Determination of serum inorganic phosphate and phosphatase. J. clin. Path. *1:* 51–59 (1957).

20 BORDIER, P.; MATRAJT, H.; HIOCO, D.; HEPNER, G. W.; THOMPSON, G. R., and BOOTH, C. C.: Subclinical vitamin D deficiency following gastric surgery, histological evidence in bone. Lancet *i:* 437–440 (1968).

21 BORDIER, P.; MIRAVET, L., and HIOCO, D.: (in press).

22 BRAMBLETT, C. A.: Pathology in the Darajani baboon. Amer. J. phys. Antropol. *26:* 331–340 (1967).

23 BRONNER, F. and LEMAIRE, R.: Comparison of calcium kinetics in man and the rat. Calc. Tissue Res. *3:* 238–248 (1969).

24 BROOKS, H. and BLAIR, W. K.: Osteomalacia of primates in captivity. Rep. N. Y. Zool. Soc. *9:* 135–175 (1905).

25 BYERS, P. and SMITH, R.: Trephine for full thickness iliac crest biopsies. Brit. med. J. 682–683 (1967).

26 CAMERON, A. H. and HILL, W. C. O.: The Arnold-Chiari malformation in a sacred baboon. J. Path. Bact. *70:* 552–554 (1955).

27 CAMPBELL, A. W. and CLELAND, J. B.: Cage paralysis in monkeys. J. Comp. Path. *32:* 95–104 (1919).

28 CHAPMAN, W. L. and ALLEN, J. R.: Multiple neoplasia in a rhesus monkey *(Macaca mulatta)*. Path. vet. *5:* 342–352 (1968).

29 CHEN, P. S. and BOSMANN, H.: Effects of vitamins D_2 and D_3 on serum calcium and phosphorus in rachitic chicks. J. Nutr. *83:* 133–139 (1964).

30 COHEN, B.: An anaesthetic method for oral surgery on monkeys; in GRAHAM-JONES Small animal anaesthesia (Pergamon Press, Oxford 1964).

31 Copp, D. H.: Parathormone, calcitonin and calcium homeostasis; in Comar and Bronner Mineral metabolism – an advanced treatise. III. Calcium physiology (Academic Press, London 1969).

32 Corson-White, E. P.: Osteomalacia in monkeys. Arch. intern. Med. *30:* 620–628 (1922).

33 Corson-White, E. P.: Osteitis deformans in monkeys. Arch. intern. Med. *30:* 790–796 (1922).

34 Cowdry, E. V. and Scott, G. H.: Effects on monkeys of small doses of a concentrated preparation of Viosterol. Arch. Path. *22* (1): 1–23 (1936).

35 Crane, J. T.; Pickering, D. E.; Van Wagenen, G., and Smyth, F. S.: Growth and metabolism in normal and thyroid ablated infant rhesus monkeys *(Macaca mulatta).* VII. Morphology of normal and thyroid ablated infant rhesus monkeys. Amer. J. Dis. Child. *87:* 708–723 (1954).

36 Dammrich, K.: Osteoporosis in young animals. Path. vet. *4:* 435–463 (1967).

37 Day, P. L.: The nutritional requirements of primates other than man. Vitamins Hormones N. Y. *2:* 71–105 (1944).

38 Deftos, L. J.; Lee, M. R., and Potts, J. J., jr.: A radioimmunoassay for thyrocalcitonin. Proc. nat. Acad. Sci., Wash. *60:* 293–299 (1968).

39 Delahunt, C. S. and Lasson, L. J.: Thalidomide syndrome in monkeys. Science *146:* 1300–1305 (1964).

40 De la Peña, A. and Goldzieher, J. W.: Clinical parameters of the normal baboon; in Vagtborg The baboon in medical research, vol. II (S. W. Foundation for Research and Education, San Antonio 1967).

41 Delsal, J. L. et Manhouri, H.: Etude comparative des dosages colorimétriques du phosphore. IV. Dosage de l'orthophosphate en présence d'esters phosphoriques (nouvelles méthodes). Bull. Soc. Chim. biol. *40:* 1623–1636 (1958).

42 De Luca, H. F.: Mechanism of action and metabolic fate of vitamin D. Vitamins Hormones N. Y. *25:* 315–367 (1967).

43 Dent, C. E.: Rickets and osteomalacia from renal tubule defects. J. Bone Jt. Surg. *34-B:* 266–274 (1952).

44 De Vries, E. J.: Determination of vitamin D in blood. J. Vitam. *15* (3): 189–194 (1969).

45 Dodds, G. S. and Cameron, H. C.: Studies on experimental rickets in rats. Amer. J. Anat. *55:* 135 (1934).

46 Dodds, G. S. and Cameron, H. C.: The healing process. Amer. J. Path. *14:* 273–296 (1938).

47 Doods, G. S. and Cameron, H. C.: The behaviour and fate of the cartilage remnants in the rachitic metaphysis. Amer. J. Path. *15:* 723–740 (1939).

48 Dodds, G. S. and Cameron, H. C.: The relation of rickets to growth with special reference to the bones. Amer. J. Path. *19:* 169–185 (1943).

49 Doyle, F. H.: Some quantitative radiological observations in primary and secondary hyperparathyroidism. Brit. J. Radiol. *39:* 161–167 (1966).

50 Dreizen, S.; Levy, B. M.; Bernick, S.; Hampton, J. K., jr., and Kraintz, L.: Studies on the biology of the periodontium of marmosets. III. Periodontal bone changes in marmosets with osteomalacia and hyperparathyroidism. Israel J. med. Sci. *3* (5): 731–738 (1967).

51 DRURY, R. A. B. and WALLINGTON, E. A.: Carleton's histological technique, 4th ed. (Oxford University Press, London 1967).

52 DU BOULAY, G. H. and CRAWFORD, M. A.: Nutritional bone disease in captive primates. Symp. zool. Soc., Lond. No. 21, pp. 223–236 (1968).

53 DUCKWORTH, W. L. H.: On the natural repairs of fractures as seen in the skeleton of anthropoid apes. J. Anat., Lond. *46:* 81–85 (1911).

54 DU MOND, F. V.: The squirrel monkey in a semi-natural environment; in ROSENBAUM and COOPER The squirrel monkey (Academic Press, London 1968).

55 FAIRBROTHER, R. W. and HURST, E. W.: Spontaneous diseases observed in 600 monkeys. J. Path. Bact. *616:* 867–873 (1932).

56 FANCONI, V. G. und GIRARDET, P.: Familiärer persistierender Phosphat-diabetes mit D-vitaminresistenter Rachitis. Helv. paediat. Acta. *7:* 14–41 (1952).

57 FIENNES, R. N. T-W-: Report of the pathologist for the year 1962. Proc. zool. Soc., Lond. *143:* 53–78 (1964).

58 FIENNES, R. N. T.-W.-: Cystic bone disease in South American primates. III. Proc. zool. Soc., Lond. *143:* 521–523 (1964).

59 FIENNES, R. N. T.-W.-: Report of the pathologist for the year 1965. J. Zool., Lond. *154:* 173–191 (1968).

60 FLEISCH, H.: Role of nucleation and inhibition in calcification. Clin. Orthop. *32:* 170–180 (1964).

61 FLOWER, W. H.: Exhibition of and remarks upon the skull of a young chimpanzee. Proc. zool. Soc., Lond.: 634–636 (1882).

62 FOSTER, G. V.; DOYLE, F. H.; BORDIER, P., and MATRAJT, H.: Effect of thyrocalcitonin on bone. Lancet *ii:* 1428–1431 (1966).

63 FOURMAN, P. and ROYER, P.: Calcium metabolism and the bone, 2nd ed. (With the collaboration of MICHAEL J. LEVELL and D. BRIAN MORGAN (Blackwell Scientific, Oxford 1968).

64 FOX, H.: Diseases in captive wild mammals and birds (Lippincott, Philadelphia 1923).

65 FOX, H.: Notes on the skeleton of a chimpanzee found in the wild with a serious chronic inflammation of the posterior extremity. Rep. Penrose Res. Lab. zool. Soc. Philad., pp. 18–25 (1937).

66 FOX, H.: Chronic arthritis in wild mammals. Trans. amer. phil. Soc. *31:* 71–149 (1939).

67 FRASER, D. R. and KODICEK, E.: Unique biosynthesis by kidney of a biologically active vitamin D metabolite. Nature, Lond. *228:* 764–766 (1970).

68 FROST, H. M.: Thickness of osteoid seams. Clin. Orthop. *25:* 175–179 (1962).

69 FROST, H. M.: Measurement of human bone formation by means of tetracycline labelling. Canad. J. Biochem. Physiol. *41:* 31–42 (1963).

70 FROST, H. M.; WILSON, R. V., and RAMSER, J. R.: Thickness of osteoid seams in some diseases of man. Clin. Orthop. *29:* 119–124 (1966).

71 GERSTENBERGER, H. J.: Rickets in monkeys *(Macaca rhesus)*. Amer. J. Dis. Child. *56:* 694 (1938).

72 GORSUCH, T. T.: Radiochemical investigations on the recovery for analysis of trace elements in organic and biological material. Analyst Lond. *84:* 135–140 (1959).

73 GRASBECK, R.: Normal values and statistics. Scand. J. clin. Lab. Invest. *21:* 193–195 (1968).

74 GRAHAM-JONES, O.: Cystic bone disease in South American primates. I. Proc. zool. Soc., Lond. *143:* 517–519 (1964).

75 GUILLOUD, N. B.; ALLMOND, B. W.; FROESCHLE, J. E., and FITZGERALD, F. L.: Paralytic poliomyelitis in laboratory primates. J. amer. vet. med. Ass. *155:* 1190–1193 (1969).

76 HAIGH, M. V. and SCOTT, A.: Some radiological and other features for assessing the age of the rhesus monkey using animals of known age. Lab. anim. Care *15:* 57–73 (1965).

77 HAM, A. W. and LEESON, T. S.: Histology, 5th ed. (Pitman Medical Publishing, London 1965).

78 HAMERTON, A. E.: Report on the deaths occurring in the society's garden during the year 1935. Proc. zool. Soc., Lond.: 659–686 (1936).

79 HAMPTON, J. K.; HAMPTON, S. H., and LANDWEHR, B. T.: Observations on a successful breeding colony of the marmoset *(Oedipomidas oedipus)*. Folia primat. *4:* 265–287 (1966).

80 HANCOX, N.: The osteoclast; in BOURNE The biochemistry and physiology of bone (Academic Press, New York 1956).

81 HANCOX, N. and BOOTHROYD, B.: Structure-function relationship in the osteoclast; in SOGGNNAES Mechanisms of hard tissue destruction, pp. 497–514 (American Association for the Advancement of Science, Washington 1963).

82 HARRIS, L. J.: The significance of vitamins in practical experience. Brit. med. J. *ii:* 367–373 (1933).

83 HARPER, R. A. and POSNER, A. S.: Measurement of non-crystalline calcium phosphate in bone mineral. Proc. Soc. exp. Biol. Med. *122:* 137–142 (1966).

84 HARRIS, R. S.; MOOR, J. R., and WANNER, R. L.: Calcium metabolism of the normal rhesus monkey. J. clin. Invest. *40:* (ii) 1766–1774 (1961).

85 HEGSTED, D. M.; MILLS, R. C.; ELVEHJEM, C. A., and HART, E. B.: Choline in the nutrition of chicks. J. biol. Chem. *138:* 459–466 (1941).

86 HENDRICKX, A. G.; AXELFOD, L. R., and CLAYBORN, L. D.: Thalidomide syndrome in baboons. Nature, Lond. *210:* 958–959 (1966).

87 HILL, W. C. O.: Spina bifida in a sacred baboon *(Papio hamadryas)*. Ceylon J. Sci. D *5:* 9–15 (1939).

88 HILL, W. C. O.: Report of the society's prosector for the year 1953. Proc. zool. Soc., Lond. *124:* 303–311 (1954).

89 HIOCO, D.; BORDIER, P.; MIRAVET, L.; DENYS, H., and TUN-CHOT, S.: Prolonged administration of calcitonin in man. Biological, isotopic and morphological effects. In: Calcitonin 1969. Proc. 2nd Int. Symp., London 1969 (Heinemann, London 1969).

90 HIRSCH, P. F. and MUNSON, P. L.: Thyrocalcitonin. Physiol. Rev. *49* (3): 548–622 (1969).

91 HOFFMAN, R. G.: Statistics in the practice of medicine. J. amer. Med. Ass. *185:* 864–873 (1963).

92 HUECKEL, H. J. and ROGERS, Q. R.: Urinary excretion of hydroxyproline-containing peptides in man, rat, hamster, dog and monkey after feeding gelatin. Comp. Biochem. Physiol. *32:* 7–16 (1970).

93 HUME, E. M.: Monkey welfare. Proc. Nutr. Soc. *16:* 87–93 (1957).

94 HUNT, R. D.; GARCIA, F. G., and HEGSTED, D. M.: Vitamin D requirements of New World primates. Fed. Proc. *25* (2006): 545 (1966).

95 Hunt, R. D.; Garcia, F. G., and Hegsted, D. M.: A comparison of vitamin D_2 and D_3 in New World primates. I. Production and regression of osteodystrophia fibrosa. Lab. Anim. Care *17* (2): 222–234 (1967a).

96 Hunt, R. D.; Garcia, F. G., and Hegsted, D. M.: Hypervitaminosis D in New World monkeys. Amer. J. clin. Nutr. *22* (3): 358–366 (1969).

97 Hunt, R. D.; Garcia, F. G.; Hegsted, D. M., and Kaplinsky, N.: Vitamin D_2 and D_3 in New World primates. Influence on calcium absorption. Science *157:* 943–945 (1967b).

98 Jeffree, G. M.: Phosphatase activity in the limb bones of monkeys *(Lagothrix humboldti)* with hyperparathyroidism. J. clin. Path. *15:* 99–111 (1962).

99 Jha, G. J.; Deo, M. G., and Ramalingaswami, V.: Bone growth in protein deficiency. A study in rhesus monkeys. Amer. J. Path. *53:* 1111–1123 (1968).

100 Jowsey, J.: Age changes in human bone. Clin. Orthop. *17:* 210–218 (1960).

101 Jowsey, J.: Studies of Haversian systems in man and some animals. J. Anat., Lond. *100:* 857–864 (1966).

102 Jowsey, J.: Quantitative microradiography, a new approach to the evaluation of metabolic bone disease. Amer. J. Med. *40:* 485–491 (1966).

103 Jowsey, J.: Age and species differences in bone. Cornell Vet. *58:* suppl., pp. 74–94 (1968).

104 Karski, J.: Morphology of parathyroid glands in *Macaca rhesus* and *Macaca cynomolgus*. Folia Morph., Warsz. *23:* 281–289 (1964).

105 Keane, B. E.; Spiegler, G., and Davis, R.: Quantitative evaluation of bone mineral by a radiographic method. J. Radiolrit. B. *32:* 162–167 (1959).

106 Kennard, M. A.: Abnormal findings in 246 consecutive autopsies on monkeys. Yale J. Biol. Med. *13:* 709–712 (1941).

107 Kent, S. P.; Vawter, G. F.; Dowben, R. M., and Benson, R. E.: Hypervitaminosis D in monkeys; a clinical and pathologic study. Amer. J. Path *34* (1): 37– (1958).

108 King, E. J. and Armstrong, A. R.: A convenient method for determining serum and bile phosphatase activity. Canad. med. Ass. J. *31:* 376–381 (1934).

109 King, E. J.; Haslewood, G. A. D., and Delory, G. E.: Micro-chemical methods of blood analysis. Lancet *i:* 886–892 (1937).

110 King, E. J.; Haslewood, G. A. D.; Delory, G. E., and Beall, D.: Micro-chemical methods of blood analysis. Lancet *i:* 207–208 (1942).

111 Kodicek, E.; Lawson, D. E. M., and Wilson, P. W.: Biological activity of a polar metabolite of vitamin D_3. Nature, Lond. *228:* 763–764 (1970).

112 Krook, L. and Barrett, R. B.: Simian bone disease – a secondary hyperparathyroidism. Cornell Vet. *52:* 459–492 (1962).

113 Lapin, B. and Yakovleva, L. A.: Comparative pathology in monkeys (Thomas, Springfield, Ill. 1963).

114 Lehner, N. D. M.; Bullock, B. C.; Clarkson, T. B., and Lofland, H. B.: Biological activities of vitamins D_2 and D_3 for growing squirrel monkeys. Lab. Anim. Care *17* (5): 483–493 (1967).

115 Lehner, N. D. M., Bullock, B. C., and Clarkson, T. B.: Squirrel monkeys as models for the study of rickets; in Vagtborg Use of non-human primates in drug evaluation. Southwest Foundation for Research and Education (University of Texas Press, Austin 1968).

116 LEHNINGER, A. L.: Mitochondria and calcium ion transport. Biochem. J. *119:* 129–138 (1970).

117 LEQUIN, R. M.; HACKENG, W. H. L.; SCHOPMAN, W., and CARE, A. D.: Comparison between the bioassay and radioimmunoassay of calcitonin in plasma of several species. J. Endocrin *45* (2): 309–310 (1969).

118 LUCAS, N. S.; HUME, E. M., and SMITH, H. H.: On the breeding of the common marmoset *(Hapale jacchus* Linn.) in captivity when irradiated with ultra violet light. Proc. zool. Soc., Lond.: 447–451 (1927).

119 LUCAS, N. S.; HUME, E. M., and HENDERSON, H.: On the breeding of the common marmoset in captivity when irradiated with ultra violet light. A ten-year family history. Proc. zool. Soc., Lond. A. *107*: 205–211 (1937).

120 LUCKEY, T. D.: A study in comparative nutrition. Comp. Biochem. Physiol. *2:* 100–124 (1961).

121 LUSTED, L. B.; PICKERING, D. E.; FISHER, D., and SMYTH, F. S.: Roentogenographic features of skeletal development in normal and thyroid ablated rhesus monkeys. Amer. J. Dis. Child. *86:* 426–435 (1953).

122 MACINTYRE, I.: Calcitonin – a general review. Calc. Tissue Res. *1:* 173–182 (1967).

123 MACINTYRE, I.: Calcitonin – a review of its discovery and an account of purification and action. Proc. roy. Soc. B. *170:* 49–60 (1968).

124 MCLEAN, F. C. and URIST, M. R.: Bone. Fundamentals of the physiology of skeletal tissue, 3rd ed. (University of Chicago Press, Chicago 1968).

125 MANN, G. V.; WATSON, P. L., and ADAMS, L.: Primate Nutrition. I. Cebus monkeys – normal values. J. Nutr. *47:* 213–224 (1952).

126 MANSON, J. D.: The lamina dura. Oral Surg. *16:* 432–438 (1963).

127 MARTIN, J. E.; COLE, W. C., and WHITNEY, R. A.: Tuberculosis of the spine (Pott's disease) in a rhesus monkey. J. amer. vet. med. Ass. *153:* 914–917 (1968).

128 MATRAJT, H. and HIOCO, D.: Solochrome cyanine R as an indicator dye of bone morphology. Stain Technol. *41:* 97–100 (1966).

129 MERTZ, W.: Chromium occurrence and function in biological systems. Physiol. Rev. *49* (2): 163–239 (1969).

130 MILCH, R. A.; RALL, D. P., and TOBIE, J. E.: Bone localization of the tetracyclines. J. nat. Cancer Inst. *19:* 87–93 (1957).

131 MILLER, C. E. and KINARD, R.: A case of generalized bone tuberculosis in a rhesus monkey. Lab. Anim. Care *14:* 264–267 (1964).

132 MOLNAR, Z.: Development of the parietal bone of young mice. I. Crystals of bone mineral in frozen dried preparations. J. Ultrastruct. Res. *3:* 39–45 (1959).

133 MOODIE, R. L.: Roentgenologic studies of Egyptian and Peruvian mummies. Mem. Field. Mus. natur. Hist. Anthrop. *3:* 1–66 (1931).

134 MORGAN, R. and FOURMAN, P.: The diagnosis of osteomalacia and osteoporosis. Brit. J. Hosp. Med. *2* (4): 901–908 (1969).

135 MOSS, D. W.; CAMPBELL, D. M.; ANAGNOSTON-KAKARAS, E. M., and KING, E. J.: Characterisation of tissue alkaline phosphatases and their partial purification by starch gel electrophoresis. Biochem. J. *81:* 444–448 (1961).

136 MOSS, D. W. and KING, E. J.: Properties of alkaline phosphatase fractions separated by starch gel electrophoresis. Biochem. J. *84:* 192–195 (1962).

137 MUNSON, P. L.: Biological assay of parathyroid hormone; in GREEP and TALMAGE The parathyroids (Thomas, Springfield 1961).

138 Napier, J. R. and Napier, P. H.: A handbook of living primates (Academic Press, London 1967).

139 Neumann, G. J.: The determination of normal ranges from routine laboratory data. Clin. Chem. *14* (10): 979–988 (1968).

140 Neuman, E. W. F. and Neuman, M. W.: The chemical dynamics of bone mineral (University of Chicago Press, Chicago 1958).

141 Nordin, B. E. C.: The pathogenesis of osteoporosis. Lancet *i:* 1011–1014 (1961).

142 Norman, A. W.: The mode of action of vitamin D. Biol. Rev. *43:* 97–137 (1968).

143 Owen, M. and Shaetar, M. R.: Uptake of ^{3}H glucosamine by osteoclasts. Nature, Lond. *220:* 1335–1336 (1968).

144 Park, E. A.: Observations on the pathology of rickets with particular reference to the changes of the cartilage-shaft junctions of the growing bones. Bull. N. Y. Acad. Med. *15:* 495–543 (1939).

145 Pautard, F. G. E.: A biomolecular survey of calcification; in Fleisch, Blackwood and Owen Calcified tissues Proc. 3rd European Symp., Davos 1965 (Springer, New York 1966).

146 Pearce, A. G. E.: Histochemistry theoretical and applied, vol. I, 3rd ed. (Churchill, London 1968).

147 Petroff, N.; Krotkina, N.; Barabadze, E.; Guelsteine, V.; Melinoff, R. et Smoilovskaya, E.: L'induction experimentale des tumeurs malignes chez les singes. Acta Un. int. Cancr. *15:* 119–123 (1959).

148 Posner, A. S.: Crystal chemistry of bone mineral. Physiol. Rev. *49* (4): 760–792 (1969).

149 Prockop, D. J.: Isotopic studies on collagen degradation and the urine excretion of hydroxyproline. J. clin. Invest. *43* (3): 453–460 (1964).

150 Pryce, J. D.: Level of haemoglobin in whole blood and red blood cells and proposed convention for defining normality. Lancet *ii:* 333–336 (1960).

151 Raisz, L. C.: Bone resorption in tissue culture factors influencing the response to parathyroid hormone. J. clin. Invest. *44:* 103–116 (1965).

152 Randall, F. E.: The skeletal and dental development and variability of the gorilla. Hum. Biol. *16:* 23–76 (1944).

153 Rasmussen, H. and Arnaud, C.: Parathyroid hormone and vitamin D; in Hioco, D. J. L'ostéomalacie (Tours 1965), pp. 221–236 (Mason, Paris 1967).

154 Rasmussen, H.; Arnaud, C.; Hawker, C., and Stedingk, M. von: The relationship between vitamin D and parathyroid hormone. J. clin. Invest. *42:* 1940–1946 (1963).

155 Reed, O. M. and Kriewaldt, F. H.: Bone dyscrasias in a baboon. J. amer. vet. med. Ass. *151:* 923–925 (1967).

156 Reynolds, J. J.: Inhibition by calcitonin of bone resorption induced *in vitro* by vitamin A. Proc. roy. Soc. *170:* 61–69 (1968).

157 Reynolds, J. J. and Minkin, C.: Bone studies *in vitro.* Use of calcitonin as a specific inhibitor to bone resorption. In: Calcitonin 1969. Proc. 2nd Int. Symp., London 1969 (Heinemann, London 1970).

158 Robinson, C. J.; Martin, T. J., and MacIntyre, I.: Phosphaturic effect of thyrocalcitonin. Lancet *ii:* 83–84 (1966).

159 Robison, R.: The possible significance of hexosephosphoric esters in ossification. Biochem. J. *17:* 286–293 (1923).

160 Romel, W. C.; Mancusa, S. J. La, and DuFrene, J. K.: Detection of serum alkaline

phosphatase isoenzymes with phenolphthalein monophosphate following cellulose acetate electrophoresis. Clin. Chem. *14:* 47–57 (1969).

161 ROSE, G. A.: Experiences with the use of interrupted carmine red and continuous chromium sesquioxide marking of human faeces with reference to calcium, phosphorus and magnesium. Gut *5:* 274–279 (1964).

162 ROSE, G. A.: Personal communication.

163 RUCH, T. C.: Diseases of laboratory primates (Saunders, Philadelphia 1959).

164 SCHACHTER, D.; KIMBERG, D. V., and SCHENKER, H.: Active transport of calcium by the intestine. Action and bioassay of vitamin D. Amer. J. Physiol. *200:* 1263–1271 (1961).

165 SCHULTZ, A. H.: Notes on diseases and healed fractures of wild apes and their bearing on the antiquity of pathological conditions in man. Bull. Hist. Med. *7* (6): 571–582 (1939).

166 SCHULTZ, A. H.: Growth and development of the orang-utan. Contrib. Embryol. Carneg. Inst. *29:* 57–110 (1941).

167 SCHULTZ, A. H.: Age changes and variability in gibbons. Amer. J. Phys. Anthrop. *2:* 1–129 (1944).

168 SCHULTZ, A. H.: The occurrence and frequency of pathological and tetralogical conditions and of twinning among non-human primates. Primatologia *1:* 965–1014 (1956).

169 SCHULTZ, A. H.: Cranial and dental variability in Colobus monkeys. Proc. zool. Soc., Lond. *130:* 79–105 (1958).

170 SCHULTZ, A. H.: Age changes and variability in skulls and teeth of Central American monkeys. Proc. zool. Soc., Lond. *133:* 337–390 (1960).

171 SELWYN TAYLOR (ed.): Bioassay. Calcitonin, chapter I. Proc. Symp. Thyrocalcitonin and 'C' Cells, London 1969 (Heinemann Medical Books, London 1970).

172 SISSONS, H. A.; HOLLEY, K. J., and HEIGHWAY, J.: Normal bone structure in relation to ostéomalacia; in HIOCO L'osteomalacie, pp. 19–31 (Masson, Paris 1967).

173 SOKOLOFF, L.; SNELL, K. C., and STEWART, H. L.: Spinal ankylosis in old rhesus monkeys; in URIST Clin. orthop. and related research, vol. 61 (Pitman Medical Publishing, Philadelphia 1960).

174 SOUTHARD, E. E.: A case of Pott's disease in the monkey. J. med. Res. *14:* 393–398 (1906).

175 STEFKO, W. H.: Materials on comparative osseous tuberculosis in monkeys. Proc. Trudy mosk. Zoopk. *1:* 304 (1940).

176 STRATES, B. and NEUMAN, W. F.: On the mechanisms of calcification. Proc. Soc. exp. Biol. Med. *97:* 688–691 (1958).

177 TERMINE, J. D. and POSNER, A. S.: Infra red analyses of rat bone. Age dependancy of amorphous and crystalline mineral fractions. Science *153:* 1523–1525 (1966).

178 TERMINE, J. D. and POSNER, A. S.: Amorphous/crystalline interrelationships in bone mineral. Calc. Tissue Res. *1:* 8–23 (1967).

179 TODD, T. W.: Observations of osteomalacia in the zoological collections of Manchester and Cleveland. Proc. zool. Soc., Lond. *2:* 855–860 (1913).

180 VAES, G.: The role of lysomes and of their enzymes in the development of bone resorption induced by parathyroid hormone; in TALMAGE and BELANGER Parathyroid hormone and thyrocalcitonin (calcitonin), pp. 318–328. Excerpta Medica International Congress Series No. 159 (1968).

181 VAGTBORG, H.: The baboon. An annotated bibliography and supplement, vol. I, 1965; vol. II, 1967; vol. III, 1969 (Southwest Foundation for Research and Education, San Antonio).

182 VAN WAGENEN, G. and CATCHPOLE, H. R.: Physical growth of the rhesus monkey *(Macaca mulatta)*. Amer. J. Phys. Anthrop. *14:* 245–251 (1956).

183 VARLEY, H.: Practical clinical chemistry, 4th ed. (Heinemann Medical Books. London 1967).

184 VAUGHAN, J. M.: The physiology of bone (Clarendon Press, Oxford 1970).

185 VITALE, J. J.; VELEZ, H.; GUZMAN, C., and CORREA, P.: Magnesium deficiency in the Cebus monkey. Circulat. Res. *12:* 642–650 (1963).

186 WEIDMANN, S. M.: Calcification in skeletal tissue. Int. Rev. connect. Tissue Res. *1:* 339–377 (1963).

187 WHITBY, L. G. and LANG, D.: Experience with the chromic oxide method of faecal marking in metabolic balance investigations on humans. J. clin. Invest. *39* (6): 854–863 (1960).

188 WIGGINS, H. S. and DAWSON, A. M.: An evaluation of unabsorbable markers in the study of fat absorption. Gut *2:* 373–376 (1961).

189 WILKINSON, J. H.: Isoenzymes (Spon, London 1965).

190 WILSON, J. G. and GAVAN, J. A.: Congenital malformations in non-human primates. Spontaneous and experimentally induced. Anat. Rec. *158:* 99–109 (1967).

191 WINTERS, R. W.; GRAHAM, J. B.; WILLIAMS, T. F.; MCFALLS, V. W., and BURNETT, C. H.: A genetic study of familial hypophosphataemia and vitamin D-resistant rickets with a review of the literature. J. Med. Cincinnati *27:* 97–141 (1958).

192 WOESSNER, J. F., jr.: The determination of hydroxyproline in tissue and protein samples containing small proportions of this amino acid. Arch. Biochem. Biophys. *93:* 440–447 (1961).

193 WOODS, J. F. and NICHOLS, G.: Collagenolytic activity in rat bone cells. Characteristics and intracellular location. J. cell. Biol. *26:* 747–757 (1965).

194 WOOTTON, I. D. P.: Micro-analysis in medical chemistry (Churchill, London 1964).

195 WOOTTON, I. D. P.; KING, E. J., and MACLEAN SMITH, J.: The quantitative approach to hospital biochemistry. Normal values and the use of biochemical determination for diagnosis and prognosis. Brit. med. Bull. *7:* 307–311 (1951).

196 YEN, P. K-J. and SHAW, J. H.: Radio-calcium uptake in the calcified tissues of rhesus monkeys at various time intervals. J. dent. Res. *42:* 168–176 (1963).

197 LAWSON, D. E. M.; FRASER, D. R.; KODICEK, E.; MORRIS, H. R., and WILLIAMS, D. H.: Identification of 1,25-dihydroxycholecalciferol, a new kidney hormone controlling calcium metabolism. Nature, Lond. *230:* 228–230 (1971).

Authors' address: Dr. M. J. M. BARKER and Dr. R. T. HERBERT, Department of Pathology, Nuffield Institute of Comparative Medicine, Zoological Society of London, Regent's Park, *London, N. W. 1* (England)

Pathology of Simian Primates, Part I, pp. 520–541 (Karger, Basel 1972)

Skeletal Muscle Pathology in Non-Human Primates

M. NELLY GOLARZ DE BOURNE and GEOFFREY H. BOURNE

Yerkes Regional Primate Research Center, Emory University, Atlanta, Ga.

Contents

I.	General	520
II.	*Macaca mulatta*	523
III.	Gorilla	526
IV.	Chimpanzee	528
V.	Java Macaques	534
VI.	Squirrel Monkeys	537
VII.	Celebes Black Ape	537
VIII.	Drill	538
IX.	Other Groups	538
X.	Discussion	538
Acknowledgment		540
References		540

I. General

The skeletal musculature has undoubtedly in recent years been one of the most studied systems in the body.

Neuromuscular diseases in man have always received great interest from investigators who have also studied the lower species in the hope of finding animal models for human muscular diseases.

Few animal species show naturally occurring myopathies; probably the best known is that found in a strain of mice in which the pathology is strikingly similar to that of human progressive muscular dystrophy [1]; even so differences exist [2] between the mouse and the human dystrophy. Spontaneous myopathic changes have also been found in the chicken, duck, hamster, cow, and lamb.

A number of other species of animals have been used for the experimental induction of myopathies. Among them are some species of non-human primates used in studies of the effects of ischemia, nerve damage, nutritional deficiencies, etc. [3, 4, 5]. However, there is no reference in this literature to the presence of any pre-existing myopathic changes and it was not stated if the animals were laboratory or wild-born.

The studies that the authors have made over the last few years have shown that myopathic changes occur to varying degrees in all species of primates examined. In most cases these occur without obvious clinical symptoms.

The material presented here deals only with naturally occurring muscle pathology, and excluded experimental induction of pathology.

The microscopic anatomy of normal muscle of non-human primates is the same as that of human skeletal muscle. The individual fibers are bound together by endomysium, bundles of fibers are surrounded by perimysium and the muscle itself is enveloped by the epimysium and covering all the muscles is the fascia. These connective tissue sheaths have the usual components of fibers and cells with few or no fat cells under normal circumstances; blood vessels and nerves of different caliber run in this connective tissue; lymphatic vessels are difficult to demonstrate. The individual muscle fiber, encased by the sarcolemma, contains myofibrils, sarcoplasm, endoplasmic reticulum and sarcosomes, but these and other components (Golgi apparatus, glycogen, fat, lysosomes) can be seen only imperfectly by light-microscopy. The nuclei, elongated in shape, are situated on the periphery of the normal muscle fiber in a subsarcolemmal position.

As in man the various components of the muscle react in several similar ways to different damaging agents, which may be infections, trauma, nutritional deficiency, nerve damage, parasites or genetic factors. Muscle will react to these stimuli by changes in the subsarcolemmal nuclei, which may shrink or increase in size with a vesicular appearance, vary in relative number, and migrate into the center of the fiber to align themselves in rows. These changes may or may not be accompanied by changes in the muscle fiber itself, such as atrophy, swelling, loss of striations, splitting, hyalinization, flocculation, degeneration, regeneration and/or phagocytosis. Fibers which are being absorbed may be replaced by proliferation of the endomysium but this may also proliferate without loss of muscle fibers. Fat cells may also develop between the fibers leading to fatty replacement or to fatty infiltration without loss of myoplasm. Generalized increase in fat stain inside the fibers in an overall pattern can also occur on occasions. There may be an accumulation of inflammatory cells, macrophages, lymphocytes, plasma cells and sometimes neutrophils in the muscle. These may be found either inside the muscle fiber in cases of degeneration and phagocytosis or in the connective tissue sheaths, sometimes in a perivascular position. Simultaneously with destruction of the muscle fibers it is possible to find active regeneration.

The normal architecture of the muscle can also be upset by the apparent longitudinal splitting of the fibers.

No specific abnormalities were observed in the intramuscular blood vessels besides the few cases of perivascular inflammatory infiltration. The innervation was only sporadically studied and no obvious abnormality was detected in the few specimens observed.

All these changes have been found to varying degrees in the muscles of the non-human primates so far studied but not necessarily all of them in any one muscle.

The different species examined were distributed as follows:

Black ape	*Cynopithecus niger*	1
Capuchin monkey	*Cebus albifrons*	2
Chimpanzee	*Pan troglodytes*	33
Galago	*Galago crassicaudatus*	2
Gorilla	*Gorilla gorilla*	3
Java macaque	*Macaca fascicularis*	12
Drill	*Mandrillus leucophaeus*	1
Orang utan	*Pongo pygmaeus*	6
Pigtail	*Macaca nemestrina*	1
Rhesus macaque	*Macaca mulatta*	53
Squirrel monkey	*Saimiri boliviensis nigriceps* or *Saimiri sciureus*	16
Potto	*Perodicticus potto*	2
Slow loris	*Nycticebus coucang*	3
Spider monkey	*Ateles geoffroyi*	2
Owl monkey	*Aotus trivirgatus*	1
Mandrill	*Mandrillus sphinx*	1

The majority of these were wild-born; a few were born in captivity. The rhesus monkey was the only species in which enough specimens were available to establish that animals born and raised under optimum laboratory *controlled* conditions show a lesser degree of pathology than wild-born imported animals.

The animals used in the studies described were mostly young adults, but 2 of the gorillas were only 5 years old. Unless otherwise stated no obvious clinical symptoms were present. Some of the muscles were obtained from biopsy on the animals and others were obtained at autopsy. In the former the muscle samples were taken routinely from the gastrocnemius and the deltoid muscles; at necropsy the flexors and extensors of the forearm, the quadriceps, pectoral muscles, and the lumbar muscles were taken in addition. The diaphragm and the tongue muscles were also studied in autopsied animals.

Since these observations were made in the course of a study of the histochemical pattern of skeletal muscle in primates, all of the muscle samples were fresh-frozen and cryostat-sectioned, but for this report we used only the preparations stained with routine histological techniques and the results of histochemical reactions will be mentioned only when relevant to the histopathological changes. The largest number of animals studied in any one species has been 53 rhesus monkeys *(Macaca mulatta)*.

II. Macaca mulatta

Individual muscle samples from these animals have shown at least one of the following pathological entities. Necrosis of individual and groups of muscle fibers with and without regeneration of the fibers found in several animals (fig. 2c and g). The regenerative process was evident by the formation of myotubes formed from myoblasts with large pale globular nuclei (fig. 1b and f). In some of the preparations faint cross striations could be detected in the myotubes (fig. 1b). In several specimens the necrotic muscle fibers were being heavily phagocytosed (fig. 3c and g), both types of lesion were accompanied by varying degrees of phagocytosis. In one specimen fine reticular fibers were extending over the degenerating fibers (fig. 3d), in the first stages of connective tissue replacement. Infiltration of phagocytes and other cells (e.g. lymphocytes) was present not only in areas where degeneration was taking place (fig. 3a and b, fig. 2c), but was also found around intra-muscular blood vessels (fig. 1g), between the muscle fasciculi (fig. 1d), and interstitially between the individual fibers themselves (fig. 1h). Fat was also seen to infiltrate the muscle in one obese animal (fig. 1h).

Some specimens showed muscle fibers with vacuoles in the center of the fiber and others narrow splits extending from the periphery of the fiber to wards the center (fig. 3f). More careful observations suggested that such splitting of the fibers might have been preceded by the formation of a central vacuole. In other specimens these splits were widened and invaded by connective tissue (fig. 3e). In some cases, however, the vacuolation of the fiber was accompanied by flocculation and invasion by phagocytes.

Several specimens showed great variation in size and shape of individual muscle fibers (fig. 2a, b, d and e). In some animals nuclear changes were seen; central migration of nuclei (fig. 2e), and rowing of nuclei were also present.

A number of fibers in one specimen showed a concentration of lactic dehydrogenase activity in the center of the fiber indicating that there may be a concentration of mitochondria in this area. It is of interest that King Engel [6] has shown the accumulation of darkly staining rod-shaped particles in the center of muscle fibers in some psychotic patients though he makes no claim that these are mitochondria.

One animal showed infection of the muscle fibers with Sarcosporidia (fig. 4a and b). These have been observed before in many mammals including monkeys [8, 9]. Sarcosporidia are protozoa belonging to the class Sporozoa. These organisms penetrate the muscle fibre and extend along it sometimes reaching a length of 1 or 2 cm, The cytoplasm breaks up into spores while the organism is increasing in size and eventually it becomes composed only of spores enclosed by a cuticle. The spores are usually sickle-shaped.

Since our muscle samples were obtained at autopsy, no further studies were made as to the extent of the infestation throughout the body musculature.

Fig. 1. a Floccular degeneration of a muscle fiber. The floccular masses stain intensely with PAS. *Macaca mulatta.* Periodic Acid-Schiff's stain (×400). *b* Muscle regeneration in an isolated animal; evidence of myotube formation, their cross striations (arrows) could be observed under the microscope by slightly varying the focus. *Macaca mulatta.* Diaphragm Muscle. Hematoxylin and eosin (×320). *c* Interstitial cellular infiltration (arrow) in an area of relative normal muscle fiber architecture. *Macaca mulatta.* Thigh Muscle. Hematoxylin and eosin (×50.4). *d* Infiltration of the perimysium with mononuclear cells, the only muscle change is the inward migration of the nuclei. *Macaca mulatta.* Thigh Muscle. Hematoxylin and eosin (×50.4). *e* From the same lesion as figure 1, flocculation with cellular infiltration and phagocytosis of a muscle fiber. *Macaca mulatta.* Hematoxylin and eosin (×160). *f* PAS positive myotubes in an area of muscle regeneration. *Macaca mulatta.* Diaphragm Muscle. Periodic Acid-Schiff's stain (×480). *g* Infiltration of mononuclear cells in a perivascular position. *Macaca mulatta.* Thigh Muscle. Hematoxylin and eosin (×50.4). *h* Extensive infiltration by adipose tissue in an isolated animal suffering from obesity. *Macaca mulatta.* Diaphragm Muscle. Oil red O stain for fat (×50.4).

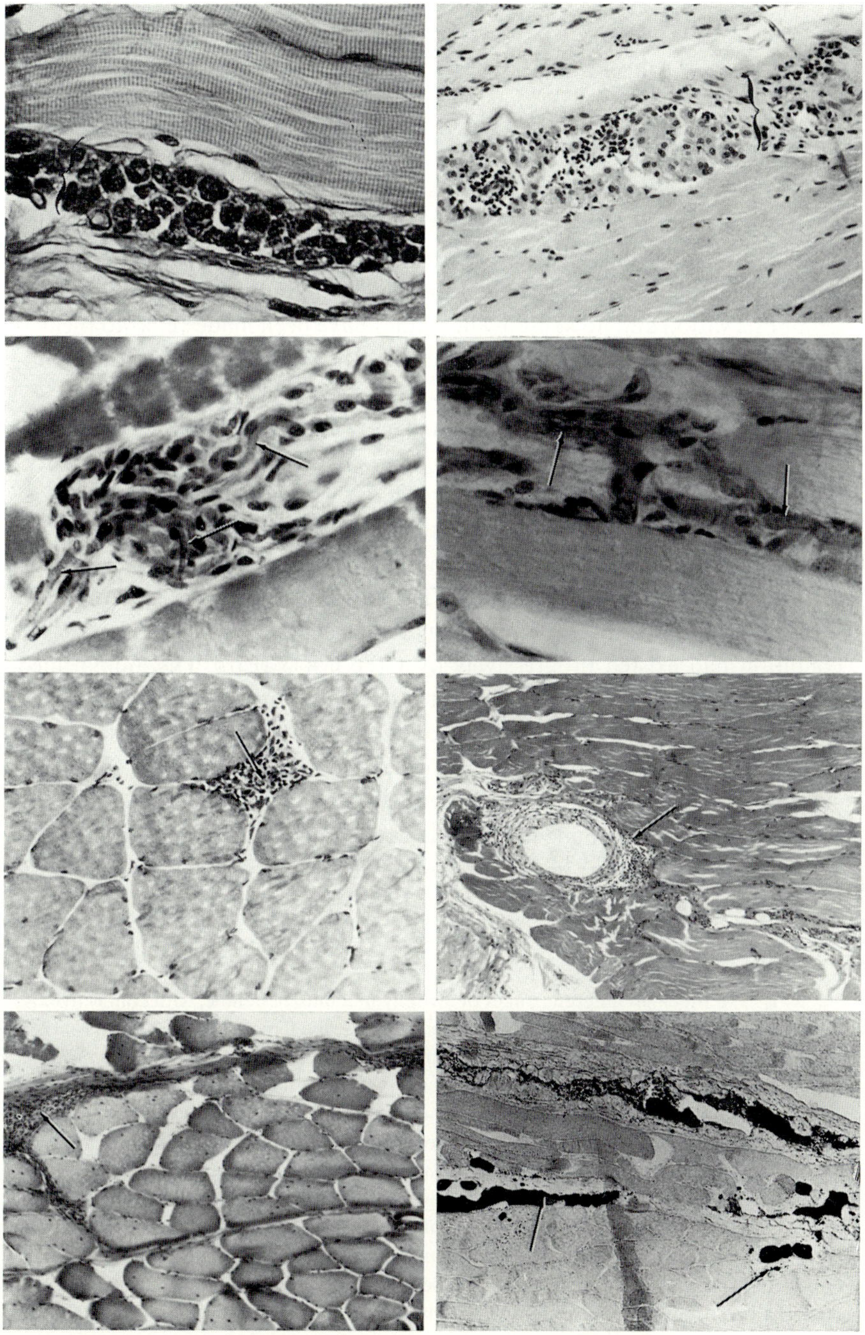

a e b f c g d h

III. Gorilla

Among the other species of non-human primates one of the 2 normal gorillas examined also showed the presence of Sarcosporidia in the biopsy materials of gastrocnemius muscle.

The sarcocysts appeared inside single muscle fibers in several fasciculi while the rest of the fibers retained normal appearance (fig. 5e). In one sarcocyst besides the presence of the sickle-shaped spores, droplets stained with Sudan black could be observed within the cuticle (fig. 5f). This is a very recent observation and no subsequent studies have been made yet as to the distribution of Sarcosporidia throughout the other muscles or as to the evolution of the lesions.

The same gorilla showed other changes in the same muscle such as heavy fatty infiltration of the perimysium (fig. 5a) and a few necrotic muscle fibers which were undergoing phagocytosis (fig. 5c). In both animals isolated foci of interstitial cell infiltration (fig. 5b) and fat deposition were seen.

A third gorilla studied was a 7-year-old female. At death she had marked atrophy of the limbs on the left side. The clinical diagnosis of her condition was that of infantile spastic hemiplegia and intractable cerebral seizures. This condition had existed progressively over the last 5 years. The animal died during surgical procedures unrelated to her condition [10]. Most of the muscles observed in the affected limbs showed extensive changes and even those fasciculi which were architecturally normal presented fibers with central nuclei (fig. 5d). The most common changes observed in the affected extremities were small fibers with rounded contours with complete loss of the normal polygonal architecture (fig. 5g). There was also extensive vacuolar degeneration

---

Fig. 2. a Fibers in different stages of atrophy in a muscle bundle, with connective tissue replacement. *Macaca mulatta.* Gastrocnemius muscle. Hematoxylin and eosin (× 126). *b* Massive atrophy with connective tissue replacement in several muscle bundles. *Macaca mulatta.* Thigh muscle. Hematoxylin and eosin (× 128). *c* Phagocytosis of a disintegrating muscle fiber. *Macaca mulatta.* Lumbar muscle. Alkaline phosphatase with hematoxylin and eosin (× 480). *d* An elongated section showing several atrophic fibers with internal nuclei and connective tissue replacement. *Macaca mulatta.* Thigh muscle. Hematoxylin and eosin (× 128). *e* In the same specimen as above, in a more transverse section, dimensional polymorphism, vestige of fiber splitting, proliferating endomysium and migration of nuclei from the subsarcolemmal position, *Macaca mulatta.* Thigh muscle. Hematoxylin and eosin (× 128). *f* Myotubes (right arrow) in a muscle fiber undergoing simultaneous necrosis and regeneration. *Macaca mulatta.* Gomori's technique for alkaline phosphatase (× 1,200).

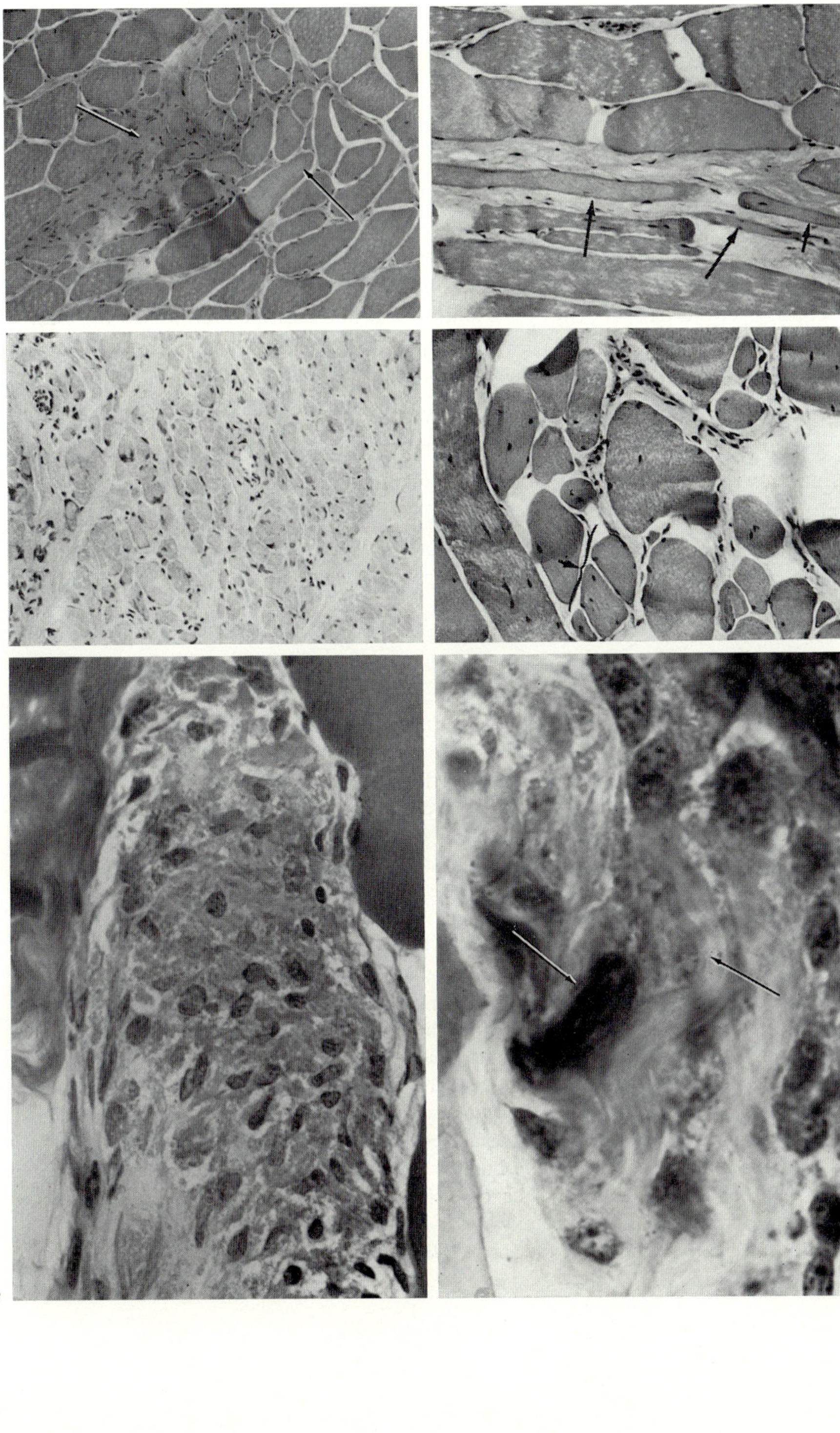

a *d*

b *e*

c *f*

without cellular infiltration or phagocytosis (fig. 6a). The contents of the fibers were in various stages of disintegration, some with the myoplasm contracted away from the sarcolemma, and even some empty sarcolemmal sheaths could be observed (fig. 5h).

IV. Chimpanzee

In muscle taken from 33 chimpanzees, the whole range of pathological changes could be seen, though not all in any one animal, and mostly in isolated foci. With the exception of one, these animals showed no clinical symptoms of muscle disease. In one case with absence of clinical symptoms, in biopsy of the deltoid muscle extensive changes were found in the form of great variation in muscle fiber diameter (fig. 6b), and there were a number of large swollen fibers interspersed with groups of small angular fibers (fig. 6b). In some of the fascicles the swollen fibers showed the beginnings of formation of fissures (fig. 6c). Sporadic necrotic muscle fibers were also found in this specimen, but central migration of muscle fiber nuclei was found in nearly all the fibers.

Fig. 3. a One muscle fiber with a 'bursting' of necrosis and phagocytosis. *Macaca mulatta.* Diaphragm muscle. Hematoxylin and eosin (×320). *b* A single muscle fiber undergoing near total destruction, at center necrosis (lower arrow), above it remnants of the muscle fiber (arrows) which can be seen also at the lower left angle. Artificially produced contraction bands at the top fiber and ATP positive blood vessel to the right. *Macaca mulatta.* Lumbar muscle. Gomori's technique for alkaline phosphatase (×224). *c* Centrally located nuclei, some atrophic fibers and, in one fiber, simultaneous hyaline and floccular degeneration (arrows). *Macaca mulatta.* Diaphragm muscle. Hematoxylin and eosin (×128). *d* In the same area as the above section, reticular fibers can be seen at right arrow in the site of connective tissue invasion of a muscle fiber still partly intact (left arrow). *Macaca mulatta.* Diaphragm muscle. Wilder's Reticulum stain (×192). *e* A muscle bundle still maintaining some normal architecture while one of the fibers is undergoing hyalinization and phagocytosis (arrow), two other fibers are undergoing invasion by the proliferating endomysium; migrated nuclei are observed in the three affected fibers. Gomori's technique for alkaline phosphatases was applied which shows a positive reaction in the capillaries and nuclei present. The horseshoe appearance of the connective tissue invasion is not common in the authors' experience. *Macaca mulatta.* Lumbar muscle. Substrate adenosine 5′triphosphate (×224). *f* Splitting of a fiber and phagocytosis. *Macaca mulatta.* Lumbar muscle. Gomori's alkaline phosphatase plus hematoxylin and eosin (+480). *g* Floccular degeneration in a single fiber in the same muscle as above. *Macaca mulatta.* Lumbar muscle. Hematoxylin and eosin (×160).

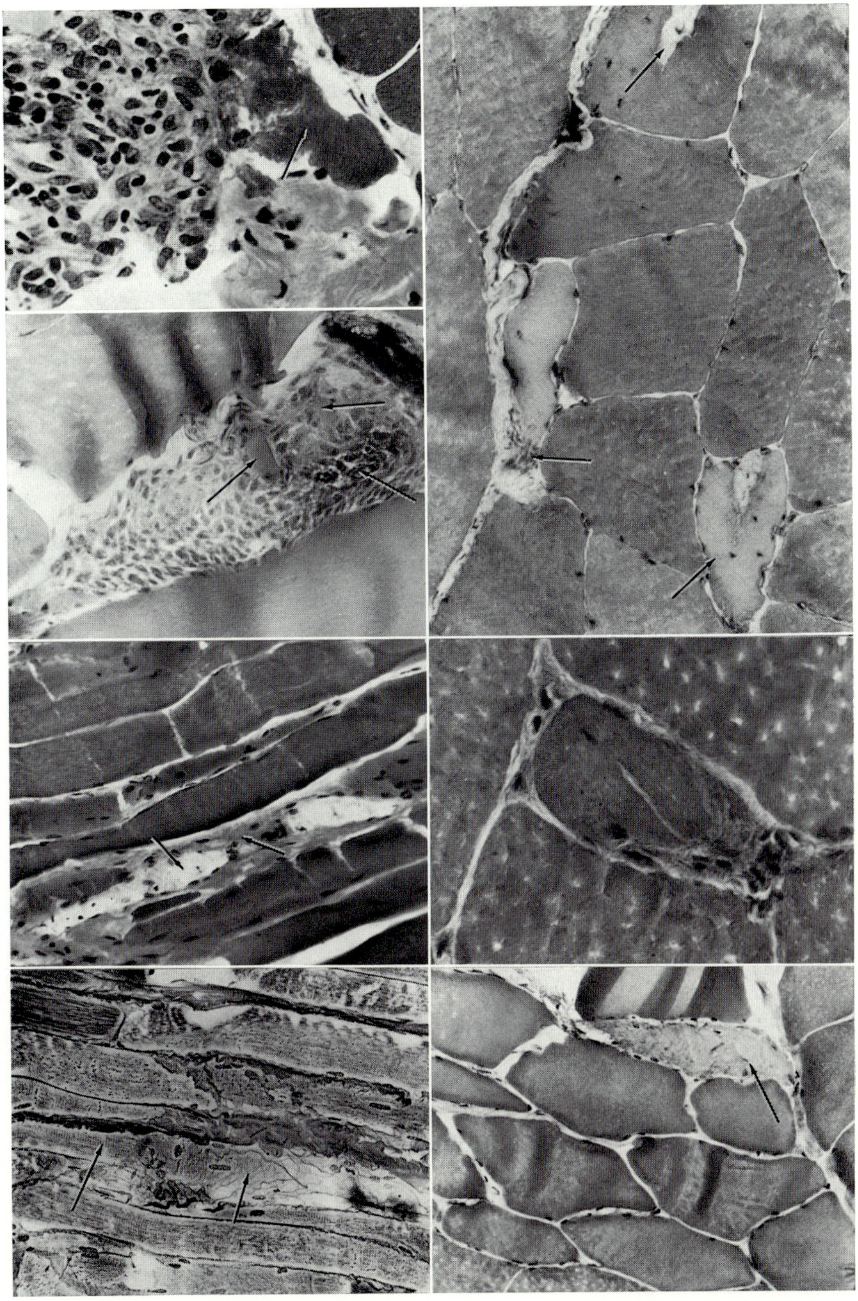

a b c d e f g

One of the chimpanzees studied in this group developed weakness of the hind legs which became progressively worse over some months and then slowly began to improve. Biopsy samples were taken from the vastus lateralis and the gastrocnemius muscles of the affected extremities.

The overall architecture of the muscle showed little change. With standard histological techniques some swollen fibers and a rare angular atrophic fiber were found (fig. 6d, 7d), but many of the fibers showed a zonal differential staining. Some other fibers were deeply fissured with endomysium penetration (fig. 7f). With histochemical techniques especially those for hexokinase and the oxidative enzymes, e. g. lactic dehydrogenase, succinic dehydrogenase and diphosphopyridine nucleotide diaphorase (fig. 6e and f), it was seen that the reaction in the zonally differentiated fibers was predominantly located in a peripheral ring around the fiber with the center giving little or no reaction (fig. 7a, b and c). There was obvious loss of normal structure in the center of many of these affected fibers (fig. 7b and c), some of which also had a centrally located nucleus. This kind of structural changes of the muscle fiber giving it a 'target' or 'targetoid' appearance is considered as a sign of denervation in some human neuromyopathies [11]. There is also a rare human muscle disease called Central Core disease which presents similar zonal irregularities in the individual muscle fibers [12]. This disease is of hereditary origin and clinically is characterized by a nonprogressive muscular weakness.

Fig. 4. a A halo of myoplasm surrounding a sarcocyst, no significant changes are observed in the adjacent fibers. *Macaca mulatta*. Arm muscle. Hematoxylin and eosin (× 50). *b* Enlarged view of the cyst; at upper arrow, the cuticle; the lower arrow points to thin trabeculae which give lobulated appearance to the cyst. Individual spores can be easily identified. *Macaca mulatta*. Arm muscle. Trichrome stain (× 1,000). *c* Interstitial cellular proliferation in an otherwise normal muscle. *Macaca fascicularis*. Hematoxylin and eosin (× 80). *d* Very atrophic fibers with great increase of muscle nuclei. *Saimiri sciureus*. Hematoxylin and eosin (× 80). *e* Necrosis and phagocytosis of an already atrophic fiber. The fenestrated appearance of the adjacent fibers (arrow) is due to ice crystals formed during the technical manipulation. *Cebus apella*. Gastrocnemius muscle. Trichrome stain (× 400). *f* Several very atrophic fibers among normal muscle fibers in the same fascicle. *Mandrillus leucophaeus*. Gastrocnemius muscle. Sudan black B stain (× 160). *g* The majority of the fibers in each fascicle show atrophy with thickening of the interstitial connective tissue. No evidence of inflammatory processes. *Mandrillus leucophaeus*. Gastrocnemius muscle. Hematoxylin and eosin (× 400). *h* An isolated muscle fiber heavily stained with Sudan black B stain for fats; several atrophic fibers are also present. *Mandrillus leucophaeus*. Gastrocnemius muscle. Trichrome stain (× 63).

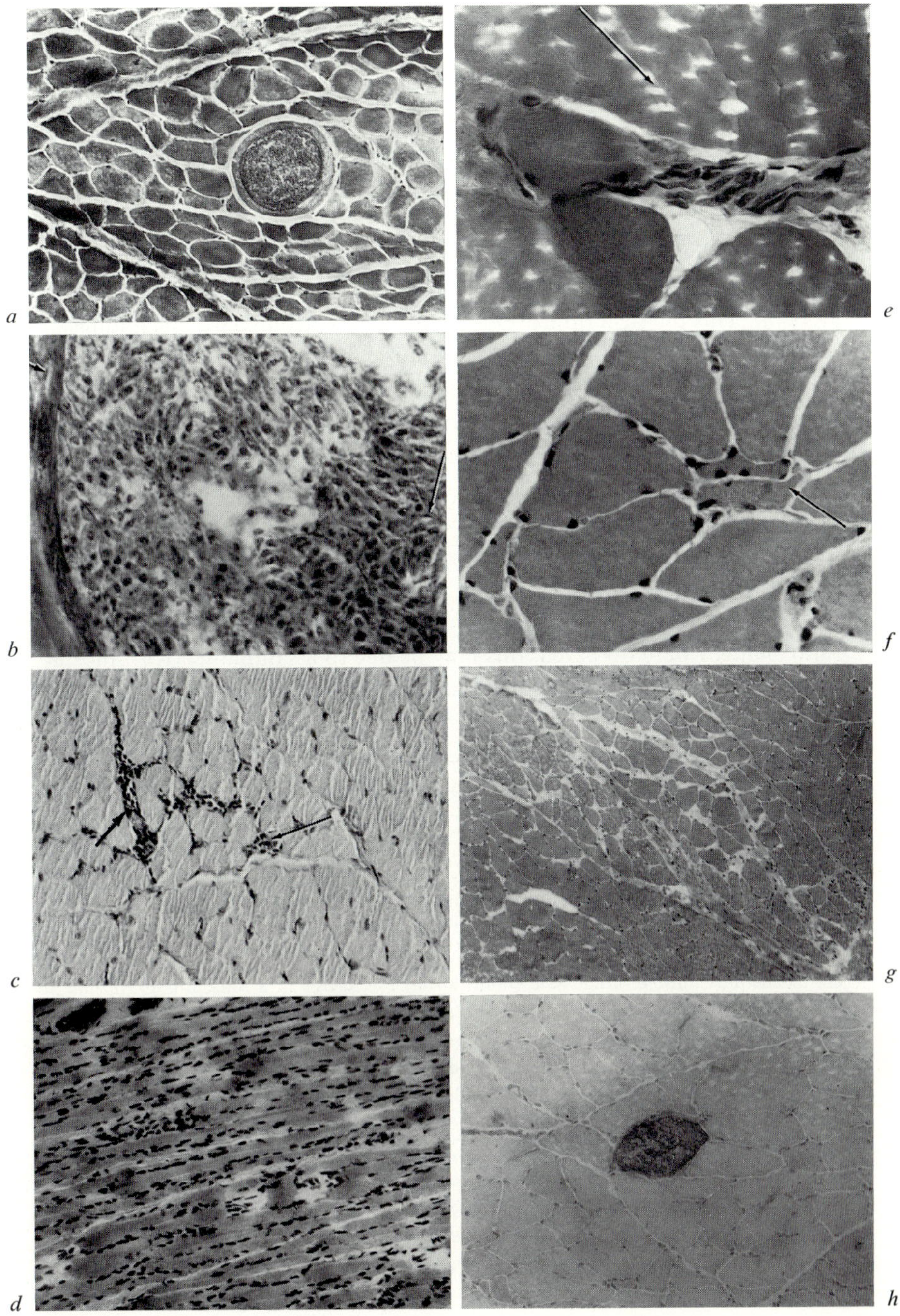
a
e
b
f
c
g
d
h

Sixty days after the original biopsy, in an attempt to make a more complete evaluation, electromyographic studies were made of the affected muscles by Dr. JOHN BASMAJIAN of the Department of Physical Medicine of Emory University, and a wider number of muscles were biopsied. The electromyograph suggested that the animal was suffering from a peripheral polyneuritis of unknown etiology although some evidence suggests a traumatic origin.

This second series of biopsies did not show any zonally differentiated fibers and it is of interest that by this time the clinical symptoms showed regression. This physical improvement plus the electromyographic findings indicated that the condition was of neuromuscular origin and in its final stages of resolution. No necrosis nor cellular infiltration or connective tissue proliferation were found in the biopsies of this animal.

Recently DUBOWITZ [13] has observed target and targetoid fibers during experiments of cross inner vation, and he contends that these 'are a feature of renervation and not of denervation' which would explain the improvement of the clinical symptoms and the simultaneous disappearance of the target fibers in our animal case.

Although electron-microscopic studies were attempted, no ultrastructural changes were observed, this may be due to a sampling error, since not all the fibers had 'targets', and the majority of the fibers had a normal staining pattern even with histochemical techniques,

We have no explanation for the appearance of fissures in muscle fibers of two chimpanzees as described above; some publications of recent years

---

Fig. 5. a Fatty infiltration of the perimysium, it is not unusual to find some fat in that position in normal muscle. *Gorilla gorilla.* Gastrocnemius muscle. Trichrome stain (×63). Interstitial cellular infiltration limited to the fibers of a muscle bundle. *Gorilla gorilla.* Gastrocnemius muscle. Trichrome stain (×160). *c* Necrosis in a muscle fiber. *Gorilla gorilla.* Gastrocnemius muscle. Trichrome stain (×1,000). *d* Centrally located nuclei in an otherwise normal muscle, in longitudinal section. *Gorilla gorilla.* Arm muscle. Trichrome stain (×160). *e* Sarcosporidia in a single fiber at arrow. *Gorilla gorilla.* Gastrocnemius muscle. Trichrome stain (×160). *f* Sarcosporidia in a muscle fiber stained for fat. Fat droplets inside sarcocyst; the sarcosporidia were stained with hematoxylin (arrow). The host muscle fiber forms a ring around the cyst. *Gorilla gorilla.* Gastrocnemius muscle. Sudan black Bystain (×1,000). *g* In a cross-section all the muscle fibers show complete loss of the usual polyzonal architecture. *Gorilla gorilla.* Thigh muscle. Hematoxylin and eosin (×160). *h* Extensive vacuolar degeneration without cellular infiltration or phagocytosis. Empty sheaths can be seen, some with remnants of myoplasm and nuclei, others at different stages of loss of contents. *Gorilla gorilla.* Thigh muscle. Hematoxylin and eosin (×63).

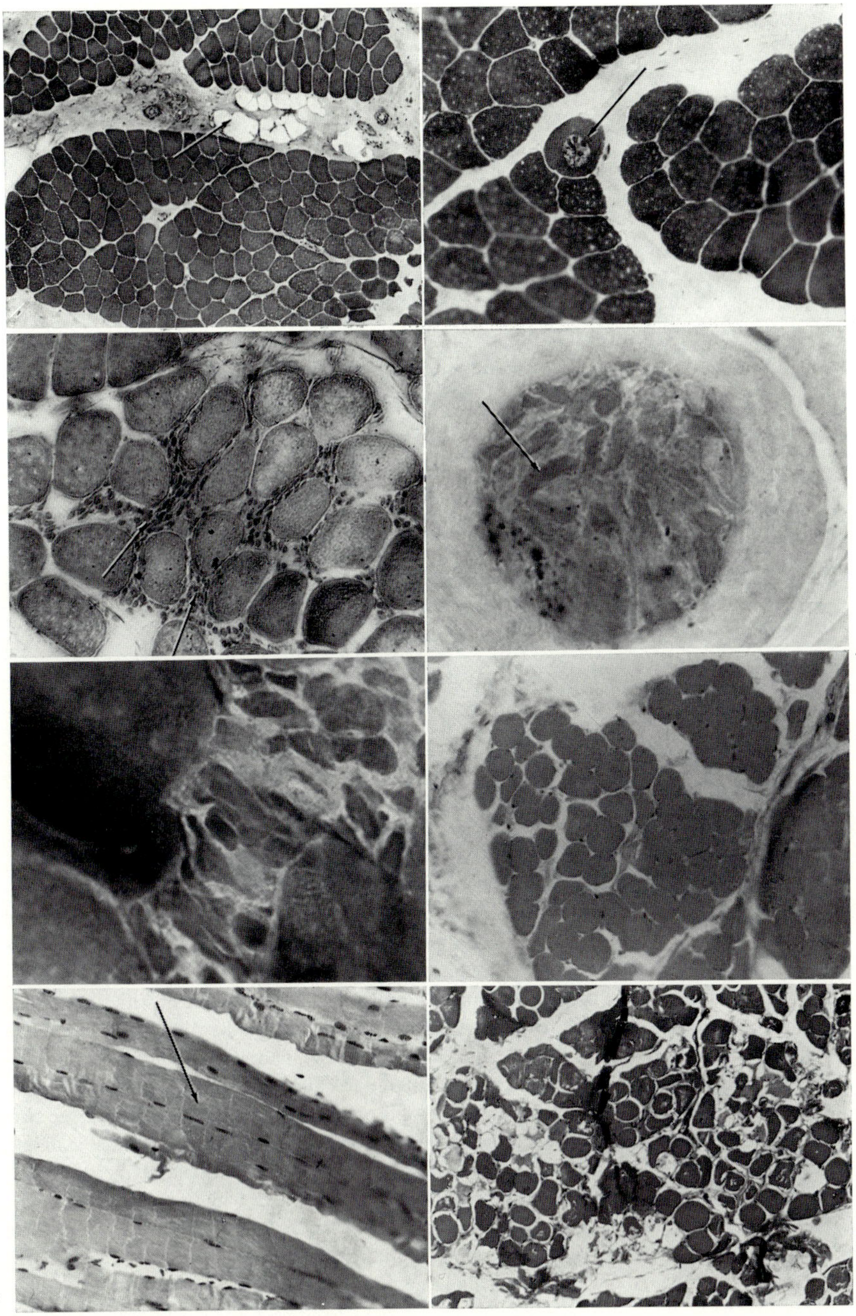

a e b f c g d h

describe fiber splitting as a result of hypertrophy following periodic strenuous work [14, 15]; however, there is no record of unusual muscular exercise in our animals.

V. Java Macaques

The initial studies by the authors were carried out on Java macaques *(Macaca fascicularis)* and it was the changes found in the muscles of these animals which led to a more extensive study of primate muscle. One of the females examined suddenly became very obese after 2 years of captivity and at autopsy following euthanasia, widespread fatty infiltration was observed throughout all the organs including the muscles. A similar type of fatty infiltration was found in 2 female *Macaca mulatta* confined in separate complete isolation for a year. In all 3 animals the fat invaded not only the endomysium, but the individual fibers also showed increased deposition of fine fatty droplets well demonstrated with Oil red 'O' and Sudan black stains. We have also received unconfirmed reports from Africa of similar changes in African green monkeys in captivity.

The other lesions seen in the muscles of the Java monkeys were focal such as interstitial cell infiltration (fig. 4c) and sporadic muscle atrophy. Central migration of muscle nuclei was also common.

Fig. 6. a Muscle fibers in several stages of vacuolar degeneration, swelling, and fragmentation. Same specimen as previous. *Gorilla gorilla.* Thigh muscle. Hematoxylin and eosin (× 160). *b* Central nuclei, groups of atrophic fibers (arrows) and swollen fibers. The muscle sample was taken from a specimen which has been receiving steroids and other immunosupressive drugs in the previous three months. Radiation treatment was also applied in the previous 30 days. *Pan troglodytes.* Deltoid muscle. Periodic Acid-Schiff's reaction (× 160). *c* Same specimen as above. Central muscle nuclei, fiber necrosis and 'splitting' can be observed in several fibers. Gomori's technique for alkaline phosphatase. *Pan troglodytes.* Deltoid muscle. Substrate adenosine diphosphate (× 400). *d* Difference of fiber size, some small angular fibers, others large and rounded, several showing differential staining. *Pan troglodytes.* Thigh muscle. Trichrome stain (× 75.6). *e* Several targetoid fibers (arrows) from the leg which showed muscular weakness in a chimpanzee with peripheral polyneuritis. *Pan troglodytes.* Thigh muscle. Reaction for DPN diaphorase (× 75.6). *f* Several muscle fibers show zonal differential staining described in the literature as 'targetoid' fibers. Same specimen as above. *Pan troglodytes.* Thigh muscle. Reaction for DPN diaphorase (× 128).

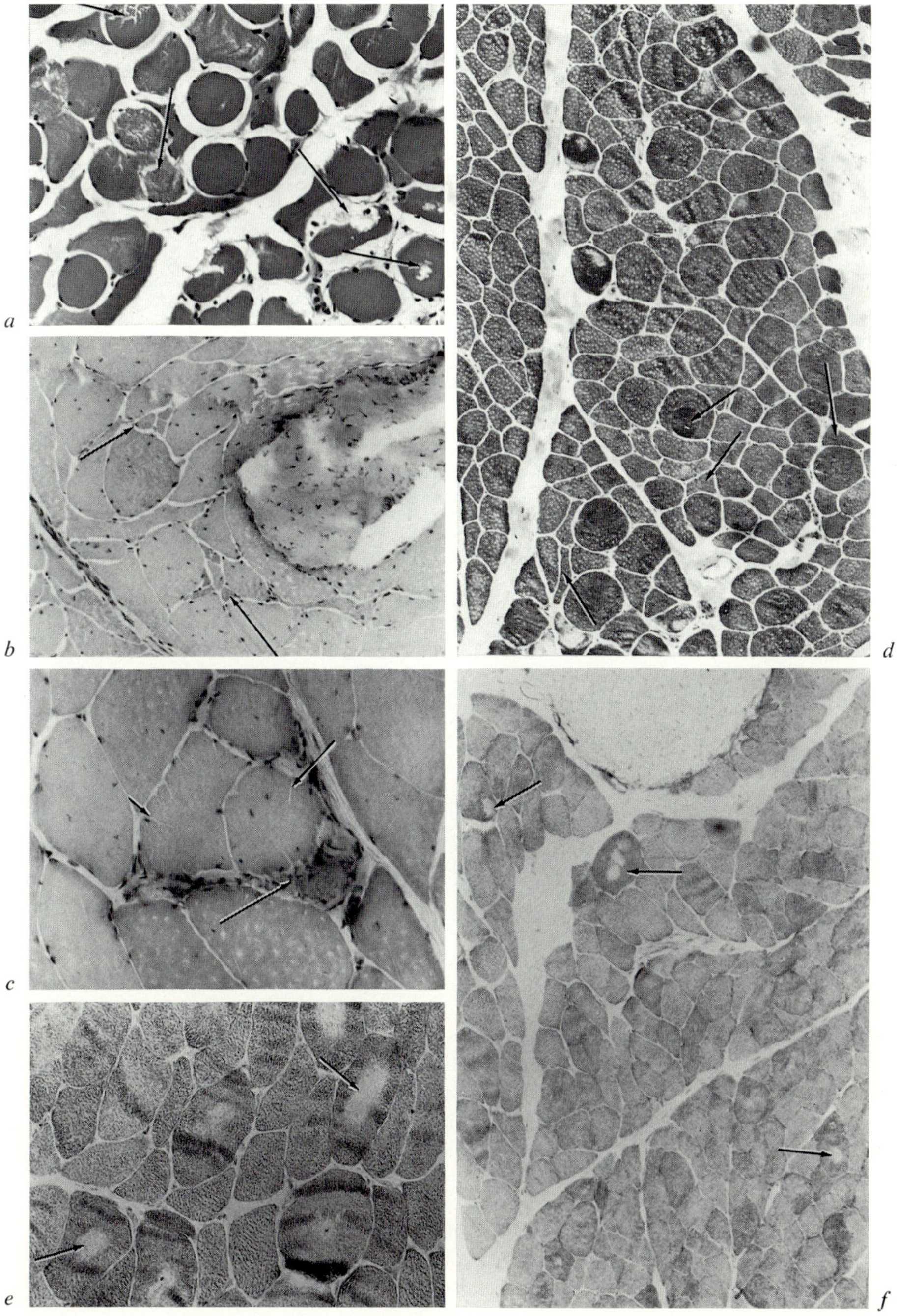

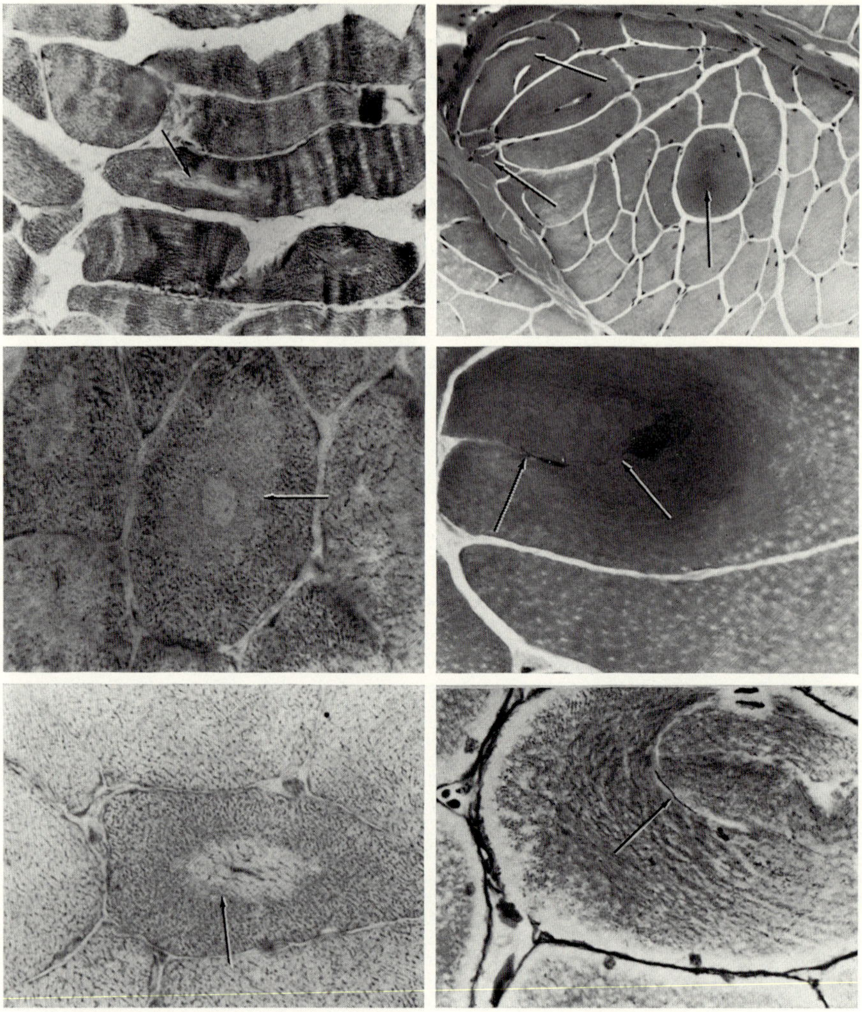

a d b e c f

Fig. 7. a Zonal differences in 'staining' (arrow) in a muscle fiber longitudinally sectioned. From a chimpanzee with polyneuritis in its terminal stages of resolution. The dark transversal bands are caused by contraction artifacts due to manipulation. *Pan troglodytes.* Thigh muscle. DPN diaphorase (× 128). *b* Three concentric zones diferentially 'staining' in a muscle fiber, the adjacent fibers show similar patterns. From the same specimen as above. *Pan troglodytes*. Thigh muscle. Reaction for DPN diaphorase (× 320). *c* From the same area as the above section, the muscle fibers show concentric differentiation when 'stained' for hexokinase demonstration. *Pan troglodytes*. Thigh muscle. Hexokinase reaction (× 320). *d* One large rounded swollen fiber and several splitting fibers occurring in the same muscle fascicle. *Pan troglodytes*. Thigh muscle. Hematoxylin and eosin (× 128). *e* Muscle fiber with a bizarre

VI. Squirrel Monkeys

In the squirrel monkey *(Saimiri sciureus* and *Saimiri boliviensis nigriceps)* incidence of muscle changes was so extensive that it is difficult to accept that many of these affected animals could function competently; other organs often show extensive pathology also. This extreme condition may be the reason why these animals are so delicate and have such a large death rate when attempts are made to acclimatize them to laboratory conditions; alternatively the muscle changes may indicate a past or present infection with a viral or other agent which makes the animals specially sensitive to stress. No other species of primate studied so far show such frequent and extensive myopathic changes.

The most striking lesion is the great relative increase of muscle nuclei (fig. 4d), their migration into the center of the fibers and their arrangement in rows, so that chains of nuclei may be seen extending for some considerable distance along the fibers. Perivascular and interstitial cellular infiltration was found in many muscle specimens, in some quite extensive.

Flocculation of the fibers occurred with or without phagocytosis; infiltration of the endomysium with fat was also observed and less often variation of muscle fiber size in individual fasciculi was seen.

VII. Celebes Black Ape

Only one Celebes black ape *(Cynopithecus niger)* was available for this study and the muscle was collected *post-mortem*. A number of the muscle fibers showed splitting and there was great disparity in muscle fiber diameter, due to some fibers being swollen; a number of others showed apparent splitting into smaller fibers. There was a moderate degree of perivascular cellular infiltration and occasional fibers showed central nuclei, in some cases with nuclear rowing.

---

structure in an apparent process of splitting, the fissure can be followed at both arrows. Same specimen as above. *Pan troglodytes*. Thigh muscle. Hematoxylin and eosin (×320). *f* In the same specimen as above, stained with Wilder's technique for reticulum fibers, a vestige of endomysium formation can be seen at arrow. *Pan troglodytes*. Thigh muscle. Wilder's Reticulum stain (×320).

VIII. Drill

One drill *(Mandrillus leucophaeus)*, originally a magnificent adult male specimen with a weight of 31 kg, developed a rapidly crippling infectious arthritis which remained in a chronic condition. The disease was first observed in June 1967, and in November 1969, the animal was found cachectic and was euthanized due to its emaciated condition (it weighed only 11.5 kg).

The muscles of the hind legs which had been first affected showed a generalized appearance of atrophy of entire fascicles (fig. 4f) or of small groups of fibers amidst normal muscle fibers, this was accompanied by interstitial connective tissue proliferation (fig. 4f). There occurred sporadically isolated large rounded fibers with a granular appearance which were heavily stained by fat stains (fig. 4h). There was little or no cellular infiltration.

It is difficult to assess how much of the lesions observed in the muscles were of a primary nature and how much were secondary to the reduced joint mobility induced by the disease and to the inanition and cachexia of the final stages.

IX. Other Groups

Other non-human primates so far examined were the capuchin monkey *(Cebus albifrons* and *Cebus apella)*, the galago *(Galago crassicaudatus)*, the pigtail macaque *(Macaca nemestrina)*, the potto *(Perodictius potto)*, the slow loris *(Nycticebus coucang)*, the spider monkey *(Ateles geoffroyi)*, the owl monkey *(Aotus trivirgatus)* and the mandrill *(Mandrillus sphinx)*. No extensive or serious muscle changes were found in these animals but they all showed sporadic isolated atrophic or necrotic muscle fibers and areas of interstitial cellular infiltration. From all these animals, however, only a limited amount of muscle tissue was available and there was a great variation in the lesions from one animal to another.

X. Discussion

With the large incidence of changes observed in the relatively small group of primates studied, it seems possible that the chance of finding congenital neuromuscular diseases in laboratory-born animals should be good. In the wild, such handicapped animals would have small chance of survival.

Even if they reached dealers they would not be suitable for export since there is a logical selection of specimens most likely to survive.

There is no evidence of a similar extensive incidence of sub-clinical muscle diseases in humans, at least in the Western countries. Such changes as have been noted have tended to occur in muscle biopsy specimens of relatives or carriers of hereditary myopathies. In old and senescent humans degenerative muscle changes are much more common but their effects may be seen clinically.

A few years ago, the authors made a study of a group of young, healthy medical students as controls for studies on human muscular dystrophy. All except one of the specimens showed a normal histological and histochemical picture. The exception showed extensive muscle changes including a number of those described in this chapter but there were no clinical symptoms. It was not established whether this student was related to any patient with a hereditary muscular disease.

We have also made studies of normal healthy rats, mice, rabbits, and guinea pigs; these were laboratory animals raised for generations under optimum control conditions. We have never observed any myopathic lesions in these animals in the absence of clinical symptoms or experimental provocation.

It is possible that some intrinsic myopathic factor may be responsible for some of the observed changes in the subhuman primates examined; they could be expressions of very early stages of a progressive disease, or a disease with a low impairment of the motor function but the majority of the cases are most likely due to traumatic damage, inadequate caging conditions, parasites, inadequate dietary conditions, or infectious diseases which have relapsed below the clinical level.

In our studies it was rare to find subhuman primates in which clinical symptoms of muscle disease were present. It is obvious that an animal with muscle weakness would not survive long in the wild.

In the absence of apparent extrinsic causes, it is difficult to establish whether focal muscle changes in a biopsy specimen are due to a specific secondary myopathy since there are no unequivocal characteristic lesions that can be detected, at least with the available methods.

We have often observed hemorrhage in muscle biopsies of human patients and monkeys; this is clearly due to manipulation prior to the biopsy. Recently a methodical study of later stages following muscle injury due to external trauma has been made and localized lesions of the muscle fibers were described similar to many of those observed in our animals [6].

Chronic malnutrition produces generalized atrophy of the muscle fibers but this is reversible when adequate diet is re-established. But some specific dietary deficiencies can produce profound pathological changes in the muscle. Of these factors the best studied is vitamin E deficiency, but other nutritional factors can also produce changes [17].

There are many parasites which can invade and damage the muscle fibers [18] and as in the case of trichinosis non-specific lesions can be widespread without evidence of the offending organism [19]. We have not yet actually observed trichina in any of our specimens.

During the course of some viral infectious diseases, occurrence of muscular lesions has also been described [20] and muscle damage has been demonstrated in rabbits after injection with sublethal doses of diphtheria toxin, and some of the changes were not completely reversible [21].

The widespread incidence of these pathological processes in subhuman primate muscle is of considerable interest and the assessment of the effects of captivity under control conditions on the animal as a whole and on the skeletal musculature in particular, merits further extensive study.

Acknowledgment

The studies conducted at the Yerkes Center reported in this chapter were carried out with the help of the National Aeronautics and Space Agency grant number NGR-11-001-016 and grant number RR-00165 of the National Institutes of Health.

We appreciate the help of Dr. Harold McClure, veterinary pathologist to the Center, for performing the autopsies on the animals used in this work and Drs. Michale Keeling and Wesley Bonner, Center veterinarians, for performing the biopsies.

This work was carried out with the technical help of Becky Floyd and Sonja Norton.

References

1 Michelson, A. M.; Russell, E. S., and Harman, J. H.: Dystrophia Muscularis; a hereditary primary myopathy in the house mouse. Proc. nat. Acad. Sci., Wash. *41:* 1079–1084 (1955).

2 Golarz, M. N. and Bourne, G. H.: Some histochemical observations on the muscles of mice with hereditary muscular dystrophy. Acta anat. *43:* 193–203 (1960).

3 Mason, K. E. and Telford, L. R.: Some manifestations of vitamin E deficiency in the monkey. Arch. Path. *43:* 363 (1947).

4 West, W. E.: Muscular dystrophy of vitamin E deficiency; in Bourne and Golarz Muscular dystrophy in man and animals (Karger, Basel 1963).

5 Dinning, J. S.: Vitamin E deficiency in the monkey. Rev. canad. Biol. *21:* 501–503 (1962).
6 Engel, W. K. and Meltzer, H.: Histochemical abnormalities of skeletal muscle in patients with acute psychoses. Science *68:* 273–276 (1970).
7 Bourne, G.: Sarcosporidia. J. roy. Soc. west. Austr. *19:* 1–8 (1932).
8 Mandour, A. M.: A report on the presence of Sarcocystis in different hosts. Trans. roy. Soc. trop. Med. Hyg. *58:* 287 (1964).
9 Zaman, V.: Sarcocystis sp. in the slow loris, *Nycticebus coucang*. Trans. roy. Soc. trop. Med. Hyg. *64:* 189–190 (1970).
10 Keener, E. B. and Guilloud, N.: Investigation of epileptic seizures in an infantile spastic hemiplegic gorilla. Proc. 2nd Int. Congr. Primat. *3:* 55–58 (Karger, Basel 1969).
11 Engel, W. K.: Muscle target fibers, a newly recognized sign of denervation. Nature, Lond. *191:* 389–390 (1961).
12 Shy, M. G.; Engel, W. K., and Wanko, T.: Central core disease; a myofibrillary and mitochondrial abnormality of muscle. Ann. intern. Med. *56:* 511–520 (1962).
13 Dubowitz, V.: Discussion of target fibers. Structural and cytochemical characteristics; in Milhorat Exploratory concepts in muscular dystrophy and related disorders. Excerpta med. Foundation, New York 1967, pp. 266–267.
14 Linge, B. van: The response of muscle to strenuous exercise. J. Bone Jt. Surg. *44B:* 71 (1962).
15 Reitsma, W.: Some structural changes in skeletal muscles of the rat after intensive training. Acta morph. neerl. scand. *3:* 229–246 (1970).
16 Hathaway, P. W.; Dahl, D. S., and Engel, W. K.: Myopathic changes produced by local trauma. Arch. Neurol., Chicago *21:* 355–357 (1969).
17 Mason, K. E.: Effect of nutritional deficiencies upon muscle; in Bourne Structure and function of muscle, vol. 3, pp. 171–207 (Academic Press, New York 1960).
18 Graham, P. C. C.: Parasitic infections; in Bourne The structure and function of muscle, vol. 3, pp. 109 (Academic Press, New York 1960).
19 Drachman, D. A. and Tuncbay, T. O.: The remote myopathy of trichinosis. Neurology *15:* 1127–1135 (1965).
20 Seway, L. C.; Enzinger, F., and Hines, H. M.: Production of dystrophic lesions in skeletal muscles of dutch rabbit by diphtheria toxin. Arch. Path. *66:* 344–351 (1958).
21 Field, E. J.: Virus infections; in Bourne The structure and function of muscle, vol. 3, pp. 85 (Academic Press, New York 1960).

Authors' address: Dr. G. H. Bourne and Dr. M. Nelly Golarz de Bourne, Yerkes Regional Primate Research Center, Emory University, *Atlanta, GA 30322* (USA)

Pathology of Simian Primates, Part I, pp. 542–638 (Karger, Basel 1972)

Diseases of the Nervous System

K. R. Brizzee

Tulane University, Delta Regional Primate Research Center, Covington, La.

Contents

Foreword by the Editor 543
Introduction 545
Congenital Abnormalities 545
  Naturally Occurring 545
  Experimentally Induced 548
    Vincristine 548
    Thalidomide 548
    Hypoxia 548
Vascular Diseases 551
Demyelinating Diseases 552
  Experimental 552
    Cyanide Poisoning 552
    Sodium Azide Poisoning 554
    Barbiturate Poisoning 554
    Hypoxia 555
    Carbon Monoxide 555
    Allergic Encephalomyelitis 556
    Lead Poisoning 558
  Naturally Occurring 558
    Confluent Leukoencephalosis and Perivascular Myelosis 558
    Isolated Papillomacular Degeneration 559
Infectious Diseases 560
  Viral 560
    Encephalitides 560
    B Virus Infection 567
    Acute Polioencephalitis 569
    Diffuse Non-Suppurative Meningitides and Meningoencephalitides 575
    Neurotropic Newcastle Infection 576
    Neurotropic Dengue 576

Lymphocytic Choriomeningitis 577
Coxsackie and Echo Viruses 577
Equine Encephalitis 580
Venezuelan Equine Encephalitis 581
Encephalitis Affecting Basal Ganglia 582
Austrian Tick-Borne Encephalitis 583
Non-Suppurative Meningitis 584
Infectious Diseases-Bacterial 587
Suppurative Meningitis 587
Tuberculosis 591
Anthrax 599
Encephalopathies Related to Dysenteric Syndromes 600
Fungal Infections 602
Torulosis (Cryptococcus) 602
Other Fungal Infections 602
Degenerative Diseases 602
Diseases of Cerebellum 602
Diseases of Basal Ganglia 604
Manganese Poisoning 604
Carbon Disulfide Poisoning 606
Spontaneous Degeneration 606
Widespread Spontaneous Degeneration of CNS 607
Kuru 609
Pigmentation and Siderosis 610
Organic Arsenical Compounds 613
Acute Amaurotic Epilepsy 613
Acute Amaurosis without Epilepsy 614
Nutritional Deficiencies 614
Experimental 614
Spontaneous Occurrence 616
Peripheral Neuropathy 617
Parasitic Infections 617
Epilepsies 618
Tumors 619
Pathological Effects of Intrathecal Injection of Spinal Anesthetic and Vasoconstrictor Drugs 619
Pathological Effects of Ionizing Radiation 620
Pathological Effects of Hyperthermia 631
References 631

Foreword by the Editor

The commonest form of nervous involvement likely to be encountered in simians of different species is paraplegia. This may be due to the following general causes:

(1) traumatic; (2) nutritional or metabolic; (3) helminth cysts or larvae; (4) non-specific degeneration; (5) infectious causes. *Trauma*. Newly imported animals frequently suffer from traumatic lesions, which may have been acquired (a) in the wild before capture; (b) at the time of capture; or (c) after capture during transit or during the period of acclimatisation to the new environment of captivity. Newly captured monkeys are often found to have old healed fractures, and it seems that they do frequently miss their hold in the wild and suffer fractures in consequence. Injuries sustained during capture usually arise from missiles used against them and these may vary from small arrows or darts to buckshot. A shot pressing on the spinal cord will easily show on X-ray.

The commonest cause of locomotor impairment in captivity arises from two causes: (1) metabolic bone disease especially in young animals (*vide* the chapter on Diseases of the Skeleton); (2) as a complication of dysentery (*vide* this chapter); (3) motor neuron degeneration following deficiencies of vitamin A or vitamins of the B group; (4) osteoporosis arising from confinement in a cage that is too small; (5) a combination of these factors; (6) simple constipation or bowel infection. It should not be difficult to eliminate these as possible causes of the trouble.

Outside of these causes, the possibility should be considered that the animal is suffering from van Bogaert's disease (leuco-encephalo-myelitis). This disease may well have an infectious cause, though its etiology has remained until now a mystery. The disease is well described here and by VAN BOGAERT and INNES [1962]. In spite of the possibility of an infectious cause, only one animal in a den or collection appears to become affected at a time. This disease may be progressive or show remissions, or even regress. The symptoms are reasonably characteristic.

When this disease is suspected and indeed in all cases in which the cause of a paraplegia cannot be determined, the animal should be handled with the greatest caution for fear that infection could be passed to persons in contact with the sick animal.

Diagnosis of the neuropathies of simians is far from simple, as will be realised from a study of the chapter that follows.

In this chapter, all the illustrations have been supplied by the Armed Forces Institute of Pathology at Bethesda and by Dr. H. SEIBOLD of the Delta Primate Center; we wish to give acknowledgment for this. We express our particular thanks to Colonel F. M. GARNER for the ready assistance he rendered to Dr. BRIZZEE.

For an account of the histological changes occurring in simian Rabies, reference should be made to the chapter on Rabies (part II) prepared by the Editor.

Introduction

Naturally occurring diseases of the nervous system in subhuman primates are less prevalent than in man. Nevertheless, neuro-pathological conditions are sufficiently common in these forms to warrant their study from a number of standpoints including their use as experimental models for the study of human disease and for comparative studies to understand better both human and animal diseases. The study of neuropathology in subhuman primates is also of importance from the standpoint of rendering experimental studies on nervous systems more valid by helping to eliminate or avoid pitfalls introduced by unrecognized disease conditions in experimental subjects.

In this brief chapter we cannot hope to cover the broad subject of neurological diseases in subhuman primates in any exhaustive fashion. It is therefore necessary to be somewhat selective as to the reports to be included, although on certain topics where data are limited or where the more significant data have become available in recent years, we have presented some appreciable amount of detail.

The sequence of subjects considered presents some problems in that certain topics (e.g. hypoxia and diseases resulting in myelin degeneration) cannot be comfortably included in a single section. However, we have presented the various topics in a sequence which we believe to be the most logical.

Congenital Abnormalities

While there is an extensive literature on congenital abnormalities in mammals, comparatively little information is available on naturally occurring anomalies in subhuman primates.

On the other hand, as will be evident from work cited below, investigations on experimentally induced birth defects, particularly as related to hypoxia, are proceeding at an accelerated pace.

Naturally Occurring

A case of spina bifida was recorded by HILL [1939] in a male baby Sacred Baboon *(P. hamadryas)*. The condition was localized and occurred without spinal hernia. The affected region was located immediately anterior to the root of the tail and measured 12 mm long by 15 mm transversely. The covering tissue was thick, soft and spongy, very vascular and thrown more or less in folds, bulging to some extent beyond the level of the surrounding skin.

The neural arches of four vertebrae immediately succeeding those attached to the iliac bone were incomplete dorsally. The covering membrane was connected by its deep surface, along the median line, with the membranes lining the vertebral canal. The spina bifida, however, involved the spinal cord as well as the meninges. The membrane covering the spina bifida represented an outer ectodermal stratum and a deeper mesodermal stratum.

The spinal cord immediately rostral to the deformity was small but had the usual outline for the sacral region. The central canal was greatly enlarged and contained blood; nerve cells were few and confined to certain areas at the periphery. A thin layer of white matter occurred at the surface of the cord and large blood spaces were present throughout the nervous substance. Caudad the spinal cord became more superficial and joined the disc of the spina bifida at a point one fourth the length of the disc from its cranial extremity. Microscopic examination of this area revealed numerous small nerves cut transversely on both sides beneath the peripheral transitional epithelial zone of the disc. In the midline the spinal cord terminated by turning dorsally to effect a junction with the central depressed area of the disc. The central canal was traced to an opening at the bottom of the depressed area.

Beneath the disc there were some nerve strands, containing large blood filled spaces between the nerve fibers. In the mid-line there was a ganglionic mass, showing well-formed nerve cells with their nuclear structure completely preserved. These were surrounded by connective tissue capsules with blood filled spaces between such elements and presumed to be either unipolar or bipolar since no processes were observed in sections. There was some evidence of a pial investment, but no evidence of dura mater or arachnoid. The whole ganglionic mass had a well-defined connective tissue capsule. No direct connection of the ganglionic mass and the spinal cord was evident.

Further study of this case by CAMERON and HILL [1955] revealed an Arnold-Chiari syndrome, in which the hindbrain appeared relatively small. A large tongue of cerebellar tissue passed caudally on the dorsal surface of the medulla and was indented posteriorly by the bony margin of the foramen magnum. The occipital horns of the lateral ventricles were slightly dilated. The lower part of the Arnold-Chiari formation extended caudally as far as the level of the neck of the first rib, a distance of 3 cm from the tip of the odontoid process, and consisted of a herniated and greatly elongated fourth ventricle covered dorsally by atrophic cerebellar tissue and ventrally by a thin membrane. The medulla oblongata was markedly elongated and extended to a level 1.5cm caudal to the tip of the odontoid process, with a hump on its dorsal surface. The pia-arachnoid investing the malformation appeared opaque and deeply congested. The cervical cord was 1.5 cm in length and its entire dorsal surface was covered by the cerebellar herniation. There was no lumbar enlargement. The first seven pairs of cervical nerve roots passed obliquely upwards, the eighth pair transversely and the thoracic and lumbar roots obliquely downward. Intense passive vascular engorgement of the involved pia-arachnoid was observed. The cavity of the fourth ventricle extended to the tip of the cerebellar herniation. The choroid plexus in this area exhibited engorged, hemorrhagic tufts, some containing fibrinous material and coarsely granular hemosiderin.

Congenital hydrocephalus of the internal type was described by Fox [1923] in a single specimen of *Cercopithecus aethiops* with the rupture of the right occipital cortex, chronic leptomeningitis and hemorrhage into the calvarium.

Lapin and Yakovleva [1963] described a multiple monstrosity in prematurely born twin monkeys *(Macaca mulatta)*. One of the twins had a cranium, about $^1/_3$ normal size for its age but no description of the histology of the nervous system was given. In one monkey multiple cysts were observed in the white matter of the brain which were presumed by the authors to be due to developmental defects.

A pronounced multiple monstrosity was also observed by these authors in a prematurely born *Macaca mulatta*. This was a nonviable monster weighing 215 grams. A cutaneous sac with a funnel shaped retraction on the anterior surface was observed, the latter corresponding to the oral cavity. Removal of the skin in the region of the cephalic sac revealed a small cranium about $^1/_3$ the normal size. The cerebrum was very small and comprised almost entirely of parietal lobes, sharply delimited from rudiments of the frontal lobes. The cerebellum was also very small and the cerebellar hemispheres had the form of two symmetrical leaflets of brain tissue. The vermis of the cerebellum was more normally developed. The brain occupied a relatively small part of the cranium and the space between was occupied by fluid. The authors expressed the opinion that it was highly probable that a connection existed between the development of the monstrosity and prolonged, almost incessantly recurring, dysenteries in the pregnant monkey during gestation as well as repeated administration of massive doses of various antibiotics. They also expressed the opinion that the development of monstrosities at the breeding farm from which this animal was obtained, was to some degree connected with avitaminosis and in particular, with a deficiency of vitamin A.

Experimentally Induced

Vincristine

COURTNEY and VALERIO [1968] reported that a monkey *(M. mulatta)* given 0.175 mg/kg vincristine sulfate intramuscularly on the twenty ninth day of gestation produced an offspring with encephalocele. In this animal there was an absence of the bony calvaria involving the frontal, parietal, and temporal bones, the gap being covered by membrane plus dura mater. At autopsy a localized area of contusion, affecting the meninges and superficial area of cerebral cortex was observed which was assumed to have resulted from trauma. In all other aspects the brain, including the pituitary, was normal.

Thalidomide

In monkeys treated with thalidomide after implantation of the ovum before formation of the fetal limbs, DELAHUNT and LASSEN [1964] described enlargement of the head, probably from internal hydrocephaly associated with a defective foramen magnum similar to the Arnold Chiari malformation in man. The portion of the occipital bone forming the posterior half of the foramen magnum was absent and the neural arches of the first and second cervical vertebrae were missing.

Hypoxia

In one investigation [WINDLE, 1963] about 130 monkeys were asphyxiated during or after birth. Asphyxiation for more than seven minutes invariably produced at least transient neurologic signs and permanent brain damage. Basic neuropathology was limited to certain brain-stem centers associated with general body sensation and hearing but not with vision. Most vulnerable were the inferior colliculus and ventrolateral group of thalamic nuclei. Nerve cells in these structures were destroyed and replaced by glial scars. Lesions were sharply circumscribed, bilaterally symmetrical and non hemorrhagic. Motor systems were spared.

Twelve to 17 minutes of placental separation or umbilical cord clamping led to asphyxial apnea, and resuscitation of the newborn monkey was necessary. In these animals the neuropathology involved additional nuclei of afferent pathways. The basic lesions of the thalamic and collicular nuclei were larger, and focal necrosis and sometimes cavitation were observed. Additional nuclear lesions appeared in the cerebellum, basal ganglia, cerebral cortex and even spinal cord of the most severely asphyxiated animals.

Symmetrical damage of bra in stem nuclei was observed in monkeys subjected to asphyxia near term [RANCK and WINDLE, 1959]. Those most consistently and severely damaged were the nucleus of the inferior colliculus, gracile and medial cuneate nuclei, roof nuclei of cerebellum, ventral posterior group of thalamic nuclei, globus pallidus, putamen and vestibular nuclei. The cerebral cortex was severely damaged in only one monkey. Lesions began with primary nerve cell and, less frequently, neuroglia cell lysis and loss. Secondary damage of myelin sheaths, and reactions of astrocytes, endothelial cells vascular adventitial cells and phagocytes were noted. A relation of lesions to vascular distribution was not apparent. Hemorrhages were seldom encountered.

In another study [FARO and WINDLE, 1967] no sign of regeneration from neonatal asphyxia was ever noted and scars of discrete primary lesions were clearly visible even after 5 years. Sections from some regions revealed widespread reduction in numbers of nerve cells without compensatory increase in scar forming cells. These were regions of transneuronal degeneration.

Monkey fetuses of different gestational ages and at term were subjected to both acute total and partial asphyxia, followed by resuscitation, to determine the patterns of brain pathology produced [MYERS, 1967]. In these animals a consistent pattern of brain pathology resulted which included bilaterally symmetrical damage in lateral portions of the posterior and lateral ventral nuclei. With slightly longer asphyxia, bilaterally symmetrical damage was seen in some specimens in subthalamic nuclei and in dorsomedial and centromedial nuclei of the thalamus.

When term fetuses were subjected to longer periods of asphyxia, the bilateral destructive changes in the thalamus in some animals extended broadly to encompass a considerable part of this area but also the putamen and deeper layers of the cortex in certain areas. Destructive changes were also observed in the lateral portions of the tegmentum and in the gray matter of the spinal cord.

Results of umbilical cord compression in fetuses of different ages revealed that more immature fetuses may sustain longer periods of total asphyxia before exhibiting damage to the brain. The earlier in gestation asphyxia occurred, the less was the tendency for damage to the less mature rostral brain structures, thus it appeared that the cortex, thalamus and basal ganglia were relatively protected in earlier fetuses. With more severe insults, the entire CNS was destroyed in some animals with only the dorsal root, autonomic ganglion cells, rod and cone cells and bipolar cells of the retina remaining. The thalamus and basal ganglia in cases of this type, exhibited a coagulatory necrosis with failure of ingrowth of viable reactive connective tissue elements. Blood vessels within such areas remained occluded by plugs of necrotic debris within and around the vessels. The brain stem below the thalamus exhibited a type of pathology in which there was preservation of a viable glial stroma with retention of the gross morphology of the tissues but with destruction of all neurons in these zones. Still another type of pathology in these cases was characterized by severe cystic degeneration of both cortex and white matter with connective tissue replacement throughout and with gitter cells or compound granular corpuscles suffusing the loose reactive collagenous stroma.

Following asphyxia of newborn monkeys ulegyria was observed, characterized by bilaterally symmetrical involvement in the paracentral region, and sulci in the parasagittal regions including the principal, arcuate, intraparietal and other sulci, with severity of involvement diminishing, the greater the distance from the central fissures. Bilaterally symmetrical involvement was also observed in the putamen, head of the caudate nucleus and lateral and posterior ventral nuclei of thalamus. The process, both in cortex and in basal nuclei, was characterized by a perivenular distribution, and there was a tendency for extension of the process between adjoining affected vessels, resulting in an overall reticulated appearance. In affected zones there was destruction of neurons with an associated astrocytic gliosis.

In newborn monkeys *(M. mulatta)* delivered by cesarean section at 153 days and 163 days of gestation and immediately asphyxiated for about 12 h, there was a widespread, abnormal accumulation of glycogen in glial cells (mainly astrocytes) of both gray and white matter [Mossakowsky *et al.*, 1968]. These changes became apparent about 10 h after asphyxia and tended to disappear after several days. An abnormal deposition of glycogen in glial cells was also demonstrated by electron microscopy. The glycogen accumulation was preceded by a marked increase in phosphorylases and UDPG-glycogen transferase activities. The latter was demonstrable in animals sacrificed 1 h after asphyxia, and abnormal activity of aminopeptidase in neuronal groups which were previously shown to be especially susceptible to asphyxia, was seen at the same time. Such neuronal groups later displayed evidence of histopathological damage. Disturbances of the blood brain barrier were demonstrable in animals sacrificed 10 h or more after asphyxia. This was characterized by selective localization of the tracer in the cellular components of the parenchyma, especially in the neurons. The changes were confined mainly to the gray matter.

In another experiment ten term monkey fetuses *(M. mulatta)* were subjected to a prolonged partial asphyxia consisting of a combination of hypoxia, hypercarbia, and a mixed respiratory and metabolic acidosis [MYERSETAL, 1969].

Seven out of ten fetuses subjected to the asphyctic insult *in utero* revealed brain swelling of variable degree, described as mild to moderate. The swelling was characterized by herniation of the cerebellar tonsils and vermis, with flattening of cerebral convolutions. The degree of brain swelling correlated best with the length of hypoxia hypercarbia and pH below 7.

Vascular Diseases

Sclerotic cerebral vessels were described in subhuman primates by LUCKE [1923]. In one animal *(M. mulatta)* less than one year of age the cerebral vessels were markedly sclerotic and atheromatous with considerable calcification of the media. Arteriosclerosis of the brain vessels was also observed in a *Papio comatus*. Associated areas of necrosis and gliosis of the brain substance were observed in both specimens. In the younger animal the intima of the vessels was relatively normal, but in the *Papio* the intima commonly contained atheromatous or calcified plaques. In two capuchin monkeys the intima was thickened and hyalinized, with the internal elastica split and fibrillated.

Four additional animals showed more or less well-marked arteriosclerosis. In a pigtailed macaque *(M. nemestrinus)* the cerebral vessels were markedly sclerotic and atheromatous, generally showing well-marked calcification involving particularly the mesial coats. In a baboon *(P. porcarius)* a similar condition was found. Necrotic and gliosed foci and atrophic, sclerosed and calcified ganglion cells were present as the result of vascular deficiency in some of the animals. In a brown cebus monkey *(C. fatuellus)* a number of small arteries in the Sylvian fissure had a moderately proliferated and hyalinized subendothelial intima, and reduplicated, split internal elastica. The mesial coats were normal. A mild, localized fibrosis of the pia was present. No foci of necrosis or gliosis were found. A few ganglion cells in cerebral cortex were atrophic. In a white throated cebus *(C. hypoleucus)* the subendothelial connective tissue of the intima in a number of pial as well as cerebral vessels was markedly thickened and hyalinized. This occasionally caused almost complete obliteration of the vascular lumens. The internal elastica was split and fibrillated, but the mesial coats were generally not involved. There was no evidence of calcification or focal necrosis. A number of atrophic and sclerotic ganglion cells were present and in places the pia was mildly thickened and fibrosed.

SCHERER [1944] observed a single case of arteriosclerosis in a presenile chimpanzee (age over 20 years) in which the systemic arteries and the basilar artery were affected.

A gorilla ('Bushman'-Chicago Zoo age 20 years +) exhibited sclerotic changes in blood vessels of the heart and nervous system [STEINER *et al.*, 1955]. The sclerotic changes were moderate and those of the cerebrum and meninges exhibited only thickening of the vessel walls. In basal ganglia the vessels showed even fewer changes.

LINDSAY and CHAIKOFF [1957] described atherosclerotic changes in the main arterial trunks of a baboon *(P. anubis)*. The arterial lesions were characterized by initial degeneration of the internal elastic membrane or inner medial elastic layer, followed by localized deposition of mucopolysaccharide substance. Intimal fibrosis was also observed with formation of elastic reticulum, and collagenous fibers, leading to the development of concentric intimal thickening or intimal plaques. Lipid infiltration of the intimal lesions was minimal and appeared to be secondary to intimal degeneration and fibrosis. Foam cells were not observed in the arteriosclerotic lesions, but small cholesterol deposits were demonstrable deep in the intimal plaques.

In young and adult monkeys *Cebus albifrons*, fed either a control or atherogenic diet for periods of up to 2 years, carotid artery atherosclerosis was more prevalent in males than females in both series, but no lesions were found in cerebral arteries in any of the animals [BULLOCK *et al.*, 1969].

Infarction of larger blood vessels of the brain (fig. 1) is occasionally seen at autopsy in subhuman primates with no evident relationship to antecedent trauma or disease conditions.

Demyelinating Diseases

Experimental

Cyanide Poisoning

In fasting animals *(M. mulatta)* which received repeated daily doses of a 0.2% solution of KCN each morning by intramuscular injection, the cere-

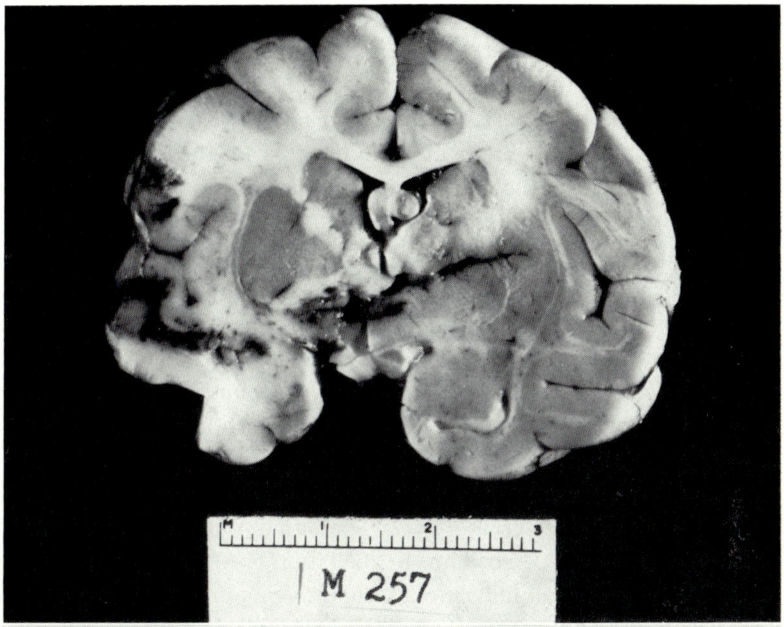

1

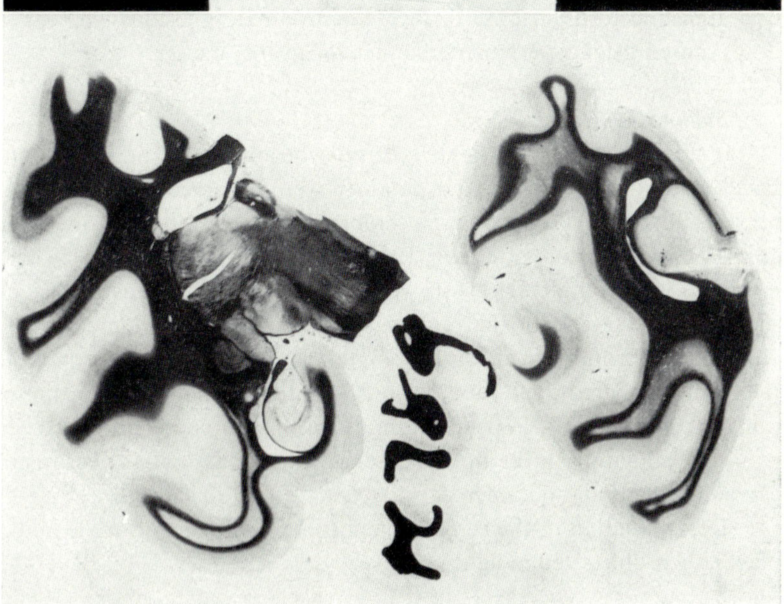

2

Fig. 1. Infarction of brain in Rhesus monkey.
Fig. 2. Demyelination following cyanide and azide poisoning ($\times 0.9$).

bral white matter in a number of subjects exhibited areas of softening which were grayish in color, and often separated from the cortex by patches or areas of normal white matter [HURST, 1940]. The lesions tended to be bilaterally symmetrical, and demyelination and necrosis were often observed in gray matter as well as white matter. In the areas of the lesions, cell packing density (mainly microglia) was commonly increased and partial demyelination was observed. Demyelination was found in occipital, temporal, parietal and frontal lobes in decreasing order of frequency. There was no tendency toward paraventricular lesions, but the corpus callosum, optic tracts and anterior commissure were common sites of lesions. In the former two, the central portions were more selectively affected. The optic nerves and chiasma commonly showed less damage. Demyelinated fiber bundles commonly extended from the centrum semiovale into the gyri. Necrotic areas, from 1 mm to 10 mm in diameter, with a rim of normally stained myelin, were seen in some cases. Similar areas of necrosis were observed in widely scattered areas of gray matter including cerebral cortex, pallidum, putamen, caudate nucleus, substantia nigra and cerebellum. In the latter, the Purkinje cells were often missing and meningeal infiltrations were sometimes observed. Prior intoxication appeared to predispose to severe demyelinating lesions whereas single doses caused the most extensive damage in the gray matter.

Sodium Azide Poisoning

In fasting monkeys which received intramuscular injections of 2% NaN_3, the caudate nucleus and putamen were observed to exhibit bilateral necrosis fairly commonly, and the cerebral cortex was often involved together with these structures [HURST, 1942]. The lesions varied from partial demyelination to complete necrosis with loss of axis cylinders as well as myelin sheaths. More severe anoxia resulted in lesions in the cerebellar cortex as well as in the cerebrum. The effect of a single dose was more lasting than that of KCN and rendered the animal more susceptible to subsequent administration of KCN or NaN_3. Repeated administration frequently produced necrosis or demyelination in the optic nerves and tracts as well as necrosis in the caudate nucleus and putamen.

Demyelination in the cerebral hemisphere following cyanide and azide poisoning is illustrated in figure 2.

Barbiturate Poisoning

Two monkeys poisoned with barbiturates (sodium pentabarbital and somifane administered intramuscularly) exhibited early changes in the white matter [HURST, 1942]. These included vacuolation in the centrum semiovale of cerebrum, but changes in cerebellum, pons, spinal cord and optic tracts were also seen. An 'over cellularity' was observed, apparently due to proliferation of both microglia and oligodendroglia. Pyknosis and karyorrhexis of some oligodendroglia was observed and a few glial cells exhibited enlarged somata.

Hypoxia[1]

A single sudden exposure to a simulated altitude of 32,000 feet for 25 min produced extensive laminar necrosis in the cerebral cortex of monkeys *(M. mulatta)* [MORRISON, 1946]. In the supragranular layers vacuolation of the cells was observed with the vacuolated cells found in clusters or zones. In the deeper layers the vacuolated type of change was of a more advanced nature with vacuolation, streakiness of cytoplasm and satellitosis.

With repeated exposures to mild hypoxia the first histologic changes occurred in the cell bodies of the cortical gray matter. This occurred at a level of about 12 or 13 volumes percent of oxygen in the blood in long repeated exposures. When the percentage of oxygen was reduced to about 10 volumes percent, and the number of exposures was increased, the white matter also became involved. On these preparations the white matter presented a pattern of demyelination in the corpus callosum, centrum ovale and the adjacent fingers of subcortical white matter which in the cases of more severe anoxia, suggested a resemblance to Schilder's disease.

Except for the lesion of the white matter, frank necrosis ordinarily occurred only after episodes of anoxia severe enough to result in respiratory arrest. The frontal lobe was most often involved and the temporal lobe least involved with such tissue changes. The cerebellum was more frequently involved than the basal ganglia, and the spinal cord and medulla were unaffected by any degree of anoxia compatible with life. In the cerebellum, long rows of Purkinje cells frequently showed large vacuoles around the periphery with the remainder of the cell tending to be somewhat hyperchromatic but otherwise in fairly good condition. In some cases, however, rows of shrunken, pyknotic, hyperchromatic cells with overstained processes were visible. The cerebellar granule cells did not appear to be affected.

Carbon Monoxide

Changes in monkeys characterized by demyelination and softening were observed by LUND [1956] in cerebral white matter and corpus callosum. Neu-

1 Effects of hypoxia in fetuses are described under 'Congenital Anomalies'.

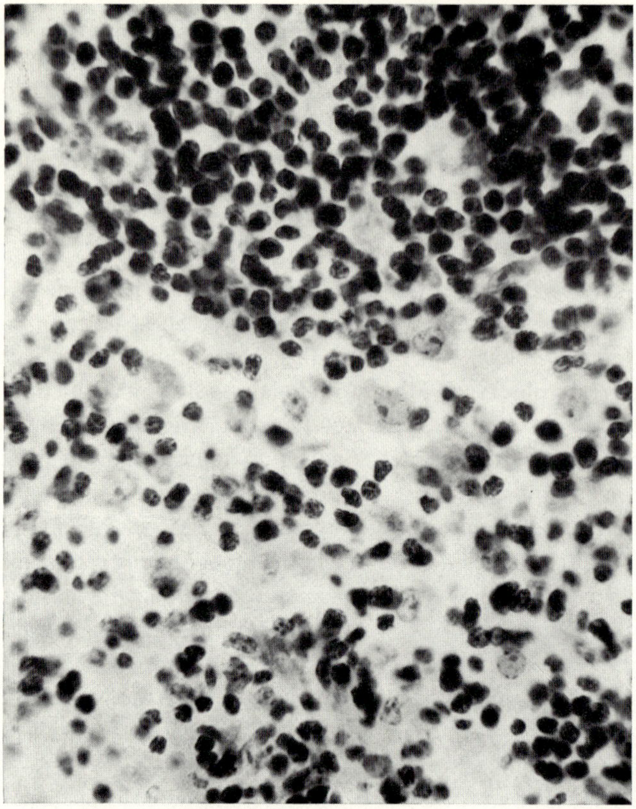

Fig. 3. Allergic encephalitis in Rhesus monkey (HE ×1026).

ronal changes were observed in hippocampus, dentate nucleus, and putamen and there was some loss of Purkinje cells in the cerebellum.

Allergic Encephalomyelitis

Injection of homologous or heterologous brain adjuvant emulsion resulted in lesions in cerebral and cerebellar white matter, pons and optic pathways [KERSTING and PETTE, 1957]. In the acute phases, the cerebrospinal fluid showed an increased white count, and decreased total protein and gamma globulin. After the acute phase there was an increase in percentage of globulin to protein. The most characteristic lesion in this condition was a perivascular myelinoclisis affecting venules, veins, capillaries, arterioles and small arteries in that sequence. Lesions in pons, and cerebellum, especially in rela-

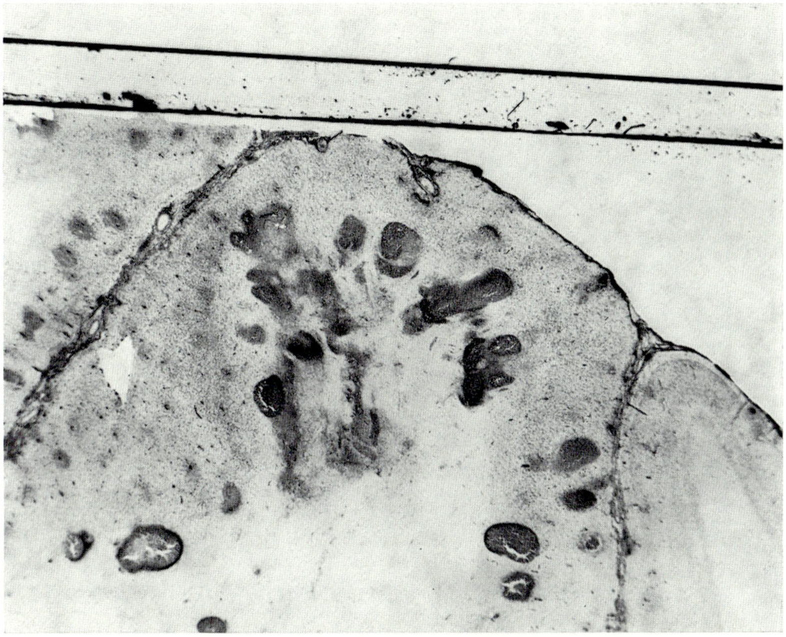

Fig. 4. Allergic encephalomyelitis. Condition similar to that described in text (× 28.5).

tion to the fourth ventricle, and in the cerebrum, were found with the white matter being more severely affected than the gray matter. In some animals alterations were observed in optic nerves, tracts and geniculate bodies.

Initially there was an increase in polymorphonuclear neutrophils, the appearance being similar to that shown in figure 3, with some eosinophils and occasionally fibrin in the vascular walls and perivascular spaces. Perivascular cuffing became extensive and perivascular areas of demyelination appeared to have spread and coalesced, with some preservation of axis cylinders. Gitter cell activity was seen, but astrocytosis was mild or moderate. Acute lesions were seen in close proximity to healed lesions, and the oldest lesions appeared as clearcut patches of demyelination with varying degrees of gliosis and preservation of some axis cylinders. The appearance of cerebral cortical gyri in a monkey *(M. mulatta)* with allergic encephalomyelitis is shown in figure 4. Subcortical demyelination in this condition is shown in fig. 5.

Lesions in the cerebrum were seldom seen in association with atrophy of cerebral gyri or cerebellar folia. Complete necrosis of all structures in the area

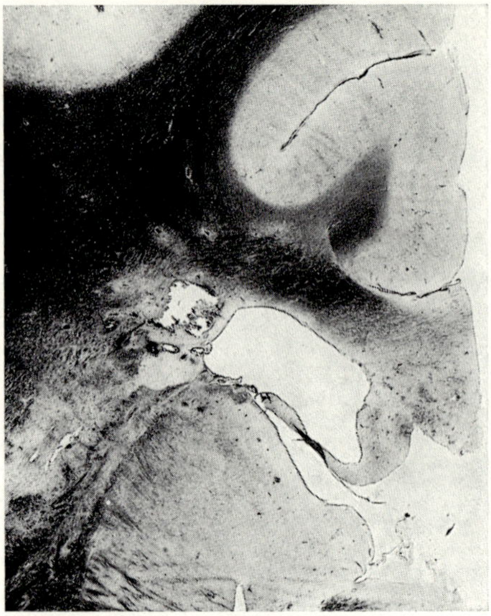

Fig. 5. Subcortical demyelination in allergic encephalomyelitis in monkey (×10).

of a lesion was seen in some cases and such lesions had the appearance of old areas of encephalomalacia.

Lead Poisoning

A description of a case of lead poisoning was given by FISHER [1954] in a gorilla but no information on neuropathological alterations were given. A similar case was described by BISHOP [1956 – Cite VAN BOGAERT and INNES, 1962] as an unpublished communication but again no information on neuropathological effects are available.

Naturally Occurring

Confluent Leukoencephalosis and Perivascular Myelosis

Optic Type

This disease has been observed mainly in adult monkeys *(M. mulatta)* of both sexes, but also in baboons, orangutans, and a lemur. The characteristic alterations were found in optic disc, and optic chiasma and extended to include the optic nerves as well as the tracts [LEVADITI *et al.*, 1930]. The chiasma was transformed into a mass of granular corpuscles and its vessels showed considerable perivascular cuffing. The lesions were variable in extent and severity, and generally were observed to begin in the deeper parts of the optic nerves and tracts, then to spread to the subpial zone.

Cerebral Type

The characteristic feature of this type, in the subhuman primates, has been perivascular demyelination of irregular extent but tending toward bilateral symmetry [van Bogaert, 1939a, b; van Bogaert and Innes, 1962]. Larger areas of demyelination were produced by coalescence of the smaller foci. These were localized in the centrum ovale but did not involve the internal capsule, short association fibers, corpus callosum, callosomarginal gyrus, orbital gyri or temporal lobe. The most severe damage has generally been seen in parietal lobes and to a lesser extent the frontal and occipital lobes. Characteristically in this condition demyelination is complete in the center of a lesion but progressively less so toward its periphery. Most reports indicated damage to axis cylinders, which were reduced in numbers. Later there were filiform foci following the direction of destroyed fibers and veins, often accompanied finally by a fibrillary gliosis.

Spinal Type

Lesions occurred mainly in the thoracic cord, with a lower incidence in the cervical and lumbar regions [van Bogaert, 1939; van Bogaert and Innes, 1962]. The borders were poorly defined and bore no relationship to the topography of the spinal tracts. They were scattered in some cases or tended to be concentrated in the deeper portions of the dorsal columns and marginal regions of the ventral columns in others. In some cases involvement of spinal nerve roots and peripheral nerves, with proliferation of neurilemma cells has been observed. The ventral roots were less affected than the dorsal roots. The three forms of this disease were often observed together, and the spinal form was seldom found without one or both of the other types. The etiology of the disease is unknown but some workers believe it may be the result of gastrointestinal disorders and dietary deficiencies or imbalances.

Isolated Papillomacular Degeneration

The characteristic finding in this disease in primates has been that of demyelination of the deeper fibers of the optic tracts and chiasma, leaving the marginal tissue unaffected. In some cases a triangular field of degeneration was seen, the apex coinciding with the central vessel and the base with the lateral surface of the nerves. The condition has been observed to occur together with generalized tuberculosis, colitis, nutritional disorders and, to a lesser extent, with cerebrovascular diseases [VAN BOGAERT and INNES, 1962].

Infectious Diseases

Viral

Encephalitides

Experimental West Nile Encephalitis

Neuronal changes in monkeys *(M. mulatta)* inoculated with this virus were seen in large pyramidal cells of cerebral cortex, granular cell neurons and Purkinje neurons of cerebellum, and neurons of hippocampus [VAN BOGAERT and INNES, 1962]. No inclusion bodies were seen but acidophil changes were observed in the affected neurons. Monkeys inoculated intracerebrally with West Nile virus in another study exhibited two forms of neuron degeneration [MANUELDIS, 1956]. In one form pinkish discoloration of the cytoplasm was observed with dissolution of Nissl substance and pyknosis of nuclei. In the other form, chromatolysis and swelling of the cytoplasm was observed together with peripheral apposition of pyknotic nuclei. Moderate to occasional marked accumulation of inflammatory cells, largely lymphocytes, was observed in the Virchow-Robin spaces. Changes similar to those described in this investigation are illustrated in figure 6. Neuronal changes, perivascular infiltrates and vascular changes were not considered as being necessarily interdependent phenomena. Astrocytes exhibited marked increases in nuclear size (fig.7) and microglia cells showed increase in size and number, and were the most common elements in cell nodules, which generally had a loose structure and were associated with blood vessels. The lesions were focal, of moderate to mild intensity and were scattered throughout the brain and spinal cord, being more prominent in the cerebellum, spinal cord and medulla oblongata but somewhat less evident in pons, thalamus, striatum-pallidum, and cerebral cortex. A focal severe meningitis was seen over the cerebellum but meninges elsewhere showed only mild changes. No characteristic lesions were seen in peripheral ganglia.

Experimental Japanese B Encephalitis

Intracranial and intraocular inoculations were found by TOGA [1957] to be effective in producing this disease, the most susceptible primates being young monkeys *(M. mulatta)*. Alterations in neurons and proliferation of

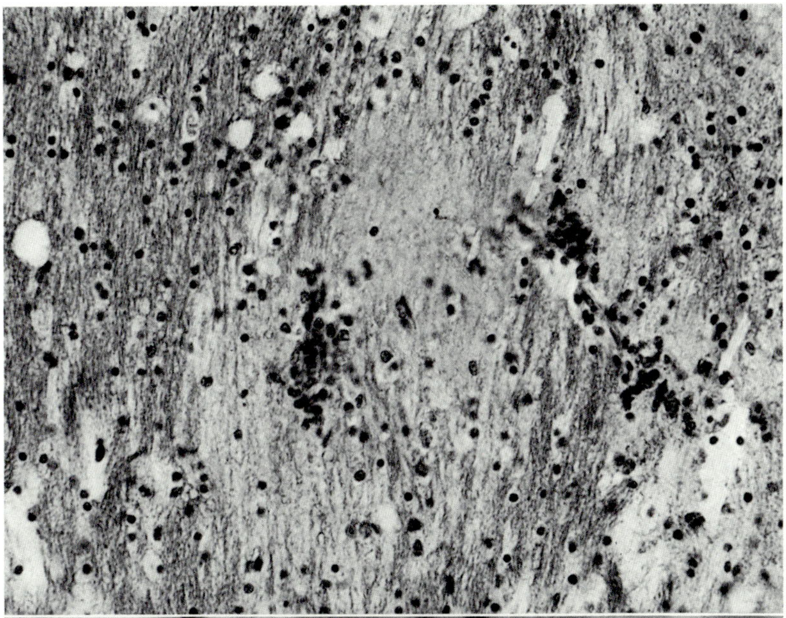

6

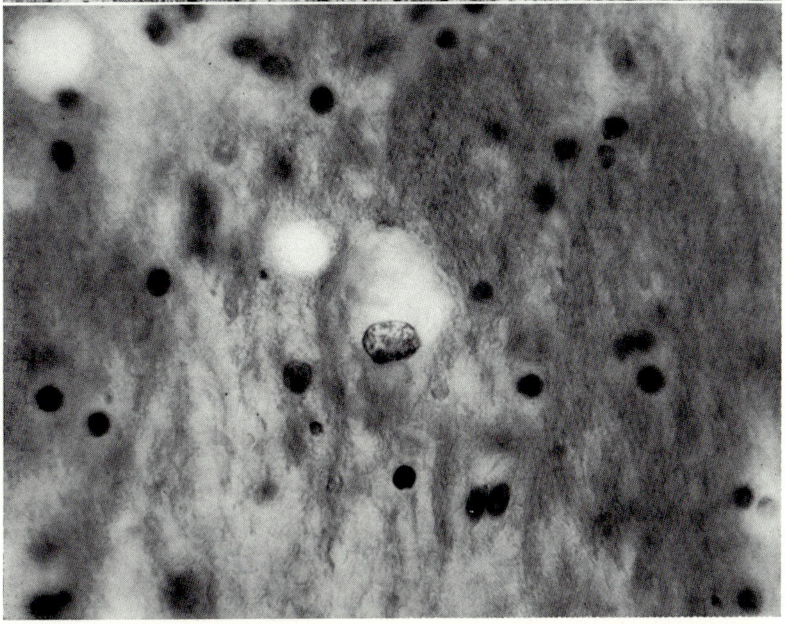

7

Fig. 6. West Nile encephalitis. Tissue destruction and accumulation of inflammatory cells in perivascular spaces (×280).

Fig. 7. Swelling of astrocytic nucleus in West Nile encephalitis (×490).

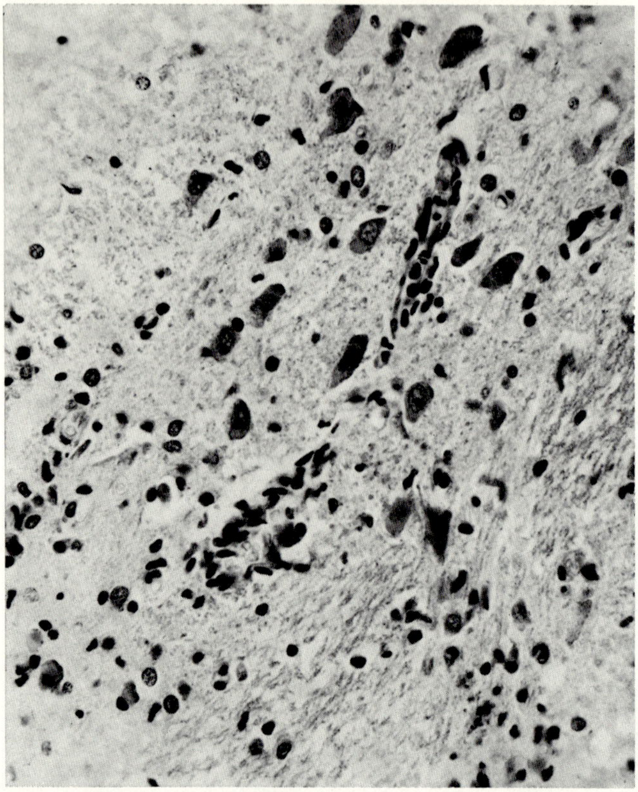

Fig. 8. Perivascular cuffing and degenerative changes in neurons of Pons in St. Louis encephalitis in Rhesus monkey, *M. mulatta* (He ×298).

glial cells together with perivascular infiltrations were observed in cerebral cortex, striatum, thalamus, substantia nigra, lamina quadrigemina, medulla oblongata, deep cerebellar nuclei and spinal cord. After intracerebral inoculation of the virus in monkeys *(M. cynomolgus)* PAYAN *et al.* [1957] observed perivascular infiltrations and glial nodules in the thalamus and brain stem reticular formation and also distributed as discrete glial nodules throughout the neuraxis. On the periphery of edematous foci, astrocytic reactions were apparent and acute degenerative changes in neurons were observed.

In this study the earliest lesions observed in an animal that was sacrificed, in the absence of all symptoms, in the 4th post inoculation day (attenuated virus) were of a purely exudative character, and were represented by cerebral edema and vascular congestion. On the 8th day the inflammatory

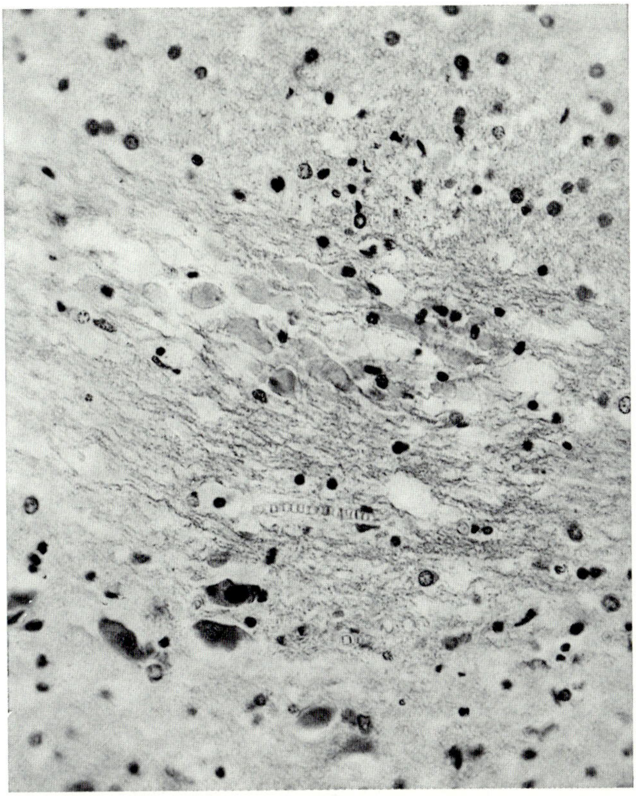

Fig.9. Rosenthal fibers and neurons exhibiting degenerative changes in Pons, St. Louis encephalitis (HE ×285).

change was characterized by perivascular infiltrations and glial nodules situated in the gray matter. These were particularly evident at the level of the thalamic formation and the reticular matter of the cerebral trunk. An animal inoculated with a virulent virus and sacrificed on the 7th day exhibited fewer perivascular infiltrations but more evident glial nodules.

In general, necrotic foci in white matter were always associated with inflammatory lesions of the gray matter.

Experimental St. Louis Encephalitis

In monkeys *(M. mulatta)* which received intracerebral and intraperitoneal inoculations of a thick brain emulsion (0.5–2.0cc) from fatal encephalitis cases, a marked perivascular round cell infiltration of the brain was ob-

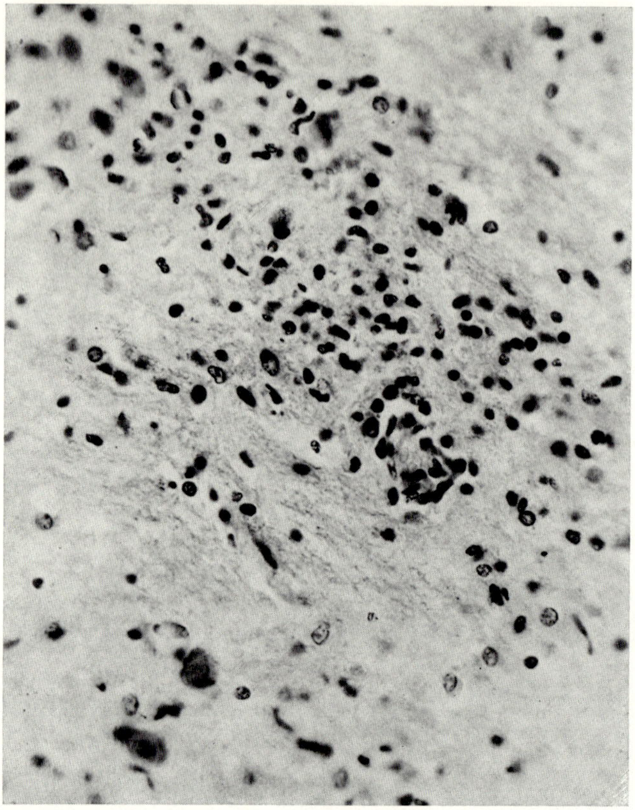

Fig. 10. Perivascular round cell infiltration in Pons, St. Louis encephalitis (HE ×275.)

served together with some nerve cell destruction scattered diffusely through the brain, bulb and spinal cord [MUCKENFUSS *et al.*, 1933].

Perivascular cuffing together with degenerative changes in neurons of Pons in monkeys *(M. mulatta)* is illustrated in figure 8 and similar changes, together with Rosenthal fibers are seen in Pons in figure 9. Perivascular round cell infiltration in Pons and in Substantia Nigra are shown in figures 10 and 11 respectively.

Experimental Inclusion Body Encephalitis

Cerebral lesions have been observed in monkeys *M. mulatta)*, following intracerebral inoculation of the virus [PELC *et al.*, 1958], in which glial infil-

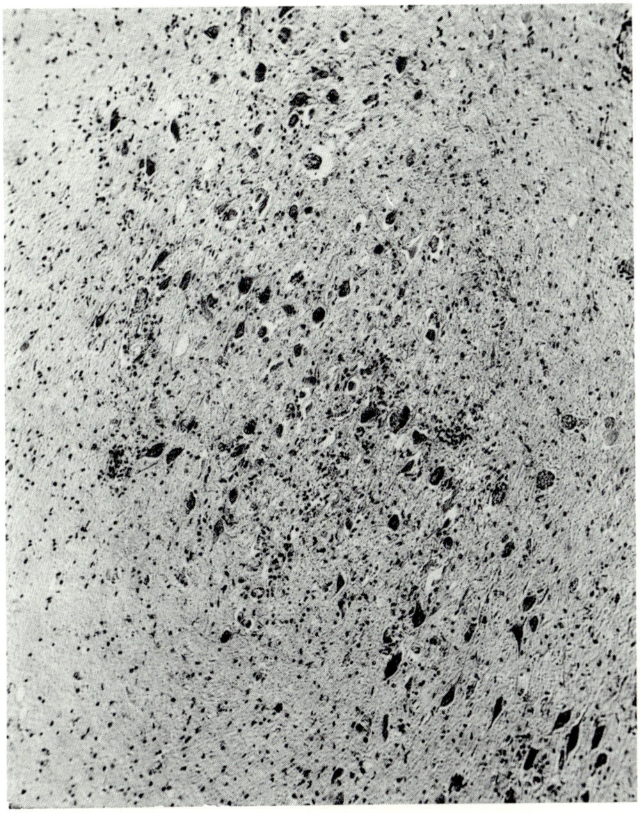

Fig. 11. Perivascular round cell infiltration in substantia nigra in St. Louis encephalitis (HE ×123).

trations were observed in the gray matter and to a lesser degree in thalamus and oculomotor nuclei. Necrotic foci with an inflammatory reaction, gitter cells and perivascular lymphocytic infiltrations were found in parts of the cortex and basal ganglia and intranuclear inclusion bodies were observed in glial cells. Inclusion bodies in brain cells of a gibbon infected with Herpes virus are illustrated in figure 12.

Corticostriate Encephalitis

In a Rhesus monkey *(M. mulatta)* BODECHTEL [1930] observed severe lymphocytic infiltrations in frontoinsular regions, with both laminar and diffuse neuronal losses and a gliosis and accumulation of red cells. In monkeys

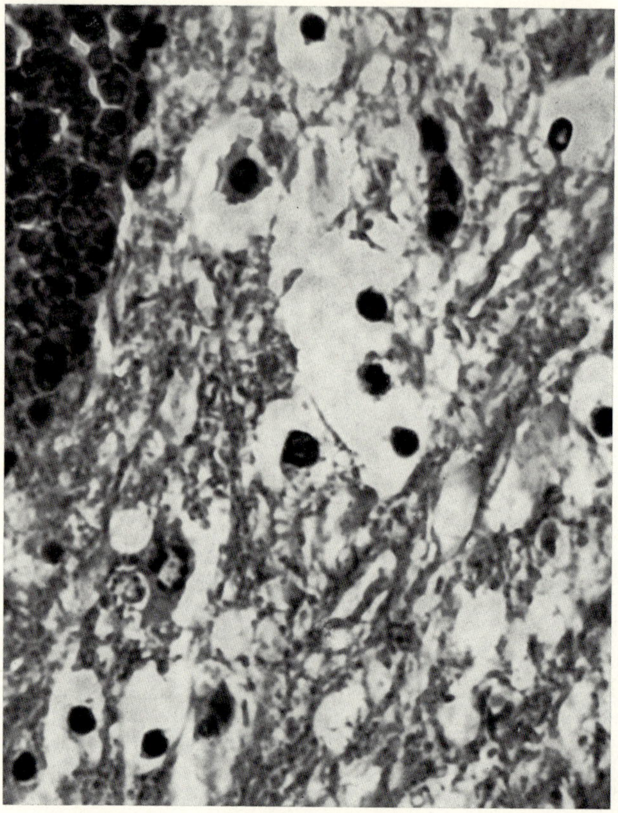

Fig. 12. Inclusion bodies in brain cells of a gibbon infected with Herpes virus (HE × 850.)

(M. mulatta) inoculated intracerebrally and intraperitoneally with a 20% suspension from the brain of an infected monkey, the principal lesion was characterized by many areas of paling from coagulation necrosis in the caudate and lenticular nuclei bilaterally [RICHTER, 1945a]. The lesions appeared as small round or oval areas but were often confluent to produce larger irregular lesions and were devoid of ganglion cells. They contained a few thickened blood vessels with some infiltrations of round cells and numbers of pyknotic glial nuclei. There was no softening or progressive glial reaction within the lesions, but many necrotic areas were surrounded by a zone of intense glial reaction with hyperplasia of all glial forms. In these zones many of the vessels were greatly thickened from proliferation of intima and adventitia and were commonly infiltrated by polymorphonuclear leukocytes and lym-

phocytes. In some areas there was a diffuse parenchymatous infiltration of exudative cells, especially PMN's. No lesions were found in the cerebral cortex, cerebellum, midbrain, pons, medulla or spinal cord. However, in the striatum and pallidum, scattered and diffuse inflammatory changes were observed which consisted of progressive reactions of glia, with rod-cell formation and perivascular and parenchymatous infiltrations of lymphocytes and leukocytes. The thalamus exhibited some diffuse increase of glia and infiltration of round cells about some blood vessels. Blood vessels of the brain except in the regions of the lesions, did not appear pathologic.

In baboons, described by VAN BOGAERT and INNES [1962], the lesion in the nervous system included a moderate lymphocytic meningitis and bilaterally symmetrical encephalitis involving the caudal part of the putamen and extending into the pallidum and internal capsule. The visual motor cortex, and frontoparietal and occipital regions showed mainly perivascular cuffings. Pial infiltrations were observed in brain stem, cerebellum and cord but these structures were otherwise, normal.

B Virus Infection

In *M. mulatta* and *Callithrix* the virus (natural cases) produced a poliomyelitis [MELNICK and BANKER, 1954], with perivascular cuffing, foci of inflammation and neuronophagia. In intracerebrally inoculated Rhesus monkeys [SABIN and HURST, 1935], an intense meningitis was observed which was most marked near the site of inoculation but also affecting the opposite hemisphere, brain stem, and cerebellum and cord. The exudate was less cellular than in cerebral vaccinia and accompanied by more acute destruction of tissue. By the 8th day, replacement of the original PMN infiltrate by mononuclears was occurring. Inclusions preceded necrosis in every type of meningeal cell. In meningeal vessels the adventitia was nearly always involved to some extent and was often wholly necrotic and heavily infiltrated with polymorphs. In less damaged vessels, marked cellular proliferation occurred. Nuclear inclusions were abundant in some cells. Extensions of the meningeal lesion downwards along the perivascular sheaths of the penetrating vessels were observed as far as the first or second cortical layer.

In one animal glial nuclei were pale and greatly swollen, usually without oxychromatic material, but sometimes showed inclusions. Microglial proliferation was observed with evident mitotic figures. In the general cortex the subpial glial cells were commonly affected, and exhibited inclusion bodies. Numerous small areas of cortex contained a few necrotic nerve-cells, while neighboring neurons and glial cells showed specific alterations, presumably chromatolytic in nature. In the dorsal spinal ganglia, occasional nerve cells were acutely necrotic, usually without signs of cellular reactions.

Following peripheral inoculation of the virus a few microglial nodules were observed in the medulla but no specific lesions could be detected.

Monkeys *(M. mulatta)* with naturally occurring B virus infection [Keeble *et al.*, 1958] did not show any involvement of cerebral cortex, thalamus, corpora quadrigemina, cervical cord or lumbar cord. Some of the monkeys, however, exhibited localized lesions of the pons and medulla and in some there was involvement of the roots of cranial nerves V and VII. Glial and lymphocytic infiltration and perivascular cuffing appeared most frequently in the adventitial tissues around the nerve roots. Chromatolytic alterations were seen in a small proportion of neurons in the facial and trigeminal nuclei, but neuronophagia was not observed. Lymphocytes were occasionally seen, but the cell reaction consisted mainly of oligodendroglia and microglia. The lesion of greatest significance in the CNS was in the nucleus and tract of the descending branch of the trigeminal nerve between the roots of origin of the facial and auditory nerves. Most of the animals showed glial infiltration of the white matter but no major neuron involvement was observed. In the medulla oblongata lesions were observed in association with the tractus solitarius and its nucleus and with the ventral spinocerebellar tract. Mild gliosis was noted in both sites, and cuffed vessels were commonly present in the surrounding tissue. Inclusions were observed in oligodendroglia especially in the area of perivascular cuffing. In general the lesions with B virus infection appear to be localized to the portions of the brainstem receiving sensory fibers from the mouth and tongue.

In contrast to the above a study carried out for the purpose of evaluating the severity of the effects of group B arboviruses in monkeys (*M. mulatta)* revealed that the thalamus, lenticulostriate complex, substantia nigra, reticular formation, vestibular nuclei, cerebellar cortex and anterior horn were more severely involved than were other CNS sites [Nathanson *et al.*, 1966a]. It was suggested that these nuclei might be used as 'indicator' centers for assessment of severity of histologic lesions. It was noted, however, that marked variation was found in the severity and distribution of lesions in monkeys inoculated with a single virus, and it was the opinion of the authors that this might constitute the major source of variance in evaluating each virus.

In a later study by these authors, in which 43 males and 35 females monkeys *(M. mulatta)* were inoculated intracerebrally with arboviruses, it was observed that nuclei of maximal susceptibility were the lenticulostriate complex, thalamus and substantia nigra in the hemispheres; cerebellar cortex

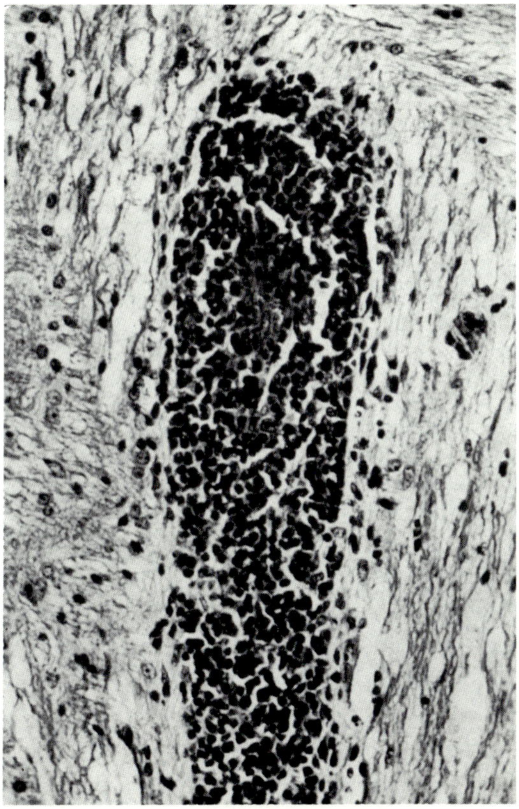

Fig. 13. Perivascular cuffing in Pons of Rhesus monkey, *M. mulatta*, in B virus infection (HE ×280).

and reticular formation in the brain stem, and the anterior horn and anterolateral white matter in the cord [NATHANSON *et al.*, 1966b].

The characteristic perivascular cuffing which commonly occurs in B (Herpes) virus infection is illustrated in figure 13. Cuffing and gliosis is shown in reticular formation of medulla oblongata in figure 14 and gliosis of the descending root of the trigeminospinal tract is shown in figure 15.

Acute Polioencephalitis

Naturally Occurring

Among the subhuman primates this condition has been found mainly in baboons. The histopathologic lesions [VAN BOGAERT and INNES, 1962] were

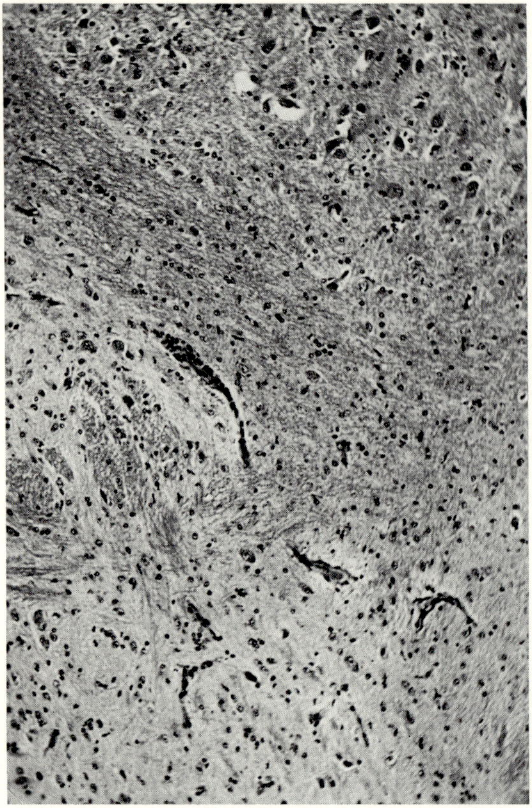

Fig. 14. Perivascular cuffing and gliosis in B virus infection in medulla oblongata in Rhesus monkey, *M. mulatta* (HE ×80).

limited to the central nervous system but neurons were not selectively involved. An acute, disseminated meningoencephalitis was observed which was mainly lymphocytic and macrophagic in character. In the brain the foci were perivascular with adjacent diffuse glial infiltrations. In the spinal cord the lesions were widely disseminated at all levels and were also observed in the medulla, reticular formation and olives. The dentate nuclei and white matter of cerebellum were severely involved. In the midbrain the tegmentum was more severly damaged than the cerebral peduncles. The hypothalamus, pallidum, internal capsule, claustrum, amygdaloid nuclei and substantia nigra were also involved. In cerebral cortex the frontoorbital regions were most severely affected, and slight lesions were present in the visual cortex.

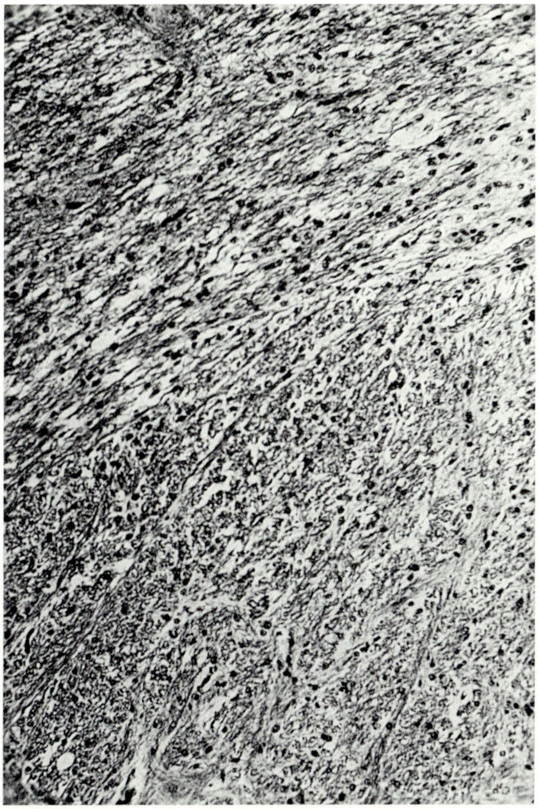

Fig. 15. Gliosis in descending root of trigemino-spinal tract in B virus infection (HE ×95).

Spontaneous cases of poliomyelitis in subhuman primates have also been reported by HOWE and CRAIGIE [1943]; BODIAN [1948, 1952]; and cases suspected to be poliomyelitis were reported by TANAKA [1939]; GOLDMAN [1935]; and LINDAU [1935].

A meningo-polioencephalitis which had a predilection for the brain stem was observed in baboons [VAN BOGAERT, 1959]. The localization and characteristics of the disease process were reminiscent of those seen in Borna disease and rabies. The pathologic picture of two of the cases was that of a very severe encephalitis of the entire brain stem, spreading anteriorly to the hypothalamus and the septal nuclei, and laterally to the claustrum. The spinal cord presented perivascular and nodular infiltrations in the posterior, more than in the anterior, gray matter with moderate involvement of the white columns.

Spontaneous meningoecephalitis was also observed by VAN BOGAERT in two baboons *(Papio hamadryas)*. The tissue changes consisted of perivascular inflammatory reactions, to a lesser extent, of glial nodules in the corpus striatum, the motor and orbital cortices and less so in the pallido-thalamic areas near the corpus striatum.

Experimental

Following a study of experimental poliomyelitis induced by intracerebral inoculation of virulent cord emulsion in 18 monkeys *(M. mulatta)* HURST [1929] concluded that poliomyelitis was an inflammatory disease of the whole nervous system with special involvement of the anterior horn neurons. The degeneration was not considered to be attributable to the accompanying interstitial inflammation. The nerve cell destruction was maximal in the cervical and lumbar regions where from ½ to ⅔ of the anterior horn cells were destroyed. The dorsal region of the cord was less severely damaged but many cells were destroyed in this area also.

Even in the less virulent infections, the destruction progressed through a sequence of changes, the first being central chromatolysis followed by complete solution of the Nissl substance. Often both the nucleus and cytoplasm swelled, became pale and the nucleus disappeared, but generally the nucleus exhibited shrinkage, stained more darkly than usual and became excentric in position. The abnormally large nucleolus was extruded in some cells. The cytoplasm became reticulated and vacuoles appeared which distorted the intracellular networks of neurofibrils. As degeneration proceeded, neuronophagia began. In addition to variable numbers of polymorphs and a few lymphocytes and glial cells, microglia gathered rapidly in these areas. These cells were involved in removal of dead neurons and even after 48 h, only a focus of these cells, the 'nodule poliomyelitique' remained to indicate its previous site.

In the medulla oblongata, perivascular infiltration in the floor of the fourth ventricle and in the substantia reticularis was always seen and was frequently more intense than in the cord. Focal and diffuse infiltrates occurred in the nervous tissue of the same regions but showed no special predilection for cranial nerve nuclei. Nerve cell damage however was never as severe as seen in the cord. Meningitis was not considered a necessary feature of the disease.

BODIAN [1948] studied the histophatological changes from the earliest time at which pathological effects became evident to late stages of recovery in 50 Rhesus monkeys killed from a few days to one year after inoculation with several 'strains' of virus of recent human origin.

In monkeys in which the onset of paralysis was abrupt, ventral root axons of motor neurons destroyed on the first day of paralysis, and probably many on the preceding day, were morphologically intact until the third day after onset of paralysis when degenerative changes were observed throughout the length of the axon. On the first and second day of paralysis axons appeared to degenerate centrifugally from the cell body for a short distance.

Even when inflammatory changes were most intense, no inflammatory cells of any kind were found anywhere along the axons of cells undergoing active phagocytosis. From this evidence it was concluded that the primary focus of virus activity was in the nerve cell body.

Cytological changes which appeared to be associated with irreversible neuron injury and were considered as indicators of necrosis included the following: (1) Complete loss of cytoplasmic basophilia. (2) Disappearance of mitochondria. (3) Breakdown of neurofibrils. (4) Severe nuclear changes, other than formation of 'inclusions' (including severe nuclear pyknosis, complete loss of nuclear basophilia, rupture of nuclear membrane or nuclear extrusion). (5) Cytoplasmic vacuolization. (6) Glossy appearance of cytoplasm, with some Nissl substance remaining at the periphery. (7) Severe cellular basophilia, usually associated with beginning neuronophagia. (8) Neuronophagia; while disappearance of mitochondria and breakdown of neurofibrils were considered to be indicators of necrosis, these elements were morphologically intact except in necrotic cells.

Axons were found to degenerate between 3 and 4 days after destruction of the cell body but usually not earlier than the third day. This was the typical time course of secondary degeneration, and indicated further that the primary damage to motorneurons was in the cell body.

The most common change in nerve cells in the earliest stage of the preparalytic period was a diffuse thinning of Nissl bodies throughout the cell body. Complete dissolution of Nissl bodies was seen only in the late preparalytic stage in many neurons. Nuclear changes in the preparalytic stage were generally not apparent until severe cytoplasmic chromatolysis had occurred and generally consisted initially of the formation of acidophilic inclusion bodies. Changes in mitochondria were not apparent in cells showing mild or even severe chromatolysis, but very few mitochondria remained when presumably irreversible change had occurred in the cell. Neurofibrillar changes were also not apparent until severe cell changes had taken place. The entire sequence of neuron changes from early injury to cell necrosis was present in the preparalytic period, although frankly necrotic cells were few in number. Neuron changes of mild or severe degree in the preparalytic period were entirely independent of inflammatory changes and occurred in regions of the gray matter of the anterior horn where no inflammatory responses had occurred. At the same time leukocytes were present in large numbers in other regions where nerve cell changes were so slight that they were considered within the normal range.

Central chromatolysis was rarely observed in the preparalytic stage although it was common in the subacute stage, and nuclear excentricity was not observed. In later stages in the preparalytic period polyblasts were often present in large numbers in addition to numerous leukocytes. Granulocytes were more numerous in the lumbar than in the cervical cord. In some cases they were present in great numbers in the gray matter and appeared to migrate into surrounding white matter.

On the first day of paralysis the predominant cellular alteration was that of diffuse chromatolysis, but many cells in various stages of necrosis and neuronophagia were present. Nuclear changes were rarely severe, but in severe chromatolysis intranuclear acidophilic inclusions were commonly seen. Basophilic nuclear inclusions were found in some cells with shrunken or highly vesicular nuclei. Neurons about to undergo neuronophagia were generally highly basophilic. The inflammatory process was more severe than on the preceding day, but with a tendency toward a reduction in granulocytes. The outstanding histopathological feature on the 2nd and 3rd days of paralysis was the scarcity of normal appearing neurons. Residual neurons in late stages of diffuse chromatolysis were observed, although most of the neurons were completely destroyed. Normal appearing granulocytes were uncommon and the principal cells of the inflammatory response were polyblasts. The earliest phases of axon degeneration in ventral root fibers were seen at this stage. Some early stages of neuronophagia were seen.

At four to six days of paralysis many cells showed aggregation of Nissl substance near the cell and nuclear membranes and there were very few neurons which did not show alterations Many cells, however, showed aggregation of Nissl substance near the cell and nuclear membranes, and many transitional stages were seen. Many ventral root axons showed severe degenerative changes, and in some cases no intact axons were visible in segments showing extensive destruction of motor neurons. In segments where many neurons remained some were normal even though most showed severe chromatolytic changes. At 7 to 12 days of paralysis the characteristic feature of neuron morphology was the predominance of central chromatolysis in remaining cells other than normal appearing. A greater proportion of normal or nearly normal neurons was found than in the preceding period. Nuclear inclusions were mainly acidophilic but many cells of otherwise normal appearance showed basophilic particles surrounding the nucleolus and usually surrounded by a shrinkage space. Lymphocytes tended to disappear in parenchymal areas in this period, but their number increased in perivascular infiltrations and cuffing was more prominent than before.

During the 3rd week after onset of paralysis, less than 10% of all neurons were abnormal in appearance. The predominant pathological stage was central chromatolysis with evidence of recovery of Nissl substance. Intranuclear inclusions were present in some neurons and the increased size of perivascular infiltrations was more than in earlier stages. From 35 to 49 days no neurons with severely abnormal appearance were found. The proportion of chromatolytic neurons was very small. Large acidophil nuclear inclusion bodies were seen in some cells. Lymphocytic perivascular infiltrations assumed the proportions of lymphatic nodules in some cases but some cord regions in the same subjects were nearly free from lymphocytes.

The morphological recovery of over 90% of surviving neurons was attained at the end of the 1st month of the disease. During the next few weeks the residual chromatolytic cells and lymphocytes were few in number. The essential disappearance of the leukocytic infiltration was roughly coincident with the recovery of most remaining neurons.

Heavily cuffed vessels persisted for at least a year in some cases. During the late convalescent period histiocytes were greatly reduced and affected regions of the anterior horn were characterized by a variable degree of gliosis, marking the terminal stage of the non-neuronal reparative process.

Statistical evidence from populations of motor neurons studied during acute and recovery periods indicated that a large proportion of such neurons, injured to the extent of severe loss of Nissl substance recovered their normal appearance.

Diffuse Non-Suppurative Meningitides and Meningoencephalitides

After intracerebral injection of cerebral biopsy substance from encephalitis cases into monkeys *(M. cynomolgus)* brain lesions were observed which were described as being comparable to those observed in human encephalitis from the standpoint of the topography of the lesions and the inflammatory pattern [PELC *et al.*, 1958]. The lesions were situated mainly in the hemispheres in the vicinity of the inoculation. Discrete lesions were seen immediately adjacent to the site but generally they tended to be scattered in the vicinity of the inoculation. The motor regions seemed to be little involved.

Despite the rather dissimilar clinical courses and symptoms observed in these conditions, the histological picture as described by VAN BOGAERT [1959] is fairly uniform. This investigator described a cellular reaction with large numbers of plasma cells, and predominance of histiocytes, or a cellular reaction which was purely lymphocytic in nature. This reaction in some cases involved large areas of the meninges overlying a sulcus or a discrete zone at the apex of a gyrus with a macroglial involvement of the molecular layer of the adjacent cortex. The parenchymal blood vessels showed a mild adventitial cuffing. This was also observed in vessels of white matter and basal ganglia. A diffuse macrogliosis sometimes involved the superficial layers of the cortex. VAN BOGAERT also reported that a few Purkinje cells may be lost in the cerebellum and in the dentate nuclei and a few cells may be missing as a result perhaps of a vascular insufficiency. The choroid plexuses were often involved.

Pericapillary infiltrations or glial nodules were seen in a few cases, in the 'structures of the base, the white matter, and the cerebellar white matter'. Occasionally areas of gliosis and neuronophagia were seen in these structures. The author felt that the absence of lesions of the gray matter of the spinal cord argued against a possible relationship with poliomyelitis.

In two baboons *(P. cynocephalus* and *P. porcarius)* the meninges were infiltrated here and there with round cells [LUCKE, 1928]. Many of the smaller vessels, particularly those in the pons, possessed jackets of several layers of lymphocytes and plasma cells, usually strictly confined to the adventitial sheaths, but occasionally extending beyond the limiting glial membrane, and occurring as irregular scattered foci in the cerebral substance. There was a mild proliferation of the cerebral glia and various ganglion cells were present but difficult to evaluate because of the possibility of postmortem degenerations or the result of the terminal disease.

Neurotropic Newcastle Infection

The vestibulocerebellar complex in infected monkeys *(M. mulatta)* exhibited severe lesions but the remainder of the brain and spinal cord revealed small lesions or no lesions [VAN BOGAERT and INNES, 1962].

Neurotropic Dengue

Severe lesions were seen in the spinal gray and substantia nigra in monkeys *(M. mulatta)* and to a lesser extent the caudate nucleus and putamen [VAN BOGAERT and INNES, 1962].

Lymphocytic Choriomeningitis

The infection in monkeys *(M. mulatta)* was characterized by an irregular, but nearly constant, lymphocytic infiltration of the choroid plexuses of the lateral ventricles, moderate, irregularly distributed lymphocyte infiltration of the leptomeninges and very few foci of cellular gliosis of vessel sheaths and spinal cord [LILLIE, 1936]. Meningeal and choroid plexus infiltrations were observed to persist for long periods after infection. In spinal root ganglia, focal lymphocyte infiltration and sheath cell proliferation were observed.

Subependymal edema and focal lymphocyte infiltration were present in a few animals while ependymal desquamation from the ventricular walls and from the villi of choroid plexuses was also seen. In the brain there were occasional small vessels mantled by lymphocytes and a few small nodes of glia cells.

Coxsackie and Echo Viruses

The pattern of involvement for Coxsackie A, Coxsackie B and Echo viruses is similar to that for poliovirus infections. The areas of CNS most damaged are spinal cord, pons and medulla [HORSTMAN and YAMADA, 1968]. Monkeys *(M. mulatta)* were inoculated intracerebrally, intraspinally and intramuscularly with virus suspensions derived from infected mouse brain and/or torso or monkey cord [HORSTMAN and MANUELDIS, 1958]. The inoculated animals all exhibited lesions in spinal cord, medulla, cerebellar nuclei, pons, motor cortex, thalamus and hypothalamus. In most animals there were also lesions in the putamen, globus pallidus, caudate nucleus and cerebral cortex, including the parietal lobes. In some the involvement of the motor cortex was more marked than that in the medulla. The lesions in the thoracic region of spinal cord were less pronounced than in cervical and lumbar areas. In the cerebellar cortex lesions were seen in the vermis in some cases, and, less frequently, in the hemipheres. In cerebral cortex the lesions were slight in the frontal, parietal, temporal and occipital lobes and consisted only of perivascular infiltrates, with a few loosely formed glial nodules. The lesions were generally characterized by varying degress of chromatolytic changes in neurons, perivascular and diffuse infiltrates and glial nodules. In the spinal cord the neuronal lesions were asymmetrical and confined to the anterior horns, but many perivascular infiltrates and glial nodules were present in the posterior horns. The majority of the altered neurons in the cord and elsewhere underwent dissolution but some were removed by phagocytosis, and typical neuronophagia was demonstrated. Both cytoplasmic and nuclear changes were observed, and acidophilic necrosis was found in nerve cells in the spinal cord, cerebellar nuclei and motor cortex. The perivascular infiltrates were predominantly mononuclear, composed chiefly of lymphocytes. In some cases in the cord, and more commonly in the pons, cerebellar nuclei and motor cortex, diffuse infiltrates composed of PMNs were seen. Phagocytosis of dead neurons by microglial and other cells was observed, and many twisted, elongated microglial cells were scattered diffusely or occasionally formed glial nodules in affected areas.

The Russian Type 4 poliovirus is Coxsackie A7 [HABEL and LOOMIS, 1957]. This virus is capable of causing polio-type lesions in the monkey CNS. American strains of Coxsackie A7 are also pathogenic for monkeys and produce histological changes in CNS of certain inoculated animals. Such tissue alterations are indistinguishable from those caused by polioviruses. Degrees of neuronal degeneration varying from chromatolysis to vacuolization of cytoplasm, nuclear pyknosis and neuronophagia were seen. Additional lesions were focal gliosis of gray matter accompanied by polymorphonuclear leukocytic and lymphocytic perivascular infiltrations. Inflammatory and glial lesions also were observed in white matter and there was some loss of neurons in the spinal cord.

In some animals the lesions were distributed in such a manner that differentiation of their infections from those caused by polioviruses was possible. The caudate nucleus, putamen or both in these animals showed lesions and some had lesions in hippocampus and occipital cortex, areas rarely involved in poliovirus infections. The nearly equal degree of involvement of the precentral (motor) and postcentral (anterior parietal) cortex was also of use in differentiating this infection from that of poliovirus infection [HABEL and LOOMIS, 1957].

Histological differences from poliovirus infections are largely a matter of degree. Neither Coxsackie A, B, nor Echo viruses induce the massive involvement commonly seen in poliomyelitis [HORSTMANN and YAMADA, 1968], the process being generally milder than in poliomyelitis infections. The Russian AB-IV strain of Coxsackie A7 produced lesions in monkeys in temporal, parietal and occipital lobes, areas rarely involved in poliomyelitis, in which cortical involvement is confined to the precentral gyrus [HORSTMANN and YAMADA, 1968; HORSTMANN and MANUELDIS, 1958; HABEL and LOOMIS, 1957].

Monkeys inoculated with Echo viruses 2, 7, 9, 14, 16, 18 exhibited inflammatory lesions in scattered foci in gray matter but occasionally in white substance (LOU and WENNER, 1962]. The lesions consisted of accumulations of leukocytes, mainly mononuclear, surrounding small and medium sized blood vessels and, less frequently, interstitial infiltration of inflammatory cells. Gliosis secondary to neuronal damage was also present. In some cases the meninges were thickened by focal fibrous tissue proliferation. Neuronal lesions were found most often following intraspinal injection of types 7, 14, and 16.

Following experimental infection of monkeys *(M. irus)* with 7 strains of Coxsackie virus, large aggregates of mononuclear inflammatory cells and extensive perivascular cuffs of lymphocytes were widely distributed through the brain and spinal cord [WENNER *et al.*, 1960]. PMN's were occasionally observed surrounding necrotic neurons. Satellitosis, neuronophagia, chromato-

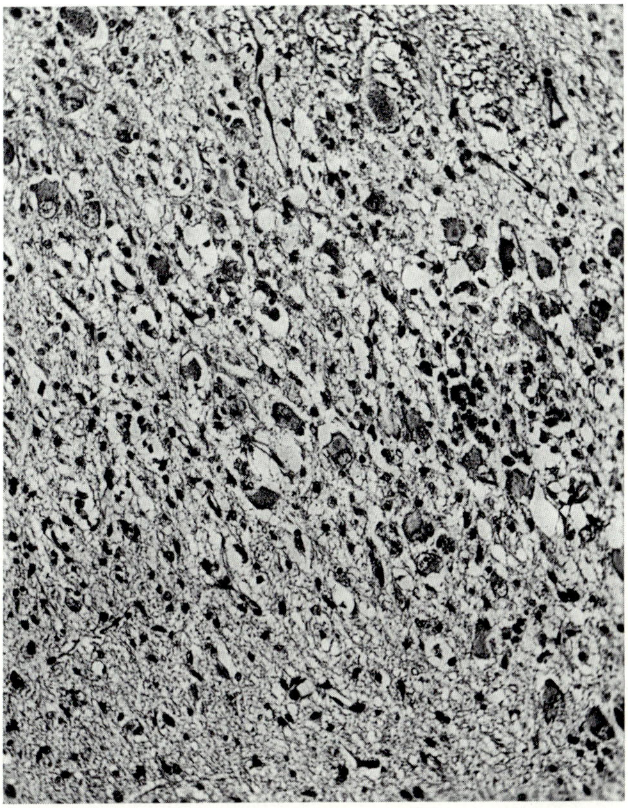

Fig. 16. Degenerative changes in neurons and round cell infiltration in Pons in monkey *(M. mulatta)* infected with Venezuelan equine encephalitis (HE ×123).

lytic changes, karyolysis and gliosis were observed. In 10 of the 13 monkeys studied both brain stem and spinal cord were involved, but in 3 the lesions were found only in pons and medulla. Although neuronal changes such as nuclear displacement, chromatolysis and occasionally karyolysis, were seen the relationship of these solitary changes to infection remained uncertain.

Equine Encephalitis

Monkeys *(M. mulatta)* which were inoculated intracerebrally with unpassed virus from a fatal human case of eastern equine encephalitis and with two unpassed mosquito isolates [NATHANSON *et al.*, 1969] exhibited severe lesions in the forebrain, especially in the lenticular nuclei, thalamus and cerebral cortex. The lesions were characterized by a very marked loss of neurons. Despite the severity of this cell loss, there was little evidence of gliosis, atrophy or cyst formation. In areas of marked neuronal loss, the major inflam-

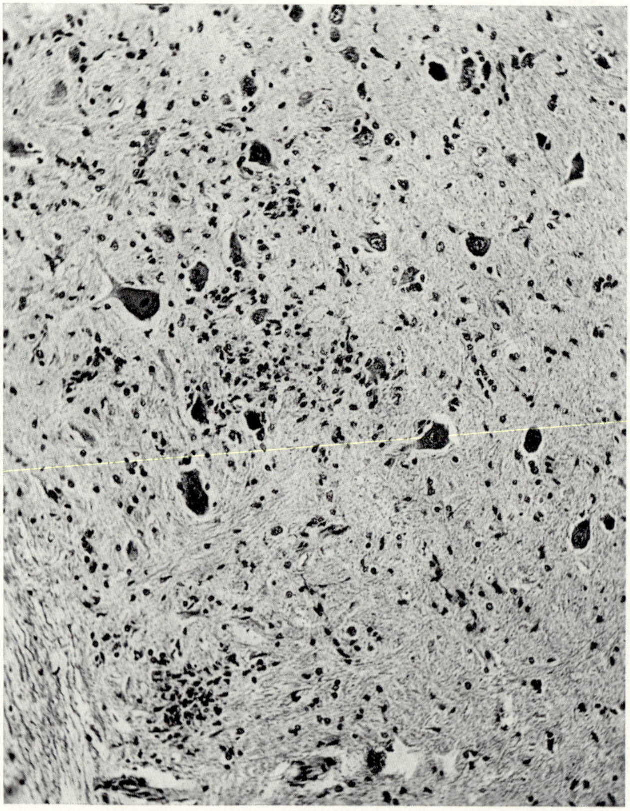

17a

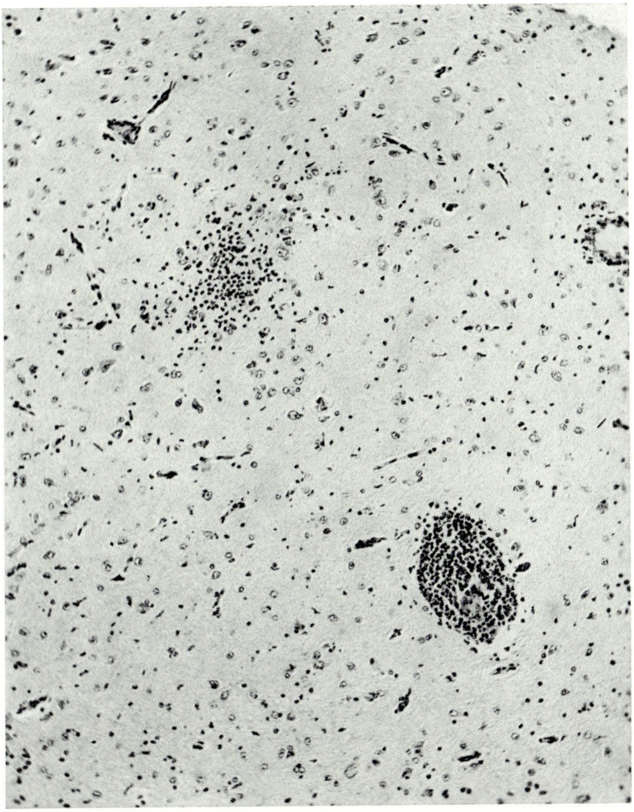

17b

Fig. 17. Degenerative changes in neurons in spinal cord (a) and round cell infiltration of spinal cord (a) and thalamus (b) in monkey *(M. mulatta)* infected with Venezuelan equine encephalitis (a) (HE ×123), (b) (H and E ×109).

matory changes were perivascular cuffs and leptomeningitis of varying severity. In many areas of CNS, including the spinal cord, only slight changes, mainly scattered small focal infiltrations and mild leptomeningitis, were observed.

Venezuelan Equine Encephalitis

Venezuelan equine encephalitis is known to occur naturally in subhuman primates but no comprehensive published descriptions are available. Degenerative changes in neurons and round cell infiltration of the pons in a rhesus monkey *(M. mulatta)* infected with this virus is shown in figure 16.

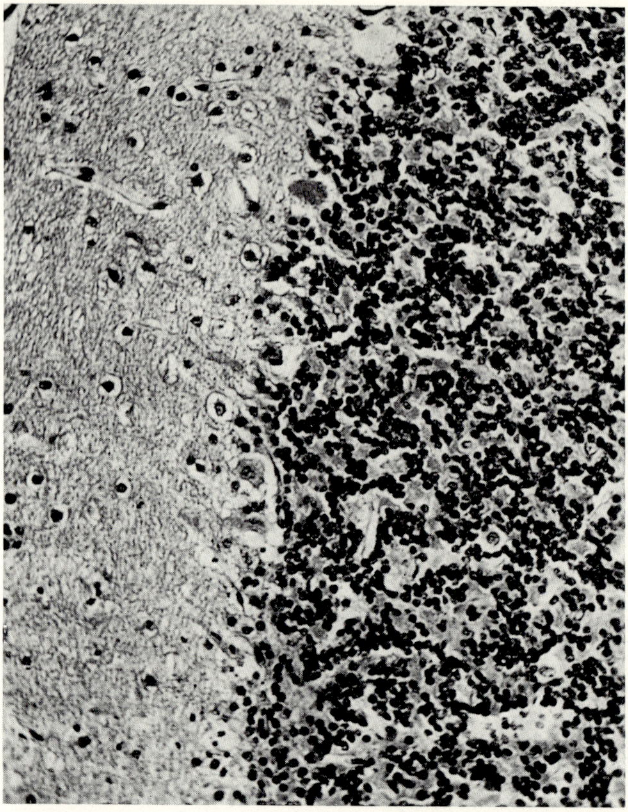

Fig. 18. Neuronal (Purkinje cell) degeneration and round cell infiltration of cerebellum in monkey *(M. mulatta)* with Venezuelan equine encephalitis (HE ×200).

Similar changes are illustrated in spinal cord (fig. 17), cerebellum (fig. 18), frontal cortex (fig. 19) and hippocampus (fig. 20).

Encephalitis Affecting Basal Ganglia

In a monkey *(M. mulatta)* which was injected with a buffered saline solution of frontal cortex tissue from a patient who had died of chronic encephalitis, an encephalitis developed characterized by many areas of 'paling' from coagulation necrosis. Such areas were strewn throughout the caudate and lenticular nuclei on both sides of the brain [RICHTER, 1945]. These areas contained only a few thickened vessels, some with infiltrations of round cells and some pyknotic glial nuclei but no ganglion cells. No softening was observed, and there was no glial reaction within the lesions, although a peripheral glial

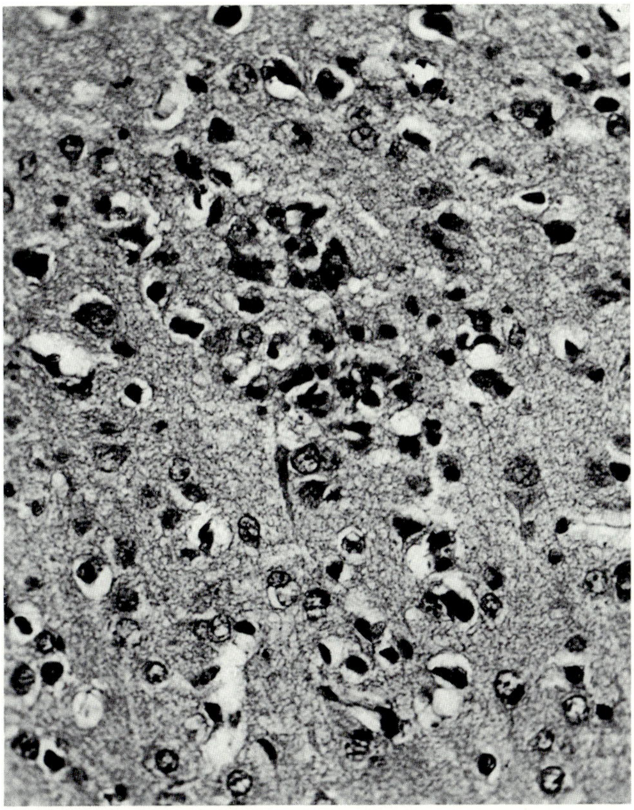

Fig. 19. Degenerative changes in neurons and round cell infiltration of frontal cortex in monkey *(M. mulatta)*, with Venezuelan equine encephalitis (HE ×285).

reaction with hyperplasia of all glial forms was seen surrounding some of the necrotic areas. In these necrotic zones many of the vessels were thickened from proliferation of intima and adventitia, and were often infiltrated both by lymphocytes and PMN's.

In the thalamus there was some diffuse increase of the glia and infiltration of round cells about a few vessels. No lesions were observed in cerebral cortex, cerebellum, brain stem or spinal cord.

Austrian Tick-Borne Encephalitis

Austrian tick-borne encephalitis is also known to occur naturally in subhuman primates but no extensive descriptions of the pathology are available. Perivascular cuffing in a subhuman primate infected with this disease (species

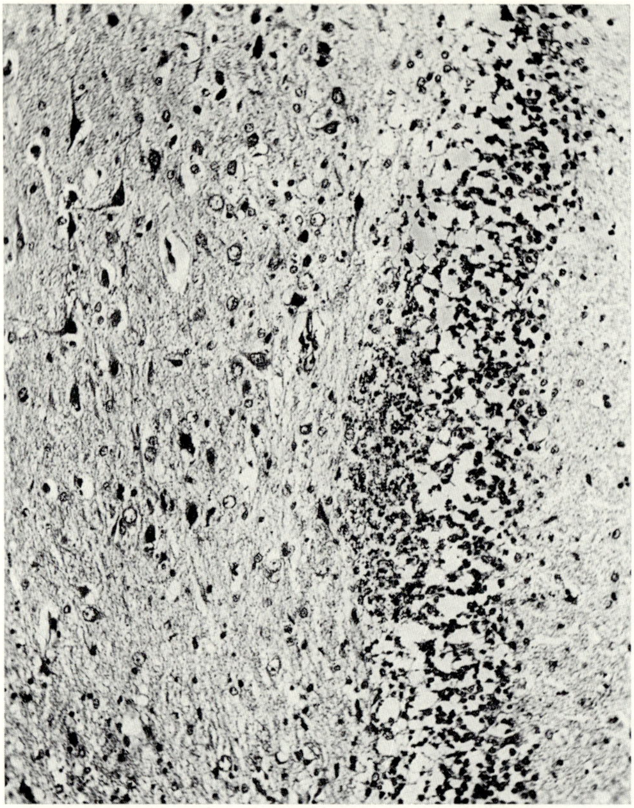

Fig. 20. Degenerative changes in neurons and round cell infiltration of hippocampus in monkey *(M. mulatta)*, infected with Venezuelan equine encephalitis (HE ×124).

unknown) and degenerative neuronal changes in cerebral cortex are illustrated in figure 21. Round cell infiltration in the thalamus of the same animal is shown in thalamus in figure 22.

Non Suppurative Meningitis

VAN BOGAERT and INNES [1962] reported, in *Erythrocebus patas*, that one group of cases, with relapsing epilepsy but no general physical disorder, exhibited plurifocal lymphocytic meningitis with scanty perivascular infiltrations in the white matter of the centrum ovale, while the brain stem and cord were normal. In another group with progressive changes in behavior and an odd gait, with deformation of bones of extremities, there was a diffuse lym-

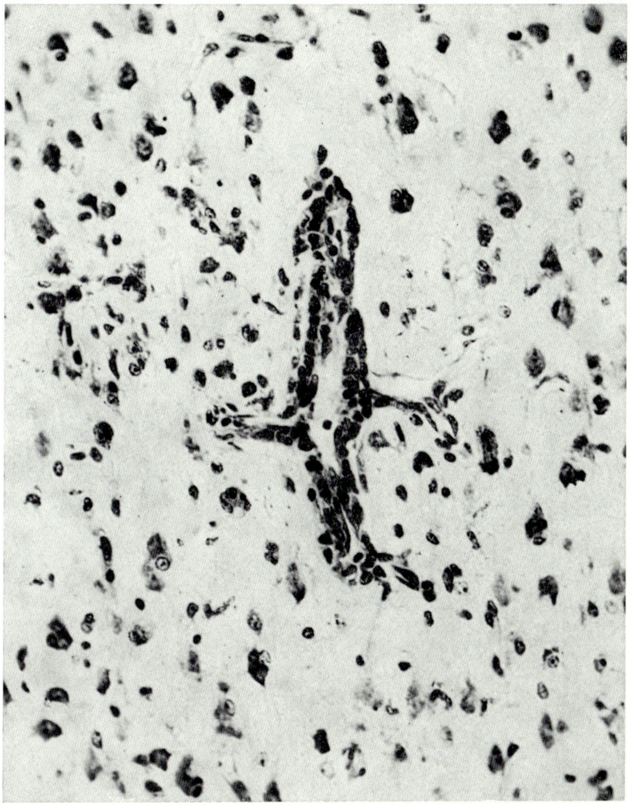

Fig. 21. Perivascular cuffing and degenerative changes in neurons in primate (species unknown) with Austrian tick-borne encephalitis (HE ×290).

phocytic meningitis with plurifocal basilar localization and a sparse perivascular glial infiltration in the thalamo-striate and orbital regions. A third group consisted of a few animals which exhibited clinical signs indicative of some characteristic severe neurologic disorder but no lesions severe enough to account for such signs.

In one member of this group the tissues exhibited focal lymphocytic-histiocytic meningitis located at the summits of gyri, around the chiasma, and scanty infiltrations were found in the basal ganglia. Macrogliosis was seen in the molecular zones and in layers I to III in the entire cerebral cortex.

The animals of all of these groups, with different clinical findings, had a somewhat similar pathological pattern. The tissue reaction was observed to

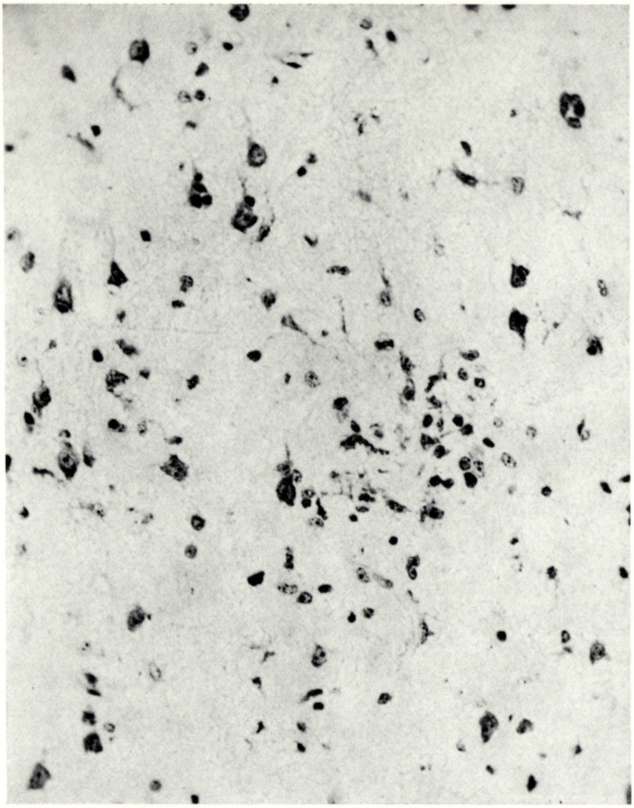

Fig. 22. Round cell infiltration in thalamus in primate (species unknown) with Austrian tick-borne encephalitis (HE ×290).

be spread out in sulci or at the summits of gyri in cerebral cortex, and in the subjacent molecular parenchyma there was a microglial infiltration. Vessels penetrating the pia had very slight cuffing and similar lesions were seen in the white matter and basal ganglia. The diffuse macrogliosis was observed to spread into the superficial cortex in some cases. Some loss of Purkinje cells and rarefaction of the dentate nuclei in cerebellum was also noted.

These authors were of the opinion that any relationship to lymphocytic choriomeningitis would be impossible to assess and noted that, in monkeys experimentally infected with the viral disease, the meningeal lesions were more severe, more diffuse and more strictly localized to the meninges than in the naturally occurring disease. They noted that the histologic picture was different from that seen in experimental B [SABIN] virus infection, in which there were also characteristic inclusion bodies and suggested that paratyphoid organisms, which have been reputed to cause meningitis in children and in pigs, may also cause the disease in monkeys, especially those with concomitant dystenteric trouble.

Infectious Diseases-Bacterial

Suppurative Meningitis

VAN BOGAERT and INNES [1962] described a case of suppurative meningitis in a monkey *(M. mulatta)* which arose from a purulent otitis and rhinitis, and exhibited a diffuse purulent meningoencephalitis extending to the tuber cinereum. Numerous perivascular infiltrations were observed in the putamen, pallidum, and chiasma. There was some infiltration into the molecular zone of cerebral cortex but no other marked involvement of the cortex. In one case, meningitis followed an acute purulent tonsillitis. Here the suppurative process was marked in the lateral recess of the fourth ventricle with ependymitis and extensions of the process into brain stem and cerebellum. In another case a Ceropithecoid contracted suppurative meningitis after cranial fracture and exhibited diffuse cerebrospinal meningitis with ependymitis and extension of the process into hippocampus and ventral pons. Laminae II and III of cerebral cortex appeared pale, presumably from ischemia.

A few cases of suppurative meningitis have been reported at the London Zoo [FIENNES, 1967] but no pathology was reported. WEINBERG [1909] reported discrete lesions in the brain, spinal cord and meninges in a female chimpanzee which had been inoculated with the syphilis organism. The pia mater was dull in appearance and the central canal contained pus.

At Delta Primate Center a monkey *(Callicebus moloch)* was observed with encephalitis and septicemia associated with a *Diplococcus pneumoniae* infection of the middle ear. At autopsy the base and cortical lobes of the brain were grayish and soft and a fibropurulent material was present over a portion of the lower part of the left hemisphere. Small necrotic foci with round cell aggregations were demonstrated and there were numerous Gram-positive organism resembling diplococci in necrotic foci and beneath the pia mater (fig. 23–25).

Another monkey of the same species with an upper respiratory infection, and bronchopneumonia and clinical symptoms of progressive paralysis

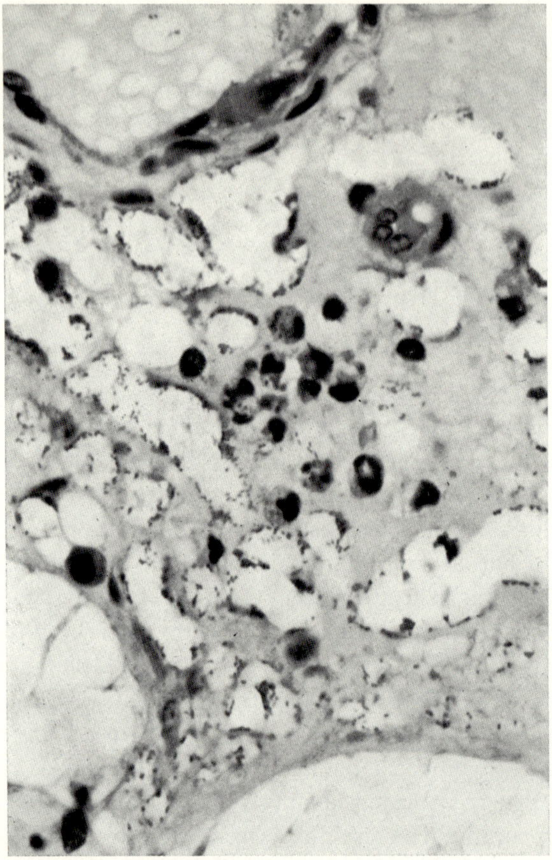

Fig. 23. Round cell aggregation in necrotic focus in cerebral cortex of monkey *(Callicebus moloch)*, with encephalitis and septicemia associated with middle ear infection. Diplococci are evident throughout the area (HE ×950).

exhibited a large abscess, which invaded and destroyed part of the internal capsule, thalamus and basal ganglia (fig. 26). The abscess contained numerous small, rod-shaped bacteria around necrotic blood vessels. The organisms were presumed to be *Pseudomonas aeruginosa* since a bacterial examination of heart blood, lung tissue and colon contents yielded cultures of these organisms.

In a vervet monkey *(Cercopithecus aethiops)* with bronchopneumonia at Delta Primate Center, microscopic examination of the brain revealed extensive suppurative leptomeningitis, with some extension of the inflammato-

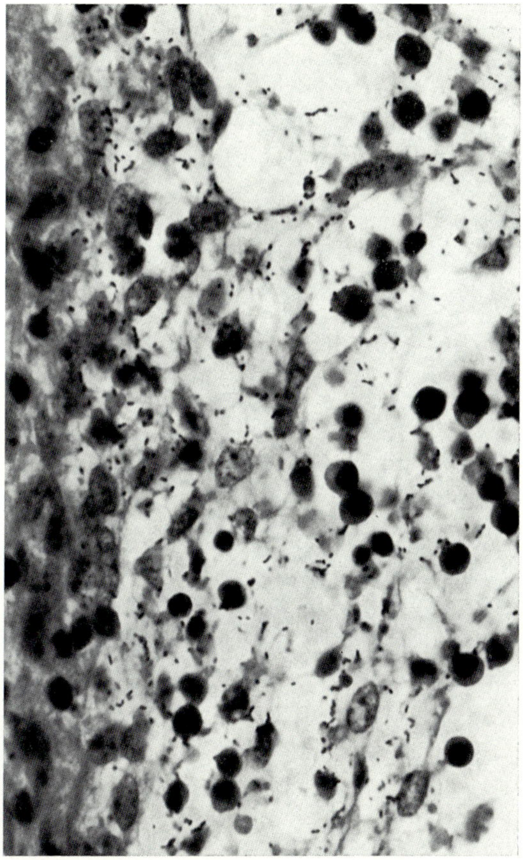

Fig. 24. Diplococci in meninges of monkey *(Callicebus moloch)*, with encephalitis and septicemia associated with a middle ear infection (Gram stain and HE ×950).

ry process into adjacent brain tissue (fig. 27). Numerous Gram-positive diplococci were present in the purulent meningeal exudate. The organism recovered from the meninges was identified as *Diplococcus pneumoniae* on the basis of morphology, Gram-staining reaction, colony characteristics on blood agar plates and optochin susceptibility.

A young patas monkey *(Erythrocebus patas)* born at Delta Primate Center developed pneumonia and expired after intensive treatment with antibiotics and oxygen therapy. The meninges of one cerebral hemisphere had a deep dark red discoloration and appeared to contain pus along some of the

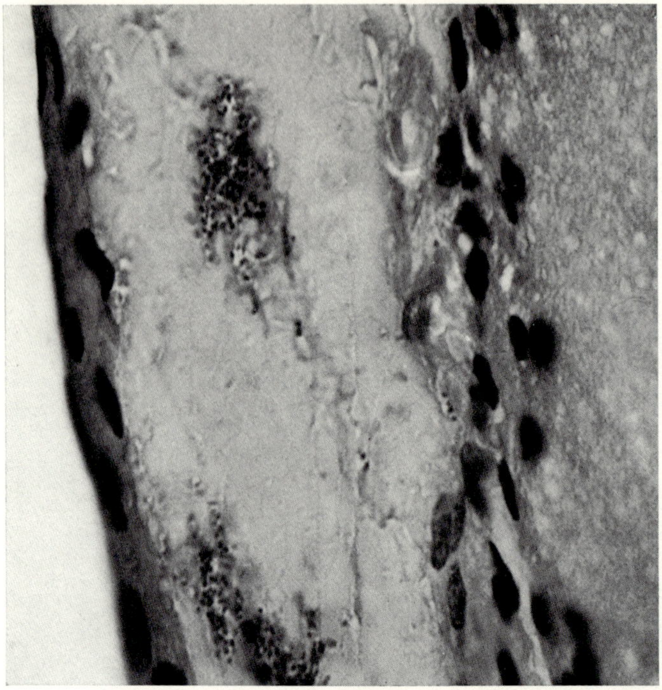

Fig. 25. Subarachnoid hemorrhage (hemolyzed) with focal accumulations of Diplococci in monkey *(Callicebus moloch)*, with encephalitis and septicemia associated with a middle ear infection (HE ×950).

sulci. Microscopic examination of the brain revealed suppurative meningitis with numerous Gram-positive diplococci in the exudate (fig. 28, 29).

Leptomeningitis was also observed in a squirrel monkey *(Saimiri sciurea)* which had received a penetrating wound to one cerebral hemisphere (fig. 30). Cultures of heart blood remained sterile during a period of 7 days incubation and no enteric pathogens were isolated from the colon contents. On microscopic examination of the brain the lesions proved to be large abscesses (fig. 31) containing numerous bacteria.

A juvenile male chimpanzee *(Pan troglodytes)* with acute bacterial endocarditis at Delta Primate Center exhibited microabscesses (fig. 32) together with small hemorrhages containing Gram-positive cocci in the brain. Bacteriologic examination of cardiac vegetations, heart blood, lung tissue and bile in this animal revealed coagulase positive *Staphylococcus aureus*. *Escherichia coli* was also isolated from the gall bladder contents.

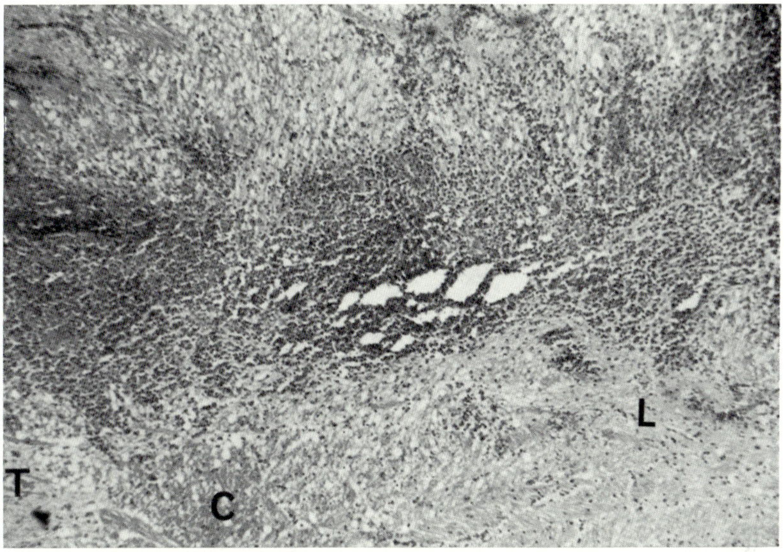

Fig. 26. Abscess in thalamus, internal capsule and basal ganglia, presumably caused by *Pseudomonas aeruginosa* in *Callicebus moloch* (Thalamus – T, Internal Capsule – C, Lentiform nucleus – L. HE ×20).

Tuberculosis

According to VAN BOGAERT and INNES [1962] the disease in monkeys is comparable only to primary infection and its early postprimary generalization as commonly seen in children. There is no phase analogous to chronic isolated tuberculosis (phthisis) of adult human beings. VOGEL [1955, Cite VAN BOGAERT and INNES, 1962] found no instance of tuberculosis of the brain in 69 *Macaca mulatta* with the disseminated disease.

A monkey *(M. mulatta)* with generalized but chiefly lymphatic tuberculosis, showed a large plaque on the external surface of the dura over the vertex where it was adherent to a yellow, fairly firm nodule about 10 mm in diameter. The nodule was spherical, and deeply imbedded in the brain surface. It was not encapsulated and there was no peripheral reactive zone. The meninges were not altered elsewhere and the blood vessels were not congested. The pia arachnoid contained no excess of fluid but the summits of the convolutions were flattened. The mass was located in the posterior frontal convolution near the longitudinal fissure, occupying nearly its entire width and penetrating about 1 cm. The core was 2 mm in diameter and completely degenerated [Fox, 1923].

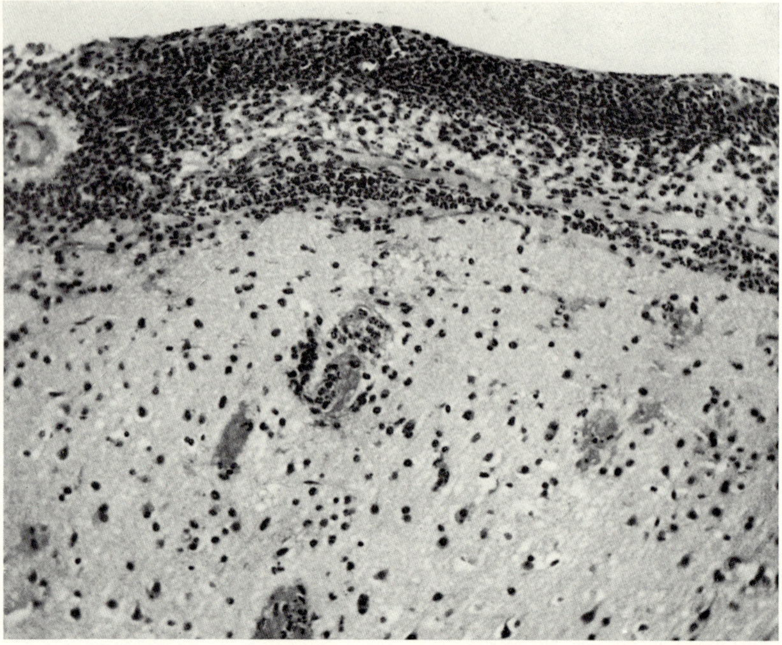

Fig. 27. Purulent meningitis in *Cercopithecus aethiops* with extension of inflammatory reaction into cerebral cortex (HE ×140).

In a drill baboon *(P. leucophaeus)* it was observed by the same author at autopsy that the dura was adherent to the left side of the brain in the prefrontal region in the upper third, and an area of softening lay beneath this point. Tubercle bacilli were found in the pus removed from the area of softening beneath the left cortex. Three foci of softening were observed in the brain, one in the left hemisphere in the prefrontal region, extending from just beneath the cortex in the upper third of the region downward almost to the base of the brain. The area was 2.5 cm in diameter. The area of softening consisted of caseated material and pus in which tubercle bacilli were present. A second area 1.5 cm in diameter was found on the right side of the brain. This consisted of a circumscribed mass of caseous material but there was no pus present. The apex of this area was just beneath the cortex in the prefrontal region on the right side, 0.5 cm in front of the central fissure and 1 cm below the superior surface of the brain. A third area of softening was observed posteriorly, in white substance, in the parieto-occipital region. This was 6 mm in diameter.

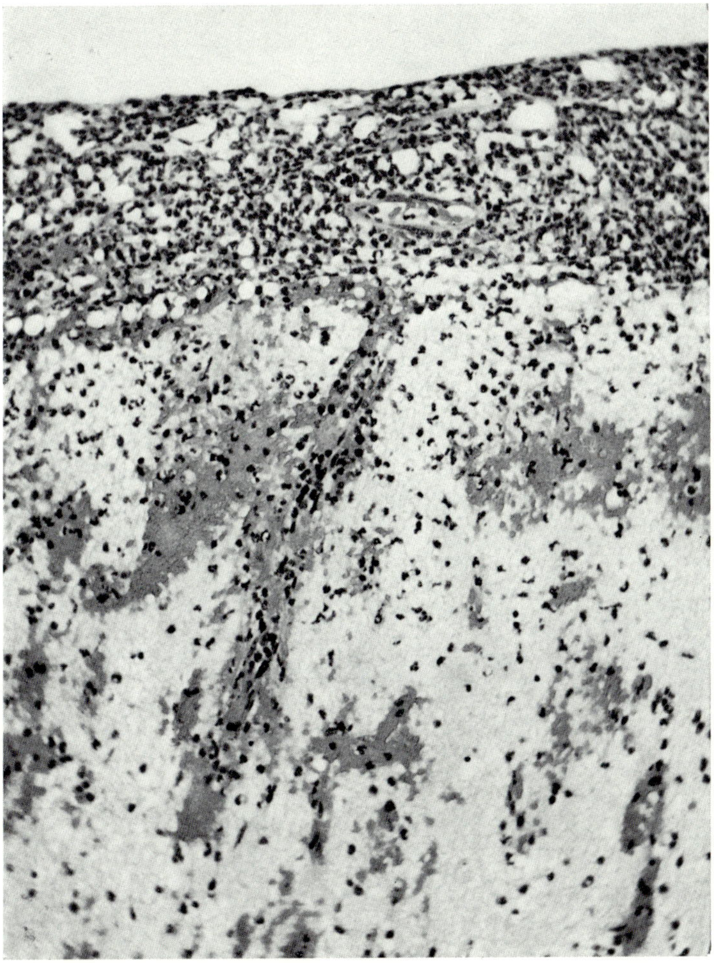

Fig. 28. Suppurative meningitis in a patas monkey *(Erythrocebus patas)*, with diplococcal pneumonia, with extension of the inflammatory process into cerebral cortex (HE ×140).

As a result of these lesions, the optic thalamus and lenticular nucleus, and the posterior limb of the internal capsule, on the left side were partly destroyed. The pyramidal tracts on the opposite side to this lesion exhibited definite degenerative changes, but these were not considered to be severe, considering the extent of damage to the internal capsule.

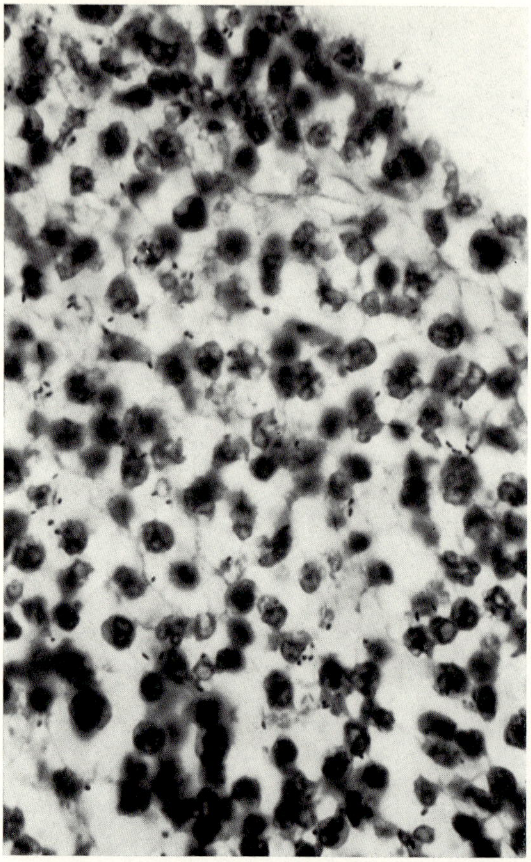

Fig. 29. Diplococcal organisms in meninges of patas monkey *(Erythrocebus patas)*, with diplococcal pneumonia (Gram stain and HE ×950).

One baboon *(P. anubis)* with tuberculosis exhibited purulent ophthalmitis [FOX, 1923]. The author was of the opinion that the tuberculous process had probably begun in the region of the optic nerve and involved the whole bulb.

SCHERER [1944] observed a higher incidence of tuberculosis, than usual when a careful search was made for solitary tubercles in the brain. This worker noted tubercles varying from the size of a pinhead to the size of a cherry stone in various regions of the brain and meninges. In some animals, diffuse meningeal involvement without tubercle formation but with hyperemia of the pia mater was observed. PFEIFER [1935, Cite VAN BOGAERT

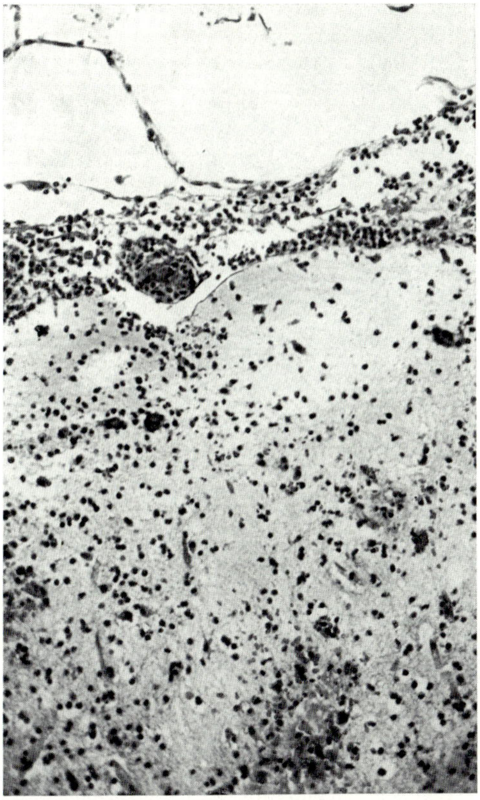

Fig. 30. Leptomeningitis with small abscesses in cerebral cortex apparently resulting from penetrating wound of cerebral hemisphere (HE ×120).

and INNES, 1962] described two types of lesions following injection of india ink into cerebral vessels in two macaques and a baboon. In one type small areas of edema were observed in relation to the tuberculous lesions, while in the second type vessel dilatation was seen.

LUCKE [1923] showed a relatively large solitary tubercle in the pons, adjacent to the sensory nucleus of the left fifth nerve of a sooty mangabey *(Cercocebus fuliginosus)*. The necrotic area was circular and about 2 mm in diameter. The central portions were crumbly and stained lightly with eosin. The periphery consisted of a thin and irregularly arranged wall of large, pale epithelioid cells, between which were found several typical tuberculous giant cells. Encircling the lesion was an indefinite ring formed by scanty

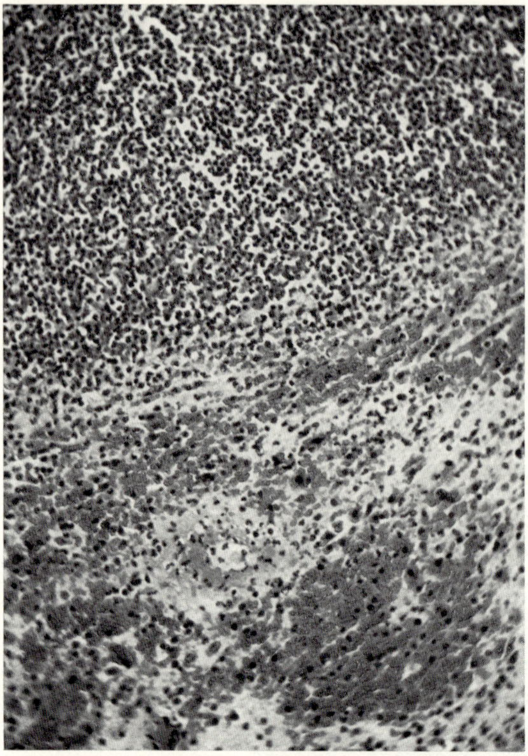

Fig. 31. Large abscess located deeply in cerebral cortex of a squirrel monkey *(Saimiri scuireus)*, and apparently resulting from an adjacent penetrating wound of cerebral cortex (HE ×60).

numbers of large and round cells. The smaller elements possessed deep staining rounded nuclei and but little cytoplasm, the larger cells had fairly abundant cytoplasm and paler nuclei; their cell bodies were indefinite and faded into the surrounding structures. Similar elements were scattered for some distance through the loose and slightly edematous nerve substance. A small blood vessel with an obliterated lumen lay just within the wall of the necrotic area. Elsewhere in the pons there were a number of very small vessels with a few round cells in their adventitial sheaths. The meninges were here and there slightly edematous and infiltrated with occasional round cells.

MARTIN *et al.* [1960] reported a tuberculous abscess at autopsy in a monkey *(M. mulatta)*. The abscess was observed to incorporate the 9th

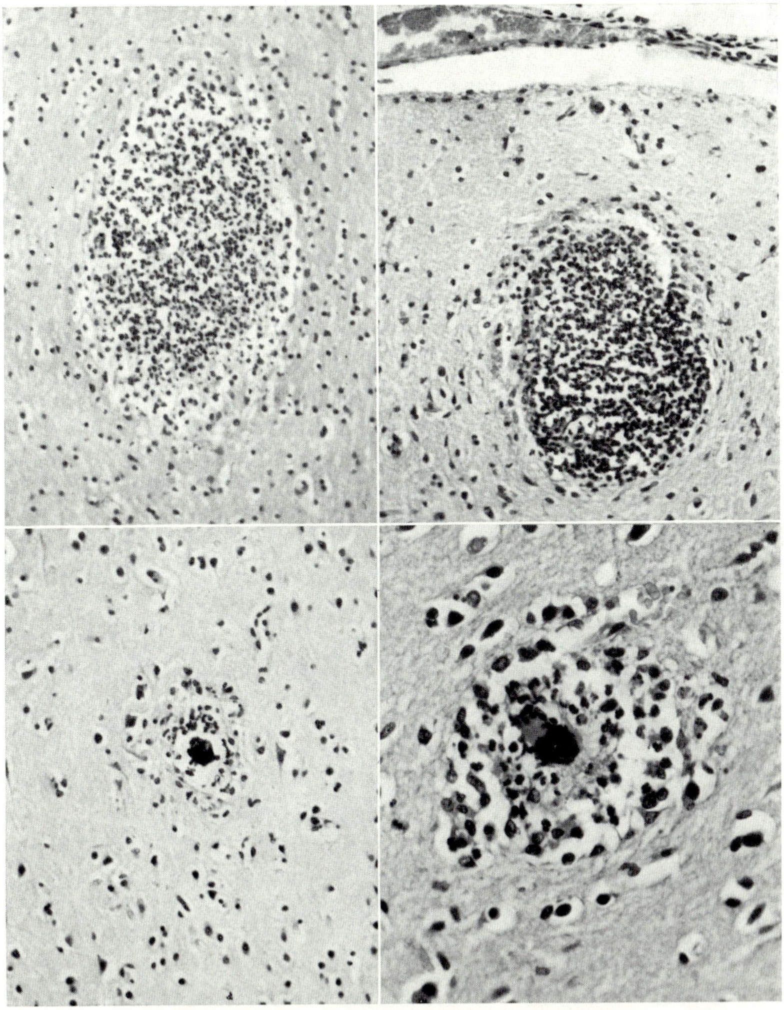

a b c d

Fig. 32. a, b, c, d. Small abscesses containing Gram-positive cocci in cerebral cortex of young male chimpanzee with acute bacterial endocarditis (a,b,c $\times$ 60, d $\times$ 160).

through 11th thoracic vertebral bodies, with resultant collapse of the 10th vertebral body and kyphosis of the thoracic spine. The spinal canal between the 9th and 11th thoracic vertebrae was narrowed, and there was malacia, with loss of all architectural characteristics of the spinal cord in this area by confluent tubercles.

FAIRBROTHER and HURST [1932] reported that involvement of the nervous system in cases of tuberculosis in primates was rare, but that in one case a focus in the bone adjacent to the temporal pole had become adherent to meninges and brain, and had invaded the latter to the depth of several millimeters. Tuberculous meningitis was never seen, but in one case numerous curious cellular foci were scattered throughout the brain, and in one a small tuberculoma was situated in the antero-lateral white matter of the lumbar cord. Distributed throughout the basal nuclei and centrum ovale, particularly numerous in the brain stem, white matter of the cerebellum and cerebellar cortex and but rarely seen in the cerebral cortex and white matter of the spinal cord, were numerous small cellular foci composed apparently of microglial cells with occasional lymphocytes. Among the former, mitoses were frequent. The larger foci showed an acellular reticulated or fibrillated central area, staining bright orange with van Gieson, with a palisade-like arrange-

33a

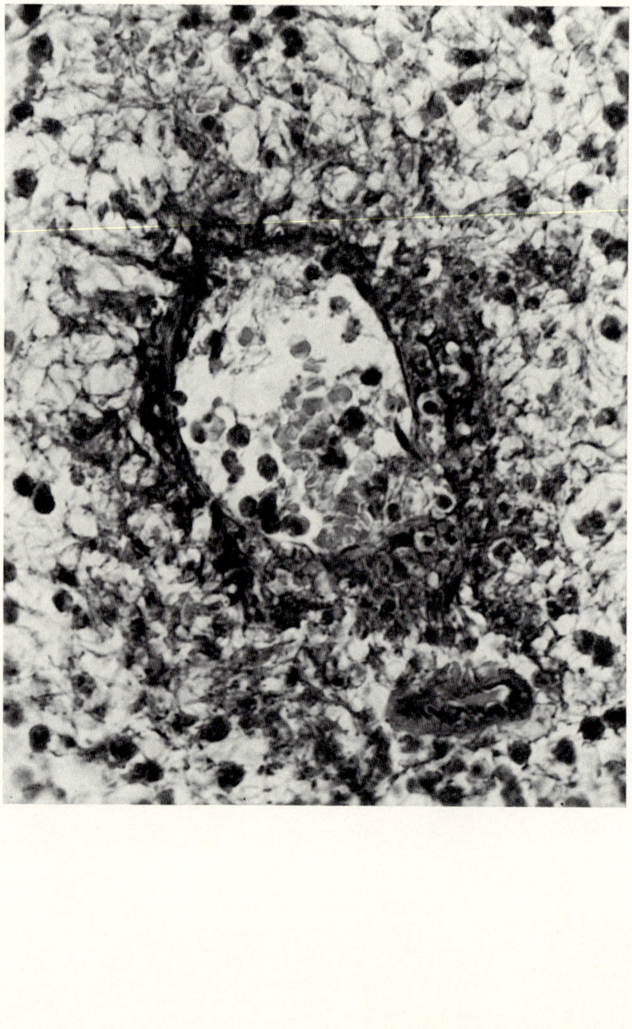

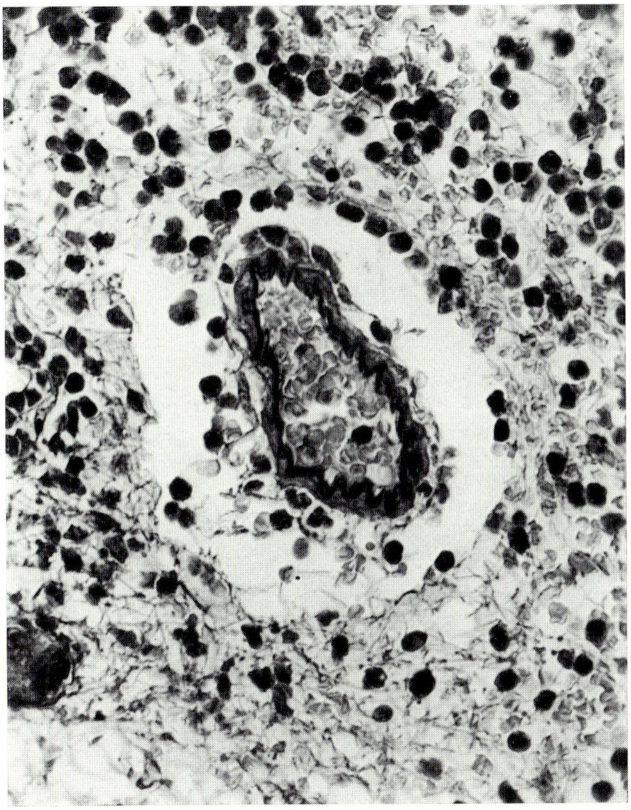

33b

Fig.33. Tissue reaction to experimental anthrax infection of brain in monkey *(M. mulatta)* (a ×546, b ×475).

ment of the bordering cells. Frequently extravasation of blood had occurred within the focus and in the surrounding tissues. Some of the foci obviously surrounded a capillary or pre-capillary; in others an occluded vessel was probably present. There was no suggestion of the existence of tuberculous meningitis, nor did foci occur in the meninges. No tubercle bacilli were observed in foci which did not show the structure of a typical tubercle.

Anthrax

No descriptions of anthrax infections involving the brain in subhuman primates are available, although it is known that the primate brain may be involved in anthrax infections. The tissue reaction to the infection in a monkey *(M. mulatta)* is illustrated in figure 33.

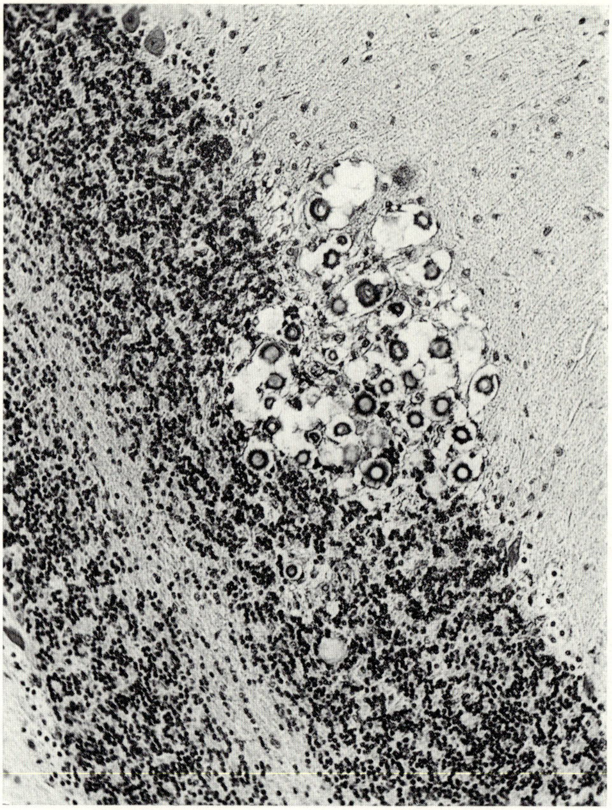

Fig. 34. Cryptococcus organisms in the cerebellum of a monkey *(M. mulatta)* (Mucocarmin ×157).

Encephalopathies Related to Dysenteric Syndromes

Histopathological findings in monkeys suffering from dysenteric conditions have included degenerative changes in papillomacular bundle, optic nerve and chiasma, inflammatory and degenerative changes in the brain and meningoencephalitis. VAN BOGAERT and INNES [1962] described a case in an orang-utan in which acute homogeneous changes were found in the neurons of the dentate nucleus, olives, and basal ganglia without any glial or mesodermal reaction. Glial nodules were observed in the subependymal regions of the third and fourth ventricles. Small microglial collections occurred in the deeper layers of the temporal and parietal cortex and in the peduncles.

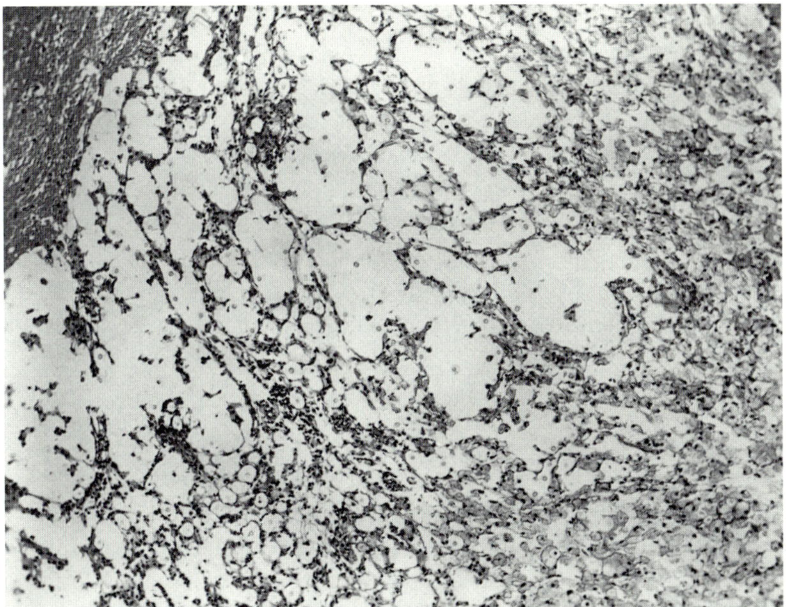

35

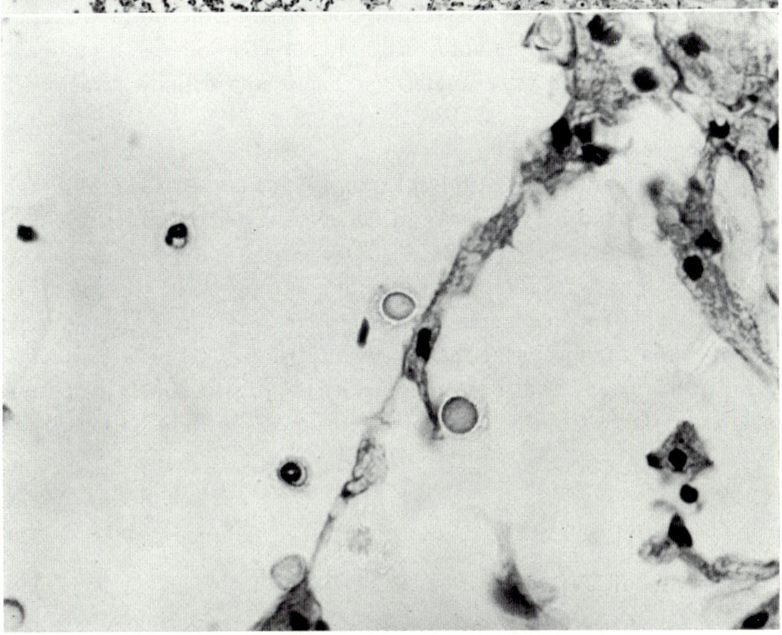

36

Fig. 35. Encephalomalacia in cerebrum of monkey *(M. mulatta)*, resulting from invasion by Cryptococcus organisms. (×110)

Fig. 36. Cryptococcus organisms in cerebrum of monkey *(M. mulatta)* (×320).

In an 8-year-old chimpanzee these workers reported homogeneous changes in the neurons of both dentate nuclei in cerebellum. Pyramidal cells of the motor cortex were also affected and a microglial reaction was seen in this area (described by SCHERER). In four specimens of *Macaca mulatta* with dysentery, homogeneous changes in neurons of the cerebellar cortex, dentate nuclei, pallidum and in the large motor neurons of the cortex were seen without necroses. It was concluded by these authors [VAN BOGAERT and INNES, 1962] that the dysenteric colitis may itself produce neuronal lesions as in the case of nutritional infantile diarrhea, and disseminated nodular glial reactions are often found together with neuronal lesions as in encephalitis.

Fungal Infections

Torulosis (Cryptococcus)

Several strains of torula, recovered from human brain lesions and from a granulomatous lesion of the knee, were each inoculated intracardially into monkeys *(M. mulatta)* [KESSEL and HOLTZWART, 1936]. Brain lesions were found to occur within the cortex and smears from the brain were positive for torula. Meningitis was not a marked characteristic in the monkey even though brain lesions were numerous. There was definite invasion of the spinal cord tissue.

Other Fungal Infections

While cases of candidiasis [CATANEI, 1925; FIENNES, 1967, YOUNG *et al.*, 1957], histoplasmosis [WALKER and SPOONER, 1966; NEGROM, 1965], coccidioidomycosis [HAMERTON, 1936; NEGROM, 1965], and actinomycosis [WEIDMANN, 1935], have been reported in subhuman primates, there is no record of nervous system involvement in any of these diseases. The presence of cryptococcus organisms in the cerebellum of a monkey *(M. mulatta)* is illustrated in figure 34 and the presence of the organism and the encephalomalacia resulting from invasion of cerebrum is shown in figures 35 and 36.

Degenerative Diseases

Diseases of Cerebellum

Exposure of monkeys to manganese aerosol [VAN BOGAERT and DALLEMAGNE, 1946] resulted in cerebellar atrophy. Using the same method VAN BOGAERT and INNES [1962] observed funicular myelosis and papillomacular

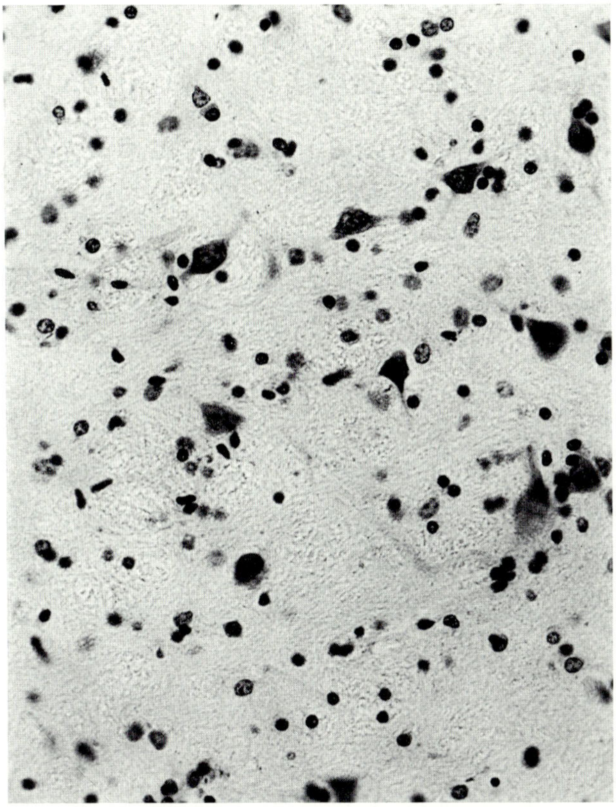

Fig. 37. Atrophic changes in neurons of globus pallidus in manganese encephalitis, in monkey *(M. mulatta)* (Cresyl violet ×290).

degeneration resembling spontaneous leukoencephalomyelosis and cerebellar cortical atrophy. The parts most severely affected in the order of intensity were the culmen, simplex lobe, quadrangular lobe, inferior semilunar lobe, uvula, median anterior lobe and deeper parts of the superior semilunar lobe. These changes included complete degeneration of the Purkinje and granule cells of the cortical layers, gliosis arising from the Bergman layer, diffuse gliosis in the white matter and deep nuclei, and degenerative lesions in the dentate nuclei, especially the dorsomedial part, and olives. The fastigial nuclei and roof were altered to a lesser extent. Myelin stains showed some pallor in the cerebellar white matter, especially near the tips of atrophic folia, but special stains did not reveal the presence of broken down lipids.

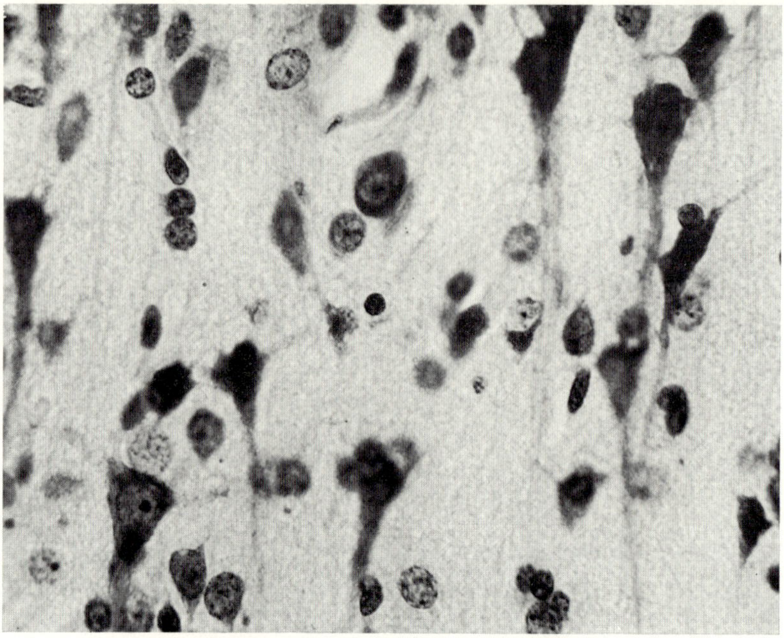

Fig. 38. Atrophic changes in neurons of cerebral cortex in manganese encephalitis in monkey *(M. mulatta)* (Cresyl violet ×855).

Naturally occurring cerebellar atrophy was reported by URBAIN *et al.* [1940] in a gelada. A bullous lesion was found in the right part of the cerebellum. Atrophy of the white matter in the opposite cerebral hemisphere was seen, especially in the temporal and frontal lobes. Lesions of the contralateral inferior olive and branchium pontis were also observed.

Diseases of Basal Ganglia

Maganese Poisoning

Six monkeys *(M. mulatta)* were subjected to gradually increasing doses of manganese chloride, administered intraperitoneally every other day from the 4th to 22nd months of life [MELLA, 1924]. Hematoxylin and eosin, cresyl violet and/or phosphotungstic acid hematoxylin stains, revealed slight, but evenly distributed chromatolytic changes in the cerebral cortex, but there was no gliosis, myelin degeneration or meningeal or vascular abnormalities.

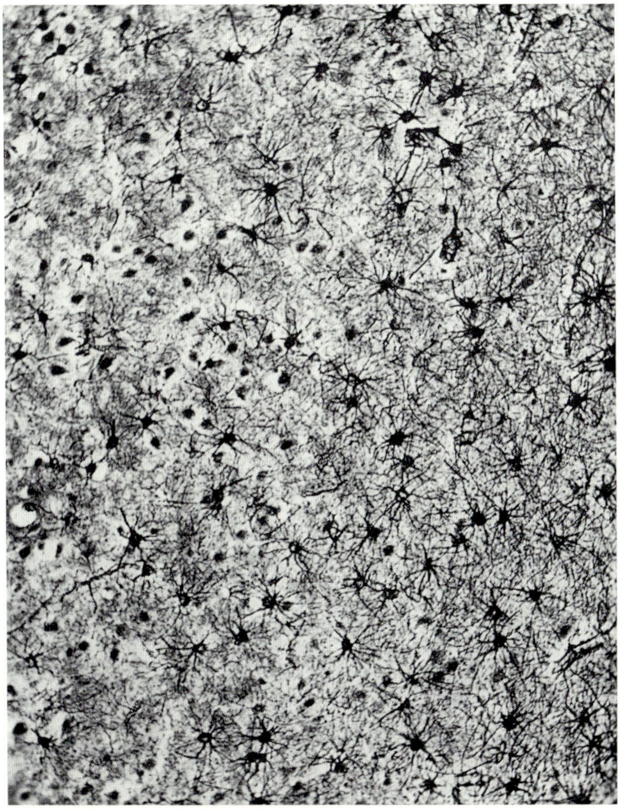

Fig.39. Astrocytosis in cerebral cortex of monkey *(M. mulatta)*, in manganese encephalitis (Cajal gold sublimate ×157).

Atrophic changes in neurons in manganese poisoning are illustrated in figure 37 and 38. In the putamen and caudate nucleus the cells were shrunken and pyknotic, often without nuclei and neuronophagia was sometimes observed. The large cells of the pallidum were swollen and vacuolated and showed excentric nuclei. In certain scattered areas gliosis was observed which was sometimes quite dense. Figure 39 illustrates the astrocytic reaction in this type of encephalitis. In Mella's investigation a loss of myelin was observed on many of the fibers connecting the caudate nucleus and putamen. In the pallidum some of the large cells were shrunken and pyknotic, others were swollen and chromatolytic, while others had excentric nuclei and vacuoles. A loss of myelin sheaths was observed in the ansa lenticularis and there were areas of gliosis in the pallidum in one brain. No definite alterations were observed in cerebellum or pons with the exception of occasional swollen ganglion cells in the latter. In these areas there was no gliosis and the blood vessels appeared normal. The spinal cord was normal and peripheral nerves showed no Marchi degeneration. The muscle fibers appeared normal and there was no evidence of nuclear proliferation. No information on tissue fixation methods was given.

A monkey treated with repeated intramuscular injections of insoluble manganese dioxide suspended in olive oil and sacrificed 14½ months after the initiation of the injection revealed diffuse and holotopistic brain alterations [PENTSCHEW *et al.*, 1963]. The holotopistic alterations were most severe in the subthalamic nucleus and medial pallidum and were characterized by a nearly complete loss of neurons and a great proliferation of bizarre glial cells, provisionally classified as 'immature' glia. In the lateral pallidum nerve cell loss was slight but the cells were atrophic. The pattern of the 'immature' glia was different from that of the medial pallidum. In the magnocellular part of the lateral habenular nucleus and in field H of Forel there was a significant increase of the glial population with the appearance of 'immature' glial cells, but without neuronal involvement. The proliferation of such abnormal glia was seen in both the gray and white matter of the cerebrum, brain stem, and cerebellum.

In the hippocampus a single lesion was observed, characterized by an accumulation of small 'immature' astrocytes with some neuronal involvement.

The most outstanding feature of manganese encephalopathy in subhuman primates as well as in man, appears to be the severe selective damage to the subthalamic nucleus and pallidum.

Carbon Disulfide Poisoning

Extensive bilateral and symmetrical necrosis of the globus pallidus and zona reticulata of the substantia nigra was observed in four monkeys *(M. mulatta)* which were subjected to carbon disulfide poisoning from chronic inhalation over periods of from 12 to 21 months [RICHTER, 1945].

Spontaneous Degeneration

A study of spontaneous degeneration in the basal ganglia revealed that Nissl staining of putamen and caudate nucleus was reduced due to loss of 'small cell elements' particularly in the central regions [RICHTER and KLUVER, 1944]. In the narrow peripheral zone these elements were present in almost normal numbers and large ganglion cells were also relatively intact, although many exhibited chromatolytic alterations. Loss of large cells was also observed in zones of complete destruction of small cells. Nearly all glia showed changes such as swelling, hyperchromatosis, and pyknosis. The putamen was affected almost in its entirety while the posterior part of the body of the caudate nucleus was almost intact. However, the tail was partially destroyed.

In addition to the above changes perivascular ring hemorrhages were seen in the lateral $^1/_2$ to $^3/_4$ of the putamen and within the claustrum and external capsule. These were observed around veins and capillaries but also, at times, in relation to small arteries.

These hemorrhages were observed bilaterally and mainly in the rostral putamen, while severe histological alterations, as noted above, were observed mainly in the striatum. Diffuse slight retrogressive changes of the neurons, generally chromatolytic in nature, were also observed. Such alterations were particularly evident in supraoptic nucleus and the nucleus posterior pedamenti lateralis.

Such mild diffuse cellular changes were also observed in cerebral cortex but without cell loss or disturbance of cellular architecture. There was some glial reaction with increase of glia nuclei in smaller projections of white matter into gyri. This was less marked centrally.

Severe localized cell destruction was observed in the entire ventral and lateral band of the Cornu Ammonis but no hemorrhages were observed. Severe retrogressive cell changes were observed in the dentate nuclei but without cell loss. The glia were increased in number or hyperplastic and myelin staining was normal.

Slight retrogressive changes in some neurons of the spinal cord were observed with more advanced changes, including vacuolation and neuronophagia in some cells. The leptomeninges were normal.

Widespread Spontaneous Degeneration of CNS

In a 22-year-old gorilla *(Gorilla gorilla)* which succumbed to acute passive congestion of the lungs of the type associated with cardiac insufficiency, many of the larger ganglion cells (right frontal pole, right parietal and right occipital) contained collections of lipochrome pigment but otherwise were normal [Steiner *et al.*, 1965]. A moderate degree of acute swelling of oligodendroglia was observed. Many of the arterioles and smaller arteries showed moderate thickening of their walls, but the vessels of the parenchyma and leptomeninges were otherwise normal. In basal ganglia the walls of the smaller arteries showed less thickening than vessels of the cortex. In the cerebellum a moderate degree of swelling of oligodendroglia was observed.

In the pons and medulla some of the larger neurons of the tegmental nuclei showed chronic degenerative changes, including peripheral displacement of Nissl substance, swollen cell bodies, or pyknotic and shrunken cell bodies, but the majority of neuronal elements were normal. The ganglion cells of the olivary nuclei contained considerable lipochrome pigment but were otherwise normal. Some of the arterioles and small arteries showed moderate thickening and hyalinization of the vessel walls. No alterations in myelin were reported.

In spinal cord moderate demyelination was seen in the medial portions of the dorsal columns. There was definite, though less marked demyelination in the midportions of both anterolateral white columns at thoracic and cervical levels. At all levels, most of the large neurons of the ventral gray columns showed marked chronic degenerative changes, consisting of diffuse chromatolysis and bulbous swelling of cell bodies, with excentric, deformed nuclei. Virtually no normal sensorimotor neurons were observed. The small neurons were more normal but some of those also showed degenerative changes. A marked reduction of ventral horn cells was observed.

In dorsal root ganglia, widespread degenerative changes were present in neurons. The earliest degenerative change was pyknosis and shrinkage of the nucleus with diffuse chromatolysis. This was followed by disappearance of the nucleus, shrinkage, vacuolation and pale staining of cytoplasm, with beginning hyperplasia of capsular cells. A few cells showed an accumulation of lipochrome pigment. Some neurons had disappeared and been replaced by solid circular nests of capsular cells.

The same type of degenerative changes were observed in the neurons of the Gasserian ganglion as were observed in the dorsal root ganglia, but the proportion of normal cells was considerably higher.

Occasional degenerating neurons were also seen in the sympathetic trunk ganglia, but the nerve fibers in the trunks appeared normal.

The sciatic nerves revealed a notable loss of myelin as well as axis cylinders, with compensatory thickening of the endoneurium, giving a picture of marked fibrosis of the nerve bundles.

To summarize the above, widespread chronic degenerative changes were observed in neurons at all levels of spinal cord and dorsal root ganglia, with secondary demyelination and disappearance of axis cylinders in the peripheral nerves and with demyelination in the ascending tracts. Only a few of the neurons of the medulla and pons were similarly affected and none of the cells of the upper brain stem, cerebellum or cerebrum showed any such atrophic or degenerative changes. The histological characteristics of the degenerating neurons ruled out the possibility that they were retrograde changes secondary to primary lesions of the peripheral nerves. There were no changes to suggest a post inflammatory state, and the lesions seemed to represent a primary degenerative process.

In a case of acute ascending paralysis observed in a chimpanzee *(Pan troglodytes)* autopsied soon after death, it was observed that the cerebrospinal fluid was slightly turbid and contained a few PMN's, and the surface of the brain and cord were congested [ADLER and CLARK, 1923]. Many of the cells of the spinal cord, medulla and cortex appeared normal and exhibited Nissl granules. A number of cells in the anterior and posterior horns, and in Clarke's column showed faintly staining protoplasm, an absence of Nissl granules and, in some cells, an excentric nucleus. Vacuoles from $2\mu m$ to $6\mu m$ appeared in the cytoplasm of some cells. The vessels were congested and small hemorrhages were found. Similar changes were observed in the medulla, where vacuolation of degenerated nerve cells was more marked than in the spinal cord. Sections of motor cortex from the limb areas showed capillary enlargement but no hemorrhages were seen, and the gross cellular changes found in the cord and medulla were not observed. No pathological changes were found in peripheral nerves.

Kuru

Post mortem examination of a chimpanzee with a Kuru-like syndrome [BECK *et al.*, 1966] revealed no signs of either cortical atrophy, raised intracranial pressure, abscess formation, circumscribing, softening or hemorrhage. The most striking change in the brain tissue was a status spongiosus, affecting to a varying degree all the gray matter of the brain. In the spongy areas there was a great loss of neurons with severe disorganization of the cytoarchitecture and pronounced proliferation and hypertrophy of astrocytes. Many of the surviving nerve cells showed cytoplasmic vacuolation and accumulation of fatty material. The hypothalamic nuclei appeared normal but with some excess of neurosecretory material in the median eminence and upper infundibular stem of the pituitary stalk. The white matter was thinned but extensive breakdown of myelin was not evident on microscopic examination. Neutral fat was present in small accumulations free in the tissue and some within compound granular corpuscles around blood vessels.

In the cerebellum and brain stem the spongy state was less developed than in the cerebral hemispheres. Loss of Purkinje cells was observed irregularly throughout the cerebellum. There was a moderate loss of granule cells, and fibrous gliosis was seen throughout the cerebellar cortex. In cerebellar white matter there was myelin breakdown with a diffuse fibrous gliosis. The degeneration was generally greater in the vermis than in the hemispheres.

In the brain stem, most of the neurons in the pontine nuclei as well as the inferior olives appeared degenerated. The nuclei of the cranial nerves, however, were better preserved. Among the sensory nuclei the medial vestibular nucleus exhibited considerable degeneration, and generalized astrocytic hypertrophy in the gray matter with mild fibrous gliosis throughout. In spinal cord only a mild fibrous gliosis was seen in the gray columns. Neither Alzheimer senile plaques nor periodic-acid/Schiff positive plaques were present in the nervous system and no inclusion bodies were observed.

Following intracerebral passage of a suspension of the brain tissue from a human patient with kuru in three chimpanzees, the brains showed extensive degeneration and vacuolation of neurons and early status spongiosus of cortical gray matter. The incubation period appeared to be about 1 year in these animals [GAJDUSEK, 1967].

Twenty three and twenty six months following inoculation of two spider monkeys *(Ateles geoffreyi)* with brain tissue from a chimpanzee *(Pan satyrus)* with experimentally induced kuru, an intense, diffuse status spongiosus of cerebral gray matter was seen with marked astroglial hypertrophy and neuronal loss with vacuolation of neurons [GAJDUSEK *et al.*, 1968].

Two female chimpanzees inoculated with brain suspensions from infected patients revealed diffuse, non-specific changes as seen in brains of kuru victims, together with the cerebellar system degeneration characteristic of the disease [GAJDUSEK *et al.*, 1966]. The spongy degenerative changes in the brain in this disease are illustrated in figure 40.

Pigmentation and Siderosis

The black pigmentation of nerve cells of substantia nigra was clearly visible in well-differentiated Nissl preparations in both the orang-utan and chimpanzee [SCHERER, 1939]. The isolated granules were easily recognizable in 20μm sections and were generally situated at the periphery of the cell with

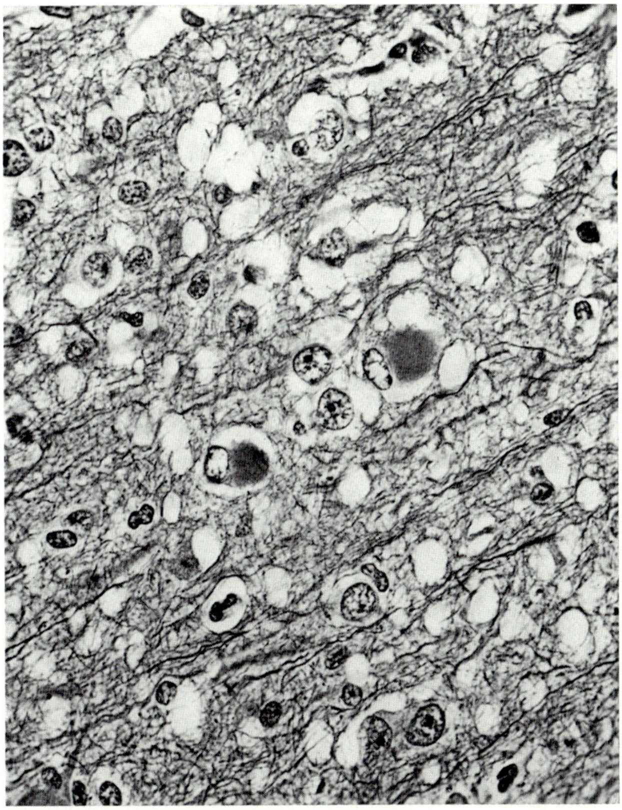

Fig. 40. Degenerative changes in brain of monkey *(M. mulatta)*, in kuru (Holmes stain) (×850).

the nucleus always clearly visible. Frequently melanin granules were scattered in cell dendrites as far as twice the length of the cell body.

In monkeys and gibbons the melanin pigmentation of the substantia nigra was less intense than in higher anthropoids, but was always present. The natural color varied from brownish to yellowish and the amount varied in different cells.

In one of three lemurs melanin pigmentation was recognized in unstained sections, but only in a few nerve cells.

There was a definite difference of pigmentation in the orang-utan brains in the adult as compared with the younger specimens (exact ages unknown). The adult showed the same degree of melanin pigmentation of cells in substantia nigra which was observed in adult chimpanzees. Ninety percent of the cells of the young animals were apparently unpigmented in unstained sections. Masson impregnation, however, showed granules in about 50% of the cells, the majority containing only a few granulations.

In chimpanzees one young adult showed less pigmentation than two older animals. In one chimpanzee over 20 years old, there was a pronounced atherosclerosis of cerebral basal arteries and coronary arteries, but no diminution of melanin pigment in nerve cells and no pigment scattered in glia cells of substantia nigra.

The locus coeruleus in chimpanzees was unpigmented except in Masson preparations. In the latter, black granules were seen in a considerable number of cells of this structure.

The melanin pigmentation of the pia melanophores was more intense in the higher apes than in man. There appeared to be no parallelism between the degree of pigmentation in pia, substantia nigra and locus coeruleus.

Large 'calcified' bodies were frequently observed in the anterior part of the globus pallidus and putamen, and occasionally in the meninges [FAIRBROTHER and HURST, 1932]. Vascular 'calcification' was much less common and usually took the form of mulberry-body formation in the artery walls of the globus pallidus and putamen, with adventitial calcification occurring occasionally.

Most of the animals affected were immature and in possession of their milk teeth.

In a male monkey *(Cercocebus torquatus* [*atys*]*)* four years of age which had been maintained on a normal diet but injected daily with 100μm estrone monobenzoate 41 days before sacrifice, calcareous deposits were observed in the globus pallidus [GLEES, 1943]. Macroscopically the membranes, vessels and gross structure of the brain appeared normal. In hematoxylin and toluidine blue sections the pallidum was crowded with deep staining patches bilaterally. The individual deposits were irregular in shape and up to 3 mm in diameter. No marked leukocytic infiltrations or glial reaction around them was noted. The deposits were occasionally intimately related to blood vessels embedded in the wall between the tunica media and tunica adventitia but in most cases were distributed in the ground substance of the nervous tissue. They were confined largely to the anterior $^2/_3$ of the pallidum and not present in any other area of the brain. The nerve cells surrounding the deposits had undergone various amounts of destruction and the nerve fiber plexus in silver sections was less dense than normal and many fibers showed degenerative changes. The deposits gave a strong positive reaction for iron (HCl and potassium ferrocyanide test).

Organic Arsenical Compounds

In monkeys (species not stated) a single toxic dose of 4 (2:4-diamino-6 pyrimidylamino) phenylarsine oxide caused hemorrhagic or, in some cases, non-hemorrhagic necroses at several sites in the central nervous system [HURST, 1959]. The most typical and commonest lesions were hemorrhagic necrosis of the external geniculate bodies and ventral part of the pons. The triazine analogue of the above compound produced similar lesions. In three of four monkeys which received multiple doses of neoarsphenamine, no nervous lesions were detected. The fourth animal exhibited many small necrotic foci. Monkeys which received repeated doses of tryparsamide showed gliosis in some portions of the cerebellar white matter and in the optic tracts. A second type of lesion, of different distribution, was characterized as a localized inflammatory lesion in parts of the walls of the meningeal and cerebral blood vessels and in the choroid plexuses.

Acute Amaurotic Epilepsy

Pathologic examination of Rhesus monkeys with this disease showed parenchymatous lesions of the vascular type (characterized by recent necrobiosis) in the cortex and in subcortical white matter [VAN BOGAERT and SCHERER, 1938; VAN BOGAERT and INNES, 1962]. These lesions were generally symmetric and localized chiefly in layers in the occipital lobes. They sometimes occurred in subcortical occipital and central regions. Marked diffuse macroglial proliferation in the molecular layer was observed with a tendency to vascular and macroglial proliferation in the deeper layers. Circumscribed vascular nodules and focal leukocytic meningeal reactions were also observed. In three cases glial complexes were observed in the molecular layer of the cerebellum, similar to those seen in cases of human epilepsy. The peripheral visual paths, spinal cord and peripheral nerves were normal. According to VAN BOGAERT and INNES [1962], this condition is characterized by diffuse macroglial proliferation in the molecular layer of the cerebral cortex, fresh degenerative vascular lesions in the subcortical cerebral white matter, especially in occipital areas and central cortex, perivascular lesions and glial nodules, cellular infiltrations of the meninges and, less constant, diffuse vascular proliferations and intracerebral inflammatory lesions. The cortical lesions in the cases described by these authors had a vascular character and a laminar or pseudolaminar distribution with the neuronal lesions appearing to be of an ischemic anture (e.g. with loss of Nissl granules and shrinkage of nuclei). Laminae III and IV were markedly damaged in every case with notably less damage in laminae II, IV, and VI. Lesions in the white matter were poorly outlined foci of demyelination, which were sometimes contiguous with the overlying cortical lesions. As seen in Nissl preparation, glial proliferation was predominant without vascular changes. Lesions of the white matter were constant in most cases but varied markedly in severity.

The diagnosis of this condition, according to these authors is based on the presence of the following: (1) blindness without lesions in the ocular fundus, (2) hemiplegia, (3) intensity of epileptic attacks, (4) symmetry of vascular lesions in occipital areas, especially the area striata, (5) milder central lesions with macrogliosis of the cerebral molecular zones, (6) endothelial proliferations in vessels of the cerebral cortex and (7) leukocytic infiltration in the meninges adjacent to the focal lesions.

Acute Amaurosis without Epilepsy

In this condition disseminated necrotic foci were seen and were numerous in cortical areas. These were accompanied by a vigorous mesenchymal reaction arising from arterioles. The foci occurred in both gray and white matter, mainly in the occipital convexity and in the interhemispheric surfaces, but somewhat less so in the temporal and lingual gyri. Laminae III, IV and V were particularly affected. Nodular vascular proliferations and masses of granulations giving a 'moth-eaten' appearance were evident and were not related to thrombosis, thrombophlebitis or angiitis obliterans. Acute amaurosis with and without epilepsy had the following in common: blindness with mydriasis, no loss of pupillary reflexes, and subsequent unilateral or bilateral hemiplegic signs. Pleocytosis of cerebrospinal fluid was found in acute amaurosis more often with epilepsy than without epilepsy. These diseases have been observed only in *Macaca mulatta* [VAN BOGAERT and INNES, 1962].

Nutritional Deficiencies

Experimental

Monkeys *(M. mulatta)* kept on a vitamin A deficient diet for six and one half months exhibited no evidence of destruction of myelin sheaths

(Marchi method), nerve fibers or ganglion cells [GRINKER and KANDEL, 1933]. Mild nonspecific alterations were found in some ganglion cells and in the endothelial lining of the smaller blood vessels.

Monkeys *(M. irus)* maintained on a vitamin E deficient diet, revealed fat deposits in Marchi preparations in the region of the pyramids, medial lemniscus and restiform bodies [WECHSLER and GLOBUS, 1942]. Smaller amounts were distributed diffusely in other parts of the pons and medulla. In one case there was a notable increase of fatty substance in the hypoglossal nuclei. A small amount of fat was observed in the subcortical regions in the motor area but none was seen in the motor cortex. In the spinal cord fat was found widely distributed throughout the white matter but was somewhat more abundant in the posterior columns and in the region of the spinocerebellar tracts. No increase in fat was noted in the anterior horn cells.

Scarlet red preparations revealed practically no fat in any of the sections of the pons, medulla, spinal cord, or peripheral nerves. A moderate amount was seen in blood vessels and an occasional fat containing compound granular cell was seen in the subcortex. Fine sprays of fat granules were seen in the motor cortex in some sections. In one monkey degenerative changes in large motor cells of the cerebral cortex were seen and some degenerative changes were observed in anterior horn cells of the lumbar region on one side of the spinal cord.

An aged control monkey showed changes in Marchi preparations similar to those observed in experimental animals.

In 7 rhesus monkeys subjected to one or more episodes of acute thiamine depletion (blood thiamine $4\gamma/100$ ml or less 2 weeks after thiamine was removed from diet [RINEHART, GREENBERG and FRIEDMAN, 1947]) bilateral symmetrical foci of accumulation of many microglia and fewer astrocytes were seen. The vascular proliferative reactions described in Wernicke's disease were occasionally seen but were not prominent. No significant lesions were seen in the spinal cord and no structural alterations were seen in peripheral nerves.

In monkeys *(M. mulatta)*, subjected to a vitamin B6 deficient diet for more than a year, the large nerve cells of the cerebral cortex showed swelling, eccentricity of nuclei and loss of Nissl granules [VICTOR and ADAMS, 1956]. No such changes were observed in the basal ganglia, subthalamus, thalamus, brain stem, cerebellum, spinal cord, dorsal root ganglia, or ventral and dorsal nerve roots. In some animals an increase in number and size of astrocytic nuclei was observed in all layers of the cerebral cortex and the most superficial parts of the subcortical white matter. There appeared to be no parallel between nerve cell and astrocytic changes. In one animal there was a loss of myelinated fibers in the lower medulla at the pyramidal decussation and throughout the spinal cord. The peripheral nerves were affected in all but one of 13 animals. In the involved nerves, the myelin had degenerated in a patchy, segmental distribution.

Several months after the termination of a diet high in vitamin D (163,000 USP units per animal, per day) 6 monkeys *(M. mulatta)* which had received high doses of X-irradiation (dose not stated) of the head expired. Necrotic areas were observed in the path of the irradiation through the brain and deposits of iron and calcium were found in such areas. Similar deposits were observed by VOGEL and PICKERING [1956] in monkeys which were exposed to focal ionizing radiation to the head in the absence of a high vitamin D diet. Other evidence of calcification was noted in the CNS.

Spontaneous Occurrence

In a colony of monkeys *(M. sylvanus)* at the Pasteur Insitute of Morocco in 1940 [LEBLOND and CHAULIN-SERVINIERE, 1942] Marchi stained tissues revealed Wallerian degeneration of myelin sheaths, with the lesions predominating in peripheral nerves, especially in the lower limbs. Degenerative changes were also seen in the facial nerve in one animal. The disease did not appear to be influenced by sex but it was evident sooner in adolescent monkeys of 3 years of age than in younger animals (1–1 ½ to 2 years of age) or adult animals.

An adult male gorilla *(Gorilla gorilla)* exhibiting anoxia, dyspnea, extreme flexion of toes and feet, easy fatigueability and locomotion almost entirely by using the arms died at the age of nearly 23 years [STEINER *et al.*, 1955]. A tentative diagnosis about 6 months prior to death by a group of physicians was 'a combination of senile changes, some arteriosclerosis, and peripheral neuritis'. The immediate cause of death was acute congestive heart failure with dyspnea, hypostatic pulmonary hyperemia and edema, and passive congestation of the liver. It was believed that the observed abnormalities, including neuropathy with related lesions in muscles, cardiopathy and changes in arteries, hemosiderosis of the spleen and testicular sclerosis may have had their basis in a nutritional deficiency. The exact nature of the deficiency was not known but it was suggested that the gorilla has a special nutritional requirement not possessed by other primates. Microscopic study of the CNS revealed that many of the larger ganglion cells of the cerebral cortex contained collections of lipochrome pigment. The brain pathology in this animal has been described above under degenerative diseases.

Peripheral Neuropathy

Experimental polyradiculitis has been produced in *Macaca mulatta* by intracerebral injection of a bacteria-free filtrate of fecal matter from a suspected poliomyelitis patient [PAPPENHEIMER *et al.*, 1951]. Some monkeys exhibited radicular lesions including degenerative changes of the spinal roots and peripheral nerves. The axis cylinders showed irregular beadings, and swellings and vacuole formation with a loss of the neurokeratin network in the myelin sheaths. The most severe lesions were observed at lumbar and sacral levels. In the spinal cord no neuronal or pathway deterioration was noted but extensive perivascular infiltrations of lymphocytes were present. The condition has been likened to the Landry-Guillian-Barre syndrome in man.

In monkeys *(M. mulatta)* which received 20 mg/kg/day of acrylamide orally for 8 weeks, then 30 mg/kg/day for an indeterminate period, conduction velocity in the sciatic nerves decreased by 22% and ulnar nerves by 20%. The affected nerves exhibited myelin degeneration and destruction of axons with the greater changes occurring in the nerves of the caudal limbs [LESWING and RIBELN, 1969].

Parasitic Infections

Localized transverse myelitic softening was observed in the guenon monkey *(Cercopithecus brazza)* in which a larva of *Porocephalus latus* was attached to the dura mater at T11-12 [SCHERER, 1944]. Granulomata in the cerebral hemispheres were observed in *Macaca mulatta* with an amoebic dysentery syndrome.

Cysts of *Cysticercus cellulosae* were observed in the subdural and subarachnoid spaces in brain substance in animals of the same species but little tissue reaction other than a slight gliosis was reported [WALKER, 1936]. Compression myelitis and secondary degeneration of the pyramidal tracts was observed in lemurs with a larval cyst *(Cysticercus bovis)* at about the 4th thoracic level [HAMERTON, 1934]. A cysticercus cyst in the brain of a rhesus monkey *(M. mulatta)* is shown in figure 41. Toxoplasma has been demon-

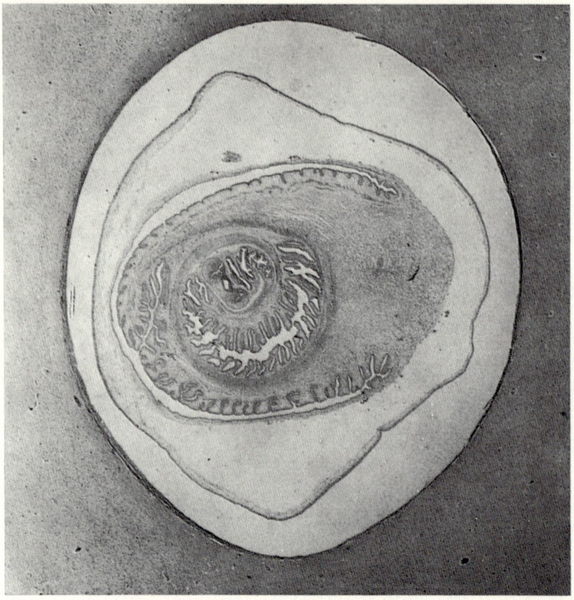

Fig. 41. Cysticercus cyst in brain of monkey *(M. mulatta)* (HE ×10).

strated in the white-faced monkey *(Cebus capicinus)*, macaque *(M. tantala)*, squirrel monkey *(Saimiri sciurea)*, baboon *(Papio sp.)* and chimpanzee [LEVINE, 1961]. SUREAU *et al.* [1962] also recorded this organism from the lemur, and POKORNY [1961] has made essentially the same observations, but no reports of neuropathological alterations incident to these organisms have been reported.

Epilepsies

In Patas monkeys epileptic crises have been observed to last for months or years [VAN BOGAERT and INNES, 1962]. These cases exhibited some selective neuronal necrosis with cellular loss in the cerebral cortex described as denticulate, disseminated and laminar in form and distribution, sometimes affecting layers II-IV and sometime III, especially in the temporooccipital regions. Layer V consistently appeared intact. The changes were described as being comparable to those described in man with different forms of anoxia and a marginal macrogliosis. The condition was characterized by demyelination subjacent to the cerebral convolutions exhibiting the necrotic changes.

Tumors

An astrocytoma in the right occipital and temporal regions as well as an intraocular tumor were reported by SELLHEIM [1936] in a baboon *(Papio anubis)*.

A Green monkey *(Cercopithecus aethiops)* at the Philadelphia Zoological Gardens exhibited a gray area 3 × 2 × 1 cm, with a softened center, in the middle of the cerebrum [FOX, 1923]. The more solid parts consisted of glia tissue, blood vessels, and degenerated cells. The author believed it might 'belong among the gliomata or gliosis' but no true gliomatous formations could be found and it was not classified by the author as a tumor or inflammation.

In a sooty mangabee at the Philadelphia Zoological Garden [FOX, 1923] which died after a distinct convulsion with semiconsciousness, a large cyst was found to occupy the posterior third of the left hemisphere. Its walls were composed of a thin (one-sixteenth inch) rim of cerebral substance and the meninges. Its contents were clear.

Pathological Effects of Intrathecal Injection of Spinal Anesthetic and Vasoconstrictor Drugs

Monkeys *(M. mulatta)* which received intrathecal injections of ephedrine sulfate in single injections of 5% solutions or multiple injections of 1.0% 1.66%, 2.5% or 5% (60 mgm/kg), solutions exhibited severe chromatolytic changes in neurons of the gray matter of the spinal cord [BRIZZEE and WU, 1951]. Single injections of 2.5% solutions or multiple injections of 1% solutions (48 mg/kg) resulted in only slight alterations in the chromidial substance. Myelin changes occurred in all monkeys which received multiple injections and in one which was sacrificed 14 days following a single injection of a 5% solution. The myelin changes were observed mainly in the posterior funiculi. Cellular alterations and myelin degeneration varied directly according to the dosage and concentration of the drug with some variation in severity at any given concentration or dose level. One monkey sacrificed 30 days after receiving a large total dose of ephedrine sulfate in multiple injections of 5% solution showed a considerable degree of recovery from chromatolytic changes.

Chromatolysis of spinal cord neurons was also observed following intrathecal injection of epinephrine (0.88 to 10.0 mg/kg) or procaine (2.2 to 3.5 mg/kg) (BRIZZEE *et al.*, 1951 a, b).

Pathological Effects of Ionizing Radiation

In 67 monkeys *(M. mulatta)* exposed to a barium 140-lanthanum 140 source in doses of from 1,000 to 30,000 R at 1,000 R/min (whole body) it was observed that an inflammatory reaction in the vessel walls dominated the pathological picture, in the CNS and its meninges [HAYMAKER *et al.*, 1958]. No vasculitis, however, was observed in the thoracic or abdominal viscera.

The CNS vasculitis had the highest incidence at doses of 10,000 R or more and was characterized by an attendant infiltration of leukocytes into the meninges, choroid plexuses and, less commonly, the brain and spinal cord. Morphological evidence of damage was commonly seen in capillaries, and endothelial cells of larger vessels often underwent regressive changes or had disappeared. Collagen fibers of veins and the elastic lamina of arteries were occasionally severely altered.

In the CNS the vasculitis was generally more marked in the gray than in the white matter. In the gray matter the cerebral cortex and basal cerebral structures exhibited greater damage than the cerebellum, brain stem, or spinal cord. The intensity of the vasculitis varied considerably within each dose group. In some cases, the cerebral cortex showed profound necrosis in association with severe vasculitis but exhibited large ischemic foci in the absence of vasculitis in others.

Of the basal structures the thalamus exhibited the highest incidence of vasculitis. The supraoptic nucleus was most often severely affected. In the brain stem, the pons and medulla oblongata commonly exhibited vasculitis. The intensity of vasculitis in the spinal cord was relatively slight but the incidence of involvement was high.

In the cerebrum, the white matter was the seat of a perivascular edematous process in some cases, and in a few the perivascular white matter was observed to have broken down after a rather long latent period. In some cases the damage to the white matter appeared primary but was usually thought to be related to the vasculitis. Alterations in glial cells were not marked. In some cases of long duration microglia had undergone some regressive changes. Astrocytic nuclei in the pallidum and occasionally in the red zone of substantia nigra had undergone enlargement, with some resembling the 'naked' glial nuclei of Alzheimer. Macrophages containing fat were occasionally seen in lesions of the white matter.

The ependymal lining of the ventricular system, occasionally showed slight proliferative changes as well as regressive changes or actual necrosis, and the wall of the ventricular system was a site of predilection of vasculitis.

There were no notable changes in neurons of the cerebrum, brain stem or spinal cord. However, granule cells of the cerebellum and basophil cells of the adenohypophysis exhibited severe pyknosis at doses of 10,000 R and higher, over the same time span. There was no clear correlation between dosage and intensity of pyknosis. The site of predilection of the pyknosis in the cerebellum was the deeper part of the granular layer, especially in the hemispheres.

In the adenohyphysis the pyknosis was more marked in the lateral areas. The pyknosis was considered to be reversible and due mainly to hypoxidoses with primary irradiation as a contributing factor in relation to the cerebellar granule cells.

Evidence for a direct radiation effect on granule cells of the cerebellum in monkeys was furnished by the work of VOGEL [1962]. Forty eight male monkeys *(M. mulatta)*, were exposed to gamma irradiation at 1,000 R/min totalling 10,000 R with head or body shielded or no shielding. Animals were killed up to 96 h.

Cerebellar granule cells promptly (2 h) exhibited pyknotic changes and inflammation of the cerebral blood vessels, leptomeninges and choroid plexuses was observed. This type of alteration is illustrated in figure 42. The lesions were not evident in monkeys exposed to 1,000 R but were equivocally present in some animals at 2,500 R. They were observed with increasing frequency after doses of 5,000 and 10,000 R.

The severity of the lesions increased rapidly within the next 8 to 24 h, then regressed, the lesions being minimal or absent after 96 h. The pathologic alterations were of the same character and degree of severity whether the head and body or only the head was exposed but did not occur when the body was exposed with the head shielded.

In granule cells of the cerebellum the nuclear diameter was markedly reduced, with condensation of intranuclear chromatic material. Narrow margins of basophilic cytoplasm which stained deeply with pyronin were seen surrounding some such nuclei. The affected cells were usually distributed haphazardly throughout all parts of the internal granular layer. Golgi and Purkinje cells were regularly spared. The pyknotic cells were more widely separated from one another than non-pyknotic ones. The appearance was that of widened intercellular spaces but with no coagulated fluid and only a light infiltration of leukocytes and macrophages.

The prompt appearance of an exudate of leukocytes was observed in all layers of the cerebral blood vessels with veins and arteries about equally involved. Those vessels of the cerebral nuclear masses were more severely af-

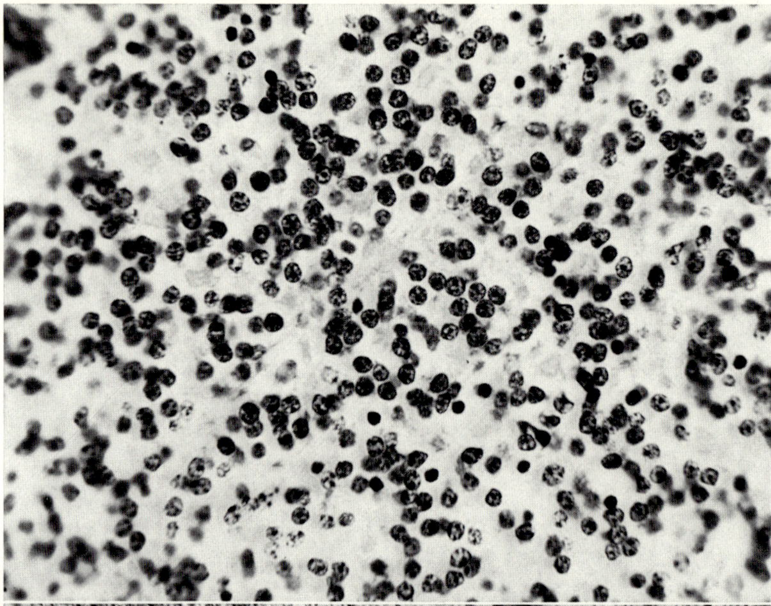

42

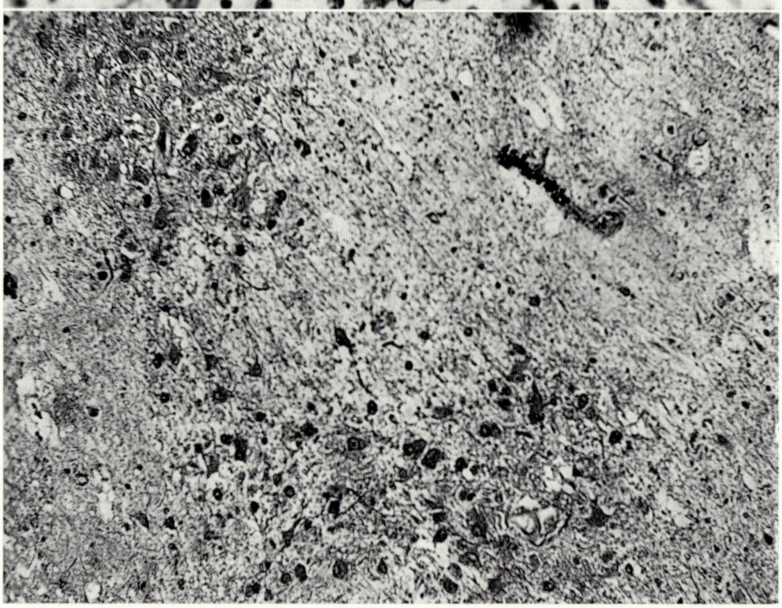

43

Fig. 42. Pyknotic granule cells in brain of subhuman primate exposed to irradiation (×940).

Fig. 43. Astrocycosis at margin of area of necrosis in subcortical white matter (HE ×157).

fected than those of the cerebral cortex while those of the brain stem cerebellum and spinal cord were similarly affected but to a lesser degree.

In the meninges an exudate was observed which commonly disseminated over the gyri and spread into the sulci. At earlier post-radiation intervals PMN's predominated followed later by decreased cellular concentrations of lymphocytes and macrophages. The inflammatory residue seen in animals killed 96h after exposure consisted of macrophages together with thin contents of cellular debris.

In the choroid plexuses, the severity of involvement and cellular composition of the inflammatory exudate was generally comparable to that of the meninges, but the choroidal epithelium was generally spared.

In another recent investigation adult monkeys *(M. mulatta)* were X-irradiated over the right side of the head at a dose of 3,500 rads (320–350 rad/min – CAVENESS *et al.* [1964]). Forty seven weeks after irradiation Nissl stained sections of cerebral cortex revealed various degrees of necrobiosis and inflammatory reactions and calcification of middle and to a lesser degree the upper cortical laminations. Golgi-Cox preparations revealed a marked loss of neurons and their dendrites. The dendritic branches that remained showed irregular thickenings but few characteristic pedunculations. Vascularity was described as poor and myelin sheaths in appropriately stained sections showed pallor, swelling and beading followed by complete disintegration and dissolution. These alterations were seen in the vicinity of, but independent of, blood vessels.

In some cortical areas, commonly near blood vessels, astrocytic hypertrophy and hyperplasia were noted and areas of necrosis surrounded by astrogliosis were seen in the subcortical white matter. This type of tissue change is illustrated in figure 43.

Electron micrographs revealed changes in fine structural constituents of the nucleus, nucleolus, nuclear membranes, and ergastoplasm including changes in pattern, shape, size of lacunae, and morphological appearance and distribution of ribonucleoprotein. Mitochondria, lysosomes, and the Golgi complex also exhibited alterations, and a large amount of osmiophilic material of varying character was noted in glia cells or irregularly dispersed within the cytoplasm of neurons, neurophil and vascular wall.

In order to evaluate the sequence of changes leading up to the tissue alterations described above these authors examined tissues at various post-irradiation intervals. In animals sacrificed 72 h post-irradiation there were no alterations in cytoarchitecture and no significant alterations in Golgi-Cox preparations. Some moderate perivascular inflammatory reactions were observed in certain areas of the leptomeninges, cortex and isolated areas of white matter. In electron micrographs a variability in distribution and pleomorphism of mitochondria was noted. Mitochondria often appeared enlarged with fragmentation of cristae and empty matrices. Marked enlargement of lacunae and variability in configuration and distribution of ribonucleoprotein in ergastoplasm and marked diffusion and enlargement of canaliculi of the Golgi complex were seen. Osmophilic structures of variable shape or size were encountered within the cytoplasm of glial cells or neuropil.

At two weeks post-irradiation minor alterations in Golgi-Cox preparations were seen but no other changes were noted.

Four weeks post-irradiation, perivascular inflammatory reactions in the cortical and subcortical regions were seen and, Sudan III preparations showed some rarefaction of the perivascular tissue associated with mononuclear inflammatory cells and some glial reaction. A granulomatous-like formation was seen in an isolated subcortical area. Its inflammatory exudate was made up mainly of mononuclear cells intermixed with some glial nuclei and a peripheral glial reaction was evident. A moderate loss of dendritic arborizations was noted, mainly basilar, and changes in the pedunculated bulbs were seen with irregular thickenings apparent along dendritic ramifications but with few spines present. Some vascular proliferation was also evident in the irradiated tissue.

Electron microscopic studies revealed grouping of glial nuclei with dense ergastoplasm. Occasional macrophages with various amounts of osmiophilic material, alternating with vesicles or vacuoles were also observed.

At 8 weeks post-irradiation the meninges showed discoloration, especially at the intergyral surface. Golgi-Cox preparations revealed a patchy loss of small dendritic ramifications in the cerebral cortex, relating especially to basilar dendrites. A loss of spines was noted along the dendrites but no change in myelin was noted.

At 12 weeks post-irradiation a neuronal rarefaction was observed in the middle and upper layers of cerebral cortex. A loss of dendritic arborizations was seen with the greatest preservation being in the apical stalks. There was a marked increase in vascularization on the right.

At 16 weeks post-irradiation a loss in basilar dendrites and alteration in pedunculated bulbs was noted. In Hortega preparations a glial reaction was suggested.

In order to gain a more precise estimate of the reduction in dendritic arborization from tissues reported in the above studies, a quantitative analysis of the processes of 48 pyramidal cells, randomly selected from irradiated and control cortex was carried out [CAVENESS *et al.*, 1965]. A significant reduction was observed in irradiated tissues in number of dendrites arising from the perikaryal surface, number of dendrites intersecting zonal boundaries at successive distances of 25 μm, in number of branching points in each zone and in total length of dendrites.

In some, though not all, monkeys dying 19 to 20 months after proton irradiation concentrically laminated and calcific bodies were observed [VAN DYKE and JANSSEN, 1963].

In monkeys *(M. mulatta)* which received 2,000 R X- or gamma irradiation at 50 R/min to the sides of the head [HAYMAKER *et al.*, 1968] the latency of lesion development was between 13 and 17 weeks. Damage occurred in white matter, cerebral cortex and hypothalamic nuclei, with the widest spectrum of lesion type in the white matter. Lesions commonly were observed in relation to altered blood vessels. Degenerated oligodendroglia containing a considerable amount of glycogen were encountered in early stages of the inception of some of the lesions. Damage to these cells was considered dependent on vascular-circulatory factors. The authors were of the opinion that the most important pathogenic factor in the development of the parenchymal lesion was vascular injury brought about by hemodynamic disturbances.

The parenchymal lesions were very diverse and differed in nature at different levels, even in the same brain. Oligodendroglia were commonly affected and were hypertrophic in regions near necrotic areas and in focal areas of tissue in between opposing altered stem vessels. In the dentate nucleus of cerebellum enlarged oligodendroglia and other reactive cells were found within linear streaks resembling infarct areas. Nerve cells in such streaks were necrotic. In tissues bordering areas of tissue necrosis many hypertrophied oligodendroglia were seen. Some lesions appeared to be related to capillary beds while others did not appear to be related to any vessel or group of vessels.

The cerebral cortex was affected to some extent in most brains but less frequently than the white matter. The most severe cortical lesions were at levels at which the white matter was heavily damaged and occasionaly lesions in cortical and white matter were coextensive.

In the hypothalamus of some animals, areas in which nerve cells had disappeared and reactive mesenchymal cells were abundant, were observed straddling the supraoptic nucleus and lateral preoptic area. A marked enlargement and distortion of larger vessels and loss of capillaries was observed in paraventricular nucleus in association with multifocal nerve cell loss and tissue calcification.

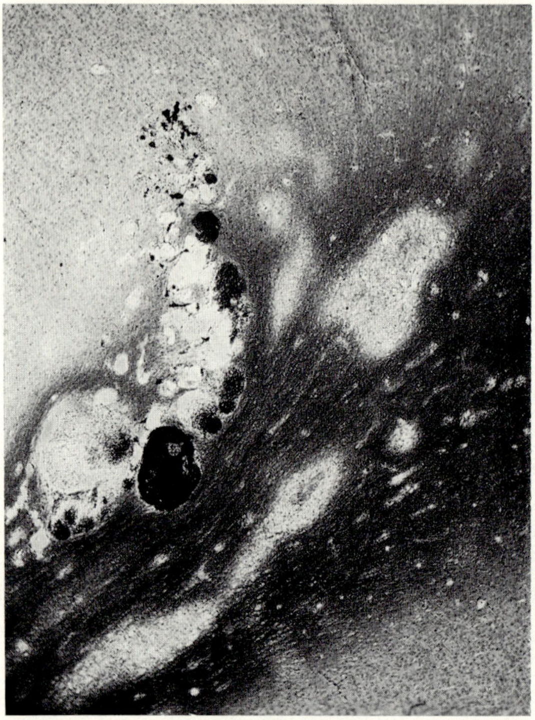

Fig.44. Demyelination resulting from ionizing radiation in brain of chimpanzee (Luxol Fast Blue ×35).

Demyelinization was regularly conspicuous in the white matter of the rostral portions of the brain of 6 monkeys *(M. mulatta)* sacrificed 14 to 22 months after exposure of the ocular regions to 850 r.e.p. of 14 MeV neutron radiation [VOGEL and PICKERING, 1956]. Demyelinization was not present in the brain of a monkey 2 months after radiation under identical conditions, or in those of 5 non-radiated animals serving as controls. Lesions of this type are illustrated in figure 44. In early lesions, the individual myelin sheaths were varicose and fragmented while the neurons, axons and glial cells remained normal in appearance. With the passage of time, the degeneration of myelin became more marked and in later stages was accompanied by a degeneration of the axis cylinders, a proliferation of astrocytes and microglia, and minor cytological changes in the oligodendroglia. The process occurred essentially without inflammation or notable changes in cerebral or meningeal blood vessels. It was clear from the observed results that neutron radiation

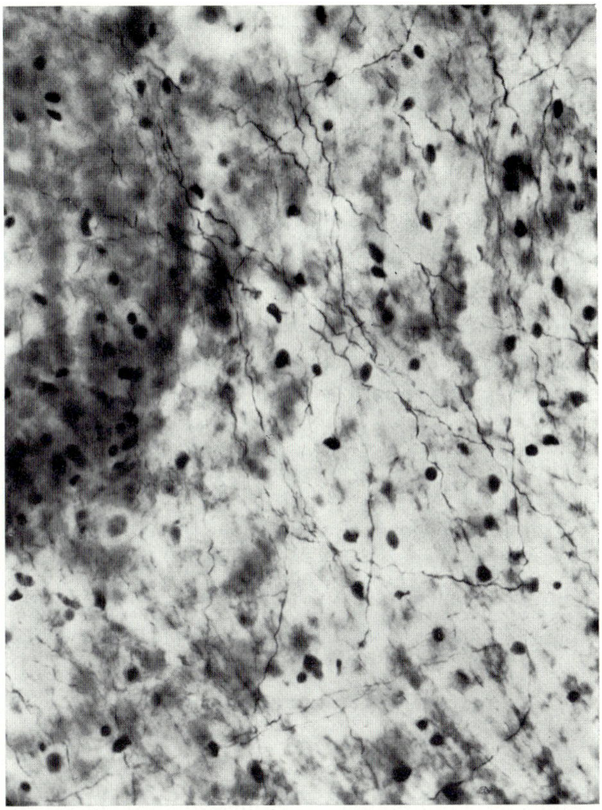

Fig. 45. Preservation of axons in area of demyelination in cerebral cortex following irradiation (Bodian Stain ×133).

had the property of destroying myelin in the living animal, and inducing changes that are notably similar to those that characterize disseminated encephalomyelitis in human beings.

In tissues stained by Cajals method, VOGEL and PICKERING [1956] observed dense plexuses of well preserved axis cylinders and many of those crossed large areas of demyelination, some showing beading and fragmentation. Tissue alterations of this type are shown in figure 45. The white matter contained many fibrillary astrocytes but the cortical gray showed little or no astrocytic gliosis. In areas of white matter that were devoid of myelin, the oligodendroglial cells were moderately decreased in number, but except for some loss and shortening and swelling of processes, were normal in appearance. The oligodendroglia in the cortex were normal in distribution and had small, rounded nuclei with fine chromatin material. The microglia were increased in number throughout the frontal lobes.

Following 3,500 R X-irradiation to the head only it was observed that an inflammatory exudate formed which was predominantly of a lympho-histiocytic character [ROIZIN, 1968]. The inflammatory reaction was generally perivenous or periarterial with peri-meso and/or endarteritis, at times leading to thrombosis. The vascular walls and more frequently the perivascular spaces and tissue sometimes contained fibrinoid material. The perivascular inflammatory exudate, generally circumscribed, also was observed to be diffuse of a granulomatous nature. At times, areas of necrosis and malacia were associated with the lesions. There appeared to be a predilection of lesions for the white matter which was more frequently noted in later stages and was then progressive. Microglia, astrocytes and to a lesser extent oligodendroglia were observed to exhibit progressive reactions followed by various metamorphic changes during the early development of the lesions. During the later stages, astrocytes especially of the fibrous type exhibited reactive proliferation and marked gliosis, especially at the periphery of necrotic areas. In the later periods the microglia and oligodendroglia appeared to undergo degenerative and regressive changes including pyknotic changes, rod cell formation and metamorphosis into argentophil bodies.

Ultrastructural studies of brain tissues from animals exposed to 3,500 R (320–350 R/min) revealed variable dilations of the canalicular system in the endoplasmic reticulum, variation in number of mitochondria in presynaptic terminals and the presence of glycogen granules in the light glial processes [ROIZIN and SCHADE, 1968]. At four weeks post radiation a larger number of astrocytes was detected on the radiated side and at 12 weeks synaptic vesicles exhibited variation in size, irregularity in shape and a decrease in numbers mainly in the radiated side.

At 20 weeks the most salient findings were the presence of variable concentrations of glycogen granules in neuroglial cells and their processes in the neuropil and around the blood vessels. Myelin sheath alterations were observed in some animals. Mitochondrial pleomorphism was observed with a tendency to congregate in groups. Various degrees of myelin alteration and disintegration were encountered, especially in areas of necrosis. In these areas various degrees of disintegration of synapses was observed.

At 44 weeks isolated synapses were observed between areas undergoing fibrillary gliosis. In areas at the periphery of necrotic regions some vessels showed marked thickness of vascular walls and were surrounded by glial proc-

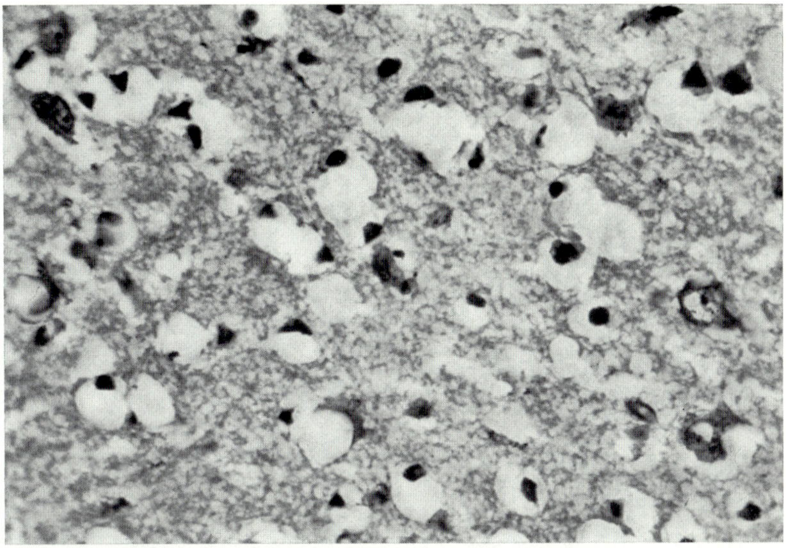

Fig. 46. Degenerative changes in neurons of cerebral cortex in monkey *(M. mulatta)*, which was sacrificed *in extremis* due to presumptive hyperthermia (heatstroke) (× 860).

esses which demonstrated a remarkable concentration of protofibrils. Some axons appeared to be undergoing myelination.

In quantitative ultrastructural studies it was found that neuron volume decreased significantly at 4 weeks post-irradiation and one animal showed a significant increase in the irradiated site. The latter was also observed in two animals sacrificed at 12 weeks and one of the 28 week group. The decrease in neuron-volume was also consistent for animals in the 12, 20, 28, and 44 week groups.

In monkeys *(M. mulatta)* which received a single exposure of above 3,000 R of 23 MeV X-rays, the entire hypothalamus was usually seriously damaged as were the cortical areas through which the radiation passes [ARNOLD, 1954]. At single doses of from 1,500 to 3,000 R the paraventricular and supraoptic nuclei were selectively damaged. From 5 to 6 months after irradiation there was a considerable deficit of cells. Many of the surviving cells had irregular outlines or appeared as bare nuclei. There was no reactive gliosis about the affected areas.

In monkeys *(M. mulatta)* exposed to 7,000 to 14,000 R of 23 MeV, acute radiation necrosis of all neural tissue was observed from 1 day to 6 weeks

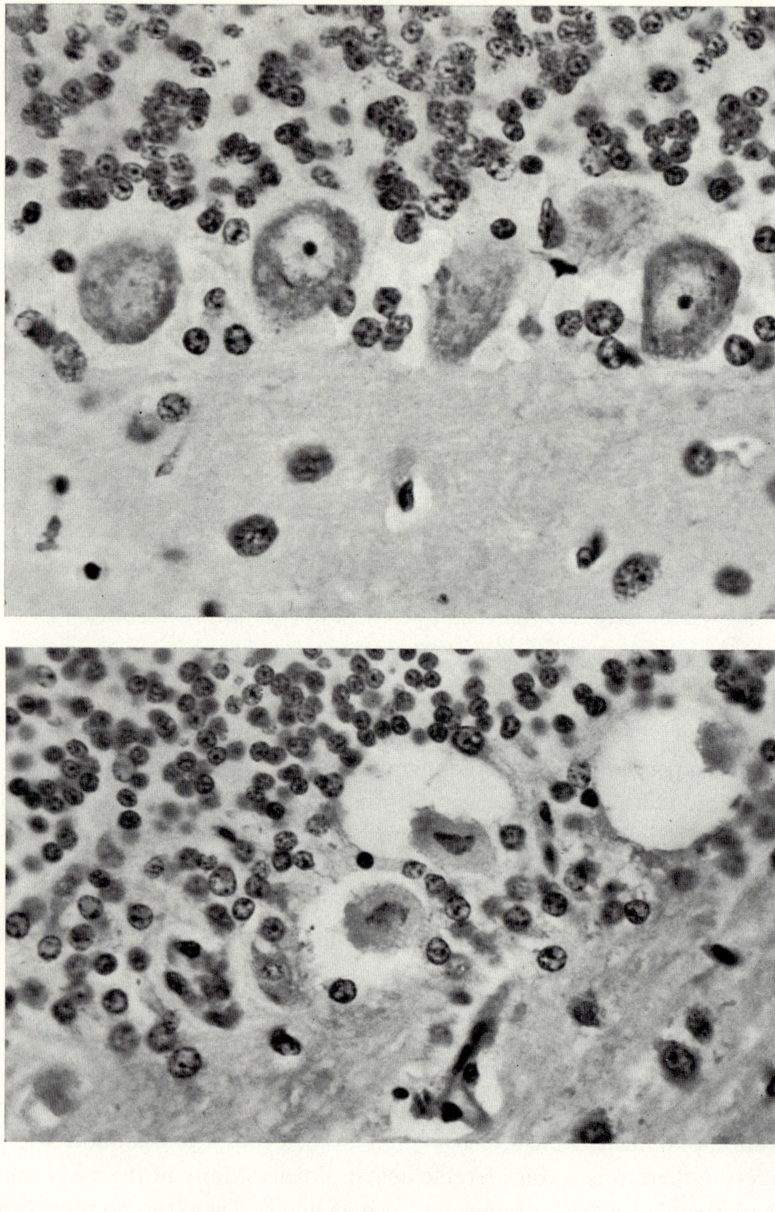

a

b

Fig. 47. Degenerative changes in Purkinje cell neurons of cerebellum in monkey *(M. mulatta)*, which was sacrificed *in extremis* due to presumptive hyperthermia (heatstroke). a. Normal Purkinje neurons, b. degenerating Purkinje neurons (×950).

post-exposure [Arnold *et al.*, 1954]. Between 6 weeks and 5 months the acute stage persisted but no gliosis was observed.

Doses of 5,000 to 7,000 R resulted in acute inflammation, hemorrhages, perivascular exudates, partial necrosis, myelin changes, and edema from one day to 6 weeks post exposure. At 6 weeks to 5 months there was partial histological recovery, and between 5 and 8 months delayed, non-selective diffuse radionecrosis was seen. Between 8 and 12 months there was vascular hyalinization with very slight gliosis, and between 12 and 24 months there was moderate to intense gliosis of radionecrotic areas with obliteration of vascular changes. At doses between 3,000 and 5,000 R acute inflammation, hemorrhages, perivascular exudates and edema were seen between 1 day and 6 weeks, but complete histological recovery was seen after 5 months. Between 5 and 8 months delayed radionecrosis was observed selectively in white matter with a 'deficient' gliosis. Between 8 and 12 months mild gliosis of radionecrotic white matter was noted with vascular alterations, and between 12 and 24 months an intense gliosis was seen. At doses from 1,500 to 3,000 R acute inflammatory exudates and edema was observed after 1 day and up to 6 weeks. Between 6 weeks and 5 months complete histological recovery occurred but mild degenerative changes were again seen in myelin between 5 and 8 months. Between 8 and 12 months delayed radionecrosis, selective for white matter was observed with a 'deficient' gliosis, and mild gliosis was observed between 12 and 24 months.

Pathological Effects of Hyperthermia

A monkey *(M. mulatta)* received at Delta Primate Center appeared to be catatonic on arrival and was treated with fluids. It gradually became weaker, however, and was sacrificed five days after its arrival at the Center. On microscopic examination, widespread degeneration of neurons in the cerebral cortex (fig. 46) and of Purkinje cell neurons of the cerebellar cortex (fig. 47) was observed. These changes were considered to be consistent with a presumption that hyperthermia (heatstroke) had been involved in the accidental death.

References

1 Adler, S. and Clark, E. J.: A case of acute ascending paralysis in a chimpanzee. Ann. trop. Med. Parasit. *17:* 299–301 (1923).

2 Arnold, A.: Effects of X-irradiation on the hypothalamus; a possible explanation for the therapeutic benefits following X-irradiation of the hypophyseal region for pituitary dysfunction. J. clin. Endocrin. *14:* 859 (1954).

3 Arnold, A.; Bailey, P., and Laughlin, J. S.: Effects of betatron radiations on the brain of primates. Neurology *4:* 165–178 (1954).

4 Beck, E.; Daniel, P. M.; Alpus, M. *et al.*: Experimental 'Kuru' in chimps. A pathological report. Lancet *ii:* 1056–1059 (1966).

5 Bischop, D. de: Un cas de saturnisme chez un gorille. Soc. Roy. Zool. d'Anvers (Unpublished communication to Int. Union Zool. Soc. Directors, Chicago 1956).

6 Bland-Sutton, J.: Tumors in animals. Jour. Anat. Physiol. *19:* 415–475 (1885).

7 Blinzinger, J. S.; Magrath, D. und Boulger, L.: Elektronenmikroskopische Beobachtungen bei experimenteller Poliomyelitis. Experientia *24:* 1095–1096 (1968).

8 Bodechtel, G.: Spontanencephalitis bei einem Affen. Z. Hyg. InfektKr. *111:* 331–336 (1930).

9 Bodian, D.: Poliomyelitis in an uninoculated rhesus monkey and in orally inoculated monkeys receiving deoxypyridoxine. Amer. J. Hyg. *48:* 87–93 (1948).

10 Bodian, D.: A reconsideration of the pathogenesis of poliomyelitis. Amer. J. Hyg. *54–55:* 414–438 (1952).

11 Bodian, D.: Amer. J. Hyg. *64:* 104 (1956).

12 Brizzee, K. R. and Wu, J. J.: Studies on the effects of intrathecal injections of ephedrine sulphate on the spinal cord. J. Neuropath. clin. Neurol. *1:* 234–249 (1951).

13 Brizzee, K. R.; Wu, J. J.; Harnagel, D. J., and Smith, S. M.: Histologic alterations in the spinal cord of the rhesus monkey produced by intrathecal injections of epinephrine, with notes on neurological symptoms following the injections. Anat. Rec. *109:* 13 (1951a).

14 Brizzee, K. R.; Harnagel, D. J., and Smith, S. M.: Histological studies on the effect of intrathecal procaine in the monkey. Anat. Rec. *109:* 104 (1951b).

15 Bullock, B. C.; Clarkson, T. B.; Lehner, N. D. M.; Lofland, H. B., jr., and St. Clair, R. W.: Atherosclerosis in *Cebus albifrons* monkeys. III. Clinical pathologic studies. Exper. and Molec. Pathology *10:* 39–62 (1969).

16 Cameron, A. H. and Hill, W. C. O.: The Arnold-Chiari malformation in a sacred baboon *(Papio hamadryas)*. J. Path. Bact. *70:* 552–555 (1955).

17 Catanei, A.: Présence d'un *Monilia* sur la langue des singes d'Algérie. C. R. Soc. Biol., Paris *93:* 92–94 (1925).

18 Caveness, W. F.; Carsten, A., and Schade, J. P.: Functional and structural alterations following X-irradiation of the cerebral cortex in the monkey. International Congress Series *100:* 784–787 (1965).

19 Caveness, W. F.; Roizin, L.; Innes, J. R. M., and Carsten, A.: Delayed effects of X-irradiation on the central nervous system of the monkey; in Haley and Snider Response of the Nervous System to Ionizing Radiation. Proc. Int. Symp. Los Angeles, Calif. Acad. Press, 448–475 (1964).

20 Courtney, K. D. and Valerio, D. A.: Teratology in the *Macaca mulatta.* Teratology *1:* 163–172 (1968).

21 Davison, C.: Disseminated demyelination of the central nervous system in monkeys and allied disorders in man. J. Neuropath. Psychopath. *14:* 227–238 (1933).

22 Delahunt, C. S. and Lassen, L. J.: Thalidomide syndrome in monkeys. Science *146:* 1300–1305 (1964).

23 FAIRBROTHER, R. W. and HURST, E. W.: Spontaneous diseases observed in 600 monkeys. J. Path. Bact. *616:* 867–873 (1932).

24 FARO, M. D. and WINDLE, W. F.: Progressive degenerative changes in brains of monkeys surviving neonatal asphyxia in brain damage in the fetus and newborn from hypoxia or asphyxia, pp. 24–27 (Ross Laboratories, Columbus 1967).

25 FELDMAN, L. A.; SHEPPARD, R. D., and BORNSTEIN, M. B.: Herpes simplex virus-host cell relationships in organized cultures of mammalian nerve tissues. J. Virol. *2:* 621–628 (1968).

26 FISHER, L. E.: Lead poisoning in a gorilla. J. amer. vet. med. Ass. *125:* 478–479 (1954).

27 FOX, H.: Disease in captive wild mammals and birds. Incidence, description, comparison *8:* (1923).

28 FIENNES, R.: Zoonoses of primates. The Epidemiology and ecology of simian diseases in relation to man (1967).

29 GAJDUSEK, D. C.; GIBBS, C. J., jr., and ALPERS, M.: Experimental transmission of a kuru-like syndrome to chimpanzees. Nature *209:* 794 (1966).

30 GAJDUSEK, D. C.; GIBBS, C. J., jr., and ALPERS, M.: Transmission and passage of experimental 'kuru' to chimpanzees. Science *155:* 212 (1967).

31 GAJDUSEK, D. C.; GIBBS, C. J., and ASHER, D. M.: Transmission of experimental kuru to the spider monkey *(Ateles geoffreyi)*. Science *162:* 693–694 (1968).

32 GLEES, R.: Siderosis of the globus pallidus in a monkey. Jour. Neurol. Neurosurg. Psychiat. *6:* 92–93 (1943).

33 GOLDMANN, K.: Spontaninfektion spinaler Kinderlähmung bei Affen. Ber. tierärztl Wschr. *51:* 497–499 (1935).

34 GRINKER, R. R. and KANDEL, E.: Experimental vitamin (A, B, B2 and B complex) deficiency. Histologic changes in the central nervous system. Arch. Neurol., Chicago *30:* 1287–1297 (1933).

35 HABEL, K. and LOOMIS, L. N.: Coxsackie A7 virus and the Russian poliomyelitis type 4. Proc. Soc. exp. Biol., N.Y. *95:* 597 (1957).

36 HAMERTON, A. E.: Report on deaths occurring in the society's gardens during 1936. Proc. zool. Soc., Lond, p. 443 (1936).

37 HAMERTON, A. E.: *Cysticercus bovis* in the spinal cord of a ringtailed lemur *(Lemur catta)*. Trans. roy. Soc. trop. Med. Hyg. *28:* 2 (1934).

38 HAYMAKER, W.; IBRAHIM, M. Z. M.; MIQUEL, J., and CALL, N.: Delayed radiation effects in the brain of monkeys exposed to X- and Y-rays. J. Neuropath. exp. Neurol. *27:* 50–79 (1968).

39 HAYMAKER, W.; NAUTA, W.; SLOPER, J. C.; LAQUEUR, G.; PICKERING, J. E., and VOGEL, S. F.: The effects of barium ^{140}, lanthanum ^{140}, (Gamma) radiation on the central nervous system and pituitary gland of Macaque monkeys. J. Neuropath. exp. Neurol. *17:* 12 (1958).

40 HILL, W. C. O.: Spina bifida in a sacred baboon *(Papio hamadryas)*. Ceylon J. Sci. (D) *5:* 9–15 (1939).

41 HORSTMANN, D. M. and MANUELIDIS, E. E.: Russian Coxsackie A-7 virus ('AB IV' strain) – neuropathogenicity and comparison with polio virus. J. Immunol. *81:* 32 (1958).

42 HORSTMANN, D. M. and YAMADA, N.: Enterovirus infections of the central nervous

system. Infections of the Nervous System. Res. Publ. ARNMD, Vol. 44, Chapter 14, p. 236. The Assoc. for research in nervous and mental diseases (1968).

43 HOWE, H. A. and CRAIGIE, J.: The question of spontaneous acquisition of poliomyelitis virus by rhesus monkeys. J. Bact. *45:* 87–88 (1943).

44 HURST, E. W.: Histology of experimental poliomyelitis. J. Path. Bact. *32:* 457–477 (1929).

45 HURST, E. W.: The lesions produced in the central nervous system by certain organic arsenical compounds. J. Path. Bact., Lond. *77:* 523–534 (1959).

46 HURST, E. W.: Experimental demyelination of the central nervous system. I. The encephalopathy produced by potassium cyanide. Austr. J. exp. Biol. med. Sci. *18:* 201–223 (1940).

47 HURST, E. W.: Experimental demyelination of the central nervous system. III. Poisoning with potassium cyanide, sodium azide, hydroxylamine, narcotics, carbon monoxide, etc., with some consideration of bilateral necrosis of basal ganglia. Austr. J. exp. Biol. med. Sci. *20:* 297–312 (1942).

48 HURST, E. W. and MCLENNAN, G. C.: A case of Schilder's encephalitis occurring in the monkeys. Rep. S. Aust. zool. Soc. *60:* 25–30 (1937).

49 KEEBLE, S. A.; CHRISTOFINIS, G. J., and WOOD, W.: Natural virus-B infection in rhesus monkeys. J. Path. Bact. *76:* 189–199 (1958).

50 KENT, S. P.; VANITER, G. F.; DOWBEN, R. M., and BENSON, R. E.: Hypervitaminosis D in monkeys *(Macaca mulatta);* a clinical and pathological study. Amer. J. Path. *34:* 37–59 (1958).

51 KENT, S. P. and PICKERING, J. E.: Neoplasms in monkeys *(Macaca mulatta);* spontaneous and radiation induced. Cancer *11:* 138–147 (1958).

52 KERSTING, G. und PETTE, E.: Zur Pathohistologie und Pathogenese der experimentellen «allergischen» Encephalomyelitis des Affen. Dtsch. Z. Nervenheilk. *176:* 387–426 (1957).

53 KESSEL, J. F. and HOLTZWART, F.: Experimental studies with torula from a knee infection in man. Amer. J. trop. Med. Hyg. *15:* 467–478 (1953).

54 LAPIN, B. A. and YAKOVLEVA, L. A.: Comparative pathology in monkeys (Thomas, Springfield 1963).

55 LEBLOND, C. P. and CHAULIN-SERVINIERE, J.: Spontaneous beriberi of the monkey as compared with experimental avitaminosis. Amer. J. Med. Sci. *203:* 100–109 (1942).

56 LEVADITI, C.; HORNUS, G. et SCHOEN, R.: I. Cécite spontanée épidémique du singe dans ses rapports avec les leuconevraxites amaurotiques humaines. Mécanisme pathogénique. Bull. l'acad. Méd. *110:* 394–406 (1933). II. Cécite spontanée du singe. Essais de transmission. C. R. Soc. Biol. *110:* 406–412 (1933).

57 LEVINE, N. D.: Protozoan parasites of domestic animals and of man. (Burgess, Minneapolis 1961).

58 LESWING, R. J. and RIBELIN, W. E.: Physiologic and pathologic changes in acrylamide neuropathy. Arch. Environ. Health *18:* 22–29 (1969).

59 LEVADITI, C.; LEPINE, P. et SCHOEN, R.: Maladie de Schilder-Foix (sclérose cérébrale centro-lobaire) spontanée chez le singe. C. R. Soc. Biol. *104:* 986–991 (1930).

60 LILLIE, R. D.: Pathologic histology of lymphocytic choriomeningitis in monkeys. Rubl. Hlth. rep. wash. *51:* 303–310 (1936).

61 LINDAU, H.: Über einen neuen Fall von spinaler Kinderlähmung bei einem Schimpansen. Tierärztl. Rdsch. *44:* 124–126 (1938).

62 LINDSAY, S. and CHAIKOFF, I. L.: Arteriosclerosis in the baboon. Naturally occurring lesions in the aorta and coronary and iliac arteries. Arch. Path. *63:* 460–471 (1957).

63 LOU, T. and WENNER, H. A.: Experimental infections with enteroviruses. V. Studies on virulence and pathogenesis in cynomolgus monkeys. Arch. ges. Virusforsch. *12:* 313 (1962).

64 LUCKE, B.: Spontaneous cerebral lesions in monkeys. Arch. Neurol. Psychiat. *10:* 212–225 (1923).

65 LUND, O. E.: Histologische Befunde bei experimentellen akuten Kohlenoxydvergiftungen. Eine Untersuchung an pervitinbehandelten Rhesusaffen. Arch. Gewerbepath. Gewerbehyg. *15:* 96–109 (1956).

66 MANUELDIS, F. E.: Neuropathology of experimental West Nile virus infection in monkeys. J. Neuropath. *15:* 448–460 (1956).

67 MARTIN, J. E.; COLE, W. C., and WHITNEY, R. A., jr.: Tuberculosis of the spine (Pott's disease) in a Rhesus monkey *(M. mulatta).* J. amer. vet. med. Ass. *153:* 914–917 (1960).

68 MELLA, H.: Experimental production of basal ganglia symptomatology in *Macaca rhesus.* Arch. Neurol., Chicago *11:* 405–417 (1924).

69 MELNICK, J. L. and BANKER, D. D.: Isolation of B virus (Herpes group) from the cerebral nervous system of a rhesus monkey. J. exp. Med. *100:* 181–194 (1954).

70 MORRISON, L. R.: Histopathological effects of anoxia on the central nervous system. Arch. Neurol., Chicago *55:* 1–34 (1946).

71 MOSSAKOWSKY, M. J.; LONG, D. M.; MYERS, R. E., and RODRIQUES DE CURET, H.: Early histochemical changes in perinatal asphyxia. J. Neuropath. exp. Neurol. *27:* 500–516 (1968).

72 MUCKENFUSS, R. S.; ARMSTRONG, C., and MCCORDOCK, H. A.: Encephalitis: Studies in experimental transmission. Pub. Hlth Rep. Wash. *48:* 1341–1343 (1933).

73 MYERS, R. E.: Patterns of perinatal brain damage in the monkey. Brain damage in the fetus and newborn from hypoxia or asphyxia, pp. 17–21 (Ross Laboratories, Columbus 1967).

74 MYERS, R. E.; BEARD, R., and ADAMSONS, K.: Brain swelling in the newborn rhesus monkey following prolonged partial asphyxia. Neurology *19:* 1012–1018 (1969).

75 NATHANSON, N.; GOLDBLATT, D.; THIND, I. S.; DAVIS, M., and PRICE, W. H.: Histological studies of the monkey neurovirulence of group B arboviruses. I. A semiquantitative grading scale. Amer. J. Epidem. *82:* 359–381 (1966a).

76 NATHANSON, N.; DAVIS, M.; THIND, I. S., and PRICE, W. H.: Histological studies of the monkey neurovirulence of group B arboviruses. II. Selection of indicator centers. Amer. J. Epidem. *84:* 524–540 (1966b).

77 NATHANSON, N.; STOLLEY, P. D., and BOOLUKOS, P. J.: Eastern equine encephalitis. Distribution of central nervous system lesions in man and Rhesus monkey. J. Comp. Path. *79:* 109–115 (1969).

78 NEGROM, P.: Histoplasmosis-diagnosis and treatment (Thomas, Springfield 1965).

79 PAPENHEIMER, A. M.; BAILEY, O. T.; CHEEVER, F. S., and DANIELS, J. B.: Experimental polyradiculitis in monkeys. J. Neuropath. exp. Neurol. *1:* 48–62 (1951).

80 PAYAN, H.; TOGA, M. et GASTANT, H.: Etude anatomique de l'encéphalite japonaise B expérimentale chez le singe. Ann. Anat. *2:* 150–181 (1957).

81 PELC, S. *et al.*: Experimental results obtained in human encephalitis, subacute type

encephalitis et inclusions, subacute sclerotic leukoencephalitis. Rev. Neurol. *98:* 3–24 (1958).

82 PENTSCHEW, A.; ENBER, F. F., and KOVATCH, R. M.: Experimental manganese encephalopathy in monkeys. J. Neuropath. exp. Neurol. *22:* 488–499 (1963).

83 PERDRAU, J. R.: Schilder's encephalitis periaxialis diffusa in a rhesus monkey. J. Path. Bact. *33:* 991–994 (1930).

84 PFEIFER, R. A.: Kreislauf und Hirntuberkulose. Untersuchung zur Pathoangio-architektonic der entzündlichen Hirnkrankheiten (Theodor Steinkopf, Dresden 1935).

85 POKORNY, J.; HUBNER, J., and ZASTERA, M.: Isolation of toxoplasma gondii from certain domestic and free-living animals in Czechoslovakia. Cs. Epidem. *10:* 323 (1961).

86 RANCK, J. B., jr. and WINDLE, W. F.: Brain damage in the monkey, *Macaca mulatta*, by asphyxia neonatorum. Exp. Neurol. *1:* 130–154 (1959).

87 RICHTER, R. B.; Encephalitis affecting the basal ganglia in monkeys. J. Neuropath. exp. Neurol. *4:* 16–26 (1945a).

88 RICHTER, R. B.: Degeneration of the basal ganglia in monkeys from chronic carbon disulfide poisoning. J. Neuropath. exp. Neurol. *4:* 324–353 (1945b).

89 RICHTER, R. B. and KLIIVER, N.: Spontaneous striatal disease in a monkey. J. Neuropath. exp. Neurol. *3:* 49–62 (1944).

90 RINEHART, J. F.; GREENBERG, L. D., and FRIEDMAN, M.: Experimental thiamine deficiency in the rhesus monkey. Amer. J. Path. *23:* 879–881 (1947).

91 ROIZIN, L. and SCHADE, J. P.: Pathogenesis of X-irradiation effects in the monkey cerebral cortex. V. Ultrastructural findings. Brain Res. *7:* 87–109 (1968).

92 ROIZIN, L.: Pathogenesis of X-irradiation effects in the monkey cerebral cortex. III. Histopathologic findings. Brain Res. *7:* 28–58 (1968).

93 SABIN, A. B. and HURST, E. W.: Studies on B virus. IV. Histopathology of the experimental disease in rhesus monkeys and rabbits. Brit. J. exp. Path. *16:* 133–148 (1935).

94 SELLHEIM, A. P.: Une tumeur cérébrale chez un singe. J. belge Neurol. Psychiat. *36:* 240–242 (1936).

95 SCHERER, H. J.: Melanin pigmentation of the substantia nigra in primates. J. comp. Neurol. *71:* 91–95 (1939).

96 SCHERER, H. J.: Funikuläre Spinalerkrankung mit schwerer Beteiligung des Grosshirnmarkes und Opticusveränderungen bei fünf Pavianen. Z. ges. Neurol. Psychiat. *141:* 212–234 (1932).

97 SCHERER, H. J.: Vergleichende Pathologie des Nervensystems der Säugetiere *8:* (Georg Thieme, Leipzig 1944).

98 SIIM, J. C.: Human toxoplasmosis (Munksgaard, Copenhagen 1960).

99 STEINER, P. E.; RASMUSSEN, T. B., and FISHER, L. E.: Neuropathy, cardiopathy, hemosiderosis and testicular atrophy in *Gorilla gorilla.* Arch. Path. *59:* 5–25 (1955).

100 SUREAU, P.; RAYNAUD, J. P.; LAPEIRE, C. et BRYGOO, E. R.: Premier isolement de *Toxoplasma gondii* à Madagascar: toxoplasmose spontanée expérimentale du *Lemur catta.* Bull. Soc. Path. exot. *55:* 357 (1962).

101 TANAKA, T.: Über spontane Poliomyelitisinfektion beim Affen. Nagasaki Igakki Zassmi *17:* 1802–1809 (1939).

102 TOGA, M.: Etude clinique, électroencéphalo-graphique et anatomique de l'encéphalite expérimentale B chez le singe (Office Universitaire de Polycopie, Aix-en-Provence 1957).

103 URBAIN, A.; REISE, W. et NOWEL, J.: Atrophie cérébelleuse observée chez un gélada *(Theropithecus gelada*, Ruppel*)*. Rev. Path. comp. *41:* 176–179 (1940–1941).

104 VAN BOGAERT, L.: Studies on spontaneous primary encephalitides in the monkey. I. A polioencephalitis with predilection for the brain stem in the baboon *(Papio hamadryas)*. II. Encephalitis with a cortico-striate predilection in the baboon. III. Diffuse non-suppurative meningitides and meningo-encephalitides with subacute and chronic course. J. Neuropath. exp. Neurol. *18:* 294–323 (1959).

105 VAN BOGAERT, L.: Sur l'atrophie papillaire bilatéral des singes (formes optique de la leucoencéphalose périvasculaire confluente) Group belge d'Etudes O. N. O. and N. C. C. R. Bruxelles Méd. 492 (1939a).

106 VAN BOGAERT, L.: Les aspects anatomiques d'une maladie démyélinisante observée chez le singe, la leucoenéphalomyélose périvasculaire confluente. Ann. Inst. Pasteur *63:* 315–366 (1939b).

107 VAN BOGAERT, L. et DALLEMAGNE, M. J.: Approche expérimentale des troubles nerveux du manganisme. Mschr. Psychiat. Neurol. *111:* 60–89 (1946).

108 VAN BOGAERT, L. et DEWULF, A.: Sur un processus nécrotisant nilaire d'origine inconnue chez *Macacus rhesus*, se présentant cliniquement sous forme d'une cécite. J. belge Neurol. Psych. *38:* 583–597 (1938).

109 VAN BOGAERT, L. and INNES, J. R. M.: Neurologic disease of apes and monkeys; in INNES and SAUNDERS Comparative Neuropathology, pp. 55–146 (Academic Press, New York/London 1962).

110 VAN BOGAERT, L. and SCHERER, H. J.: Acute amaurotic epilepsy in *Macacus rhesus*. Arch. Neurol., Chicago *40:* 521 (1938).

111 VAN DYKE, D. and JANSSEN, P.: J. Neurosurg. *20:* 289 (1963).

112 VICTOR, M. and ADAMS, R. D.: The neuropathology of experimental vitamin B6 deficiency in monkeys. Amer. J. clin. Nutr. *4:* 346–353 (1956).

113 VOGEL, F. S. and PICKERING, J. E.: Demyelination induced in the brains of monkeys by means of fast neutrons. Pathogenesis of the lesion and comparison with the lesions of multiple sclerosis and Schilder's disease. J. exp. Med. *104:* 435–442 (1956).

114 VOGEL, F. S.: Tuberculosis in the *Macaca mulatta* and in the chimpanzee. Lecture at the Armed Forces Institute of Pathology, Washington (1955).

115 VOGEL, F. S.: In response of nervous system to ionizing radiation. Proc. Int. Symposium (Acad. Press, Chicago 1962).

116 WALKER, J. and SPOONER, E. T. C.: Natural infection of the African baboon *(Papio papio)* with the large-cell form of histoplasma. J. Path. Bact. *80:* 436–438 (1960).

117 WALKER, A. E.: Cysticercosis cellulosae in the monkey. J. Comp. Path. *49:* 141–145 (1936).

118 WECHSLER, I. S. and GLOBUS, J. H.: Histological observations on the nervous system of monkeys experimentally deprived of vitamin E. J. Neuropath *1:* 430–437 (1942).

119 WEIDMANN, F. D.: Dermatosis of monkeys and apes: 9th Inter. Congr. Dermatol. Deliberations *1:* 600–606 (1935).

120 WEINBERG, M.: Un cas de méningite cérébrospinale chez le chimpanzé. Bull. Soc. Path. exot. *2:* 202–204 (1909).

121 WENNER, H. A.; LOU, T., and KAMITSUKA, P. S.: Experimental infections with coxsackie viruses. Arch. ges. Virusforsch. *10:* 426 (1960).

122 WINDLE, W. F.: Neuropathology of certain forms of mental retardation. Science *140:* 1186, 1189 (1963).

123 Young, R. J.; Fremming, B. D.; Benson, R. E., and Harris, M. D.: Care and management of a *Macaca mulatta* monkey colony. Proc. anim. care panel *7:* 67–82 (1957).

124 Zaman, V. and Goh, T. K.: Isolation of *Toxoplasma gondii* from the slow loris *Nycticebus concang*. Ann. trop. Med. Parasit. *62:* 52–53 (1968).

Author's address: Dr. Kenneth R. Brizzee, Tulane University, Delta Regional Primate Research Center, *Covington, LA 70433* (USA)

Pathology of Simian Primates, Part I, pp. 639–670 (Karger, Basel 1972)

Diseases of the Oral Cavity

B. COHEN and W. H. BOWEN

Department of Dental Science, Royal College of Surgeons of England, London

Contents

General 639
Abnormalities in Post-Natal Development 639
Abnormalities in Morphology 641
Injuries to the Teeth 649
Malnutrition 652
Infectious Diseases 658
Tumours 669
References 670

General

The content of this chapter unless otherwise stated is based on observations made on between three and four hundred living specimens of *Macaca irus* housed in a colony maintained for the purposes of dental research, as well as some five hundred autopsy examinations of animals of the same species used in a vaccine-testing programme. The material includes thirty monkeys born in the colony.

Abnormalities in Post-Natal Development

Abnormalities in the number of teeth, or of their morphology, are rarely seen, but variations in the age at which teeth erupt cover a wide time scale. Since the presence or absence of teeth in both the deciduous and permanent

Table Ia. Sequence and age of eruption of deciduous teeth in *Macaca irus*

Sequence	Teeth	Age of eruption in weeks			Number of teeth in sample
		Average	Minimum	Maximum	
1st	Lower central incisors	2.6	0	5	36
2nd	Upper central incisors	2.9	0	6	36
3rd	Lower lateral incisors	5.2	1	7	36
4th	Upper lateral incisors	6.7	2	9	36
5th	Lower canines	10.3	4	14	36
6th	Upper first molars	10.3	7	14	35
7th	Lower first molars	10.5	7	13	40
8th	Upper canines	10.7	4	13	32
9th	Lower second molars	25.1	16	34	36
10th	Upper second molars	27.0	16	41	36

dentitions is frequently used as a yardstick for age estimation, it is important to recognise that the degree of variation renders such calculations of doubtful value.

Table Ia and Ib indicate the disparity in times of tooth eruption observed in 27 animals born in captivity. It will be appreciated that age estimation based on data of this sort is at best only approximate. At worst it approaches guesswork as, for example, during the period following eruption of the first permanent molars; as long as twelve months may then elapse during which no further tooth eruption takes place.

Recently BOWEN and KOCH [1] have evolved a method which enables an accurate estimate to be made of the age of an animal up to four years. Thus far, however, the method has been applied to *M. irus* only. It is probable that the principles of the method can be applied to most primates. The procedure entails the taking of two dental radiographs, one of the mandibular molar-premolar area and the other of the maxillary incisor area. From these radiographs it is possible to determine whether permanent teeth in these areas have started to mineralise (fig. 1) or whether crowns of certain teeth are completely mineralised (fig. 2).

A simple chart (fig. 3) has been drawn up to facilitate the use of clinical and radiographical information in estimating the age of an animal. The use of this chart may be illustrated as follows. If clinical examination reveals that all first permanent molars and all permanent incisors have erupted, the animal falls within the age range of 26–37 months. If the radiographic examina-

Table Ib. Sequence and age of eruption of permanent teeth (not including 3rd molars) in *Macaca irus*

Sequence	Teeth	Age of eruption in months			Number of teeth in sample
		Average	Minimum	Maximum	
1st	Lower first molars	16.1	14	18	32
2nd	Upper first molars	17.5	14	20	30
3rd	Lower central incisors	28.8	23	33	20
4th	Upper central incisors	28.9	23	34	21
5th	Lower lateral incisors	30.5	25	34	13
6th	Upper lateral incisors	32.2	26	37	13
7th	Lower canines	42.3	36	49	6
8th	Upper first premolars	43.2	37	46	6
9th	Lower first premolars	44.2	40	46	6
10th	Lower second premolars	44.2	40	46	6
11th	Lower second molars	44.2	40	46	6
12th	Upper second premolars	45.5	39	49	3
13th	Upper second molars	46.0	40	49	3
14th	Upper canines	47.3	40	51	3

tion then shows crowns of the lower second permanent molars to be completely mineralised and those of the lower second premolars to be incompletely mineralised, the estimate of age can be narrowed to 26–28 months.

Although this method gives an accurate assessment of age, useful conclusions can also be drawn by merely observing the number of teeth present. For example, the presence of 28 teeth, irrespective of whether they are all permanent or a mixture of deciduous and permanent, has been found to be a reasonable indicator of sexual maturity and is a quick and useful guide to the selection of females for breeding.

Abnormalities in Morphology

1. General. A wide range of observations in dental morphology has been described in human and animal dentitions, but many are of interest as rarities rather than of practical importance. Abnormalities in tooth development are certainly uncommon in *M. irus*, although the examination of one thousand dentitions is scarcely sufficient to form any sort of assessment as to their overall prevalence.

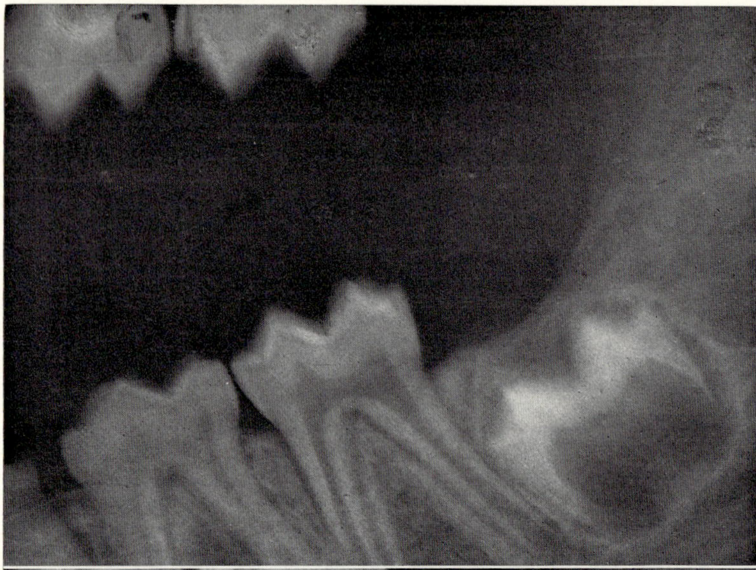

1

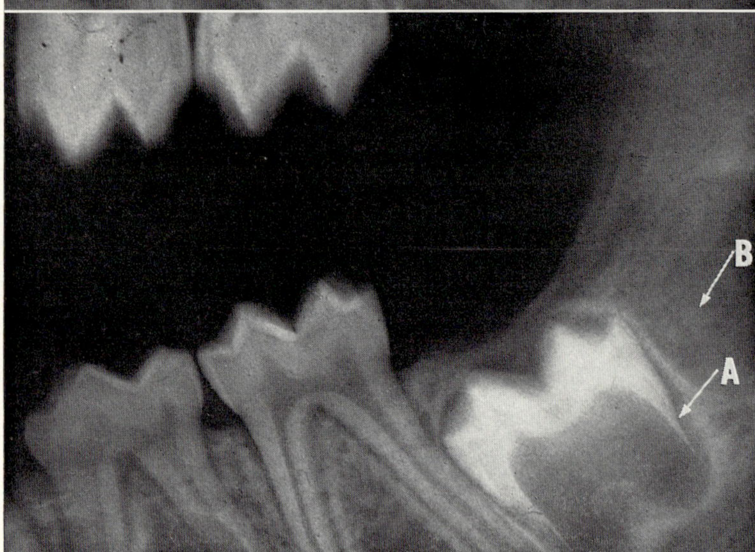

2

Fig. 1. Radiographs of lower first permanent molar of 10-month-old monkey. Tooth crown is not completed.

Fig. 2. Radiograph from the same monkey as in figure 1 at 11 months of age. A indicates the change from convexity to concavity of the proximal surface (= crown completed). B indicates the first sign of mineralisation of the lower second permanent molar.

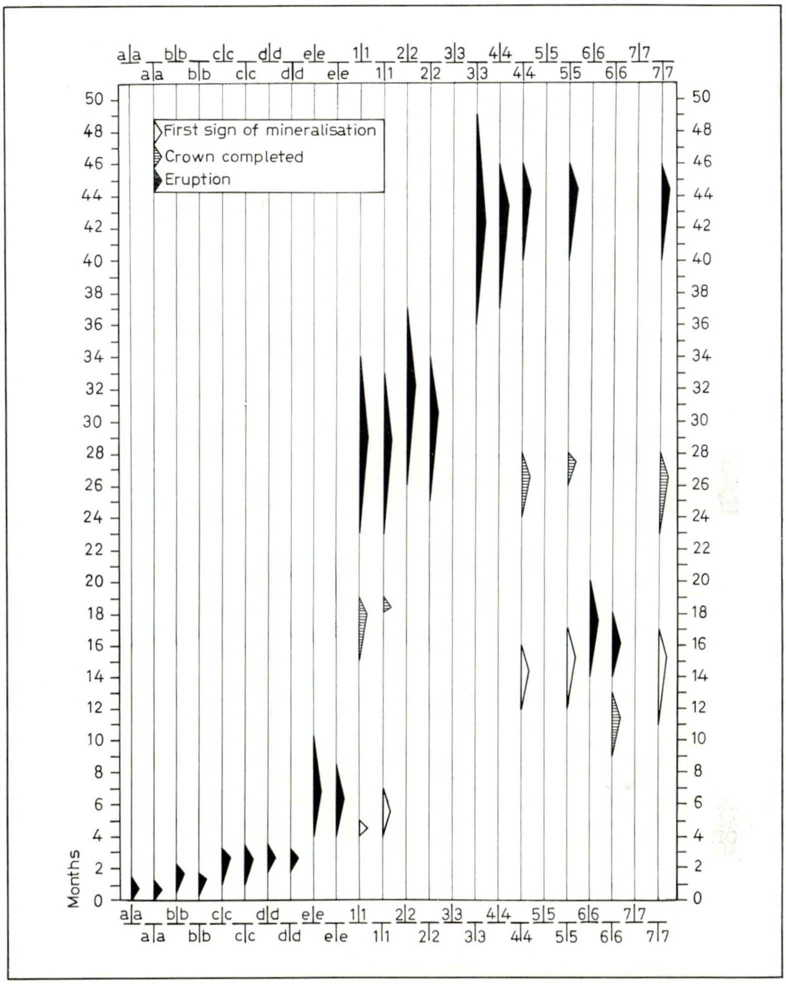

Fig. 3. Chart summarising radiological and eruption data for age estimation.

2. *Malposition of teeth* in the dental arch has been seen only in those instances where there has been premature loss of deciduous predecessors (fig. 4) or where removal of permanent teeth has led to drifting of those adjacent, as in figure 5. In usual circumstances, where all the teeth are present, the dental arches of both the deciduous and permanent dentitions are astonishingly regular. It is not possible to predict with certainty the type of malposition which will ensue following premature tooth loss. The animal depicted in

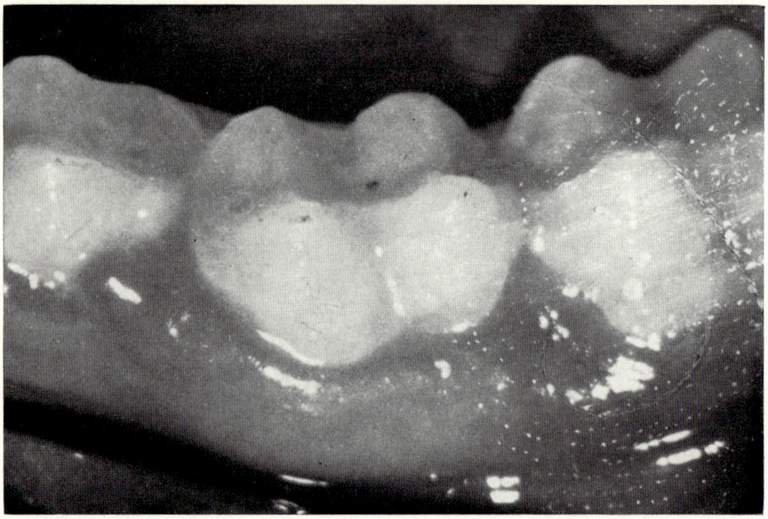

Fig. 4. Crowded permanent molars following early loss of deciduous teeth in the same quadrant.

figure 4 exhibited crowding of the mandibular permanent teeth following extraction of deciduous molars on the same side, but similar extractions in other animals have not resulted in malposition. The loss of permanent incisors in the animal shown in figure 5 has resulted in the tipping mesially of the lateral incisors abutting upon this abnormal diastema. More severe malocclusion has been seen in instances where normally opposing teeth are missing, as is exemplified by the disturbed dentition of the animal shown in figure 6.

3. *Fusion of teeth* may take several forms in the human dentition. These are known by a variety of terms indicating the degree and type of union of the dental tissues. In the case of complete fusion the union of two tooth germs results in a tooth of twice the normal size. Where a single tooth germ undergoes division as a result of an abnormal invagination, the anomaly known as gemination ensues in which two teeth are incompletely separated as, for example, when two crowns share a single root. Two separate and completely formed teeth may become fused by the amalgamation of their roots with newly deposited cementum, a phenomenon known as concrescence. The fusion of a lower canine and lateral incisor is shown in figure 7. The union is incomplete, but a partial fusion of the crowns has occurred, indicating that

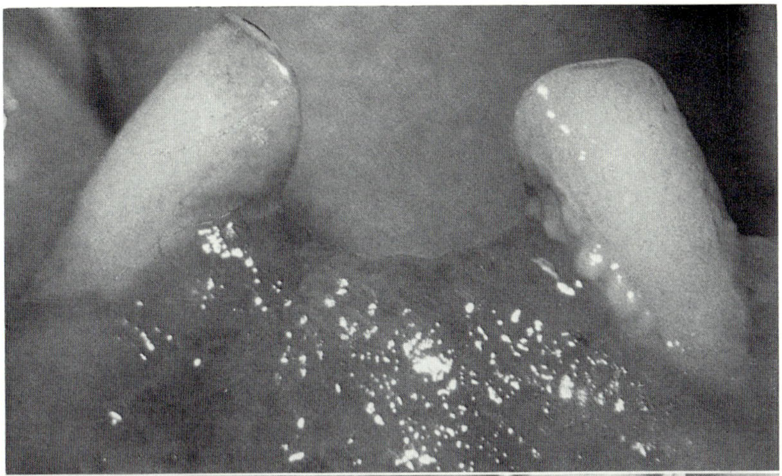

5

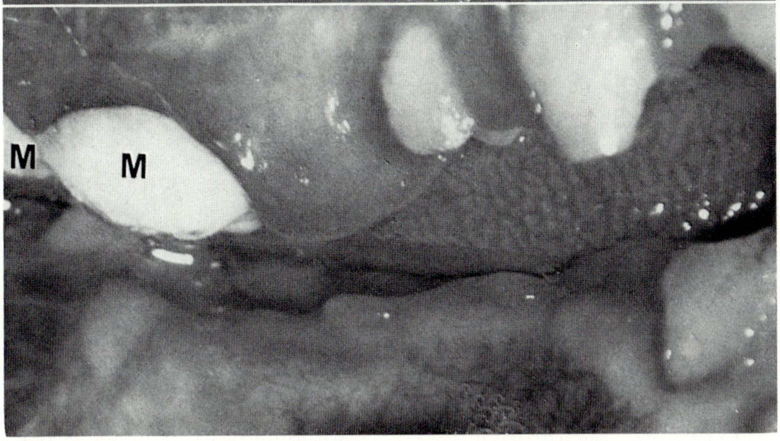

6

Fig. 5. Lateral incisors tipped towards the midline across the space left by the loss of the central incisors.

Fig. 6. Loss of mandibular cheek teeth has resulted in downward migration of the unopposed upper molars (M).

the anomaly originated early in tooth development, and certainly before the completion of amelogenesis.

4. *Enamel hypoplasia* is a condition in which a disturbance in enamel formation results in the appearance of defects of the tooth surface. Such a disturbance may occur during the development of the enamel matrix or at a

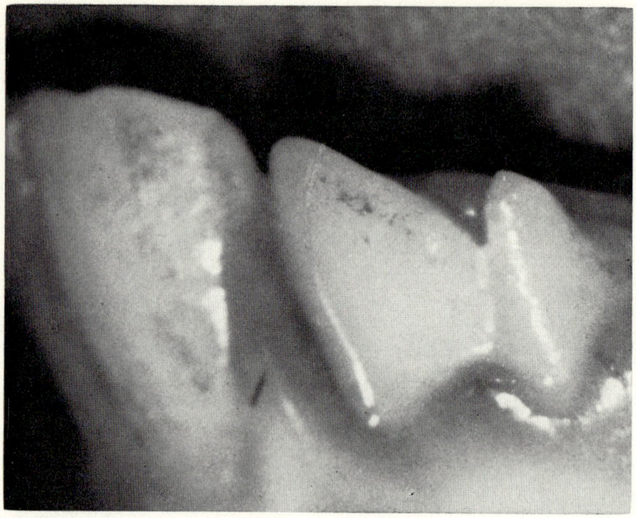

Fig. 7. Fusion of incisor and canine.

later stage during calcification and maturation. A variety of different factors may be responsible for similar imperfections in the finally formed tooth. Nutritional deficiencies are an obvious cause of enamel hypoplasia; severe infections during the formative period, the ingestion of certain chemicals, and local trauma are equally possibilities to be considered. Although the enamel defect gives little or no indication of the underlying cause, the area of the tooth or teeth affected often provides testimony for chronological identification of the insult. It is worth remembering that even in cases of severe maternal nutritional insufficiency the enamel of the infant may be free of outward blemishes; it is remarkable that the developing tooth appears to have first call on resources of maternal or foetal calcium even when severe calcium deprivation exists. Consequently it is not surprising that hypoplasia of enamel in the decidous dentition is rarely seen.

Figure 8 shows a case of enamel hypoplasia in the incisor teeth of an animal brought into captivity at the age of eleven months which suggests that the defect may be a reflection of dietary disturbance at the time of capture. Hypoplastic enamel in the incisors of a rhesus monkey is shown in figure 9, the sharp demarcation of the defect pointing to a disturbance during the latter half of the first year of life.

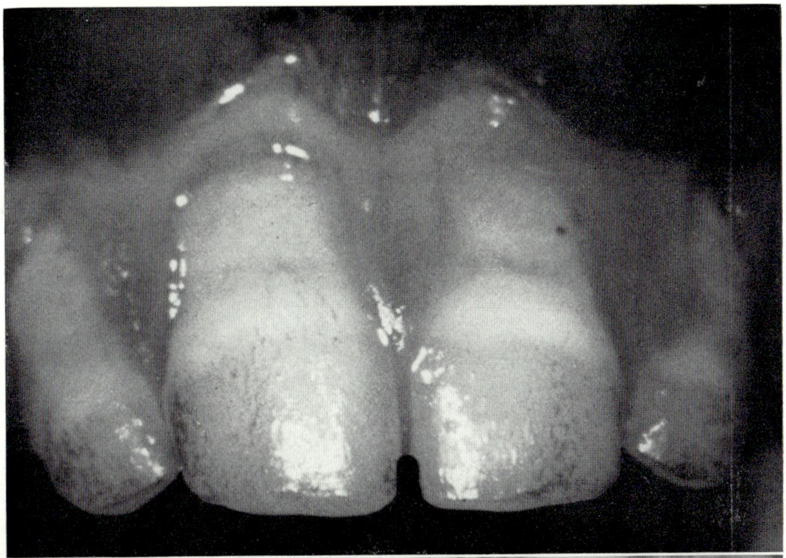

8

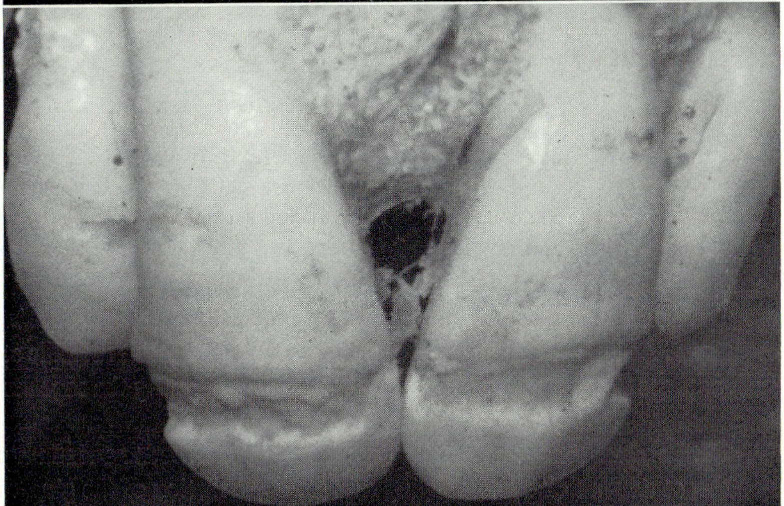

9

Fig. 8. Enamel hypoplasia in upper incisors.

Fig. 9. Severe hypoplasia; from the skull of a specimen of *M. mulatta* in the Odontological Museum of the Royal College of Surgeons of England.

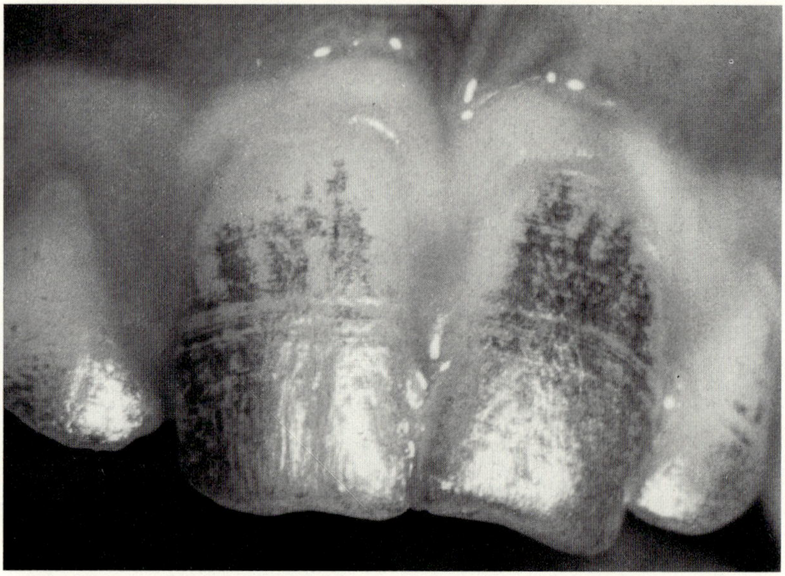

Fig. 10. Metallic stains on incisor teeth.

5. *Pigmentation* of teeth may, rarely, be endogenous in origin but much more commonly arises from exogenous causes. Intrinsic discolouration of the tooth substance may be derived from blood pigments in cases of haemolytic disease and, if confined to limited areas of the teeth, indicates the period during which the adverse effect was in operation. Exogenous pigmentation is usually bacterial in origin, both orange and green stains of this sort being encountered. Animals housed in metal cages frequently exhibit metallic stains, such as those shown in figure 10, which result from gnawing the cage mesh.

6. *Abnormalities of the soft tissues* can usually be traced to an inflammatory state, as a result of which varying degrees of distortion of the gingival architecture may be seen. Figure 11 exemplifies chronic hyperplastic gingivitis. Abnormalities which are not inflammatory in origin are uncommon as, for example, the rarity illustrated in figure 12. In the mouth of this young animal small vesicular protuberances were present over a wide area of the oral mucosa including the cheeks, the buccal sulcus, the attached gingiva and the soft palate. A biopsy of the lesion revealed the presence of abundant subepithelial follicles of lymphoid tissue, resembling accessory tonsillar aggregations (fig. 13).

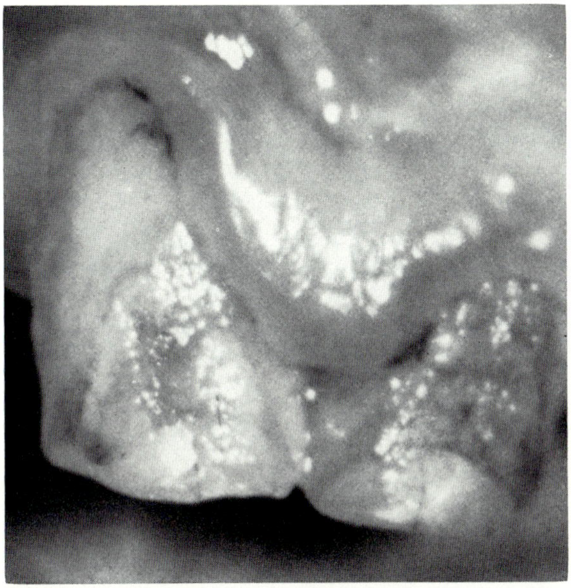

Fig. 11. Hyperplastic gingivitis. There is a gross enlargement of the labial interdenta papilla and the free gingival margin has a rolled edge.

A minor deviation from the normal soft tissue morphology is sometimes seen in the presence of an enlarged frenum between the maxillary central incisors.

Pigmentation of the soft tissues is frequently seen, resembling the patchy melanotic discolouration of the gingivae and cheeks common in many human races.

A curious example of soft tissue abnormality is that shown in figure 14. In this monkey the lower lip is completely atonic, a feature first noticed when the animal arrived in the colony aged a little more than one year; it has persisted, unchanged, for two years subsequently. This lack of lip-seal appears to have provoked a degree of mandibular hyperplasia in the mental region, but the health of the unnaturally exposed mucosa is unimpaired and the occlusal relationship of upper and lower incisor teeth is normal.

Injuries to the Teeth

1. Broken and chipped teeth are commonly seen in adult animals brought into captivity. The teeth of *M. irus* are exceedingly secure in their attachment

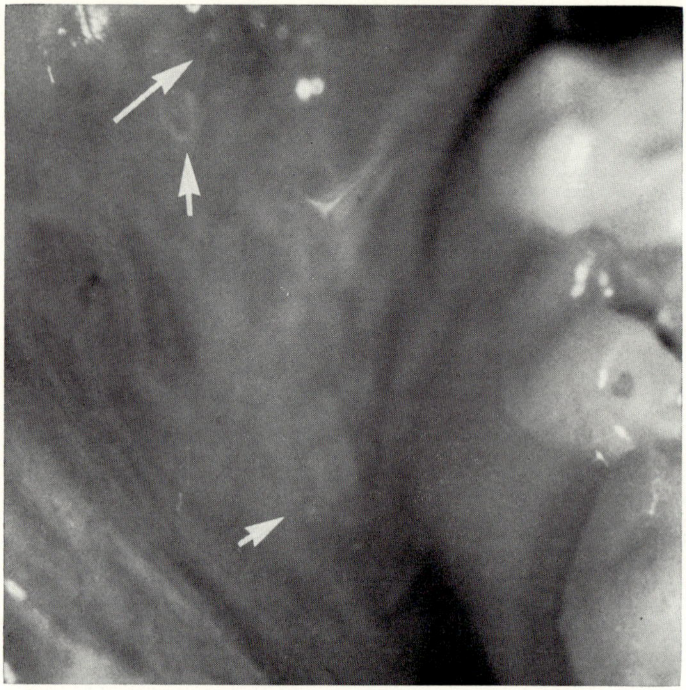

Fig. 12. Accessory lymphoid tissue in the form of multiple raised, irregular swellings of the mucosa.

to bone, so that severe trauma is more likely to fracture the tooth crown than to result in loss of the complete tooth.

Chipped teeth are occasionally the cause of food or debris becoming impacted in the abnormal space. There is little doubt that damage brought about in this way has been mistakenly described as dental caries in the past, and there are many early authors who have speculated on the difference in distribution of carious lesions between man and monkey. The production of true caries in captive monkeys (see below) has lent weight to the belief that most of the so-called caries described in wild monkeys is in fact an aftermath of local trauma or attrition and is not dental caries in the accepted sense of the term. Figure 15 is a photograph of a skull in which tooth destruction was primarily traumatic in origin although no doubt followed by food stagnation and caries-like deterioration.

2. Attrition is a physiological process in which long usage and the consumption of gritty foods result in the gradual wearing away of tooth sur-

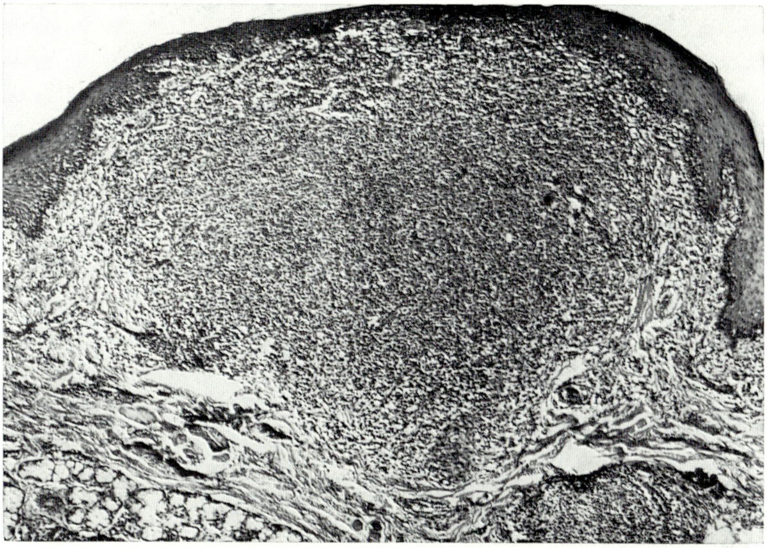

Fig. 13. Microscopic appearance of one of the follicles shown in figure 12. Haematoxylin and eosin ×55.

faces. It is, of course, especially prevalent in wild animals. As wear proceeds the surface of the crown becomes progressively more flattened; as enamel wears away and dentine becomes exposed the lustrous white appearance of the crown is replaced by the dull yellow hue of dentine. This form of wear brings in its train a gradual alteration in the relationship of adjacent as well as opposing teeth; the overall effect is to reduce the self-cleansing action of mastication and to favour the accumulation of debris between and around the teeth. Increasing attrition stimulates the dentine-forming cells of the pulp chamber to lay down secondary dentine. In this way, the pulp chamber becomes progressively smaller and the external loss of calcified substance is compensated for by the internal deposition of new dentine; unless attrition is unusually rapid, the pulp chamber is thus protected from exposure.

Examples of attrition are to be seen in most older animals, as is shown in figure 15. Too much reliance should not be placed upon attrition as an aid to the estimation of age since to a large extent the wear of teeth is influenced by the composition of diet. However, the degree of attrition seen in third molars provides the basis for an approximation.

3. *Abrasion* is a pathological effect produced by the exposure of teeth to abnormal wear. Characteristically this type of wear is localised to particular

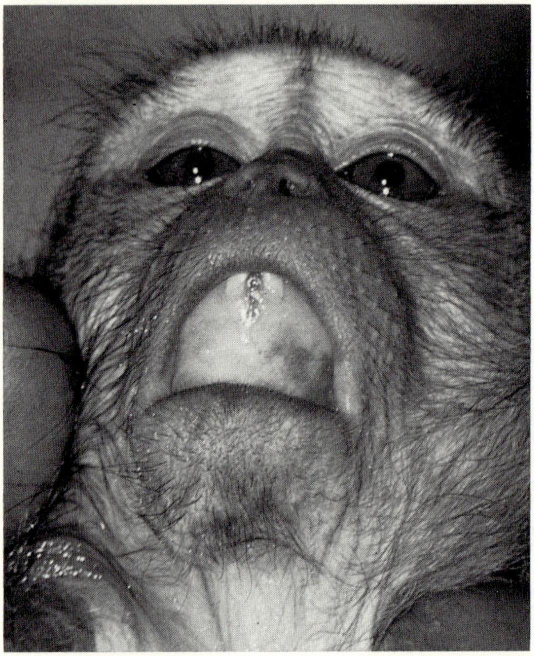

Fig. 14. Completely flaccid lower lip accompanied by overgrowth of the mental region.

areas. Caged animals which develop a habit of gnawing at the bars of the cage are likely to abrade their teeth in this fashion. Facets of wear can be seen on the incisal edges of the teeth shown in figure 16, metallic deposits are present on the lower teeth, and it is possible that the chipped tooth may have been damaged in the same way, although this is not commonly seen.

Injuries to the soft tissues of the mouth are usually a consequence of fighting. *Lacerations* of the lips and cheeks are sometimes seen in these instances. Figure 17 is an example of an injury to the tongue of an aggressive male possibly self-inflicted in the course of a fight.

Malnutrition

Effects of malnutrition upon the teeth are, as mentioned above, rare in deciduous teeth. They are however, occasionally discernible at microscopic level in the teeth of young animals which have suffered dietary deficiencies

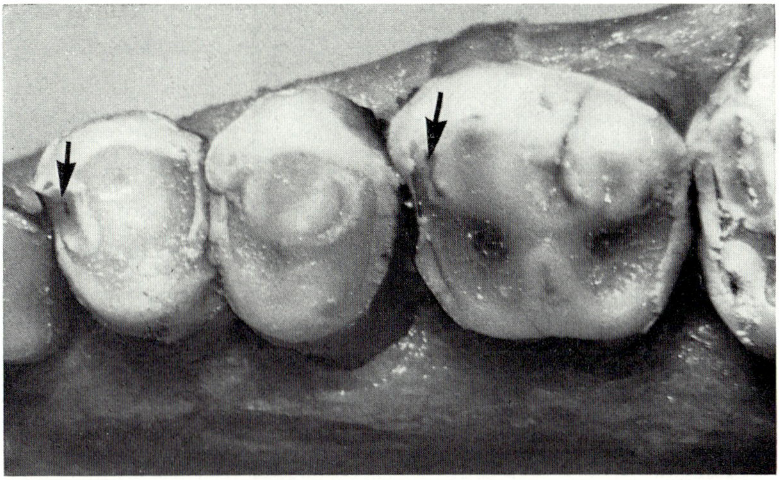

Fig. 15. The occlusal surfaces of the cheek teeth have been worn flat, and dentine, darker than enamel, has become exposed. Arrows denote chipped and worn areas mimicking the appearance of carious destruction.

post-natally. Figure 18 illustrates such a case; in this instance a young animal, brought into captivity soon after birth, failed to thrive. Sections of the teeth provide chronological evidence of malnutrition in the disordered root formation; the crown of this tooth, which formed *in utero*, exhibited no such derangement of structure.

A characteristic deformity assumed to arise from dietary deficiency is frequently seen in young animals brought in from the wild state. This takes the form of what at first sight appears to be an over-development of the premaxilla. The anterior segment of the upper jaw is broader than normal and the vault of the palate is flattened. The alveolar ridge as a whole, but most markedly in the premaxillary region, has a rolled and rounded contour (fig. 19).

The evidence in support of the belief that this condition is a manifestation of dietary deficiency is derived, in the first instance, from the observation that the phenomenon seems to be most common in animals which are undersized and underweight; but more important, from its gradual disappearance in animals which thrive on the nutritionally balanced diet provided in captivity.

A complex and contradictory literature has been built up on the subject of bone disease in monkeys. RUCH [2] has reviewed the controversies arising

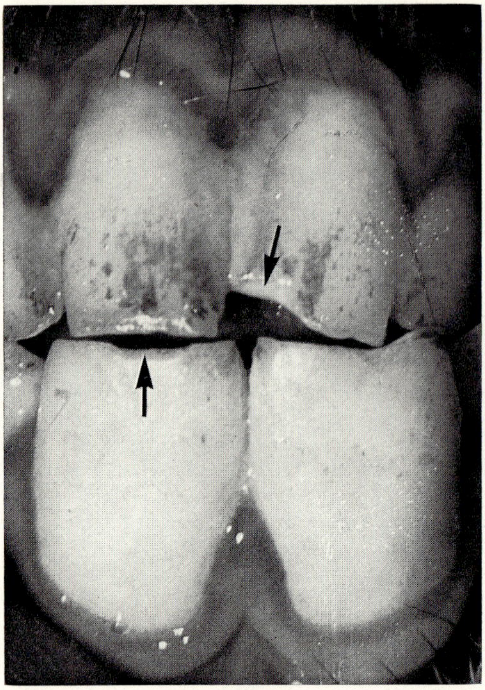

Fig. 16. Chipped tooth and facets of wear (arrowed) on incisor teeth.

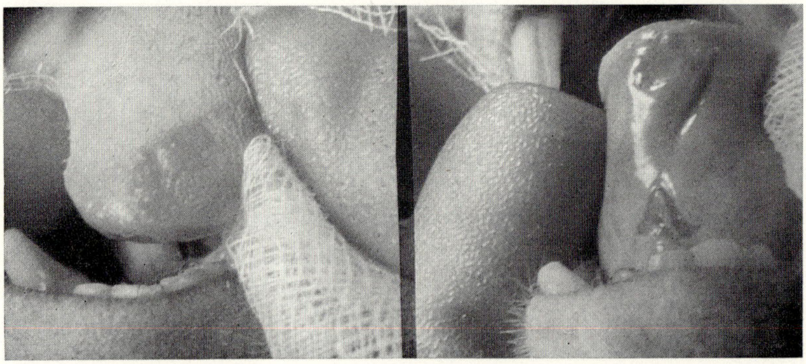

Fig. 17. Laceration on ventral surface of tongue; the resultant haematoma is visible on the dorsal aspect.

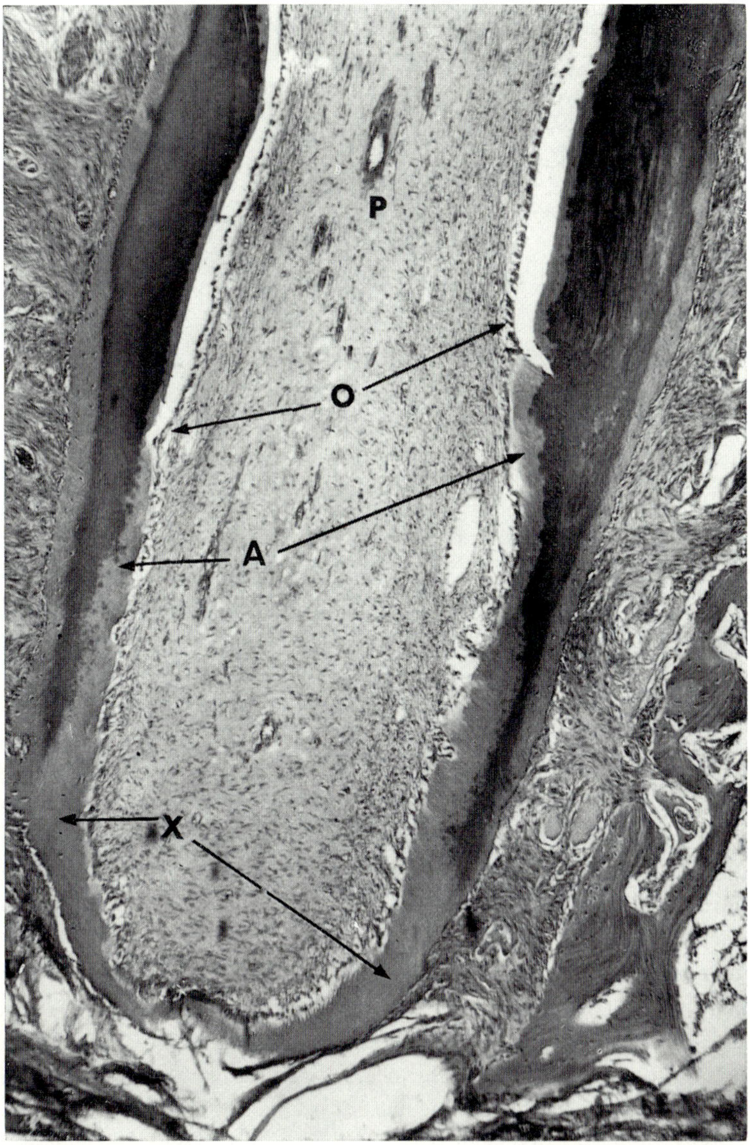

Fig. 18. Disordered root formation in a severely malnourished young monkey; P = pulp chamber, O = extremity of normal odontoblast layer, A = abnormally wide predentine, X = irregular non-calcifying matrix. Haematoxylin and eosin ×50.

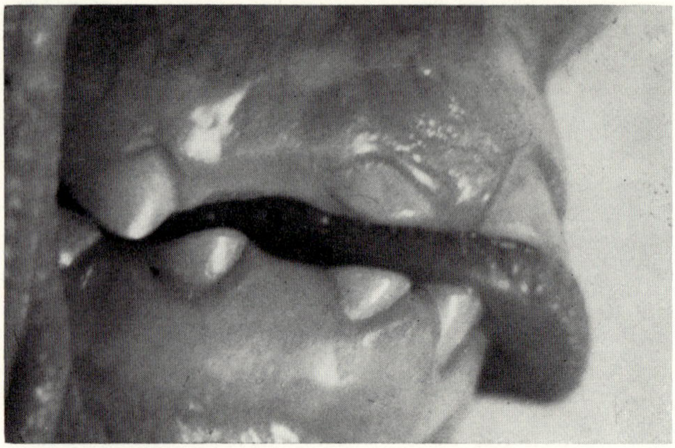

Fig. 19. Profile of malnourished young animal with lips retracted showing bulbous, rounded contour of the tooth-supporting alveolar bone.

from many reports of Paget's disease and von Recklinghausen's disease in primates. Our own observations suggest that the apparent overgrowth of the premaxilla is in fact attributable to undue plasticity of bone, and is in all probability a rachitic manifestation.

A biopsy taken from one of the affected animals shows coarse bone, markedly basophilic, in which trabecular pattern, size and distribution of osteocytes, and the disposition of osteoclasts and osteoblasts all deviate from the normal (fig. 20). The microscopic appearance of bone in an animal maintained on an improved diet, fortified with a supplement of Vitamin D, shows (fig. 21) that extensive remodelling has taken place and that new trabeculae of normal appearance have come to enclose the previous poorly-formed bone by incremental apposition. Two years after the biopsy shown in figure 22 a further biopsy from the same animal, in which the premaxilla has since attained normal contour, shows that stigmata of the pristine osteodystrophy can still be detected at a microscopic level. Although the overall histological structure, like the clinical appearance, is normal, high magnification shows that original poorly-formed bone has not been entirely remodelled.

Within the soft tissues of the mouth, manifestations of deficiency states are entirely similar to those described in humans. *Glossitis* is to be expected; the tongue may be swollen, the lateral borders being indented by impressions of the cheek teeth, it is often abnormally smooth, and it has a magenta colour instead of the normal pink appearance so characteristic of health. *Bleeding* of the gingivae may be severe, and if protracted may be followed by intractable

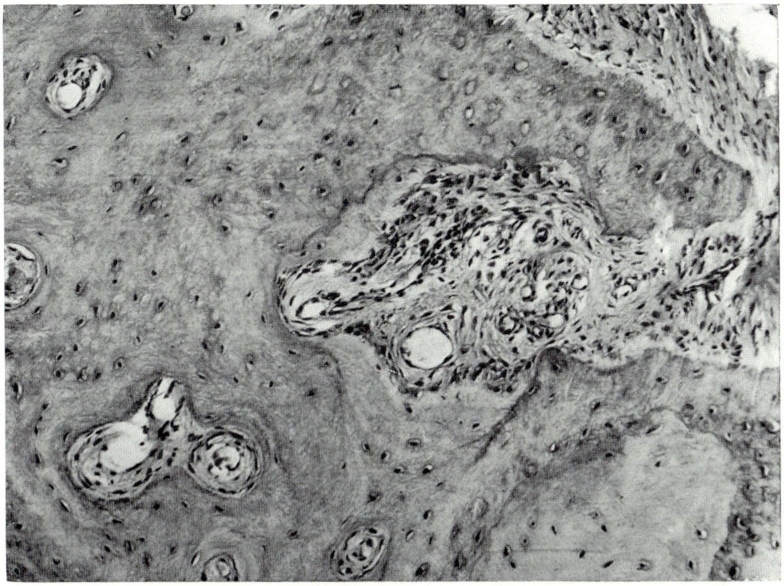

Fig. 20. Biopsy showing poorly mineralised coarse bone. Haematoxylin and eosin ×90.

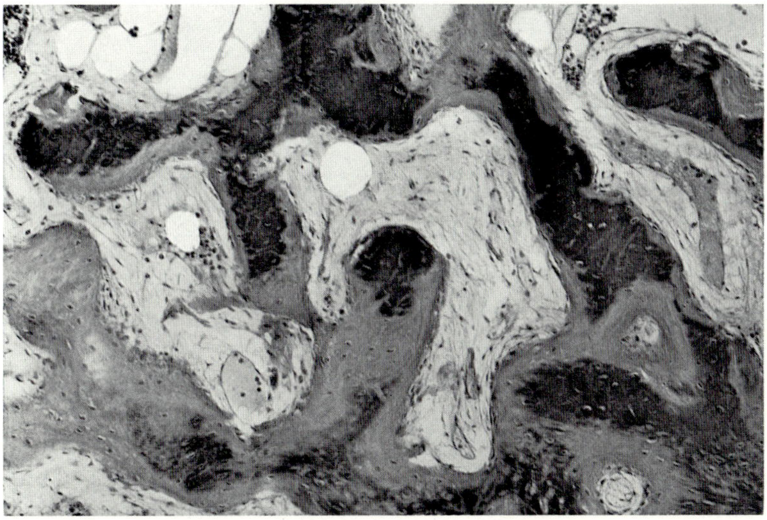

Fig. 21. Biopsy of bone from an animal that had been brought into captivity in a state similar to that shown in figure 20. Improved nutrition in captivity is reflected in the apposition of new, normal bone to the pre-existing coarse and poorly mineralised trabeculae.

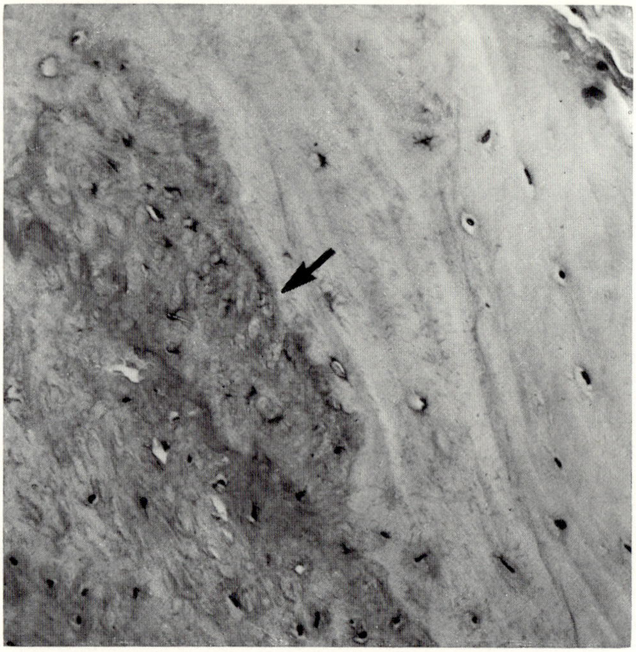

Fig. 22. Biopsy taken two years after that shown in figure 20. Arrow denotes previously formed irregular bone.

periodontal disease (see below) long after the nutritional deficiency has been corrected; the same conditions that predispose to gingival haemorrhage give rise to petechial haemorrhages elsewhere in the body, the eyelids often being conspicuously affected. *Cheilitis* has been observed in malnourished animals, but the fissuring soon disappears when nutrition is improved.

Infectious Diseases

1. Dental Caries. Although there are reports in the literature on the occurrence of dental caries in animals brought in from the wild this disease is rarely observed in recently captured monkeys. The present authors have studied some 400 such animals of the species *M. irus* and in no instance were carious lesions found. Early carious lesions which appear as faint opaque spots on the enamel surface can be detected clinically only with the aid of a good light and by means of a sharp metal probe. These lesions as a rule devel-

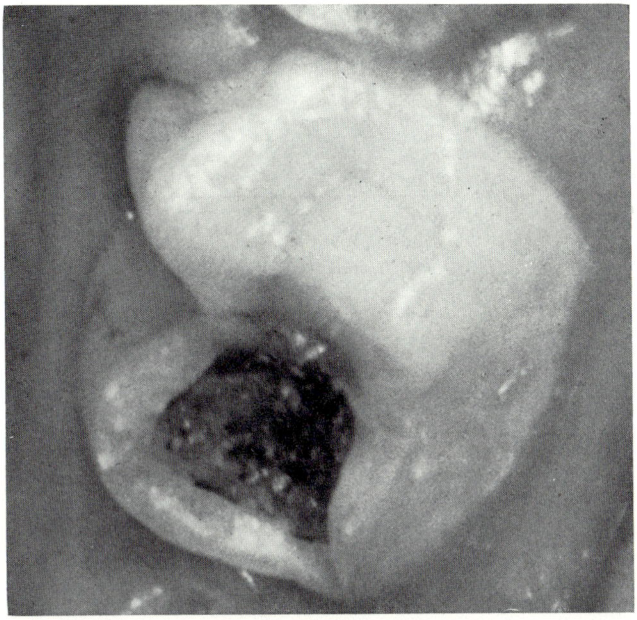

Fig. 23. Large carious cavity in the occlusal surface of a first molar.

op in pits and fissures of molar teeth, and in approximal areas of incisors and molars. On clinical examination the probe 'sticks' in the enamel lesion. As caries progresses the lesion spreads into dentine and the opaque white area breaks down to form an open cavity (fig. 23). All phases of the developing lesion can be observed radiographically (fig. 25) and in this respect, as well as histologically (fig. 26), caries in experimental monkeys is identical with that seen in humans.

COLYER [3] examined the skulls of a large number of primates covering a considerable range of species. The percentage of animals with caries varied from nil in *Mandrillus*, *Erythrocebus*, and *Allenopithecus* to 10% in *Cryopithecus*. Six of 400 macaques which COLYER examined were found to have dental caries. He also observed that although the overall prevalence was low, caries occurred more frequently in Old World monkeys than in New World animals.

Caries was found by SCHULTZ [4] in some of the skulls of 252 wild *M. mulatta*. The disease was not observed in the young animals, but 2.1% of the adult group and 13.7% of the aged group had dental caries.

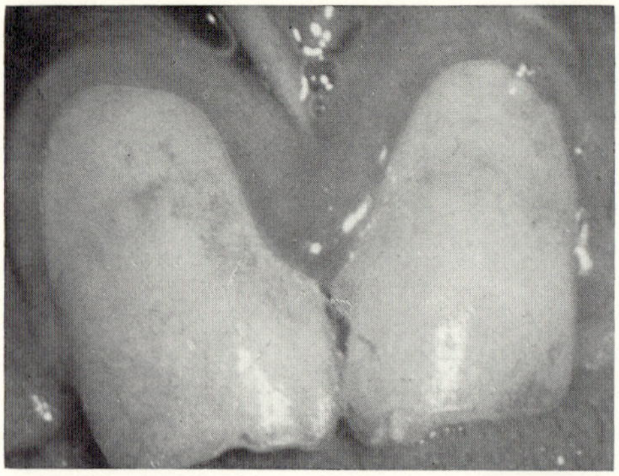

Fig. 24. Carious lesions in upper central incisors.

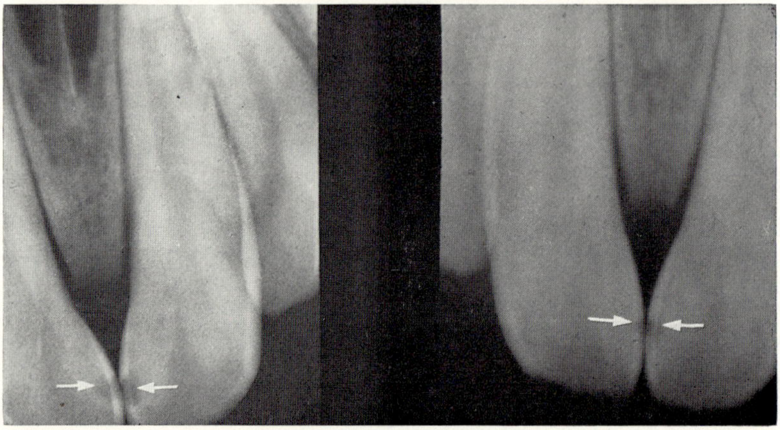

Fig. 25. Radiographs of the lesions shown in figure 24. Arrows denote the interstitial caries.

Out of 2,700 animals of the species *Cercopithecus aethiops* [5] only 20 were observed to have dental decay. Most of the 37 cavities detected were small.

One of the outstanding features in COLYER's study was that caries was commoner in incisors than in other teeth. This is also in marked contrast to the situation which prevails in animals being used for experimental purposes [6–8]. Of all the carious teeth found in wild monkeys, 66% were incisors, 12.9% were premolars and 20.9% were molars. The overwhelming majority (94.2%) of the cavities were found interproximally and the remainder were

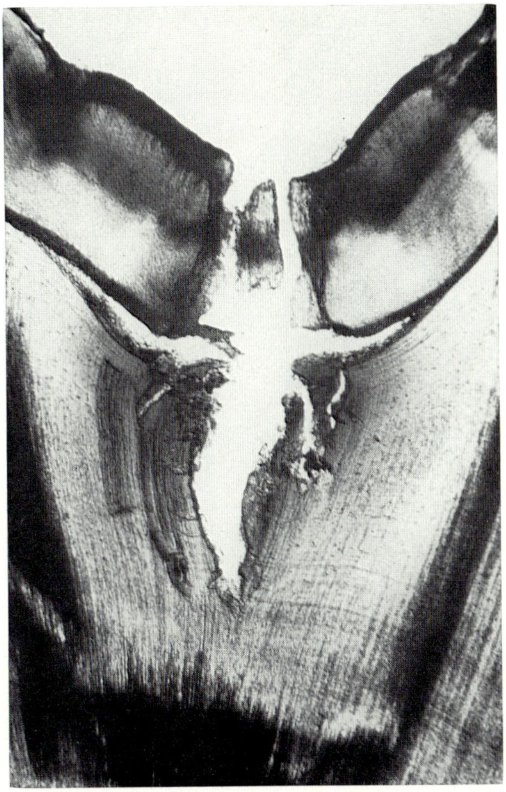

Fig. 26. Ground section showing a carious lesion which has penetrated enamel vertically and is spreading vertically and laterally into the dentine.

occlusal cavities. As explained previously in discussing the appearance of figure 15 this unique distribution of caries can no doubt be explained by the considerable amount of attrition found on the occlusal surfaces of the teeth of mature wild monkeys.

Caries is more prevalent in animals which have been in captivity for some time than in wild animals. COLYER [3] observed lesions in more than 70 of 915 Cercopithecinae he examined. Not all the species within the group were affected equally, prevalence ranging from 20% in *Erythrocebus* to 2.6% in the *Cercocebus*.

A substantial amount of caries activity was reported by SCHULTZ [4] in a group of 290 captive *M. mulatta*, lesions being found in almost 25% of the adult animals. On the basis of histological examination of teeth AUSKAPS and

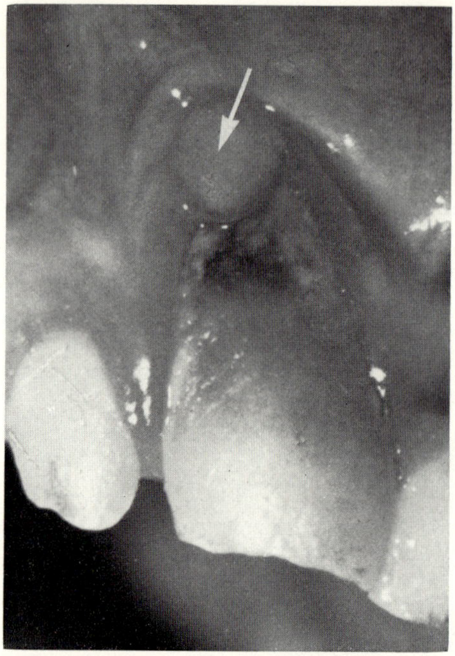

Fig. 27. Arrow denotes an alveolar abscess pointing on the gingival surface, a lesion referred to colloquially as a 'gumboil'.

SHAW [9] detected caries in 12 of 17 *M. philippinensis.* None of the lesions was detectable by radiological examination.

The distribution of caries in captive animals contrasts markedly with that occurring in wild animals. Molars and premolars are the teeth most often affected; almost 90% of all cavities found in captive monkeys by COLYER [10] occurred in molar teeth.

Considering the close similarity of monkeys to man it is remarkable how infrequently they have been used in the experimental investigation of dental caries. The methods used to induce the disease experimentally in these animals have been described by COHEN and BOWEN [7] and BOWEN [6, 8]. Animals which are infected by mouth with a dextran forming streptococcus, *Strept. mutans* [8] and fed a diet rich in readily fermentable carbohydrate, develop dental caries rapidly.

It has been found that the age at which the animals are exposed to cariogenic influences is of importance in determining susceptibility to the disease. Newly erupted teeth are more susceptible to attack than teeth which have

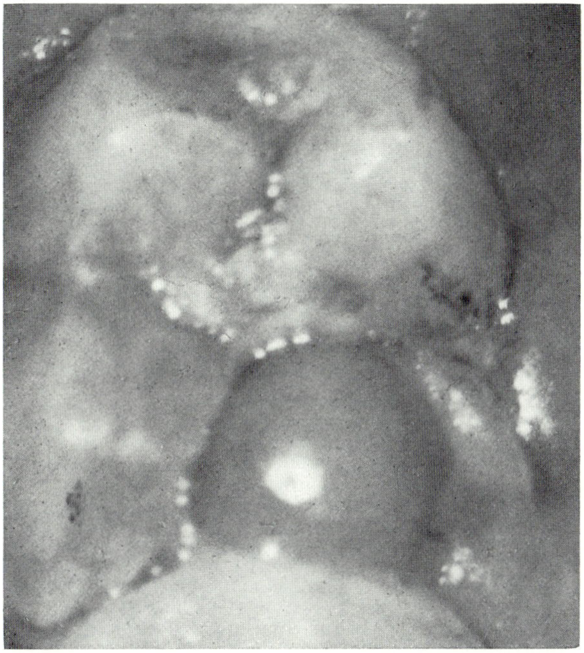

Fig. 28. Large carious cavity in an upper molar tooth, into which a gingival polyp has proliferated.

been present in the mouth for some time. Teeth which erupt into the cariogenic environment or have been present for less than 6 months before being exposed to a cariogenic environment develop clinical caries in about 8 months. In contrast, teeth which have been present for more than 1 year before being exposed rarely become carious, and when they do the lesions take about 14 months to develop.

The local complications of dental caries extend beyond the loss of tooth substance, or even the tooth itself. Exposure of the pulp chamber results in infection spreading down the root canal, and abscess formation within the alveolar bone may ensue (fig. 27). This in turn may lead to extensive suppuration with cellulitis of surrounding soft tissue as well as spreading osteomyelitis. There have been several reports of such bone lesions in the skulls of wild animals.

An occasional consequence of cavitation in a tooth is the unimpeded proliferation of contiguous soft tissue. Gingival polyps of this sort, an example of which is shown in figure 28, should not be mistaken for neoplasms.

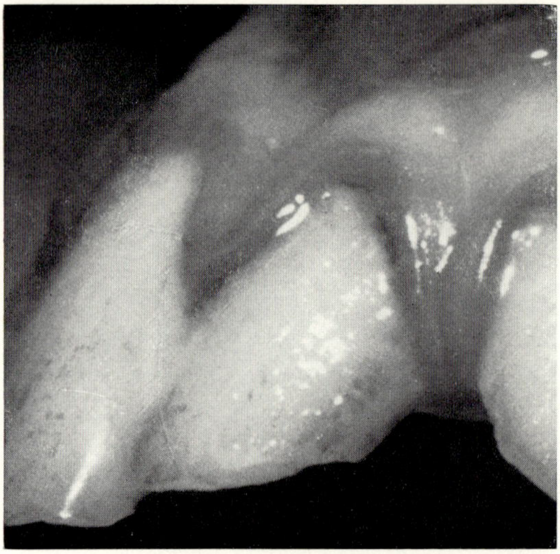

Fig. 29. Inflammation of the gingival margin.

2. Gingivitis. Inflammatory conditions of the gingival margin are present to a minor degree in most monkeys (fig. 16 and 29) just as they are rarely absent in the human mouth. Marginal gingivitis is usually associated with deposits of salivary calculus at the necks of the teeth, a condition illustrated in figure 30. Although the inflammatory state is frequently attributed to the irritant effect of these stagnant deposits of calcium salts, there is evidence to show that the causal relationship may be reciprocal for the teeth may become encrusted in this manner following prolonged gingivitis (fig. 31).

Not all accumulations at the gingival margin are calcified, nor are all of salivary origin. Food debris, desquamated cells, precipitated protein and aggregations of bacteria may all contribute to soft debris known as *materia alba*, and an adherent film known as dental plaque is normally present to greater or lesser extent in all animals.

Uncomplicated gingivitis undergoes resolution following the removal of the irritant, but when the condition becomes chronic it will usually progress until the periodontal ligament is affected, resulting in the condition known as periodontal disease.

3. Periodontal disease is a generic term applied to lesions of the tissues supporting the teeth as a result of which the teeth become so loose, and their surroundings so septic, that they are eventually exfoliated. Pyor-

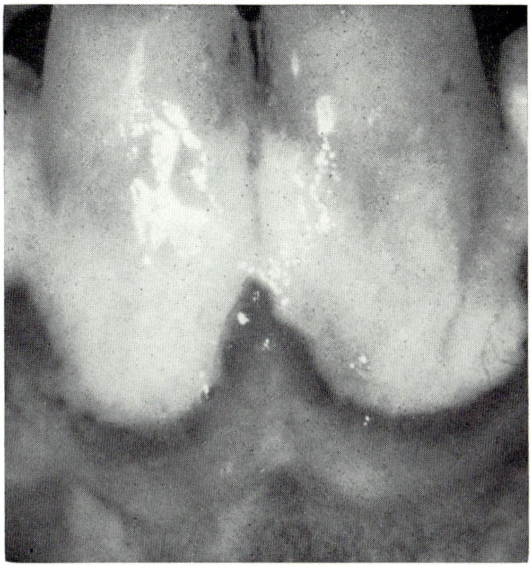

Fig. 30. Marked accumulation of calculus around the necks of incisor teeth.

rhoea, a term which has fallen into disuse, is descriptive of the terminal stage of this condition, when pus oozes from the gingival crevice surrounding the tooth.

It must be appreciated that the relationship of the gingival tissues to the teeth which they invest normally changes with age. In the period after tooth eruption is completed, the cervical border of the enamel remains covered by the marginal gingiva. At maturity the level of the gingiva has receded until the entire crown is exposed, and further recession takes place with age so that cementum as well as enamel is laid bare. This process (colloquially referred to as 'getting long in the tooth') is known as physiological recession, and is regarded as normal; it should not be mistaken for the gradual dissolution of the supporting structures which occurs in periodontal disease.

The chief feature distinguishing periodontal disease is the development of pockets around the teeth. These pockets arise from abnormal deepening of the sulcus between the teeth and their investing soft tissues. In order for such separation to occur the detachment of epithelium from the tooth, and of periodontal fibres from cementum, must take place. Many theories have been advanced to explain the initiation of this process, but since its progression is heavily influenced by bacterial agencies the condition is usually classified as an infectious disease. The development of an abnormal space between tooth

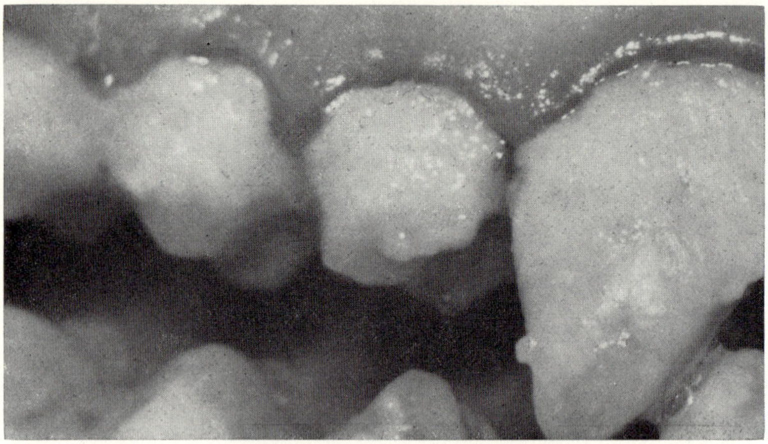

Fig. 31. Cheek teeth encrusted with calculus in an animal with advanced periodontal disease.

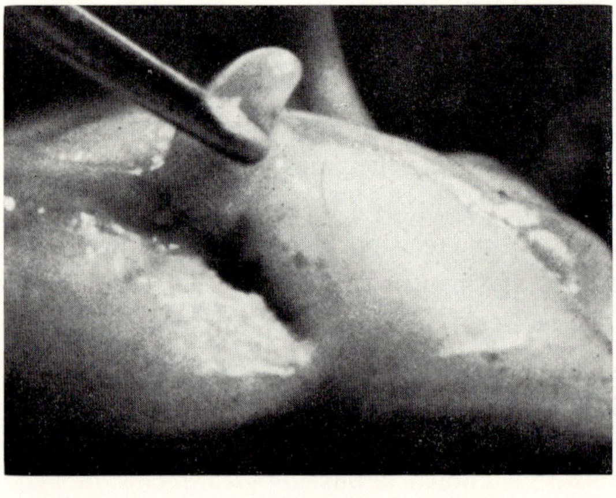

Fig. 32. Early periodontal pocket.

and gingiva inevitably leads to stagnation of debris and favours bacterial proliferation. The dissolution of periodontal fibres is a lengthy process and bone comes to be infected and contaminated by fluids from the mouth only after the condition of chronic ulceration has been in existence for a prolonged period – for years rather than months.

Periodontal disease usually begins, and is usually most marked, in the interdental regions. With the gradual dissolution of the fibres comprising the periodontal ligament, contamination spreads to other parts of the tooth peri-

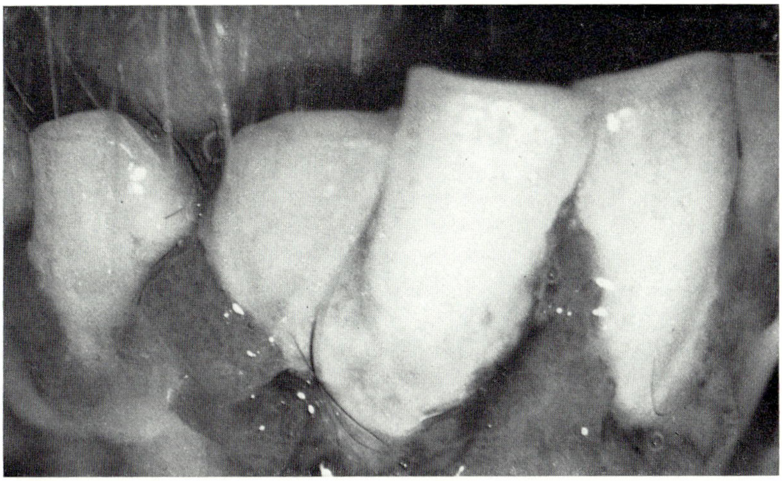

Fig. 33. Advanced periodontal disease marked by immobility of teeth and deep interdental pockets.

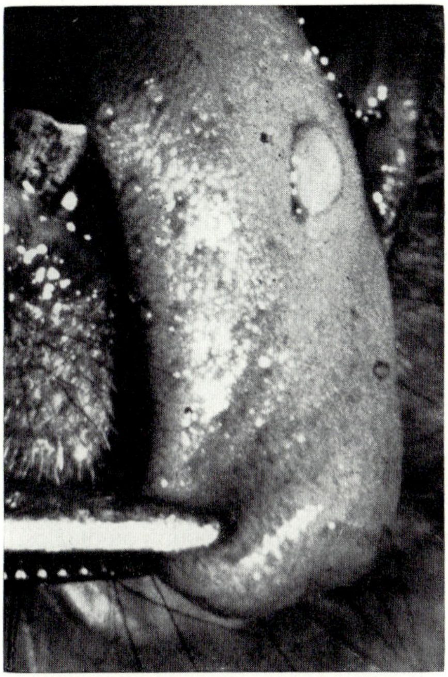

Fig. 34. Shallow, sharply outlined ulcer on dorsum of tongue.

Fig. 35. Pigmented warty outgrowth on margin of lip.

meter, and with the spread in depth bone undergoes resorption, and mobility of the affected teeth becomes more marked. The condition may be localised to separate areas of the same mouth.

An early stage in periodontal disease may be detected when the papilla which fills the embrasure between adjacent teeth can be deflected by light pressure with a fine instrument. This state is depicted in the illustration of an early pocket between the central and lateral incisors of a monkey in figure 32. Later stages (fig. 33) are more easily recognised as the gingival outline exhibits marked swelling and irregularity and as the teeth become mobile. Bleeding from the gingivae is an important sign at all stages of the disease and is sometimes so marked that the saliva is bloodstained.

4. Herpes B Infection (Herpesvirus simiae). Animals coming from the wild may be infected with monkey B virus *(Herpesvirus simiae)*. This infection is comparatively harmless to monkeys and may manifest itself as a series of vesicular eruptions around the muco-cutaneous junction on the lips. These vesicles frequently rupture and a reddish-brown scab may form. Shallow ulcerations, usually covered by a yellowish plaque and sharply delineated from the surrounding tissues are frequently observed on the tongue (fig. 33).

A high proportion of animals coming from the wild have virus B antibodies without displaying overt signs of the disease. There is some cross reac-

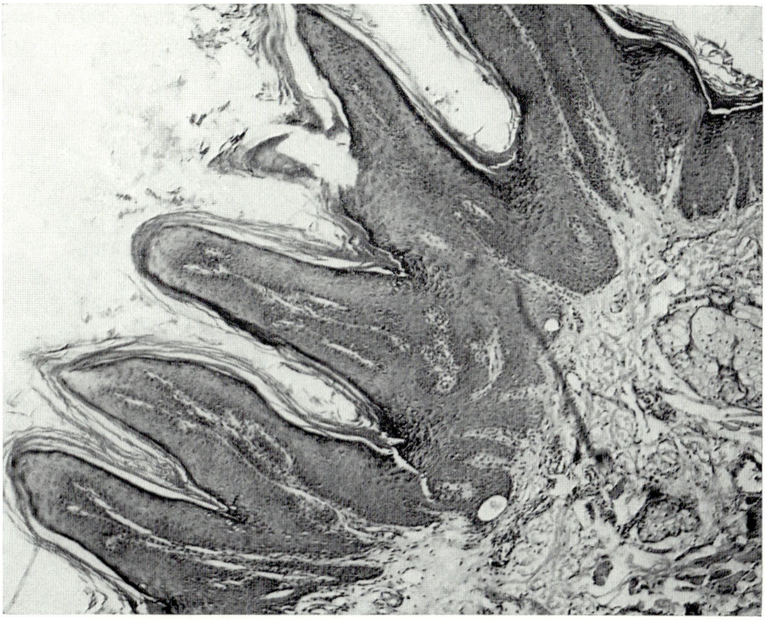

Fig. 36. Microscopic appearance of lesion depicted in figure 35.

tion between antibodies against *Herpes simplex* and B virus, but there is no convincing evidence that a high titre of *H simplex* antibodies in humans gives protection against infection by B virus. Because of the danger of this virus to humans it is wise to assume that all animals are carriers and to take precautions to ensure that humans do not become infected. Many laboratories destroy all recently captured animals exhibiting ulceration in and around the mouth.

Tumours

Neoplastic disease in the mouth, either spontaneous or induced, has rarely been reported in non-human primates. In the studies from which the content of this chapter is largely derived, the only new growth observed was a pigmented papilloma situated at the muco-cutaneous junction of the upper lip (fig. 35). The fact that similar warty outgrowths have been observed on skin sites in several other animals in the colony suggests that the lesion may be infective in origin and also that it arises from skin rather than mucous membrane; even so, this is of interest since skin tumours are also exceedingly uncommon in monkeys. The microscopic appearance of this lesion is typically papillomatous and notable for the abundance of melanocytes (fig. 36).

References

1 Bowen, W. H. and Koch, G.: Laboratory animals *4:* 113–123 (1970).

2 Ruch, T. C.: Diseases of laboratory primates, pp. 472–479 (1959).

3 Colyer, F.: Four lectures on abnormal conditions of the teeth of animals and their relation to similar conditions in man. Dent. Board U.K. XI (1931).

4 Schultz, A. H.: Eruption and decay of permanent teeth in primates. Amer. J. phys. Anthropol. *19:* 489–581 (1935).

5 Ockerse, T. and Jager, C. L. de: Dental caries produced in the vervet monkey. Brit. dent. J. *102:* 93–96 (1957).

6 Bowen, W. H.: A bacteriological study of experimental dental caries in monkeys. Int. dent. J. *15:* 12–53 (1965).

7 Cohen, B. and Bowen, W. H.: Dental caries in experimental monkeys. Brit. dent. J. *121:* 269–276 (1966).

8 Bowen, W. H.: The induction of rampant dental caries in monkeys *(M. irus)*. Caries Res. *3:* 227–237 (1969).

9 Auskaps, A. M. and Shaw, J. H.: Studies on the dentition of the cynomolgus monkey. J. dent. Res. *36:* 432–436 (1957).

10 Colyer, F.: Variations and diseases of the teeth of animals (Bale and Danielson, London 1936).

Authors' address: Dr. B. Cohen and Dr. W. H. Bowen, Department of Dental Science, Royal College of Surgeons of England, Lincoln's Inn Fields, *London, W.C.2* (England)

Pathology of Simian Primates, Part I, pp. 671–710 (Karger, Basel 1972)

The Respiratory and Alimentary Systems

R. N. T-W-Fiennes, B. A. Lapin, E. K. Dzhikidze
and L. A. Yakovleva

Department of Pathology, Nuffield Institute of Comparative Medicine, Zoological Society of London, London

Contents

I. General Survey ... 671
No Pathogen Involved ... 672
Opportunist Pathogens ... 675
Primary Pathogens ... 677
II. Gastro-Intestinal Diseases of Simian Primates ... 681
1. Bacillary Dysentery ... 681
2. Salmonellosis ... 685
3. Enterocolitis Caused by the Organisms of the Proteus Group ... 688
4. Amoebiasis ... 689
5. Rare Conditions of the Alimentary Tract ... 689
6. Bezoars ... 690
7. Appendicitis ... 691
8. Hernia ... 693
9. Gallstone Disease ... 693
III. Diseases of the Respiratory Tract ... 694
1. Pneumonia ... 694
2. Lung Mites ... 701
3. Pulmonary Nocardiosis ... 704
References ... 704

I. General Survey
(R. N. T-W-Fiennes)

In the assessment of alimentary and respiratory diseases, certain clear-cut distinctions require to be made. The first is between recently captured and imported animals, which have not yet passed through a period of quarantine and acclimatisation, and animals which are part of an established colony. The second important distinction is between conditions solely attributable to some pathogen, and those which are either not due to a living pathogen or in which the pathogen's activities are secondary to some other underlying causes.

We can thus distinguish three types of syndrome: (1) in which no pathogen is involved; (2) in which pathogens are involved secondarily; (3) in which pathogens are the primary cause of the disease. Opportunist organisms are usually those which affect the alimentary or respiratory systems, though dermatophytes may also behave in this way.

No Pathogen Involved

Conditions in which no pathogens are involved may be congenital, or due to trauma, nutritional defects, or stress.

Congenital defects. Congenital defects may be either genetic or acquired *in utero* prior to parturition. Congenital defects generally are described in another chapter by ADOLPH SCHULTZ. Two defects only are worthy of mention in relation to the present survey.

A small number of cases of varices of the oesophagus with atresia of the gastric end have been found in South American Primates. This may lead to difficulties of deglutition, especially when some infective condition of the oesophagus or stomach is also involved.

The second condition to be noticed is that of diaphragmatic hernia which has been seen by the author on more than one occasion in squirrel monkeys *(Saimiri sciureus)*. This condition has appeared to be due to a congenital defect of the diaphragm rather than to trauma. This opinion is based on the site of the cleft in the diaphragm, at the oesophageal hiatus between the two diaphragmatic leaves.

Traumatic conditions. Traumatic conditions may be acquired at the time of capture, during transit, or as a result of handling or fighting with cagemates after arrival. With many primate species, a high stress component may accompany trauma, especially in animals already suffering from the stress of capture and transit. Thus death may follow comparatively minor injuries.

The commonest injuries encountered are penetrating wounds of the thorax. In South American species, buckshot is commonly found in the pleura, together with fractures of one or more ribs. Diagnosis, whether in the living or the dead animal, is easily made by X-ray. In a gibbon, examined *post-mortem* by the author, a penetrating suppurative wound of the left chest was found, which contained the broken end of a small arrow or dart made of bamboo. Such wounds may lead to pleuritis and the death of the animal.

A further common form of trauma in primates follows careless dosage with nutritional supplements or medicaments. This causes 'inhalation pneumonia', which occurs all too frequently. This accident is easily diagnosed *post-mortem* by the presence of the offending substance in the bronchioles, and in histological preparations by the peribronchial inflammatory reaction around fluid or debris.

Nutritional causes. Nutritional causes of alimentary troubles may be due to failure to feed, to depressed appetite, to the presence of foreign bodies which are swallowed, to food that is of poor or unsuitable quality or mouldy, or to actual deficiencies.

Failure to feed may be due to a number of quite simple causes, which may be easily overcome and avoided. Many newly-acquired monkeys, particularly of the South American species, have been acquired by the slaughter of the mother when the young animal is either unweaned or only partially weaned. In such cases, the animal may be physically unable to eat the food provided, and would need to be hand fed. Without the individual and personal attention that could only be given in a home, there is little hope for such animals and they usually die of inanition.

Another common form of food deprivation arises when the other cage inmates prevent the subordinate animal from getting access to the food. In such a case, the deprived animal also may grab what he can and feed too quickly with consequent digestive disturbance and colic. Monkeys, newly-arrived, frequently suffer from sub-clinical infections of one type or another and at this time fail to feed, and find difficulty in resuming the habit. It is, therefore, good practice with newly-arrived animals to add especially palatable foods, such as grapes, to the standard diet.

Foreign bodies are not commonly swallowed by monkeys, because of their meticulous habits of picking over their food and examining it before it is eaten. Depraved appetites, too, are rare, though less commonly encountered than failure to feed. The only trouble of this nature encountered at the London Zoo has been with nocturnal prosimians, such as galagos and pottos. Possibly, because of boredom at night when nobody is about, they tend to lick their fur and cases have not been infrequent when the stomach has been completely filled with a cast of impacted fur. Bezoars are discussed below by Lapin *et al.*

In most Primate colonies today, pelleted diets are given, and the standard high-protein monkey chows are excellent. It must, however, be remembered that different groups of monkeys have different requirements. Leaf-eating species, such as colobus and presbytis, have a high requirement of vegetable fibre; to a lesser extent this is true of those primates, of which the natural diet is largely vegetarian, such as chimpanzee and gorilla. To these animals, supplements of high fibre vegetable foods, such as cabbage, need to be given if the intestinal tract is to be maintained in good working order. Some South American monkeys have a high protein requirement, which is often supplied by meal-worms fed on oatmeal; such meal-worms are virtually bags of oatmeal and cannot provide the protein required. It appears that most New World primates must be provided with very large supplements of vitamin D3, whereas Old World primates require vitamin D2; in an animal that acquires rickets, the rib cage is narrowed and cannot properly be expanded; animals so affected usually die eventually of pneumonia.

A frequent cause of colic is stale or mouldy food. Such food is easily given, where management of food distribution is inadequate. Pellets at the bottom of the sack easily become mouldy if the sack becomes wet because left outside or when the food store is hosed. It is a common practice, where sacks are emptied into food bins, to refill the bin before the pellets at the bottom have been used. When they are eventually used, they may be very stale indeed, and probably also mouldy. Bins should never be refilled until the old pellets have been removed, and the bin has been properly cleaned and sterilized.

Any oil-containing foods are an especial danger, if not fresh. Foods that are rancid, in which the oil contains more than 8% at most of free fatty acid, such as groundnuts, can readily cause fatal duodenitis. It must be remembered that fats continue to deteriorate slowly even at refrigeration temperatures; deterioration in hot places is very rapid indeed.

The effects of actual dietary deficiencies on the alimentary and respiratory systems are complex. In vitamin A deficiency, the integument and the lining cells of the upper respiratory tract suffer metaplastic changes. Thus the animal is more exposed to respiratory infections. Vitamins of the B-complex are important to many enzyme reactions, and when absent the skin, the haemopoietic tissues, the central nervous system, the peripheral nerves, the liver and the bowel may all be affected. B deficiencies often occur secondarily to the use of antibiotics, which may destroy a portion of the bowel micro-flora responsible for its synthesis. As was shown by the writer [34a] some groups of monkeys (Callicebus spp.) may have an unusually high requirement of these vitamins and in their absence respiratory and alimentary troubles ensue. The effect of B deficiencies on the bowel are probably to injure absorption as a result of a sprue-like condition. It is suspected that such conditions may interfere with calcium absorption and thus exacerbate deficiencies or imbalances of calcium or deficiency of vitamin D.

It is evident from what has been said that attention must be paid to supplementing the excellent pelleted foods available on the market with any special additions required by different primate groups. Failure to do so will lead to avoidable alimentary or respiratory troubles either of a mechanical or an infective nature.

Opportunist Pathogens

Newly-acquired primates of all groups harbour or may acquire many pathogens, which are harmless under good conditions. Such pathogens are: (1) those they carry in the wild; (2) those they acquire in the land of their origin from their captors; (3) those they acquire from contacts with humans or simians in their ultimate place of captivity. Such pathogens may be viral, bacterial, fungal, protozoal or helminthic.

Of the viruses, which animals normally carry, those of the herpes and cytomegalovirus groups are undoubtedly the most important. They may either set up disease of their own account, usually septicaemic with pneumonia as the most important feature, or they may weaken the animal's resistance so that opportunist organisms become active. Such organisms usually cause pneumonia and are pneumococci, klebsiella, bordetella, streptococci and staphylococci (fig. 1a and 1b).

Of the viruses, which animals may acquire, the most important are probably infectious hepatitis, measles and pox.

Of the bacteria, which may become active, the most important are probably pneumococci, meningococci, klebsiella, bordetella, streptococci, staphylococci, salmonella, shigella, and other alimentary pathogens consisting of serotypes of *Escherichia coli*, proteus, and paracolon groups. One may also mention *Pasteurella pseudotuberculosis*, which is not uncommon as a bowel pathogen in simians and may also come to infect the lungs and other organs. *Pseudomonas aeruginosa* may also be isolated from intestinal contents. It does not appear to be pathogenic in the bowel, but may enter the body proper to cause nephritis and septicaemia.

Fungi may become active as a result of stressed conditions or because of removal of competition by other pathogens, following the use of antibiotics. The organism mostly concerned is probably candida, which may normally be an inactive constituent of the flora of the oral mucosa. Occasional cases

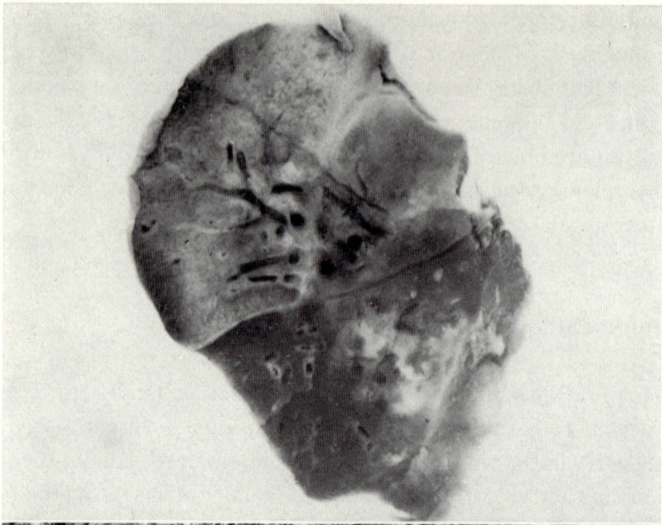

a

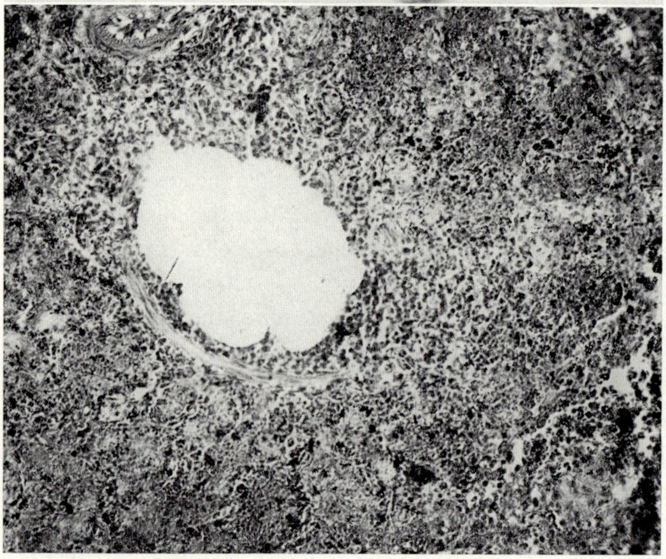

b

Fig. 1. a Membranous pneumonia in *Macaca mulatta. b* Broncho-pneumonia lesion of bronchi in *Papio hamadryas.*

are also encountered of mucormycosis of the large bowel; one such case was described by FIENNES [34].

Amongst the protozoal infections of the bowel, which may get out of hand in the stressed or unhealthy animal, are *Entamoea bhistolytica, Tricho-*

monas spp., *Coccidia*, *Giardia lamblia* and *Balantidium coli* (fig. 2a, b, c). Amongst the helminthic, the most serious are *Oesophagostomum* in the great apes, and *Strongyloides* in monkeys. The oesophagostomes live in nodules beneath the bowel mucosa, and are thus protected from the action of anthelmintics. Under conditions in captivity, they readily build up to serious numbers and can cause the death of gorillas and orang-utans. *Strongyloides* can undergo either an external or an internal reproduction cycle, so that its numbers can build up to serious proportions. The adult worm burrows into the wall of the intestine and can cause perforation with consequent peritonitis (fig. 3a, b, c). Serious parasites in some South American species are the thorny-headed worms (Acanthocephala), which often cause bowel obstruction and death.

Primary Pathogens

The most important of the primary pathogens, which cause disease of the respiratory and alimentary systems in imported primates, is undoubtedly tuberculosis. Human-type tuberculosis in Old World species causes a rapidly fatal form of phthisis recalling that of human children. The bovine type develops more slowly and may be regressive. Often lesions are found in the bowel, and the disease may develop from a calcified mediastinal gland years after the primary infection. New World monkeys are far less susceptible to human-type tuberculosis. In them it may take either pulmonary or alimentary forms and the course may be prolonged.

Primary pneumonias due to pneumococci are common, particularly in Old World primates and especially in chimpanzees. These respond rapidly to the correct antibiotic, so long as some other primary cause is not responsible. All groups of primates are susceptible to pseudomyxovirus, but not to the myxoviruses themselves, nor to rhinoviruses. Either shigellae or salmonellae may act as primary bowel pathogens but, as stated, these can also be dormant pathogens and only cause disease secondarily to some other disturbing influence.

Stress. The effects of stress are primarily on the nervous, endocrine and circulatory systems. In stressed animals there are, however, prominent lesions in the alimentary and respiratory systems. Shock and stress are hypotensive conditions with impaired blood return to the right side of the heart. Animals dying in the acute stage (shock) show abdominal hyperaemia, together with anaemia and collapse of the lungs, sometimes with localised haemorrhages. The liver is hyperaemic and shows superficially a nutmeg appearance. The spleen may be enlarged, but more usually is normal in size and appearance. There are haemorrhages and focal necroses in the cortices of the adrenal glands. Very frequently, there is haemorrhage into the lumen of the small intestine. These changes are secondary to the failure of forward move-

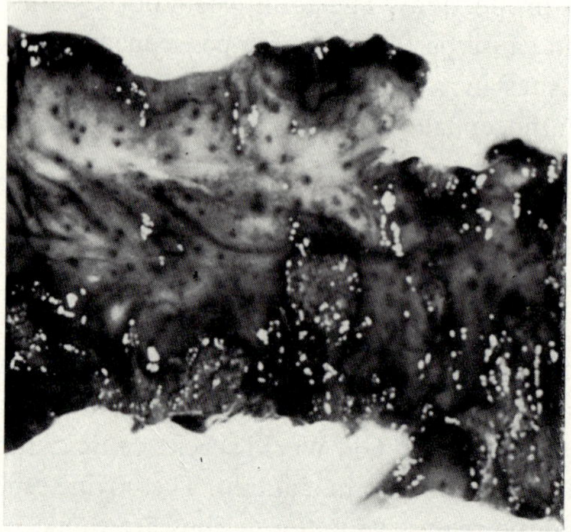

2a

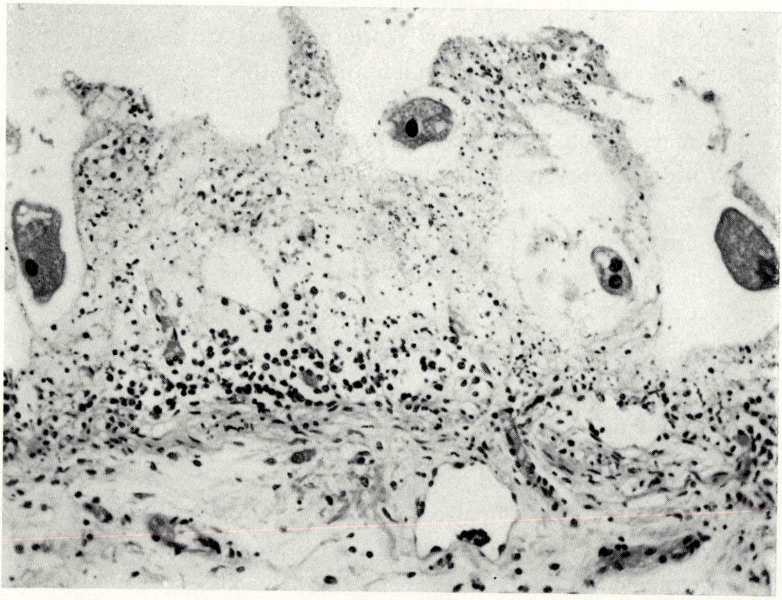

2b

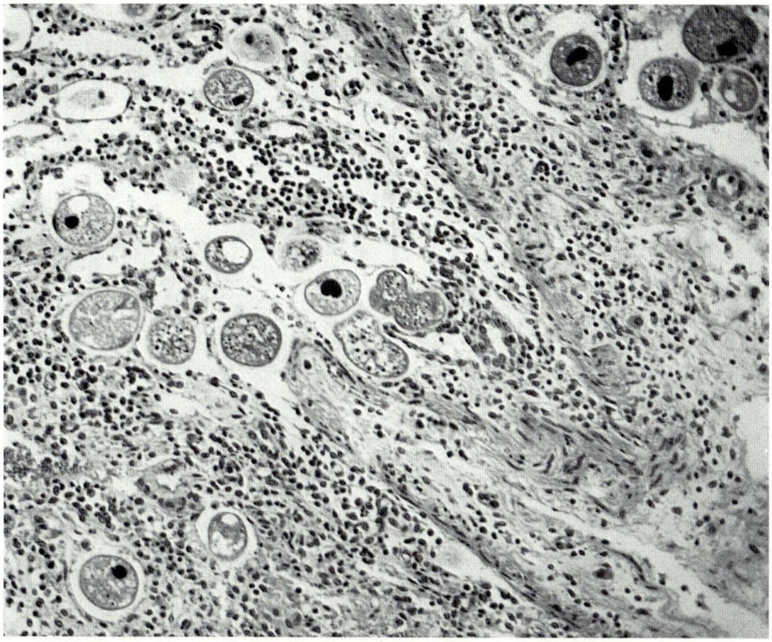

2c

Fig. 2. a Chimpanzee. Colon, infected with Balantidium. *b* Callicebus. Balantidiasis, with ulcerative caecitis. *c* Chimpanzee. Balantidiasis. Necrosis of colon.

ment of blood by way of the posterior vena cava to the right heart [FIENNES and DU BOULAY, in press].

In the case of an animal that is more chronically hypotensive (i.e. stressed) the pathological picture is similar but less marked. In addition, ulceration of the gastric pylorus or duodenum or both may be found. In the early stages, the pylorus is found to be covered with a thick coat of viscid adherent mucus, beneath which the pyloric mucosa shows early signs of erosion.

Diagnosis. It is often difficult to find the true cause of deaths from alimentary and respiratory causes in non-human primates. Where a distinct primary pathogen is involved, cultures of sputum, faeces or rectal swabs, should serve to determine the cause of the trouble and appropriate steps can be taken. When the pathogens are secondary to some other primary cause, clinician and pathologist should consult together to eliminate possible causes of the trouble one by one, bearing in mind the various factors that may be involved.

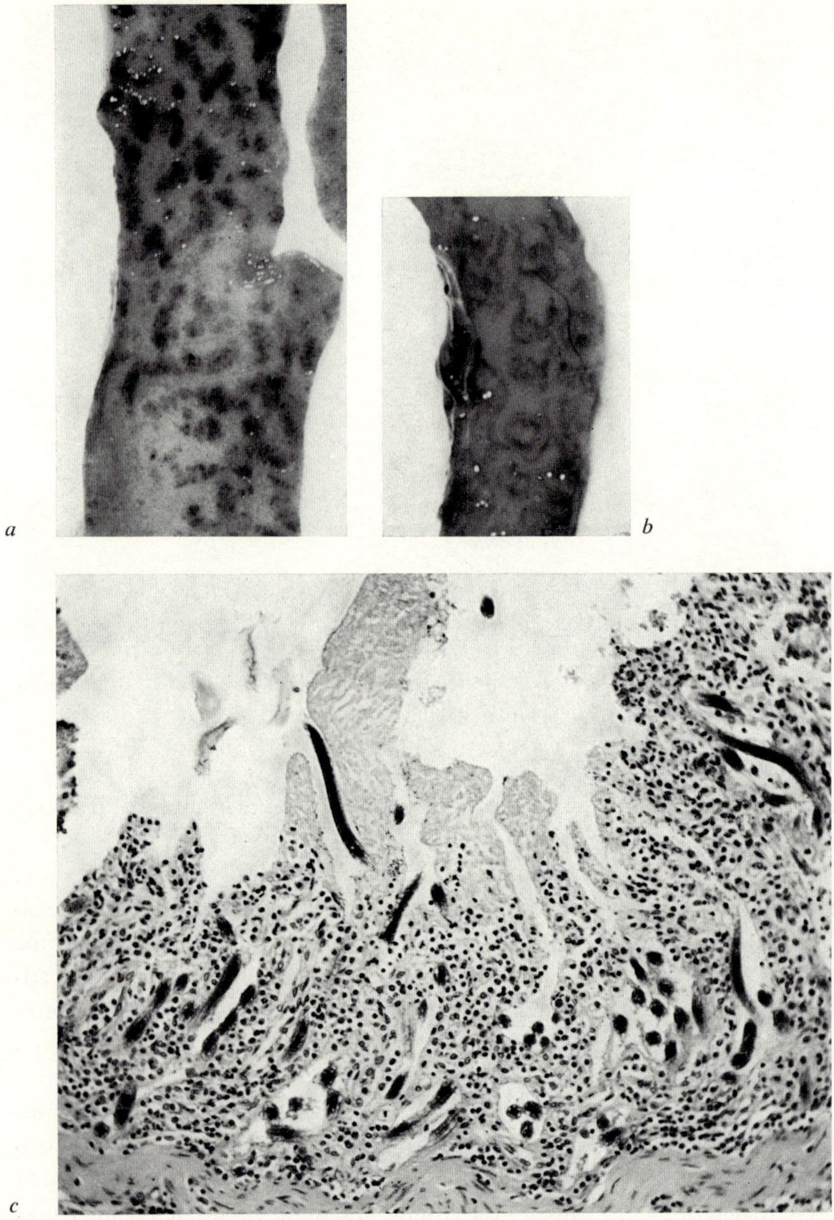

Fig. 3. *a* Chimpanzee, Strongyloidiasis. Colon. *b* Chimpanzee, Strongyloidiasis. Colon. *c* Chimpanzee, Strongyloidiasis. Colon.

It is of especial importance to remember that many primate diseases may be dangerous to man and every possible precaution must be taken in handling both living and dead material.

II. Gastro-Intestinal Diseases of Simian Primates

(B. A. Lapin, E. K. Dzhikidze and L. A. Yakovleva)

Gastro-intestinal disease in simians may be due to viral, bacterial, protozoal or helminthic causes. The viral nature of these diseases still lacks definite proof, though viruses have been isolated from the intestinal contents of monkeys and apes [102].

Unusually frequent passage of faeces may also be the result of non-specific infections of the gastro-intestinal tract [87]. Diarrhoea of non-infectious origin may be caused by change of diet or emotional stress.

1. Bacillary Dysentery

Bacillary dysentery is a common disease of primates living in captivity [1, 8, 10, 11, 14, 19, 28, 32, 44, 92, 93, 103, 107]. It occurs frequently in galagos, vervets, patas, macaques and baboons [28, 34, 38, 45, 73, 76, 83, 94, 96, 108]. The apes and rhesus macaques *(Macaca mulatta)* are the most susceptible [21, 24, 35, 76, 105, 107]. *Papio hamadryas*, on the other hand, is relatively resistant [29, 76, 107].

i) Etiology. The agents which cause bacillary dysentery in simians are of the same subgroups and serotypes of shigellae, as cause dysentery in humans, these being *Shigella flexneri*, *S. sonnei*, *S. schmitzii* and *S. dysenteriae* [1, 3, 22, 25, 54, 66, 67, 90, 94, 95, 106, 110]. *S. dysenteriae* has not so far been isolated from monkeys, but a single record exists of dysentery due to it in chimpanzees [34].

The sources of infection for simians are human patients and carriers, infected food, sick animals and asymptomatic carriers. Bacillary dysentery does not occur in wild-living simians [18, 68, 77, 106]. Even so, cases may be observed in simians very soon after capture, when the animals are housed in captivity; in such cases various serotypes of shigella may be isolated from the faeces [77]. While it is evident that monkeys become infected very quickly, the incidence of the disease amongst monkeys born or living in captivity for any length of time is low, in spite of the wide spread of carrier animals. According to evidence at the Sukhumi colony, the incidence of the disease varies from 0.38 to 5.7%. In newly-arrived monkeys, however, the incidence during the early months of climatic adaptation may be very high and mortality 30–80% [41, 76]. In animals adapted to climate and conditions of captivity, dysentery occurs most frequently in early spring, autumn and winter. In newly-imported monkeys, outbreaks occur at any season of the year, but soon after arrival in the colony.

ii) Symptoms. The clinical manifestations, as in man, are variable. Studies at Sukhumi, based on more than 750 cases [39, 40, 41], show that the course may be divided into chronic and acute forms. The chronic disease may be mild, moderate or severe.

In the mild form, general toxaemic effects are absent. Faeces are passed infrequently, but contain blood and mucus. General health is not obviously impaired, and the duration is only three to five days.

With the moderate form there is a mild degree of toxaemia, anaemia and frequent passage of faeces containing blood and mucus, and sometimes pus. Unless treatment is given, the disease is progressive and ends in death.

The severe form of dysentery is accompanied by pronounced toxaemia. The general state of health deteriorates rapidly, and anorexia and prostration develop. At first, faeces are passed frequently and contain blood and mucus; later, they become purulent and resemble meat slops. Body temperature is subnormal and signs of desiccation develop. Untreated animals die in two to three days.

A very mild form also occurs, in which the faeces are usually rather softer than usual. In such cases, rectoscopy [40] and bacteriological examination are most important for diagnosis. Sometimes too, the acute form passes into the chronic, manifested as a relapsing or chronic debilitating disease. The tendency to a chronic course is seen most frequently in newly-imported animals or those exhausted from experiments [41].

In infants, dysentery usually resembles ordinary or toxic dyspepsia, in which severe manifestations of general intoxication develop prior to the faeces becoming liquid.

Typical cases of sanguinolent-mucous colitis occur most frequently in apes and macaques. They are observed less frequently in *Cercopithecus aethiops* (vervet) and *Erythrocebus patas*, only rarely in baboons. In baboons, the course of dysentery is usually atypical with the prevalence of ordinary dyspepsia or gastro-entero-colitis [29, 39].

Apart from clinically affected animals, many monkeys become so-called 'healthy' carriers. According to the Sukhumi data, numbers in the colony harbouring shigellae vary widely from 16.8 to 56.8%, even though the incidence of disease is low. Carriers occur amongst all species, macaques, baboons, vervets, patas and apes. The carrier state takes two forms: (1) a transitional form, in which bacilli are excreted for short periods only; (2) the chronic form, in which bacilli are excreted indefinitely. Amongst short term carriers, there are no pathological lesions of the gastro-intestinal tract. Amongst the chronic carriers, dysentery bacilli are excreted in great numbers and quite constantly.

Careful, clinical observation, together with rect-oro-manoscopy and pathological studies, have shown that chronic carriers are in fact dysenteric, though in a state of relative compensation [26]. In such animals, the mucous membrane of the intestine is inflamed and ulcerated. Transition from the asymptomatic carrier state to chronic dysentery is accompanied by an increase of excretion of bacilli in the faeces, 10^7 or more [106].

iii) Pathology. Pathomorphological alterations in simian dysentery show a great similarity to those in man [44, 89, 116]. Lesions are predominantly localised in the large intestine. The intestine is generally distended especially near the caecum, and the distensions alternate with areas of contraction. In its mildest form, there is a desquamative mucous catarrh. The mucosa of the large intestine is unevenly swollen and pink with occasional haemorrhages; the folds are thickened. Histologically, oedema is seen together with a marked thickening of the mucosa and evidence of desquamation and necrosis of the epithelium. The crypts are generally enlarged and overflowing with mucus. The connective tissue stroma of the mucous membrane is infiltrated by cellular elements of unusually large size. The blood vessels of the mucosa and submucosa are dilated. Occasional haemorrhages are seen (fig. 4a).

In more severe cases, the process acquires the character of diphtheritic colitis (fig).5. The mucous membrane of the large intestine shows scattered islets of dirty grey film adhering to it. Histologically, there are deep necrotic patches, which reach down at times to the submucosa. There is also oedema of the submucosa, and infiltration of the tunica musculorum mucosae by leucocytes and histiocytes (fig. 6 and 7).

Sometimes, but more rarely, the necrotic process in the mucosa is predominant, and the cellular infiltration less marked. Such a process occurs in infants and has the character of 'follicular colitis'.

In cases of dysentery, ulcerative colitis has also been described (fig. 4b) in the literature. Ulcers are located some distance from the lymph nodules, showing erosive areas of varying size and depth. Ulcers sometimes penetrate to the submucosa. Morphologically, chronic, dysenteric colitis in monkeys is characterized by polypous and ulcerative lesions in the intestinal mucosa [89]. This form of dysentery is very rare, and usually observed in animals over five years of age.

Such are the lesions found in the large intestine. In the small intestine, usually in the lower portion and the ileo-caecal region, catarrhal changes are manifest. The lymph nodules show moderate hyperaemia, haemorrhage and oedema, and leucocyte infiltration.

In the spleen, there is moderate oedema, hyperaemia and sometimes haemorrhages.

The liver, kidneys and myocardium show changes of an infectious and toxic character.

In general, catarrhal colitis is most commonly found amongst both climatically adapted and newly-arrived animals. Diphtheritic forms occur in

4a

4b 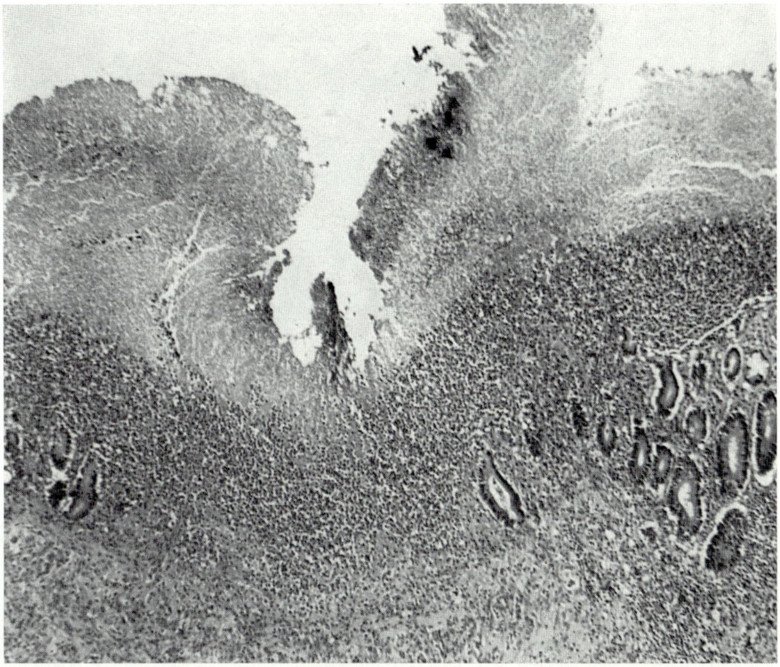

Fig. 4. a Bacillary Dysentery in *Macaca mulatta. b* Bacillary Dysentery in *Macaca mulatta.*

only 8% of both groups [76, 78]. Thus the morphological and clinical manifestations of bacillary dysentery closely resemble those in man. The course is generally more severe in monkeys and the death rate, 30 to 60%, is high. In man, lesions are mostly found in the lower division of the large intestine – the sigmoid colon and the rectum. In simians, lesions are generally localised in the area of the caecum and ascending colon [76].

2. Salmonellosis

i) Etiology. Bacteria belonging to various types of salmonella are widespread in nature, particularly among wild and laboratory rodents. It is possible that many outbreaks in simians originate with the rodent fauna. The serotypes involved are variable, but belong chiefly to groups B, C, D and E. In groups B and D, *S. typhimurium*, *S. stanley* and *S. enteritidis* are the most pathogenic [2, 30, 88, 98]. The usual sources of infection are food products

Fig. 5. Diphtheritic colitis in *Macaca mulatta* with dysentery.

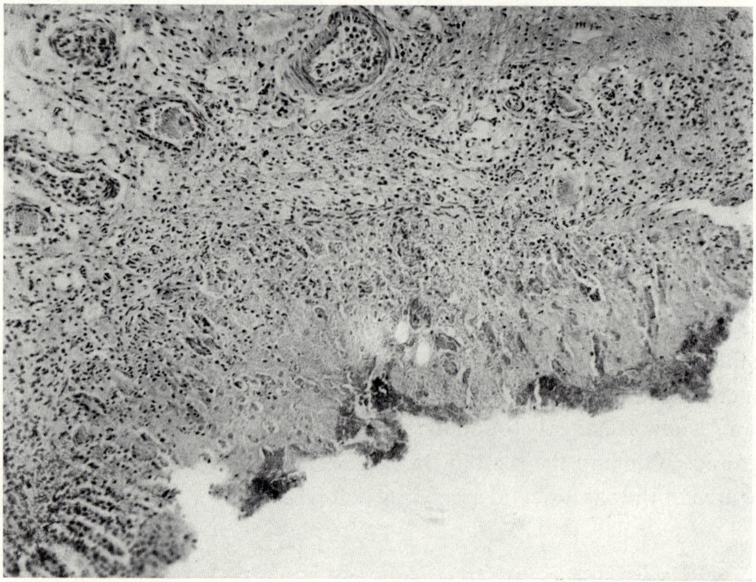

Fig. 6. Edema and necrosis in dysenteric colon mucosa (staining with hematoxylin eosin ×60).

and water, possibly contaminated with rodent or bird faeces, healthy carriers, and sick animals. Its occurrence is usually sporadic, but epidemics sometimes occur [2, 20, 25, 46, 63, 70, 84, 113]. The numbers of healthy carrier simians in colonies have been variously estimated from 0.5 to 60% [7, 12,

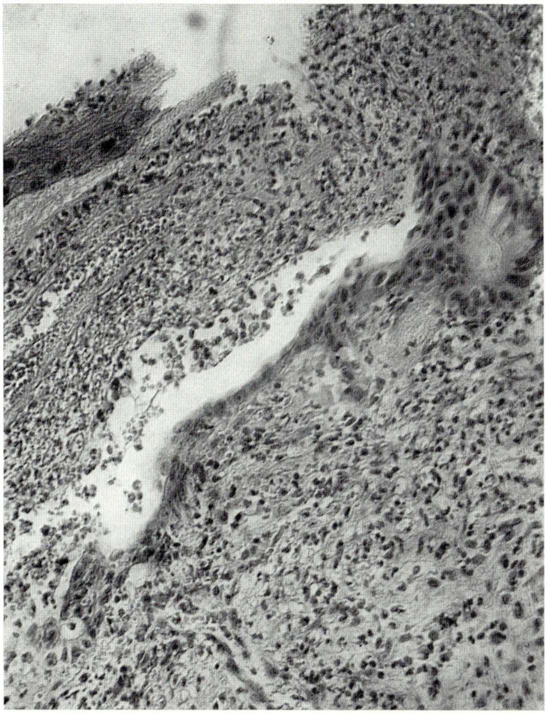

Fig. 7. Ulcer of the epithelium in *Macaca mulatta* with dysentery (Van Gieson ×300 staining).

30, 36, 37, 47, 53, 99, 101]. The carrier state most commonly occurs in apes, macaques, baboons and capuchins. Under Sukhumi conditions, salmonellosis has its highest incidence in cold weather, winter and spring [7]. (In other areas, such as Great Britain, outbreaks occur more commonly in the summer months, Ed.)

ii) Symptoms. The disease has an acute beginning, characterized by anorexia, prostration, diarrhoea without mucus and blood, watery stools with pellets of faeces resembling peas, and hypothermia in the more severe cases. Sometimes, the skin shows a widespread rosy rash [30].

iii) Pathology. Macroscopical changes have the character of catarrhal enteritis or gastro-entero-colitis with hyperplasia of the Peyer's patches, lymph nodules, and mesenteric lymph glands [76]. The causative agent can be isolated from the blood, spleen, liver, lymph nodes, bone marrow and intestinal contents.

Microscopically, there is hyperplasia of pale-staining histiocytes, sometimes with the formation of fine granulomas. Hyperplasia of the splenic follicles is commonly found. However, catarrhal changes of the stomach and large intestine are not a general feature of the disease. Ulcerative and diphtheritic erosions of the mucosa of the large intestine frequently occur, a form of the disease described in *P. hamadryas* [2, 30, 113]. In addition, specific focal bronchopneumonia is a common complication, and salmonellae can be isolated from lung tissue.

3. Enterocolitis Caused by the Organisms of the Proteus Group

Proteus group organisms are relatively pathogenic, though they may be isolated from the intestines of healthy monkeys.

i) Proteus vulgaris is often associated with epizootic outbreaks of haemorrhagic gastro-entero-colitis in weak and newly-arrived animals [111, 112]. The onset is acute and the condition associated with severe general symptoms including anorexia, prostration and diarrhoea with mucus and blood in the faeces which often have a foetid smell. The body temperature is reduced and the animals die within 3 to 5 days. *P. vulgaris* can be isolated in large numbers from the blood, parenchymatous organs, and intestinal contents.

The autopsy reveals sharp swelling of the intestinal segments, the lumen being filled with liquid contents. The inflammation of the bowel is catarrhal-/haemorrhagic. In most cases there is accompanying severe dystrophy of parenchymatous organs. The chief site of lesions with these organisms lies in the stomach and upper bowel, in contrast to what occurs with shigellae. Haemorrhages are also more pronounced than in the true dysenteries. In contrast to the true dysenteries also, proteus never penetrates into the epithelial cells of the intestine, though they may be surrounded by them [79].

ii) Proteus morgani also is relatively pathogenic, causing 'summer diarrhoea' in children and a number of animals, including simians [48, 50, 76, 80, 81, 86].

The disease is accompanied clinically by general sluggishness, vomiting, flatulence, and the development of a gastro-enteritis lasting from two to five days. *P. morgani* may be isolated from the intestinal contents and from the parenchymatous organs of moribund animals.

The disease is characterized morphologically by catarrhal desquamative gastro-enteritis and distinct hyperplasia of Peyer's patches and lymphoid nodules of the intestine. There is also hyperplasia of the spleen and mesenteric lymph nodes.

4. Amoebiasis

A number of species of amoeba may be found in the intestines of various monkey species. The most significant is *Entamoeba histolytica* which is pathogenic for both simians and man. Asymptomatic carriers of vegetative forms and cysts by *E. histolytica* are widespread amongst Asian macaques [13, 16, 17, 64, 74, 96]. There are variations of resistance in different species of primate, the macaques being more resistant, New World species very sensitive. Tissue invasion of the large intestine is commonly observed.

The disease is characterized by frequent passage of faeces containing blood and mucus, the external symptoms being similar to those of bacillary dysentery; thus bacteriological diagnosis is important. Lesions are found predominantly in the caecum and colon. The mucous membrane shows catarrhal inflammation with a varying number of ulcers of different sizes.

The ulcers are covered with a white/yellowish coating, but the bottom may be clear. The margins are raised. Also, the presence of small follicular abscesses is typical of simian amoebiasis [114] (fig. 8a, b, c). Vegetative forms of *E. histolytica* can be detected in smears and blood scrapings from the ulcers. In simian primates, in contrast to human, pathogenic amoebae do not penetrate into the submucosa and muscular layers of the intestine. This suggests that *E. histolytica* is less pathogenic to simians [96] than to humans.

5. Rare Conditions of the Alimentary Tract

Oesophagitis. Inflammation of the oesophageal mucosa is very rare in simians as an independent disease. Amongst the vast morphological material at Sukhumi, only one single case has been observed, and it has never been diagnosed in a living animal. The lesions in the organ may be haemorrhagic or rarely diphtheritic.

(Two cases of oesophageal varices were seen in South American species at the Delta Primate Center, during 1968. A number of cases of candidiasis of the oesophagus were also encountered in New World species (fig. 9a, b) Ed.).

Alimentary Tract. Degenerative conditions of the alimentary tract affecting the muscular layer and inflammatory lesions caused by bacterial infection have been described by BARTLAKOWSKI [9].

6. Bezoars

Solid formations varying in form and size are sometimes found in captive monkeys. They may consist of hair and plant debris held together by mucus (fig. 10). The inflammatory process in the stomach is chronic and is accompanied by coarse reorganization of the mucous membrane with the growth of crypts into the submucosa (fig. 11) [76].

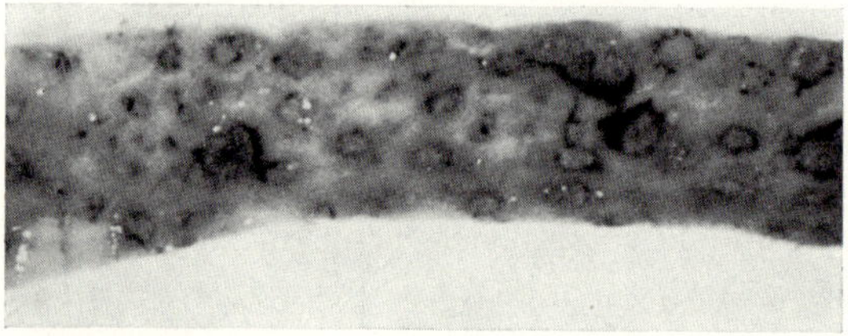

8a

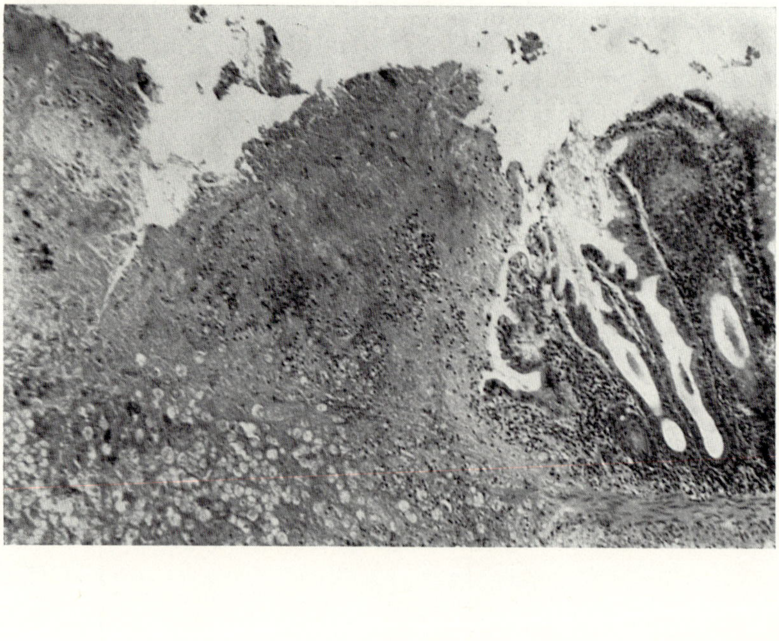

8b

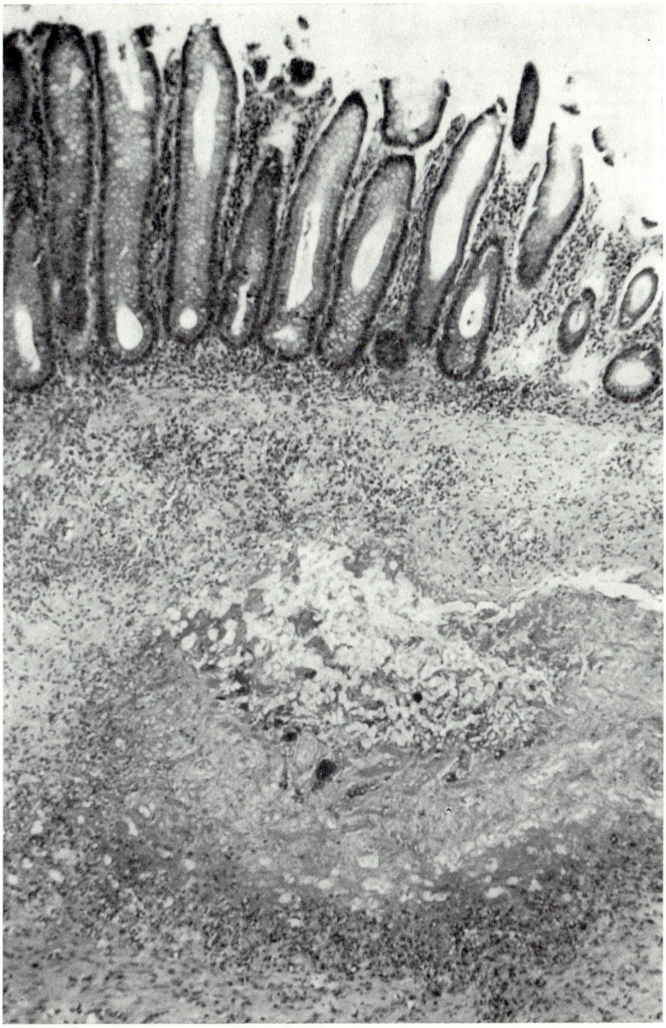

8c

Fig. 8. a Callicebus. Colon with acute amoebiasis. *b* Callicebus. Colon. Amoebic ulceration. *c* Callicebus. Colon. Amoebiasis: submucoid abscess.

7. Appendicitis

Acute inflammation of the appendix has only been described in anthropoids: chimpanzee, orang-utan, gorilla and gibbon [51, 55, 72, 108, 109]. Appendicitis is a common cause of death in captive gibbons. Death is usually

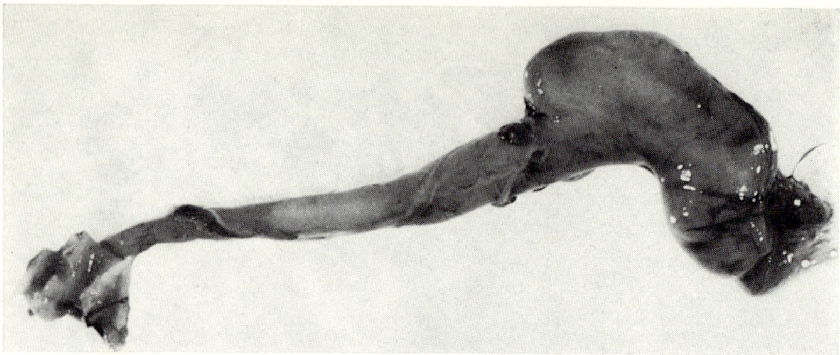

a

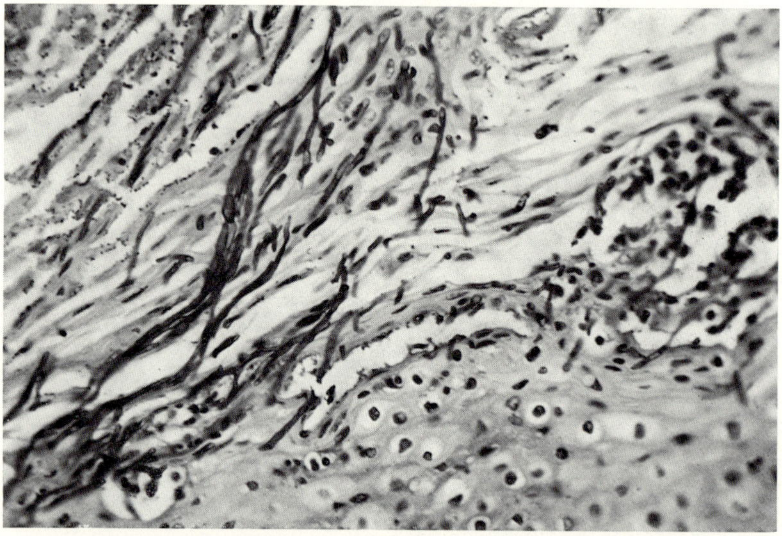

b

Fig. 9. a Callicebus. Oesophagus and stomach. The oesophagus is much dilated due to candidiasis. *b* Callicebus. Oesophagus, Candidiasis.

from secondary peritonitis and septicaemia [55]. It is difficult to diagnose appendicitis in live animals. The morphological examination shows an acute haemorrhagic, ulcerative-haemorrhagic or purulent inflammation of the appendix. The development of local peritonitis and adhesive processes often takes place in the disease. The caecum, the ileum and the colon are sometimes also involved in the process.

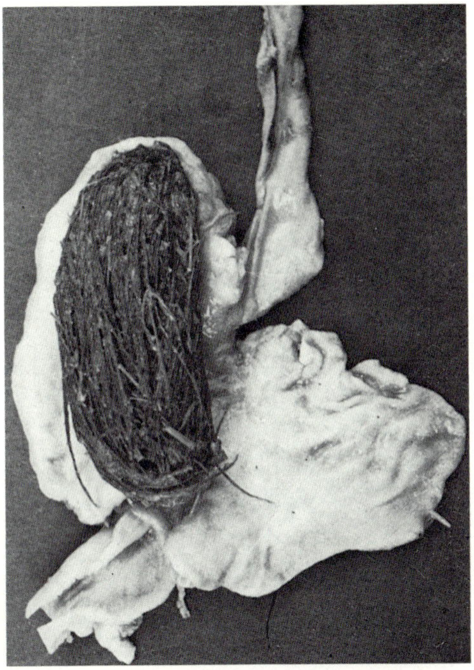

Fig. 10. Bezoar in the stomach.

8. Hernia

This disease is rare in monkeys. Only one case of strangulated inguinal hernia was recorded in a 6-year-old baboon at the Sukhumi colony during 40 years. Strangulated hernia was also described in other monkey species (mangabey, mandril, orang-utan, *Macaca nemestrina*, Mona.) [33, 34, 91, 96, 98, 104].

9. Gallstone Disease

The formation of cholesterol or cholesterol-lime stones in the gallbladder has been described in Asian and African primates *Cercopithecus aethiops*, Macaques, *Papio hamadryas*, *P. anubis*, orang-utans [76, 96].

Asymptomatic presence of stones is more often detected in adult monkeys (fig. 12). Death from gallstone disease is rather rare. In only 1 of the 14 cases observed at the Sukhumi colony gallstone disease was the direct cause

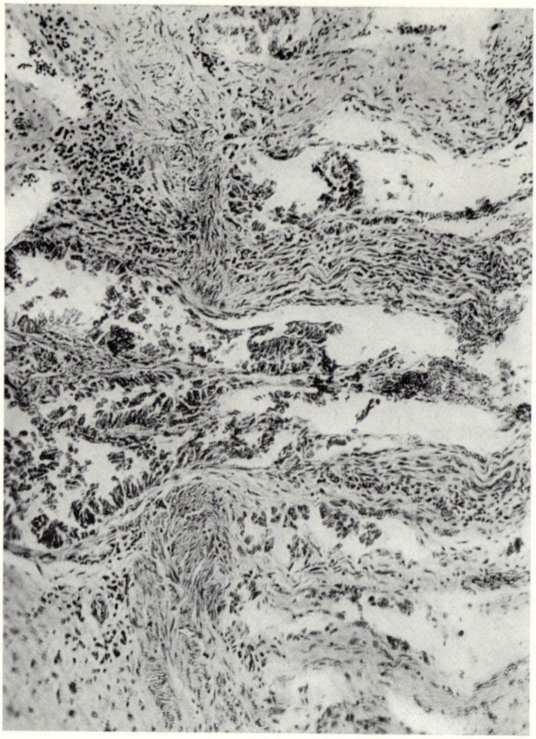

Fig. 11. Chronic gastritis with ingrowth of crypts in the submucosa.

of death. Morphological changes may not be detected in the wall of the gallbladder containing stones. Rarely the gallbladder is sharply distended and shows haemorrhages. Acute cholecystitis develops in the obturation of the bladder caused by the presence of a stone in it [49].

III. Diseases of the Respiratory Tract
(B. A. LAPIN, L. A. YAKOVLEVA and E. K. DZHIKIDZE)

1. Pneumonia

i) Etiology. Pneumonia is a common disease in all groups of simian primates, Asian, African and South American. It accounts for the death of some 20–50% of the animals in captivity [33, 58, 69, 76, 97, 115]. Different species, however, show differing susceptibility [82, 97, 115].

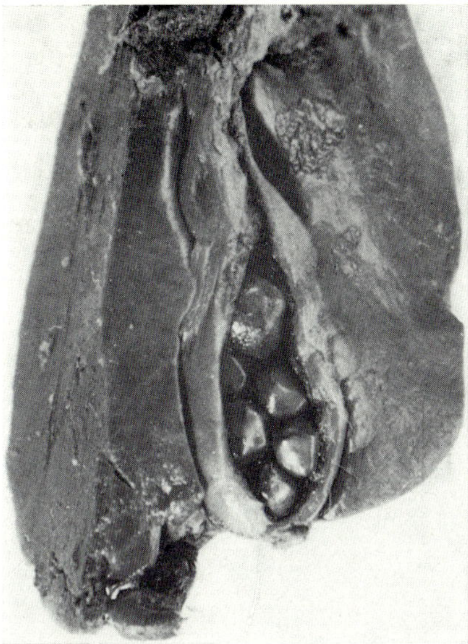

Fig. 12. Faceted calculi in gall bladder.

In the great majority of cases, pneumonia develops as a complication of some other disease, but it may also occur as a primary infection especially in infants up to 6 months of age. Amongst climatically adapted animals, the incidence of pneumonia is usually sporadic, though small epidemics sometimes occur. Amongst newly-arrived monkeys, pneumonia commonly occurs during the first 3 to 4 weeks. In all groups, the commonest cause is one of the types of pneumococcus *(Diplococcus pneumoniae)* [76]; according to Sukhumi records, this organism is responsible for 95% of cases (fig. 13a, b, c, d). Other organisms involved are: α and β - haemolytic streptococci, staphylococci and micrococci, *Haemophilus influenzae*, *Bordetella bronchiseptica* (fig. 14a, b), *Pasteurella* and *Klebsiella pneumoniae* [4–6, 33, 42, 61, 62, 76, 85, 97].

Symptoms. Coughing, sneezing, and rhinitis are some symptoms of pneumonia [100]. There are general disorders, by which the condition may be recognised. Such are: cyanosis of the upper eyelids and round the mouth; short breath and dilatation of the nostrils; the heart beat becomes more frequent; the red cell sedimentation rate is increased. Mucoid or muco-purulent discharges from the nose are usually seen and the temperature is elevated. The duration of the disease is usually from 7 to 12 days.

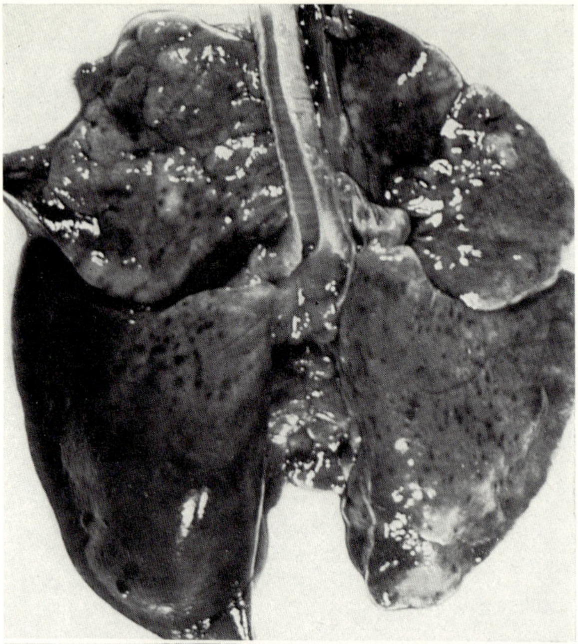

13a

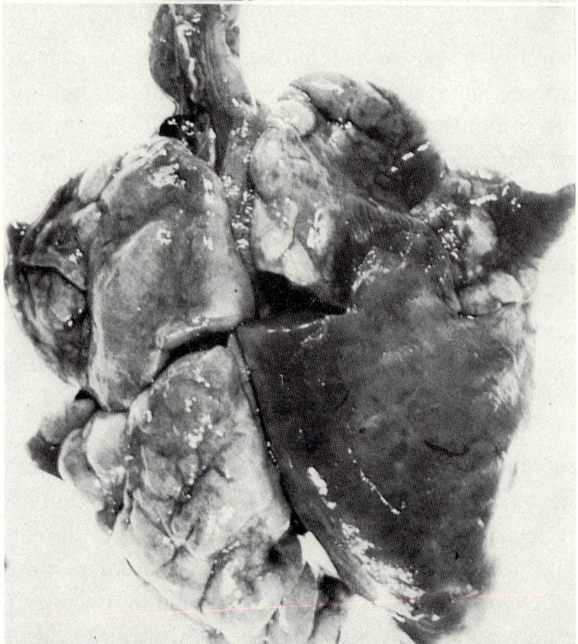

13b

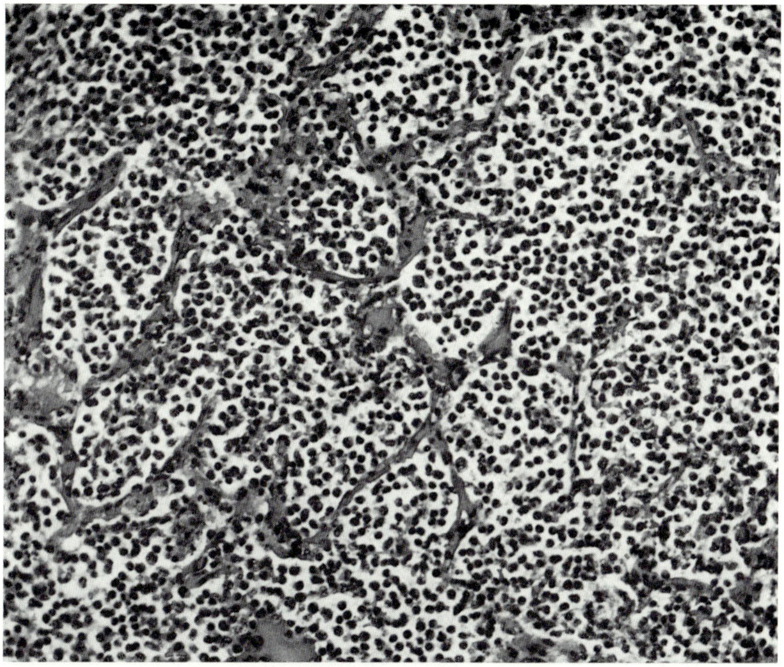

13c

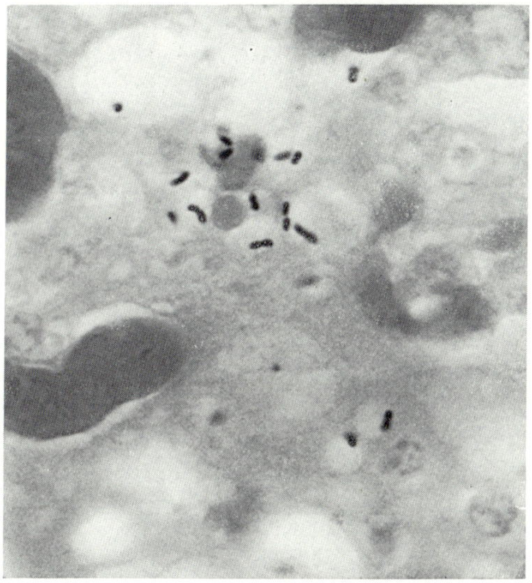

13d

Fig. 13. a Patas. Lungs. Pneumococcal pneumonia. *b* Patas. Lungs. Pneumococcal pneumonia. *c* Patas. Pneumococcal pneumonia. *d* Patas. Pneumococcal pneumonia.

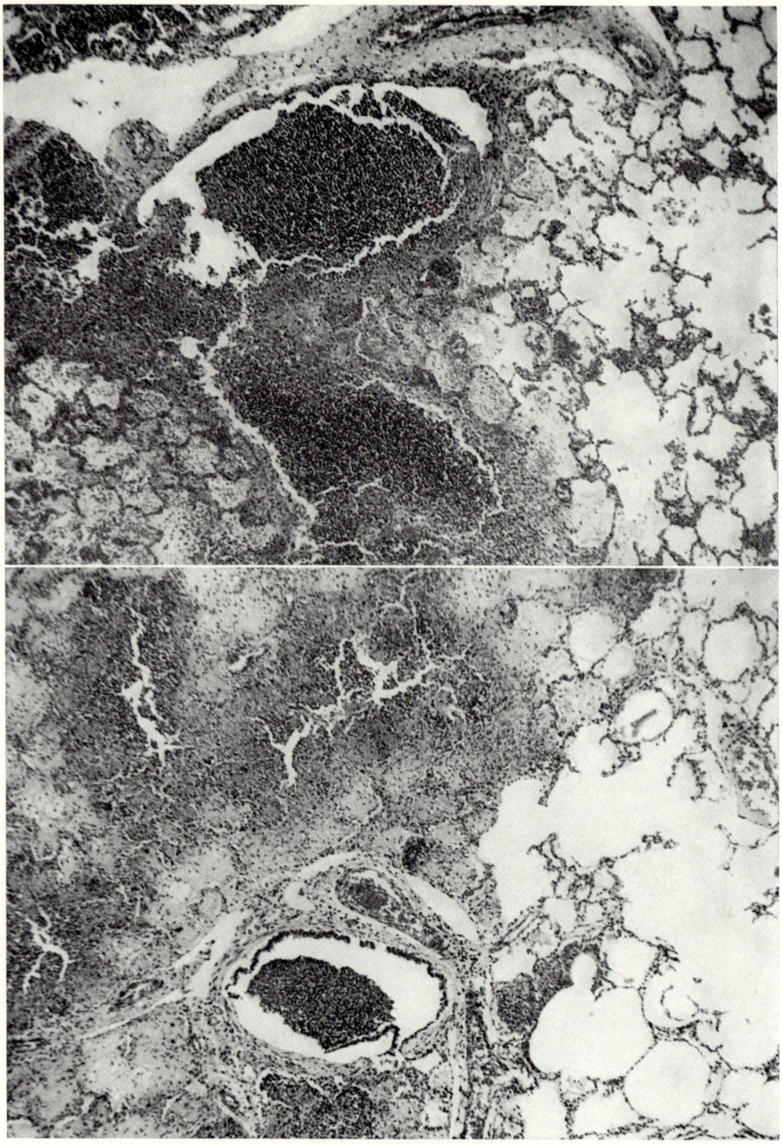

a

b

Fig. 14. a Callicebus. Bronchopneumonia attributed to Bordetella sp. *b* Callicebus. Bronchopneumonia attributed to Bordetella sp.

ii) Diagnosis. Percussion of the chest is liable to give misleading results, though a shortened note may be found in most cases over the affected portion of the lung. When auscultation is used, a change in the character of the respirations may be noted, from a sharply relaxed vesicular sound to bronchial noises, consisting of dry, mixed and rarely crepitant rales. Roentgenologically, the affected areas show increased opacity and the appearance of shadows.

iii) Pathology. The majority of pneumonias (86%) are bilateral. Initial inflammatory changes are usually detectable around the bronchi, particularly in the areas near the hilus. The pleura often becomes involved also. Histologically, the alveolar exudate is usually purulent and consists predominantly of neutrophil leucocytes mixed with erythrocytes, desquamated epithelial cells, and a small quantity of free fibrin. Both free and phagocytised diplococci may also be seen in the exudate. The bronchi are usually also involved in a desquamative or ulcerative bronchitis. There are also numerous atalectases. In cases of interstitial pneumonia the alveolar septa appear very thickened with the lumina of the alveoli constricted. The bronchial tree may also show oedema and infiltration by cellular elements.

Broncho-pneumonia may develop as a complication of enteritis and colitis. In infants up to 6 months, it occurs as a primary infection [76, 57].

In croupous pneumonia, the condition is usually bilateral. The lower lobes of the lungs are mostly affected. The affected lobe is increased in volume and thickened (fig. 15a). In section, the tissue is homogeneously thick and fine-grained. Histologically, the exudate is found to contain fibrin and leucocytes, sometimes with a large admixture of erythrocytes (fig. 16). The exudate at the periphery of the inflammatory focus is serous, and may also contain numerous macrophages. The bronchi, too, become involved in the process, and contain exudate similar to that of the alveoli. In the pleural cavities, exudate is present containing fibrin flakes. The inflammatory process extends from the areas of the lung near the hilus towards the periphery. Oedematous fluids containing bacteria in the periphery of the hepatized zone indicate a progressive process [60].

Pneumonias caused by *Klebsiella pneumoniae* are characterized by the development of numerous foci of muco/leucoytic inflammation, which have a tendency to fuse. The result of this coalescence is that many lobes become affected and may be totally consolidated. Gram-negative diplobacilli are found in the exudate [59, 60].

Atypical pneumonias, with giant cell infiltration of the alveolar and

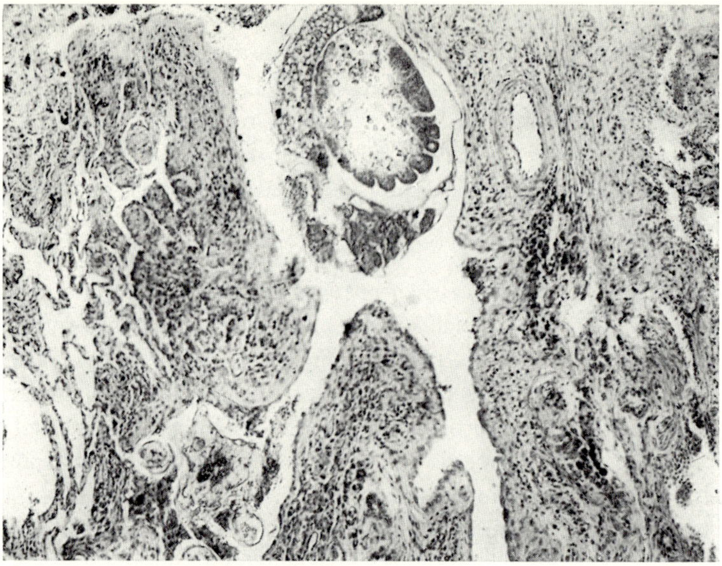

a

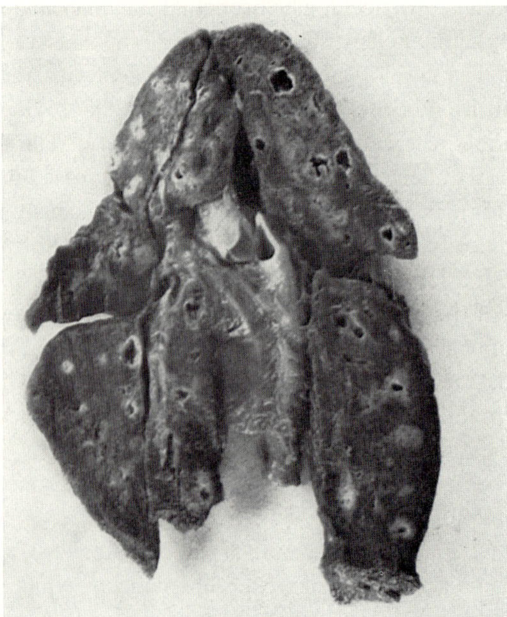

b

Fig. 15. a Lung infection with mites (pneumonyssus), causing bronchitis (staining with hemotoxylin eosin ×150). *b* Lung infection with mites (pneumonyssus), causing bronchitis (staining with hemotoxylin eosin ×150).

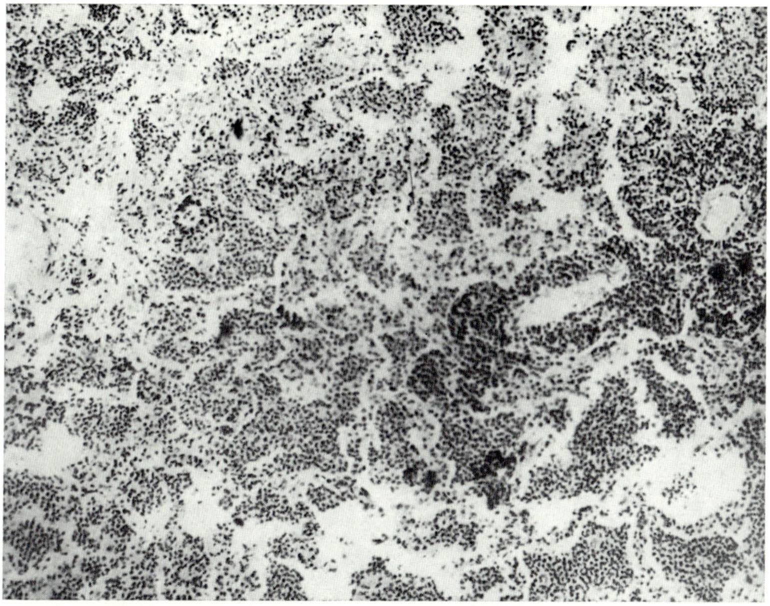

Fig. 16. Pneumonia. Leucocytic exudate (staining with hemotoxylin eosin ×150).

bronchial epithelium, are sometimes also found in monkeys. In these, intracytoplasmic and intranuclear oxyphilic inclusions can be detected. These simian pneumonias resemble human viral pneumonias.

During recovery, gradual resorption of the exudate takes place. However, in some cases the exudate becomes organised, and the lung tissue is 'carnified' [76].

Complications of pneumonia occur somewhat rarely. Such are: gangrene; lung abscesses; serous pleuritis; pneumococcal meningitis; meningo-encephalitis; pericarditis; myocarditis; and septicaemia [52, 76, 78, 96].

2. Lung Mites

Mite infestation of the lungs is widespread amongst simian primates [71, 75, 76, 96]. (See also the section on Lung Mites by HALL and INNES, part II.) The species most commonly affected are: *Macaca mulatta, M. irus, M. nemestrina, Cercopithecus aethiops, Hamadryas gelada, Papio hamadryas* and the chimpanzees [15, 43, 76] (fig. 17a, b).

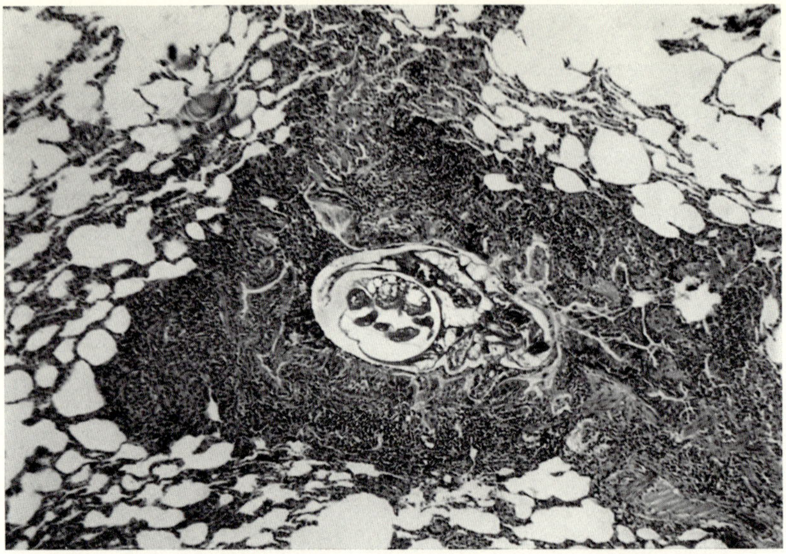

a

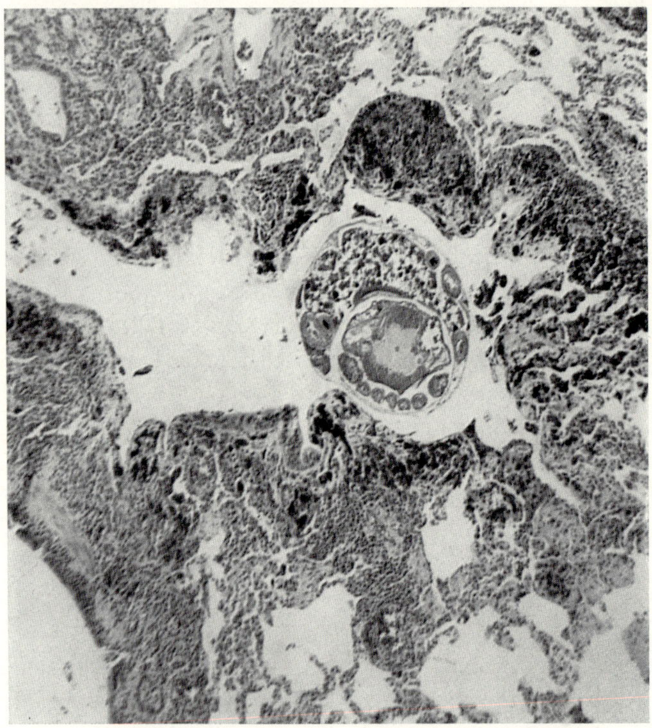

b

Fig. 17. a Talapoin. Lung mites in bronchioles and peribronchiolar reaction. *b Macaca mulatta.* Lung mites in alveolus of lung.

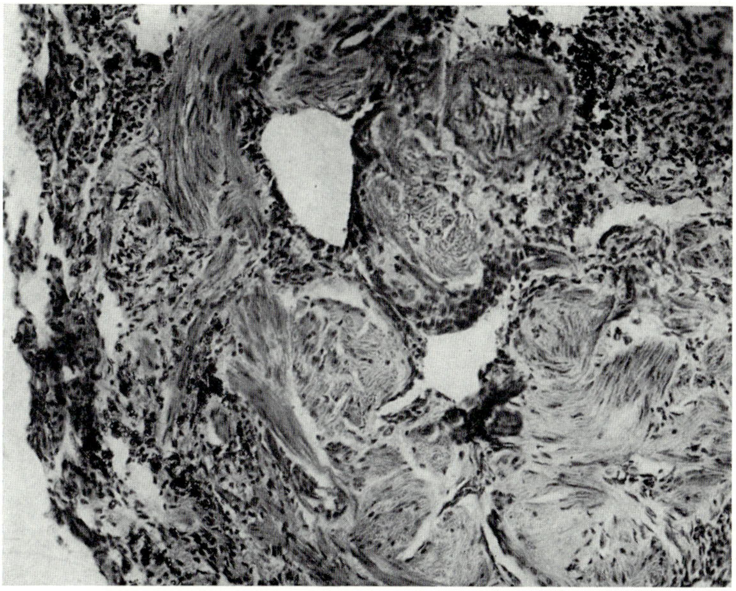

Fig. 18. Muscular cirrhosis of lung tissue in bronchitis due to lung mites.

In the great majority of cases, mites can be detected in the lungs of newly-caught animals, but they may occur also in animals that have been kept in captivity for a long time [76]. The percentage affected of the most susceptible groups varies widely between 46 to 100% [32, 68, 76]. The disease is caused by the mites *Pneumonyssus simicola.* Both nymphs and adults are found in the tissues of the lungs; they are whitish and 0.3 to 0.5 mm in length. The mites invade the bronchial tree, and become located mostly beneath the pleura. Rare specimens may be found in the alveoli. The mechanism of transmission has not been ascertained.

The disease is usually relatively asymptomatic, though often attended by chronic cough. When the lungs are seriously invaded, dyspnoea and cardiopulmonary insufficiency may develop.

The appearances of affected lungs are fairly typical. The surface is seen to contain numerous yellowish and greyish-white foci, the size of a pinhead; these are situated beneath the pleura and resemble miliary tubercles. When the colour is greyish-white, the focus is a fresh one. In cross-section, the foci consist of cavities with rigid and rather thick walls, from which the parasite can be extracted fairly easily with a teasing needle. Surrounding the parasite and covering the walls of the cavity completely or partly are greyish, friable, masses of detritus (fig. 15b). In these cavities, can often be seen an opening into an adjacent bronchiole, so that the lesions are in reality 'bronchiectasis'. Mites may also be present in the bronchioles, chiefly in the terminal, subpleural branches. In these cases, there is severe destructive bronchiolitis and bronchiectasis with muscular cirrhosis (fig. 15a and 18). Where the process is far advanced, there is a close resemblance to non-specific tuberculosis [76]. Localisation of mites in the alveoli, which is much less common, is accompanied by emphysematous dilatation of those alveoli which contain mites, and the appearance of inflammatory foci.

3. Pulmonary Nocardiosis

Pulmonary nocardiosis has been described in *Macaca mulatta* [65]. The authors have studied also two cases. The disease consists of an extensive necrotic or caseous pneumonia, which affects one or more lobes of the lungs. Usually, the lymph nodes of the trachea and bronchi, and the liver and spleen are also involved. Direct smears from the affected portion of a lung show fine Gram-positive, branching or thread-like, filaments, which have a tendency to fragmentation. Nocardia is also frequently acid-fast.

References

1 ABRAMOVA, E. V.: Spontaneous dysentery in monkeys. Trud. Sukhumi Biol. Sta. *1:* 244–275 (1949).

2 AKSENOVA, A. S. and HUTSISHVILI, L. M.: Breslau parathyroid infection in monkeys. Tr. Sukhumsk. Biol. stantsii AMN SSSR. *1949:* 270–287.

3 AKSENOVA, A. S.: Dysentery Newcastle bacteria in monkeys. Vopr. infect. pathol. v experimente na obezjanah. Sukhumi *1968:* 21–24.

4 AKBROIT, E. Y.; IVANOV, M. T.; KUKSOVA, M. I., and DANKO, L. V.: Characterization of pneumonia in macaques imported to Sukhumi monkey nursery from India and Vietnam. Vopr. physiol. exper. pathol. i. radiobiol., pp. 273–276 (Sukhumi 1965).

5 AKBROIT, E. Y.: Microflora of monkey and human respiratory tracts. Vopr. physiol. exper. pathol. i. radiobiol. Sukhumi, pp. 264–267.

6 AKBROIT, E. Y.; KUKSOVA, M. I., and DANKO, L. V.: Etiology, clinic and blood picture in pneumonia in green monkeys. Vopr. physiol. exp. patol. i. radiobiol. Sukhumi (1965).

7 AKSENOVA, A. S.: Paratyphoid carriers in monkeys. Vopr. inf. patologii v experimente na obesjanah. Sukhumi, pp. 159–166 (1958).

8 Bach, F. W.; Fülscher, I. und Harnack, S.: Über eine durch Affen verursachte Ruhr-Epidemie. Veröffentlicht aus dem Geb. der Med. *34:* Suppl. 2, pp. 3–13 (1931).

9 Bartlakowski, J.: Speiseröhrenerkrankungen bei Affen. Virchows Arch. *277:* 670–672 (1930).

10 Baumann, F. B.: A note of the spontaneous occurrence of bacillary dysentery in monkey. Philippine J. Sci. *5:* 481 (1910).

11 Bernhard, G. und Markoff, W. N.: Über Modifikationen bei Bakterien. Zbl. Bakt. *65:* 1–4 (1912).

12 Bokkenheuser, V.: Salmonellae and shigellae in gall bladders from man and monkey. Sth afr. J. Med. Sci. *27:* suppl. 3, pp. 60–64 (1962).

13 Bond, V. P.; Bostick, W.; Hansen, E. L., and Anderson, H. H.: Pathologic study of natural amebic infection in macaques. Amer. J. trop. Med. Hyg. *26:* suppl. 5, pp. 625–629 (1946).

14 Bowman, F. B.: A note on the spontaneous occurrence of bacillary dysentery in monkeys. Philippin. J. Sci. *135:* 481–484 (1910).

15 Born, E. und Zöliger-Müller, B.: Beiträge zur vergleichenden Pathologie. I. Über die Lungenakariasis des Rhesusaffen *(Macaca mulatta)*. Zbl. allg. Path. path. Anat. *94:* 7–8 und 340–345 (1956).

16 Burova, L. F.: Experimental study of histolytica-like amebas in lower monkeys. Arkh. biol. nauk. *55:* suppl. 3, pp. 66–67 (1939).

17 Burova, L. F.: Intestinal protozoa in lower monkeys. Vopr. kraevoyi parasitologii. *1938:* 309–322.

18 Carpenter, K. P. and Cooke, E. R. W.: An attempt to find shigellae in wild primates. J. Comp. Path. *75:* 201–204 (1965).

19 Cass, I. S.: Enteric infection in monkeys. Proc. Anim. Care *3:* 14–22 (1952).

20 Christoph, H. J.: Salmonellose bei Schimpansen. Zool. Gart., Lpz. *29:* 58–62 (1964).

21 Cockburn, T. A.: Intestinal infection of primates. Trans. roy. Soc. trop. Med. Hyg. *42:* 509–510 (1949).

22 Cook, R.: The naturally occurring diseases of laboratory animals and practical measures for control of these diseases. J. Med. Lab. Technol. *11:* 30–35 (1953).

23 Cruickshank, J. C.: Bacillary dysentery in laboratory monkeys. Monthly Bull. Min. Hlth Publ. *9:* 277–279 (1950).

24 David, H. und Schirl, A.: Über Ruhrerkrankungen bei Menschenaffen. Zool. Bakt. *144:* 43–44 (1939).

25 Dobberstein, Y.: Über die häufigsten Todesursachen der in zoologischen Gärten gehaltenen Tiere. Med. Klinik *32:* 311–315 (1936).

26 Dzhikidze, E. K.: Dysentery carriage in monkeys. Kand. dis., Sukhumi (1954).

27 Dzhikidze, E. K.: Experimental study of the nature and regularity of dysentery carriage; in Vopr. inf. pathologii v experimente na obesjanah., Sukhumi *1958:* 7–19.

28 Dzhikidze, E. K.: Shigella dysenteriae carriage in monkeys. Zh. Mikrobiol. Epid. Immunol. *27:* suppl., pp. 44–48 (1956).

29 Dzhikidze, E. K.; Gvasava, I. S., and Kavtaradze, K. N.: Peculiarity of the dysentery carriage and clinical course in *Papio hamadryas;* in Vopr. physiol. exp. pathologii i radiobiol., Sukhumi *1965:* 215–222.

30 Dzhikidze, E. K. and Stasilevich, Z. K.: Analysis of salmonella carriage in mon-

keys for the period of 8 years. Vopr. physiol. i. experim. pathologii., Sukhumi 1968: 289–296.

31 EDWARDS, J. B.: Strangulated inguinal hernia in a monkey. Vet. Rec. *61:* 387–388 (1949).

32 FAIRBROTHER, R. W. and HURST, E. W.: Spontaneous diseases observed in 600 monkeys. J. Path. Bact. *35:* 867–873 (1932).

33 FEGLEY, H. C.: Mass transport of monkeys with antibacterial substances; in Care and diseases of the research monkey *85:* 3 (1960).

34 FIENNES, R.: Zoonoses of primates (Weidenfeld & Nicolson, London; Cornell University Press, Ithaca 1967).

34a FIENNES, R. N. T-W-: in BALNER and BEVERIDGE Infections and immuno-suppression in subhuman primates, pp. 151–152 (Munksgaard, Copenhagen 1970).

35 FRATTA, I. and MANN, P. H.: Bacteriophage typing of staphylococci isolated from various species of domestic and laboratory animals. Canad. J. comp. Med. *24:* 270–272 (1960).

36 GALTON, M. M.; MITCHELL, B. B.; CLARK, G., and RIESEN, A. H.: Enteric infections in chimpanzee and spider monkeys with special reference to a sulfadiazin resistent shigella. J. infect. Dis. *83:* suppl. 2, pp. 147–154 (1948).

37 GARNER, A. and MORALES, C.: Occurrence of salmonella and shigella organisms in monkeys transported from India to Randolph Air Force Base. USAF School of Aviation Medicine. Rep. No. 56–140 (1956).

38 GOLTENBOTH, R.: Ein Ausbruch von bakterieller Ruhr bei Menschenaffen. Kleintier-Praxis *12:* 79–82 (1967). – Biol. Abstr. *49:* 30591 (1968).

39 GVASAVA, I. S.: Clinical peculiarities of monkey dysentery; in Vopr. inf, pathologii v experimente na obesjanah., Sukhumi *1958:* 31–38.

40 GVASAVA, I. S.: Clinic, diagnostic and antibiotic therapy of monkey bacterial dysentery. Kand. dis., Sukhumi (1959).

41 GVASAVA, I. S. and KAVTARADZE, K. N.: Dysenteric clinic and bacteriology in newly-imported monkeys. Tr. Inst. Exp. Pathol. Ther. AMN SSSR Sukhumi *1965:* 224.

42 GRAVES, J. L.: Bordetella bronchiseptica isolated from a total case of bronchopneumonia in an African green monkey. Lab. Anim. Care *18:* 405–406 (1968).

43 GRZIMEK, B.: Tod durch Zungenmilben bei einem Schimpansen. Der Zoologische Garten *18:* 5–6, 249 (1951).

44 GRÜHNBAUM, A.: Some experiments on enterica scarlet fever and measles in the chimpanzee. Brit. Med. J. *i:* 817 (1904).

45 GUPTA, B. M.: Fatal flexner bacillus infection in anthropoidea ape *(Hylobates hoolock)*. Ind. Med. Gaz. *1933:* 138–139.

46 HAAS, R. und PETERSON, K. F.: Shigella und Salmonella Befunde bei Affen. Med. Klin. *43:* 1947–1949 (1959).

47 HABERMANN, R. T. and WILLIAMS, F. P.: Diseases seen at necropsy of 708 *Macaca mulatta* (rhesus monkey) and *Macaca philippineun* (Cynomolgus monkey). Amer. J. vet. Res. *18:* 419–426 (1957).

48 HAMERTON, A. E.: Report of the Society's Prosector for the year 1928. Proc. zool. Soc., Lond. *vi:* 49–59 (1929).

49 HAMERTON, A. E.: Report on the deaths occurring in the Society's Gardens during the year 1930. Proc. zool. Soc., Lond. *101:* 11, 527–555 (1931).

50 Hamerton, A. E.: Report on the deaths occurring in the Society's Gardens during the year 1933. Proc. zool. Soc., Lond. *104:* 11, 389–422 (1934).

51 Hamerton, A. E.: Report on the deaths occurring in the Society's Gardens during 1939–1940. Proc. zool. Soc., Lond. *3:* 151–185 (1941).

52 Hamerton, A. E.: Report on the deaths occurring in the Society's Gardens during the year 1944. Proc. zool. Soc., Lond. *3–4:* 1946).

53 Hardy, A. V.: Problems in the control of infectious diseases in a large colony of primates. Proc. Anim. Care Panel *5:* 16–21 (1954).

54 Hardy, P. H.; Lindsey, J. R., and Melby, E. C.: Transferable multiple drug resistance in enterobacteria from non-human primates. Johns Hopk. Hosp. Bull. *123:* 29–37 (1968).

55 Hill, W. C. O.: Appendicitis in gibbons. Proc. zool. Soc., Lond. *124:* 13–16 (1954).

56 Ivanov, M. T.: Pathomorphologic study of untreated croupous pneumonia in macaques. Arch. Path. *3:* 33–39 (1962).

57 Ivanov, M. T.: The morphology of focal pneumonia in monkeys. Arch. Path. *24:* suppl. 12, pp. 39–44 (1962).

58 Ivanov, M. T.: Pneumonia in monkeys (morphologic studies). Autoref. dissert. kand., Sukhumi (1964).

59 Ivanov, M. T.: On the pathologic anatomy of Friedlander's pneumonia in rhesus monkeys. Arch. Path. *28:* suppl. 2, pp. 76–78 (1966).

60 Ivanov, M. T. and Akbroit, E. Y.: Pathomorphologic characters of green monkey pneumonia. Vopr. physiol. exper. pathol. i radiobiol., Sukhumi *1965:* 280–282.

61 Ivanov, M. T. and Aksenova, A. C.: Pathologic anatomy and bacteriology of monkey pneumonia; in Mater. konf. po biol. u. pathol. obesjan Tes. dokl. *68:* 1961).

62 Ivanov, M. T.; Aksenova, A. S., and Danko, L. V.: On the etiology and pathogenesis of monkey pneumonia. Mater. konf. Inst. Exp. Pathol. i Ther. AMN SSSR, Sukhumi, 6 str. (1963).

63 Jansen, Y.: Salmonellose *(S. typhimurium)* infection bei Affen. Infekt. Kr. Haustiere *56:* 135–144 (1940).

64 Johnson, C. M.: Observations on natural infections of *Entamoeba histolytica* in ateles and rhesus monkeys. Amer. J. Hop. Med. *21:* 49–61 (1941).

65 Jones, A. M. and Wyand, D. S.: Pulmonary nocardiosis in the rhesus monkey. Path. Vet. *3:* 588–600 (1966).

66 Kavtaradze, K. N.: Characterization of dysentery cultures, isolated from imported monkeys. Tes. dokl. nauchn. konf. Inst. Exp. Pathol. i Ther. AMN SSSR *38:* (1963).

67 Kavtaradze, K. N. and Gasparian, T. S.: Characteristics of dysentery carriage in monkeys according to the materials of monkey nursery for the recent 10 years (1951–1961); in Mater. konf. po biolog. i pathol. obezjan Tes. dokl. Sukhumi, *1961:* 17–18.

68 Kalter, S. S.; Kuntz, R. E.; Myers, B. J.; Eugster, A. K.; Rodriguez, A. R.; Benke, M., and Walter, G. V.: The collection of biomedical specimens from Baboons *(Papio* sp.*)* Kenya, 1966. Primates *9:* 123–139 (1968).

69 Kennard, M. A.: Abnormal findings in 246 consecutive autopsies on monkeys. Yale J. Biol. Med. *13:* 701–712 (1941).

70 Kent, T. H.; Formal, S. B., and Labrec, E. H.: Salmonella gastroenteritis in sus monkeys. Arch. Path. *82:* 2, 272–279 (1966).

71 KIRSCH, E.: Pathologische Lungenveränderungen beim Rhesusaffen, hervorgerufen durch *Pneumonyssus simicola* (acarina). Z. Parasitenk. *14:* 626–636 (1950).

72 KOCH, W.: Bericht über das Ergebnis der Obduktion des Gorilla Bobby des Zoologischen Gartens zu Berlin. Ein Beitrag zur vergleichenden Konstitutionspathologie. Veröff. Konst. Wehrpath. *9:* supp. 3, p. 36 (1937).

73 KOCH, W. und DEIMED, B.: Über Krankheiten der Menschenaffen (1952).

74 KUNTZ, R. E.; MYERS, B. J., and VICE, T. E.: Intestinal protozoa and parasites of the gelada baboon *(Theropithecus gelada* KÜPPEL, 1835). Proc. Helm. Sci., Wash. (1967).

75 LAPIN, B. A.: Tick bronchitis in monkeys. Arch. Path. *18:* 4, 57–61 (1956).

76 LAPIN, B. A. and YAKOVLEVA, L. A.: Assays of comparative pathology of monkeys (1960).

77 LAPIN, B. A.; DZHIKIDZE, A. K., and YAKOVLEVA, L. A.: Diseases of monkeys in natural conditions of Vietnam (Moskva, 1965).

78 LAPIN, B. A.; NORKINA, L. N.; CHERKOVICH, G. M.; YAKOVLEVA, L. A.; KUKSOVA, M. I.; ALEKSEEVA, L. V.; FUFACHEVA, A. A., and STARTSEV, V. T.: Monkey as an object of medical and biological experiments, Sukhumi (1963).

79 LEVENBOOK, I. S. and VORONIN, E. S.: On the morphology of acute intestinal diseases of rhesus monkeys caused by *Proteus vulgaris*. Arch. Path. *27:* 7, 53–58 (1965).

80 LOVELL, R.: Discussion on monkeys and human disease. Proc. roy. Soc. Med. *22:* 820–823 (1929).

81 LOVELL, R.: Isolation of *Bacterium morgani* from mammals, birds and reptiles at Zoological Gardens. J. Path. Bact. *32:* 79 (1929).

82 LUND, E. and PETERSON, K. B.: Pneumococci in monkeys. Acta path. microbiol. scand. *45:* 309–318 (1959).

83 MATSUOKA, K.; KISHI, T., and YAMADA, H.: Shigella harbored by apes and monkeys and treatment of shigellosis in them. J. jap. vet. med. Ass. *15:* 230–233 (1962).

84 MACGAUGHEY, Y.; SCHMIDT, E.; VELAUDAPILLAE, T., and WEINMAN, A.: *Salmonella typhimurium* in young elephants and chimpanzees. Vet. Rec. *65:* 431–432 (1953).

85 MONTANO, A.; TOMES, S. F., and OLIVELRA e SILVA: Contagious bronchopneumonia caused by *Bordetella bronchi septica* in imported monkeys; comparative histology with human findings. Med. contemp. *84:* 303–315. – Abstract: Biol. Abstr. *49:* 41354 (1968).

86 MORGAN, H.: Upon the bacteriology of the summer diarrhea of infants. Brit. Med. J. *11:* 16 (1907).

87 MUTO, T.; UEDA, K., and NAKAGAWA, M.: A spontaneous outbreak of dysentery-like disease among experimental monkeys. J. jap. vet. med. Ass. *16:* 267–269 (1963).

88 NEFEDIEVA, N. P. and AKSENOVA, A. S.: Study of salmonella antigenic structure isolated from monkey intestine; in Vopr. infect. parathol. in experim. na. obesjanah. Sukhumi *1958:* 167–171.

89 OGAWA, H.; TAKAHASHI, R.; HONJO, S.; TANASAKA, M.; FUJIWARA, T.; ANDO, K.; NAKAGAWA, M.; MUTO, T., and IMAIRUMI, K.: Shigellosis in Cynomolgus monkeys *(Macaca irus)*. III. Histopathological studies on natural and experimental Shigellosis. Jap. J. Med. Sci. *17:* 6, 321–332 (1964).

90 PINKERTON, M. E.: Shigellosis in the baboon *(Papio* sp.*)*. Lab. Anim. Care *18:* 1, 11–21 (1968).

91 Ratcliffe, H. L.: Left inguinal hernia in a pig-tailed macaque *(Macacus nemestrinus)*. Rep. Lab. comp. Path., Philad. *1930:* 51–52.

92 Ravaut, F. et Dopter, C.: Une épidémie de dysenterie bacillaire chez des macaques. Bull. Soc. Path. exot. *1909:* 2, 17–20.

93 Rewell, R. E. and Bridges, R. E.: An outbreak of *Shigella flexneri* infection among rhesus monkeys. Mon. Bull. Min. Hlth. Lab. Serv. *7:* 25–29 (1948).

94 Rewell, R. E. and Bridges, R. F.: Dysentery in monkeys. Brit. Med. J. *2:* 876 (1949).

95 Rewell, R. E.: Outbreak of *Shigella schmitzii* infection in men and apes. Lancet *256:* 220–221 (1949).

96 Ruch, T.: Diseases of laboratory primates (Philadelphia, London 1959).

97 Sauer, R. M.: The roles of infections and noninfectious diseases in monkey health; in Care and diseases of the research monkey. *85:* 3 (1960).

98 Scott, H. H.: Report on the deaths occurring in the Society's gardens during the year 1927. Proc. Zool. Soc., Lond. *1928:* suppl. 1, pp. 81–119.

99 Schneider, N. J.; Prather, E. C.; Lewis, A. L., and Scatterday, Y. E.: Enteric bacteriological studies in a large colony of primates; in Care and diseases of the research monkey. *1960:* 935–941.

100 Simavonian, V. G. and Bochrova, D. N.: On the clinical and roentgenological picture of lower monkey pneumonia; in Vopr. physiol. exper. pathol. i radiobiol., pp. 268–272 (Sukhumi 1965).

101 Stasilevich, Z. K.: Experimental salmonellosis and colienteritis in monkeys. Kand. dissert., Sukhumi (1961).

102 Soike, K. F.; Coulston, F.; Day, P.; Deibel, R., and Plager, H.: Viruses of the alimentary tract of chimpanzees. Exp. molecul. Path. *7:* 3, 259–303 (1967).

103 Stadie, H.: Etwas über Erkrankungen von Affen. Zoologischer Garten Leipzig. *3:* 55–58 (1930).

104 Straus, W. L.: The thoracic and abdominal viscera of primates, with special reference to the orang-utan. Proc. amer. Phil. Soc. *76:* 1–85 (1936).

105 Summers, G. A. C. and Linton, A. H.: *Shigella flexneri* enzootic in captive rhesus monkeys. Brit. Med. J. *4882:* 283–285 (1954). – Exc. Med. *8:* 137 (1955).

106 Tanaka, M.; Honjo, S.; Fujiwara, T.; Hagiwara, T.; Ogawa, H., and Imaizumi, K. Shigellosis in cynomolgus monkeys *(Macaca irus)*. I. Epidemiological surveys on Shigella infection rate. Jap. J. Med. Sci. *17:* 5, 259–265 (1964).

107 Triotskiy, V.; Dzhikidze, A. K.; Tumanian, M. A., and Aksenova, A. S.: Dysentery in monkey as experimental and biological model of human dysentery; in Teor. i. pract. vopr. med. i. biol. v. exp. na. obesjanah., Medgiz *1956:* 139–158.

108 Weinberg, M.: Un cas d'appendicite chez le gorilla. Bull. Soc. Path. exot. *1:* 556–560 (1908).

109 Weinberg, M.: Appendicite et vers intestinaux chez le chimpanzé. C. R. Soc. Biol., Paris *60:* 661–662 (1906).

110 Victorova, O. N. and Aksenova, A. S.: The study of Schmitz-like strains, isolated from monkey. Tr. Sukhumi Biol. Stantsii AMN SSSR, Moskva *1949:* 264–269.

111 Voronin, E. S. and Gavrilov, V. I.: The study of the proteus role in etiology of acute gastrointestinal diseases in monkey; in Poliomyel. i drugie enterovirusn. inf. M. *1963:* 243–244.

112 VORONIN, E. S. and GAVRILOV, V. I.: The study of the role of *Proteus vulgaris* in etiology of acute gastrointestinal diseases in monkeys, ZHMAI *6:* 102–106 (1964).

113 TSVETAEVA, N. P.: Diseases of the animals of the Moscow Zoo; in Tr. Moskovsk. zooparka t. 2, vip. 2, M. 1941, 53–54, 57.

114 VICKERS, J. H.: Gastrointestinal diseases of primates; in KIRN Current veterinary therapy. III. Small animal practice (Saunders, Philadelphia 1968).

115 VICKERS, J. H.: Respiratory diseases of primates; in KIRN Current veterinary therapy. III. Small animal practice, pp. 398–400 (1968).

116 YAKOVLEVA, L. A.: On the pathoanatomical manifestations of dysentery in monkeys; in Vopr. infect. pathol. v. experimente na obezjanah, Sukhumi *1958:* 39–54.

Authors' addresses: R. N. T-W-FIENNES, M. A., M.R.C.V.S., Head, Department of Pathology, Nuffield Institute of Comparative Medicine, Zoological Society of London, Regent's Park, *London, N.W. 1* (England); Dr. B. A. LAPIN, Dr. E. K. DZHIKIDZE and Dr. L. A. YAKOVLEVA, Institute of Experimental Pathology and Therapy, USSR Academy of Medical Sciences, *Sukhumi* (USSR)

Pathology of Simian Primates, Part I, pp. 711–755 (Karger, Basel 1972)

Liver Function and Disease in Primates

S. M. KRUCKENBERG, C. E. CORNELIUS and J. E. COOK

College of Veterinary Medicine, Kansas State University, Manhattan, Kan.

Contents

Introduction ... 711
Classification of Liver Function Tests ... 712
Tests Measuring Hepatic Uptake, Conjugation and Excretion of Certain Organic Anions in Health and Disease ... 713
- A. Bilirubin ... 713
- B. BSP Clearance ... 713
- C. Indocyanine Green (ICG) ... 718

Tests Dependent Upon Elevations of Serum Enzyme Activities Due to Hepatic Necrosis or Cholestasis ... 718
- A. Serum Glutamic Pyruvic Transaminase (SGPT) and Serum Glutamic Oxaloacetic Transaminase (SGOT) ... 718
- B. Serum Alkaline Phosphatase (SAP) ... 719
- C. Miscellaneous Enzymes ... 719

Miscellaneous Tests ... 725
- A. Serum Protein ... 725
- B. Prothrombin Time ... 730
- C. Thymol Turbidity ... 730
- D. Cephalin-Cholesterol Flocculation ... 730

Compendium of Hepatic Diseases: Spontaneous and Experimental ... 732
- A. Viral Diseases ... 732
- B. Bacterial Diseases ... 736
- C. Nutritional Deficiency Diseases ... 738
- D. Parasites ... 740
- E. Miscellaneous ... 743

References ... 748

Introduction

The multiplicity of hepatic disorders in free-ranging and captive nonhuman primates has necessitated a more thorough understanding of techniques by which investigators can assess liver function. The diagnosis of liver disorders in nonhuman primates has only recently been facilitated by more thorough studies in biomedical research using various species as experimental animals.

The innumerable functions of the liver and their measurement magnifies the complex nature of the organ. A meaningful impression of the functional status of the organ cannot be provided by a single function test. Unfortunately no single test can be relied upon to differentiate totally the various forms of icterus in primates: prehepatic or hemolytic, hepatocellular and obstructive. Various attempts to relate functional failure with structural changes have been disappointing since more than one anatomical unit of liver can be involved: (1) parenchymal mass; (2) canalicular, cholangiolar and extrahepatic biliary system; (3) vascular, hemodynamic unit; and (4) the reticuloendothelial system [1].

The dissociation of liver function continually challenges the clinical investigator. The loss of certain hepatic functions to a greater degree than others amplifies this complexity.

The use of a battery of liver function tests will aid in: (1) detection of both primary and secondary hepatic disorders; (2) differentiation of the type of icterus or anemia; (3) prognostication of the disorder; and (4) investigation of the metabolic observations occurring in specific investigations.

The intent of this chapter is to present a brief introduction to: (1) the measurement of hepatic function in primates and (2) the hepatic diseases occurring spontaneously and produced experimentally.

Classification of Liver Function Tests

Limited data available on liver function tests in nonhuman primates suggests an arbitrary classification as follows:

1) Tests measuring hepatic uptake, conjugation and excretion of certain organic anions – bilirubin, sulfobromophthalein (BSP) and indocyanine green (ICG).

2) Tests dependent upon the measurement of serum enzymes – serum glutamic pyruvic transaminase (SGPT), serum glutamic oxaloacetic transaminase (SGOT), serum alkaline phosphatase (SAP), leucine aminopeptidase (LAP) and lactic acid dehydrogenase (LDH).

3) Miscellaneous tests such as serum protein electrophoresis, prothrombin time, flocculation tests, etc.

Tests Measuring Hepatic Uptake, Conjugation and Excretion of Certain Organic Anions in Health and Disease

A. Bilirubin

Normal values for serum bilirubin concentrations in various primates are presented in table I.

Studies on the transmission of human viral hepatitis to marmoset monkeys by Deinhardt *et al.* [2, 1962] revealed hyperbilirubinemia occurred in only 20% of the affected animals and was not a sensitive indicator of this experimental hepatopathy. Following the inoculations of squirrel monkeys with *Leptospira icterohaemorrhagiae* organisms, total serum bilirubin levels were elevated 10.4 mg/100 ml by the tenth day with 50% direct-reacting pigment [3]. Veno-occlusive disease in monkeys which received monocrotaline intraperitoneally was characterized by marked elevations of serum bilirubin over 1.0 mg/100 ml (averaged 2.8 mg/100 ml) in 8 of 18 animals. All animals which survived exhibited serum bilirubin levels below 1.0 mg/100 ml [4]. A mild nutritional Laennec's cirrhosis in the monkey, characterized by fatty metamorphosis and fibrosis, was not accompanied by clinical jaundice, and serum bilirubin values did not increase above normal levels [5]. No differences have been observed in the serum bilirubin levels in morphine-addicted rhesus monkeys as are questionably reported to be present in man [6].

In experimental obstructive jaundice total serum bilirubin levels reached a mean of 15.5 mg/100 ml and bilirubin was excreted in the urine [7]. Conjugated bilirubin levels also became elevated in some cases of yellow fever [8].

Limited studies and clinical observations on nonhuman primates suggest some similarities to humans in the interpretation of serum bilirubin levels during episodes of hemolytic, hepatic and obstructive jaundice. However, since serum bilirubin levels reportedly approximate 25% that of man, significant elevations may still not surpass normal human values [9].

B. BSP Clearance

BSP clearance and retention values for various nonhuman primates are shown in table II.

Deinhardt *et al.* [2] reported increased BSP retention in chimpanzees in experiments with human viral hepatitis. Retentions up to 15% were observed at 40 min. Rhesus monkeys have been studied extensively using BSP to mea-

Table I. Normal levels of serum bilirubin in various nonhuman primates

Nonhuman primate	Total bilirubin (mg/100 ml)			Conjugated bilirubin (direct reacting, mg/100 ml)			Condition	Reference
	Mean	SD	Range	Mean	SD	Range		
Chimpanzee	–	–	0.0–0.4	–	–	0.0–0.1	Normal	2
Chimpanzee	–	–	up to 1.5	–	–	up to 0.5	Infectious hepatitis	2
Chimpanzee	0.17	0.05	–	–	–	–	Normal	36
Papio								
African	–	–	–	0.30 ± 0.17	–	–	Normal	37
Domestic	–	–	–	0.23 ± 0.17	–	–	Normal	37
Cross	–	–	–	0.23 ± 0.14	–	–	Normal	37
Nairobi	–	–	–	0.34 ± 0.17	–	–	Normal	37
Male	–	–	–	0.24 ± 0.17	–	–	Normal	37
Female	–	–	–	0.24 ± 0.14	–	–	Normal	37
Papio (male)	A sudden rise in indirect bilirubin and associated fall in prothrombin levels suggested that obstruction of bile ducts due to bile duct hyperplasia in pyridoxine deficient baboons was present							38
Macaca mulatta	An increase in serum bilirubin occurred in monkeys on a diet that produced nutritional cirrhosis							39
M. mulatta	Serum bilirubin levels remained normal during experimental schistosomiasis in monkeys							32
M. mulatta	–	–	–	0.35	–	–	Normal	10
M. mulatta	–	–	–	0.83	–	–	Veno-occlusive disease	10
M. mulatta	0.38[1] + 0.28–		–	0.8[1] ± 0.28	–	–	Normal	28
M. mulatta and *M. irus*	0.10 + 0.33–		0.08–0.18	–	–	–	Normal	34
	No change in serum bilirubin in monkeys inoculated with materials from men with infectious hepatitis and mononucleosis							34

M. mulatta	0.28	–	0.1–0.6	0.2	–	0.1–0.3	Nutritional cirrhosis	5
M. mulatta	15.5	–	10.8–18.3	–	–	–	Experimental obstructive jaundice	7
M. mulatta	–	–	–	–	–	0–0.2	Normal	8
M. mulatta	–	–	–	–	–	0.2–10.0	Experimental yellow fever	8
M. mulatta	0.15	0.15	–	–	–	–	Normal	6
M. mulatta	0.125	0.13	–	–	–	–	Morphine addicted	6
Saimiri sciureus	0.5	–	–	0.2	–	–	Normal	3
S. sciureus	10.4	–	–	5.4	–	–	Experimental infection with *Leptospirae icterohaemorrhagiae*	3
S. sciureus	0.2	0.3	0.0–1.9	–	–	–	Normal	40
Alouatta caraya	0.75	0.06	–	0.17	0.03	–	Hepatic pigmentation	12
Marmoset	0.20	–	–	0.10	–	0.0–0.3	Normal	41
Marmoset	Apparent increase in serum bilirubin in 20% of animals inoculated with human hepatitis tissues							41
Marmoset								
Tamarins	0.209	–	–	0.102	–	0.00–2.11	Normal	29
Cotton tops	0.200	–	–	0.106	–	0.00–2.07	Normal	29
Marmoset								
Tamarins	0.209	–	–	0.102	–	0.00–2.11	Normal	16
Cotton tops	0.200	–	–	0.106	–	0.00–2.07	Normal	16

1 As reported, but could be in error in reporting.

Table II. BSP clearance and retention in various nonhuman primate species

Species	Percent retention 5 mg/kg	Fractional clearance %/min	Condition	Reference
Chimpanzee	2–8%/20 min	–	Normal	2
Chimpanzee	0–3%/40 min	–	Normal	2
Chimpanzee	up to 43%/20 min	–	Experimental hepatitis	2
Chimpanzee	up to 15%/40 min	–	Experimental hepatitis	2
Chimpanzee	4.7±2.6%/20 min	–	Normal	36
Macaca mulatta	0.9–3.1%/45 min	–	Normal	5
M. mulatta	3.7–8.9%/45 min	–	Cirrhosis (6 months)	5
M. mulatta	21–46%/45 min	–	Cirrhosis (16 months)	5
M. mulatta	0.7±1.8%/30 min	–	Normal	42
M. mulatta	2.95±2.95%/45 min	–	Normal	6
M. mulatta	2.20±1.41%/45 min	–	Morphine addicted	6
M. mulatta	–	17.6±4.2%/min	Normal	28
M. mulatta	5%/30 min	–	Normal	10
M. mulatta	35–65%/30 min	–	Veno-occlusive disease	10
M. mulatta	1.58%/45 min	–	Experimental hepatitis	34
M. mulatta	5.7%(2–10)/30 min	–	Normal	
M. mulatta	4.7%(0–8)/45 min	–	Normal	43
M. mulatta	7.2%(3–16)/30 min	–	Normothermic anesthetized	43
M. mulatta	2.0%(2)/45 min	–	Normothermic anesthetized	43
M. mulatta	17.47%(11.3–29)/30 min	–	Hypothermic anesthetized	43
M. mulatta	8.71%(5–15)/45 min	–	Hypothermic anesthetized	43
M. mulatta	0–10%/30 min	–	Normal	8
M. mulatta	up to 80%/30 min	–	Experimental yellow fever	8
M. mulatta	0.71±0.13%/30 min	–	Normal	44
M. mulatta	1.6%(0–3.6)/30 min	–	Normal	14

Table II (continued)

Species	Percent retention 5 mg/kg	Fractional clearance %/min	Condition	Reference
M. mulatta	L-asparginase, derived from *E. coli*, will cause hepatic necrosis and BSP retention but it was also found that caloric restriction will also cause a prompt retention of BSP			14
M. mulatta	0.25%(0–0.9)/30 min	–	Normal	32
M. mulatta	4.77(0.14–15.30)/ 30 min	–	Experimental schistosomiasis	32
Cercopithecus aethiops				
Male	5.7±3.3% (1.5–19.7)/30 min	–	Normal	46
Female	5.5±6.4% (1.3–33.0)/30 min	–	Normal	46
Alouatta caraya	–	27 (17–46)	Normal	12
Cebus sp.	–	43±7.47%/ 5 min	Normal	47
Cebus sp.	–	81.3±8.21%/ 5 min	Normal	47

sure hepatic excretory function; conditions have included nutritional cirrhosis [5] and veno-occlusive disease [10] with BSP retentions up to 45 and 65%, respectively. In wholebody hypothermic studies on anesthetized rhesus monkeys, a significant elevation in blood ammonia levels occurred after blood administration via gastric tube; however, BSP retention at 45 min after injection was increased by 70% in 2 animals 4 h after the beginning of the experiment [11]. In experimental fever, retentions up to 80% at 30 min have been reported [8].

BSP clearance studies on free-ranging howler monkeys with hepatic pigmentation did not suggest that an excretory defect for such organic anions exists as in Dubin-Johnson syndrome in man and mutant Corriedale sheep [12]. In 2 howler monkeys with pigmented livers, the relative hepatic storage (S) and transport maxima (Tm) for BSP were calculated by the infusion method of Wheeler *et al.* [13]. This method allows for the independent calculation of S and Tm without any concern about hepatic blood flow or hepatic vein BSP concentrations. Tm's for 2 free-ranging howler monkeys were 0.89 and 2.3 mg/min/10 kg body weight; S's were 19.0 and 11.5 mg

BSP stored/mg/100 ml of plasma/10 kg, respectively. Although the Tm and S values varied considerably in the 2 howler monkeys, they were both in the same range reported for normal man and dog.

C. Indocyanine Green (ICG)

The plasma clearance rate of ICG was recently determined by VOGIN *et al.* [14, 1966] in rhesus monkeys. Half-life values for ICG plasma disappearance in females averaged 1.9 ± 0.11 min (fractional clearance, %/min: 36.5 ± 2.6). In male monkeys, half-life values were 1.3 ± 0.13 min (52.9 ± 4.8%/min). The slopes of the clearance curves were observed to be exponentially linear throughout a sampling period of 12 min. Pentobarbital anesthesia delayed clearance, possibly from a lowering of the body temperature. The biliary recovery of ICG was complete by 5 h after injection with maximal excretion within 2 h. The arteriohepatic-venous extraction ratio of dye ranged from 21 to 41%.

Tests Dependent Upon Elevations of Serum Enzyme Activities Due to Hepatic Necrosis or Cholestasis

A. Serum Glutamic Pyruvic Transaminase (SGPT) and Serum Glutamic Oxaloacetic Transaminase (SGOT)

Normal values for SGPT and SGOT activities in various primates are presented in table III. SGPT is apparently specific for hepatic necrosis since the distribution of this enzyme is confined primarily to liver tissue. Other enzymes such as GOT, isocitric dehydrogenase (ICD) and lactic dehydrogenase (LDH) are present in high concentrations in most tissues of mammals [15] and their activities can be elevated by major necrosis present in any tissues in the body. GOT, ICD and LDH activities in the serum can be used to assess the degree of hepatic necrosis if the liver is known to be the involved organ and other tissues can be excluded.

Elevated transaminase levels have been utilized in a multitude of experimental studies in various primates. These include: human viral hepatitis research in squirrel monkeys [3]; a case of hepatic hemangioma in a rhesus monkey [17]; veno-occlusive hepatic disease in rhesus monkeys [4]; malarial infections with associated hepatic necrosis in baboons [18]; elevations in

SGPT and SGOT due to restraint as compared to anesthesia in chimpanzees [19]; carbon tetrachloride intoxication in rhesus monkeys [20]; and elevations in transaminases from pyridoxine administration to monkeys [21]. Of special interest was a study by MARSH *et al.* [22, 1955] which showed that 4 mg of pyridoxine per day was necessary to maintain maximum transaminase levels in the blood of rhesus monkeys.

B. Serum Alkaline Phosphatase (SAP)

Values for SAP activities in various primates can be found in table IV.

Since elevation in the SAP level can occur from osteoclastic activity of the bone in such conditions as excessively rapid growth, rickets, osteomalacia, osteogenic sarcoma and secondary hyperparathyroidism, the interpretation of SAP activity in relation to the liver presupposes that these other conditions do not exist. KROOK and BARRET [23] reported a 16-fold increase in SAP of cebus monkeys with simian bone disease.

Elevations in the SAP activity associated with liver disease are usually due either to intrahepatic cholestasis occurring in necrosis or fibrosis or to obstruction of the extrahepatic bile duct.

C. Miscellaneous Enzymes

Recently, a multitude of enzymes have been measured to assess hepatic necrosis in primates. To date, the SGPT test remains the enzyme of choice due to its specificity for liver, ease of measurement and past experience in its use.

Values for serum isocitric dehydrogenase (SICD) are presented in table V. Elevations in the serum activity of isocitric dehydrogenase have been reported in experimental hepatitis in patas monkeys, chimpanzees [24] and in marmosets [16].

Values for lactic acid dehydrogenase are reported in table VI [41].

In one study [16] the isoenzymes of LDH were determined by electrophoretic separation in control and human viral hepatitis infected marmosets during the acute stage of the disease. At the time of acute liver injury, the most rapidly moving isoenzyme, which is predominantly in the liver, was increased both in its percentage of the total serum LDH activity as well as in absolute amounts.

Table III. Serum glutamic oxaloacetic transaminase and glutamic pyruvic transaminase levels for various nonhuman primate species

Species	SGOT	Units	SGPT	Units	Condition	Reference
Chimpanzee	20.3±4.9	Sigma-Frankel	16.9±6.2	Sigma-Frankel	Normal	36
Chimpanzee	570	Sigma-Frankel	480	Sigma-Frankel	Hepatitis	24
Chimpanzee	30.7±18.2	Sigma-Frankel	19.6±16.8	Sigma-Frankel	Experimental malaria	18
Chimpanzee	21.5±8.7	Sigma-Frankel	12.9±7	Sigma-Frankel	Normal	18
Chimpanzee	27.90±14.35	Reitman-Frankel, DADE modification	21.70±14.09	Reitman-Frankel, DADE modification	Normal adult	48
Chimpanzee	31.43±4.02	Reitman-Frankel, DADE modification	31.64±3.06	Reitman-Frankel, DADE modification	Normal juvenile	48
Chimpanzee	10–35	Karmen	–	–	Normal	2
Chimpanzee	up to 246	Karmen	–	–	Experimental hepatitis	2
Chimpanzee	47.5±13.8	Reitman-Frankel	36.9±16.4	Reitman-Frankel	Normal	49
Chimpanzee	80.9±36.0	Reitman-Frankel	152±107.0	Reitman-Frankel	Acute hepatitis	49
Chimpanzee	51.3±21.0	Reitman-Frankel	37.4±15.0	Reitman-Frankel	Normal after hepatitis	49
Chimpanzee	59.8	Reitman-Frankel	40.6	Reitman-Frankel	Physical restraint	19
Chimpanzee	43.69	Reitman-Frankel	30.13	Reitman-Frankel	Phencyclidine restraint	19
Chimpanzee	12	International	10	International	Normal	50
Chimpanzee	23–69	Reitman-Frankel, DADE modification	24–59	Reitman-Frankel, DADE modification	Normal	51
Baboon	51±5.4	Reitman-Frankel, DADE modification	33.25±7.27	Reitman-Frankel, DADE modification	Normal	48
African	34±8.9	Reitman-Frankel, DADE modification	18±12	Reitman-Frankel, DADE modification	Normal	37
Domestic	30±11	Reitman-Frankel, DADE modification	17±7	Reitman-Frankel, DADE modification	Normal	37

Cross	41±8.2	Reitman-Frankel, DADE modification	22±5	Reitman-Frankel, DADE modification	Normal	37
Nairobi	10±4	Reitman-Frankel, DADE modification	29±11	Reitman-Frankel, DADE modification	Normal	37
Male	–	–	17.7±5.9	Reitman-Frankel, DADE modification	Normal	37
Female	-	–	22.8±9.3	Reitman-Frankel, DADE modification	Normal	37
Cercopithecus aethiops						
Male	61±9.8	Sigma-Frankel	38±9.2	Sigma-Frankel	Normal	46
Female	54±9.9	Sigma-Frankel	47±13.3	Sigma-Frankel	Normal	46
Erythrocebus patas	20	Reitman-Frankel	15	Reitman-Frankel	Normal	24
Erythrocebus patas	160	Reitman-Frankel	355	Reitman-Frankel	Experimental hepatitis	24
Macaca mulatta	27±6.5	1	13.1+5	2	Normal	28
M. mulatta	74 (53–126)	3	–	–	Normal	5
M. mulatta	38	Karmen	–	–	Normal	52
M. mulatta	56.7	Karmen	–	–	Vibration stress	52
M. mulatta	35.7	Karmen	–	–	Noise and confinement stress	52
M. mulatta	258	Karmen	–	–	Acceleration stress	52
M. mulatta	97	Karmen	–	–	Heat stress	52
M. mulatta	42 (31–58	Sigma-Frankel	30 (22–45)	Sigma-Frankel	Normal	20

1 Trans Ac Kit, Warner-Chilcott, Morris Plains, NJ.
2 Wroblewski and LaDue.
3 Method not given in article.

Table III (continued)

Species	SGOT	Units	SGPT	Units	Condition	Reference
M. mulatta	49 (38–65)	Sigma-Frankel	130 (39–250)	Sigma-Frankel	CCl_4 poisoning	20
M. mulatta	33.3 (32–38	Sigma-Frankel	–	–	Normal	53
M. mulatta	72	Sigma-Frankel	–	–	Increased muscular activity	53
M. mulatta	465[4]	Sigma-Frankel	–	–	Normal	30
M. mulatta	1619[4]	Sigma-Frankel	–	–	Aflatoxin poisoning	30
M. mulatta	20.2±11.8	Reitman-Frankel	–	–	Normal	42
M. mulatta	128	Reitman-Frankel, DADE modification	76	Reitman-Frankel, DADE modification	Hepatic hemangioendothelioma	17
M. mulatta	No change in SGOT or SGPT levels in animals inoculated with *Plasmodium cynomolgi*					54
M. mulatta	28.75[5]	–	23.25[6]	–	Normal	27
M. mulatta	20.4±9.9	Karmen	37.9±19.1	Karmen	Normal	6
M. mulatta	28.9±10.8	Karmen	23.3±8.00	Karmen	Morphine addicted	6
M. mulatta	Administration of pyridoxine in daily doses of 10- to 15 mg for 4 to 6 weeks will regularly cause a 30 to 100% increase in transaminase activity of whole blood					21
M. mulatta						
Male	39.7±11	Reitman-Frankel, DADE modification	27.5±6.6	Reitman-Frankel, DADE modification	Normal	55
Female	41.0±11.8	Reitman-Frankel, DADE modification	28.2±7.6	Reitman-Frankel, DADE modification	Normal	55

Monkey (genus or species not given)	Direct relationship between pyridoxine intake and transaminase levels					22
Macaca speciosa	238	Reitman-Frankel	–	–	Veno-occlusive disease	56
Saimiri sciureus	70	Reitman-Frankel	–	–	Normal	3
S. sciureus	146	Reitman-Frankel	–	–	Leptospirosis	3
S. sciureus	138.1 ± 61.7	Reitman-Frankel	117.2 ± 74.9	Reitman-Frankel	Normal	40
Marmoset						
Tamarins	98.7 (48–204)	Karmen	–	–	Normal	29
Cotton tops	113.3 (62–106)	Karmen	–	–	Normal	29
Marmoset	105 (48–204)	Karmen	–	–	Normal	16
Marmoset	Identical values as reference 29 above					16
Marmoset	143 ± 32.8	Reitman-Frankel, DADE modification	20.2 ± 8.25	Reitman-Frankel, DADE modification	Normal	48
Alouatta caraya	786 ± 10	Karmen	35 ± 08	Karmen	Hepatic pigmentation	12
Monkey	12 (10–13)	Karmen	11 (4–18)	Karmen	Normal	57
Monkey	12 (2–30)	Reitman-Frankel	12 (8–15)	Reitman-Frankel	Normal	57

4 μg of keto acid formed, expressed as pyruvic acid per ml per h.
5 Trans Ac Kit, Warner-Chilcott, Morris Plains, N.J.
6 Wroblewski and LaDue.

Table IV. Serum alkaline phosphatase levels for various nonhuman primates

Species	Average	Units	Condition	Reference
Macaca mulatta	16	Bessey, Lowry and Brock	Normal	27
M. mulatta	20.2 (18–24)	King-Armstrong	Normal	5
M. mulatta female	2.5±1.3	Sigma	Normal	42
M. mulatta	9.5±5.15	Sigma	Normal	28
M. mulatta	17.46 (8.2–25.1)	Bessey, Lowry and Brock	Normal	4
M. mulatta	29 (20–57)	Klein-Babson-Read	Normal	20
M. mulatta	2.42±0.14	Sigma	Normal	44
M. mulatta	6.7	Micrograms of inorganic phosphate liberated per h	Normal	30
M. mulatta	632±25	International units = micromoles substrate liberated per min per ml of serum, J 25°C	Normal	26
M. mulatta	711±52		Normal	26
M. mulatta	797±46		Protein deficiency	26
M. mulatta	(2.5–20.3)	Bodansky	Normal	58
Papio				
African	15.4±11.1	Sigma	Normal	37
Domestic	15.3±11.0	Sigma	Normal	37
Cross	44.7±7.5	Sigma	Normal	37
Nairobi	21.1±7.4	Sigma	Normal	37
Males	32.0±20.2	Sigma	Normal	37
Females	22.0±14.3	Sigma	Normal	37
Nairobi				
Group I	9.3±8.6	Sigma	Normal	37
Group II	12.8±11.5	Sigma	Normal	37
Cercopithecus aethiops				
Males	30 (10.3–111.0)	Bodansky	Normal	46
Females	12 (4.5–28.5)	Bodansky	Normal	46
Saimiri sciureus	21.8 (6–49)	Bessey, Lowry and Brock	Normal	40
Alouatta caraya	Test conducted but no results given		Pigmented liver	12
Marmoset				
Tamarins	29.78 (11.7–75.9)	Bessey, Lowry and Brock	Normal	29
Cotton tops	11.93 (3.6–39,9)	Bessey, Lowry and Brock	Normal	29

Table IV (continued)

Species	Average	Units	Condition	Reference
Chimpanzee	13.6±4.8	Shinowara	Normal	36
Lagothrix sp.	45.7	Bodansky	Simian bone disease	23
Cebus sp.	80.7	Bodansky	Simian bone disease	23
Cebus sp.	5.0	Bodansky	Normal	23

In contrast to the above work is the report of an increase in the electrophoretically slowest LDH (fraction 5) in hepatic necrosis induced by carbon tetrachloride poisoning. An increase in total LDH levels in the serum is reported in X-irradiation in rhesus monkeys [25].

Leucine aminopeptidase (LAP) levels have been reported in normal rhesus monkeys and monkeys on a protein deficient diet [26]. The range of these monkeys was from 82 to 110 international units (IU) and was not affected by dietary protein deficiency. Two other reports of LAP have given normal levels of 193 [27] and 218±18 units [28] using the Goldbarg and Rutenburg method.

Serum cholinesterase levels have been reported for chimpanzees [2] (0.9–1.35 Michel units or delta pH/h) and for marmosets 0.56 (0.17–1.89) Michel units [16 and 29]. The value of this enzyme in assessing liver function appears to be limited, although it has been thoroughly investigated in nonhuman primates in connection with liver function and pathology.

Other serum enzymes that have been used in human and veterinary medicine in assessing liver function and pathology and that require further investigation in nonhuman primates are serum arginase, ornithine carbamyl transferase (OCT) and sorbital dehydrogenase (SD).

Miscellaneous Tests

A. Serum Proteins

Quantitative and qualitative alterations in serum proteins are frequently observed in liver dysfunctions. Total protein, albumin, globulins, albumin:

Table V. Serum isocitric dehydrogenase (SICD) levels in nonhuman primates

Species	Units	SD	Range	Condition	Reference
Macaca mulatta	12.75[1]		6–21	Normal	27
M. mulatta	326[2]	179	–	Normal	6
M. mulatta	250[2]	103	–	Morphine addicted	6
M. mulatta	Increase in SICD as cirrhosis of dietary origin developed, but no actual values given			Cirrhosis	39
Erythrocebus patas	300	–	–	Normal	24
E. patas	1620	–	–	Experimental hepatitis	24
Chimpanzee	2544	–	–	Hepatitis	24
Marmoset					
Tamarins	624.3[3]	–	274–1,421	Normal	29
Cotton tops	813.2[3]	–	458–1,466	Normal	29
Tamarins	624.3[3]	–	274–1,421	Normal	29
Cotton tops	7,000[3]	–	–	Experimental hepatitis	16

1 AMADOR, E. and WACKER, W. E. C.: Enzymatic methods used for diagnosis. In Glick methods for biochemical analysis, vol. 13, pp. 286–287 (Interscience Publishers, New York 1965).
2 BOWERS, G. N., jr.: Measurement of isocitric dehydrogenase activity in body fluids. Clin. Chem. *5:* 509 (1959).
3 WOLFSON and WILLIAMS-ASHAM: Isocitric and 6-phosphogluconic dehydrogenase in human blood serum. Proc. Soc. exp. Biol., N.Y. *96:* 231–234 (1957).

globulin ratio (A:G ratio) and quantitative fractionation of the serum proteins by electrophoresis are excellent aids in determining and differentiating hepatic disorders. Normal values and values found in various diseases are presented in table VII.

Total serum protein values generally fall into the range of 6.0–8.5 g/100 ml. Plasma protein values are slightly higher since the protein removed in the clot is not measured with total serum protein. Normal marmosets have a total serum protein level of 5.01 g/100 ml [16]. Hypoproteinemia is reported in veno-occlusive disease [4 and 10], cirrhosis of the liver [5] and aflatoxin-induced cirrhosis [30].

Table VI. Lactic dehydrogenase values for various nonhuman primates

Species	Units	Condition	Reference
Macaca irus	Increase in LDH5 (electrophoretically slowest) in experimental carbon tetrachloride poisoning		59
Papio			
African	469 ± 62	Normal	37
Domestic	589 ± 185	Normal	37
Cross	627 ± 166	Normal	37
Nairobi	504 ± 202	Normal	37
Male	372 ± 114	Normal	37
Female	431 ± 184	Normal	37
Nairobi			
Group I	593 ± 130	Normal	37
Group II	984 ± 247	Normal	37
Macaca mulatta	376 ± 175[1]	Normal	25
M. mulatta	1,129 ± 225[1]	X-irradiation (24 h)	25
M. mulatta	457 ± 17[2]	Normal	28
M. mulatta	449[1]	Normal	27
Saimiri sciureus	381 ± 110	Normal	40
Marmoset	146.4[3]	Normal	41
Marmoset	1,115.4[3]	Hepatitis	41
Marmoset	Increase in the fastest moving LDH band in experimental hepatitis		41

1 Sigma units.
2 Wacker method.
3 International Units/l.

Albumin is synthesized only in the liver and, therefore, a decreased synthesis will be reflected in blood levels of albumin. Other factors such as inadequate protein intake, excessive protein breakdown and loss of albumin and proteins in conditions such as nephritis, severe hemorrhage, etc. must be taken into consideration when interpreting results of blood chemistry studies. Hypoalbuminemia is seen in chronic liver disease. The half-life of albumin has been reported to be 15 days, therefore, the drop in albumin levels probably will not be evident if liver disease of an acute nature is present [35].

Albumin: globulin ratios (A: G ratio) are valuable aids in assessing liver function but should not be used without also viewing total protein, total albumin and total globulin levels. Albumins and globulins are independent

Table VII. Normal values for serum proteins, albumin, globulin and A:G ratio

Species	Total proteins	Albumin	Globulin	A:G Ratio	Condition	Reference
Chimpanzee						
Adult	7.40±0.59	4.81	2.59	1.86±0.60	Normal	48
Juvenile	8.06±0.18	4.59	3.47	1.32±0.12	Normal	48
Chimpanzee	6.72±0.74	2.71±0.35	4.01	0.70±0.14	Normal	18
Chimpanzee	6.50±0.84	2.51±0.44	3.99	0.64±0.10	Experimental malaria	18
Chimpanzee						
Preadolescent male	7.2	3.9	3.3	1.18	Normal	36
Preadolescent female	7.0	3.8	3.2	1.18	Normal	36
Postadolescent male	7.1	4.3	2.8	1.53	Normal	36
Postadolescent female	7.4	4.4	3.0	1.46	Normal	36
Papio sp.	6.32±0.63	3.72	2.60	1.41±0.10	Normal	48
Papio						
African	6.65±0.44	3.34±0.61	3.34	1.0±0.35	Normal	37
Domestic	6.68±0.34	3.30±0.56	2 54	1.3±0.35	Normal	37
Cross	6.19±1.66	3.41±0.92	2.84	1.2±0.18	Normal	37
Nairobi	6.23±0.53	2.57±0.49	3.61	0.71±0.23	Normal	37
Male	6.59±0.41	3.50±0.39	3.01	1.16±0.20	Normal	37
Female	6.68±0.32	3.10±0.59	3.40	0.91±0.28	Normal	37
Nairobi						
Group I	7.3±0.54	3.4±0.45	3.8±0.63	0.9±0.24	Normal	37
Group II	7.7±0.45	3.7±0.62	3.9±0.82	0.9±0.27	Normal	37
Macaca mulatta	7.7 (6.9–8.3)	4.6 (4.0–5.1)	3.2	1.43	Normal	27
M. mulatta	6.76 (6.41–7.78)	4.2 (3.5–5.2)	2.57 (1.73–3.38)	1.78 (1.2–3.0)	Normal	5
M. mulatta	6.66 (3.66–8.66)	2.88 (2.0–3.8)	3.78 (1.26–6.66)	0.74 (0.3–1.9)	Cirrhosis, 6 months	5
M. mulatta	4.77 (4.24–5.72)	2.82 (2.2–3.4)	4.49 (4.00–5.50)	0.64 (0.4–0.8)	Cirrhosis, 16 months	5
M. mulatta	7.8±0.78	4.9±0.46	2.9±0.64	1.92±0.79	Normal	28
M. mulatta	7.4	4.59	2.81	1.63	Normal	10
M. mulatta	4.8	0.96	3.84	0.25	Veno-occlusive disease	10
M. mulatta	No significant change from control values (values not given) CCl_4, NO_2 and ozone toxicosis					20

Table VII (continued)

Species	Total proteins	Albumin	Globulin	A:G Ratio	Condition	Reference
M. mulatta	8.43	5.64	2.79	2.07	Normal	30
M. mulatta	5.90	2.80	3.10	0.93	Aflatoxin	30
M. mulatta	8.25 (8.1–8.5)	4.38 (3.8–5.0)	3.87 (3.2–4.6)	1.13 (0.85–1.56)	Normal	32
M. mulatta	7.25 (6.85–7.65)	3.25 (2.50–3.85)	4.00 (3.15–5.15)	0.81 (0.49–1.22)	Experimental schistosomiasis	32
M. mulatta	–	–	–	2.20	Normal	31
M. mulatta	No values for total protein given			0.45	Pregnancy	31
M. mulatta	Proteins remained normal (no actual values given)				Normal	54
M. mulatta	Proteins remained normal (no actual values given)				Experimental malaria	54
M. mulatta	7.4±0.2	4.9±0.1	2.5±0.2	1.99±0.15	Normal (24% protein diet)	26
M. mulatta	6.7±0.2	4.0±0.2	2.7±0.1	1.49±0.10	Normal (17% protein diet)	26
M. mulatta	6.0±0.1	3.3	2.7±0.1	1.22±0.27	Protein deficiency (3.5% protein diet)	26
M. speciosa	7.2	4.54	1.94	2.34	Normal	4
M. speciosa	4.8	1.34	3.45	0.39	Veno-occlusive disease	4
Marmoset	7.20±0.70	3.71±1.01	3.49	1.11±0.63	Normal	48
Marmoset	–	(1.9–5.3)	3.7(2.3–6.1)	–	Normal	29
Marmoset White lipped Tamarins and Cotton tops	5.01±0.92	(1.9–5.3)	3.7(2.3–6.1)	–	Normal	16

variables, so A:G ratio will not reflect whether or not the concentrations of albumin, globulin or both are abnormal. Values of A:G ratios from 0.70 to 2.34 are reported as normal (table VII). Although total serum protein, total albumin, or total globulin values were not given, the A:G ratio changed from 2.20 to 0.45 during pregnancy in rhesus monkeys and then returned to normal after parturition [31]. Altered ratios are also reported for experimental malaria in chimpanzees [18], cirrhosis of the liver [5], veno-occlusive disease [10], aflatoxin-induced cirrhosis [29], and experimental schistosomiasis [32].

B. Prothrombin Time

Prothrombin is known to be synthesized in the liver and is vitamin K dependent. Prolonged prothrombin time may be a consequence of deficient vitamin K absorption (because of absence of bile salts in the intestines, resulting in decreased plasma concentration of prothrombin) or a result of impaired synthesis by the hepatic cells. Vitamin K administered by a parenteral route will result in a normal prothrombin time within 24 h, unless there is impaired hepatic synthesis. Intestinal disorders, such as diarrhea, may alter absorption and synthesis of vitamin K in the intestinal tract and cause a prolonged prothrombin time [33].

Normal values and values in vitamin K deficiency, cirrhosis and veno-occlusive disease are given in table VIII. At the present time this test is of little value in assessing liver function in nonhuman primates, because obstruction of the bile ducts is extremely rare and experimental evidence demonstrates that other tests are much more sensitive for determining liver function.

C. Thymol Turbidity

The thymol turbidity test is based on the fact that some degree of turbidity occurs when serum is mixed with a saturated solution of thymol. In cases of hepatitis, this turbidity increases and can be measured with a spectrophotometer. Values (normal and abnormal) are listed in table IX. Elevated values are seen in experimental hepatitis in chimpanzees [2] and in rhesus monkeys [34]. In human medicine this test is normal in biliary obstruction and positive in hepatocellular disease [35]. More work needs to be done with this test to establish its true value in liver diseases of nonhuman primates.

D. Cephalin-Cholesterol Flocculation

This test is used in human medicine to differentiate between obstructive and hepatogenous jaundice [35]. In general the test will be positive in hepato-

Table VIII. Prothrombin times in nonhuman primates

Species	Seconds	SD[4]	Range	Condition	Reference
Macaca mulatta					
Plasma	13.18	2.54	(10.5–18.5)	Normal[1]	60
Serum	54.12	23.57	(19.1–180)	Normal[2]	60
M. mulatta	12.0[3]	0.3	–	Normal	61
M. mulatta	–	23.0	(20.0–24.8)	Vitamin K deficiency	62
M. mulatta	–	16.2	(12.6–19.5)	Normal	62
M. mulatta	12[3]	–	–	Normal	10
M. mulatta	21	–	(15–23)	Veno-occlusive disease	10
M. mulatta	13.2	–	(12.7–14)	Cirrhosis	5
M. nemestrina	12.6	0.3	–	Normal	61
M. fuscata	11.6	0.2	–	Normal	61
M. radiata	12.6	0.2	–	Normal	61
L. fulvus	12.6	0.3	–	Normal	61
Cebus niger	14.0	0.6	–	Normal	61
Papio sp.	Fall in prothrombin in blood of baboons on a pyridoxine deficient diet				38
M. speciosa	10.9	0.2	–	Normal	61
M. speciosa	12	–	–	Normal	56
M. speciosa	Up to 25	–	–	Veno-occlusive disease	56

1 Sharpio, S. and Weiner, M.: Coagulation, Thrombosis and Dicumoral (Brooklyn Medical Press, New York 1949).
2 Weiner, N. and Wald, N.: Proc. Soc. exp. Biol., N.Y. *80:* 8 (1952).
3 Quick test.
4 Standard deviation.

cellular destruction and normal with obstructive jaundice. A number of false positives and false negatives have been observed in man.

Test results have been reported in normal chimpanzees (average 2.2+ on a 0–4+ scale) [36], normal and experimental human hepatitis in chimpanzees (0–4+ controls and 0–4+ in attempted hepatitis) [2] and in normal rhesus monkeys following the unsuccessful inoculation of human hepatitis material [34]. The test appears at present to be of little value in the diagnosis of hepatopathies in nonhuman primates.

Table IX. Thymol turbidity levels in various nonhuman primates

Species	Levels	Units	Condition	Reference
Chimpanzee	0.2–3.0	Shank-Hoagland	Normal	2
Chimpanzee	Up to 8.1	Shank-Hoagland	Experimental hepatitis	2
Macaca mulatta	0.93±40 (.25–2.00)	Not given	Normal	34
M. mulatta	2.0 (2.0–3.25)	Not given	Experimental hepatitis	34
M. mulatta	0.51±0.30	Shank-Hoagland	Normal	6
M. mulatta	0.54±0.20	Shank-Hoagland	Morphine addicted	6
M. mulatta	0.27±0.19	Shank-Hoagland	Normal	28
Alouatta caraya	2.1±0.51	McLagan	Hepatic pigmentation	12
Marmoset				
Tamarins	0.41 (0.1–1.6)	McLagan	Normal	29
Cotton tops	0.24 (0.1–0.9)	McLagan	Normal	29
Tamarins	0.41 (0.1–1.6)	McLagan	Normal	16
Cotton tops	0.24 (0.1–0.9)	McLagan	Normal	16

Compendium of Hepatic Diseases: Spontaneous and Experimental

A. Viral Diseases

Yellow fever (natural outbreak) [63]

Alouatta seniculus insulanus (red howler). The livers were pale, smooth and of normal size with a fine mottling of the surface due to necrosis. All livers had a distinct yellow tinge. Yellow fever virus was recovered from each of the 7 monkeys described. Councilman bodies were numerous. Necrosis of the liver parenchymal cells was most severe in the midzonal area. Intranuclear inclusions were seen in 2 of the livers.

Yellow fever (natural outbreak) [64]

Alouatta seniculus insulanus (red howler). The Councilman bodies in hepatic cells, which stain red with hematoxylin and eosin, were studied with the aid of the electron microscope. The bodies were well circumscribed, but had no membranes surrounding them or their internal vacuoles. They probably represented condensed cytoplasmic material. No yellow fever virus particles or lipid droplets were found in the Councilman bodies.

Experimental yellow fever [65]

Macaca mulatta (rhesus). Electron microscopy studies revealed increased density of the hepatic cell nuclear sap resulting from the passage of granules of ribonucleoprotein to the cytoplasm. In the early stages of the disease most of the virus was present in the perinuclear region, but later the cytoplasm became packed with viral particles.

The Torres inclusion body was a degenerative change within the nucleus and consisted of enlargement and distortion of the nucleolus and aggregation of amorphous material.

In the cytoplasm the main degenerative changes were swelling and distortion of the mitochondria, partial or complete loss of the membranous elements of the endoplasmic reticulum and accumulation of lipid droplets. Later the entire cell became distorted and the bile canaliculi were lost through rupture of the liver plates.

Experimental yellow fever [66]

Cercopithecus aethiops tantalus (green monkey) and *Erythrocebus patas patas* (patas monkey). Changes in ribonucleoprotein in the nucleus started at 2–6 days after inoculation. This material accumulated in the cytoplasm. Focal areas of congestion and fatty changes were seen in the lobular midzones. The cells did not proceed to necrosis as was the case in *Macaca mulatta.*

B-virus (natural infection) (Herpesvirus simiae) [67]

Macaca mulatta (rhesus). In the livers of 7 out of 12 animals studied, lesions ranged from diffuse infiltration of polymorphonuclear and mononuclear cells into the periportal connective tissue to scattered foci around the central veins. The nuclei were pyknotic or karyorrhectic with pale acidophilic cytoplasm components. Intranuclear inclusions were found in the areas of extensive inflammatory reaction.

B-virus (natural infection) [68]

Macaca mulatta (rhesus). Intranuclear inclusion bodies in hepatic cells were reported.

B-virus and concurrent simian giant-cell pneumonia infection [69]

Macaca mulatta (rhesus). Cloudy swelling of the hepatic parenchymal cells was described. Purplish cytoplasmic bodies were observed in the bile duct epithelium. Some of these inclusions measured up to 6 μm in diameter and may have actually represented degenerating nuclei.

Inclusion body hepatitis [70]

Marmoset. Acidophilic intranuclear inclusion bodies were present. Multiple circumscribed degenerative foci were seen. The hepatic cells were swollen; the cytoplasm was extremely pale, granular and occasionally vacuolated. This most likely was an outbreak of *Herpes T.*

Spontaneous Herpes T and Herpes M [71]

Aotus trivirigatus (owl monkey). Viral isolates were made and the lesions in the livers of 6 animals were described. Numerous necrotic foci up to 5 mm in diameter were seen. There were 2 types of intranuclear inclusion bodies present. In one type there was margination of the nuclear chromatin with a halo around the acidophilic to basophilic inclusion. Most of these inclusions did not stain by the Feulgen technique. The second type of inclusion completely filled the nucleus. The halo was absent, but there was margination of the chromatin. These latter inclusions were amphoteric and most of them stained by the Feulgen technique. Intranuclear inclusions of both types were also found in the bile duct epithelium but not in the gallbladder.

Herpes T (natural infection) [72]

Aotus trivirigatus (owl monkey). The liver was congested with necrotic foci randomly distributed through the parenchyma. Lesions ranged from coagulative necrosis to complete dissolution of the cellular elements with neutrophilic infiltration. Nuclear changes included karyorrhexis and margination of the chromatin with pale purple intranuclear inclusions which tended to fill the entire nucleus of the hepatic cell. Multinucleated hepatic cells with intranuclear inclusions were also seen.

Experimental Herpes simplex infection [73]

Aotus trivirigatus (owl monkey). Areas of focal necrosis were prominent in the peripheral and midzonal areas of the liver. Typical COWDRY type A and B inclusions were seen in the nuclei in these areas.

Herpesvirus hominis (Herpes simplex) infection (natural infection) [74]

Aotus trivirigatus (owl monkey). Tissues from 24 monkeys were observed. Grossly, focal areas of necrosis appeared as gray-white, and sometimes reddish, foci. Histopathologic lesions were similar to those described for *Herpesvirus T* infection in owl monkeys.

Experimental hepatitis (human type) and mononucleosis [34]

Macaca mulatta (rhesus) and *M. irus* (cynomolgus). Experimental inoculation of 16 rhesus and 4 cynomolgus monkeys with blood from human cases of hepatitis and mononucleosis did not result in clinically apparent disease. Total bilirubin, thymol turbidity, cephalin cholesterol flocculation and alkaline phosphatase remained essentially normal after inoculation.

Experimental hepatitis (human type) [*120*]

Erythrocebus patas patas (patas). Parenchymal changes consisted of occasional small foci of liver cells showing a slight increase in nuclear and nucleolar size with a concurrent increase in nucleolar and cytoplasmic ribonucleoprotein. Necrosis, inflammatory cells and swelling of connective tissue fibroblasts were seen. Baboons, drills and mangabeys did not develop hepatitis when injected with the infective material. No clinical evidence of jaundice was seen.

Experimental hepatitis (human type) [*24*]

Erythrocebus patas patas (patas) and chimpanzees. Histopathologic changes in the liver and changes in SGOT, SGPT and SICD suggested that 1 patas monkey developed hepatitis following inoculation with materials from a case of human hepatitis. Ballooning of the liver cells, monocytic infiltrates within the sinusoids as well as in the stroma of the portal canals, formation of acidophilic bodies, apparent increase in the number of Kupffer cells, and distortion of the cell cords associated with focal collapse and condensation of reticulum fibers were seen. There was no lipofuscin pigment in the Kupffer cells as has been reported in pathologic studies in hepatitis in man. SGOT increased from a mean normal value of 20 units to 160 units, SGPT from 15 units to 355 units, and SICD from 300 units to 1,620 units. The serum appeared icteric, although bilirubin levels were not determined. The other 8 patas monkeys and 4 chimpanzees apparently did not develop hepatitis, since histopathologic changes or SGOT, SGPT and SICD changes did not occur.

Spontaneous viral hepatitis [*24*]

Chimpanzee. Histopathologic findings were those of a generalized glandular transformation of liver cord cells showing an occasional central vein lumen. Marked destruction of liver cell cords was seen, accompanied by monocytic infiltrates within the sinuses and infiltrates of monocytes and polymorphonuclear leukocytes within the stroma of the portal canals. Similar lesions have been reported in infectious hepatitis of people in Ghana.

Experimental hepatitis (human type) [*16*]

Saguinus fuscicollis, S. nigricollis, S. oedipus (marmosets). Elevations of SGOT, SICD and serum bilirubin occurred following inoculation of marmosets with serum and blood samples from human patients with hepatitis. Histopathologic findings were similar, but differed in certain respects from hepatic changes observed in human hepatitis.

Experimental hepatitis (human type) [*41*]

Saguinus fuscicollis, S. nigricollis, S. oedipus (marmosets). SGOT, SICD, and serum bilirubin became elevated. The fastest moving LDH isoenzyme increased in percentage of total LDH activity. Focal necrosis and portal inflammation with bile duct proliferation were seen.

Chimpanzee and/or human hepatitis (adenovirus infection) [*75*]

Chimpanzee. The isolation of 7 adenoviruses from chimpanzees which had been specifically associated with spontaneous or experimental hepatitis was discussed. Descriptions of pathologic changes were not given.

Infectious hepatitis (nonhuman primate type) [*49*]

Chimpanzee. In the acute phase, SGOT and SGPT levels increased. SGOT and SGPT became normal in subsiding hepatitis. Liver biopsies from 9 animals showed pathological changes indicating viral hepatitis. A correlation between the lesions and transaminase levels was shown.

Experimental Kyasanur Forest disease [*119*]

Macaca radiata (bonneted monkey). Eight *Macaca radiata* monkeys were exposed intravenously to KFD virus. Four of the animals were non-immune and 4 had been vaccinated with KFD vaccine. Two of the non-immune animals died and were necropsied. The livers were fatty and occasional areas of focal necrosis were observed. Some lymphocytic infiltrates were evident in the portal areas. The Kupffer cells contained numerous red cells and apparent nuclear fragments of leukocytes.

B. Bacterial Diseases

Acute dysentery [*76*]

Macaca mulatta (rhesus). Adiposis of the liver occurred in some cases. Perivascular infiltration with round cells was seen.

Experimental leptospirosis [*3*]

Saimiri sciureus (squirrel). Of 5 species of monkeys inoculated with *Leptospira interrogans,* serotype *icterohemorrhagiae* 4 species were susceptible;

however, only the squirrel monkey developed severe clinical signs of the disease. Areas of necrosis and bacteria were seen on histopathologic examination.

Leptospirosis, L. interrogans, serotype icterohemorrhagiae [122]

In 1968 an outbreak of leptospirosis occurred in a colony of barbary apes. The livers were enlarged, yellow in color and friable. On microscopic examination there was dissociation of the hepatocytes. Leptospira were observed in silver-stained liver tissue sections.

Experimental Salmonella typhosa infection [77]

Chimpanzee. Gross and microscopic lesions of the liver and other organs of 37 animals were evaluated. The lesions resembled those observed in mild typhoid fever in man.

Experimental Tyzzer's disease [78 and 79]

Monkeys (no genus or species given). TAKAGAKI could not demonstrate the organism or evidence of pathologic change on microscopic examination of the livers of monkeys following experimental inoculation. But Nivens reported Tyzzer's disease in the liver of 1 monkey. This organism may be stained with Giemsa or fuchsin.

Experimental anthrax [80]

Macaca mulatta (rhesus). Changes in the livers of 10 immature animals were described and varied from congestion and edema to degeneration and necrosis. There were widespread focal areas of degeneration and destruction of hepatic cells, frequently associated with mild inflammatory reactions and invariably with septicemia.

Shigellosis [81]

Macaca irus (cynomolgus). This report was concerned with 31 natural cases and 16 experimental ones. Most of the cases were due to *Shigella flexneri* [2a], followed by *S. sonnei*, *S. flexneri* [1a] and *S. dysenteriae* [2]. Vacuolation of the liver cells occurred in the central areas with fatty droplets in the peripheral zones. A slight increase in numbers of polymorphonuclear leukocytes and monocytes were seen in the sinusoids. Kupffer cells were prominent in some cases.

Tuberculosis [82]

Forty percent of 274 cases of tuberculosis had liver involvement.

Vibrio cholerae infection [83]

Erythrocebus patas patas (patas). Sixteen patas monkeys were studied. The wall of the gallbladder showed edema with foci of lymphocytes, histiocytes, plasma cells and polymorphonuclear leukocytes in the muscularis. Occasional necrosis was seen.

C. Nutritional Deficiency Diseases

Nutritional cirrhosis [39]

Macaca mulatta (rhesus). Eight animals were studied. Elevated levels of SGOT, SGPT, SICD, serum cholesterol and, eventually, serum bilirubin occurred. Fatty metamorphosis, pseudoxanthoma cells, coarsely granular acidophilic hepatocytes, and Mallory bodies were seen. Pseudolobulation without efferent veins and which replaced normal lobules was defined as cirrhosis.

Experimental nutritional cirrhosis [39]

Macaca mulatta (rhesus). Six males and 6 females were fed a low protein diet rich in fat and deficient in choline. Males were most severely affected. Increased levels of SGOT, SGPT, SICD, serum cholesterol and serum bilirubin were seen. Progressive fatty metamorphosis was followed by a replacement of signet ring fat cells with granular acidophilic hepatocytes, progressive formation of Mallory bodies, and appearance of reactive cellular infiltrates about necrobiotic liver cells. Pseudolobules developed at about 8 months on the diet. The liver cells around the efferent veins exhibited severe degenerative changes leading to cytolysis and fibrosis.

Riboflavin and pyridoxine deficiencies [38]

Papio sp. (baboon). Fatty changes were seen in the 4 pyridoxine-deficient animals. There was a remarkable variation in the sizes of the liver cells and their nuclei, as well as in the number of nuclei per cell. Multinucleated liver cells with up to fifteen nuclei were observed. Mitotic figures seemed to be unusually frequent in both liver and bile duct epithelium. A sudden rise in indirect bilirubin and an associated fall in prothrombin level were seen in the pyridoxine deficient group. No fatty changes or other significant pathology was seen in the 4 riboflavin-deficient animals.

Vitamin B_6 deficiency (pyridoxine) [21]

Macaca mulatta (rhesus). Forty animals were placed on a pyridoxine-deficient diet. SGOT and SGPT levels increased or decreased in direct relationship with dietary levels of pyridoxine. Cirrhosis was sometimes seen with fatty metamorphosis.

Pyridoxine deficiency [22]

Monkey (no genus or species given). Decreased dietary pyridoxine levels decreased transaminase levels in the serum.

Vitamin K deficiency [62]

Macaca mulatta (rhesus). Prothrombin time was prolonged. No pathologic changes were observed in the liver. Apparently some vitamin K was synthesized by the intestinal microflora and absorbed.

Choline deficiency [84]

Papio sp. (baboon). Two males were fed a choline-deficient diet for 2 months. Centrolobular fatty changes were reversible when the animals were placed on a normal diet. Liver function tests remained normal throughout the experiment.

Experimental hyperhistidinemia [85]

Macaca mulatta (rhesus). Four infant rhesus monkeys developed a persistent lipemia after being placed on the diet. Hepatomegaly with increased glycogen deposition was seen. Only slight fatty infiltration occurred and fibrosis was absent. Examination by electron microscopy failed to reveal a specific structural analysis. There was no correlation between the hyperlipemia and the amount of fat in the liver.

Kwashiorkor-like synodrome [86]

Cercopithecus aethiops (African green). Eight growing animals were fed a diet composed entirely of unenriched yellow cornmeal. Enlargement of the liver occurred as a result of fatty infiltration which began in the periportal areas. Cells were filled with vacuoles which stained with oil red 0 in frozen sections and proved to be lipid.

Protein dietary levels and aflatoxin induced liver injury [87]

Macaca mulatta (rhesus). Thirteen animals were divided into 2 groups which received 500 μg and 100 μg of aflatoxin respectively. The livers were fatty with a finely granular surface. Generalized fatty changes and proliferation of the ductular cells occurred. High protein levels in the diet decreased the amount of change in the liver when low levels of aflatoxin were fed, but not when higher levels were fed.

D. Parasites

Hepatocystis kochi infection [88]

Cercopithecus aethiops (African green). An incidence survey was made from over 1,000 imported animals. The incidence ranged as high as 40% of those examined from Kenya and Tanganyika down to 2% average from Ethiopia and Somalia. *Hepatocystis kochi* developed in the parenchymal cells, progressing from minute round bodies of chromatin to spheres with peripheral nuclei. The nuclei later became scattered throughout the cytoplasm. Vacuolation occurred, and fluid accumulated in large central vacuoles which expanded to form merocysts. The merozoites collected in enormous numbers in the cyst walls, which eventually ruptured. Scars formed and could be seen on the surface of the liver as fibrotic spots.

Experimental Plasmodium ovale infection [89]

Pan troglodytes (chimpanzee). One splenectomized immature animal and a second intact one were used for the study. The intact chimpanzee displayed a natural resistance to the blood forms; the splenectomized chimpanzee was susceptible. The tissue phase was not affected by natural resistance as was the blood form. The parenchymal cells that contained the parasites were greatly enlarged. The host cell cytoplasm was pushed to the periphery, making a wall between the parasite and the host cell envelope, and was piled up particularly in the vicinity of the host cell nucleus. The host cell nucleus was hypertrophied. The host cell nucleus frequently stained less strongly than the nucleus of an uninfected cell. Multiplication of nucleoli within the host cell nucleus was not seen. The host cell nucleus was not flattened and the parasite was forced to make accomodation for it.

Experimental Plasmodium cynomolgi infection [90]

Macaca mulatta (rhesus). Tissue forms of *P. cynomolgi* were reported in detail at various intervals following inoculation. The earliest forms were detected at 5 days in hepatic cells which were both vacuolated and enlarged.

Experimental Entamoeba histolytica infection [91]

Macaca mulatta (rhesus). Four young animals were experimentally infected. In no case was *E. histolytica* found in the host tissues. However, amoebae were found in the colon contents.

A fatal case of amebiasis [92]

Chimpanzee. A 3-year-old chimpanzee died suddenly. Three months prior to death *E. histolytica* trophozoites were first identified in the stool.

The liver was markedly enlarged. Numerous discrete abscesses were evident upon gross examination. Entamoeba were identified within the walls of the abscesses and in some of the contents.

Schistosoma mansoni infection [93]

Macaca mulatta (rhesus), *M. irus* (cynomolgus), *M. speciosa* (stump-tailed), *Cebus apella* (cebus), *Ateles geoffroyi* (spider), *Papio anubis* (baboon), *Pan satyrus* (chimpanzee), *Callithrix aurita* (marmoset), *Saimiri sciureus* (squirrel), and *Tupaia* sp. (tree shrew). All species became infected. Parasite ova and pseudotubercles were found in the liver. Granulomas were generally of moderate size and healing took place with minimal distortion of the liver. Only the chimpanzee showed moderate portal fibrosis and development of portocaval collateral anastomoses.

Experimental schistosomiasis [32]

Macaca mulatta (rhesus). Four young monkeys were infected. Blood samples and liver biopsies were taken at 6, 17, 26 and 40 weeks following infection. Serum albumin decreased, followed by an increase in globulin. BSP retention increased by 6 weeks but returned to normal by 40 weeks.

Acute hepatitis was characterized by increased eosinophils, parenchymal inflammation and foreign body granulomas at 6 weeks following infection. Chronic changes characterized by hepatic fibrosis and portal hypertension were not obtained.

Fasciola hepatica [94]

Macaca mulatta (rhesus). One hundred rhesus monkeys were examined. Five percent incidence was reported from Benzubik and Furmaga (1959).

Clonorchis sinensis infection [95]

Macaca phillapensis (cynomolgus). *Clonorchis sinensis* infection was found in 3 of 60 monkeys examined. They evoked a slight dilation of the bile ducts.

Athesmia sp. infection [96]

Cebus albifrons (cebus). *Athesmia* sp., probably *A. foxi*, were found in the bile ducts of a cinnamon ringtail monkey. The liver had a marked increase in fibroblastic cells, macrophages, and lymphocytes in many periportal areas. The epithelial lining of the bile ducts containing the parasites was partially or completely destroyed, leaving the trematodes in direct contact with the periportal connective tissue.

Athesmia foxi infection [97]

Saimiri sciureus (squirrel). No gross pathology was seen in the 2 monkeys infected. Histologically, moderate cholangiectasis, partial biliary obstruction and mild biliary stasis involving bile canaliculi were observed. Pericholangitis was not seen. No trematodes were found in the gallbladder.

Athesmia foxi infection [98]

Cebus alfibrons (cebus). The tissues of 16 monkeys were examined. Gross pathologic changes were not discernible. Microscopically flukes could be seen in the bile ducts. Upon microscopic examination the portal triads contained slightly increased numbers of lymphocytes and monocytes. Ducts containing flukes were dilated.

Trematodes: Athesmia foxi, Platynosomum marmoseti, and P. amazonensis infection [99]

Saguinus fuscicollis and *S. nigricollis* (tamarins). Four hundred and fifty-five tamarins were necropsied. *A. foxi* and *P. amazonensis* were observed in the gallbladder and bile ducts of about 8% of the animals examined. *P. marmoseti* were found in 2 animals.

Capillaria hepatica infection [94]

Monkeys (no genus or species given). In this review article general comments pertaining to susceptibility and pathogenesis were included. The adult parasite migrated through the liver and laid numerous eggs. The eggs and adult parasite when present could be seen microscopically.

Capillaria hepatica infection [95]

Macaca mulatta (rhesus). Rhesus monkeys were examined. Segments of capillarids and masses of ova were found in the liver of 1 rhesus monkey.

Echinococcus granulosus infection [100]

Mandrillus leucophaeus (drill baboon). A 9-year-old female drill baboon appeared in good health until 2 weeks prior to death. Only 10% of the functional hepatic parenchyma remained due to 2 massive echinococcus cysts. Many central veins were dilated and early centrolobular degenerative changes were seen. A few inflammatory cells were observed in the portal areas.

Hydatid disease [101]

Papio cynocephalus (baboon). A 3–5-year-old female baboon undergoing a physical examination was found to have numerous small abdominal masses. She succumbed within 18 months. One large hydatid cyst was found within the liver parenchyma.

Hydatid disease [*102*]

Macaca mulatta (rhesus). Small white cysts were found in the liver of 2 monkeys. The type of cysts indicated an *Echinococcus granulosus* infection. Three previous references were also reviewed.

Visceral larva migrans [*103*]

Macaca mulatta (rhesus). Sixty-nine percent of *Toxocara canis* larvae given to a monkey were recovered by peptic digestion of the liver 4 months following oral administration of the parasites. Two animals which were infected several years earlier were biopsied periodically and the livers still included viable larvae.

Pentastomid infection [*104*]

Miopithecus talapoin (talapoin) and *Cercopithecus hamlyni* (HAMLYN's guenon monkey). A young adult male died suddenly. Large numbers of larvae were imbedded in the liver. A second monkey, a young *Cercopithecus hamlyni*, presented similar gross changes. There were no inflammatory changes associated with the parasites.

E. Miscellaneous

Cytoplasmic inclusions [*105*]

Monkey (genus and species not given). Cytoplasmic inclusions of unknown etiology were described. These bodies were predominantly spherical; however, some were rod-shaped and filamentous. The spherical bodies were up to 5 μm in size. The bodies lay in a vacuole, were numerous and could coalesce. These inclusions were stained acidophilic with hematoxylin and eosin. With Laidlaw's method for staining inclusion bodies, the spherical bodies stained intensely crimson. The significance of these bodies was unknown.

Hepatic hemangioendothelioma [*17*]

Macaca mulatta (rhesus). A sexually immature male rhesus which had been superficially irradiated with 800 rad of 55 Mev protons developed severe dypsnea 2 years post irradiation. Roentgenograms revealed an enlarged liver. A diagnosis of hemangioendothelioma was made upon microscopic examination.

Aflatoxin [30]

Macaca mulatta (rhesus). Six animals were used in the investigation. Biopsies were taken at weekly intervals. SGOT levels increased two-fold after 2 weeks of aflatoxin's being fed at the rate of 1 mg daily.

At the end of 4 weeks on the diet, fat content of the liver was elevated, serum albumin had fallen significantly, and the A:G ratio was reduced. Serum bilirubin levels increased more than four-fold, alkaline phosphatase levels doubled and SGOT levels increased four-fold. Fatty changes appeared first in the periportal areas. Biliary fibrosis developed over a three-week period. In general the lesions were similar to those reported for ducklings.

Experimental Veno-occlusive disease [4]

Macaca speciosa (stumptailed). Eighteen experimental animals were each given 0.125 g of monocrotaline per kilogram of body weight intraperitoneally. Biopsies were taken at stated time intervals. Decrease in total serum protein by more than 50% was seen. There was an absolute increase in globulin and an absolute decrease in albumin. Serum bilirubin increased and later returned to normal. SGOT levels became elevated and averaged 238 units in those monkeys that died.

The endothelium of the hepatic veins became fragmented and edematous. Obliteration of the lumens occurred due to collection of blood cells, fibrin, macrophages and proteinaceous fluids. Some occluded vessels were entirely replaced by fibrous connective tissue while others became canalized.

Electron microscopy study of the liver after exposure to a pure oxygen atmosphere [106]

Macaca mulatta (rhesus). Serial biopsies were taken from 2 monkeys over a period of 22 days exposure. Initially an increase in smooth endoplasmic reticulum, at the expense of the rough form, and glycogen were seen in the livers of monkeys exposed to a pure oxygen atmosphere. At the end of 2 weeks in this environment, however, these changes were no longer evident. These studies were conducted with the aid of an electron microscope. No changes were noted with light microscopy.

Methyl alcohol poisoning [107]

Macaca mulatta (rhesus). Forty-one young rhesus were exposed to oral, inhalation and skin absorption of methyl alcohol. Parenchymatous degeneration with focal necrosis was seen.

Aflatoxin B_1 [108]

Macaca mulatta (rhesus), *M. irus* (cynomolgus), *Cercopithecus aethiops* (African green). Sixty-two animals were given the toxin orally or intraperitoneally. New-born and infant monkeys were found most susceptible. No liver tumors were observed after 6 months to 2 years on aflatoxin, but hepatic changes similar to the pre-neoplastic period seen in monkeys fed nitrosodiethylamine (DENA) did occur. Centrolobular necrosis was seen in those animals dying during the first 2 weeks of aflatoxin treatment.

Marked proliferation of bile ducts within the portal areas occurred and the intralobular bile canaliculi were often plugged with inspissated bile. Males were more severely affected than females.

Metastasizing carcinoma of the gallbladder [109]

Macaca mulatta (rhesus). A 10-year-old female rhesus developed a spontaneous metastasizing carcinoma of the gallbladder. The neoplasm also involved the right superior lobe of the liver, lesser omentum, portal vein and common bile duct. Metastases were found in the lungs, peritoneum, and pancreas. The tumor consisted mostly of cuboidal cells that lined irregular cystic spaces. Mitotic figures were numerous.

O'GARA in a review of the literature cited 4 tumors of the liver and gallbladder in nonhuman primates. One was that of a hepatic cell adenoma of liver reported by ZVETAEVA in 1941 in a *Presbytis entellus* monkey. The other 3 tumors cited involved only the gallbladder.

Granulocytic leukemia X-irradiation [110]

Macaca mulatta (rhesus). One rhesus monkey developed spontaneous granulocytic leukemia following exposure to 400 Mev proton irradiation. In the liver there were diffuse infiltrations of neoplastic cells, with heavy collars of neoplastic cells around the hepatic and portal vessels.

Whole body X-irradiation [111]

Macaca mulatta (rhesus). Ten adult female rhesus monkeys were exposed to 325–650R. No significant changes in liver lipids or histopathology were seen.

X-irradiation [112]

Saimiri sciureus (squirrel). Thirty-six monkeys were given 800 R of total body X-irradiation. Well circumscribed hemorrhagic foci were scattered throughout the livers. The foci resembled cavernous dilations of pre-existing sinusoids when parenchymal cells in the area were destroyed. Foci of bacteria were present in the hemorrhagic areas, in sinusoidal spaces and in the lumens of a few small vessels. Inflammatory cells were absent. Parenchymal cells which were not affected by the extravasated blood and bacteria seemed normal. In hepatic tissue specially stained for iron, considerable deposition of Prussian blue-positive material was found in the cytoplasm of hepatic and Kupffer cells.

X-irradiation [25]

Macaca mulatta (rhesus). One-hundred four males and females were given graded single exposures of 2 Mev X-rays. An increase in LDH levels was observed for the first 4 days post-irradiation. SAP and LAP remained normal for the 90-day observation period.

Pigmentary liver disease [113]

Alouatta seniculus (howler). The livers of 290 howler monkeys were studied histologically for this report. The livers examined were macroscopically a darker-than-normal brown to dark blue-green. Some of the discolored livers had nodules with normal hepatic coloration. Seventy-eight percent of the animals had coarse, granular, amorphous brown pigment in the liver cells when observed with light microscopy. The pigment was occasionally seen in the epithelium of the bile ducts. Some multinucleated hepatocytes were seen in the heavily pigmented livers. The pigment gave negative results for bilirubin and iron, but was moderately positive with Sudan IV and strongly positive with Sudan black and with the PAS reaction; it was also argentophilic. The granules were isotropic, exhibited a dull brown fluorescence and appeared to be lipofuscin. The pigment was also observed in Kupffer cells. Unidentified nematodes were found in the small bile ducts in 2 animals.

Cholelithiasis [76]

Marmoset. Hamerton reported cholecystitis in a green marmoset and listed the cause of death as obstruction of common bile duct by calculi.

Cholelithiasis [76]

Marmoset. Two marmosets, 11 macaques, 4 baboons and 1 gelada each had concretions in the gallbladder which ranged in size from small grains to large calculi.

Amyloidosis [114]

Saimiri sciureus (squirrel). Amyloidosis was seen in the liver of a 2-year-old squirrel monkey. Hematoxylin and eosin stain of the liver showed accumulations of acidophilic amorphous material assumed to be amyloid. When tissue sections were stained with Congo red and viewed with polarized light, a greenish birefringence was seen.

Hemosiderosis [*115*]

Gorilla gorilla (gorilla). A 22-year-old male succumbed after a 7 month illness characterized primarily by paresis. A large amount of coarse brown pigment present in the liver cord cells gave a strongly positive Prussian blue reaction for iron and negative Hotchkiss reaction for lipofucsin. This pigment was minimal in Kupffer cells and not observed in the bile duct epithelium, although small deposits were seen in the connective tissue of the portal triads. Foamy cells were seen in the mucosa of the gallbladder. These changes were degenerative in nature and are usually associated with chronic passive congestion.

Experimental coccidioidomycosis [*116*]

Macaca mulatta (rhesus). Forty-six animals were evaluated. Microscopic lesions appeared in the liver as early as 9 days after exposure and consisted primarily of epithelioid and multinucleated giant cells. Neutrophils and cryptococcus spherules were also seen. An excellent detailed description of the pulmonary changes was presented.

Cold and heat stress [*117*]

Saimiri sciureus (squirrel). Enlargement of the liver was seen in cold-exposed monkeys, whereas there was a distinct decrease in growth in the heat-exposed group. In the heat exposed group there was clear cut decrease in ratio of liver weight to total body weight and true diminution in active hepatic tissue. The number of animals observed was not listed.

Physical restraint [*51*]

Chimpanzee. A pronounced rise in SGOT and SGPT occurred during 24 h of physical restraint; levels returned to normal following the experiment. No changes occurred in SAP and total serum protein during restraint.

Phencyclidine hydrochloride treatment and physical restraint [*19*]

Chimpanzee. The SGOT and SGPT levels of 6 animals were higher in physical restraint when compared with chemorestraint (phencyclidine hydrochloride).

Estrogen inactivation [*121*]

Macaca mulatta (rhesus). Normal BSP retention and cephalin-cholesterol flocculation tests were seen in 4 ovarioectomized females following implantation of estrone into the spleen. Estrogens are not inactivated by the liver of the monkey as compared to rodents which completely inactivate estrone implanted into the spleen.

Experimental obstructive jaundice [*7*]

Macaca mulatta (rhesus). The serum bilirubin levels of 5 animals were elevated. There was an increase in urinary bilirubin excretion following ligation of the common bile duct.

Pregnancy [*31*]

Macaca mulatta (rhesus). The average albumin to globulin ratio of 42 monkeys changed from 2.2:1 to 1:2.2 during pregnancy.

Myocarditis [*118*]

Macaca mulatta (rhesus). Twenty-three 1 year old clinically normal rhesus were studied. Eighteen of the animals had microscopic evidence of myocarditis. Virtually all of the animals had infiltrations of the liver portal areas with lymphocytes and plasma cells. Multinucleated (3–15 nuclei) hepatocytes were seen in 7 of the monkeys. Accompanying these latter changes were congestion and edema of the sinusoids with separation of the individual parenchymal cells. The multinucleated cells were randomly distributed and they were not related to any zone of the lobules.

References

1 LICHTMAN, S. S.: Diseases of the liver, gallbladder and bile ducts; vol. 1 (Len and Lebiger, Philadelphia 1953).

2 DEINHARDT, F.; COURTOIS, G.; DHERTE, P.; OSTERRIETH, P.; NINANE, G.; HENLE, G., and HENLE, W.: Studies of liver function tests in chimpanzees after inoculation with human infectious hepatitis virus. Amer. J. Hyg. *75:* 311–321 (1962).

3 MINETTE, H. P. and SHAFFER, M. F.: Experimental leptospirosis in monkeys. Amer. Soc. trop. med. Hyg. *17:* 202–212 (1968).

4 ALLEN, J. R.; CARSTENS, L. A., and KATAGIRI, G. J.: Hepatic veins of monkeys with veno-occlusive disease. Arch. Path. *87:* 279–289 (1969).

5 GAISFORD, W. D. and ZUIDEMA, G. D.: Nutritional Laennec's cirrhosis in the *Macaca mulatta* monkey. J. surg. Res. *5:* 220–235 (1965).

6 Brooks, F. P.; Deneau, G. A.; Potter, H. P., jr.; Reinhold, J. G., and Norris, R. F.: Liver function tests in morphine-addicted and in nonaddicted rhesus monkeys. Gastroenterology *44:* 287–290 (1963).

7 Cameron, J. L.; Pulaski, E. J.; Abei, T., and Iber, F. L.: Metabolism and excretion of bilirubin-C^{14} in experimental obstructive jaundice. Ann. Surg. *163:* 330–338 (1966).

8 Wakeman, A. M. and Morrell, C. A.: Chemistry and metabolism in experimental yellow fever in *Macacus rhesus* monkeys. Arch. intern. Med. *50:* 876–833 (1932).

9 Schmid, R.; Hammaker, L., and Axelrod, J.: The enzymatic formation of bilirubin glucuronide. Arch. Biochem. *70:* 285–288 (1957).

10 Allen, J. R. and Carstens, L. A.: Veno-occlusive disease in rhesus monkeys. Amer. J. vet. Res. *29:* 1681–1694 (1968).

11 Zuidema, G. D.; Gaisford, W. D.; Kowalczyk, R. S., and Wolfman, E. F., jr.: Whole-body hypothermia in ammonia intoxication. Effect on monkeys with portacaval shunts. Arch. Surg., Chicago *87:* 578–582 (1963).

12 Katz, S.; Gilardoni, A.; Genovese, N.; Wikinski, R. W.; Cornelius, C. E., and Malinow, M. R.: Liver function studies in free ranging howler monkeys with hepatic pigmentation. Lab. anim. Care *18:* 626–630 (1968).

13 Wheeler, H. O.; Meltzer, J. K., and Bradley, S. E.: Biliary transport and hepatic storage of sulphobromophthalein sodium in the anesthetized dog, in normal man and patients with hepatic diseases. J. clin. Invest. *39:* 1131–1144 (1960).

14 Vogin, E. E.; Moreno, O. M.; Brodie, D. A., and Mattis, P. A.: Indocyanine green clearance in the rhesus monkey *(Macaca mulatta)*. J. appl. Physiol. *21:* 1880–1882 (1966).

15 Cornelius, C. E.; Bishop, J.; Sartzer, J., and Rhode, E. A.: Serum and tissue transaminase activities in domestic animals. Cornell Vet. *49:* 116–126 (1959).

16 Deinhardt, F. and Deinhardt, J.: The use of platyrrhine monkeys in medical research. Symp. zool. Soc., Lond. *17:* 127–152 (1966).

17 Woodruff, J. M. and Johnson, D. K.: Hepatic hemangioendothelioma in a rhesus monkey. Path. Vet. *5:* 327–332 (1968).

18 Sadun, E. H.; Williams, J. S., and Martin, L. K.: Serum biochemical changes in malarial infections in man, chimpanzees and mice. Milit. Med. *131:* 1094–1106 (1966).

19 Krushak, D. H. and Hartwell, W. V.: Effect of phencyclidine on serum transaminase in chimpanzees. J. amer. vet. med. Ass. *153:* 866–867 (1968).

20 Thomas, A. A.: Low ambient pressure environments and toxicity. Arch. environm. Hlth *11:* 316–322 (1965).

21 Rinehart, J. F. and Greenberg, L. D.: Vitamin B_6 deficiency in the rhesus monkey. Amer. J. clin. Nutr. *4:* 318–328 (1956).

22 Marsh, M. E.; Greenberg, L. D., and Rinehart, J. F.: The relationship between pyridoxine ingestion and transaminase activity. I. Blood hemolysates. J. Nutr. *56:* 115 (1955).

23 Krook, L. and Barrett, R. B.: Simian bone disease; a secondary hyperparathyroidism. Cornell Vet. *52:* 459–492 (1962).

24 Smetana, H. F.: Experimental and spontaneous viral hepatitis in primates. Lab. Invest. *14:* 1366–1374 (1965).

25 DALRYMPLE, G. V.; LINDSAY, I. R., and GHIDONI, J. J.: The effect of 2 mev X-rays on whole-body irradiated primates; the radiations of space. I. USAF School of Aerospace Medicine, Aerospace Medical Division, SAM-TR-65-*9:* 1–25 (1965).

26 ORDY, J. M.; SAMORAJSKI, T.; ZIMMERMAN, R. R., and RADY, P. M.: Effects of postnatal protein deficiency on weight gain, serum proteins, enzymes, cholesterol, and liver ultrastructure in a subhuman primate *(Macaca mulatta)*. Amer. J. Path. *48:* 769–791 (1966).

27 ELTRINGHAM, J. R.; ANDERSON, D. R., and DEHART, R. L.: Effect of Azothioprine and Amethopterin on secondary disease in the rhesus monkey. USAF School of Aerospace Medicine, Aerospace Medical Division, SAM-TR-67-*94:* 1–14 (1967).

28 ANDERSON, D. R.: Normal values for clinical blood chemistry tests of the *Macaca mulatta* monkey. Amer. J. vet. Res. *27:* 1484–1489 (1966).

29 HOLMES, A. W.; PASSOVOY, M., and CAPPS, R. B.: Marmosets as laboratory animals. Lab. anim. Care *17:* 41–47 (1967).

30 TULPULE, P. G., MADHAVAN, T. V., and GOPALAN, C.: Effect of feeding aflatoxin to young monkeys. Lancet *i:* 962–963 (1964).

31 ALLEN, J. R. and SIEGFRIED, L. M.: Hematologic alterations in pregnant rhesus monkeys. Lab. anim. Care *16:* 465–471 (1966).

32 BRUCE, J. I.; WARREN, K. S., and SADUN, E. H.: Observations on the pathophysiology of *Schistosomiasis* in monkeys. Exp. Parasitology *13:* 194–198 (1963).

33 STEE, E. W. VAN: Some observations on the clinical management of the chimpanzee. USAF School of Aerospace Medicine, Aerospace Medical Division, SAM-TDR-64-*45:* 1–44 (1964).

34 EVANS, A. S.; EVANS, B. K., and STURTZ, V.: Standards for hepatic and hematologic tests in monkeys. Observations during experiments with hepatitis and mononucleosis. Proc. Soc. exp. Biol., N.Y. *82:* 437–440 (1953).

35 MILLER, S. E.: A textbook of clinical pathology, pp. 1–947 (Williams and Wilkins, Baltimore 1966).

36 WISECUP, W. G.; HODSON, H. H.; HANLY, W. C., and FELTS, P. E.: Baseline blood levels of the chimpanzee *(Pan troglodytes)*. Liver function tests. Amer. J. vet. Res. *30:* 955–962 (1969).

37 PENA, A. DEL LA and GOLDZIEHER, J. W.: Clinical parameters of the normal baboon. Baboon med. Res. *2:* 379–389 (1967).

38 FOY, H.; GILLMAN, T.; KONDI, A., and PRESTON, J. K.: Hepatic injuries in riboflavin and pyridoxine deficient baboons-possible relations to aflatoxin induced hepatic cirrhosis and carcinoma in Africans. Nature *212:* 150–153 (1966).

39 SMETANA, H. F. and GREER, W. E.: Experimental study of the histogenesis of nutritional cirrhosis in primates. Recent Advanc. Gastroenterology *3:* 431–437 (1967); Biol. Abstr. *49:* (1968).

40 NEW, A. E.: in ROSENBLUM and COOPER The squirrel monkey, pp. 418–419 (Academic Press, New York 1968).

41 DEINHARDT, F.; HOLMES, A. W.; CAPPS, R. B., and POPPER, H.: Studies on the transmission of human viral hepatitis to marmoset monkeys. I. Transmission of disease serial passages, and description of liver lesions. J. exp. Med. *125:* 673–688 (1967).

42 KING, T. O. and GARGUS, J. L.: Normal blood values of the adult female monkey *(Macaca mulatta)*. Lab. anim. Care *17:* 391–396 (1967).

43 Zuidema, G. D.; Gaisford, W. D.; Kowalczyk, R. S.; Wolfman, E. F., jr., and Arbor, A.: Whole-body hypothermia in ammonia intoxication. Effect on monkeys with portacaval shunts. Arch. Surg. *87:* 62–66 (1963).

44 King, T. O.: Normal blood values of the adult female *Macaca mulatta* monkey. Toxico, appl. Pharmacol. *6:* 351 (1964).

45 Rakieten, N.; Gordon, B.; Cooney, D.; Davis, R. D., and Rall, D. P.: Nutritional aspects of the fatty liver induced in the rhesus monkey *(Macaca mulatta)* by *escherichia coli L-asparaginase*. Proc. amer. Ass. Cancer Res. *10:* 71 (1969).

46 Pridgen, W. A.: Values for blood constituents of the african green monkey *(Cercopithecus aethiops)*. Lab. anim. Care *17:* 463–468 (1967).

47 Mann, G. V.; Watson, P. L., and Adams, L.: Primate nutrition. J. Nutr. *47:* 213–224 (1952).

48 Burns, K. F.; Ferguson, F. G., and Hampton, S. H.: Compendium of normal blood values for baboons, chimpanzees, and marmosets. Amer. J. clin. Path. *48:* 484–494 (1967).

49 Hartwell, W. V.; Kimbrough, R. D., and Love, G. J.: Serum transaminase activities related to pathologic changes of liver in chimpanzees. Amer. J. vet. Res. *29:* 1449–1452 (1968).

50 Rahm, U.: Observations during chimpanzee captures in the Congo: Neue Ergebnisse der Primatologie, 195–207 (1967).

51 Caraway, B. L.: Physiologic baseline studies of zoologic specimens, physiologic values of the normal immature chimpanzee under restraint. USAF School of Aerospace Medicine, Aerospace Medical Division, SAM-TR-66-*100:* 1–24 (1966).

52 Cope, F. W. and Polis, B. D.: Increased plasma glutamic-oxaloacetic transaminase activity in monkeys due to nonspecific stress effect. Aviation Med. *30:* 90–96 (1959).

53 Robinson, F. R.; Gisler, D. B., and Dixon, D. F., jr.: Factors influencing normal SGOT levels in the rhesus monkey. Lab. anim. Care *14:* 275–282 (1964).

54 Kisliuk, R. L.; Friedkin, M.; Reid, V.; Crawford, E. J.; Schmidt, L. H.; Rossan, R.; Lewis, D.; Harrison, J., and Sullivan, R.: The toxicity of tetrahydrohomofolic acid in normal and malaria-infected monkeys. J. path. exper. Therapeutics, *159:* 416–421 (1968).

55 Robinson, F. R. and Ziegler, R. F.: Clinical laboratory data derived from 102 *Macaca mulatta*. Lab. anim. Care *18:* 50–68 (1968).

56 Allen, J. R. and Carstens, L. A.: Sequential ultrastructural changes in the hepatic vessels of monkeys with veno-occlusive disease. Amer. J. Path. *13:* 52A (1968).

57 Zimmerman, H. J.; Schwartz, M. A.; Boley, L. E., and West, M.: Comparative serum enzymology. J. Lab. clin. Med. *66:* 961–972 (1965).

58 Asatiani, V. S.; Pichkhaya, T. P., and Ageeva, A. K.: The blood chemistry of the lower monkeys. Proc. Soc. exp. Biol., N.Y. *47:* 203–206 (1959).

59 Wieme, R. J. and Maercke, Y. van: The fifth (electrophoretically slowest) serum lactic dehydrogenase as an index of liver injury. Ann. N.Y. Acad. Sci. *94:* 898–911 (1961).

60 Krise, G. M.: Normal blood picture of the *Macaca mulatta* monkey. J. appl. Physiol. *12:* 482–484 (1958).

61 Seaman, A. J. and Malinow, M. R.: Blood clotting in nonhuman primates. Lab. anim. Care *18:* 80–84 (1968).

62 HILL, R. B.; SCHENDEL, H. E.; RAMA RAO, P. B., and JOHNSON, B. C.: Effect of dietary vitamin K on the nutritional status of the rhesus monkey. J. Nutr. *84:* 259–264 (1964).

63 ANDERSON, C. R. and DOWNS, W. G.: The isolation of yellow fever virus from the livers of naturally infected red howler monkeys. Amer. J. trop. med. Hyg. *4:* 662–664 (1955).

64 BARUCH, E.; CARBONELL, L., and WEIBEL, J.: Fine structure of councilman bodies in the liver of *Aluatta seniculus seniculus L* infected with yellow fever virus. Exp. cell Res. *29:* 50–53 (1963).

65 BEARCROFT, W. G. C.: Electron-microscope studies on the liver cells of yellow fever-infected rhesus monkeys. J. path. Bact. *80:* 421–426 (1960).

66 BEARCROFT, W. G. C.: Studies on the livers of yellow fever-infected african monkeys. J. path. Bact. *83:* 49–58 (1962).

67 KEEBLE, S. A.; CHRISTOFINIS, G. J., and WOOD, W.: Natural virus-B infection in rhesus monkeys. J. path. Bact. *76:* 189–201 (1958).

68 PIERCE, E. C.; PIERCE, J. D., and HULL, R. N.: B virus: its current significance, description and diagnosis of a fatal human infection. Amer. J. Hyg. *68:* 242–250 (1958).

69 SOTO, P. J. and DEAUVILLE, G. A.: Spontaneous simian giant-cell pneumonia with coexistent B virus infection. Amer. J. vet. Res. *25:* 793–804 (1964).

70 SAUER, R. M. and BISHOP, R. W.: Inclusion body hepatitis in marmosets. Lab. anim. Care *13:* 790–792 (1963).

71 HUNT, R. D. and MELENDEZ, L. V.: Spontaneous *Herpes-T* infection in the owl monkey *(Aotus trivirigatus)*. Path. Vet. *3:* 1–26 (1966).

72 EMMONS, R. W.; GRIBBLE, D. H., and LENNETTE, E. H.: Natural fatal infection of an owl monkey *(Aotus trivirigatus)* with *Herpes-T* virus. J. infect. Dis. *118:* 153–159 (1968).

73 KATZIN, D. S.; CONNOR, J. D.; WILSON, L. A., and SEXTON, R. S.: Experimental Herpes simplex infection in the owl monkey. Proc. Soc. exp. Biol., N.Y. *125:* 391–398 (1967).

74 MELENDEZ, L. V.; ESPANA, C.; HUNT, R. D.; DANIEL, M. D., and GARCIA, F. G.: Natural Herpes simplex infection in the owl monkey *(Aotus trivirigatus)*. Lab. anim. Care *19:* 38–45 (1969).

75 HILLIS, W. D.; HOLMES, A. W., and DAVISON, V.: Serologic characterization of adenoviruses isolated from chimpanzees associated with viral hepatitis. Proc. Soc. exp. Biol., N.Y. *129:* 366–369 (1968).

76 LAPIN, B. A. and YAKOVLEVA, L. A.: Comparative pathology in monkeys (Thomas, Springfield 1963).

77 EDSALL, G.; GAINES, S.; LANDY, M.; TIGERTT, W. D.; SPRINZ, H.; TRAPANI, R. J.; MANDEL, A. D., and BENENSON, A. S.: Studies on infection and immunity in experimental typhoid fever. I. Typhoid fever in chimpanzees orally infected with *Salmonella typhosa*. J. exp. Med. *112:* 143–166 (1960); Prelim. rep. Fed. Proc. *15:* 614 (1956).

78 TAKAGAKI, Y.; ITO, M.; NAIKI, M.; FUJIWARA, K.; OKUGI, M.; MAEJIMA, K., and TAJIMA, Y.: Experimental Tyzzer's disease in different species of laboratory animals. Jap. J. exp. Med. *36:* 519–534 (1966).

79 NIVEN, J. S.: Tyzzer's disease in laboratory animals. Z. Versuchstierk. *10:* 168–174 (1968).

80 Berdjis, C. C.; Gleiser, C. A., and Hartman, H. A.: Experimental parenteral anthrax in *Macaca mulatta*. Brit. J. exp. Path. *44:* 101–115 (1963).

81 Ogawa, H.; Takahashi, R.; Honjo, S.; Takasaka, M.; Fujiwara, T.; Ando, K.; Nakagawa, M.; Muto, T., and Imaizumi, K.: Shigellosis in cynomolgus monkeys *(Macaca irus)*. III. Histopathological studies on natural and experimental Shigellosis. Jap. J. med. sci. Biol. *17:* 321–332 (1964).

82 Francis, J.: Tuberculosis in animals and man (Cassell, London 1958).

83 Greer, W. E.; Jiricka, Z., and Felsenfeld, O.: Gallbladder damage and prolonged excretion of cholera vibrios in *Erythrocebus patas* (32738). Proc. Soc. exp. Biol., N.Y. *127:* 551–555 (1968).

84 Hoffbauer, F. W.; Frederich, W., and Zaki, F. G.: Choline deficiency in baboon and rat compared. Arch. Path. *79:* 364–369 (1965).

85 Kerr, G. R.; Wolf, R. C., and Waisman, H. A.: A disorder of lipid metabolism associated with experimental hyperhistidinemia in *Macaca mulatta*. Symp. zool. Soc., Lond. *17:* 371–392 (1966).

86 Follis, R. H., jr.: A kwashiorkor-like syndrome observed in monkeys fed maize. Proc. Soc. exp. Biol., N.Y. *96:* 523–528 (1957).

87 Madhavan, T. V.; Rao, K. S., and Tulpule, P. G.: Effect of dietary protein level on susceptibility of monkeys to aflatoxin liver injury. Indian J. med. Res. *53:* 984–989 (1965).

88 Vickers, J. H.: *Hepatocystis kochi* in *Cercopithecus* monkeys. J. amer. vet. med. Ass. *149:* 906–908 (1966).

89 Bray, R. S.: Studies on malaria in chimpanzees. IV. *Plasmodium ovale*. Amer. J. trop. med. Hyg. *6:* 638–645 (1957).

90 Hawking, F.; Perry, W. L. M., and Thurston, J. P.: Tissue forms of a malaria parasite, *Plasmodium cynomolgi*. Lancet *i:* 783–789 (1948).

91 Miller, M. J.: The experimental infection of *Macaca mulatta* with human strains of *Entamoeba histolytica*. Amer. J. trop. Med. *1:* 417–428 (1952).

92 Fremming, B. D.; Vogel, F. S.; Young, R. J., and Bensen, R. E.: A fatal case of amebiasis with liver abscesses and ulcerative colitis in a chimpanzee. J. amer. vet. med. Ass. *126:* 406–407 (1955).

93 Sadun, E. H.; Lichtenberg, F. von, and Bruce, J. I.: Susceptibility and comparative pathology of ten species of primates exposed to infection with *Schistosoma mansoni*. Amer. J. trop. med. Hyg. *15:* 705–718 (1966).

94 Graham, G.: Parasitism in monkeys. Ann. N.Y. Acad. Sci. *85:* 842–860 (1960).

95 Habermann, R. T. and Williams, F. P.: Diseases seen at necropsy of 708 *Macaca mulatta* (rhesus monkey) and *Macaca philippinensis* (cynomolgus monkey). Amer. J. vet. Res. *18:* 419–426 (1957).

96 Ewing, S. A.; Helland, D. R.; Anthony, H. D., and Leipold, H. W.: Occurrence of *Athesmia* sp. in the cinnamon ringtail monkey, *Cebus albifrons*. Lab. anim. Care *18:* 488–492 (1968).

97 Bostrom, R. E. and Slaughter, L. J.: Trematode *(Athesmia foxi)* infection in two squirrel monkeys *(Saimiri sciureus)*. Lab. anim. Care *18:* 493–495 (1968).

98 Garner, E.; Hemrick, F., and Rudiger, H.: Multiple helminth infections in cinnamon-ringtail monkeys *(Cebus albifrons)*. Lab. anim. Care *17:* 310–315 (1967).

99 Cosgrove, G. E.; Nelson, B., and Gengozian, N.: Helminth parasites of the tamarin, *Saguinus fuscicollis*. Lab. anim. Care *18:* 654–656 (1968).

100 POWERS, R. D.; PRICE, R. A.; MATTLIN, R. H., and HOUK, R. P.: Echinococcosis in a drill baboon. J. amer. vet. med. Ass. *149:* 902–905 (1966).

101 CROSBY, W. M.; IVEY, M. H.; SHAFFER, W. L., and HOLMES, D. D.: Echinococcus cysts in the savannah baboon. Amer. Ass. lab. anim. Sci. *18:* 395–398 (1968).

102 HEALY, G. R. and HAYES, N. R.: Hydatid disease in rhesus monkey. J. Parasit. *49:* 837 (1963).

103 BEAVER, P. C.: Toxocarosis (visceral larva migrans) in relation to tropical eosinophilia. Bull. Soc. Path. exot. *55:* 555–576 (1963).

104 HEUSCHELE, W. P.: Internal parasitism of monkeys with the pentastomid, *Armillifer armillatus.* J. amer. vet. med. Ass. *139:* 911–912 (1961).

105 PAPPENHEIMER, A. M. and HAWTHORNE, J. J.: Certain cytoplasmic inclusions of liver cells. Amer. J. Path. *12:* 625–634 (1936).

106 SCHAFFNER, F.; ROBERTS, D. K.; GINN, F. L., and ULVEDAL, F.: Electron microscopy of monkey liver after exposure of animals to pure oxygen atmosphere. Proc. Soc. exp. Biol., N.Y. *121:* 1200–1203 (1966).

107 SCOTT, E.; HELZ, M. K., and MCCORD, C. P.: The histopathology of methyl alcohol poisoning. Amer. J. clin. Path. *3:* 311–319 (1933).

108 O'GARA, R. W.: Early and late toxic effects in the livers of monkeys receiving aflatoxin B_1 and N-nitrosodiethylamine (DENA). Proc. amer. Ass. Cancer Res. *9:* 217 (1968).

109 O'GARA, R. W.; KELLY, M. G., and KERBER, W. T.: A spontaneous metastasizing carcinoma of the gallbladder in a rhesus monkey. Folia primat. *6:* 284–291 (1967).

110 SIEGAL, A. M.; CASEY, H. W.; BOWMAN, R. W., and TRAYNOR, J. E.: Leukemia in a rhesus monkey *(Macaca mulatta)* following exposure to whole-body proton irradiation. J. Hematology *32:* 989–996 (1968).

111 CONIGLIO, J. G.; DARBY, W. J.; EFNER, J. A.; FLEMING, J., and HUDSON, G. W.: Effect of total-body X-irradiation on fat balance and liver lipids in the rhesus monkey. Amer. J. Physiol. *184:* 113–118 (1956).

112 ALLEN, J. R. and CARSTENS, L. A.: Response of *Saimiri sciureus* monkeys to total body X-irradiation: Clinical signs and pathologic changes. Amer. J. vet. Res. *29:* 2179–2187 (1968).

113 MARUFFO, C. A.; MALINOW, M. R.; DEPAOLI, J. R., and KATZ, S.: Pigmentary liver disease in howler monkeys. Amer. J. Path. *49:* 445–456 (1966).

114 BANKS, K. L. and BULLOCK, B. C.: Naturally occurring secondary amyloidosis of a squirrel monkey *(Saimiri sciureus).* J. amer. vet. med. Ass. *151:* 839–842 (1967).

115 STEINER, P. E.; RASMUSSEN, T. B., and FISHER, L. E.: Neuropathy, cardiopathy, hemosiderosis and testicular atrophy in *Gorilla gorilla.* Arch. Path. *59:* 5–25 (1955).

116 BLUNDELL, G. P.; CASTLEBERRY, M. W.; LOWE, E. P., and CONVERSE, J. L.: The pathology of *Coccidioides immitis* in the *Macaca mulatta.* Amer. J. Path. *39:* 613–630 (1961).

117 CHAFFEE, R. R. J.; ALLEN, J. R.; BREWER, M.; HORVATH, S. M.; MASON, C., and SMITH, R. E.: Cellular physiology of cold- and heat-exposed squirrel monkeys *(Saimiri sciurea).* J. appl. Physiol. *21:* 151–157 (1966).

118 SOTO, P. J.; BEALL, F. A.; NAKAMURA, R. M., and KUPFERBERG, L. L.: Myocarditis in rhesus monkeys. Arch. Path. *78:* 681–690 (1964).

119 WEBB, H. E. and CHATTERJEA, J. B.: Clinico-pathological observations on monkeys

infected with Kyasanur Forest disease virus, with special reference to the haemopoietic system. Brit. J. Haemat. *8:* 401–413 (1962).

120 Bearcroft, W. G. C.: Hepatitis in african monkeys. Nature *197:* 806–807 (1963).

121 Hooker, C. W.; Drill, V. A., and Pfeiffer, C. A.: Failure of the liver of the monkey to inactivate estrogens *in vivo*. Proc. Soc. exp. Biol., N. Y. *65:* 192–194 (1947).

122 Shive, R. J.; Green, S. S.; Evans, L. B., and Garner, F. M.: Leptospirosis in barbary apes *(Macaca sylvana)*. J. amer. vet. med. Ass. *155:* 1176–1178 (1969).

Authors' address: Dr. Samuel M. Kruckenberg, Dr. Charles E. Cornelius and Dr. James E. Cook, College of Veterinary Medicine, Kansas State University, *Manhattan Kans.* (USA)

Pathology of Simian Primates, Part I, pp. 756–808 (Karger, Basel 1972)

The Cardiovascular System[1]

W. P. McNulty and M. R. Malinow

Oregon Regional Primate Research Center, Beaverton, Ore. and University of Oregon Medical School, Portland, Ore.

Contents

General ... 757
Heart ... 758
  Endocardium and Valves ... 758
  Myocardium ... 761
    Diseases of Abnormal Depositions ... 761
    Nutritional Deficiency Diseases ... 762
      Protein Deficiency ... 762
      Thiamine Deficiency ... 763
    Myocarditis ... 764
      Bacteria ... 764
      Viruses ... 764
      Rickettsia ... 764
      Parasites ... 765
      Unknown Etiology ... 765
    Circulatory Insufficiency ... 767
    Cardiac Hypertrophy ... 774
    Congenital Heart Disease ... 775
    Neoplasms ... 777
  Pericardium ... 777
Blood Vessels ... 779
  Arteries ... 779
    Atherosclerosis ... 779
      Prevalence ... 779
      Morphology, Progression and Complications ... 780
      Induction ... 788
    Hypertensive Arteriopathy ... 792

1 Publication No. 447 from the Oregon Regional Primate Research Center, supported in part by Grant No. FR 00163 from the National Institutes of Health, Grant No. HE 12694-02 from the Public Health Service, and the John C. Higgins Memorial Foundation.

Medial Calcification 795
Necrotizing Arteritis (Periarteritis nodosa) 796
Aneurysms 796
Veins 799
References 801

General

Clinically apparent cardiovascular disease is uncommon in non-human primates, perhaps because their life expectancy is short. The true age of most captive adult primates and of free-ranging simians is unknown, and therefore the potential life span of almost all species has not been established. Furthermore extended autopsy series are available only for rhesus *(Macaca mulatta)* and cynomolgus *(Macaca irus)* monkeys [22, 42, 57, 64], for howler monkeys *(Alouatta caraya)* [75, 85], for a limited number of chimpanzees [58] and baboons *(Papio* sp.*)* [62], and for miscellaneous primates in the animal reports of the Philadelphia and London zoos. The dearth of data precludes discussion of the incidence by species of naturally occurring heart disease and generally restricts us to qualitative description and comparison with human conditions.

On the other hand, subclinical alterations in the heart and vessels, which are potentially of great interest for experimental exploitation, are not at all rare. However, since the diagnosis of minimal disease depends on tissue sample size, diagnostic criteria, and the interest of the observer, the reported incidences of spontaneous arteriosclerosis and chronic myocarditis (to be discussed below) vary widely. The highest reported incidences for these minor abnormalities are probably more accurate, but it should be borne in mind that there is little evidence that the health or longevity of the animals is in any way affected.

Cardiovascular pathology in primates has been reviewed by RUCH [113], LAPIN and YAKOVLEVA [64] and FINLAYSON [24] who refer to numerous case reports from the Philadelphia and London zoos, many of which are, however, merely mentioned by species and diagnosis. Since comparison of incidences by species is not yet feasible (and it makes little biologic sense to lump together the heterogeneous group of non-human primates), only those references that provide informative description will be included here.

Investigation of induced disease, particularly atherosclerosis, has become quite active; we will consider experimental morbid anatomy together with the naturally occurring.

Accurate identification of species is important in a comparative study, but unfortunately, taxonomy is a field of many disagreements. Not only do experts differ on identification, but names for readily recognizable species change with changing evolutionary formulations. Fortunately there is reasonable consensus for some species commonly used in investigations, and we will use the common name for rhesus *(Macaca mulatta)*, cynomolgus *(Macaca irus)*, squirrel *(Saimiri sciureus)*, and howler *(Alouatta caraya)* monkeys. In other cases, we will use the identification – common, generic, or genus-and-species – given by the original authors.

Heart

Endocardium and Valves

The normal histology of the cardiac valves of monkeys, as well as the changes ascribed to aging, have been described by Lapin and Yakovleva [64].

The incidence of valvular lesions, other than age-related, is apparently low: 6 of over 1,200 simians studied by Lapin and Yakovleva [64], 2 of over 1,000 reported by Roberts [110], 2 of 280 free-ranging howler monkeys studied by Maruffo and Malinow [84], and 1 of 940 examinations of the hearts of various primates at the Oregon Regional Primate Research Center (ORPRC). These incidences should be regarded as minimum estimates, since routine post-mortems do not always include careful inspection of the valves. For example, the 6 cases of Lapin and Yakovleva were found among only 40 hearts selected for detailed examination.

Of these 6 cases, 3 were classified as acute; in one vegetations were found on the mitral valve, and in two the mitral valve was infiltrated with leukocytes. Roberts [110] reported in detail acute nonulcerative inflammation of the mitral and aortic valves of a rhesus monkey, probably with compromise of competence, associated with myocarditis and pericarditis (fig. 1). Chrisp and McLeod [15] found a right atrial endocardial vegetation in a rhesus monkey with endophthalmitis and septicemia *(Aerobacter aerogenes)*. The vegetation was near the tip of a catheter, and the valves were not involved. The single case of acute vegetative mitral valvulitis seen at ORPRC in an infant rhesus, age 4 months, was associated with abscesses in the myocardium and kidney.

Chronic mitral valvulitis, with vegetations on the valve and mural endocardium, occurred in a Pluto cercopitheque *(C. leucocampyx)* [116]; sclero-

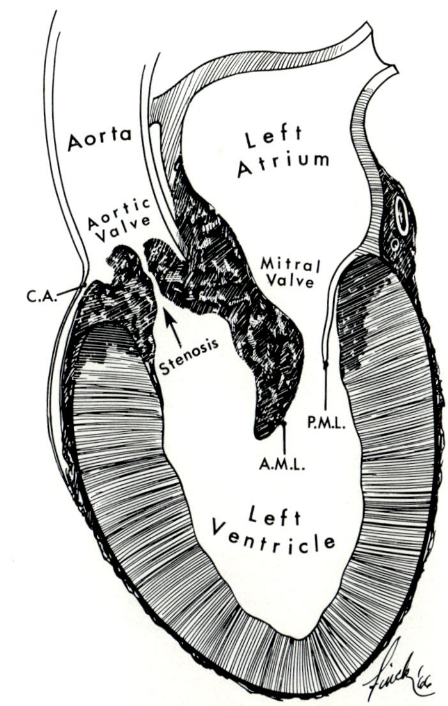

Fig. 1. Rhesus monkey. Diagram of heart with acute valvulitis. The aortic valvular cusps and the anterior mitral leaflet (AML) are markedly thickened, and the aortic valve is stenotic. The posterior mitral leaflet (PML) is normal. The inflammatory process (dark shading) also involved the atrial wall, left ventricular wall, and pericardium. Fig. 7 in ROBERTS and INNES [110], reprinted with permission of the American Heart Journal.

sis and deformation of the mitral valve, with stenosis and insufficiency, in a 'monkey' and a baboon [64]; and chronic mitral and aortic valvulitis in a female Barbary ape *(Macaca sylvana)* [24]. Thickening and deformation of the mitral valves of baboons, associated with a uniquely high incidence of amyloidosis, was reported by GILLMAN and GILBERT [36]. ROBERTS [110] described scarring of the mitral valve in a rhesus monkey that strongly resembled human rheumatic valvulitis; the valve was stenotic and incompetent, the atria were dilated and the ventricles hypertrophied, and the liver was chronically congested (fig. 2). Among 280 wild howler monkeys, two had a thickened mitral valve with narrowing of the orifice. There was no significant fu-

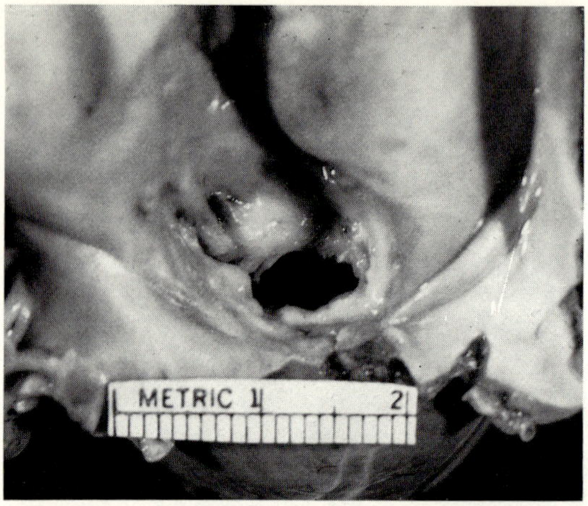

a

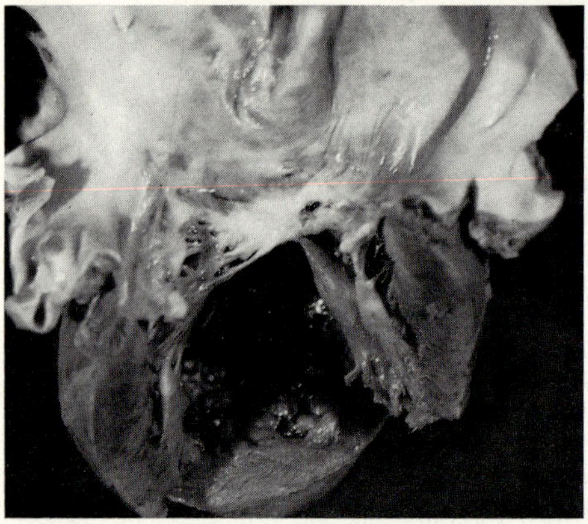

b

Fig. 2. Rhesus monkey, adult female. Chronic mitral valvulitis. (a) Mitral valve as seen from above, before it was opened. The orifice of the valve is more or less fixed; the valve is both stenotic and incompetent. (b) Left heart after opening through the posterior leaflet of the mitral valve. The left atrium is extremely dilated, thin walled, and free of clot. Just above the mitral valve the atrial endocardium is thickened, probably a 'jet lesion'. The leaflets of the mitral valve are thickened, particularly at the margins, and the chordae tendinae are short and thick. From fig. 3 in Roberts and Innes [110], reprinted with permission of the American Heart Journal.

sion of the cusps and the left atrium was normal in size and shape. The other valves were unaltered [84].

The etiology of all these types of valvular disease remains obscure. In several instances, the acute conditions were said to have occurred in 'septic' animals, but identification of organisms was not recorded. In some cases, the descriptions of chronic mitral valvulitis tantalizingly resemble the findings in human rheumatic disease, but in no case have Aschoff bodies been observed. However, these lesions are often very difficult to find in chronic, inactive, but well-documented human rheumatic fever.

Experimental attempts to produce rheumatic carditis in monkeys by infection with streptococcus have been unsuccessful [137], but the numbers of animals and species used have been small, and even in the human population only a small percentage is afflicted. Whether non-human primates can serve as models for rheumatic fever, as well as for other 'collagen' diseases that affect the heart and valves, remains unanswered.

Chronic valvulitis in man, usually ascribed to rheumatic fever, may sometimes be caused by a virus. Verrucous mitral and aortic valvulitis and cicatricial mitral stenosis were found in cynomolgus monkeys experimentally infected with Coxsackie B virus, and viral antigen was identified in the damaged valves as long as 200 days after infection [134].

Myocardium

Diseases of Abnormal Depositions

Intracellular fatty metamorphosis is common in the myocardium of squirrel monkeys on diets high in butter and cholesterol. Highly refractile granules, stainable with oil-red-O, appear in the muscle fibers throughout the ventricles, leading in advanced cases to obliteration of cross-striations and nuclei (fig. 3). Fatty infiltration, or the appearance of adipose cells in sheets and strands between bundles of muscle fibers, is frequent in obese animals; no functional significance is known. In the heart of a 15-year-old overfed chimpanzee *(Pan satyrus)* that died unexpectedly, excessive fat was found in the epicardium and myocardium, involving even the atrio-ventricular bundle [50]. Death was thought to be due to heart failure, but confirmatory clinical or pathological evidence was not given.

In older non-human primates, lipofuscin pigmentation ('wear and tear pigment') collects in the customary perinuclear location in myocardial cells (fig. 4).

Amyloidosis of the heart was an unexpected finding in autopsies of 3 sexually mature female baboons *(Papio ursinus)* from a colony of 50 [36]. Besides the damage to the mitral valve mentioned above, amyloid was deposited diffusely in connective tissue throughout the myocardium, particularly

Fig. 3. Squirrel monkey, adult female. Fatty metamorphosis in the myocardium. High fat diet. The globules of lipid are linearly arranged between the myofibrils. Frozen section, oil red 0 and hematoxylin (×250).

near blood vessels. In fact, it was patchily distributed throughout the entire vascular tree, including liver, kidneys, and spleen. Since the animals had been clinically well except for disturbances in menstrual cycles and terminal diarrhea, the authors suggested that these were examples of primary amyloidosis.

Nutritional Deficiency Diseases

Although myocardial lesions thought to be due to nutritional inadequacy have been seen in primates held for some time in captivity, few reports deal with cardiac damage associated with specific single deficiencies. The difficulty of ascertaining whether single deficiencies do exist, the likelihood that the effects of combined deficiencies are more than simply additive, and the rather nonspecific nature of the lesions described, all contribute to the nebulous picture of nutritional cardiac disease.

Protein Deficiency

Chauhan *et al.* [13] fed 48 young male rhesus monkeys divided into 2 groups a semipurified diet differing in the amount of casein (0% and 15%). Lesions were observed 5 to 12 weeks after the start of the deficient diet. In the

Fig. 4. Rhesus monkey, adult female. Lipofuscin pigment deposits near ends of nuclei in muscle cells of left ventricle. Hematoxylin and eosin (×400).

test animals, the heart was smaller than in the control monkeys and appeared flabby and pale. Light microscopy showed atrophy of the myocardial muscle fibers and occasional areas of myolysis. In groups of muscle fibers, the striations were hazy or not visualized; in some areas, the intercalated discs were unusually prominent. After 10 weeks of the deficient diet, interfibrillar fibrosis appeared. Pathological changes, however, were generally more advanced in the skeletal muscles than in the myocardium. In a similar study, RACELA *et al.* [103] found only atrophy; ultrastructurally the numbers of myofilaments and mitochondria were reduced here and there.

Thiamine Deficiency

Experimental thiamine deficiency was induced in 7 rhesus monkeys by RINEHART and GREENBERG [109]. The heart appeared dilated, focal necrosis of heart muscle was observed by light microscopy, and 'hydropic' swelling of subendocardial muscle fibers of the conduction system was described. In 'African monkeys' a diet of polished rice supplemented with ascorbic acid caused swelling of myocardial fibers with loss of striations as well as focal necrosis unattended by inflammatory reaction [26]. The changes were reversed with supplementary thiamine.

Myocarditis

Bacteria

Abscesses may be present in septic states sometimes associated with acute valvulitis[2] [110]. Since the organism has almost never been identified, the assumption of bacterial etiology has rested on histopathology and associated clinical findings. Tuberculous granulomas in the myocardium have been reported [96], but diptheritic myocarditis has not been described.

Viruses

To establish a viral etiology on histopathologic grounds alone is nearly impossible, unless pathognomonic inclusion bodies can be found. Many cases of myocarditis of unknown etiology (described below) may be virus-caused, but the anatomical changes are so nonspecific that unless characteristic clinical syndromes or antibody rises can be recognized or viruses isolated even a presumptive diagnosis of viral disease is unwarranted.

Myocarditis with the inclusions typical for salivary gland inclusion virus has been seen in a chimpanzee [143]; adrenals and salivary glands were also affected.

Focal chronic myocarditis has been seen in cynomolgus hearts 50 to 200 days after inoculation of Coxsackie B virus [134], but according to the published descriptions, the lesions do not differ substantially from those of 'spontaneous' focal chronic myocarditis found in many simian hearts (see below). Acute diffuse myocarditis caused by Coxsackie virus, such as occurs in human infants, has not been reported in monkeys.

Members of the encephalomyocarditis group of viruses, originally found in rodents, were isolated from a chimpanzee in Florida [115], and a squirrel monkey and two night monkeys *(Aotus trivirgatus)* in Columbia [111] succumbed with severe myocarditis. The virus was also found in man, a rhesus monkey, mosquitoes, and a mongoose in Uganda [21]. Thus the distribution appears to be world-wide, and rodents may be the reservoir. The gross and microscopic changes in the fatal cases were severe but not specific. Typically the heart was dilated, flabby, and pale and microscopically showed diffuse lymphocytic infiltrate and necrosis of muscle (fig. 5); marked pleural and pericardial effusions were also present.

Rickettsia

Monkeys are susceptible to typhus [7] and Rocky Mountain spotted fever [148], but studies of specific morbid anatomy of the heart have not been made.

2 Unpublished.

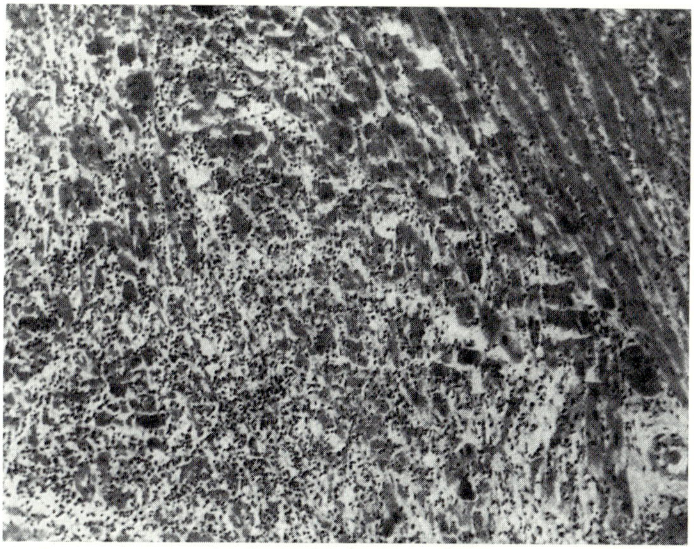

Fig. 5. Gibbon, adult male. Myocarditis due to encephalomyocarditis virus. There is widespread involvement with areas of necrosis. Hematoxylin and eosin stain (× 110). Fig. 1 in Schmidt [115] reprinted with permission of the American Journal of Pathology.

Parasites

Leishmanial forms, presumably of *Trypanosoma cruzi*, were found in nests of muscle fibers of a *Cebus albifrons* (fig. 8), which showed a conduction abnormality by electrocardiography during life and multifocal chronic myocarditis at autopsy [10]. Filariasis is common in squirrel monkeys; occasionally we have seen in the myocardium small foci of necrosis and reaction associated with microfilaria.

Toxoplasmosis occurs spontaneously in several species of New World monkeys [8, 90] and in prosimians [97, 152]; Old World monkeys, though relatively resistant, can sometimes be infected experimentally [19]. Myocardial focal necrosis with demonstrable parasites are seen (fig. 7).

In the ORPRC series, sarcosporidiosis was noted in a lesser bushbaby *(Galago senegalensis)* and a rhesus monkey (fig. 8); the infestation was slight and almost certainly of no clinical significance. Echinococcosis and cysticercosis of the heart have not been reported nor seen in our collection.

Unknown Etiology

Most of the diagnoses of myocarditis made at autopsy fall into this category; the finding is usually unexpected and of uncertain clinical importance.

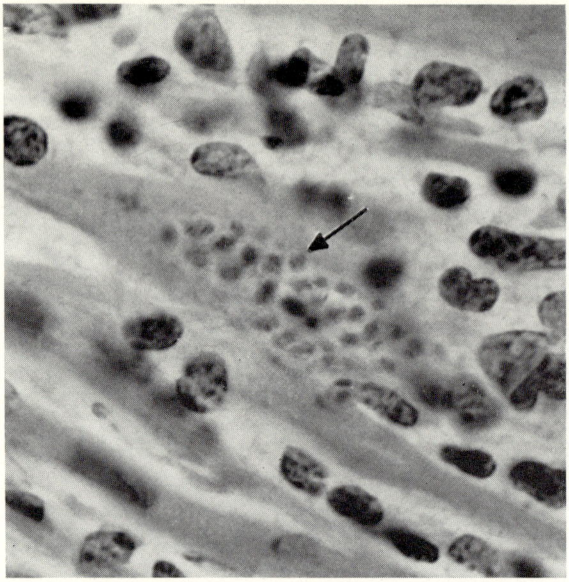

Fig. 6. Cebus albifrons, adult female. Nest of leishmanial forms of *T. cruzi* in a cardiac muscle fiber. Small, rodlike kinetoplast is visible in 1 cell (arrow). Hematoxylin and eosin stain (× 1,000). Fig. 3 in Bullock *et al.* [10], reprinted with permission of the Journal of the American Veterinary Medical Association.

Focal collections of lymphocytes and sometimes mononuclear cells are commonly found in random perivascular locations in the myocardium, sometimes with various degrees of alteration of the myocytes and increase in connective tissue (fig. 9). The incidence appears to vary with the diagnostic threshold of the observer. Fox [27] listed an incidence of 2.3% of 498 simians and none in 86 prosimians; the nature of the changes was not described. Eighteen of 36 rhesus hearts, carefully examined as controls in studies of experimental rheumatic fever, had lesions like those described above, two with focal changes in muscle cells and connective tissue [137]. Nine of 13 rhesus monkeys used as controls for experimental anaphylactic shock also had interstitial lymphocytic infiltrates in the myocardium without associated myocellular necrosis [98]. Similar findings have been noted in 26 of 336 rhesus hearts and in 13 of 363 hearts from other primates at ORPRC; since no systematic search for the condition has been made, these figures are undoubtedly low.

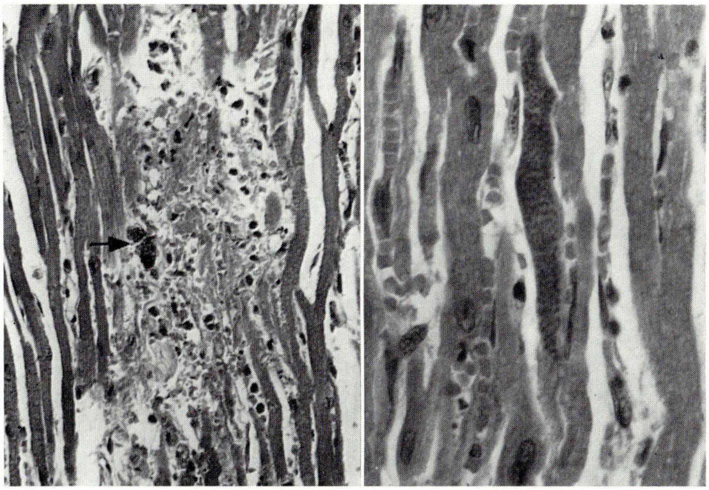

a *b*

Fig. 7. Squirrel monkeys. Cardiac toxoplasmosis. (a) Focus of necrotic myocarditis (interventricular septum) containing toxoplasma aggregates (arrow). Hematoxylin and eosin stain (×290). (b) Intracytoplasmic colony of toxoplasma within sarcoplasm of right ventricular myocardial fiber. Hematoxylin and eosin stain (×640). Fig. 5 and 6 in McKissick *et al.* [90], reprinted with permission of Pathologia Veterinaria.

The cause of 'chronic interstitial myocarditis' is probably not bacteria or a virus of the encephalomyocarditis group [121]. Usually clinical disability is absent; in Soto's [121] study, 17 rhesus monkeys with various degrees of myocarditis had no electrocardiographic abnormalities. Among instances of bundle branch block in rhesus monkeys Malinow [12] found 2 cases with scattered foci of myocarditis, some in the region of the conduction fibers (fig. 10), but no definite correlation could be established. The condition is not necessarily related to the laboratory environment, since 8% of 200 free-ranging howler monkeys had focal chronic myocarditis without myocellular necrosis in all cases but one. There was no discernible correlation with age, but the incidence was somewhat higher in females [84].

Circulatory Insufficiency

Small myocardial scars and diffuse fibrosis in human hearts are usually associated with coronary insufficiency. The same relationship can be seen in simians: a male chimpanzee *(Pan troglodytes)*, exhibited for 458 months at the Philadelphia Zoo, had atherosclerosis of the smaller coronary arteries and many scattered scars in the myocardium [106]. Usually, however, small

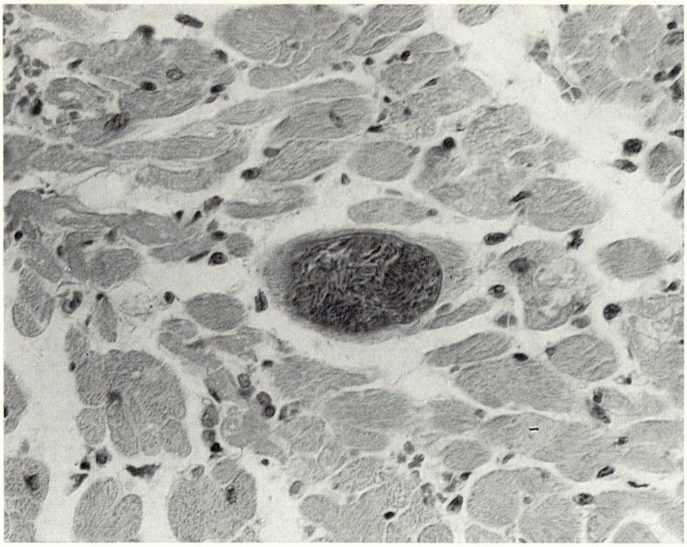

Fig. 8. Rhesus monkey, adult female. Sarcocystis infestation in a myocardial fiber. Hematoxylin and eosin stain (×400).

scars are found in hearts without discernible arterial disease, and a scar gives little clue of the original insult. One can be confident only that necrosis of the muscle has occurred; possibly those instances of 'chronic focal myocarditis' that are accompanied by myocellular alterations account for many of these small scars.

Large scars without associated arterial disease have also been noted in a Kenya baboon [88] and a *Macaca lapunder (nemestrina)* [64]. Extensive necrosis with fatty metamorphosis, unexplained by arterial occlusion, occurs sometimes in squirrel monkeys on high-fat diets[3] [102] (fig. 11). The etiology in these cases is unknown.

Infarcts caused by observed coronary occlusion are rare; only five cases have been reported for non-human primates on a regular diet. A black lemur *(Lemur macaco)* suffered a fatal infarct of the apical portion of the left ventricle, associated with atherosclerosis and obesity [45]. Both coronary arteries were partly obstructed with atheromatous deposits in a white-collared mangabey *(Cercocebus collaris)* with a myocardial infarct [50]. This animal, which had been in the zoo 7 years, represents the only case so far reported in

3 Unpublished.

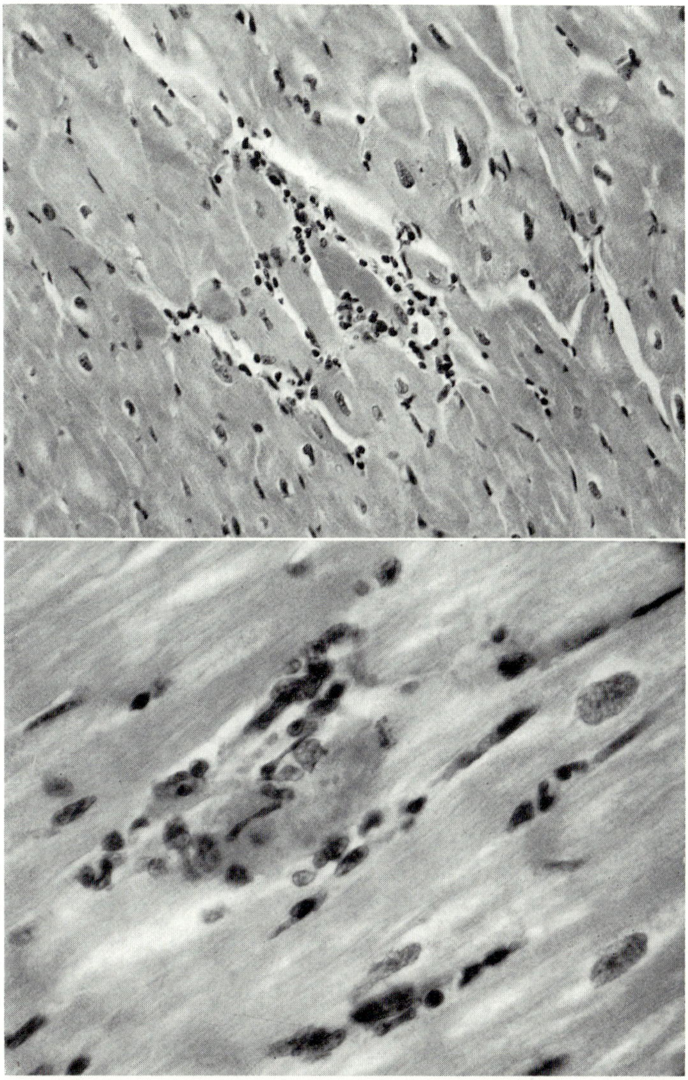

a

b

Fig. 9. Rhesus monkeys. Chronic myocarditis of unknown etiology. (a) Focus of myocarditis, with slight degree of inflammation around darkly staining necrotic myofiber. Hematoxylin and eosin stain (×288). (b) Focus of myocarditis showing necrotic myofiber with surrounding polymorphonuclear neutrophils and mononuclear cells. Hematoxylin and eosin stain (×721). Courtesy of Dr. D. B. Hackel.

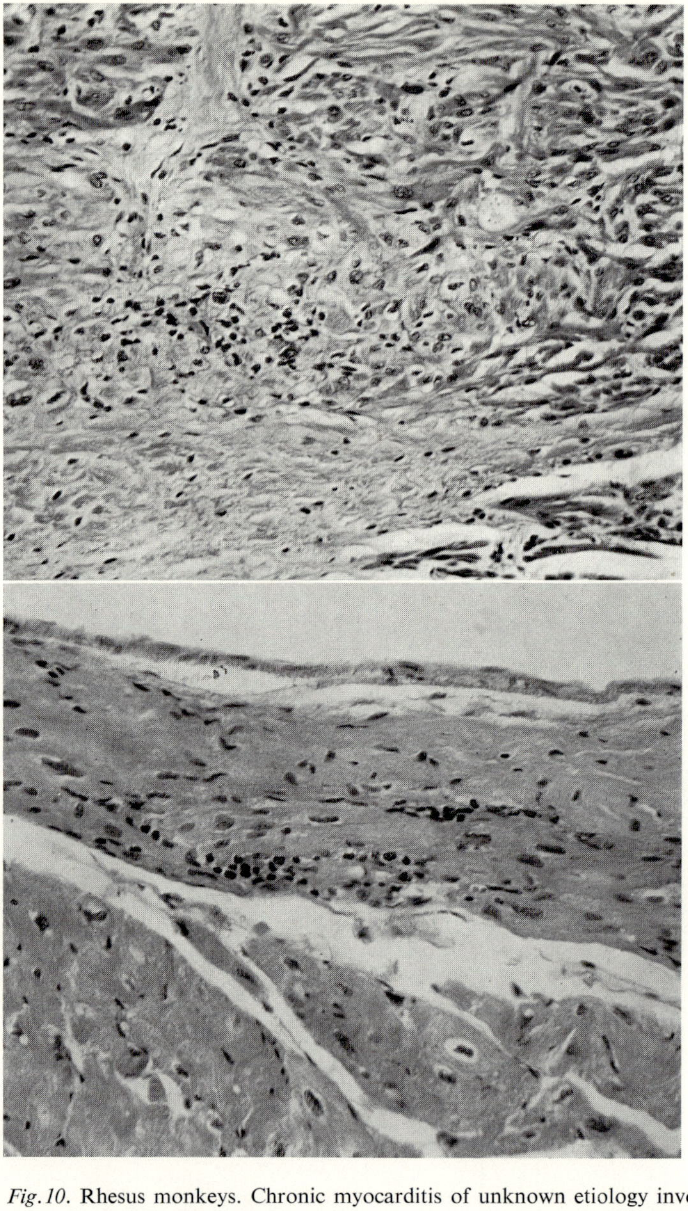

a

b

Fig. 10. Rhesus monkeys. Chronic myocarditis of unknown etiology involving the conduction system. (a) Slight degree of inflammation involving edge of atrio-ventricular node; AV node is above and central fibrous body below. Hematoxylin and eosin stain ($\times 288$). (b) Slight chronic inflammatory infiltrate involving left bundle (left ventricular cavity above). Hematoxylin and eosin stain ($\times 288$). Courtesy of Dr. D. B. Hackel.

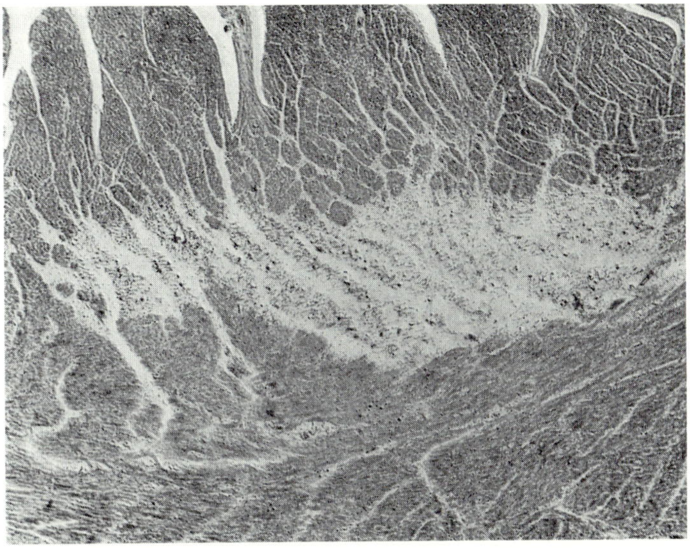

Fig. 11. Squirrel monkey, adult male. Necrosis of the left ventricular myocardium. High fat diet. Frozen section, oil red 0 stain (×25).

an Old World monkey not given a special diet. An extensive hemorrhagic infarct of the myocardium of the auricular-ventricular groove, with thrombosis of the coronary sinus, was found in a concolor gibbon *(Hylobates concolor);* both coronary orifices were choked with atheromatous deposits [52]. Old and recent posterior wall infarcts, together with considerable intimal thickening of coronary arteries, occurred in a chimpanzee, although an occlusion could not be demonstrated [80]. A male *Gorilla gorilla,* dying after 34 years in captivity, had an anterior left ventricular transmural infarct with an endocardial thrombus and probable embolus in the superior mesenteric artery [105]. The larger coronary arteries contained small intimal plaques; significant arteriosclerosis was limited to branches of less than one millimeter in diameter. This appears to be the only reported death from complications of occlusive atherosclerosis in an ape fed a low-fat diet throughout its entire adult life.

Myocardial infarction was first mechanically induced and the electrocardiogram studied by DeWaart *et al.* [144] in open-chested Javanese monkeys. Using previously implanted snares, Malinow *et al.*[4] followed electrocardiographic, pressure, and flow changes after sudden obstruction of the an-

4 Unpublished.

terior descending coronary artery in conscious, unrestrained rhesus monkeys. Extensive myocardial infarcts were produced in the septum and anterior wall of the left ventricle (fig. 12), and the electrocardiogram showed the typical pattern and evolution of anteroseptal infarction as seen in the human. Grayson and Irvine [39] followed the changes in blood flow in the myocardium after open-chested ligation of the anterior descending coronary artery in Patas and West African green monkeys and in baboons.

Myocardial infarction resulting from experimentally induced atherosclerosis of the coronary arteries has been difficult to achieve. Taylor *et al.* [135] reported the first and only known instance of fatal myocardial infarct resulting from diet-induced hypercholesterolemia in a rhesus monkey. Healed, healing, and acute infarcts were seen in the inner three-quarters of the left ventricular wall, the septum, the papillary muscles, and a small portion of the posterior right ventricular wall and adjacent papillary muscles. The lumens of all major coronary arteries were markedly reduced, and the left circumflex branch had three areas of occlusion with old recanalization and recent laminated thrombus (fig. 13). The arterial media were markedly

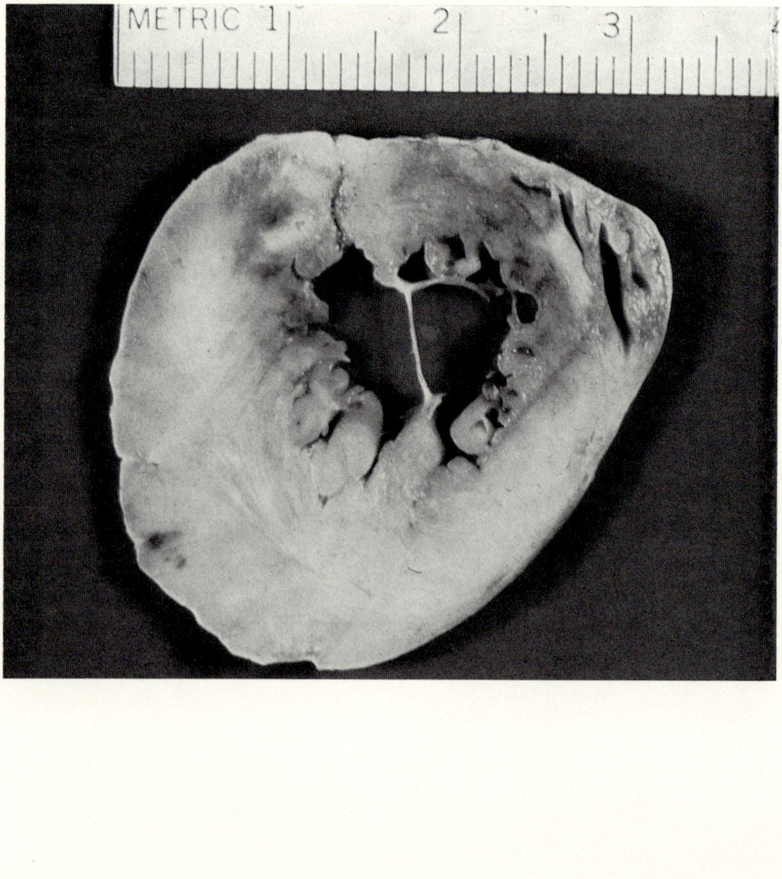

12a

12b

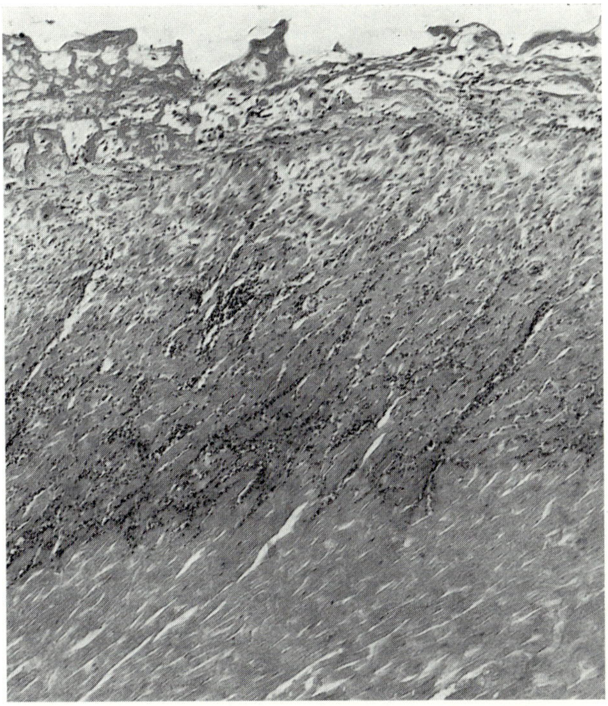

Fig. 12. Rhesus monkeys, adult. Myocardial infarcts following ligation of the proximal left anterior descending coronary artery. (a) Male, 4 days post-ligation. Mid-ventricle cross-section of the heart, showing discoloration of the anterior left ventricular wall and most of the interventricular septum. (b) Female, 2 days post-ligation. Section of anterior left ventricle showing total infarction (bottom), a peripheral band of edema and leukocytic infiltrate (middle), and a rim of viable muscle cells just beneath the pericardium (top). Fibrinous exudate adheres to the pericardium. Hematoxylin and eosin stain (×63).

damaged, with reactive fibrosis and calcification. Two changes commonly associated with coronary occlusions in man, intramural hemorrhages and ulceration of atheromatous plaques, did not occur.

Three of 14 chimpanzees on a 12 month diet containing 0.5% cholesterol and 45% of calories in corn oil or coconut oil had fibrosis of the myocardium [3]. In one, a diffuse scar was present in the left ventricle but no fresh necrosis. Intramural lipid deposits in the coronary arteries were marked. In the second, focal fibrosis and necrosis were present, and gross lesions in the arteries were less extensive. Scattered foci of fibrosis occurred in the third case, but the coronary arteries were completely free of lesions. As the authors

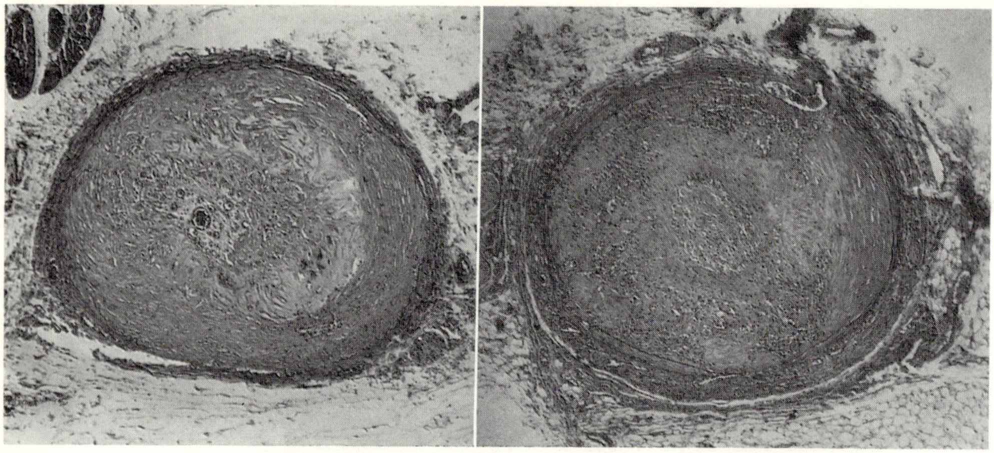

a b

Fig. 13. Rhesus monkey, adult female. Induced atherosclerosis of the coronary arteries, with thrombosis and occlusion. (a) Multiple channels of recanalization in well-organized occlusive lesion involving channel which had already been markedly narrowed by severe atherosclerosis. Hematoxylin and eosin stain. (b) Laminated thrombus occluding lumen which had been previously reduced to less than 10% of its original size. Hematoxylin and eosin stain. Fig. 8 and 9 in Taylor *et al.* [135], reprinted with permission of the Archives of Pathology.

caution, causal relationship between coronary insufficiency and myocardial damage was difficult to establish.

Cardiac Hypertrophy

Standards for heart weights in non-human primates have not yet been generally established. The relative heart weight in 200 free-ranging howler monkeys was around 0.4% [123]. A value of 0.5% was found in 155 ORPRC squirrel monkeys, except for animals weighing less than 100 g (0.7%) and over 700 g (0.4%). In 60 rhesus monkeys the relative heart weight was 0.56% in the weight range 1 to 2 kg, 0.46%, in 2 to 3 kg, and 0.4% above 3 kg [32]. Analysis of heart weights in 350 rhesus monkeys at ORPRC gave a value of about 0.6% for infant rhesus monkeys under 1 kg and a slowly decreasing value of 0.5 to 0.4% in the body weight range of 1 to 10 kg. In this series, 20 animals had heart weights at least half again the average value. None of these bore associated post-mortem or clinical evidence of hypertension or other cardiovascular disease; 6 were newly-arrived animals dying with pneumonia or intractable diarrhea, probably with acute weight loss, and 5 were pregnant.

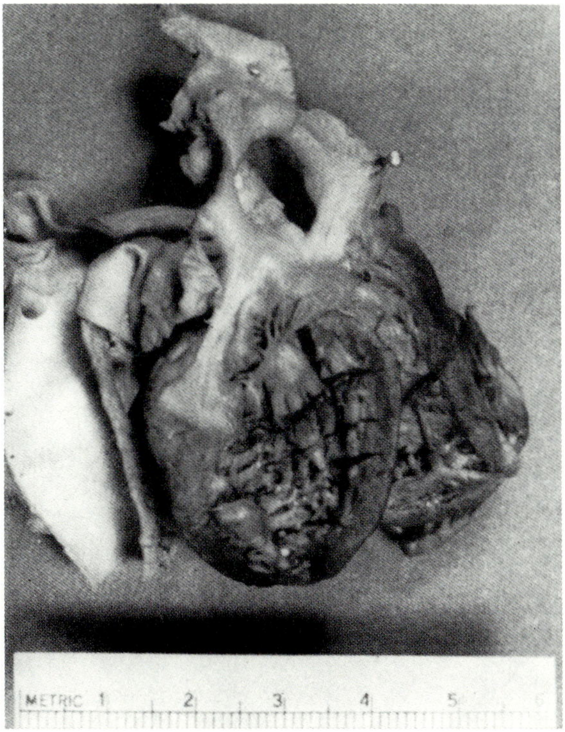

Fig. 14. Pan satyrus, immature male. Heart opened on left side, showing large interatrial septal defect. Fig. 1 in HACKEL *et al.* [43], reprinted with permission of Laboratory Investigation.

LAPIN and YAKOVLEVA [64] studied 6 rhesus monkeys and 8 baboons with clinical hypertension, some over a period of years. Hypertrophy of myocardial fibers was described, but measurements of heart size were not given. Cardiac hypertrophy and dilatation, associated with unilateral hydronephrosis, were present in a wild-shot *Cercopithecus ascanius schmidtii* [44]. The simian heart could be expected to enlarge in response to hypertension, anemia, arteriovenous shunts, and chronic pulmonary disease, but quantitative observations have been scanty.

Congenital Heart Disease

Few cases of congenital anomalies of the heart have been described in monkeys (table I; fig. 14 and 15). Except for the last 2 cases believed to have been induced by a derivative of thalidomide they have, of course, been inci-

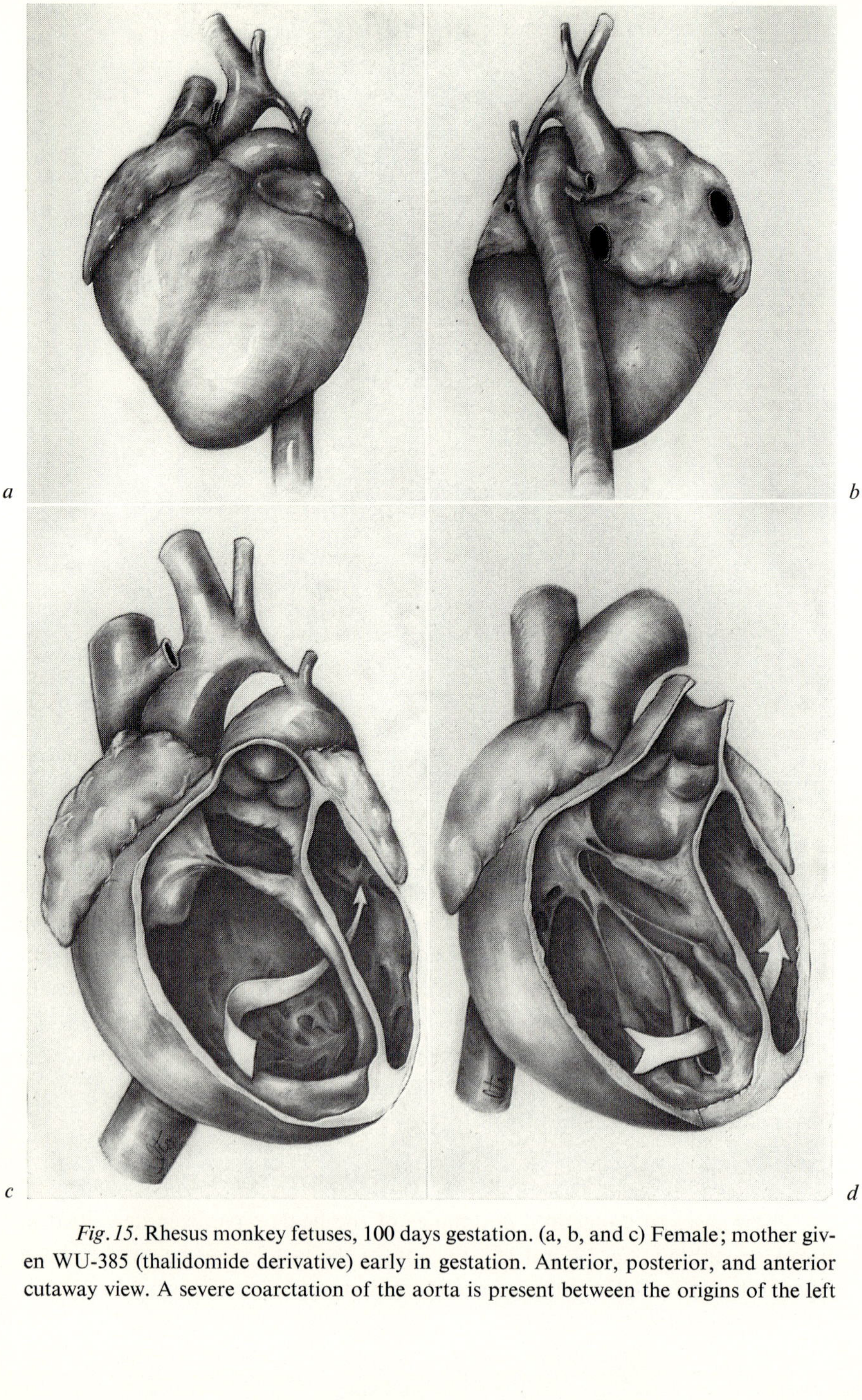

Fig. 15. Rhesus monkey fetuses, 100 days gestation. (a, b, and c) Female; mother given WU-385 (thalidomide derivative) early in gestation. Anterior, posterior, and anterior cutaway view. A severe coarctation of the aorta is present between the origins of the left

dental findings. The spectrum of anomalies is similar to that seen in humans. The described defects in case No. 8 are consistent with the tetralogy of Fallot, associated with aortic stenosis.

Neoplasms

Myocardial neoplasms, either primary or metastatic, have not been reported in non-human primates.

Pericardium

Most reported cases of pericarditis not obviously related to a primary myocarditis have been incidental findings of fibrous plaques, adhesions, and lymphocytic accumulations of unknown cause and probably of no clinical significance. The incidence is uncertain since often such lesions are probably not thought worthy of mention. In 498 primate autopsies, Fox [27] listed 18 diagnoses of 'pericarditis' without further comment. MARUFFO and MALINOW [84] found changes of the sort described above in 19 of 280 free-ranging howler monkeys.

However, pericarditis of a more serious nature does occur. Beta hemolytic streptococci were isolated from a serofibrinous pericardial exudate in an orang-utan with streptococcal sepsis [100]. We have found unclassified streptococci in an adhesive fibrinous pericarditis in a female dwarf galago *(Galago demidovii)* without any other detectable focus of infection. Serous effusions were seen in rhesus monkeys with pulmonary tuberculosis [96], though acid-fast bacilli in the fluid were not reported. Most cases have been of unknown etiology: serous pericarditis and myocarditis in a young tantalus monkey *(Cercopithecus tantalus)* [52], fibrinous pericarditis in a female Barbary ape *(Macaca sylvana)* [24], adhesive pericarditis in a female mandrill *(Mandrillus sphinx)* [24], dry pericarditis in an old female *Macaca sylvana* [23], and a chronic adhesive and calcifying, though nonconstrictive, pericarditis in a howler monkey [84].

Of particular interest have been epizootics of unknown cause. Six rhesus monkeys in a single shipment died with purulent pericarditis at the Breslau

---

common carotid and left subclavian arteries. The ductus arteriosus provides a straight-through communication between the pulmonary artery and descending aorta. The interventricular septum is almost totally absent, and both major arteries lead from the common chamber. (d) Male; mother given thalidomide early in gestation. Anterior cutaway view. An interventricular septal defect is present, low in the septum.

Table I. Congenital cardiac anomalies in non-human primates

Cardiac anomaly	Associated findings	Species, sex, age	Reference
1. Interatrial septal defect (fossa ovalis)	None	Chimpanzee, male, 2 years	43
2. High interventricular septal defect (0.5 cm). Right ventricular hypertrophy	Cyanosis	Rhesus, 3 years	64
3. Interventricular septal defect		Rhesus, stillborn	64
4. Enlarged patent ductus arteriosus, right atrial and ventricular hypertrophy		Rhesus, stillborn	64
5. Enlarged patent ductus arteriosus; aortic hypoplasia, aortic and pulmonary valvular deformities		Hamadryad baboon, stillborn	64
6. Pulmonic stenosis, right atrial and ventricular hypertrophy	Visceral congestion, ascites, and pericardial effusion	Rhesus, 1½ years	64
7. Interventricular septal defect, overriding aorta, aortic and pulmonic stenosis	Fetal lobulation of kidneys, anomalies in digits	Hamadryad baboon, stillborn	64
8. Interventricular septal defect, mitral and tricuspid valvular deformities, dilatation of pulmonary artery, left ventricular hypoplasia, right ventricular hypertrophy		Hamadryad baboon, 7 days	64
9. Left superior vena cava, with partial pulmonary venous drainage into caval system		*Pongo pygmaeus*	12
10. Atrial septal defect (2 cm), cardiac hypertrophy, pulmonary arterial dilatation		*Cercopithecus ascanius schmidtii*, female, adult	24
11. Persistent left vena cava, patent ductus arteriosus, probe-patent foramen ovale, right and left ventricular hypertrophy	Chronic passive congestion of liver, hypertrophy and arteritis of small pulmonary arteries	Rhesus, male, 3 years, 5 months	31
12. Large interventricular septal defect, infantile coarctation of aorta, enlarged ductus arteriosus, persistent left vena cava		Rhesus, female, 100 days gestation, thalidomide treatment	91
13. Low small interventricular septal defect		Rhesus, male, 100 days gestation, thalidomide treatment	91

Zoo [122]. A thickened adherent pericardium was seen in 2 rhesus monkeys in a group of experimental animals that had been afflicted with acute respiratory illness at the start of a study of vitamin E deficiency [25]. In Edinburgh 8 chimpanzees *(Anthropopithecus troglodytes)*, aged 3½ months to 11 years, died in less than one year with varying patterns of serous, fibrinous, and fibrous pleuritis and pericarditis with pericardial fluid accumulations of from a few ml to more than a liter. The bacteriological findings were not consistent, and two attempts to isolate virus were unsuccessful [4]. The clinical course differed from that described in chimpanzees for pericarditis and myocarditis caused by encephalomyocarditis virus.

Uremic pericarditis as such has not been identified. A congenital imperfection of the pericardium with herniation of the left ventricle occurred in an orang-utan *(Pongo pygmaeus)* [20], and a pericardial neoplasm, either a primary mesothelioma or metastatic adenocarcinoma with undetected primary site, was found in a *Cebus apella* [56].

Blood Vessels

Arteries

Atherosclerosis

Whether non-human primates are adequate experimental models for the study of human atherosclerosis depends on the answers to 3 kinds of questions. (1) Does intimal disease occur spontaneously in the arteries of simians; if so, is it common and is the incidence affected by captivity? (2) Is it morphologically similar to or identical with human atherosclerosis? Is it progressive, sex-related, and associated with sequelae in heart, brain, kidney, and extremities? (3) Can it be experimentally manipulated by diet, environment, and drugs?

Prevalence

Arterial intimal disease occurs in many species of wild and captive primates; the reports have been reviewed frequently in the past decade [16, 17, 24, 67, 68, 70, 127, 133, 141]. It occurs in all genera (table II), provided various criteria are accepted: (1) gross intimal sudanophilia; (2) deposition of stainable lipid either intra- or extracellularly in intima and inner media; (3) gross intimal plaques; (4) microscopic intimal thickening, with or without alteration in elastic lamella; and (5) classical intimal atheroma with fibrous cap and necrotic depths containing free lipid and cholesterol.

Although the evidence is ample that atherosclerosis occurs throughout the order of primates, most reports are based on one or several animals; only a few give figures for the prevalence by species, wild and captive (table III).

The figures for wild animals are based only on the observations made on series of primates killed in the field. The overall incidence of aortic sudanophilia in adults is 50% or more in all species examined, except *Cercopithecus ascanius* and *Macaca mulatta*.

Data for captive animals have come from zoos and collections for experimental laboratories. Although the numbers of examinations for single species are small and the information is quite heterogeneous (differences in duration and nature of captivity and of diet, variations in diagnostic criteria), we can reasonably conclude that atherosclerosis of the aorta is qualitatively about as common in wild as in captive primates. Furthermore, no species so far examined has been totally free from the disease.

Although some authors contend that atherosclerosis is more severe in captive than in wild animals, reliable comparison is difficult in the absence of accurate estimates of the age of free-ranging primates. Many animals in the wild probably do not grow as old as the most elderly and most atherosclerotic individuals in captive collections.

Morphology, Progression and Complications

The answers to the second group of questions are crucial, but they can hardly be given with confidence until the pathogenesis of the human disease itself is much better understood. The essential elements of human atherosclerosis are lipid accumulation and cellular proliferation, yet after decades of controversy we simply do not know whether the initial lesion is a cellular alteration caused by accumulation of lipid or an accumulation of lipid in areas of change from other causes. We are ignorant because human atherosclerosis is largely inaccessible for pathogenetic studies. Thus our special interest in non-human primates has a bootstrap character: we could better compare primate and human disease if we more fully understood the process in man, but on the other hand, we can only hope to learn about human disease through appropriate experimental models.

The numerous case reports and surveys (table II and III) as well as investigations of induced atherosclerosis (see following) have included anatomical descriptions of the lesions, ranging in detail from a mapping of gross intimal sudanophilia (fig. 16) to careful histologic analysis. The points of similarity to human atherosclerosis are frequently stressed; no author has declared for qualitative morphologic differences.

Table II. Natural atherosclerosis in non-human primates. Occurrence by species

Species	Reference	Species	Reference
Prosimians		Old World monkeys	
Lemur macaco	45	*Macaca mulatta*	14, 53, 68, 150
Lemur speck	140	*Macaca nemestrina*	68
Lemur sp.	24	*Macaca maura*	68
Galago crassicaudatus	47	*Macaca sylvana*	68
Galago sp.	24	*Macaca*, mixed sp.	24
		Cynopithecus niger	68
New World monkeys		*Cercocebus fuliginosus*	50
Callimico sp.	64, 24	*Cercocebus collaris*	50
Callicebus sp.	2	*Cercocebus albigena*	130
Aotus sp.	104	*Cercocebus alterimus*	68, 140
Alouatta caraya	71, 73	*Cercocebus nictitans*	68, 129
Cebus hypoleucus	69	*Cercopithecus ascanius*	130, 140
Cebus fatuellus	69	*Cercopithecus aethiops*	41, 130, 140
Cebus albifrons	68	*Cercopithecus talapoin*	140
Cebus 'capuchin'	64	*Cercopithecus brazzae*	140
Saimiri sciureus	92, 94, 107	*Cercopithecus nictitans*	24
Ateles paniscus	68	*Cercopithecus mona*	24
Ateles geoffroyi verrucosus	24	*Papio porcarius*	69
Ateles sp.	17	*Papio papio*	140
Lagothrix humboldtii	51, 105	*Papio cynocephalus*	41
Lagothrix sp.	2, 24	*Papio doguera*	66
		Papio anubis/doguera	88
Apes		*Papio ursinus*	35, 59
Hylobates concolor	52	*Papio hamadryas*	46, 47, 48, 49
Hylobates lar pileatus	68	*Papio hamadryas/anubis*, hybrid	24
Hylobates lar	24, 105		
Pongo pygmaeus	24, 104	*Papio* sp.	29
		Mandrillus sphinx	68
Pan troglodytes	50, 106, 125	*Theropithecus*	24
Pan satyrus	24, 68, 138	*Colobus angolensis*	140
Pan schweinfurthii	139, 140		
Pan sp.	3, 30, 80, 114, 131		
Gorilla gorilla	68, 105, 126, 129		
Gorilla beringei	24		

Listing includes diagnosis of arterio- or atherosclerosis by any criterion; some reports are for wild and some for captive animals on 'regular' diets (animals receiving atherogenic diets not included); some are for single animals, others are for series.

Table III. Natural atherosclerosis in non-human primates. Prevalence by species

Species	Prevalence	Criteria	Reference
Free ranging			
Alouatta caraya	209/287 (aorta)	Gross sudanophilia	71
	113/287 (aorta)	Gross intimal fibrosis	
Alouatta caraya	50%/143 (coronary)	Microscopic intimal thickening	77
Saimiri sciureus	97/220 (aorta)	Gross sudanophilia	94
Macaca mulatta	23/150 (aorta)	Gross fatty streaks or gross fibrosis	14
Cercocebus albigena	30/49 (aorta)	Gross sudanophilia	130
Cercopithecus ascanius	5/64 (aorta)	Gross sudanophilia	130
Papio doguera/anubis	70/163 (aorta)	Gross sudanophilia	88
	75/155 (coronary)	Intimal thickening, microscopic	
Papio sp.	14/23 (aorta)	Gross sudanophilia	54
	8/23 (aorta)	Fibrous plaques	
Pan troglodytes	2/5 (aorta)	Sudanophilia	139
	5/5 (coronary)	Intimal thickening	
Captive			
Galago sp.	2/11 (aorta)	Fatty streaking	24
Callimico sp.	6/47 (aorta)	Gross plaques	64
Callimico sp.	9/23 (aorta)	Fatty streaking	24
Saimiri sciureus	72/74 (aorta)	Microscopic atheromata	92
	17/58 (coronary)	Microscopic atheromata	
Cebus albifrons	3/12 (aorta)	Microscopic atheromata	11
	2/24 (coronary)	Microscopic atheromata	
Macaca (mixed sp.)	10/762 (aorta)	Gross plaques	64
Papio ursinus	19/31 (aorta)	Microscopic fat or fibrosis	37
Papio ursinus	31/44 (coronary)	Microscopic intimal fibrosis	35
Papio cynocephalus			
immature	2/54 (aorta)	Gross sudanophilia	41
mature	11/15 (aorta)	Gross sudanophilia	
	2/15 (coronary)	Fibroelastic plaques	
Papio (mixed sp.)	22/156 (aorta)	Gross sudanophilia or plaques	64
Cercopithecus aethiops	32/82 (aorta)	Gross sudanophilia	41
Pan satyrus	6/9 (aorta)	Fatty streaking	24
Pan satyrus	13/13 (aorta)	Microscopic intimal thickening	68
	12/13 (coronary)	Microscopic intimal thickening	
Pan sp.	16/17 (aorta)	Gross sudanophilia	3
Pan sp.	11/14 (aorta)	Gross sudanophilia	131

Fig. 16. Squirrel monkey, adult female. Aortic sudanophilia. High fat diet. Arch of the aorta on the right.

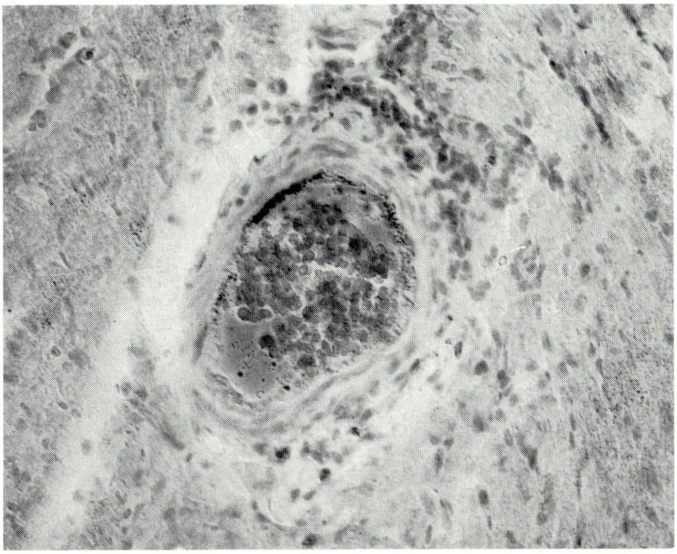

Fig. 17. Squirrel monkey, adult female. Small droplets of fat along the intima and internal elastic membrane of a coronary artery of a squirrel monkey fed a high fat diet. Frozen section, oil red 0 stain ($\times 250$).

The minimal recognizable lesions have consisted of either lipid droplet accumulation in the intima, along the internal elastica (fig. 17), or in the cells of the inner media; or intimal thickening with increased cellularity, accumulation of acid mucopolysaccharides, and formation of elastic and collagen fibers without lipid deposition (fig. 18); or a combination of both processes. The mature lesion has shown the classic pattern of an intimal plaque with fibrous cap, deep necrosis with masses of free lipid and sometimes calcification, and extensive fragmentation and reduplication of elastic fibers (fig. 19). Side-by-side comparisons by single observers of material from various species carry special weight in the argument for a unity of the disease in all pri-

Fig. 18. Rhesus monkey, adult female. Aortic intimal thickening. There is some reduplication of the internal elastica, and numerous fine elastic fibers are present in the plaque Verhoeff-van Gieson stain (× 70).

mates. In a detailed light microscopic study of aortic and coronary lesions, Andrus and Portman [2] concluded that except for extent and progression, there were no essential differences in spontaneous disease in woolly monkeys and chimpanzees, between spontaneous and diet-induced disease both in these species and in cebus and squirrel monkeys, or between atherosclerosis in any of these animals and man. A study of the hearts and aortas of 67 primates from 17 species of New and Old World monkeys and apes, all of which were captive and many of which were over 10 years old, brought Lindsay and Chaikoff [68] to the same conclusion. However, they believed that human and induced simian atherosclerosis show more lipid deposition than the spontaneous variety in monkeys.

Detailed histologic descriptions have been provided in one-species studies of aortas and coronary arteries in howler monkeys [73, 77], wild and captive baboons [35, 37, 66, 88], *Cebus albifrons* [11], squirrel monkeys [92, 94], rhesus monkeys [117, 136], and chimpanzees [3, 131].

Ultrastructural changes in early aortic intimal lesions appear to be the same in spontaneous and induced atherosclerosis in rhesus monkeys [117], in baboons [34] (fig. 20), in squirrel monkeys [86, 87], and in man [6, 33]. The

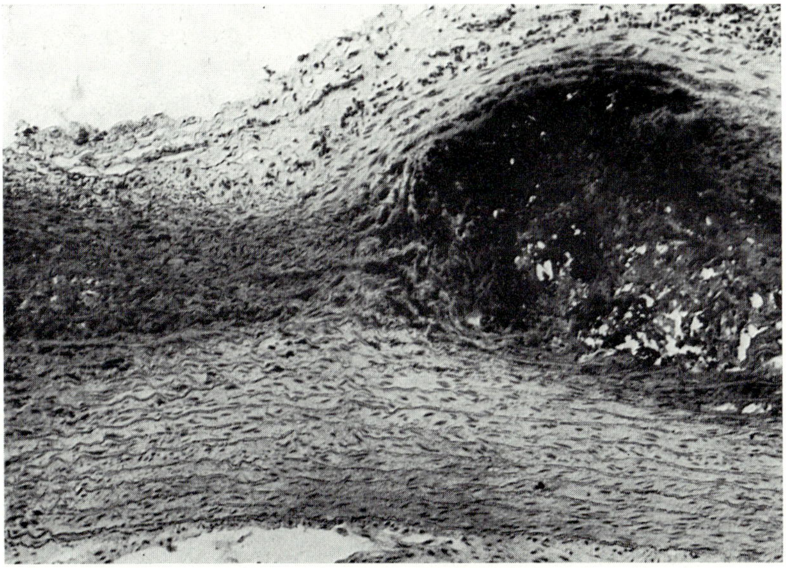

Fig. 19. Rhesus monkey. Mature atheroma in thoracic aorta of an animal fed a diet high in fat and cholesterol. Besides considerable intimal proliferation, fibrosis, intense calcification, cholesterol slits and necrosis of the plaque are also present. Frozen section, oil red 0 stain (× 100). Fig. 9 in WISSLER [147], reprinted with permission of the publisher, S. Karger.

key migratory, proliferative, and early lipidophilic element in all is the smooth muscle cell, although the origin of the foam cell is still in doubt.

How simple intimal thickening with lipid deposition relates to frank atherosclerosis has not yet been resolved by studies of primate vessels. LAPIN and YAKOVLEVA [64] carefully distinguished such thickening as a 'physiologic' development or aging process in macaques. MARUFFO and MALINOW [81] also found intimal thickening in nonatherosclerotic areas of the aortas of old howler monkeys and ascribed it to age. But ANDRUS and PORTMAN [2] and LINDSAY and CHAIKOFF [68] believed that fibromusculoelastic intimal thickening without lipid is the earliest change in atherosclerosis.

Although atherosclerosis seems morphologically the same in all primates, the susceptibility of different species is not uniform, and the differences do not correlate with taxonomy. The majority of wild baboons [88], *Cercocebus albigena* [130], and *Cercopithecus aethiops* [41] have aortic sudanophilia, whereas few *Cercopithecus ascanius* [130] and macaques [14, 64] do. Yet cellular intimal thickening in the aorta and coronaries is common in rhe-

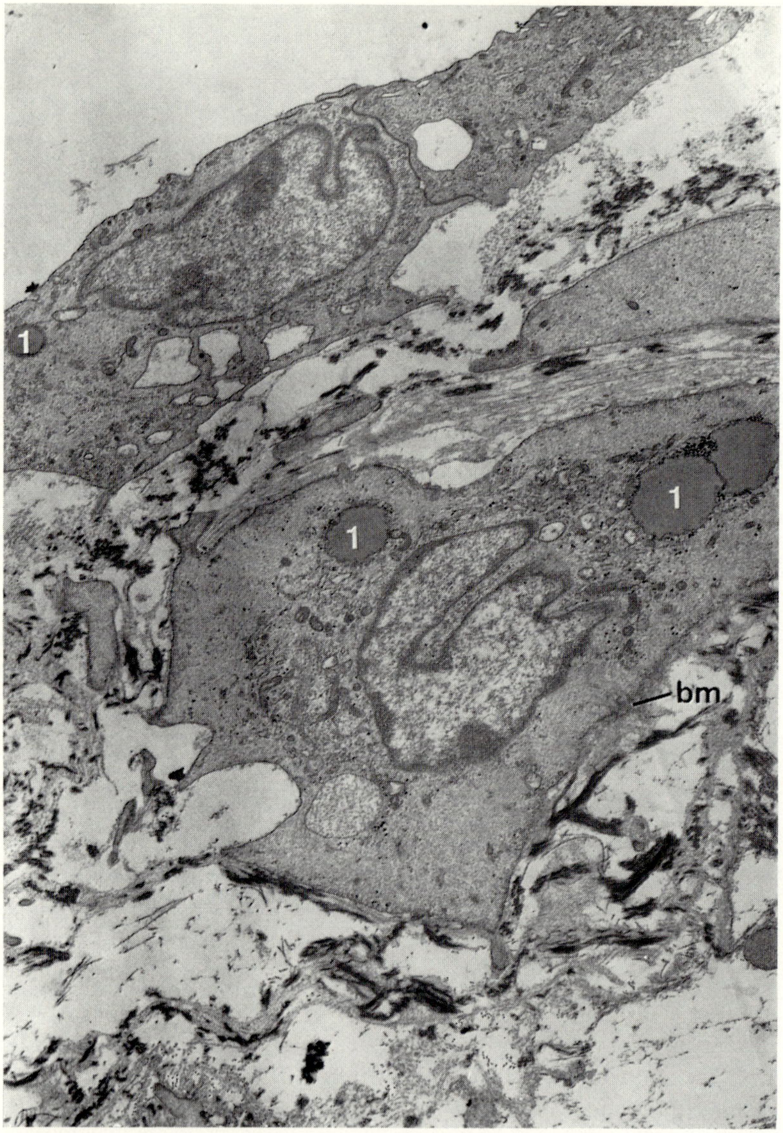

Fig. 20. Papio cynocephalus, young adult male. Sudanophilic intimal thickening in aorta, experimental diet. Endothelial cell has lipid inclusion (1) in cytoplasm similar to those in smooth-muscle cell at center. Smooth-muscle cell shows a limiting basement membrane (bm), numerous pinocytotic vesicles along the plasma membrane, myofilaments in the cytoplasm, and numerous granules of glycogen about the lipid inclusions (× 14,080). Fig. 2 in Geer *et al.* [34], reprinted with permission of the American Journal of Pathology.

sus monkeys[5] [14, 64]. Among New World monkeys, natural aortic sudanophilia is common in squirrel [94] and howler monkeys [73] but uncommon in *Cebus albifrons* [11, 17]; but both cebus and squirrel monkeys respond to atherogenic diet [11, 74, 78, 93, 146].

Furthermore, individual predisposition to atherosclerosis is probably not normally distributed even among animals of one species, sex, and age. When a semiquantitative method was adopted for scoring coronary arterial sudanophilia in squirrel monkeys fed a semipurified, high cholesterol, high fat diet, a distinctly asymmetric distribution of scores was obtained, which appeared to be at least bimodal[6]. Genetic studies are desirable and may be feasible in the growing number of primate breeding colonies.

The prevalence and severity of atherosclerosis correlates with age in baboons [41, 88], macaques [64], squirrel monkeys [94], howler monkeys [73], mixed Old and New World monkeys and apes [68]. Arteriosclerosis in zoological collections, reviewed by FINLAYSON [24], has occurred mostly in elderly animals.

However, sex differences have not been found, except for a suggestively higher incidence of aortic sudanophilia in wild adult female squirrel monkeys [94] and a definite male preponderance in wild howler monkeys [73].

The familiar consequences of atherosclerosis in the human heart, brain, kidneys, abdominal viscera, and extremities are rare in non-human primates. Only five spontaneous myocardial infarcts have been reported as noted above, and no unambiguous spontaneous arteriosclerotic changes in the brain or other organs have been described. Still the assumption that peripheral arteries are largely unaffected in monkeys and apes is unwarranted. The disease was more severe in coronary and cerebral arteries than in the aorta in 2 captive chimpanzees and one gorilla [128] (fig. 21). Sudanophilia and plaques are common in the cerebral arteries of captive chimpanzees fed a 'natural' diet [3, 125]. Fibroelastic intimal thickening in cerebral vessels occurs in a small percentage of macaques [14, 64], although none was found in a large series of wild adult baboons with a high incidence of aortic sudanophilia [88].

Other systemic arteries have also been involved: subclavian, iliac, carotid, and abdominal arteries in chimpanzees [3] and renal arteries in baboons [49, 50] and rhesus monkeys[7] [14].

5 Unpublished.
6 Unpublished.
7 Unpublished.

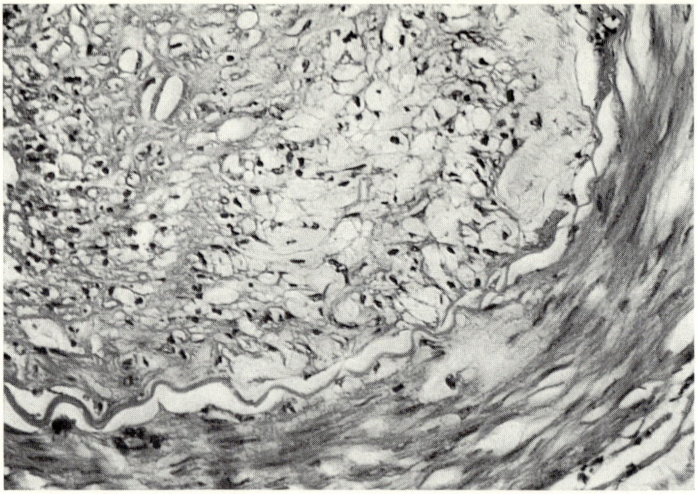

Fig. 21. Pan troglodytes, female, 9 years old. Section through a large intimal plaque in the basilar artery. The inner portion of the lesion contains a large clump of cells whose cytoplasms are greatly distended with lipid. Toward the lumen (upper left) the plaque contains more collagen, although oil red 0 stains of adjacent sections revealed considerable lipid in this region also. Hematoxylin and eosin stain (×128). Fig. 2 in Stout and Lemmon [128], reprinted with permission of Experimental and Molecular Pathology.

Induction

Intimal thickening and changes in elastica occur in the vessels of infant rhesus monkeys just as in man, but attempts to modify these processes by varying the type of fat in the formula during the first year of life have been unsuccessful [40, 99]. Nevertheless, prolonged elevated intake of dietary cholesterol and fat, particularly animal fat, has uniformly raised serum cholesterol and induced or aggravated aortic and coronary atherosclerosis in adults of every primate species tested. Among New World monkeys, experiments have been conducted mainly with cebus and squirrel monkeys. The results have been comparable, even though spontaneous atherosclerosis is unusual in cebus and common in squirrel monkeys (table III).

Using a diet adequate in choline, high in cholesterol and corn oil, and poor in sulfur-containing amino acids, Mann [78] induced hypercholesterolemia and atherosclerotic plaques in the aorta and major branches and coronaries of *Cebus fatuellus.* Microscopically, the plaques were composed chiefly of foam cells, with little fibrosis or change in elastica. As the author cautioned, applications to human disease must be limited since there is no reason to suspect such dietary deficiency to be common in man.

WISSLER [146] found that a high cholesterol, high fat diet was atherogenic in *Cebus albifrons* if the fat was supplied as butter or coconut oil, but not corn oil. BULLOCK *et al.* [11] reported that an atherogenic diet caused quantitatively more aortic sudanophilia in old than in young *Cebus albifrons*. Using essentially the basic semisynthetic diet of MANN with supplements of corn oil, coconut oil, and cholesterol, PORTMAN and ANDRUS [101] compared capuchins *(Cebus albifrons)*, woolly monkeys *(Lagothrix lagotricha)*, and squirrel monkeys for diet-induced susceptibility to aortic sudanophilia in 6 months. Squirrel monkeys were the most susceptible, cebus monkeys the least; coconut oil was more atherogenic than corn oil. In squirrel and woolly monkeys, the extent of aortic sudanophilia correlated well with serum cholesterol.

In squirrel monkeys hypercholesterolemia resulted from a mixed diet supplemented with cholesterol, fat, or both for 11 months [74]. Aortic atherosclerosis was increased by all lipid additions, but coronary sudanophilia was increased only when cholesterol was added to the diet. Butter aggravated this effect even though not preventing hypercholesterolemia; margarine seemed to be protective. In later experiments with a semisynthetic diet supplemented by cholesterol and butter, more extensive changes occurred in the aorta and coronary arteries in 4 months than in 11 months in the first series [76]. In this later series, intramural coronary arteries were more severely afflicted with intimal and medial deposits of lipid than the larger subepicardial vessels.

Young and old squirrel monkeys did not differ in the extent of aortic atherosclerosis induced by the addition of cholesterol to the diet for 12 months, but coronary arterial disease was more severe in the old animals. Supplemental dietary protein increased the atheromatous response to cholesterol in both groups [93].

Among Old World monkeys, only baboons, rhesus and cynomolgus monkeys have been tested. Investigating pyridoxine deficiency in rhesus monkeys, RINEHART and GREENBERG [108] unexpectedly induced lesions morphologically resembling atherosclerosis in man. MUSHETT and EMERSON [95] confirmed their results in the rhesus monkey, KUZUYA [62] in the Japanese macaque. However, evidence is lacking that this deficiency ordinarily contributes to atherosclerosis in either man or monkeys.

An early attempt to cause disease in 2 young rhesus monkeys with modest supplemental cholesterol for 1½ years was unsuccessful [55]. However, MANN and ANDRUS [79] found massive atheromatosis in an adult rhesus male on a 3½ year diet containing high levels of cholesterol and fat supplied as egg yolk and corn oil. After the third month, the level of blood cholesterol was around 1,000 mg%; by the 28th month, cutaneous xanthomata had appeared. At autopsy diffuse sudanophilia and confluent plaques were found in the aorta and major branches; the coronary arteries were strikingly narrowed, and isolated intimal lesions were found as far distally as the circle of Willis and the radial and dorsalis pedis arteries. Microscopically the intimal lesions were typical atheromata.

This single animal was not atypical; Taylor [136] caused moderate to severe atherosclerosis of the aorta and branches and coronaries in 15 rhesus monkeys fed high butter and cholesterol diets for 5 years. The serum cholesterol remained above 250 mg%, ranging to much higher levels. Two animals with levels consistently over 500 mg% developed cutaneous and tendinous xanthomata and exhibited the most severe atherosclerosis at autopsy. One of them suffered a fatal myocardial infarction (as noted above) with atheromatous occlusion of the major coronary arteries (fig. 13). The earliest changes consisted simply of stainable extracellular lipid in the intima and along the elastica. Progressively more advanced lesions showed phagocytosis of the lipids, reactive fibrosis, and finally mature atheromatous plaque formation with central necrosis, massive free lipid deposits, hyalinization, vascularization of the intima and media, and even calcification and thrombosis. Control animals, like rhesus on a normal diet, showed only minimal thickening of the aortic intima.

Kramsch and Hollander [61] found that dietary atherosclerosis showed a marked predilection for the proximal segments of coronary arteries in cynomolgus monkeys. After 12 months on a diet containing cholesterol and cottonseed oil, young adult males had hypercholesterolemia and gross atherosclerosis in the coronary arteries but not in the aorta or its branches. Morphologically the coronary lesions consisted of (1) intimal thickening with changes in the elastic fibers and accumulation of acid mucopolysaccharides but not of stainable fat; (2) the same with extra- and intracellular lipid; (3) markedly swollen intima with alterations in elastic fibers and foam cell accumulation, but little or no acid mucopolysaccharides. Microscopically, the first 2 types of lesion were seen in controls that had no gross atherosclerosis.

The authors stressed that the cerebral vessels were free from change, but in preliminary experiments Malinow[8] observed early intimal plaques in the cerebral vessels of cynomolgus monkeys fed a semipurified diet supplemented with cholesterol and butter.

8 Unpublished.

Fatty streaks in the aorta of baboons, which have a high incidence of naturally occurring aortic sudanophilia, resulted when they were fed dietary cholesterol [41, 132]; however, the coronary arteries were only slightly affected.

ANDRUS *et al.* [3] aggravated aortic and coronary atherosclerosis in chimpanzees with high dietary fat and cholesterol. Corn oil alone was ineffective in these animals as in other species, whereas coconut oil and cholesterol, separately or additively, raised serum cholesterol and accentuated the disease. In accordance with their earlier results [2], the authors concluded that the lesions were morphologically indistinguishable from those in New World monkeys, although atheromata in the epicardial coronary arteries were more frequent and severe in the apes.

'Stress' is difficult to define, let alone measure, but it may influence the course of atherosclerosis in primates. LAPIN and YAKOVLEVA [64] believed that vascular disease was more marked in animals under prolonged psychic stress, and RATCLIFFE [104] attributed the rising incidence of arteriosclerosis in a zoo population, including primates, in part to increased crowding. ST. CLAIR *et al.* [124] found in 30 squirrel monkeys that the average serum cholesterol, which was 116 mg% soon after trapping, had risen to 194 mg% when they arrived at the laboratory. When the animals were placed in large semi-secluded cages, the level fell to 139 mg% but rose again to 164 mg% when they were moved to noisier, more crowded cages.

LANG [63] found higher serum cholesterol and more coronary atherosclerosis in squirrel monkeys subjected to psychic stress and electric shock for 25 months than in controls fed the same moderately high fat, high cholesterol diet. VLODAVER [142] reported intimal thickening in the coronary arteries of young adult rhesus monkeys after only a few weeks of captivity in closely confining cages, whereas no such lesions were present in freshly imported animals.

Whether spontaneous or induced atherosclerosis can be modified by drugs has not been widely explored. Wild adult female squirrel monkeys have been reported somewhat more prone to aortic sudanophilia than males [94], but manipulation of the female endocrine status by spaying or administering progesterone and estrogen, singly or in combinations, did not affect the arterial changes induced by an atherogenic diet[9].

Hypothyroidism in man is accompanied by hypercholesterolemia and may accelerate atherosclerosis. When hypothyroidism was induced in rhesus

9 Unpublished.

monkeys with ^{131}I, severe atheromatosis of the aorta, renal arteries, and coronary arteries was seen after 1 to 2 years of a diet high in fat and very high (5 to 10%) in cholesterol [151]. The hypothyroid state did not accelerate the development of atherosclerosis but did cause higher levels of cholesterol. Furthermore, the extent of the disease did not seem to differ from that seen by Taylor [136] in euthyroid monkeys on a similar diet for about the same time. Scott *et al.* [118] found more advanced atheromata in the aortas of rhesus monkeys receiving propylthiouracil and sodium cholate in a high fat, high cholesterol diet; but since the animals not receiving the antithyroid drug also received much less cholesterol, the significance of suppressing thyroid activity was not clear.

Malinow[10] saw no effect of d-thyroxine on diet-induced atherosclerosis in squirrel monkeys. In parallel experiments, pyridinolcarbamate [76], reportedly protective against experimental atherosclerosis in rabbits [120], and chondroitin sulfate A were also ineffective.

Comments: With increasing age, most primates, including man, are affected to some extent with atherosclerosis. Any dietary regimen that elevates blood cholesterol also accelerates the disease in most though not all animals. The wide variation in individual susceptibility is a serious hindrance to experimental work. Monkeys are comparatively expensive, increasingly rare, and difficult to manage, and hence their use in large numbers to neutralize individual variations is impractical. Methods of estimating the susceptibility of each animal early in the course of an experiment are being developed[11] [18], and such methods may themselves shed light on pathogenesis.

Species also differ in susceptibility and vary in pathogenesis. For example, stainable lipid in minimal lesions accumulates more in New World than in Old World simians (except for baboons), and the small intramural coronary arteries are the first to be affected in New World monkeys, whereas the proximal epicardial segments are primarily involved in man and probably in macaques.

Finally, data on the progression of atherosclerosis in captive, aging, and 'normally' fed monkeys have not yet been systematically gathered. Those data that are available have come almost entirely from zoos and the Sukhumi Station in Russia; because of the pressure for use, animals in experimental colonies seldom grow old enough to exhibit advanced atherosclerosis.

Hypertensive Arteriopathy

Little is known about the anatomical changes that occur with spontaneous hypertension in the simian vascular tree. A study of 6 rhesus monkeys and 8 baboons with hypertension, caged for long periods of time or experi-

10 Unpublished.

11 Unpublished.

a 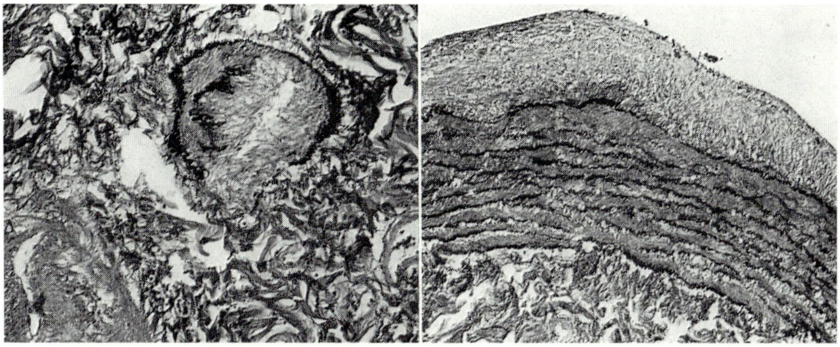 *b*

Fig. 22. Rhesus monkey, adult male. Pulmonary arterial changes due to lung mites *(Pneumonyssus simicola)*. (a) Medium-size pulmonary arteries in a bronchial ray, adjacent to a focus of infestation. The media and intima are distorted and hyperplastic, and much of the elastica is destroyed. Verhoeff-van Gieson stain (× 100). (b) Intimal thickening and interruption in the internal elastica in a large pulmonary artery, remote from foci of infestation. Verhoeff-van Gieson stain (× 100).

mentally made neurotic, showed thickening and smudging of the arteriole walls in the kidney and brain. The kidneys showed patchy and diffuse atrophy of the tubules, scarring, and glomerular fibrosis. One rhesus monkey died with a massive right intracerebral hemorrhage, which was attributed to hypertension [64].

In macaques severely infested with lung mites *(Pneumonyssus simicola)*, marked arterial intimal thickening, arterial tortuosity, and arteriolar medial hypertrophy occurs in the pulmonary circuit [148] (fig. 22). Pulmonary hypertension is probably a complication of protracted and severe infestation with these parasites, but neither pulmonary arterial pressure nor the relative weight of the right ventricle has yet been measured.

Experimental hypertension, induced by some variant of renal arterial constriction, has been followed by arteriolar changes in the eye and brain. Nine months after bilateral renal artery constriction in a macaque, KEYES and GOLDBLATT [60] found in the eyegrounds bilateral localized areas of edema and copper- or silver-wire arteries with segmental occlusion. Fifteen months after the operation, the monkey died of acute enteritis; the blood pressure had been 300/220 mm Hg during the last six months. The retinal arterioles showed reduplication of the elastica and adventitial thickening; the choroidal arteries had more advanced changes in the media plus hyaline degeneration of the intima.

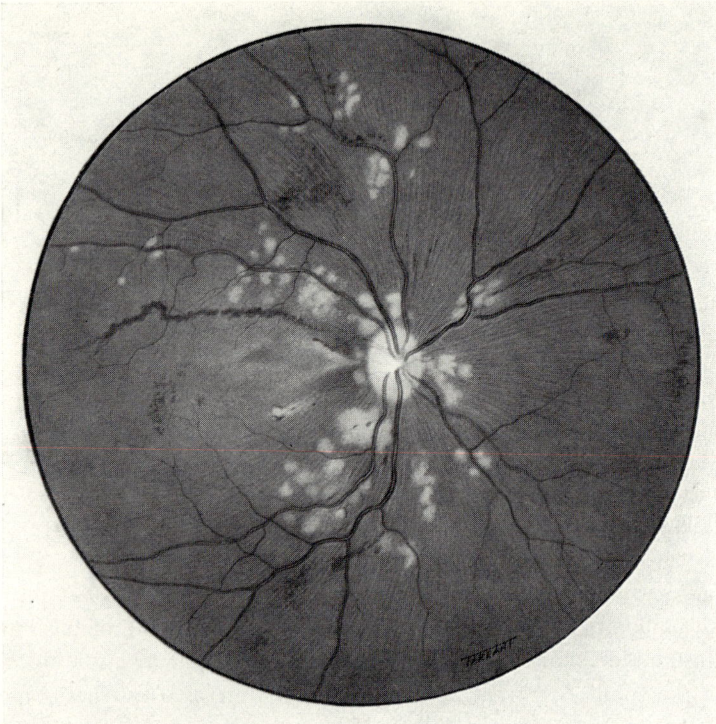

Fig. 23. Cynomolgus monkey. Drawing of the fundus of the right eye in experimental renal hypertension, showing neuroretinal edema, numerous cotton-wool spots, linear and punctate hemorrhages, and arterial narrowing. Fig. 1 in Ashton *et al.* [5], reprinted with permission of the Transactions of the Ophthalmological Societies of the United Kingdom.

Ashton *et al.* [5] reproduced the full complex of human malignant hypertensive retinopathy – neuroretinal edema, cotton-wool spots, hemorrhages, and arteriolar narrowing – in a cynomolgus monkey with surgical renal arterial stricture and hypertension of 6 weeks duration (fig. 23). Surprisingly, the systolic pressure reached only 170, though it rose from a normal level of 120 in 8 days. In studies with both light- and electron-microscopy, the choroidal vessels were unaffected, whereas the retina was edematous and partly detached, the axons were edematous, and the arterioles were segmentally necrotic. No vascular changes were found in other organs, including the brain, whereas Rodda and Denny-Brown [112] described segmental spasm and frank arteriolonecrosis in both the cortical and meningeal vessels of cynomolgus monkeys made hypertensive with Goldblatt clips.

The effect of hypertension on atherogenesis was tested by McGill *et al.* [89] in rhesus monkeys made hypertensive by constriction of the renal arteries. The animals were fed a mixture containing 60 mg cholesterol (supplied by horsemeat and lard) per day, substantially at the level of the average American diet. Two types of lesions developed in the aorta: fatty streaks primarily in the descending thoracic aorta and fibrous plaques more commonly in the abdominal portion. Histologically, these arterial lesions differed in several respects from those found in man. They contained more interstitial mucous material and less fat and fibrous tissue. None of the complications of human lesions (calcification, vascularization, hemorrhage, necrosis, or thrombosis) were found in the monkeys, though it should be noted that the maximum exposure of any animal was only 20 months.

Medial Calcification

Medial calcification has been reported infrequently. After toxic amounts of vitamin D had been accidentally administered to 558 rhesus monkeys, 9 of the 24 animals that died between the 55th and 88th day after the start of the high vitamin D diet showed myocardial necrosis [59]. The intramyocardial vessels were often mineralized, sometimes partially necrotic and occasionally thrombosed. Mineral deposits were seen in the aortic media along the elastic fibers and along the internal elastica in other exposed animals of the group. Fine calcific granules were deposited in the aortic media of a mona monkey *(Cercopithecus mona)* and of a male thick-tailed bushbaby *(Galago crassicaudatus)* that had been in captivity for 9 years [24]. In a brown lemur *(Lemur fulvus)*, the aortic ring was calcified, and many small, hard, white nodules were present in the lower thoracic and abdominal aorta. These nodules consisted of foci of lysed muscle cells and calcium encrustations on elastica lamellae [24].

In a case of extensive calcific disease of the aorta in a free-ranging howler monkey the entire aorta and proximal portions of the branches of the arch and of the renal and mesenteric arteries were firm and inelastic with a rugose intimal surface [82]. Although the intima was diffusely fibrous, no sudanophilic material was grossly detectable, and atheromata were not found microscopically. So much calcium was present in the media that the structure was obliterated, but none was seen in the subendothelium (fig. 24). In the less severely affected areas, calcification appeared to begin on the elastic lamellae, including the internal elastica. Large amounts of PAS-positive material were near the calcified masses as well as in the calcium-free areas of the aorta and main branches.

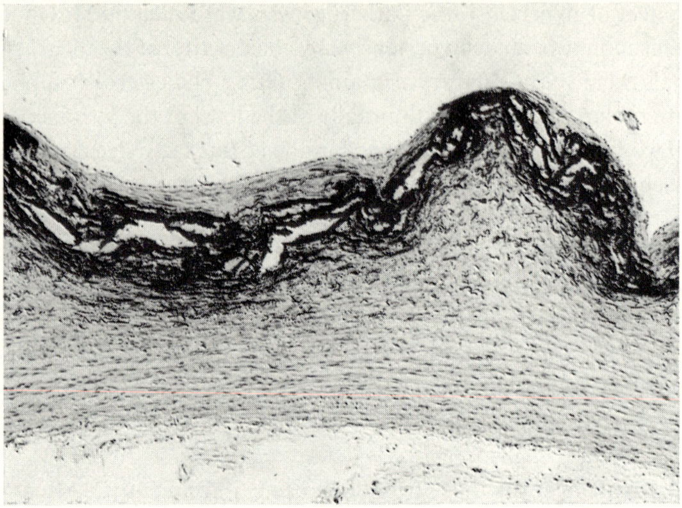

Fig. 24. Howler monkey, adult male. Marked deposition of calcium in the internal third of the media of the aorta, with preferential disposition in the vicinity of elastic lamellae. von Kossa stain (×27).

Necrotizing Arteritis (Periarteritis nodosa)

Arterial lesions called periarteritis nodosa were seen in the viscera of 4 unilaterally nephrectomized rhesus monkeys treated for 80 days with 2-methyl-9-fluorocortisol, a potent synthetic mineralocorticoid [119]. The microscopic descriptions were meager, however, and the published microphotographs showed chiefly intimal thickening rather than the classic changes of periarteritis described in man. The arterial changes may well have been those of malignant hypertension; in fact, elevated blood pressure and severe nephrosclerosis were seen in the animals.

Occasionally necrotizing, inflammatory changes are seen in the small coronary arteries of squirrel monkeys given a semipurified diet high in cholesterol and butter[12] (fig. 25). The pathogenesis and relation to the experimental diet are unknown.

Aneurysms

Syphilis is not natural in primates other than man [113], and abdominal aortic atherosclerosis of the devastating severity so common in man has not been seen in monkeys or apes; hence aneurysms from these causes have not

12 Unpublished.

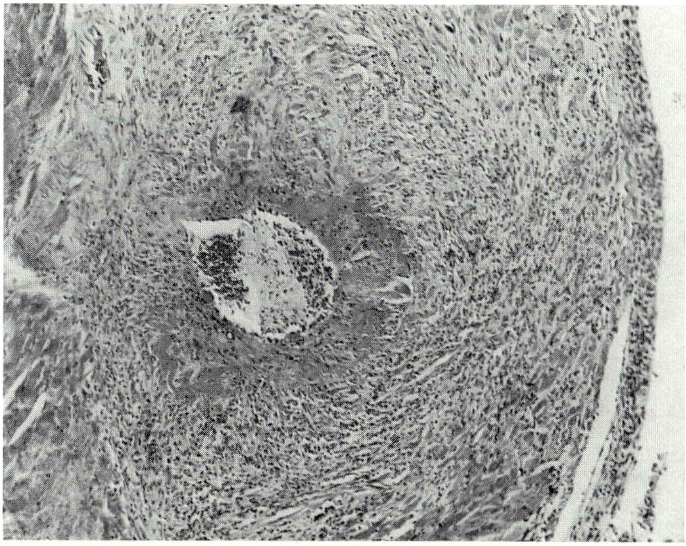

Fig. 25. Squirrel monkey, adult male. High cholesterol – high fat diet. Fibrinoid necrosis of the wall of an intramyocardial coronary artery, with surrounding acute inflammatory cellular reaction. Hematoxylin and eosin (× 100).

been reported. A probably mycotic aneurysm occurred in a red monkey *(Erythrocebus patas)* captive in the Philadelphia Zoo for 14 years [28]. It was circumscribed, partly calcified, and located just above the aortic valve. Gram-positive bacteria were present in the thrombus overlying an aneurysm with an extensively and chronically inflamed wall in the ascending aorta of a vervet et *(Cercopithecus aethiops)* monkey [24].

Saccular aneurysms associated with medial defects at the forks of cerebral vessels have been found in chimpanzees [3, 125] (fig. 26). One animal died from subarachnoid hemorrhage from the rupture of one of the sacs. Considerable atherosclerosis was present in the cerebral arteries, but no definite relationship to the aneurysms could be established. As in the common human lesion, the arterial media and internal elastic lamina ended abruptly at the entrance to the sac.

A dissecting aneurysm in the lower thoracic aorta of a squirrel monkey was noted in passing by MIDDLETON *et al.* [92], but histologic details were not given. A 'double-barrelled aorta' extending from the second intercostal arch to the coeliac axis was found in a male howler monkey obtained in the wild [83] (fig. 27). The internal surface was rugose in the false lumen, which was

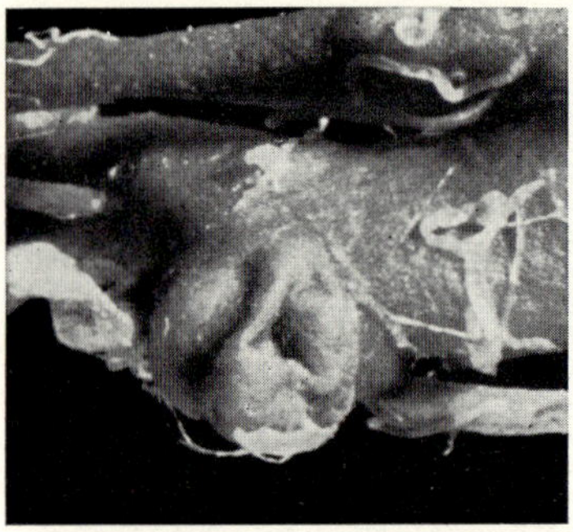

a

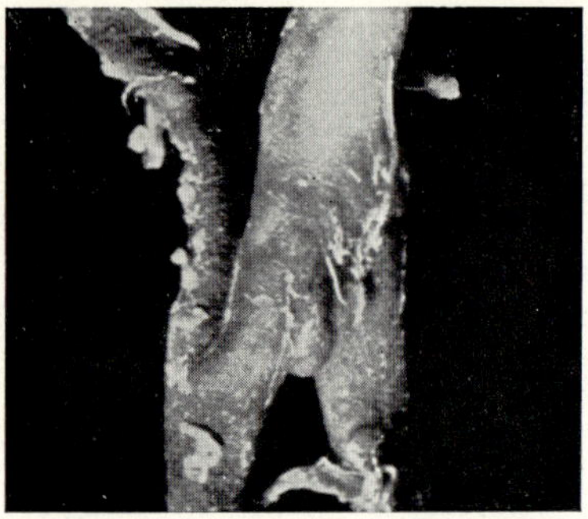

b

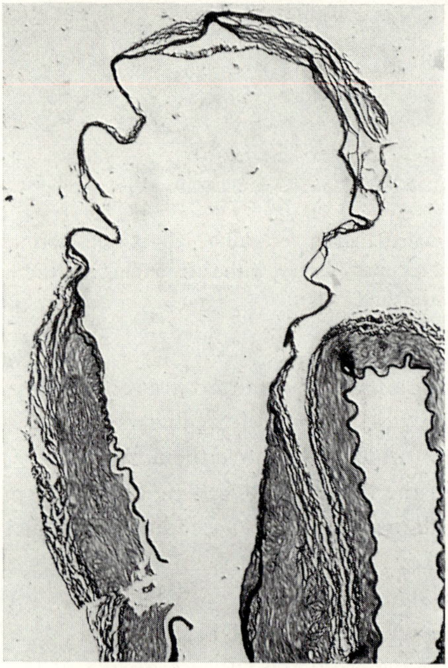

c

Fig. 26. (a and b) *Pan troglodytes*, female, 8 years old. Small aneurysm with collapsed sac, arising from the internal carotid artery, (a), and small aneurysm in fork of anterior communicating artery and the left anterior cerebral artery, (b). Fig. 1 and 2 in Stehbens [125], reprinted with permission of the Journal of Pathology and Bacteriology. (c). Chimpanzee, female, approximately 18 years old. Right middle cerebral artery with marked attenuation of the wall, which over wide areas consists solely of a thin collagenous membrane. Fig. 20 in Andrus *et al.* [3], reprinted with permission of the publisher, S. Karger.

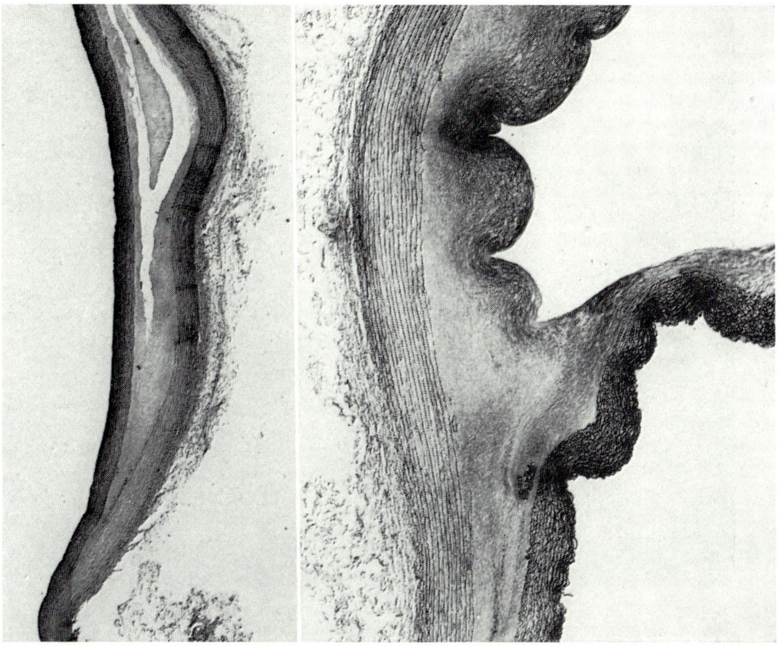

a b

Fig. 27. Howler monkey, adult male. Dissecting aneurysm of the descending aorta. (a) Edge of the aneurysm, showing split of the media to form a false lumen. True lumen to the left. Verhoeff-van Gieson stain (× 11). (b) Another level, showing a thick layer of fibrous tissue containing many small elastic fibers lining the false lumen. True lumen to the right. Verhoeff-van Gieson stain (× 34).

bridged by fibrous strands. Foci of basophilic, PAS-positive material were seen in the media proximal to the dissection, with fragmentation of the elastic fibers and necrosis of the muscle cells, whereas both walls of the false channel showed minimal alterations except for calcification of the elastic lamellae. An atherosclerotic plaque was seen in the arch, unrelated to the dissection, and the subintima of the true lumen of the descending aorta showed minimal fibrosis. The authors believed the changes met the criteria of classic 'idiopathic cystic necrosis'.

Veins

Spontaneous disease of the systemic veins, such as thrombophlebitis, varicosities, hemorrhoids, and arteriovenous fistulae has not been reported.

a 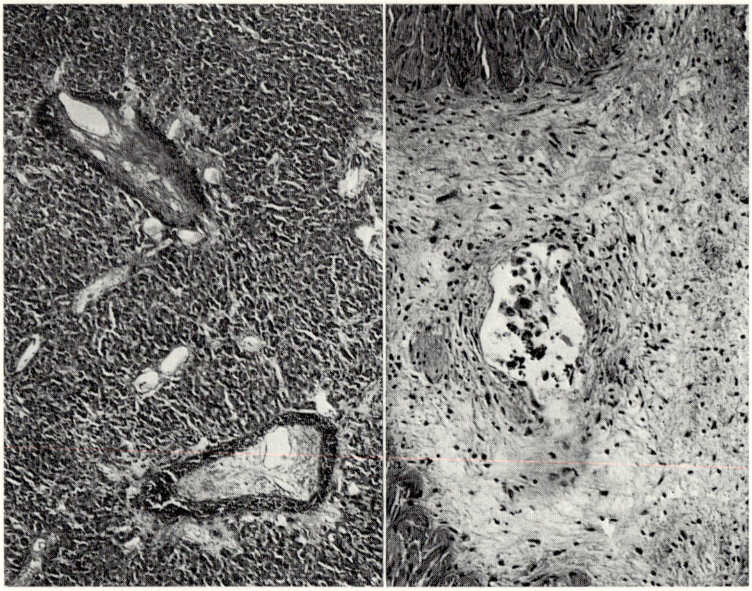 b

Fig. 28. Macaca speciosa. Veno-occlusive disease of the liver due to *Crotolaria* alkaloids. (a) Central veins three weeks after a single injection, which were probably completely occluded, show recanalization of loose edematous fibrous tissue and a peripheral ring of dense collagen. Hematoxylin and eosin stain (×163). (b) After 6 months and four injections, the structure of the original vessel has been replaced by irregular fibrous tissue with a single small lumen, either residual or a recanalization. Hematoxylin and eosin stain (×410). Courtesy of Dr. J. R. Allen.

Since chronic liver disease is also uncommon, naturally occurring alterations in the veins of the portal system have not been found. Experimental damage to the liver and abdominal great veins has, however, produced expected changes. Two monkeys infected with cercariae of *Schistosoma mansoni* developed granulomatous periphlebitis and fibrosis around small and medium-sized intrahepatic portal veins, in reaction to both ova and dead worms; and histologic evidence of portal hypertension was present in the spleens of both monkeys [38]. After the portal vein was ligated in 6 monkeys 60 to 90 days after ligation of the supradiaphragmatic inferior vena cava and azygos veins, esophageal varices and prominent superficial abdominal veins developed in 4 of 6 monkeys [65]. 'Cavernous transformation' occurred in 4 of 13 portacaval shunts in rhesus monkeys, the transformation apparently representing enlargement of the collaterals rather than recanalization of a thrombus [145].

Ingestion by West Indian natives of 'bush tea' prepared from members of the leguminous *Crotolaria* genus sometimes results in obliterative endophlebitis in small and medium hepatic veins together with centrilobular necrosis and eventual cirrhosis and portal hypertension [9]. ALLEN [1] produced similar changes in *Macaca speciosa* by both oral and intraperitoneal administration of monocrotaline, extracted from *Crotolaria spectabilis*. The primary lesion was believed to be damage to the endothelium of the central veins, with ensuing edema, hemorrhage, and thrombosis leading finally to fibrosis and occlusion of the vessels (fig. 28).

References

1 ALLEN, J. R.; CARSTENS, L. A., and KATAGIRI, G. J.: Hepatic veins of monkeys with veno-occlusive disease. Arch. Path. *87:* 279–289 (1969).

2 ANDRUS, S. B. and PORTMAN, O. W.: Comparative studies of spontaneous and experimental atherosclerosis in primates. Symp. zool. Soc., Lond. *17:* 161–177 (1966).

3 ANDRUS, S. B.; PORTMAN, O. W., and RIOPELLE, A. J.: Comparative studies of spontaneous and experimental atherosclerosis in primates. II. Lesions in chimpanzees including myocardial infarction and cerebral aneurysms. Progr. biochem. Pharmacol., vol. 4, pp. 393–419 (Karger, Basel/New York 1968).

4 APPLEBY, E. C. and FRASER, G.: Six fatal cases of non-specific pericarditis in the chimpanzee. J. Path. Bact. *84:* 245–248 (1962).

5 ASHTON, N.; PELTIER, S., and GARNER, A.: Experimental hypertensive retinopathy in the monkey. Trans. ophthal. Soc. U.K. *88:* 167–186 (1969).

6 BALIS, J. U.; HAUST, M. D., and MORE, R. H.: Electron-microscopic studies in human atherosclerosis; cellular elements in aortic fatty streaks. Exp. molec. Path. *3:* 511–525 (1964).

7 BARKER, L. F.; PALMER, A. E.; KIRSCHSTEIN, R. L.; HOPPS, H. E., and PATT, J. K.: Experimental infection with *Rickettsia prowaseki* in rhesus and vervet monkeys. J. infect. Dis. *117:* 285–291 (1967).

8 BERNIRSCHKE, K. and RICHART, R.: Spontaneous acute toxoplasmosis in a marmoset monkey. Amer. J. trop. Med. Hyg. *9:* 269–273 (1969).

9 BRAS, G.; JELLIFFE, D. B., and STUART, K. L.: Veno-occlusive disease of the liver with nonportal type of cirrhosis, occurring in Jamaica. Arch. Path. *57:* 285–300 (1954).

10 BULLOCK, B. C.; WOLF, R. H., and CLARKSON, T. B.: Myocarditis associated with trypanosomiasis in a cebus monkey *(Cebus albifrons)*. J. Amer. vet. med. Ass. *151:* 920–922 (1967).

11 BULLOCK, B. C.; CLARKSON, T. B.; LEHNER, N. D. M.; LOFLAND, H. B., jr., and ST.CLAIR, R. W.: Atherosclerosis in *Cebus albifrons* monkeys. III. Clinical and pathologic studies. Exp. molec. Path. *10:* 39–62 (1969).

12 CHASE, R. E. and DEGARIS, C. F.: Anomalies of venae cavae superiores in an orang. Amer. J. phys. Anthrop. *24:* 61–65 (1938).

13 CHAUHAN, S.; NAYAK, N. C., and RAMALINGASWAMI, V.: The heart and skeletal mus-

cle in experimental protein malnutrition in rhesus monkeys. J. Path. Bact. *90:* 301–309 (1965).

14 Chawla, K. K.; Murthy, C. D. S.; Chakavarti, R. N., and Chhuttani, P. N.: Arteriosclerosis and thrombosis in wild rhesus monkeys. Amer. Heart J. *73:* 85–91 (1967).

15 Chrisp, C. E. and MacLeod, I. F.: Bilateral endogenous endophthalmitis in a rhesus monkey *(Macaca mulatta)* with bacterial *(Aerobacter aerogenes)* endocarditis and septicemia. J. ped. Opthal. *6:* 166–169 (1969).

16 Clarkson, T. B.: Spontaneous atherosclerosis in subhuman primates; in Roberts and Straus Comparative atherosclerosis, pp. 211–214 (Harper and Row, New York 1965).

17 Clarkson, T. B.; Bullock, B. C., and Lehner, N. D. M.: Pathologic characteristics of atherosclerosis in New World monkeys. Progr. biochem. Pharmacol., vol. 4, pp. 420–428 (Karger, Basel/New York 1968).

18 Clarkson, T. B.: Personal communication (1970).

19 Cowen, D. and Wolf, A.: Toxoplasmosis in the monkey; acute fatal infection experimentally produced in a young *Macaca mulatta.* J. infect. Dis. *77:* 144–157 (1945).

20 DeGaris, C. F.: Pericardial patency and partial ectocardia in a newborn orang-utan. Anat. Rec. *59:* 69–82 (1934).

21 Dick, G. W. A.: Mengo encephalomyelitis virus; pathogenicity for animals and physical properties. Brit. J. exp. Path. *29:* 559–577 (1948).

22 Fairbrother, R. W. and Hurst, E. W.: Spontaneous diseases observed in 600 monkeys. J. Path. Bact. *35:* 867–873 (1932).

23 Fiennes, R. N. T-W-: The Zoological Society of London, report of the Society's pathologist for the year 1963. J. Zool. *148:* 341–362 (1962).

24 Finlayson, R.: Spontaneous arterial disease in exotic animals. J. Zool. *147:* 239–343 (1965).

25 Fitch, C. D. and Dinning, J. S.: Vitamin E deficiency in the monkey. V. Estimated requirements and the influence of fat deficiency and antioxidants on the syndrome. J. Nutr. *79:* 69–78 (1963).

26 Follis, R. H., jr.: Experimental thiamine deficiency. Proc. conf. on beriberi, endemic goiter, and hypervitaminosis A, Princeton 1958. Fed. Proc. *17:* suppl. 2, pp. 23–31 (1958).

27 Fox, H.: Disease in captive wild mammals and birds, pp. 48–82 (Lippincott, Philadelphia 1923).

28 Fox, H.: Special subjects of pathological interest. Rep. Lab. Mus. comp. Path. zool. Soc., Phil. *1927:* 26–27.

29 Fox, H.: Arteriosclerosis in lower mammals and birds; its relation to the disease in man; in Cowdry Arteriosclerosis; a survey of the problem, pp. 155–157 (McMillan, New York 1933).

30 Fox, H.: Some comments on arteriosclerosis in wild mammals and birds. Bull. N.Y. Acad. Med. *15:* 748–756 (1939).

31 Freigang, B. and Knobil, E.: Patent ductus arteriosus with pulmonary hypertension and arteritis in a rhesus monkey. Yale J. Biol. Med. *40:* 239–242 (1967).

32 Fremming, B. D.; Benson, R. E., and Young, R. J.: Weights of organs in sixty-six male *Macaca mulatta* monkeys. J. appl. Physiol. *8:* 155–158 (1955).

33 GEER, J. C.: Fine structure of human aortic intimal thickening and fatty streaks. Lab. Invest. *14:* 1764–83 (1965).

34 GEER, J. C.; CATSULIS, C.; McGILL, H. C., and Strong, J. P.: Fine structure of baboon aortic fatty streak. Amer. J. Path. *52:* 265–286 (1968).

35 GILBERT, C. and GILLMAN, J.: Structural modifications in the coronary artery of the baboon *(Papio ursinus)* with special reference to age and endocrine status. Sth afr. J. med. Sci. *25:* 59–70 (1960).

36 GILLMAN, J. and GILBERT, C.: Primary amyloidosis in the baboon *(Papio ursinus);* its relationship to macromolecular syndromes and diseases of the connective tissue. Acta med. scand. *152:* suppl. 306, pp. 155–189 (1955).

37 GILLMAN, J. and GILBERT, C.: Atherosis in the baboon *(Papio ursinus);* its pathogenesis and etiology. Exp. Med. Surg. *15:* 181–221 (1957).

38 GILLMAN, T.: Venous obstruction in the pathogenesis of hepatic bilharziasis; a preliminary report of comparative findings in rats, monkeys and man. Ann. trop. Med. Parasit. *51:* 409–416 (1957).

39 GRAYSON, J. and IRVINE, M.: Myocardial infarction in the monkey; studies on the collateral circulation after acute coronary occlusion. Cardiovasc. Res. *2:* 170–178 (1968).

40 GREENBERG, L. D. and GONZALES, I. E.: Influence of fatty acid composition of infant formulas on development of arteriosclerosis. Exp. molec. Path. *10:* 240–249 (1969).

41 GRESHAM, G. A.; HOWARD, A. N.; MCQUEEN, J., and BOWYER, D. E.: Atherosclerosis in primates. Brit. J. exp. Path. *46:* 94–103 (1965).

42 HABERMANN, R. T. and WILLIAMS, F. P., jr.: Diseases seen at necropsy of 708 *Macaca mulatta* (rhesus monkey) and *Macaca philippinensis* (cynomolgus monkey). Amer. J. vet. Res. *18:* 419–426 (1957).

43 HACKEL, D. B.; KINNEY, T. D., and WENDT, W.: Pathologic lesions in captive wild animals. I. Interatrial septal defect in a chimpanzee. II. Adenocarcinoma of thyroid in a bear. III. Toxoplasmosis in a kangaroo. Lab. Invest. *2:* 154–163 (1953).

44 HADDOW, A. J.: Field and laboratory studies on an African monkey, *Cercopithecus ascanius schmidtii* Matschi. Proc. zool. Soc., Lond. *122:* 297–394 (1952–53).

45 HAMERTON, A. E.: Report on the deaths occurring in the society's gardens during the year 1930. Proc. zool. Soc., Lond. *1931:* 527–555.

46 HAMERTON, A. E.: Report on the deaths occurring in the society's gardens during the year 1934. Proc. zool. Soc., Lond. *1935:* 443–474.

47 HAMERTON, A. E.: Report on the deaths occurring in the society's gardens during the year 1935. Proc. zool. Soc., Lond. *1936:* 659–686.

48 HAMERTON, A. E.: Report on the deaths occurring in the society's gardens during the year 1936. Proc. zool. Soc., Lond. *107B:* 443–473 (1937).

49 HAMERTON, A. E.: Report on the deaths occurring in the society's gardens during the year 1937. Proc. zool. Soc., Lond. *108B:* 489–526 (1938).

50 HAMERTON, A. E.: Report on the deaths occurring in the society's gardens during the years 1939–1940. Proc. zool. Soc., Lond. *111B:* 151–185 (1941).

51 HAMERTON, A. E.: Report on the deaths in the society's gardens during 1942. Proc. zool. Soc., Lond. *113B:* 149–160 (1943).

52 HAMERTON, A. E.: Report on the deaths occurring in the society's gardens during the year 1944. Proc. zool. Soc., Lond. *115:* 371–384 (1945–46).

53 HIGGINBOTHAM, F. H. and VINCENT, K.: Histochemical study of young rhesus aorta. Anat. Rec. *160:* 365–366 (1968).

54 HOWARD, A. N.; GRESHAM, G. A.; BOWYER, D. E., and LINDGREN, F. T.: Aortic and coronary atherosclerosis in baboons. Progr. biochem. Pharmacol., vol. 4, pp. 438–444 (Karger, Basel/New York 1968).

55 HUEPER, W. C.: Experimental studies in cardiovascular pathology. XIV. Experimental atheromatosis in *Macacus rhesus* monkeys. Amer. J. Path. *22:* 1287–1291 (1946).

56 IGLESIAS, R. and LIPSCHÜTZ, A.: Effects of prolonged estrogen administration in female New World monkeys, with observations on pericardial neoplasm. J. Endocrin. *5:* 88–98 (1947–48).

57 KENNARD, M. A.: Abnormal findings in 246 consecutive autopsies on monkeys. Yale J. Biol. Med. *13:* 701–712 (1941).

58 KENNARD, M. A. and WILLNER, M. D.: Findings at autopsies of seventy anthropoid apes. Endocrinology, Springf. *28:* 967–976 (1941).

59 KENT, S. P.; VAWTER, G. F.; DOWBEN, R. N., and BENSON, R. E.: Hypervitaminosis D in monkeys; a clinical and pathologic study. Amer. J. Path. *34:* 37–59 (1958).

60 KEYES, J. E. L. and GOLDBLATT, H.: Experimental hypertension. VIII. Vascular changes in the eyes. Arch. Ophthal., Chicago *20:* 812–828 (1938).

61 KRAMSCH, D. M. and HOLLANDER, W.: Occlusive atherosclerotic disease of the coronary arteries in monkey *(Macaca irus)* induced by diet. Exp. molec. Path. *9:* 1–22 (1968).

62 KUZUYA, H.: Arteriosclerosis in pyridoxine-deficient monkeys. Primates *2:* 99 (1959).

63 LANG, C. M.: Effects of psychic stress on atherosclerosis in the squirrel monkey *(Saimiri sciureus)*. Proc. Soc. exp. Biol. N.Y. *126:* 30–34 (1967).

64 LAPIN, B. A. and YAKOVLEVA, L. A.: Noninfectious diseases, diseases of the cardiovascular system; in Comparative pathology in monkeys, pp. 132–176 (Thomas, Springfield 1963).

65 LAUFMAN, H.; BERNARD, V.; ROACH, H. D., and CHAMPLAIN, G.: Experimental production of esophageal varices in the *Macaca rhesus*. Surg. Gynec. Obstet. *110:* 451–456 (1960).

66 LINDSAY, S. and CHAIKOFF, I. L.: Arteriosclerosis in the baboon. Arch. Path. *63:* 460–471 (1957).

67 LINDSAY, S. and CHAIKOFF, I. L.: Naturally occurring arteriosclerosis in animals: a comparison with experimentally induced lesions; in SANDLER and BOURNE Atherosclerosis and its origin, pp. 349–437 (Academic Press, New York 1963).

68 LINDSAY, A. and CHAIKOFF, I. L.: Naturally occurring arteriosclerosis in nonhuman primates. J. Atheroscler. Res. *6:* 36–61 (1966).

69 LUCKE, B.: Spontaneous cerebral lesions in monkeys; their significance in experimental pathology. Arch. Neural Psychiat. *10:* 212–225 (1923).

70 MALINOW, M. R.: Atherosclerosis in subhuman primates. Folia primat. *3:* 277–300 (1965).

71 MALINOW, M. R. and MARUFFO, C. A.: Aortic atherosclerosis in free-ranging howler monkeys *(Alouatta caraya)*. Nature, Lond. *206:* 948–949 (1965).

72 MALINOW, M. R.: An electrocardiographic study of *Macaca mulatta*. Folia primat. *4:* 51–65 (1966).

73 Malinow, M. R. and Maruffo, C. A.: Naturally occurring atherosclerosis in howler monkeys *(Alouatta caraya)*. J. Atheroscler. Res. *6:* 368–380 (1966).

74 Malinow, M. R.; Maruffo, C. A., and Perley, A. M.: Experimental atherosclerosis in squirrel monkeys *(Saimiri sciurea)*. J. Path. Bact. *92:* 491–510 (1966).

75 Malinow, M. R. and Maruffo, C. A.: Arterial pathology; in Malinow Biology of the howler monkey. Bibl. primat *7:* 175–189 (Karger, Basel/New York 1968).

76 Malinow, M. R.; Perley, A., and McLaughlin, P.: The effect of pyridinolcarbamate on aortic and coronary atherosclerosis in squirrel monkeys *(Saimiri sciurea)*. J. Atheroscler. Res. *8:* 455–461 (1968).

77 Malinow, M. R. and Storvick, C. A.: Spontaneous coronary lesions in howler monkeys *(Alouatta caraya)*. J. Atheroscler. Res. *8:* 421–431 (1968).

78 Mann, G. V.; Andrus, S. B.; McNally, A., and Stare, F. J.: Experimental atherosclerosis in cebus monkeys. J. exp. Med. *98:* 195–217 (1953).

79 Mann, G. V. and Andrus, S. B.: Xanthomatosis and atherosclerosis produced by diet in an adult rhesus monkey. J. Lab. clin. Med. *48:* 533–550 (1956).

80 Manning, G. W.: Coronary disease in the ape. Amer. Heart J. *23:* 719–724 (1942).

81 Maruffo, C. A. and Malinow, M. R.: Postnatal changes in the aorta of howler monkeys *(Alouatta caraya)*. J. Geront. *21:* 119–123 (1966).

82 Maruffo, C. A. and Malinow, M. R.: Aortic mediocalcinosis in a howler monkey *(Alouatta caraya)*. J. Path. Bact. *92:* 236–238 (1966).

83 Maruffo, C. A. and Malinow, M. R.: Dissecting aneurysm of the aorta in a howler monkey *(Alouatta caraya)*. J. Path. Bact. *92:* 567–570 (1966).

84 Maruffo, C. A. and Malinow, M. R.: Pathologic changes of the heart in free-ranging howler monkeys (*Alouatta caraya*, Humboldt, 1812). Amer. J. vet. Res. *28:* 237–243 (1967).

85 Maruffo, C. A.: General pathology; in Malinow Biology of the howler monkey. Bibl. primat. *7:* 151–174 (Karger, Basel/New York 1968).

86 McCombs, H. L.; Hayes, K. C., and McGandy, R. B.: Ultrastructure of coronary artery atherosclerosis in the monkey. Fed. Proc. *28:* 448 (1969).

87 McCombs, H. L.; Zook, B. C., and McGandy, R. B.: Fine structure of spontaneous atherosclerosis of the aorta in the squirrel monkey. Amer. J. Path. *55:* 235–252 (1969).

88 McGill, H. C., jr.; Strong, J. P.; Holman, R. L., and Werthessen, N. T.: Arterial lesions in the Kenya baboon. Circulat. Res. *8:* 670–679 (1960).

89 McGill, H. C., jr.; Frank, M. H., and Geer, J. C.: Aortic lesions in hypertensive monkeys. Arch. Path. *71:* 96–102 (1961).

90 McKissick, G. E.; Ratcliffe, H. L., and Koestner, A.: Enzootic toxoplasmosis in caged squirrel monkeys *Saimiri sciureus*. Path. vet. *5:* 538–560 (1968).

91 McNulty, W. P., jr. and Wuest, H. M.: Thalidomide, teratogeny and structure: teratogenic action of WU-385 on the rhesus. Teratology *2:* 265 (1969).

92 Middleton, C. C.; Clarkson, T. B.; Lofland, H. B., and Prichard, R. W.: Atherosclerosis in the squirrel monkey. Arch. Path. *78:* 16–23 (1964).

93 Middleton, C. C.; Clarkson, T. B.; Lofland, H. B., and Prichard, R. W.: Diet and atherosclerosis of squirrel monkeys. Arch. Path. *83:* 145–153 (1967).

94 Middleton, C. C.; Rosal, J.; Clarkson, T. B.; Newman, W. P., and McGill, H. C., jr.: Arterial lesions in squirrel monkeys. Arch. Path. *83:* 352–358 (1967).

95 Mushett, C. W. and Emerson, G. A.: Arteriosclerosis in pyridoxine-deficient monkeys and dogs. Fed. Proc. *15:* 526 (1956).

96 Nair, C. P. and Ray, A. P.: Observations on the incidence and types of tuberculosis in rhesus monkeys on autopsy studies. Indian J. Malariol. *9:* 185–189 (1955).

97 Nigi, H. and Itakura, C.: Spontaneous toxoplasmosis in *Lemur catta.* Primates *9:* 155–160 (1968).

98 Nishimura, E. Y.; Patterson, R.; Fink, J. N., and Wennemark, J. N.: Myocardial lesions associated with experimental passive transfer of immediate type hypersensitivity. Lab. Invest. *15:* 1269–1278 (1966).

99 Pickering, D. E.; Fisher, D. A.; Perley, A.; Basinger, G. M., and Moon, H. D.: Influence of dietary fatty acids on serum lipids. Amer. J. Dis. Child. *102:* 42–51 (1961).

100 Pilot, I.: A mucoid encapsulated hemolytic streptococcus in fatal sepsis of an orang-utan. J. infect. Dis. *61:* 220–221 (1937).

101 Portman, O. W. and Andrus, S. A.: Comparative evaluation of three species of New World monkeys for studies of dietary factors, tissue lipids, and atherogenesis. J. Nutr. *87:* 429–438 (1965).

102 Prichard, R. W. and Bullock, B. C.: Spontaneous myocardial necrosis in a squirrel monkey. Fed. Proc. *24:* 310 (1965).

103 Racela, A. S., jr.; Grady, H. J.; Higginson, J., and Svoboda, D. J.: Protein deficiency in rhesus monkeys. Amer. J. Path. *49:* 419–443 (1966).

104 Ratcliffe, H. L.; Yerasimides, T. G., and Elliott, G. A.: Changes in the character and location of arterial lesions in mammals and birds in the Philadelphia zoological garden. Circulation *21:* 730–738 (1960).

105 Ratcliffe, H. L.: Rep. Penrose Res. Lab. zool. Soc. Philadelphia *1961:* 12–13.

106 Ratcliffe, H. L.: Rep. Penrose Res. Lab. zool. Soc. Philadelphia *1962:* 19–20.

107 Rewell, R. E.: Uterine fibromyomas and bilateral ovarian granulosa-cell tumours in a senile squirrel monkey, *Saimiri sciurea.* J. Path. Bact. *68:* 291–293 (1954).

108 Rinehart, J. F. and Greenberg, L. D.: Arteriosclerotic lesions in pyridoxine-deficient monkeys. Amer. J. Path. *25:* 481–491 (1949).

109 Rinehart, J. F. and Greenberg, L. D.: Effect of experimental thiamine deficiency on the heart of the rhesus monkey. Arch. Path. *48:* 89–95 (1949).

110 Roberts, W. C. and Innes, J. R. M.: Valvular heart disease in the monkey *(Macaca mulatta).* Amer. Heart J. *72:* 206–213 (1966).

111 Roca-Garcia, M. and Sanmartin-Barberi, C.: The isolation of encephalomyocarditis virus from *Aotus* monkeys. Amer. J. trop. Med. Hyg. *6:* 840–852 (1957).

112 Rodda, R. and Denny-Brown, D.: The cerebral arterioles in experimental hypertension. II. The development of arteriolonecrosis. Amer. J. Path. *49:*365–381 (1966).

113 Ruch, T. C.: Diseases of laboratory primates, pp. 352–379 (Saunders, Philadelphia 1959).

114 Scherer, H. J.: Vergleichende Pathologie des Nervensystems der Säugetiere, p. 108 (Thieme, Leipzig 1944).

115 Schmidt, E. C. H.: Virus myocarditis; pathologic and experimental studies. Amer. J. Path. *24:* 97–117 (1948).

116 Scott, H.: Report on the deaths occurring in the society's gardens during the year 1926. Proc. zool. Soc., Lond. *1927:* 173–198.

117 Scott, R. F.; Morrison, E. S.; Jarmolych, J.; Nam, S. C.; Kroms, M., and Couls-

TON, F.: Experimental atherosclerosis in rhesus monkeys. I. Gross and light microscopy features and lipid values in serum and aorta. Exp. molec. Path. *7:* 11–33 (1967).

118 SCOTT, R. F.; JONES, R.; DAOUD, A. S.; ZUMBO, O.; COULSTON, F., and THOMAS, W. A.: Experimental atherosclerosis in rhesus monkeys. II. Cellular elements of proliferative lesions and possible role of cytoplasmic degeneration in pathogenesis as studied by electron microscopy. Exp. molec. Path. *7:* 34–57 (1967).

119 SELYE, H. and BOIS, P.: The hormonal production of nephrosclerosis and periarteritis nodosa in the primate. Brit. med. J. *i:* 183–186 (1957).

120 SHIMAMOTO, T.; NUMANO, T., and FUJITA, T.: Atherosclerosis-inhibiting effect of an antibradykinin agent, pyridinolcarbamate. Amer. Heart J. *71:* 216–227 (1966).

121 SOTO, P. J., jr.; BEALL, F. A.; NAKAMURA, R. M., and KUPFERBERG, L. L.: Myocarditis in rhesus monkeys. Arch. Path. *78:* 681–690 (1964).

122 STADIE, R.: Etwas über Erkrankungen von Affen. Mit einem Nachwort von HONIGMANN, Breslau. Zool. Gart., Lpz., N.F. *3:* 55–58 (1930); cit. RUCH Diseases of laboratory primates, p. 356 (Saunders, Philadelphia 1959).

123 STAHL, W. R.; MALINOW, M. R.; MARUFFO, C. A.; POPE, B. L., and DEPAOLI, R.: Growth and age estimation of howler monkeys; in MALINOW Biology of the howler monkey, Bibl. primat. *7:* 59–80 (Karger, Basel/New York 1968).

124 ST.CLAIR, R. W.; MACNINTCH, J. E.; MIDDLETON, C. C.; CLARKSON, T. B., and LOFLAND, H. B.: Changes in serum cholesterol levels of squirrel monkeys during importation and acclimation. Lab. Invest. *16:* 828–832 (1967).

125 STEHBENS, W. E.: Cerebral aneurysms of animals other than man. J. Path. Bact. *86:* 161–168 (1963).

126 STEINER, P. E.; RASMUSSEN, T. B., and FISHER, L. E.: Neuropathy, cardiopathy, hemosiderosis, and testicular atrophy in *Gorilla gorilla.* Arch. Path. *59:* 5–25 (1955).

127 STOUT, C. and GROOVER, M. E., jr.: Spontaneous versus experimental atherosclerosis. Ann. N.Y. Acad. Sci. *162:* 89–98 (1969).

128 STOUT, C. and LEMMON, W. B.: Predominant coronary and cerebral atherosclerosis in captive nonhuman primates. Exp. molec. Path. *10:* 312–322 (1969).

129 STRONG, J. P.: Arterial lesions in primates; in ROBERTS and STRAUSS Comparative atherosclerosis, pp. 244–252 (Harper and Row, New York 1965).

130 STRONG, J. P. and TAPPEN, N. C.: Naturally occurring arterial lesions in African monkeys. Arch. Path. *79:* 199–205 (1965).

131 STRONG, J. P.; KRITCHEVSKY, D., and MARTINEZ, R. D.: Sudanophilic aortic lesions in chimpanzees. Arch. Path. *81:* 544–547 (1966).

132 STRONG, J. P. and MCGILL, H. C., jr.: Diet and experimental atherosclerosis in baboons. Amer. J. Path. *50:* 669–690 (1967).

133 STRONG, J. P.; EGGEN, D. A.; NEWMAN, W. P., III, and MARTINEZ, R. D.: Naturally occurring and experimental atherosclerosis in primates. Ann. N.Y. Acad. Sci. *149:* 882–894 (1968).

134 SUN, S. C.; SOHAL, R. S.; BURCH, G. E.; CHU, K. C., and COLCOLOUGH, H. L.: Coxsackie virus B4 pancarditis in cynomolgus monkeys resembling rheumatic heart lesions. Brit. J. exp. Path. *48:* 655–661 (1967).

135 TAYLOR, C. B.; PATTON, D. E., and COX, G. E.: Atherosclerosis in rhesus monkeys. VI. Fatal myocardial infarction in a monkey fed fat and cholesterol. Arch. Path. *76:* 404–412 (1963).

136 TAYLOR, C. B.: Experimental pathology of atherosclerosis and myocardial infarction

in rhesus monkeys. Meth. Achievem. exp. Path., vol. 1, pp. 355–384 (Karger, Basel/New York 1966).

137 Vanace, P. W.: Experimental streptococcal infection in the rhesus monkey. Ann. N.Y. Acad. Sci. *85:* 910–930 (1960).

138 Vastesaeger, M. M.; Gillot, P. et Parmentier, R.: L'athérosclérose coronarienne chez les vertébrés supérieurs vivant en jardin zoologique. Acta Cardiol *15:* 12–30 (1960).

139 Vastesaeger, M. M. and Delcourt, R.: Spontaneous atherosclerosis and diet in captive animals. Nutr. Diet. *3:* 174–188 (1961).

140 Vastesaeger, M. M.; Vercruysse, J. et Delcourt, R.: Tendance spécifique à l'atherosclérose et lipides alimentaires chez les animaux sauvages. Bull. Soc. roy. zool., Anvers *27:* 3–32 (1963).

141 Vastesaeger, M. and Delcourt, R.: Some aspects of spontaneous and experimental cardiovascular disease in Old World monkeys and Pongidae. Symp. zool. Soc., Lond. *17:* 179–194 (1966).

142 Vlodaver, Z.; Medalie, J., and Neufeld, H. N.: Coronary arteries in immature monkeys; preliminary report of relationships to activity and diet. J. Atheroscler. Res. *8:* 923–933 (1968).

143 Vogel, F. S. and Pinkerton, H.: Spontaneous salivary gland virus disease in chimpanzees. Arch. Path. *60:* 281–285 (1955).

144 Waart, A. de; Storm, C. J., and Loumans, A. K. J.: Ligation of the coronary arteries in Javanese monkeys. I. Introduction, general experimental results, especially the changes in the ventricular electrocardiogram. Amer. Heart J. *11:* 676–704 (1936).

145 Williams, A. O. and Johnston, G. W.: Cavernous transformation of the portal vein in rhesus monkeys. J. Path. Bact. *90:* 613–618 (1965).

146 Wissler, R. W.; Frazier, L. E.; Hughes, R. H., and Rasmussen, R. A.: Atherogenesis in the cebus monkey. I. A comparison of three food fats under controlled dietary conditions. Arch. Path. *74:* 312–322 (1962).

147 Wissler, R. W.: Recent progress in studies of experimental primate atherosclerosis. Progr. biochem. Pharmacol., vol. 4, pp. 378–392 (Karger, Basel/New York 1968).

148 Wolf, G. L.; Cole, C. R.; Carlisle, H. N., and Saslaw, S.: The pathogenesis of Rocky Mountain spotted fever in monkeys, infected by inhalation. Arch. Path. *84:* 486–494 (1967).

149 Woodard, J. C.: Acarous *(Pneumonyssus simicola)* arteritis in rhesus monkeys. J. amer. vet. med. Ass. *153:* 905–909 (1968).

150 Yakovleva, L. A.: A case of atherosclerosis in a macaque rhesus monkey; in Utkin Theoretical and practical problems of medicine and biology in experiments on monkeys, pp. 165–170 (Pergamon, New York 1960).

151 Younger, R. K.; Scott, H. W.; Butts, W. H., and Stephenson, S. E., jr.: Rapid production of experimental hypercholesterolemia and atherosclerosis in the rhesus monkey; comparison of five dietary regimens. J. surg. Res. *9:* 263–271 (1969).

152 Zaman, V. and Goh, T. K.: Isolation of *Toxoplasma gondii* from the slow loris, *Nycticebus coucang*. Ann. trop. Med. Parasit. *62:* 52–53 (1968).

Authors' address: Dr. W. P. McNulty and Dr. M. R. Malinow, Oregon Regional Primate Research Center, 505 N.W. 185th Avenue, *Beaverton, OR 97005* (USA)

Pathology of Simian Primates, Part I, pp. 809–820 (Karger, Basel 1972)

Blood Formation Disturbances in Monkeys

B. A. LAPIN and L. A. YAKOVLEVA

Institute of Experimental Pathology and Therapy, USSR Academy of Medical Sciences, Sukhumi

Contents

General 809
Disturbances of White Cell Production 809
The Effects of Radiation 815
Disturbances of Red Cell Production 817
References 819

General

This chapter includes the description of pathologic changes in the blood forming system appearing primarily as an independent disease as well as secondarily in the form of a complication of certain pathologic states. Such proliferative processes, affecting the blood forming system, as leukosis, lymphogranulomatosis, plasma cell neoplasms are included. The chapter does not contain material on local tumors.

Disturbances of White Cell Production

The true blood diseases connected with the development of proliferative processes are very rare in monkeys. The first described leukemia cases in a baboon and African green monkey [20 and 3] cited by T. RUCH [1958] in the chapter dealing with lymphocythemia remain obscure due to negligible changes in the peripheral blood (taking into consideration large variability of blood indices in healthy monkeys), slight spleen involvement and absence of

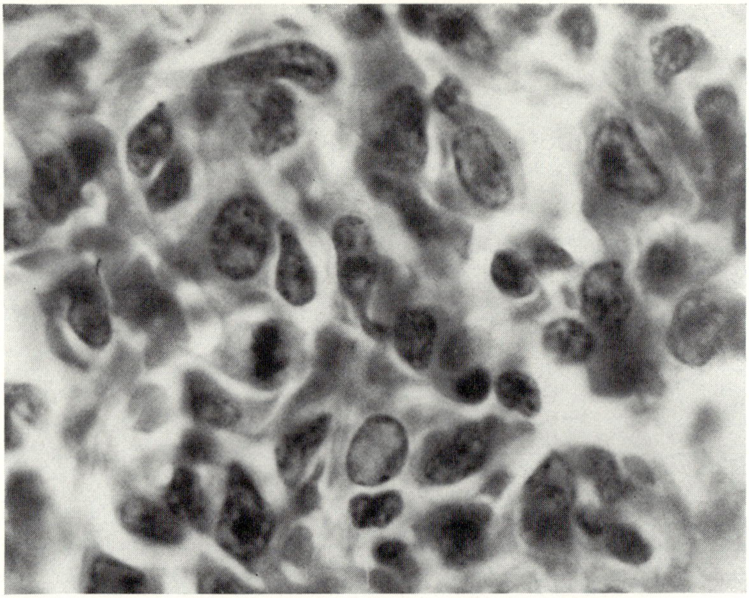

Fig. 1. Proliferation of reticular cells in spleen in reticulosis (Green monkey). HE × 900.

histopathological studies of viscera. Single cases of blood diseases are known in some monkeys having been in captivity for a long time.

There is a mention of the development of HODGKIN's disease in a baboon [12] as well as lymphogranulomatosis in a lion marmoset and of 4 cases of lympholeukemia in green monkeys, HUMBOLDT's woolly monkey and Senegal galago [9, 10, 11]. A number of cases of lymphatic leukemia and lymphoma with disseminated lesions of the lymphatic system and viscera were described in Stair's monkey, rhesus monkey and sooty mangabey [8, 13], [14 (cited by 18)]. The possibility of lymphatic leukemia development in gibbons with the involvement of different lymph node groups and viscera is also known [21, 6, 5, 19]. In the abovementioned studies the authors diagnosed the pathologic processes described by them as lymphoma, lymphosarcoma, lymphatic leukemia, but the patho-anatomical description (even only macroscopic in some cases) allowed us to treat this disease as leukemia. The description of a myeloid leukemia case in monkeys (genus and species not indicated) is also known [22]. 3 cases of proliferative processes in the blood forming system of adult green monkeys were registered in a troop of different species of monkeys (2,000 animals) during 15 years of observation

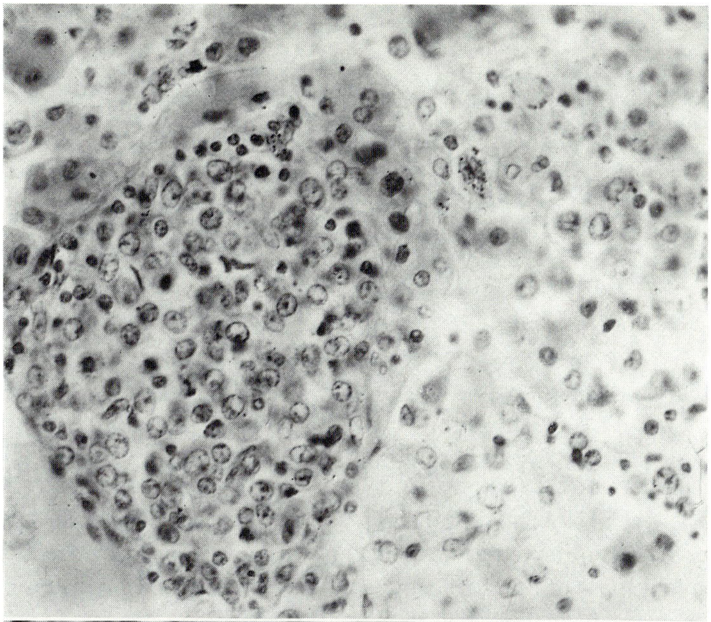

2

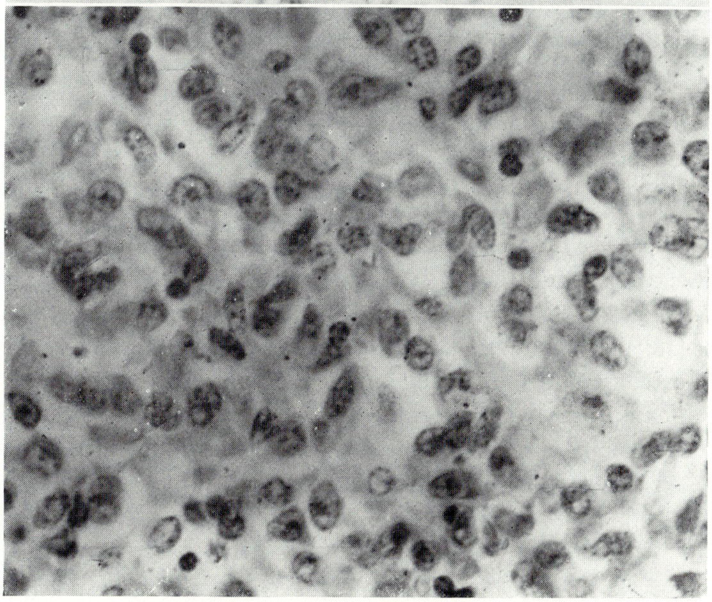

3

Fig. 2. Proliferation of reticular cells in liver in reticulosis (Green monkeys). HE × 300.

Fig. 3. Proliferation of reticular cells in lymph node in reticulosis (Green monkey). HE × 600.

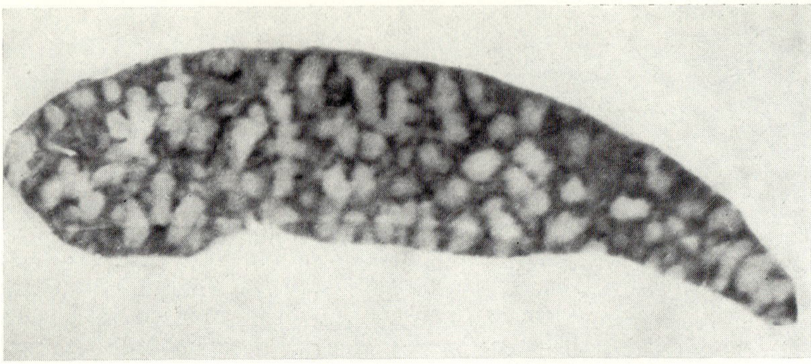

Fig. 4. Proliferation in spleen in malignant lymphogranulomatosis (Green monkey).

in the Sukhumi monkey colony [16, 17]. One was lymphogranulomatosis and two cases were of leukemia. Both cases of leukemia (haemocytoblastosis and reticulosis) were observed to be aleukemic. They were characterized by systemic involvement of different groups of lymph nodes and by the lesions of spleen, bone marrow and other viscera.

Histological findings showed proliferation of stem cell elements in spleen (fig. 1), bone marrow, liver (fig. 2), lymph nodes (fig. 3), kidneys and in some other organs.

In animals with lymphogranulomatosis the lesions of the spleen (fig. 4) and lymph nodes were the result of specific infiltration of these organs with young cell elements such as lymphocytes, reticulocytes (fig. 5), eosinophils and STERNBERG cells-like giant cells (fig. 6). Infiltration with these cells was also observed in lungs [fig. 7], bone marrow, and kidneys. Among blood forming system diseases, plasma cell neoplasms in green monkeys [17] and gibbons [18, 19] with multiple bone lesions resembling those in man in the same disease, are known to have been mentioned twice.

A disease of the bone system [25] accompanied by an increased bone fracture and proliferation of fibro-reticular and osteoid tissue at the site of lesions was also described in green monkeys.

Fig. 5. Proliferation of reticular cells in malignant lymphogranulomatosis (Green monkey). HE × 160.

Fig. 6. Giant cell in malignant lymphogranulomatosis. HE × 160.

Fig. 7. Lung infiltration in malignant lymphogranulomatosis (Green monkey). HE × 60.

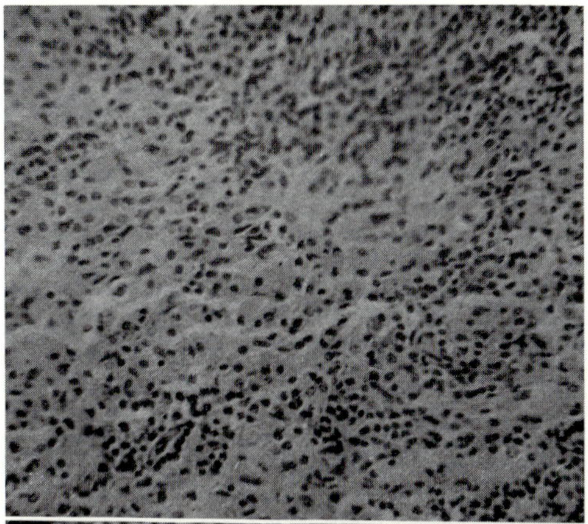

5

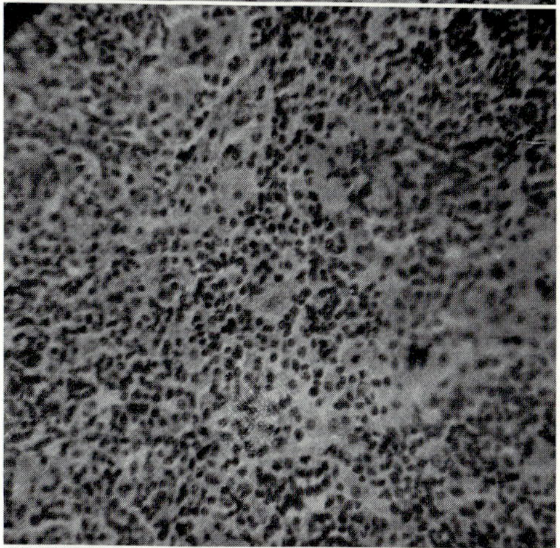

6

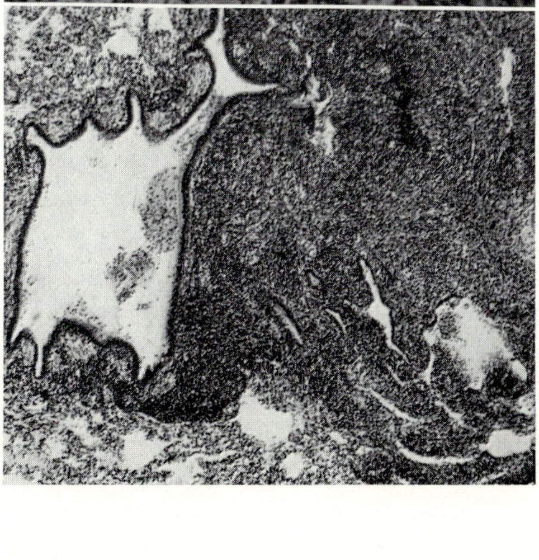

7

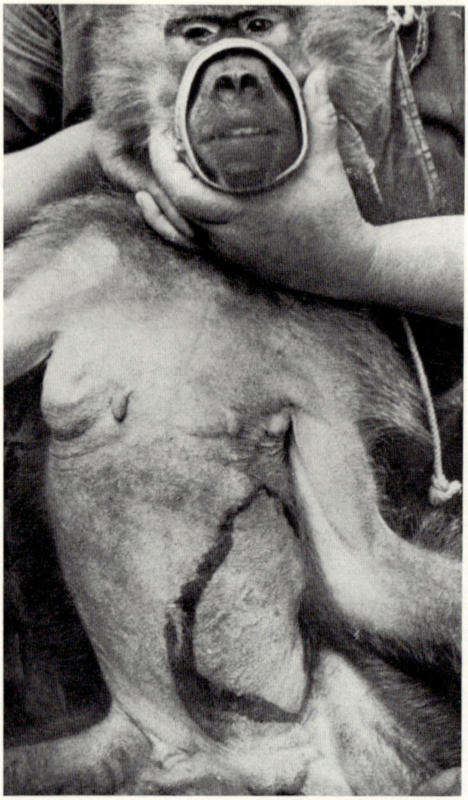

Fig. 8. A leukemic baboon with sharp splenomegaly (the contours of the spleen are outlined on the abdominal wall) and enlarged peripheral lymph nodes.

21 mortal cases resulting from leukosis-reticulosis [26, 27] were recorded among adult hamadryas baboons of the Sukhumi colony which were in contact with monkeys inoculated with human leukemic blood. The disease was accompanied by pronounced splenomegaly and generalized lymph node hyperplasia (fig. 8) and also the involvement of the liver (fig. 9), kidneys (fig. 10), lungs and bone marrow. Proliferation of immature reticular cells (fig. 11) was observed in these organs histologically. The incidence of the disease made 3.5% to the total stock of adult animals of these species (600 individuals).

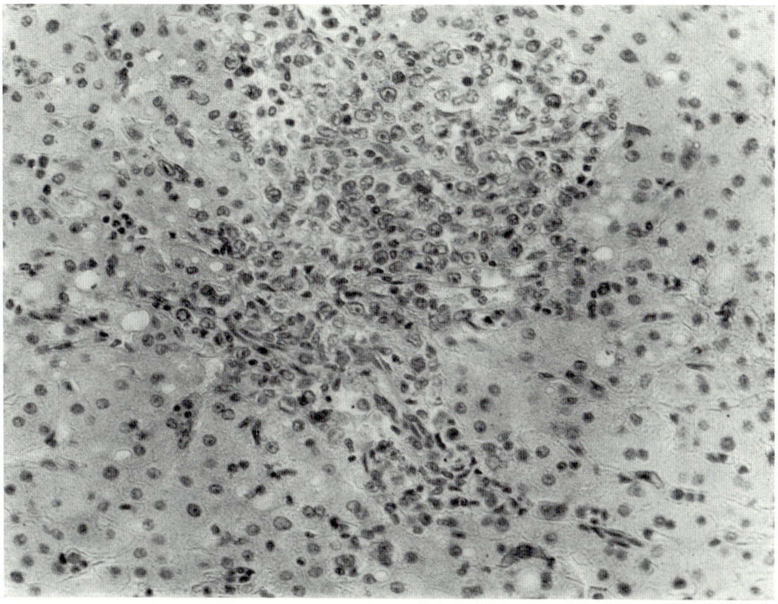

Fig. 9. Proliferation of immature reticular cells in the liver in reticulosis. HE × 600.

The Effects of Radiation

In addition to spontaneous development of proliferative processes in the blood forming system of monkeys the possibility of leukemia induction by ionizing radiation is also known. However, in contrast to the information regarding some rodents, single thought massive radiation seems to be insufficient for inducing leukemia in monkeys. Apparently in this connection there are no literary data on monkey leukemia as a result of single massive radiation in spite of the fact that in various countries large numbers of monkeys have been used in model studies of the mechanisms of action of ionizing radiation.

Long-term observations (more than 10 years) at the Sukhumi breeding station of a group of monkeys (200 animals, rhesus in the main) exposed to X- or γ-radiation in sublethal or minimal lethal doses did not reveal leukemia in these animals.

Prolonged chronic irradiation in small doses is apparently more effective in regard to the induction of monkey leukemia [7]. At the Sukhumi

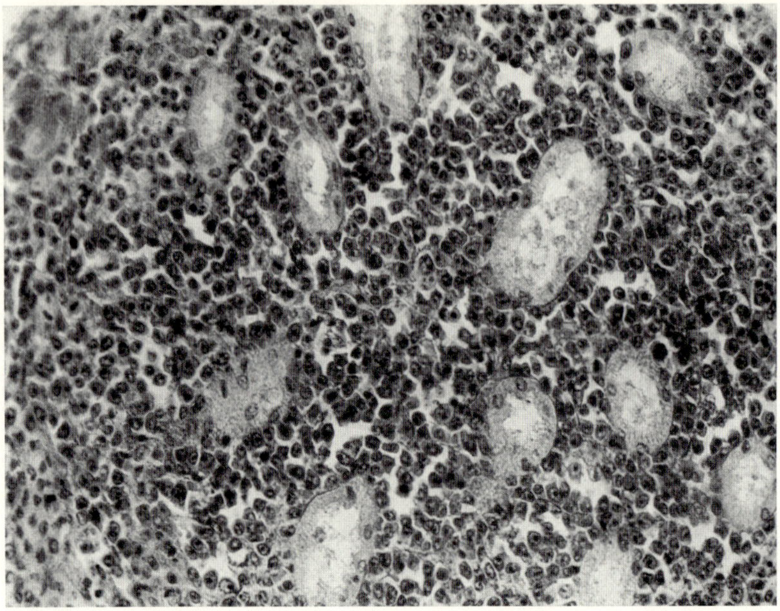

Fig. 10. Intensive infiltration of kidney tissue by malignant reticular cells. HE × 600.

breeding station one of 4 hamadryas baboons exposed to daily γ-radiation at a dose of 6 r for several years developed haemocytoblastosis. Radiation was discontinued as soon as the animal had received a total dose of 2,800 r. Haemocytoblastosis developed on the background of full recovery of blood formation after radiation. At the time of the animal's death its peripheral blood showed moderate leukocytosis (22,700 cells in 1 mm^3) and a great number of haemocytoblasts (35% cells of leukogramma). Haemocytoblasts accounted for 84% of bone marrow cells (fig. 12). The process was accompanied by a pronounced megaloblastic type of anemia, a haemorrhagic diathesis and extensive proliferation of undifferentiated cell elements in the lymph nodes, spleen, liver, thymus gland tissue, kidney and dura mater.

The second case of leukemia was revealed in one of 4 rhesus monkeys exposed to long-term neutron radiation. Myeloid leukemia with a large number of cell forms in the peripheral blood appeared after the animal was exposed to a total dose of 680 r [28].

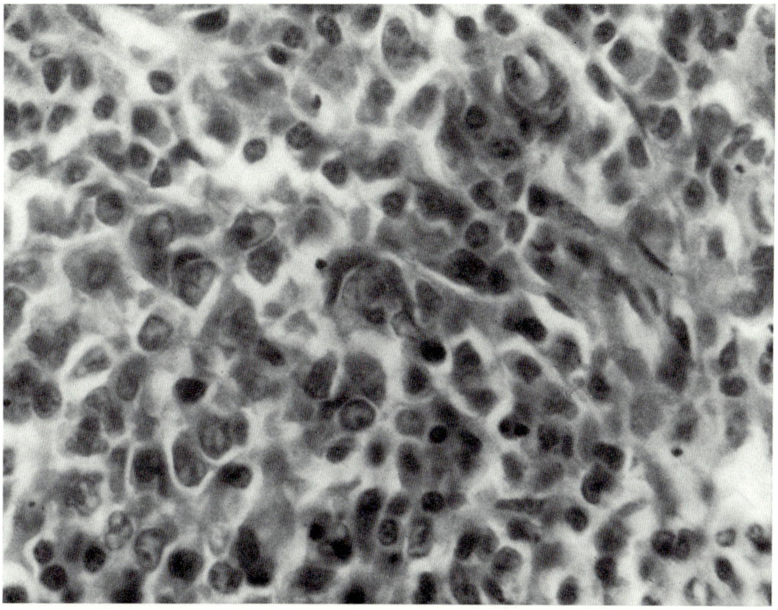

Fig. 11. Proliferation of immature reticular cells in the spleen in reticulosis in a hamadryas baboon. HE × 1,200.

Disturbances of Red Cell Production

Disturbances of red-cell production are rather often observed in monkeys; they are usually of a secondary character, and either alimentary or the result of complications from infectious diseases and helminth invasions in captive monkeys [1, 2, 23]. These secondary anemias are accompanied by changes in leukogramma according to the main process [23]. Literary data dealing with this question showed that, equally with infectious diseases (dysentery in particular) and helminth invasion, alimentary avitaminoses especially as a result of insufficiency of vitamin B_2, B_6, pantothenic and folic acids, take a special place in the development of anemic conditions in monkeys. West [24] reported on the development of anemia as a result of D_1, BII, B_2, B_6 hypovitaminosis and decrease of pantothenic and folic acids in diet.

The possibility of anemia developing in monkeys after protein decrease in their diet is known [4]. In monkeys with prolonged serious anemia, foci of extramedullary blood formation can appear (fig. 13) [17].

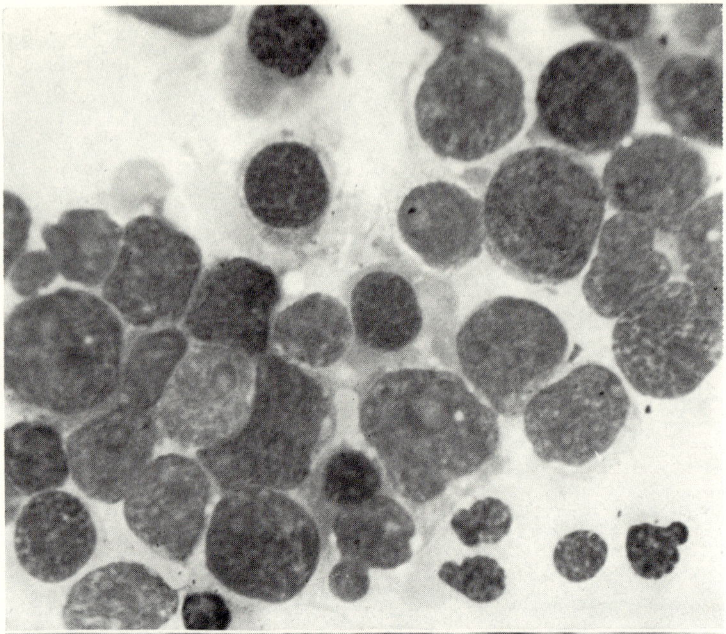

12

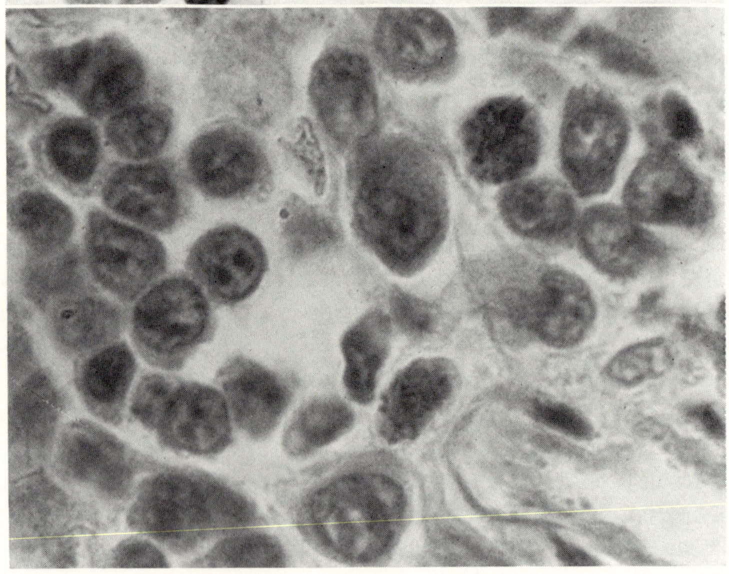

13

Fig. 12. Proliferation of undifferentiated cells in bone marrow in haemocytoblastosis (*P. hamadryas*). HE × 1,500.

Fig. 13. The foci of extramedullary haemopoiesis in anemia. HE × 1,200.

While analyzing the blood picture of newly imported macaca monkeys, M. J. KUKSOVA [15] revealed more pronounced hypochromic anemia in dysenteric animals after antibiotic treatment. In these animals red-cell counts could reach 3,000,000 in 1 mm^3 of blood, and haemoglobin values fell as low as 36 to 50 percent. Single normoblasts were observed among nuclear blood forms. In the bone marrow of these animals either depression of red series (decrease in cells of erythroid series up to 2 to 19 percent from 27 to 30 percent in normal state) or arrest of erythroblast maturation was observed. Erythroblasts made up 50% of cells in bone marrow. Parallel with the changes in the red series, in infectious diseases monkeys developed a shift to the left in the leukocyte formula with neutrophil young cell forms (neutrophils being staff-like for the most part, metamyelocytes and myelocytes being less in number).

As far as the two most common monkey infections is concerned, changes in the white cells are more pronounced in pneumonia than in dysentery.

References

1 AKBROIT, E. Y.; IVANOV, M. T.; KUKSOVA, M. I., and DAN'KO, L. V.: Character of pneumonia in *M. mulatta* imported to Sukhumi monkey colony from India and Viet-Nam. Vopr. physiol., exper. patol. i radiol. Sukhumi 273 (1965).

2 AKBROIT, E. Y.; KUKSOVA, M. I., and DAN'KO, L. V.: Etiology, clinic and blood picture in pneumonia in green monkeys. Vopr. physiol., exp. patol. i radiol. Sukhumi 277 (1965).

3 CORSON-WHITE, E. P.: Chronic lymphatic leukemia in a green monkey *(Cercopithecus callitrichus)*. Rep. Lab. Comp. Path. Philad. *57*: 40–44 (1929) and Arch. Path. *8:* 1019 (1929).

4 DE LA IGLESIA, F. A.; PORTA, E. A., and HARTRATT, P.: Effect of dietary protein levels of the *Saimiri sciureus*. Exp. Path. *7:* 182–195 (1967).

5 DEPAOLI, A. and GARNER, F. M.: Acute lymphocytic leukemia in whitecheeked gibbon *(Hylobates concolor)*. Cancer Res. *28:* 2559–2561 (1968).

6 DIGIACOMO, R. F.: Burkitt's lymphoma in a white-handed gibbon *(Hylobates lar)*. Cancer Res. *27:* 1178–1179 (1967).

7 DZHIKIDZE, E. K. and YAKOVLEVA, L. A.: A case of leukemia in a monkey exposed to prolonged X-ray irradiation. Probl. hematol. i pereliv. krovi *9:* 58–62 (1965).

8 HAMERTON, A. E.: Primary degeneration of the spinal cord in monkeys, a study in comparative pathology. Brain *65:* 193–204 (1942).

9 HILL, O.: Report of the Society's Prosector for the Year 1951. Proc. zool. Soc., Lond. *122:* 515–533 (1952).

10 HILL, O.: Report of the Society's Prosector for the Year 1952. Proc. zool. Soc. Lond. *123:* 227–251 (1953).

11 HILL, O.: Report of the Society's Prosector for the Year 1953. Proc. zool. Soc. Lond. *124:* 303–311 (1954).

12 KELLY, A. L.: Annual report of the hospital and research committee. Zoonooses Res. *21:* 7 (1948).

13 KENT, S. P. and PICKERING, J. E.: Neoplasms in monkeys *(Macaca mulatta):* Spontaneous and irradiation induced. Cancer *11:* 138–147 (1958).

14 KLÜVER: Personal communication (cited by C. H. LINGEMAN).

15 KUKSOVA, M. I.: Some changes in blood count in imported *M. mulatta* during treatment of dysentery with antibiotics. Vopr. physiol., exper. patol. i radiol. Sukhumi 242 (1965).

16 LAPIN, B. A.: A case of lymphogranuloma in a green monkey; in UTKIN Theoretical and practical problems of medicine and biology in experimentation on monkeys (Medgiz, Moscow 1956).

17 LAPIN, B. A. und YAKOVLEVA, L. A.: Vergleichende Pathologie der Affen. Jena 123 (1964).

18 LINGEMAN, C. H.: Spontaneous hematopoietic neoplasms of nonhuman primates, review, case report, and comparative studies. National Cancer Institute Monograph *32:* 157–170 (1969).

19 LINGEMAN, C. H.: Plasma cell neoplasms of man and animals. National Cancer Institute Monograph *32:* 303–311 (1969).

20 MASSAGLIA, A. C.: Leukemia in the monkeys; Note on a syndrome in a monkey *(Cynocephalus sphinx)* similar to that of chronic lymphatic leukemia in man. Lancet *i:* 1056–1057 (1923).

21 NEWBERNE, J. W. and ROBINSON, V. B.: Spontaneous tumours in primates – a report of two cases with notes on the apparent low incidence of neoplasms. Amer. Res. *21:* 150–155 (1960).

22 OSHIMA, F.: Über die Geschwülste bei wilden Tieren. Gann *31:* 220–223 (1937).

23 RUCH, T. C.: Diseases of laboratory primates (Philadelphia, London 1959).

24 WEST, E. S.: Dietary requirements of subhuman primates. Primate News *6:* 4–7 (1968).

25 WRIGHT, D. H. and BELL, T. M.: Bone disease in African green monkeys. Lancet *7366:* 969–970 (1964).

26 YAKOVLEVA, L. A.; LAPIN, B. A.; KUKSOVA, M. I.; MARKALAN, D. S., and LOMOVA, T. M.: Material on the leukosis/reticulosis among hamadryas baboons and attempts of its transplantation on monkeys. Proceedings of the symposium on Diffusion, virology and immunology of leukoses among man and animals, p. 155–164 (Riga 1970).

27 YAKOVLEVA, L. A.; LAPIN, B. A.; ASANOV, S. S.; LOMOVA, T. M.; GVASAVA, I. S.; KOKOSHA, L. V.; AGRBA, V. S., and KUKSOVA, M. I.: The onset of leukosis/reticulosis among hamadryas baboons of the Sukhumi breeding (in press).

28 ZALUSKY, R.; GHIDONI, Z. Y.; MCKINLEY, Y.; LEFFINGWELL, T. Y., and MELVILLE, G. S.: Leukemia in the rhesus monkeys *(Macaca mulatta)* exposed to whole body neutron irradiation. Radiat. Res. *25:* 410–416 (1965).

Authors' address: Dr. B. A. LAPIN and Dr. L. A. YAKOVLEVA, Institute of Experimental Pathology and Therapy, USSR Academy of Medical Sciences, *Sukhumi* (USSR)

Pathology of Simian Primates, Part I, pp. 821–840 (Karger, Basel 1972)

The Urinary System

J. A. Roberts

Tulane University, Delta Regional Primate Research Center, Covington, La.

Contents

Anatomy ... 821
  Kidney ... 821
  Ureter, Bladder, Urethra ... 822
Renal Physiology ... 823
Ureteral Physiology ... 825
Natural Pathology ... 826
  Congenital Anomalies ... 826
  Obstruction ... 827
  Neoplasm ... 828
  Urolithiasis ... 828
  Vascular Disease ... 829
  Chronic Renal Disease ... 830
Urinary Tract Infection ... 831
  Bacterial ... 831
  Specific Infections ... 836
References ... 838

Anatomy

Kidney

The urinary tract in all species studied has been found to be similar to that of the human except for two minor anatomic variations: the left kidney lies lower in the abdomen than the right as opposed to human anatomy in which the right kidney lies lower, and the subhuman primate has a unipapillary kidney as opposed to the human multipapillary kidney. The phylogenet-

Table I. Comparative structural parameters of the kidney

Glomerular Volume (mm^3/Gm kidney)		Length of cortical nephrons (mm)		
		Proximal tubule	Loop henle	Distal tubule
M. Mulatta	49.7	3.6–9.6	0.9–1.9	1.4–6.6
Man	29.4	7.2–23.1	0–5.6	1.6–4.2

From RYTAND [2] and BENNETT [3].

ic similarity of all primates at first makes the single papilla an unexpected finding. Although most lower mammals have a unipapillary kidney, such dissimilar animals as the beaver, pig, ox and elephant all have a multipapillary kidney similar to that of man. OLIVER [1], states that the formation of the multipapillary kidney must represent a mutational change in some species which led to an increase in structural efficiency. The balance between adequate kidney function and increasing body size was met in other species such as the subhuman primate by adaptive growth changes in the nephron itself. An example of extreme adaptive growth change in nephron length (loop of Henle) is the desert rat, the longer nephron allowing greater absorption of water than in most mammals.

A greater efficiency of man's nephron is suggested by the finding of a smaller glomerulus [2], but since the total renal mass per unit of surface area is the same in all mammals, the tubular length is greater in man than monkey [3] (table I). Although comparisons based on surface area have been shown by SMITH [4] to be more valid, in the absence of surface area values, STAHLS [5] extension of the allometric technique of BRODY [6] to primates will yield a reasonable estimation of total renal weight (in g) by the formula: $6.3\ M^{0.87}$ where M (body weight in kg) is known.

Ureter, Bladder, Urethra

The microscopic anatomy of the urinary tract is the same as in man. The mucosa of the ureter, bladder and urethra to the verumontanum in the male and to just below the bladder neck in the female, consists of several layers of transitional cells. The distal urethra in male and female is covered by stratified or pseudostratified columnar epithelium except for its most distal extent which is squamous epithelium [7].

Table II. Comparative parameters of renal function

Species	Glomerular filtration rate	Effective renal plasma flow	Filtration fraction	Hydropenic U/P osmolality	Maximum concentrating ability
	ml/min/kg			milliosmols/liter	
Man	1.97	10.0	0.184	3.89	1,400
Dog	4.29	13.5	0.317		2,300
M. mulatta	1.96	8.06	0.243	2.66	
M. speciosa	3.29	15.1	0.218	2.27	1,191

From Smith [4], Odell [10] and Bennett [3].

There is no limiting basement membrane beneath the mucosa of the bladder and ureter other than a reticulin network which merges with ground substance and connective and elastic tissue in the submucosa. In the ureter this is surrounded by two layers of smooth muscle, a circular and longitudinal layer and an external rather loose connective tissue adventitia [8]. The bladder muscle has not been dissected but generally consists of decussating smooth muscle fibers. In at least three species, *Papio papio*, *Macaca speciosa* and *Macaca irus* unusual intracytoplasmic round eosinophillic bodies of unknown function are seen in the transitional cells of the ureter, bladder and urethra (fig. 1).

Renal Physiology

Renal function, especially regarding concentrating ability and filtration fraction, has been found to be more similar to the human than other experimental animals (table II). Determination of renal clearances has been found to yield reproducible results during tranquilization as is necessary to handle adult animals. Standard clearances have been done [9] as well as isotopic clearances by the intradermal injection of isotope with epinephrine to maintain a constant serum level [10]. While reported in the dog, constant infusion of radioisotopes to determine clearances has not been used in the primate [11].

Stop-flow studies of renal function have not been reported in the primate, but micropuncture has been found to be possible in *M. mulatta* [3]. The advantages of micropuncture studies in the monkey besides the phyloge-

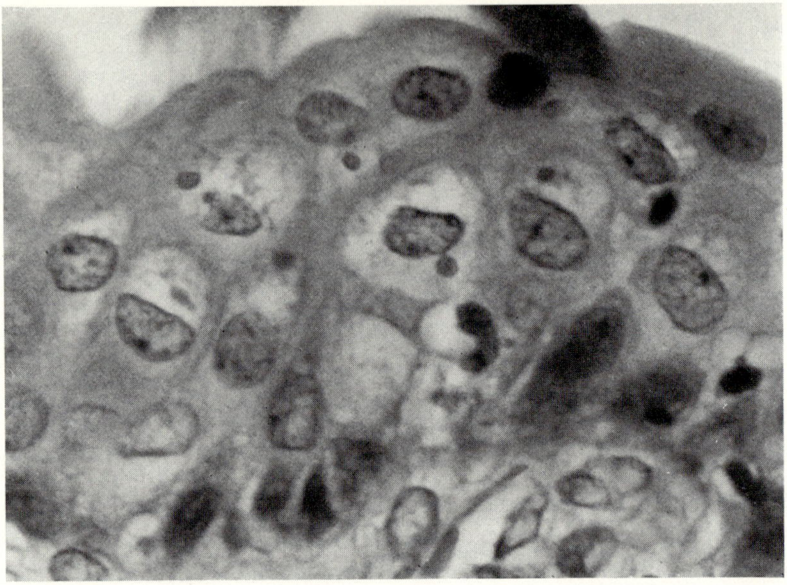

Fig. 1. Many of the transitional cells of the bladder of *Papio papio* contain homogeneous eosinophilic round to oval bodies within their cytoplasm. The largest are those in cells lining the lumen. Hemotoxylin and eosin ×1,800.

netic similarity to man are: more of the proximal tubule is available for puncture than either the dog or rat; the rat absorbs more water in the distal tubule than either the monkey or dog; and the rat does not respond to some pharmacologic agents such as diuretics as do both the dog and monkey. Finally and most important, a strong inference can be made from micropuncture studies that the nephron of the monkey has the potential to generate the single effect, that process fundamental to the countercurrent mechanism for the concentration of urine, since it has been shown that in cortical nephrons at least, sodium is reabsorbed in hypertonic proportions in the Loop of Henle.

Studies of urinary concentration and the response to osmotic diuresis [12] in *Saimiri sciureus* again reveals the similarity of monkey and man, for in both, unlike the dog, positive free water clearance does not typically occur. Thus in primates complete osmotic equilibration occurs in the distal tubules and collecting duct. The response to sustained osmotic diuresis in both man and monkey is a decrease in the tubular clearance of water unlike the rat where it rises progressively in response to such diuresis.

The existence of erythropoietin (a humoral factor involved in the control of erythropoiesis) has been confirmed in several primate species (*M. mulatta* [13], *Tamarinus nigricollis* [14] and the chimpanzee [15]). The kidney is thought to be the principal site of production, but bilateral nephrectomized tamarins continued to produce erythropoietin when subjected to hypoxia. Hydronephrosis was found not to be associated with an erythropoietin response unlike renal ischemia.

Marked variation in urine volume is noted in a compilation of daily measurements [16] and urine output seems high when considered on a weight basis. Undoubtedly such measures should be reported on the basis of surface area rather than weight. Urine volume will reflect variation in diet from dry to fresh food, and since the primate most frequently maintains the upright posture with the male often voiding out the side of a cage, special metabolism cages with solid sides are necessary for accurate collection. Accurate recording of fluid intake is possible by the use of a nipple constructed so that the animal must suck the fluid to obtain it [17].

Ureteral Physiology

That ureteral physiology may differ in a comparison of the primate to lower mammals is suggested by the fact that hydronephrosis of pregnancy occurs at the same frequency and with the same behavior in both subhuman primates and man [18]. Studies in normal *M. mulatta* females utilizing the cystoscopic introduction of non-obstructing ureteral catheters (3–4 Fr) connected to pressure transducers have revealed that peristaltic activity and pressure readings are similar to both those in the dog and man [19] (table III). Ureteral pressure was seen to be maximal in the lower ureter as in man [20], and the effect of diuresis in simulating obstruction of the ureter was confirmed as

Table III. Comparative ureteral physiology (intact bladder-upper ureter)

	Peristalsis/min	Pressure – mm Hg	
		Baseline	Amplitude
Dog [Boyarsky]	9 (2–20)	10 (Maximum)	13 (8–44)
M. mulatta	5 (2–10)	7 (5–12)	7 (2–20)
Human [Sala]	3 (2–6)	7 (3–15)	16 (1–20)

From Roberts [19].

shown in the dog [21]. The effect of pharmacologic agents on the ureter have not been studied but the tranquilizer used (Innovar-Vet) does not seem to effect the ureter unlike sodium pentobarbital or halothane anesthesia which markedly depress ureteral activity, at least when these agents are used for anesthesia following Innovar-Vet.

Natural Pathology

Congenital Anomalies

Several congenital anomalies have been recorded [22–26] as would be expected considering the high frequency of urinary tract anomalies in man. These include congenital absence or aplasia, fused kidneys, polycystic kidneys, and aberrant renal artery with hydronephrosis.

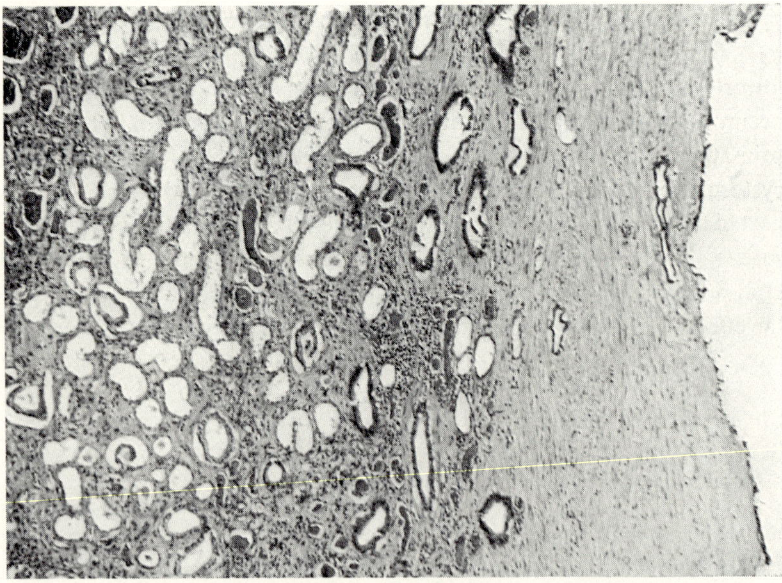

Fig. 2. Hydronephrosis secondary to ureteral obstruction by a leiomyomatous uterus in a chimpanzee. Interstitial fibrosis and dilatation of collecting tubules with hyaline casts. Hematoxylin and eosin ×60.

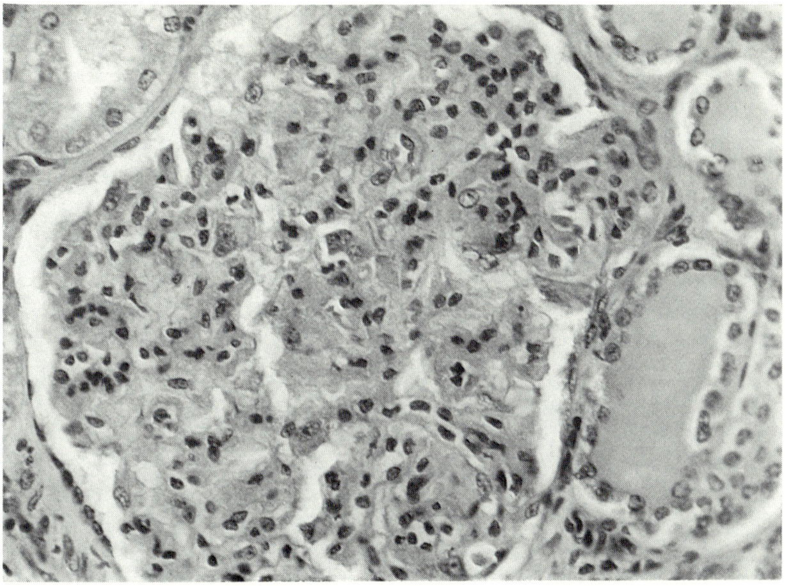

Fig. 3. Glomerulonephritis or 'the toxemic lesion' in a chimpanzee dying during pregnancy. The glomerulus is ischemic with turgid endothelial cells and decrease in capillary lumens. Hematoxylin and eosin ×400.

Obstruction

Obstructive disease is unusual, but that called hypoplasia of the kidney [23], by description would more likely represent ureteral obstruction or vesicoureteral reflux leading to hydronephrosis and secondary renal atrophy. One instance of ureteropelvic obstruction from an aberrant artery is reported [26]. The sequela of bladder neck obstructive disease is described only once in the literature [Ruch, p. 465]. An additional case of uremia from either phimosis or neurogenic bladder is described in Lapin [p. 91] in an *M. mulatta* with spinal tuberculosis and paraplegia.

In an aged female chimpanzee dying at this center, very large leiomyomata of the uterus apparently caused bilateral ureteral obstruction with hydronephrosis as is reported in the human female (fig. 2). Urinalysis should be a part of prenatal care, since recently at this center a chimpanzee died during pregnancy with glomerular changes similar to those reported in the human dying of toxemia of pregnancy (fig. 3). A *M. mulatta* is described elsewhere

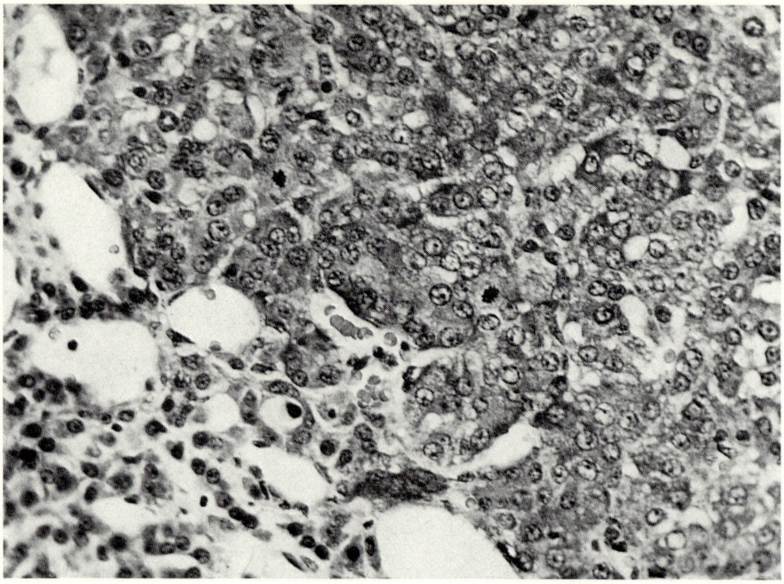

Fig. 4. Pheochromocytoma in *Macaca irus*. The neoplastic cells of the adrenal medulla have large vesicular nuclei. Two mitotic figures are seen. Hematoxylin and eosin ×400.

to have died of arteriolonephrosclerosis following chronic nephrosis during the third pregnancy [Ruch, p. 464].

Neoplasm

Neoplasms are rare. Although one pheochromocytoma of the left adrenal of a Macaca irus has been found at this institution (fig. 4), no tumors have been seen in the urinary tract. Renal carcinoma [27] is the tumor reported in the literature. An unusual familial occurrence of renal carcinoma in *M. mulatta* was not completely studied before all the animals died [28].

Urolithiasis

Urolithiasis has been described twice in subhuman primate species [29]. Nephrocalcinosis, however, has been shown to naturally occur in 86% of a series of squirrel monkeys *(Saimiri sciureus)* [30]. Studies of calcium metabo-

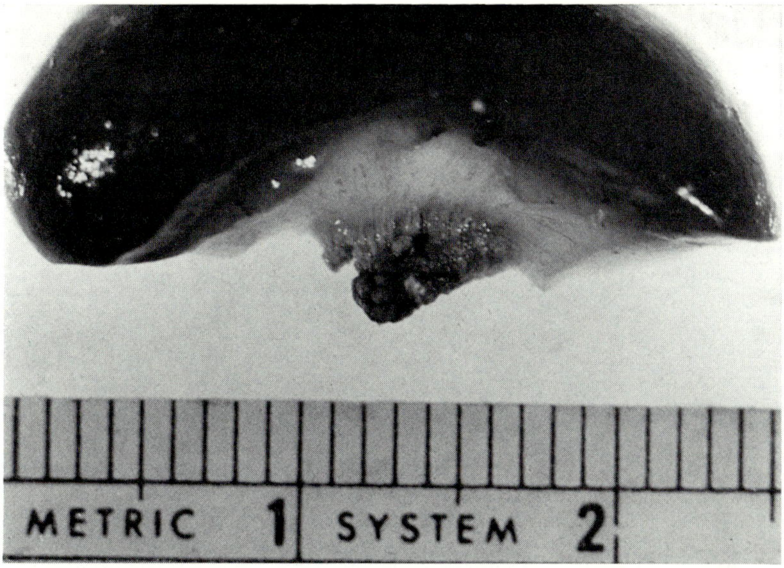

Fig. 5. Experimental papillary tip stone from feeding of oxamide to *Saimiri sciureus*. In addition to the calculus attached to the papilla, streaks in the papilla are oxamide in collecting tubules.

lism suggest that the monkey is more similar to man than rat or dog, and would thus be expected to be a more suitable animal for the study of stone disease [31]. Not only is the percentage of injected Ca [45] excreted in the monkey (11%) compared closely to man (8%) but the ratio of urine to fecal excretion 2/1 in each. Attempts to form oxalate stone by producing oxaluria from Vitamin B^6 deficiency [32] and the feeding of ethylene glycol [33], both standard techniques in the rat, have to date been unsuccessful in the production of stones, although oxaluria occurs. Preliminary experiments at this institution have shown that the feeding of oxamide, the diamide of oxalic acid, does produce papillary tip calcification or actual calculi in the chimpanzee, *Macaca speciosa*, and *Saimiri sciureus* (fig. 5).

Vascular Disease

Although atherosclerosis occurs naturally in several primate species, renal lesions associated with it have not been reported. In those dietary studies in which atherosclerosis has been produced changes in the renal arteries are

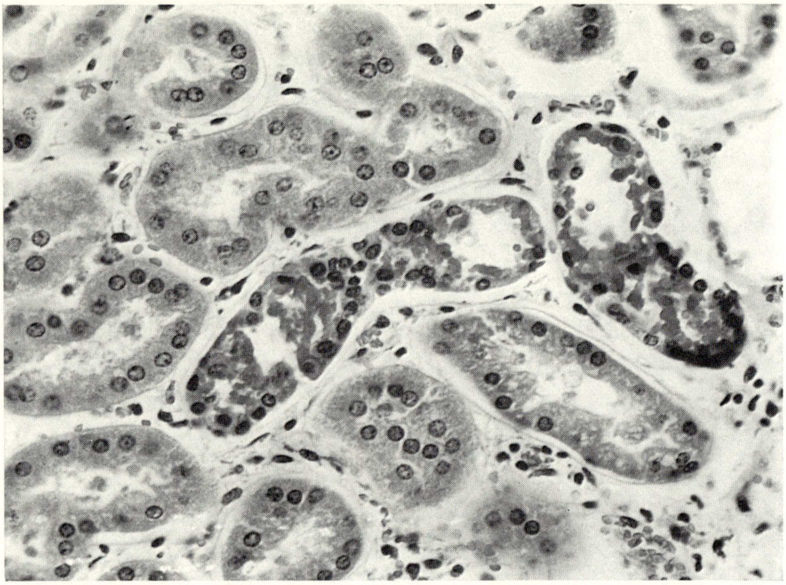

Fig. 6. Acute tubular necrosis or 'lower nephron nephrosis' following trauma in *Macaca irus*. Some tubules show necrosis, tubulorrhexis, and contain debris with the staining reaction of hemoglobin. Hematoxylin and eosin ×400.

not mentioned. Hypertension secondary to a renal arterial lesion is reported only once, but small vessel disease of the kidney is reported in animals with long standing hypertension by LAPIN [p. 134]. Renal hypertension has been produced by the surgical production of renal ischemia in *Macaca mulatta* [34] following the initial work of GOLDBLATT [35] which included primates.

Chronic Renal Disease

The parenchymal disease which occurs most frequently is tubular, for acute tubular necrosis and lower nephron nephrosis is often seen in animals dying following shock from trauma or exposure (fig. 6).

Glomerulitis suggestive of the changes seen in human toxemia of pregnancy was seen in a chimpanzee dying during pregnancy [36]. Acute glomerulonephritis has been described [37], as has subacute proliferative glomerulonephritis causing the nephrotic syndrome [38]. A squirrel monkey dying here shows the microscopic appearance of acute glomerulonephritis (fig. 7).

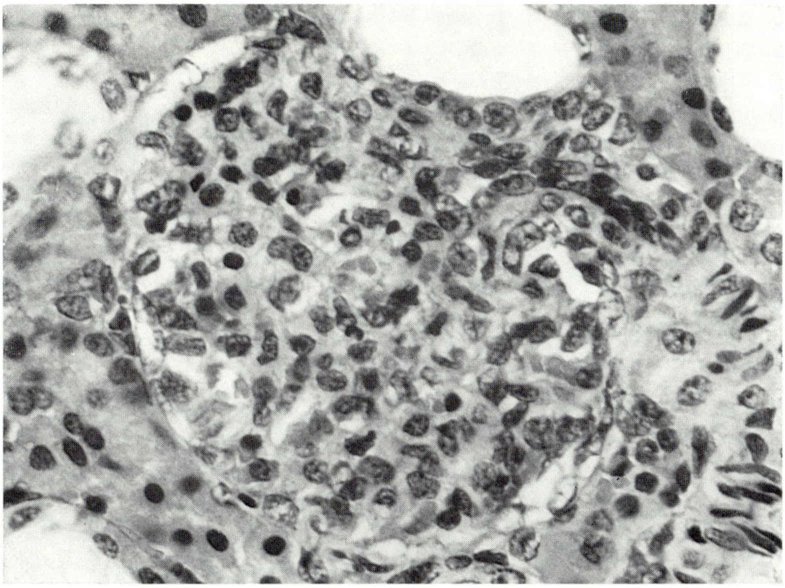

Fig. 7. Acute glomerulonephritis in *Saimiri sciureus* characterized by hypercellularity of all glomeruli with normal tubules. Hematoxylin and eosin ×400.

Lesions suggestive of acute glomerulonephritis have been produced in monkeys infected with influenza virus in combination with Type C streptococcus; [39] and in animals given antiglomerular antibodies developed in the rabbit, Masugi nephritis has been produced [40].

Ethylene glycol has been shown to cause the typical nephrotoxicity seen in the human [33] (fig. 8) and RUCH mentions nephrotoxicity from mercuric chloride, potassium bichloride and uranium nitrate [p. 463].

Urinary Tract Infection

Bacterial

Urinary tract infection is probably more common than realized. Naturally occurring infection from both *E. coli* and *Proteus mirabilis* has been reported in the squirrel monkey [41]. While RUCH states that renal inflammation or 'nephritis' is a commonly listed finding in reports of deaths from zoos, FOX [p. 274] reported that nephritis was more common in other mammals.

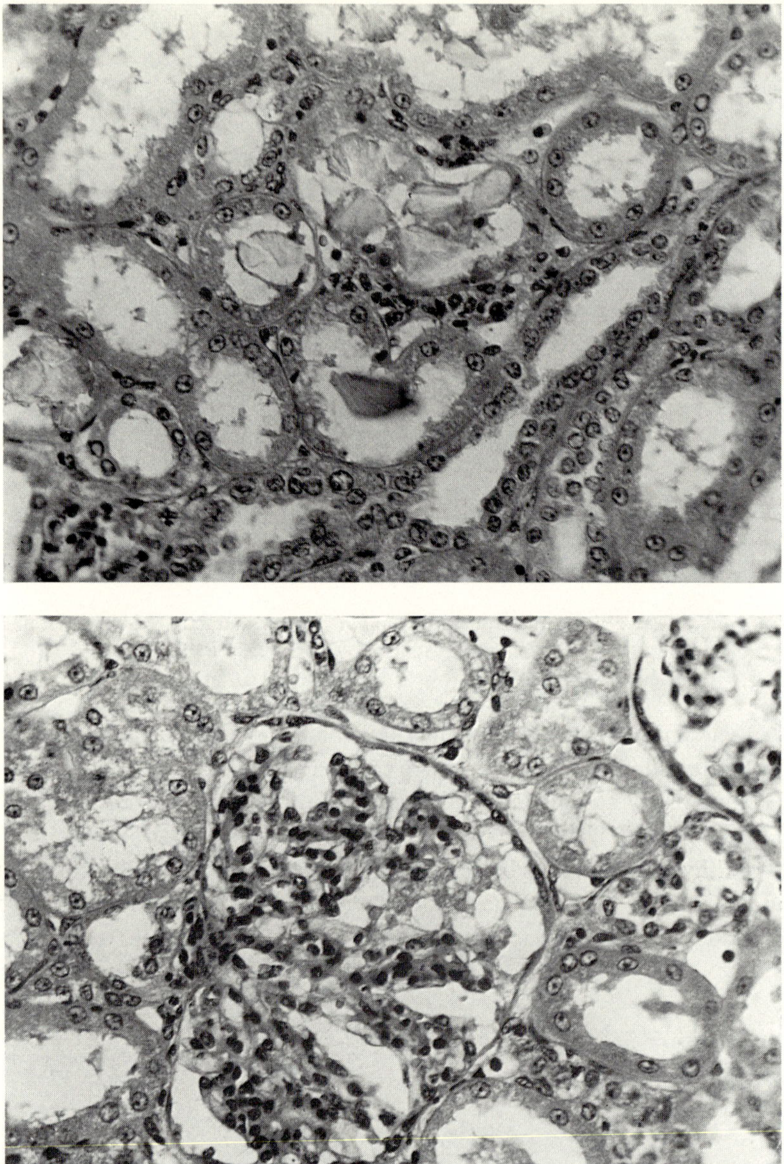

a

b

Fig. 8. Nephrotoxicity of ethylene glycol. *a* Calcium oxalate crystals within proximalrenal tubules and necrosis of contiguous epithelium in *Macaca mulatta.* Hematoxylin and eosin × 600. *b* A glomerulus with protein precipitate in the capsular space and adhesions of glomerular tufts to Bowman's capsule. Hematoxylin and eosin × 600.

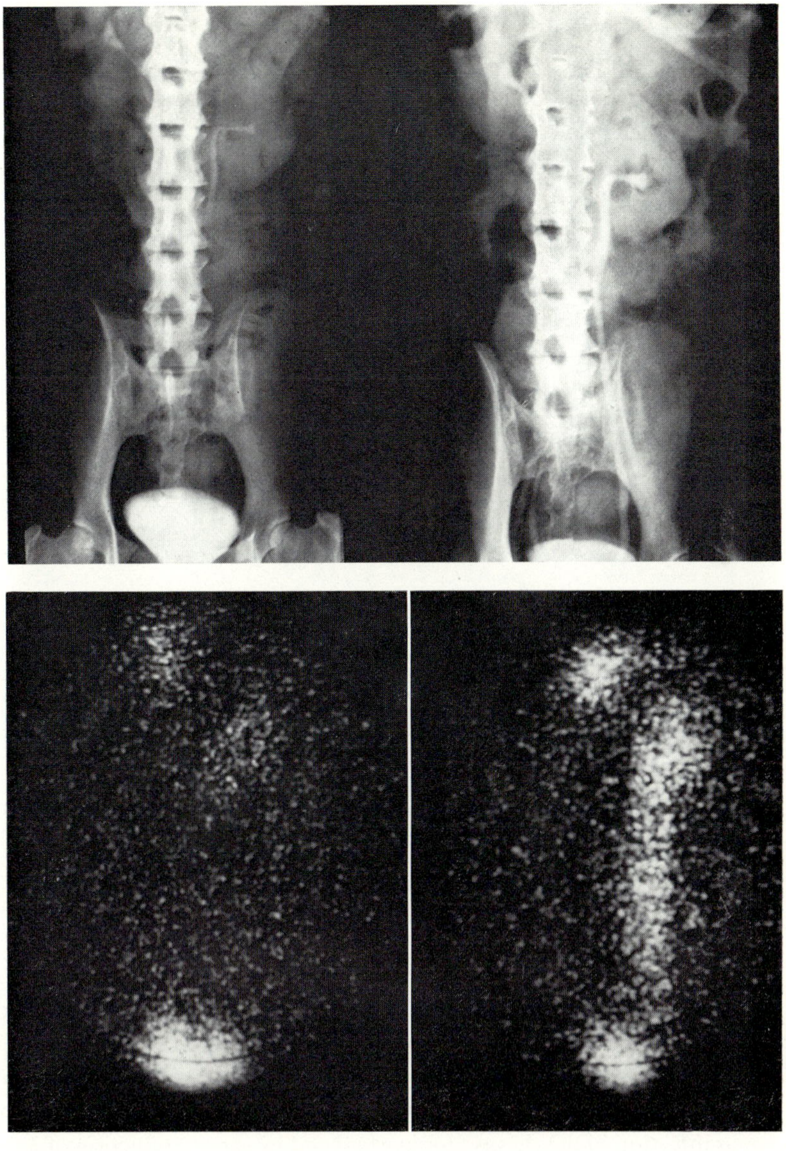

a

b

Fig. 9. A spontaneous coliform urinary tract infection in *Macaca speciosa*. *a* Excretory urogram depicting normal (on the left) and left ureteral dilation during infection (on the right). *b* ^{131}I Hippuran scintiphoto comparing normal (on the left) to left ureteral hold-up of isotope (on the right).

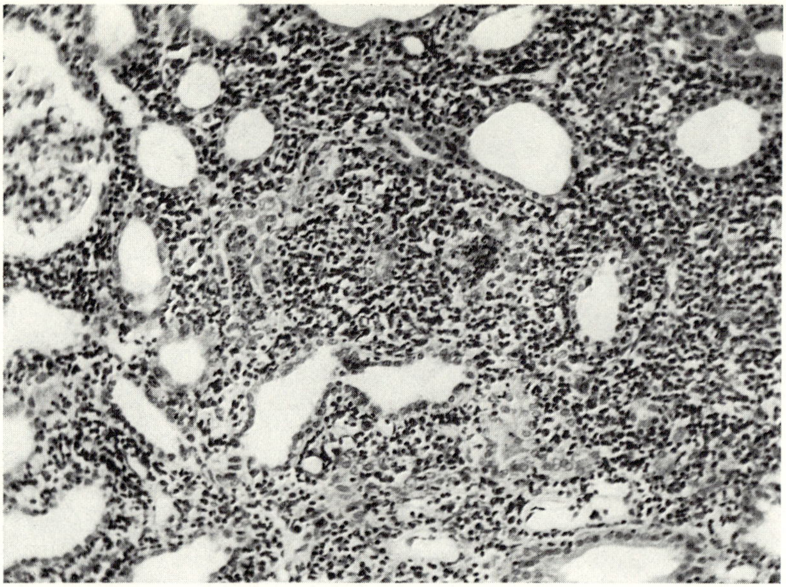

Fig. 10. Pyelonephritis in a macaque with marked interstitial cellular infiltrate and two intratubular pus casts. Hematoxylin and eosin ×200.

Lapin [pp. 9–11] reports no renal inflammatory disease at Sukhumi out of over 1,200 autopsies. Fiennes [42] study of the deaths at the London Zoo over the last 15 years (750 primates) showed a very low incidence of renal pathology with a maximum of 4.2% in old world monkeys and only 1.25% in cebids.

An occasional spontaneous infection has been seen in our laboratories in both *Macaca mulatta* and *speciosa* females. In three animals abnormal ureteral function producing temporary hydroureter has been found with *E. coli* and *Proteus mirabilis* infection (fig. 9). This is thought to be secondary to the endotoxin produced by coliform bacilli, which has been found in the dog to cause decreased peristaltic activity [43]. Experimentally pyelonephritis has been produced by ureteral ligation followed by hematogenous *Staphylococcus aureus* infection in the rhesus monkey [44] and by the introduction of *Escherichia coli* into the bladder of marmosets *(Callithrix jacchus)* [45].

Acute pyelonephritis is suggested by the finding of pyuria and bacilluria in the animal manifesting the illness only by lassitude and decreased food intake progressing to total inactivity except when aroused (fig. 10). A leukocytosis is present but fever unusual. The disease has infrequently been de-

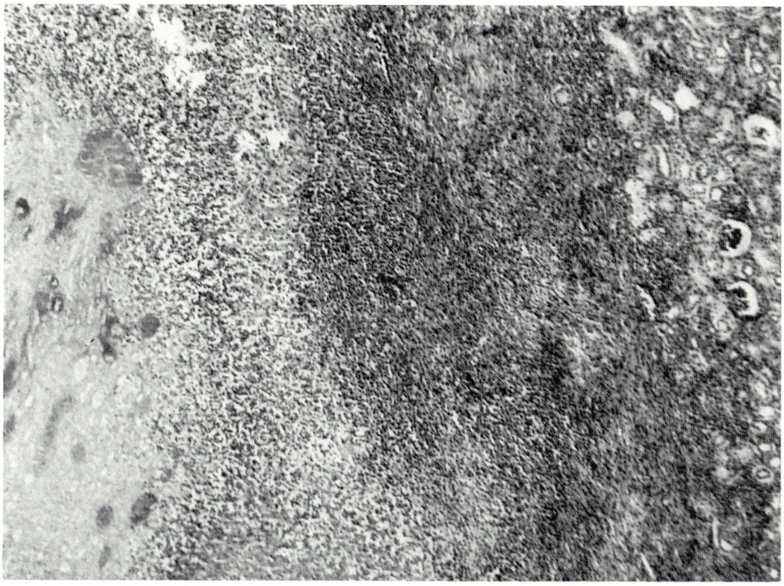

Fig. 11. The edge of a septic renal infarct in *Callicebus moloch.* To the left cellular ghosts are present within infarcted tissue, bordered by inflammatory cellular infiltration and normal tissue. Hematoxylin and eosin ×60.

scribed, but pyelonephritis led to death from septicemia in one *M. mulatta* at this Center, and an orang-utan reported by FIENNES, in both cases due to *Pseudomonas aeruginosa.* In addition, a septic renal infarct in a squirrel monkey has been seen here (fig. 11). Parenteral medication is necessary in the acute case, long-term medication being difficult unless the antibiotic is relatively tasteless in which case it may be mixed with the food or water. Recently, long term administration by surgical implantation into the subcutaneous tissues of antibacterials in silastic has been successful [46].

Organisms of any number obtained by catheterization are probably significant. However, because organisms are found as a natural inhabitant of the urethra (at least in man) and preparation for aseptic catheterization is difficult in the monkey, colony count of less than 1,000 colonies/ml is considered a contaminant and this value in repeated studies of *Macaca mulatta* has proved to be reliable. Aseptic catheterization is important since introduction of unusual organisms can lead to infection and, on occasion, fatal complication. A female *Macaca mulatta* was inadvertently infected with *Staphylococcus aureus* and succumbed within 48 hours of gangrenous cystitis (fig. 12).

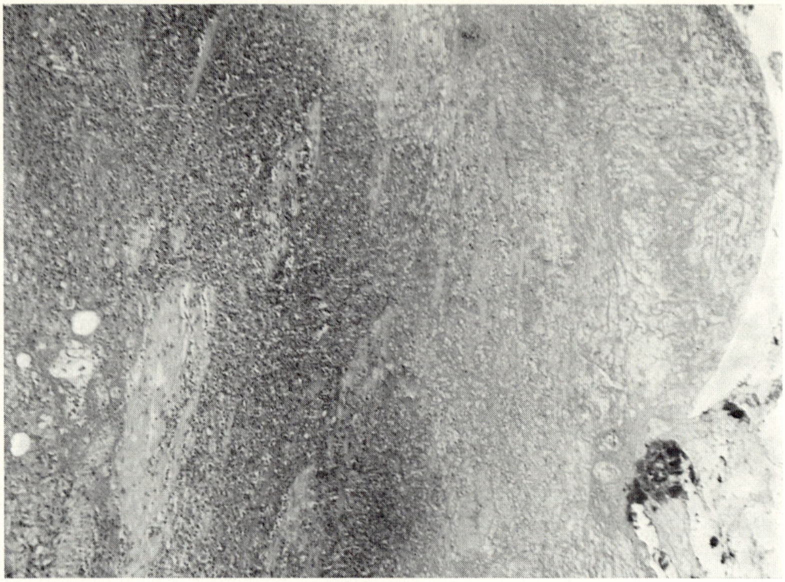

Fig. 12. Gangrenous cystitis from *Staphylococcus aureus* in *Macaca mulatta.* The mucosa (right) is absent with necrosis of submucosa and most of the muscularis. The marked inflammatory cellular response extended into the adventitia. Hematoxylin and eosin ×60.

The male urethra is easily catheterized in all primate species, but the trauma of catheterization with introduction of urethral organisms into the prostatic urethra may cause prostatitis (fig. 13). This infrequently occurs with a single catheterization. If it does and is untreated it may lead to epididymitis. In the female, after careful preparation with antiseptics, the urethra can easily be catheterized although in some species this requires a nasal speculum to visualize an intravaginal meatus. In animals even as small as the squirrel monkey, a convenient catheter is a disposable polyethylene infant feeding tube of 8-Fr size. Most macaque female urethras will easily contain 10 to 12-Fr instrument, and can be dilated with impunity to 18-Fr.

Specific Infections

Renal tuberculosis is commonly seen in primates dying of the disease as terminally it consists of a miliary form. The primary focus is uncommonly

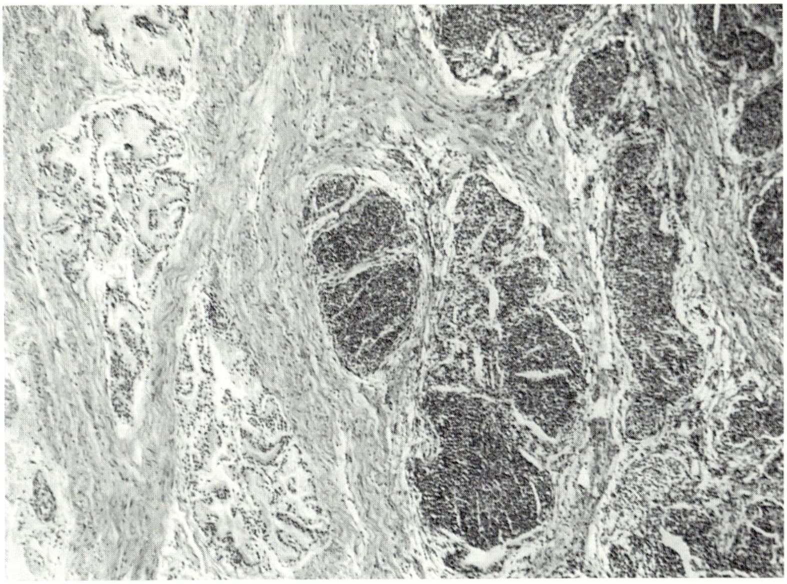

Fig. 13. Acute prostatitis in a chimpanzee. Normal prostatic glands on the left with nearby glands showing destruction of epithelium and polymorphonuclear cells but little interstitial infiltration. Hemotoxylin and eosin ×60.

the kidney; lung, liver, and spleen all being more frequent. The disease usually leads to death within 24 weeks of its onset, so long standing renal tuberculosis would not be expected. Of 498 primates studied by Fox [p. 486], 192 had tuberculosis, and of those 84 showed involvement of the kidney.

The following are briefly mentioned since further information and references may be found in Fiennes 'Zoonoses of Primates': Weil's disease (leptospirosis) probably occurs naturally, but has not been reported since 1927 when an epizootic caused the death of a large colony of chimpanzees with grossly affected kidneys. A naturally occurring significant antibody titre has been described in wild primates. While antibodies to *Treponema pertenue* (yaws) have been commonly found in some African monkeys, lesions due to the organism have not been described, though actual recovery of organisms is now reported. *Schistosoma haematobium* has been described in African baboons, and the question of their being a carrier of the disease raised. Cytomegalic inclusion disease represents the only renal pathology known to occur from viral infection in the monkey, though no renal involvement has been described in simians. Whether it is a renal disease rather than a finding in an animal with a viral disease excreting the virus via the urine, has not been proved.

References

1 OLIVER, J.: Nephrons and kidneys (Harper & Row, New York 1968).

2 RYTAND, D. A.: The number and size of mammalian glomeruli as related to kidney and to body weight, with methods for their enumeration and measurement, Amer. J. Anat. *62:* 507 (1937–38).

3 BENNETT, C. M.; BRENNER, B. M., and BERLINER, R. W.: Micropuncture study of nephron function in the rhesus monkey. J. clin. Invest. *47:* 203 (1968).

4 SMITH, H. W.: The kidney (Oxford University Press, New York 1964).

5 STAHL, W. R.: Organ weights in primates and other animals. Science *150:* 1039 (1965).

6 BRODY, S.: Bioenergetics and Growth (Reinhold, New York 1945).

7 ROBERTS, J. A. and SEIBOLD, H. R.: The histology of the primate urethra (accepted for publication in Folia Primatologica).

8 ROBERTS, J. A. and WEXLER, N.: Comparative anatomy of the mammalian ureter (in preparation).

9 PICKERING, D. E. and SUSSMAN, H. H.: Renal Function Studies in Monkeys *(Macaca mulatta)*. Amer. J. vet. Res. *23:* 667 (1962).

10 ODELL, R. M.; RADWIN, H. M.; BERNSTEIN, L. H., and SCHLEGEL, J. U.: Renal function in *Macaca speciosa*. J. appl. Physiol. *24:* 366 (1968).

11 OESTER, A.; WOLF, H., and MADSEN, P. O.: Double isotope technique in renal function testing in dogs. Invest. Urol. *6:* 387 (1969).

12 SELKURT, E. E. and WATHER, R. L.: Renal concentrating mechanism of the squirrel monkey. Amer. J. Physiol. *213:* 191 (1967).

13 MELBY, E. C.; MURPHY, G. P., and MIRAND, E. A.: Erythropoietin levels in response to ureteral occlusion or reinsertion in the rhesus monkey. Invest. Urol. *3:* 92 (1965).

14 MIRAND, E. A.: Studies on erythropoietin in tamarins. Invest. Urol. *2:* 579 (1965).

15 MURPHY, G. P.; MIRAND, E. A.; WADE, J. C., and MELBY, E. D.: Erythropoietin response to renal stress in the chimpanzee. Invest. Urol. *5:* 234 (1967).

16 MORROW, A. C. and TERRY, M. W.: Urine volume in nonhuman primates. A tabulation from the literature (Primate Information Center, Regional Primate Research Center at the University of Washington, Seattle 1968).

17 POVAR, M. L. and SALK, R.: A device for measuring fluid intake by primates. J. exp. Anal. Behav. *8:* 187 (1965).

18 WAGENEN, G. VAN and JENKINS, R. H.: Pyeloureteral dilation in successive pregnancies. J. Urol. *49:* 228 (1943).

19 ROBERTS, J. A.; CLAYTON, T., and MATHER, F.: Ureteral physiology in Macaques.

20 KIIL, F.: The function of the ureter and renal pelvis. (Saunders, Philadelphia 1957).

21 BOYARSKY, S. and MARTINEZ, J.: Ureteral peristaltic pressures in dogs with changing urine flows. J. Urol. *87:* 26 (1962).

22 SCOTT, H. H. and CAMB, H.: Congenital malformations of the kidney. Proc. zool. Soc., Lond. *2:* 1529 (1925).

23 Hill, W. C. O.: Congenital abnormalities of the urinary tract in primates. Folia primat. *2:* 111 (1964).
24 David, G. F. X. and Ramaswami, L. S.: Unilateral hypoplasia of the kidney of a female langur. Folia. primat. *5:* 312 (1967).
25 Maruffo, C. A. and Cramer, D. L.: Congenital renal malformations in monkeys. Folia. primat. *5:* 305 (1967).
26 Haddow, C. J.: Field and laboratory studies on an African monkey, *Cercopithecus ascanius* schmidti Matschie. Proc. zool. Soc., Lond. *122:* 297 (1952).
27 Chapman, W. L., jr. and Allen, J. R.: Multiple Neoplasia in a rhesus monkey. Path. Vet. *5:* 347 (1962).
28 Ratcliffe, J. L.: Familial occurrence of renal carcinoma in rhesus monkeys. Amer. J. Path. *16:* 619 (1940).
29 Pryor, W. H., jr.; Chang, C., and Roulston, G. L.: Urolithiasis in a taiwan monkey *(Macaca cyclopis)*. A literature review and case report. Lab. Animal Care *19:* 862 (1969).
30 Drach, G. W. and Bowen, J.: Nephrocalcinosis in the squirrel monkey. Surg. Forum *17:* 501 (1966).
31 Harris, R. S.; Moor, J. R., and Wanner, R. L.: Calcium metabolism of the normal rhesus monkey. J. Clin. Invest. *40:* 1766 (1966).
32 Stone, F. J.; Andrus, S. B., and Portman, O. W.: Primates in Medical Research with special reference to new world monkeys; in Pickering Proceedings of a Conference on Research with Primates, p. 59 (Tektronix Foundation, Beaverton 1963).
33 Roberts, J. A. and Seibold, H. R.: Ethylene glycol toxicity in the monkey, Toxicol appl. Pharmacol. *15:* 624 (1969).
34 Smith, C. C. and Ansevin, A.: Experimental Hypertension in the Rhesus monkey. Proc. Soc. exp. Biol., N.Y. *96:* 432 (1957).
35 Goldblatt, H.: Studies on Experimental Hypertension. III. The production of persistent hypertension in monkeys (macaque) by renal ischemia. J. Exp. Med. *65:* 671 (1937).
36 Stout, C. and Lemmon, W. B.: Glomerular capillary endothelial swelling in a pregnant chimpanzee. Amer. J. Obstet. Gynec. *105:* 212 (1969).
37 Fiennes, R. N.: Report of the pathologist for the year 1960. Proc. zool. Soc. Lond *137:* 173 (1961).
38 Feldman, D. B. and Bree, M. M.: The nephrotic syndrome associated with glomerulonephritis in a rhesus monkey *(Macaca mulatta)*. J. Amer. vet. med. Ass. *155:* 1249 (1969).
39 Wilson, H. E.; Saslaw, S.; Doan, C. A.; Woolpert, O. C., and Schwab, J. L.: Reactions of monkeys to experimental mixed influenza and streptococcus infections. J. exp. Med. *85:* 199 (1947).
40 Markowitz, A. S.; Clasen, R.; Nidus, B. D., and Ainis, H.: Streptococcal Related Glomerulonephritis. J. Immunol. *98:* 161 (1967).
41 Clarkson, B.; Bullock, B. D.; Lehner, N. D. M., and Feldner, M. A.: The Squirrel monkey as a laboratory animal, in 'Animal Models for Biomedical Research' Publication 1594 (National Academy of Sciences, Washington 1968).
42 Fiennes, R. N.: Unpublished data.
43 Teague, N. and Boyarsky, S.: The effect of coliform bacilli upon ureteral peristalsis. Invest. Urol. *5:* 423 (1968).

44 DATTA, S. K. and CHAKRAVARTI, R. N.: Experimental pyelonephritis in the rhesus monkey. Nature *209:* 90 (1966).

45 ROCHA, H.; DA SILVA TELES, E., and BRITO, E.: Studies on experimental bacteremia and pyelonephritis in the marmoset *(Callithrix jacchus)*. Proc. Soc. exp. Biol., N.Y. *129:* 506 (1968).

46 MANGELSON, N. L. and COCKETT, A. T. K.: Long-term antibiotic therapy using silicone rubber as a carrier. Surg. Forum *28:* 531 (1967).

Author's address: Dr. JAMES A. ROBERTS, Tulane University, Delta Regional Primate Research Center, *Covington, LA 70433* (USA)

Pathology of Simian Primates, Part I, pp. 841–877 (Karger, Basel 1972)

The Female Reproductive System[1]

D. C. KRAEMER and N. C. VERA CRUZ

Southwest Foundation for Research and Education, San Antonio, Tex.

Contents

I. Anatomy 842
A. External Genitalia 842
B. Internal Genitalia 843
C. Mammary Glands 845
II. Physiology 846
A. Puberty 846
B. Breeding Season 846
C. Birth Season 849
D. Sexual Receptivity (Estrus) 849
E. Sexual (Menstrual) Cycle 850
1. Hormone Patterns 850
2. Ovarian Changes 851
3. Endometrial Changes 852
4. Changes in Cervical Mucus 852
5. Vaginal Changes 853
6. Perineal (Sex Skin) Changes 853
7. Changes in the External Genitalia 854
F. Gestation 854
G. Parturition 856
H. Lactation 856
III. Pathology 857
A. Prenatal and Neonatal Mortality 857
1. Embryo Mortality 857
2. Congenital Malformations 858
3. Placental Abnormalities 858

1 Preparation of this chapter and the unpublished work therein was supported in part by US Public Health Service Research Grant FR-GM00 451 from the Institute of General Medical Sciences.

4. Abortions and Stillbirths 858
5. Dystocia 859
6. Neonatal Mortality 859
B. Noninfectious Gynecologic Pathology 859
1. Ovaries 859
2. Uterus and Fallopian Tubes 860
3. Vagina and External Genitalia 861
4. Mammary Glands 861
C. Infectious Gynecologic Pathology 862
1. Viral 862
2. Mycoplasma 864
3. Bacterial Infections 864
4. Mycotic Infections 866
5. Protozoan Infections 867
6. Parasitic Helminth Infections 867
References 868

I. Anatomy

A. External Genitalia

The external genitalia of the nonhuman primate female are the vestibule of the vagina, the clitoris, and the labia. In some species, notably those belonging to the family Cercopithecidae and Pongidae, the perineum or sex skin is well-developed and is also classified as an external genital organ.

The perineum is a thickened, pigmented and highly vascularized area of the skin surrounding the anus and the external genitalia. It is bounded dorsally by the base of the tail, ventrolaterally by the ischial callosities and ventrally by the pelvic arch. It undergoes cyclical changes in color, size and conformation in response to ovarian hormones; the changes vary among species, individuals and different cycles in an individual. The perineum is well-developed in *Theropithecus*, *Mandrillus*, *Papio*, *Pan*, *Cercocebus*, *Miopithecus talapoin*, *Cynopithecus niger*, and *Macaca nemestrina*. Two patterns of sex skin turgescence have been observed. The more common pattern is a smooth, unlobulated swelling rarely extending laterally beyond the ischial callosities; less common is a lobulated pattern with the sex skin extending laterally into the hairless buttocks and even into the flanks.

The vestibule or the external opening of the vagina is a funnel-shaped, moderately shallow recess ventral to the anus. The urethra enters into the vestibule ventral to the vaginal opening and is surrounded, at least in baboons and rhesus monkeys, by 2 papillary folds.

The clitoris consists of erectile tissue covered with stratified squamous epithelium; it is located in the ventral commissure of the vestibule. While it is a prominent structure in all nonhuman primate species, it is most prominent in *Loris*, *Nycticebus*, *Perodicticus*, and *Galago*. The clitoris in these prosimians is covered with a cutaneous prepuce and the urethra opens at its apex. The clitoris in *Saimiri*, *Cebus*, and *Ateles* is elongated and pendulous. However, the urethra does not tunnel the clitoris but opens into the vestibule [1, 2]. In the species which exhibit sex skin changes, the clitoris is often masked by the clitoral lobes of the perineum during maximal turgescence.

Surrounding the vestibule and the clitoris are poorly developed longitudinal folds, the labia, which in most species are the labia minora. Labia majora are absent in most species of nonhuman primates except in *Lemur*, *Saguinus*, and *Lagothrix*. In Catarrhine monkeys [3] the labia majora are also generally lacking but in some instances rudimentary labia are present. Between the labia is a fissure, the rima pudendi. In species with cyclic perineal changes, the rima appears as an irregular shallow slit with puckered edges during sex skin quiescence but becomes a deep slit with smooth walls extending from the anus to the clitoris during sex skin turgescence.

B. Internal Genitalia

The internal genitalia of female nonhuman primates consist of the ovaries, the uterus, the Fallopian tubes or oviducts, and the vagina.

The ovaries are paired organs located on each side of the uterus. They vary in size and shape among and even within species. They are small, laterally compressed bodies in *Tupaia*, relatively large and ovoid in Platyrrhine monkeys; ovoid to globular and occasionally prismatic in *Cercopithecus;* flattened, bean-shaped bodies with a rectilinear attached region, a highly convexed free margin, and 2 pointed poles in *Erythrocebus* [3] and oval or fusiform in *M. mulatta* [4]. They lie in a peritoneal ovarian fossa in most prosimians *(Lemur*, *Loris*, and *Galago)* [5]. The ovaries are attached to the dorsal aspect of the broad ligament by a short mesovarium, to the uterus by the ovarian ligament, and to the Fallopian tube by the ovarian fimbria. In *Ateles*, *Lagothrix*, and *Callicebus* the superior mesosalpinx along the cranial border of the tube partly shelters the ovaries, and in Catarrhine monkeys, the ovaries are overshadowed cranially by a hood-like fold derived from the anterior mesosalpinx [6].

The Fallopian tube or oviduct is a thick-walled, convoluted tube which originates from the lateral margin of the uterus at the junction of the uterine fundus and the uterine body. It is divided into 3 portions: (1) the isthmus or the constricted one third portion proximal to the uterus, a portion of which courses through the visceral peritoneum of the uterus; (2) the ampulla which is the intermediate dilated portion that curls over the ovary, and (3) the infundibulum which possesses membranous uterine and ovarian fimbriae of about equal length. In *Tarsius*, the oviducts end in a single extended fimbria which overlaps the ovary. The tubes are markedly coiled in most Platyrrhine monkeys but they are relatively straight in *Saimiri* and *Callithrix*. In the great apes *(Hylobates, Pan, Pongo, Gorilla)* the Fallopian tubes are very similar to those in women although the fimbriae are generally longer and the uterine and ovarian fimbriae are poorly demarcated [7]. The slit-like abdominal opening into the oviduct is considerably larger than the opening into the uterus. The oviduct is supported by the inferior mesosalpinx as well as a superior mesosalpinx. The latter is absent in the great apes [4].

The uterus of nonhuman primates is of the simplex type except in *Tupaia*, *Lemur*, and *Tarsius* which possess a bicornuate uterus with relatively long uterine horns. The uterus is located in the pelvic cavity and is roughly subdivided into: (1) the fundus, which lies cranial to the utero-tubal junction; (2) the body, which is a muscular pear-shaped structure containing most of the cavity; (3) the isthmus, which is the narrow caudal portion of the uterus, and (4) the cervix, which leads out into the vagina. In general, the uterus of Platyrrhine and Catarrhine monkeys and the great apes is similar in structure although considerable variation occurs in size and shape. For example, the uterus of *Ateles* is approximately the same size as in *Pongo* and *Hylobates*, while the uterus of *Cebus* is much smaller than that of *M. mulatta*. The fundus of the uterus in Platyrrhine and Catarrhine monkeys is globular. In *Callithrix*, the uterine fundus has 2 lateral projections and a median groove on the ventral surface suggesting a bicornuate structure which on sectioning proves to be simplex. The uterus of *Gorilla* is much larger than that of *Pan* while the uteri of *Pongo* and *Hylobates* are similar in size. In the great apes *(Pan, Pongo,* and *Hylobates)* the fundus is flattened dorso-ventrally as in man, but in *Gorilla* it is globular as in Catarrhine monkeys.

The cervix also varies structurally as well as in dimensions among the nonhuman primates. Generally, it is more developed in lower monkeys than in the great apes. In *M. mulatta* the cervical canal is bent in two almost right-angle turns, with the proximal curve convex towards the ventral aspect and the distal curve towards the dorsal aspect of the uterus. An S-shaped curve is sometimes also found in the baboon [8]. The external cervical os is slit-like. The dorsal lip of the cervix is usually thicker than the ventral lip. In the *Goril-*

la the cervix is large and exceeds that of humans. Other anthropoid apes, however, have small cervixes, especially the *Hylobates* in which the cervical portion protruding within the vaginal vault is hardly noticeable. A similar condition is found in *Aotus*. In other Platyrrhine and Catarrhine monkeys, however, this structure projects into the vagina, resulting in a marked vaginal fornix or 'portio'.

The vagina is a thick-walled muscular structure which extends caudally from the uterus to the vestibule. In the New World monkeys, the mucous membrane is elaborately folded in longitudinal and transverse ridges marked with raised papillae. It is constricted near the vestibule and dilated towards its medial portion. In *Gorilla* and *Hylobates*, the vaginal epithelium is arranged in low transverse folds. These folds are absent in *Pan* and *Pongo* [9].

C. Mammary Glands

The anatomy of the mammary glands of anthropoids, Platyrrhines and Catarrhines has recently been reviewed [10]. In general there is 1 pair of pectoral mammary glands. The number and position varies more in the prosimians [11], which may have 1 or 2 pectoral pairs with or without an abdominal pair.

The mammary areas in some species are unique. In *Loris* and *Nycticebus*, 'lacteal tracts' of approximately 5×1 cm are formed during lactation when some chest hair is lost. In *Theropithecus gelada*, the mammary area is a large bare patch which is surrounded by a necklace of red fleshy lumps [12]. The nipples in *Cercopithecus aethiops* and *T. gelada* are close together allowing the infant to nurse both simultaneously. They are far apart in *Alouatta*, *Lagothrix*, and some *Callithrix* and *Pithecia*. *Galago senegalensis* has 1 pair of mammary glands with 3 pairs of nipples spaced from clavicle to groin [13]. The nipples of *M. mulatta* (9–12 mm long $\times$ 5–6 mm wide) have 2 to 10 ducts and vary in color from pink to bright red [14]; those of *Pan* (15×14 mm) are black with 8 or 9 ducts [15]; those of *Pongo pygmaeus* (40 mm long), and of *G. senegalensis* (4 mm long) have 2 ducts and of *Nycticebus coucang* (7–10 mm long) have 7 to 10 ducts [16].

The mammary glands of adult nonhuman primates are richly vascularized flat sheets or discs of tissue which are so thin beneath the skin that they are hardly palpable in nonpregnant females or even in early pregnancy. In *M. mulatta* the glands may extend laterally into the axillary regions and caudally into the costal borders [17]. Those of the *Cercopithecus sabaeus* measure 20–45 cm^2; those of *Pan* are approximately 50 cm^2 [15]. The lobules of alveoli are supported by connective tissue stroma and are connected with collection ducts radiating from the nipples. The glands become enlarged and the alveoli are distended after parturition.

The mammary glands of *P. cynocephalus* [18] and *Macaca irus* [19] histologically resemble those of *M. mulatta* [14]. The mammary glands of lactating *G. senegalensis* differ histologically from those of *M. mulatta* in that striated muscle fibers are found between the lobules. This is not the case in the glands of *Nycticebus coucang* [16].

II. Physiology

A. Puberty

The age of females at puberty varies markedly among the species of nonhuman primates, ranging from 6 months in *Tupaia* to 10 years in *Hylobates* and in *Pan* (table I). In general, puberty is attained early in prosimians (0.5 to 1.5 years) and much later in the anthropoid apes (5–10 years). Platyrrhine monkeys attain sexual maturity in 15 to 18 months while Catarrhine monkeys become sexually mature at 3 to 5 years of age. Puberty is attained before mature body weight is reached. Among the several factors that possibly influence the age at puberty are environment, nutrition and selection. Thus, pubertal age might differ between 'captive' and 'wild' populations. Puberty may be accelerated by controlled environmental and nutritional conditions as well as by the added selection pressure for early breeding imposed on a captive population.

In some species *(M. mulatta, Papio,* and *Pan)*, the onset of puberty is generally marked by initiation of cyclical turgescence of the sex skin and initial menstrual bleeding (menarche). Menarche is also accompanied by certain physical changes such as the opening of the vaginal orifice *(Galago)*, full canalization of the vagina *(Cercopithecus aethiops)*, vulvar swelling *(Erythrocebus patas)*, and cervical mucus secretion *(Macaca sinica)* [20]. Mammary gland development also accompanies the onset of sexual maturity.

Following menarche there is generally a period of adolescent sterility characterized by menstrual irregularities, anovulatory cycles and high abortion rates [20, 21].

B. Breeding Season

Many species of nonhuman primates are polyestrous and breed throughout the year (table I). The *Lemur* [22], *Loris* [23], *Saimiri, Galago, Sa-*

Table I. The age at puberty, breeding season, gestation period and menstrual cycle duration of nonhuman primates (compiled from the literature)

Common name	Scientific name	Age at puberty, years	Breeding season	Gestation period, days	Menstrual cycle duration, days
Prosimians					
Tree shrew	*Tupaia spp.*	0.5	all year	41–50	–
Lemur	*Lemur spp.*	1.5	once/twice a year	120–135	–
Loris	*Loris sp.*	–	twice a year	160–174	–
	Nycticebus spp.	–	all year	193	37–54
Potto	*Perodicticus spp.*	–	–	170	34–47
Galago (Bush Baby)	*Galago spp.*	1.5	April-Oct.	120–146	19–39
Tarsier	*Tarsius sp.*		all year	180	23–24
New World monkeys (Platyrrhine)					
Tamarin	*Saguinus spp.*	1.5	February	140–145	–
Marmoset	*Callithrix spp.*	1.2	all year	140	–
Squirrel	*Saimiri sciureus*	–	July-Aug.	168–182	24–25
Night owl	*Aotus trivirgatus*	–	–	140–180	–
Titi	*Callicebus sp.*	–	all year	–	–
Capuchin/ringtail	*Cebus spp.*	–	all year	180	15–20
Spider	*Ateles spp.*	–	–	139	24–27
Woolly	*Lagothrix lagotricha*	4	–	130–150	–
Howler	*Alouatta spp.*	–	–	140	–
Old World monkeys (Catarrhine)					
Vervet/ African green	*Cercopithecus aethiops*	–	–	180–213	30
Guenon	*C. mitis*	–	–	–	–
Talapoin	*Miopithecus talapoin*	–	all year	196	–
Mangabey	*Cercocebus spp.*			–	29–33
Patas	*Erythrocebus patas*	3–4	all year	170	30
Macaques:					
Barbary ape	*Macaca sylvana*	–	–	210	27–33
Cynomolgus/ crabeating	*M. fascicularis/irus*	4.3	all year	153–179	28
Japanese	*M. fuscata*	3.5	–	170–180	28
Rhesus	*M. mulatta*	3.5	all year	146–180	28
Pig-tailed	*M. nemestrina*	4	all year	162–186	31
Bonnet	*M. radiata*	3–4	–	153–169	31

Table I (continued)

Common name	Scientific name	Age at puberty, years	Breeding season	Gestation period, days	Menstrual cycle duration days
Stump tailed	*M. speciosa*	–	–	–	–
Celebes	*Cynopithecus niger*	–	–	155–175	31
Silvered langur	*Presbytis cristatus*	3.5	–	168	30
Hanuman langur	*P. entellus*	–	April-Aug.	196	30
Gelada baboon	*Theropithecus sp.*	5	all year	–	–
Mandrill	*Mandrillus spp.*	–	–	245	–
Savannah baboon	*Papio sp.*	3–4	all year	154–183	35
Great apes					
Gorilla	*Gorilla gorilla*	5–8	–	251–289	30–31
Chimpanzee	*Pan troglodytes*	6–10	all year	216–260	35
Orangutan	*Pongo pygmaeus*	–	–	275	29
Gibbon	*Hylobates spp.*	5–10	–	210	–
Siamang	*Symphalangus syndactylus*	–	–	230–235	20–43

guinus, and *Presbytis* [20, 21, 24] have seasonal breeding patterns. Such seasonality appears to be a product of adaptation to the environment. Photoperiod length, temperature, and rainfall, aside from their direct effects on the physiological reactions of the animals, also control to a large extent the quality and quantity of vegetation and thereby indirectly influence breeding patterns. Examples of the effects of environment on the breeding season are seen in the Madagascar lemur *(Microcebus murinus)* in which the mating season changes from the normal of October to February in Madagascar to the period of March to August when transported to the northern hemisphere [25]. Free-ranging *M. mulatta* in Puerto Rico [26, 27, 28] and *M. fuscata* in Takasakiyama, Japan [27] are reportedly only seasonally polyestrous. Although there are insufficient data on the mating season of wild chimpanzees, it has been observed that they frequently breed from August to November [29]. However, under captive conditions where the diet and the environment are controlled, no restricted seasonal breeding seasons are observed in either the rhesus or the chimpanzee. Likewise, captive baboons show no seasonal variation in sexual activity although higher conception rates occur in early spring and late summer [21].

In a survey of the reproductive cycles in a variety of nonhuman primate species [27] it was concluded that, in general, the timing of the mating season and the resulting conceptions is associated with decreasing day length, high levels of precipitation, and favorable diet. Decreasing daylight was similarly implicated as a trigger to the onset of the mating season in Japanese macaques and Barbary apes [30].

C. Birth Season

Species with a restricted mating season naturally have a similarly restricted period when most of the births occur. In the rhesus colony in Puerto Rico, where the mating season is from August to December, the birth season is from February to June with the peak between March and May [27]. In northern India, free-ranging rhesus monkeys have a birth peak in April with a few births occurring in September [31]. Although births occur throughout the year in caged baboons, the majority of births occur during the fall and winter months, with peaks in October and again in January [32].

In the *Saimiri*, which exhibit marked reproductive seasonality, the birth season is confined to a period of 2 to 3 months (February–April) in their natural habitat. When a colony of *Saimiri* was transferred from Peru to Florida there was a definite and well-defined shift in the birth season of the colony to April–June in the 1st year, May–July in the 2nd year and June–August in the 3rd and subsequent years [33]. Such shifts in birth season are directly related to shifts in sexual activity due to changes in the environment and possibly to the organization and integration of the colony.

D. Sexual Receptivity (Estrus)

The duration of sexual receptivity or estrus is species-dependent. In some species, estrus is exhibited by the females only during a limited interval within the menstrual cycle. The duration of estrus is 2 days in the *Lemur* and *Perodictus*, 3 to 4 days in *Pongo pygmaeus*, 5 to 7 days in *Presbytis entellus* and *Papio cynocephalus*, 9 days in *M. mulatta* and *M. fuscata* and 11 days in *M. fascicularis*. The female chimpanzee *(Pan)* shows sexual interest only during the period of maximum perineal swelling although copulation is sometimes observed even when the sex skin is deturgesced. The gibbon *(Hylobates)* female engages in copulation throughout the menstrual cycle and even during pregnancy.

Post-partum estrus is apparently rare in the nonhuman primate. In the *Galago*, *post-partum* estrus occurs only when the neonate is lost [23]. Free-ranging lactating *M. mulatta* resume their sexual activities later than non-lactating females [28] and in the spider monkey *(Ateles)*, ovulation ceases completely during lactation [34] while in the howler monkey *(Alouatta)*, ovulation is suppressed only during early lactation.

E. Sexual (Menstrual) Cycle

Nonhuman primates may undergo periodic intervals of sexual activity either throughout the year or only during certain seasons of the year. The duration of the sexual cycle in various nonhuman primate species is shown in table I. The length of the cycle varies between species, among individuals of the same species and also within individuals. During the cycle changes occur in the perineum, external genitalia, vaginal and cervical mucosa, endometrium, ovaries, and oviducts. In some species *(Cercopithecidae*, *Pongidae)* the cycle features a period of overt menstrual bleeding as in the human female. The cycle and the changes that occur therein are mediated and controlled by gonadotropic (FSH and LH) and gonadal (estrogenic and progestational) hormones. The cycle generally consists of 2 main phases of approximately equal length: (1) the follicular phase, at the height of which females are generally sexually receptive, and (2) the luteal phase. Ovulation occurs between these two phases. The luteal phase is followed by menstruation in certain species.

1. Hormone Patterns

Comparative aspects of reproductive endocrinology in nonhuman primates remain, for the most part, undefined. Studies have been limited mainly to *Macaca* and *Papio* with a few isolated studies in *Pan* and *Gorilla*. The recent advances in technology, which permit measurement of plasma levels of gonadotropic and steroid hormones, will undoubtedly bring about rapid expansion of knowledge in this field.

Basically, the patterns of secretion of the gonadotropic and steroid hormones in the nonhuman primates which have been studied are similar to those in man, although some significant quantitative and qualitative differences have been found. Urinary excretion of follicle stimulating hormone (FSH) in the chimpanzee is highest during the mid-follicular phase of the cycle [35]. The biological assay used in this study failed to detect any rise in FSH around the time of ovulation as has been found in the human female.

Plasma luteinizing hormone (LH) patterns are basically similar in the cycles of rhesus monkeys and baboons. There appears to be only 1 major peak during the cycle. It occurs around mid-cycle, apparently preceding ovulation [36, 37]. Although nonhuman primates are responsive to human gonadotropins, there are apparently some differences in the hormone molecules since there is a lack of immunologic cross-reactivity between simian LH and antisera produced to human LH or chorionic gonadotropin (HCG) [36].

Plasma progesterone and 'progestin' (a semiquantitative estimate of combined progesterone and 17α-hydroxypregn-4-ene 3,20 dione) levels have been determined in the rhesus monkey and baboon sexual cycles. There is a mid-cycle rise in 'progestin' which apparently precedes ovulation and then a drop to basal levels often occurs, probably on the day of ovulation. In the baboon, progesterone and of course 'progestins' both rise following ovulation to a broad peak which is reached on approximately day 7 post-ovulation; they then return to basal levels 12–14 days post-ovulation [37, 38]. In the rhesus monkey, the progesterone level may increase slightly 1–2 days preceding ovulation, in contrast to the baboon. In other respects the pattern is similar except that the plasma levels are slightly lower than in the baboon [36, 39, 40].

In the baboon urinary excretion of estrogens attains a mid-cycle peak which coincides with the plasma LH and 'progestin' peaks mentioned earlier [37]. The broad peak of urinary estrogen secretion seen during the luteal phase of the human menstrual cycle was not observed in the baboon cycle.

2. Ovarian Changes

Cyclic changes in ovarian morphology have been described extensively in *P. hamadryas*, *P. porcarius* [41], and in *Macaca mulatta* [42].

Resting follicles in both baboons and rhesus monkeys measure approximately 1 mm in diameter. Follicular enlargement commences when the corpus luteum begins to degenerate. A wave of follicular growth occurs at the start of the follicular phase. Atresia of the follicles occurs at all times during follicular development, consequently, only a few preovulatory follicles mature. Usually a single follicle ripens, reaching a diameter of approximately 6 mm at ovulation which occurs around the middle of the menstrual cycle. Ovulation may occur alternately from the 2 ovaries or repeatedly from a single ovary. The cavity of the ruptured follicle measures approximately 4 mm and is filled with fluid containing leukocytes, fibrin and fibroblasts. The granulosa cells begin to swell and show signs of transition into luteal cells in approximately 2 days. Full morphological development of the corpus luteum occurs from about day 8 to day 12 after ovulation. Luteal cell degeneration takes place soon after maximum maturity of the corpus luteum is attained. Two types of auxiliary luteal bodies have been observed in *M. mulatta*. Aberrant corpora lutea are believed to be remains of previously active corpora lutea which did not degenerate in the usual manner. Accessory corpora lutea are luteinized follicles which have not ovulated [42, 43, 44].

3. Endometrial Changes

The cyclic changes of the endometrium of *M. mulatta* [45, 46] and the Hamadryas baboon *(P. hamadryas)* [41] have been studied extensively.

The endometrium of nonhuman primates responds to gonadal hormones in a manner similar to that of other mammalian species which exhibit menstrual or estrous cycles. During the estrogenic phase, increased vascularization and proliferation of the endometrium occur but the endometrial glands remain relatively uncoiled. During the progestational phase, the endometrium thickens, the surface epithelium changes from low to high columnar and the endometrial glands become coiled and secrete actively. In the late luteal phase the uterine epithelium degenerates. Cellular desquamation, bleeding and endometrial destruction are characteristic of the late luteal phase although the degree to which they occur varies with species. Menstrual bleeding lasting 3–4 days on the average is seen most commonly in Cercopithecidae and Pongidae. In most Cebidae and green monkeys only the superficial lining of the endometrium undergoes degeneration, and only minute bleeding is observed.

The physiological and hormonal mechanisms of menstruation in the nonhuman primate have been reviewed recently [20].

4. Changes in Cervical Mucus

The fluid and cellular contents of the cervical canal and glandular crypts undergo cyclical qualitative and quantitative changes during the nonhuman primate menstrual cycle. Like the vaginal epithelium, the stratified cervical epithelium lining the mouth and distal 3–4 mm of the cervix also change during the cycle. The amount of mucus in the cervical canal is greatest during the preovulatory and mid-luteal phases of the cycle. Erythrocytes and cellular debris are found in abundance in the cervical mucus during menstruation. A characteristic cervical mucus with a disagreeable odor is discharged from the vagina during estrus in *M. sinica* and *M. radiata*. A sanguineous discharge of cervical origin is also seen in the *Loris* during the proestrus phase of the cycle.

A characteristic cyclic variation in the quality of the cervical mucus is seen in *M. mulatta* [47]. Maximal fern-pattern formation was noted on the day preceding ovulation and a drastic regression occurs on the subsequent day. Similar patterns are apparent in *P. cynocephalus* (unpublished data).

5. Vaginal Changes

The vaginas of some nonhuman primates undergo cyclical cytological changes. In the baboon, vaginal cytological changes are correlated with the perineal cycle [48]. This correlation was not seen in *Saimiri sciureus* [49] but is evident in the rhesus and other species. In the baboon [50], the menstrual phase of the cycle is characterized by the occurrence of erythrocytes, leukocytes, epithelial cells, mucus and cellular debris in the vaginal smears. Red blood cells disappear and only a few leukocytes remain in the postmenstrual phase. During this phase, there is an increase in small and intermediate cells and superficial cells. Parabasal cells are seen occasionally. A sharp increase in leukocytes, superficial cells and mucus characterizes the preovulatory phase while the ovulatory phase is marked by an abundance of superficial epithelial cells and the absence of mucus and leukocytes. The return of leukocytes and mucus marks the postovulatory phase. Clumping, curling and folding of the epithelial cells are noticeable. The luteal phase is characterized by a drop in the karyopyknotic index and an increase in leukocytes and mucus. Lastly, the premenstrual phase is typified by an increase in superficial cells and a decrease in mucus. Bacteria and cellular debris are also more plentiful.

6. Perineal (Sex Skin) Changes

In many species of nonhuman primates (Section I,A), the appearance of the perineum undergoes cyclical changes during the menstrual cycle. Sex skin swelling is most pronounced in baboons *(Papio)* and in the pig-tailed macaque *(M. nemestrina)*. In the baboon the perineal cycle has been correlated with the time of ovulation; this occurs approximately 3 days prior to sex skin deturgescence [51]. The perineal cycle in the baboon has 2 phases: (1) a turgescent, and (2) a deturgescent phase. The turgescent phase starts shortly after menstruation and coincides with the preovulatory or follicular phase of the cycle. During this phase the sex skin changes in color from dull pink to bright red, and starts to swell. When the skin is maximally swollen, it is fully distended, devoid of wrinkles and has a smooth, shiny appearance. The deturgescent phase begins with a loss of color and a decrease in size and turgidity of the swelling, with a corresponding increase in wrinkles. The swelling gradually subsides. This phase coincides with the postovulatory or luteal stage of the cycle.

In the pregnant baboon, the perineum generally remains quiescent throughout the gestation period although the color often changes to a deeper red, similar to that of the hairless skin area around the external genitalia.

7. Changes in the External Genitalia

In some species, such as the lorisoid primates, in which the labia of the vulva are more developed visible external swelling of the labia occurs. In the Potto *(Perodicticus)*, the 'superficial labia' are widely spread and flushed tesserae appear on each side while the 'deep labia' are separated and the vagina is open [23]. The vaginal orifice of *G. senegalensis* not only shows periodic opening and closure during the sex cycle but also changes considerably in size. Marked changes in the external genitalia are also observed in *Miopithecus talapoin.*

F. Gestation

The gestation periods of various nonhuman primates are shown in table I. The length of gestation varies from species to species with the majority falling within 140–170 days. *Tupaia* has a remarkably short gestation period of 41–50 days in contrast to the long periods observed in *M. sylvana* (210 days), *Mandrillus* (245 days) and the great apes (210–289 days).

Chorionic gonadotropin levels have been measured in the blood and urine of pregnant *M. mulatta* [52, 53]. Monkey chorionic gonadotropin (MCG) is detected in the urine as early as day 12 of pregnancy in some individuals. Maximum levels of urinary excretion are reached between days 18 and 28; declining to nondetectable levels around day 38. MCG is detectable in serum between days 18 and 35 of pregnancy in this species. Chorionic gonadotropin was detected in the urine of a *Gorilla gorilla* from the 8th (first sample taken) to the 35th week of pregnancy. Peak excretion occurred in the consecutive samples collected at weeks 16 and 23 [54].

The pattern of plasma progesterone concentrations during gestation has been determined for *M. mulatta* [40]. There is a sharp rise in plasma progesterone on days 9–11 of pregnancy to approximately 8 ng/ml which is about twice the level of the normal cycle peak. This rise is short-lived, returning to 2–3 ng/ml by days 19–23 of pregnancy; it occurs at a time when the progesterone level would be decreasing in a normal nonpregnant cycle. Another peak lasting approximately 15 days and reaching a level of about 5 ng/ml begins around day 29–33 of pregnancy. The plasma progesterone concentration then declines to approximately 1 ng/ml on the 40th to 50th day of pregnancy, then rises to 2–4 ng/ml for the remainder of gestation.

In *P. cynocephalus* the plasma 'progestin' level rises to approximately 50 ng/ml between days 35 and 50 of pregnancy. This is approximately 5 times the level of the nonpregnant cycle peak. The 'progestin' level then generally fluctuates between 20 and 40 ng/ml during the remainder of pregnancy [38].

Comparative aspects of the importance of pregnanediol as an excretion product of progesterone in nonhuman primates remain unclear although several points have been established. Pregnanediol is detectable in *P. cynocephalus* urine at least during the last 2 months of pregnancy during which time it is excreted at a relatively constant level of approximately 100 μg/24 h [55]. The major urinary metabolites of exogenously administered 4-^{14}C-progesterone in the nonpregnant *P. cynocephalus* are pregnanediols and androsterone [175]. Small amounts of pregnanediol have also been isolated from the urine of pregnant chimpanzees [56]. Although pregnanediol was not detected in the urine of pregnant 'monkeys' (presumably rhesus) [57], later studies revealed the presence of pregnanediol in urine of both pregnant rhesus monkeys and rhesus monkeys injected with large doses of progesterone [176]. However, quantitative importance of this metabolite was not reported. Urinary androsterone was estimated to account for 5–19% of exogenously administered progesterone in the nonpregnant pig-tail monkey [177].

Estrone is the predominant urinary estrogen excreted during pregnancy in *P. cynocephalus* [55] and *M. mulatta* [58], with estradiol and estriol being excreted in much lower quantities. In *Gorilla gorilla* [59] and *Pan troglodytes* [60] estriol is predominant in the urine during pregnancy.

Although plasma 'progestin' determinations show promise of being a useful means of pregnancy diagnosis as early as days 12–15 of pregnancy, the reliability of this method has not yet been established. Pregnancy can be detected in most nonhuman primates after day 25 of gestation either by rectal digital palpation or transabdominally.

Another sign of pregnancy in the baboon and rhesus is the coloration of the hairless skin area around the external genitalia; this usually turns deep red during pregnancy. However, this change generally occurs later in pregnancy than the detectable changes in uterine size and tonicity and also varies between individuals. There are also certain observable behavioral changes in most species during pregnancy. The animals tend to become increasingly quiet and gentle and sexual receptivity is reduced although males will attempt to copulate.

G. Parturition

The signs of approaching parturition have been described recently in nonhuman primates [21]. Impending parturition is usually preceded by changes in sleep, feeding and watering patterns, frequency of urination, and other activities. In the chimpanzee the descent of the abdominal bulge, which corresponds to descent of the fetal head into the pelvic inlet, occurs about 1 week prior to parturition. Extrusion of the cervical plug, which consists of a blood-tinged mucus exudate, is by far the most reliable sign of approaching labor.

Normally, the fetus in all nonhuman primates develops in a breech presentation. Version or turning occurs late in pregnancy; from 23 to 65 days prior to parturition in *Saimiri*, 4 to 80 days in *M. mulatta* and as late as within the last 24 h in *P. cynocephalus*. Compared to man, the frequency of breech presentation is higher in the nonhuman primates (12% in the rhesus, 16% in the baboons, and 4% in man). Usually the fetal head faces the mother's symphysis pubis in most nonhuman primates [21].

A number of behavioral characteristics have been observed in females in labor: squatting and raising of outstretched arms in baboons; short, quick running, or running followed by squatting and grunting in chimpanzees; and freezing into unusual positions, and staring into space in the rhesus and in the squirrel monkey. Labor and parturition usually occur at night in the rhesus, squirrel monkey and baboon. It is not uncommon for labor contractions to stop if parturition has not occurred by sunrise and to resume again at sundown. This nocturnal birth pattern also occurs in the wild and is probably significant for survival. The chimpanzee is an exception to this nocturnal birth pattern. It is common for nonhuman primate mothers to eat the placenta, and in many instances the stillborn infants as well, shortly after delivery.

H. Lactation

The physiology of the rhesus monkey mammary gland has been described extensively [14] and the scanty data known for other species have been reviewed recently [10]. Basically, the development of the mammary gland and the initiation and maintenance of lactation in the nonhuman primate appears to be under the same hormonal control as in other mammals. Ovariectomy or hypophysectomy after the first month of pregnancy does not affect mammary development or initiation of lactation in *M. mulatta* [14, 61]. However, hypophysectomy prevents maintenance of lactation [61]. The length of lactation varies with each species, ranging from about 2 months in some prosimians to approximately 2 years in the great apes. Nutritional factors, separation or death of the infant, age and size of the mother and, of course, pathological conditions of the mammary glands are some of the factors that affect the length of lactation. Involution of the mammary glands generally occurs after cessation of lactation.

III. Pathology

Several reviews cover various aspects of pathology of reproduction in nonhuman primates [62, 63, 64, 65]. The present review covers pathologic conditions of the female and her offspring. The etiologies of many of the conditions described are not clearly understood. It is, therefore, difficult to make clear-cut distinctions between infectious and non-infectious diseases. This is particularly true in the case of neoplasms. Only those neoplasms which result from experimental inoculation of viruses have been included in the 'infectious' section although it is recognized that others may one day be found to be caused by infectious agents.

A. Prenatal and Neonatal Mortality

1. Embryo Mortality

The incidence of early embryonic mortality varies little between species of nonhuman primates. In *P. cynocephalus*, approximately 30 and 15% of the preimplantation [66] and postimplantation [67] embryos, respectively, are morphologically abnormal. A similar incidence of abnormal preimplantation embryos was observed in *M. mulatta* [66] and 1 of 2 preimplantation sites was found to be abnormal in *P. troglodytes* [68]. The defective segmented ova are characterized by necrotic blastomeres, distorted or malpositioned embryonic or trophoblastic blastomeres, and distorted morulae or blastulae. In the postimplantation embryo, degeneration is marked by cellular pyknosis in the embryonic disk, the placenta and associated membranes. There are no descriptions of embryonic mortality in *P. cynocephalus* from the 4th to the 7th week of pregnancy (period of organogenesis) [64]. However, abnormal embryos during this stage of pregnancy are seen in *M. mulatta* [69].

A number of factors are thought to cause defects in ova and early embryonic mortality. In *P. cynocephalus*, mating early in or at the optimum time of the ovulatory phase of the cycle results in a higher conception rate than mating late in the ovulatory stage of the cycle. This may be due to aging of ova either before or after ovulation. Similarly, aged sperm cells probably produce more defective zygotes than those which fertilize ova soon after attaining fertilizing capacity.

2. Congenital Malformations

Developmental malformations recorded in the nonhuman primate include anencephaly, hydrocephaly, anophthalmia, diaphragmatic hernia, facial cutaneous angioma, pigmental nevus, subdural hydroma, congenital blindness, hermaphroditism, agenesis of the genitalia, renal malformations, cleft lip, cleft palate, polydactylism, micromelia, amelia and other skeletal defects [70–74]. The incidence of specific spontaneous congenital malformations is 0.5% or less [63, 75, 76, 77].

3. Placental Abnormalities

Placental malformation occurs infrequently in nonhuman primates. In one study, 3 double placentas, 1 circumvallate placenta and 1 placenta with a membraneous insertion of the umbilical cord were found out of approximately 200 cases [64]. Two cases of placenta praevia have been reported [62]. In 165 consecutive baboon placentas, only one was observed to have a single umbilical artery [78]. A more common occurrence is variation in the transverse communicating artery. In species with bidiscoid placenta, variations in lobe size also occur.

4. Abortions and Stillbirths

The incidence of abortions and stillbirths varies with species, parity, season, caging conditions, degree of handling, geographic location and other factors, including infectious diseases. It is highest in primigravid animals, in animals late in their reproductive life, in animals kept in cages rather than in pens, and in animals born in captivity and during the winter months [63].

The abortion rate in nonhuman primates ranges from 4 to 16%. It is approximately 4% in baboons [76]. In a colony of baboons and macaques in which the abortion rate decreased from 14.1 to 4.6% over a 5-year period [26], the decline was attributed to avoidance of stresses such as restraints and changing of caging conditions.

The stillbirth rate is approximately 15% in the rhesus [79], 12.3% in a colony consisting mainly of *M. mulatta*, 14.6% in green marmosets, 13.6% in baboons, 5% in *C. atys*, *M. mulatta*, and *M. radiata*, and 20% in *M. nemestrina* [63, 64].

5. Dystocia

Dystocia related to abnormal size relationships, positional abnormalities and uterine inertia are known to occur, although at a low rate, in nonhuman primates. One case reported in a 35-year-old *P. troglodytes* was due to cervical blockage by a uterine fibroid tumor [80]. Deformity as a result of dystocia was observed in a Barbary ape *(M. sylvana)* and in a pigtailed macaque *(M. nemestrina)* [62].

The most common positional abnormality in nonhuman primates is facial presentation, followed by posterior presentation. Other types of presentation generally occur in conjunction with premature births and abortions.

6. Neonatal Mortality

Prematurity is a major cause of neonatal mortality. Deaths among premature infants, generally in their first weeks of life, are often related to immature respiratory systems [81]. Agalactia, cannibalism and maternal abandonment are other common causes of early infant mortality. A hemolytic disease resembling erythroblastosis fetalis occurs in marmosets [82]. Megaloblastic anemia of pregnancy is common in captive squirrel monkeys [83].

B. Noninfectious Gynecologic Pathology

1. Ovaries

The various types of ovarian pathology which have been described in nonhuman primates are listed in table II. The most frequently observed pathological condition is the cystic follicle. The follicle cyst in the baboon [21] as well as the benign cystic teratoma and the cavernous hemangioma in the rhesus monkey [84] were observed in pregnant animals. It is not known whether these conditions developed after conception or whether these conditions failed to inhibit conception. Of 8 rhesus monkeys having follicle cysts, 4 were classified as fertile while 3 exhibited secondary and 1 primary infertility [85]. Parovarian cysts, observed in the same study, were associated more frequently with infertility. Three females with parovarian cysts experienced primary infertility, 5 secondary infertility and only 1 was classified in the fertile group. It is not known whether this indicates that the parovarian cyst is a cause of, a sequel to infertility or a coincidental finding. The types of ovarian tumors found in nonhuman primates are quite similar to those occurring in human subjects [84]. Infantile ovaries observed in an adult baboon were associated with an entire reproductive tract which was much smaller than normal [21].

Table II. Ovarian pathology in nonhuman primates

Pathological conditions	Species	Number of cases	Reference No.
Infantile ovaries	*P. cynocephalus*	1	70
Infantile ovaries	*M. mulatta*	1	86
Aberrant ovary	*Papio*	1	87
Aberrant ovary	*M. mulatta*	1	88
Parovarian cysts	*M. mulatta*	9	85
Mineral deposits	'monkeys'	common	62
Simple cyst	*M. mulatta*	1	89
Simple cyst	*M. philippinensis*	1	62
Follicle cysts	various	6	63
Follicle cysts	*M. mulatta*	8	85
Follicle cysts	*P. cynocephalus*	1	21
Corpus luteum cysts	*M. mulatta*	1	85
Follicle hematomas	*M. mulatta*	2	85
Benign neoplastic tumors:			
Serous cystadenoma	*M. mulatta*	1	90
Serous cystadenoma	*M. mulatta*	1	85
Papillary serous cystadenoma	*M. mulatta*	1	84
Papillary serous cystadenoma	*M. mulatta*	1	91
Benign cystic teratoma	*M. mulatta*	1	91
Benign cystic teratoma	*M. mulatta*	1	84
Brenner tumor	*A. paniscus*	1	92
Cavernous hemangioma	*M. mulatta*	1	84
Adenoma	(not indicated)	1	93
Malignant tumors:			
Adenocarcinoma	*M. irus*	1	94
Undifferentiated	*M. philippinensis*	1	95
Granulosa cell	*S. sciurea*	1	96
Granulosa cell	*M. mulatta*	1	97
Sarcoma	*P. anubis*	1	63

2. Uterus and Fallopian Tubes

Five cases of cervical dysplasia have been observed in *M. mulatta* [85]. Cervical and uterine prolapse occur infrequently in the nonhuman primate. In 1,693 recorded births of monkeys at the Sukhumi breeding farm, only 1 case of prolapsed uterus was observed [63]. Also, only 1 case of uterine prolapse out of 1,190 rhesus births was noted [64]. Uterine torsion has been recorded in a baboon [63]. The same investigators also recorded 3 uterine ruptures in 715 births. The ruptures occurred at the cervico-uterine junction. Uterine ruptures in *M. mulatta* have also occurred along previous Caesarian section scars, and spontaneous ruptures have been seen in *M. nemestrina* and *Cercocebus atys* [64].

Four cases of ectopic pregnancy have been reported [63, 81]. Two of these cases were in *M. mulatta*, one in *P. hamadryas* and the 4th, a tubal pregnancy, was in a 'green marmoset'.

The first reported case of ectopic endometrium (endometriosis) in the nonhuman primate was described in an adult *M. mulatta* [98]. This was followed by a report of proliferating endometrial gandular tissue on the surface of the ovaries in an adult *M. irus* [99]. Several other cases of endometriosis in *M. mulatta* have been reported. One had been thyroidectomized and received propylthiouracil [88] and another was associated with endocrine abnormalities consequent to hypothalamic degeneration following a temporal lobectomy procedure [100]. Endometriosis was also found in 2 monkeys receiving dienestrol [62] and in a rhesus monkey with a history of irradiation but no abdominal surgery [101]. In a study involving 63 *M. mulatta* females with various types of gynecologic pathology, 21 (33%) had endometriosis [102].

Neoplasms of the uterus and uterine cervix which have been reported in nonhuman primates are listed in table III.

3. Vagina and External Genitalia

Two cases of vaginal papillomatosis in *M. mulatta* and 1 case of vaginal leukoplakia in *M. talapoin* have been described [62]. Vaginal hyperkeratosis and papillomatosis were also observed in 'monkeys' under various experimental conditions which may have resulted in vitamin A deficiency [63]; however, a direct causal relationship was not established. A case of primary infertility of a *M. mulatta* female was reported in which clitoromegaly was the only remarkable abnormality [85]. The sex skin is a frequent site of lacerations and was the site of a squamous cell carcinoma in a 19-year-old *P. hamadryas*.

4. Mammary Glands

Mammary adenocarcinoma has been described in 2 elderly multiparous *M. mulatta* [62]. Spontaneous mammary cancer in monkeys has also been reported [80, 103–107]. A well-differentiated spontaneous primary tumor of the mammary gland has been seen in *Pongo pygmaeus* [108]. Metastases were found throughout the body of the animal, including kidney, axillary lymph nodes, lungs, pleura, pericardium, liver, bladder, uterus, thyroid and the air

Table III. Neoplasms of the uterus and uterine cervix in nonhuman primates

Types of Tumor	Species	Number of cases	Reference No.
Uterus:			
Polyp	*P. hamadryas*	1	63
Polyp	*M. mulatta*	9	85
Papillomatous excrescences	*M. mulatta*	1	109
Undifferentiated carcinoma	*M. fascicularis*	1	95
Cystadenoma	*P. cynocephalus*	1	110, 111, 112
Adenomyoma	*P. cynocephalus*	1	113
Myoma	'Monkeys'	2	114, 115
Myoma	*M. mulatta*	3	85
Leiomyofibromatous nodule	*Lemur macaco*	1	116
Hemangioma	*M. mulatta*	1	85
Hamartoma	*Saguinus oedipus*	1	117
Uterine Cervix:			
Papilloma	*M. sinica*	1	112
Polyp	*Macaca spp.*	8	63
Adenoma	*M. mulatta*	1	118
Adenoma	*M. fascicularis*	1	106
Carcinoma	*M. mulatta*	3	85, 119, 120
Carcinoma	*M. sinica*	1	121

sac. The occurrence of supernumerary nipples and gynecomastia in the nonhuman primate has been reviewed recently [10].

C. Infectious Gynecologic Pathology

1. Viral

a) Rubella

Common laboratory primates, *Erythrocebus patas* [122, 123], *Macaca mulatta* [124, 125], and *Papio cynocephalus* [76] are susceptible to wild and attenuated strains of rubella virus.

M. mulatta and *M. fascicularis*, experimentally infected with rubella virus at various stages of pregnancy, show high incidences of fetal size reduction, fetal mortality, stillbirths and abortions in addition to otic, cutaneous, chorionic, osseous and vascular abnormalities. The virus affects the products of conception and often involves the placenta. Fetal abnormality, mortality and abortions are more frequent when animals are infected early in pregnancy [124, 125, 126].

A marked increase in the incidence of stillbirths and abortions was observed in pregnant *P. cynocephalus* experimentally infected with rubella virus either by intravenous inoculation or by exposure to inoculated animals [76].

b) Cytomegalovirus

As early as 1935, extranuclear inclusion bodies suggestive of cytomegalovirus were seen in the cebus monkey *(C. fatuellas)* [127]. Cytomegaloviruses (CMV, cytomegalic inclusion virus, salivary gland virus) are classified among herpes viruses. A comprehensive survey of primate sera for antibodies to viruses of man and simians reveals that herpes-type viruses are present in a wide variety of nonhuman primate species as well as man [128, 129]. CMV are reportedly highly species-specific [129, 120] and do not, in general, cross-react with CMV from other species. However, a cross-reaction between a vervet CMV and human strains has been reported [131]. An attempt to produce CMV infections in pregnant *M. mulatta* was unsuccessful [132].

c) Other viruses

M. mulatta fetuses are susceptible to tumor formation following inoculation *in utero* with Rous sarcoma virus [133]. Intravenous inoculation appears to cause a higher incidence of abortion and embryo mortality than does intramuscular injection. Some fetuses survive to term and the tumors tend to regress spontaneously after birth. Adenovirus type 12 given to pregnant *P. cynocephalus* and *M. mulatta* by amnioinjection does not produce tumors but does induce abortions in the former species [134, 135]. Trachoma and inclusion conjunctivitis (TRIC) viruses from cervixes of infected women cause trachoma in infant baboons. TRIC also causes trachoma in infant *M. cyclopis* and cervicitis in the pregnant animals. The experimental cervical infections, however, do not produce eye lesions in the offspring. Interestingly, there is increased susceptibility of these offspring to eye inoculations [136]. Many other viruses (mumps, herpes-simplex, rubeola, vaccinia, western equine encephalitis, chickenpox, smallpox, coxsackie B) have been implicated in reproductive failure in man, laboratory rodents, and certain domestic animals, but none of these viruses have yet been implicated or studied in the nonhuman primate. However, as interest broadens in this area, the role of viruses in reproduction will be further elucidated. Viral surveys [128, 129, 137] are major contributions to progress in this area. The findings of a strain of virus (CV 52) belonging to the Para 2 group of viruses in the testes of a chimpanzee is noteworthy, although its implications relative to reproduction are not understood [128].

2. Mycoplasma

Antibodies to mycoplasma *(M. pneumoniae)* occur in the sera of many species of nonhuman primates [129, 138]. Results of a comprehensive survey of chimpanzees also reveal a high incidence of antibodies to PPLO [139]. Mycoplasma organisms, predominantly *M. orale II*, *M. salivarium*, and *M. hominis I*, have been recovered from conditioned, pregnant *M. mulatta*. While no signs of the disease were observed in these monkeys, it is not certain that these species of mycoplasma are entirely nonpathogenic under all conditions [140]. Several species of mycoplasmas, predominantly *M. orale II*, *M. salivarium* and *M. laidlawii*, were also isolated from *C aethiops* within 3 weeks of capture [141]. The recovery of these organisms from animals soon after capture suggests that they may have been present in the 'wild' population. Mycoplasma isolated from various orifices (throat, nose, rectum, and vagina) of nonhuman primates *(P. cynocephalus*, *C. sabaeus*, *M. mulatta*, *P. troglodytes)* have some antigens in common with *M. pneumoniae* [138]. The isolation of mycoplasma from 2 aborted *P. troglodytes* fetuses is also noteworthy. However, a causal relationship between mycoplasmas and nonhuman primate abortions remains in question.

3. Bacterial Infections

a) Brucellosis

Although this is a world-wide infectious disease affecting a wide variety of mammalian species including man, brucellosis is not prevalent in nonhuman primates. Positive serology to *B. abortus* was observed in 8 of 72 *P. cynocephalus*, in 2 of 17 *P. troglodytes*, and in 3 of 17 *P. pygmaeus*. No brucella antibodies were found in *M. mulatta*, *Cercopithecus pygerythrus* or *G. gorilla* [142].

b) Vibriosis

As with many reproductive disorders, very little has been published on vibriosis in laboratory primates. Serological studies on 212 *P. papio* revealed 192 animals to have antibodies agglutinating *V. fetus* [143]. At the same time, screening for brucella antibodies on the sera of the same animals turned up negative results. Vibrio organisms have been isolated from the abdominal fluid of an adult female *M. mulatta* which exhibited signs of lethargy, weight loss and slight chronic diarrhea [97]. The clinical features observed and the frequency of occurrence in the nonhuman primate appear to be completely different from those observed in other animal species.

c) Leptospirosis

Leptospira pomona, which has been implicated in reproductive failure in domestic animals, has not been linked, so far, to disorders of reproduction in the nonhuman primate. However, another species of *Leptospira* have been suspected of being involved in nonhuman primate reproductive disorder. In 1965 to 1966, an extensive epidemic of *L. ballum* infection occurred at the baboon colony at the Southwest Foundation for Research and Education [144]. Associated with the outbreak was a significant rise in the incidence of abortions and stillbirths among pregnant *P. cynocephalus*. The females that aborted showed mild leucocytosis, and approximately 25% of the group showed monocytosis. Culture of samples from the aborted fetuses, however, failed to show leptospira organisms; thus, there is no direct evidence to implicate leptospira with the abortions.

d) Puerperal Sepsis

An early report describes the occurrence of puerperal sepsis causing peritonitis and septicemia in a baboon [145]. Puerperal sepsis was also observed in a chimpanzee; streptococci were seen in the vaginal smears at approximately 5½ months *ante-partum* [146]. When the animal died 15 days *post-partum*, autopsy revealed pronounced uterine infection. Two cases of septic abortions in *Galago senegalensis*, caused by an unidentified gram-negative bacillus, have recently been described [147].

e) Staphylococcal Infections

Staphylococcus aureus has been implicated as a pathogen in cases of abortions and stillbirths in *M. mulatta* but no causative relationship has been established [97]. Staphylococcal infections are infrequently found in laboratory primates. For instance, in a colony of 1,500 animals, only 1 animal, a patas monkey, was observed to be infected with *S. aureus* [148]. At necropsy, abscesses were found in the lungs, liver, spleen and mesentery of the small intestine, and the ureters were swollen and filled with purulent material. *S. aureus* and *Actinomyces sp.* have also been isolated from a rhesus monkey with acute pyometra and hemopurulent vaginal discharge [149].

f) Syphilis

Apparently, simians are not naturally infected with *Treponema pallidum* [62]. Experimentally, however, various nonhuman primate species have been infected with the disease. A mandrill [150] and a sacred baboon [151] inoculated with human infective material developed syphilitic lesions. The occurrence of a large, crusted, late syphiloderm was reported in an owl monkey *(Aotus trivirgatus)* 8 months after intravenous infection with virulent *T. pallidum* expressed from a human penile chancre [152]. The development of uveitis and the presence of spirochetes in the aqueous humor and in the skin lesions attest to the generalized syphilitic involvement. Histopathology of the lesions was considered 'consistent with *Condyloma latum*' [153]. *P. troglodytes* has also been experimentally infected with *T. pallidum* [154]. One chimpanzee was inoculated in the skin of the back with darkfield-positive serous material from a human syphilitic lesion. From the lesion that developed, 5 more chimpanzees, 2–4 years old, were inoculated with the infectious material into the skin at the base of the penis or into the left lip of the vulva. Three of these animals developed darkfield-positive lesions and the sera from the infected animals were reactive to various serological tests for syphilis.

The results of a serological survey of 250 *P. troglodytes* in the United States suggest that there is an enzootic treponematosis in captive chimpanzees in this country. Of the animals tested, 10% were reactive in the Venereal Disease Research Laboratory slide test, 16.4% were reactive in the *T. pallidum* immobilization test, and 15.6% were reactive in the fluorescent treponemal antibody-absorption test [155].

4. Mycotic Infections

a) Candidiasis

Candida was found to be the most frequently occurring yeast in the vagina of *P. cynocephalus* [156]. It has also been reported that *P. cynocephalus* is more susceptible to experimental candidiasis than *C. pygerythrus* [157]. Intravenous inoculation of *P. cynocephalus* with 5×10^6 cells of *C. albicans* isolated from a case of human vaginitis caused death 5 days post-inoculation. General dissemination of the infection throughout the major organs was noted at necropsy of *P. cynocephalus*, whereas no significant signs of infection were seen in the inoculated *C. pygerythrus*.

b) Actinomycosis

Actinomyces sp. was isolated from a *M. mulatta* which had acute pyometra [149]. However, *S. aureus* was also isolated and these 2 organisms were suspected as the probable etiologic agents.

5. Protozoan Infections

a) Trichomoniasis

The flagelated protozoan, *Trichomonas fetus* causes fetal death, abortion, pyometra and eventual sterility in many species of animals. It does not survive in the vagina of common laboratory monkeys [62]. Although the evidence is very scanty from a comparative standpoint, species lower in the primate series appear to be receptive hosts and harbor this organism. Trichomonads resembling *T. fetus* were found in a tarsier *(Tarsius carbonarius)* which aborted before the infection was detected [158].

Another species of trichomonad *(T. vaginalis)* thrives in the human and nonhuman primate vagina [159, 160, 161]. Neither in spontaneously nor in experimentally infected monkeys could gross pathologic lesions be attributed to this protozoan. *T. hominis* normally inhabits the intestines of man and monkeys and causes infection of the vagina of *M. mulatta* when artificially inoculated [162]. Other experiments, however, failed to produce any infection of the reproductive tract [160, 163].

b) Toxoplasmosis

Toxoplasma gondii antibodies have been found in *Papio sp.* [164]. However, there is no evidence linking *Toxoplasma* with reproductive disorders in nonhuman primates. Nevertheless, since some reproductive failures in man have been attributed to this disease, its possible effects on nonhuman primate reproduction warrant investigation.

6. Parasitic Helminth Infections

a) Schistosomiasis

Nonhuman primates are susceptible to schistosoma infections [165]. *C. aethiops* [166] and *Papio sp.* [167, 168, 169] were naturally infected with *S. mansoni. Cercocebus sp.* was found to be infected with a schistosome '*Bilharzia magna*' [170], which is neither *S. mansoni* nor *S. haematobium* [171]. *S. haematobium* was seen in 1 of 15 Kenya baboons. Eggs of the same species of schistosomes were found in the bowel in 1 of 20 *C. aethiops* [165]. The same investigator also reported the presence of *S. mattheei* in 4 of 15 Kenya baboons.

Experimental exposure of 10 primate species to *S. mansoni* reveals that all are susceptible to schistosomes, although the type, intensity, and course of infection varies considerably [172]. In *M. mulatta*, *M. fascicularis*, and *M. speciosa*, there is a high rate of recovery of worms from the tissues following exposure but a tendency for gradual self-cure. In *P. anubis* and *P. troglodytes* worm recovery is lower but the egg excretion rate is maintained for long periods. In *Callithrix aurita, Saimiri sciureus,* and *Tupaia sp.*, the egg recovery rate is low and the infestation tends to be aborted at an early stage. In *Ateles geoffroyi* and *Cebus apella,* the number of parasites recovered is relatively low. While eggs of the schistosome were found in all the organs studied (lung, liver, intestines, spleen, and kidney), apparently no attempts were made to recover eggs from the genital tract. However, eggs of *S. mansoni* have been found in the genital tract of a female *P. cynocephalus* [173]. Even though there are no data regarding the influence of schistosomes on reproduction of nonhuman primates, it is apparent that heavy deposition of eggs in the reproductive organs could impair fertility.

b) Filariasis

An isolated case of a filarid worm *(Tetrapetalonema digitata)* infestation has been reported in 2 adult gibbons *in habitat* [174]. The ovaries were found to be invaded by the parasite.

References

1 HILL, W. C. O.: Primates. Comparative anatomy and taxonomy, vol. 4 (University Press, Edinburgh 1960).

2 HILL, W. C. O.: Primates. Comparative anatomy and taxonomy, vol. 5 (University Press, Edinburgh 1962).

3 HILL, W. C. O.: Primates. Comparative anatomy and taxonomy, vol. 6 (Wiley, New York 1966).

4 WISLOCKI, G. B.: The reproductive systems; in HARTMAN and STRAUS The anatomy of the rhesus monkey, p. 231 (Williams & Wilkins, Baltimore 1933).

5 HILL, W. C. O.: Primates. Comparative anatomy and taxonomy, vol. 2 (University Press, Edinburgh 1955).

6 HILL, W. C. O.: Primates. Comparative anatomy and taxonomy, vol. 3 (University Press, Edinburgh 1957).

7 ECKSTEIN, P. and ZUCKERMAN, S.: Morphology of the reproductive tract; in PARKES Marshall's physiology of reproduction, vol. 1 (Longmans, London 1960).

8 HENDRICKX, A. G. and KRAEMER, D. C.: Methods used in collecting and studying embryos; in HENDRICKX Embryology of the baboon (University of Chicago Press, Chicago 1971).

9 DEMPSEY, E. W.: The structure of the reproductive tract in the female gibbon. Amer. J. Anat. *67:* 229 (1940).

10 BUSS, D. H.: Mammary glands and lactation; in HAFEZ Comparative reproduction of nonhuman primates (Thomas, Springfield, in press).

11 SCHULTZ, A. H.: The number of young at a birth and the number of nipples in primates. Amer. J. phys. Anthropol. *6:* 1 (1948).

12 HILL, W. C. O.: Primates. Comparative anatomy and taxonomy, vol. 8 (Wiley-Interscience, New York 1970).

13 BUTLER, H.: Seasonal breeding of the Senegal Galago *(Galago senegalensis senegalensis)* in the Nuba Mountains, Republic of the Sudan. Folia primat. *5:* 165 (1967).

14 SPEERT, H.: The normal and experimental development of the mammary gland of the rhesus monkey, with some pathological correlations. Contrib. Embryol. Carneg. Inst. *32:* 9 (1948).

15 HILL, W. C. O.: The external genitalia of the female chimpanzee; with observations on the mammary apparatus. Proc. zool. Soc., Lond *121:* 133 (1951).

16 AHMED, M. M. U. and KANAGASUNTHERAM, R.: A note on the mammary glands in the lesser bush baby *(Galago senegalensis senegalensis)*. Acta anat. *60:* 253 (1965).

17 STEWART, T. D.: The skin and its appendages; in HARTMAN and STRAUS Anatomy of the rhesus monkey, pp. 34–35 (Williams & Wilkins, Baltimore 1933).

18 BUSS, D. H.: (Personal communication).

19 ZUCKERMAN, S.: The menstrual cycle of primates. IV. Observations on the lactation period. Proc. zool. Soc. Lond. : 593 (1931).

20 HAFEZ, E. S. E.: Reproductive cycles; in HAFEZ Comparative reproduction of nonhuman primates (Thomas, Springfield, in press).

21 HENDRICKX, A. G. and KRAEMER, D. C.: Primates; in HAFEZ Reproduction and breeding techniques for laboratory animals, pp. 316–335 (Léa & Febiger, Philadelphia 1970).

22 JOLLY, A.: Lemur behavior. A Madagascar field study (University of Chicago Press, Chicago 1966).

23 MANLEY, G. H.: Reproduction in lorisoid primates; in Comparative biology of reproduction in mammals. Symp. Zoological Society of London, No. 15 (Academic Press, London 1966).

24 NAPIER, J. R. and NAPIER, P. H.: A handbook of living primates (Academic Press, London 1968).

25 PETTER-ROUSSEAUX, A.: Reproductive physiology and behaviour of the Lemuroidea; in BUETTNER-JANUSCH Evolutionary and genetic biology of primates, vol. II, pp. 91–132 (Academic Press, New York 1964).

26 KOFORD, C. B.: Population dynamics of rhesus monkeys on Cayo Santiago; in DEVORE Primate behavior, pp. 160–174 (Holt, Rinehart & Winston, New York 1965).

27 LANCASTER, J. B. and LEE, R. B.: The annual reproductive cycle in monkeys and apes; in DEVORE Primate behavior, pp. 486–513 (Holt, Rinehart & Winston, New York 1965).

28 VANDENBERGH, J. G. and VESSEY, S.: Seasonal breeding of free-ranging rhesus monkeys and related ecological factors. J. Reprod. Fertil. *15:* 71 (1968).

29 GOODALL, J.: Chimpanzees in the Gombe Stream Reserve; in DEVORE Primate behavior, pp. 425–473 (Holt, Rinehart & Winston, New York 1965).

30 MACROBERTS, M. H. and MACROBERTS, B. R.: The annual reproductive cycle of the Barbary ape *(Macaca sylvana)* in Gibraltar. Amer. J. phys. Anthropol *25:* 299 (1966).

31 SOUTHWICK, C. H.; BEG, M. A., and SIDDIQI, H. R.: A population survey of rhesus monkeys in northern India. II. Transportation routes and forest areas. Ecology *42:* 669 (1961).

32 HENDRICKX, A. G. and KRIEWALDT, F. H.: Reproductive parameters of the baboon. Lab. Anim. Care *18:* 36 (1968).
33 DUMOND, F. V.: The squirrel monkey in a seminatural environment; in ROSENBLUM and COOPER The squirrel monkey, pp. 87–145 (Academic Press, New York 1968).
34 DEMPSEY, E. W.: The reproductive cycle of New World monkeys. Amer. J. Anat. *3:* 381 (1939).
35 MCARTHUR, J. W. and PERLEY, R.: Urinary gonadotropin excretion by infrahuman primates. Endocrinology *84:* 508 (1969).
36 KIRTON, K. T.; NISWENDER, G. G.; MIDGLEY, A. R., jr.; JAFFEE, R. B., and FORBES, A. D.: Serum luteinizing hormone and progesterone concentration during the menstrual cycle of the rhesus monkey. J. Clin. Endocrin. Metab. *30:* 105 (1970).
37 STEVENS, V. C.; SPARKS, S. J., and POWELL, J. E.: Levels of estrogens, progestogens and luteinizing hormone during the menstrual cycle of the baboon. Endocrinology *87:* 658 (1970).
38 KRAEMER, D. C.; PENA, A. DE LA, and VERA CRUZ, N. C.: Plasma levels of progesterone and progestin during the sexual cycle and pregnancy of the baboon *(Papio cynocephalus)* (in press).
39 JOHANSSON, E. D. B.; NEILL, J. D., and KNOBIL, E.: Periovulatory progesterone concentration in the peripheral plasma of the rhesus monkey with a methodologic note on the detection of ovulation. Endocrinology *82:* 143 (1968).
40 NEILL, J. D.; JOHANSSON, E. D. B., and KNOBIL, E.: Patterns of circulating progesterone concentrations during the fertile menstrual cycle and the remainder of gestation in the resus monkey. Endocrinology *84:* 45 (1969).
41 ZUCKERMAN, S. and PARKES, A. S.: The menstrual cycle of the primates. Proc. zool. Soc. Lond. *1:* 139 (1932).
42 KOERING, M. J.: Cyclic changes in ovarian morphology during the menstrual cycle in *Macaca mulatta*. Amer. J. Anat. *126:* 73 (1969).
43 CORNER, G. W.; BARTELMEZ, G. W., and HARTMAN, C. G.: On normal and aberrant corpora lutea of the rhesus monkey. Amer. J. Anat. *59:* 433 (1936).
44 CORNER, G. W.: Accessory corpora lutea in the ovary of monkey *Macaca rhesus*. Ann. Fac. Med. Montevideo *25:* 553 (1940).
45 BARTELMEZ, G. W.: (With the collaboration of CORNER, G. W. and HARTMAN, C. G.) Cyclic changes in the endometrium of the rhesus monkey *(Macaca mulatta)*. Contrib. Embryol. Carneg. Inst. *34:* 99 (1950).
46 BENSLEY, C. M.: Cyclic fluctuations in the rate of epithelial mitosis in the endometrium of the rhesus monkey. Contrib. Embryol. Carneg. Inst. *34:* 87 (1950).
47 DAVID, A. and MASTROIANNI, L., jr.: Cervical mucus arborization in the rhesus monkey. J. Reprod. Fertil. *17:* 495 (1968).
48 HENDRICKX, A. G.: The menstrual cycle of the baboon as determined by the vaginal smear, vaginal biopsy, and perineal swelling; in VAGTBORG The baboon in medical research, vol. II (University of Texas Press, Austin 1967).
49 CLEWE, T. H.: Observations on reproduction of squirrel monkeys in captivity. J. Reprod. Fertil. suppl. *6:* 151 (1969).
50 HENDRICKX, A. G. and KRAEMER, D. C.: Reproduction; in HENDRICKX Embryology of the baboon (University of Chicago Press, Chicago 1971).
51 HENDRICKX, A. G. and KRAEMER, D. C.: Observations on the menstrual cycle, opti-

mum mating time and pre-implantation embryo of the baboon, *Papio anubis* and *Papio cynocephalus*. J. Reprod. Fertil. suppl. *6:*, p. 119 (1969).

52 ARSLAN, M.; MEYER, R. K., and WOLF, R. C.: Chorionic gonadotrophin in the blood and urine of pregnant rhesus monkeys *(Macaca rhesus)*. Proc. Soc. exp. Biol. Med. *125:* 349 (1967).

53 TULLNER, W. W.: Urinary chorionic gonadotropin excretion in the monkey *(Macaca mulatta)* – early phase. Endocrinology *82:* 874 (1968).

54 TULLNER, W. W. and GRAY, C. W.: Chorionic gonadotropin excretion during pregnancy in a gorilla. Proc. Soc. exp. Biol. Med. *128:* 954 (1968).

55 MERKATZ, I. R. and BELING, C. G.: Urinary excretion of oestrogens and pregnanediol in the pregnant baboon. J. Reprod. Fertil. suppl. *6:* 129 (1969).

56 FISH, W. R.; DORFMAN, R. I., and YOUNG, W. C.: Metabolism of steroid hormones. III. The isolation of pregnanediol-3(*a*),20(*a*) from the urine of pregnant chimpanzees. J. biol. Chem. *143:* 715 (1942).

57 WESTPHAL, U. and BUXTON, L. L.: Failure to recover pregnanediol glucuronidate in monkeys, rabbits and cats. Proc. Soc. exp. Biol. Med. *42:* 749 (1939).

58 HOPPER, B. R. and TULLNER, W. W.: Urinary estrogen excretion patterns in pregnant rhesus monkeys. Steroids *9:* 517 (1967).

59 HOPPER, B. R.; TULLNER, W. W., and GRAY, C. W.: Urinary estrogen excretion during pregnancy in a gorilla *(Gorilla gorilla)*. Proc. Soc. exp. Biol. Med. *129:* 213 (1968).

60 ELMADJIAN, F. and FORCHIELLI, E.: Characterization of hormonal steroids of the chimpanzee: changes observed in adrenal cortical function during simulated and actual space flight. Proc. 1st Int. Congr. on Hormonal Steroids, vol. 2, p. 535 (Academic Press, New York 1965).

61 AGATE, F. J.: The growth and secretory activity of the mammary glands of the pregnant rhesus monkey *(Macaca mulatta)* following hypophysectomy. Amer. J. Anat. *90:* 257 (1952).

62 RUCH, T. C.: Diseases of laboratory primates (Saunders, Philadelphia 1959).

63 LAPIN, B. A. and YAKOVLEVA, L. A.: Comparative pathology in monkeys (Thomas, Springfield 1963).

64 HENDRICKX, A. G. and GILES-NELSON, V.: Reproductive failure; in HAFEZ Comparative reproduction of nonhuman primates (Thomas, Springfield, in press).

65 KRAEMER, D. C. and VERA CRUZ, N. C.: Infectious diseases influencing reproduction; in HAFEZ Comparative reproduction of nonhuman primates (Thomas, Springfield, in press).

66 HENDRICKX, A. G. and KRAEMER, D. C.: Preimplantation stages of baboon embryos *(Papio sp.)*. Anat. Rec. *160:* 181 (1968).

67 HENDRICKX, A. G.; HOUSTON, M. L.; KRAEMER, D. C.; GASSER, R. F., and BOLLERT, J. A.: Embryology of the baboon (University of Chicago Press, Chicago 1971).

68 HEUSER, C. H.: The chimpanzee ovum in the early stages of implantation. J. Morph. *66:* 155 (1940).

69 PADGET, D. H.: Spina bifida and embryonic neuroschisis. A causal relationship. Johns Hopk. Med. J. *123:* 233 (1968).

70 HENDRICKX, A. G. and GASSER, R. F.: A description of a diaphragmatic hernia in a sixteen week baboon fetus *(Papio sp.)* Folia primat. *7:* 66 (1967).

71 KOFORD, C. B.; FRABER, P. A., and WINDLE, W. F.: Twins and teratisms in rhesus monkeys. Folia primat. *4:* 221 (1966).

72 MARUFFO, C. A. and CRAMER, D. L.: Congenital renal malformations in monkeys. Folia primat. *5:* 305 (1967).

73 SCHULTZ, A. H.: The occurrence and frequency of pathological and teratological conditions and of twinning among nonhuman primates. Folia primat. *1:* 965 (1956).

74 SULLIVAN, D. J. and DROBECK, H. P.: True hermaphrodism in a rhesus monkey. Folia primat. *4:* 309 (1966).

75 WILSON, J. G. and GAVAN, J. A.: Congenital malformations in nonhuman primates. Spontaneous and experimentally induced. Anat. Rec. *158:* 99 (1967).

76 HENDRICKX, A. G.: Teratological findings in a baboon colony. Proc. Conf. on Nonhuman Primate Toxicology, pp. 120–123. Food and Drug Administration, Washington, D. C. (1966).

77 COURTNEY, K. D. and VALERIO, D. A.: Teratology in the *Macaca mulatta.* Teratology *1:* 163 (1968).

78 HOUSTON, M. L. and HENDRICKX, A. G.: Observations on the vasculature of the baboon placenta *(Papio sp.)* with special reference to the transverse communicating artery. Folia primat. *9:* 68 (1968).

79 VALERIO, D. A.; PALLOTTA, A. J., and COURTNEY, K. D.: Experiences in large scale breeding of simians for medical experimentation. Ann. N. Y. Acad. Sci. *162:* 282 (1969).

80 GUILLOUD, N. B.: The breeding of apes. Experience at the Yerkes Regional Primate Research Center and a brief review of the literature. Ann. N. Y. Acad. Sci. *162:* 297 (1969).

81 PICKERING, D. E.: The laboratory-confined Mulatta macaque monkey. Reproduction characteristics. Proc. Conf. on Nonhuman Primate Toxicology, pp.107–113. Food and Drug Administration, Washington, D.C. (1966).

82 GENGOZIAN, N.: Marmosets. Their potential in experimental medicine. Ann. N. Y. Acad. Sci. *162:* 336 (1969).

83 ROSENBLUM, L. A.: Some aspects of female reproductive physiology in the squirrel monkey; in ROSENBLUM and COOPER The squirrel monkey, pp. 147–170 (Academic Press, New York 1968).

84 MARTIN, C. B., jr.; MISENHIMER, H. R., and RAMSEY, E. M.: Ovarian tumors in rhesus monkeys *(Macaca mulatta).* Report of three cases. Lab. Anim. Care *20:* 686 (1970).

85 DIGIACOMO, R. F. and MCCANN, T. O.: Gynecologic pathology in the *Macaca mulatta*, pt 1. Amer. J. Obstet. Gynec. *108:* 538 (1970).

86 VAN WAGENEN, G.: Mating in relation to pregnancy in the monkey. Yale J. Biol. Med. *17:* 745 (1945).

87 CULINER, A.: Physiological activity of a single follicle occurring in an aberrant baboon ovary. Nature, Lond. *157:* 443 (1946).

88 KROHN, P. L.: Endometriosis and supernumerary ectopic ovarian tissue in a rhesus monkey. J. Obstet. Gynaec. brit. Cwlth. *58:* 430 (1951).

89 VADOVA, A. V. i GEL'SHTEIN, V. I.: Dlite'lnoe primenonie senestrola pri spontannem rake molochnoi lezy u obez'iany. (Prolonged use of synestrol in spontaneous mammary carcinoma of monkey.) Vop. Onkol. *2:* 391 (1956).

90 Flinn, R. M.: Serous cystadenoma of the ovary in a rhesus monkey. J. Path. Bact. *94:* 451 (1967).
91 Crews, L. M.; Kerber, W. T., and Feinman, H. T.: An ovarian tumor of dual nature in a rhesus monkey. Path. vet. *4:* 157 (1967).
92 Chiba, T.: On tumors of the wild animals. Gann *50:* suppl., p. 300 (1959).
93 Lombard, L. S. and Witte, E. J.: Frequency and types of tumors in mammals and birds of the Philadelphia Zoological Garden. Cancer Res. *19:* 127 (1959).
94 Fox, H.: Mortality and matters of pathological interest. Rep. Penrose Res. Lab., pp. 13–17 (1937).
95 Newberne, J. W. and Robinson, J. B.: Spontaneous tumors in primates. A report of two cases with notes on the apparent low incidence of neoplasms in subhuman primates. Amer. J. vet. Res. *21:* 150 (1960).
96 Rewell, R. E.: Report of the Society's pathologist for the year 1949. Proc. zool. Soc. Lond. *120:* 485 (1950).
97 Valerio, D. A.; Miller, R. L.; Innes, J. R. M.; Courtney, K. D.; Pallotta, A. J., and Guttmacher, R. M.: *Macaca mulatta.* Management of a breeding colony (Academic Press, New York 1969).
98 Fraser, A. D.: Ectopic endometrium in a *Macacus rhesus.* J. Obstet. Gynaec. brit. Cwlth. *36:* 590 (1929).
99 Joachimovits, R.: Spontane endometriode Drusenwurcherung im over bei Mensch und Affe. Abh. Gynäk. *54:* 1419 (1930).
100 Kluver, H. and Bartelmez, G. W.: Endometriosis in a rhesus monkey. Surg. Gynec. Obstet. *92:* 650 (1951).
101 McClure, H. M.; Graham, C. E., and Guilloud, N. B.: Widespread endometriosis in a rhesus monkey *(Macaca mulatta).* Proc. 2nd. Int. Congr. Primat. *3:* 155 (1969).
102 McCann, T. O. and Myers, R. E.: Endometriosis in rhesus monkeys. Amer. J. Obstet. Gynec. *106:* 516 (1970).
103 Chopra, H. C.: Electron microscopic detection of an oncogenic-type virus in a monkey breast tumor. (Abstr.) Proc. amer. Ass. Cancer Res. *11:* 16 (1970).
104 Mason, M. M.; Baker, J. R., and Ilievski, V. R.: Histopathology of a spontaneous tumor in a *Macaca mulatta* in which RNA virus particles were found. (Abstr.) Proc. amer. Ass. Cancer Res. *11:* 53 (1970).
105 Shimkin, M. B. and Shope, R. E.: Some observations on cancer research in the Soviet Union. Cancer Res. *16:* 915 (1956).
106 Vadova, A. V. and Gel'shtein, V. I.: Spontaneous tumors in Catarrhine monkeys according to the data obtained in the monkey colony of the Sukhumi medico-biological section; in Utkin Theoretical and practical questions of medicine and biology in experiments on monkeys (Pergamon Press, New York 1960).
107 Wadsworth, J. R.: Tumors in zoo animals. Animaland *20:* 2 (1953).
108 Broch, M.: Carcinoma solidum simplex mammae bei einem Orangutan *(Pongo pygmaeus).* Zbl. allg. Path. path. Anat. *109:* 474 (1966).
109 Hill, W. C. O.: Report on the Society's prosector for the year 1953. Proc. zool. Soc. Lond. *124:* 303 (1954).
110 Ratcliffe, H. L.: Tumors in captive primates. Report of two cases. Amer. J. Pat. *8:* 117 (1932).
111 Ratcliffe, H. L.: Cancer in animals. Trans. Coll. Physicians Philad. *54:* 152 (1932).

112 RATCLIFFE, H. L.: Incidence and nature of tumors in captive wild animals and birds. Amer. J. Cancer *17:* 116 (1933).
113 HAMERTON, A. E.: Report on the deaths occurring in the Society's gardens during 1932. Proc. zool. Soc. Lond. *1:* 451 (1933).
114 BLAND, S.: (1885) cited by RUCH in Diseases of laboratory primates (Saunders, Philadelphia 1959).
115 OSHIMA, F.: Tumors in wild animals. Gann *31:* 220 (1937).
116 FOX, H.: Diseases in captive wild mammals and birds. Incidence, description, comparison (Lippincott, Philadelphia 1923).
117 MORELAND, A. F. and WOODARD, J. C.: A spontaneous uterine tumor in a New World primate *Saguinus (Oedipomidas) oedipus*. Path. vet. *5:* 193 (1968).
118 GRIFFITH, P. and HULSE, F.: (Unpublished observations.) Cited by RUCH in Diseases of laboratory primates, p. 552 (Saunders, Philadelphia 1959).
119 HISAW, F. L. and HISAW, F. L., jr.: Spontaneous carcinoma of the cervix uteri in a monkey *(Macaca mulatta)*. Cancer *11:* 810 (1958).
120 STERNBERG, S. S.: Carcinoma *in situ* of the cervix in a monkey *(Macaca mulatta)*. Amer. J. Obstet. Gynec. *82:* 96 (1961).
121 FOX, H.: Mortality and matters of pathological interest, pp. 14–16. Rep. Penrose Res. Lab (1936).
122 SIGURDARDOTIR, B.; GIVAN, K. F.; ROZEE, K. R., and RHODES, A. J.: Association of virus with cases of Rubella studied in Toronto: Propagation of the agent and transmission to monkeys. Canad. med. Ass. J. *88:* 128 (1963).
123 DRAPER, C. C. and LAURENCE, G. D.: Susceptibility of *Erythrocebus patas* monkeys to Rubella virus. J. med. Microbiol. *2:* 249 (1969).
124 PARKMAN, P. D.; PHILLIPS, P. E., and MEYER, H. H., jr.: Experimental rubella virus infection in pregnant monkeys. Amer. J. Dis. Child. *110:* 390 (1965).
125 SEVER, J. L.; MEIER, G. W.; WINDLE, W. F.; SCHIFF, G. M.; MONIF, G. R., and FABIYI, A.: Experimental rubella in pregnant rhesus monkeys. J. infect. Dis. *116:* 21 (1966).
126 DELAHUNT, C. S.: Rubella-induced cataracts in monkeys. Lancet *7441:* 825 (1966).
127 COWDRY, E. V. and SCOTT, G. H.: Nuclear inclusions suggestive of virus action in the salivary glands of the monkey, *Cebus fatuellas*. Amer. J. Path. *11:* 647 (1935).
128 KALTER, S. S.; RATNER, J.; KALTER, G. V.; RODRIGUEZ, A. R., and KIM, C. S.: A survey of primate sera for antibodies to viruses of human and simian origin. J. Epidemiology *86:* 552 (1967).
129 KALTER, S. S.: Nonhuman primates in viral research. Ann. N. Y. Acad. Sci. *162:* 499 (1969).
130 MEDEARIS, D. N., jr.: Comparative aspects of reproductive failure induced in mammals by viruses; in BENIRSCHKE Comparative aspects of reproductive failure, pp. 333–349 (Springer, New York 1967).
131 BLACK, P. H.; HARTLEY, J. W., and ROWE, W. P.: Isolation of a cytomegalovirus from African green monkey. Proc. Soc. exp. Biol. Med. *112:* 601 (1963).
132 MEDEARIS, D. N.: Mouse cytomegalovirus. III. Attempts to produce intrauterine infections. Amer. J. Hyg. *80:* 113 (1964).
133 BERMAN, L. D.; COTES, P. M., and SIMONS, P. J.: Tumors and immunological response in rhesus monkeys born after *in utero* inoculation with Rous sarcoma virus. J. Nat. Cancer Inst. *39:* 119 (1967).

134 COTES, P. M.: Discussion of the role of viruses in the etiology of leukemia in man and other animals. Proc. Symp. Some Rec. Dev. Comp. Med., zool. Soc. Lond., 1965.

135 KALTER, S. S.; RATNER, I. A.; BRITTON, H. A.; VICE, T. E.; EUGSTER, A. K., and RODRIGUEZ, A. R.: Inoculations of baboons *(Papio)* with adenovirus type 12. The effect of thymectomy, splenectomy, immunopressive drug (Imuran), radiation and amnioinjection; in VAGTBORG The baboon in medical research, vol. 2, pp. 683–695 (University of Texas Press, Austin 1967).

136 ALEXANDER, E. R. and CHIANG, W. T.: Infection of pregnant monkeys and their offspring with TRIC agents. Amer. J. Ophthal. *63:* 1145 (1967).

137 GAJDUSEK, D. C.; ROGERS, N. G.; BASNIGHT, M.; GIBBS, C. J., jr., and ALPERS, M.: Transmission experiments with kuru in chimpanzees and the isolation of latent viruses from the explanted tissues of affected animals. Ann. N. Y. Acad. Sci. *162:* 529 (1969).

138 HUTCHISON, V. E.; PINKERTON, M. E., and KALTER, S. S.: Incidence of mycoplasma in nonhuman primates. Lab. Anim. Care *20:* 914 (1970).

139 KALTER, S. S. and GUILLOUD, N. B.: Chimpanzee viruses; in BOURNE The chimpanzee, vol. 2, pp. 361–389 (Karger, Basel 1970).

140 MADDEN, D. L.; HILDEBRANDT, R. J.; MONIF, G. R. G.; LONDON, W. T.; SEVER, J. L., and MCCULLOUGH, N. B.: The isolation and identification of mycoplasma from *Macaca mulatta*. Lab. Anim. Care *20:* 467 (1970).

141 MADDEN, D. L.; HILDEBRANDT, R. J.; MONIF, G. R. G.; LONDON, W. T.; MCCULLOUGH, N. B., and SEVER, J. L.: The isolation and identification of *Mycoplasma* from *Cercopithecus aethiops*. Lab. Anim. Care *20:* 471 (1970).

142 KALTER, S. S.; AL-DOORY, Y.; KUNTZ, R. E., and PINKERTON, M. E.: Infectious diseases associated with the use of primates; in VAGTBORG Use of nonhuman primates in drug evaluation. A symposium, pp. 505–538 (University of Texas Press, Austin 1968).

143 JACOTOT, H. et VALLEE, A.: Elements d'enquête sérologique sur l'infection par *Vibrio foetus* en France. Ann. Inst. Past. *98:* 601 (1960).

144 FEAR, F. A.; PINKERTON, M. E.; CLINE, J. A.; KRIEWALDT, F., and KALTER, S. S.: A leptospirosis outbreak in a baboon *(Papio sp.)* colony. Lab. Anim. Care *18:* 22 (1968).

145 BROOKS, H. and BLAIR, W. R.: Annual report of the medical department of the Zoological Park for 1903. Rep. N. Y. zool. Soc. *8:* 104 (1903).

146 TINKLEPAUGH, O. L.: Parturition and puerperal sepsis in a chimpanzee. Anat. Rec. *53:* 193 (1932).

147 BUTLER, H.: Septic abortion in the Senegal bush baby *(Galago senegalensis senegalensis)*. Folia primat. *13:* 207 (1970).

148 VICKERS, J. H.: Generalized staphylococcic infection in a monkey. J. amer. vet. med. Ass. *141:* 256 (1962).

149 LANG, C. M. and BENJAMIN, S. A.: Acute pyometra in a rhesus monkey *(Macaca mulatta)*. J. amer. vet. med. Ass. *155:* 1156 (1969).

150 ZABOLOTNYI, D. K.: On experimental syphilis in baboons. Arkh. biol. nauk *11:* 155 (1904).

151 MECHNIKOV, I. I. and RUCH, E.: Studies on experimental syphilis. Report 3. Ann. Inst. Past. *18:* 657 (1904).

152 ELSAS, F. J.; SMITH, J. L.; ISRAEL, C. W., and GAGER, W. E.: Late syphilis in the primate. Brit. J. vener. Dis. *44:* 267 (1968).

153 SMITH, J. L.: Correspondence. Syphilis in the owl monkey. Brit. J. vener. Dis. *45:* 170 (1969).

154 BROWN, W. J.; KUHN, U. S. G.; TOLLIVER, E. A., and NORINS, L. C.: Experimental syphilis in the chimpanzee. Immunoglobulin class of early antibodies reactive with *Treponema pallidum*. Brit. J. vener. Dis. *46:* 198 (1970).

155 KUHN, U. S. G.; BROWN, W. J., and FALCONE, V. H.: (1968) WHO/VDT/RES/68.137 cited by BROWN *et al.:* Experimental syphilis in the chimpanzee. Immunoglobulin class of early antibodies reactive with *Treponema pallidum*. Brit. J. vener. Dis. *46:* 198 (1970).

156 AL-DOORY, Y.; KALTER, S. S., and FREDERICKSON, M.: The mycoflora of the subhuman primate. II. The flora of the rectum and vagina of the baboon in captivity. Mycopath. Mycologia Appl. *31:* 332 (1967).

157 AL-DOORY, Y.: An immune factor in baboon anti-*Candida* serum. Sabouraudia (in press).

158 HILL, W. C. O.; PORTER, A., and SOUTHWICK, M. D.: The natural history, endoparasites and psuedoparasites of the tarsiers *(Tarsius carbonarius)* recently living in the Society's menagerie. Proc. zool. Soc. Lond. *122:* 79 (1952).

159 TRUSSELL, R. E.: *Trichomonas vaginalis* and trichomoniasis (Thomas, Springfield 1947).

160 TRUSSELL, R. E. and MCNUTT, S. H.: Animal inoculations with pure cultures of *Trichomonas vaginalis* and *Trichomonas foetus*. J. infect. Dis. *69:* 18 (1941).

161 VASQUEZ-COLET, A. and TUBANGUI, M.: The identity and incidence of a flagellated parasite found in human urine. J. Phillip. Islands Med. Ass. *16:* 231 (1936).

162 HEGNER, R.: Infections of the vagina of rhesus monkeys with *Trichomonas hominis* from man. J. Parasit. *20:* 247 (1934).

163 KESSEL, J. F. and GAFFORD, J. A.: Observations on the pathology of *Trichomonas vaginalis* and on vaginal implants with *Trichomonas vaginalis* and *Trichomonas intestinalis*. Amer. J. Obstet. Gynec. *39:* 1005 (1940).

164 ARCHER, J. F. and BEVERLEY, J. K. A.: Immunofluorescence in the diagnosis of ovine abortion due to *Toxoplasma gondii*. Vet. Rec. *78:* 369 (1966).

165 NELSON, G. S.: Schistosome infections as zoonoses in Africa. Trans. roy. Soc. trop. Med. Hyg. *54:* 301 (1960).

166 CAMERON, T. W. M.: A new definitive host for *Schistosoma mansoni*. J. Helminth. *6:* 219 (1928).

167 MCQUAY, R. M.: Susceptibility of a Louisiana species of *Tropicorbis* to infection with *Schistosoma mansoni*. Exp. Parasit. *1:* 184 (1952).

168 MILLER, J. M.: *Papio doguera* (dog face baboon), a primate reservoir host of *Schistosoma mansoni* in East Africa. Trans. roy. Soc. trop. Med. Hyg. *54:* 44 (1960).

169 STRONG, J. P.; MCGILL, H. C., and MILLER, J. M.: *Schistosoma mansoni* in the Kenya baboon. Amer. J. trop. Med. Hyg. *10:* 25 (1961).

170 COBBOLD, T. S.: On some new forms of entozoa. Trans. Linn. Soc. Lond., zool. *22:* 363 (1859).

171 LEIPER, R. T.: Report on the results of the Bilharzia mission in Egypt, 1915. J. roy. Army med. Corps *30:* 235 (1918).

172 SADUN, E. H.; LICHTENBERG, F. VON, and BRUCE, J. I.: Susceptibility and comparative pathology of ten species of primates exposed to infection with *Schistosoma mansoni*. J. trop. Med. Hyg. *15:* 705 (1966).
173 KUNTZ, R. and MYERS, B. J.: (Personal communications).
174 SCHULTZ, A. H.: Notes on diseases and healed fractures of wild apes, and their bearing on the antiquity of pathological conditions in man. Bull. Hist. Med. *7:* 571 (1939).
175 GOLDZIEHER, J. W. and AXELROD, L. R.: Urinary metabolites of 4-^{14}C-progesterone in the baboon *(Papio spp.)*. Gener. comp. Endocr. *13:* 201 (1969).
176 CHAMBERLAIN, J.; KNIGHTS, B. A., and THOMAS, G. H.: A system of analysis by gas chromatography of 17α- and 17β-pregnane-3,20-diols and their identification as metabolites of progesterone in man, the monkey, rabbit and guinea pig. J. Endocrinol. *28:* 235 (1964).
177 JEFFERY, J. D'A.: Metabolism of progesterone in the pig-tail monkey *(Macaca nemestrina)*. J. Endocrinol. *34:* 387 (1966).

Authors' address: Dr. D. C. KRAEMER and Dr. N. C. VERA CRUZ, Southwest Foundation for Research and Education, *San Antonio, Tex.* (USA)

Pathology of Simian Primates, Part I, pp. 878–888 (Karger, Basel 1972)

The Male Reproductive System

J. A. Roberts

Tulane University, Delta Regional Primate Research Center, Covington, La.

Contents

I. Anatomy and Physiology 878
1. A Male Sexual Cycle 878
2. The Testis 879
3. The Semen 880
4. The External Genitalia 880
5. The Accessory Glands 880
II. Pathology 882
1. Infertility 882
2. Cryptorchidism 884
3. Inflammatory Disease 884
4. Neoplasm 884
5. Miscellaneous 886
General References 887
References 887

I. Anatomy and Physiology

1. A Male Sexual Cycle

It has been thought that the existence of uninterrupted sexual activity in primates sets them apart from other mammals [1]. Recently, however, it has been shown that not only seasonal variation in sexual activity but also in the size of the testes [2] and in the spermatogenic cycle [3] occurs in free ranging *Macaca mulatta* in the colony on Cayo Santiago, Puerto Rico and *Cercopithecus aethiops* on St. Kitts, West Indies [4]. In addition, field studies of several macaque species, show that copulation ceases during the birth season of four to seven months [5]. More recent observations of reproduction in *Saimi-*

ri sciureus in a seminatural state and in the field show that the male undergoes an annual cycle [6]. During the spermatogenic phase a change in secondary sexual characteristics called the 'fatted' condition occurs in the squirrel monkey. This may be due to climatological changes as the 'fatted' condition was found at a different time of year in a Miami colony of monkeys as opposed to those in their natural South American environment. Thus good evidence for seasonality of reproduction in the male has been found for both old and new world monkeys. It has been suggested, however, that in animals raised in the laboratory, this seasonality is not marked [7]. These facts are mentioned to show that care must be taken in the interpretation of results of studies of spermatogenesis in these species, bearing in mind the seasonality of spermatogenesis.

2. The Testis

The testes after puberty are usually scrotal, but in some tree shrews *(Ptilocercus lowii)* and some lemurs *(Loris tardigradus, Perodicticus potto,* and *Daubentonia)* the testes descend only seasonally. The testes are usually retropenial, but are found to the side of the penis or anterior to it in *Perodicticus potto* and the gibbon *(Hylobates)*. The testis is suspended by a long mesorchium in most primate species. It is said that there is no definite dartos muscle but a strong cremasteric is present and most primates retain the ability to carry the testes in the inguinal position. The testes of prosimians, monkeys and the chimpanzee are larger than those of man on a weight basis. The processus vaginalis remains open in monkeys, but in all anthropoids excepting the gibbon *(Hylobates)* it is closed. Comparative studies of the embryology of human and monkey testes have been done by van Wagenen [8].

The process of spermatogenesis as shown by the PAS technique is quite similar in all mammalian species [9]. Electron microscopy reveals no condensation at the apex of the subacrosomal space in the mature spermatozoa of man as is seen in monkeys and other mammals [10]. The subacrosomal space is small and vacuoles within the substance of the nucleus are found in human and chimpanzee *(Pan troglodytes)* spermatozoa [11] (fig. 1). The highly heterogeneous population of spermatozoa normally found in studies of human ejaculates by MacLeod [12] are in sharp contrast to the uniform appearance of spermatozoa from other mammals including the nonhuman primate. This makes experimental evaluation of agents responsible for infertility difficult in man. The use of the nonhuman primate for such studies might lead to more reliable data especially if the chimpanzee were used.

3. The Semen

Studies of primate semen have been mainly descriptive, the techniques of electroejaculation for collection of such semen samples being described by MASTROIANNI [13], WEISBROTH [14], ROUSSEL [15], and VALERIO [16]. A technical problem in such studies is that of coagulation of the semen specimens. Liquefaction of monkey semen has been found to be incomplete after many hours of observation unlike that of human semen which normally liquefies completely within twenty minutes of ejaculation. VAN WAGENEN has shown that the coagulating gland of *M. mulatta*, the cranial lobe of the prostate, is responsible for this characteristic [17]. ROUSSEL states that treatment with trypsin solution will liquefy this coagulum and that the coagulum contains at least as high a concentration of spermatozoa as does the fluid portions of the ejaculate [18]. Surgical excision of the cranial lobe in *Macaca mulatta* is possible and is a practical method of obtaining liquid semen for studies [19].

4. The External Genitalia

In most mammals the penis is slung in a cutaneous sheath along the mid-ventral line of the abdomen. However, a pendulous scrotum and penis are characteristic of primates (but not confined to them, as shown by the bear). A sex skin is noted in most old world monkeys but not commonly in new world monkeys. An os penis (baculum) of varying size is present in most primates but is said to be absent in *Tupaiidae*, and in spider *(Ateles)* and woolly *(Lagothrix)* monkeys.

5. The Accessory Glands

Both intra-urethral and periurethral glands (Littre) are found in primates as are large bulbo-urethral glands [20]. The prostate gland is of interest in that it can be divided in most nonhuman primates into a cranial and caudal lobe, the cranial lobe being homologous to the middle lobe and posterior commissure and the caudal lobe homologous to the lateral lobes of man. The

---

Fig. 1. Spermatozoa of *Pan troglodytes*, mid sagital section with a small subacrosomal space and nuclear vacuoles. Courtesy of L. E. FRANKLIN. Fixed in glutaraldehyde followed by osmium tetroxide, stained with uranyl magnesium acetate and lead citrate. × 39,500.

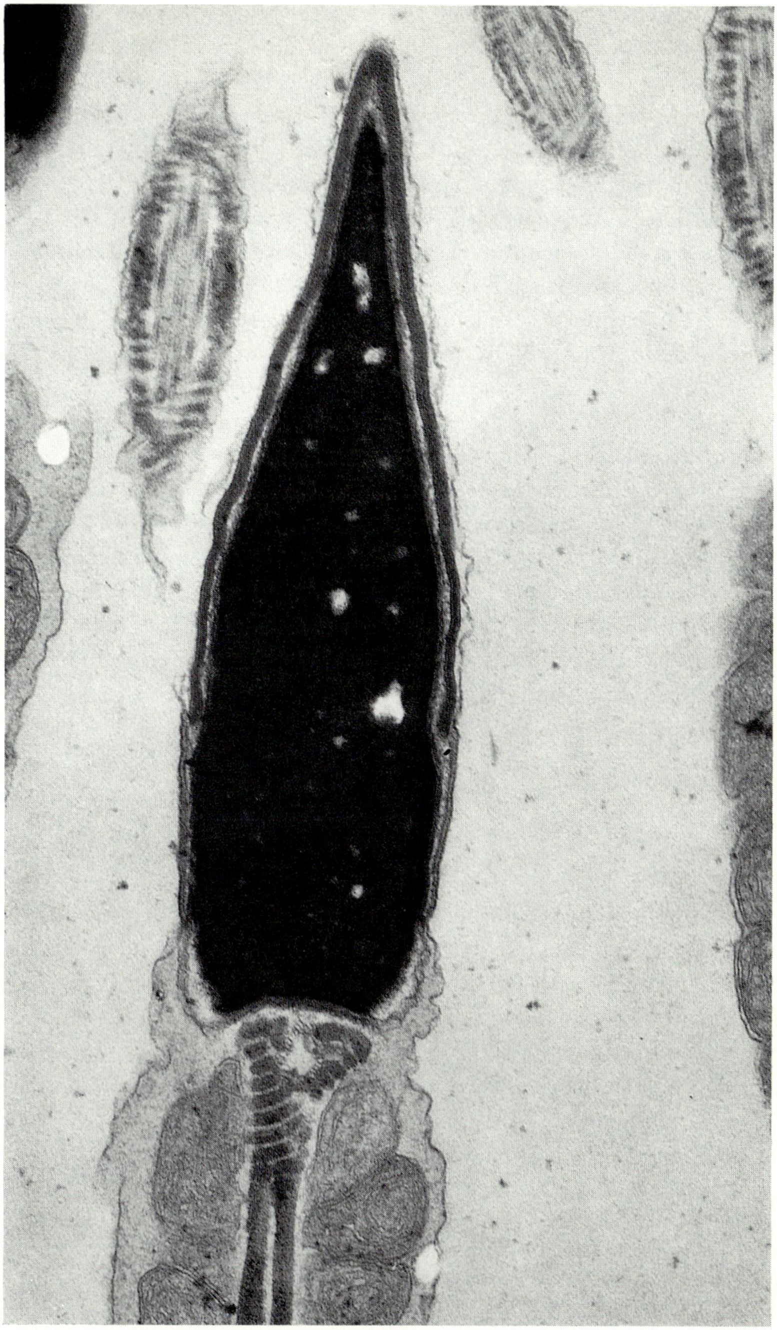

lobes differ histologically and also functionally, for VAN WAGENEN showed in *Macaca mulatta* that the cranial lobe of the prostate contributed that secretion necessary for coagulation of the semen [17]. Two species have been shown to have single lobed prostates, *Pongo pygmaeus* and a marmoset *(Hapale jacchus)*. The prostate of primates has a high acid phosphatase content like that of man, the amount of acid phosphatase being associated with functional activity of the gland, and thus with maturity (see MANN, p. 183). It is of interest that of the accessory reproductive glands the only gland found in all orders of mammals excluding monotremes is the prostate. A new approach to the study of prostatic function is reported in the baboon *(Papio ursinus)* utilizing radioisotopes of zinc since the concentration of zinc in the prostate is high [21].

The prostatic utricle is absent in *Lemuroidea* but is present in other primates. It contains no glands in macaques and *Papio* but in many species of *Tupaiidae* [22], in *Cercopithecus aethiops*, *Cercopithecus mona*, *Simias*, and *Pan troglodytes*, a definite uterus masculinus with glands is said to occur.

The seminal vesicle is absent in *Daubentonia*. An ampulla of the vas deferens is seen in *Tupaiidae*, the langur *(Simias)* and in the gibbon *(Hylobates)*. The vas deferens and ducts of the seminal vesicles join in a short common ejaculatory duct, however, in old world monkeys, and in *Tarsiidae*, *Galago*, *Propithecus* and marmosets [23].

II. Pathology

1. Infertility

A small temperature gradient from the abdomen to the scrotum has been noted in primates like that of dog and man. Thus, cryptorchidism might be expected to be associated with a decreased fertility rate. It is of interest that in lower species such as the tree shrew the testes are carried in the scrotum only during the breeding season. In addition, in most primates the testes are highly retractile and often are carried in the inguinal region. Thus the

---

Fig. 2. a Prostatitis in *Saimiri sciureus*. Note interstitial lymphocytic infiltration and polymorphonuclear cells within glandular lumen. H and E × 480. *b* Prostatitis in *Pan troglodytes*. Interstitial infiltrate is minimal. There is marked polymorphonuclear infiltration with tissue destruction of some glands, while adjacent glands are normal. H and E × 80.

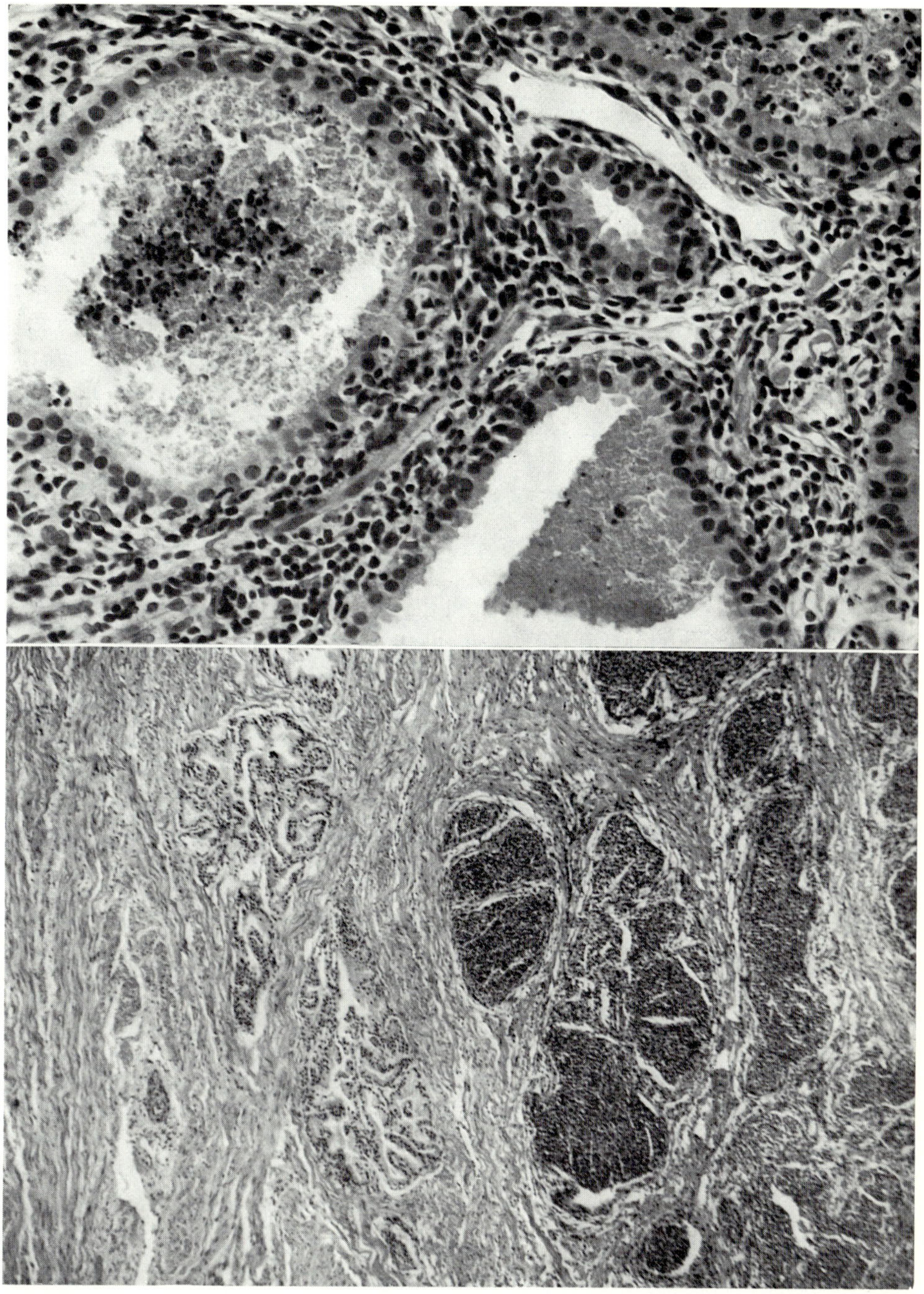

a

b

changes in spermatogenesis that have been noted in the seasonal breeding of macaques, *Saimiri sciureus* and *Cercopithecus aethiops* may be related to the position of the testes, although such correlations have not been made. One famous case of infertility is that of the 'Bushman' who died at a Chicago zoo. This gorilla's testes showed sclerosis and loss of the germinal epithelium. It was associated with cardiopathy, neuropathy and hemosiderosis and was probably due to a nutritional deficiency [24].

2. Cryptorchidism

Anomalous position of the testes is not widely reported, but in wild gibbons *(Hylobates)* the incidence of unilateral cryptorchidism was found to be 5% in adults [25]. RUCH reports unilateral cryptorchidism in seven anthropoids and one monkey (p. 460).

3. Inflammatory Disease

Although not naturally reported the *Treponema pallidum* has been used to infect the urethra of the owl monkey as an experimental model of the disease [26]. Prostatitis or urethritis is uncommon in primates but of many species of primates at this Center, prostatitis, as defined by inflammatory cells within the prostatic glands and interstitium occurred once in *Pan troglodytes* and *Saimiri sciureus* (fig. 2). The chimpanzee had previously been catheterized for experimental purposes, however.

FOX reports three cases of genital tuberculosis (p. 498), and one primary genital tuberculosis (p. 315), all in *Macaca fuscatus*.

4. Neoplasm

A benign prostatic adenoma has been reported in *Lemur catta* and glandular hyperplasia in *Macaca mulatta* both by FOX (p. 314). In addition, a single adenoma was noted in a *Callicebus moloch* at autopsy at this institution and is illustrated in figure 3. Carcinoma of the prostate is rare in mammals

---

Fig. 3. Caudal lobe of prostate, *Callicebus moloch. a* Normal glands, some distended with secretion. H and E × 80. *b* Adenoma with cystic hyperplasia and a small area of fibromuscular hyperplasia. H and E × 80.

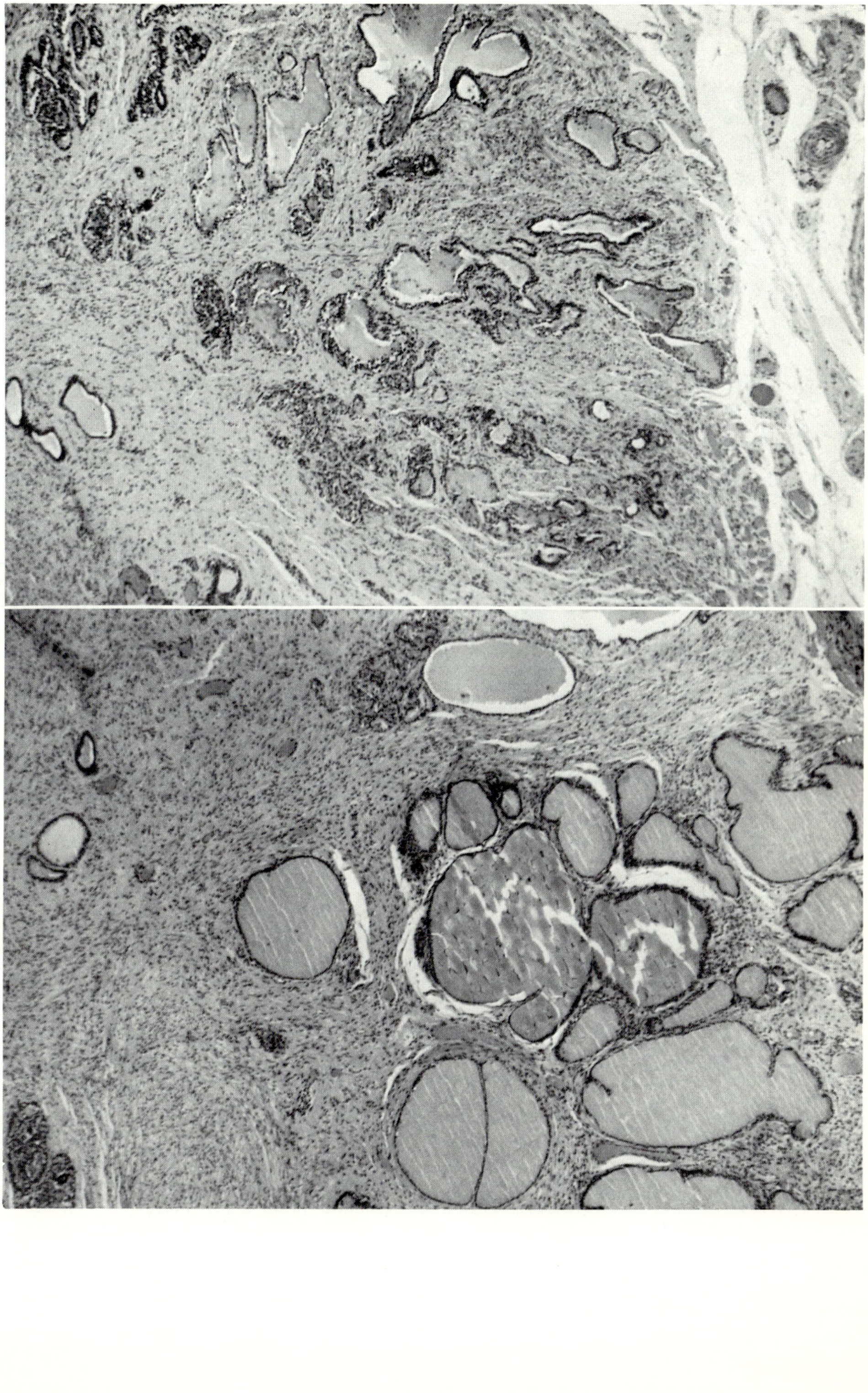

a

b

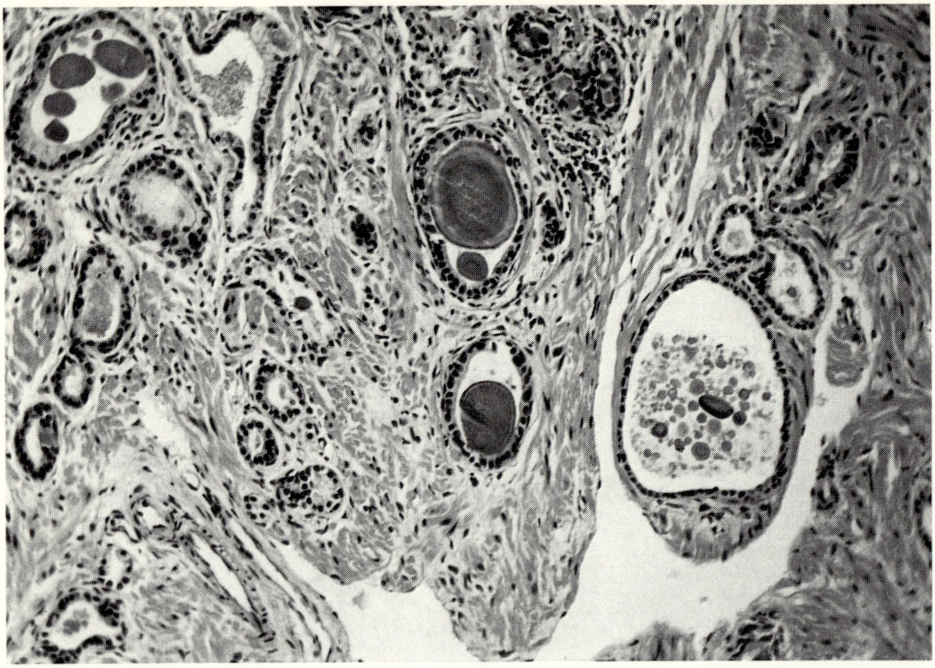

Fig. 4. Prostate of *Papio papio* containing corpora amylacia. They vary in size and show typical laminations. H and E ×240.

other than man. This may well be because carcinoma of the prostate, at least in man, is associated with increased age and most laboratory animals do not reach an extreme age. One carcinoma of the prostate was reported by ENGLE in an aged *Macaca mulatta* [27].

5. Miscellaneous

Calculi of the prostate have not been reported in the nonhuman primate but corpora amylacia are common in at least *Papio papio* and *Pan troglodytes* (fig. 4).

General References

FOX, H.: Diseases in captive wild mammals and birds (Lippincott, Philadelphia 1923).
RUCH, T. C.: Diseases of laboratory primates. (Saunders, London/Philadelphia 1967).

PARKES, A. S.: Marshall's physiology of reproduction. Vol. I, chapt. 2 (Little and Brown, Boston 1965).
HILL, W. C.: Primates. (University Press, Edinburgh 1953).
PRICE, D.: Comparative aspects of development and structure in the prostate. Nat. Cancer Inst. Monograph *12:* 1 (1963).
MANN, T.: The biochemistry of semen and of the male reproductive Tract (Wiley, New York 1964).

References

1 ZUCKERMAN, S.: The social life of monkeys and apes (Rontledge and Pail, London 1932).
2 SADE, D. S.: Seasonal cycle in size of testes of free-ranging *Macaca mulatta.* Folia primat. *2:* 171 (1964).
3 CONWAY, C. H. and SADE, D. S.: The seasonal spermatogenic cycle in free-ranging rhesus monkeys. Folia primat. *3:* 1 (1965).
4 CONWAY, C. H. and SADE, D. S.: Annual testis cycle of the green monkey *(Cercopithecus aethiops)* on St. Kitts, West Indies. J. Mammal. *50:* 833 (1969).
5 LANCASTER, S. B. and LEE, R. B.: The annual reproductive cycle in monkeys and apes. From 'Primate Behavior, Field Studies of Monkeys and Apes' (Holt, Rinehart and Winston, New York 1965).
6 MOND, F. V. DU and HUTCHINSON, T. C.: Squirrel monkey reproduction: the 'fatted' male phenomenon and seasonal spermatogenesis. Science *158:* 1067 (1967).
7 WAGENEN, G. VAN: Fertility of the colony born male macaque. Folia primat. *5:* 241 (1967).
8 WAGENEN, G. VAN and SIMPSON, M. E.: Embryology of the ovary and testis *Homo sapiens* and *Macaca mulatta* (Yale Univ. Press. New Haven 1965).
9 CLERMONT, Y. and LEBLOND, C. P.: Spermiogenesis of man, monkey, ram and other mammals as shown by the 'Periodic acid-Schiff' technique. Amer. J. Anat. *96:* 229 (1955).
10 BEDFORD, J. M.: Observations on the fine structure of spermatozoa of the Bush Baby *(Galago senegalensis)*, the african green monkey *(Cercopithecus aethiops)* and man. Amer. J. Anat. *121:* 443 (1967).
11 FRANKLIN, L. E.: Personal communication.
12 MACLEOD, J.: The male factor in fertility and infertility. Fertil. Steril. *1:* 347 (1950).
13 MASTROIANNI, L., jr. and MANSON, W. A., jr.: Collection of monkey semen by electroejaculation. Proc. Soc. exp. Biol., Med. *112:* 1025 (1963).
14 WEISBROTH, S. and YOUNG, F. A.: The collection of primate semen by electroejaculation. Fertil Steril. *16:* 229 (1965).
15 ROUSSEL, J. D. and AUSTIN, C. R.: Improved electro-ejaculation of primates. J. Inst. Animal Techn. *19:* 22 (1968).
16 VALERIO, D. A.; ELLIS, E. B.; CLARK, M. L., and THOMPSON, G. E.: Collection of semen from macaques by electroejaculation. Lab. Anim. Care *19:* 250 (1909).
17 WAGENEN, G. VAN: The Coagulating function of the cranial lobe of the prostate in the monkey. Anat. Rec. *66:* 411 (1936).

18 Roussel, J. D. and Austin, C. R.: Enzymic liquefaction of primate semen. Int. J. fertil. *12:* 288 (1967).

19 Greer, W. E.; Roussel, J. D., and Austin, C. R.: Prevention of coagulation in monkey semen by surgery. J. Reprod. Fertil. *15:* 153 (1968).

20 Roberts, J. A. and Seibold, H. R.: The histology of the primate urethra. Accepted for publication in Folia primat. (1970).

21 Schooness, R.; DeKlerk, J. N., and Murphy, G. P.: Anatomy, radioisotopic blood flow and glandular secretory activity of the baboon prostate. Baboons in organ transplantation: Supplement to the Sth afr. med. J. (1968).

22 Alcala, J. R. and Conway, C. H.: The gross and microscopic anatomy of the uterus masculinus of tree shrews. Folia primat. *9:* 216 (1968).

23 Beattie, J.: The anatomy of the common marmoset. Proc. zool. Soc., Lond. *1927:* 593 (1927).

24 Steiner, P. E.; Rasmussen, T. B., and Fisher, L. E.: Neuropathy, cardiopathy, hemosiderosis and testicular atrophy in *Gorilla gorilla*. Amer. med. Ass. Arch. Path. *59:* 5 (1955).

25 Schultz, A. H.: The relative weight of the testes in primates. Anat. Rec. *72:* 387 (1938).

26 Elsos, F. J.; Smith, J. L.; Israel, C. W., and Goger, W. E.: Syphilis in the owl monkey. Brit. J. vener. Dis. *44:* 267 (1968).

27 Engle, E. T. and Stout, A. P.: Spontaneous primary carcinoma of the prostate in a monkey *(Macaca mulatta)*. Amer. J. Cancer *39:* 334 (1940).

Author's address: Dr. James A. Roberts, Tulane University, Delta Regional Primate Research Center, *Covington, LA 70433* (USA)

Subject Index

Abortion and stillbirth 159, 160, 858
Acanthocephala 677
Acrocephaly 162
Actinomycosis 602, 865, 866
Activity rhythms in relation to feeding 71, 72
Adeno-virus 863
ADP role in platelet aggregation 318
Adrenal gland
  tumours of 214–216, 408
  structure and pathology 407–409
Aflatoxin 204, 225, 226, 739, 744, 745
Agenesis of genitalia 858
Albinism 184
Alimentary canal
  morphology in relation
    to feeding habits 72–75
Alimentary system 671–694
  tumours of 209–214
  gastric ulceration 334, 335
  caecum as site of immunocyte regulation 381
  auto-allergic phenomena 391, 392
  bacillary dysentery 681–685
  salmonellosis 685–688
  enterocolitis due to proteus 688, 689
  amoebiasis 689, 690
Allenopithecus 27
Allergy and anaphylaxis 366–371
  as cause of skin
    diseases 420
  effect on heart 766
Amaurosis acute 613, 614
Amelia, hemimelia, phocomelia, micromelia 172, 858
Amoebiasis 675, 689, 690, 740
Amyloidoses
  of liver 746, 747
  of heart 761, 762
Amyotrophic lateral sclerosis 4
Anaemia 817–819
  nutritional 241, 251, 253–255
Anencephaly 161, 858
Aneurysms 796–799
Angioma 858
Anophthalmia 858
Anorexia and failure to feed 673
Anotia (hypogenesis of ear auricles) 471
Anthrax 599, 737
Anubis baboon 31, 32
Aotus (night monkey, douroucouli) 17, 18, 48
Appendicitis 691, 692
Arnold-Chiari syndrome 470, 546
Arsenical poisoning 613
Ascaris
  allergic responses to 369–371
  role in auto-immune disease 385
Ascorbic acid 255, 257, 258
Asphyxia, neonatal 548, 549
Assam macaque *(M. assamensis)* 24
Ateles (spider monkeys) 18, 52, 53

Atherosclerosis and arteriosclerosis 5, 6, 767, 768, 779–792
association with hypothyroidism 407
of cerebral vessels 551, 552
medial calcification 795
of kidneys 829–830
in pyridoxine deficiency 252, 253
Athesmia 741, 742
Atlas, abnormalities of 167
Austrian tick-borne encephalitis 583, 584
Autoimmune diseases 380–398
etiology 381–385
sites of immunocyte regulation 381, 382
auto-antibodies 384
induced by microbial agents 385, 386
rheumatoid diseases 386–388
Sjögren's syndrome 388
lupus erythematosus 388, 389
scleroderma and related lesions 389, 390
thyroiditis 390
allergic gastritis and pernicious anaemia 391
auto-allergic phenomena of the gut 391, 392
diseases of the liver 392
renal diseases 393
heart 393
myasthenia gravis 393, 394
endocrine glands 394
ophthalmic conditions 394
central nervous system 395
Aye-aye (Daubentoniidae) 14

Baboons 18, 19, 31–33
Balantidium 677
Baldness, common 420
Barbary ape *(M. sylvanus)* 26
Barbiturate poisoning 554, 555
Bezoars 673, 690
Biotin 255, 256, 422
Black ape (Cynopithecus) 22, 23, 26
Blindness 428, 429
amaurotic, with and without epilepsy 614
congenital 858
Blood 80–93
volume 80
haematocrit 80
coagulation 80–83, 318–330
cells 83–88
sedimentation 87
serum proteins 89–97, 342–379
antigenic properties of 93–95, 342–379
physico-chemical and amino acid composition 96
serum proteins
genetic polymorphism 96, 97
non-protein components 97–104
oxygen dissociation curve 122, 123
fibrinolysis, platelets 82, 83, 318–330
Blood formation 809–820
disturbances of white cell production 809–814
disturbances of red cell production 817–819
production of erythropoietin by kidneys 825
Blood groups 270–317
cross-reaction of human and simian 271–273
human ABO system 273–275
ABO system of apes 275–278
ABO system of Old World monkeys 278, 279
ABO system of New World monkeys 279–281
Lewis types 282–284
M-N types 284–288
Rh system 288–292
the I-i factors 292, 293
Xga blood factor 293
simian-type blood groups 296–302
chimpanzees 296, 297
gibbons 297, 298
Celebes black apes 300
baboons 300, 301
rhesus 301, 302
Blood pressure 117–121
Bone marrow 88, 89
Bone structure and physiology 479–491
Bonnet macaque *(M. sinica)* 26
Bordetella
role in auto-immune diseases 385
in respiratory disease 675, 695

Borna disease 571
Brucellosis 864
Bursa of Fabricius of birds' function in immunogenesis 381
B. virus (herpesvirus simiae) infection 567–569, 733, 734

Cage paralysis 442, 443
Calcium, deficiency of 259, 260, 445 et seq.
Calcitonin 436, 438, 488–490
Callicebus (titi monkeys) 17, 50, 51
Callimico 43
Callitrichidae 17, 43–48
Callithrix (marmosets) 43–48
Candidiasis 602, 675, 690, 691, 866
Capillaria 742
Capuchins (Cebus) 17, 18, 49, 50
Carbon monoxide poisoning 555, 556
Carcinogens, chemical 216–230
Cardiovascular system 756–808
  endocardium and heart valves 758–761
  myocardium 761–777
  nutritional disease 762, 763
  myocarditis 764–767
  circulatory insufficiency 767–774
  cardiac hypertrophy 774, 775
  congenital heart disease 775–777
  neoplasms 777
  pericardium 777–779
  arteries 779–792
  hypertensive arteriopathy 792–795
  medial clacification 795
  necrotizing arteritis (periarteritis nodosa) 796
  aneurysms 796–799
  veins 799–801
Caries, dental 658–663
Cataract 429
  associated with diabetes mellitus 404
Catarrhini 18–20
Cebidae 17, 18
Cebuella 43
Cebus (capuchins) 17, 18, 49, 50
Central core disease 530
Cephalhaematoma 472
Cercocebus (mangabeys) 18, 19, 31–33
Cercopithecoidea 18
Cercopithecus 19, 20, 28–30
Cervical ribs 168, 169
Cheilitis 658
Cheirogaleidae (Lemuriformes) 14
Chimerism in marmosets 364
Chimpanzee 20, 39–42
Cholelithiasis (gallstones) 693, 694, 746
Choline deficiency 256, 257, 739
Chromosomal abnormalities 160
Circulation 104–121
  pulse rate 104–108
  electro-cardiogram 108–117
Cirrhosis of liver 738
Cleft foot 174
Cleft lip 161
Cleft palate 161, 470, 858
Clonorchis 741
Club foot 174
Coccidia 677
Coccidioidomycosis 602, 747
Coccygeal foveas 169
Coccygeal tubercles 169
Colobus 18, 38, 39
Congenital abnormalities 158–189, 470–472, 545–548, 858
  of kidneys 826
Conjunctivitis 428
Corneal ulceration 428, 429
Coronary insufficiency 5, 767–774
  association with hypothyroidism 407
Corynebacterium
  role in auto-immune disease 385
Coxsackie viruses
  nervous disease caused by 577–580
  association with disease ot the heart 761, 764
Crab-eating macaque *(M. fascicularis)* 25
Craniotabes 437, 438
Crohn's disease 392
Cryptococcus 602
Cryptorchidism 184, 884
Cyanide poisoning 552–554
Cycasin 227, 228
Cynomolgus (crab-eating macaque) 25
Cynopithecus 22, 23, 26
Cysticercus, nervous system 617

Cytomegalic inclusion disease (salivary gland inclusion virus)
heart 764
urinary tract 837
reproductive tract 863

Daubentoniidae (Lemuriformes) 14
Dengue 576
Dermatitis, nutritional (vide skin, diseases of) 250–253
Detached retina 429
Diabetes mellitus 403, 404
Diets, natural 58–77
Digestive system
normal values 123–133
saliva, composition and secretion 123–127
gastric juice, composition and secretion 127–132
gastric movements 132
small intestine, movements and secretion 132, 133
Diplococcus (Pneumococcus) pneumoniae 588–590, 675, 695
Disuse atrophy 472–474
Douroucouli *(Aotus)* 12, 18, 48
Drill (baboons) 18, 19, 31–33
Dwarf lemurs (Cheirogaleidae) 14
Dysentery
bacillary 681–685
encephalopathies resulting from 600–602
Dystokia 859

Echinocuccus cysts
role in auto-immune disease 385
of liver 742, 743
Echinococcus granulosus 742, 743
Echo viruses, nervous disease caused by 577–580
Eczema, infantile 422
Encephalocoele 548
Encephalomyelitis, allergic 395
Encephalo-myocarditis viruses 764, 779
Endocrine system 399–413
feedback mechanisms 399–400
classification 400–401
hypothalamo-hypophyseal axis 402, 403
pituitary tumours 403
pancreas 403, 404
pancreatic tumours 403
parathyroid 404, 405
thyroid 405–407
adrenal glands 407–409
sex hormones 850–855
Entamoeba histolytica 675, 689, 740, 741
Enterobacteriaceae in auto-immune diseases 385
Epididymitis 836
Epilepsy, acute amaurotic 613, 614, 618
Equine encephalitis 580–582
Erythrocebus (patas monkey) 20, 27, 28
Escherichia 675
role in auto-immune diseases 385
role in urinary tract diseases 831, 834
Essential fatty acids 243, 421, 422

Fasciola 741
Fats, nutritional requirements 243
Feeding rabits 60–66
Felty's syndrome 387
Filariasis 765, 868
Folic acid 255
Fontanellar bones 165
Foreign bodies in alimentary canal 673
Fractures 471, 472
Franklin's disease 344
Frostbite 421
Fungi, *vide also* Candida, Mucormycosis, Histoplasmosis
role in auto-immune disease 385

Gallstones (cholelithiasis) 693, 694, 746
Gammapathies, idiopathic, monoclonal 344
Gelada (Theropithecus) 33–35
Gestation 85, 855
Giardia 677
Gibbons 20
Gingival polyps 663
Gingivitis, bleeding of gums 648, 649, 656–658, 664
Glossitis 656
Gorilla 20
Goundou 437

Green monkey *(Cercopithecus)* 19, 20, 28–30
Grivet *(Cercopithecus)* 19, 20, 28–30
Growth hormone 402, 403
Guenons *(Cercopithecus)* 19, 20, 28–30
Guinea baboon 31
Guinea worm 425
Gynaecomastia (supernumerary nipples) 161, 184, 861, 862

Haemocytoblastosis 816
Haemophilia 328
Haemophilus 695
Haemosiderosis 747
Hamadryas baboon 18, 19, 31–33
Harelip 161
Hashimoto's disease (chronic thyroiditis) 390
Heat stroke (hyperthermia) 631
Hemiatrophy of face, palate, tongue with other deformities 161
Hemimelia 471
Hepatitis of infectious origin 732–743
Hepatitis, infectious (specific disease) 4, 675
Hepatitis, nutritional 242, 243
Hepatocystis 740
Hermaphroditism 184, 858
Hernia
  congenital 184
  diaphragmatic 672, 858
  strangulated inguinal 693
Herpes 424, 425, 668, 669, 733, 734
Herpes virus group 567–569, 733, 734
  H. simiae 863
Histoplasmosis 602
Hodgkins disease 206, 207, 230, 810
Hominoidea 20, 39–42
Hookworms 425
Hormones, *vide* Endocrine system
Hydatid disease 742, 743
Hydrocephalus 161, 471, 547, 858
Hydroma, subdural 858
Hydronephrosis 826, 827
Hylobates (gibbons) 20
Hypergammaglobulinaemia 382
Hyperhistidinaemia 739
Hypertension 4, 5
  relation to stress 334
  hypertensive arteriopathy 792–795
  secondary to renal-arterial lesions 830

Immunology (auto-immune diseases) 6, 7, 342–379, 381–398
  antibody production 352–359
  nutritional aspects of 241, 358
  immunoglobulins 343–352
  transplacental transfer of 361
  transfer in lactation 361
  foetal production of 361–363
  in newborn 363, 364
  hypersensitivity, *vide* Allergy and Anaphylaxis
Impetigo 424
Implantation, faulty 160
Inca bone formation 163
Inclusion body encephalitis 564, 565
Inclusion body hepatitis 734, 743
Infertility 882, 883

Japanese B encephalitis 560–563
Japanese monkey *(M. fuscata)* 26
Jaundice 748

Klebsiella 675, 695
Kuru 4, 386, 609, 610
Kwashiorkor, *vide* Protein deficiency 241, 242, 739
Kyasanur forest disease 736

Lagothrix (woolly monkey) 18, 51, 52
Langur *(Presbytis)* 18, 19, 39
Larva migrans (visceral) 743
Lead poisoning 558
Lectins 274
Leishmaniasis 425
Lemuriformes 12–15
Leontiasis ossea 437, 438
Leontopithecus *(Leontideus)* 47–48
Leprosy, role of leprosy bacilli in autoimmune disease 385
Leptospirosis 736, 737, 836, 865
Leuco-encephalo-myelitis (van Bogaert's disease) 544, 558, 559

Leukaemia 204–209, 230, 810–814
Lion-tail macaque *(M. silenus)* 26
Listeria, role in auto-immune disease 385
Liver and biliary system 711–755
  tumours of 201–204
  disease and the clotting system 328
  auto-immune disease of 387, 392
  liver function tests 712–732
  infectious diseases 732–738
  nutritional diseases 738, 739
  parasitic diseases 740–743
Lobster claw 174
Lorisiformes 13–15
Lung mites 701–704
Lupus erythematosus 383, 387, 389
Lymphocythemia 809, 810
Lymphocytic choriomeningitis 577
Lymphogranulomatosis 810

Macaques 19, 20–27
Macroglobulinaemia, Waldenstròm's 344
Magnesium deficiency 260
Malaria
  effect on clotting systems 328
  latent, activation by stress 335
  as auto-immune disease 386
  liver 740
Malbrouk *(Cercopithecus)* 19, 20, 28–30
Mammary glands 861, 862
Mandrill 18, 19, 31–33
Mangabeys 18, 19, 31–33
Manganese, lesions of nervous system due to 602–604
Mange 423
Marmosets 17, 31–48
Measles 424, 425, 675
Meningitis 424, 428, 587–590
Meningococci 675
Methyl alcohol poisoning 744
Metopism 163
Micrococci 695
Micromelia 858
Middle-ear disease 424, 428
Miopithecus (talapoin) 35–38
Molluscum contagiosum 424
Mongolism 160
Monsters 161, 162, 546–548
Moor monkey *(Cynopithecus)* 22, 23, 26
Mouse lemurs (Cheirogaleidae) 14
Mucormycosis 675
Muscular dystrophy 248, 249, 520
Muscular system 520–541
Myasthenia gravis 393, 394
Mycobacterium, role in auto-immune disease 385
Mycoplasma 864
  role in auto-immune disease 385
Myelocoele 471
Myeloma, multiple 344
Myocardial Infarction 337, 758–762, 764–767
Myocarditis 767–775
  liver changes in 748
  'chronic interstitial' 767
  'chronic focal' 767
Myxovirus 677, 831

Nasalis (proboscis monkey) 39
Neoplasms (all organs, see also below) 190–238
  as auto-immune response 382
  pituitary tumours 403
  of the pancreas 403
  of the adrenal 407, 408, 828
  of the nervous system 619
  of the mouth 669, 670
  of the liver 745
  of the heart 777
  of the pericardium 779
  leiomyoma of uterus causing obstruction of the ureters 827
  of the urinary system 828
  of the genital system 862, 884, 885
  of the parathyroids 438
  of the skeleton 474, 475
Nephritis 831–838
Nephrocalcinosis 828, 829
Nervous system 542–638
  electro-encephalography 138–146
  hypoxia 548–551, 555
  atherosclerosis and arteriosclerosis 551, 552
infarction 552
  encephalomyelitis (allergic) 556–558

confluent leukoencephalosis and perivascular myelosis 558, 559
isolated papillomacular degeneration 559, 560
viral encephalitides 560–587
diffuse non-suppurative meningitides and meningoencephalitides 575, 576
encephalitis affecting basal ganglia 582, 583
non-suppurative meningitis 584–587
suppurative meningitis 587–590
encephalopathies related to dysenteries 600–602
fungal infections 602
degenerative diseases 602–609
cerebellum, diseases of 602–604
basal ganglia, diseases of, widespread spontaneous degeneration 607–609
pigmentation and siderosis 610–612
nutritional deficiencies 614–616
parasitic infections 617, 618
epilepsies 618
tumours 619
ionizing radiation 620–631
effects of hypertension 793
Neuroses 4, 5
Nevus 858
Newcastle disease 576
Niacin (nicotinic acid) 254
Night monkey *(Aotus)* 17, 18, 48
Nocardiosis 704, 705
Nutrition 239–269
skin diseases 421
nervous system 614–616
teeth 652–658
liver 738, 739
heart 762, 763
arteries 788–792

Oesophagostomum 677
Oesophagus, varices 672
oesophagitis 689, 690
Olfactory and tactile senses 429
Oligodactyly 174
Olive baboon 31, 32
Oncology 7–9, 190–238
Orang utan 20
Osteitis deformans (Paget's disease) 437, 438
Osteitis fibrosa cystica (Von Recklinghausen's disease) 438
Osteoarthritis 476–479
Osteodystrophia fibrosa 438, 439
Osteodystrophy, renal 439
Osteogenesis imperfecta 441, 442
Osteomalacia 244–248, 437, 456–462
Osteomyelitis 475, 476, 663
Osteoporosis 462–464
Ovaries, pathology 859, 860

Paget's disease (osteitis deformans) 437, 438
Pancreas, tumours of 204
Pantothenic acid 256, 422
Papio (baboons) 18, 19, 31–33
Paracolon bacilli 675
Paraplegia (hemiplegia etc.) 544, 609, 614
nutritional 253, 254
Parathyroid glands 404, 405, 444, 467–469, 490
Parathyroid tumours 438
Parturition 856
Pasteurella
in auto-immune disease 385
in respiratory disease 695
Patas *(Erythrocebus)* 20, 27, 28
Pellagra 422
Pentastomida 743
Periarteritis nodosa (necrotizing arteritis) 796
Pericarditis 758, 777–779
Periodontal disease (pyorrhoea) 664–669
Peripheral neuropathy 617
Pernicious anaemia 391
Peromelia 172, 470
Pharmacology 6, 414, 415
Phimosis 827
Phocomelia 471
Phylogeny and classification 11–57
tree shrews 11, 12
Prosimii 12–16
Tarsiiformes 16
Anthropoidea 16–42
Platyrrhini 17–18

taxonomy of macaques 20–27
taxonomy of patas 27, 28
taxonomy of vervets 28–33
taxonomy of baboons 31–33
taxonomy of geladas 33–35
taxonomy of talapoins 35–38
taxonomy of Colobinae 38–39
taxonomy of chimpanzees 39–42
taxonomy of marmosets 42–48
taxonomy of other South American monkeys 48–53
classification table 54,55
Phytic acid, role in skeletal disease 449
Pigmentary disease
liver 746, 747
heart 761, 762
Pigtail macaque *(M. nemestrina)* 23
Plagiocephaly 164, 165
Plasmodium
role in auto-immune disease 385
experimental infection in liver 740
Pneumococcus (Diplococcus) 588–590, 675, 695
Pneumonia
inhalation 673
infectious 675–677, 694–701
Pongidae 10
Poisons and toxins 414, 415
Polioencephalitis 569–575
Poliomyelitis 569–575
Polydactylism 161, 172–174
Polyoma, as auto-immune disease 386
Porocephalus 617
Pox 424, 675
Pregnancy toxaemia 827, 828
Prenatal and neonatal mortality 857–859
Presbytis (langurs) 18, 19, 39
Proboscis monkey *(Nasalis)* 39
Prostatis 836, 885, 886
Protein deficiency, *vide also* Kwashiorkor 241–243
liver 739
heart 762, 763
skeleton 464–466
Proteus 688, 689
role in urinary tract diseases 831, 834
Pseudomonas 588
role in auto-immune disease 385
as cause of middle-ear disease 424, 428
in meningitis 588
in urinary tract diseases 835
Pseudomyxovirus 677
Pseudotuberculosis 675
Psoriasis 420
Puberty 846–850
Puerperal sepsis 865
Pulse
rate 104–108
wave, velocity of 120, 121
Pyelonephritis 834, 835
Pygathrix (snub nose monkey) 39
Pyorrhoea, *vide* gingivitis (see also Part II for association with *Entamoeba gingivalis)* 648, 649, 656–658, 664
Pyridoxine (vitamin B 6) 252, 253, 422, 615
and liver 738, 739

Rabies 571
Radiation
tumours due to 193, 220, 221
effects on nervous system 620–631
effects on liver 745, 746
effects on blood-forming organs 815, 816
von Recklinghausen's disease 438
Reproductive system, tumours of 139–198
Reproductive system, female 841–877
anatomy 842–845
physiology 846–857
pathology 857–867
Reproductive system, male 878–888
anatomy and physiology 878–880
pathology 882–886
Reproductive cycles (a) male (b) female 850–854, 878, 879
Respiratory system
normal values 121–123
gaseous exchange and respiratory quotient 133, 134
tumours of 204
chronic bronchitis (auto-immune) 387
diseases of 694–704
Reticulosis 812, 813
Retinopathy – hypertensive 794
Rhesus *(M. mulatta)* 24, 25

Rheumatoid diseases 386–388
Rhinopithecus, *vide* snub nose monkey 39
Rhinovirus 677
Riboflavin 250–252
  and liver 738, 739
Rickets 244–248, 437, 447–456
  teeth 653–656
  congenital 449, 450
  vitamin D resistant 447, 439, 440
Rickettsia 764
Ringworm 419, 423, 424
Rocky Mountain spotted fever 764
Rous sarcoma virus 863
Rubella 471, 862, 863
Russell's viper venom 323

Sacred baboon 31
Saguinus 44–47
Saimiri (squirrel monkey) 18, 49
Salivary gland inclusion virus (cytomegalic inclusion disease) 764, 837, 863
Salmonella 765, 677, 685–688
  skin lesions 424
  liver lesions 737
Sarcoptes 423
Sarcosporidia 524–526, 765
Scabies 420
Scaphocephaly 165
Schilder's disease 555
Schistosoma
  allergic responses to 368
  hepatic infections 741, 800
  urinary tract 837
  reproduction system 867, 868
Schizophrenia 395
Scleroderma 389, 390
Scoliosis, lumbar, congenital 169
Scurvy 257, 258, 466
Sero-primatology 302–311
  chimpanzees 305–310
  gibbons 310, 311
Sex linkage, Xga blood factor 293
Sexual swellings 19, 25, 842–854
Shigella 675, 677, 681–685
  liver lesions 737
Siamang (gibbons) 20
Sjögren's syndrome 388
Skeletal system 433–519
  tumours of 191–193, 474, 475
  endocrine diseases 405, 438, 439, 467–470
  protein deficiency, effect on 242, 464–466
  vitamin C 251, 258, 466
  vitamin D 439, 440, 445–467
  infectious conditions 475, 476
  diseases of joints 476–479
Skin and appendages and subcutaneous tissues 198–201, 418–426
  nutritional diseases 250 et seq.
Snub nose monkey *(Pygathrix* or *Rhinopithecus)* 39
Special senses 428, 429
Spider monkeys (Ateles) 18, 52, 53
Spina bifida 470, 471, 545, 546
Spleen, accessory 184
Sprue 391
Squirrel monkey (Saimiri) 18, 49
Staphylococcus
  in respiratory diseases 675, 695
  in meningitis 590
  in urinary tract disease 834
  in reproductive system 865
St. Louis encephalitis 563, 564
Streptococci
  role in auto-immune disease 385
  role in skin disease 424
  role in dental caries 662
  role in respiratory diseases 675, 695
  role in rheumatoid cardiac disease 761
  role in nephritis 831
Stress 331–341
  effect on bone marrow values 89
  effect on blood sugar and cholesterol 99, 336
  effect on gastric juice secretion 131, 336
  causes of 331, 332
  metabolic parameters 332, 333, 336–340
  clinical signs 334, 335
  gastric ulceration caused by 334, 335, 409
  effect on immunological responses 335, 336
  atherosclerosis associated with 337

cortico-steroid secretion in 336
endocrine factors 338, 339
effects on growth hormone levels 402
effects on adrenal and corticosteroid output 408, 409
effects on alimentary and respiratory tracts 677–679
effects on liver 747
effects on arteries 791
Strongyloides 423, 677
Stumptail macaque *(M. arctoides)* 21, 22
Swamp monkey *(Allenopithecus)* 27
Symphalangus (Siamang, gibbons) 20
Syndactyly 172
Syphilis 866

Talapoin *(Miopithecus)* 35–38
Tamarins 44–47
Tantalus monkey 28, 29
Teeth and gums 639, 670
supernumary 160, 161
abnormal development 177–184, 641–649
congenital absence of 177–184
displacement of 177–184
diseases of 639–670
gingivitis 648, 649, 664–668
injuries 649–652
malnutrition, effects of 257, 258, 652–658
infectious diseases 658, 659
tumours 669, 670
Temperature, body 134–137
Teratology 158–189
Thalidomide 160, 471, 548
Theropithecus (Gelada) 18, 19, 31–33
Thiamine (vitamin B 1) 249, 250
effect on heart 763
Thrombocytopenia 328
Thymus function in immunogenesis 381
Thyroxine
skeletal disease 405–407, 469, 470
foetal requirements and production 406
influence on cholesterol and atherosclerosis 791, 792
Tibetan macaque *(M. thibetana)* 21, 26
Titi *(Callicebus)* 17, 50, 51
Tolerogenesis 26, 364–366
Toque macaque 21, 26
Torulosis *(Cryptococcus)* 602
Toxocara
allergic responses to 369
larva migrans 743
Toxoplasmosis 765, 867
Trachoma, virus 863
Transferrins 97
Trauma 470–472, 544, 672
Tree shrews (Tupaioidea) 11, 12
Treponematosis 437, 837, 866
Trichinosis 540
Trichomonas 675, 867
Trichophyton 423
Trichosoma 425
Trypanosoma cruzi 765
Tuberculosis 677
role of tubercle bacilli in auto-immune disease 385
of nervous system 591–599
of liver 737
of heart 764
of the skeleton 475, 827
of urinary tract 836, 837
of genital system 884
Tumours, *vide also* Neoplasms 190–238
Twinning 185
in marmosets 364, 365
Typhus 764
Tyzzer's disease 737

Ulcerative colitis 392
Urethritis 884
Urinary system 821–840
renal function, normal 137, 138
cystic kidneys- congenital 184
renal agenesia 184
tumours of 192, 213–216
auto-immune diseases 393
hypertension 793–795
anatomy 821–823
physiology 823–826
urinary output 823, 824
natural pathology 826–831
obstructive disease 827, 828
vascular disease 829, 830
chronic renal disease 830, 831

urinary tract infections 831–838
role of kidneys in skeletal disease 439
Urolithiasis 828, 829
Uterus and fallopian tubes 860, 861

Valvulitis
mitral 758, 759
rheumatic 761
van Bogaert's disease, *vide* Leuco-encephalo-myelitis
Venezuelan equine encephalitis 581, 582
Veno-occlusive disease 744
Vertebrae, malformations of 167–171
Vervets *(Cercopithecus)* 19, 20, 28–30
Vibriosis 864, 865
Viruses, oncogenic 7–9
Vitamin A 243, 244
deficiency in pregnancy 160
diseases of skin 421, 422
and monsters 548
and nervous system 614, 615
and alimentary system 674
eye lesions 243, 244
Vitamin B group 249–256
skin diseases 421, 422
nervous diseases 615, 616
alimentary system 674
heart 763
Vitamin B 12 253, 254
role in pernicious anaemia 391
Vitamin C 257, 258
Vitamin D 244–248, 447–456
effect on nervous system 616
alimentary system 674
hypervitaminosis D 246–248, 466, 467, 795
Vitamin E 248, 249
and muscle pathology 540
and nervous system 615
Vitamin K, deficiency of 249, 328
and liver 739

Warts 424
Weil's disease, *vide* Leptospirosis
Wernicke's disease 250
West Nile encephalitis 560
Woolly monkeys (Lagothrix) 18, 51, 52
Wounds, *vide* Trauma

Yaba and Yaba-like disease 424
Yaws, *vide* Treponematosis
Yellow fever 732, 733

Zinc, deficiency of 259
Zygodactyly (webbing) 177

Author Index

Aaberg, T. M. 430
Abei, T. 749
Abramova, E. V. 704
Abrams, R. L. 402, 409
Abruzzo, L. J. L. 396
Adachi, K. 198, 223, 231, 426, 428
Adams, C. K. 430
Adams, L. 516, 751
Adams, R. D. 269, 615, 637
Adamson, R. H. 200, 223, 230, 231, 233
Adamsons, K. 635
Adelsberger, L. 314
Ades, H. W. 429
Adey, W. R. 154
Adler, S. 609, 631
Adlersberg, D. 148
Agate, F. J. 871
Agieva, A. K. 147, 751
Ahmed, M. M. U. 869
Ainis, H. 849
Akbroit, E. Y. 704, 707, 819
Aksenova, A. S. 704, 707, 708, 709
Alagille, D. 329
Alarie, Y. 147
Alcala, J. R. 888
Al-Doory, Y. 426, 875
Alekseeva, 152, 708
Alepa, F. P. 146, 311, 374, 387, 395
Alexander, E. R. 875
Alexander, M. 266
Allegra, F. 428
Allen, E. 193, 198, 222, 235, 236
Allen, J. C. 373
Allen, J. R. 511, 749, 750, 751, 754, 801, 839
Allison, A. C. 374
Allmond, B. W. 514
Alpers, M. 396, 632, 633, 875
Altman, R. L. 138, 146
Amador, E. 726
Amkraut, A. A. 379
Anagnoston-Kakaras, E. M. 516
Anast, C. 510
Anderson, C. R. 300, 301, 314, 752
Anderson, D. R. 750
Anderson, E. T. 146
Anderson, H. H. 705
Anderson, J. M. 377
Anderson, J. R. 395
Anderson, S. 58, 76
Ando, K. 708, 753
Andrada, E. C. 391, 395, 413
Andrada, F. P. 390, 391, 395
Andrada, J. A. 413
André, A. 276, 277, 311
Andreeva, A. V. 89, 152
Andrewes, C. H. 235
Andrus, S. B. 9, 223, 231, 261, 265, 266, 268, 407, 409, 411, 784, 785, 789, 791, 801, 805, 806, 839
Annenkov, G. A. 92, 94, 97, 146, 155

Ansevin, A. 839
Anthony, H. D. 753
Antonov, A. M. 215, 231
Appleby, E. C. 801
Appley, M. H. 332, 340
Arao, T. 426, 427
Arbeshan, C. E. 367, 374, 378, 379, 428
Arbor, A. 751
Argur, H. 396, 411
Arhangelskaja, N. A. 146
Archer, J. F. 876
Arita, L. 426
Armstrong, A. R. 508, 515
Armstrong, C. 635
Armstrong, J. A. 235
Armstrong, M. L. 261
Arnaud, C. 489–490, 510, 517
Arndt, H. J. 405, 409
Arnold, A. 631, 633
Arnstein, P. 432
Aronson, S. B. 397
Arslan, M. 871
Arterberry, J. D. 388, 396
Asakura, S. 233
Asatiani, V. S. 146, 751
Asher, D. M. 633
Asherson, G. L. 384, 385, 388, 395
Ashley Montagu, M. F. 459, 510
Ashmore, J. 339, 341
Ashton N. 794, 801
Atta, A. G. 105, 109, 113, 114, 147
Attinger, F. 128, 129, 130, 148
Aurbach, G. D. 511
Auskaps, A. M. 670
Austen, K. F. 378
Austin, C. R. 887, 888
Avioli, L. V. 440, 507, 510
Axelrod, L. R. 514, 749, 877

Bach, F. W. 705
Bagg, H. J. 192, 231
Baidakova, Z. L. 10
Bailey, O. T. 635
Bailey, P. 632
Baird, C. A. 510
Baireuther, R. 119, 149
Baker, J. R. 873
Baldwin, M. 315
Balis, J. U. 801
Baillieux, R. E. 375
Balner, H. 302, 312, 315
Bangham, D. R. 360, 361, 377
Banker, D. D. 567, 635
Banks, K. L. 754
Bannerjee, C. M. 123, 147
Barabadze, E. M. 9, 10, 191, 208, 220, 221, 234–235, 236
Barker, L. F. 801
Barney, G. A. 261, 265
Barnett, E. V. 397
Barnicot, N. 147
Barnstein, N. J. **153**
Barrett, R. 147
Barrett, R. B. 265, 405, 411, 437, 438, 515, 719, 749
Barrow, E. M. 329
Barrow, M. V. 160, 186
Bartelmez, G. W. 870, 873
Bartlakowski, J. 690, 705
Barton, M. N. 266
Baruch, E. 752
Basinger, G. M. 9, 806
Basnight, M. 875
Bassi, R. 510
Bateson, W. 172, 177, 181, 187
Batson, J. S. 377
Bauchop, T. 74, 76
Baud, C. A. 510
Bauer, J. A. 411
Baum, J. 379
Baumann, F. B. 705
Beall, D. 515
Beall, F. A. 754, 807
Bearcroft, W. G. C. 5, 9, 198, 231, 752, 755
Beard, M. E. J. 88, 151, 264
Beard, R. 635
Bearn, A. 153
Beasley, J. W. 510
Beattie, J. 888
Beaver, P. C. 754
Beck, E. 609, 632
Beck, J. S. 390, 395
Beck, P. 404, 409

Becker, R. O. 511
Bedford, J. M. 887
Beer, B. 412
Beg, M. A. 869
Behrman, K. E. 123, 147
Beischer, D. E. 105, 147
Belanger, L. F. 511
Beling, C. G. 871
Belkania, 116, 117
Bell, M. 426
Bell, N. 147
Benacerraf, B. 375
Benda, R. 426
Bender, M. A. 146
Benenson, A. S. 752
Benfield, V. 30
Benirschke, K. 364, 377, 801
Benke, M. 707
Benjamin, S. A. 875
Bennejeant, C. 177, 187
Bennett, C. M. 822, 838
Bennett, D. G. 431
Bensley, C. M. 870
Benson, R. E. 264, 511, 515, 638, 753, 802, 804
Berdjis, C. C. 252, 261, 753
Bereznay, Y. 311
Berg, E. W. 511
Berghe, L. Van den 84, 85, 86, 88, 148
Berliner, R. W. 838
Berman, L. D. 365, 377, 874
Bernard, J. 222, 231, 314
Bernard, V. 804
Bernhard, G. 705
Bernick, S. 512
Bernier, G. M. 395
Bernstein, F. 274, 311
Bernstein, L. H. 139, 140, 153, 838
Berson, S. A. 511
Bert, J. 148, 432
Bertram, U. 388, 395
Best, C. H. 269
Beutner, E. H. 395
Beverley, J. K. A. 876
Bienenstock, J. 372
Biggs, R. 330
Bilimoria, H. S. 269
Biltz, R. M. 511
Bischop, D. De 632
Bishop, J. 749
Bishop, R. W. 752
Black, P. A. 874
Blackwell, L. H. 86, 153
Blair, W. K. 442, 443, 511, 875
Blakemore, C. 430
Blanco, S. 409
Bland Sutton, J. 195, 197, 199, 237, 403, 413, 435, 438, 449, 450, 511
Blinzinger, J. S. 632
Blitstein, J. 84, 86, 148
Blix, S. 330
Blokhin, N. M. 199, 231
Blough, D. S. 432
Blundell, G. P. 754
Bochrova, D. N. 709
Bodansky, A. 508, 511
Bodechtel, G. 565, 632
Bodian, D. 571, 572, 632
Bogaert, L. van, 544, 558, 559, 560, 567, 569, 571, 572, 576, 584, 587, 591, 600, 602, 613, 618, 637
Bois, P. 807
Boitnott, J. D. 267
Boitrub, B. A. 152
Bokkenheuser, V. 705
Bolano, C. R. 428
Bolau, H. 172, 187
Boley, L. R. 751
Bollert, J. A. 871
Bollman, Y. 156
Bock, L. 161, 186
Bond, V. P. 705
Bongiovanni, A. 412
Bonne, C. 198, 211, 213, 222, 231
Boolukos, P. J. 635
Booth, A. H. 62, 64, 76
Boothroyd, B. 514
Bordier, P. 440, 511, 513, 514
Boren, J. J. 341
Born, E. 705
Bornstein, M. B. 633
Borutau, H. 105, 148
Boshaar, E. 314
Bosmann, H. 511

Bossak, E. 148
Bostick, W. 705
Bostrom, R. E. 753
Botkin, C. C. 233
Bouffard, B. 405, 412
Boulay, G. H. Du 261, 329, 445, 449, 513
Boulger, L. 632
Bourlière 76
Bourne, G. H. 189, 540, 541
Bowen, J. A. 261, 839
Bowen, N. H. 125, 148, 640, 662, 670
Bowers, G. M., jr. 726
Bowman, F. B. 705
Bowman, R. W. 237, 754
Bowyer, D. E. 9, 264, 803, 804
Boyarsky, S. 838, 839
Boycott, B. B. 431
Boyer, S. H. 311
Brack, M. 197, 231
Bradley, S. E. 749
Brady, J. V. 335, 336, 340, 411, 412
Bragg, W. C. 429
Brain, P. 315
Brambell, F. W. R. 360, 376
Bramblett, C. A. 56, 511
Brancato, G. J. 315
Brandom, W. F. 241, 261, 374, 381, 395
Brannon, R. B. 269
Bras, G. 801
Bray, R. S. 753
Brecher, G. 330
Bree, M. M. 839
Bree, P. J. H. van 33, 56
Breedis, C. 234
Brenner, B. M. 838
Brewer, M. 754
Bridges, R. E. 709
Bristow, J. D. 115, 148
Britton, H. A. 156, 875
Brito, E. 840
Brizee, K. R. 619, 632
Brobeck, J. R. 403–409, 410
Broch, M. 873
Brodie, D. A. 128, 129, 131, 148, 749
Brody, S. 822, 838
Brom, A. J. 430
Bronner, F. 511
Brooks, F. P. 128, 129, 130, 131, 148, 155
Brooks, H. 442, 443, 511
Brown, G. M. 402, 403, 409
Brown, K. N. 374
Brown, M. H. 377
Brown, P. C. 394, 395
Brown, W. J. 876
Brownhill, L. E. 377
Bruce, J. I. 378, 750, 753, 877
Bruchea, J. M. 430
Bruhn, J. M. 134, 148
Bruno, D. 9
Brunschwig, A. 209, 212, 234, 237
Bryan, A. H. 430
Brygoo, E. R. 202, 231, 636
Buescher, E. L. 339, 340
Buettner-Janusch, J. 33, 56, 311
Bullock, B. C. 258, 261, 265, 515, 552, 632, 754, 789, 801, 802, 806, 839
Bunina, T. L. 10
Bunting, C. H. 206, 229, 231
Burch, G. E. 807
Burch, P. R. J. 380, 383, 396
Burkholder, P. M. 397
Burnet, F. M. 380, 382, 383, 384, 385, 392, 396, 397
Burnett, C. H. 519
Burnham, L. J. 312
Burns, K. F. 148, 751
Burova, L. F. 705
Burrows, W. 375
Burtin, B. 397
Burwell, R. G. 380, 381, 383, 396
Buss, D. H. 868, 869
Bütler, C. 387, 396
Butler, H. 869, 875
Butnev, Y. P. 148
Butts, W. H. 269, 413, 808
Buxton, L. L. 871
Byers, P. 498, 511

Cabrera, A. 49, 56
Cadigan, F. O., jr. 315
Cahan, A. 312

Call, M. 633
Camb, H. 838
Cameron, A. H. 511, 546, 632
Cameron, H. C. 455, 512
Cameron, J. L. 749
Cameron, T. W. M. 876
Cammock, E. E. 412
Campbell, A. W. 511
Campbell, C. E. M. 431
Campbell, D. H. 379
Campbell, D. M. 442, 516
Campbell, F. P. 431
Camus, J. 402, 409
Candela, P. B. 276, 277, 311, 315
Cannon, M. D. 153
Capps, R. B. 9, 151, 750
Caraway, B. L. 751
Carbonara, A. O. 374
Carbone, P. P. 233
Carbonell, L. 752
Care, A. D. 516
Carlisle, H. N. 808
Carlson, A. J. 409
Carpenter, K. P. 705
Carpenter, M. L. 431
Carr, C. D. 429
Carsten, A. 632
Carstens, L. A. 748, 751, 754, 801
Casey, H. W. 237, 754
Cass, I. S. 705
Castleberry, M. W. 754
Catanei, A. 602, 632
Catchpole, H. R. 503, 519
Catsulis, C. 803
Caulfield, A. H. W. 366, 377
Cave, A. J. E. 429
Caveness, W. F. 141, 148, 623, 625, 632
Celano, M. J. 312
Ceppellini, R. 311
Chachibaya, I. A. 237
Chaffee, R. R. J. 754
Chaikoff, I. L. 552, 635, 785, 804
Chakravarti, R. N. 132, 154, 802, 840
Chalmers, J. N. M. 264
Chamberlain, J. 877
Champlain, G. 804
Chang, C. 839
Chang, S. 341
Chapman, W. L. 197, 199, 231, 511, 839
Charipper, H. A. 153
Charles-Dominique, P. 61, 64, 65, 76
Chase, R. E. 188, 801
Chatterjea, J. B. 754
Chauhan, S. 762, 801
Chaulin-Serviniere, J. 265, 616, 634
Chazov, E. I. 9
Cheever, F. S. 635
Chen, P. S. 511
Chernigovskaya 82
Cherkovich, G. M. (vide also Tscherkowitsch) 105, 106, 108, 121, 135, 136, 141, 148, 149, 152, 155, 341, 708
Chez, R. A. 149, 412
Chhutani, P. N. 154, 802
Chiang, W. T. 875
Chiba, T. 873
Chick, H. 261
Chien Shu 150
Cho, C. T. 428
Chopra, H. C. 873
Chorazyma, H. 429
Chrisp, C. E. 758, 802
Christeller, E. 405, 409
Christian, C. L. 389, 396
Christie, R. J. 161, 186
Christofinis, G. J. 634, 752
Christoph, H. J. 705
Chu, K. C. 807
Chutna, J. 397
Citters, R. L. van 80, 104, 106, 109, 115, 119, 120, 156
Clark, E. J. 609, 631
Clark, G. 706
Clark, M. L. 887
Clark, P. F. 269
Clarke, R. W. 155
Clarkson, T. B. 258, 261, 265, 266, 413 515, 632, 801, 807, 839
Clasen, R. 839
Clayborn, L. D. 514
Cleland, J. B. 442, 511
Clermont, Y. 887

Clewe, T. H. 402, 410, 870
Cline, J. A. 875
Cobbold, T. S. 876
Cockburn, T. A. 705
Cockett, A. T. K. 840
Cohen, B. S. 88, 149, 495, 511, 662, 670
Cohen, D. W. 192, 210, 212, 231
Cohen, L. 315
Cohen, S. 346, 373, 376
Colcolough, H. L. 807
Cole, C. R. 808
Cole, W. C. 516, 635
Collomb, H. 148
Colyer, F. 177, 178, 180, 181, 182, 183, 187, 660, 661, 662, 670
Condie, R. M. 380, 396
Coniglio, J. G. 754
Connor, J. D. 752
Connor, W. E. 261
Conrad, D. G. 340
Conrad, M. 330
Converse, J. L. 754
Conway, C. H. 887, 888
Conzelman, G. M., jr. 223, 231
Coogan, P. S. 431
Cook, R. 705
Cooke, E. R. W. 705
Coombs, A. M. 384, 396
Coombs, R. R. 396
Cooney, D. 751
Cooper, R. W. 200, 223, 230, 231
Cooper, T. 156
Cooperman, J. M. 251, 261
Cope, F. W. 751
Copp, D. H. 490, 512
Cordy, D. R. 405, 410
Cornelius, C. E. 749
Corner, G. W. 870
Correa, P. 269
Corson-White, E. P. 210, 231, 405, 410, 437, 458, 512, 819
Cosgrove, G. E. 753
Cotes, P. M. 361, 363, 365, 377, 874, 875
Coulston, F. 9, 268, 709, 806, 807
Counts, M. A. 262
Courtney, K. D. 186, 548, 632, 872, 873
Courtois, G. 311, 748
Cowdry, E. V. 246, 262, 403, 410, 466, 503, 512
Cowen, D. 802
Cowey, A. 430, 431
Cowles, C. A. 374
Cox, G. C. 262
Cox, G. E. 9, 268, 413, 807
Cox, L. G. 262
Craigie, J. 571, 634
Cramer, D. L. 839, 872
Cran, J. A. 212, 231, 426
Crane, J. T. 512
Crawford, E. J. 751
Crawford, M. A. 261, 329, 445, 449, 513
Crews, L. M. 197, 232, 873
Cronkite, E. P. 330
Crook, J. H. 34, 56
Crosby, W. M. 754
Crout, D. W. 105, 155, 231
Cruickshank, J. C. 705
Cueto, J. 262
Cugell, D. W. 122, 123, 153
Culiner, D. 872
Cunningham, W. F. 208, 232
Currie, R. A. 262
Cuthbertson, W. F. J. 204, 226, 232

Dack, G. M. 264
Dahl, D. S. 541
Dallemagne, M. J. 637
Dalrymple, G. V. 750
Damian, R. T. 347, 373, 374
Dammrich, K. 464, 512
Dance, N. 147
Dandelot, P. 28, 56
Danes, L. 426
Daniel, M. D. 752
Daniel, P. M. 632
Daniels, J. B. 635
Danko, L. V. 704, 707, 819
Daoud, A. S. 267, 807
Darby, W. J. 261, 262, 265, 754
Dart, R. 76

Datta, J. K. 410, 840
Daughaday, W. H. 409
Daumy, O. M. 316
David, A. 870
David, G. F. X. 839
David, H. 705
Davis, D. 76
Davis, J. H. 313, 316
Davis, L. J. 150
Davis, M. 635
Davis, R. 515
Davis, R. A. 128, 129, 130, 131, 155
Davis, R. D. 751
Davis, R. T. 432
Davison, C. 632
Davison, V. 752
Dawe, C. J. 203, 232, 234
Dawson, A. M. 495, 519
Dawson, C. A. 432
Day, P. L. 149, 240, 262, 446, 512, 709
Deauville, G. A. 752
Dede, J. A. 236
Deeds, F. 377
Deftos, L. 512
Degaris, C. F. 188, 801, 802
De Hart, R. L. 750
Deibel, R. 709
Deicher, N. 314
Deimed, B. 708
Deinhardt, F. 4, 9, 407, 410, 713, 748, 749
Deinhardt, J. B. 410, 749
Deklerk, J. N. 888
Delahunt, C. S. 512, 548, 632, 874
Delcourt, R. 808
De La Peña 503, 512
De Leon, W. 208, 232
Delory, E. G. 515
Delsal, J. L. 506, 512
De Luca, H. F. 490, 512
Demis, D. J. 426
Dempsey, E. W. 868, 80
Deneau, G. A. 749
Dennis, J. 375
Dennis, L. H. 330
Denny-Brown, D. 794, 806
Denson, K. W. 330
Dent, C. E. 439, 512
Denys, H. 514
Deo, M. G. 262, 266, 268, 515
Depaoli, A. 210, 232, 754, 807
Derwelis, S. K. 426
De Simone, A. R. 396
Desmedt, J. E. 429
Destombes, P. 231
Devine, J. 410
Devore, I. 66, 76
De Vries, E. J. 509, 512
Dewson, J. H. 429
Dewulf, A. 637
Dhariwal, A. 412
Dherte, P. 748
Diakonova, S. M. 144, 156
Diaz-Rivera, R. S. 88, 155
Dick, G. W. A. 802
Diemel, B. 405, 411
Digiacomo, R. F. 210, 232, 872
Dikovenko, E. A. 152
Dillingham, L. A. 189, 237
Dinning, J. S. 248, 262, 266,541, 802
Dische, F. 330
Dittmer, D. 138, 146
Dixon, D. F., jr. 751
Doan, C. A. 839
Dobberstein, Y. 705
Dobelle, W. H. 431
Dobson, J. F. 237
Dobson, R. L. 232
Dobson, R. M. 426
Dodds, G. S. 455, 512
Doebbler, T. K. 390, 396, 397
Doenhoff, A. von 330
Dolling, J. E. 431
Dolnick, C. H. 427
Domaniewska-Soblzak, K. 313
Doniach, D. 392, 396
Donohoo, J. S. 267
Dopter, C. 709
Dorfman, R. I. 871
Dougherty, S. F. 377
Douglas, J. D. 426
Dowben, R. M. 264, 512, 515, 804
Downs, W. G. 752

Doyle, F. H. 512, 513
Drach, G. W. 839
Drachman, D. A. 541
Draper, C. C. 874
Dray, S. 152
Dreizen, S. 125, 149, 262, 426, 462, 512
Drewer, D. E. 429
Dreyer, W. J. 373
Dreyfus, P. M. 262
Drobeck, H. P. 188, 872
Drill, V. A. 755
Drury, R. A. B. 513
Du Bois, E. L. 388, 396
Dubowitz, U. 541
Duckworth, W. L. H. 513
Dufrene, J. K. 517
Duke, B. O. 426
Dumond, F. V. 513, 870, 887
Dumonde, D. C. 380, 384, 385, 390, 393, 396
Dupuy, M. E. 312
Dvizhkov, P. P. 232
Dyer, H. M. 228, 232
Dyke, D. van 625, 637
Dzhikidze, E. K. 152, 221, 232, 705, 708, 709, 819

Eastoc, J. E. 125, 148
Eberlein, W. 412
Eberl-Rothe, G. 149
Eckstein, D. 868
Edelman, G. M. 347
Edsall, G. 752
Edwards, J. B. 206
Edwards, J. B. 206
Efner, J. A. 754
Eggen, D. A. 807
Egozcue, J. 160, 186
Eichelberger, J. W. 330
Eide, P. 377
Eidelman, S. 391, 396
Elliot, M. 312
Elliot, M. W. 197, 232
Elliot, O. S. 232
Ellis, C. M. 431
Ellis, E. B. 887
Elmadjian, F. 871
Elsas, F. J. 876, 888
Eltringham, J. R. 750
Elvehjem, C. A. 255, 261, 265, 268, 269, 514
Emerson, G. A. 789, 806
Emmons, R. W. 752
Enber, F. F. 636
Engel, G. L. 392, 396
Engel, W. K. 541
Engle, E. T. 193, 197, 232, 886, 888
Engstrom, H. 429
Enzinger, F. 541
Epstein, A. 373
Epstein, H. 339
Erikson, L. B. 149
Espana, C. 752
Eugster, A. K. 707, 875
Evans, A. S. 750
Evans, B. K. 750
Evans, C. A. 233
Evans, L. B. 755
Ewing, S. A. 753
Eyquem, A. 391, 397

Fabiyi, A. 874
Fahey, J. L. 372, 374
Fairbrother, R. W. 513, 598, 612, 633, 706, 802
Falcone, V. H. 876
Fanconi, V. G. 439, 513
Farber, P. A. 186
Faro, M. D. 559, 633
Farr, R. S. 355, 356, 362, 375, 378
Farrer, D. N. 431
Fazal, H. 267
Fear, F. A. 875
Federlin, K. 394, 396
Fegley, H. C. 706
Feinberg, A. R. 378, 426
Feinberg, S. M. 367, 378, 426
Feinman, H. T. 232, 873
Feldman, A. Z. 137, 149
Feldman, D. B. 839
Feldman, H. A. 137, 153
Feldman, J. 315
Feldman, L. A. 633
Feldner, M. A. 621, 839

Felsenfeld, A. D. 374, 376
Felsenfeld, O. 261, 269, 336, 340, 358, 374, 375, 376, 383, 392, 395, 753
Felts, P. E. 750
Ferguson, F. G. 148, 751
Field, E. J. 541
Fiennes, R. N. T-W- 262, 332, 334, 341, 414, 513, 587, 602, 633, 675, 706, 802, 835, 837, 839
Filer, L. J. 249, 263
Fillios, L. C. 231, 409
Findley, J. 76
Fineg, J. 313
Fink, J. N. 379, 806
Fink, M. A. 379
Finkeldey, W. 213, 232
Finkelstein, M. S. 375
Finlayson, R. 757, 787, 802
Finstad, J. 373
Fischer, W. 301, 311
Fischer-Williams, M. 144, 149
Fish, W. R. 871
Fisher, C. W. R. 402, 410, 411
Fisher, D. A. 9, 516, 866
Fisher, L. E. 558, 633, 636, 754, 807, 888
Fisher, N. 312
Fisk, R. T. 311
Fitch, C. D. 266, 802
Fitzgerald, F. L. 514
Fitzgerald, G. 261
Flanders, L. E. III. 231
Flatt, R. E. 426
Fleisch, H. 513
Fleischhacker, H. 189
Fleishman, R. W. 149
Fleming, J. 754
Flinn, R. M. 197, 232, 263, 873
Flower, W. H. 470, 513
Follis, R. H. 240, 245, 263, 753, 802
Fooden, J. 21, 22, 56, 62, 65, 71, 73, 76, 149
Foord, R. G. 31
Forbes, A. D. 870
Forchielli, E. 871
Ford, D. M. 186, 427, 428
Formal, S. B. 707
Forssman, J. 272, 311
Forsyth, R. P. 116, 118, 119, 149
Foster, G. V. 513
Fourman, P. 513, 516
Fox, H. 32, 33, 195, 196, 197, 199, 202, 212, 213, 215, 232, 233, 403, 407, 410, 476, 477, 513, 547, 594, 619, 633, 777, 802, 873, 874, 886
Foy, H. 251, 263, 750
Fraber, P. A. 872
Fradkin, R. 160, 172, 186
Francis, J. 753
Frank, M. H. 805
Franklin, L. E. 887
Franzl, R. E. 336, 341
Fraser, A. D. 195, 233, 873
Fraser, D. R. 440, 491, 513
Fraser, G. 801
Fraser, H. F. 263, 268
Fratta, I. 706
Frazier, L. E. 10, 269
Frederich, W. 753
Frederickson, M. 876
Freedman, L. 188
Freeman, M. J. 375
Freigang, B. 802
Fremming, B. D. 511, 638, 753, 802
French, S. D. 129, 149
Friedkin, M. 751
Friedman, M. 267, 615, 636
Froeschle, J. E. 514
Frost, H. M. 481, 485, 513
Fudenberg, H. 345, 349, 373, 375
Fufacheva, A. A. 105, 106, 149, 152, 708
Fujita, T. 807
Fujiwara, K. 752
Fujiwara, T. 135, 151, 264, 708, 709, 753
Fukuda, N. 233
Fülscher, I. 705
Furuya, Y. 174, 187

Gadekar, K. 233
Gafford, J. A. 876
Gager, W. E. 876
Gaines, S. 752
Gaisford, W. D. 263, 748, 751

Gajdusek, D. C. 9, 386, 396, 610, 633, 875
Gallardo, F. O. E. De 106, 142, 149
Galton, M. M. 706
Ganla, . 153
Garcia, F. G. 247, 264, 514, 515, 752
Garcia, J. F. 403, 410
Gardashian, A. H. 10
Gargus, J. L. 151, 750
Garner, A. 706, 801
Garner, E. 753
Garner, F. M. 210, 232, 755
Gasparian, T. S. 707
Gasser, R. F. 188, 871
Gastant, H. 635
Gautier-Hion, A. 35, 56, 66, 70, 76
Gavan, J. A. 519, 872
Gavin, J. 311
Gavrilov, V. I. 709
Geager, F. 153
Geer, J. C. 803, 805
Gelb, A. G. 312
Gelshtein, V. I. 9, 192, 196, 197, 213, 215, 236, 413, 517, 872, 873
Gendon, U. Z. 149
Gengozian, N. 280, 311, 364, 372, 753, 872
Genovese, M. 749
Gergen, J. A. 144, 150
Gershoff, S. N. 359, 376
Gerstenberger, H. J. 245, 263, 454, 513
Geschikter, C. F. 193, 233
Geschwind, I. I. 403, 410
Getz, G. S. 269
Ghidoni, J. J. 238, 750, 820
Giacometti, L. 427
Gibbs, C. J., jr. 396, 633, 875
Gibbs, G. E. 404, 410
Gifford, H. 410
Gilardoni, A. 749
Gilbert, C. 210, 233, 759, 803
Gilbert, E. 262
Giles-Nelson, . 871
Gill, T. J. 359, 376
Gillman, J. 210, 233, 759, 803
Gillman, T. 750, 803
Gillot, P. 8
Gilmore, L. I. 411
Ginn, F. L. 754
Ginzton, L. L. 267
Girard, J. P. 379
Girard, P. 378
Girardet, P. 439, 513
Gisler, D. B. 751
Gitlin, D. 233
Givan, K. F. 874
Glaisher, 233
Glees, R. 612, 633
Gleiser, C. A. 753
Glickstein, M. 431
Globus, J. H. 249, 269, 615, 637
Glynn, L. E. 380, 395, 396
Godelard, J. 265
Goger, W. E. 888
Goh, T. K. 638, 808
Golarz, M. N. 540
Goldblatt, D. 635
Goldblatt, H. 793, 804, 830, 839
Goldman, E. A. 57
Goldman, H. M. 192, 210, 212, 231
Goldmann, K. 571, 633
Goldschmidt, W. 187
Goldsmith, E. I. 414
Goldstein, E. B. 431
Goldstein, M. 339
Goldzieher, J. W. 503, 512, 750, 877
Goltenboth, R. 706
Gonzales, I. E. 803
Good, R. A. 380, 381, 396
Goodall, J. 68, 69, 76, 869
Goodfriend, L. 378
Goodlow, R. J. 233
Goodman, M. 150, 153, 272, 311, 372
Goodrich, J. S. 125, 149
Gopalan, C. 9, 105, 109, 114, 150, 155, 242, 263, 266, 750
Gordon, B. 751
Gordon, E. 156
Gordon, E. B. 276, 313, 315, 316, 317
Gorlin, 210
Gorsuch T. T. 506, 513
Goslee, D. G. 377
Goss, L. H. 276, 277, 311, 315

Gottlieb, A. J. 312
Gowey, A. 429
Gowgiel, J. M. 192, 221, 233
Grabar, . 372, 397
Grady, H. J. 266, 806
Graham, C. E. 873
Graham, E. S. 431
Graham, G. 753
Graham, J. B. 329, 519
Graham, P. C. C. 541
Gras, J. 383, 396
Grasbeck, R. 502, 513
Graves, J. L. 706
Gray, C. W. 871
Gray, K. G. 395
Gray, W. R. 373
Graybiel, A. 108, 150
Grayson, J. 772, 803
Green, K. F. 431
Green, R. G. 229, 233
Green, S. S. 755
Greenberg, L. D. 154, 251, 252, 261, 263, 266, 267, 269, 615, 636, 749, 763, 789, 803
Greene, C. T. 377
Greene, F. C. 378
Greene, N. D. 373, 374
Greene, T. W. 378
Greer, W. E. 261, 268, 269, 340, 374, 375, 376, 378, 383, 392, 395, 396, 750, 753
Gregersen, M. J. 150
Gresham, G. A. 9, 264, 804
Grey, H. M. 373, 374
Grey, R. M. 236
Gribble, D. H. 752
Grigorova, O. P. 150
Grindlay, Y. 156
Grinker, R. B. 263, 615, 633
Groover, M. E. 104, 105, 150, 807
Grover, A. D. 267
Groves, C. P. 20
Grubb, R. 312, 373
Grzimek, B. 706
Guenter, C. A. 116, 122, 123, 150
Guillon, J. C. 231
Guillon, R. 432
Guilloud, N. B. 159, 186, 376, 514, 541, 872, 873, 875
Gulati, S. M. 427
Gupta, B. M. 706
Guthrie, C. B. 153, 313
Guttmacher, R. M. 186, 873
Guzman, C. 269
Gvasava, I. S. 705, 706

Haagensen, C. D. 232
Haas, R. 706
Habel, K. 578, 633
Habermann, R. T. 706, 753, 803
Hackel, D. B. 803
Hackeng, W. W. L. 516
Haddow, A. J. 803
Haddow, C. J. 839
Hafez, E. S. E. 869
Hagemenas, F. 186
Hagiwara, T. 709
Haglin, J. J. 105, 150
Haigh, M. V. 514
Halberg, P. 388, 395
Hall, A. S. 88, 150
Hall, K. 66, 76
Ham, A. W. 452, 514
Hamerton, A. E. 195, 210, 212, 213, 233, 405, 410, 438, 443, 514, 602, 617, 633, 706, 707, 803, 819, 874
Hamilton, A. 265
Hamilton, H. C. 403, 410
Hamlin, R. L. 109, 113, 115, 150
Hammaker, L. 749
Hampai, A. 379
Hamper, J. 312
Hampton, J. K. Jr. 377, 462, 512, 514
Hampton, J. W. 82, 83, 150, 330
Hampton, S. H. 148, 377, 462, 514, 751
Hance, A. J. 432
Hancox, N. 514
Hanlon, C. R. 156
Handa, J. 116, 133, 152
Hanly, C. 315
Hanly, W. C. 750
Hansen, E. L. 705
Hansen, H. A. 139, 140, 156
Hansen, L. T. 133, 153

Harden, A. 263, 264
Hardisty, R. M. 330
Hardy, A. Y. 707
Hardy, P. H. 707
Harkens, H. N. 412
Harman, J. H. 540
Harnack, S. 705
Harnagel, D. J. 632
Harper, R. A. 486, 514
Harris, L. J. 264, 509, 512
Harris, M. D. 638
Harris, R. L. 378
Harris, R. S. 264, 514, 839
Harrison, J. 751
Hart, E. B. 514
Hartley, J. W. 874
Hartman, C. G. 159, 161, 186, 194, 233, 870
Hartman, F. A. 410
Hartman, H. A. 753
Hartratt, P. 819
Hartwell, W. V. 749, 751
Hasabov, G. A. 144, 150
Haslewood, G. A. D. 515
Hathaway, P. W. 541
Hatta, J. van, 427
Haust, M. D. 801
Hawker, 517
Hawkey, C. 150, 329, 330
Hawking, F. 753
Hawkins, J. 430
Hay, R. 428
Hayama, S. 151
Hayden, R. O. 116, 119, 151
Hayes, N. R. 754
Haymaker, W. 620, 625, 633
Healey, L. A. 139, 140, 154
Healy, G. R. 754
Hediger, H. 189
Hedman, S. 397
Heerden, P. D. 151
Heffner, H. E. 430
Heffner, R. R., jr. 375
Hegner, R. 876
Hegsted, D. M. 9, 247, 264, 514, 515
Heighway, J. 518
Heinrichs, L. 189
Helland, D. R. 753
Helling, L. S. 431
Helz, M. K. 754
Hemrick, F. 753
Henderson, D. A. 426
Henderson, H. 516
Hendrick, A. G. 188, 514, 868, 869, 870, 871, 872
Henle, G. 748
Henle, W. 748
Heremans, J. F. 344, 372, 373, 374
Hernandez-Morales, F. 88, 155
Herrman, G. R. 105, 109, 113, 114, 151
Hershkovitz, P. 43, 56
Hess, M. 387, 396
Hetler, R. A. 264
Heuglin, M. T. 35, 56
Heuschele, W. P. 754
Heuser, C. H. 871
Hickman, R. L. 378
Hideo, Nigi (vide also Nigi) 151
Higginbotham, F. H. 804
Higginson, J. 266, 806
Hildebrandt, R. J. 875
Hill, C. W. 336, 341, 359, 364, 374, 376
Hill, R. B. 264, 752
Hill, W. C. O. 28, 37, 39, 40, 41, 43, 56, 58, 61, 75, 77, 161, 169, 186, 187, 188, 195, 233, 279, 281, 305, 314, 471, 511, 545, 546, 632, 707, 819, 820, 839, 868, 869, 873, 887
Hillis, W. D. 752
Hillyer, G. V. 376
Hinchcliffe, R. 430
Hines, H. M. 541
Hioco, D. 511, 514, 516
Hirokaua, M. 237
Hirose, Y. 312
Hirsch, P. F. 490, 514
Hirschfeld, J. 96, 151
Hirszfeld, H. 303, 312
Hirszfeld, L. 303, 312
Hisaw, F. L. 193, 196, 233, 874
Hisaw, F. L., jr. 874
Hitchcock, C. R. 105, 150

Hladik, C. 73, 77
Hobbs, K. R. 376, 377
Hodes, H. L. 139, 155
Hodgkinson, R. G. 430
Hodson, H. H. 750
Hoffbauer, F. W. 264, 753
Hoffman, H. A. 312
Hoffman, R. G. 502, 514
Holborow, E. J. 380, 395, 396
Hollander, W. 264, 790, 804
Holley, K. J. 518
Holman, R. L. 805
Holmes, A. W. 9, 151, 750, 752
Holmes, D. D. 754
Holmes, R. H. 150
Holmes, R. L. 410
Holtzwart, F. 602, 634
Honjo, S. 135, 151, 258, 264, 708, 709, 753
Hood, L. 346, 373
Hooff, J. A. R. A. M., van, 37, 56
Hooker, C. W. 755
Hopper, B. R. 871
Hopps, H. E. 801
Horger, E. O. 412
Hornus, G. 634
Horstmann, D. M. 577, 578, 633
Horvath, S. M. 754
Houba, U. 514
Houk, R. P. 754
Houser, R. G. 408, 410
Houston, M. L. 871, 872
Howard, A. N. 9, 264, 803, 804
Howard, R. J. 365, 377
Howe, N. A. 571, 634
Howell, M. 76, 329
Hrdlička, A. 165, 187
Hsu, H. F. 378
Hsu, K. C. 397
Hsu, S. L. Y. 378
Hu, F. 427
Hubel, D. H. 431
Hubner, J. 636
Hudson, G. W. 754
Hueckel, H. J. 507, 514
Huehus, E. K. 147
Hueper, W. C. 9, 804
Hughes, R. H. 10, 269
Hull, E. W. 203, 233
Hull, R. N. 752
Hume, E. M. 261, 454, 514, 516
Hunt, R. D. 247, 264, 450, 467, 509, 514, 515, 752
Huntley, C. C. 398
Hurst, E. W. 513, 554, 555, 567, 572, 598, 612, 613, 633, 634, 636, 706, 802
Huser, H. J. 88, 151, 153, 264, 313
Hutchinson, D. L. 149
Hutchinson, T. C. 887
Hutchinson, W. 195, 233
Hutchison, V. E. 875
Hutsishvili, C. M. 704
Hynes, J. D. 391, 395

Iber, F. L. 349
Ibrahim, M. Z. M. 633
Igarashi, M. 430
Iggo, A. 427
Iglesia, F. A. De La 819
Iglesias, R. 199, 233, 804
Ilievski, V. R. 873
Imaizumi, K. 151, 264, 708, 709, 753
Ingle, D. 339
Ingram, D. G. 396
Ingram, G. I. C. 329, 410
Inman, M. 330
Innes, J. R. M. 186, 238, 544, 559, 560, 567, 569, 576, 584, 587, 591, 600, 602, 613, 618, 632, 637, 806, 873
Inoye, M. 151
Irvine, M. 772, 803
Isa, A. M. 376
Ishizaka, K. 350, 371, 374, 378
Ishizaka, T. 350, 371, 374, 378
Israel, C. W. 876, 888
Itakura, C. 151, 806
Itani, J. 172, 187
Ito, K. 374
Ito, M. 752
Ivanov, M. T. 704, 707, 819
Ivey, M. H. 754
Ivy, A. C. 129, 154
Izawa, K. 67, 77

Jacobson, E. D. 129, 130, 131, 155

Jacotot, H. 875
Jacox, R. F. 397
Jaffee, R. B. 870
Jager, C. L. De 670
James, W. 59, 77
Jamieson, M. F. 198, 199, 231
Jansen, Y. 707
Janssen, P. J. 625, 637
Janota, M. 264
Jarmolych, J. 268, 806
Jeffree, G. M. 515
Jeffrey, J. D'A 877
Jelliffe, D. B. 801
Jenkins, R. H. 838
Jha, G. J. 464, 515
Jiřička, J. 375
Jiřička, Z. 753
Joachimovits, R. 873
Johansson, E. D. B. 870
Johansson, S. G. O. 376
Johnson, B. C. 264, 752
Johnson, C. M. 707
Johnson, D. K. 202, 238, 749
Johnson, G. L. 431
Johnson, T. A. 412
Johnson, L. G. 430
Johnston, G. S. 139, 140, 151
Johnston, G. W. 808
Jolly, A. 869
Jolly, C. J. 27, 57, 147
Jones, A. E. 430
Jones, A. M. 707
Jones, C. 59, 62, 64, 65, 66, 67, 69, 77
Jones, E. S. 265
Jones, O. G.- 263, 451, 514
Jones, R. 267, 807
Jones, V. E. 378
Jordon, R. E. 390, 395
Josso, F. 329
Jowsey, J. 515
Joy, R. M. 432
Jung. G. 196, 233

Kabat, E. A. 312
Kaboya, S. 432
Kado, R. T. 154
Kadur, B. M. 427
Kagan, I. G. 376
Kaiser, G. C. 156
Kalmikoa, K. P. 133, 154
Kalter, S. S. 122, 137, 153, 156, 373, 374, 376, 707, 874
Kaminer, B. 104, 105, 109, 114, 151
Kamitsuka, P. S. 637
Kanagasuntheram, R. 869
Kandel, E. 263, 615, 633
Kano, K. 187, 396
Kaplinsky, M. 515
Kar, A. B. 406, 410
Karel, L. 151
Karski, J. 515
Kasai, G. J. 375
Katagiri, G. J. 748, 801
Katz, S. 412, 749, 754
Katzberg, A. A. 405, 410
Katzin, D. S. 752
Katzin, E. M. 312
Kavtaradze, K. N. 705, 706, 707
Kawamura, R. **9**
Keane, B. E. 515
Keeble, S. A. 568, 634, 752
Keefe, J. R. 432
Keener, E. B. 541
Kellogg, R. 57
Kellog, W. H. 430
Kelly, A. L. 403, 410, 820
Kelly, E. 265
Kelly, M. G. 200, 203, 210, 222, 223, 230, 231, 232, 233, 234, 235, 754
Kende, M. 376
Kennard, M. A. 213, 234, 403, 410, 707, 804
Kenshalo, 427
Kent, J. H. 378, 707
Kent, S. P. 191, 199, 210, 213, 215, 220, 234, 264, 401, 403, 407, 411, 466, 515, 634, 804
Kenyon, A. J. 398
Kerber, W. T. 232, 234, 235, 427, 754, 873
Kerr, G. R. 411, 412, 753
Kersting, G. 556, 634
Kessel, J. F. 602, 634, 876
Kettel, L. J. 122, 123, 153

Keyes, J. E. L. 793, 804
Kill, I. F. 838
Killam, E. K. 431
Killam, K. F. 431, 432
Killander, J. 372
Kim, C. S. 874
Kim, D. 243, 264
Kimberg, D. V. 518
Kimbrough, R. D. 213, 234, 751
Kinard, R. 475, 516
King, C. T. G. 186
King, E. J. 508, 515, 516, 519
King, T. 151
King, T. O. 751
Kinney, T. D. 803
Kirsch, E. 708
Kirschstein, R. L. 801
Kirtley, B. 375
Kirton, K. T. 870
Kishi, T. 708
Kisliuk, R. L. 751
Kite, J. H., jr. 390, 395, 396, 397, 406, 411
Klembovsky, A. I. 9
Klinkhart, C. 301, 311
Klüver, H. 209, 212, 234, 237, 873
Kluver, N. 606, 636
Knepton, J. C. 105, 147
Knezevic, A. 88, 150
Knights, B. A. 877
Knobil, E. 402, 403, 411, 412, 802, 870
Knouff, R. A. 410
Knowles, F. G. W. 407, 411
Koch, G. 640, 670
Koch, W. 403, 405, 411, 708
Kodicek, E. 440, 491, 513, 515
Koering, M. J. 870
Koestner, A. 805
Koford, C. B. 160, 161, 186, 869, 872
Kokaya, G. I. 105, 108, 109, 110, 113, 114, 151
Kondi, A. 263, 750
Konovalov, N. V. 10
Kontaxis, N. E. 412
Kopec, A. C. 3, 13
Korinek, J. 153
Kornevitz, G. V. 124, 155
Kostuchenko, L. V. 151
Kovacs, B. A. 378
Kovatch, R. M. 636
Kowalczyk, R. S. 749, 751
Kraemer, D. C. 868, 869, 870, 871
Kraintz, L. 512
Krakower, C. 232
Kramer, K. L. 377
Kramsch, D. M. 264, 790, 804
Kratochvil, C. H. 313, 316
Kreig, H. 62, 77
Kriewaldt, F. H. 451, 517, 870, 875
Krise, G. M. 84, 85, 86, 151, 751
Kritchevsky, D. 807
Krohn, P. L. 264, 872
Kroms, M. 9, 268, 806
Kronberger, H. 195, 234
Kronvall, G. J. 373
Krook, L. 265, 405, 411, 437, 438, 515, 719, 749
Krotkina, N. A. 9, 191, 193, 208, 212, 216, 220, 222, 234, 236, 517
Krupe, M. 314
Krushak, D. H. 749
Kuhn, V. S. G. 187, 876
Kuksova, M. I. 84, 85, 86, 87, 88, 89, 90, 91, 151, 152, 156, 708, 820
Kulkarn, A. 150
Kumar, T. C. A. 403, 411
Kummer, H. 57
Kunkel, H. G. 373, 374, 380, 384, 387, 389, 396
Kuntz, R. E. 376, 707, 708, 875, 877
Kunz, H. W. 376
Kupferberg, L. L. 754, 807
Kurland, L. T. 234
Kuttner, A. 376
Kuzuya, H. 789, 804

Labrec, E. H. 707
Lacey, J. I. 332, 341
Lagutina, N. I. 124, 125, 152
Lakin, J. D. 352, 354, 356, 357, 374, 375
Lancaster, J. B. 869, 887
Landon, J. C. 238, 377, 431

Landsteiner, K. 271, 272, 273, 275, 278, 280, 284, 288, 293, 300, 312
Landwehr, B. T. 514
Landy, M. 752
Lang, C. M. 337, 341, 404, 411, 791, 804, 875
Lang, D. 494, 519
Langer, B. 510
Langston, W. C. 149, 262
Lannoy, C. W. De 105, 110, 114, 152
Lapeire, C. 636
Lapin, B. A. 119, 152, 155, 159, 160, 186, 192, 195, 196, 199, 210, 213, 214, 222, 234, 334, 337, 341, 474, 515, 547, 634, 708, 752, 757, 758, 775, 791, 820, 827, 830, 834, 871
Laqueur, G. L. 227, 234, 633
Lasry, J. E. 109, 119, 156
Lassen, L. J. 512, 548, 632
Laties, A. M. 431
Laufman, H. 804
Laughlin, J. S. 632
Laurence, G. D. 874
Laursen, A. C. 226, 232
Lavender, J. F. 341
Lawson, D. E. M. 515
Layton, L. L. 366, 377, 378
Leach, E. H. 266
Leary, G. A. 432
Lease, J. G. 255, 256, 265
Leblond, C. P. 265, 616, 634, 887
Ledouble, F. 161, 186
Lee, F. 378, 426
Lee, M. R. 512
Lee, R. B. 869, 887
Lee, S. 377, 378
Lees, M. H. 116, 118, 119, 122, 133, 152
Leeson, T. S. 452, 514
Leffingwell, T. P. 238, 820
Lehner, N. D. M. 245, 258, 261, 265, 439, 498, 516, 632, 801, 802, 839
Lehninger, A. L. 487, 516
Leinfelder, J. T. 431
Leiper, R. T. 876
Leipold, H. W. 753
Lemaire, R. 511
Lemmon W. B. 807, 839
Lendrum, F. C. 193, 233
Lennes, G. 311
Lennette, E. H. 375, 752
Lepine, P. 634
Leppi, T. J. 152
Lequin, R. M. 516
Leswing, R. J. 634
Levaditi, C. 558, 634
Levaditi, J. 231
Levenbook, I. S. 708
Leventhal, C. M. 137, 149
Levine, M. D. 634
Levine, P. 284, 293, 300, 312
Levitt, M. R. 139, 165
Levy, B. M. 125, 149, 200, 216, 223, 234, 262, 426, 512
Lewis, A. L. 709
Lewis, D. 751
Lewis, J. P. 146
Lewis, U. J. 268
Liat, L. 61, 77
Lichtenberg, E. von, 378, 753, 877
Lichter, E. 152
Lichtman, S. S. 748
Liebman, P. A. 431
Lillie, R. D. 577, 634
Lindau, H. 571, 634
Lindeman, H. H. 429
Lindgren, F. T. 264, 750, 804
Lindsay, I. R. 750
Lindsay, S. 552, 635, 784, 804
Lindsey, J. R. 707
Linge, B. van, 541
Linton, A. H. 709
Lipschütz, A. 199, 233, 804
Litwin, S. 152
Lodder, J. 231
Loew, E. R. 268
Lofland, H. B. 261, 265, 266, 413, 632, 801, 807
Loghem, E. 152
Loghem, J. J. van, 386, 397
Lombard, L. S. 202, 234, 407, 411, 873
London, W. T. 875
Long, D. M. 635

Longcope, W. T. 208, 234
Longmire, R. Z. 129, 149
Loomis, L. N. 578, 633
Lou, T. 578, 635, 637
Loumans, A. K. J. 808
Love, G. J. 751
Lovell, R. 708
Lowe, C. U. 265
Lowe, E. P. 754
Luca, H. F. De 510
Lucas, C. C. 269
Lucas, N. S. 454, 516
Lucke, B. 198, 199, 229, 234, 551, 576, 595, 635
Luckey, . 516
Lüdke, H. 235
Luker, M. F. 374
Lund, J. 510
Lund, O. E. 635, 708
Luntz, M. H. 394, 396
Lushbaugh, C. C. 222, 235
Lusted, L. B. 516
Luther, M. 190, 235

Macapinlac, M. P. 261, 265
Machado, A. de B. 38, 57
Machida, H. 427
Macfarlane, R. G. 330
MacGaughey, Y. 708
Machener, R. 430
MacIntyre, I. 490, 516, 517
Mack, R. S. 510
Mackay, I. R. 380, 382, 383, 384, 388, 392, 396, 397
Mackenzie, J. C. 427
Mackler, B. 347, 373
Maclean, P. D. 48, 57, 150
Macleod, I. F. 758, 802
Macleod, J. 879, 887
Macnintch, J. E. 265, 413, 807
Macpherson, J. 330
MacRoberts, B. R. 869
MacRoberts, M. H. 869
Madden, D. L. 875
Maddock, S. 152
Madhavani, T. V. 750, 753
Madsen, P. O. 838
Maejima, K. 752
Maercke, Y. van, 751
Magakyan, G. O. 118, 152
Magoon, H. W. 402, 411
Magrath, D. 632
Mahl, C. T. 127, 152
Mahno, G. 428
Mahon, R. C. J. 430
Makarchenko, A. F. 152
Malinovsky, C. V. 124, 125, 152
Malinov, M. R. 9, 80, 81, 82, 105, 106, 108, 109, 110, 114, 115, 116, 118, 119, 121, 133, 134, 138, 146, 152, 154, 155, 197, 235, 265, 329, 749, 751, 754, 758, 771, 777, 790, 792, 804, 805, 807
Malley, A. 347, 370, 373, 379, 427
Manalo-Estrella, P. 268
Mancini, G. 374
Mancusa, S. J. La, 517
Mandel, A. D. 752
Mandour, A. M. 541
Mangelson, N. L. 840
Manhouri, H. 506, 512
Manley, G. H. 869
Mann, G. V. 9, 251, 265, 407, 411, 516, 751, 288, 789, 805
Mann, J. D. 312
Mann, P. H. 706
Mann, T. 887
Manning, G. W. 805
Manson, W. A. 516, 887
Manueldis, F. E. 560, 577, 578, 633
Marassero, F. 144, 153
Marchoux, E. 405, 411
Margolis, J. 330
Margolis, S. 267
Markoff, W. H. 705
Markova, A. Y. 431
Markowitz, A. S. 839
Marks, L. B. 431
Marsh, J. T. 36, 341
Marsh, M. E. 719, 749
Marsh, W. L. 312
Marshall, R. W. 128, 129, 131, 148
Marsland, W. P. 150
Martensson, L. 373

Martin, C. B. jr.
Martin, J. E. 475, 516, 596, 635
Martin, L. K. 749
Martin, R. D. 12, 57
Martin, T. J. 517
Martinez, J. 838
Martinez, R. D. 807
Martucci, R. 76
Maruffo, C. A. 9, 188, 197, 202, 215, 235, 265, 266, 407, 411, 754, 758, 777, 804, 805, 807, 839, 872
Mason, C. 754
Mason, D. E. 265
Mason, J. W. 129, 130, 131, 155, 333, 339, 341, 408, 411, 412
Mason, K. E. 263, 265, 541
Mason, M. M. 873
Masouredis, S. P. 297, 312
Massaglia, A. C. 210, 235, 820
Massopust, L. C. 432
Masterton, B. 430
Mastroianni, L., jr. 870, 880, 887
Matrajt, H. 511, 513, 516
Matsuoka, K. 708
Matthews, C. 82, 83, 150, 329
Mattis, P. A. 749
Mattlin, R. H. 754
Maurer, P. H. 354
May, C. D. 265, 266, 268
Mbaya, V. 263, 375
Mechnikov, I. I. 875
Medalie, J. 808
Medearis, D. N., jr. 874
Medin, D. L. 432
Mehta, U. C. 266
Meier, G. W. 874
Meissner, W. A. 407, 413
Melby, E. C. 707, 738
Melendez, L. V. 752
Mella, H. 604, 635
Melmon, K. L. 149
Melnick, J. L. 567, 635
Melnikov, R. A. 9, 191, 221, 235, 236
Meltzer, H. 541
Meltzer, J. K. 749
Melville, G. S. 238, 820
Menzie, C. 329
Mergenhagen, S. E. 377
Merkatz, I. R. 871
Mertz, W. 495, 516
Mesnil, F. 405, 411
Metzgar, R. S. 394, 397
Metzger, H. 373
Meyer, H. H., jr .874
Meyer, J. S. 116, 133, 152
Meyer, R. K. 874
Meyer, V. 402, 411, 412
Michelson, A. M. 540
Mickelsen, O. 234
Middleton, C. C. 266, 413, 797, 805, 807
Midgley, A. R., jr. 870
Milch, R. A. 516
Miller, C. E. 475, 516
Miller, C. O. 414, 415
Miller, C. P., jr. 272, 275, 280, 312
Miller, E. G., jr. 268
Miller, J. M. 430, 876
Miller, M. J. 753
Miller, R. L. 186, 873
Miller, S. E. 750
Mills, R. C. 514
Milstein, C. 376
Mims, V. 262
Minden, P. 378
Minette, H. P. 748
Minkin, C. 517
Mintz, D. H. 404, 412
Miquel, J. 633
Mirand, E. A. 838
Miravet, L. 511, 514
Misenhimer, H. R. 872
Mitchell, B. B. 706
Miyamoto, T. 122, 123, 153
Mollison, T. 272, 312, 313
Molnar, Z. 486, 516
Monif, G. R. G. 874, 875
Montagna, W. 231, 232, 427, 428
Montano, A. 708
Moodie, R. L. 479, 516
Moody, D. B. 430
Moon, H. D. 9, 263, 269, 806
Moor, J. R. 264, 514, 839
Moore, J. W.

Moor-Jankowsky, J. 88, 153, 156, 276, 279, 283, 299, 313, 315, 316, 317
Morales, C. 706
More, R. H. 801
Moreland, J. T. 432
Moreno, O. M. 749
Morgan, H. 708
Morgan, R. 516
Morgan, W. D. 232
Morrell, C. A. 749
Morris, G. C. R. 125, 148
Morrison, L. R. 555, 635
Morrison, E. S. 268, 806
Morrison, P. 133, 135, 153
Morrow, A. C. 138, 153, 838
Moss, D. W. 516
Moskov, F. S. 426
Mossakowsky, M. J. 550, 635
Mougey, E. H. 411
Mourant, A. E. 303, 313
Movis, H. J. 129, 149
Muckenfuss, R. S. 564, 635
Muir, A. R. 427
Mukha, C. V. 154
Munroe, J. S. 229, 235
Munson, P. L. 151
Murphy, G. P. 151, 838, 888
Murphy, R. 9
Murthy, C. D. S. 802
Muskett, C. W. 266, 806
Muto, T. 708, 753
Myers, B. J. 707, 708, 877
Myers, J. H. 153
Myers, R. E. 549, 635, 873
Mysorekar, N. R. 154
McAlpine, K. 208, 232
McArthur, J. W. 870
McCaffree, 150
McCall, K. B. 250, 261, 265, 269
McCann, J. O. 872, 873
McCann, S. M. 412
McCarrison, R. 213, 235
McCarty, C. 156
McClean, F. C. 516
McClure, H. M. 873
McCoombs, H. L. 805
McCord, C. P. 754
McCordock, H. A. 735
McCoy, F. W. 410
McCullough, N. B. 875
McCune, C. L. 379
McFalls, Y. W. 519
McGandy, R. B. 805
McGerity, J. L. 378
McGill, H. C. 268, 795, 803, 805, 807, 876
McHugh, P. R. 341
McKinley, J. 238, 820
McKissick, G. E. 805
McLaughlin, P. 805
McLennan, G. C. 634
McMaster, P. D. 336, 341
McNally, A. 265, 411, 805
McNulty, W. P., jr. 805
McNutt, S. H. 876
McQuay, R. M. 876
McQueen, J. 9, 803
McRitchie, E. 314
McVean, G. W. 431

Nagle, J. P. 408, 412
Naiki, M. 752
Nair, C. 806
Nakagawa, M. 708, 753
Nakagawa, S. 233
Nakamura, R. M. 754, 807
Nam, S. C. 268, 806
Napier, J. 57, 58, 60, 61, 74, 77, 343, 372, 517, 869
Napier, P. 57, 58, 60, 61, 66, 74, 77, 343, 372, 517, 869
Naquet, R. 149, 132
Narasimhan 153
Nastuk, E. L. 397
Nathanson, N. 568, 580, 635
Nauta, W. 633
Nayak, N. C. 801
Nefedieva, N. P. 708
Neff, K. 128, 129, 130, 148
Negrom, P. 602, 635
Neill, J. D. 870
Nelson, B. 753
Nelson, G. S. 876
Nesturh, M. F. 150

Neufeld, H. H. 808
Neutze, J. 149
Neuman, E. W. F. 486, 517, 518
Neuman, N. W. 486, 517
Neumann, G. J. 502, 517
New, A. E. 750
Newberne, I. W. 195, 210, 235, 401, 407, 412, 873
Newman, W. P. III. 805, 807
Ney, R. 379
Nichols, G. 519
Nichols, R. E. 199, 235
Nidus, B. D. 839
Nies, A. S. 149
Nigi, H. (vide also Hideo) 153, 806
Ninane, G. 311, 748
Nisjuda, T. 67, 77
Nishimura, E. Y. 806
Niswender, G. G. 870
Niven, J. S. F. 229, 235, 752
Noades, J. 311
Noback, C. V. 210, 235, 432
Noel, P. 397
Noguchi, Y. 153
Nordin, B. E. C. 517
Norins, L. C. 876
Norkina, L. N. 152, 708
Norman, A. W. 440, 490, 509, 517
Normansell, D. E. 373
Norris, R. 749
Nossal, G. J. V. 380, 383, 397
Notkins, A. L. 377
Nouvel, J. 637
Noyes, W. F. 200, 223, 235
Numano, T. 807
Nuttall, G. H. F. 272, 314
Nyhus, L. M. 412

O'Conner, P. 197, 213, 235
Ockerse, T. 670
Odell, R. M. 139, 153, 823, 828
Oester, A. 838
O'Gara, R. W. 198, 222, 230, 233, 234, 235
Ogawa, H. 264, 708, 709, 753
Ogilvie, B. M. 378
Okugi, M. 753
Oliphant, M. C. 105, 154
Oliveira, L. F. De 431
Oliveira e Silva, 708
Oliver, J. 838
Orange, R. P. 378
Ordy, J. M. 266, 432, 750
Orlova, S. D. 153
Oshima, F. 195, 210, 235, 874
Osserman, K. E. 397
Osterrieth, P. M. 311, 748
Ottensooser, F. 286, 314
Ottow, B. 169, 187
Overholser, M. D. 193, 222, 235
Overman, R. P. 153
Owen, M. 517
Owen, R. D. 300, 301, 314
Oxnard, C. E. 254, 263, 264, 266

Pace, H. 133, 153
Padget, D. H. 871
Padmavathi, S. 105, 109, 114, 150, 242, 263
Pagnic, A. 144, 153
Palotta, A. J. 153, 186, 872, 873
Palmer, A. E. 801
Paluska, E. 153
Paparella, M. M. 430
Papermaster, D. S. 376
Pappenheimer, A. M. 617, 635, 754
Parer, J. T. 116, 118, 119, 122, 133, 152
Park, E. A. 517
Parker, M. L. 409
Parker, W. 99, 153
Parkes, A. S. 870, 887
Parkman, P. D. 874
Parshall, C. J., jr. 375, 377
Parson, H. T. 265
Passovoy, C. M. 410, 750
Paterson, P. Y. 380, 395, 397
Patt, J. K. 801
Patterson, R. 122, 123, 147, 369, 370, 375, 379
Patton, D. E. 9, 268, 807
Patton, N. M. 426
Pautard, F. G. E. 517
Payan, H. 562, 635

Payton, W. T. 233
Pearce, A. G. E. 517
Pearlman, F. 379
Pearson, K. 187
Pearson, W. M. 261, 265
Pechet, L. 329
Pelc, S. 574, 575, 635
Pellegrino, E. D. 511
Peltier, S. 807
Pena, A. Del La 750, 870
Pennington, L. L. 412
Pentschew, A. 606, 636
Perdrau, J. R. 636
Pereira, H. G.
Perkins, E. M. 53, 57, 186, 426, 427, 428
Perley, A. M. 9, 265, 805, 806
Perley, R. 870
Perry, W. L. M. 753
Peters, H. R. 288, 316
Peterson, K. F. 708
Petran, E. 268
Petrov, N. M. 9, 191, 220, 236, 517
Pette, E. 556, 634
Petter-Rousseaux, A. 869
Pettit, A. 192, 196, 236, 405, 412
Pezet, 192, 236
Pfeifer, R. A. 594, 636
Pfeiffer, C. A. 194, 198, 222, 236, 755
Phillips, P. E. 874
Phillips, P. H. 268
Picard, J. 344, 345, 346, 372
Picard, P. 153
Pichkhaya, T. P. 751
Pickering, D. E. 9, 138, 139, 140, 153, 160, 186, 191, 199, 210, 213, 215, 220, 234, 403, 406, 411, 412, 512, 516, 616, 626, 627, 633, 634, 637
Pierce, E. C. 752
Pierce, J. D. 752
Pike, R. M. 375
Pilot, I. 806
Pinkerton, H. 808
Pinkerton, M. E. 376, 708, 875
Pinto, J. S. 232, 426
Pirofsky, B. 380, 397
Pitney, W. R. 269
Plager, H. 709
Plentl, A. A. 213, 236
Plimmer, W. G. 199, 215, 236
Podvalnaya, M. J. 237
Podliachouk, L. 314
Pogosyants, Y. Y. 231
Pohlaman, R. 410
Pokorna, Z. 397
Pokorny, J. 618, 636
Polis, B. D. 751
Polihs, E. 411
Poncet, M. 149
Ponder, E. J. 153
Pope, B. L. 807
Popper, H. 9, 750
Poppen, K. J. 252, 266
Porta, E. A. 819
Porter, A. 876
Porter, F. S. 266
Porter, R. P. 377
Porter, R. R. 346, 364, 373
Porter, R. W. 129, 149, 340
Portman, O. W. 9, 240, 261, 266, 268, 784, 785, 801, 806, 839
Posner, A. S. 485, 486, 487, 488, 514, 517, 518
Possovoy, M. 146, 151
Postnikova, Z. A. 9, 236
Potter, H. P., jr. 749
Potts, J. J. 511, 512
Poulik, E. 150, 153
Poulik, M. D. 345, 346, 347, 348, 373
Povar, M. L. 838
Powell, J. E. 870
Powers, R. D. 754
Prather, E. C. 709
Pratt, D. A. H. 226, 232
Prehn, R. T. 237
Prendergast, R. A. 375
Preston, J. K. 750
Price, D. 887
Price, R. A. 754
Price, W. H. 635
Prichard, R. W. 266, 805, 806
Prichod'ko, W. 153
Pridgen, W. A. 751

Prine, J. R. 426
Prockop, D. J. 507, 510, 517
Proehl, E. C. 266
Prokop, O. 314
Prou-Wartelle, O. 329
Pruzansky, J. J. 374, 375, 379
Pryce, J. D. **517**
Pryor, W. H., jr. **839**
Psemicka, P. 428
Pucheran, M. 35, 57
Pulaski, E. J. 749
Putterson, R. 153
Putz, Z. 314
Pychkaya, T. P. 147
Pye, A. 430

Quick, A. J. 329
Quinlivan, W. L. G. 360, 376

Race, R. R. 293, 311, 314
Racela, A. S. 266, 806
Radwin, H. M. 139, 140, 153, 838
Rady, D. M. 266, 750
Raeva, N. V. 88, 156
Rahm, U. 38, 57, 172, 187, 751
Raisz, L. C. 517
Rakieten, M. 751
Rall, D. P. 516, 751
Rama Rao, P. B. 264, 752
Ramalingaswami, V. 244, 262, 266, 268, 515, 801
Ramanathan, K. 9
Ramaswami, L. S. 839
Ramirez de Arellano, M. T. R. De 141, 149, 154
Ramser, J. T. 513
Ramsey, E. M. 872
Ranck, J. B., jr. 549, 636
Randall, F. E. 181, 188, 517
Ranizla, R. 430
Rankow, R. M. 192, 236
Ranson, S. W. 410, 411
Rao, M. V. R. 240, 267, 753
Rapp, W. 391, 397
Rasmussen, A. F., jr. 269, 341
Rasmussen, H. 510, 517
Rasmussen, R. A. 10, 269
Rasmussen, T. B. 636, 754, 807, 888
Ratcliffe, H. 192, 195, 197, 199, 202, 212, 215, 234, 236, 267, 403, 412, 709, 791, 805, 806, 873, 874
Ratcliffe, J. L. 839
Ratner, B. 376
Ratner, I. A. 122, 137, 156
Ratner, J. J. 376, 875
Ratnoff, O. D. 329
Ravant, F. 709
Rawson, R. A. 150
Ray, A. P. 806
Raynaud, J. P. 636
Reed, O. M. 451, 517
Reese, W. H. 233, 427
Register, U. D. 268
Rehlmann, D. F. 153
Reichlin, S. 409
Reid, R. T. 378
Reid, V. 751
Reinhold, J. G. 749
Reis, D. J. 105, 154
Reise, W. 637
Reisman, R. E. 379, 428
Reite, M. L. 154
Reitsma, W. 541
Rekeys, E. 150
Remane, A. 163, 186
Renold, A. E. 339, 341
Retief, C. P. 151
Rewell, R. E. 195, 196, 236, 709, 806, 873
Reyes, C. 208, 232
Reynolds, J. J. 517
Reynolds, L. B. 122, 123, 153
Reynolds, P. C. 430
Rhode, E. A. 749
Rhodes, A. J. 874
Rhodes, J. M. 154
Ribelin, W. E. 634
Richards, W. H. G. 330
Richart, R. 801
Riche, D. 149
Richter, R. B. 566, 582, 606, 636
Ridley, D. 330
Ridley, F. 128, 129, 130, 131, 148
Riesen, A. H. 706

Riesenfeld, A. 188
Rigdon, R. H. 109, 114, 154
Rinehart, J. F. 154, 250, 253, 261, 263, 266, 267, 615, 636, 749, 763, 789, 806
Riopelle, A. J. 9, 261, 317, 801
Riper, D. C. van, 426
Ripley, S. 39, 57
Ripps, H. 431, 432
Ritchie, L. S. 376
Ritter, H. 314
Rittner, C. 154
Roach, A. M. 231, 409
Roach, H. D. 804
Roberts, D. K. 754
Roberts, J. 268, 758
Roberts, J. A. 825, 838, 839, 888
Roberts, M. 379
Roberts, W. C. 806
Robinson, C. J. 517
Robinson, F. R. 109, 113, 150, 751
Robinson, J. A. 412
Robinson, U. B. 195, 210, 235, 401, 407, 412, 873
Robison, R. 486, 517
Robson, J. G. 431
Roca-Garcia, M. 806
Rocha, H. 840
Rodda, R. 794, 806
Rodgers, F. C. 244, 267
Rodman, N. F. 329
Rodriguez, A. R. 707, 874
Rodriques Decuret, H. 635
Rogers, C. 314
Rogers, N. G. 875
Rogers, Q. R. 507, 514
Roizin, L. 628, 632
Romel, W. C. 517
Root, A. 412
Roques, R. 405, 412
Rosal, J. 805
Rose, B. 378
Rose, G. A. 495, 518
Rose, N. R. 378, 390, 395, 396, 397, 411, 413
Rosen, S. 138, 155
Rosenblum, L. A. 17, 255, 267, 872
Rosin, S. 314
Rossan, R. 751
Rossen, R. D. 374
Rothman, S. S. 128, 129, 130, 131, 155
Roulston, G. L. 839
Roussel, J. D. 880, 887
Roussy, G. 402, 409
Rowe, D. A. 372
Rowe, W. P. 874
Rowell, N. R. 395
Royer, P. 513
Rozee, K. R. 874
Ruch, E. 875
Ruch, T. C. 190, 196, 197, 236, 240, 244, 249, 252, 267, 329, 334, 341, 401, 404, 408, 412, 414, 415, 437, 443, 518, 653, 670, 709, 757, 806, 809, 820, 827, 828, 831, 871, 884, 886
Rudiger, H. 753
Ruegamer, W. R. 261
Ruffie, J. 303, 314
Rumery, R. E. 263
Ruppell, E. 57
Ruskin, A. 109, 114, 154
Russell, E. S. 540
Rutherford, R. B. 267
Ruttkay-Nedecky, I. 115, 148, 149, 154
Rychlikova, M. 397
Rytand, D. A. 822, 838

Sabater Pi, I. J. 67, 70, 77, 186, 188
Sabin, A. B. 267, 567, 636
Sade, D. S. 887
Sadun, E. H. 378, 749, 753, 877
Saiger, G. L. 150
Saint Martin, J. 391, 397
Salk, J. E. 198, 229, 237
Salk, R. 838
Samorajski, T. 266, 432, 750
Sander, P. C. 375
Sanders, B. G. 373
Sandground, J. H. 211, 213, 229, 231
Sanger, R. 293, 311, 314
Sanmartin-Barberi, C. 806
Sargent, T. D. 431
Sarich, V. 154
Sartzer, J. 749

Sasaki, H. 314
Sasaki, T. 212, 237
Saslaw, S. 808, 839
Sauer, R. M. 709, 752
Scatterday, Y. E. 709
Schaar, F. 265, 268
Schachter, D. 509, 518
Schachter, J. 432
Schade, J. P. 628, 632, 636
Schaffner, F. 754
Schalch, D. S. 409
Schally, A. V. 411
Schendel, H. W. 264, 752
Schendorf, J. G. 129, 154
Schenker, H. 518
Schepers, G. W. K. 204, 216, 223, 237
Scherer, H. J. 552, 594, 610, 613, 617, 636, 637, 806
Schiff, F. 314
Schiff, G. M. 874
Schirl, A. 705
Schlegel, J. H. 139, 140, 153, 838
Schluederberg, A. 375
Schmey, M. 212, 237
Schmid, R. 749
Schmidt, E. C. H. 708, 806
Schmidt, Ira, G. 408, 412
Schmidt, L. H. 751
Schmidt, N. J. 357, 375
Schmitt, J. 274, 276, 314
Schneider, N. J. 187, 709
Schoen, R. 634
Schoenbechler, M. J. 378
Schooness, R. 888
Schopman, W. 516
Schrier, A. M. 437
Schultz, A. H. 161, 162, 164, 165, 167, 168, 169, 170, 171, 175, 176, 178, 179, 181, 186, 187, 188, 470, 518, 659, 661, 670, 868, 872, 877, 888
Schumacker, H. R. 428
Schurman, E. A. 195, 237
Schwab, J. L. 839
Schwartz, M. A. 751
Schwarz, E. 41, 57
Scott, A. 514
Scott, E. 754
Scott, G. H. 246, 262, 410, 466, 503, 512
Scott, H. H. 199, 215, 237, 709, 806, 838
Scott, H. W., jr. 269, 403, 413, 808
Scott, R. F. 9, 268, 792, 806, 807
Seal, U. S. 373
Seaman, A. J. 81, 82, 154, 329, 751
Sear, H. 150
Sebor, J. 267
Seegar, B. C. 397
Sehgal, A. K. 132, 154
Seibold, H. R. 838, 839, 888
Sela, M. 376
Selenka, E. 181, 188
Seljeskog, E. L. 105, 150
Selkurt, E. E. 139, 154, 838
Sellers, T. 186
Sellheim, A. P. 199, 237, 619, 636
Selye, H. 332, 335, 341, 807
Serbanescu, T. 432
Sever, J. L. 874, 875
Seway, L. C. 541
Sexton, R. S. 752
Shaetar, M. R. 517
Shaffer, A. N. 432
Shaffer, M. F. 748
Shaffer, W. L. 754
Shaposhnikov, A. I. 98, 154
Sharpio, S. 731
Shaw, J. C. M. 181, 188
Shaw, J. H. 268, 519, 670
Shcherbakova, O. P. 122, 133, 134, 135, 154
Shell, N. F. 316
Sheppard, R. D. 633
Shestopalova, S. K. 124, 125, 154
Shimamoto, T. 807
Shimkin, M. B. 873
Shin, Y. 187
Shive, R. J. 755
Shope, R. E. 873
Shriver, J. E. 431, 432
Shukers, C. F. 149, 262
Shuster, J. 345, 348, 375
Shuster, Y. 152
Shy, M. G. 541

Sibal, L. R. 352, 357, 374
Siddiqi, H. R. 869
Sidley, M. A. 432
Siegal, A. M. 221, 237, 754
Sigurdardotir, B. 874
Silberschmidt, K. 286, 314
Silva Teles, E. Da 840
Silverstein, A. M. 357, 362, 375, 376, 377
Simavomian, V. G. 709
Simoes, J. 133, 135, 153
Simons, P. J. 377
Simpson, G. G. 13, 57
Simpson, J. A. 393, 397
Simpson, M. E. 887
Sinisterra, L. 9
Siskind, G. W. 375
Sissons, H. A. 518
Sitbenthal, J. 510
Sivadon, P. 395, 397
Skeith, M. D. 139, 140, 154
Skelton, F. R. 397
Slaughter, L. J. 753
Slim, J. C. 636
Slonim, A. D. 122, 133, 134, 135, 154
Sloper, S. C. 633
Smetana, H. F. 268, 749, 750
Smith, C. C. 105, 139, 140, 155, 156, 839
Smith, C. P. 128, 130, 131, 155
Smith, C. R. 109, 113, 150
Smith, F. G. 149
Smith, G. P. 336, 341
Smith, G. S. 403, 412
Smith, H. H. 516
Smith, H. W. 140, 155, 822, 823, 838
Smith, J. H. 378
Smith, J. L. 876, 888
Smith, M. J. 519
Smith, R. 498, 511
Smith, R. E. 754
Smith, S. M. 632
Smith, V. S. 150
Smith, W. E. 237
Smith, W. K. 137, 149
Smith, W. T. 254, 266
Smithers, S. R. 378
Smyth, F. S. 512, 516
Smoilovskaya, E. Y. 9, 10, 193, 220, 221, 237
Snell, K. C. 237, 518
Snow, C. E. 188
Snyder, R. L. 392, 397
Sobel, M. V. 397
Socha, W. W. 317
Soderberg, O. 397
Sohal, R. S. 807
Soike, K. F. 709
Sokoloff, L. 478, 518
Solopaev, B. P. 155
Solopaeva, I. M. 155
Soloviova, G. A. 155
Sonmitt, I. 314
Sonnenwirth, A. C. 429
Sood, S. K. 262, 266, 268
Soto, P. J. 752, 754, 807
Soulier, J. P. 329
Southard, E. E. 518
Southwick, C. H. 869
Southwick, M. D. 876
Spadaro, J. A. 511
Sparks, S. J. 870
Spector, W. S. 134, 155
Speert, H. 869
Sperling, H. G. 432
Spicer, S. S. 152
Spiegler, G. 515
Spier, R. 397
Spooner, E. T. C. 602, 637
Springer, G. F. 314
Springer, K. 231
Sprinz, H. 752
Sridastava, H. K. 406, 410
Srikantia, S. 9, 155
Sriramachari, S. 266
Stadie, R. 807
Stahl, W. R. 115, 121, 138, 155
Stanworth, D. R. 373, 377
Starck, 187
Stare, F. 9, 265, 268, 411, 805
Stark, L. G. 431, 432
Startsev, V. G. 127, 129, 130, 131, 133, 152, 155, 708
Stasilevich, Z. K. 705, 709

Stavitsky, A. B. 375
St. Clair, R. W. 261, 265, 404, 413, 632, 791, 807
Stebbins, R. S. 266
Stebbins, W. C. 430
Steblay, R. W. 393, 397
Stedingk, M. von, 517
Stee, E. W. van, 750
Steffek, A. J. 186
Stefko, W. H. 518
Stehbens, W. E. 807
Steinberg, A. G. 373, 395, 397
Steiner, P. E. 209, 212, 237, 552, 607, 616, 636, 754, 807, 888
Steinmuller, D. 237
Stephenson, S. E., jr. 269, 413, 808
Sternberg, S. S. 196, 237
Sterzl, J. 376
Stevens, V. C. 870
Stewart, C. T. 265
Stewart, H. L. 208, 237, 518
Stewart, M. J. 208, 237
Stewart, T. D. 869
Stiehm, E. R. 375
Stolley, P. D. 635
Stollnitz, F. 432
Stone, F. J. 839
Stone, M. J. 373
Stoop, J. W. 375
Storm, C. J. 109, 114, 156, 808
Storvick, C. A. 805
Stout, A. P. 197, 232, 888
Stout, C. 392, 397, 807, 839
Strates, B. 518
Straus, H. W. 366, 367
Straus, W. L., jr. 188, 709
Strauss, A. J. L. 394, 397
Strauss, R. 268
Streff, G. M. 231
Strong, J. P. 268, 803, 805, 807, 876
Struhsaker, T. T. 430
Stuart, K. L. 801
Sturtz, V. 750
Suarez, R. M. 88, 155
Sugiura, K. 430
Sugiura, S.
Sullivan, D. J. 188, 872
Sullivan, R. 751
Summers, G. A. C. 709
Sun, S. C. 807
Sundberg, R. D. 265, 268
Sunderland, D. A. 237
Sureau, P. 618, 636
Sussman, H. H. 139, 140, 153, 838
Suzuki, A. 67, 77
Suzuki, Y. 151
Svartz, H. 388, 397
Svedberg, A. 152
Svoboda, D. J. 266, 806
Sweet, A. Y. 139, 155
Switzer, J. W. 88, 155
Symons, C. 82, 150, 329

Tada, T. 374
Tajen, N. 262
Tajima, Y. 752
Takagaki, Y. 737, 752
Takahashi, R. 708, 753
Takasaka, M. 151, 264, 708, 753
Takemura, N. 151
Talbut, C. H. 379
Tallar, J. A. 427
Talmage, D. W. 354, 375
Tan, E. M. 380, 384, 387, 389, 396
Tanaka, M. 709
Tanaka, T. 153, 571, 636
Tanner, K. N. 426
Tappan, D. V. 268
Tappen, M. C. 807
Tatoyan, S. H. 106, 108, 110, 155
Taylor, B. 268
Taylor, C. B. 9, 262, 268, 407, 413, 790, 792, 807
Taylor, E. D. 411
Taylor, R. L. 376
Taylor, S. 518
Teague, N. 839
Tee, D. E. H. 376
Teichner, W. H. 332, 341
Telford, L. R. 265, 540
Tenenhouse, A. M. 510
Termine, J. D. 486, 518
Terry, M. W. 138, 153, 838
Terry, R. J. 376, 378

Terry, W. D. 311, 374
Teschan, P. 138, 155
Themann, H. 407, 413
Thind, I. S. 635
Thomas, A. A. 749
Thomas, G. H. 877
Thomas, W. A. 807
Thompson, G. E. 887
Thorington, R. 59, 62, 77
Thurston, E. 430
Thurston, J. P. 753
Tigges, J. 432
Tigges, M. 432
Tigertt, W. D. 752
Tinklepaugh, O. L. 875
Tolliver, E. A. 876
Törquist, G. 432
Tilden, E. B. 268
Tippett, P. 311, 312
Tischer, C. F. 138, 155
Titrud, L. A. 233
Tobie, J. E. 516
Todd, T. W. 457, 518
Toga, M. 560, 635, 636
Tokuda, F. 187
Tolliver, G. A. 411, 412
Tolson, W. W. 411, 412
Tomasi, T. B. 372
Tomes, S. F. 708
Tomlinson, G. H., jr. 258, 268
Topping, N. H. 268
Torres, I. 266
Torrey, J. C. 208, 237
Trapani, R. J. 752
Traynor, J. E. 237, 754
Triotskiy, V. 709
Trobaugh, F. E. 146
Trueharrt, R. E. 413
Trumbull, R. 332, 340
Trussell, R. E. 876
Tscherkowitsch, G. M. 115, 154
Tsvetaeva, N. P. 710
Tubangui, M. 876
Tufling, M. R. 510
Tullner, W. W. 871
Tulpule, P. G. 750, 753
Tumanian, M. A. 709
Tuncbay, T. O. 541
Tun-Chot, S. 514
Turner, B. C. 431
Turner, R. G. 268

Ueda, K. 708
Ugolev, A. M. 124, 125, 155
Uhlenbruck, G. 314
Uhr, J. W. 375
Ulmer, F. A., jr. 268
Ulvedal, F. 754
Uno, N. 428
Unterharnscheidt, F. 199, 237
Urbain, A. 604, 637
Urist, M. R. 516
Urmancheeva, T. G. 138, 144, 150, 155, 156
Usacheva, I. N. 88, 156
Ushkalov, A. F. 9
Usizima, H. 237
Uspensky, S. S. 156
Usunina, L. A. 148
Utkin, I. A. 87, 156

Vadova, A. V. 9, 10, 193, 196, 197, 213, 215, 236, 237, 238, 413, 872, 873
Vaes, G. 518
Vagtborg, H. 519
Valen, L. van, 54
Valentine, R. G. 235
Valerio, D. A. 172, 186, 216, 221, 238, 548, 632, 872, 873, 880, 887
Vallee, A. 875
Valois, R. L. De, 431
Vanace, P. W. 105, 109, 113, 147, 808
Vandebroek, G. 344, 372
Vandenberg, J. G. 869
Vannier, W. E. 374
Vargas, L. 193, 238
Varley, H. 519
Vasilyev, Y. M. 231
Vasquez-Colet, A. 876
Vastersaeger, M. M. 808
Vaughan, H. G. 432
Vaughan, J. H. 387, 397
Vaughan, J. M. 490, 508, 519

Vawter, D. W. 264
Vawter, G. F. 515, 804
Velaudapillae, T. 708
Velez, H. 269, 519
Vera Cruz, N. C. 870, 871
Verbitskiy, M. Sh. 89, 152
Vercruysse, J. 808
Verder, E. 268
Vesselinovitch, D. 269
Vessey, S. 869
Vice, T. E. 122, 156, 708, 875
Vickers, J. H. 710, 753, 875
Victor, M. 262, 269, 615, 637
Victorova, O. N. 709
Vincent, K. 804
Vitale, J. J. 269, 519
Vlodaver, Z. 791, 808
Vogel, S. F. 616, 621, 626, 627, 633, 637
Vogin, E. E. 749
Vojtiskova, M. 394, 397
Volkova, E. M. 42, 68
Volkova, L. S. 89, 152
Voller, A. 330
Voronin, E. S. 708, 709
Voronoff, S. 197, 213, 238
Vorwald, A. J. 204, 216, 223, 238

Waart, A. De, 109, 114, 156, 771, 808
Wacker, W. E. C. 726
Wade, J. C. 838
Wadsworth, J. R. 199, 202, 238, 873
Wagenen, G. van, 193, 238, 503, 512, 519, 838, 872, 879, 887
Wagner, R. 134, 152
Wahle, H. 135, 156
Wahlin, J. H. 262
Waisman, H. A. 250, 251, 255, 261, 265, 269
Wakeman, A. M. 749
Waksman, B. H. 397
Wald, N. 731
Walford, R. L. 383, 398
Walker, A. E. 617, 637
Walker, E. 58, 64, 77
Walker, J. G. 392, 396, 602, 637
Walker, M. D. 203, 234, 238
Wallach, J. D. 428
Wallington, E. A. 513
Walter, G. V. 707
Wang, A. C. 349, 373
Wang, C. H. I. 148
Wang-Peng, J. 232
Wanko, T. 541
Wanner, R. L. 264, 514, 839
Ward, E. N. 198, 229, 236
Warner, E. D. 261
Warren, K. S. 750
Warren, S. 407, 413
Warwick, R. A. 192, 238
Wasbhurn, S. 66, 76
Waters, E. T. 377
Wathen, R. L. 139, 154
Wather, R. L. 838
Watkins, W. M. 314
Watson, D. L. 265
Watson, P. L. 516, 751
Weale, R. P. 432
Webb, H. E. 754
Wechsler, I. S. 249, 269, 615, 637
Weibel, J. 752
Weidmann, F. D. 602, 637
Weidmann, S. M. 485, 519
Weigel, W. W. 139, 140, 156
Weil, A. 234
Weinberg, M. 587, 637, 709
Weiner, M. 731
Weinman, A. 708
Weisbroth, S. 880. 887
Weiszer, I. 369, 379
Wells, S. 404, 409
Wemyss, C. 156
Wendt, W. 803
Wennemark, J. H. 379, 806
Wenner, H. A. 578, 635, 637
Wenner, W. L. 428
Werthessen, N. T. 805
West, E. S. 820
West, M. 751
West, W. E. 540
Weston, R. E. 122, 151
Westphal, U. 871
Wexler, I. B. 317
Wexler, N. 838

Wheeler, H. O. 749
Wherry, F. W. 412
Whitby, L. G. 495, 519
White, R. 156
Whiting, M. G. 234
Whitney, R. A. 516, 635
Wicher, K. 374, 378, 379, 428
Wieme, R. J. 756
Wiener, A. S. 153, 156, 276, 277, 278, 279, 282, 283, 285, 288, 289, 298, 299, 303, 311, 312, 313, 314, 315, 316, 317
Wiesel, T. N. 431
Wiggins, H. S. 519
Wikinski, R. W. 749
Wilber, Ch. G. 105, 110, 156
Wilgram, G. F. 256, 257, 269
Wilkinson, J. H. 519
Williams, A. H. 105, 109, 113, 114, 151
Williams, A. O. 808
Williams, C. A., jr. 372
Williams, F. P. 706, 753, 803
Williams, J. S. 749
Williams, R. C. 373, 386, 398
Williams, T. F. 510, 519
Williams-Asham, 726
Willman, V. L. 156
Willner, M. D. 403, 410, 804
Wills, L. 269
Wilson, A. 154
Wilson, C. A. 752
Wilson, K. E. 269, 839
Wilson, J. A. 397
Wilson, J. G. 160, 172, 186, 471, 519, 872
Wilson, P. W. 515
Wilson, R. B. 410
Wilson, R. V. 513
Windle, W. F. 186, 229, 235, 548, 549, 633, 636, 637, 872, 874
Winters, R. W. 439, 519
Wisecup, W. G. 317
Wiseman, R. A. 379
Wislocki, G. B. 364, 377, 868
Wissler, R. W. 10, 269, 789, 808
Witebsky, E. 396, 397, 413
Witte, E. J. 202, 234, 407, 411, 873
Witzgird, J. P. 253, 269
Woessner, J. F., jr. 507, 519
Wolf, A. 802
Wolf, G. L. 808
Wolf, H. 838
Wolf, P. 330
Wolf, R. C. 753, 871
Wolf, R. H. 242, 269, 801
Wolff, S. M. 374
Wolfman, E. F. 749, 751
Wolfson, 726
Wolin, L. R. 432
Wood, W. 634, 752
Woodard, J. C. 808, 874
Woodruff, A. W. 379
Woodruff, J. M. 202, 238, 749
Woods, J. F. 519
Woolard, M. 147
Woolpert, O. C. 839
Wootton, I. D. P. 502, 504, 519
Wu, J. J. 619, 632
Wuest, H. M. 805
Wyand, D. S. 707
Wyler, F. 149

Yakovleva, L. A. 152, 159, 160, 186, 192, 195, 196, 197, 199, 210, 213, 221, 222, 232, 234, 334, 341, 474, 515, 547, 634, 708, 710, 752, 757, 758, 775, 785, 791, 804, 808, 819, 820, 871
Yalow, R. S. 51
Yamada, H. 708
Yamada, M. 577, 578, 633
Yamasawa, S. 231, 426
Yamanaka, E. 378
Yates, J. L. 206, 229, 231
Yen, P. K. J. 519
Yerkes, R. M. 159, 186
Yoshiba, K. 68, 69, 77
Young, F. A. 432, 887
Young, R. J. 511, 602, 638, 753, 802
Young, W. C. 871
Young, W. J. 311
Younger, R. K. 269, 407, 413, 808
Yu, P. N. G. 263
Zabolotnyi, D. K. 875

Zaki, F. G. 264, 753
Zalusky, R. 208, 221, 238, 820
Zaman, V. 541, 638, 808
Zastera, M. 636
Zdrodovskaya, E. P. 154
Zegers, B. J. 375
Ziegler, R. F. 751
Zilber, L. A. 10
Zilva, S. S. 263, 264
Zimmerman, H. J. 751
Zimmerman, R. R. 266, 750
Zlatin, R. S. 152
Zöliger-Müller, B. 705
Zook, B. C. 805
Zuckerman, S. 193, 212, 238, 868, 869, 870, 887
Zuidema, G. D. 263, 267, 749, 751
Zumbo, O. 267, 807
Zvetaeva, N. P. 202, 232, 238
Zweifel, F. 189
Zyl, J. A. van 151
Zyl, J. J. W. van 151